Volume 2 of 2 Engine Maintenance Manual
LD 465-1 / LD 465-1C / LT 465-1C
LDS-465-1A / LDS 465-2
Engines
TM 9-2815-210-34-2-2
for
2.5 Ton 6x6 M44A2 Truck Series
aka
Deuce and a Half

April 1981

edited by
Brian Greul

The M44A2 series of military trucks is affectionately known as the deuce and a half, or simply the deuce. These ubiquitous trucks were first produced during WWII and General Eisenhower credited them as being one of the most important US Army vehicles. This book is volume 2 of 2 for the engine maintenance manual for the LD-465 series of diesel engines. It is being printed for enthusiasts, restores, and collectors who may wish to own a quality paperback copy. The editor has endeavored to minimize changes, but the following changes were made: Feedback forms are omitted, the fonts have been updated for printing purposes on modern equipment.

Should you have suggestions or feedback on ways to improve this book please send email to Books@OcotilloPress.com

---

Edited 2021 Ocotillo Press
ISBN 978-1-954285-48-4

Printed in the United States of America

Ocotillo Press
Houston, TX 77017
Books@OcotilloPress.com

TECHNICAL MANUAL
VOLUME 2 OF 2
PART 2 OF 2

MAINTENANCE

DIRECT SUPPORT AND GENERAL SUPPORT LEVEL

ENGINE ASSEMBLY, DIESEL (MULTIFUEL): NATURALLY ASPIRATED OR TURBOCHARGED, FUEL-INJECTED, WATER-COOLED, 6-CYLINDER

MODELS: LD-465-1, NSN 2815-00-239-5824; LD-465-1C, NSN 2815-00-134-4830; LDT-465-1C, NSN 2815-00-103-2642; LDS-465-1, NSN 2815-00-075-0087; LDS-465-1A, NSN 2815-00-239-5819; AND LDS-465-2, NSN 2815-00-808-8011 AND CLUTCHES

NOTE:
THE STYLE OF THIS TM IS EXPERIMENTAL. IT IS BEING TRIED BY THE ARMY ONLY ON A LIMITED BASIS

DEPARTMENTS OF THE ARMY AND THE AIR FORCE
APRIL 1981

WARNING

EXHAUST GASES CAN BE DEADLY

Exposure to exhaust gases produces symptoms of headache, dizziness, loss of muscular control, apparent drowsiness, and coma . Permanent brain damage or death can result from severe exposure .

Carbon monoxide occurs in the exhaust fumes of internal combustion engines, and becomes dangerously concentrated under conditions of inadequate ventilation. The following precautions must be observed to insure the safety of personnel whenever engine is operated.

Do not operate engine in an enclosed area unless it is adequately ventilated.

Be alert at all times during engine operation for exhaust odors and exposure symptoms. If either are present, immediately ventilate work area. If symptoms persist, remove affected personnel from area and treat as follows: expose to fresh air; keep warm; do not permit physical exercise; if necessary, administer artificial respiration.

If exposed, seek prompt medical attention for possible delayed onset of acute lung congestion. Administer oxygen if available .

The best defense against exhaust gas poisoning is adequate ventilation,

WARNING

SERIOUS OR FATAL INJURY TO PERSONNEL MAY RESULT
IF THE FOLLOWING INSTRUCTIONS ARE NOT COMPLIED WITH.

Dry cleaning solvent is flammable. Do not use near an open flame. Keep a fire extinguisher nearby when solvent is used. Use only in well-ventilated places. Failure to do this may result in injury to personnel and damage to equipment.

Do not use more than 30 psi of air pressure for drying parts. Eye shields must be worn when using compressed air. Eye injury can occur if eye shields are not used.

Do not use dip tank unless you are experienced at it. Chemicals in dip tank can cause serious burns to personnel .

Do not use less than six universal mounting arms to support cylinder block on maintenance stand . Using less than six arms may cause cylinder block to fall, causing damage to equipment and injury to personnel.

Use goggles, rubber gloves, and rubber apron when cleaning parts in carbon removing compound . Make sure there is enough ventilation. Be careful not to breathe in fumes or let compound touch skin. If compound is splashed on skin, flush with water and wash with alcohol, preferably 2 to 3 percent.

Carbon removing solvent is flammable. Do not use near an open flame. Keep a fire extinguisher nearby when solvent is used. Use only in well-ventilated places. Failure to do this may result in injury to personnel and damage to equipment.

Face shields and gloves must be worn when using heated cleaning compound. Work only in well-ventilated area. Do not breathe in steam. Failure to use proper precautions may result in injury to personnel.

Rocker arm assembly is under spring tension. Be careful when taking off retaining ring or parts of rocker arm assembly will fly off, causing injury to personnel.

Hydraulic pump oil reservoir cover is spring loaded. Be careful when taking off cover bolt so cover does not fly off. Failure to do so may cause injury to personnel or damage to equipment.

The voltage output of the ignition unit is strong enough to cause a dangerous electrical shock. Do not touch any bare wires or connections during operation. Refe r to decal on unit.

Crankshaft is very heavy. Do not let fingers or hand get between crankshaft and engine block. Do not let crankshaft bump engine block. Failure to use prope r caution may result in damage to equipment and injury to personnel.

Flywheel is very heavy. It should be lifted into place by two persons. After flywheel is in position it must be held firmly until mounting bolts are put in. Failure to do so may cause injury to personnel and or damage to equipment.

Pressure plate assembly is heavy. It must be held firmly up and against flywheel until mounting screws are put in. If it is not held up, it will fall and cause injury to personnel and damage to equipment.

Starter weighs fifty pounds. Be careful to hold it up firmly when taking it off mounting studs . Starter could fall and cause injury, to personnel and damage to equipment.

Make sure all dynamometer power switches are set to OFF position. Test equipment can cause an electrical shock and injure personnel.

During dynamometer test,be alert at all times during engine operation for exhaust odors and exposure symptoms . If either are present, immediately ventilate work area. Permanent brain damage or death can result from severe exposure.

Use hearing protection during dynamometer test. The test is very noisy and the noise can cause damage to your hearing.

Smoking, sparks or open flame are not allowed within 50 feet of work area when working on fuel system. Fuel is highly flammable and can explode, causing injury to personnel and damage to equipment.

Wear welder's gloves when handling hot parts. Failure to do so will result in serious burns .

The high fuel pressure used during fuel injector nozzle and holder assemblies tests and checks can puncture the skin and cause blood poisoning and possible death. Keep hands away from nozzle.

* TM 9-2815-210-34-2-2
T.O. 38G1-48-12-2-2

TECHNICAL MANUAL
NO. 9-2815 -210-34-2-2
TECHNICAL ORDER
NO. 38G1-48-12-2-2

DEPARTMENTS OF THE ARMY
AND
THE AIR FORCE
Washington, DC, 3 April 1981

TECHNICAL MANUAL

VOLUME 2 OF 2

PART 2 OF 2

MAINTENANCE

DIRECT SUPPORT AND GENERAL SUPPORT LEVEL

ENGINE ASSEMBLY, DIESEL (MULTIFUEL): NATURALLY ASPIRATED OR TURBOCHARGED, FUEL-INJECTED, WATER-COOLED, 6-CYLINDER

MODELS: LD-465-1 , NSN 2815-00-239-5824; LD-465-1C, NSN 2815-00-134-4830; LDT-465-1C, NSN 2815-00-103-2642; LDS-465-1 , NSN 2815-00-075-0087; LDS-465-1A, NSN 2815-00-239-5819; AND LDS-465-2, NSN 2815-00-808-8011 AND CLUTCHES

REPORTING OF ERRORS AND RECOMMENDING IMPROVEMENTS

You can help improve this manual. If you find any mistakes or if you know of a way to improve the procedure, please let us know. Mail your letter, DA Form 2028 (Recommended Changes to Publications and Blank Forms), or DA Form 2028-2 located in the back of this manual direct to: Commander, US Army Tank-Automotive Command, ATTN: DRSTA-MB, Warren, Michigan 48090. A reply will be furnished to you.

TABLE OF CONTENTS

* This manual, together with TM 9-2815-210-34-1, 3 April 1981; and TM 9-2815-210-34-2-1, 3 April 1981, supersedes TM 9-2815-210-34, 16 June 1978.

TABLE OF CONTENTS-CON T

TABLE OF CONTENTS-CONT

TABLE OF CONTENTS-CON T

TABLE OF CONTENTS-CON T

TABLE OF CONTENTS-CONT

TABLE OF CONTENTS-CONT

TABLE OF CONTENTS-CON T

TABLE OF CONTENTS-CONT

TABLE OF CONTENTS-CON T

TABLE OF CONTENTS-CONT

TABLE OF CONTENTS-CON T

TABLE OF CONTENTS-CON T

LIST OF ILLUSTRATIONS

LIST OF TABLES

LIST OF TABLES - CONT

LIST OF TABLES - CONT

CHAPTER 4

CLEANING, INSPECTION, AND REPAIR

Section I. SCOPE

4-1. EQUIPMENT ITEMS COVERED. This chapter gives instructions for cleaning, inspection, and repair of engine accessories and engine components.

4-2. EQUIPMENT ITEMS NOT COVERED. All equipment items are covered in this

Section II. GENERAL CLEANING, INSPECTION, AND REPAIR PROCEDURES

TOOLS: No special tools required

SUPPLIES: Solvent, dry cleaning, type II (SD-2), Fed. Spec P-D-680
Compressed air source, 30 psi max
Stiff wire
Penetrant kit, MIL-I-25135

PERSONNEL: One

EQUIPMENT CONDITION: Engine disassembled.

4-3. CLEANING.

a. Clean all parts before inspection, after repair, and before assembly.

b. Clean all bearing cones and cups. Refer to Inspection, Care, and Maintenante of Antifriction Bearings, TM 9-214.

WARNING

Dry cleaning solvent is flammable. Do not use near an open flame. Keep a fire extinguisher nearby when solvent is used. Use only in well-ventilated places. Failure to do this may result in injury to personnel and damage to equipment.

Do not use more than 30 psi of *air* pressure for drying parts. Eye shields must be worn when using compressed air. Eye injury can occur if eye shields are not used.

CAUTION

When scraping gasket material from surface of parts, be careful not to scratch or gouge metal surfaces.

c. Clean all other parts with solvent. Scrape all gasket material from surface of parts. Rinse parts in clean solvent and dry with compressed air.

d. Make sure all oil passages are open. Open clogged passages with compressed air or by working a stiff wire back and forth. Flush with solvent.

c. After cleaning, all parts should be covered or wrapped in plastic or paper to protect them from dust and dirt.

4-4. INSPECTION.

Check all bearing cones and cups. Refer to Inspection, Care, and Maintenance of Antifriction Bearings, TM 9-214.

b. Small chips, burrs or scratches in castings that do not go into screw holes or openings can be repaired.

c. Check all mating flanges on housings and supports for warpage with a straight edge or on a surface plate. Check mating flanges for discoloring. Discoloring of flanges will mean that oil has been leaking which may mean that flanges are warped.

d. Check aluminum castings for cracks using penetrant kit. If dye penetrant inspection equipment is not available, use a magnifying glass and a strong light.

e. Check all other type castings for cracks with magnetic particle inspection equipment. If magnetic particle inspection equipment is not available, use a magnifying glass and a strong light.

NOTE

There are no wear limits on gear teeth. Good judgement is needed to know if parts are good.

f. Check all gears for cracks, worn teeth, and pitted surfaces.

g. Check all threaded parts for worn or stripped threads.

4-5. REPAIR.

a. Smooth out any chips, burrs, or scratches on castings, gears, and shafts.

b. Weld cracks and small holes in castings. Refer to TM 9-237.

c. Drill out any bolts or studs broken off in tapped holes.

d. Drill out threaded holes that are stripped or out-of-round to the next larger size and retap them. When putting engine together, use a bolt or stud the size of the newly tapped hole.

e. Tell the machine shop to mill warped castings.

Section III. CLEANING, INSPECTION, AND REPAIR OF ENGINE COMPONENTS AND ACCESSORIES

TOOLS: Valve seat replacing tool, pn 11642007
Exhaust valve seat replacer, pn 11642006
Hand reamer, pn 12254220
Air compressor pulley adjusting wrench, pn 10935288
Fuel passages cleaning brush, pn ST 876

SUPPLIES: Dry cleaning solvent, type II (SD-2), Fed. Spec P-D-680
Air compressor unloader valve gasket
Air compressor intake air cleaner gasket
Air compressor air discharge manifold gasket

SUPPLIES: (cont)
Croucus cloth
Lubricating oil, ICE, OE/HDO 10, MIL-L-2104
Compressed air source, 30 psi max
Prussian blue, MIL-P-30501
Sealing compound, type II, MIL-S-45180
Clean lint-free cloth
Copper tube stoc k
Flexible non-metallic tube stock
Oil gage rod preformed packing
Oil drain plug gaskets (2)
Oil pump pressure relief valve shims
Penetrant kit, MIL-I-25135
Magnifying glas s
Breather tube adapter gasket
Breather valve sprin g
Breather diaphragm assembly (2)
Oil filter post gasket
Oil cooler and filter housing gaskets (2)
Fuel filter body gaskets (2)
Fuel filter relief valve preformed packing
Fuel filter element
Plastic tube stock
Hydraulic pump preformed packing
Hydraulic pump manifold gasket
Hydraulic pump adapter gasket
Silicone lubricant, MIL-L-25681
Carbon removing solvent, MIL-C-25107
Thermostat seal
Copper tube, 2 feet long
Teflon tape
Cylinder head water manifold hoses (4)
Exhaust manifold seal rings (6)
Turbocharger exhaust elbow gaske t
Cotter pins (as req'd)
Hydraulic pump oil reservoir gasket
Hydraulic pump oil reservoir preformed packing
Hydraulic pump oil reservoir body gasket
Hydraulic pump oil reservoir filter element
Diesel fuel, VV-F-800
Flame heater pipe plugs (2)
Flame heater fuel line filter
Timing gear cover seal
Wood V-blocks (2)

PERSONNEL: Tw o

EQUIPMENT CONDITION: Engine disassembled.

4-6. CRANKCASE ASSEMBLY, CRANKSHAFT, AND CAMSHAFT.

a. Removal of Pipe Plugs and Inspection of Expansion Plugs.

FRAME 1

1. Take out five pipe plugs (1).

NOTE

Coolant leakage around expansion plugs (2) will show up as rust, scale or stain on side of crankcase (3) below plug .

2. Check that three expansion plugs (2) are not damaged and that there are no signs of coolant leakage. Mark damaged or leaking plugs for removal.

GO TO FRAME 2

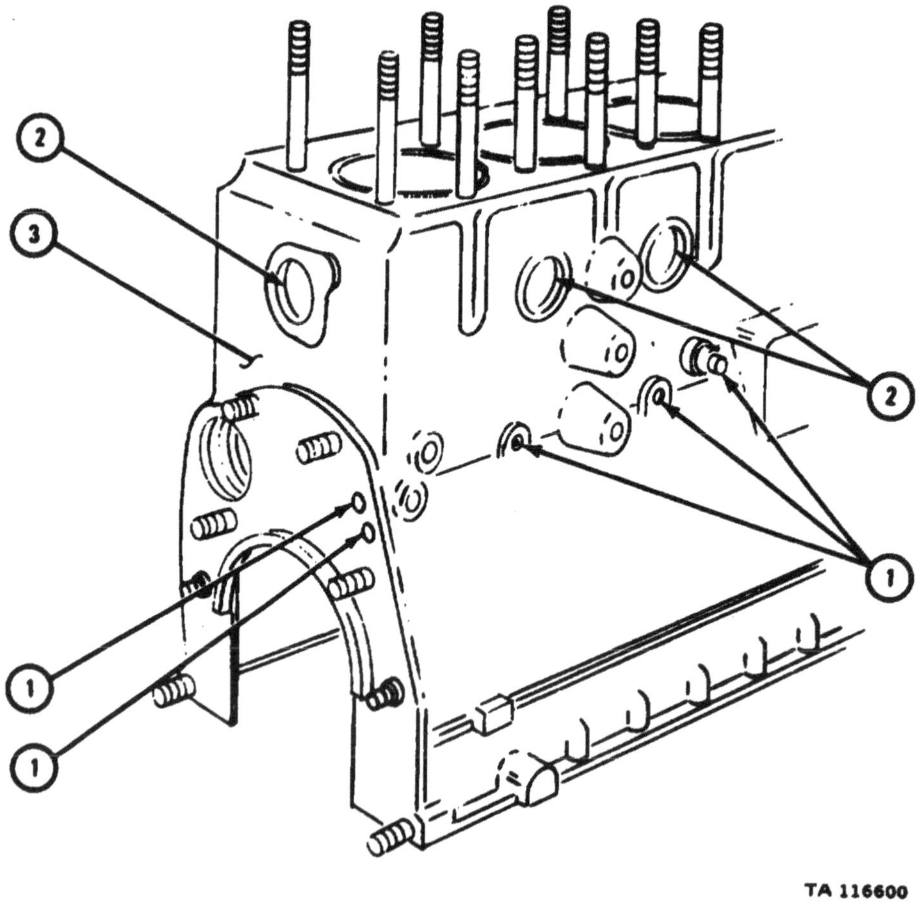

FRAME 2

1. Takeout six pipe plugs (1) .

NOTE

Coolant leakage around expansion plugs (2) will show up as rust, scale or stain on side of crankcase (3) below plug .

2. Check that two expansion plugs (2) are not damaged and that there are no signs of coolant leakage. Mark damaged or leaking plugs for removal.

GO TO FRAME 3

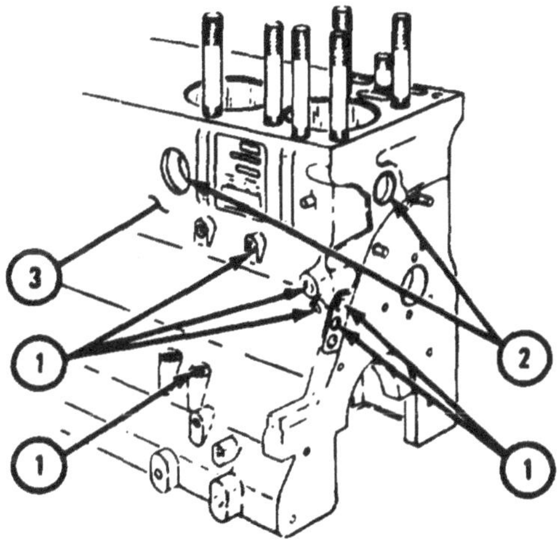

FRAME 3

NOTE

There is an expansion plug (1) behind identification plate (2) . Identification plate should not be taken off unless there are signs of coolant leakage under it.

Coolant leakage around expansion plug (1) will show up as rust, scale or stain on side of crankcase (3) below expansion plug .

1. Check that there is no sign of leakage below identification plate (2).
2. If there are signs of leakage below identification plate (2), take it off using hammer and cold chisel.
3. If identification plate (2) was taken off, mark expansion plug (1) for removal.

END OF TASK

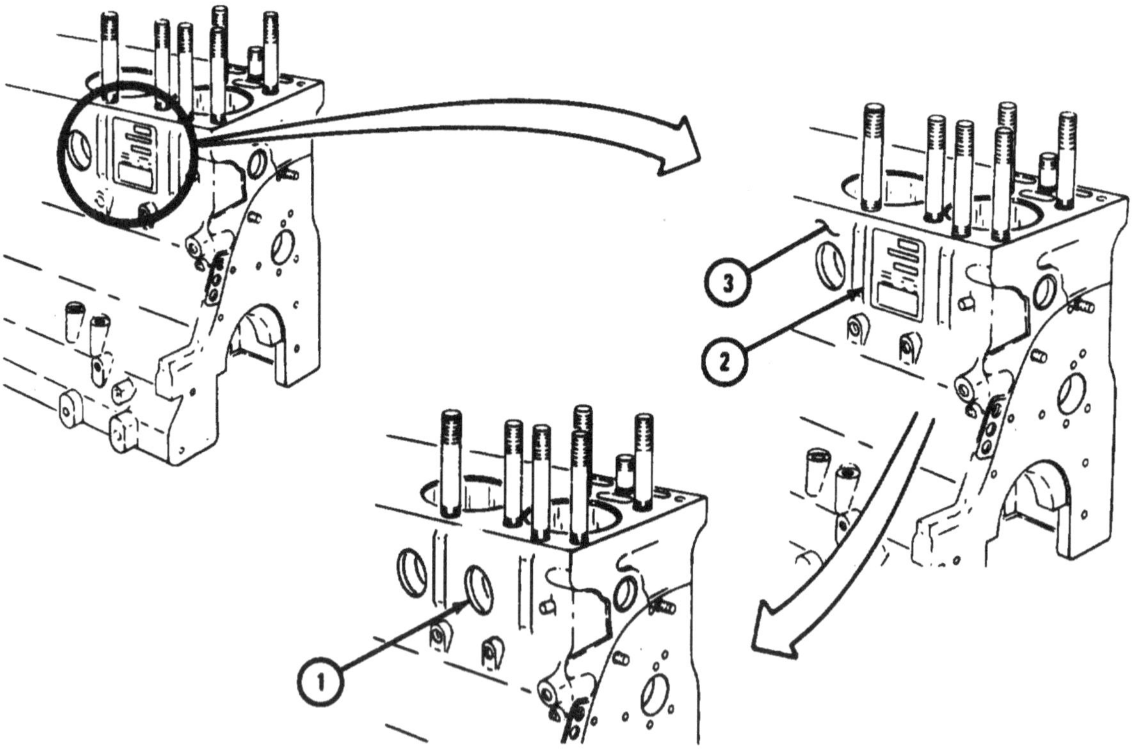

b. Removal of Expansion Plugs and Crankcase to Cylinder Head Studs.

FRAME 1

NOTE

Take out only expansion plugs (1) which were marked for removal.

This task is shown for one expansion plug (1). This task is the same for all expansion plugs marked for removal.

1. Drill a hole in center of expansion plug (1) marked for removal.
2. Using prybar, take out and throw away expansion plug (1).

GO TO FRAME 2

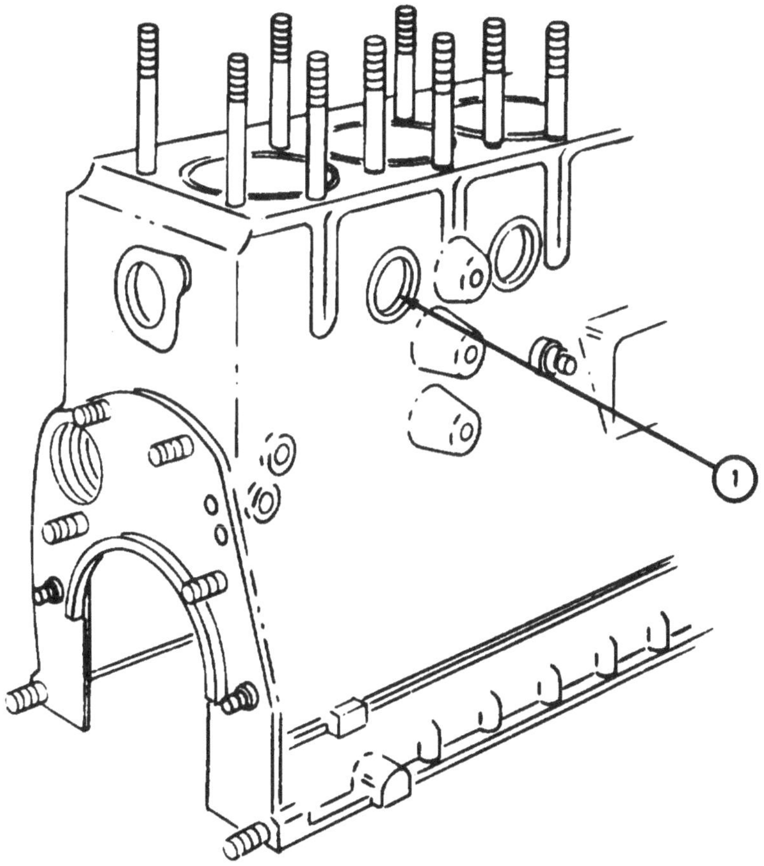

TA 116603

FRAME 2

NOTE

Cylinder sleeves (1) are usually thrown away, but cylinder sleeves that are almost new may be used again. Mark cylinder sleeves to be used again so that they will be put back in the same position in the same crankcase (2) cylinder bore.

1. Check if hone pattern can be seen in ring travel area in each of six cylinder sleeves (1) as shown. If hone pattern can be seen and there is no groove worn at top of ring travel area, mark cylinder sleeve and crankcase (2) with scribe mark (3) as shown. If hone pattern cannot be seen, or if a groove is worn at top of ring travel area, mark cylinder sleeve to be thrown away.
2. Using stud remover, take out all crankcase to cylinder head studs (4).

GO TO FRAME 3

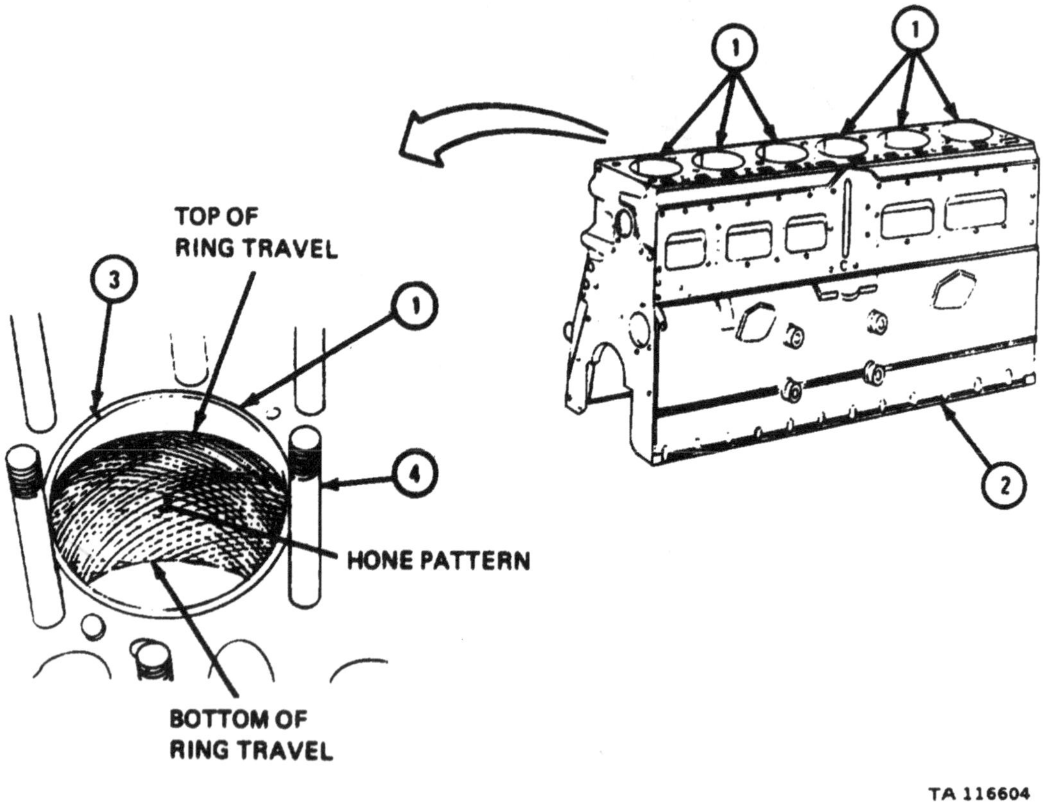

FRAME 3

NOTE

Cylinder sleeves (1) are called by numbers one to six counting from front to rear of crankcase (2).

1. Using cylinder sleeve remover and replacer kit, take out six cylinder sleeves (1). Tag cylinder sleeves marked for reuse with their number .
2. Threw away cylinder sleeves (1) that were not marked for reuse.

END OF TASK

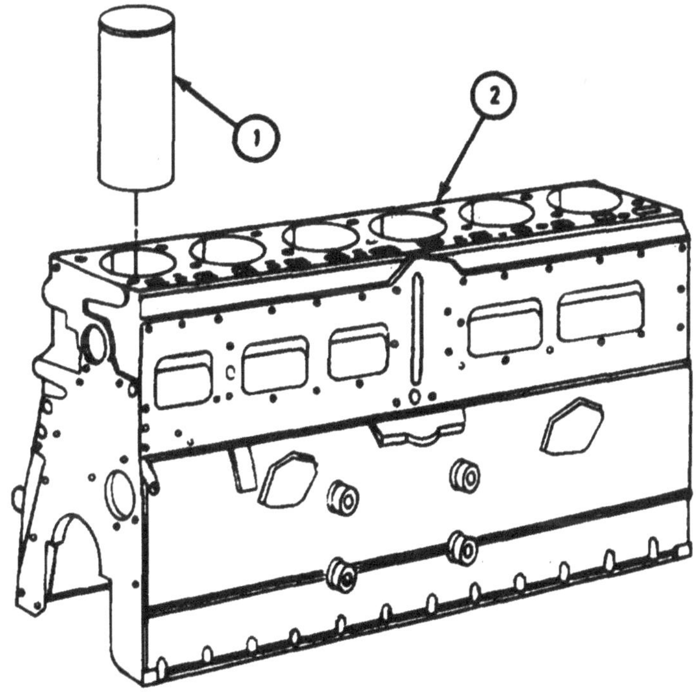

TA 116605

c. Cleaning.

FRAME 1

WARNING

Do not use dip tank (2) unless you are experienced at it. Chemicals in dip tank can cause serious burns to personnel.

1. Using crane, lower crankcase (1) into dip tank (2). Clean crankcase in dip tank.
2. Lift crankcase (1) out of dip tank (2).
3. Mount crankcase (1) on maintenance stand. Refer to para 3-31.

WARNING

Dry cleaning solvent is flammable. Do not use near an open flame. Keep a fire extinguisher nearby when solvent is used. Use only in well-ventilated places. Failure to do this may result in injury to personnel and damage to equipment.

Eye shields must be worn when using compressed air. Eye injury can occur if eye shields are not used.

4. Using fuel passage cleaning brush, clean oil passages in crankcase (1). Flush passages with dry cleaning solvent. Blow passages dry with compressed air.

GO TO FRAME 2

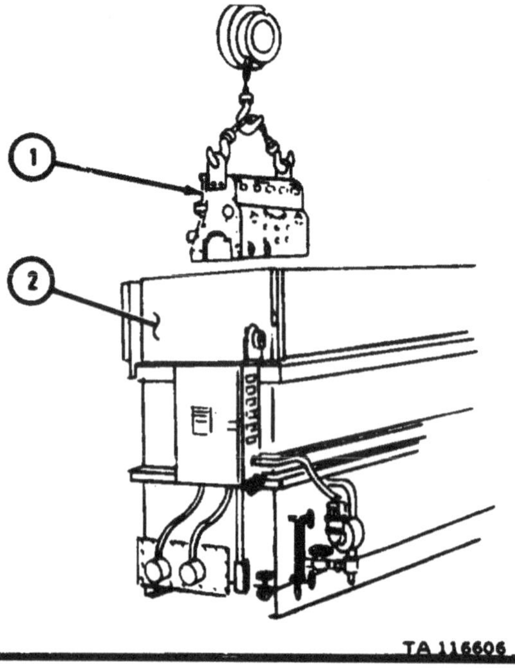

FRAME 2

WARNING

Dry cleaning solvent is flammable. Do not use near an open flame. Keep a fire extinguisher nearby when solvent is used. Use only in well-ventilated dates. Failure to do this may result in injury to personnel and damage to equipment.

CAUTION

It is easy to damage the equipment if you do not know what you are doing. Do not try to do this task unles s you are experienced at it, or you have an experienced person with you .

1. Clean 14 main bearing halves (1), seven main bearing caps (2), 14 screws (3), and 14 washers (4) with dry cleaning solvent. Use stiff bristle brush or wooden scraper to clean off sludge and gum in bearing grooves and oil holes. Dry with clean cloth.

GO TO FRAME 3

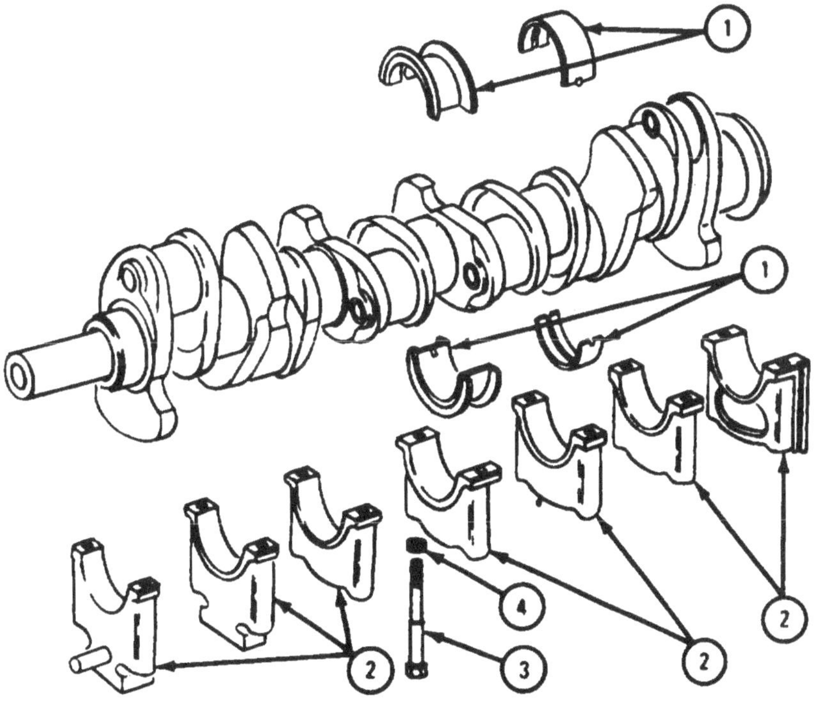

NOTE

CLEAN ONLY THOSE PARTS WHICH ARE CALLED OUT. PARTS WITHOUT CALLOUTS ARE SHOWN ONLY FOR REFERENCE PURPOSES.

TA 116607

FRAME 3

WARNING

Dry cleaning solvent is flammable. Do not use near an open flame. Keep a fire extinguisher nearby when solvent is used . Use only in well-ventilated places. Failure to do this may result in injury to personnel and damage to equipment.

1. Clean six piston cooling nozzles (1) and six nozzle mounting screws (2) with dry cleaning solvent .

WARNING

Eye shields must be worn when using compressed air. Eye injury can occur if eye shields are not used.

2. Clean oil passages in six piston cooling nozzles (1) using brass wire probes. Blow compressed air through oil passages.

GO TO FRAME 4

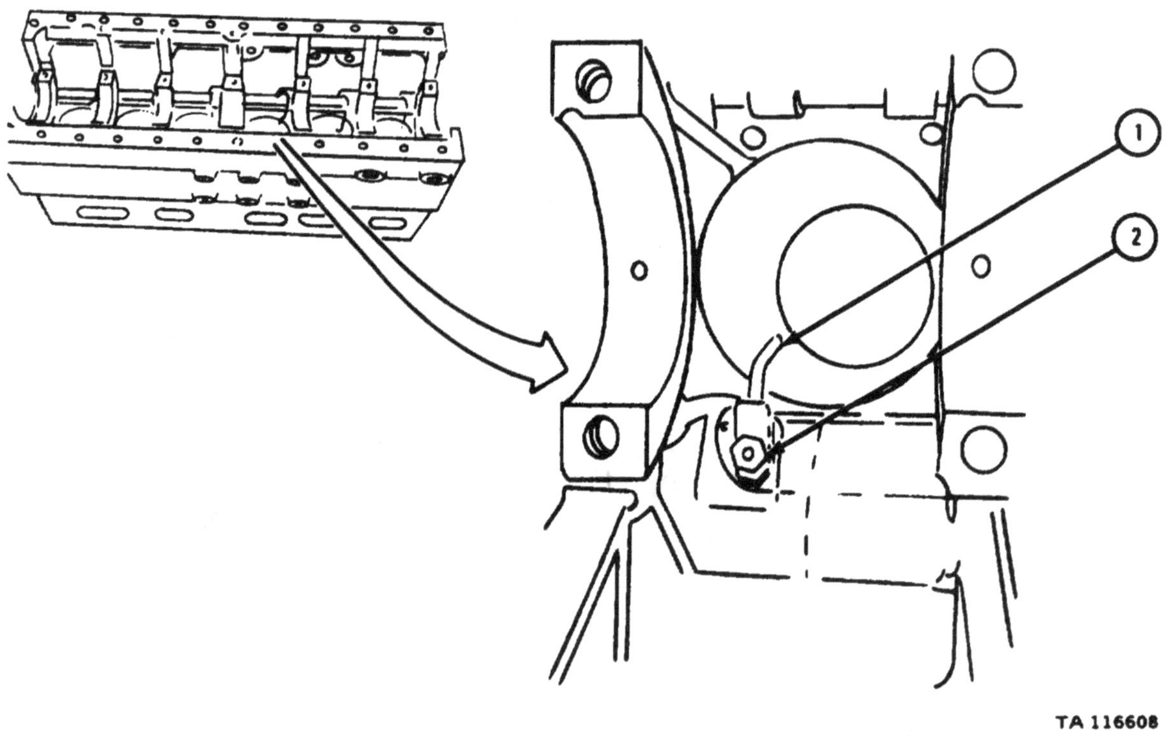

FRAME 4

CAUTION

Be careful not to nick or scratch journals on crankshaft (1) . If crankshaft journals are damaged, engine will not run properly.

1. Take out crankshaft gear woodruff key (2) .

WARNING

Dry cleaning solvent is flammable. Do not use near an open flame. Keep a fire extinguisher nearby when solvent is used. Use only in well-ventilated places. Failure to do this may result in injury to personnel and damage to equipment.

2. Clean crankshaft (1) with dry cleaning solvent. Clean off sludge and gum using stiff brush . Dry with clean cloth.

WARNING

Eye shields must be worn when using compressed air. Eye injury can occur if eye shields are not used.

3. Using fuel passage cleaning brush, clean oil passages in crankshaft (1). Flush passages with dry cleaning solvent and blow dry with compressed air.
4. Put crankshaft gear woodruff key (2) back in crankshaft (1) so it will not be lost.

GO TO FRAME 5

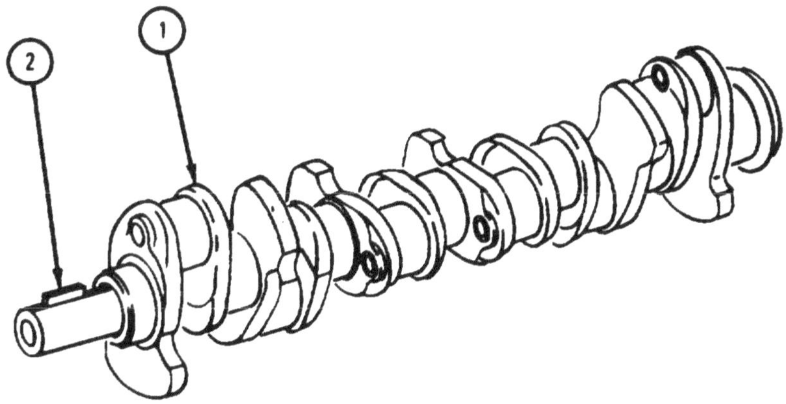

TA 116609

FRAME 5

CAUTION

Be careful not to nick or scratch camshaft journals (1). If camshaft journals are damaged, engine will not run properly .

1. Take out camshaft gear woodruff key (2).

WARNING

Dry cleaning solvent is flammable. Do not use near an open flame. Keep a fire extinguisher nearby when solvent is used. Use only in well-ventilated places. Failure to do this may result in injury to personnel and damage to equipment.

2. Clean camshaft (3) with dry cleaning solvent. Clean off sludge and gum using stiff brush . Dry with clean cloth.
3. Clean camshaft thrust plate (4), camshaft gear woodruff key (2), two camshaft thrust plate screws (5), and lockwashers (6), and camshaft gear locknut (7) with dry cleaning solvent. Dry with clean cloth .
4. Clean camshaft gear (8) with dry cleaning solvent. Clean off sludge and gum deposits with stiff brush. Dry with clean cloth .
5. Put camshaft gear woodruff key (2) back in camshaft (3) so it will not be lost.

GO TO FRAME 6

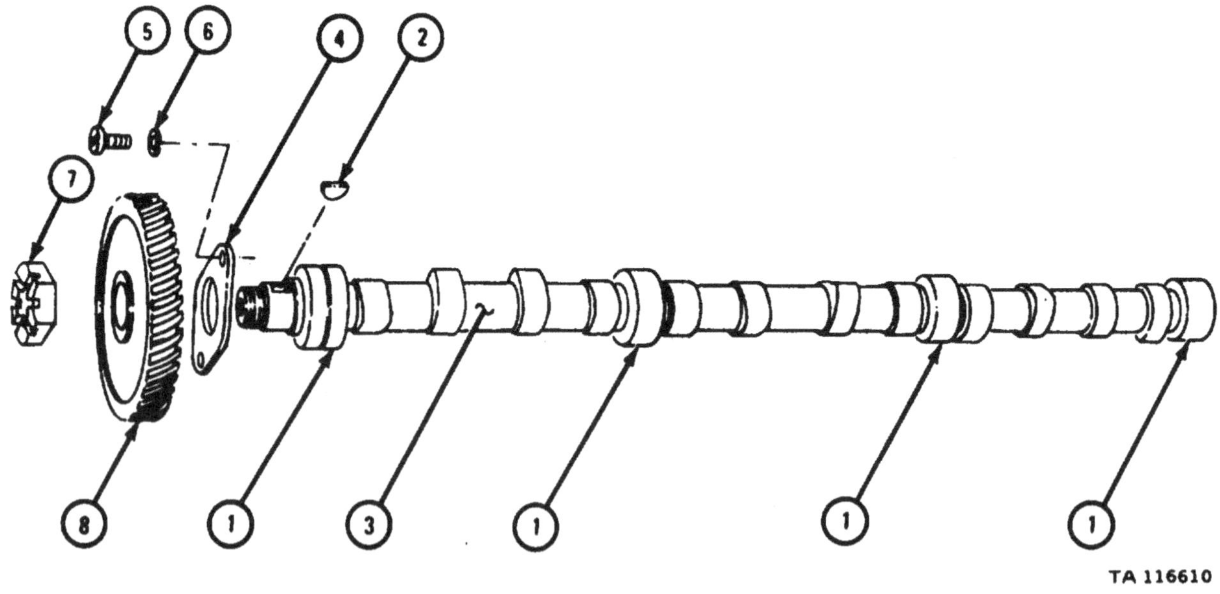

FRAME 6

CAUTION

Cylinder sleeves (1) are made of thin metal. They are easy to break and scratch. Never scrape off carbon deposits using a tool with sharp corners.

NOTE

Do this frame only for cylinder sleeves (1) marked for reuse. If no cylinder sleeves were marked for reuse, go to frame 7.

1. Clean carbon deposits off cylinder sleeves (1) after they are out of crankcase (2).

WARNING

Dry cleaning solvent is flammable. Do not use near an open flame. Keep a fire extinguisher nearby when solvent is used. Use only in well-ventilated places. Failure to do this may result in injury to personnel and damage to equipment.

2. Clean cylinder sleeves (1) with dry cleaning solvent. Dry with clean cloth.

GO TO FRAME 7

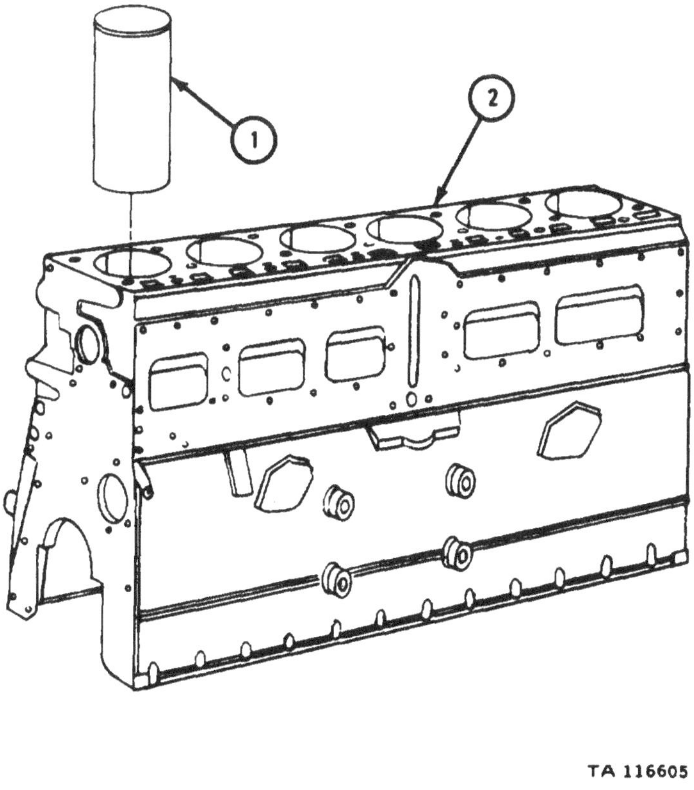

TA 116605

FRAME 7

WARNING

Dry cleaning solvent is flammable. Do not use near an open flame. Keep a fire extinguisher nearby when solvent is used. Use only in well-ventilated places. Failure to do this may result in injury to personnel and damage to equipment.

1. Clean 12 valve tappets (1) in dry cleaning solvent. Clean off sludge and gum with stiff brush .

WARNING

Eye shields must be worn when using compressed air. Eye injury can occur if eye shields are not used.

2. Clean oil passage (2) using fuel passage cleaning brush. Flush with dry cleaning solvent . Blow oil passage dry with compressed air.
3. Dry out side of 12 valve tappets (1) with clean cloth.

END OF TASK

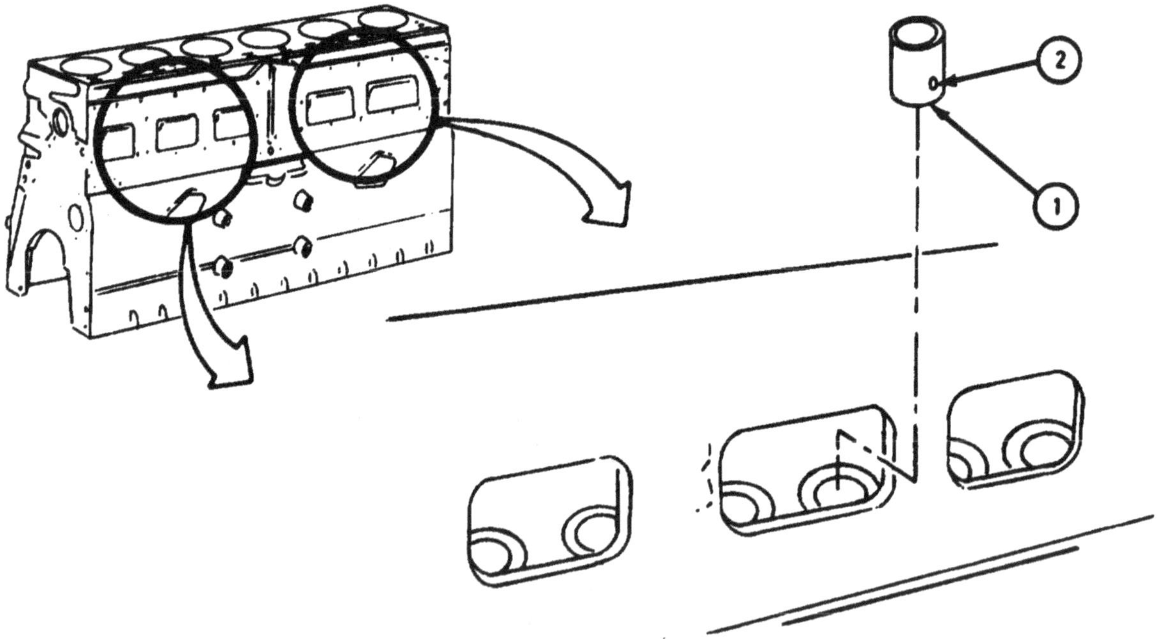

TA 116611

d. Inspection and Repair of Crankcase and Crankshaft.

CAUTION

It is easy to damage the equipment if you do not know what you are doing. Do not try to do this task unless you are experienced at it or you have an experienced person with you.

FRAME 1

NOTE

Main bearing caps (1) and thrust main bearing cap (2) are line bored with crankcase (3). If main bearing cap or thrust bearing cap is cracked, the whole crankcase assembly and all bearing caps must be thrown away.

1. Inspect crankcase assembly (3), six main bearing caps (1), and thrust main bearing cap (2). Refer to para 4-4.
2. Repair minor damage to crankcase assembly (3), six main bearing caps (1), and main thrust bearing cap (2). Refer to para 4-5.

GO TO FRAME 2

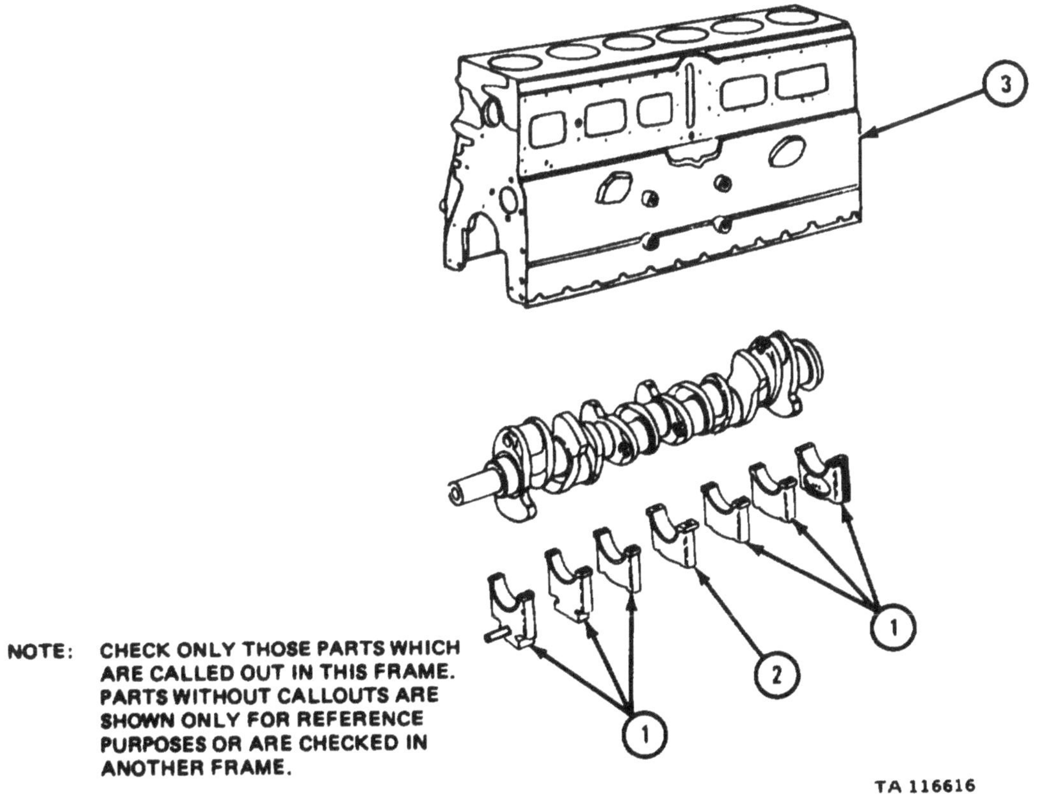
3
2
1
1
NOTE: CHECK ONLY THOSE PARTS WHICH ARE CALLED OUT IN THIS FRAME. PARTS WITHOUT CALLOUTS ARE SHOWN ONLY FOR REFERENCE PURPOSES OR ARE CHECKED IN ANOTHER FRAME.
TA 116616

FRAME 2

NOTE

Readings must be within limits given in table 4-1. If readings are not within given limits, throw away part and get a new one.

1. Check that oil pump idler gear shaft (1) is not cracked or broken. Check that oil pump idler gear shaft is not scored. Check that oil pump idler gear shaft does not have small bits of metal stuck in it.
2. Measure out side diameter of oil pump idler gear shaft (1).
3. If oil pump idler gear shaft (1) is damaged or worn, using 2-ton hydraulic press, press it out of front main bearing cap (2) .

IF OIL PUMP IDLER GEAR SHAFT (1) WAS PRESSED OUT, GO TO FRAME 3.
IF OIL PUMP IDLER GEAR SHAFT (1) WAS NOT PRESSED OUT, GO TO FRAME 5

NOTE: CHECK ONLY THOSE PARTS WHICH ARE CALLED OUT IN THIS FRAME. PARTS WITHOUT CALLOUTS ARE SHOWN ONLY FOR REFERENCE PURPOSES OR ARE CHECKED IN ANOTHER FRAME.

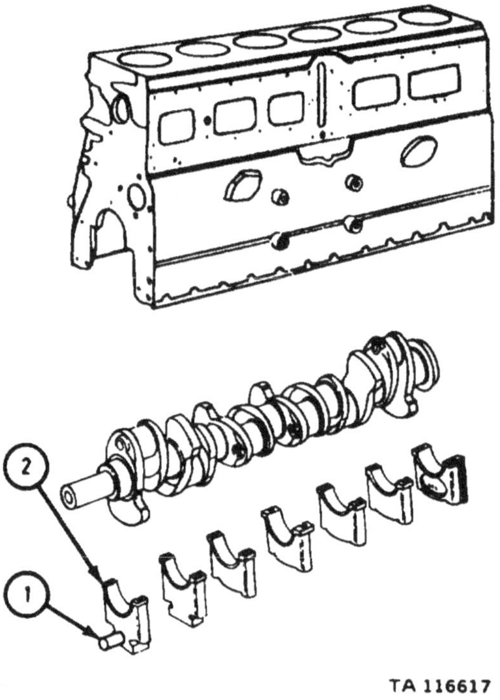

Table 4-1. Oil Pump Idler Gear Shaft Wear Limits

Index Number	Item/Point of Measurement	Size and Fit of New Parts (inches)	Wear Limits (inches)
1	Oil pump idler gear shaft	0.7869 to 0.7873	0.7867

FRAME 3

1. Measure diameter of bore (1) in front main bearing cap (2). Write down this measurement.

NOTE

New oil pump idler gear shaft (3) is different from early model oil pump idler gear shaft (4). New oil pump idler gear shaft has a shoulder (5) as shown.

2. Measure diameter of new oil pump idler gear shaft (3).

3. Fit of new oil pump idler gear shaft (3) in bore of front main bearing cap (2) must be 0.001 to 0.0025 inch.

GO TO FRAME 4

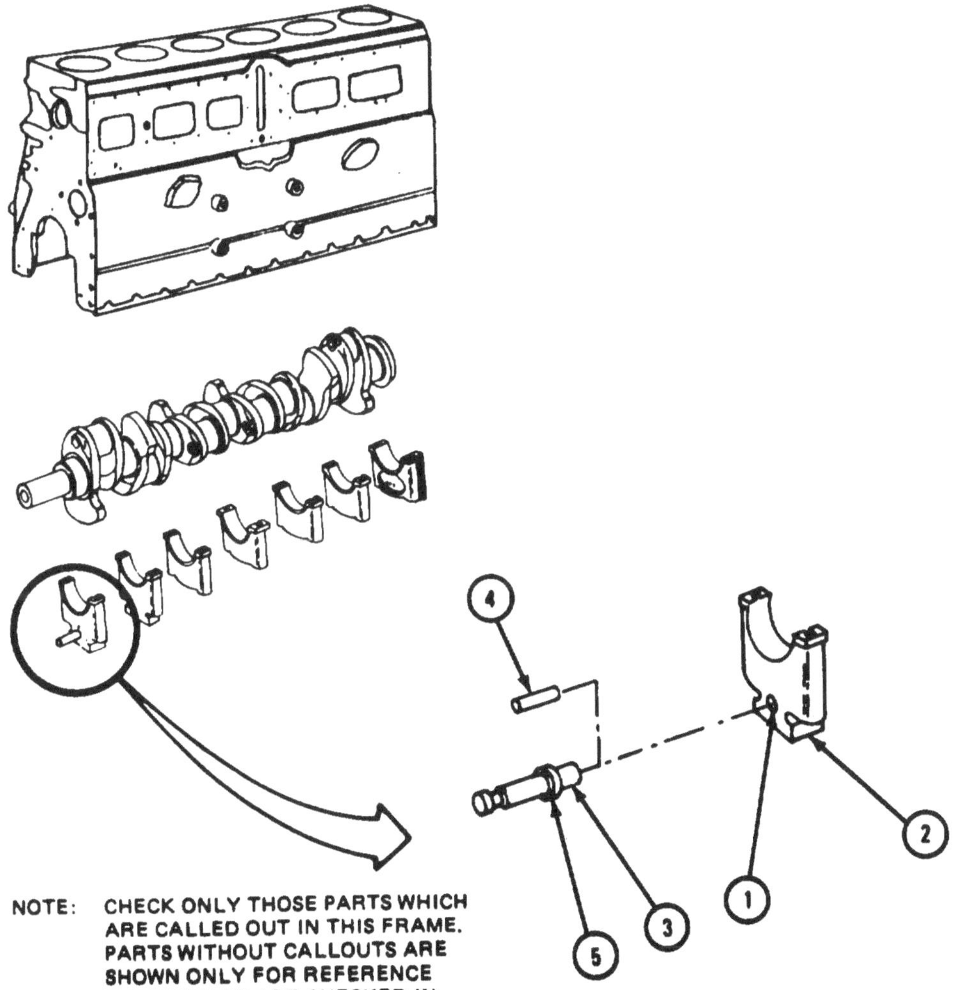

NOTE: CHECK ONLY THOSE PARTS WHICH ARE CALLED OUT IN THIS FRAME. PARTS WITHOUT CALLOUTS ARE SHOWN ONLY FOR REFERENCE PURPOSES OR ARE CHECKED IN ANOTHER FRAME.

TA 116618

FRAME 4

NOTE

Six main bearing caps (1 and 2) and main thrust bearing cap (3) are line bored with crankcase (4). If an y main bearing cap is damaged, crankcase and all seven main bearing caps must be thrown away.

1. If fit of new oil pump idler gear shaft (5) in bore of front main bearing cap (1) is not 0.001 to 0.0025 inch, throw away crankcase assembly (4) and all seven main bearing caps (1, 2, and 3) and get new ones. Then go to frame 5.

2. If fit of new oil pump idler gear shaft (5) in bore of front main bearing cap (1) is 0.0010 to 0.0025 inch, using 2-ton hydraulic press, press new oil pump idler gear shaft into bore in front main bearing cap until shoulder (6) touches face of front main bearing cap. Ring groove (7) in new oil pump idler gear shaft must face out as shown.

GO TO FRAME 5

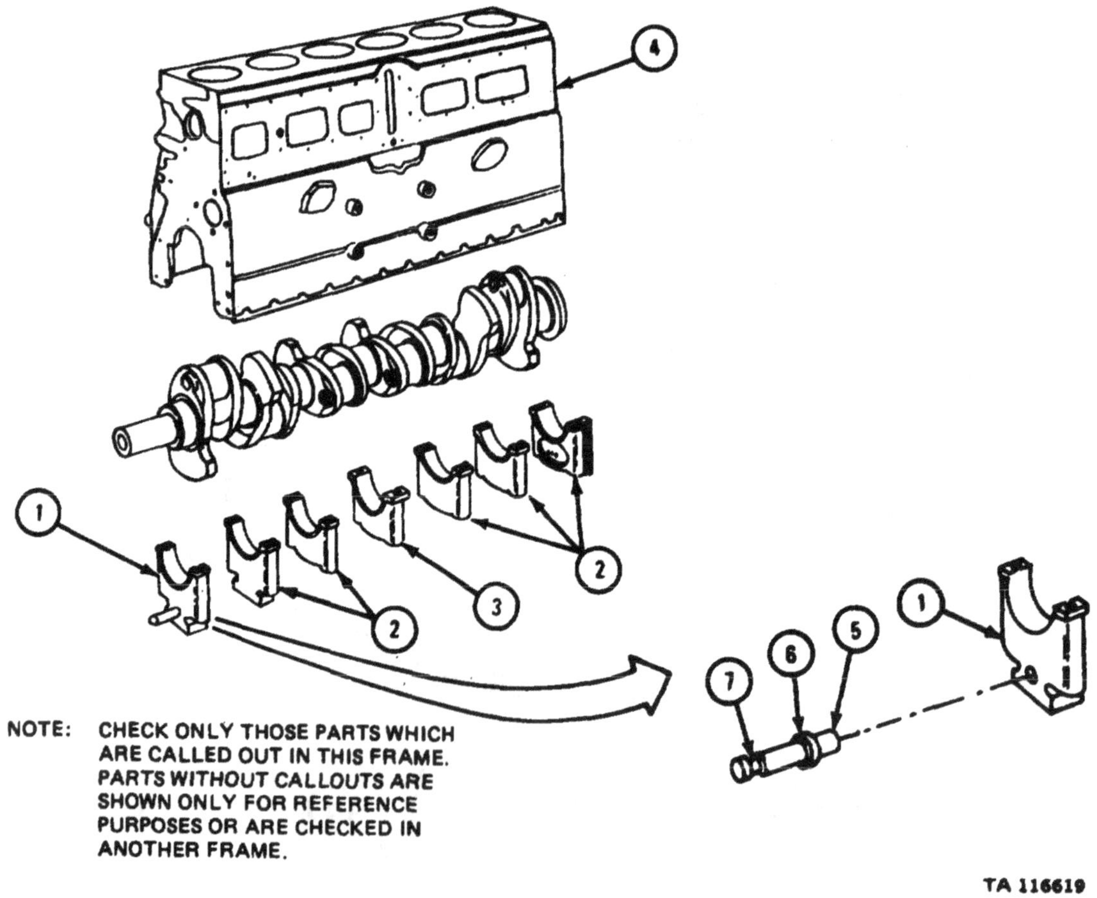
4
1
2
3
2
1
5
6
7
NOTE: CHECK ONLY THOSE PARTS WHICH ARE CALLED OUT IN THIS FRAME. PARTS WITHOUT CALLOUTS ARE SHOWN ONLY FOR REFERENCE PURPOSES OR ARE CHECKED IN ANOTHER FRAME.
TA 116619

FRAME 5

WARNING

Dry cleaning solvent is flammable. Do not use near an open flame. Keep a fire extinguisher nearby when solvent is used. Use only in well-ventilated places. Failure to do this may result in injury to personnel and damage to equipment.

1. Check that six piston oil sprayer nozzles (1) have no dents, nicks, scratches or burrs. Fix small nicks or burrs with a fine mill file or crocus cloth dipped in dry cleaning solvent. If piston oil sprayer nozzle is badly damaged, throw it away and get a new one.
2. Check that oil passage in six piston oil sprayer nozzles (1) is clear. If oil passage is stopped up, clean nozzle opening with wire probe and cleaning solvent. Clear nozzle with compressed air.
3. If oil passage is blocked because piston oil sprayer nozzle (1) is dented or bent, throw away piston oil sprayer nozzle and get a new one. Nozzle openings should be 0.070 inch on engine LDS 465-2 and 0.060 inch for all other engines.
4. Check that 14 main bearing cap bolts (2) do not have damaged threads, damaged heads, nicks or burrs. Throw away damaged main bearing cap bolts and get new ones.
5. Check that 14 main bearing cap washers (3) are not bent or damaged. Throw away damaged main bearing cap washers and get new ones.

GO TO FRAME 6

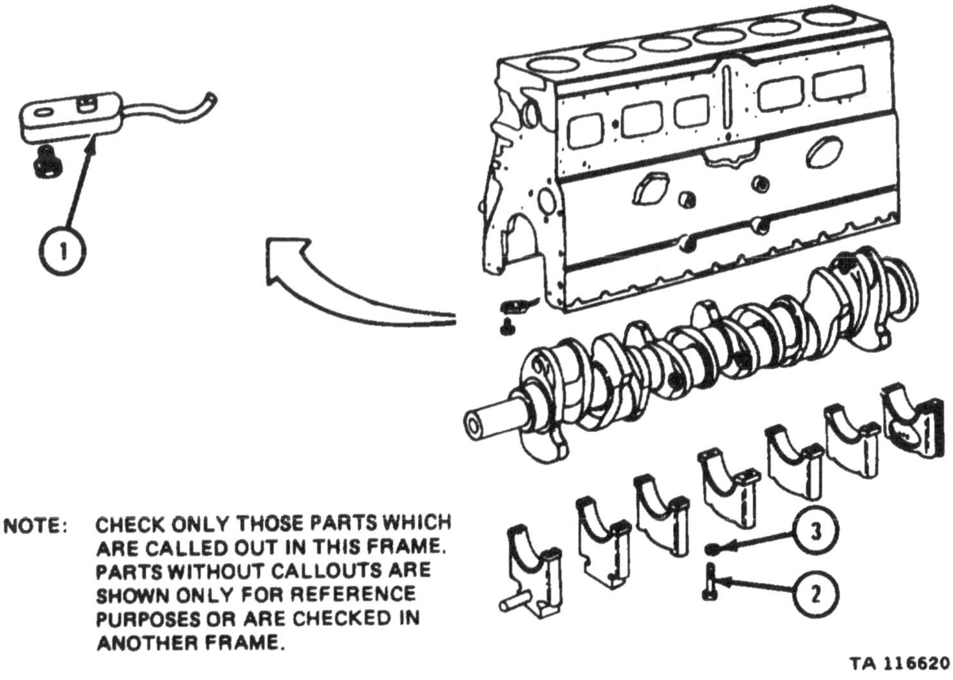
1
3
2
NOTE: CHECK ONLY THOSE PARTS WHICH ARE CALLED OUT IN THIS FRAME. PARTS WITHOUT CALLOUTS ARE SHOWN ONLY FOR REFERENCE PURPOSES OR ARE CHECKED IN ANOTHER FRAME.
TA 116620

FRAME 6

WARNING

Dry cleaning solvent is flammable. Do not use near an open flame. Keep a fire extinguisher nearby when solvent is used. Use only in well-ventilated places. Failure to do this may result in injury to personnel and damage to equipment.

1. Check that dowel pins (1) have no nicks, scratches or burrs. Fix minor burrs with a fine mill file or crocus cloth dipped in cleaning solvent.
2. Check that dowel pins (1) are not cracked, broken or loose.
3. If either of two dowel pins (1) is cracked, broken or loose, pull it out using dowel pin puller.
4. Using soft-faced hammer, tap in new dowel pin (1) until dowel pin bottoms in hole.
5. Check that all crankcase-to-cylinder-head studs (2) are not stripped, cracked, or broken.
6. If crankcase-to-cylinder-head studs (2) are damaged, get new ones.

GO TO FRAME 7

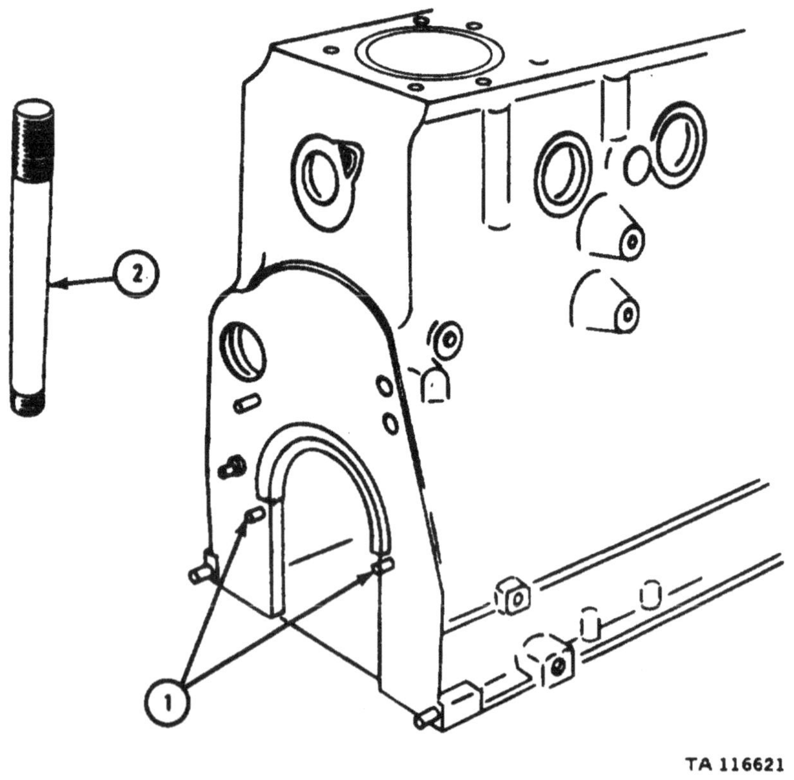

FRAME 7

WARNING

Dry cleaning solvent is flammable. Do not use near an open flame. Keep a fire extinguisher nearby when solvent is used . Use only in well-ventilated places. Failure to do this may result in injury to personnel and damage to equipment.

1. Check that dowel pins (1) have no nicks, scratches or burrs. Fix minor burrs with a fine mill file or crocus cloth dipped in cleaning solvent.
2. Check that dowel pins (1) are not cracked, broken or loose.
3. If either of two dowel pins (1) is cracked, broken or loose, pull it out using dowel pin puller.
4. Using soft-faced hammer, tap in new dowel pin (1) until dowel pin bottoms in hole.

GO TO FRAME 8

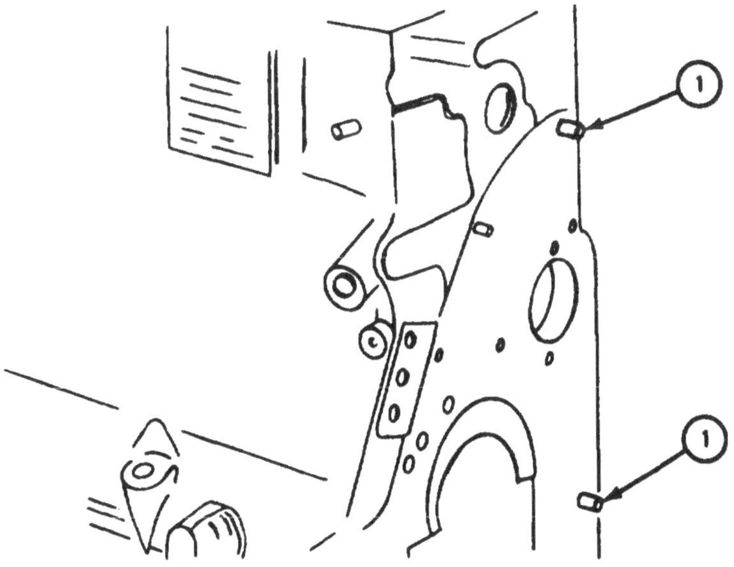

TA 116622

FRAME 8

1. Check that crankshaft gear (1) is not cracked. If crankshaft gear is cracked, throw it away and get a new one.
2. Check that teeth of crankshaft gear (1) are not worn or pitted. If crankshaft gear has pitted or badly worn teeth, throw it away and get a new one.

WARNING

Dry cleaning solvent is flammable. Do not use near an open flame. Keep a fire extinguisher nearby when solvent is used. Use only in well-ventilated places. Failure to do this may result in injury to personnel and damage to equipment.

3. Check that teeth of crankshaft gear (1) have no burrs or sharp raised metal edges. Fix small burrs and raised metal edges with a fine mill file or crocus cloth dipped in dry cleaning solvent.
4. Check that threads in two puller holes (2) in crankshaft gear (1) are not damaged. Fix minor thread damage with a tap. If threads are badly damaged, throw away crankshaft gear and get a new one.

GO TO FRAME 9

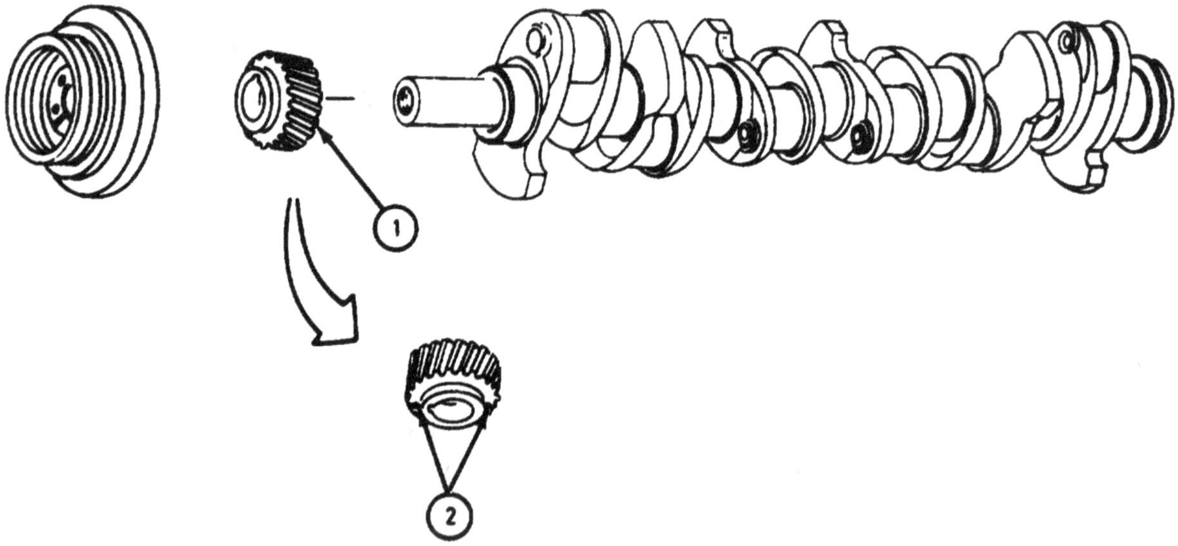

NOTE: CHECK ONLY THOSE PARTS WHICH ARE CALLED OUT IN THIS FRAME. PARTS WITHOUT CALLOUTS ARE SHOWN ONLY FOR REFERENCE PURPOSES OR ARE CHECKED IN ANOTHER FRAME.

TA 116623

FRAME 9

1. Check that crankshaft damper and pulley (1) is not cracked. If crankshaft damper and pulley is cracked, throw it away and get a new one.
2. Check that rubber insert (2) in crankshaft damper and pulley (1) is not cracked. If rubber insert is cracked, throw away damper and pulley and get a new one.

NOTE

Front and rear halves of damper and pulley (1) are held together by rubber insert (2).

3. Check that rubber insert (2) has not pulled away from front or rear half of crankshaft damper and pulley (1). If rubber insert has pulled away from front or rear half of crankshaft damper and pulley, throw away crankshaft damper and pulley and get a new one.
4. Hold rear half of crankshaft damper and pulley (1) on floor with your feet. Try to turn front half of crankshaft damper and pulley. If front half of crankshaft damper and pulley is loose, throw away crankshaft damper and pulley and get a new one.

GO TO FRAME 10

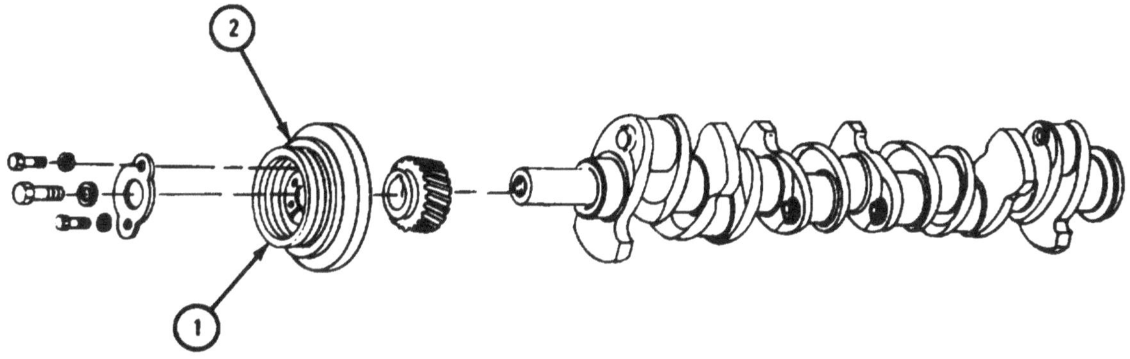

NOTE: CHECK ONLY THOSE PARTS WHICH ARE CALLED OUT IN THIS FRAME. PARTS WITHOUT CALLOUTS ARE SHOWN ONLY FOR REFERENCE PURPOSES OR ARE CHECKED IN ANOTHER FRAME.

TA 116624

FRAME 10

WARNING

Dry cleaning solvent is flammable. Do not use near an open flame. Keep a fire extinguisher nearby when solvent is used. Use only in well-ventilated places. Failure to do this may result in injury to personnel and damage to equipment.

1. Check that crankshaft damper and pulley (1) has no nicks, scratches or burrs. Fix small nicks, scratches, and burrs with a fine mill file or crocus cloth dipped in dry cleaning solvent.
2. Take out keyway seal (2) and check that keyway in crankshaft damper and pulley (1) has no burrs or cracks. Fix small burrs with a fine mill file. If keyway is cracked or badly damaged, throw away crankshaft damper and pulley and get a new one.
3. Put back keyway seal (2) so it is not lost.
4. Check that threads in two puller screw holes (3) are not damaged. Fix minor thread damage with a tap. If threads are badly damaged, throw away crankshaft damper and pulley (1) and get a new one.
5. Check that lockplate (4) is not cracked or broken. If it is cracked or broken, throw it away and get a new one.
6. Check that three screws (5) and three washers (6) are not damaged. If they are damaged, throw them away and get new ones.

IF WORKING ON ENGINE LDS-465-2, GO TO FRAME 11.
IF WORKING ON ANY ENGINE EXCEPT LDS-465-2, GO TO FRAME 12

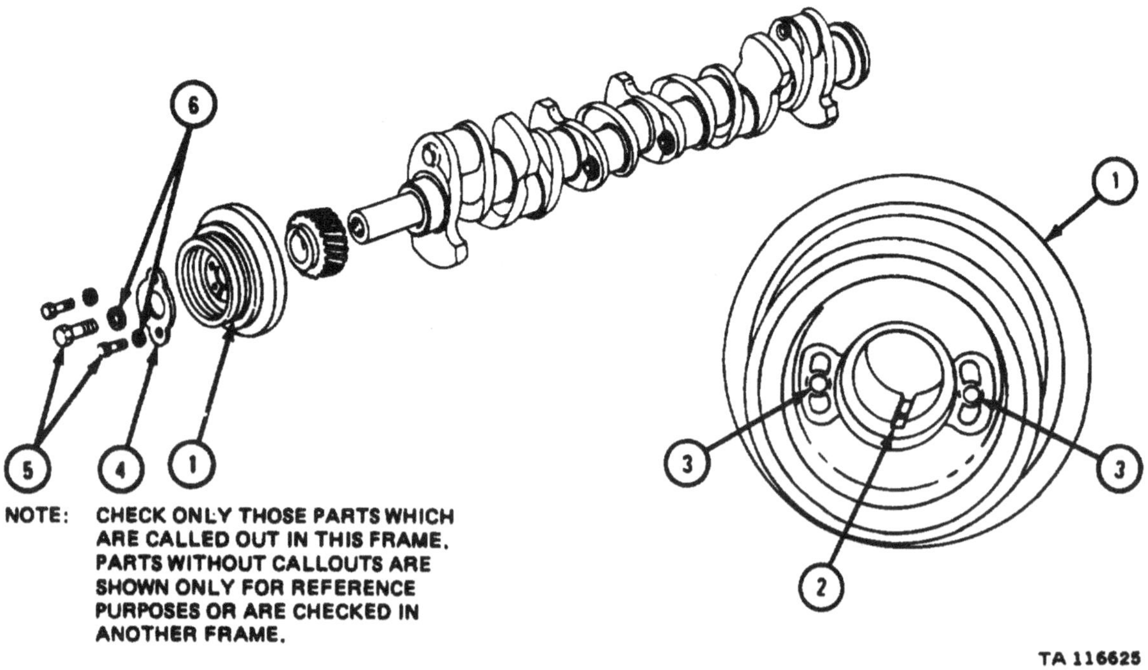

NOTE: CHECK ONLY THOSE PARTS WHICH ARE CALLED OUT IN THIS FRAME. PARTS WITHOUT CALLOUTS ARE SHOWN ONLY FOR REFERENCE PURPOSES OR ARE CHECKED IN ANOTHER FRAME.

FRAME 11

NOTE

Dirt and liquid deflector (1) is on engine LDS-465-2 only.

1. Check that dirt and liquid deflector (1) in back of crankshaft damper and pulley (2) is not loose or damaged.
2. If dirt and liquid deflector (1) is loose or damaged, throw away crankshaft damper and pulley assembly (2) and get a new one.

GO TO FRAME 12

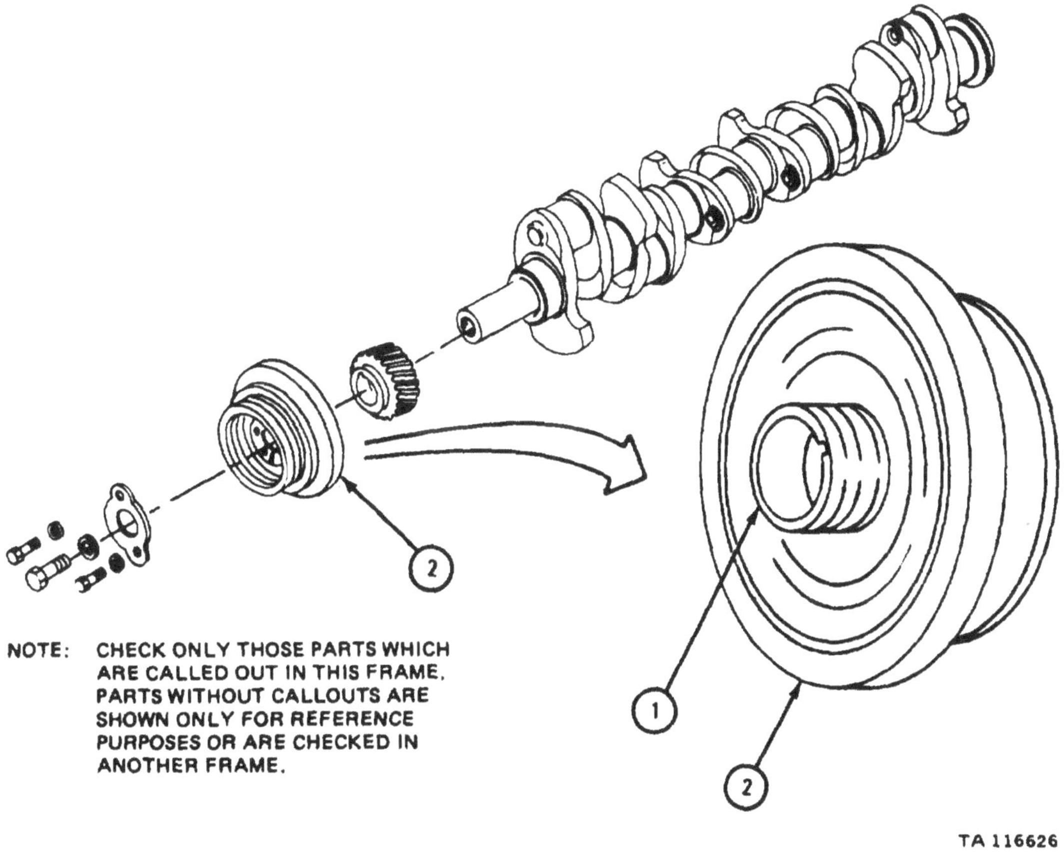

FRAME 12

CAUTION

It is easy to damage the equipment if you do not know what you are doing. Do not try to do this task unless you are experienced at it or you have an experienced person with you.

1. Use magnetic particle inspection equipment or use strong light and magnifying glass to check that crankshaft (1) is not cracked. Look extra closely around oil holes. Look extra closely at corners where crankshaft journals (2) meet crankshaft thrust faces (3) and where crankshaft cheeks (4) meet crankshaft thrust faces.
2. If crankshaft (1) is cracked, get a new one and go to frame 21.

WARNING

Dry cleaning solvent is flammable. Do not use near an open flame. Keep a fire extinguisher nearby when solvent is used. Use only in well-ventilated places. Failure to do this may result in injury to personnel and damage to equipment.

3. Check that crankshaft journals (2) are not deeply nicked, burned, grooved, scuffed or discolored from overheating. Fix very minor nicks and scratches with a crocus cloth dipped in dry cleaning solvent.
4. If crankshaft journals (2) are deeply nicked, burned, grooved, scuffed or discolored from overheating, get a new crankshaft (1).

GO TO FRAME 13

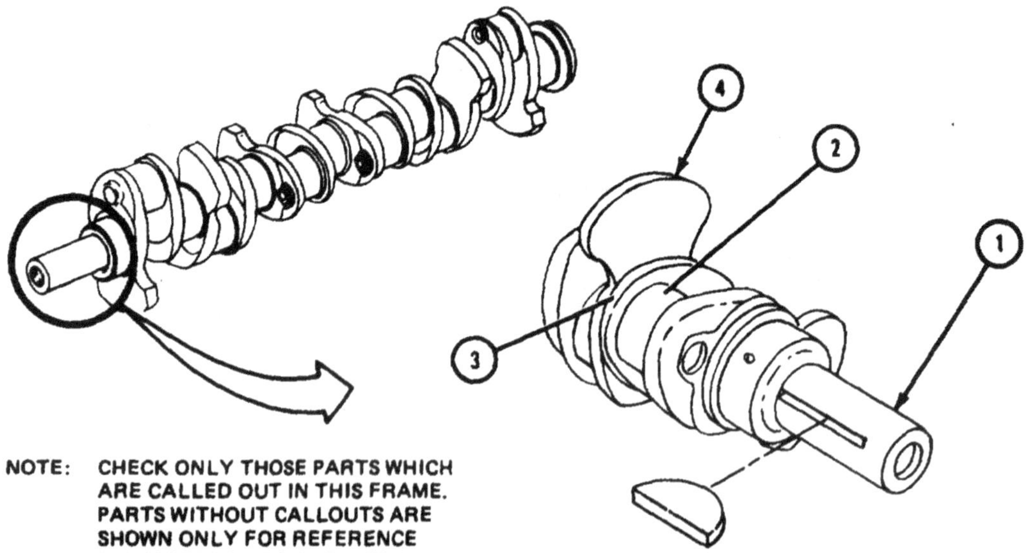

NOTE: CHECK ONLY THOSE PARTS WHICH ARE CALLED OUT IN THIS FRAME. PARTS WITHOUT CALLOUTS ARE SHOWN ONLY FOR REFERENCE PURPOSES OR ARE CHECKED IN ANOTHER FRAME.

TA 116627

FRAME 13

1. Check that crankshaft gear key (1) has no nicks. Fix minor nicks with a fine oil stone. Get a new key if it has any deep nicks.
2. Check that crankshaft gear key (1) fits snugly in keyway in crankshaft (2).
3. If crankshaft gear key (1) is loose, threw it away and put in a new one.
4. Check that crankshaft thrust faces (3) have no grooves, scuffing o r discoloration. If any thrust face is grooved, scuffed or discolored, ge t a new crankshaft.

GO TO FRAME 14

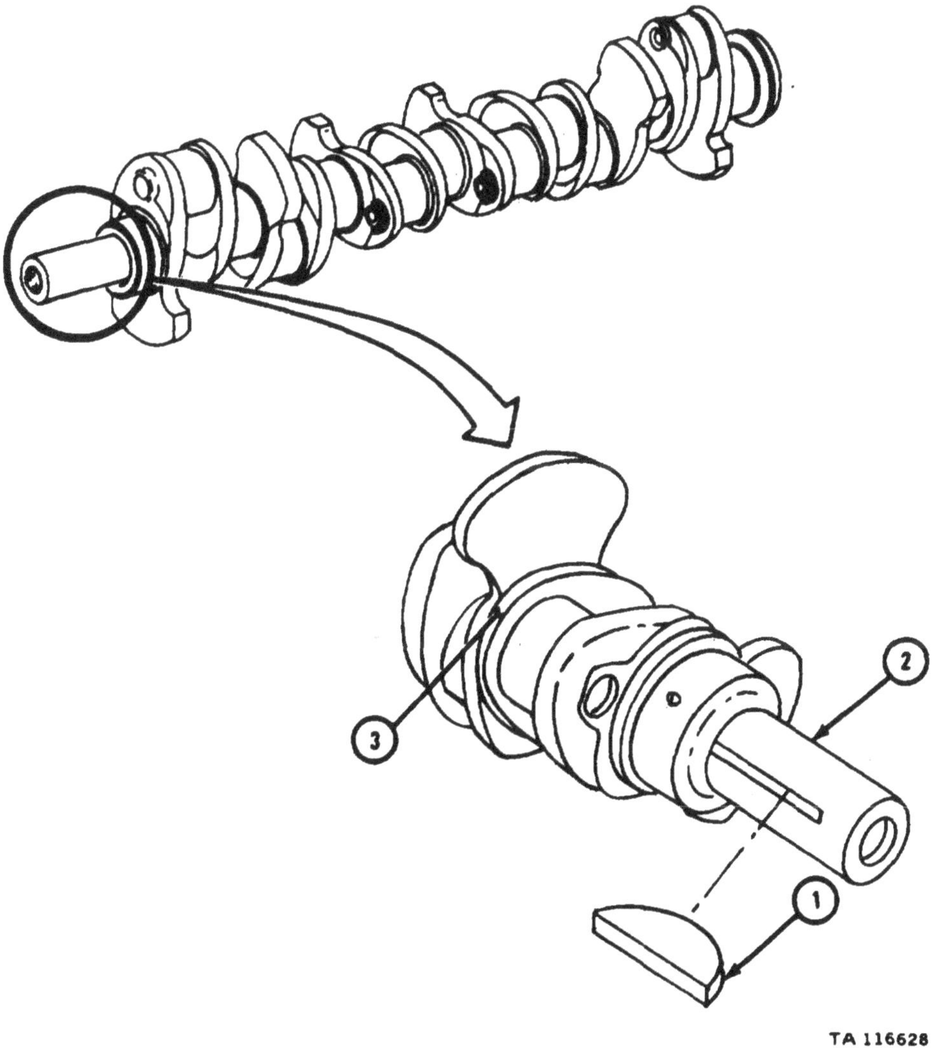

FRAME 14

CAUTION

It is easy to damage the equipment if you do not know what you are doing. Do not try to do this task unless you are experienced at it or you have an experienced person with you.

NOTE

Rear oil seal tends to wear, a groove in crankshaft flange (1) .

1. Check to see if there is a groove in crankshaft flange (1). If there is, set up dial indicator (2) as shown. For best reading, pull crankshaft away from dial indicator.
2. If groove in crankshaft flange (1) is more than 0.005 inch deep, get a new crankshaft (3) . Then go to frame 21.
3. If there is no groove or groove is less than 0.005 inch deep, check crankshaft flange (1) for minor nicks, scratches or burrs .

WARNING

Dry cleaning solvent is flammable. Do not use near an open flame. Keep a fire extinguisher nearby when solvent is used. Use only in well-ventilated places. Failure to do this may result in injury to personnel and damage to equipment.

4. If crankshaft flange (1) has minor nicks, scratches or burrs, smooth it with crocus cloth dipped in cleaning solvent.

GO TO FRAME 15

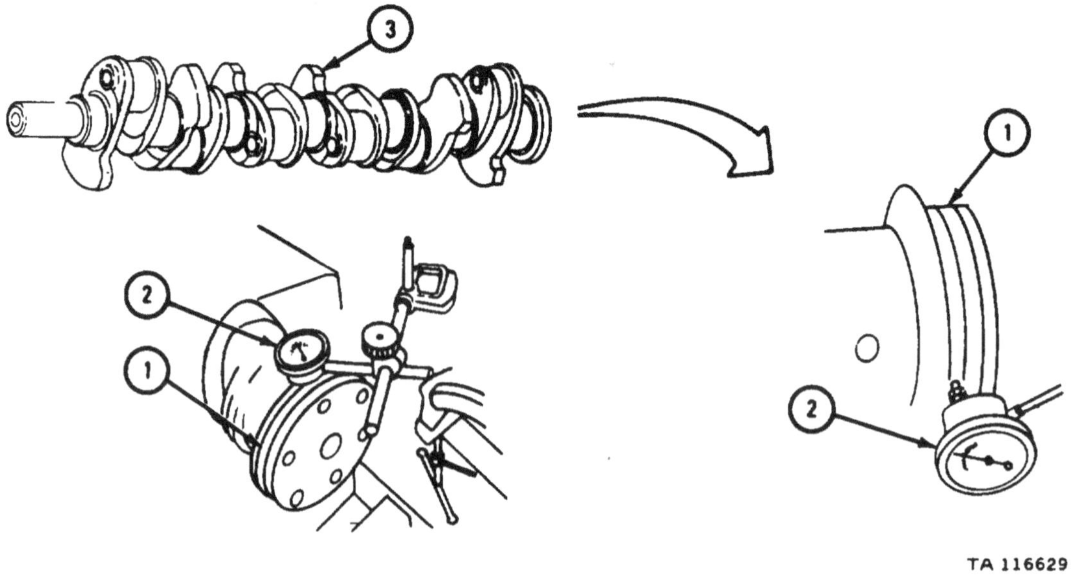

FRAME 15

CAUTION

It is easy to damage the equipment if you do not know what you are doing. Do not try to do this task unless you are experienced at it or you have an experienced person with you .

1. Check that crankshaft pilot bearing (1) is not cracked, loose or discolored from overheating . If it is cracked, loose or discolored from overheating, go to frame 16.

NOTE

Readings must be within limits given in table 4-2.

2. Measure inside diameter (2) of crankshaft pilot bearing (1).

IF READINGS ARE NOT WITHIN GIVEN LIMITS, GO TO FRAME 16.
IF READINGS ARE WITHIN GIVEN LIMITS, GO TO FRAME 18.

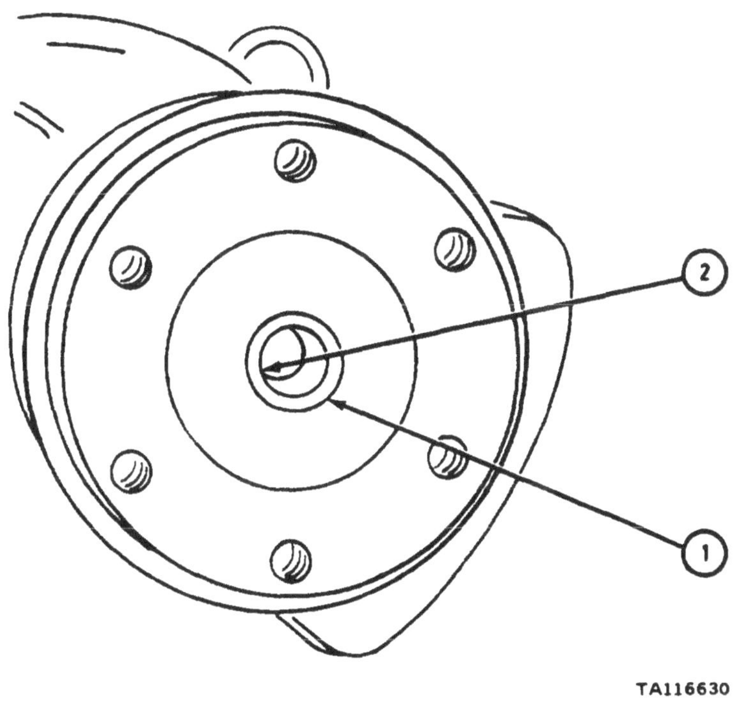

Table 4-2. Crankshaft Bushing Type Pilot Bearing Wear Limits

Index Number	Item /Point of Measurement	Size and Fit of New Parts (inches)	Wear Limits (inches)
2	Crankshaft pilot bearing inside diameter	0.7490 to 0.7505	0.7530

FRAME 16

1. Using mechanical puller (1), take out pilot bearing (2) .

GO TO FRAME 17

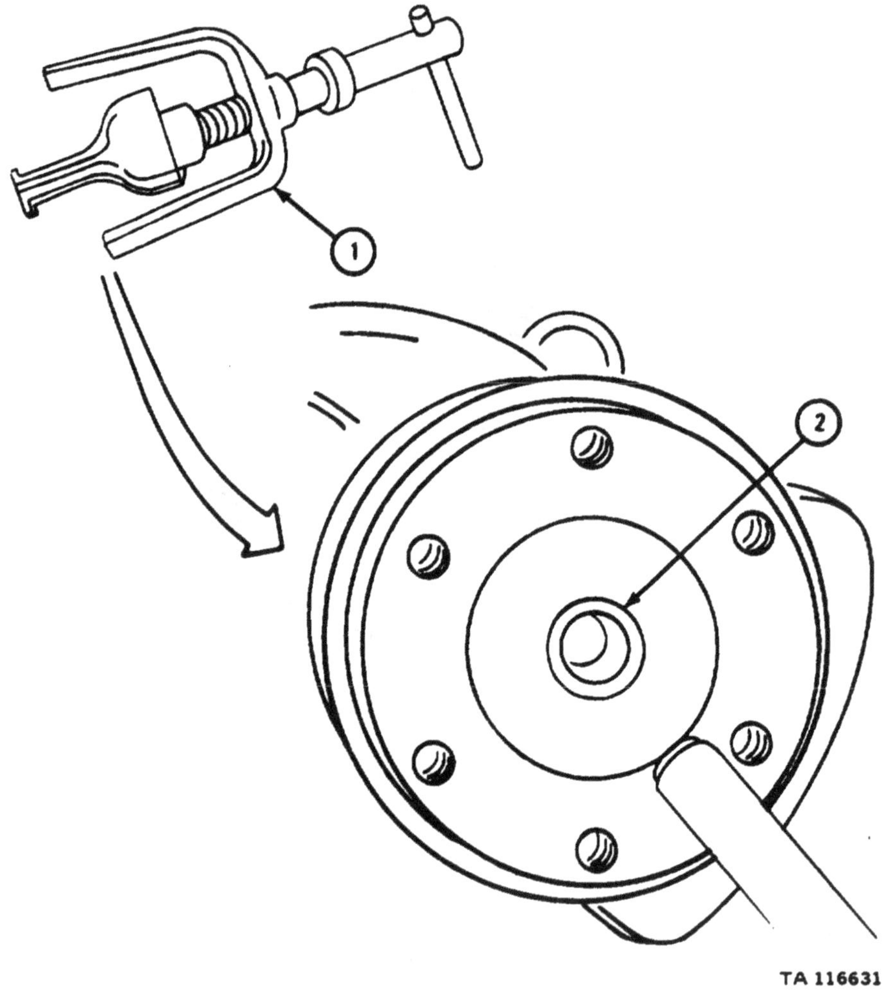

FRAME 17

NOTE

Readings must be within limits given in table 4-3. The letter T shows a tight fit. If readings are not withi n given limits, throw away part and get a new one.

1. Using gage (1) (internal measuring set) and outside micrometer (2), measure crankshaft pilot bearing inside diameter (3).
2. Using outside micrometer (2), measure crankshaft pilot bearing outside diameter (4) .
3. Using gage (1) and outside micrometer (2), measure inside diameter of crankshaft pilot bearing bore (5) .
4. Measure fit of crankshaft pilot bearing outside diameter (4) in crankshaft pilot bearing bore (5) .
5. Using heavy hammer and bearing installing tool with a pilot diameter of 0.7503 inch, put in new crankshaft pilot bearing (3).

GO TO FRAME 18

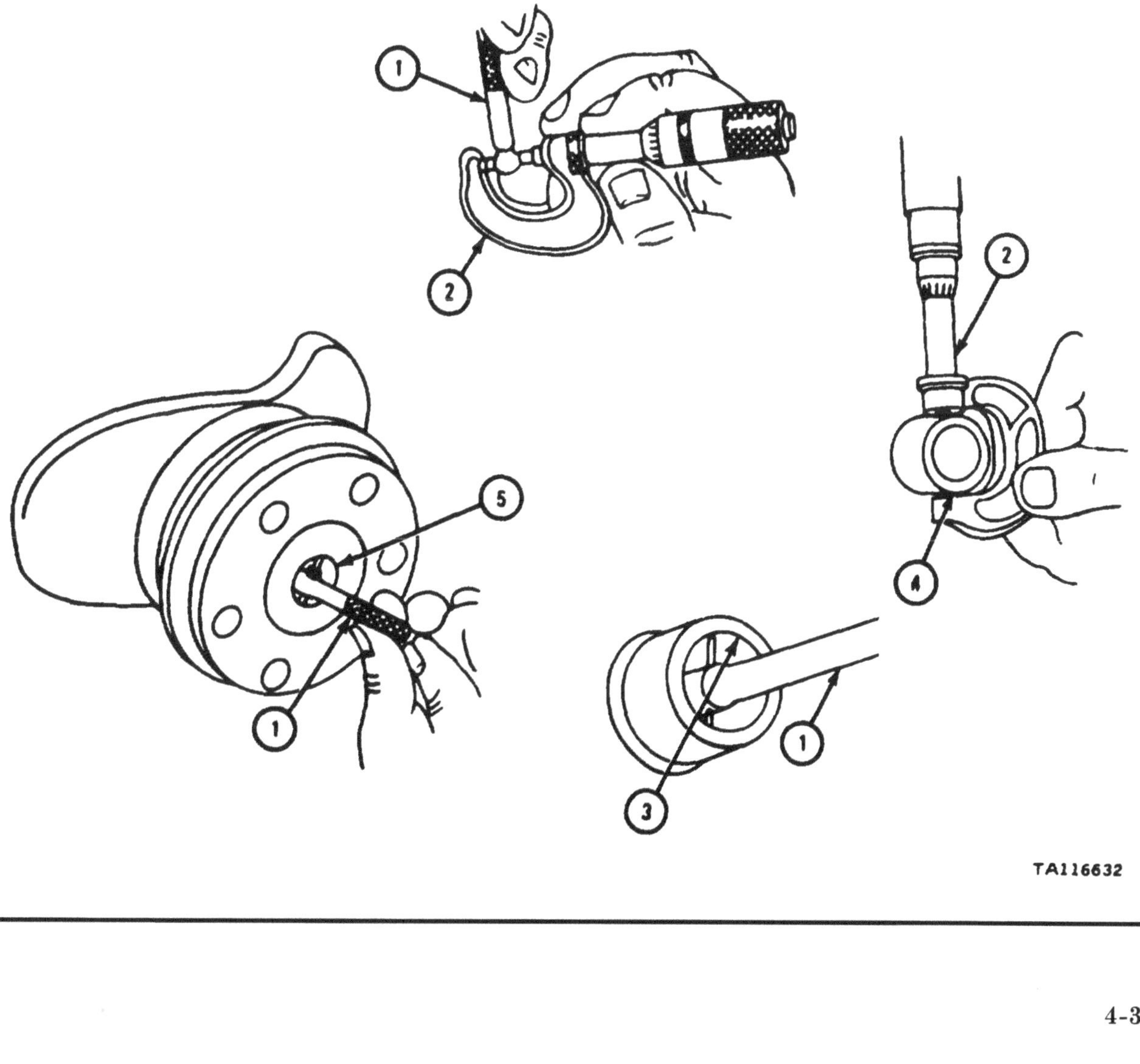

Table 4-3. Crankshaft Pilot Bearing and Bore Wear Limits

Index Number	Item /Point of Measurement	Size and Fit of New Parts (inches)	Wear Limits (inches)
3	Crankshaft pilot bearing inside diameter	0.7490 to 0.7505	0.7530
4	Crankshaft pilot bearing out side diameter	1.0030 to 1.0040	None
5	Crankshaft pilot bearing bore inside diameter	0.9990 to 1.0000	1.0010
4 and 5	Fit of crankshaft pilot bearing in crankshaft pilot bearing bore	0.0030T to 0.0050T	0.0020T to 0.0030T

FRAME 18

CAUTION

Do not use metal V-blocks to hold up crankshaft (1). Metal V-blocks will scratch crankshaft:

NOTE

Crankshaft (1) has seven main bearing journals. The y are numbered one to seven counting from front to rear of crankshaft .

1. Set crankshaft (1) down on wooden V-blocks (2). V-blocks should hold up end main bearing journals as shown.
2. Hold dial indicator base (3) firmly on floor or workbench with weights.
3. Rest dial indicator pointer (4) on number one main bearing journal (5). Set dial indicator (6) to zero.
4. Turn crankshaft (1) one full turn or more in either direction while watching dial indicator dial (6). Stop turning crankshaft when you are sure dial indicator pointer (4) rests where biggest runout reading is seen.
5. Set dial indicator dial (6) to zero again.

GO TO FRAME 19

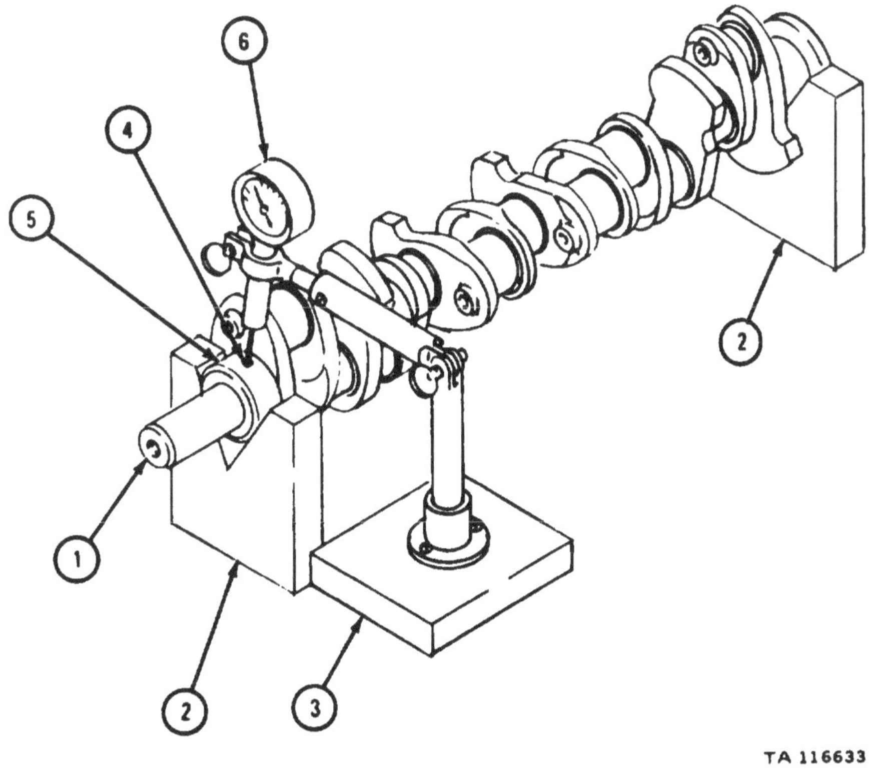

FRAME 19

NOTE

Crankshaft (1) has seven main bearing journals. They are numbered one to seven counting from front to rear of crankshaft.

1. Turn crankshaft (1) one full turn or more in either direction and note biggest runout reading seen on dial indicator dial (2). This is total indicated runout for this journal. Also note the number of the journal.
2. Do steps 2 through 5 of frame 18 and step 1 of this frame again for other six bearing journals (3 through 8).
3. Total indicated runout must not be more than 0.001 inch for any journal (3 through 9) except for number four journal (5). Total indicated runout must not be more than 0.004 inch for number four journal.
4. If total indicated runout for any journal (3 through 9) is more than limits given in step 3, get a new crankshaft (1).

GO TO FRAME 20

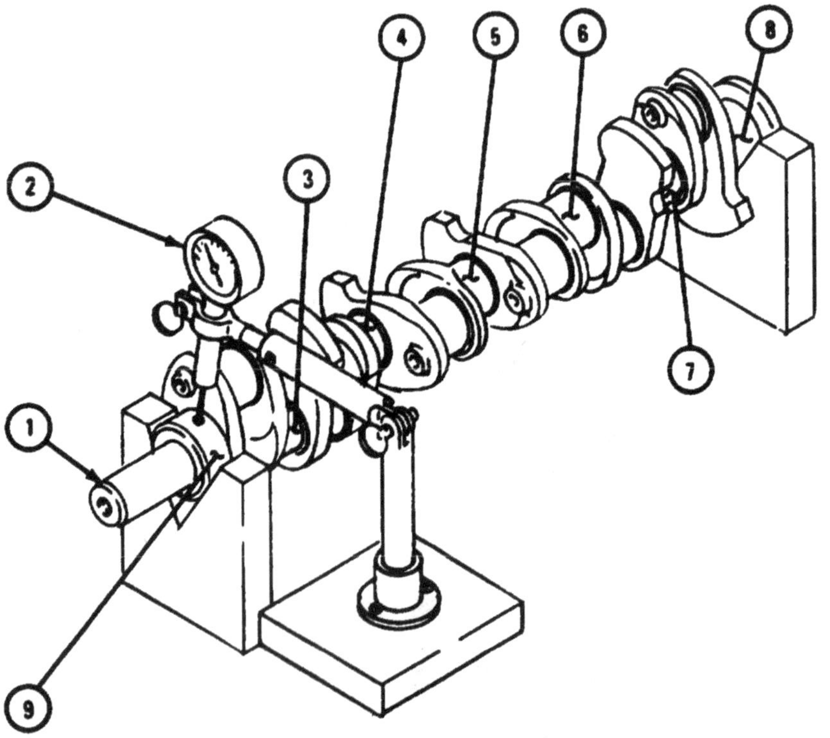

FRAME 20

CAUTION

Always support full length of crankshaft (1) to keep it from warping.

NOTE

Crankshaft (1) has seven main bearing journals. They are numbered from one to seven counting from front to rear.

1. Using micrometer, measure diameter of number one main bearing journal (2). Take measurements as close to front and rear end of main bearing journal as possible. Note measurements.
2. Note the smallest of two measurements taken in step 1 and number of journal that it was taken on (to be used in frames 21, 22, 23, and 24).
3. Do steps 1 and 2 again for other six main bearing journals (3 through 8).

GO TO FRAME 21

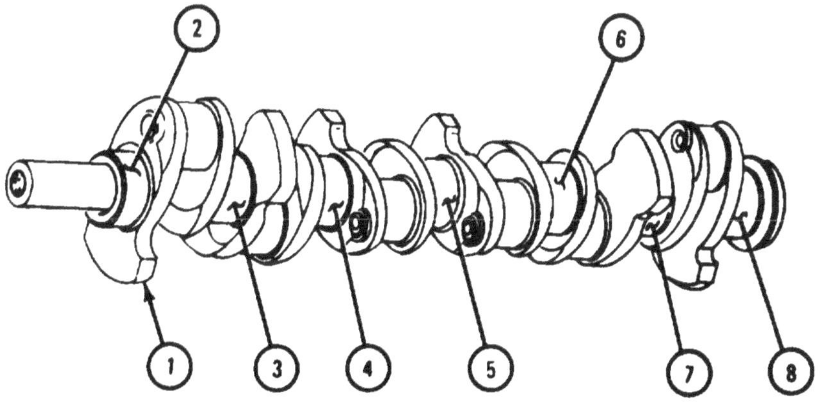

TA 116635

FRAME 21

NOTE

Four different crankshafts can be used in these engines. The diameters of main bearing journals and connecting rod journals are where they are different. The fou r crankshafts are early model standard size, late model standard size, 0.0100-inch undersize, and 0.0200-inch undersize.

1. Look at flywheel mounting face (1) of crankshaft (2). If no number is stamped here, this is a standard size crankshaft. Note this for future use.
2. Look at flywheel mounting face (1) of crankshaft (2). If the number 0.010 or 0.0100 is stamped here, this is a 0.0100-inch undersize crankshaft. Note this for future use.
3. Look at flywheel mounting face (1) of crankshaft (2). If the number 0.020 or 0.0200 is stamped here, this is a 0.0200-inch undersize crankshaft. Note this for future use.

IF WORKING ON A STANDARD SIZE CRANKSHAFT, GO TO FRAME 22.
IF WORKING ON A 0.0100-INCH UNDERSIZE CRANKSHAFT, GO TO FRAME 23.
IF WORKING ON A 0.0200-INCH UNDERSIZE CRANKSHAFT, GO TO FRAME 24

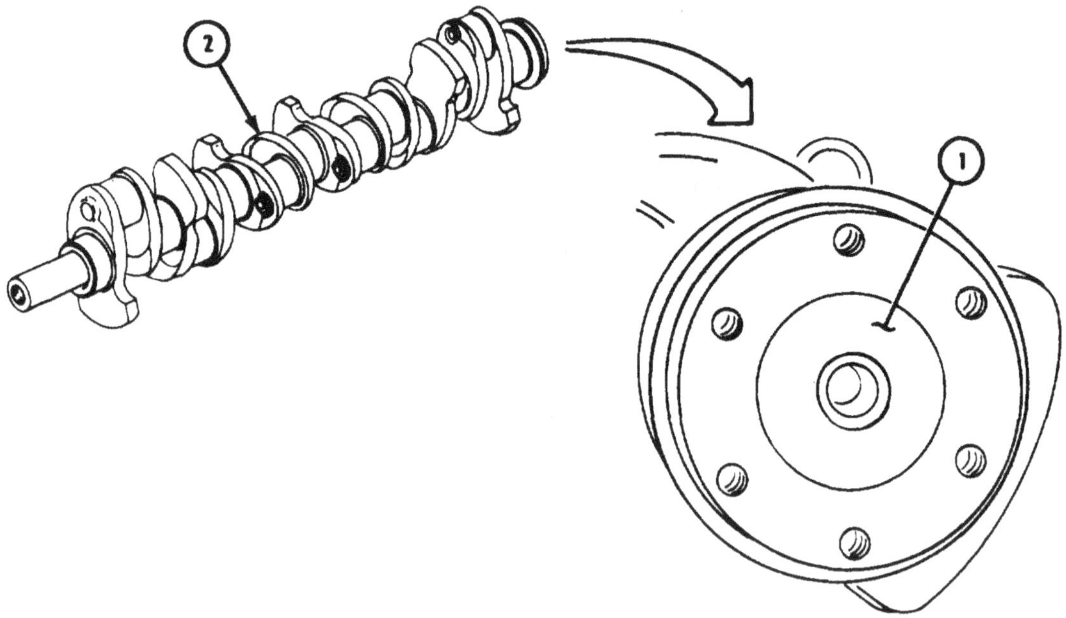

TA 116636

FRAME 22

1. Check that outside diameter measurements for seven main bearing journals (1 through 7) noted in step 3 of frame 20 are within limits given in table 4-4.

2. If any outside diameter measurement for seven main bearing journals (1 through 7) is not within limits given in table 4-4, get a new crankshaft (8).

GO TO FRAME 25

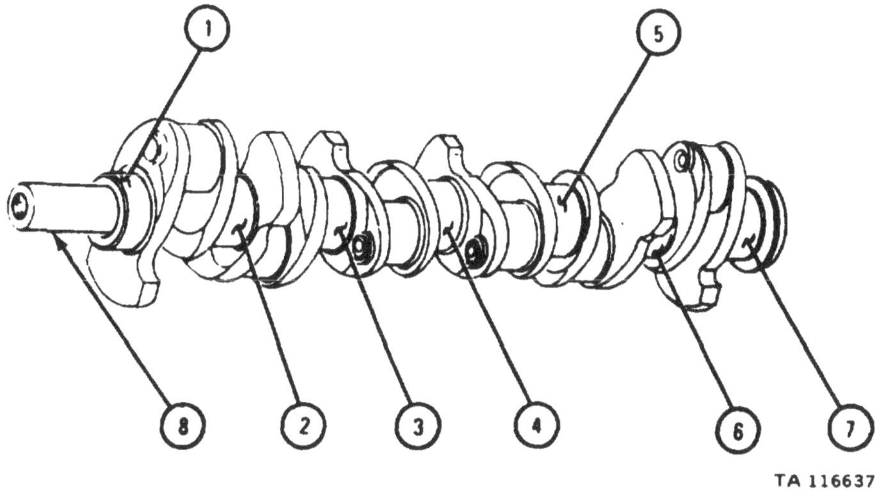

Table 4-4. Standard Size Crankshaft Main Bearing Journals Wear Limits

Index Number	Item /Point of Measurement	Size and Fit of New Parts (inches)	Wear Limits (inches)
1 through 7	Main bearing journals outside diameter		
	Late model	3.6210 to 3.6220	3.6190
	Early model	3.6200 to 3.6210	3.6190

FRAME 23

1. Check that outside diameter measurements for seven main bearing journals (1 through 7) noted in step 3 of frame 20 are within limits given in table 4-5.
2. If any outside diameter measurements for seven main bearing journals (1 through 7) is not within limits given in table 4-5, get a new crankshaft (8).

GO TO FRAME 25

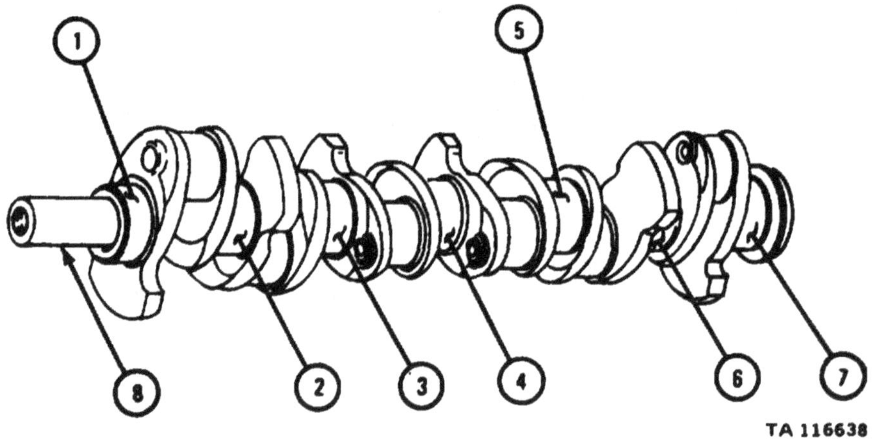

Table 4-5. 0.0100-Inch Undersize Crankshaft Main Bearing Journals Wear Limits

Index Number	Item /Point of Measurement	Size and Fit of New Parts (inches)	Wear Limits (inches)
1 through 7	Main bearing journals outside diameter	3.6110 to 3.6120	3.6090

FRAME 24

1. Check that outside diameter measurements for seven main bearing journals (1 through 7) noted in step 3 of frame 20 are within limits given in table 4-6.
2. If any outside diameter measurements for seven main bearing journals (1 through 7) is not within limits given in table 4-6, get a new crankshaft (8).

GO TO FRAME 25

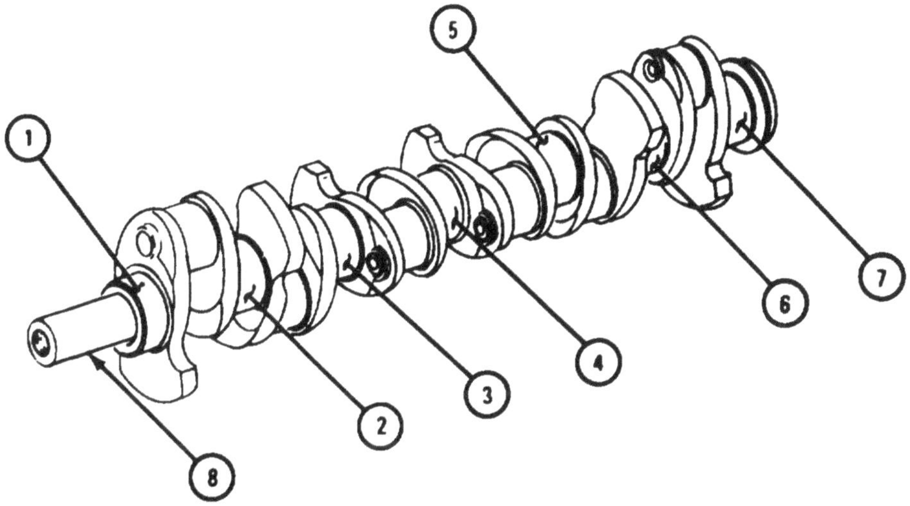

Table 4-6. 0.0200-Inch Undersize Crankshaft Main Bearing Journals Wear Limits

Index Number	Item /Point of Measurement	Size and Fit of New Parts (inches)	Wear Limit (inches)
1 through 7	Main bearing journals outside diameter	3.6010 to 3.6020	3.5990

FRAME 25

NOTE

Crankshaft (1) has six connecting rod journals. They are numbered one to six, counting from front to rear.

1. Using micrometer, measure diameter of number one connecting rod journal (2). Take measurements as close to front and rear ends of connecting rod journal as possible . Note measurement.
2. Turn micrometer 1/4 turn around number one connecting rod journal (2). Do step 1 again with micrometer in this new position. Note measurement.
3. Subtract the smaller of the two measurements taken in steps 1 and 2 from the larger of two measurements. If difference is more than 0.0010 inch, journal (2) is out of round. Get a new crankshaft (1).
4. Note the smallest of the four measurements taken in steps 1 and 2 and the number of connecting rod journal (2) it was taken on.
5. Do steps 1 through 3 again for other five connecting rod journals (3 through 7).

GO TO FRAME 26

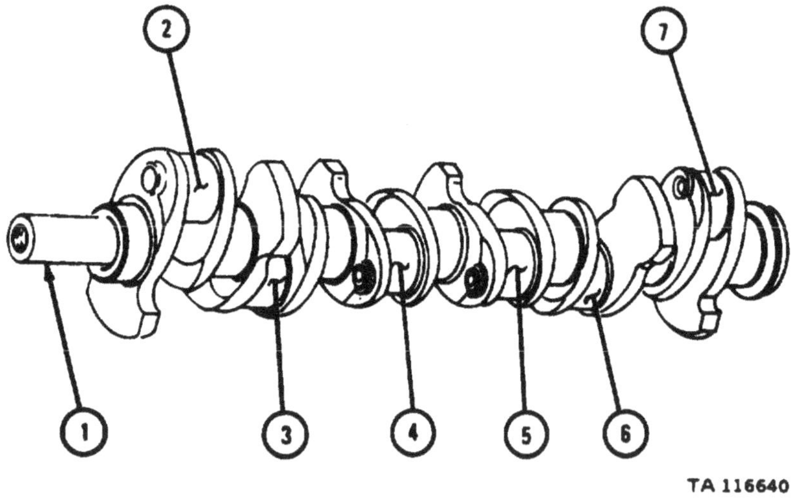

FRAME 26

1. Do frame 21 again to check size of crankshaft (1) you are working on.

IF WORKING ON STANDARD SIZE CRANKSHAFT, GO TO FRAME 27.
IF WORKING ON 0.010-INCH UNDERSIZE CRANKSHAFT, GO TO FRAME 28.
IF WORKING ON 0.0200-INCH UNDERSIZE CRANKSHAFT, GO TO FRAME 29

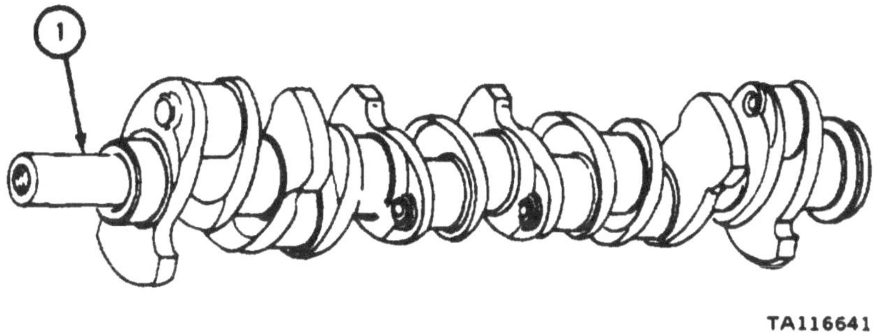

FRAME 27

1. Check that outside diameter measurements for six connecting rod journals (1 through 6) noted in step 5 of frame 25 are within limits given in table 4-7.
2. If any outside diameter measurements for six connecting rod journals (1 through 6) is not within limits given in table 4-7, get a new crankshaft (7).

GO TO FRAME 30

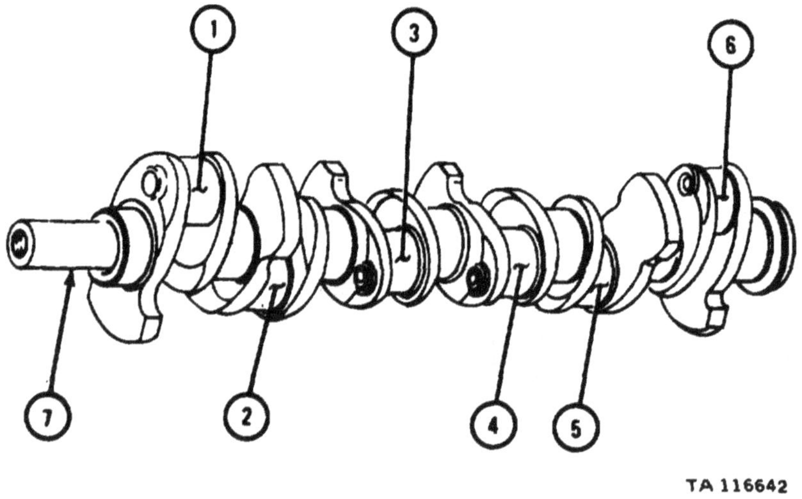

Table 4-7. Standard Size Crankshaft Connecting Rod Journals Wear Limits

Index Number	Item/Point of Measurement	Size and Fit of New Parts (inches)	Wear Limits (inches)
1 through 6	Concocting rod journa 1 outside diameter		
	Late model	2.9970 to 2.9980	2.9950
	• Early model	2.9960 to 2.9970	2.9950

FRAME 28

1. Check that outside diameter measurements for six, connecting rod journals (1 through 6) noted in step 5 of frame 25 are within limits given in table 4-8.
2. If any outside diameter measurements for six connecting rod journals (1 through 6) is not within limits given in table 4-8, get a new crankshaft (7).

GO TO FRAME 30

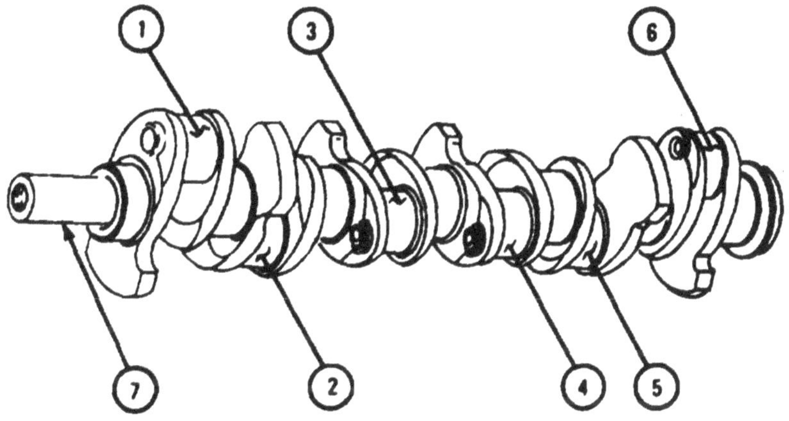

Table 4-8. 0.0100-Inch Undersize Crankshaft Connecting Rod Journals Wear Limits

Index Number	Item/Point of Measurement	Size and Fit of New Parts (inches)	Wear Limit (inches)
2 through 6	Connecting rod journal outside diameter	2.9870 to 2.9880	2.9850

FRAME 29

1. Check that outside diameter measurements for six connecting rod journals (1 through 6) noted in step 5 of frame 25 are within limits given in table 4-9.
2. If any outside diameter measurements for six connecting rod journals (1 through 6) is not within limits given in table 4-9, get a new crankshaft (7).

GO TO FRAME 30

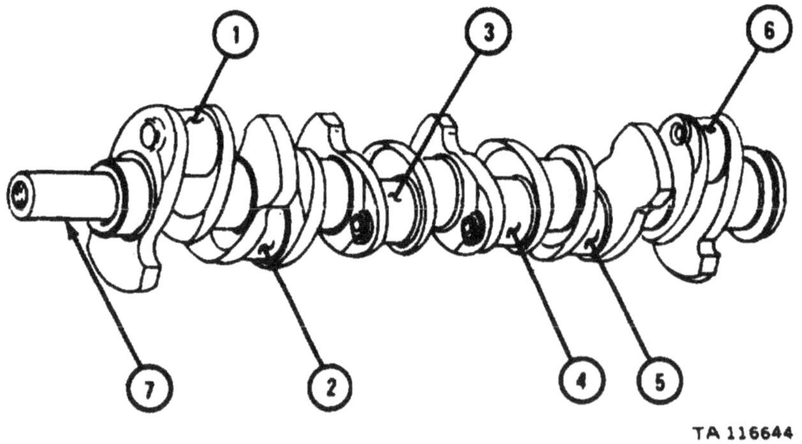

Table 4-9. 0.0200-Inch Undersize Crankshaft Connecting Rod Journals Wear Limits

Index Number	Item/Point of Measurement	Size and Fit of New Parts (inches)	Wear Limit (inches)
1 through 6	Connecting rod journal out side diameter	2.9770 to 2.9780	2.9750

FRAME 30

NOTE

Readings must be within limits given in table 4-10. The letter L shows a loose fit and the letter T shows a tight fit. If readings are not within given limits, throw away part and get a new one.

There are six main bearing journals on crankshaft. They are called by numbers one to six, counting from front to rear of crankshaft. The number four main bearing journal (3) is also called the main thrust bearing journal.

1. using gage (internal measuring set) (1) and micrometer (2), measure width of main thrust bearing journal (3).
2. Using outside micrometer (2), measure outside diameter of crankshaft pulley end (4) near rear (5) of crankshaft pulley end.
3. Using gage (1) and micrometer (2), measure inside diameter (6) of crankshaft gear (7).
4. Measure fit of crankshaft gear (7) on rear (5) of crankshaft pulley end (4).
5. Using gage (1) and micrometer (2), measure diameter of bore (8) in crankshaft damper and pulley (9).
6. Using micrometer (2), measure outside diameter of crankshaft pulley end (4) at front (10) of crankshaft pulley end.
7. Measure fit of crankshaft damper and pulley (9) on front (10) of crankshaft pulley end (4).

GO TO FRAME 31

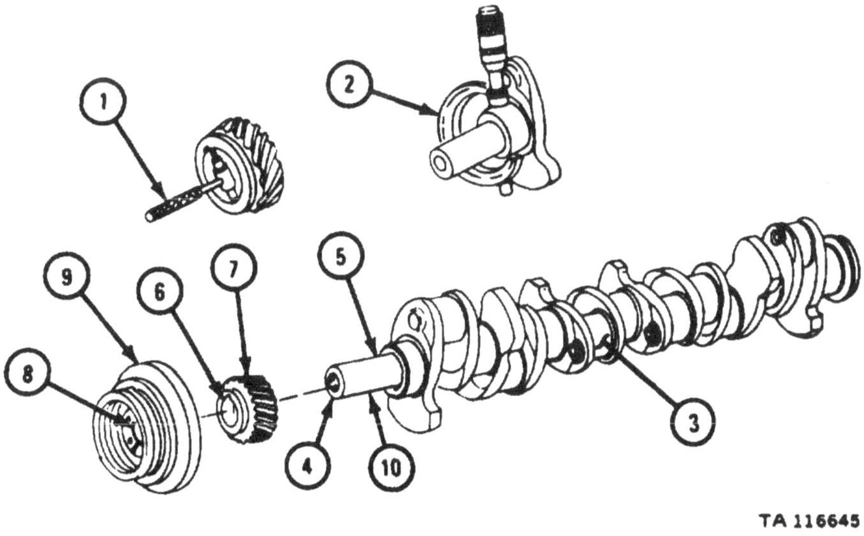

Table 4-10. Crankshaft Thrust Bearing Journal, Crankshaft Pulley End, Crankshaft Gear, and Crankshaft Damper and Pulley Wear Limits

Index Number	Item/Point of Measurement	Size and Fit of New Parts (inches)	Wear Limit (inches)
3	Crankshaft main thrust bearing journal width	2.1370 to 2.1390	2.1420
5	Crankshaft pulley end rear outside diameter	2.2492 to 2.2498	None
6	Crankshaft gear inside diameter	2.2495 to 2.2500	None
7 and 5	Fit of crankshaft gear on rear of crankshaft pulley end	0.0003T to 0.0008L	None
8	Crankshaft damper and pulley bore diameter		
	Late model	2.2478 to 2.2484	None
	Early model	2.2487 to 2.2494	None
10	Crankshaft pulley end front outside diameter	2.2487 to 2.2492	None
9 and 10	Fit of crankshaft damper and pulley on front of crankshaft pulley end		
	Late model	0.0003T to 0.0014T	None
	Early model	0.007L to 0.005T	0.0014T

FRAME 31

NOTE

Four different crankshafts (2) can be used in these engines. They differ in the diameters of the main bearing journals and connecting rod journals. The four crankshafts are : early model standard size, late model standard size, 0.0100-inch undersize, and 0.0200-inch undersize.

1. Look at flywheel mounting face (1) of crankshaft (2). If no number is stamped here, this is a standard size crankshaft. Note this for future use.
2. Look at flywheel mounting face (1) of crankshaft (2). If the number 0.010 or 0.0100 is stamped here, this is a 0.0100-inch undersize crankshaft. Note this for future use.
3. Look at flywheel mounting face (1) of crankshaft (2). If the number 0.020 or 0.0200 is stamped here, this is a 0.0200-inch undersize crankshaft. Note this for future use.

IF CRANKSHAFT (2) WAS CHANGED DURING THIS TASK, GO TO FRAME 32.
IF CRANKSHAFT (2) WAS NOT CHANGED DURING THIS TASK, GO TO FRAME 34

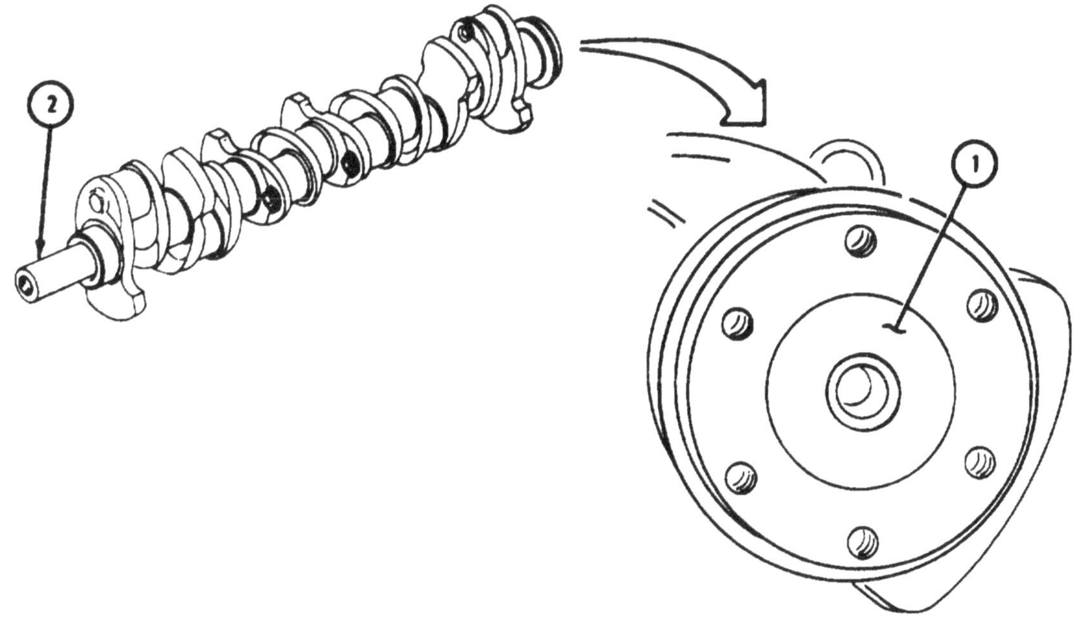

FRAME 32

NOTE

Size of crankshaft (1) taken out of engine during engine disassembly was noted in frame 21. Size of crankshaft you are working on now was noted in frame 31.

1. Check if crankshaft (1) you are working on now is same size as crankshaft taken out of engine during engine disassembly.

IF CRANKSHAFT YOU ARE WORKING ON NOW IS DIFFERENT FROM CRANKSHAFT TAKEN OUT OF ENGINE, GO TO FRAME 33.
IF CRANKSHAFT YOU ARE WORKING ON NOW IS SAME AS CRANKSHAFT TAKEN OUT OF ENGINE, GO TO FRAME 34

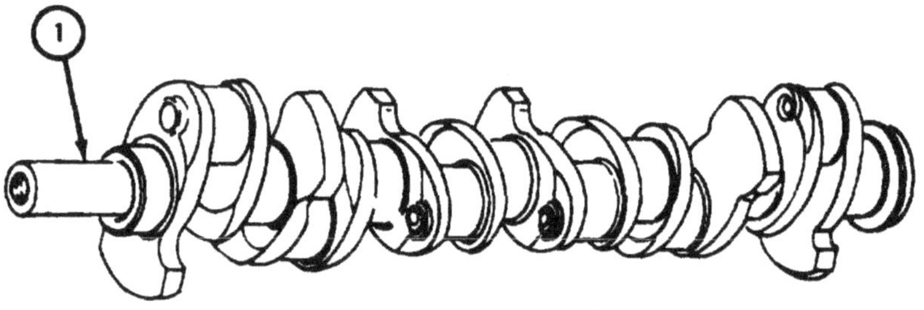

FRAME 33

1. Throw away all 14 main bearing halves (1).

NOTE

Size of crankshaft (2) you are working on now was noted in frame 31.

2. If you have an early or late model standard size crankshaft (2), get a new standard size main bearing kit.
3. If you have a 0.0100-undersize crankshaft (2), get a new 0.0100-undersize main bearing kit.
4. If You have a 0.0200-undersize crankshaft (2), get a new 0.0200-undersize main bearing kit.

GO TO FRAME 35

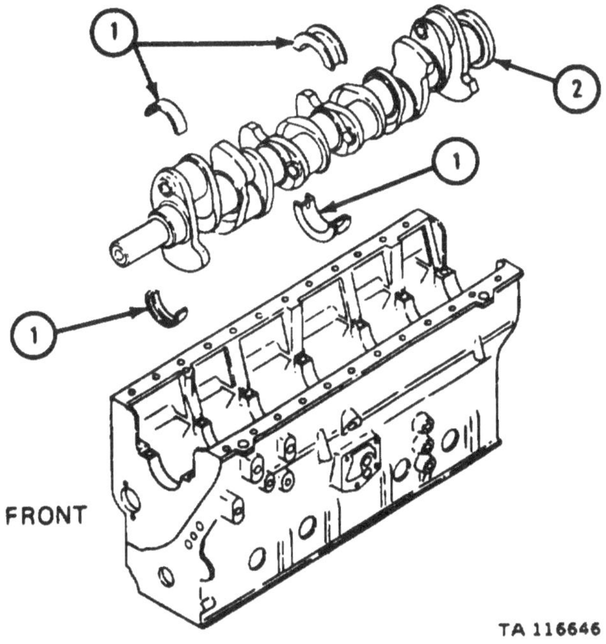

FRAME 34

NOTE

If silver-coated plating is worn off bearing half (1), brass color of bearing underneath plating coat will be seen .

1. Check that none of the 14 main bearing halves (1) is bent, scored or pitted. Check that silver-coated plating is not worn off any of 14 main bearing halves.
2. If any main bearing half (1) is damaged or worn, go to frame 33.

GO TO FRAME 35

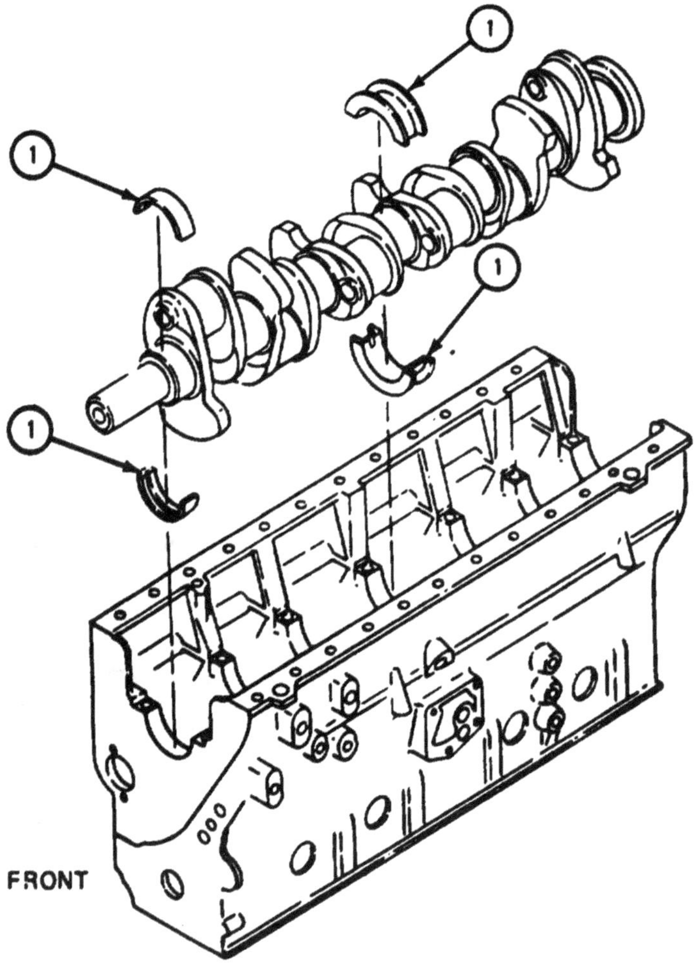

CHECK ONLY THOSE PARTS WHICH ARE CALLED OUT. PARTS WITHOUT CALLOUTS ARE SHOWN ONLY FOR REFERENCE PURPOSES OR ARE CHECKED IN ANOTHER FRAME.

TA 116647

FRAME 35

NOTE

Main thrust bearing halves (1 and 2) are different from other main bearing halves (3). Main thrust bearing halves have a flange (4) on each side.

1. Measure inside width A of main thrust bearing half (2) between flanges (4). If reading is not within limits given in table 4-11, throw away both main bearing halves (1 and 2) and all 12 main bearing halves (3).
2. Measure width of main thrust bearing cap (5). If reading is not within limits given in table 4-11, throw away crankcase (6) and all seven main bearing caps (5 and 7) and get a new crankcase.

NOTE

New crankcase (6) comes with new bearing caps (5 and 7).

3. Measure fit of main thrust bearing half (2) on main thrust bearing cap (5). If reading is not within limits given in table 4-11, do steps 1 and 2 again. If reading is still not within limits, throw away crankcase (6) and all seven main bearing caps (5 and 7) and get a new crankcase.

IF BEARING HALVES WERE NOT THROWN AWAY, GO TO FRAME 36.
IF BEARING HALVES WERE THROWN AWAY, GO TO FRAME 38

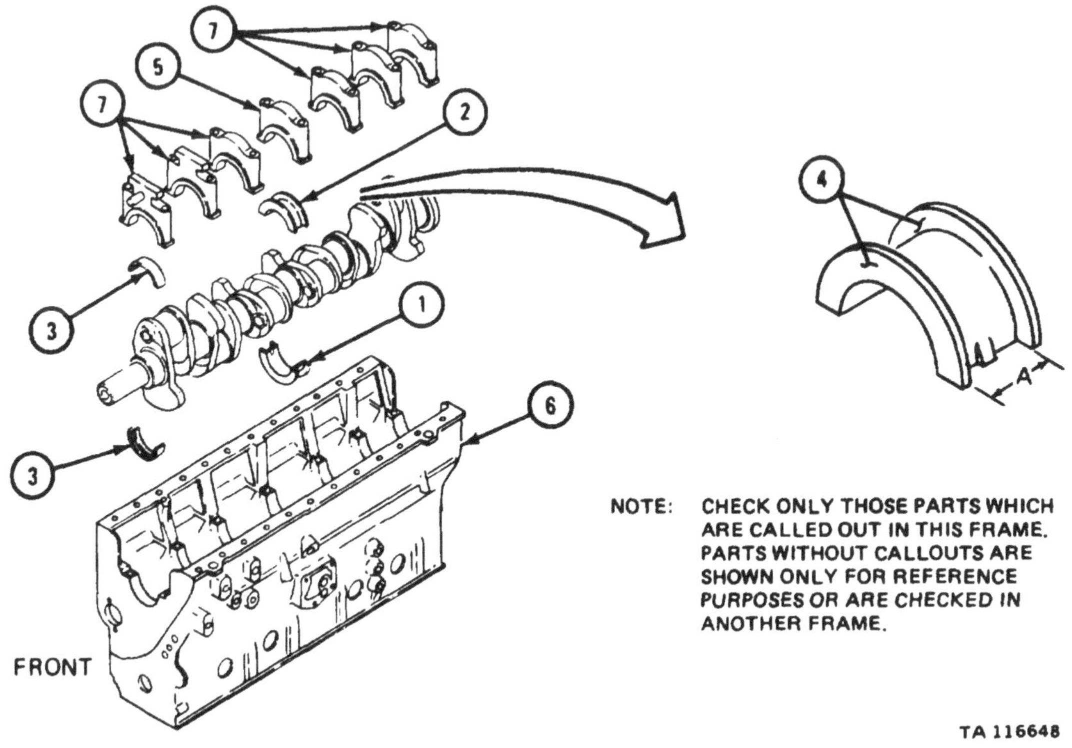
7
5
7
2
4
A
3
1
6
3
FRONT
NOTE: CHECK ONLY THOSE PARTS WHICH ARE CALLED OUT IN THIS FRAME. PARTS WITHOUT CALLOUTS ARE SHOWN ONLY FOR REFERENCE PURPOSES OR ARE CHECKED IN ANOTHER FRAME.
TA 116648

Table 4-11. Main Thrust Bearing Half Wear Limits

Index Number	Item/Point of Measurement	Size and Fit of New Parts (inches)	Wear Limit (inches)
2	Main thrust bearing half inside width A	1.8010 to 1.8045	None
5	Main thrust bearing cap width	1.7970 to 1.7990	None
2 and 5	Pit of main thrust bearing half on main thrust bearing cap	0.0020 to 0.0075	None

FRAME 36

NOTE

Main thrust bearing halves (1 and 2) are different from other main bearing halves (3). Main thrust bearing halves have a flange (4) on each side.

1. Measure inside width A of main thrust bearing half (1) between flanges (4). If reading is not within limits given in table 4-12, throw away both main thrust bearing halves (1 and 2) and all 12 main bearing halves (3).
2. Measure width of thrust bearing saddle (5). If reading is not within limits given in table 4-12, throw away crankcase (6) and seven main bearing caps (7) and get new ones.

NOTE

New crankcase (6) comes with new bearing caps (7).

3. Measure fit of main thrust bearing half (1) on thrust bearing saddle (5). If reading is not within limits given-in table 4-12, do steps 1 and 2 again. If reading is still not within limits, throw away crankcase (6), seven main bearing caps (7), both main thrust bearing halves (1 and 2), and all 12 main bearing halves (3), and get a new crankcase .

IF BEARING HALVES WERE NOT THROWN AWAY, GO TO FRAME 37.
IF BEARING HALVES WERE THROWN AWAY, GO TO FRAME 38

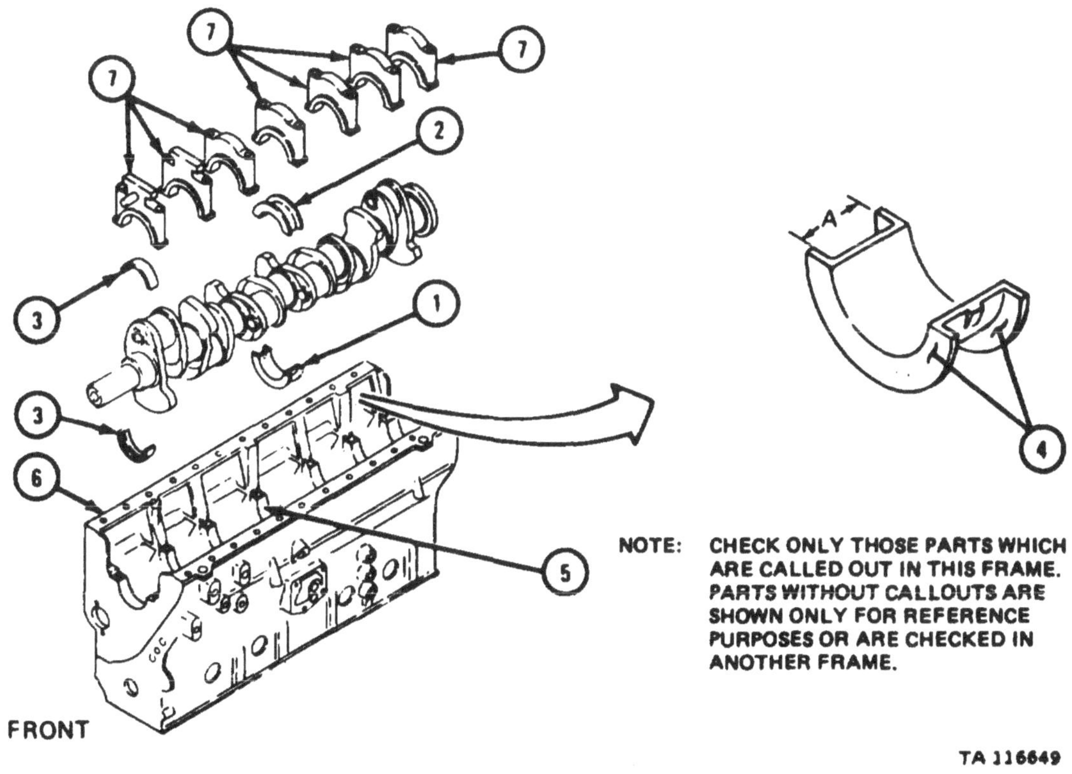
7
7
7
2
A
3
1
3
6
4
5
FRONT
NOTE: CHECK ONLY THOSE PARTS WHICH ARE CALLED OUT IN THIS FRAME. PARTS WITHOUT CALLOUTS ARE SHOWN ONLY FOR REFERENCE PURPOSES OR ARE CHECKED IN ANOTHER FRAME.
TA 116649

Table 4-12. Main Thrust Bearing Half and Saddle Wear Limits

Index Number	Item/Point of Measurement	Size and Fit of New Parts (inches)	Wear Limit (inches)
1	Main thrust bearing half inside width A	1.8010 to 1.8045	None
5	Main thrust bearing saddle	1.7970 to 1.7990	None
1 and 5	Fit of main thrust bearing half on main thrust bearing saddle	0.0020L to 0.0075L	None

FRAME 37

NOTE

Main thrust bearing halves (1) are different from other main bearing halves (2). Main thrust bearing halves have a flange (3) on each side.

Readings must be within limits given in table 4-13. The letter L shows a loose fit and the letter T shows a tight fit. If readings are not within given limits, throw away main thrust bearing halves (1) and 12 main bearing halves (2). Do not get new ones at this time.

1. Measure width of main thrust bearing journal (4).
2. Measure outside width A of both thrust bearing halves (2).
3. Measure end play between outside width A of each thrust bearing half (2) and width of thrust main bearing journal (4).
4. Measure thickness of main thrust bearing flanges (3) for both main thrust bearing halves (1).

IF MAIN BEARING HALVES WERE THROWN AWAY, GO TO FRAME 38.
IF MAIN BEARING HALVES WERE NOT THROWN AWAY, GO TO FRAME 39

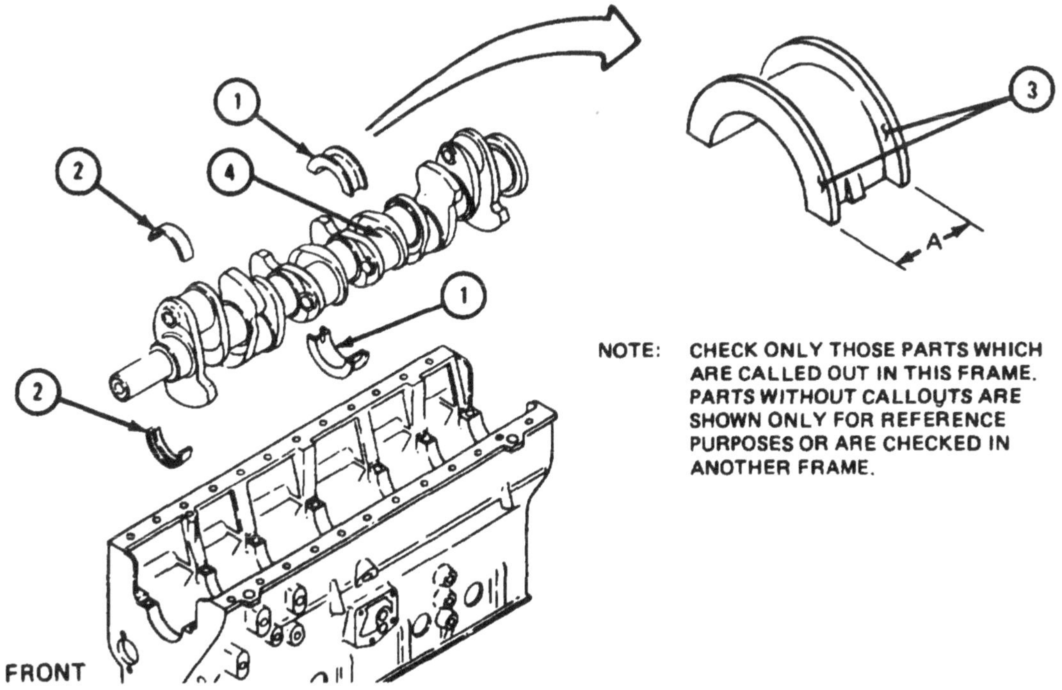

NOTE: CHECK ONLY THOSE PARTS WHICH ARE CALLED OUT IN THIS FRAME. PARTS WITHOUT CALLOUTS ARE SHOWN ONLY FOR REFERENCE PURPOSES OR ARE CHECKED IN ANOTHER FRAME.

TA 116650

Table 4-13. Crankshaft Main Thrust Bearings and Journal Wear Limits

Index Number	Item/Point of Measurement	Size and Pit of New Parts (inches)	Wear Limits (inches)
2	Thrust bearing half outside width A	2.1240 to 2.1270	2.1200
3	Main thrust bearing half flange width	0.1597 to 0.1630	0.1570
4	Thrust main bearing journal width	2.1370 to 2.1390	2.1420
4 and 2	End play between thrust bearing half and main bearing journal	0.0100L to 0.0150L	0.0220L

FRAME 38

NOTE

Four different crankshafts (1) can be used in these engines. They differ in the diameters of the main bearing journals and connecting rod journals. The four crankshafts are : early model standard size, late model standard size, 0.0100-inch undersize and 0.0200-inch undersize .

1. If you have any early or late model standard size crankshaft (1), get a new standard size main bearing kit.
2. If you have a 0.0100-undersize crankshaft (1), get a new 0.0100 undersize main bearing kit.
3. If you have a 0.0200-undersize crankshaft (1), get a new 0.0200-undersize m&i bearing kit .

IF WORKING ON STANDARD SIZE CRANKSHAFT, GO TO FRAME 39.
IF WORKING ON 0.0100-INCH UNDERSIZE CRANKSHAFT, GO TO FRAME 40.
IF WORKING ON 0.0200-INCH UNDERSIZE CRANKSHAFT, GO TO FRAME 41

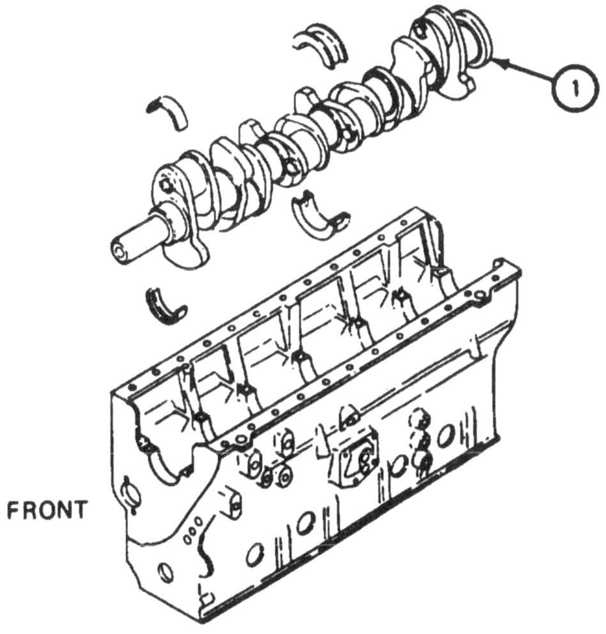

NOTE

CHECK ONLY THOSE PARTS WHICH ARE CALLED OUT IN THIS FRAME. PARTS WITHOUT CALLOUTS ARE SHOWN ONLY FOR REFERENCE PURPOSES OR ARE CHECKED IN ANOTHER FRAME.

TA 116646

FRAME 39

NOTE

Main thrust bearing halves (1) are different from other main bearing halves (2). Main thrust bearing halves have a flange (3) on each side. Wear limits are same for both kinds of main bearing half.

Readings must be within limits given in table 4-14. If readings are not within given limits, throw away all 14 main bearing halves (1 and 2) and get a new standard size main bearing kit.

1. Measure thickness of all 14 main bearing halves (1 and 2).
2. If main bearing halves (1 and 2) were changed, do step 1 again for new main bearing halves.

GO TO FRAME 42

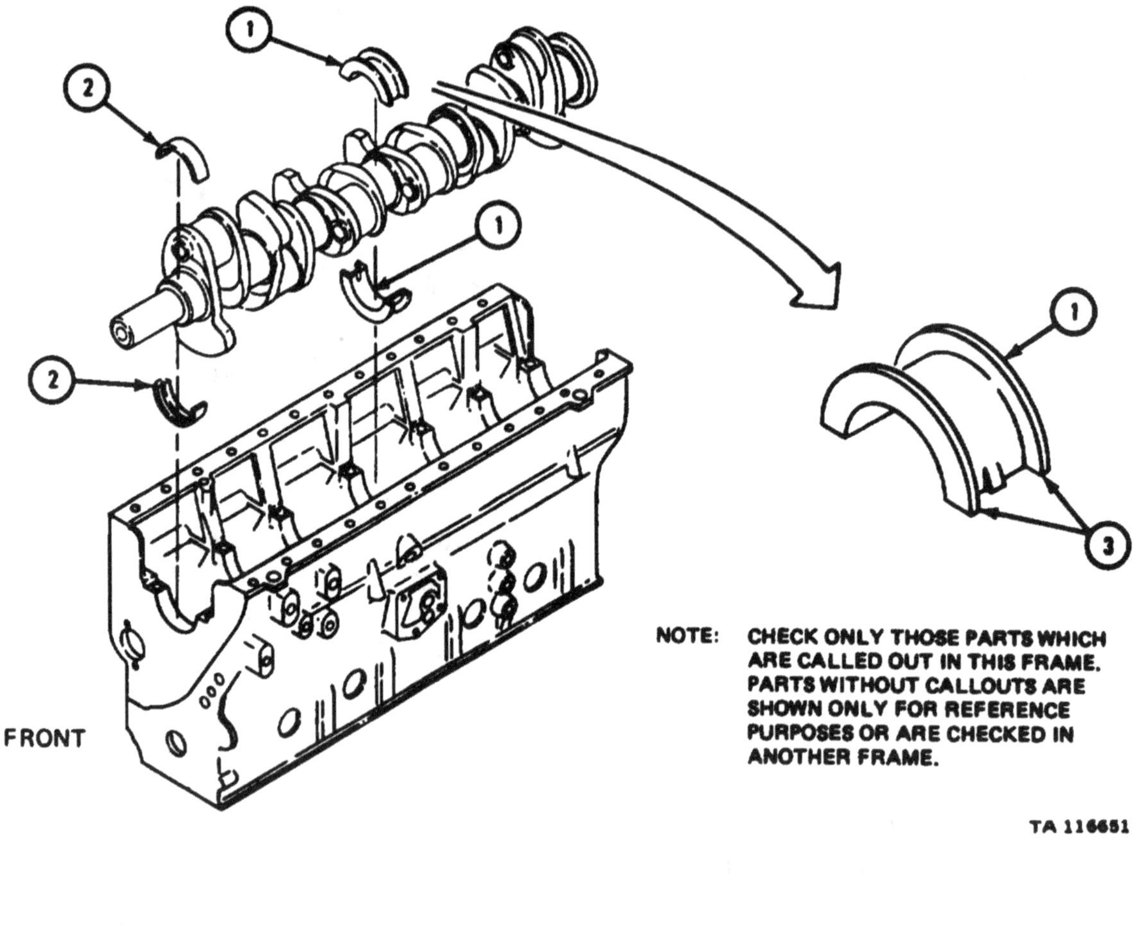
1
2
1
2
1
3
FRONT
NOTE: CHECK ONLY THOSE PARTS WHICH ARE CALLED OUT IN THIS FRAME. PARTS WITHOUT CALLOUTS ARE SHOWN ONLY FOR REFERENCE PURPOSES OR ARE CHECKED IN ANOTHER FRAME.
TA 116651

Table 4-14. Standard Size Crankshaft Main Bearing Half Wear Limits

Index Number	Item/Point of Measurement	Size and Fit of New Parts (inches)	Wear Limit (inches)
1	Main bearing half thickness	0.1554 to 0.1559	0.1544
2	Main thrust bearing half thickness	0.1554 to 0.1559	0.1544

FRAME 40

NOTE

Main thrust bearing halves (1) are different from other main bearing halves (2). Main thrust bearing halves have a flange (3) on each side. Wear limits are same for both kinds of main bearing halves.

Readings must be within limits given in table 4-15. If readings are not within given limits, throw away all 14 main bearing halves (1 and 2) and get a new 0.0100-undersize main bearing kit.

1. Measure thickness of all 14 main bearing halves (1 and 2).
2. If main bearing halves (1 and 2) were changed, do step 1 again for new main bearing halves.

GO TO FRAME 42

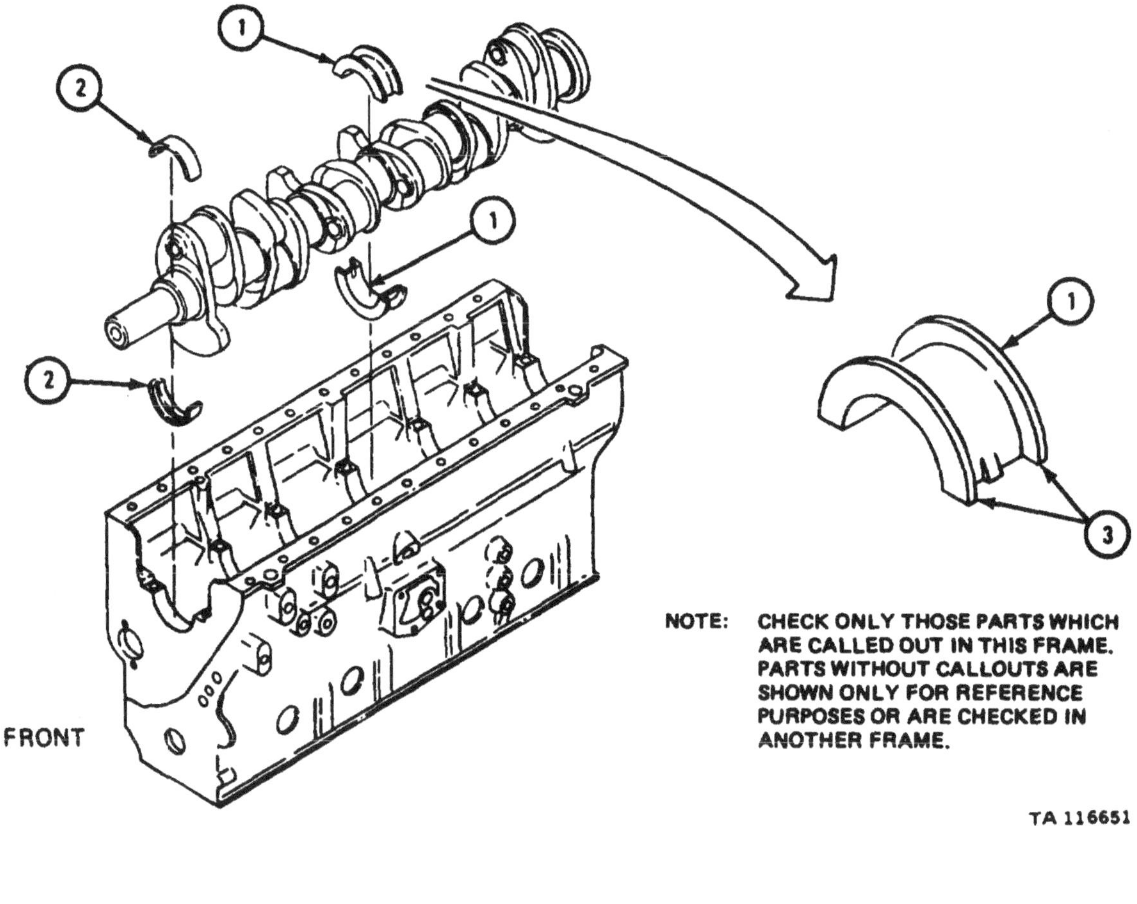
1
2
1
2
1
3
FRONT
NOTE: CHECK ONLY THOSE PARTS WHICH ARE CALLED OUT IN THIS FRAME. PARTS WITHOUT CALLOUTS ARE SHOWN ONLY FOR REFERENCE PURPOSES OR ARE CHECKED IN ANOTHER FRAME.
TA 116651

Table 4-15. 0.0100-Inch Undersize Crankshaft Main Bearing Half Wear Limits

Index Number	Item /Point of Measurement	Size and Fit of New Parts (inches)	Wear Limit (inches)
1	Main bearing half thickness	0.1604 to 0.1609	0.1594
2	Thrust main bearing half thickness	0.1604 to 0.1609	0.1594

FRAME 41

NOTE

Main thrust bearing halves (1) are different from other main bearing halves (2). Main thrust bearing halves have a flange (3) on each side. Wear limits are same for both kinds of main bearing half.

Readings must be within limits given in table 4-16. If readings are not within given limits, throw away all 14 main bearing halves (1 and 2) and get a new 0. 0200-undersize main bearing kit.

1. Measure thickness of all 14 main bearing halves (1 and 2).
2. If main bearing halves (1 and 2) were changed, do step 1 again for new main bearing halves .

GO TO FRAME 42

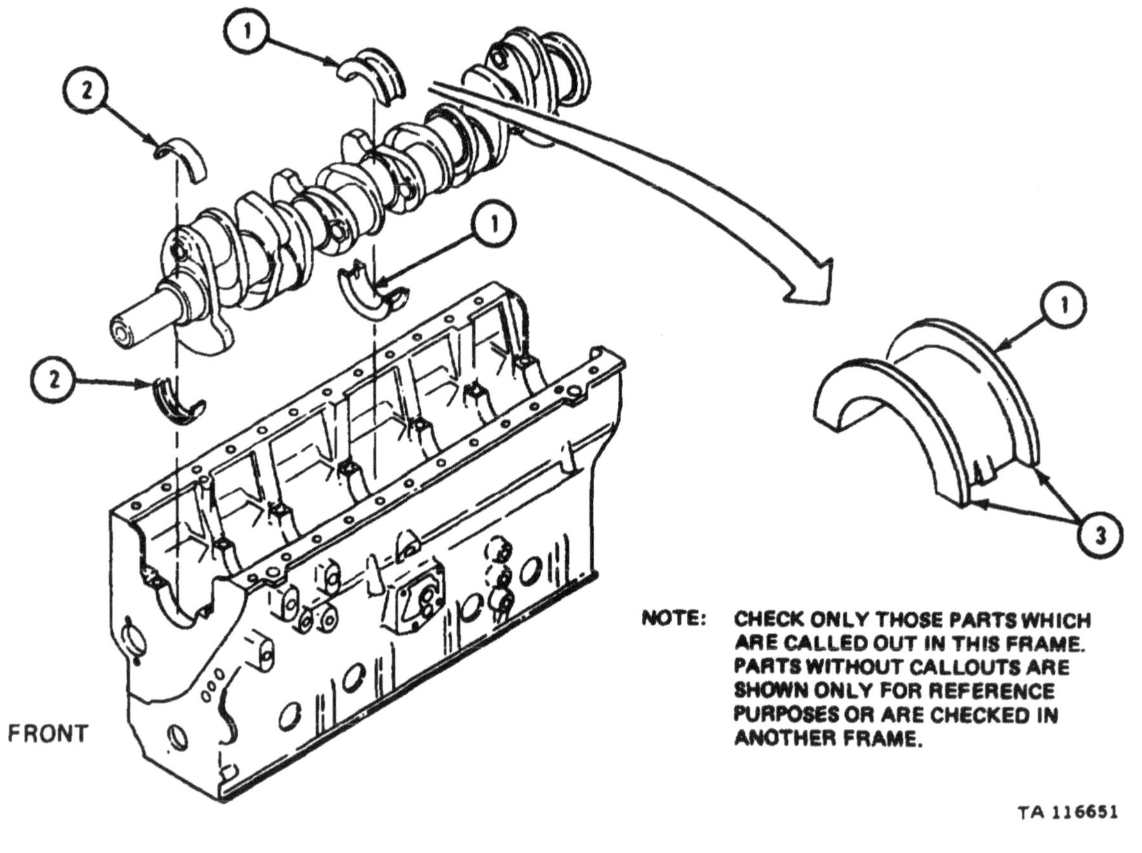
1
2
1
2
1
3
FRONT
NOTE: CHECK ONLY THOSE PARTS WHICH ARE CALLED OUT IN THIS FRAME. PARTS WITHOUT CALLOUTS ARE SHOWN ONLY FOR REFERENCE PURPOSES OR ARE CHECKED IN ANOTHER FRAME.
TA 116651

Table 4-16. 0. 0200-Inch Undersize Crankshaft Main Bearing Half Wear Limits

Index Number	Item /Point *of* Measurement	Size and Pit of New Parts (inches)	Wear Limit (inches)
1	Main bearing half thickness	0.1654 to 0.1659	0.1644
2	Thrust main bearing half thickness	0.1654 to 0.1659	0.1644

FRAME 42

NOTE

Main bearing caps (1) and bearing saddles (2) are called by numbers one to seven counting from front to rear of crankcase (3).

The number one main bearing cap (1) has oil pump idler gear shaft (4).

1. Put number one main bearing cap (1) on number one bearing saddle (2) with oil pump idler gear shaft (4) facing out as shown.
2. Coat two bolts (5) and washers (6) lightly with engine lubricating oil. Put in and hand tighten two bolts and washers.
3. Using torque wrench, tighten two bolts (5) to 80 to 90 pound-feet.
4. Using torque wrench, tighten two bolts (5) to 115 to 120 pound-feet.

GO TO FRAME 43

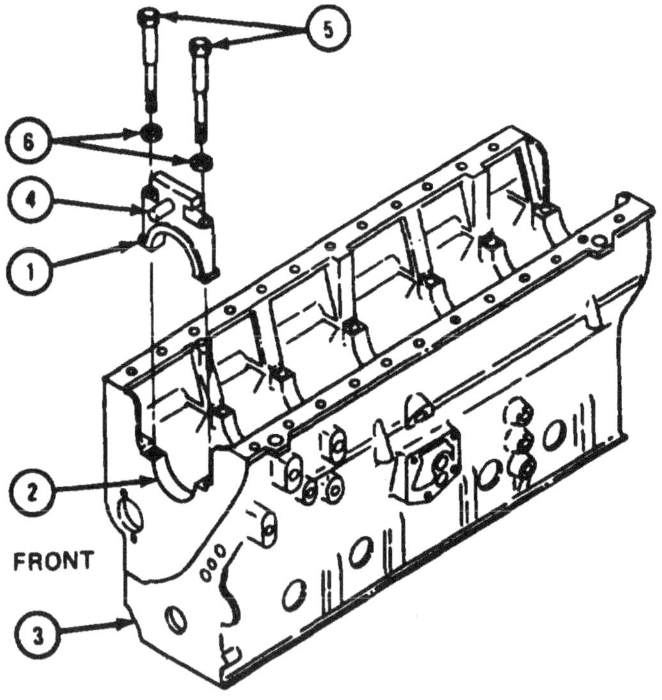

TA 116652

FRAME 43

NOTE

Seven main bearing caps (1) are line bored with crankcase (2). If any main bearing cap is damaged, crankcase and all bearing caps must be thrown away.

Main bearing bores are called by numbers one to seven counting from front to rear of engine.

1. Using dial indicator, measure inside diameter of number one bearing bore (3). Inside diameter must be 3.9370 to 3.9380 inches.
2. If inside diameter of number one main bearing bore (3) is not 3.9370 to 3.9380 inches, throw away crankcase (2) and get a new one.

NOTE

New crankcase (2) will come with new bearing caps (1).

3. **If diameter of number one main bearing bore (3) is 3.9370 to 3.9380 inches, take out two screws and washers (4) and take off number one main bearing cap (1).**

GO TO FRAME 44

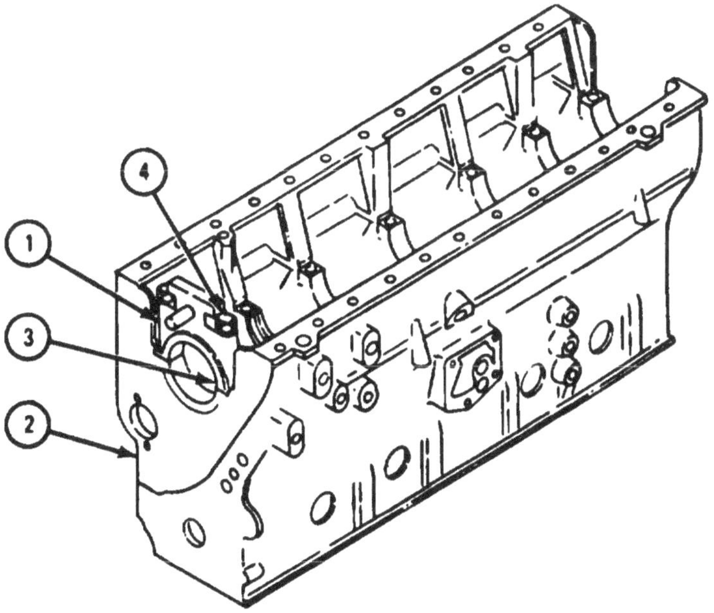

TA 116653

FRAME 4 4

NOTE

Main bearing caps (1) and bearing saddles (2) are called by numbers one to seven counting from front to rear of crankcase (3).

Number two main bearing cap (1) has a flat mounting surface (4) for mounting oil pump.

1. Put number two main bearing cap (1) on number two bearing saddle (2). Mounting holes in number two main bearing cap must line up with screw holes in number two bearing saddle.
2. Coat two bolts (5) and washers (6) lightly with engine lubricating oil, Put in and hand tighten two bolts and washers.
3. Using torque wrench, tighten two bolts (5) to 80 to 90 pound-feet.
4. Using torque wrench, tighten two bolts (5) to 115 to 120 pound-feet.

GO TO FRAME 45

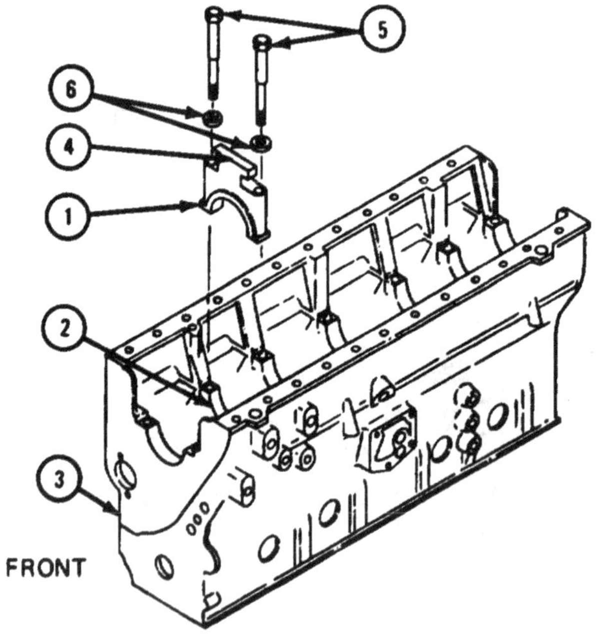

TA 116654

FRAME 4 5

NOTE

Seven main bearing caps (1) are line bored with crankcase (2). If any main bearing cap is damaged, the crankcase and all main bearing caps must be thrown away.

Bearing bores (3) are called by numbers one to seven counting from front to rear of crankcase (2).

1. Using dial indicator, measure inside diameter of number two bearing bore (3). Inside diameter must be 3.9370 to 3.9380 inches.
2. If inside diameter of number two main bearing bore (3) is not 3.9370 to 3.9380 inches, throw away crankcase (2) and get a new one.

NOTE

New crankcase (2) will come with new bearing caps (1).

3. **If diameter of number** *two* **main bearing bore (3) is 3.9370 to 3.9380 inches take out two bolts and washers (4), and take off number two main bearing cap (1).**

GO TO FRAME 46

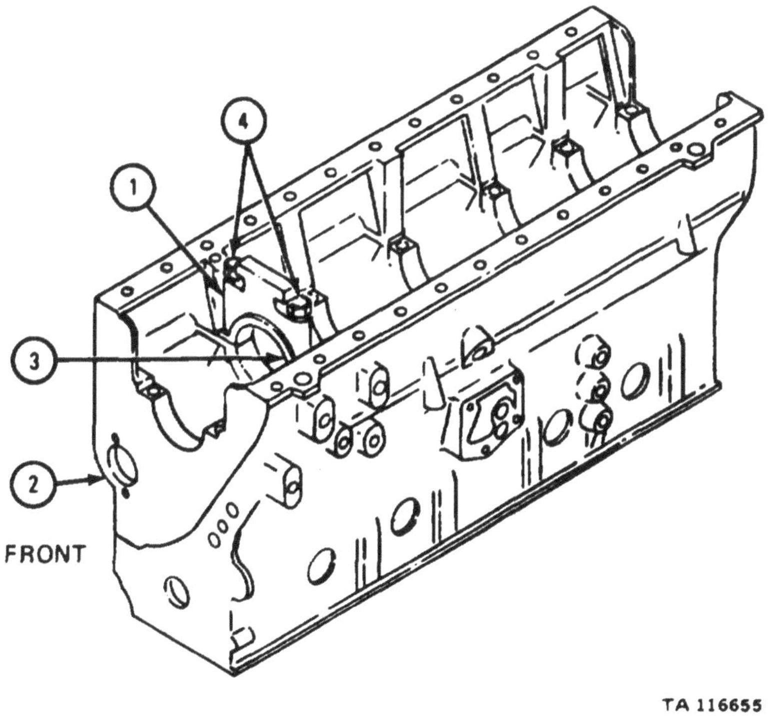

FRAME 46

NOTE

Seven main bearing caps (1) and seven bearing saddles (2) are called by numbers one to seven counting from front to rear of crankcase (3).

The number three main bearing cap (1) is marked with the number 3.

1. Put number three main bearing cap (1) on number three bearing saddle (2). Mounting holes in number three main bearing cap must line up with screw holes in number three bearing saddle.
2. Coat two bolts (4) and washers (5) lightly with engine lubricating oil. Put in and hand tighten two bolts and washers.
3. Using torque wrench, tighten two bolts (4) to 80 to 90 pound-feet.
4. Using torque wrench, tighten two bolts (4) to 115 to 120 pound-feet.

GO TO FRAME 47

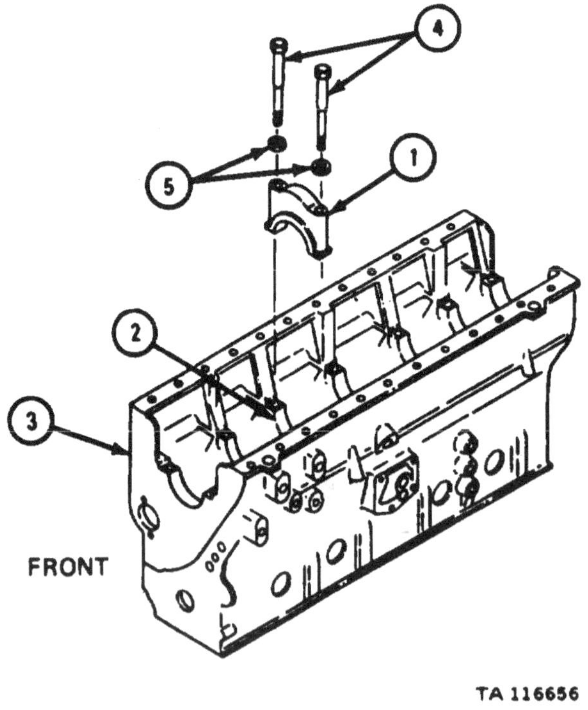

FRAME 47

NOTE

Seven main bearing caps (1) are line bored with crankcase (2). If any main bearing cap is damaged, crankcase and all main bearing caps must be thrown away.

Main bearing bores are called by numbers one to seven counting from front to rear of crankcase (2).

1. Using dial indicator, measure inside diameter of number three bearing bore (3). Inside diameter must be 3.9370 to 3.9380 inches.
2. If inside diameter of number three bearing bore (3) is not 3.9370 to 3.9380 inches, throw away crankcase (2) and get a new one.

NOTE

New crankcase (2) will come with new main *bearing* caps (1).

3. **If diameter of number three main bearing bore (3) is 3.9370 to 3.9380 inches, take out two bolts and washers (4) and take off number three main bearing cap (1).**

GO TO FRAME 48

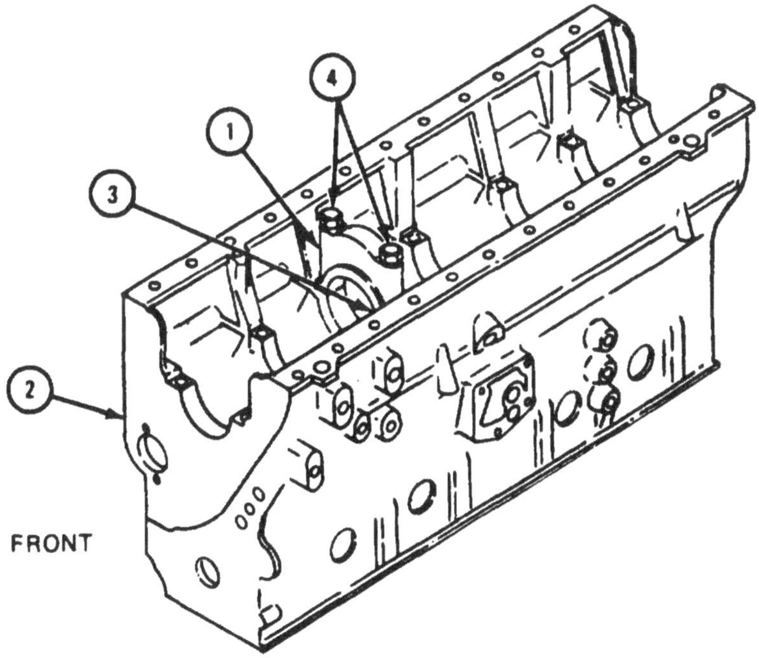

FRAME 48

NOTE

Main bearing caps (1) and bearing saddles (2) are called by numbers one to seven counting from front to rear of crankcase (3).

Number four main bearing cap (1) is called main thrust bearing cap. It is wider than other main bearing caps. Number four bearing saddle (2) is also wider than other bearing saddles.

1. Put main thrust bearing cap (1) on number four bearing saddle (2). Mountin g holes in main thrust bearing cap must line up with screw holes in number four earing saddle .
2. Coat two bolts (4) and washers (5) lightly with engine lubricating oil. Put in and hand tighten two bolts.
3. Using torque wrench, tighten two bolts (4) to 85 to 90 pound-feet.
4. Using torque wrench, tighten two bolts (4) to 115 to 120 pound-feet.

GO TO FRAME 49

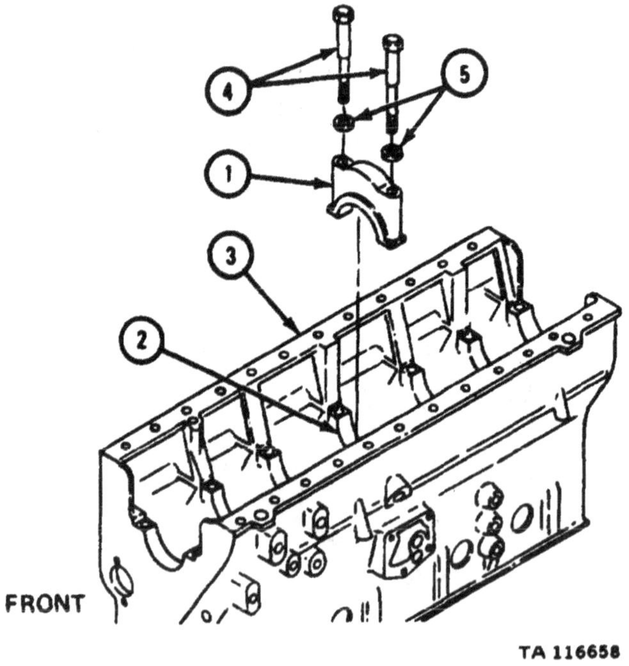

FRAME 49

NOTE

Seven main bearing caps (1) are line bored with crankcase (2). If any main bearing cap is damaged, crankcase and all main bearing caps must be thrown away.

Main bearing bores are called by numbers one to seven counting from front to rear of crankcase (2).

1. Using dial indicator, measure inside diameter of number four bearing bore (3). Inside diameter must be 3.9370 to 3.9380 inches.
2. If inside diameter of number four bearing bore (3) is not 3.9370 to 3.9380 inches, throw away crankcase (2) and get a new one.

NOTE

New crankcase (2) will come with new main bearing caps (1).

3. If diameter of number four main bearing bore (3) is 3.9370 to 3.9380 inches, take out two bolts and washers (4) and take off main thrust bearing cap (1).

GO TO FRAME 50

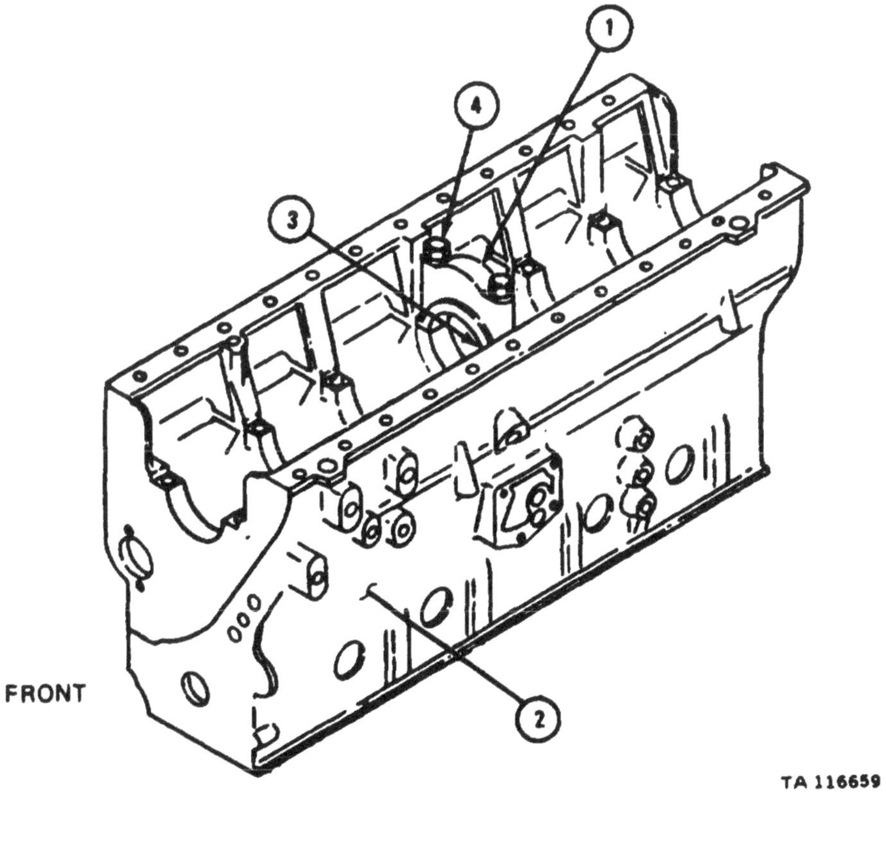

FRAME 50

NOTE

Main bearing caps (1) and bearing saddles (2) are called by numbers one to seven counting from front to rear of crankcase (3).

Number five main bearing cap (1) is marked with the number 5.

1. Put number five main bearing cap (1) on number five bearing saddle (2). Mounting holes in number five main bearing cap must line up with screw holes in number five bearing saddle.
2. Coat two bolts (4) and washers (5) lightly with engine lubricating oil. Put in and hand tighten two bolts and washers.
3. Using torque wrench, tighten two bolts (4) to 80 to 90 pound-feet.
4. Using torque wrench, tighten two bolts (4) to 115 to 120 pound-feet.

GO TO FRAME 51

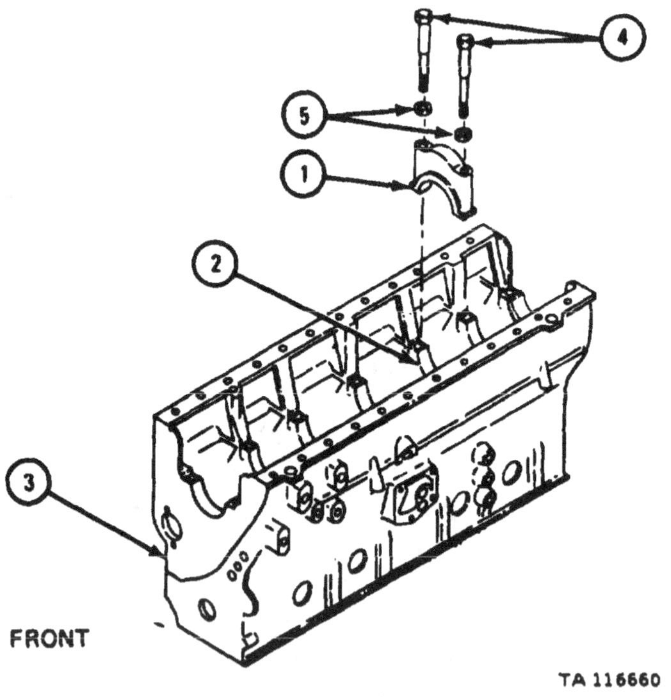

FRAME 51

NOTE

Seven main bearing caps (1) are line bored with crankcase (2). If any main bearing cap is damaged, crankcase and all main bearing caps must be thrown away.

Main bearing bores are called by numbers one to seven counting from front to rear of crankcase (2).

1. Using dial indicator, measure inside diameter of number five bearing bore (3). Inside diameter must be 3.9370 to 3.9380 inches.
2. If inside diameter of number five bearing bore (3) is not 3.9370 to 3.9380 inches, throw away crankcase (2) and get a new one.

NOTE

New crankcase (2) will come with new main bearing caps (1).

3. If diameter of number five main bearing bore (3) is 3.9370 to 3.9380 inches, take out two bolts and washers (4) and take off number five main bearing cap (1) .

GO TO FRAME 52

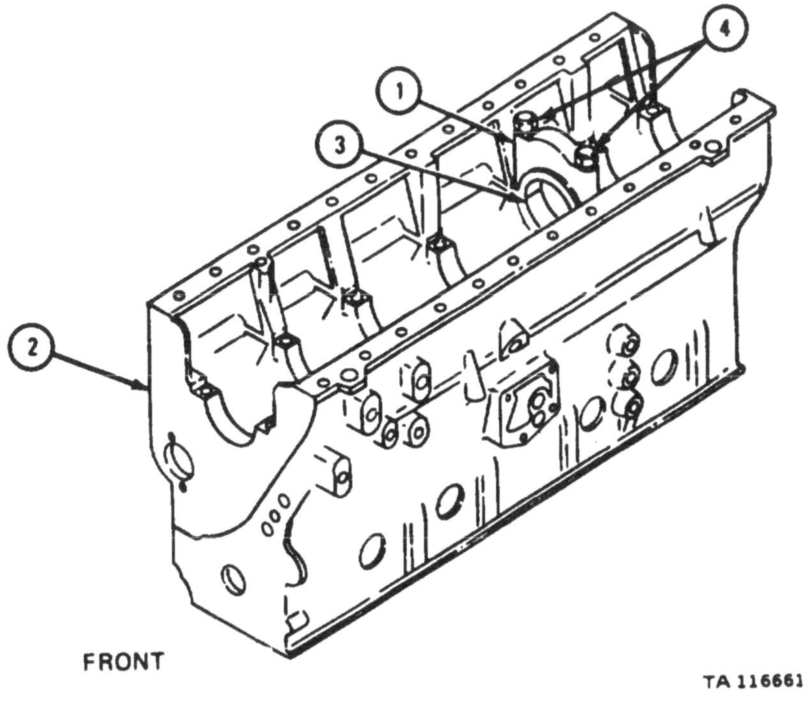

FRAME 52

NOTE

Main bearing caps (1) and bearing saddles (2) are called by numbers one to seven counting from front to rear of crankcase (3).

1. Put number six main bearing cap (1) on number six bearing saddle (2). Mounting holes in number six main bearing cap must line up with screw holes in number six bearing saddle.
2. Coat two bolts (4) and washers (5) lightly with engine lubricating oil. Put in and hand tighten two bolts and washers.
3. Using torque wrench, tighten two bolts (4) to 80 to 90 pound-feet.
4. Using torque wrench, tighten two bolts (4) to 115 to 120 pound-feet.

GO TO FRAME 53

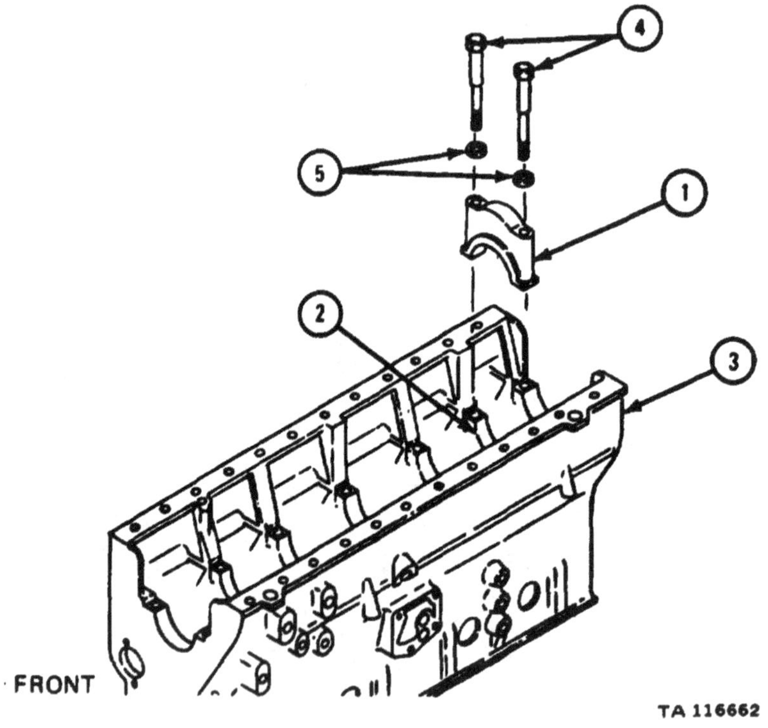

FRAME 53

NOTE

Seven main bearing caps (1) are line bored with crankcase (2). If any main bearing cap is damaged, crankcase and all main bearing caps must be thrown away.

Main bearing bores are called by numbers one to seven counting from front to rear of crankcase (2).

1. Using dial indicator, measure inside diameter of number six bearing bore (3). Inside diameter must be 3.9370 to 3.9380 inches.
2. If inside diameter of number six bearing bore (3) is not 3.9370 to 3.9380 inches, throw away crankcase (2) and get a new one.

NOTE

New crankcase (2) will come with new main bearing caps (1).

3. If diameter of number six main bearing bore (3) is 3.9370 to 3.9380 inches, take out two bolts and washers (4) and take off number three main bearing cap (1) .

GO TO FRAME 54

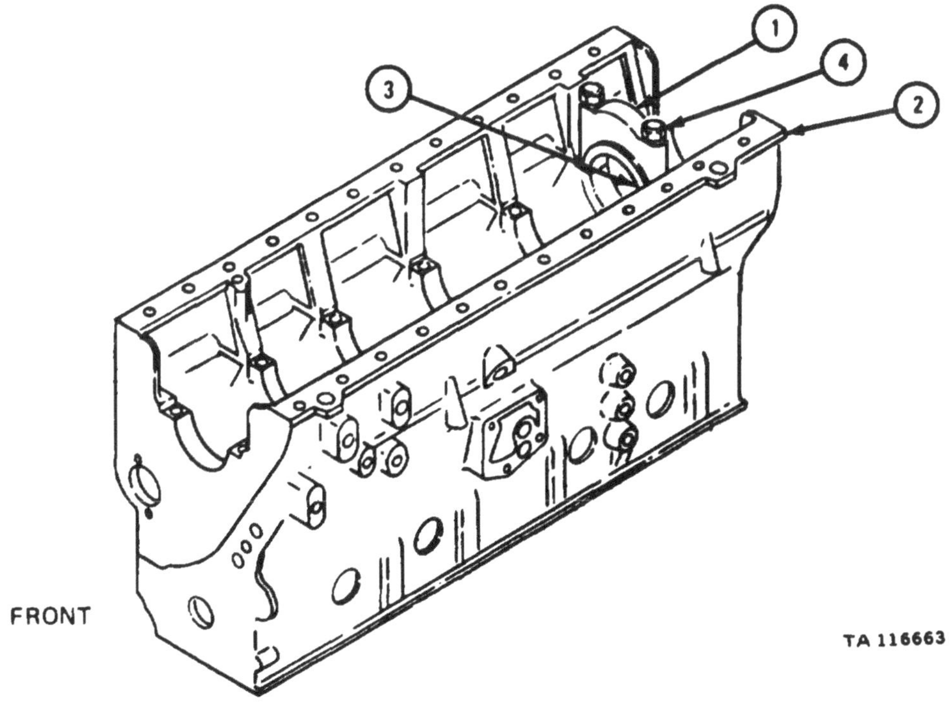

FRAME 5 4

NOTE

Main bearing caps (1) and bearing saddles (2) are called by numbers one to seven counting from front to rear of crankcase (3).

The number seven main bearing cap (1) has a slot on each side for rubber oil seals.

1. Put number seven main bearing cap (1) on number seven bearing saddle (2). Mounting holes in number seven main bearing cap must line up with screw holes in number seven bearing saddle.
2. Put number seven main bearing cap (1) on number seven bearing saddle (2).
3. Coat two bolts (4) and washers (5) lightly with engine lubricating oil. Put in and hand tighten two bolts and washers.
4. Using torque wrench, tighten two bolts (4) to 80 to 90 pound-feet.
5. Using torque wrench, tighten two bolts (4) to 115 to 120 pound-feet.

GO TO FRAME 55

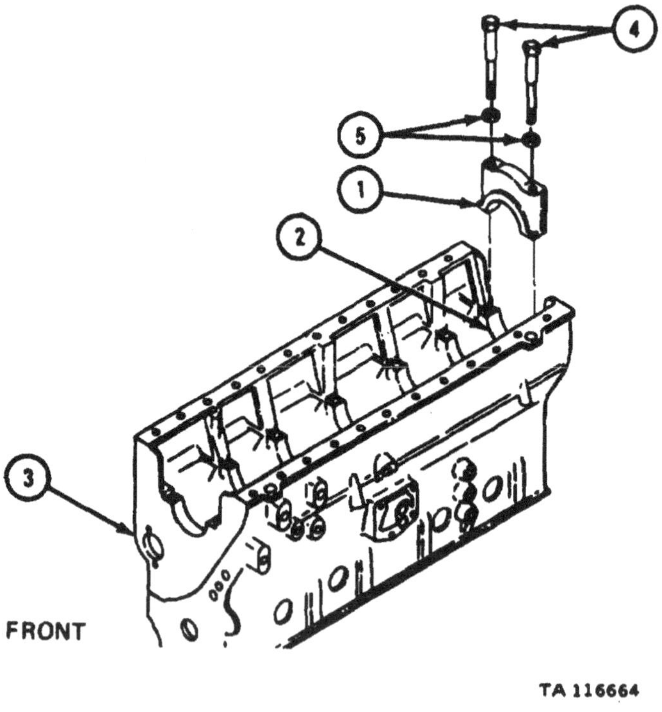

FRAME 55

NOTE

Seven main bearing caps (1) are line bored with crankcase (2). If any main bearing cap is damaged, crankcase and all main bearing caps must be thrown away.

Main bearing bores are called by numbers one to seven counting from front to rear of crankcase (2).

1. Using dial indicator, measure inside diameter of number seven bearing bore (3). Inside diameter must be 3.9370 to 3.9380 inches.
2. If inside diameter of number seven bearing bore is not 3.9370 to 3.9380 inches, throw away crankcase (2) and get a new one.

NOTE

New crankcase (2) will come with new main bearing caps (1).

3. If diameter of number seven main bearing bore (3) is 3.9370 to 3.9380 inches, take out two bolts and washers (4), and take off number seven main bearing cap (1).

GO TO FRAME 56

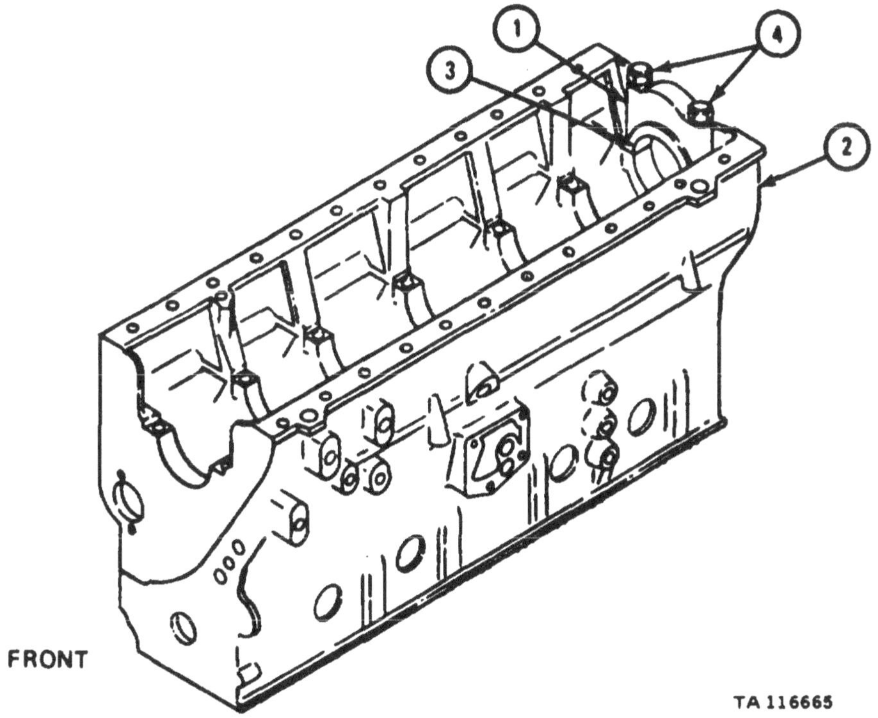

FRAME 56

NOTE

All seven upper main bearing halves (1 and 2) have an alinement tab (3) off center as shown. All seven lower main bearing halves (4 and 5) have alinement tab (6) in center as shown.

1. Spread a thin coat of Prussian blue on surface of upper main thrust bearing half (2) which rests on main thrust bearing saddle (7).
2. Put upper main thrust bearing half (2) on thrust main bearing saddle (7). Alinement tab (3) must fit in notch (8) in thrust main bearing saddle.

GO TO FRAME 57

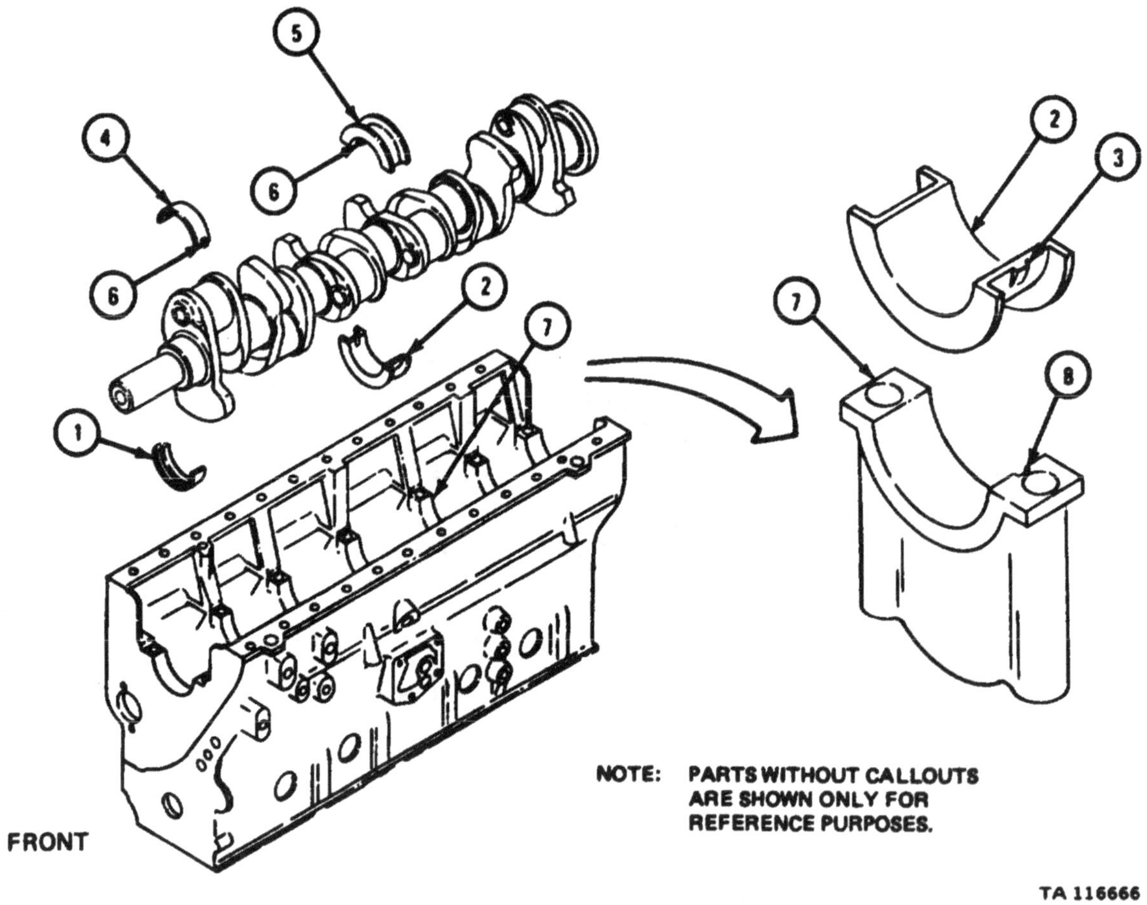

FRAME 57

NOTE

All seven upper main bearing halves (1 and 2) have an alinement tab (3) off center as shown. All seven lower main bearing halves (4 and 5) have alinement tab (6) in center as shown.

Number four upper main bearing half (2) is called upper main thrust bearing half and was already put on.

If main bearing halves (1, 2, 4, and 5) taken out of engine were not thrown away, they must be used in this task. Main bearing halves and main bearing saddles (7) are called by numbers one to seven counting from front to rear of crankcase (8). Used main bearin g halves are marked with a number in alinement tabs (3 and 6).

1. Spread a thin coat of Prussian blue on surface of number one upper main bearing half (1) which rests on number one main bearing saddle (7).
2. Put number one upper main bearing half (1) on number one main bearing saddle (7) . Alinement tab (3) must fit in notch (9) in number one main bearing saddle .
3. Do steps 1 and 2 again for all other upper main bearing halves (1).

GO TO FRAME 58

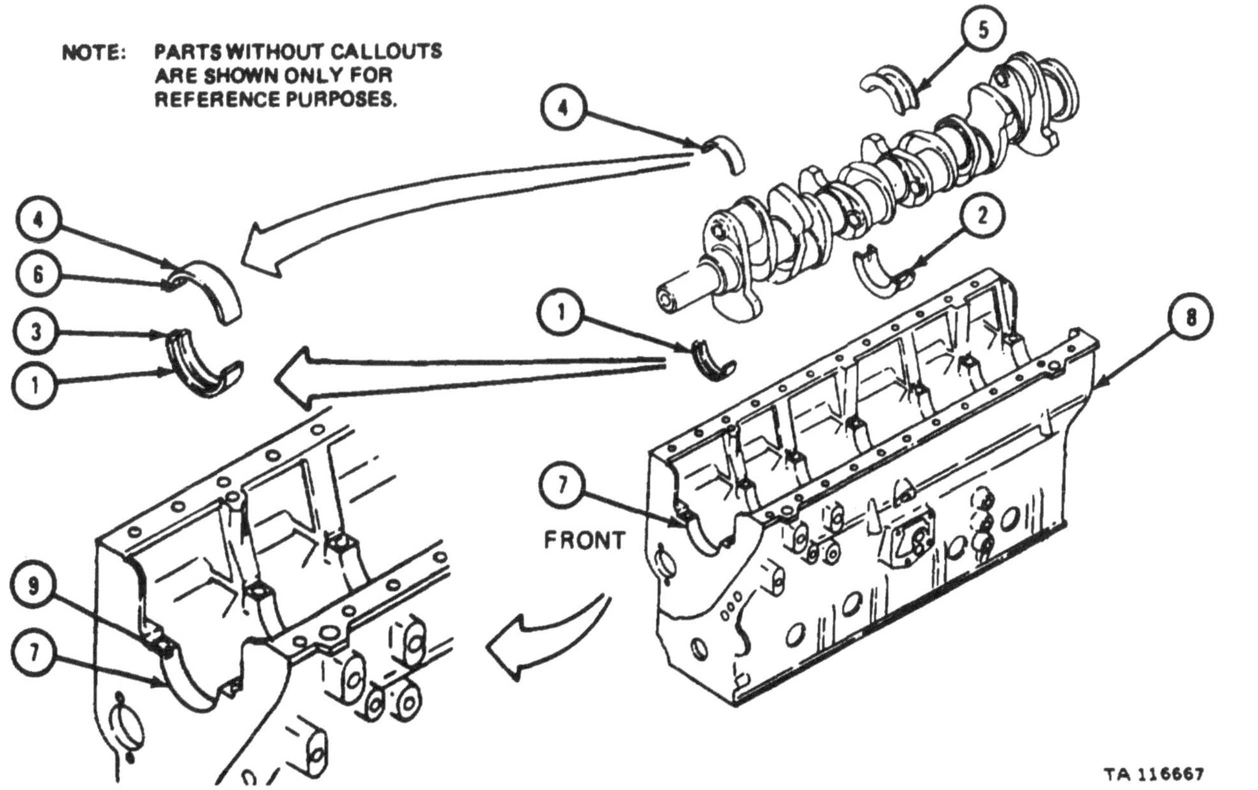

FRAME 58

NOTE

All seven lower main bearing halves (1) have an alinement tab (2) at center as shown.

Main bearing halves (1) and main bearing caps (3) are called by numbers one to seven counting from front to rear of crankcase (4) . Used main bearing halves are marked with a number in alinement tab (2).

Number one main bearing cap (3) has oil pump idler gear shaft (5) .

1. Spread a thin coat of Prussian blue on surface of number one lower main bearing half (1) which rests against number one main bearing cap (3).
2. Hold number one lower main bearing half (1) on number one main bearing cap (3) . Alinement tab (2) must fit in notch (6) in number one main bearing cap.
3. Put a light coat of engine lubricating oil on edges (7) of number one main bearing-cap (3).

GO TO FRAME 59

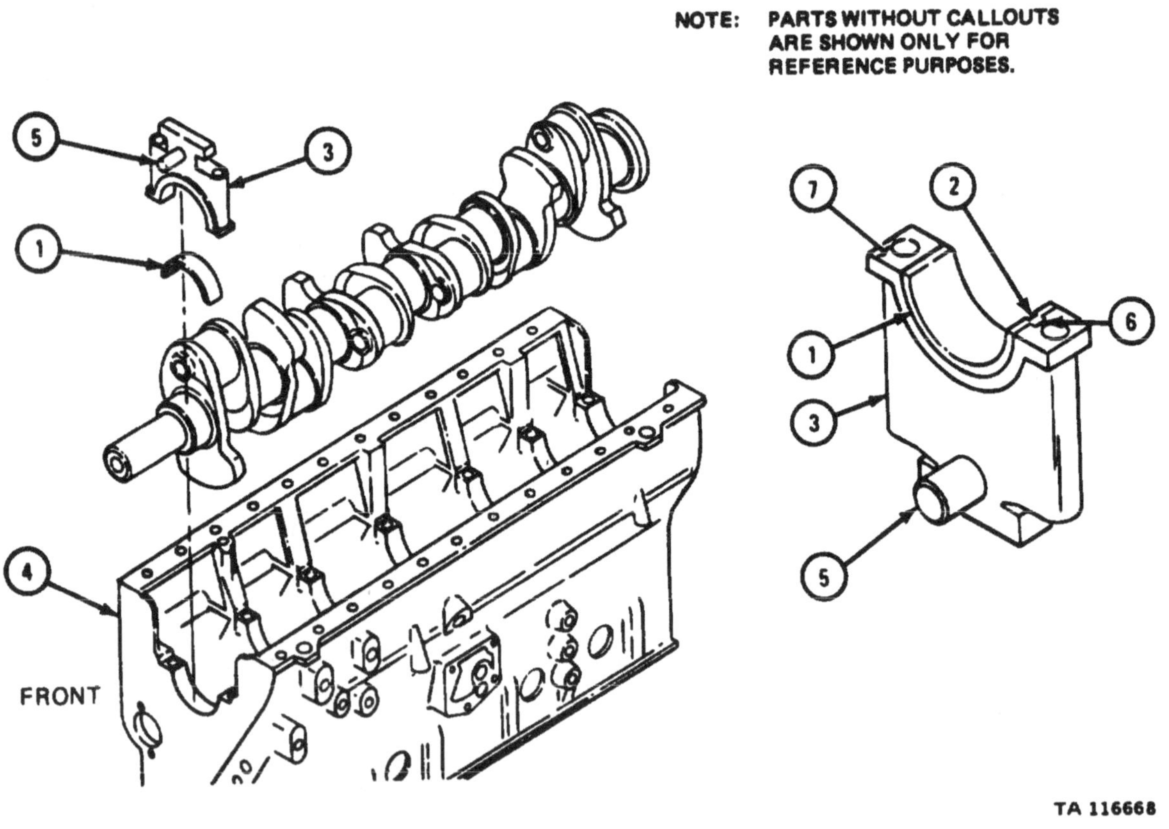

FRAME 59

NOTE

Main bearing caps with bearing halves (1) and main bearing saddles (2) are called by numbers one to seven counting from front to rear of crankcase (3).

The number one main bearing cap (1) has oil pump idler gear shaft (4) .

1. Put number one main bearing cap with bearing half (1) on number one bearing saddle (2) . Oil pump idler gear shaft (4) should face out.
2. Coat two bolts (5) and washers (6) lightly with engine lubricating oil. Put in and hand tighten two bolts and washers.
3. Using torque wrench, tighten two bolts (5) to 80 to 90 pound-feet.
4. Using torque wrench, tighten two bolts (5) to 115 to 120 pound-feet.

GO TO FRAME 60

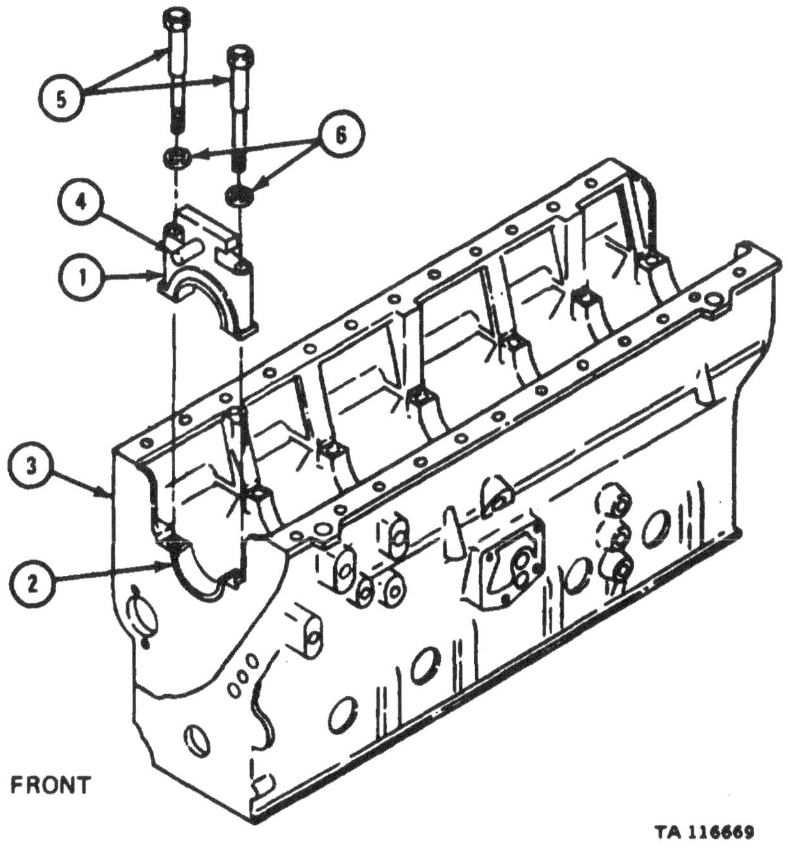

FRAME 60

NOTE

All seven lower main bearing halves (1) have an alinement tab (2) at center as shown.

Main bearing halves (1) and main bearing caps (3) are called by numbers one to seven counting from front to rear of crankcase (4) . Used main bearing halves are marked with a number in alinement tab (2).

Number two main bearing cap (3) has a flat mounting surface (5) for mounting oil pump.

1. Spread a thin coat of Prussian blue on surface of number two lower main bearing half (1) which rests against number two bearing cap (3).
2. Hold number two lower main bearing half (1) on number two main bearing cap (3) . Alinement tab (2) must fit in notch (6) in number two main bearing cap.
3. Put a light coat of engine lubricating oil on edges (7) of number two main bearing cap (3) .

GO TO FRAME 61

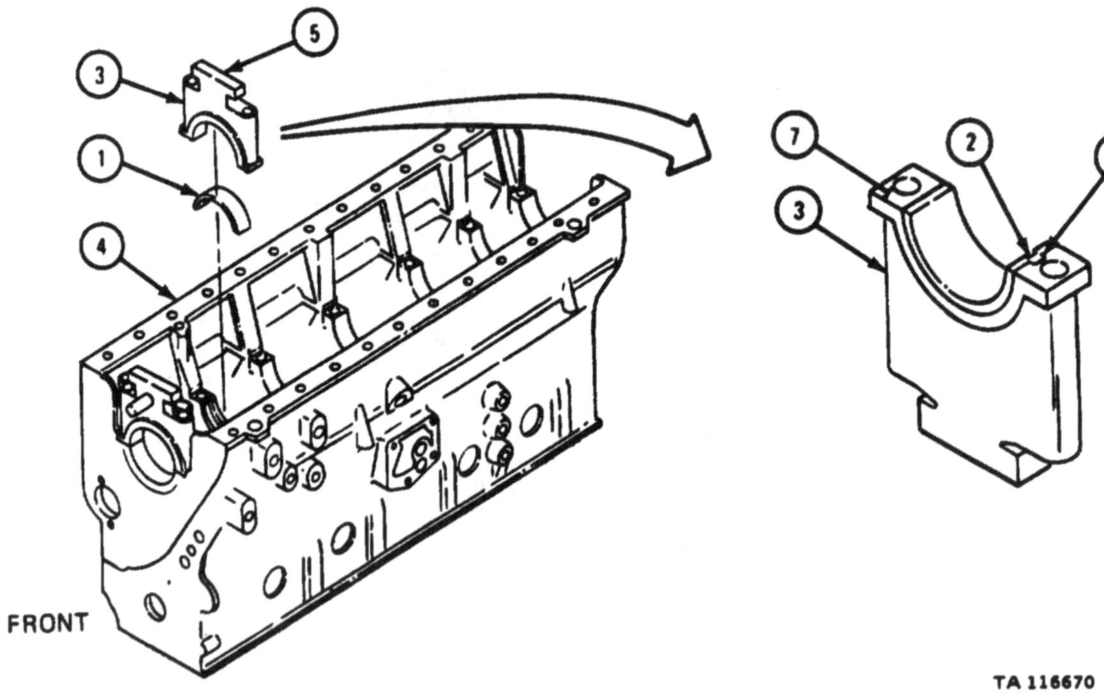

FRAME 61

NOTE

Main bearing caps (1) with bearing halves (2) and main bearing saddles (3) are called by numbers one to seven counting from front to rear of crankcase (4).

Number two main bearing cap (1) has a flat mounting surface (5) for mounting the oil pump.

1. Put number two main bearing cap (1) with bearing half (2) on number two main bearing saddle (3). Mounting holes in number two main bearing cap must line up with screw holes in number two bearing saddle.
2. Coat two bolts (6) and washers (7) lightly with engine lubricating oil. Put in and hand tighten two bolts and washers.
3. Using torque wrench, tighten two bolts (6) to 80 to 90 pound-feet.
4. Using torque wrench, tighten two bolts (6) to 115 to 120 pound-feet.

GO TO FRAME 62

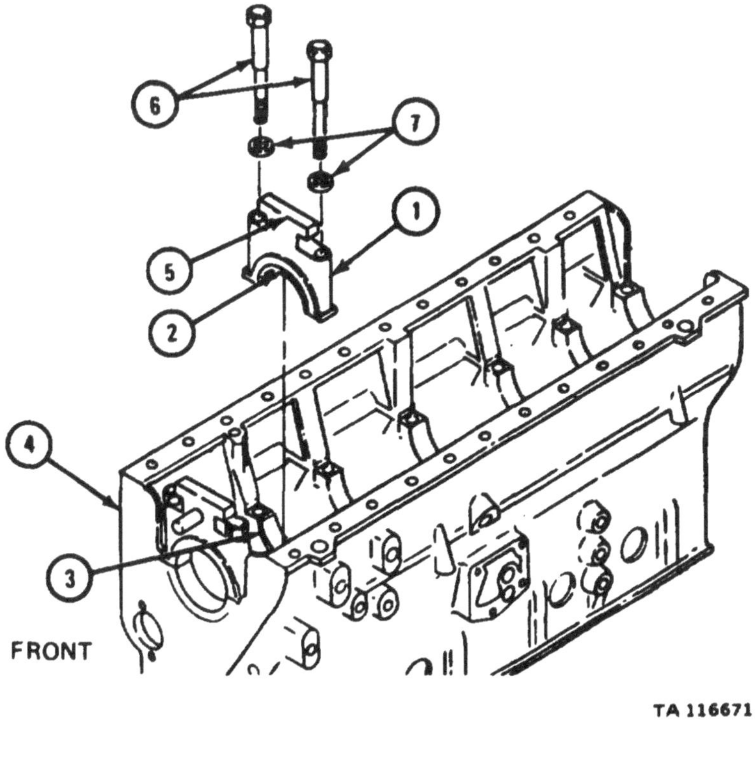

TA 116671

FRAME 62

NOTE

All seven lower main bearing halves (1) have an alinement tab (2) at center as shown.

Main bearing halves (1) and main bearing caps (3) are called by numbers one to seven counting from front to rear of crankcase (4) . Used main bearing halves are marked with a number in alinement tab (2).

The number three, number five, and number six main bearing caps look the same. They are marked with their number .

1. Spread a thin coat of Prussian blue on surface of number three lower main bearing half (1) which rests against number three main bearing cap (3).
2. Hold number three lower main bearing half (1) on number three main bearing cap (3) . Alinement tab (2) must fit into notch (5) in number three main bearing cap .
3. Put a light coat of engine lubricating oil on edges (6) of number three main bearing cap (3) .
4. Do this frame again for number five lower main bearing half and number six lower main bearing half.

GO TO FRAME 63

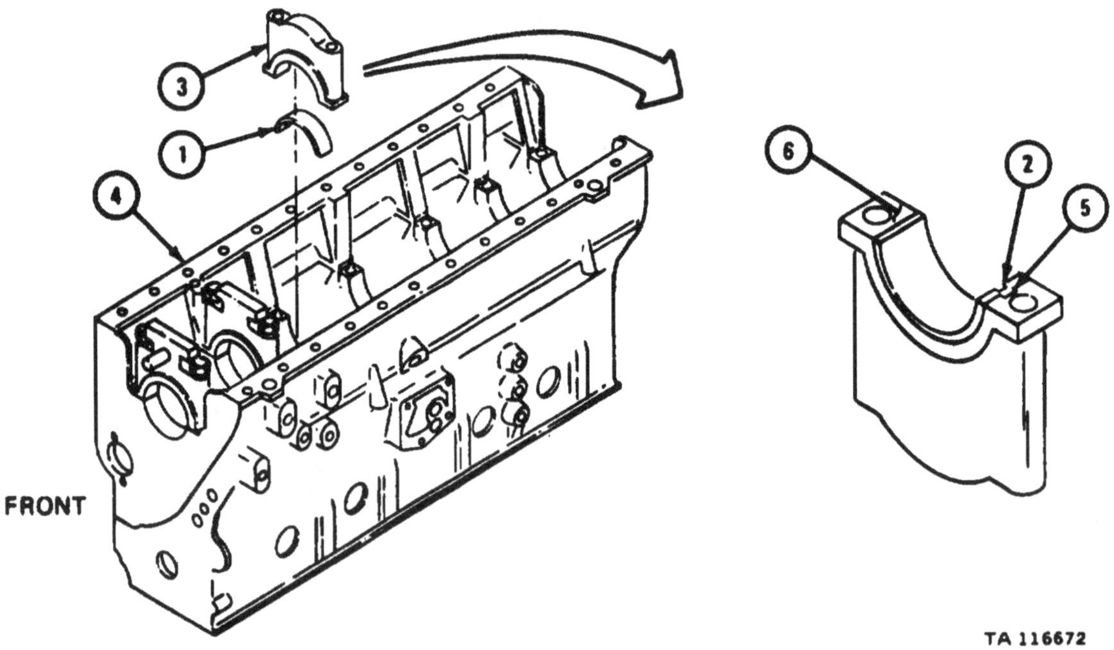

FRAME 63

NOTE

Main bearing caps (1) with bearing halves (2) and main bearing saddles (3) are called by numbers one to seven counting from front to rear of crankcase (4).

Number three, number five, and number six main bearing caps (1) look the same. They are marked with thei r number.

1. Put number three main bearing cap (1) with bearing half (2) on number three main bearing saddle (3). Mounting holes in number three main bearing cap must line up with screw holes in number three main bearing saddle.
2. Coat two bolts (5) and washers (6) lightly with engine lubricating oil. Put in and hand tighten two bolts and washers.
3. Using torque wrench, tighten two bolts (5) to 80 to 90 pound-feet.
4. Using torque wrench, tighten two bolts (5) to 115 to 120 pound-feet.
5. Do this frame again for number five main bearing cap with bearing half and number six main bearing cap with bearing half.

GO TO FRAME 64

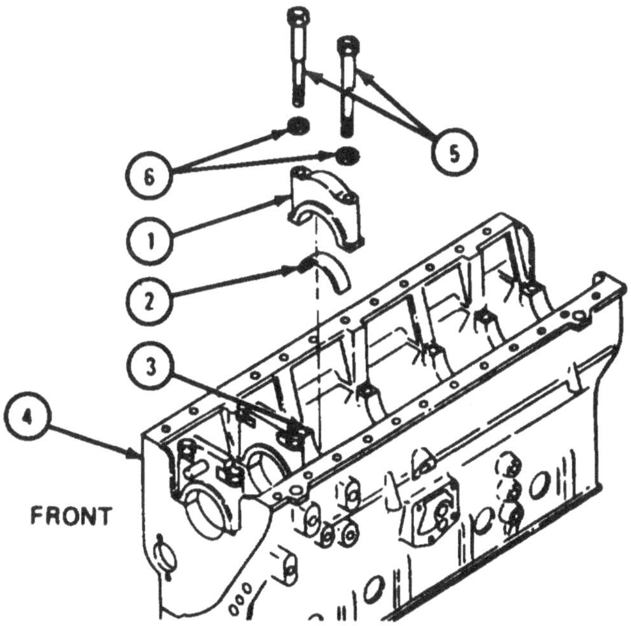

TA 116673

FRAME 64

NOTE

All seven lower main bearing halves (1) have an alinement tab (2) at center as shown.

Main bearing halves (1) and main bearing caps (3) are called by numbers one to seven counting from front to rear of crankcase (4) . Number four main bearing half and main bearing cap are called main thrust bearing half and main thrust bearing cap.

Lower main thrust bearing half (1) has a flange (5) on each side. Main thrust bearing cap (3) is wider than other main thrust bearing caps.

1. Spread a thin coat of Prussian blue on surface of lower main thrust bearing half (1) which rests against main thrust bearing cap (3).
2. Hold main thrust bearing half (1) on main thrust bearing cap (3). Alinement tab (2) must fit in notch (6) in main thrust bearing cap.
3. Put a light coat of engine lubricating oil on edges (7) of main thrust bearing cap (3) .

GO TO FRAME 65

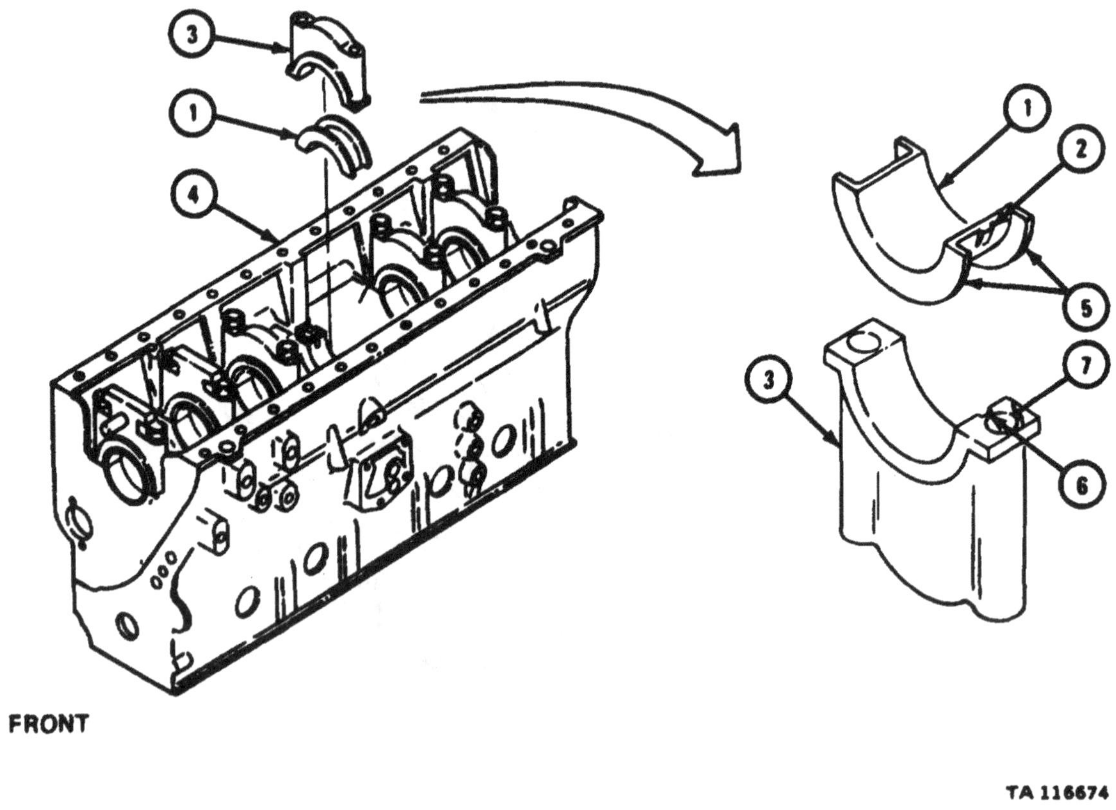

FRAME 65

NOTE

Main bearing caps (1) with bearing halves (2) and main bearing saddles (3) are called by numbers one to seven counting from front to rear of crankcase (4).

Number four main bearing cap (1) is called main thrust bearing cap. It is wider than other main bering caps .

Number four main bearing saddle (3) is called main thrust bearing saddle. It is wider than other main bearin g saddles.

1. Put main thrust bearing cap (1) with bearing half (2) on main thrust bearing saddle (3). Mounting holes in main thrust bearing cap must line up with mounting holes in main thrust bearing saddle.
2. Coat two bolts (5) and washers (6) lightly with engine lubricating oil. Put in and hand tighten two bolts and washers.
3. Using torque wrench, tighten two bolts (5) to 80 to 90 pound-feet.
4. Using torque wrench, tighten two bolts (5) to 115 to 120 pound-feet.

GO TO FRAME 66

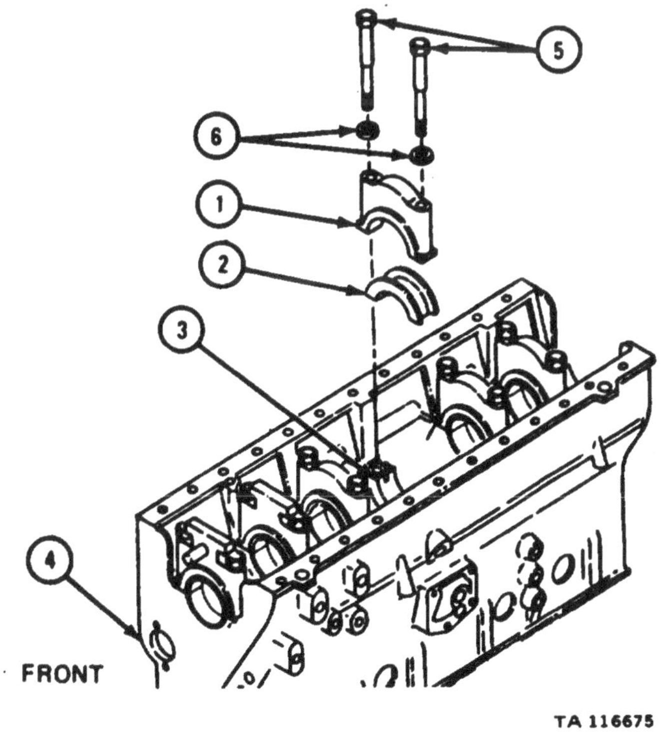

FRAME 66

NOTE

All seven lower main bearing halves (1) have an alinement tab (2) at center as shown.

Main bearing halves (1) and main bearing caps (3) are called by numbers one to seven counting from front to rear of crankcase (4). Used main bearing halves are marked with a number in alinement tab (2).

The number seven main bearing cap (3) has a slot (5) on each side for rubber oil seals.

1. Spread a thin coat of Prussian blue on surface of number seven lower q ain bearing half (1) which rests against number seven main bearing cap (3).
2. Hold number seven lower main bearing half (1) on number seven main bearing cap (3). Aline tab (2) must fit in notch (6) in number seven main bearing cap .
3. Put a light coat of engine lubricating oil *on* edges (7) of number seven main bearing cap (3) .

GO TO FRAME 67

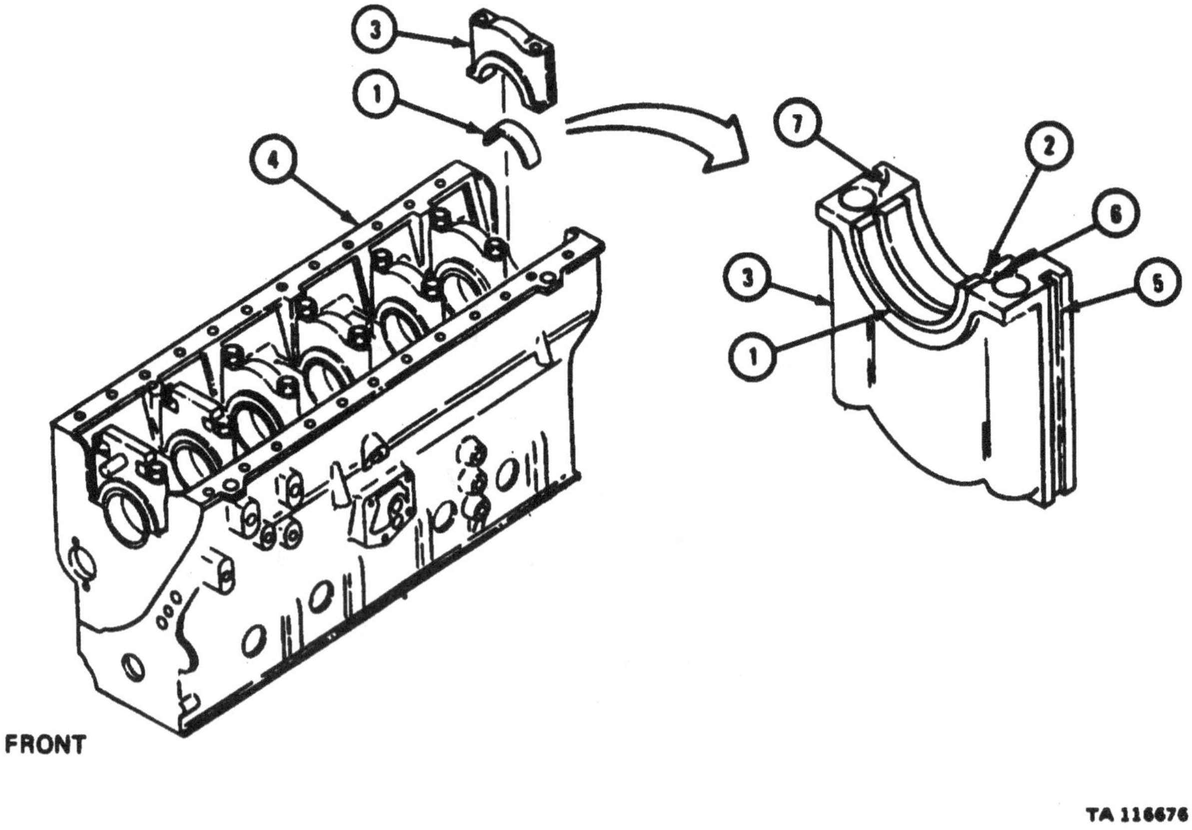

FRAME 67

NOTE

Main bearing caps (1) with bearing halves (2) and main bearing saddles (3) are called by numbers one to seven counting from front to rear of crankcase (4).

Number seven main bearing cap has a slot (5) on each side for rubber oil seal.

1. Put number seven main bearing cap (1) with bearing half (2) on number seven main bearing saddle (3). Mounting holes in number seven main bearing cap must line up with screw holes in number seven bearing saddle.
2. Coat two bolts (6) and washers (7) lightly with engine lubricating oil. Put in and hand tighten two bolts and washers.
3. Using torque wrench, tighten two bolts (6) to 80 to 90 pound-feet.
4. Using torque wrench, tighten two bolts (6) to 115 to 120 pound-feet.

GO TO FRAME 68

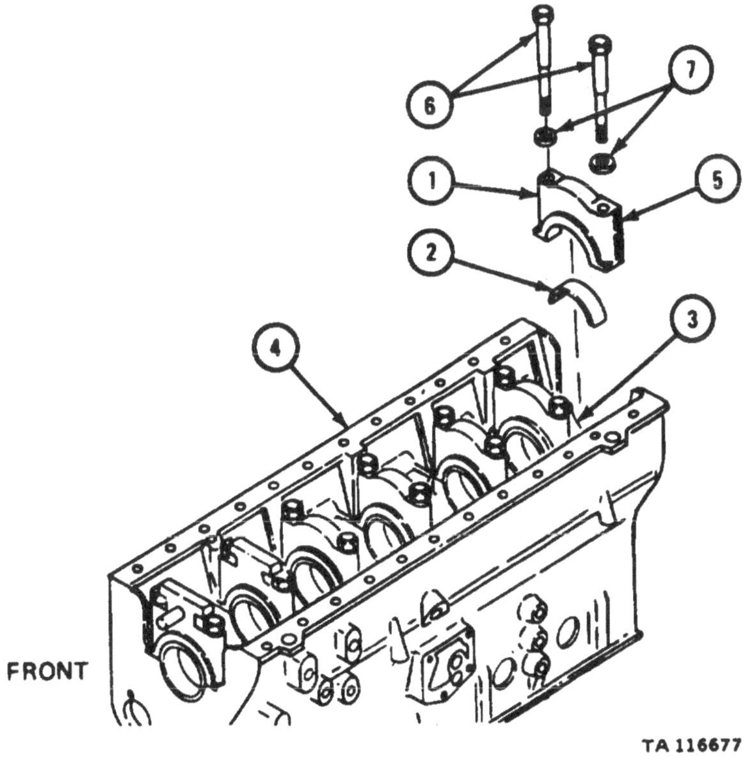

FRAME 68

NOTE

Four different crankshafts (1) can be used in these engines. They differ in the diameters of the main bearing journals and connecting rod journals. The four crankshafts are early model standard size, late model standard size, 0.0100-inch undersize, and 0.0200-inch undersize.

Size of crankshaft (1) you are now working on was noted in frame 31.

1. Check size of crankshaft (1) noted in frame 31.

IF WORKING ON STANDARD SIZE CRANKSHAFT, GO TO FRAME 69.
IF WORKING ON 0.100-INCH UNDERSIZE CRANKSHAFT, GO TO FRAME 71.
IF WORKING ON 0.0200-INCH UNDERSIZE CRANKSHAFT, GO TO FRAME 73

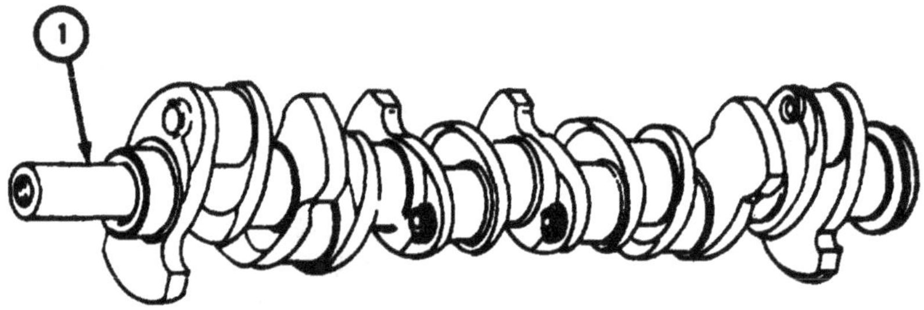

TA116641

FRAME 69

NOTE

Main bearings (1) are called by numbers one to seven counting from front to rear of crankcase (2) .

Readings must be within limits given in table 4-17.

1. Measure inside diameter of number one main bearing (1). Note this measurement
2. Subtract measurement of crankshaft number one main bearing journal diameter noted in step 3 of frame 20 from measurement taken in step 1 of this frame. This is oil clearance between number one main bearing (1) and journal. Note this measurement.
3. Do steps 1 and 2 again for number two through number seven main bearings (1).

GO TO FRAME 70

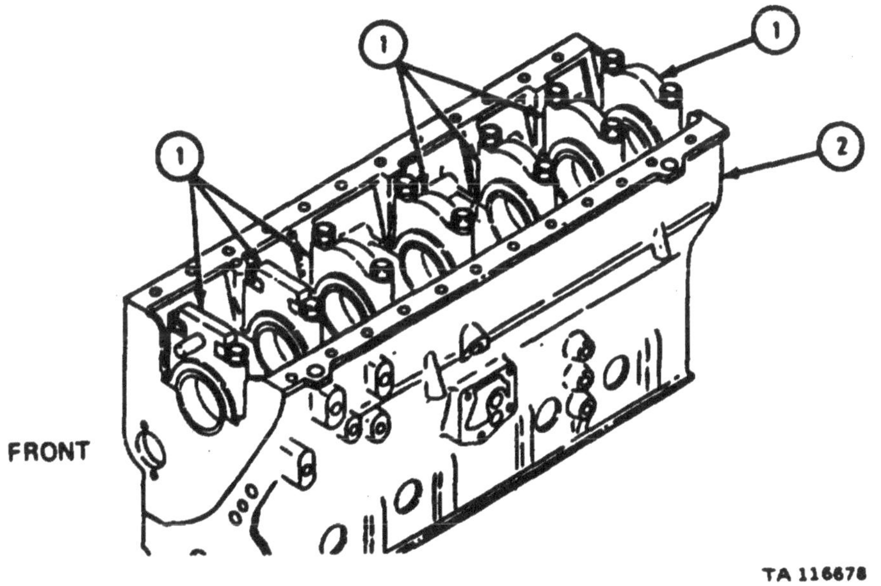

Table 4-17. Standard Size Crankshaft Main Bearings and Oil Clearance Wear Limits

Index Number	Item/Point of Measurement	Size and Fit of New Parts (inches)	Wear Limit (inches)
1	Main bearings inside diameter	3.6252 to 3.6272	3.6292
	Oil clearance between crankshaft main journals and main bearings		
	Late model	0.0032 to 0.0062	0.0080
	Early model	0.0042 to 0.0072	0.0080

FRAME 70

1. If any measurement taken in frame 69 is not within limits given in table 4-17, skip step 2 and do step 3.
2. Count the number of main bearing oil clearance measurements done in step 2 of frame 69 that are more than 0.0062 inch. If four or more oil clearance measurements are more than 0.0062 inch, do step 3. If three or fewer oil clearance measurements are more than 0.0062 inch, go to frame 75.
3. Take out two screws and washers (1) from each of seven main bearing caps (2),
4. Take off seven main bearing caps (2). Take out and throw away all 14 main bearing halves (3) and get a new standard size main bearing kit.

NOTE

If main bearing halves (3) were changed, do frames 56 to 70 again for new main bearing halves.

GO TO FRAME 75

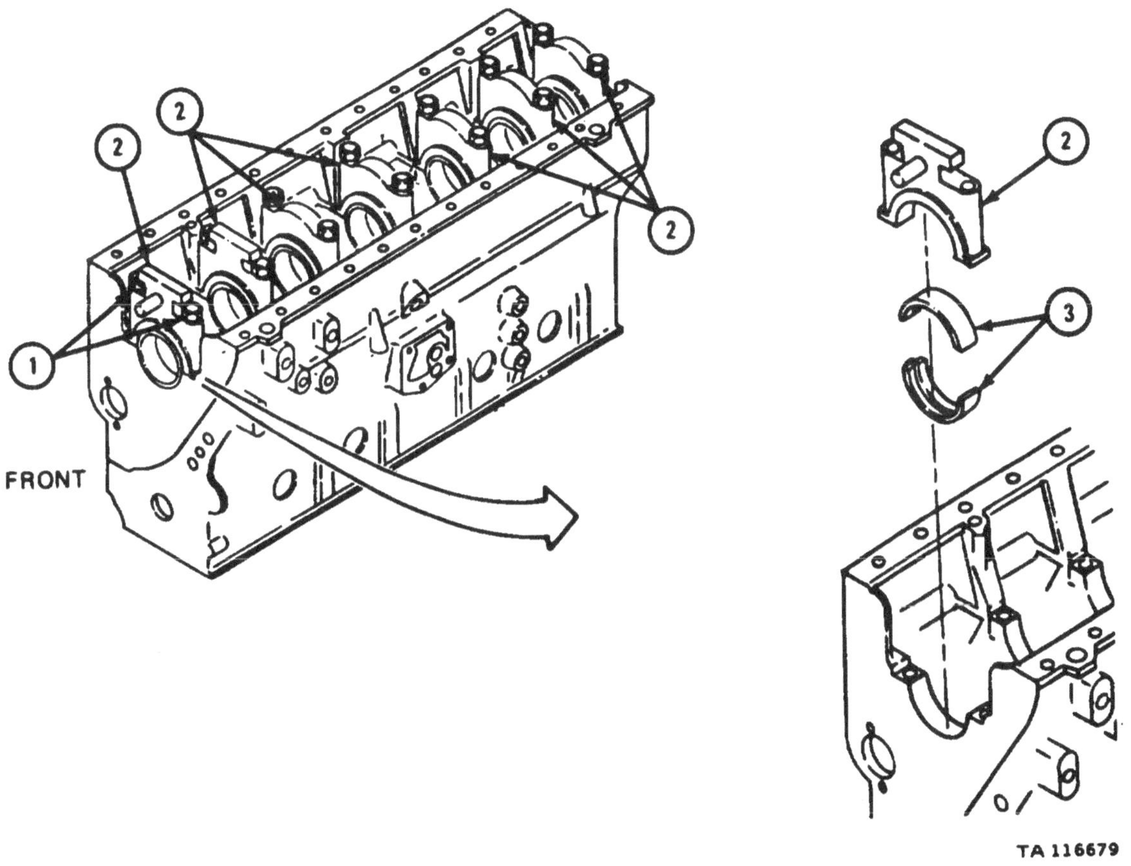

FRAME 71

NOTE

Readings must be within limits given in table 4-18.

Main bearings (1) are called by numbers one to seven counting from front to rear of crankcase (2).

1. Measure inside diameter of number one main bearing (1). Note this measurement.
2. Subtract measurement of crankshaft number one main bearing journal diameter noted in step 3 of frame 20 from measurement taken in step 1 of this frame. This is oil clearance between number one main bearing and journal. Note this measurement.
3. Do steps 1 and 2 again for number two through number seven main bearings (1)

GO TO FRAME 72

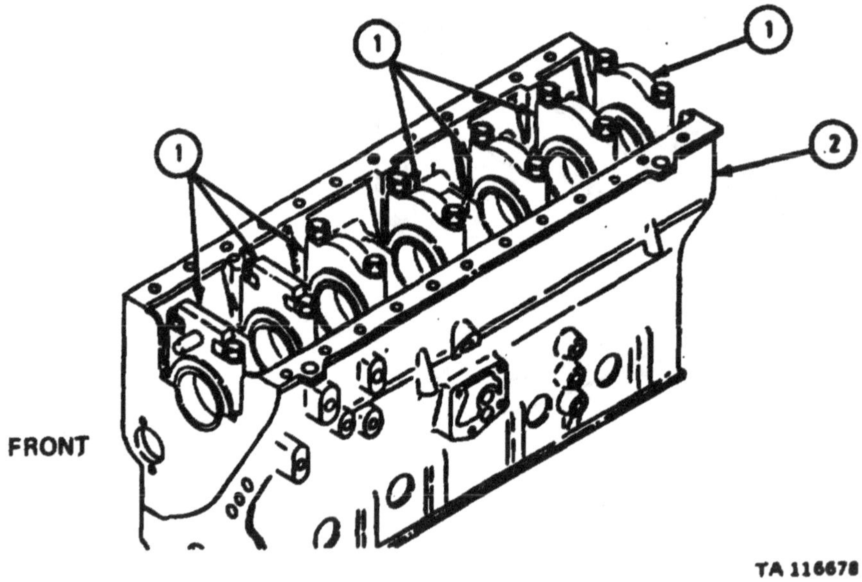

Table 4-18. 0.0100-Inch Undersize Crankshaft Main Bearings and Oil Clearance Wear Limits

Index Number	Item/Point of Measurement	Size and Fit of New Parts (inches)	Wear Limit (inches)
1	Main bearings inside diameter	3.6152 to 3.6172	3.6192
1	Oil clearance between crankshaft q ain journals and main bearings	0.0032 to 0.0062	0.0080

FRAME 72

1. If any measurement taken in frame 71 is not within limits given in table 4-18, skip step 2 and do step 3.

2. Count the number of main bearing oil clearance measurements done in step 2 of frame 71 that are more than 0.0062 inch. If four or more oil clearance measurements are more than 0.0062 inch, do step 3. If three or fewer oil clearance measurements are more than 0.0062 inch, go to frame 75.

3. Take out two screws and washers (1) from each of seven main bearing caps (2) .

4. Take off seven main bearing caps (2). Take out and throw away all 14 main bearing halves (3) and get a new 0.0020-inch undersize main bearing kit.

NOTE

If main bearing halves (3) were changed, do frames 56 to 72 again for new main bearing halves.

GO TO FRAME 75

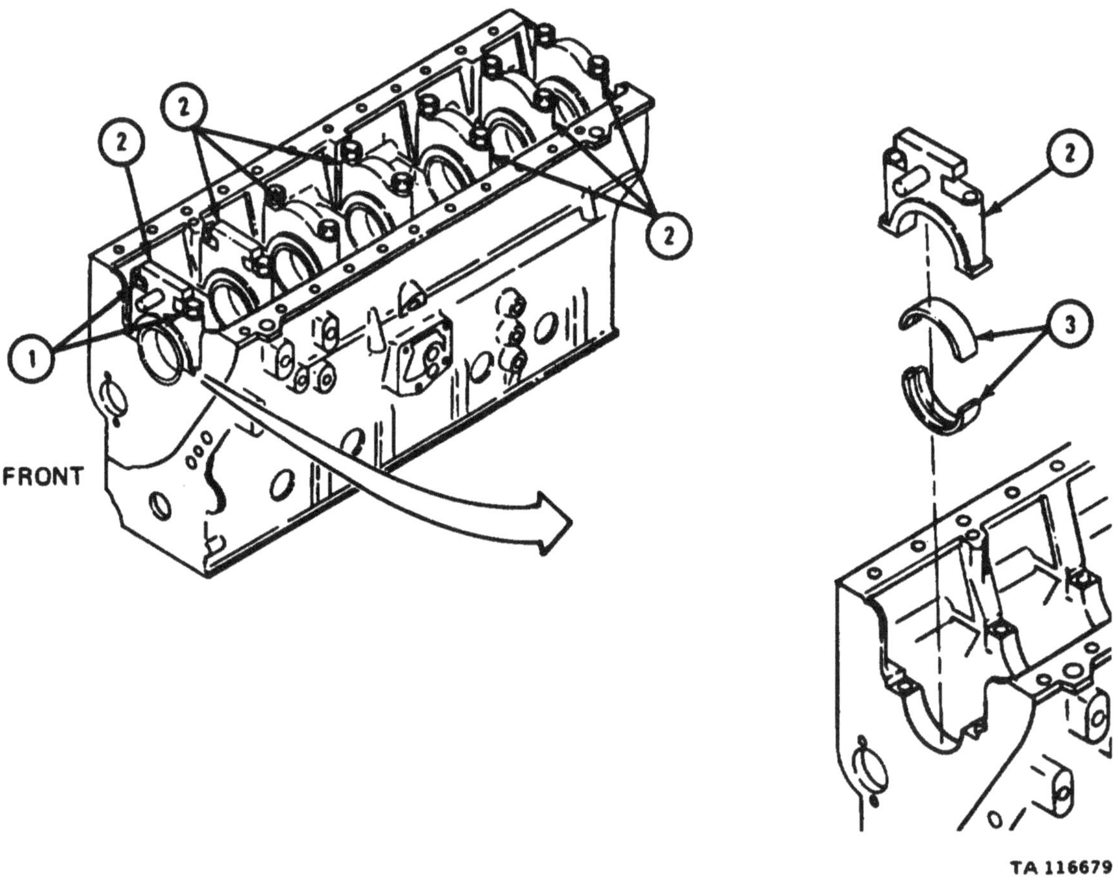

FRAME 73

NOTE

Readings must be within limits given in table 4-19.

Main bearings (1) are called by numbers one to seven counting from front to rear of crankcase (2).

1. Measure inside diameter of number one main bearing (1). Note this measurement.
2. Subtract measurement of crankshaft number one main bearing journal diameter noted in step 3 of frame 20 from measurement taken in step 1 of this frame. This is oil clearance between number one main bearing and journal. Note this measurement.
3. Do steps 1 and 2 again for number two through number seven main bearings (1),

GO TO FRAME 74

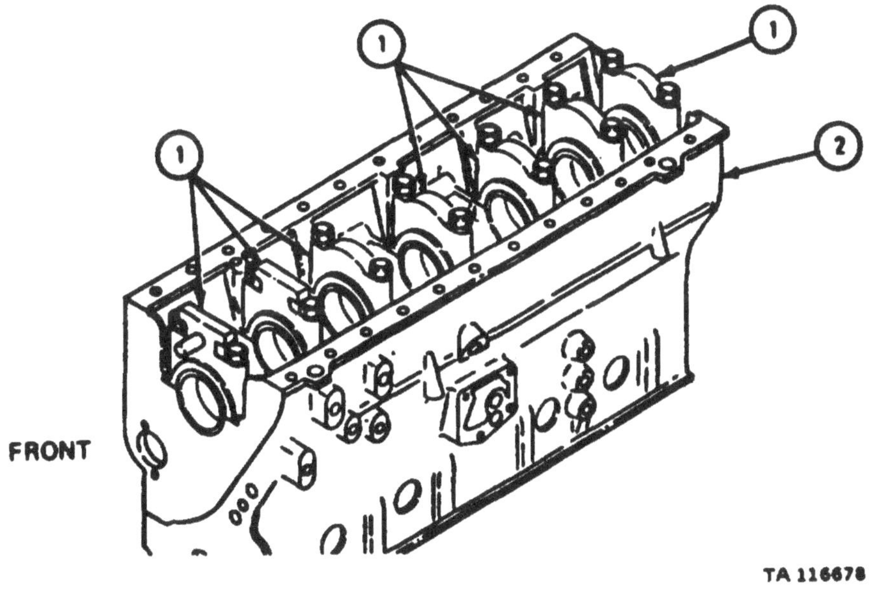

Table 4-19. 0.0200-Inch Undersize Crankshaft Main Bearings and Oil Clearance Wear Limits

Index Number	Item/Point of Measurement	Size and Pit of New Parts (inches)	Wear Limit (inches)
1	Main bearings inside diameter	3.6052 to 3.6072	3.6092
1	Oil clearance between crankshaft main journals and main bearings	0.0032 to 0.0062	0.0080

FRAME 74

1. If any measurement taken in frame 73 is not within limits given in table 4-19, skip step 2 and do step 3.

2. Count the number of main bearing oil clearance measurements done in step 2 of frame 73 that are more than 0.0062 inch. If four or more oil clearance measurements are more than 0.0062 inch, do step 3. If three or fewer oil clearance measurements are more than 0.0062 inch, go to frame 75.

3. Take out two screws and washers (1) from each of seven main bearing caps (2).

4. Take off seven main bearing caps (2). Take out and throw away all 14 main bearing halves (3) and get a new 0.0200-inch undersize main bearing kit.

NOTE

If main bearing halves (3) were changed, do frames 56 to 74 again for new main bearing halves.

GO TO FRAME 75

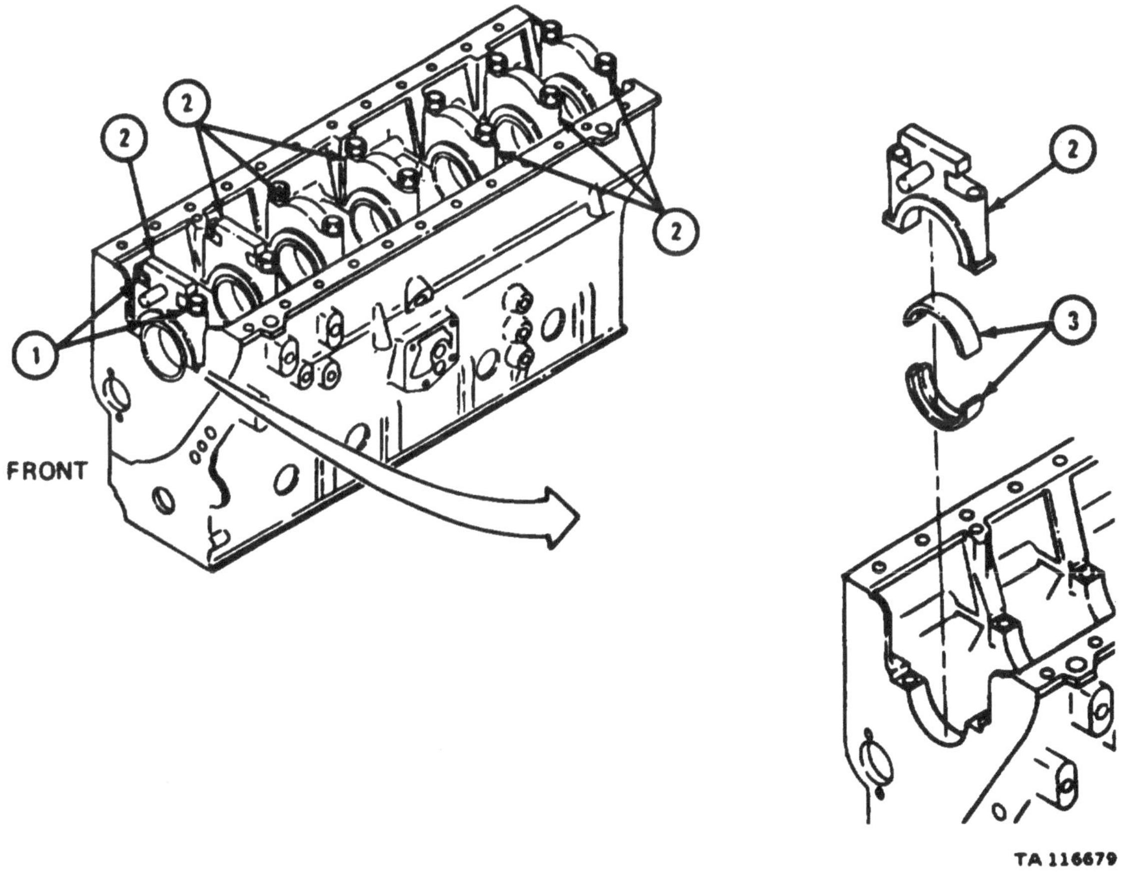

FRAME 75

NOTE

Main bearing caps (1), main bearing halves (2), and main bearing saddles (3) are called by numbers one to seven counting from front to rear of crankcase (4).

1. Take out two bolts and washers (5) and take out number one main bearing cap (1). Take off two main bearing halves (2).
2. Check that 75 percent of surface (6) of number one main bearing cap (1) is coated with Prussian blue. If less than 75 percent of surface of number one main bearing cap is coated with prussian blue, main bearing halves (2) are worn out.
3. Check that 75 percent of surface of number one main bearing saddle (3) is coated with Prussian blue. If less than 75 percent of surface of number one main bearing cap is coated with Prussian blue, main bearing halves (2) are worn out.
4. Do steps 1, 2, and 3 again for all other main bearing caps (1).

IF ANY MAIN BEARING HALVES ARE WORN OUT, GO TO FRAME 76.
IF NO MAIN BEARING HALVES ARE WORN OUT, END OF TASK

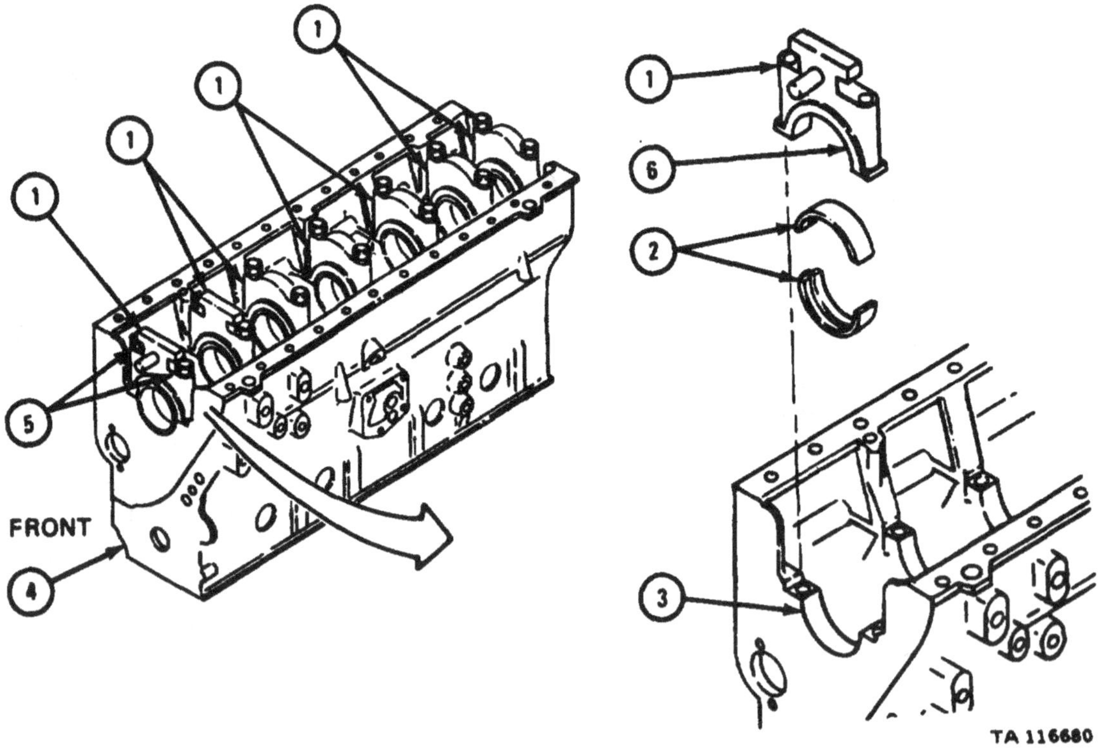

FRAME 76

NOTE

Different crankshafts (1) use different sizes of main bearing halves (2) .

Size of crankshaft (1) was noted in frame 31.

1. Throw away all 14 main bearing halves (2).
2. Check size of crankshaft noted in frame 31.
3. For early and late model standard size crankshafts (1), get a new standard size main bearing kit.
4. For 0.0100-inch undersize crankshaft s (1), get a new 0.0100-inch undersize main bearing kit.
5. For 0.0200-inch undersize crankshaft s (1), get a new 0.0200-inch undersize main bearing kit.

END OF TASK

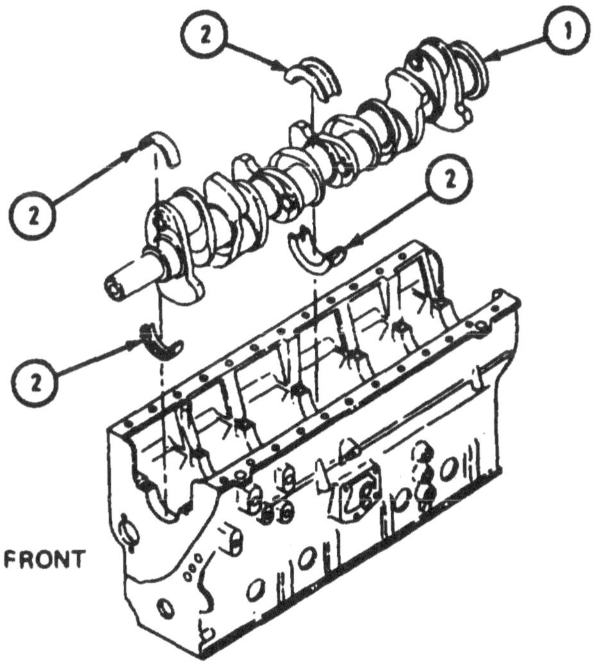

TA 116681

e. Inspection and Repair of Camshaft and Related Camshaft Parts.

CAUTION

It is easy to damage the equipment if you do not know what you are doing. Do not try to do this task unless you are experienced at it, or you have an experienced person with you.

FRAME 1

1. Using strong light and magnifying glass, check that camshaft (1) has no cracks. If camshaft is cracked, throw it away and get a new one.
2. Check that camshaft lobes (2) and bearing surfaces (3) are not worn, scuffed or scored . If camshaft lobes or bearing surfaces are badly worn, scuffed or scored, throw away camshaft (1) and get a new one.
3. Check that camshaft threads (4) are not stripped or damaged. If camshaft threads are stripped or damaged, throw away camshaft and get a new one.

WARNING

Dry cleaning solvent is flammable. Do not use near an open flame. Keep a fire extinguisher nearby when solvent is used . Use only in well-ventilated places. Failure to do this may result in injury to personnel and damage to equipment.

4. Check that camshaft (1) has no small nicks, scratches or burrs. Fix small nicks, scratches, and burrs using crocus cloth and dry cleaning solvent, or a fine mill file.

GO TO FRAME 2

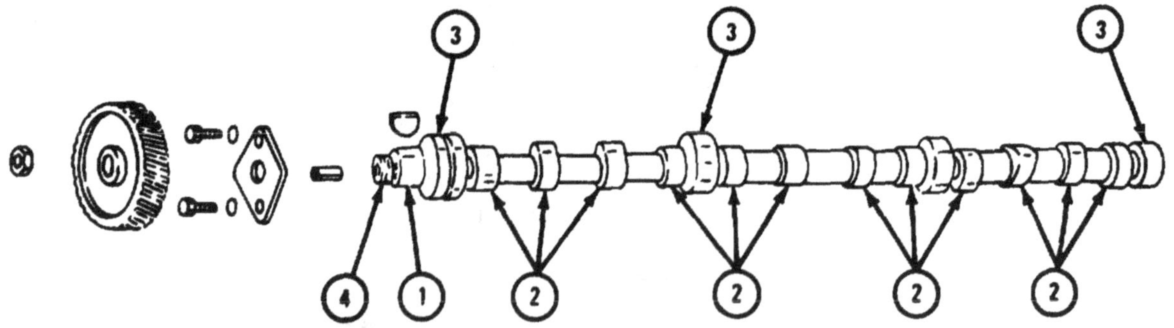

NOTE: CHECK ONLY THOSE PARTS WHICH ARE CALLED OUT IN THIS FRAME. PARTS WITHOUT CALLOUTS ARE SHOWN ONLY FOR REFERENCE PURPOSES OR ARE CHECKED IN ANOTHER FRAME.

TA 118784

FRAME 2

1. Check that woodruff key (1) is not bent, cracked or chipped. If woodruff key is bent, cracked or chipped, throw it away and get a new one.

WARNING

Dry cleaning solvent is flammable. Do not use near an open flame. Keep a fire extinguisher nearby when solvent is used. Use only in well-ventilated places. Failure to do this may result. in injury to personnel and damage to equipment.

2. Check that woodruff key (1) has no small nicks, scratches or burrs. Fix small nicks, scratches, or burrs with a fine mill file or crocus cloth dipped in dry cleaning solvent.
3. Check that woodruff key (1) fits snugly in keyway (2). If it does not, throw it away and get a new one.
4. If woodruff key (1) was changed, check that new woodruff key fits snugly in keyway (2). If it does not, throw away camshaft (3) and get a new one.
5. Check that keyway (2) is not badly damaged. If it is, throw away camshaft (3) and get a new one.

GO TO FRAME 3

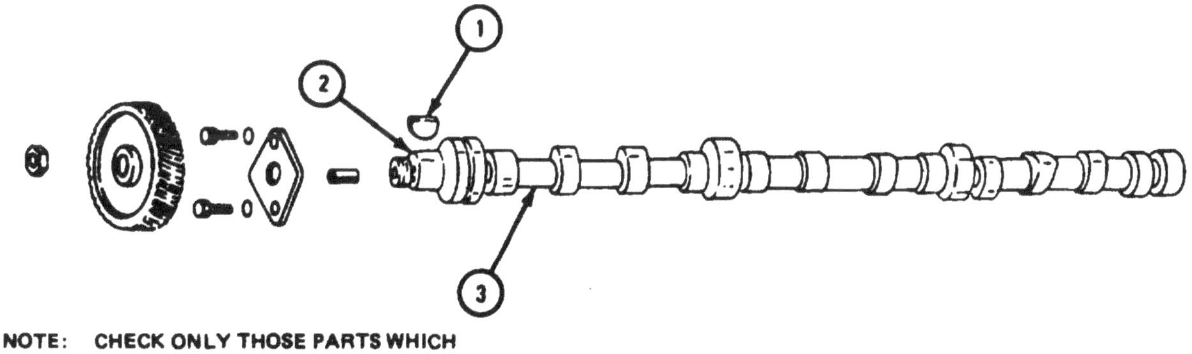

NOTE: CHECK ONLY THOSE PARTS WHICH ARE CALLED OUT IN THIS FRAME. PARTS WITHOUT CALLOUTS ARE SHOWN ONLY FOR REFERENCE PURPOSES OR ARE CHECKED IN ANOTHER FRAME.

TA 118785

FRAME 3

NOTE

Tachometer drive sleeve (1) is inside front of camshaft (2). It is shown out of camshaft here for reference purposes only .

1. Check that tachometer drive sleeve (1) is not cracked or loose. Check that slot in tachometer drive sleeve does not have rounded edges.
2. If tachometer drive sleeve (1) is cracked or loose or if slot in tachometer drive sleeve is rounded out, take out tachometer drive sleeve and press in a new one.
3. Check that camshaft gear nut (3) and two camshaft thrust plate screws (4) do not have damaged threads and are not rounded out. If part has damaged threads or is rounded out, throw it away and get a new one.
4. Check that two camshaft thrust plate lockwashers (5) are not cracked or damaged. If lockwashers are damaged, throw them away and get new ones.

GO TO FRAME 4

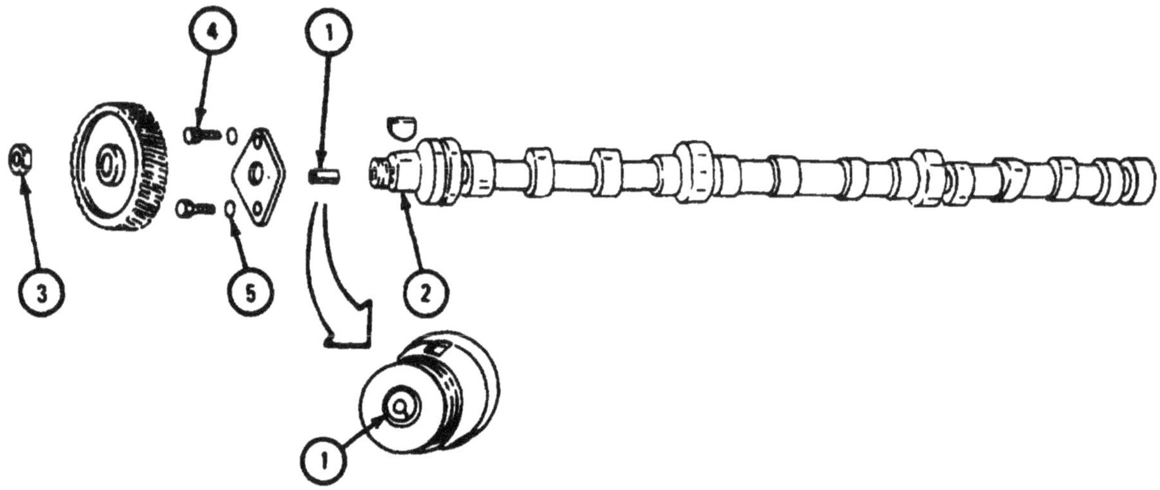

NOTE: CHECK ONLY THOSE PARTS WHICH ARE CALLED OUT IN THIS FRAME. PARTS WITHOUT CALLOUTS ARE SHOWN ONLY FOR REFERENCE PURPOSES OR ARE CHECKED IN ANOTHER FRAME.

TA 118786

FRAME 4

1. Check that neither face of camshaft thrust plate (1) is worn thin in places. Check that neither face of camshaft thrust plate has grooves worn in it.
2. If either face of camshaft thrust plate (1) is groooved or badly worn down, throw it away and get a new one.

WARNING

Dry cleaning solvent is flammable. Do not use near an open flame. Keep a fire extinguisher nearby when solvent is used. Use only in well-ventilated places. Failure to do this may result in injury to personnel and damage to equipment.

3. Check that neither face of camshaft thrust plate (1) has small nicks, scratches or burrs. To fix small nicks, scratches, and burrs, put crocus cloth soaked in dry cleaning solvent on a flat surface. Then rub face of camshaft thrust plate against crocus cloth.
4. Measure thickness of camshaft thrust plate (1) near hole in center. Reading must be within limits given in table 4-20. If readings are not within given limits, throw away camshaft thrust plate and get a new one.

GO TO FRAME 5

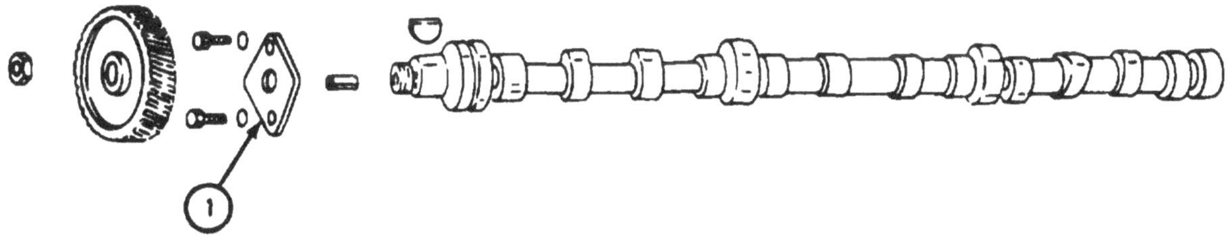

NOTE: CHECK ONLY THOSE PARTS WHICH ARE CALLED OUT IN THIS FRAME. PARTS WITHOUT CALLOUTS ARE SHOWN ONLY FOR REFERENCE PURPOSES OR ARE CHECKED IN ANOTHER FRAME.

TA 118787

Table 4-20. Camshaft Thrust Plate Wear Limits

Index Number	Item/Point of Measurement	Size and Fit of New Parts (inches)	Wear Limit (inches)
1	Camshaft thrust plate thickness	0.2400 to 0.2420	0.2380

FRAME 5

1. Check that camshaft gear (1) is not cracked. If camshaft gear is cracked, throw it away and get a new one.
2. Check that teeth of camshaft gear (1) are not cracked, pitted, chipped or worn. If they are, throw away camshaft gear and get a new one.
3. Check that teeth and keyway of camshaft gear (1) have no small nicks or burrs. Fix minor nicks and burrs with a fine mill file.
4. Check that threads for camshaft gear puller in two small holes (2) are not damaged. Fix minor thread damage with a tap.

GO TO FRAME 6

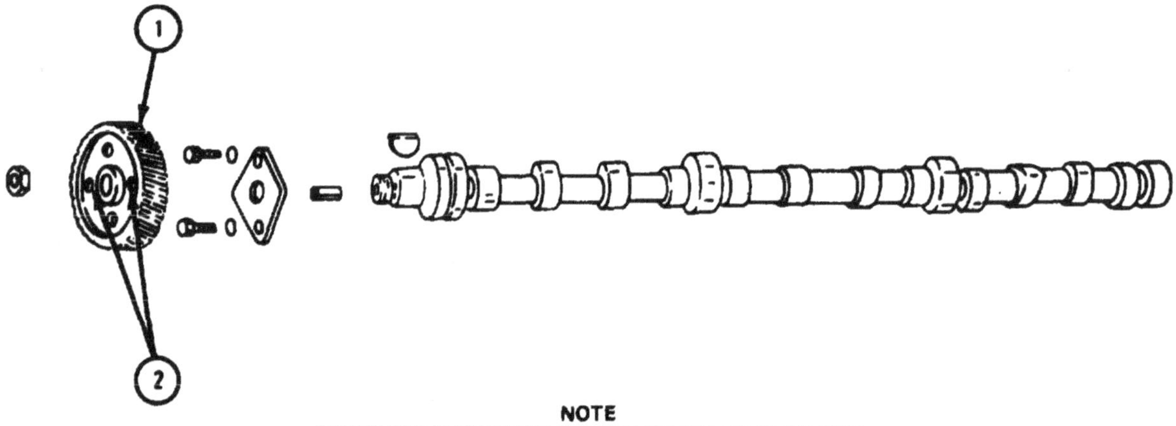

NOTE
CHECK ONLY THOSE PARTS WHICH ARE CALLED OUT IN THIS FRAME. PARTS WITHOUT CALLOUTS ARE SHOWN ONLY FOR REFERENCE PURPOSES OR ARE CHECKED IN ANOTHER FRAME.

TA 118788

FRAME 6

NOTE

Readings must be within limits given in table 4-21. The letter T shows a tight fit. If readings are not within given limits, throw away part and get a new one.

1. Measure inside diameter (1) of camshaft gear (2).
2. Measure outside diameter (3) of front end of camshaft (4).
3. Measure fit of camshaft gear (2) on camshaft (4).

GO TO FRAME 7

NOTE: CHECK ONLY THOSE PARTS WHICH ARE CALLED OUT IN THIS FRAME. PARTS WITHOUT CALLOUTS ARE SHOWN ONLY FOR REFERENCE PURPOSES OR ARE CHECKED IN ANOTHER FRAME.

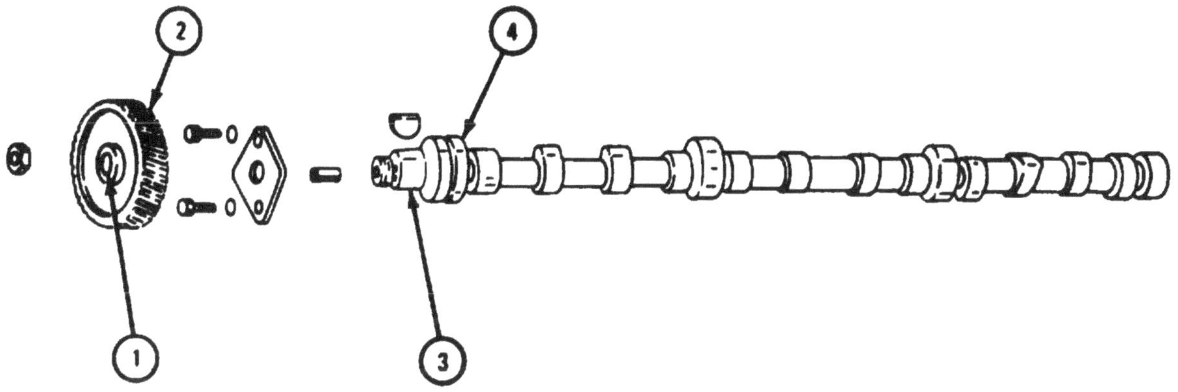

TA 118789

Table 4-21. Camshaft Gear Wear Limits

Index Number	Item/Point of Measurement	Size and Fit of New Parts (inches)	Wear Limit (inches)
1	Camshaft gear inside diameter	1.3744 to 1.3799	None
3	Camshaft front end out side diameter	1.3752 to 1.3758	None
2 and 4	Fit of camshaft gear on camshaft	0.0003T to 0.0014T	None

FRAME 7

NOTE

Readings must be within limits given in table 4-22. If readings are not within given limits, throw away part and get a new one.

If camshaft is thrown away, do steps 1 through 4 again for new camshaft.

1. Measure outside diameter of rear camshaft journal (1). Note this measurement.
2. Measure outside diameter of rear intermediate camshaft journal (2). Note this measurement.
3. Measure outside diameter of front intermediate camshaft journal (3). Note this measurement.
4. Measure outside diameter of front camshaft journal (4). Note this measurement.

GO TO FRAME 8

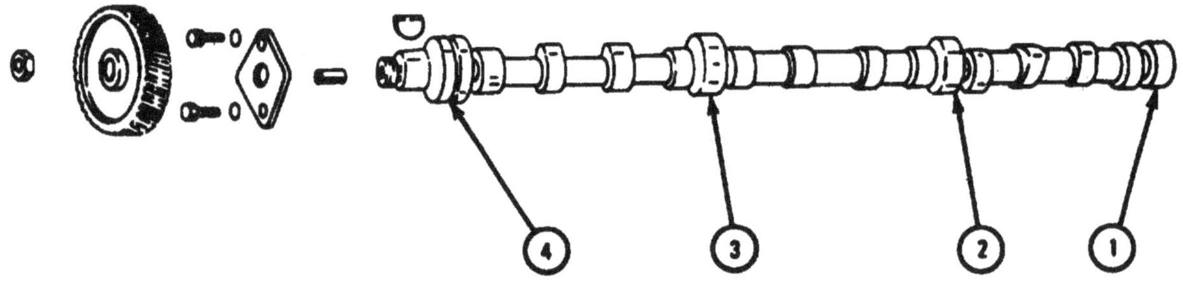

NOTE: CHECK ONLY THOSE PARTS WHICH ARE CALLED OUT IN THIS FRAME. PARTS WITHOUT CALLOUTS ARE SHOWN ONLY FOR REFERENCE PURPOSES OR ARE CHECKED IN ANOTHER FRAME.

Table 4-22. Camshaft Journals Wear Limits

Index Number	Item/Point of Measurement	Size and Fit of New Parts (inches)	Wear Limit (inches)
1	Rear camshaft journal outside diameter	1.9410 to 1.9920	1.9890
2	Rear intermediate camshaft journal outside diameter	2.3035 to 2.3045	2.3015
3	Front intermediate camshaft journal outside diameter	2.3660 to 2.3670	2.3640
4	Front camshaft journal outside diameter	2.4285 to 2.4295	2.4265

FRAME 8

CAUTION

An oil transfer liner (1) is found in crankcase web (2). It is there to pipe oil through crankcase web into main oil gallery in crankcase (3). Do not take out this oil transfer liner .

1. Check that rear camshaft bearing (4), rear intermediate camshaft bearing (5), front intermediate camshaft bearing (6), and front camshaft bearing (7) are not pitted or worn or discolored from overheating.
2. If any camshaft bearing (4, 5, 6, or 7) is pitted or worn or discolored from overheating, take them all out using camshaft bushing remover and replacer kit. Then go to frame 10 .
3. Check that rear camshaft bearing (4), rear intermediate camshaft bearing (5), front intermediate camshaft bearing (6), and front camshaft bearing (7) have no nicks, burrs or cracks .
4. If any camshaft bearing (4, 5, 6, or 7) has nicks, burrs or cracks, take them all out using camshaft bushing remover and replacer kit. Then go to frame 10.

GO TO FRAME 9

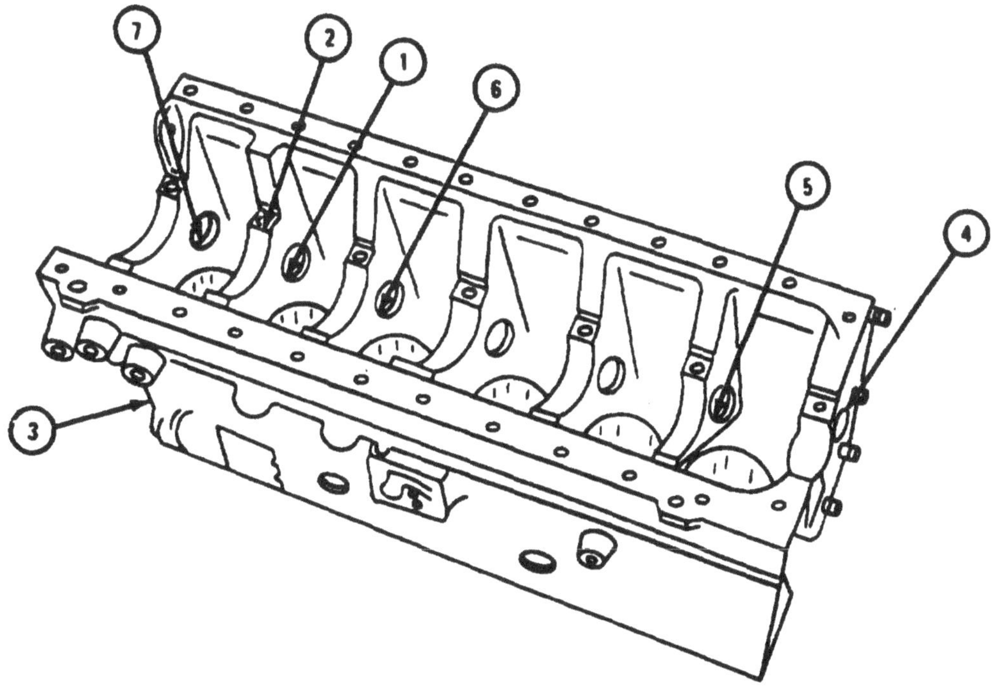

TA 118791

FRAME 9

NOTE

Readings must be within limits given in table 4-23. If readings are more than given limits, take out all four camshaft bearings (1 to 4) using camshaft bushing remover and replacer kit. Then go to frame 10.

1. Measure inside diameter of camshaft rear bearing (1). Note measurement.
2. Measure inside diameter of camshaft rear intermediate camshaft bearing (2). Note measurement.
3. Measure inside diameter of camshaft front intermediate camshaft bearing (3). Note measurement.
4. Measure inside diameter of camshaft front bearing (4). Note measurement.

GO TO FRAME 13

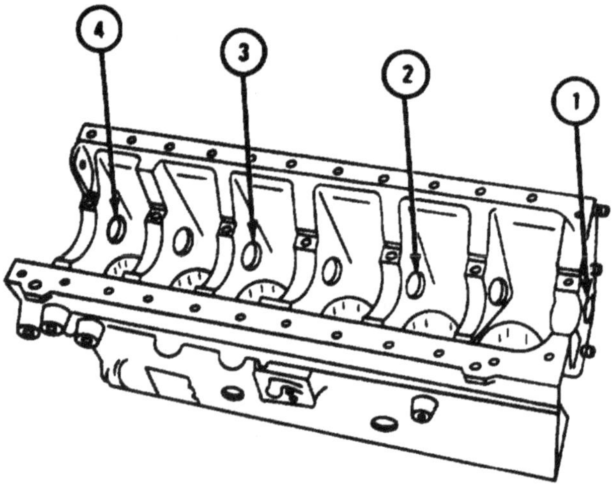

Table 4-23. Camshaft Bearing Wear Limits

Index Number	Item/Point of Measurement	Size and Fit of New Parts (inches)	Wear Limit (inches)
1	Camshaft rear bearing inside diameter	1.9940 to 1.9950	1.9970
2	Camshaft rear intermediate bearing inside diameter	2.3065 to 2.3075	2.3095
3	Camshaft front intermediate bearing inside diameter	2.3690 to 2.3700	2.3720
4	Camshaft front bearing inside diameter	2.4315 to 2.4325	2.4345

FRAME 10

NOTE

Readings must be within limits given in table 4-24. If readings are not within given limits, throw away crankcase (1) and get a new one.

If crankcase (1) is thrown away, get a new camshaft bearing kit. New crankcase does not come with camshaft bearings.

1. Measure camshaft rear bearing bore (2).
2. Measure camshaft rear intermediate bearing bore (3).
3. Measure camshaft front intermediate bearing bore (4).
4. Measure camshaft front bearing bore (5).

GO TO FRAME 11

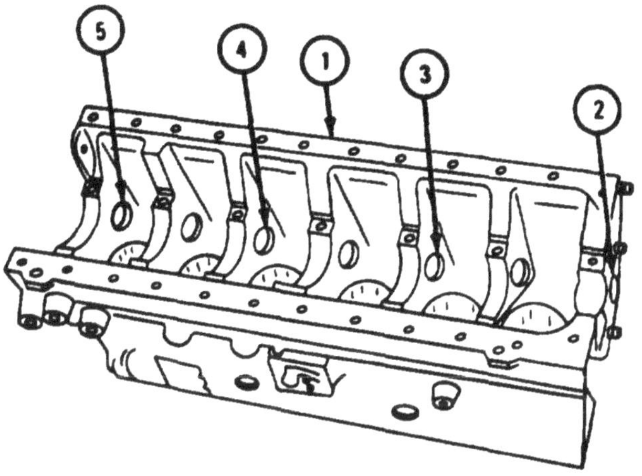

Table 4-24. Camshaft Bearing Bore Wear Limits

Index Number	Item/Point of Measurement	Size and Fit of New Parts (inches)	Wear Limit (inches)
2	Camshaft rear bearing bore inside diameter	2.1245 to 2.1255	None
3	Camshaft rear intermediate bearing bore inside diameter	2.4270 to 2.4280	None
4	Camshaft front intermediate bearing bore inside diameter	2.4995 to 2.5005	None
5	Camshaft front bearing bore inside diameter	2.5620 to 2.5630	None

FRAME 11

NOTE

There are two oil holes (1) in camshaft rear bearing bore (2). There are two oil holes (3) in rear camshaft bearing (4) .

Oil holes (3) are a little bit closer to one end of rear camshaft bearing (4) than to other end. Rear camshaft bearing must be put in rear crankcase web (5) so that end of rear camshaft bearing with oil holes closest to it faces front of crankcase, as shown.

1. Mark rear crankcase web (5) with chalk or grease pencil next to oil holes (1) in camshaft rear bearing bore (2) .

CAUTION

Both oil holes (1) in camshaft rear bearing bore (2) must line up with both oil holes (3) in rear camshaft bearing (4). Alinement of oil holes need not be perfect, but more than half the area of oil holes (3) must line up with oil holes (1). If this is not done, oil flow will be blocked and engine will be damaged.

2. Using camshaft bushing remover and replacer kit, put in new rear camshaft bearing (4). Use chalk mark on crankcase web (5) to help line up oil holes (3) with oil holes (1).

GO TO FRAME 12

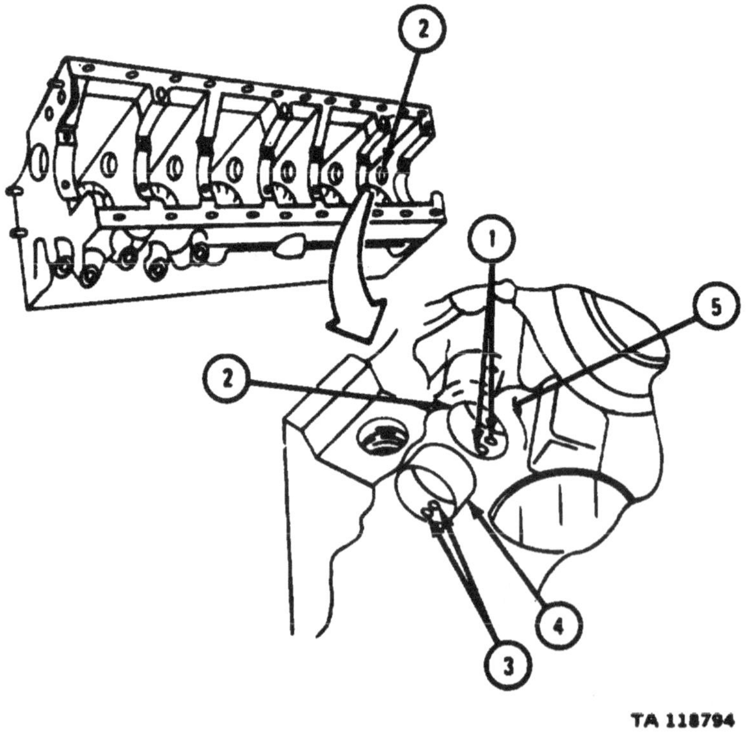

FRAME 12

NOTE

There is one oil hole (1) in camshaft front bearing bore (2), camshaft front intermediate bearhg bore (3), and camshaft rear intermediate bearing bore (4). There is one oil hole in front camshaft bearing (5), front intermediate camshaft bearing (6) and rear intermediate camshaft bearing (7).

1. Mark three crankcase webs (8) with chalk or grease pencil next to oil hole (1) in camshaft front bearing bore (2), camshaft front intermediate bearing bore (3), and camshaft rear intermediate bearing bore (4).

CAUTION

Oil holes (1) in camshaft bearing bores (2, 3, and 4) must line up with oil holes in camshaft bearings (5, 6, and 7). If oil holes do not line up, oil flow will be blocked and camshaft bearings and camshaft will be damaged.

NOTE

Oil hole in front camshaft bearing (5), front intermediate camshaft bearing (6), and rear intermediate camshaft bearing (7) is exactly at center of bearing. Because of this, camshaft bearings (5, 6, and 7) can be put in facing either way.

2. Using camshaft bushing remover and replacer kit, put in new front camshaft bearing (5), front intermediate camshaft bearing (6), and rear intermediate camshaft bearing (7). Use chalk marks on crankcase webs (8) to help line up oil hole in camshaft bearing bores (2, 3, and 4) with oil hole in camshaft bearings (5, 6, and 7).

3. Do frame 9 again.

GO TO FRAME 13

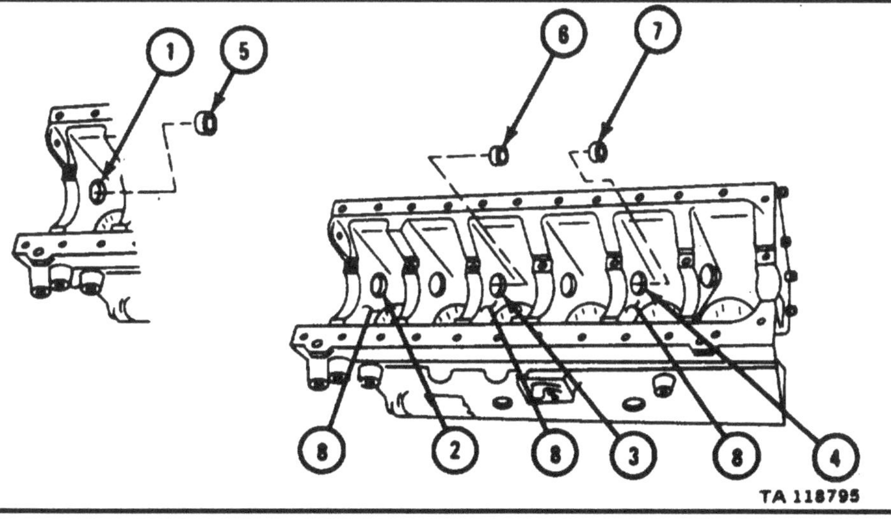

FRAME 13

NOTE

Readings must be within limits given in table 4-25. If readings are not within given limits, throw away part and get a new one.

1. Subtract measurement of outside diameter of rear camshaft journal (1) noted in step 1 of frame 7 from measurement of inside diameter of rear camshaft bearing (2) noted in step 1 of frame 9.
2. Subtract measurement of outside diameter of rear intermediate camshaft journal (3) noted in step 2 of frame 7 from measurement of inside diameter of rear intermediate camshaft bearing (4) noted in step 2 of frame 9.

GO TO FRAME 14

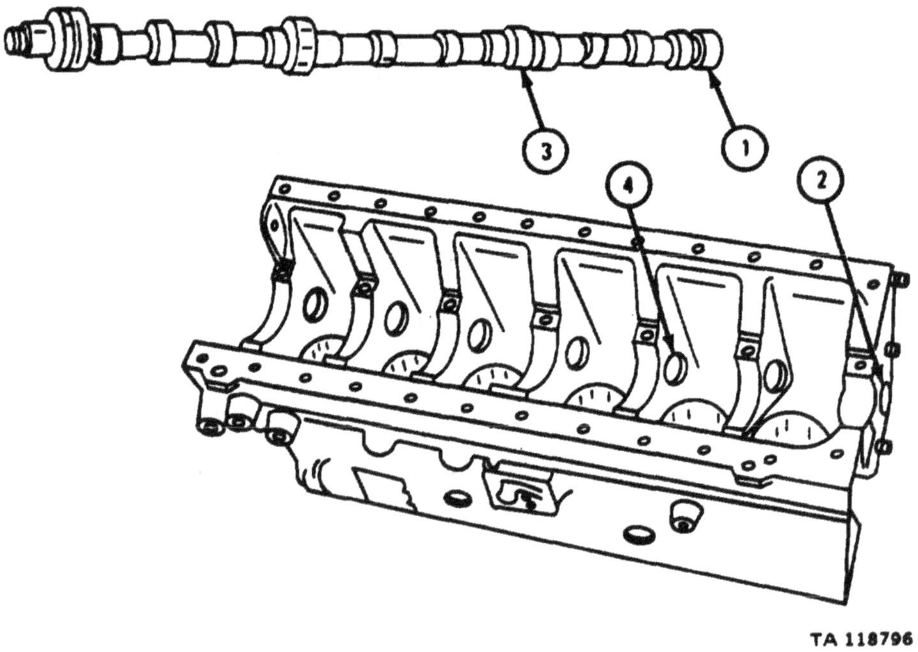

Table 4-25. Rear Camshaft Journals and Camshaft Bearings Wear Limits

Index Number	Item/Point of Measurement	Size and Fit of New Parts (inches)	Wear Limit (inches)
1 and 2	Fit of rear camshaft journal in rear camshaft bearing	0.0020 to 0.0040	0.0060 max
3 and 4	Fit of rear intermediate camshaft journal in rear intermediate camshaft bearin g	0.0020 to 0.0040	0.0060 max

FRAME 14

NOTE

Readings must be within limits given in table 4-26. If readings are not within given limits, throw away part and get a new one.

1. Subtract measurement of outside diameter of front intermediate camshaft journal (1) noted in step 3 of frame 7 from measurement of inside diameter of front intermediate camshaft bearing (2) noted in step 3 of frame 9.
2. Subtract measurement of outside diameter of front camshaft journal (3) noted in step 4 of frame 7 from measurement of inside diameter of front camshaft bearing (4) noted in step 4 of frame 9.

END OF TASK

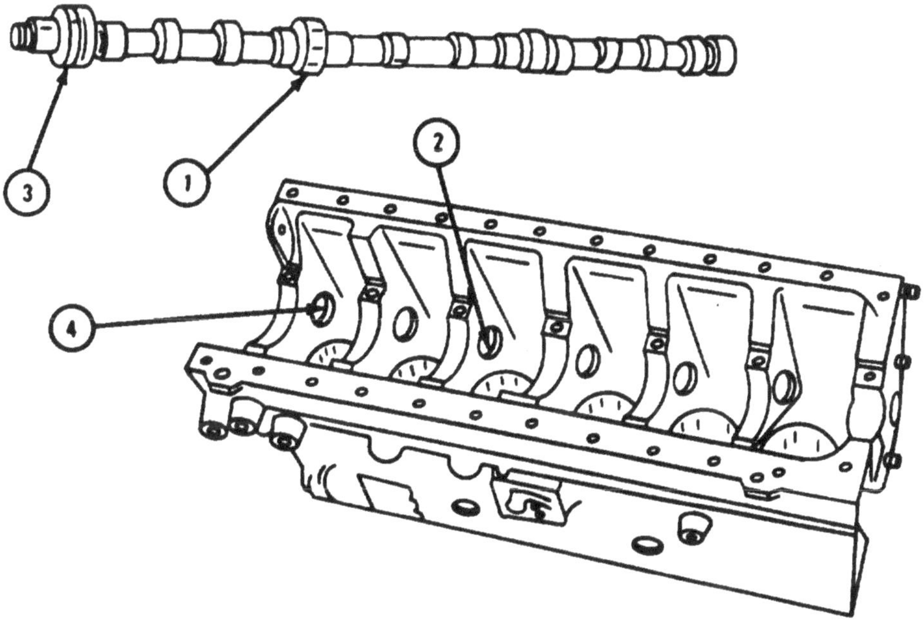

Table 4-26. Front Camshaft Journals and Camshaft Bearings Wear Limits

Index Number	Item/Point of Measurement	Size and Fit of New Parts (inches)	Wear Limit (inches)
1 and 2	Fit of front intermediate camshaft journal in front intermediate camshaft bearin g	0.0020 to 0.0040	0.0060 max
3 and 4	Fit of front camshaft journal in front camshaft bearing	0.0020 to 0.0040	0.0060 max

f. Inspection and Repair of Valve Tappets.

FRAME 1

1. Turn crankcase (2) right side up .
2. Check that 12 valve tappet bores (1) have no nicks, scratches or burrs and that they are not scored. Fix small nicks, scratches and burrs with a fine mill rat tail file or crocus cloth.

WARNING

Dry cleaning solvent is flammable. Do not use near an open flame. Keep a fire extinguisher nearby when solvent is used . Use only in well-ventilated places. Failure to do this may result in injury to personnel and damage to equipment.

3. Check that 12 valve tappets (3) have no nicks, scratches or burrs and that they are not scored . Fix small nicks, scratches, burrs, and scoring with a fine mill file or crocus cloth dipped in dry cleaning solvent. If any tappet is badly scored or damaged, throw it away and get a new one.

NOTE

Valve tappets (3) that are discolored from overheating will look blue.

4. Check that 12 valve tappets (3) are not discolored from overheating. If any valve tappet is discolored from overheating, throw it away and get a new one.
5. Check that pushrod sockets in 12 tappets (3) are not damaged. If pushrod socket in any valve tappet is damaged, throw away valve tappet and get a new one.
6. Check that bottom of 12 valve tappets (3) shows an even wear pattern all the way around . If bottom of any tappet is worn or scuffed more on one side, throw it away and get a new one.

GO TO FRAME 2

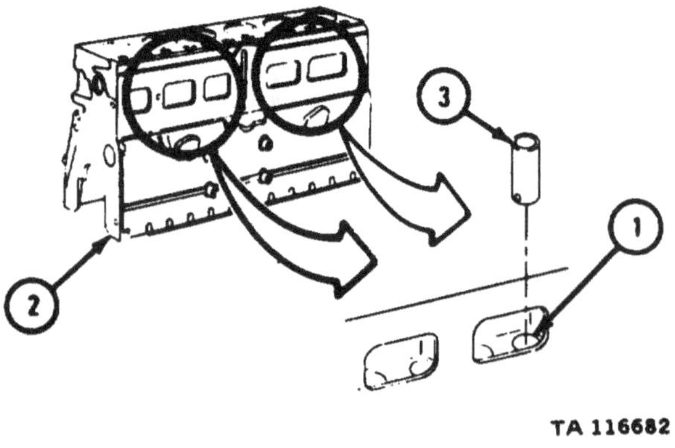

FRAME 2

NOTE

Readings must be within limits given in table 4-27. The letter L shows a loose fit. If readings are no t within given limits, throw away part and get a new one.

Valve tappet bores (1) and valve tappets (2) are called by numbers one to 12 counting from front to rear of crankcase. Valve tappets were tagged with numbers during engine disassembly .

Some early model engines used valve tappets with curved bottoms. Late model engines use valve tappets with flat bottoms.

1. Check whether valve tappets (2) have flat or curved bottoms. Use wear limits in table 4-27 for valve tappets you are working on.
2. Measure inside diameter of 12 tappet bores (1) and outside diameter of 12 valve tappets (2) .
3. Measure fit of each of 12 valve tappets (2) in proper valve tappet bores (1).

END OF TASK

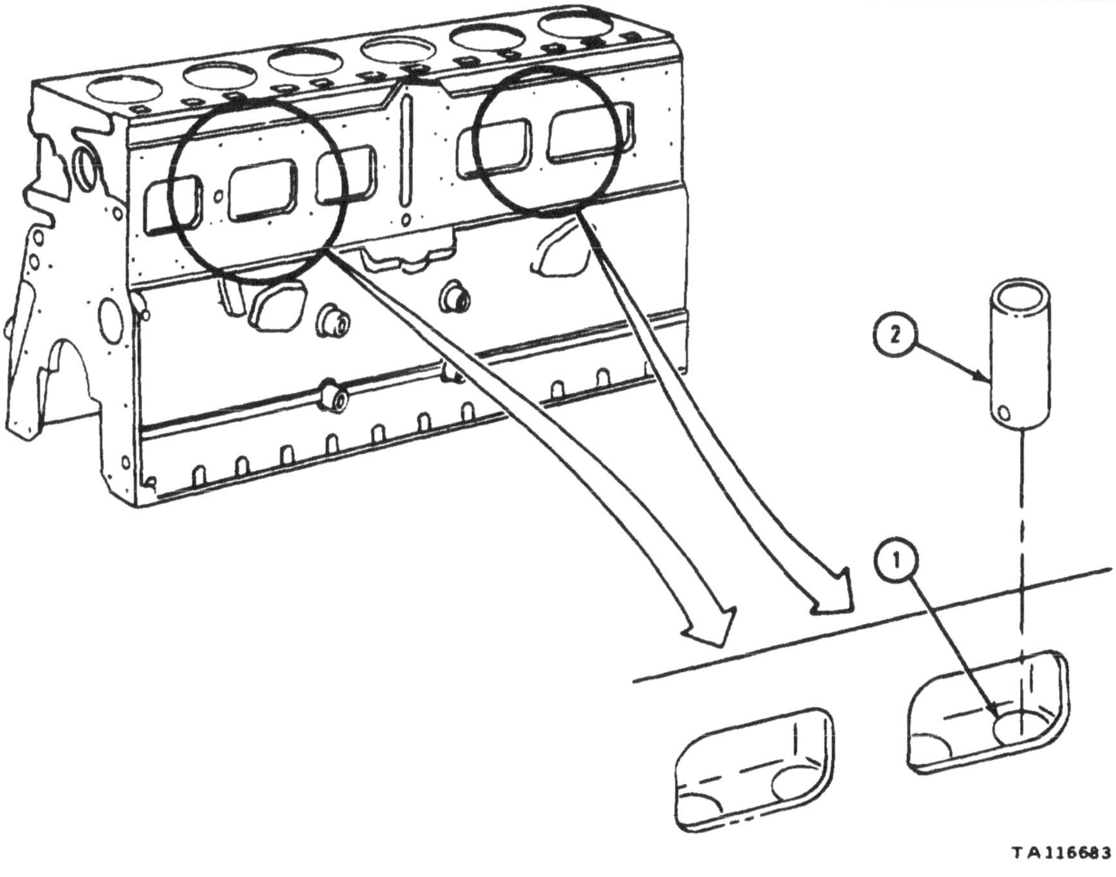

Table 4-27. Valve Tappet Bores and Valve Tappet Wear Limits

Index Number	Item/Point of Measurement	Size and Fit of New Parts (inches)	Wear Limit (inches)
1	Valve tappet bore inside diameter	1.2513 to 1.2523	1.2540
2	Valve tappets with flat bottoms out side diameter	1.2475 to 1.2480	1.2455
2	Valve tappets with curved bottoms outside diameter	1.2485 to 1.2490	1.2465
1 and 2	Fit of valve tappets with flat bottoms in tappet bore	0.0033L to 0.0048L	0.0080L max
1 and 2	Fit of valve tappets with curved bottoms in tappet bore	0.0023L to 0.0038L	0.0070L max

g. Inspection and Repair of Cylinder Bores and Cylinder Sleeves.

CAUTION

It is easy to damage the equipment if you do not know what you are doing. Do not try to do this task unless you are experienced at it, or you have an experienced person with you.

FRAME 1

WARNING

Dry cleaning solvent is flammable. Do not use near an open flame. Keep a fire extinguisher nearby whe n solvent is used. Use only in well-ventilated places . Failure to do this may result in injury to personnel and damage to equipment.

1. Check that 6 counterbores (1) at top of crankcase (2) have no burrs. Fi x small burrs with a crocus cloth dipped in dry cleaning solvent.

GO TO FRAME 2

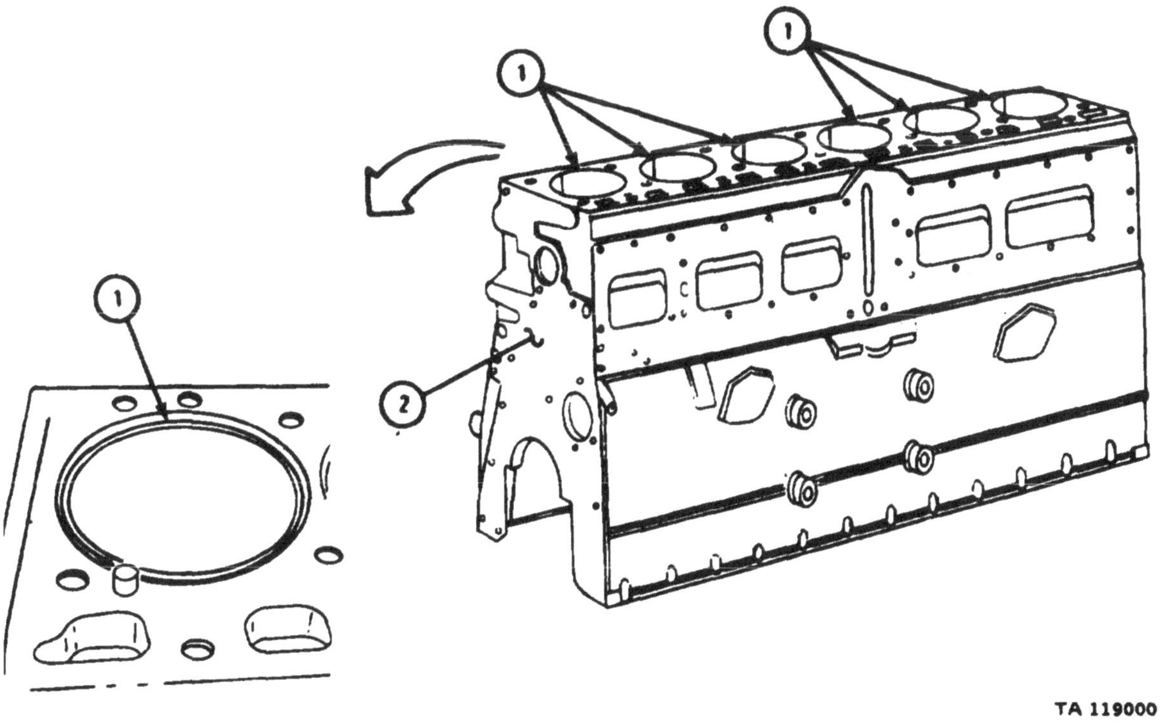

FRAME 2

NOTE

Cylinder bores (2) in crankcase (1) are machined to two different sizes. Cylinder sleeves are also machined to two different sizes to fit different size cylinder bores. The two sizes of cylinder bores and cylinder sleeves are called "Y" and "Z" .

All six cylinder bores (2) are not always the same size.

Cylinder bores (2) and cylinder sleeves are called by numbers one to six counting from front to rear of crankcase.

1. Look at left side of crankcase (1) opposite each cylinder bore (2) about one inch below top of crankcase. The letter "Y" opposite a cylinder bore means this is a Y-size cylinder bore. The letter "Z" opposite a cylinder bore means this is a Z-size cylinder bore.

NOTE

Readings must be within limits given in table 4-28. If readings are not within given limits, throw away part and get a new one.

2. Measure inside diameter of all six cylinder bores (2). Note this measurement and the number cylinder it was taken from.

GO TO FRAME 3

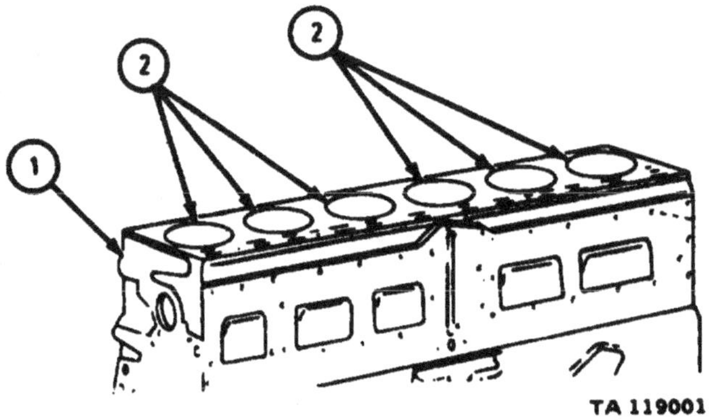

Table 4-28. Cylinder Bore Wear Limits

Index Number	Item/Point of Measurement	Size and Fit of New Parts (inches)	Wear Limit (inches)
2	Y-size cylinder bore inside diameter	4.7516 to 4.7520	None
2	Z-size cylinder bore inside diameter	4.7511 to 407515	None

FRAME 3

CAUTION

Cylinder sleeves (1) are easy to break. Handle them very carefully.

NOTE

Y-size cylinder sleeves (1) are marked with the letter "Y" on outside of sleeve about one inch below top of sleeve. Z-size cylinder sleeves (1) are marked with the letter "Z" on outside of sleeve about one inch below top of sleeve.

1. Get new cylinder sleeves (1) for old cylinder sleeves thrown away during preparation of crankcase for cleaning. Get Y-size cylinder sleeves for Y-size cylinder bores (2). Get Z-size cylinder sleeves for Z-size cylinder bores (2).
2. Check that six cylinder sleeves (1) are not cracked, nicked, scratched or scored. Check that six cylinder sleeves have no small bits of metal stuck in surface. If any cylinder sleeve is cracked, nicked, scratched or scored or has small bits of metal stuck in surface, throw it away and get a new one.

Go TO FRAME 4

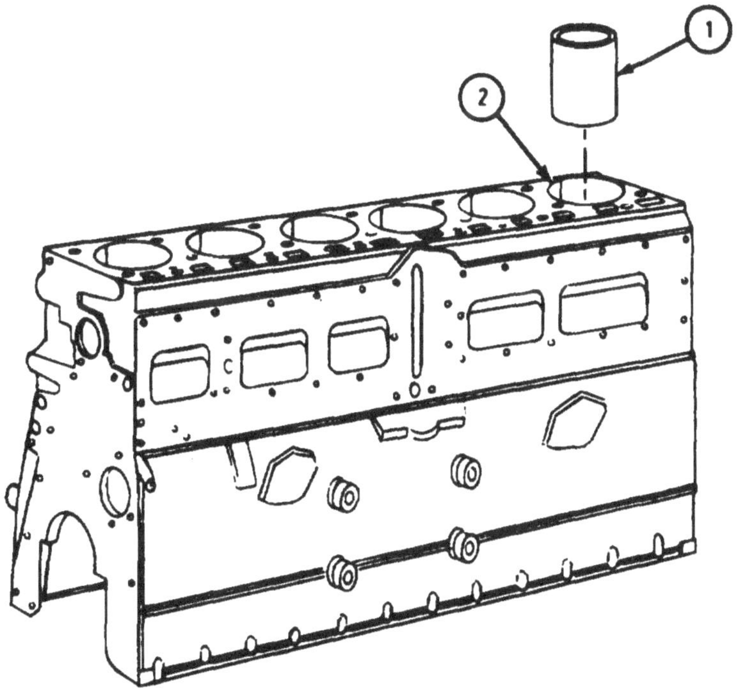

TA 119002

FRAME 4

NOTE

No special tools are needed to put in cylinder sleeves (1). If old cylinder sleeves are used again, they must be put back in the same cylinder bore (2) they were taken from.

1. Coat six cylinder sleeves (1) with engine lubricating oil and put one into each of six cylinder bores (1) . Y-size cylinder sleeves must go in Y-size cylinder bores. Z-size cylinder sleeves must go in Z-size cylinder bores.
2. Using block gage, measure and note height of top of each cylinder sleeve (1) from top surface of crankcase (3) . Cylinder sleeve is properly seated when this height is 0.0005 to 0.005 inch. If height is more than 0.005 inch, push cylinder sleeve further in .
3. Subtract biggest and smallest measurement done on three cylinder sleeves (1) closest to front of crankcase (3) . Difference should not be more than 0.002 inch. If difference is bigger, push highest cylinder sleeves further down and check heights again .
4. Do step 3 again for the three cylinder sleeves (1) closest to rear of crankcase (3).

GO TO FRAME 5

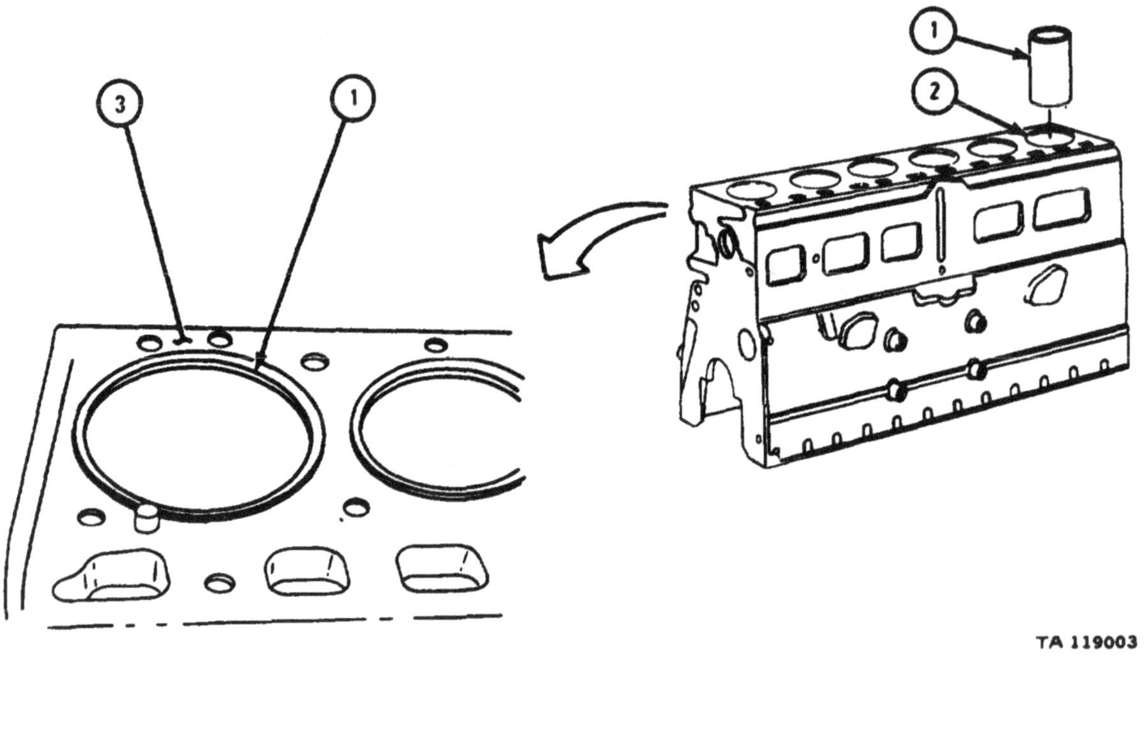

FRAME 5

NOTE

Readings must be within limits givien in table 4-29. The letter L shows a loose fit. If readings are not within given limits, throw away part and get a new one.

Cylinder sleeves (1) can be a little bit out-of-round before they are put in cylinder bores in crankcase (2). Because of this, it is not practical to measure cylinder sleeve diameter when cylinder sleeves are out of cylinder bores. Outside diameter of cylinder sleeves and fit of cylinder sleeves in cylinder bores is given in table 4-29 for reference purposes only.

1. Measure inside diameter of six cylinder sleeves (1) inside crankcase (2). Take measurement one inch from top of cylinder sleeve.

2. Turn crankcase (2) upside down.

END OF TASK

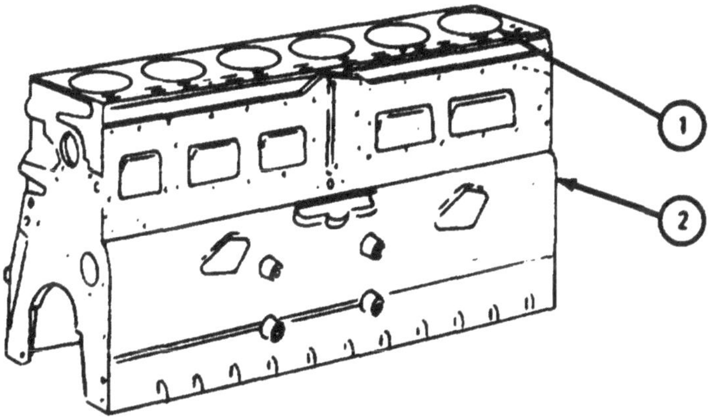

Table 4-29. Cylinder Sleeve Wear Limits

Index Number	Item/Point of Measurement	Size and Fit of New Parts (inches)	Wear Limit (inches)
1	Cylinder sleeve outside diameter	4.7500 to 4.7510	None
1 and 2	Fit of cylinder sleeve in bore of crankcase	0.0005L to 0.0015L	None
1	Cylinder sleeve inside diameter (cylinder sleeve in crankcase)	4.5630 to 4.5645	4.5665

h. Replacement of Expansions Plugs, Pipe Plugs, and Crankcase to Cylinder Head Studs.

FRAME 1

1. If expansion plug (1) behind identification plate (2) was taken out, coat new expansion plug with sealer. Then knock in new expansion plug using mallet and short brass rod until they are about 1/8-inch below surface of crankcase (3).
2. Line up four holes in identification plate (2) with mounting holes for identification plate in crankcase (3). Using hammer, tap in four straight pins (4).
3. Put in crankcase to cylinder head studs (5).

GO TO FRAME 2

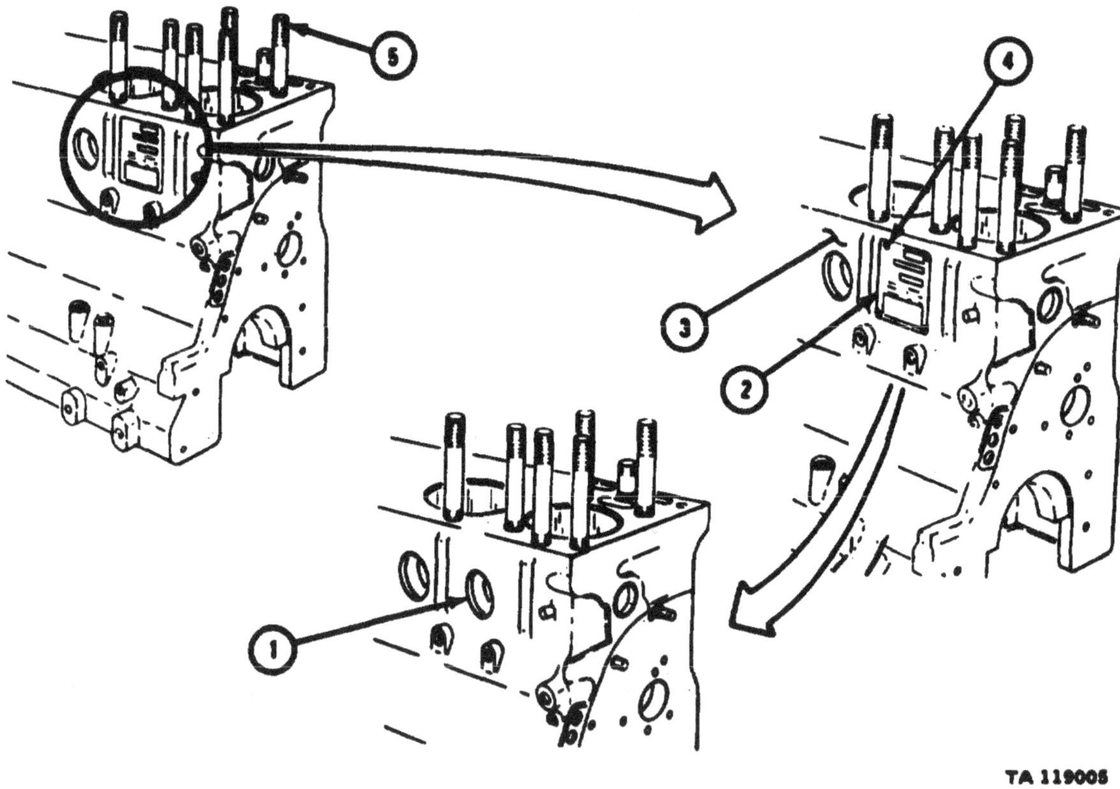

FRAME 2

1. If expansion plugs (1 or 2) were taken out, coat new expansion plugs with sealer. Then knock in new expansion plugs using mallet and short brass rod until expansion plugs are about 1/8 inch below surface of crankcase (3).

2. Put in six pipe plugs (4) .

GO TO FRAME 3

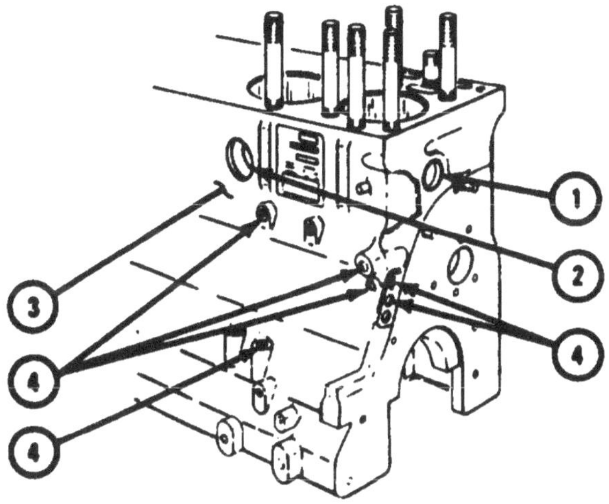

FRAME 3

1. If expansion plugs (1, 2, or 3) were taken out, coat new expansion plugs with sealer. Then knock in new expansion plugs using mallet and short brass rod until they are about 1/8 inch below surface of crankcase (4).

2. Put in five pipe plugs (5) .

END OF TASK

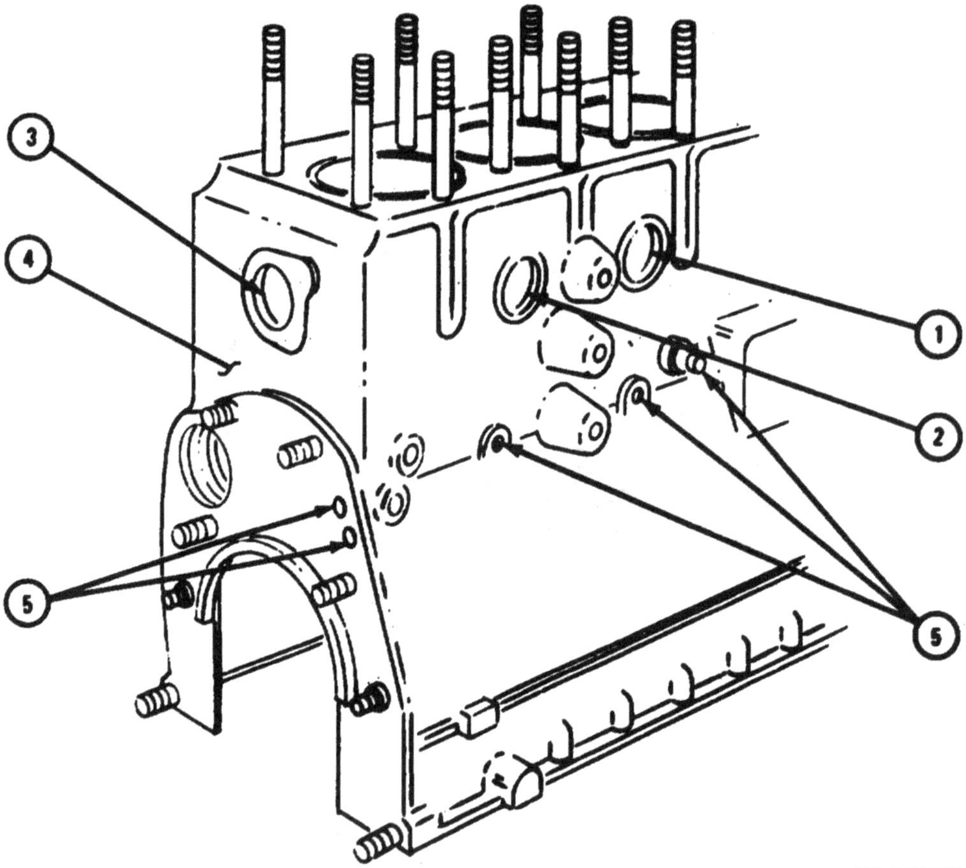

4-7. PISTONS, CONNECTING RODS, AND RELATED PARTS.

a. Cleaning.

WARNING

Use goggles, rubber gloves, and rubber apron when cleaning parts in carbon removing compound. Make sure there is enough ventilation. Be careful not to breathe in fumes or let compound touch skin. If compound is splashed on skin, flush with water and wash with alcohol, preferably 2 to 3 percent.

FRAME 1

1. Clean pistons (1), rings (2), piston pins (3), and retaining rings (4) by soaking in carbon removing compound.
2. Clean piston ring grooves (5) with stiff bristle brush or fiber scraper. Do not scratch or gouge ring grooves.
3. Clean oil holes in lower ring groove of piston (1) with No. 23 (0.154-inch) drill

GO TO FRAME 2

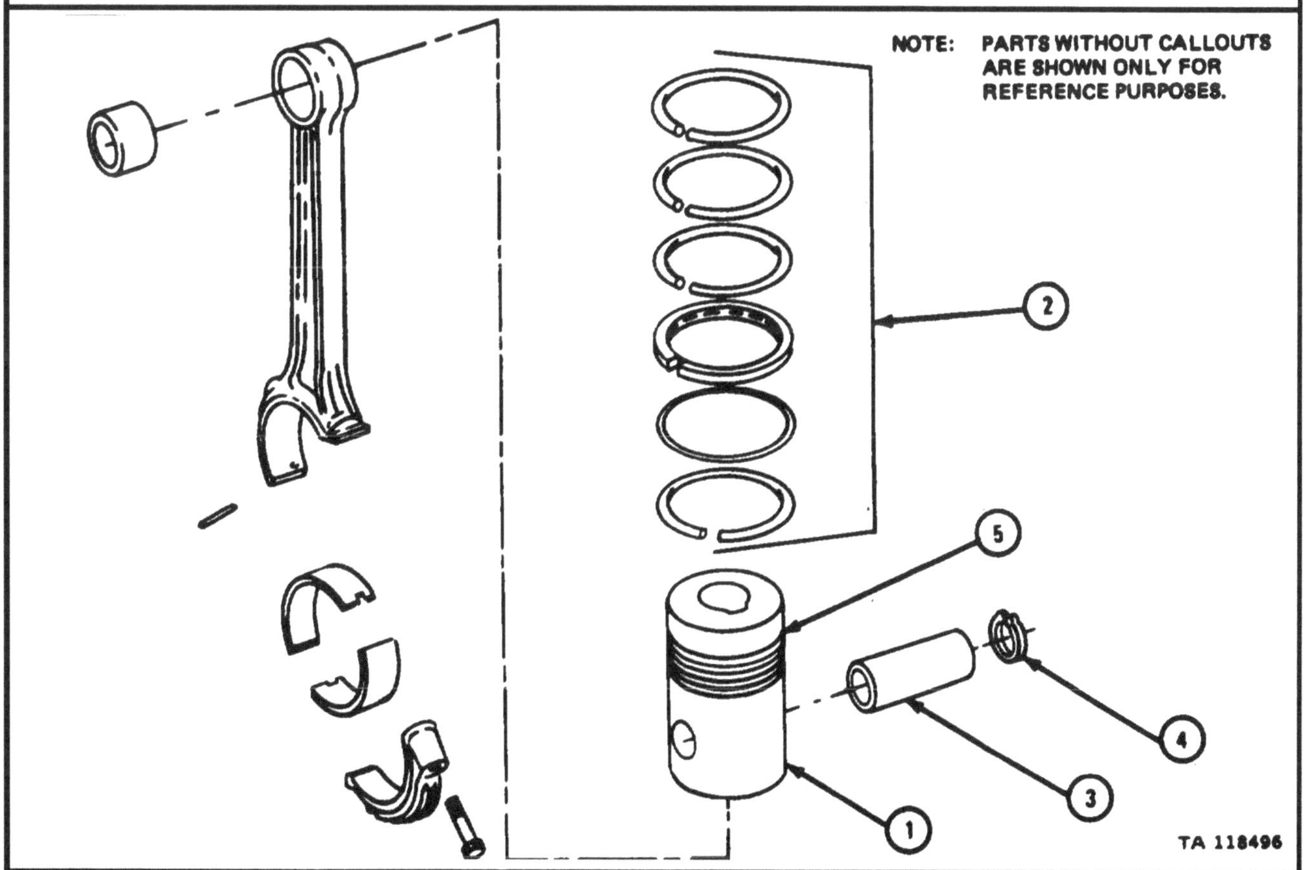
NOTE: PARTS WITHOUT CALLOUTS ARE SHOWN ONLY FOR REFERENCE PURPOSES.
1
2
3
4
5
TA 118496

FRAME 2

1. Clean carbon from cavity (1) in piston heads. Use crocus cloth .

WARNING

Dry cleaning solvent is flammable. Do not use near an open flame. Keep a fire extinguisher nearby when solvent is used . Use only in well-ventilated places. Failure to do this may result in injury to personnel and damage to equipment.

2. Clean carbon from piston pins (2) and bore (3). Use crocus cloth dipped in dry cleaning solvent .
3. Clean connecting rod bearing halves (4) and pin spring (5), Use clot h dipped in dry cleaning solvent. Do not take off bearing identification marking.
4. Clean connecting rod piston pin bearing sleeve (6). Use crocus cloth dipped in dry cleaning solvent .
5. Clean connecting rod (7) and connecting rod cap (8). Refer to para 4-3 .

END OF TASK

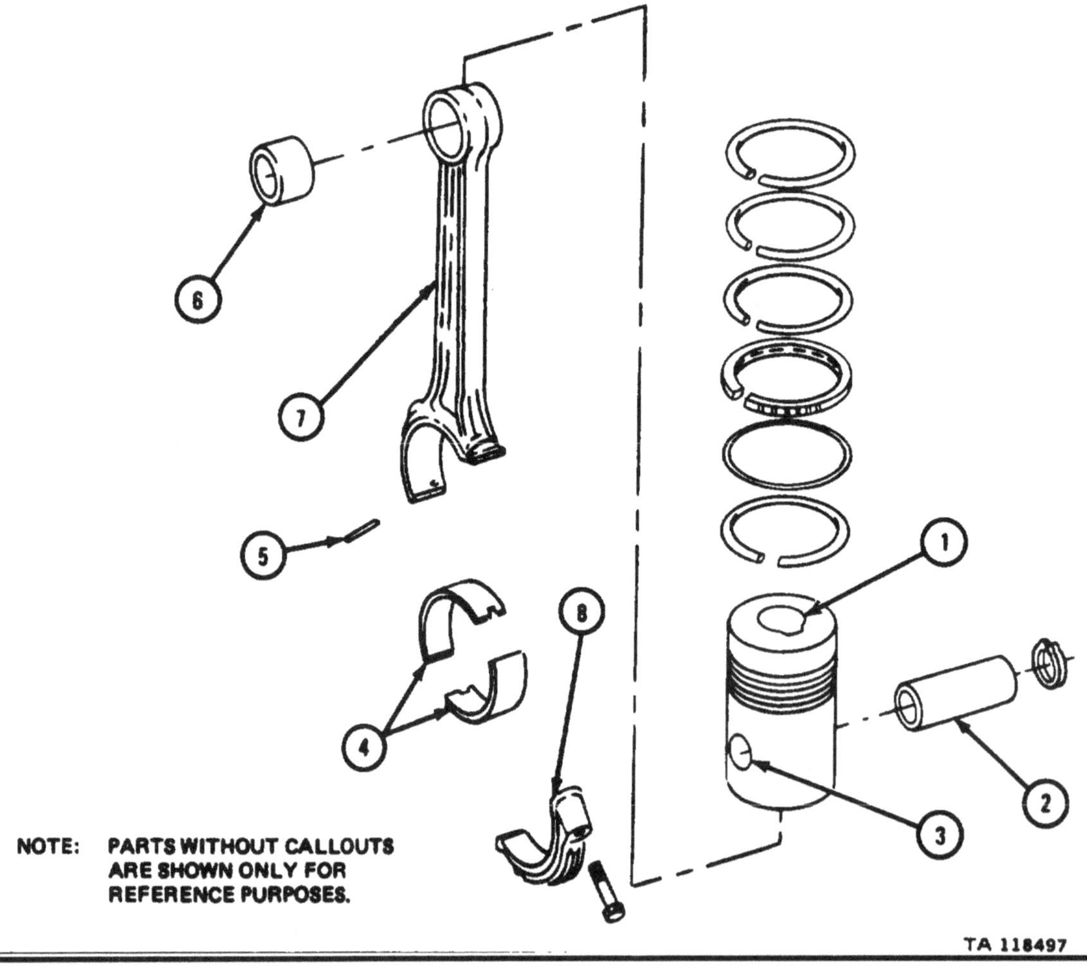

b. Inspection and Repair.

FRAME 1

1. Check that pistons (1) are not burned, scored, scuffed or have any broken ring grooves, deep scratches or signs of metal pick-up. If any piston has such damage, get a new one.
2. Check that spherical combustion chamber (2) of piston (1) is not burned or distorted and does not have cracks that are more than 0.020 inch wide or 5/16 inch in length. Fix minor damage with a fine mill file. If more repair is needed, get a new piston.
3. Check that piston pin bore (3) measurement is between 1.6251 and 1.6263 inches. If piston pin bore is oversized, get a new piston.

GO TO FRAME 2

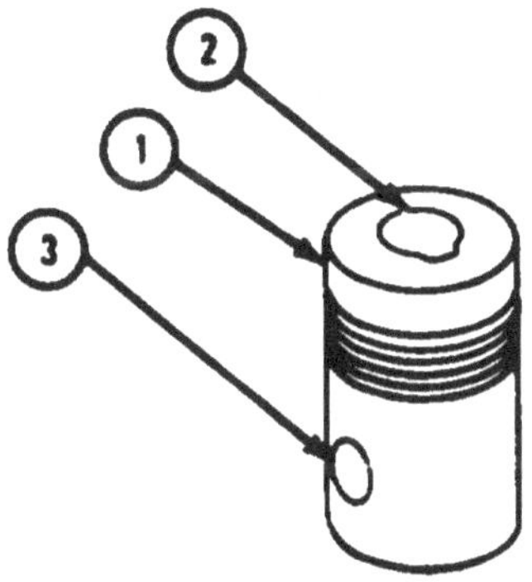

TA 118498

FRAME 2

1. Check that piston rings (1) are not scuffed, scored, chipped or have chrome plating worn through to bare metal. Check closely those rings that may have been sticking in ring grooves . If any rings are damaged in any way, get new ones.

WARNING

Dry cleaning solvent is flammable. Do not use near an open flame. Keep a fire extinguisher nearby when solvent is used . Use only in well-ventilated places. Failure to do this may result in injury to personnel and damage to equipment.

2. Check that piston pin (2) is not cracked, nicked or scratched. Fix mino r damage with a fine mill file or crocus cloth dipped in solvent. If more repair is needed on any pin, get a new one.

3. Check that retainer rings (3) are not cracked or bent. If any retainer ring is damaged in any way, get a new one.

GO TO FRAME 3

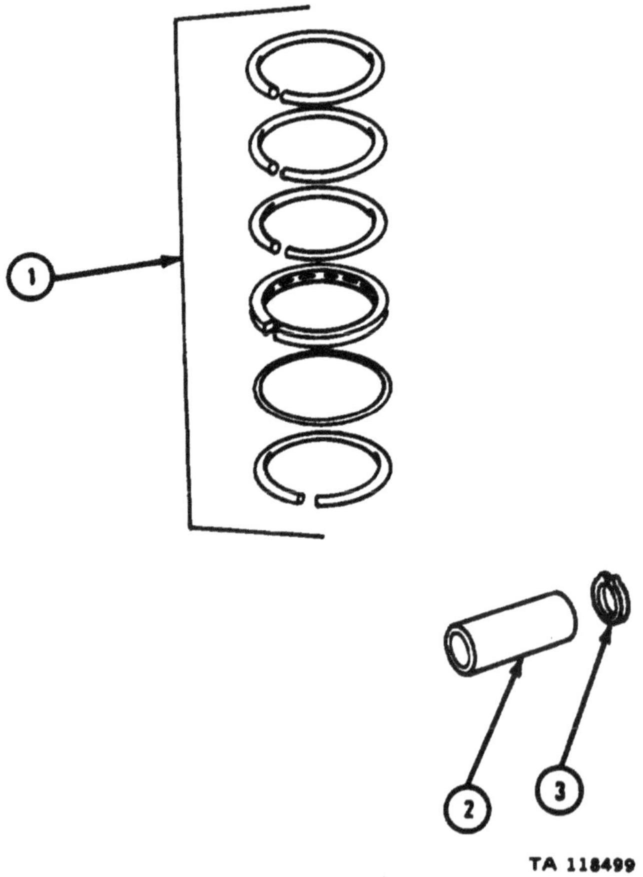

FRAME 3

1. Check that connecting rod bearing halves (1) are not nicked, cracked, scored or showing signs of burns. If bearing halves are damaged in any way, get new ones.
2. Check that connecting rods (2), caps (3), and bolts (4) are not damaged in any way. Fix minor damage to connecting rod and cap with a fine mill file and fix minor thread damage to bolts with a thread chaser. If more repair is needed for any part, get a new one.
3. Check that pin springs (5) are not damaged in any way. If any pin spring is damaged, get a new one.
4. Check that piston pin bearings (6) are not damaged in any way. If any piston pin bearing is damaged, get a new one.

END OF TASK

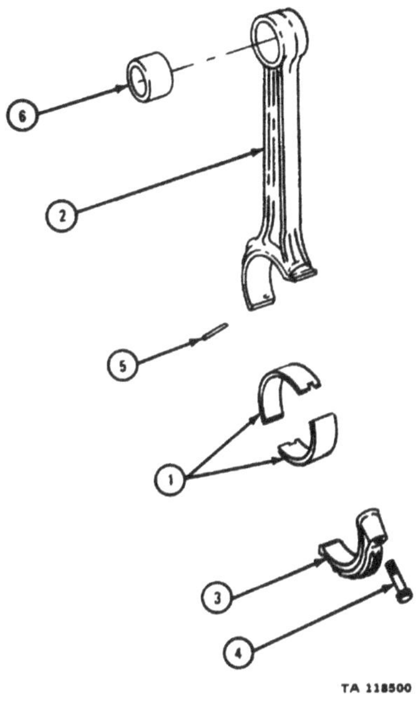

4-8. CRANKSHAFT REAR OIL SEAL HOUSING.

a. Cleaning. There are no special cleaning procedures needed. Refer to cleaning procedures given in para 4-3.

b. Inspection and Repair.

FRAME 1

WARNING

Dry cleaning solvent is flammable. Do not use near an open flame. Keep a fire extinguisher nearby when solvent is used. Use only in well-ventilated places. Failure to do this may result in injury to personnel and damage to equipment.

1. Check that rear oil seal housing (1) has no nicks, burrs, scratches or cracks. Take out minor nicks, burrs or scratches with a fine mill file or crocus cloth dipped in dry cleaning solvent. If more repair is needed get a new part.

GO TO FRAME 2

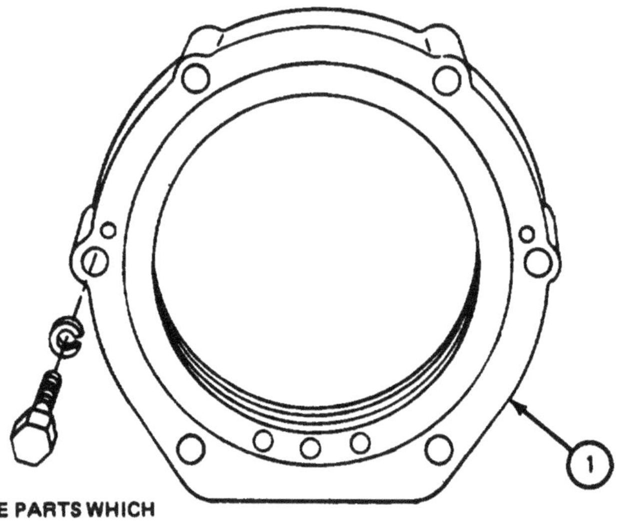

NOTE: CHECK ONLY THOSE PARTS WHICH ARE CALLED OUT IN THIS FRAME. PARTS WITHOUT CALLOUTS ARE SHOWN ONLY FOR REFERENCE PURPOSES OR ARE CHECKED IN ANOTHER FRAME.

TA 087739

FRAME 2

1. Check that six screws (1) are not stripped or damaged. If any screw is damaged, get a new screw.
2. If rear oil seal housing (2) is not damaged and can be used again, check whether it has one oil drain hole (3) or three oil drain holes (3 and 4).
3. If real oil seal housing (2) has only one oil drain hole (3), tell machine shop to drill two more drain holes (4). See Figure 4-1.

END OF TASK

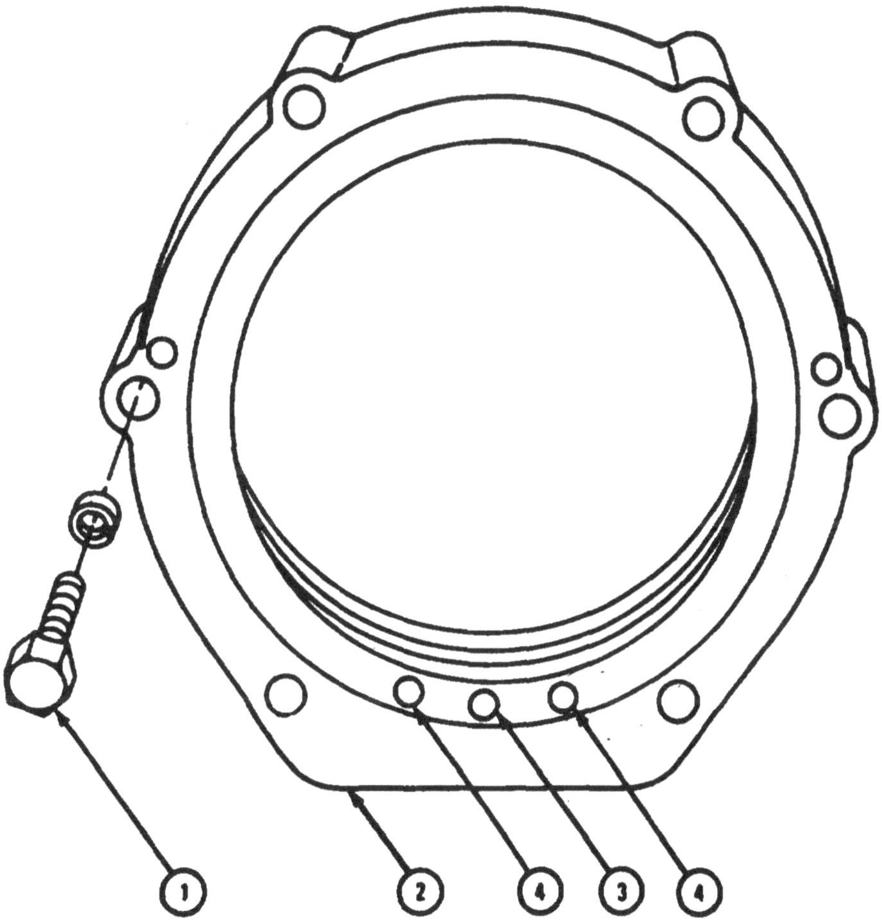

NOTE: CHECK ONLY THOSE PARTS WHICH ARE CALLED OUT IN THIS FRAME. PARTS WITHOUT CALLOUTS ARE SHOWN ONLY FOR REFERENCE PURPOSES OR ARE CHECKED IN ANOTHER FRAME.

TA 067728

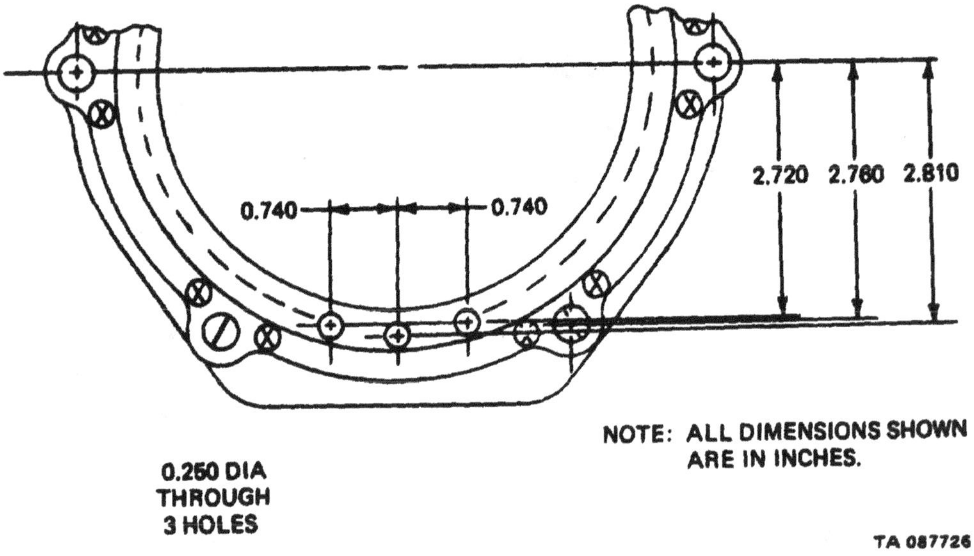

Figure 4-1. Rear Oil Seal Housing Drain Hole Machine Shop Instructions.

4-9. AIR COMPRESSOR SUPPORT.

a. Clesning.

FRAME 1

1. Unscrew and takeout pipe plug (1).

WARNING

Dry cleaning solvent is flammable. Do not use near an open flame. Keep a fire extinguisher nearby when solvent is used. Use only in well-ventilated places. Failure to do this may result in injury to personnel and damage to equipment.

2. Clean support (2) with dry cleaning solvent. Dry with clean cloth.
3. Clean oil passages (3) with a brass wire probe.
4. Wash oil passages (3) by flushing with dry cleaning solvent.

WARNING

Eye shields must be worn when using compressed air. Eye injury can occur if eye shields are not used.

5. Dry oil passages (3) by blowing them out with dry, compressed air.

END OF TASK

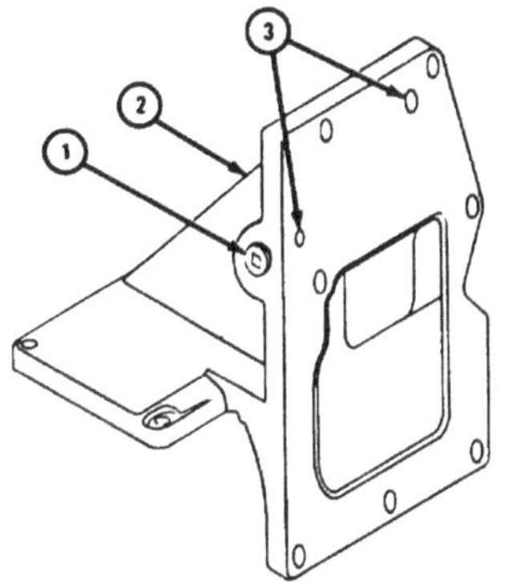

TA 113595

Inspection and Repair .

FRAME 1

1. Check that all threaded parts are not stripped or damaged in any other way. Fix minor thread damage with thread chaser or tap. If threads are badly damaged, get a new part.
2. Check that gasket surfaces (1 and 2) have no nicks, scratches or burrs. Repair small nicks, scratches or burrs with a fine mill file or crocus cloth.
3. Check that support (3) is not cracked or damaged in any other way. If support is damaged, get a new support.
4. Put in pipe plug (4) .

END OF TASK

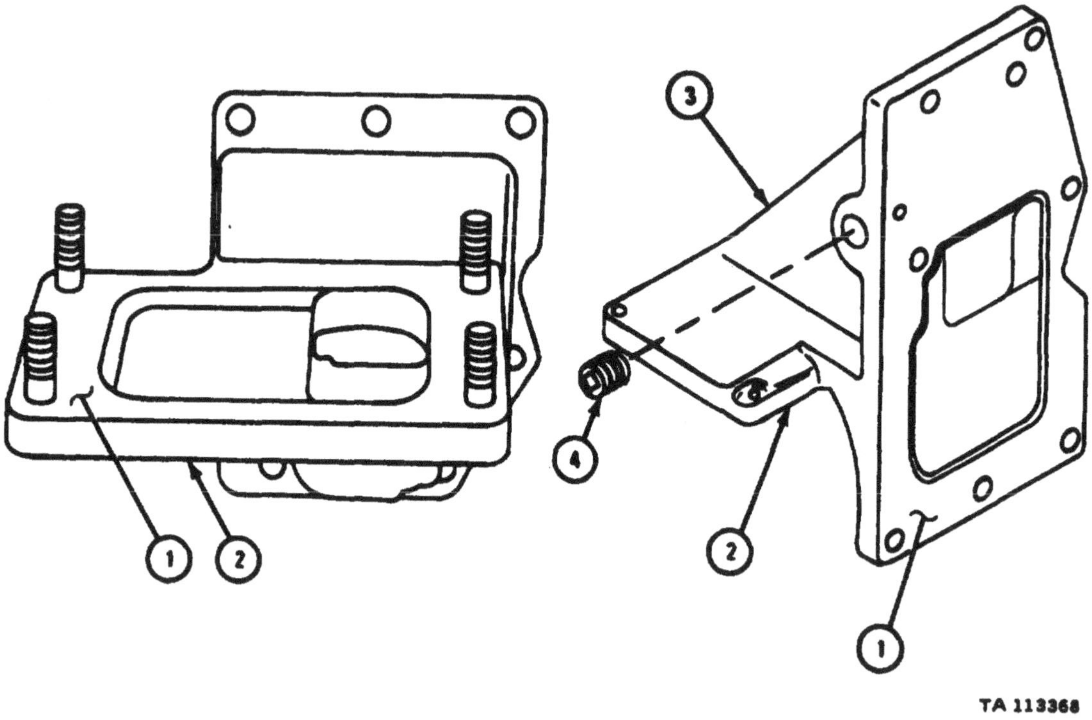

4-10. TAPPET CHAMBER COVER.

a. Cleanin g There are no special cleaning procedures needed. Refer to cleaning procedures given in para 4-3.

b. <u>Inspection and Repair.</u>

FRAME 1

1. Check that cover (1) is not bent, dented, torn, cracked or damaged in any other way. If cover is damaged, get a new one.
2. Check that cover (1) has no minor nicks, scratches or burrs. Fix small nicks, scratches or burrs with a fine mill file or crocus cloth. If more repair is needed, get a new part.

END OF TASK

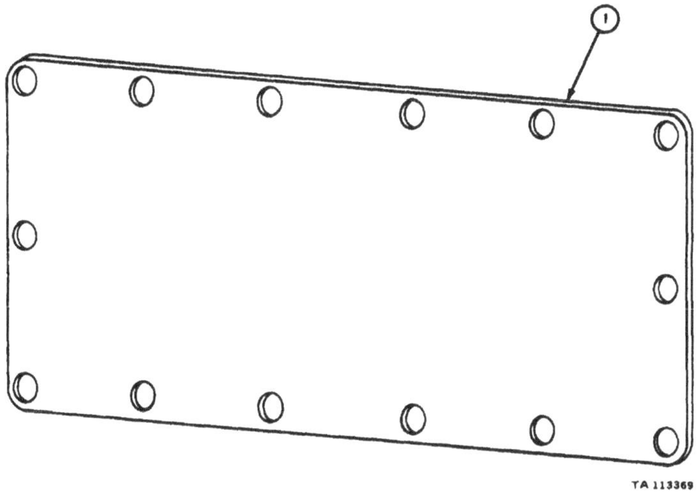

4-11. FUEL INJECTOR PUMP ADAPTER.

a. Cleaning. There are no special cleaning procedures needed. Refer to cleaning procedures given in para 4-3.

b. Inspection and Repair.

FRAME 1

1. Check that two studs (1) are not loose, stripped or damaged in any other way. Fix minor thread damage with a thread chaser. If more repair is needed, get a new stud.
2. Check that adapter (2) is not cracked and that stud holes (3) are not stripped or out-of-round. If adapter is damaged, get a new one.

END OF TASK

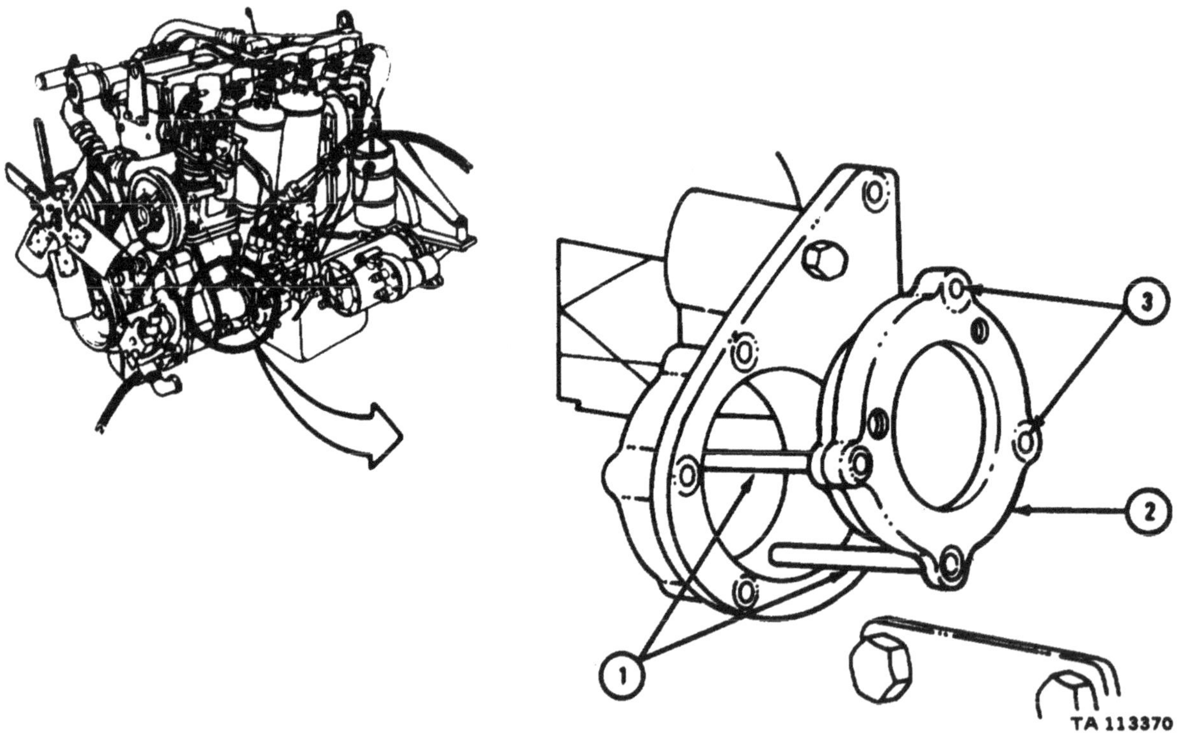

4-12. ENGINE FRONT PLATE ASSEMBLY.

a. Cleaning. There are no special cleaning procedures needed. Refer to cleaning procedures given in para 4-3.

b. Inspection and Repair.

FRAME 1

1. Check that two studs (1) in rear of plate (2) are not stripped or damaged in any other way. If studs are damaged, get new ones.
2. Check that plate (2) is not cracked, bent or damaged in any other way. If plate is damaged, get a new plate.
3. Check that plate (2) has no small nicks, scratches or burrs. Repair small nicks, scratches or burrs with crocus cloth. If more repair is needed, get a new part.

END OF TASK

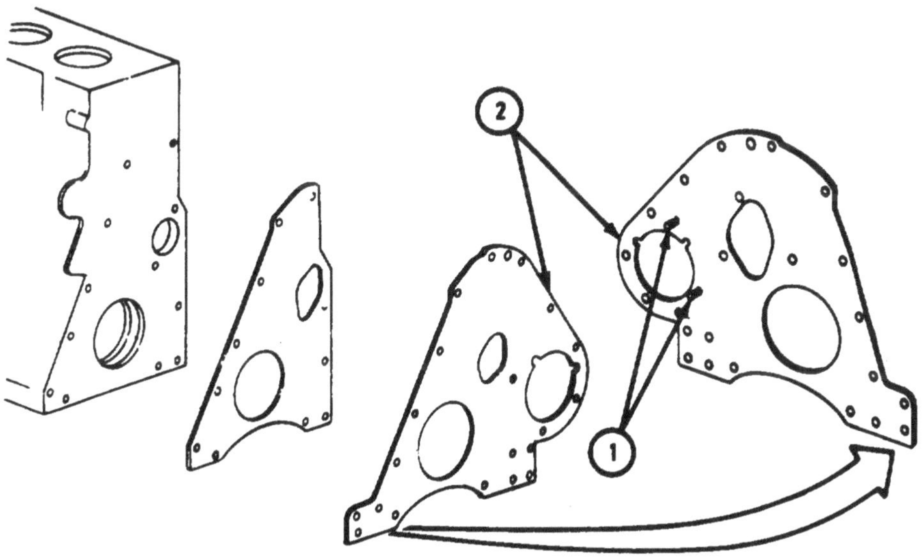

NOTE: CHECK ONLY THOSE PARTS WHICH ARE CALLED OUT IN THIS FRAME. PARTS WITHOUT CALLOUTS ARE SHOWN ONLY FOR REFERENCE PURPOSES OR ARE CHECKED IN ANOTHER FRAME.

TA 113671

4-13 OIL PUMP IDLER GEAR.

a. Cleaning For cleaning procedures for idler gear, refer to para 4-3. For cleaning procedures for bearing, refer to TM 9-214 .

b. Inspection and Repair .

FRAME 1

1. Check that idler gear (1) has no damaged teeth or scored bearing surfaces (2). Take off minor burrs from gear teeth with fine mill file. If damaged beyond repair, get a new idler gear.
2. Check that idler gear bearing bore (2) is not worn. If worn beyond limits given in table 4-4, get a new idler gear.
3. Check that bearing (3) is not damaged. Refer to TM 9-214 and table 4-4. If bearing is damaged or worn, get a new one.
4. Check that retainer ring (4) is not damaged or bent. If part is damaged or bent, get a new one.

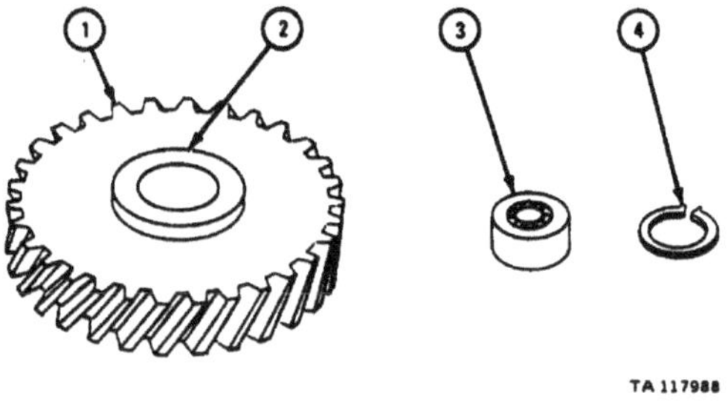

4-14. COOLANT DRAIN COCK AND OIL PRESSURE TRANSMITTER.

a. Cleaning.

FRAME 1

WARNING

Dry cleaning solvent is flammable. Do not use near an open flame. Keep a fire extinguisher nearby when solvent is used, Use only in well-ventilated places. Failure to do this may result in injury to personnel and damage to equipment.

1. Clean oil pressure transmitter (1) with dry cleaning solvent. Dry with clean cloth.
2. Clean pipe adapter (2), drain cock (3), and two elbows (4) with brass wire probe.
3. Wash pipe adapter (2), drain cock (3), and two elbows (4) by flushing with dry cleaning solvent .

WARNING

Eye shields must be worn when using compressed air. Eye injury can occur if eye shields are not used.

4. Dry pipe adapter (2), drain cock (3), and two elbows (4) by blowing them out with dry, compressed air .

END OF TASK

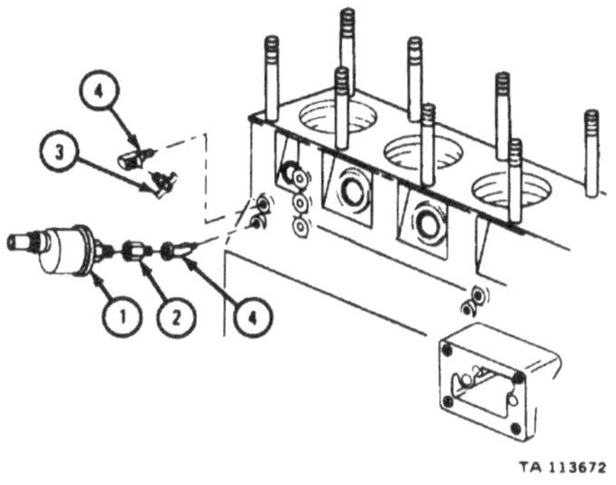

b. Inspection and Repair.

FRAME 1

1. Check that drain cock (1), elbow (2), oil pressure transmitter (3), pipe adapter (4), and elbow (5) have no damaged threads. If threads are damaged, get a new part.
2. Check that drain cock (1) is not corroded and turns freely in elbow (2). If damaged, get a new part.
3. Check that oil pressure transmitter (3) is not leaking around seams. If damaged, get a new part.
4. Check that pipe adapter (4), elbow (2), and elbow (5) are not bent or damaged in any way. If part is damaged, get a new one.

END OF TASK

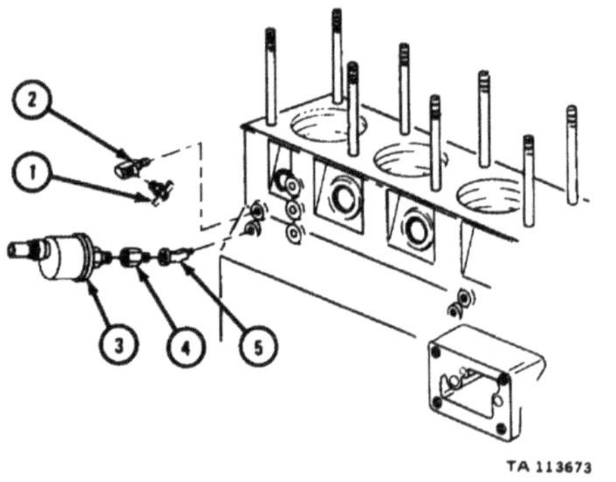

4-15. OIL GAGE ROD, ROD SUPPORT, AND SUPPORT BRACKET.

a. Cleaning.

FRAME 1

WARNING

Dry cleaning solvent is flammable. Do not use near an open flame. Keep a fire extinguisher nearby when solvent is used. Use only in well-ventilated places. Failure to do this may result in injury to personnel and damage to equipment.

Eye shields must be worn when using compressed air. Eye injury can occur if eye shields are not used.

1. Clean inside of rod support (1) using brass wire probe. Flush inside of rod support with dry cleaning solvent. Blow inside of rod support dry with compressed air.
2. Clean outside of rod support (1) with dry cleaning solvent. Clean off sludge and gum using a stiff brush. Dry with clean cloth.
3. Take preformed packing (2) off oil gage rod (3) and throw it away.
4. Clean oil gage rod (3), support bracket (4), screw (5), nut (6), and washer (7) with dry cleaning solvent. Dry with clean cloth.

END OF TASK

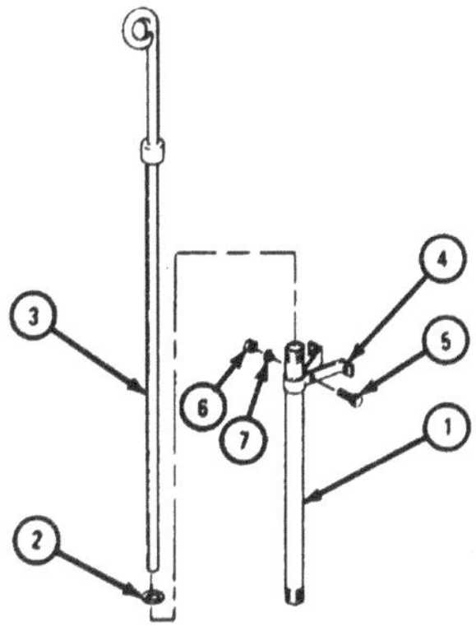

b. Inspection and Repair.

FRAME 1

1. Check that oil gage rod support (1) is not bent or cracked and does not have damaged threads . Fix minor thread damage with a thread chaser. If more repair is needed, get a new oil gage rod support.
2. Check that oil gage rod (2) is not bent or cracked and does not have damaged threads. If oil gage rod is bent, cracked or has damaged threads, get a new one.
3. Check that support bracket (3) is not cracked or bent. If it is, get a new one.
4. Check that screw (4), nut (5), and washer (6) are not damaged. If any par t is damaged, get a new one.
5. Put preformed packing (7) on oil gage rod (2) and slide it all the way up.

END OF TASK

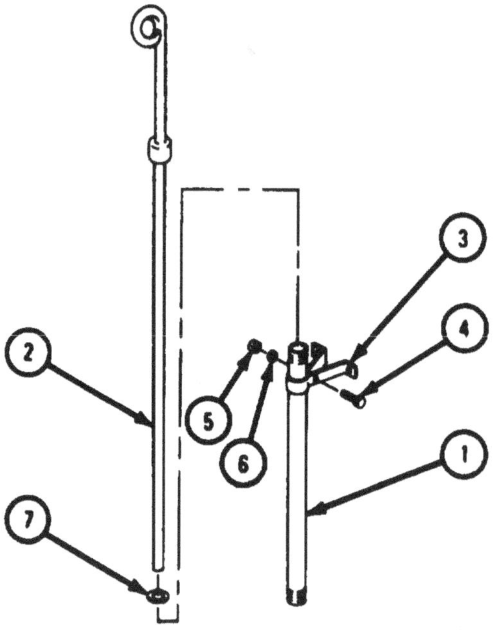

TA 118506

4-16. **OIL PAN** .

Disassembly.

FRAME 1

NOTE

If working on engine LDS-465-1, do step 2.

1. Takeout two drain plugs (1) and two gaskets (2). Throw away gaskets .
2. Take out fuel injection pump oil drain elbow (3) or oil drain plug (4).

END OF TASK

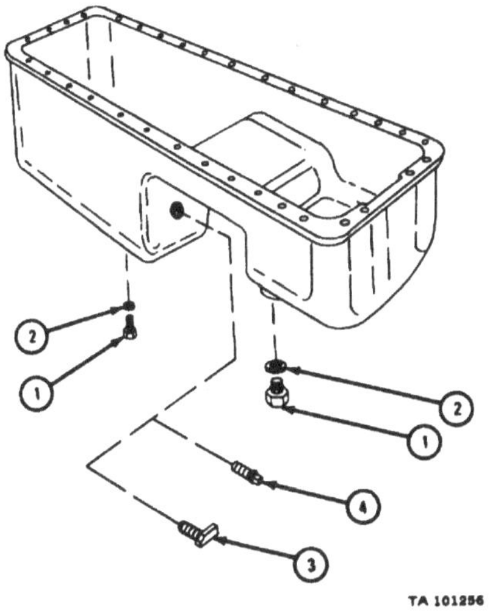

b. Cleaning. There are no special cleaning procedures needed. Refer to cleaning procedures given in para 4-3.

c. Inspection and Repair.

FRAME 1

1. Check that 30 screws (1) are not stripped or damaged in any other way. If any screw is damaged, get a new one. Check that 30 lockwashers (2) are not chipped, cracked, or bent. If any lockwasher is damaged, get a new one.
2. Check that two drain plugs (3) are not stripped or damaged in any other way. Fix minor thread damage with a thread chaser. If more repair is needed, get a new drain plug.
3. If working on engine LDS-465-1, check that drain plug (4) or fuel injection pump oil drain elbow (5) are not stripped or damaged in any other way. Fix minor thread damage with a thread chaser. If more repairs are needed, get a new part.

GO TO FRAME 2

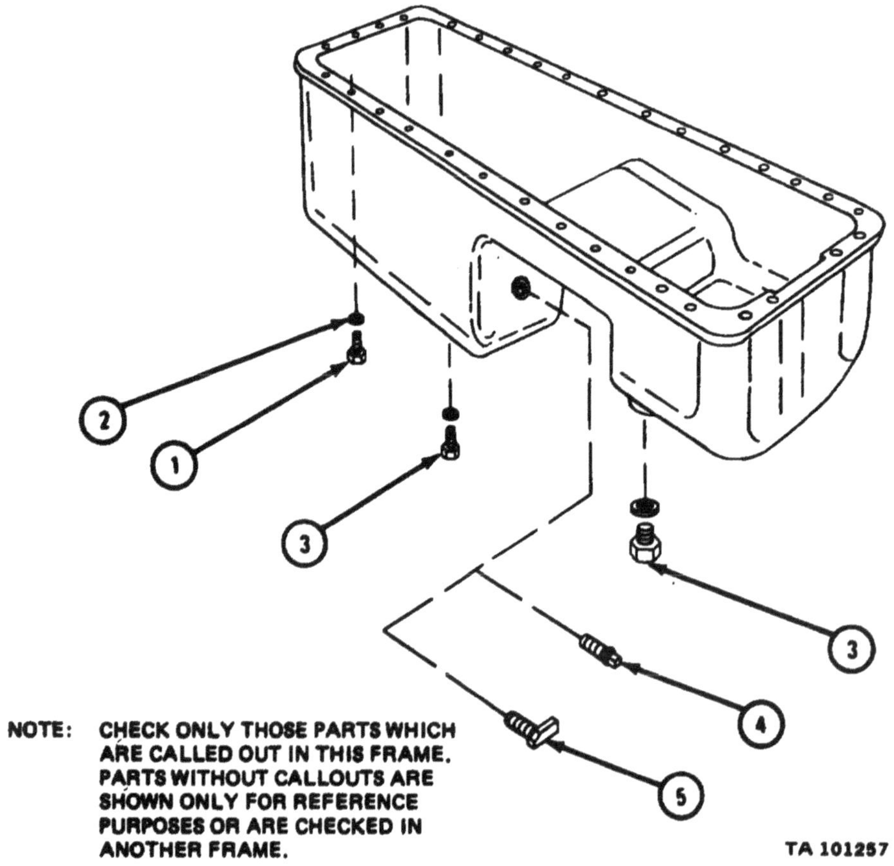
2
1
3
3
4
5
NOTE: CHECK ONLY THOSE PARTS WHICH ARE CALLED OUT IN THIS FRAME. PARTS WITHOUT CALLOUTS ARE SHOWN ONLY FOR REFERENCE PURPOSES OR ARE CHECKED IN ANOTHER FRAME.
TA 101257

FRAME 2

1. Check that threads in drain plug holes in oil pan (1) are not damaged. Fix minor thread damage with a tap. If more repair is needed, get a new oil pan.
2. Check that gasket surface of oil pan (1) has no nicks or burrs. Take out minor nicks or burrs with a fine mill file. If more repair is needed, get a new oil pan.
3. Check that oil pan (1) has no cracks or dents. To fix dents, refer to FM 43-2. To fix cracks, refer to TM 9-237 .

END OF TASK

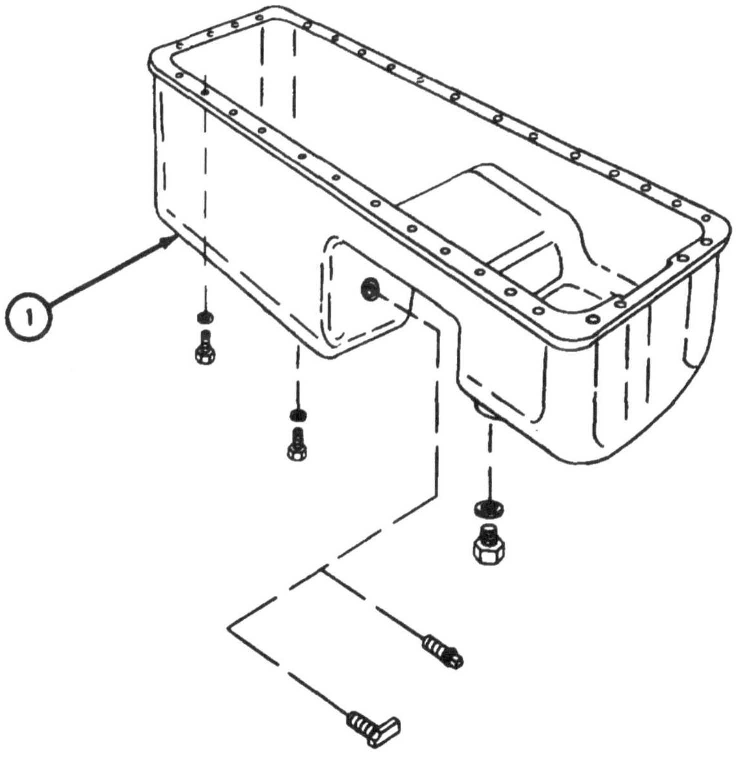

NOTE: CHECK ONLY THOSE PARTS WHICH ARE CALLED OUT IN THIS FRAME. PARTS WITHOUT CALLOUTS ARE SHOWN ONLY FOR REFERENCE PURPOSES OR APE CHECKED IN ANOTHER FRAME.

TA 101258

d. Assembly.

FRAME 1

1. Put in two drain plugs (1) with two gaskets (2).
2. If working on engine LDS-465-1, put fuel injection pump oil drain elbow (3) or fuel injection pump oil drain plug (4) in oil pan (5).

END OF TASK

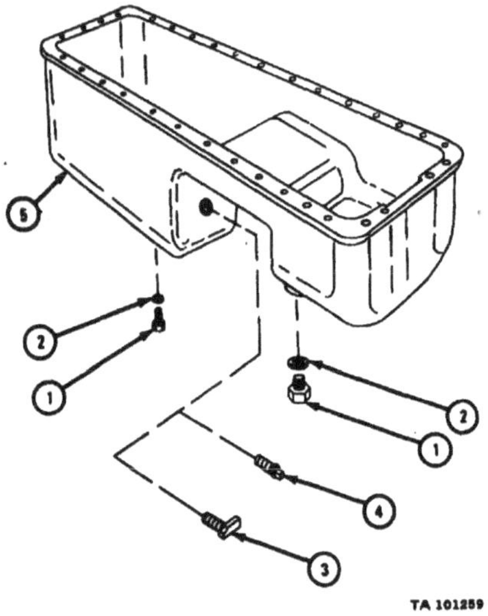

4-17. OIL PUMP ASSEMBLY AND OIL PUMP TUBES.

a. Disassembly.

FRAME 1

1. Put oil pump assembly (1) in vise with soft jaw caps as shown.
2. Put pin punch into oil pump outlet hole (2) to keep oil pump impellers from turning.

GO TO FRAME 2

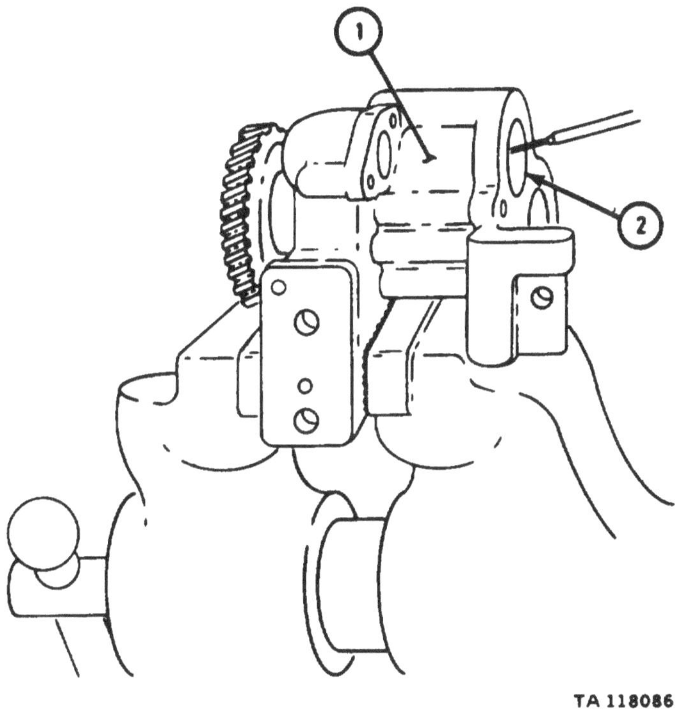

FRAME 2

1. On all engines except LDS-465-2, take out and throw away cotter pin (1) and take off slotted nut (2).
2. On engine LDS-465-2, take off locknut (3) and flat washer (4) .
3. Using universal puller (5), take off oil pump drive gear (6) .
4. Take oil pump (7) out of vise .

GO TO FRAME 3

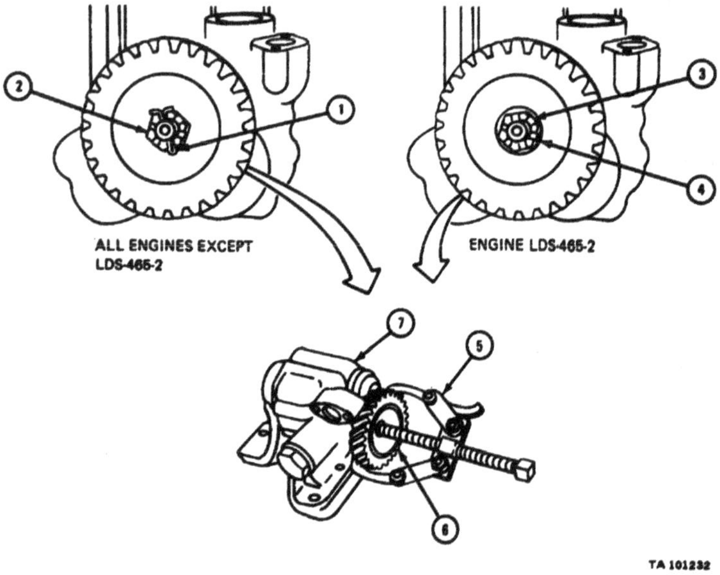

FRAME 3

CAUTION

Pressure relief valve threaded cap (2) or retaining pin (7) are under spring tension. Be very careful when taking them out or parts inside oil pump (1) will be lost.

NOTE

Two types of oil pump (1) have been used on these engines. Early model pump uses a threaded cap (2) to hold pressure relief valve spring (4) and plunger (5) inside pump. Late model pump uses a retaining pin (7) and either a washer (8), or one or two types of spring retainer (9) to hold pressure relief valve spring (4) and plunger (5) inside pump.

Shims (6) are used to set pressure relief valve. Number of shims inside pressure relief valve will be different on different pumps (1). Some pumps have no shims.

1. On early model pumps (1), *unscrew* and take out threaded cap (2). If there is a copper gasket (3) under threaded cap, take it out and throw it away. Take out pressure relief valve spring (4), plunger (5), and shims (6). Tag shims so same number of shims will be put back.
2. On late model pumps (1) take out retaining pin. (7), washer (8) or spring retainer (9), plunger (5), pressure relief valve spring (4), and shims (6). If pump has a washer instead of spring retainer, throw away washer. Tag shims so the same number of shims will be put back.

IF WORKING ON ANY ENGINE EXCEPT LDS-465-2, GO TO FRAME 4.
IF WORKING ON ENGINE LDS-465-2, GO TO FRAME 5

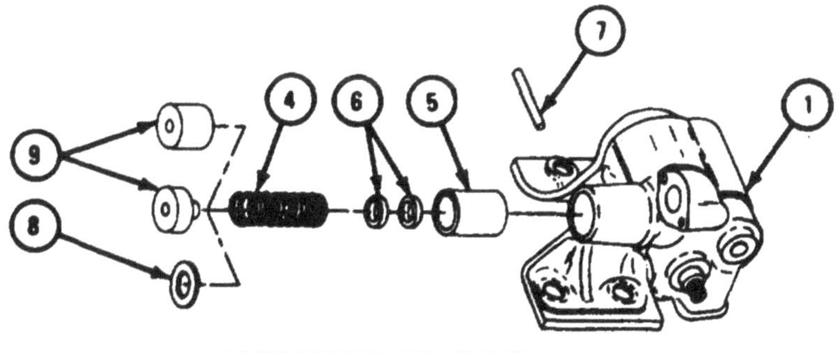

LATE MODEL OIL PUMP

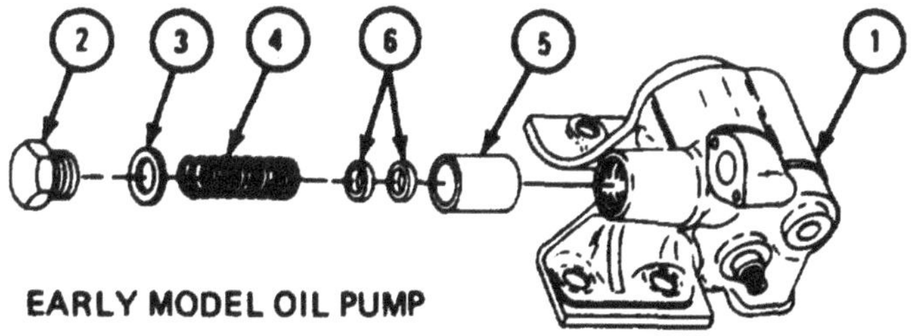

EARLY MODEL OIL PUMP

TA 101233

FRAME 4

1. Takeout two screws and lockwashers (1) , and take off scavenge oil pump cover (2) .
2. Take out scavenge oil pump drive impeller (3) and driven impeller (4).
3. Take scavenge oil pump housing (5) apart from pressure oil pump housing (6).
4. Take out pressure oil pump drive shaft impeller (7). Take pressure oil pump driven impeller (8) off driven impeller shaft (9).
5. Take square key (10) out of shaft of pressure oil pump drive shaft impeller (7) .

END OF TASK

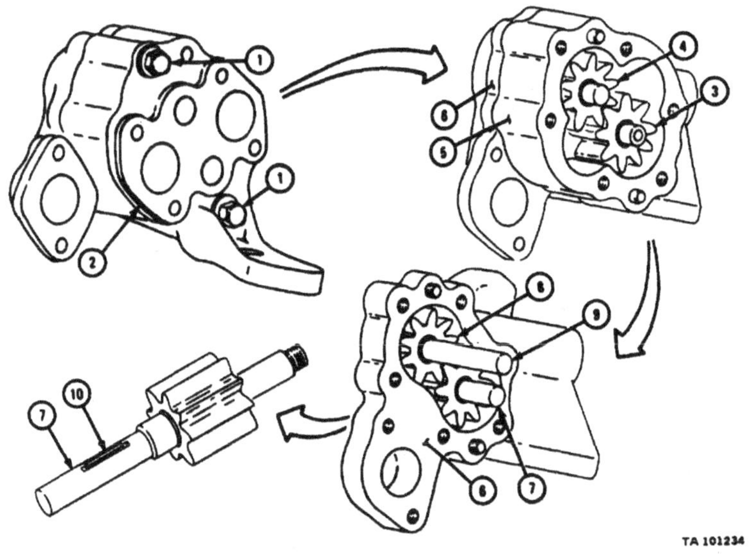

FRAME 5

1. Takeout two screws and lockwashers (1) and takeoff scavenge oil pump cover (2) .
2. Takeout scavenge oil pump driven impeller (3) .
3. Take scavenge oil pump housing (4), apart from pressure oil pump housing (5) .
4. Take pressure oil pump driven impeller (6) off shaft (7).

END OF TASK

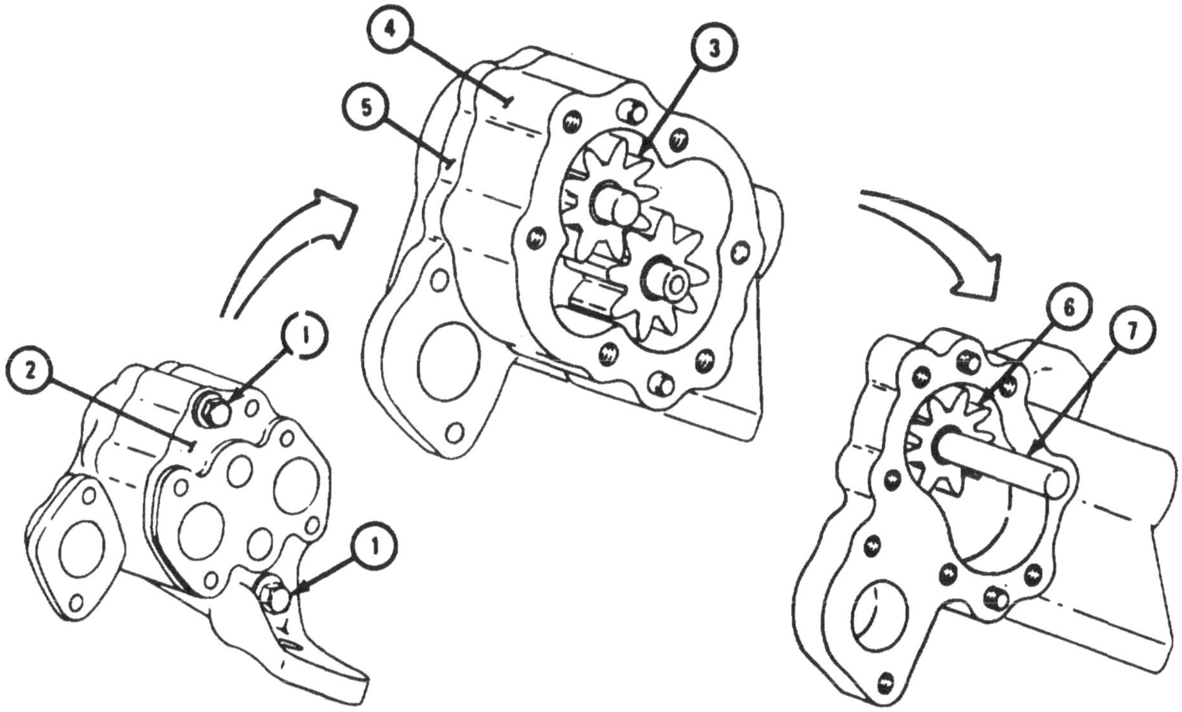

TA 101235

b. Cleaning.

(1) Oil pump tubes .

(a) Engines LD-465-1, LD-465-1C, and LDT-465-1C .

FRAME 1

CAUTION

Be careful not to damage screens (1) diring cleaning with brass wire probe.

1. Clean three oil pump tubes (2) with a brass wire probe. Use a wire brush to take off carbon and sludge.

WARNING

Dry cleaning solvent is flammable. Do not use near an open flame. Keep a fire extinguisher nearby when solvent is used. Use only in well-ventilated places . Failure to do this may result in injury to personnel and damage to equipment.

2. Wash three oil pump tubes (2) by flushing with dry cleaning solvent.

WARNING

Eye shields must be worn when using compressed air. Eye injury can occur if eye shields are not used.

3. Dry three oil pump tubes (2) by blowing them out with compressed air.
4. Clean bracket (3) and flange (4) with dry cleaning solvent. Use a stiff brus h to take off sludge and gum deposits.

END OF TASK

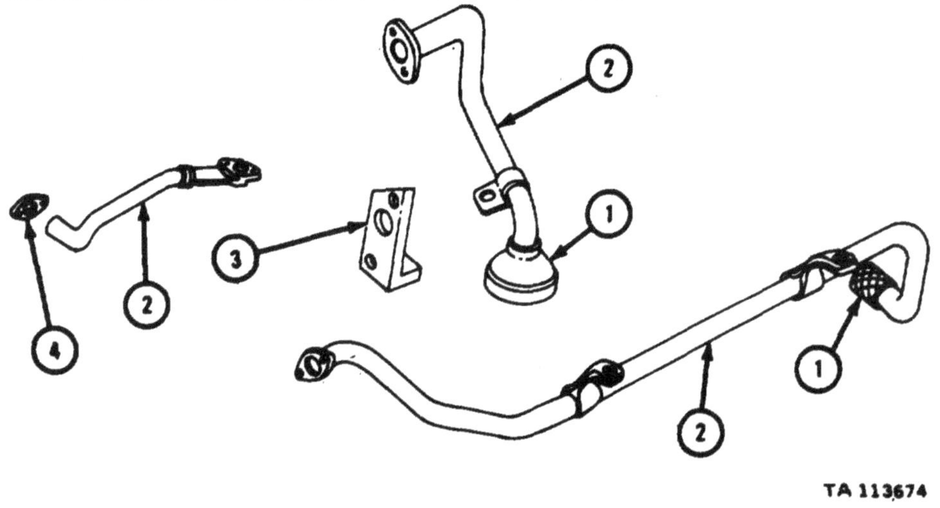

(b) Engines LDS-465-1, LDS-465-1A, and LDS-465-2

FRAME 1

CAUTION

Be careful not to damage screens (1) during cleaning with brass wire probe.

1. Clean four oil pump tubes (2) with a brass wire probe. Use a wire brush to take out carbon and sludge.

WARNING

Dry cleaning solvent is flammable. Do not use near an open flame. Keep a fire extinguisher nearby when solven t is used. Use only in well-ventilated places. Failure to do this may result in injury to personnel and damage to equipment.

2. Wash four oil pump tubes (2) by flushing with dry cleaning solvent.

WARNING

Eye shields must be worn when using compressed air. Eye injury can occur if eye shields are not used.

3. Dry four oil pump tubes (2) by blowing them out with compressed sir.

4. Clean brace (3), flange (4), and four clamps (5) using dry cleaning solvent . Use a stiff brush to take off sludge and gum deposits .

END OF TASK

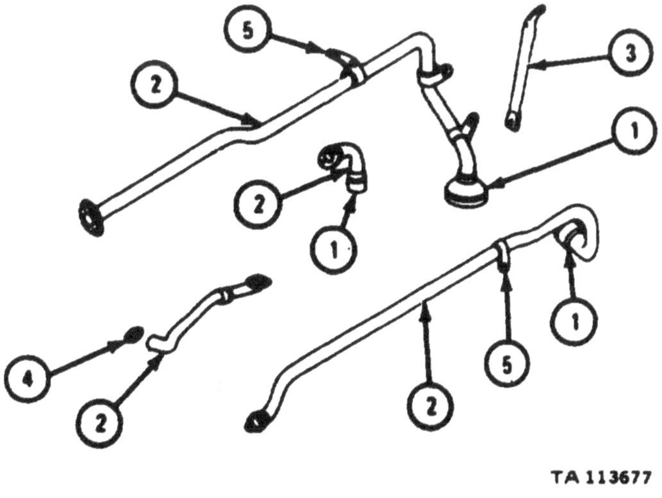

(2) Oil pump assembly. There are no special cleaning procedures needed. Refer to cleaning procedures given in para 4-3.

Inspection and Repair.

FRAME 1

NOTE

On engine LDS-465-2, impeller (1) and impeller (2) are pressed together on shaft of impeller (1). They should not be pulled apart.

1. Check that scavenge oil pump cover (3), scavenge oil pump housing (4), and pressure oil pump housing (5) are not worn, scored or damaged. If these parts are worn, scored or damaged, get a new oil pump.

WARNING

Dry cleaning solvent is flammable. Do not use near an open flame. Keep a fire extinguisher nearby when solvent is used. Use only in well-ventilated places. Failure to do this may result in injury to personnel and damage to equipment.

2. Check that machined surfaces of scavenge oil pump cover (3), scavenge oil pump housing (4), and pressure oil pump housing (5) do not have nicks, burrs or raised metal. Fix any nicks burrs or raised metal with a fine mill file or crocus cloth dipped in dry cleaning solvent.
3. Check that four impellers (1, 2, 6, and 7) are not badly worn, scored or damaged. If impellers are worn, scored or damaged, get a new oil pump.
4. Check that four impellers (1, 2, 6, and 7) do not have nicks, burrs, or raised metal. Fix small nicks, burrs or raised metal with a fine mill file or crocus cloth dipped in dry cleaning solvent.

NOTE

Engine LDS-465-2 does not have bushing (9).

5. Check that two drive impeller bushings (8) and bushing (9) do not have nicks, burrs or raised metal. Fix small nicks burrs or raised metal with a fine mill rat tail file or crocus cloth dipped in dry cleaning solvent.

GO TO FRAME 2

NOTE

CHECK ONLY THOSE PARTS WHICH ARE CALLED OUT IN THIS FRAME. PARTS WITHOUT CALLOUTS ARE SHOWN ONLY FOR REFERENCE PURPOSES OR ARE CHECKED IN ANOTHER FRAME.

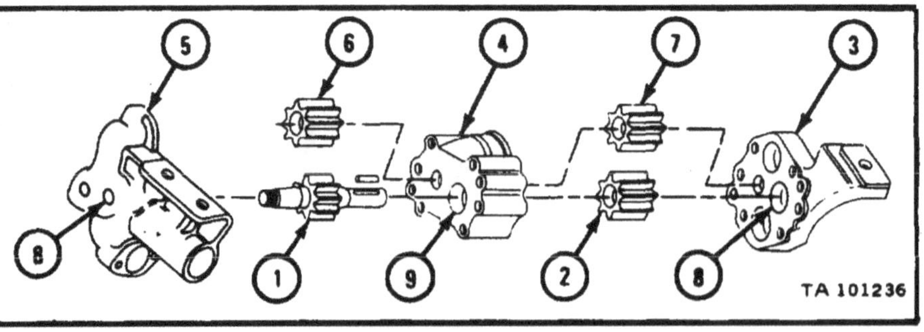

FRAME-1

1. Check if pump (1) uses threaded cap (2) to hold in pressure relief valve spring (3) and plunger (4) . If pump uses a threaded cap, this is early model pump.

2. Check if pump (1) uses retaining pin (5) and either washer (6) or one of two types of spring retainer (7) to hold in pressure relief valve spring (3) and plunger (4) . If pump uses retaining pin and washer or spring retainer, this is late model pump.

IF WORKING ON EARLY MODEL PUMP, GO TO FRAME 3.
IF WORKING ON LATE MODEL PUMP, GO TO FRAME 4

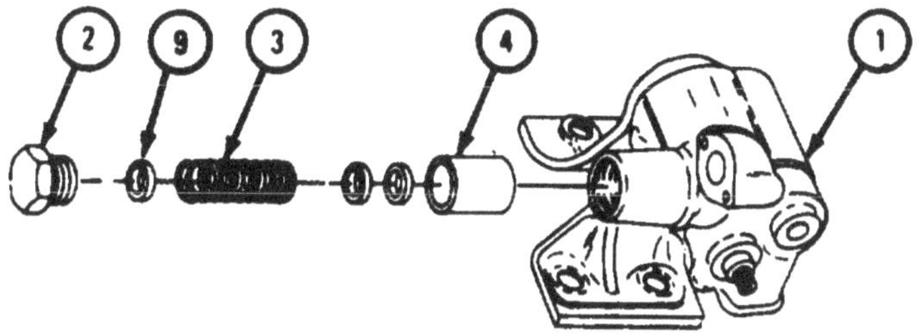

EARLY MODEL OIL PUMP

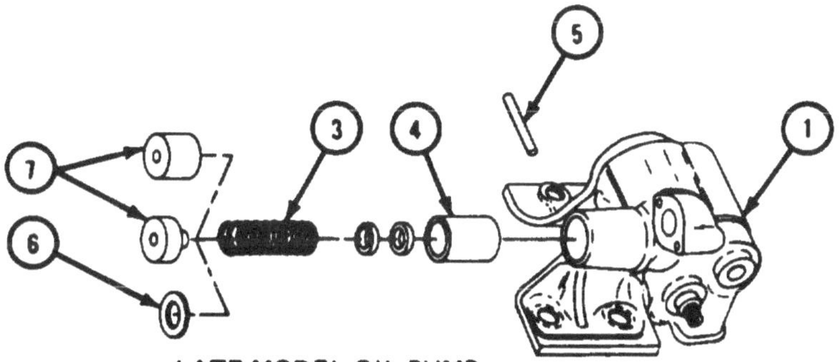

LATE MODEL OIL PUMP

NOTE: CHECK ONLY THOSE PARTS WHICH ARE CALLED OUT IN THIS FRAME. PARTS WITHOUT CALLOUTS ARE SHOWN ONLY FOR REFERENCE PURPOSES OR ARE CHECKED IN ANOTHER FRAME.

TA 101237

FRAME 3

1. Check that threaded cap (1) is not cracked and does not have damaged threads. Fix minor thread damage with a thread chaser. If more repair is needed, get a new oil pump.

NOTE

Shims (2) are used to set pressure relief valve. The number of shims may differ from pump to pump. Some pumps may have no shims.

Pressure relief valve spring (3) comes in a parts kit which has new shims (2). If you get a new parts kit, save all the shims even if pump had no shims, Shims will be used later during pump test and adjustment. Parts kit also has a new plunger (4) and a type of spring retainer (5) with no threads. This plunger and spring retainer are for late model pumps and should not be used on early model pumps.

2. Check that pressure relief valve spring (3) is not cracked or damaged. If it is cracked or damaged, throw it away and get a new one.

WARNING

Dry cleaning solvent is flammable. Do not use near an open flame. Keep a fire extinguisher nearby when solvent is used. Use only in well-ventilated places. Failure to do this may result in injury to personnel and damage to equipment.

3. Check that plunger (4) does not have cracks or burrs. Fix small burrs with a fine mill file or crocus cloth dipped in dry cleaning solvent. If more repair is needed or if plunger is cracked, throw it away and get a new one.

END OF TASK

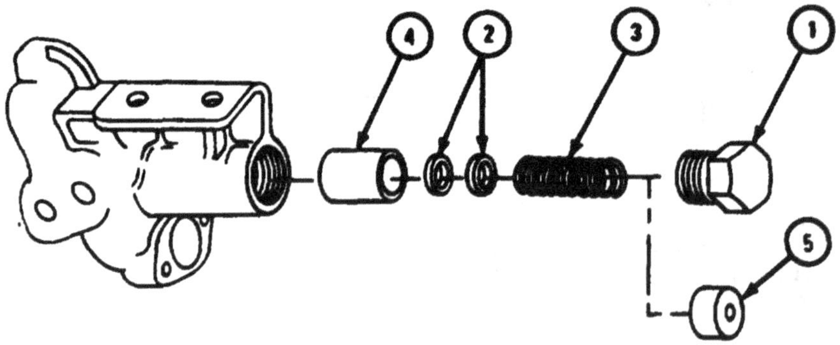

TA 101238

FRAME 4

NOTE

Shims (1) are used to set pressure relief valve. The number of shims may differ from pump to pump. Some pumps may have no shims. Save all shims in new parts kit, even if pump you are working on had no shims. Shims will be used later for pump test and adjustment.

Relief valve spring (2) and plunger (3) come in a parts kit. The kit also has a new retaining pin (4), one of two types of spring retainer (5), and shims (1). If any of these parts are damaged, all should be changed. If kit has two plungers, the longer one is for this pump.

1. Some early model oil pumps used washer (6) instead of one of two types of spring retainer (5) to hold in relief valve plunger (3) and spring (2). If working on pump with washer, threw away washer and get a parts kit with new spring retainer, retaining pin (4), plunger, spring, and shims (1).

WARNING

Dry cleaning solvent is flammable. Do not use near an open flame. Keep a fire extinguisher nearby when solvent is used. Use only in well-ventilated places. Failure to do this may result in injury to personnel and damage to equipment.

2. Check that spring retainer (5), spring (2), and plunger (3) have no cracks or burrs. Fix small burrs with a fine mill file or crocus cloth dipped in dry cleaning solvent.
3. If spring retainer (5), spring (2) or plunger (3) are cracked or badly damaged, throw away spring retainer, spring, plunger, retaining pin (4), and shims (1) and get new ones.

IF WORKING ON ENGINES LD-465-1, LD-465-1C OR LDT-465-1C , GO TO FRAME 5.
IF WORKING ON ENGINES LDS-465-1, LDS-465-1A OR LDS-465-2, GO TO
FRAME 7

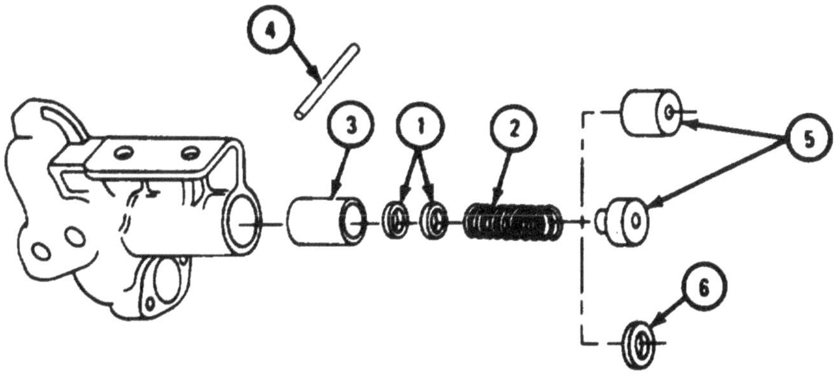

TA 101239

FRAME 5

1. Check that three oil tubes (1) do not have dents, cracks, loose flanges or flattened areas, and that screens (2) are not cracked, torn or loose or damaged in any other way. If tubes or screens are damaged, get a new tube.
2. Check that mounting flanges (3) and tube flange (4) are not warped. Fix minor warpage by working flange across a sheet of crocus cloth held tightly on a flat surface. If more repair is needed, get a new part.

GO TO FRAME 6

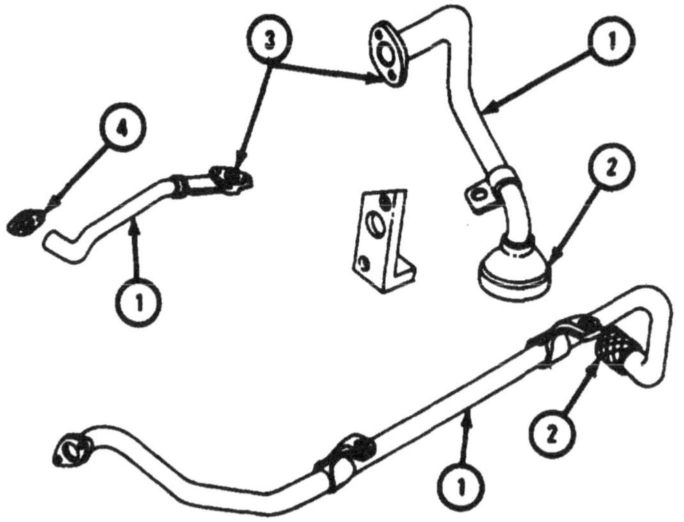

NOTE: CHECK ONLY THOSE PARTS WHICH ARE CALLED OUT IN THIS FRAME. PARTS WITHOUT CALLOUTS ARE SHOWN ONLY FOR REFERENCE PURPOSES OR ARE CHECKED IN ANOTHER FRAME.

TA 113675

FRAME 6

1. Check that three clamps (1) are not bent or cracked. If clamps are damaged, get a new oil tube (2).

2. Check that bracket (3) is not bent or cracked. If bracket is damaged, ge t a new bracket.

FOR PUMPS ON ALL ENGINES EXCEPT LDS-465-2, GO TO FRAME 9.
FOR PUMPS ON ENGINE LDS-465-2, GO TO FRAME 17

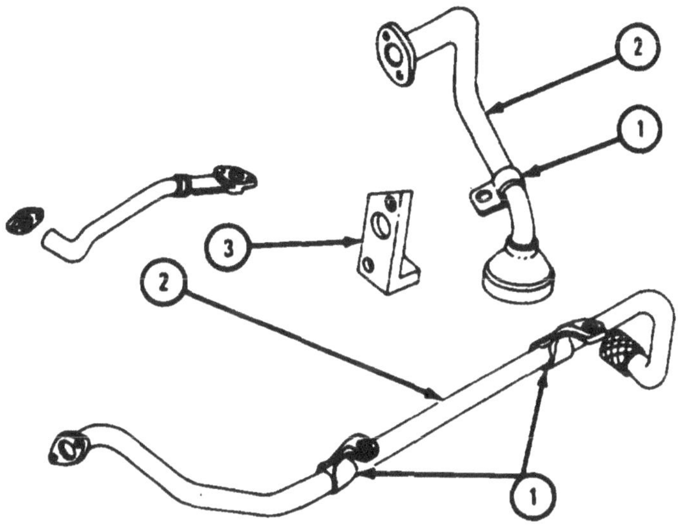

NOTE
CHECK ONLY THOSE PARTS WHICH ARE CALLED OUT IN THIS FRAME. PARTS WITHOUT CALLOUTS ARE SHOWN ONLY FOR REFERENCE PURPOSES OR ARE CHECKED IN ANOTHER FRAME.

TA 113676

FRAME 7

1. Check that four oil tubes (1) do not have dents, cracks, loose flanges or flattened areas, and that screen (2) is not cracked, torn or loose or damaged in any other way. If tubes or screen are damaged, get a new tube.
2. Check that mounting flanges (3) and tube flange (4) are not warped. Fix minor warpage by working flange across a sheet of crocus cloth held tightly on a flat surface. If more repair is needed, get a new part.

GO TO FRAME 8

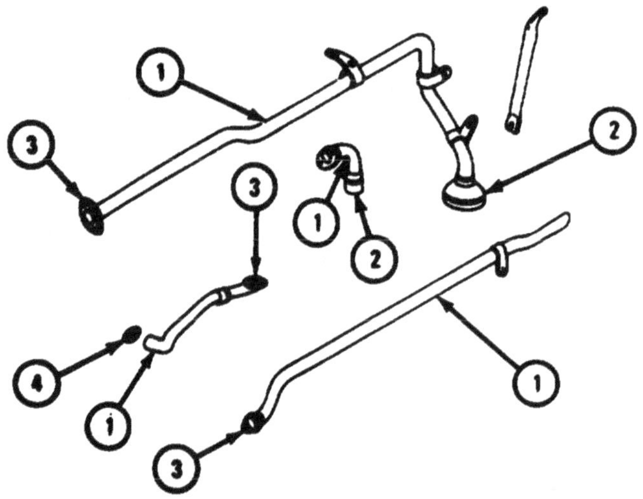

NOTE: CHECK ONLY THOSE PARTS WHICH ARE CALLED OUT IN THIS FRAME. PARTS WITHOUT CALLOUTS ARE SHOWN ONLY FOR REFERENCE PURPOSES OR ARE CHECKED IN ANOTHER FRAME.

TA 113679

FRAME 8

1. Check that four clamps (1) are not bent or cracked. If clamps are damaged, get new ones.
2. Check that brace (2) is not bent or cracked. If brace is damaged, get a new brace .

FOR PUMPS ON ALL ENGINES EXCEPT LDS-465-2, GO TO FRAME 9.
FOR PUMPS ON ENGINE LDS-465-2, GO TO FRAME 17

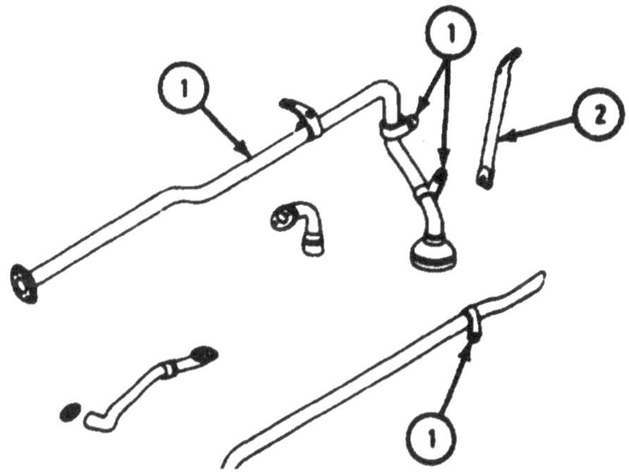

NOTE

CHECK ONLY THOSE PARTS WHICH ARE CALLED OUT IN THIS FRAME. PARTS WITHOUT CALLOUTS ARE SHOWN ONLY FOR REFERENCE PURPOSES OR ARE CHECKED IN ANOTHER FRAME.

TA 113678

FRAME 9

1. Check if pump (1) uses a threaded cap (2) to hold in relief valve spring (3) and plunger (4) . If pump uses a threaded cap (2), this is early model pump.
2. Check if pump (1) uses a retaining pin (5) and one of two types of spring retainer (6) to hold in pressure relief valve spring (3) and plunger (4). If pump uses retaining pin and spring retainer, this is late model pump.

IF WORKING ON EARLY MODEL PUMP, GO TO FRAME 10.
IF WORKING ON LATE MODEL PUMP, GO TO FRAME 11

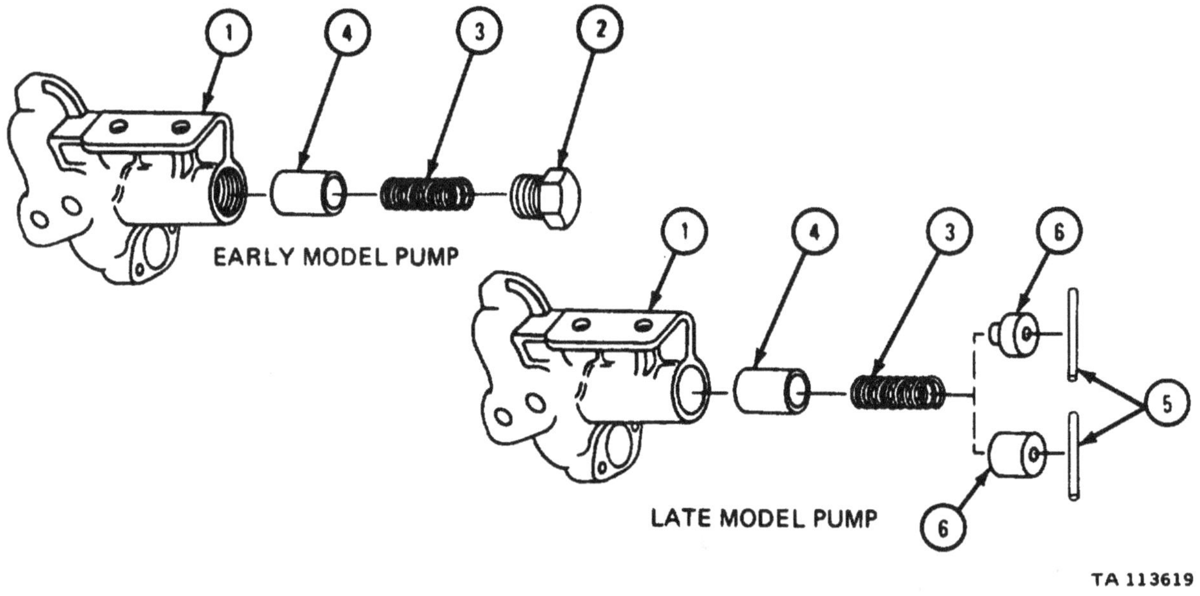

FRAME 10

NOTE

Readings must be within limits given in table 4-30. The letter L shows a loose fit. If readings are not within given limits, throw away part and get a new one.

Shims (4) are used .to adjust pressure relief valve. The number of shims may differ from pump to pump. Some pumps may have no shims.

Pressure relief valve spring (1) comes in a parts kit which has new shims (4). If you get a new parts kit, save all the shims even if pump had no shims. Shims will be used later during pump test and adjustment. Parts kit also contains a new plunger (2) and a type of spring retainer (5) with no threads. This plunger and spring retainer are for late model pumps and should not be used on early model pumps.

1. Measure free length of relief valve spring (1). Measure load needed to squeeze relief valve spring (1) to 2.100 inches. Measure maximum solid height of relief valve spring (1) .
2. Measure outside diameter of pressure relief valve plunger (2).
3. Measure inside diameter of pressure relief valve bore (3). If reading is not within limits given in table 4-30, get a new oil pump.
4. Measure fit of relief valve plunger (2) in pressure relief valve bore (3). If reading is not within limits given in table 4-30, get a new oil pump.

GO TO FRAME 11

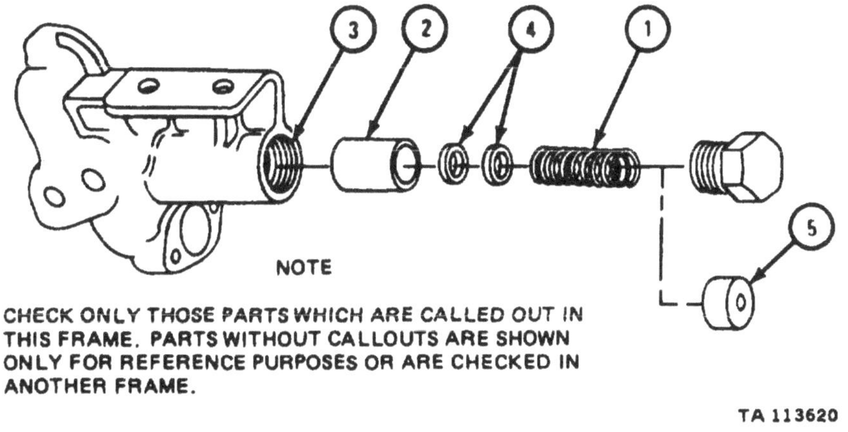
3
2
4
1
5
NOTE
CHECK ONLY THOSE PARTS WHICH ARE CALLED OUT IN THIS FRAME. PARTS WITHOUT CALLOUTS ARE SHOWN ONLY FOR REFERENCE PURPOSES OR ARE CHECKED IN ANOTHER FRAME.
TA 113620

Table 4-30. Early Model Oil Pump Pressure Relief Valve Wear Limits

Index Number	Item/Point of Measurement	Size and Fit of New Parts (inches unless otherwise noted)	Wear Limits (inches)
1	Relief valve spring free length	2.3400	None
1	Relief valve spring load at 2.100 inches	45.0 pounds to 55.0 pound s	None
1	Relief valve spring maximum solid height	1.8800	None
2	Pressure relief plunger outside diameter	0.8110 to 0.8120	0.8105
3	Pressure relief valve bore inside diameter	0.8140 to 0.8150	0.8160
2 and 3	Fit of relief valve plunger in relief valve bore	0. 0020L to 0. 0040L	0. 0060L

FRAME 11

NOTE

Readings must be within limits given in table 4-31. The letter L shows a loose fit. If readings are not within given limits, throw away part and get anew one.

Relief valve spring (1) and plunger (2) come together in a kit. Kit also has a new retaining pin (4), one of two types of spring retainer (5), and shims (6). If any of these parts are damaged, all should be changed. If kit has two plungers, longer one is for this pump.

If you get a new parts kit, save all shims (6) in the kit, even if pump had no shims. Shims will be used later for pump test and adjustment.

1. Measure free length of relief valve spring (1).
2. Measure load needed to squeeze relief valve spring (1) down to 2.100 inches.
3. Measure maximum solid height of relief valve spring (1).
4. Measure outside diameter of relief valve plunger (2).
5. Measure inside diameter of relief valve bore (3). If reading is not within limits given in table 4-31, get a new oil pump.
6. Measure fit of relief valve plunger (2) in relief valve bore (3). If reading is not within limits given in table 4-31, get a new oil pump.

GO TO FRAME 12

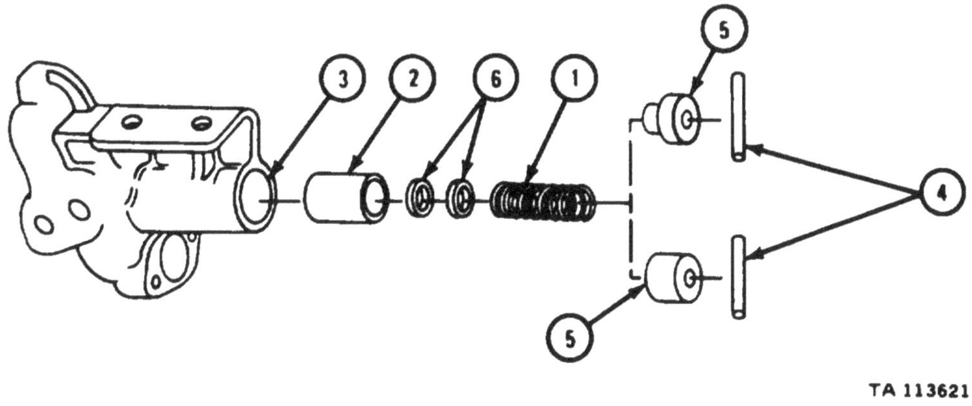

Table 4-31. Late Model Oil Pump Pressure Relief Valve Wear Limits

Index Number	Item/Point of Measurement	Size and Fit of New Parts (inches unless otherwise noted)	Wear Limit (tithes)
1	Relief valve spring free length	2.0000	None
1	Relief valve spring load at 1.8100-inches	47.0 to 57 pounds	None
1	Relief valve spring maximum solid height	1.6000	None
2	Relief valve bore outside diameter	0.8120 to 0.8125	0.8100
3	Relief valve bore inside diameter	0.8140 to 0.8150	0.8160
2 and 3	Fit of relief valve plunger in relief valve bore	0.0015L to 0.0030L	0.0060L

FRAME 1 2

NOTE

Readings must be within limits given in table 4-32. The letter L shows a loose fit. If readings are not within given limits, get a new oil pump.

1. Measure inside diameter of impeller drive shaft bushing (1).
2. Measure outside diameter (2) of impeller drive shaft.
3. Measure fit of impeller drive shaft diameter (2) in impeller drive shaft bushing (1).
4. Measure inside diameter of impeller drive shaft bushing (3).
5. Measure outside diameter (4) of impeller drive shaft.
6. Measure fit of impeller drive shaft diameter (4) in impeller drive shaft bushing (3).

GO TO FRAME 13

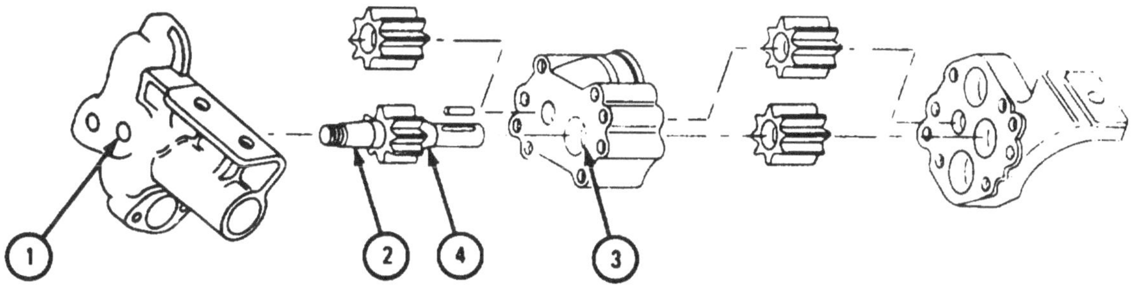

NOTE: CHECK ONLY THOSE PARTS WHICH ARE CALLED OUT IN THIS FRAME. PARTS WITHOUT CALLOUTS ARE SHOWN ONLY FOR REFERENCE PURPOSES OR ARE CHECKED IN ANOTHER FRAME.

TA 113622

Table 4-32. Oil Pump Impeller Shaft and Bushing Type Bearings Wear Limits

Index Number	Item /Point of Measurement	Size and Fit of New Parts (inches)	Wear Limits (inches)
1	Impeller drive shaft bushing inside diameter	0.6910 to 0.6920	0.6940
2	Impeller drive shaft outside diameter	0.6890 to 0.6900	0.6880
1 and 2	Fit of impeller drive shaft in impeller drive shaft bushing	0.0010L to 0.0030L	0. 0050L
3	Impeller drive shaft bushing inside diameter	0.6910 to 0.6920	0.6940
4	Impeller drive shaft outside diameter	0.6890 to 0.6900	0.6880
3 and 4	Fit of impeller drive shaft in impeller drive shaft bushing	0. 0010L to 0.0030L	0.0050L

FRAME 13

NOTE

Readings must be within limits given in table 4-33. The letter L shows a loose fit. If readings are not within given limits, get a new oil pump.

1. Measure outside diameter (1) where impeller drive shaft fits into impeller drive shaft bushing (2) .
2. Measure inside diameter of impeller drive shaft bushing (2).
3. Measure fit of impeller drive shaft outside diameter (1) in impeller drive shaft bushing (2) .

GO TO FRAME 14

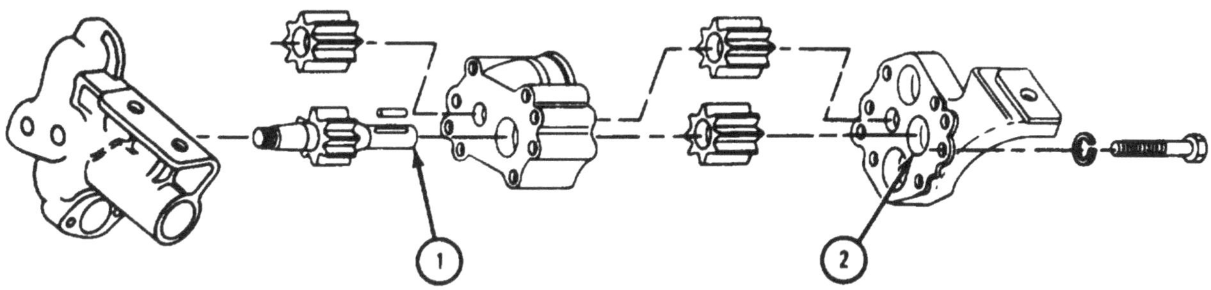

NOTE
CHECK ONLY THOSE PARTS WHICH ARE CALLED OUT IN THIS FRAME. PARTS WITHOUT CALLOUTS ARE SHOWN ONLY FOR REFERENCE PURPOSES OR ARE CHECKED IN ANOTHER FRAME.

Table 4-33. Oil Pump Impeller Drive Shaft and Bushing Wear Limits

Index Number	Item /Point of Measurement	Size and Fit of New Parts (inches)	Wear Limit (inches)
1	Impeller drive shaft outside diameter	0.4985 to 0.4990	0.4975
2	Impeller drive shaft bushing inside diameter	0.4995 to 0.5005	0.5020
1 and 2	Fit of impeller drive shaft in impeller drive shaft bushing	0.005L to 0.0020L	0.0050L

FRAME 14

NOTE

Readings must be within limits given in table 4-34. The letter L shows a loose fit. If readings are not within given limits, get a new oil pump.

1. Measure inside diameter of driven impeller bushing (1).
2. Measure outside diameter of driven impeller shaft (2).
3. Measure fit of driven impeller shaft (2) in driven impeller bushing (1).
4. Measure depth of pressure pump driven impeller bore (3).
5. Measure length A of pressure pump driven impeller (4).
6. Measure clearance between pressure pump driven impeller (4) and top of pressure pump drive impeller bore (3).

GO TO FRAME 15

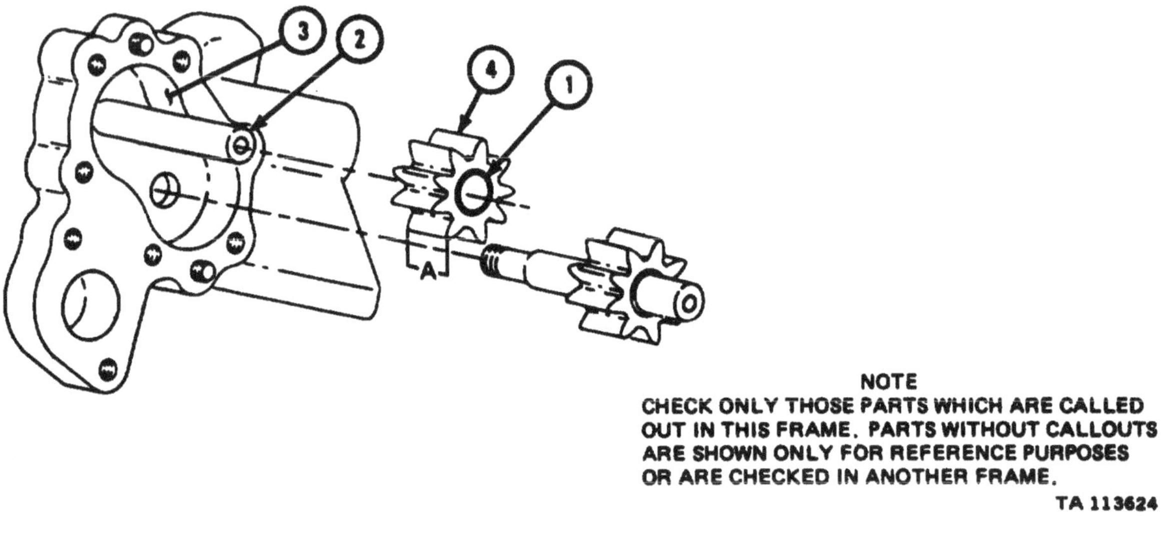
3
2
4
1
A
NOTE
CHECK ONLY THOSE PARTS WHICH ARE CALLED OUT IN THIS FRAME. PARTS WITHOUT CALLOUTS ARE SHOWN ONLY FOR REFERENCE PURPOSES OR ARE CHECKED IN ANOTHER FRAME.
TA 113624

Table 4-34. Oil Pump Driven Impeller, Impeller Bore, Impeller Shaft, and Impeller Bushing Wear Limits

Index Number	Item /Point of Measurement	Size and Fit of New Parts (inches)	Wear Limit (inches)
1	Driven impeller bushing inside diameter	0.5035 to 0.5045	0.5065
2	Driven impeller shaft outside diameter	0.5020 to 0.5025	0.5020
1 and 2	Fit of driven impeller shaft in driven impeller bushing	0.0010L to 0.0025L	0.0050L
3	Pressure pump driven impeller bore dept h	1.3770 to 1.3790	1.3820
4	Pressure pump driven impeller length A	1.3735 to 1.3750	1.3750
3 and 4	Clearance between pressure pump driven impeller and top of impeller bore	0.0020L to 0.0055L	0.0080L max

FRAME 15

NOTE

Readings must be within limits given in table 4-35. The letter L shows a loose fit. If readings are not within given limits, get a new oil pump.

1. Measure depth of pressure pump drive impeller bore (1).
2. Measure length A of pressure pump drive impeller (2).
3. Measure clearance between pressure pump drive impeller (2) and top of pressure pump drive impeller bore (1).

GO TO FRAME 16

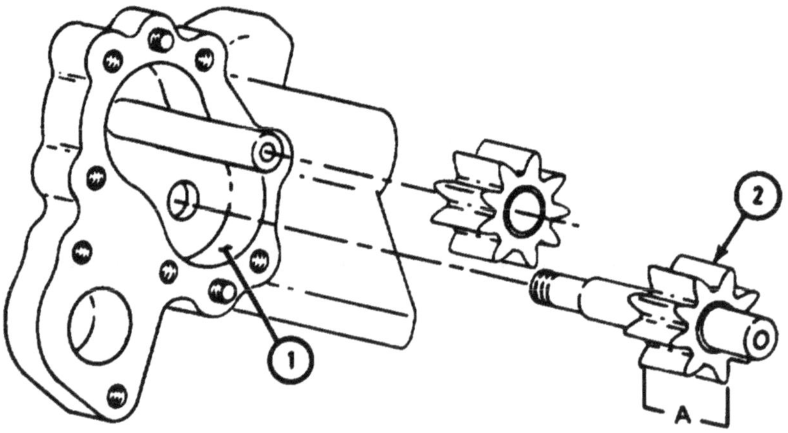

NOTE: CHECK ONLY THOSE PARTS WHICH ARE CALLED OUT IN THIS FRAME. PARTS WITHOUT CALLOUTS ARE SHOWN ONLY FOR REFERENCE PURPOSES OR ARE CHECKED IN ANOTHER FRAME.

TA 113625

Table 4-35. Oil Pump Drive Impeller and Impeller Bore Wear Limits

Index Number	Item /Point of Measurement	Size and Fit of New Parts (inches)	Wear Limit (inches)
1	Pressure pump drive impeller bore dept h	1.3770 to 1.3790	1.3820
2	Pressure pump drive impeller length A	1.3735 to 1.3750	1.3705
1 and 2	Clearance between pressure pump drive impeller and top of drive impeller bore	0.0020L to 0.0055L	0.0020L max

FRAME 1 6

NOTE

Readings must be within limits given in table 4-36. The letter L shows a loose fit. If readings are not within given limits, get a new oil pump.

1. Measure depth of scavenge pump driven impeller bore (1).
2. Measure length A of scavenge pump driven impeller (2).
3. Measure clearance between scavenge pump driven impeller (2) and top of scavenge pump driven impeller bore (1).
4. Measure depth of scavenge pump drive impeller bore (3).
5. Measure length B of scavenge pump drive impeller (4).
6. Measure clearance between scavenge pump drive impeller (4) and top of scavenge pump drive impeller bore (3).

END OF TASK

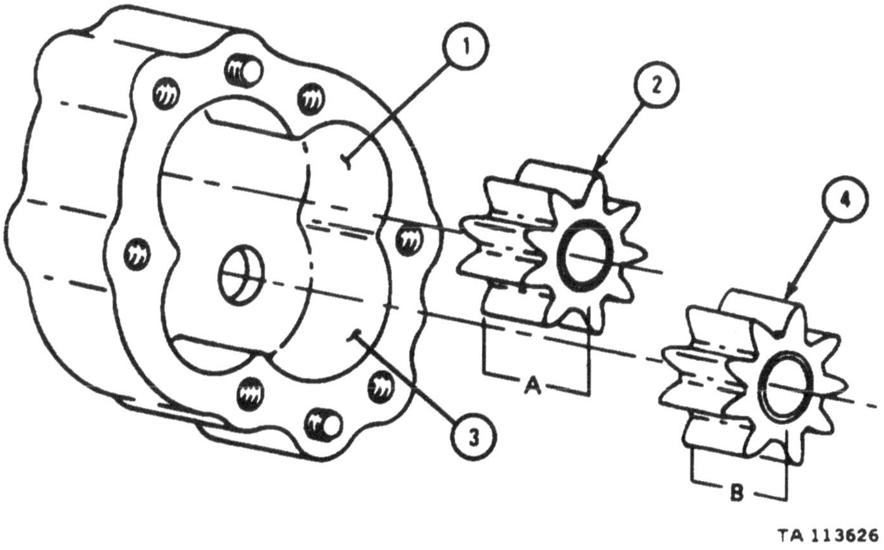

Table 4-36. Scavenge Oil Pump Impellers and Impeller Bores Wear Limits

Index Number	Item /Point of Measurement	Size and Fit of New Parts (inches)	Wear Limit (inches)
1	Scavenge pump driven impeller bore dept h	1.6270 to 1.6290	1.6330
2	Scavenge pump driven impeller length A	1.6235 to 1.6250	1.6205
1 and 2	Clearance between scavenge pump driven impeller and top of driven impeller bore	0.0020L to 0.0055L	0.0080L max
3	Scavenge pump drive impeller bore dept h	1.6270 to 1.6290	1.6330
4	Scavenge pump drive impeller length B	1.6235 to 1.6250	1.6205
3 and 4	Clearance between scavenge pump drive impeller and top of drive impeller bore	0.0020L to 0.0055L	0.0080L max

FRAME 17

NOTE

Readings must be within limits given in table 4-37. The letter L shows a loose fit. If readings are not within given limits, throw away part and get a new one.

Relief valve spring (1) and relief valve plunger (2) come in parts kit. Kit also has new retaining pin (4), one of two types of spring retainer (5), and several shims (6). If any of these parts is damaged, all should be changed. Either type of spring retainer (5) can be used on this pump . If kit has two plungers, the longer one is for this pump.

Save all shims (6) even if pump you are working on has no shims. Shims will be used later during pump test and adjustment.

1. Measure free length of relief valve spring (1). Measure load needed to squeeze relief valve spring (1) down to 1.8100 inch. Measure maximum solid height of relief valve spring (1).
2. Measure outside diameter of relief valve plunger (2).
3. Measure inside diameter of relief valve bore (3). If reading is not within limits given in table 4-37, get a new oil pump.
4. Measure fit of relief valve plunger (2) in relief valve bore (3). If reading is not within limits given in table 4-37, get a new oil pump.

GO TO FRAME 18

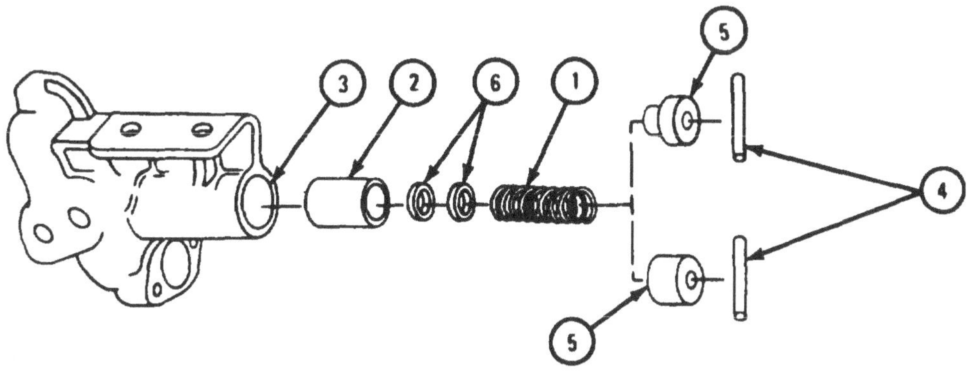

Table 4-37. Engine LDS-465-2 Oil Pump Pressure Relief Valve Wear Limits

Index Number	Item /Point of Measurement	Size and Fit of New Parts (inches unless otherwise noted)	Wear Limit (inches)
1	Relief valve spring free length	2.0000	None
1	Relief valve spring load at 1.8100-inches	47.0 to 57 pounds	None
1	Relief valve spring maximum solid height	1.6000	None
2	Relief valve plunger outside diameter	0.8120 to 0.8125	0.8105
3	Relief valve bore inside diameter	0.8140 to 0.8150	0.8165
2 and 3	Fit of relief valve plunger in relief valve bore	0.0015L to 0.0030L	0.0060L

FRAME 18

NOTE

Readings must be within limits given in table 4-38. The letter L shows a loose fit. If readings are not within given limits, get a new oil pump.

1. Measure inside diameter of impeller drive shaft bushing (1).
2. Measure outside diameter of impeller drive shaft .
3. Measure fit of impeller drive shaft diameter (2) in impeller drive shaft bushing (1).
4. Measure inside diameter of impeller drive shaft bushing (3).
5. Measure outside diameter (4) at other end of impeller drive shaft.
6. Measure fit of impeller drive shaft diameter (4) in impeller drive shaft bushing (3).

GO TO FRAME 19

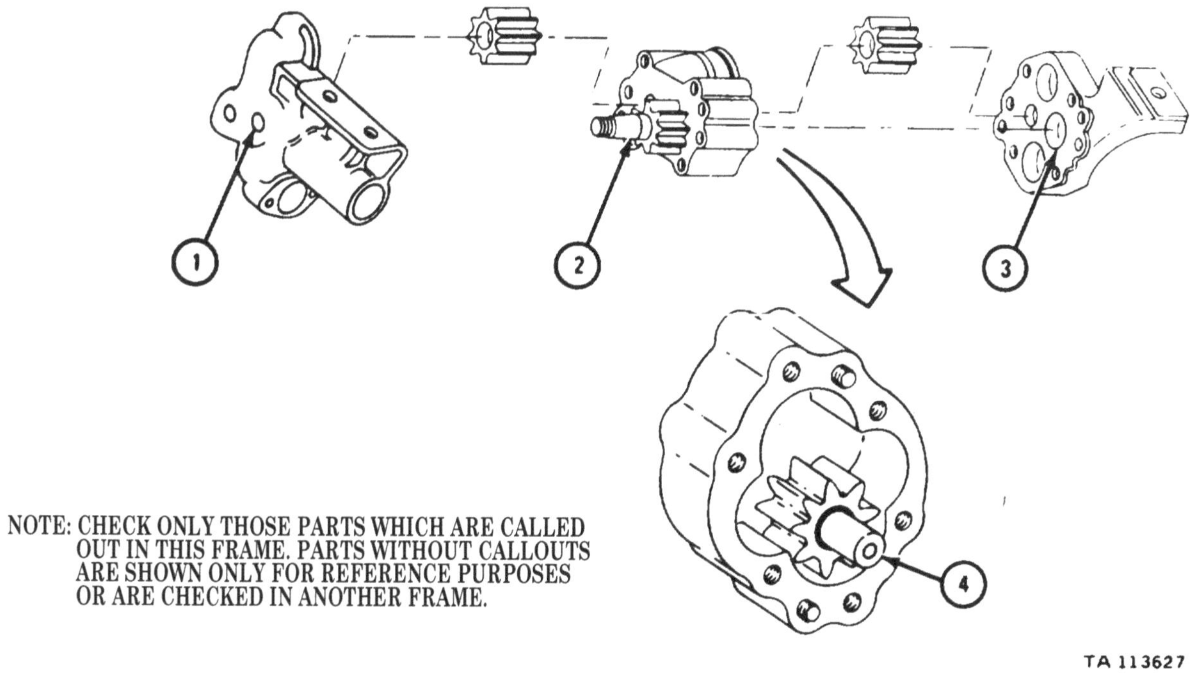
1
2
3
4
NOTE: CHECK ONLY THOSE PARTS WHICH ARE CALLED OUT IN THIS FRAME. PARTS WITHOUT CALLOUTS ARE SHOWN ONLY FOR REFERENCE PURPOSES OR ARE CHECKED IN ANOTHER FRAME.
TA 113627

Table 4-38. Engine LDS-465-2 Oil Pump Impeller Drive Shaft and Bushings Wear Limit s

Number	Item /Point of Measurement	Size and Fit of New Parts (inches)	Wear Limit (inches)
1	Impeller drive shaft bushing inside diameter	0.6910 to 0.6920	0.6940
2	Impeller drive shaft outside diameter	0.6890 to 0.6894	0.6880
land 2	Fit of impeller drive shaft in impeller drive shaft bushing	0.0016L to 0.0030L	0.0055L
3	Impeller drive shaft bushing inside diameter	0.6910 to 0.6920	0.6940
4	Impeller drive shaft outside diameter	0.6890 to 0.6894	0.6880
3 and 4	Fit of impeller drive shaft in impeller drive shaft bushing	0.0016L to 0.0030L	0.0055L

FRAME 19

NOTE

Readings must be within limits given in table 4-39. The letter L shows a loose fit. If readings are not within given limits, get a new oil pump.

1. Measure inside diameter of pressure pump driven impeller shaft bore (1).
2. Measure outside diameter of driven impeller shaft (2).
3. Measure fit of driven impeller shaft (2) in pressure pump driven impeller shaft bore (1) .
4. Measure inside diameter of scavenge pump driven impeller shaft bore (3).
5. Measure fit of driven impeller shaft (2) in scavenge pump driven impeller shaft bore (3) .

GO TO FRAME 20

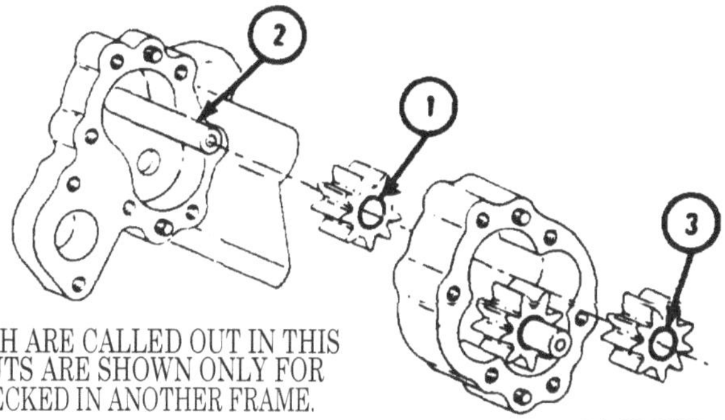

NOTE: CHECK ONLY THOSE PARTS WHICH ARE CALLED OUT IN THIS FRAME, PARTS WITHOUT CALLOUTS ARE SHOWN ONLY FOR REFERENCE PURPOSES OR ARE CHECKED IN ANOTHER FRAME.

TA 113628

Table 4-39. Engine LDS-465-2 Oil Pump Driven Impeller Shaft Wear Limits

Index Number	Item/Point of Measurement	Size and Fit of New Parts (inches)	Wear Limit (inches)
1	Pressure pump driven impeller shaft bore inside diameter	0.6877 to 0.6883	0.6898
2	Driven impeller shaft outside diameter	0.6853 to 0.6858	0.6843
1 and 2	Fit of driven impeller shaft in pressure pump driven impeller bore	0.0019L to 0.0030L	0.0055L
3	Scavenge pump driven impeller shaft bore inside diameter	0.6877 to 0.6883	0.6898
2 and 3	Fit of driven impeller shaft in scavenge pump driven impeller shaft bore	0.019L to 0.0030L	0.0055L

FRAME 20

NOTE

Readings must be within limits given in table 4-40. The letter L shows a loose fit. If readings are not within given limits, get a new oil pump.

1. Measure depth of pressure pump driven impeller bore (1).
2. Measure length A of pressure pump driven impeller (2).
3. Measure clearance between pressure pump driven impeller (2) and top of pressure pump driven impeller bore (1) .

GO TO FRAME 21

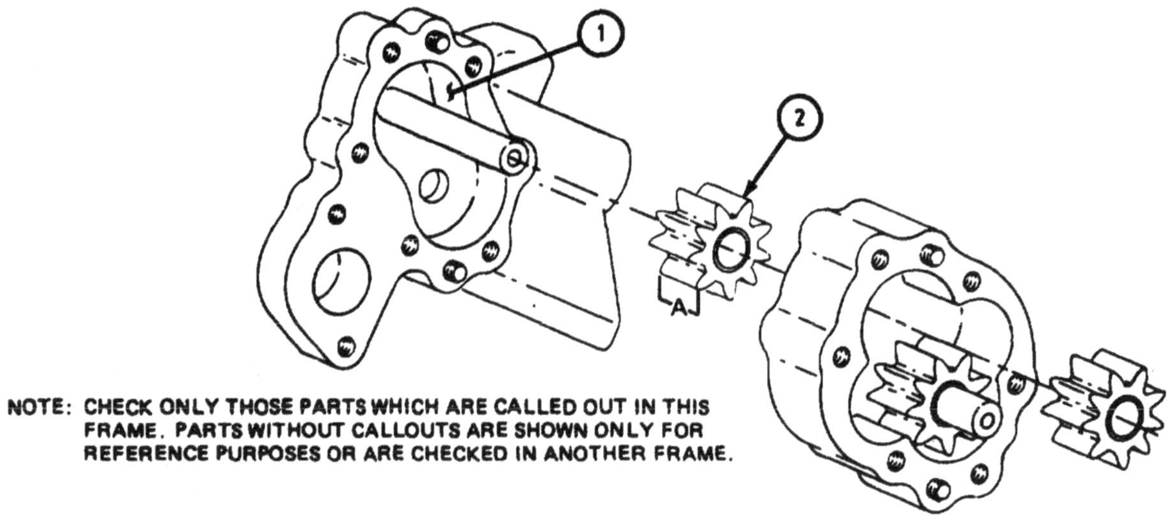
1
2
A
NOTE: CHECK ONLY THOSE PARTS WHICH ARE CALLED OUT IN THIS FRAME. PARTS WITHOUT CALLOUTS ARE SHOWN ONLY FOR REFERENCE PURPOSES OR ARE CHECKED IN ANOTHER FRAME.
TA 113629

Table 4-40. Engine LDS-465-2 Pressure Pump Driven Impeller and Driven Impeller Bore Wear Limits

Number	Item/Point of Measurement	Size and Fit of New Parts (inches)	Wear Limit (inches)
1	Pressure pump driven impeller bore dept h	1.5600 to 1.5620	1.5635
2	Pressure pump driven impeller length	1.5565 to 1.5580	1.5560
1 and 2	Clearance between pressure pump driven impeller and top of impeller bore	0.0020L to 0.0055L	0.0075L max

FRAME 21

NOTE

Readings must be within limits given in table 4-41. The letter L shows a loose fit. If readings are not within given limits, get a new oil pump.

1. Measure depth of pressure pump drive impeller bore (1).
2. Measure length A of pressure pump drive impeller (2).
3. Subtract measurement in step 1 from measurement in step 2. Result is clearance between pressure pump drive impeller (2) and top of pressure pump drive impeller bore (1).

GO TO FRAME 22

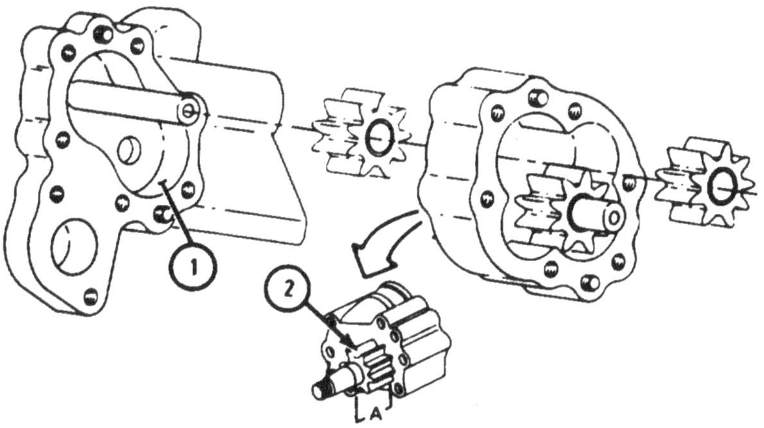

NOTE: CHECK ONLY THOSE PARTS WHICH ARE CALLED OUT IN THIS FRAME. PARTS WITHOUT CALLOUTS ARE SHOWN ONLY FOR REFERENCE PURPOSES OR ARE CHECKED IN ANOTHER FRAME.

TA 113630

Table 4-41. Engine LDS-465-2 Pressure Pump Drive Impeller and Drive Impeller Bore Wear Limits

Index Number	Item /Point of Measurement	Size and Fit of New Parts (inches)	Wear Limit (inches)
1	Pressure pump drive impeller bore dept h	1.5600 to 1.5620	1.5635
2	Pressure pump drive impeller length A	1.5565 to 1.5580	1.5560
1 and 2	Clearance between pressure pump drive impeller and top of drive impeller bore	0.0020L to 0.0055L	0.0075L max

FRAME 2 2

NOTE

Readings must be within limits given in table 4-42. The letter L shows a loose fit. If readings are not within given limits, get a new oil pump.

1. Measure depth of scavenge pump driven impeller bore (1).
2. Measure length A of scavenge pump driven impeller (2).
3. Measure clearance between scavenge pump driven impeller (2) and top of scavenge pump impeller bore (1).

GO TO FRAME 23

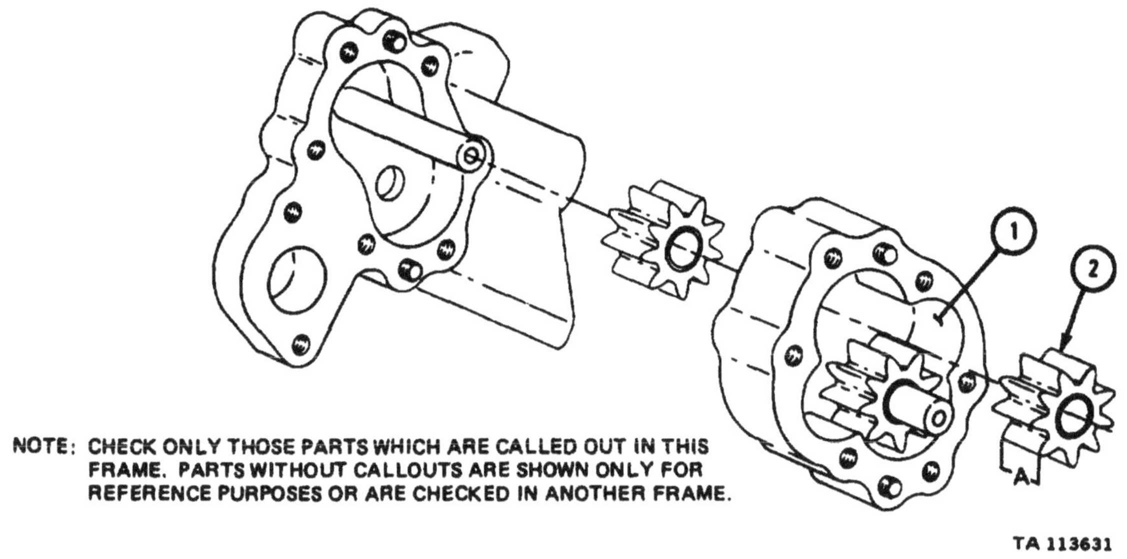
1
2
A
NOTE: CHECK ONLY THOSE PARTS WHICH ARE CALLED OUT IN THIS FRAME. PARTS WITHOUT CALLOUTS ARE SHOWN ONLY FOR REFERENCE PURPOSES OR ARE CHECKED IN ANOTHER FRAME.
TA 113631

Table 4-42. Engine LDS-465-2 Scavenge Pump Driven Impeller and Impeller Bore Wear Limits

Number	Item /Point of Measurement	Size and Fit of New Parts (inches)	Wear Limit (inches)
1	Scavenge pump driven impeller bore dept h	1.5600 to 1.5620	1.5635
2	Scavenge pump driven impeller length A	1.5565 to 1.5580	1.5560
1 and 2	Clearance between scavenge pump driven impeller and top of impeller bore	0.0020L to 0.0055L	0.0075L max

FRAME 23

NOTE

Readings must be within limits given in table 4-43. The letter L shows a loose fit. If readings are not within given limits, get a new oil pump.

1. Measure depth of scavenge pump drive impeller bore (1).
2. Measure length A of scavenge pump drive impeller (2).
3. Measure clearance between scavenge pump drive impeller (2) and top of scavenge pump drive impeller bor e (1).

END OF TASK

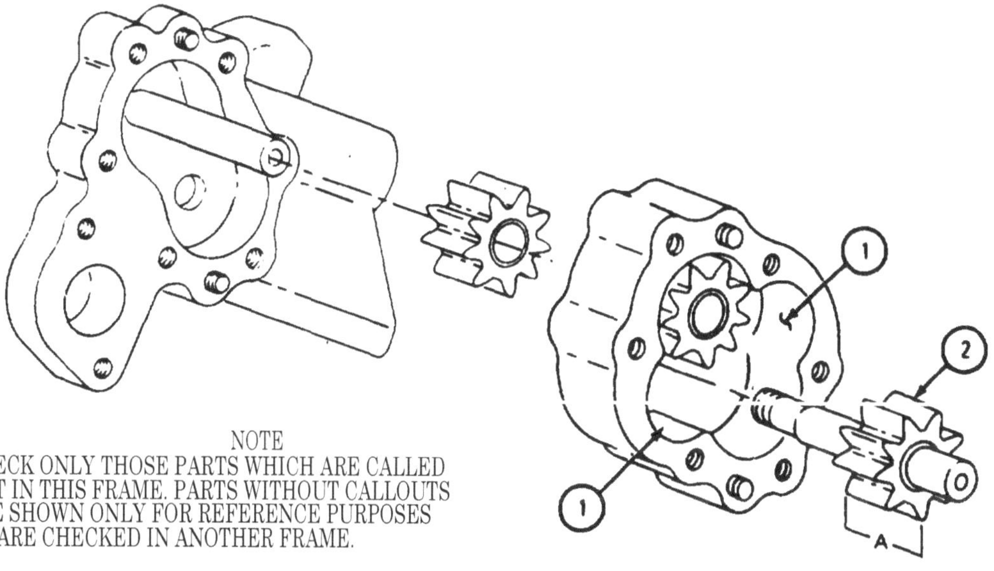

NOTE
CHECK ONLY THOSE PARTS WHICH ARE CALLED OUT IN THIS FRAME. PARTS WITHOUT CALLOUTS ARE SHOWN ONLY FOR REFERENCE PURPOSES OR ARE CHECKED IN ANOTHER FRAME.

Table 4-43. Engine LDS-465-2 Scavenge Pump Drive Impeller and Impeller Bore Wear Limits

Index Number	Item /Point of Measurement	Size and Fit of New Parts (inches)	Wear Limit (inches)
1	Scavenge pump drive impeller bore dept h	1.5600 to 1.5620	1.5635
2	Scavenge pump drive impeller length	1.5565 to 1.5580	1.5560
1 and 2	Clearance between scavenge pump impeller and impeller bore	0.0020L to 0.0055L	0.0075L max

d. Assembly.

NOTE

If working on engine LDS-465-2, go to frame 3. Put a light coat of lubricating oil on shafts and on all bushings in pressure pump.

FRAME 1

1. Put a light coat of lubricating oil on shafts (1 and 2) and on bushing (3).
2. Put square key (4) in slot of drive impeller shaft (2).
3. Put drive impeller shaft (2) in pressure pump housing (5) as shown. Threaded end of drive impeller shaft should fit through bushing (3) in drive impeller housing.
4. Slide pressure pump driven impeller (6) over driven impeller shaft (1) and into pressure pump housing (5) as shown.

GO TO FRAME 2

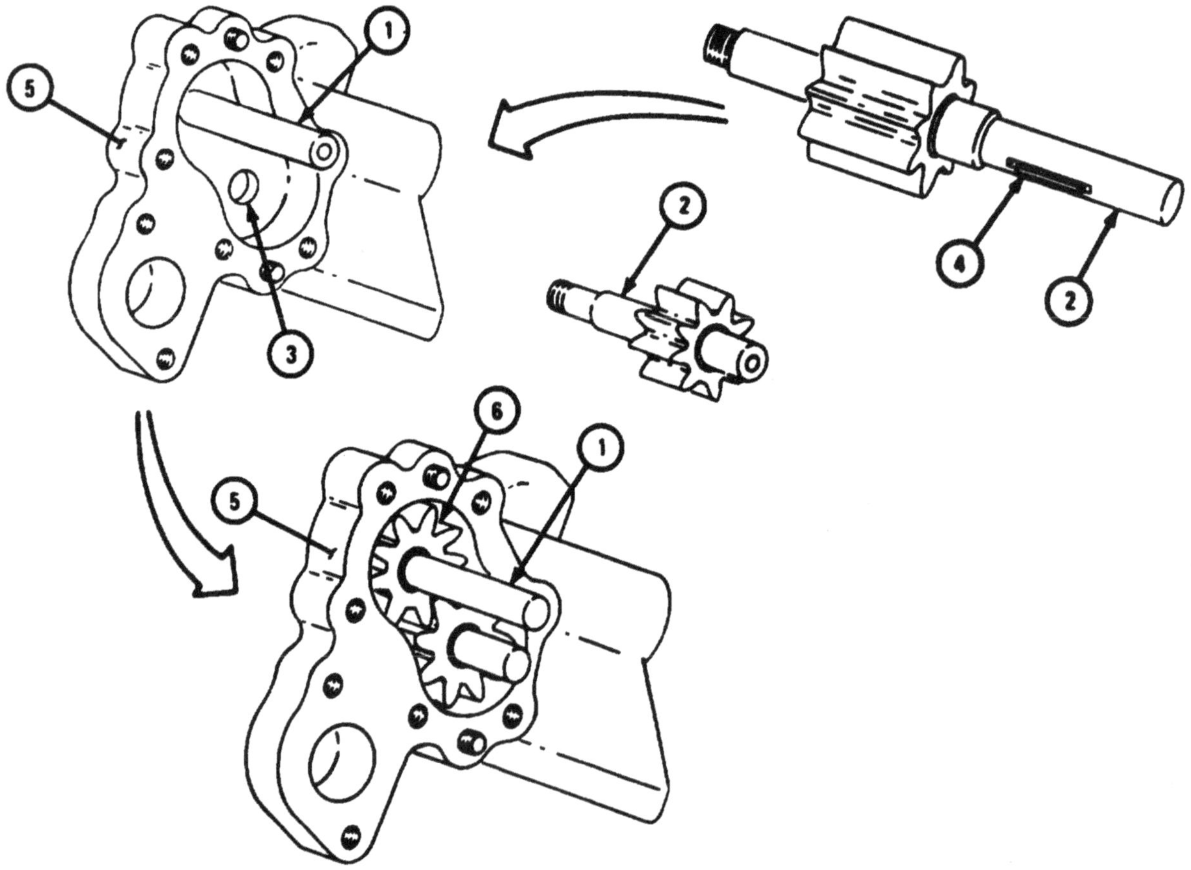

FRAME 2

1. Put scavenge pump housing (1) on pressure pump housing (2) as shown.
2. Slide scavenge pump driven impeller (3) over driven impeller shaft (4) and into scavenge pump housing (1) as shown.
3. Slide scavenge pump drive impeller (5) over drive impeller shaft (6) and into scavenge pump housing (1) as shown.
4. Put scavenge pump cover (7) on scavenge pump housing (1) as shown. Put in two screws (8) and lockwashers (9) .

GO TO FRAME 5

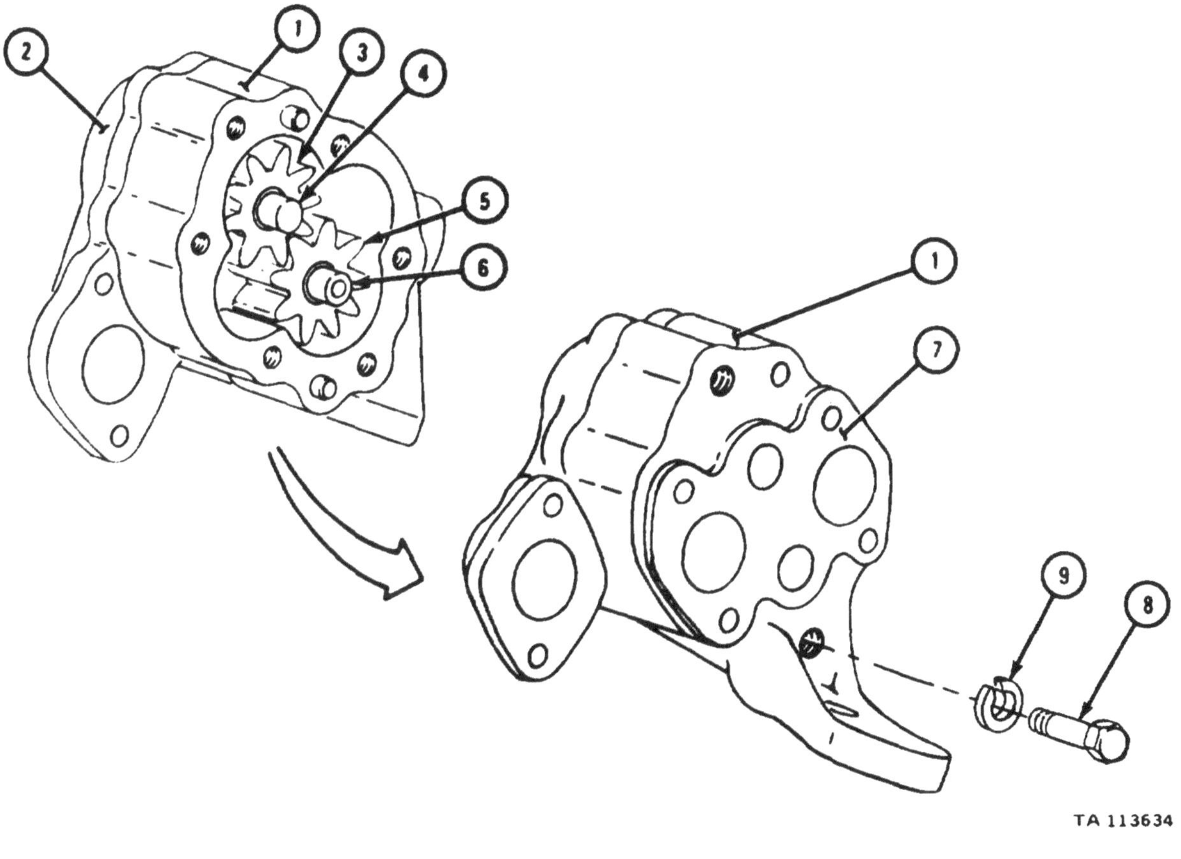

FRAME 3

1. Put a light coat of lubricating oil on shafts (1, 2, and 3) and in all bushings in pressure pump housing (4), bushing in driven impeller (5), and bushing in scavenge pump housing (6).
2. Slide pressure pump driven impeller (5) over driven impeller shaft (1) and into pressure pump housing (4) as shown.

NOTE

Drive impeller and drive impeller shaft (2) do not come apart from scavenge pump housing (6).

3. Put scavenge pump housing (6) on pressure pump housing (4) so that threaded end of drive impeller shaft (2) fits through bushing (7).
4. Slide scavenge pump drive impeller (8) over driven impeller shaft (1) and into scavenge pump housing (6) as shown.

GO TO FRAME 4

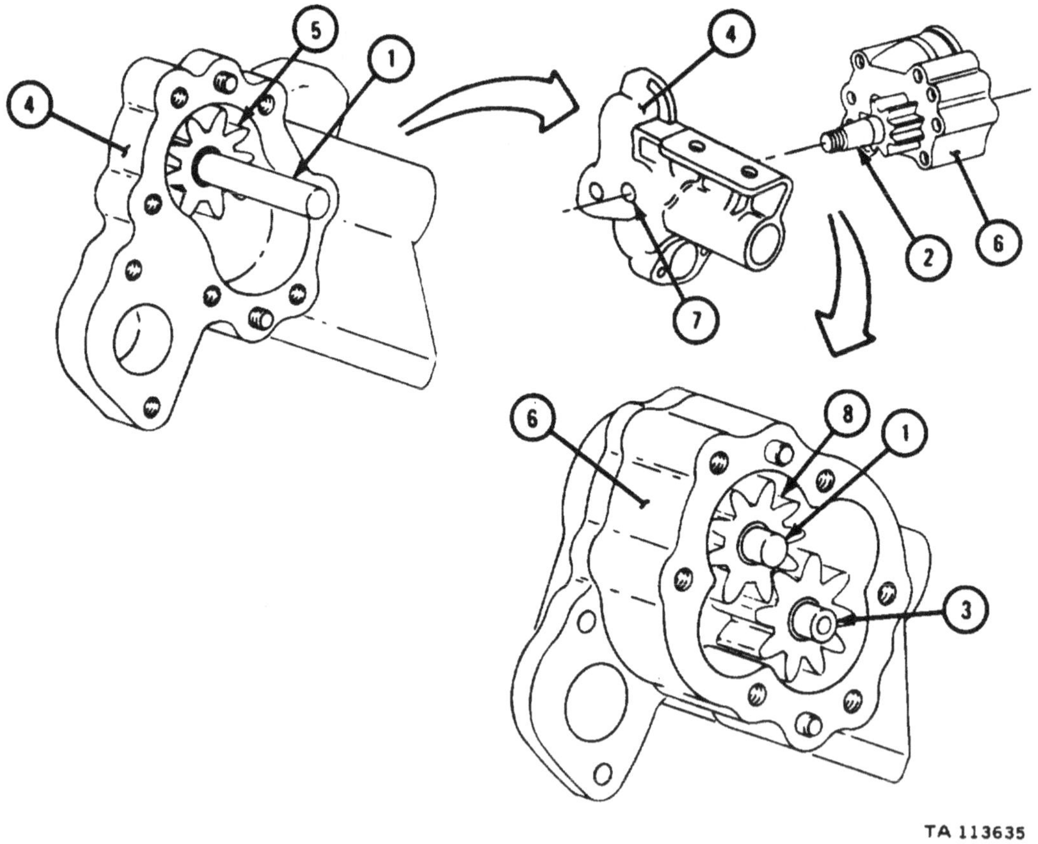

FRAME 4

1. Put scavenge pump cover (1) on scavenge pump housing (2) as shown.
2. Put in two screws (3) and lockwashers (4).

GO TO FRAME 5

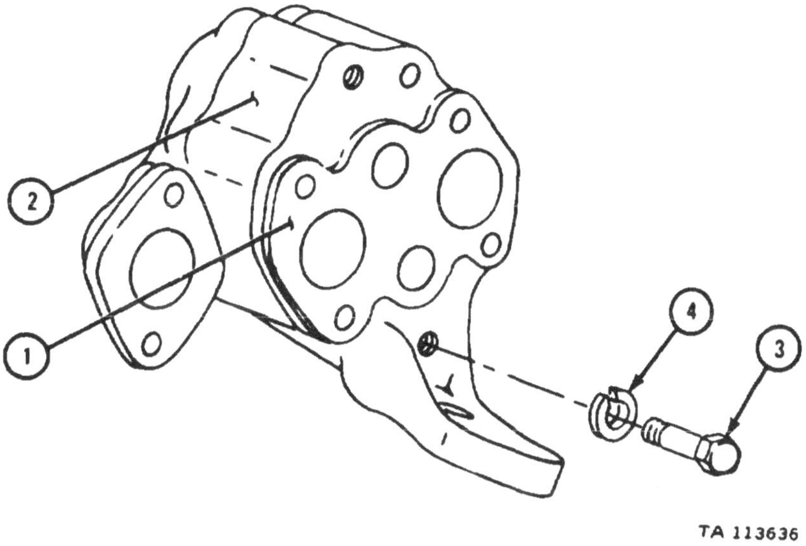

FRAME 5

NOTE

Two types of oil pumps (1) have been used on these engines. Early model pump uses threaded cap (2) to hold pressure relief valve spring (3) and plunger (4) inside pump. Late model pump uses retaining pin (5) and one of two types of spring retainer (6) to hold pressure relief valve spring and plunger inside pump .

1. Coat plunger (4) and plunger bore in oil pump (1) with lubricating oil. Put in plunger.

NOTE

Number of shims (7) will differ from pump to pump. Some pumps may have used no shims. Put back noted number of shims.

2. Put in shims (7).
3. Put in spring (3) .
4. On early model pumps (1), screw in plug (2).
5. On late model pumps (1), put in spring retainer (6) and retaining pin (5).

NOTE

You may have to use an arbor press to push spring retainer (6) in far enough to let retaining pin (5) fit in pump (1) .

GO TO FRAME 6

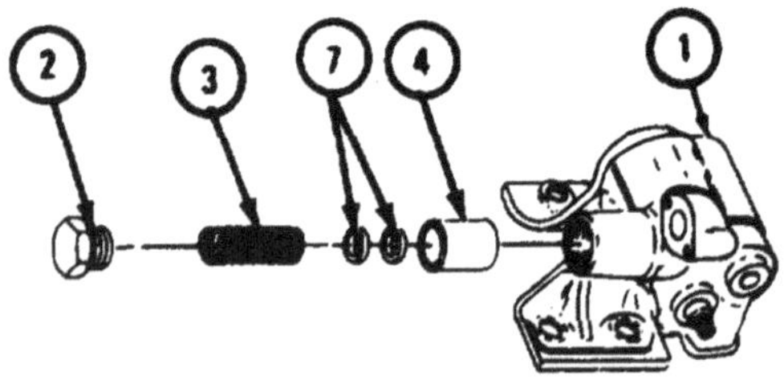

EARLY MODEL OIL PUMP

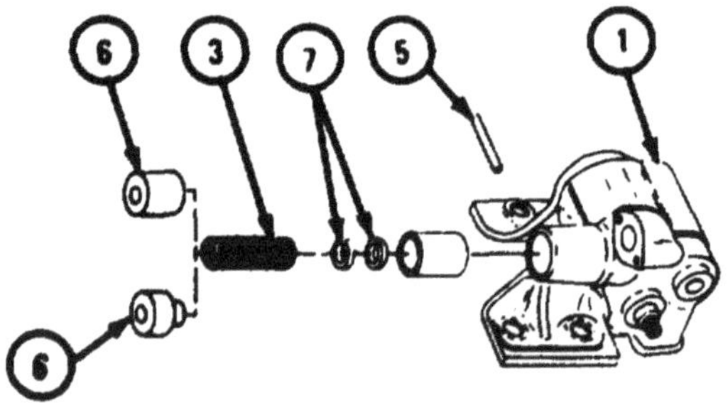

LATE MODEL OIL PUMP

TA 113637

FRAME 6

1. Put oil pump assembly (1) in vise with soft jaw caps.
2. Put pin punch into scavenge oil pump outlet hole (2) to keep pump impeller from turning .

GO TO FRAME 7

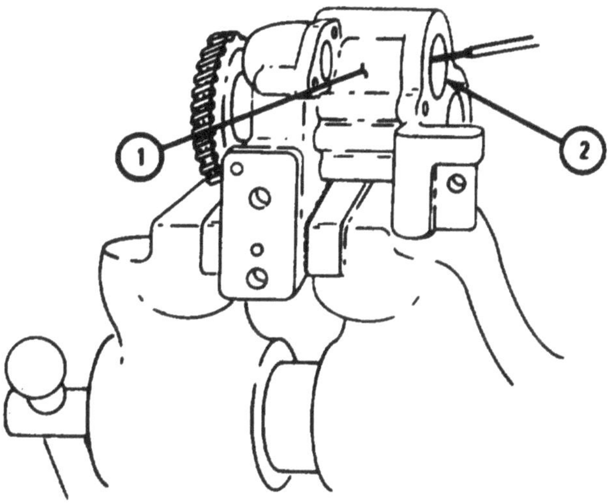

FRAME 7

1. Put drive gear (1) on oil pump shaft (2).
2. On all engines except LDS-465-2, put on slotted nut (3). Using torque wrench, tighten nut to 44 to 46 pound-feet. Put in cotter pin (4).
3. On engine LDS-465-2, put on flat washer (5) and locknut (6). Using torque wrench, tighten locknut to 44 to 46 pound-feet.
4. Take oil pump assembly (7) out of vise.

END OF TASK

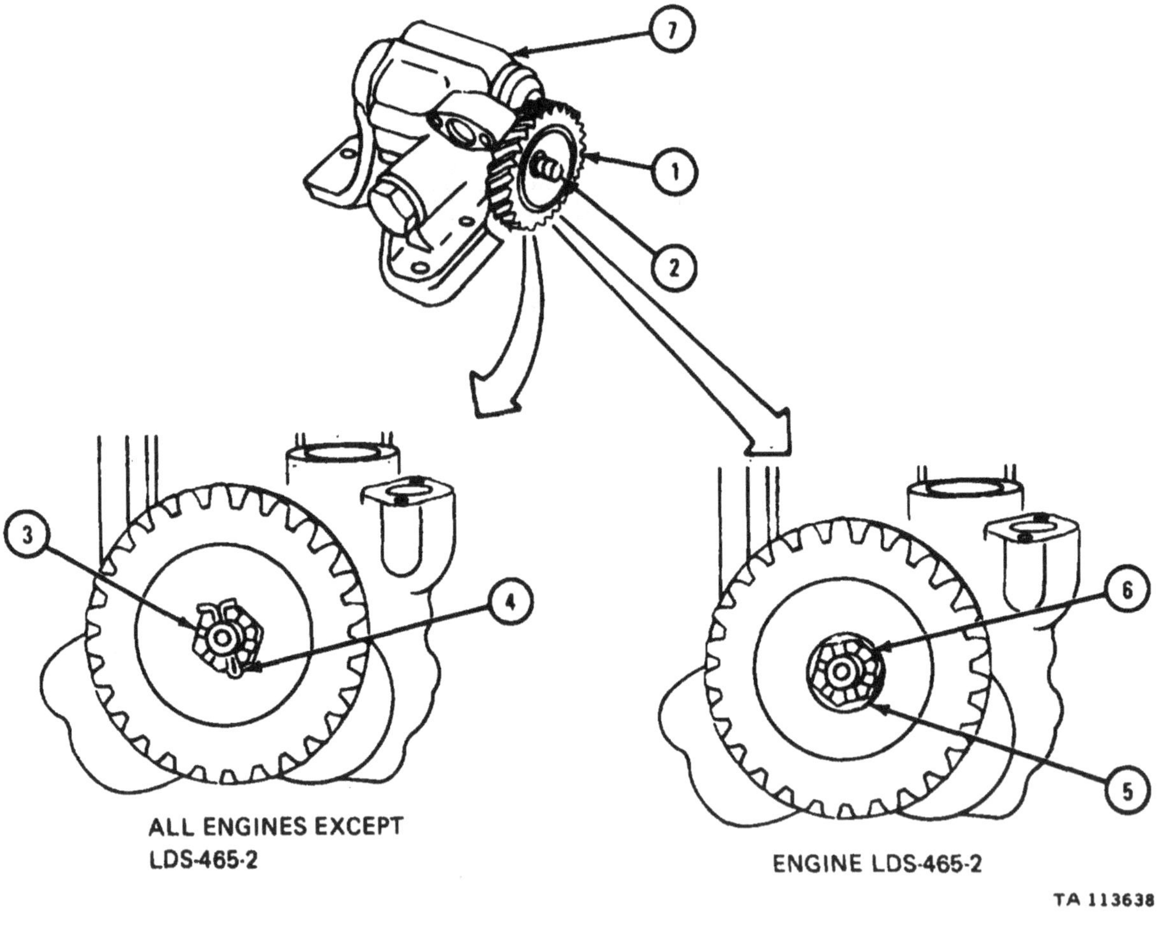

e. Adjustment and Test.

(1) Pressure relief valve adjustment .

NOTE

Oil pump has two halves. Front half of pump behind drive shaft is called pressure pump. Pressure pump pumps oil from deep sump of engine oil pan to engine main oil gallery. Rear half of pump is called scavenge pump. Scavenge pump pumps oil from shallow sump in engine oil pan to deep sump in engine oil pan.

(a) Pour some engine lubricating oil into scavenge pump so scavenge pump impellers will not be damaged during test.

(b) Hook oil pump to test bench as shown in figure 4-2. Join oil inlet hose and oil outlet hose to pressure half of pump. Check that you can see pressure relief valve hole in pump.

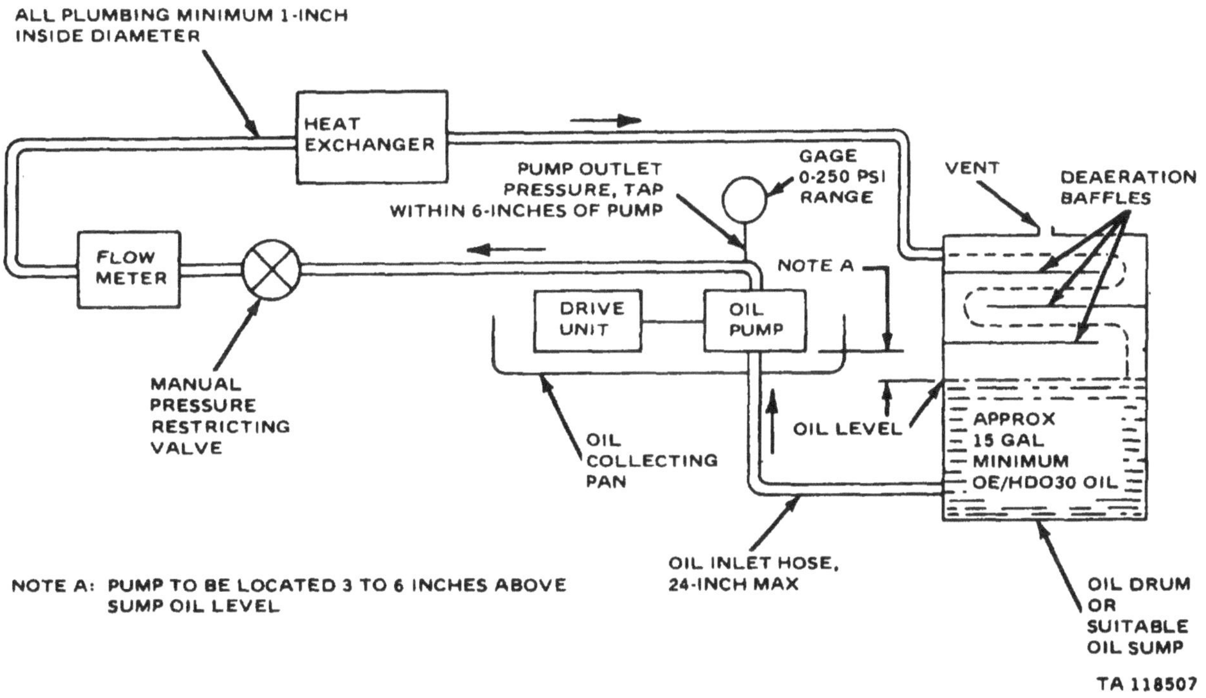

Figure 4-2. Typical Oil Pump Test Setup.

(c) Open pressure restrictor valve all the way. Set drive unit to turn pump at 1,000 to 2,000 rpm.

(d) Slowly close restrictor valve while watching pressure gage and pressure relief valve hole in pump. Check that oil starts to squirt out of pressure relief valve hole when pressure gage reads between 140 and 225 pounds per square inch.

(e) If oil does not start to squirt out of pressure relief valve hole when pressure gage reads between 140 and 225 psi, stop test. Take apart pressure relief valve. Refer to para 4-17a, frame 4. Add or take out shims shown in figure 4-3 to change relief valve opening pressure. Do steps (c) and (d) again .

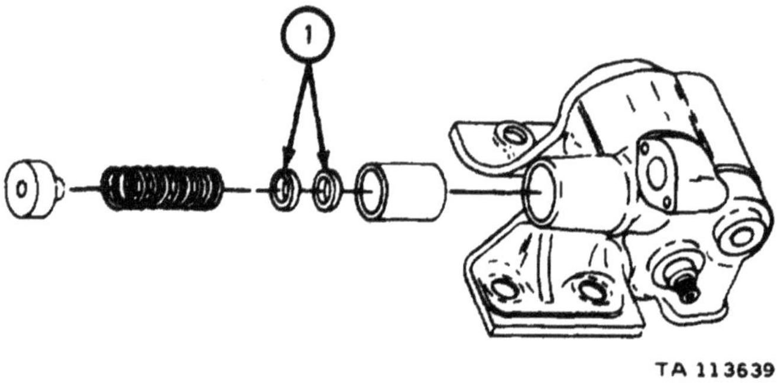

Figure 4-3. Pressure Relief Valve Shims.

(2) Flow test .

NOTE

Pressure relief adjustment must be done before flow test. Refer to para 4-17e (1).

Oil pump has two halves. Front half of pump behind drive shaft is called pressure pump. Rear half of pump is called scavenge pump.

(a) If scavenge pump has not already been oiled, pour some engine lubricating oil into scavenge pump so scavenge pump impellers will not be damaged during test .

(b) Hook oil pump to test bench as shown in figure 4-2.

(c) Open pressure restrictor valve. Start drive unit, Run pump until air is out of system and oil temperature holds steady at 200°F to 210°F for all engines except LDS-465-2 and 220°F to 240°F for engine LDS-465-2.

(d) Work drive unit to speed up oil pump to 3,420 rpm. Slowly close restrictor valve until pressure gage reads 110 psi.

(e) For all engines except LDS-465-2, check that flow meter reads 22 gallons per minute (gpm) or more. If flow meter reads less than 22 gpm, get a new oil pump.

(f) For engine LDS-465-2, check that flow meter reads 26 gpm or more. If flow meter reads less than 26 gpm, get a new oil pump.

4-18. OIL PRESSURE REGULATOR VALVE HOUSING ASSEMBLY.

a. Disassembly.

FRAME 1

NOTE

Two types of oil pressure regulator (1) have been used on these engines. Disassembly procedure is the same for both except as noted.

1. Unscrew and take out two plugs (2 and 3), two gaskets (4), spring (5), spring (6), and two plungers (7) .
2. If working on early model oil pressure regulator, take out pipe plug (8).
3. If working on late model oil pressure regulator, take out pipe plug (9).

END OF TASK

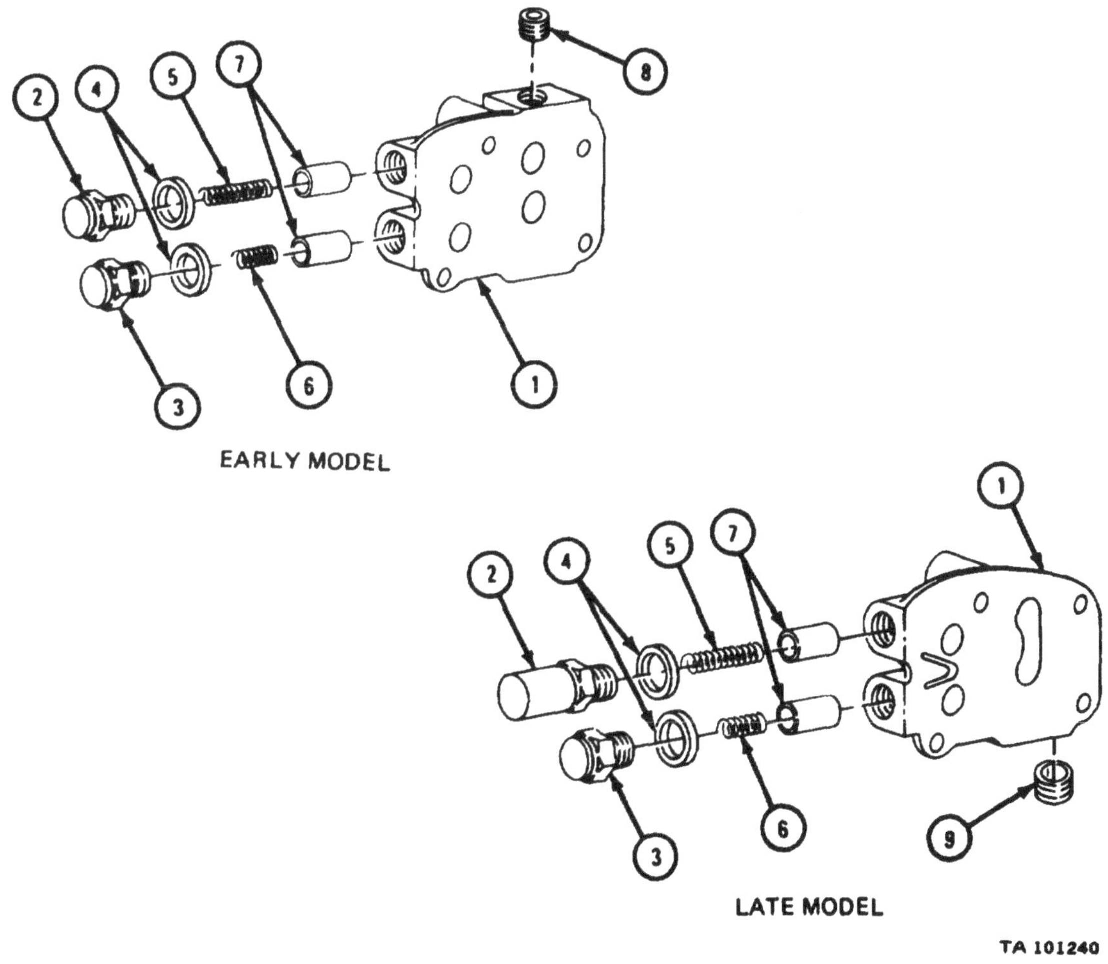

b. Cleaning.

WARNING

Dry cleaning solvent is flammable. Do not use near an open flame. Keep a fire extinguisher nearby when solvent is used. Use only in well-ventilated places. Failure to do this may result in injury to personnel and damage to equipment.

(1) Clean all parts in dry cleaning solvent.

(2) Clean off sludge and gum using a stiff brush.

WARNING

Eye shields must be worn when using compressed air. Eye injury can occur if eye shields are not used.

(3) Blow all oil passages dry with compressed air. Dry outside of oil pressure regulator housing and parts that were taken from inside housing with clean cloth.

c. Inspection and Repair.

FRAME 1

NOTE

Two types of oil pressure regulator (1) have been used on these engines. This task is the same for both except as noted.

1. Check that housing (1) is not cracked. If housing is cracked, throw it away and get a new one.

WARNING

Dry cleaning solvent is flammable. Do not use near an open flame. Keep a fire extinguisher nearby when solvent is used. Use only in well-ventilated places. Failure to do this may result in injury to personnel and damage to equipment.

2. Check that mounting surface of housing (1) has no nicks, scratches or burrs. Fix nicks, scratches, and burrs with a fine mill file or crocus cloth dipped in dry cleaning solvent.
3. Check that oil passages (2) are clean and free from dirt. Clean out blocked or dirty oil passages. Refer to para 4-3.
4. Check that two plunger bores (3) are not scored or worn. If plunger bores are scored or worn, throw away housing and get a new one.
5. Check that threaded holes in housing (1) are not damaged. Fix minor thread damage with a tap.

GO TO FRAME 2

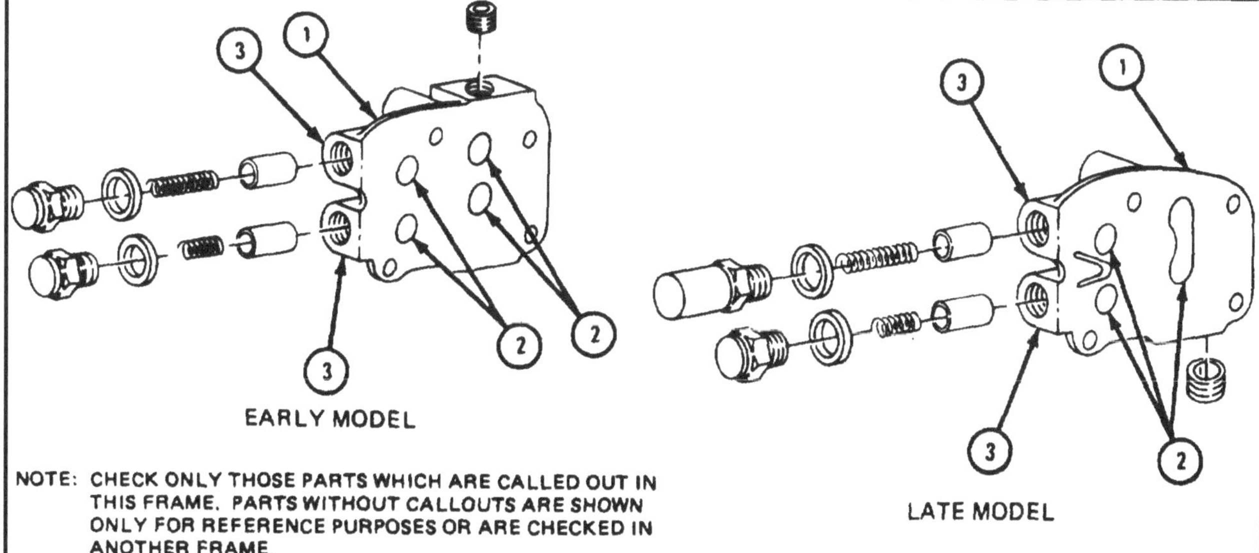

NOTE: CHECK ONLY THOSE PARTS WHICH ARE CALLED OUT IN THIS FRAME. PARTS WITHOUT CALLOUTS ARE SHOWN ONLY FOR REFERENCE PURPOSES OR ARE CHECKED IN ANOTHER FRAME.

FRAME 2

NOTE

Two types of oil pressure regulator have been used on these engines. This task is the same for both except as noted.

On early model oil pressure regulators, if upper plug (1) or upper spring (2) are found to be damaged, change both parts. Only late model plug (3) and spring (4) are now available.

1. On early model pressure regulators, check that threads on four mounting screws (5), pipe plug (6), plug (1), and plug (7) are not damaged. Fi x minor thread damage with a thread chaser.
2. On early model pressure regulators, check that plugs (1 and 7) are not cracked. Check that springs (2 and 8) are not cracked or worn. Throw away cracked plugs and cracked or worn springs and get new ones.
3. On late model pressure regulators, check that threads on four mounting screws (5), pipe plug (6) and plug (9) are not damaged. Fix minor thread damage with a thread chaser.
4. On late model pressure regulators, check that plugs (3 and 9) are not cracked. Check that spring (4) and spring (8) are not cracked or worn. Throw away cracked plugs and cracked or worn springs and get new ones.
5. Check that two plungers (10) are not cracked. If plunger is cracked , throw it away and get a new one.

WARNING

Dry cleaning solvent is flammable, Do not use near an open flame. Keep a fire extinguisher nearby when solvent is used. Use only in well-ventilated places. Failure to do this may result in injury to personnel and damage to equipment.

6. Check that two plungers (10) do not have nicks or burrs, Fix small nicks and burrs with a fine mill file or crocus cloth dipped in dry cleaning solvent.

GO TO FRAME 3

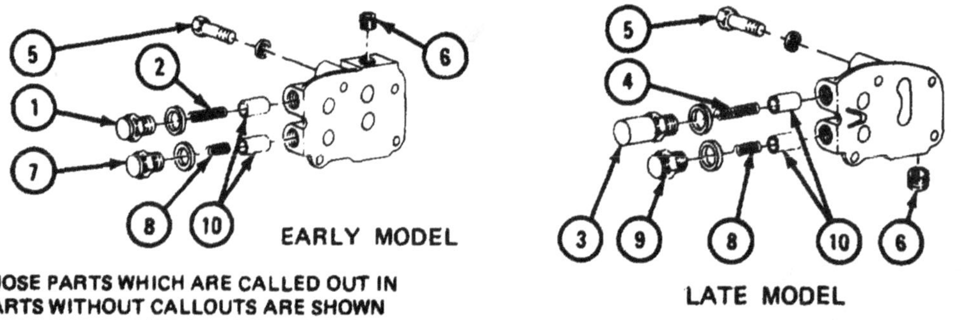

NOTE: CHECK ONLY THOSE PARTS WHICH ARE CALLED OUT IN THIS FRAME. PARTS WITHOUT CALLOUTS ARE SHOWN ONLY FOR REFERENCE PURPOSES OR ARE CHECKED IN ANOTHER FRAME.

TA 101242

FRAME 3

1. Check upper plug (1 or 2). If pressure regulator has a longer upper plug (2), this is a late model regulator. If pressure regulator has a short uppe r plug (1), this is an early model regulator.

IF WORKING ON AN EARLY MODEL PRESSURE REGULATOR, GO TO FRAME 4.
IF WORKING ON A LATE MODEL PRESSURE REGULATOR, GO TO FRAME 5

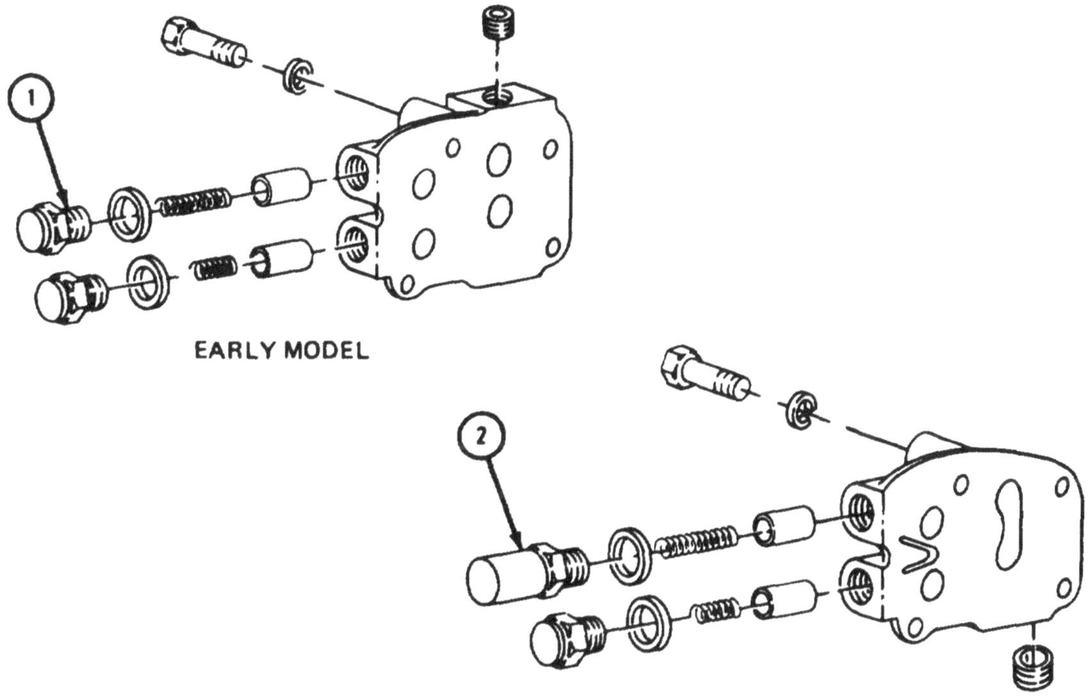

NOTE: CHECK ONLY THOSE PARTS WHICH ARE CALLED OUT IN THIS FRAME. PARTS WITHOUT CALLOUTS ARE SHOWN ONLY FOR REFERENCE PURPOSES OR ARE CHECKED IN ANOTHER FRAME.

TA 101243

FRAME 4

NOTE

Readings must be within limits given in table 4-44. If readings are not within given limits, throw away part and get a new one.

Early model oil pressure regulator valve plunger spring (1) is no longer available. If this spring is damaged, you will have to put in late model oil pressure regulator valve plunger spring, which is longer. To make room for longer spring, you must also throw away early model plug (2) and put in longer late model plug (3).

1. Measure free length of oil pressure regulator valve plunger spring (1).
2. Using spring tester, measure load needed to squeeze spring (1) down to 1.750 inches.
3. Using spring tester, measure solid height of spring (1).

GO TO FRAME 5

NOTE:
CHECK ONLY THOSE PARTS WHICH ARE CALLED OUT IN THIS FRAME. PARTS WITHOUT CALLOUTS ARE SHOWN ONLY FOR REFERENCE PURPOSES OR ARE CHECKED IN ANOTHER FRAME.

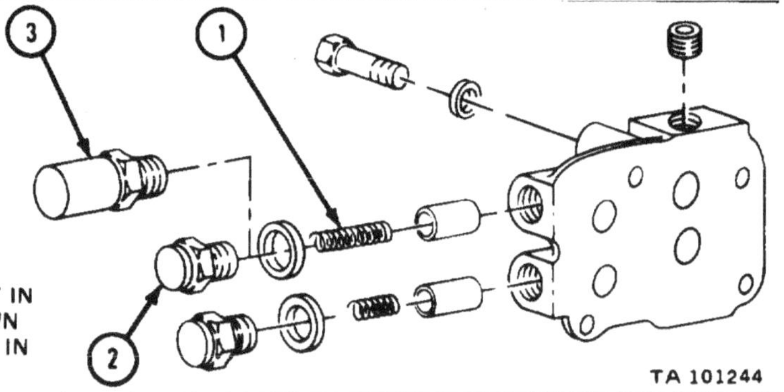

Table 4-44. Early Model Oil Pressure Regulator Valve Plunger Spring Wear Limits

Index Number	Point of Measurement	Size and Fit of New Parts (inches unless otherwise noted)	Wear Limit (inches)
1	Oil pressure regulator valve plunger spring free length	2.2800	None
1	Oil pressure regulator valve plunger spring load at 1.750 inches	24.45 to 30.45 pounds	None
1	Oil pressure regulator valve plunger spring maximum solid height	1.4990	None

FRAME 5

NOTE

Readings must be within limits given in table 4-45. If readings are not within given limits, throw away par t and get a new one.

1. Measure free length of oil pressure regulator valve plunger spring (1).
2. Using spring tester, measure load needed to squeeze spring (1) down to 3.440 inches.
3. Using spring tester, measure solid height of spring (1).

GO TO FRAME 6

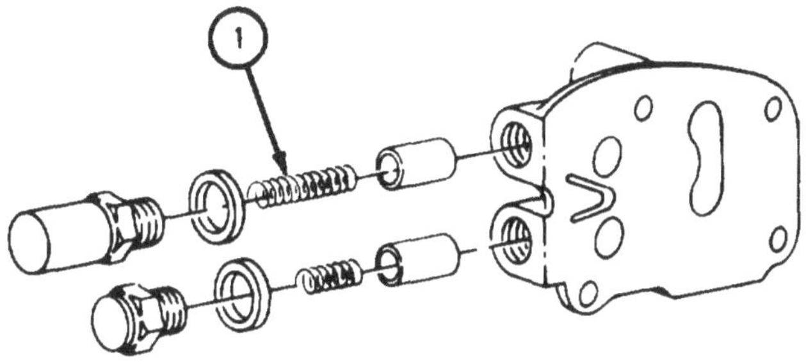

NOTE: CHECK ONLY THOSE PARTS WHICH ARE CALLED OUT IN THIS FRAME. PARTS WITHOUT CALLOUTS ARE SHOWN ONLY FOR REFERENCE PURPOSES OR ARE CHECKED IN ANOTHER FRAME.

TA 101245

Table 4-45. Late Model Oil Pressure Regulator Valve Plunger Spring Wear Limits

Index Number	Point of Measurement	Size and Fit of New Parts (inches unles s otherwise noted)	Wear Limits (inches)
1	Oil pressure regulator valve plunger spring free lengt h	4.6500	None
1	Oil pressure regulator valve plunger spring load at 3.440 inches	24.60 to 28.60 pounds	None
1	Oil pressure regulator valve plunger spring maximum solid height	2.8700	None

FRAME 6

NOTE

Readings must be within limits given in table 4-46. The letter L shows a loose fit. If readings are not within given limits, throw away part and get a new one.

1. Measure free length of piston cooling relief valve plunger spring (1).
2. Using spring tester, measure load needed to squeeze spring (1) down to 1.380 inches.
3. Using spring tester, measure solid height of spring (1).
4. Measure outside diameter of two plungers (2).
5. Measure inside diameter of two plunger bores (3).
6. Measure fit of two plungers (2) in plunger bores (3).

END OF TASK

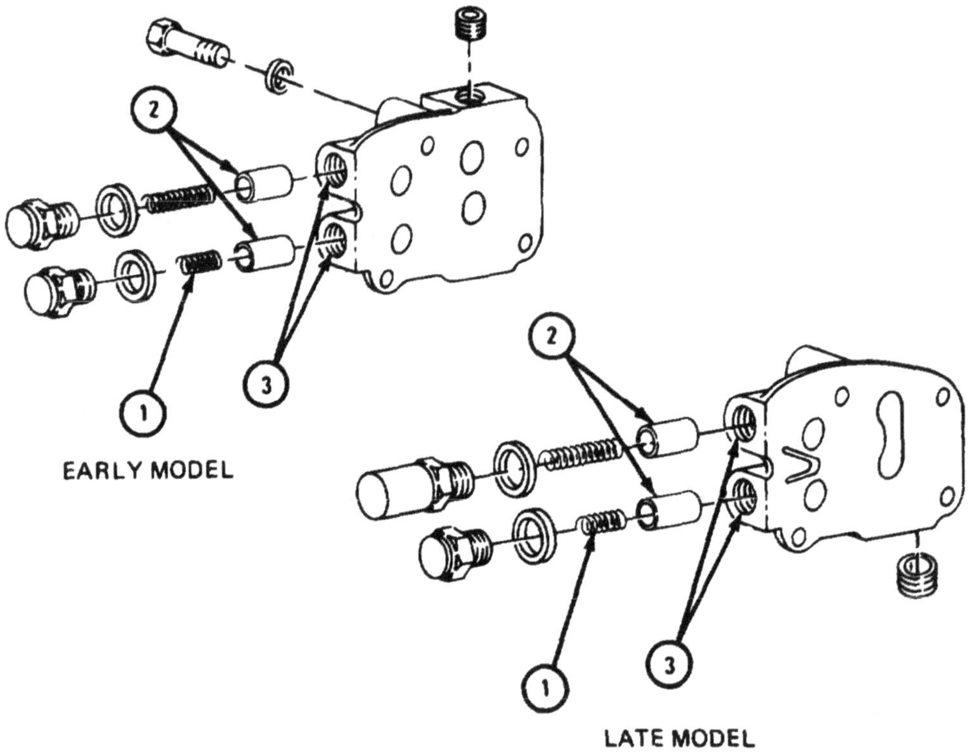

NOTE: CHECK ONLY THOSE PARTS WHICH ARE CALLED OUT IN THIS FRAME. PARTS WITHOUT CALLOUTS ARE SHOWN ONLY FOR REFERENCE PURPOSES OR ARE CHECKED IN ANOTHER FRAME.

TA 101246

Table 4-46. Oil Pressure Regulator Valve Housing Wear Limits

Index Number	Item/Point of Measurement	Size and Fit of New Parts (inches unless otherwise noted)	Wear Limit (inches)
1	Piston cooling relief valve plunger spring free length	1.6200	None
1	Piston cooling relief valve plunger spring load at 1.380 inches	7.58 to 7.72 pounds	None
1	Piston cooling relief valve spring maximum solid height	0.9500	None
2	Plunger outside diameter	0.8000 to 0.8010	0.7995
3	Plunger bore inside diameter	0.8030 to 0.8040	0.8050
2 and 3	Fit of plunger in plunger bore	0.0020L to 0.0040L	0.0050L

d. Assembly.

FRAME 1

CAUTION

Upper spring (1) is longer than lower spring (2) on both models of regulator (3). Be careful not to mix them up or regulator will not work properly.

NOTE

Two types of oil pressure regulator (3) have been used on these engines. Assembly procedure is the same for both except as noted.

Upper plug (4) and lower plug (5) are the same size on early model regulator (3).

1. Put in two plungers (6), upper spring (1), lower spring (2), two gaskets (7), upper plug (4), and lower plug (5).
2. On early model regulators (3), put in pipe plug (8).
3. On late model regulators (3), put in pipe plug (9).

END OF TASK

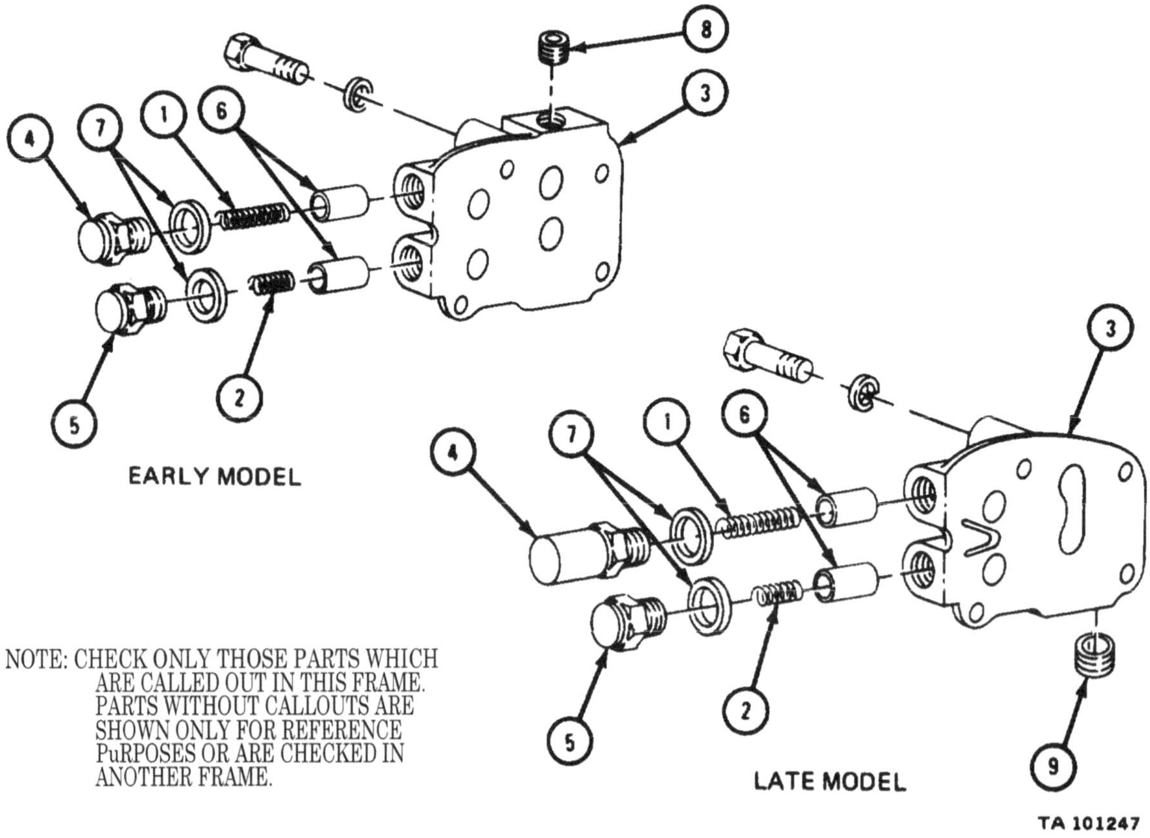
EARLY MODEL
LATE MODEL
NOTE: CHECK ONLY THOSE PARTS WHICH ARE CALLED OUT IN THIS FRAME. PARTS WITHOUT CALLOUTS ARE SHOWN ONLY FOR REFERENCE PuRPOSES OR ARE CHECKED IN ANOTHER FRAME.
TA 101247

4-19. CYLINDER HEAD ASSEMBLIES.

NOTE

This task is the same for the front and rear cylinder head assemblies (1). It must be done for both cylinder head assemblies.

Intake valves are bigger than exhaust valves.

a. Disassembly.

FRAME 1

1. Using valve spring compressor, press in outer valve spring (2) .
2. Take out two exhaust valve spring retainer locks (3).

CAUTION

Be careful not to let valve spring compressor slip off valve spring retainer (4) when easing off valve spring tension . Outer valve spring (2) and valve spring retainer can pop out and be lost.

3. Slowly work valve spring compressor to ease off tension on outer valve spring (2) .
4. Take off valve spring retainer (4) .
5. Take out outer valve spring (2) and inner valve spring (5).
6. Take out and tag exhaust valve (6) and valve rotor (7).
7. Do steps 1 through 6 again and take out and tag two more exhaust valves, valve rotors, and springs (8) and three intake valves, valve rotors, and springs (9) .

GO TO FRAME 2

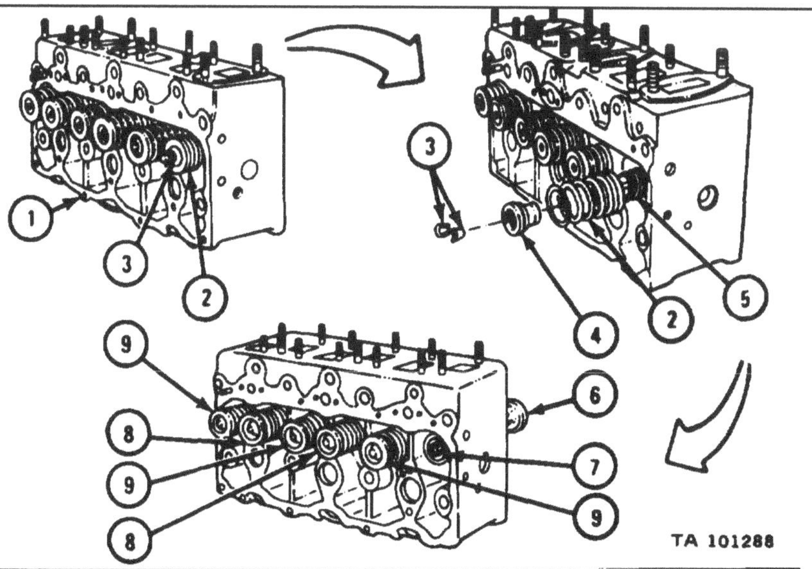

FRAME 2

1. Take out two pipe plugs (1 and 2).

NOTE

A rust colored stain on cylinder head (3) below expansion plugs (4, 5, 6, and 7) means there is coolant leakage.

Some cylinder heads (3) do not have three expansion plugs (4) .

2. Check that three expansion plugs (4), three expansion plugs (5), expansio n plug (6), and expansion plug (7) show no signs of coolant leakage.

3. If any expansion plug (4, 5, 6 or 7) shows signs of coolant leakage, punch hole in plug and pry it out. Put in new plug using brass rod and hammer.

END OF TASK

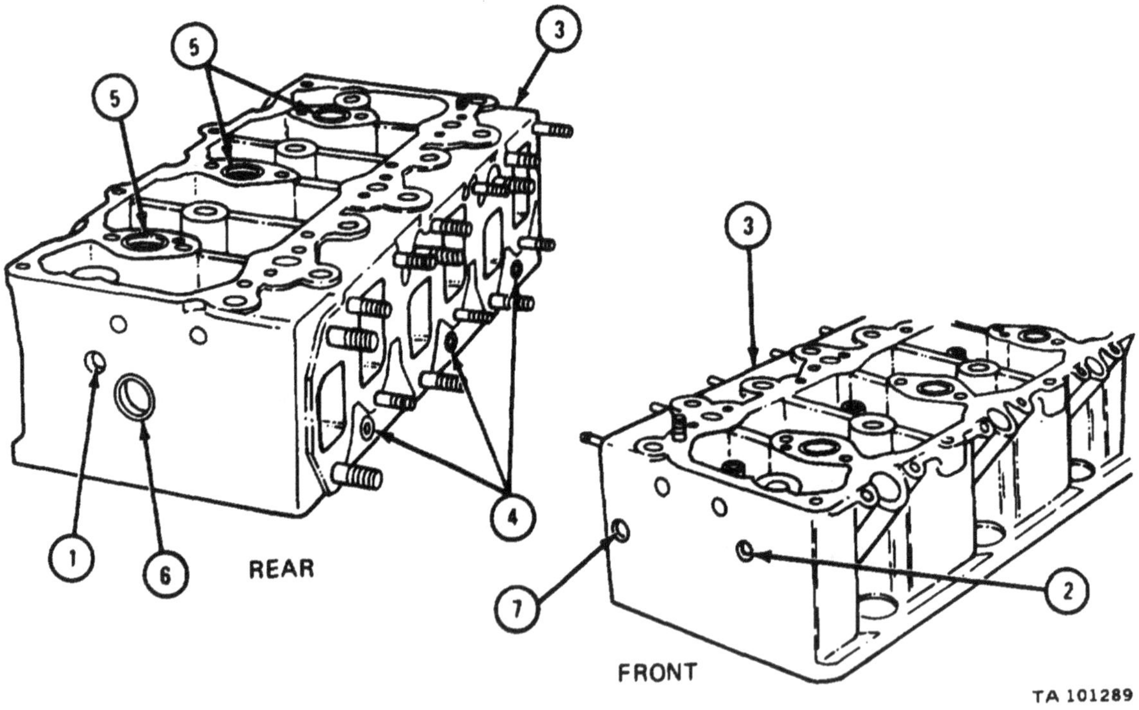

b. Cleaning.

FRAME 1

WARNING

Dry cleaning solvent is flammable. Do not use near an open flame. Keep a fire extinguisher nearby when solvent is used. Use only in well-ventilated places. Failure to do this may result in injury to personnel and damage to equipment.

CAUTION

Be careful not to nick or scratch stems or seating surface of six valves (1) . Be careful not to nick or scratch six cylinder head valve seats (2) or cylinder head gasket face (3) .

1. Clean cylinder head (4), six valves (1) , six inner valve springs (5), six outer valve springs (6), and six valve rotors (7) with solvent. Clean sludge, gum, and carbon deposits using a scraper or a stiff wire brush. Dry with clean cloth.
2. Clean carbon from six fuel injector nozzle seats (8) using stiff brush and carbon removing compound .

WARNING

Eye shields must be worn when using compressed air. Eye injury can occur if eye shields are not used.

3. Clean all oil passages in cylinder head (4) with wire brush. Flush oil passages with dry-cleaning solvent and blow dry with compressed air.

END OF TASK

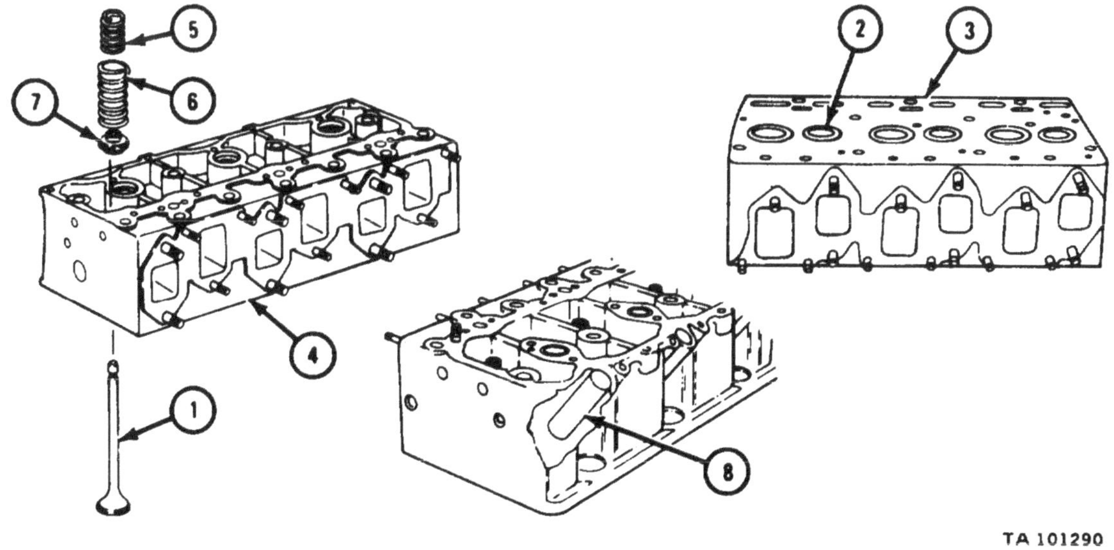

c. Inspection and Repair.

CAUTION

It is easy to damage the equipment if you do not know what you are doing. Do not try to do this task unless you are experienced or you have an experienced person with you .

NOTE

If cylinder head is damaged or worn and must be changed, new cylinder head will come assembled with new valves, valve springs, and rocker arms and rocker arm shaft.

(1) Cylinder head .

FRAME 1

NOTE

If there are only small cracks in cylinder head (1) between intake valve seat inserts (2) and exhaust valve seat inserts (3), the cylinder head need not be thrown away. If cracks show signs of coolant leakage, then throw cylinder head away and get a new one.

1. Using dye penetrant kit and magnifying glass, check that cylinder head (1) has no cracks around studs and pipe plug openings, rear corners, and bends of intake and exhaust passages. If cylinder head is cracked, get a new one.

GO TO FRAME 2

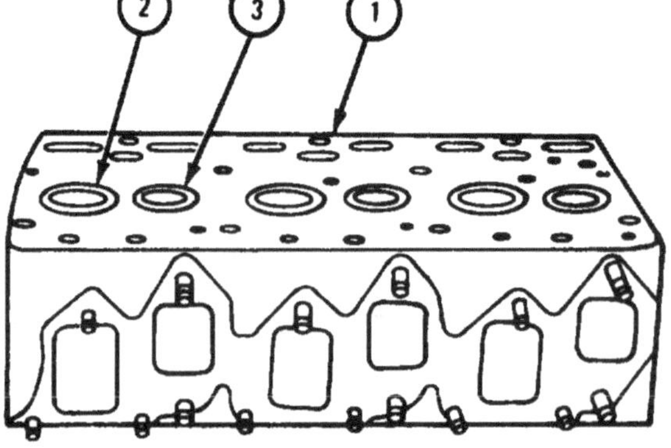

TA 101291

FRAME 2

1. Put in pipe plug (1).
2. Send cylinder head (2) to machine shop. Tell machine shop to make a sealing plate (3) as shown in figure 4-4.
3. Cut gasket (4) to fit sealing plate (3). Pu gasket and sealing plate (3) on bottom of cylinder head (2), alining holes. Bolt sealing plate and gasket to cylinder head as shown.

GO TO FRAME 3

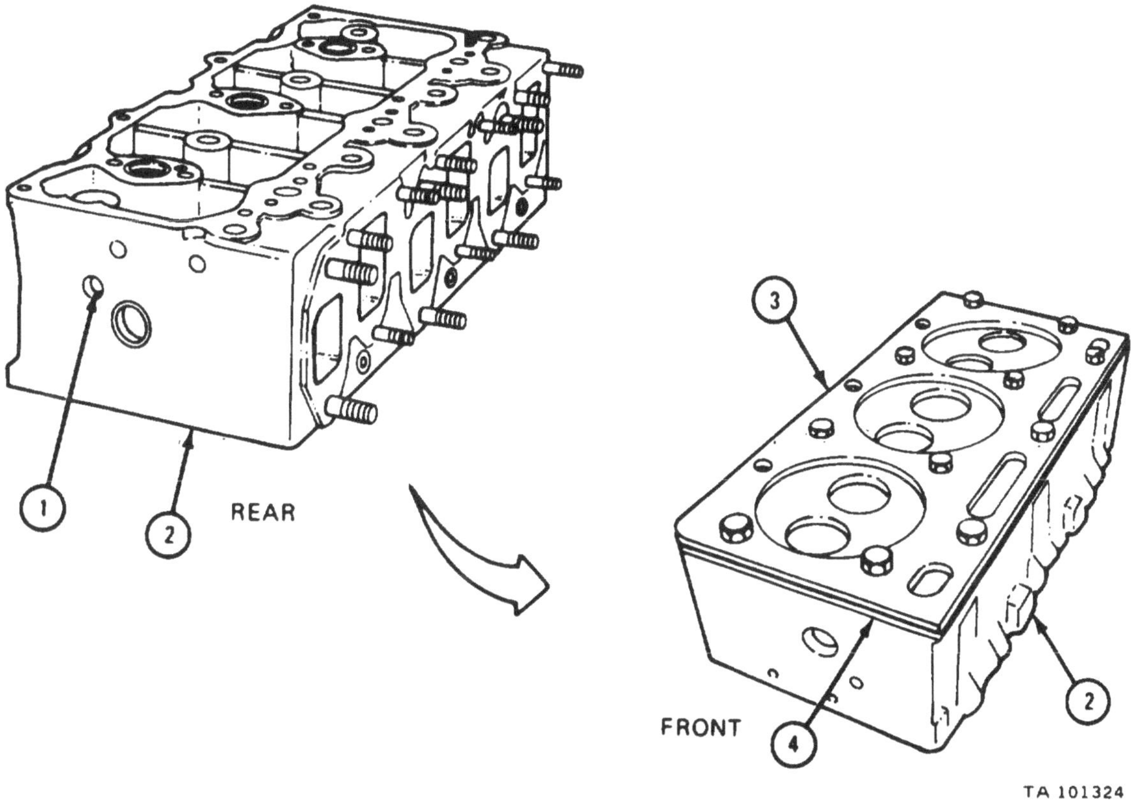

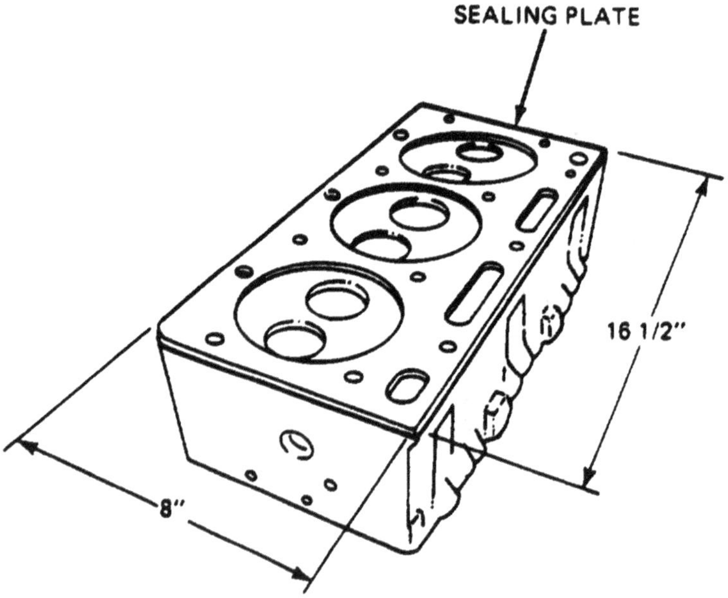

NOTE

Make plate of 1/2-inch thick common stock. Use cylinder head as a template to pick up holes. Large round holes should clear valve ports in cylinder head as shown. Plate should cover water passages in cylinder head.

Figure 4-4. Cylinder Head Sealing Plate Fabrication Instructions.

FRAME 3

1. Screw lifting eye (1) into rocker arm mounting hole at center of cylinder head (2) as shown .
2. Screw air line adapter (3) into cylinder head (2) as shown. Join air line (4) to air line adapter (3) as shown. Join other end of air line (4) to hose connector (5) on pressure tank (6) .
3. Join pressure tank (6) to compressed air source.
4. Set pressure tank (6) to put 50 psi of pressure to cylinder head (2).

GO TO FRAME 4

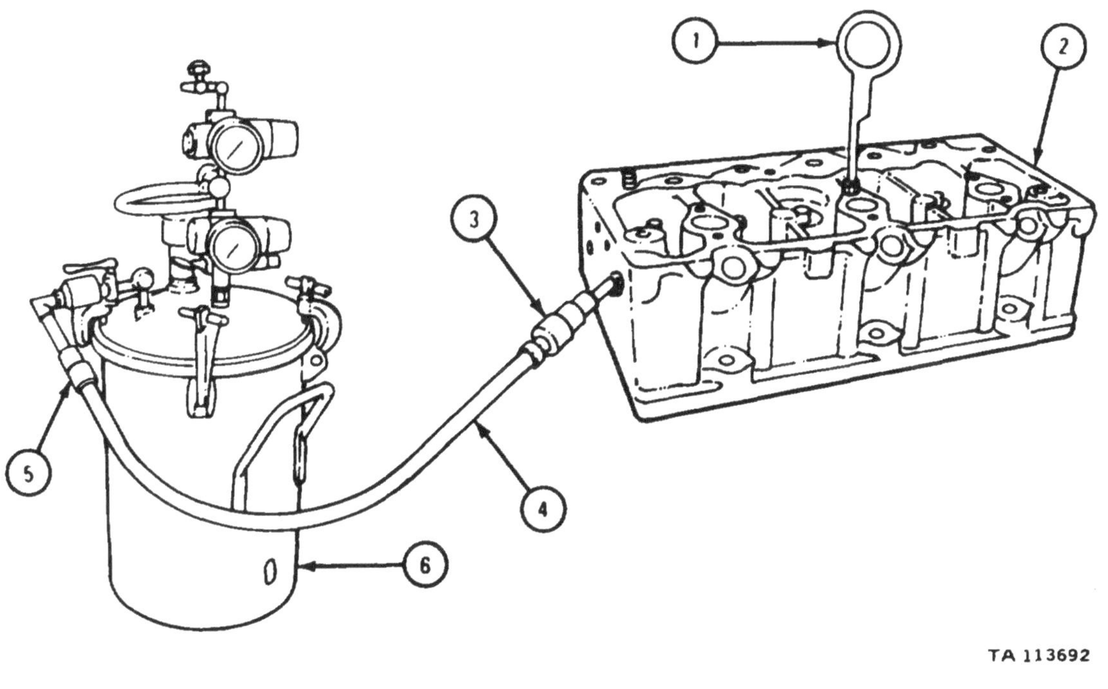

FRAME 4

1. Using chain hoist (1), put cylinder head (2) with air line still hooked up into tank of water (3) heated to 120°F.

NOTE

If cylinder head (2) leaks, air bubbles will be seen coming out of cylinder head when it is under water.

2. Check that cylinder head (2) does not leak. If leaks come from expansion plugs in cylinder head, take out leaking expansion plugs and put in new ones. Refer to para 4-19a. If leaks come from any other place, get a new cylinder head .

3. Using chain hoist (1), lift cylinder head (2) out of tank of water (3). Take off air line (4). Unscrew and take out air line adapter (5) and put in pipe plug (6) .

GO TO FRAME 5

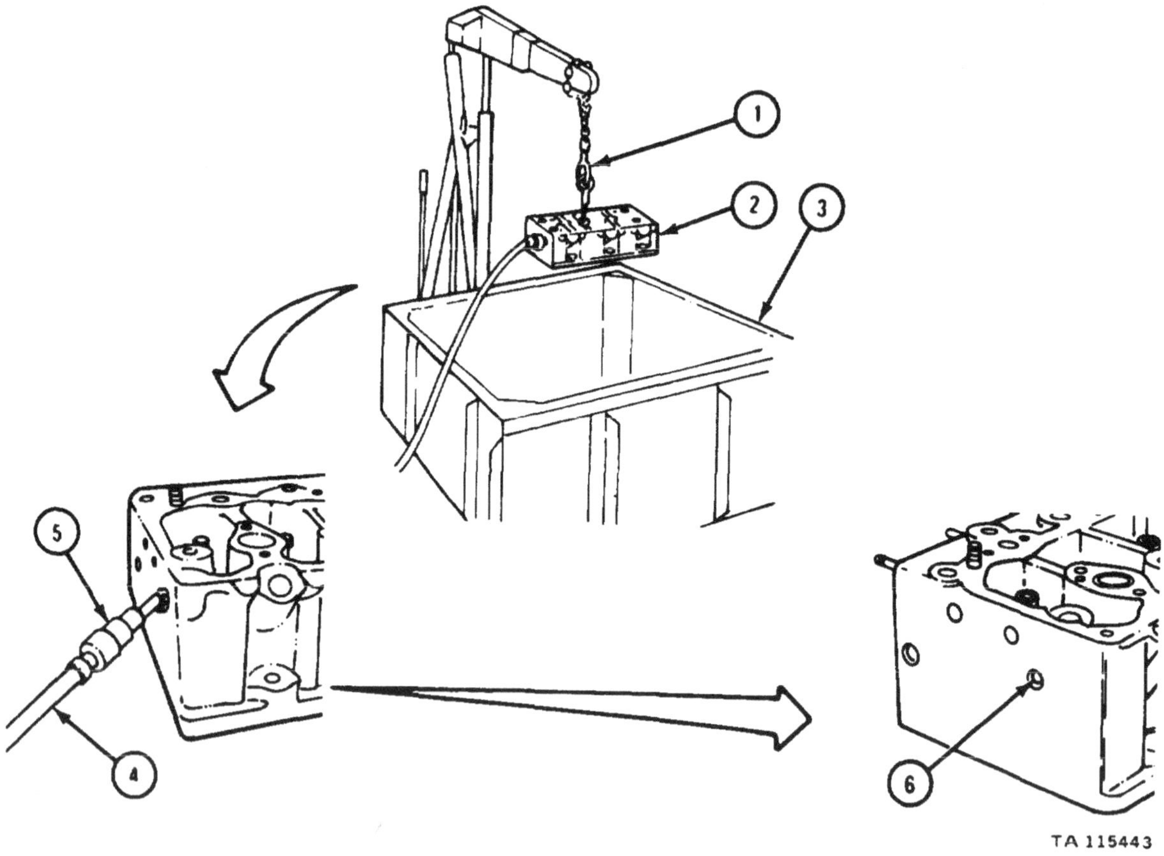

FRAME 5

1. Check that studs (1) are not bent, chipped or cracked and do not have damaged threads. Fix minor thread damage with thread chaser. If more repair is needed, take out stud and put in new one.
2. Check all threaded holes in cylinder head (2). Fix minor thread damage with tap. If threads cannot be fixed with tap, put in threaded insert.

WARNING

Dry cleaning solvent is flammable. Do not use near an open flame. Keep a fire extinguisher nearby when solvent is used. Use only in well-ventilated places. Failure to do this may result in injury to personnel and damage to equipment.

3. Check that gasket surface of cylinder head (2) has no burrs, scratche s or gouges. If cylinder head gasket surface is deeply scratched or gouged , get a new cylinder head. Fix small burrs and scratches with a fine mill file or crocus cloth dipped in dry cleaning solvent.
4. Using dial indicator, check that gasket surface of cylinder head (2) has no scratches or gouges deeper than 0.001 inch. If any scratch or gouge i s deeper than 0.001 inch, tag cylinder head for resurfacing by machine shop.

NOTE

Cylinder head (2) may have been resurfaced before. If cylinder head was resurfaced before, amount of metal machined off will be stamped on front of cylinder head about 1/2 inch from gasket surface of cylinder head. No more than 0.005 inch can be machined off. If 0.005 has already been machined off and cylinder head needs resurfacing g, get a new cylinder head.

GO TO FRAME 6

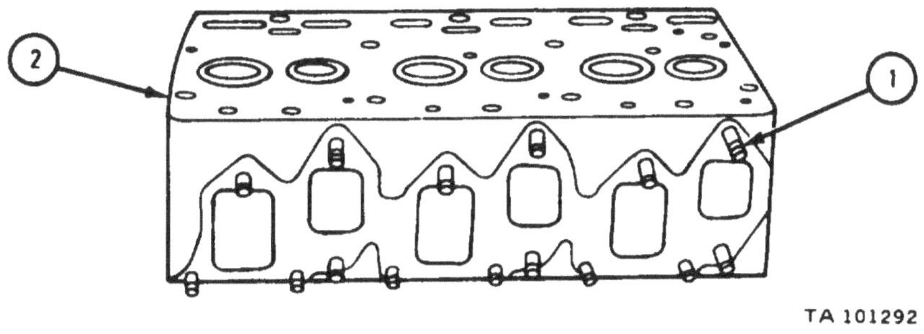

FRAME 6

1. Lay straight edge (1) along line A on gasket surface of cylinder head (2). Using 0.005inch feeler gage, check that gasket surface of cylinder head (2) is not more than 0.005 inch out-of-flat lengthwise. If gasket surface is more than 0.005 inch out-of-flat lengthwise, get a new cylinder head.
2. Lay straight edge (1) along line B on gasket surface of cylinder head (2). Do step 1 again.
3. Lay straight edge (1) along line C on gasket surface of cylinder head (2). Do step 1 again.

GO TO FRAME 7

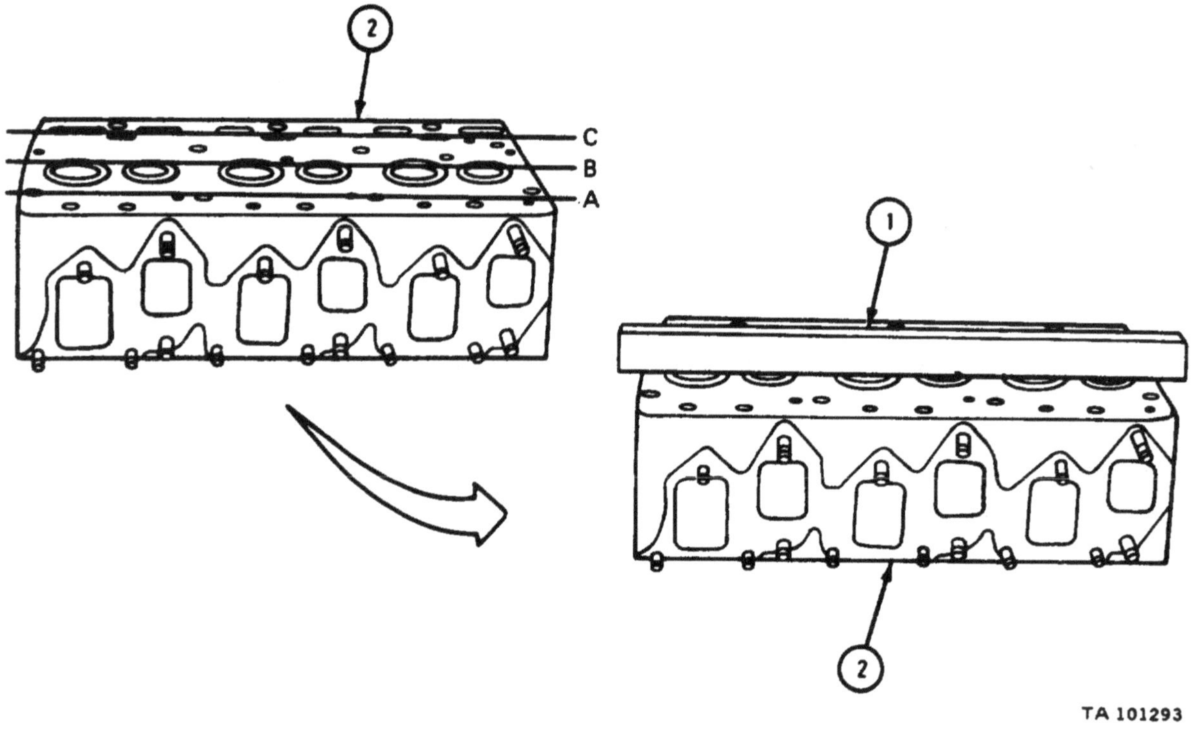

FRAME 7

1. Lay a straight edge (1) along line A on gasket surface of cylinder head (2). Using feeler gage, check that gasket surface of cylinder head is not out-of-flat crosswise .
2. If gasket surface of cylinder head (2) is more than 0.003 inch out-of-flat crosswise, get a new cylinder head. If gasket surface is out-of-flat by les s than 0.003 inch crosswise, tag cylinder head for resurfacing by reaching shop.

NOTE

Cylinder head (2) may have been resurfaced before. If cylinder head was resurfaced before, amount of metal machined off will be stamped on front of cylinder head about 1/2 inch from gasket surface of cylinder head. No more than 0.005 inch can be machined off. If 0.005 inch has already been machined off and cylinder head needs resurfacing, get a new cylinder head .

3. Lay straight edge (1) along line B on gasket surface of cylinder head (2). Do steps 1 and 2 again.
4. Lay straight edge (1) along line C on gasket surface of cylinder head (2). Do steps 1 and 2 again.
5. Lay straight edge (1) along line D on gasket surface of cylinder head (2). Do steps 1 and 2 again.

GO TO FRAME 8

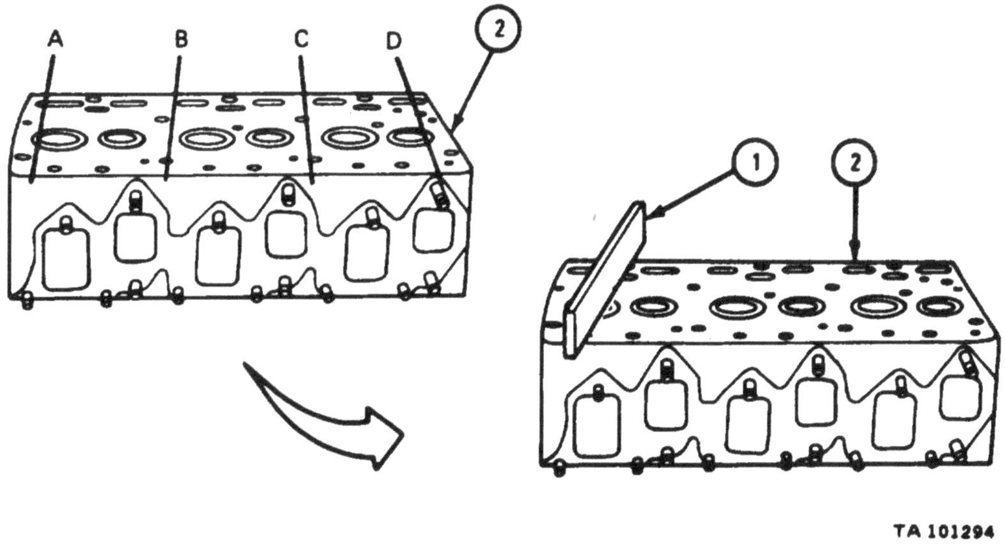

FRAME 8

NOTE

Intake valve seats (1) are bigger than exhaust valve seats.

1. Check that three intake valve seats (1) are not cracked or loose. If any intake valve seat is cracked or loose, take it out using engine valve seat knockout tool .
2. Check that three intake valve seats (1) are not burned or pitted. If any intake valve seat is burned or pitted, take it out using engine valve seat knockout tool .
3. Set cylinder head (2) on side as shown. Coat face if a new intake valve (3) with Prussian blue. Put new intake valve in cylinder head so that it rests against any intake valve seat (1).
4. Turn new intake valve (3) 1/2-turn in either direction.
5. Push new intake valve (3) straight out of cylinder head (2). Check for an even coating of Prussian blue all the way around intake valve seat (1) just checked. If there is not an even coating, tag intake valve seat just checked for grinding by machine shop.
6. Do steps 3 through 5 again for other two intake valve seats (1).

GO TO FRAME 9

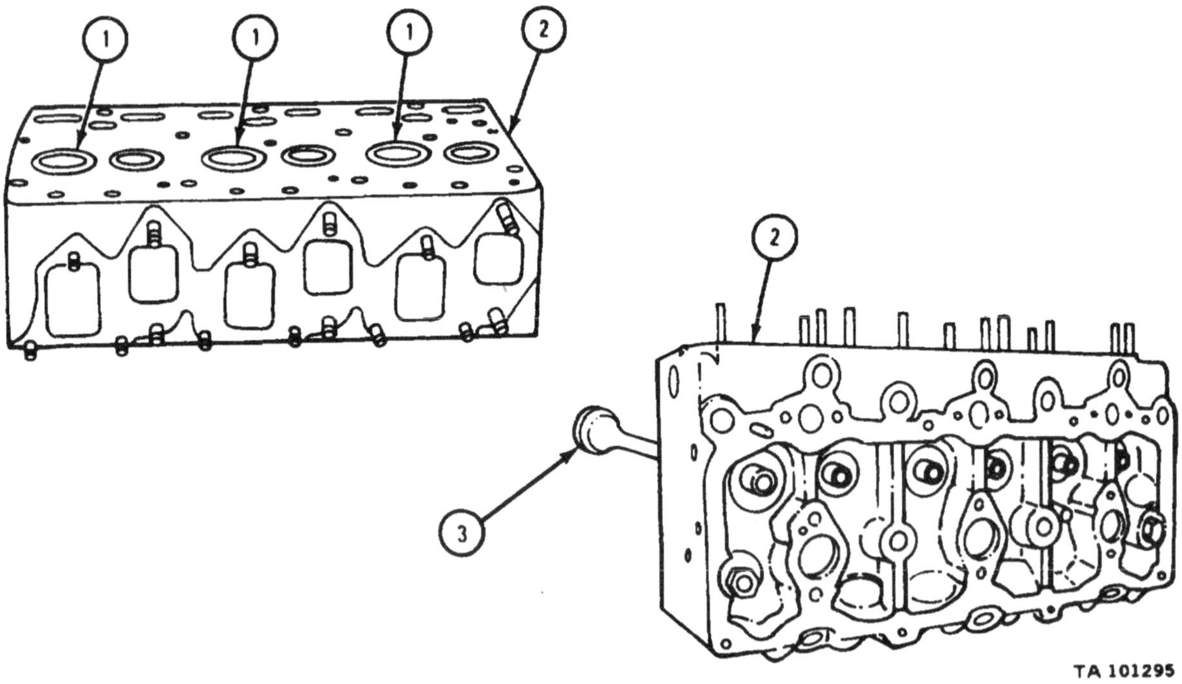

FRAME 9

1. Check that three exhaust valve seats (1) are not cracked or loose. If any exhaust valve seat is cracked or loose, take it out using engine valve seat knockout tool.
2. Check that three exhaust valve seats (1) are not burned or pitted. If any exhaust valve seat is burned or pitted, take it out using engine valve seat knockout tool .
3. Set cylinder head (2) on side as shown. Coat face of a new intake valve (3) with Prussian blue. Put new exhaust valve in cylinder head so that it rests against any exhaust valve seat (1).
4. Turn new exhaust valve (3) 1/2-turn in either direction .
5. Push new exhaust valve (3) straight out of cylinder head (2). Check for a n even coating of Prussian blue all the way around the exhaust valve seat (1) just checked. If coating is not even, tag exhaust valve seat just checked for grinding by machine shop.
6. Do steps 3 through 5 again for other two exhaust valve seats (1).

GO TO FRAME 10

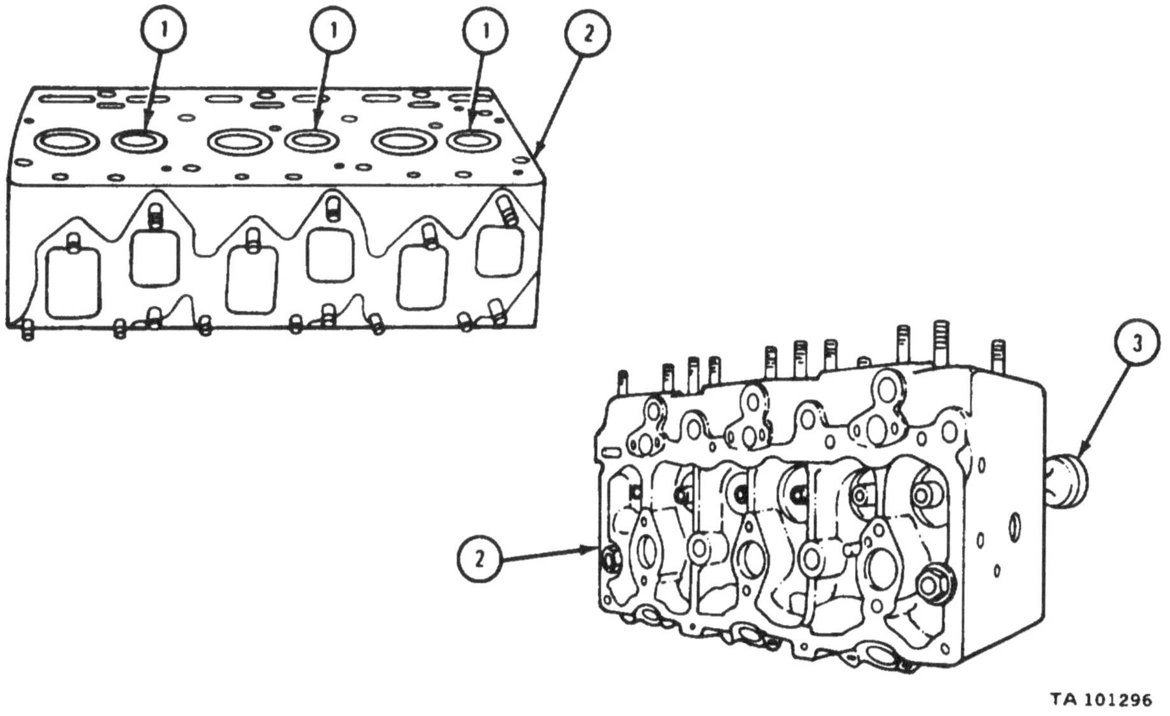

FRAME 10

NOTE

Do this frame only if cylinder head (2) was tagged for resurfacing or if any intake valve seat (1) was tagged to be machined.

Do step 1 only if cylinder head (2) was tagged for resurfacing. If cylinder head was not tagged, do steps 2 through 5.

1. Using engine valve seat knockout tool, take out all intake valve seats (1) not tagged to be machined. Tag each intake valve seat taken out "front intake," "center intake," or "rear intake," so it can be put back in the same place. Do not take out intake valve seats tagged to be machined.

NOTE

Two sizes of intake valve seat (1) are used in these engines. Cylinder head (2) can use a combination of standard and 0.100-inch oversize intake valve seats.

2. Measure inside diameter of intake valve seat counterbore (3) for each intake valve seat (1) taken out.
3. If diameter of intake valve seat counterbore (3) is 2.0580 to 2.0595 inches, this is standard size intake valve seat counterbore. If diameter is 2.0680 to 2.0695 inches, this is 0.0100-inch oversize intake valve seat counterbore. If diameter is more than 2.0695 inches, get a new cylinder head (2).
4. If inside diameter of intake valve seat counterbore is more than 2.0595 inches but less than 2.0680 inches, tag this intake valve seat counterbore to be machined to 0.100-inch oversize by machine shop.
5. Check that intake valve seat counterbore (3) is not damaged. If 0.0100-inch oversize intake seat counterbore is damaged, get a new cylinder head (2). If a standard size intake valve seat counterbore is damaged, tag it to be machined to 0.100-inch oversize by machine shop.

GO TO FRAME 11

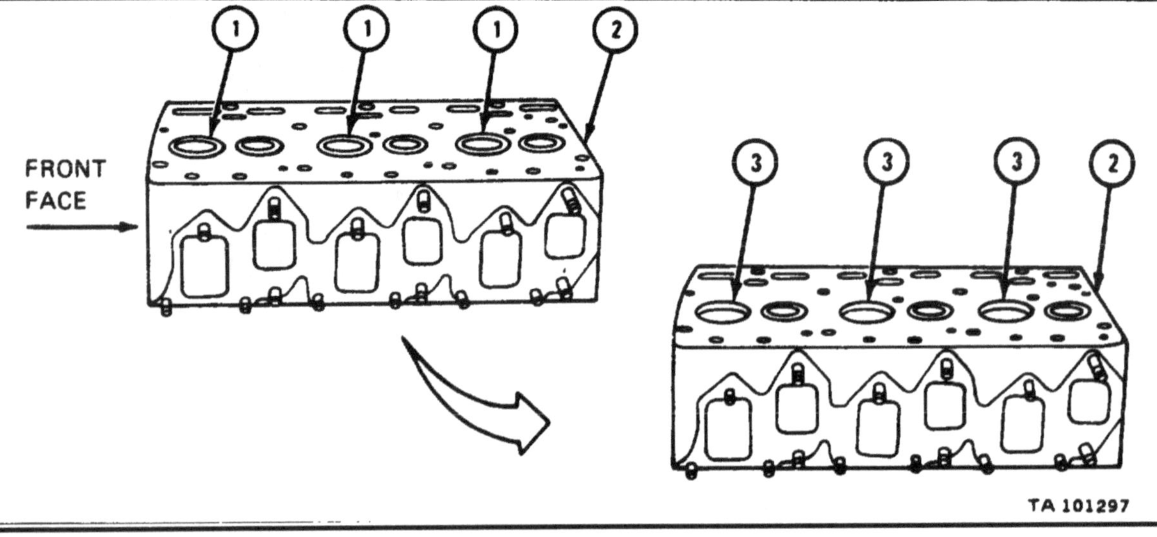

FRAME 11

NOTE

Do this frame only if cylinder head (2) was tagged for resurfacing or if any exhaust valve seat (1) was tagged to be machined.

Do step 1 only if cylinder head (2) was tagged for resurfacing. If cylinder head was not tagged, do steps 2 thorough 5.

1. Using engine valve seat knockout tool, take out exhaust valve seats (1) not tagged to machined. Tag each exhaust valve seat taken out "fron t exhaust", "center exhaust" or "rear exhaust" so it can be put back in the same place. Do not take out exhaust valve seats tagged to be machined.

NOTE

Two sizes of exhaust valve seats (1) are used in these engines. Cylinder head (2) can use a combination of standard and 0.0100-inch oversize exhaust valve seats.

2. Measure inside diameter of exhaust valve seat counterbore (3) for each exhaust valve seat (1) taken out.
3. If diameter of exhaust valve seat counterbore (3) is 1.8080 to 1.8095 inches, this is a standard size exhaust valve seat counterbore. If diameter is 1.8180 to 1.8195 inches, this is a 0.0100-inch oversize exhaust valve seat counterbore. If diameter is more than 1.8195 inches, get a new cylinder head (2).
4. If inside diameter of exhaust valve seat counterbore (3) is more than 1.8095 inches but smaller than 1.8180 inches, tag this exhaust valve seat counterbore to be machined to 0.0100-inch oversize by machine shop.
5. Check that exhaust valve seat counterbore (3) is not damaged. If a 0.0100-inch oversize intake valve seat counterbore is damaged, get a new cylinder head (2). If a standard size intake valve seat counterbore is damaged, tag it to be machined to 0.0100-inch oversize by machine shop.

GO TO FRAME 12

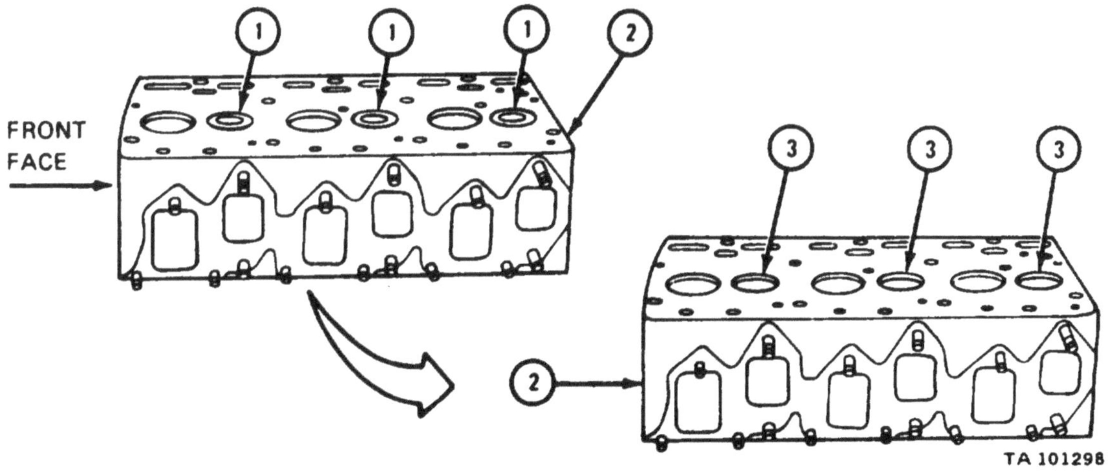

FRAME 12

NOTE

Intake valve guides and exhaust valve guides are the same.

1. Check that six valve guides (1) are not cracked, worn, scuffed or loose.
2. Check that inside diameter of six valve guides (1) is not more than 0.4420 inch. Diameter should be measured 1/8 to 1/4 inch from each end of valve guide. If any valve guide is worn more than given limit, do step 3.
3. If any valve guide (1) is damaged, worn or loose, mount cylinder head (2) in holding fixture and turn cylinder head so that bottom faces up as shown. Drive out valve guide using soft faced hammer and valve guide remover (3). Throw valve guide away, and get a new one.

IF ANY VALVE GUIDE WAS TAKEN OUT, GO TO FRAME 13.
IF NO VALVE GUIDE WAS TAKEN OUT, GO TO FRAME 16.

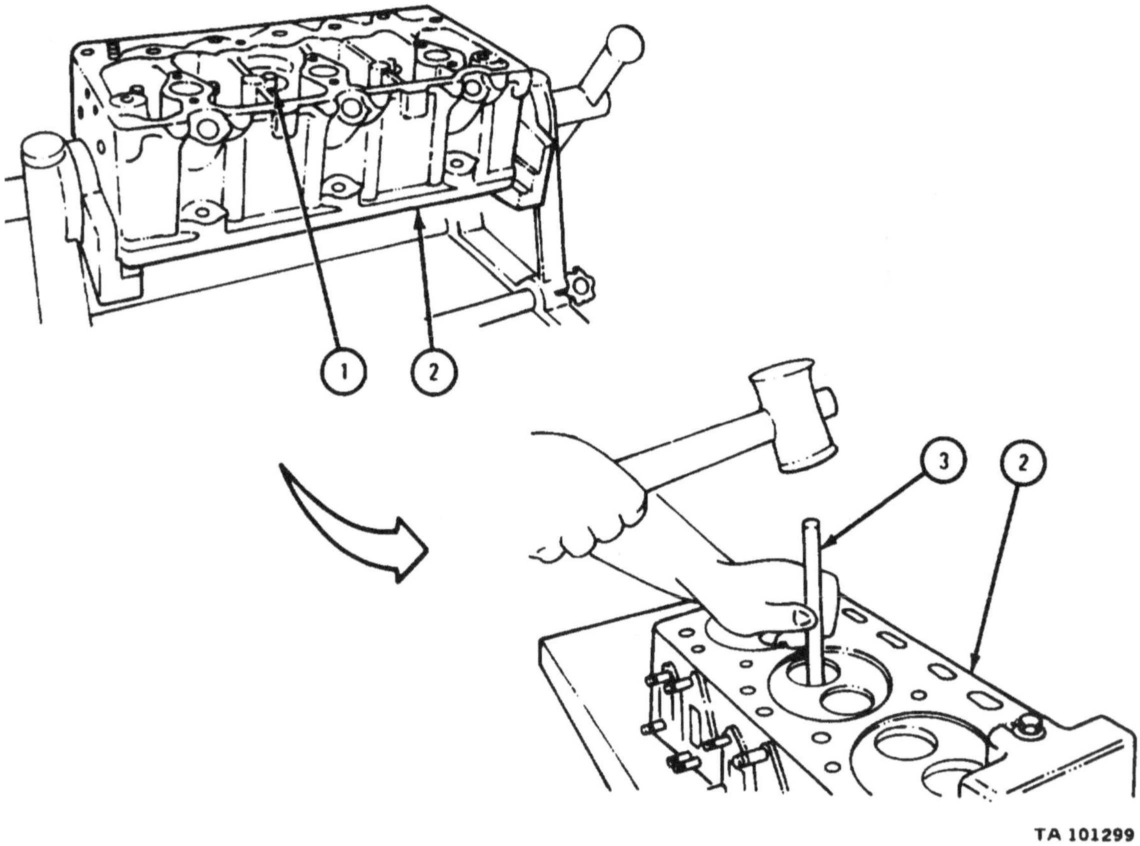

FRAME 13

NOTE

Readings must be within limits given in table 4-47. The letter T shows a tight fit. If readings are not within given limits, throw away part and get a new one.

1. Measure inside diameter of valve guide bores in cylinder head (1).

NOTE

Early model valve guides (2) are not stocked. Limits given in table 4-47 for early model valve guides are for reference purposes only. Use limits for late model valve guides .

2. Measure outside diameter of new valve guides (2).
3. Measure fit of each new valve guide (2) in its valve guide bore in cylinder head (1) .

GO TO FRAME 14

NOTE
CHECK ONLY THOSE PARTS WHICH ARE CALLED OUT IN THIS FRAME. PARTS WITHOUT CALLOUTS ARE SHOWN ONLY FOR REFERENCE PURPOSES OR ARE CHECKED IN ANOTHER FRAME.

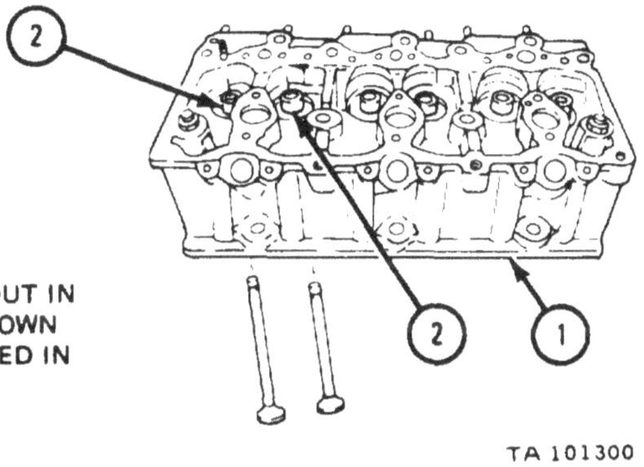

Table 4-47. Valve Guides and Valve Guide Bores Wear Limits

Index Number	Item/Point of Measurement	Size and Fit of New Parts (inches)	Wear Limit (inches)
1	Valve guide bore in cylinder head inside diameter	0.6865 to 0.6875	None
2	Valve guide outside diameter		
	Early model	0.6880 to 0.6890	None
	Late model	0.6880 to 0.6885	None
1 and 2	Fit of valve guides in cylinder head bore s		
	Early model	0.005T to 0.0025T	None
	Late model	0.0005T to 0.0020T	None

FRAME 14

NOTE

Cylinder head (1) should be mounted in holding fixture as shown.

1. Line up new valve guide (2) with valve guide bore in cylinder head (1).
2. Using soft-faced hammer and valve guide replacer, drive new valve guide (2) into valve guide bore in cylinder head (1).

GO TO FRAME 15

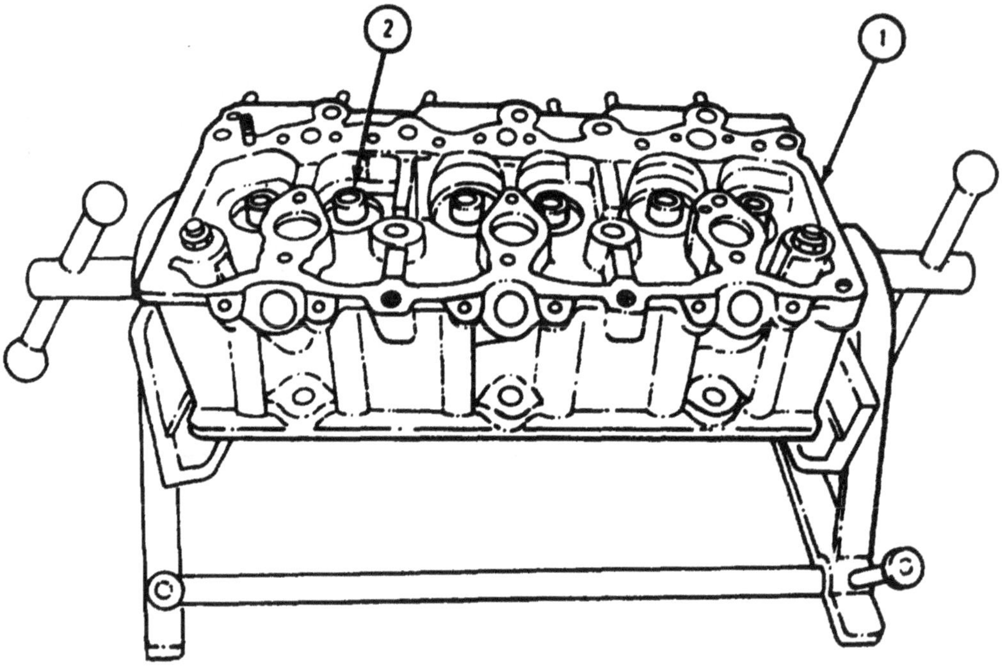

TA 101301

FRAME 15

CAUTION

Do not turn reamer (1) to the left while reaming valve guides (2). Never force reamer into valve guide. Damage to valve guide will result.

NOTE

Do this frame only for valve guides (2) that have been changed.

1. Slowly turn 0.4392 to 0.4390-inch external diameter reamer (1) to the right while pushing reamer lightly into valve guide (2) until reamer is all the way through valve guide .
2. While turning reamer (1) to the right, pull it out of valve guide (2).
3. Measure inside diameter of valve guide (2) after reaming. If inside diameter of valve guide (2) is less than 0.4388 inch, do steps 1 and 2 again. Make sure you are using right size reamer.
4. If inside diameter of valve guide (2) is more than 0.4394 inch, knock it out. Refer to frame 12. Get a new valve guide and put it in. Refer to frame 14. Then do steps 1, 2 and 3 again with right size reamer.

GO TO FRAME 16

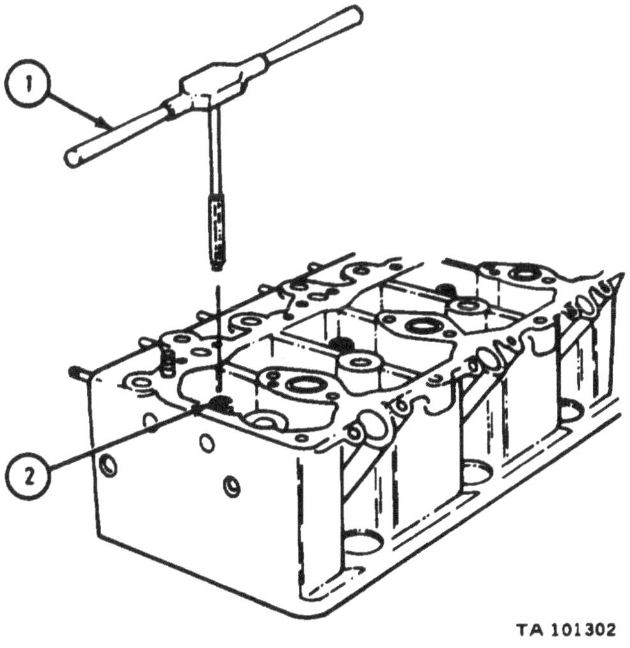

FRAME 16

WARNING

Dry cleaning solvent is flammable. Do not use near an open flame. Keep a fire extinguisher nearby when solvent is used. Use only in well-ventilated places. Failure to do this may result in injury to personnel and damage to equipment.

NOTE

Gasket surface of cylinder head (1) was already checked for nicks, scratches, and burrs. Do not try to fix gasket surface in this frame.

1. Check that cylinder head (1) has no small nicks, scratches or burrs. Fi x small nicks, scratches, or burrs with fine mill file or crocus cloth dipped in dry cleaning solvent .
2. Screw tap into all threaded holes in cylinder head (1) to clean threads.

IF CYLINDER HEAD (1) OR ANY PART OF CYLINDER HEAD WAS TAGGED FOR MACHINING , GO TO FRAME 17.
IF CYLINDER HEAD (1) WAS NOT TAGGED FOR MACHINING, GO TO FRAME 21

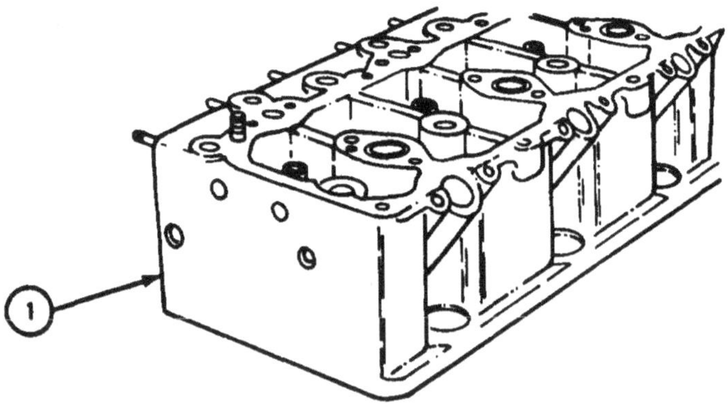

TA 101303

FRAME 17

1. Send cylinder head (1) to machine shop.

NOTE

Intake valve seat inserts (2) and exhaust valve seat inserts (3) were taken out of cylinder head (1) before head was sent to machine shop unless valve seat inserts need machining .

Intake valve seat inserts (2) are bigger than exhaust valve seat inserts (3) .

2. In machine shop, grind any intake valve seat inserts (2) and any exhaust valve seat inserts (3) tagged for machining at a 45 degree angle as shown in figure 4-5. Mark valve seat inserts for replacement if they are worn more than limits shown in figure 4-5.
3. In machine shop, if cylinder head (1) is tagged for resurfacing, take out intake valve seat inserts (2) using engine valve seat knockout tool. Ta g intake valve seat inserts that were not marked to throw away so they can be put back in the same place.
4. In machine shop, if cylinder head (1) is tagged for resurfacing, take out exhaust valve seat inserts (3) using engine valve seat knockout tool. Ta g exhaust valve seat inserts that were not marked to throw away so they can be put back in the same place.
5. In machine shop, check that intake valve seat counterbores and exhaust valve seat inserts (2 and 3) are not damaged. Mark damaged valve seat counterbores for machining.
6. In machine shop, do frame 10 for intake valve seat counterbores under any intake valve se-at inserts (2) just taken out. Do frame 11 again for exhaust valve seat counterbores under any exhaust valve seat inserts (3) just taken out.

GO TO FRAME 18

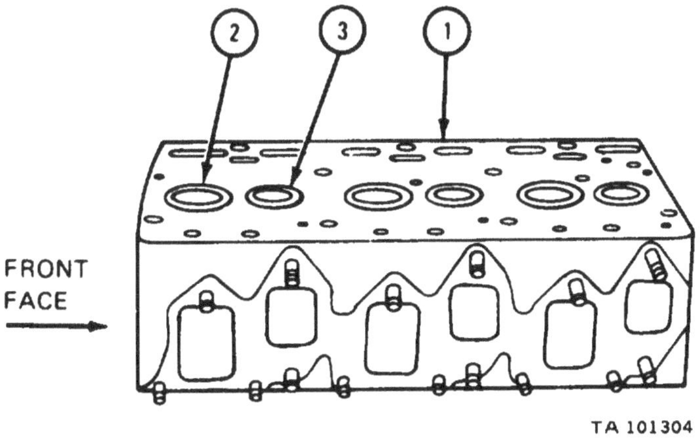

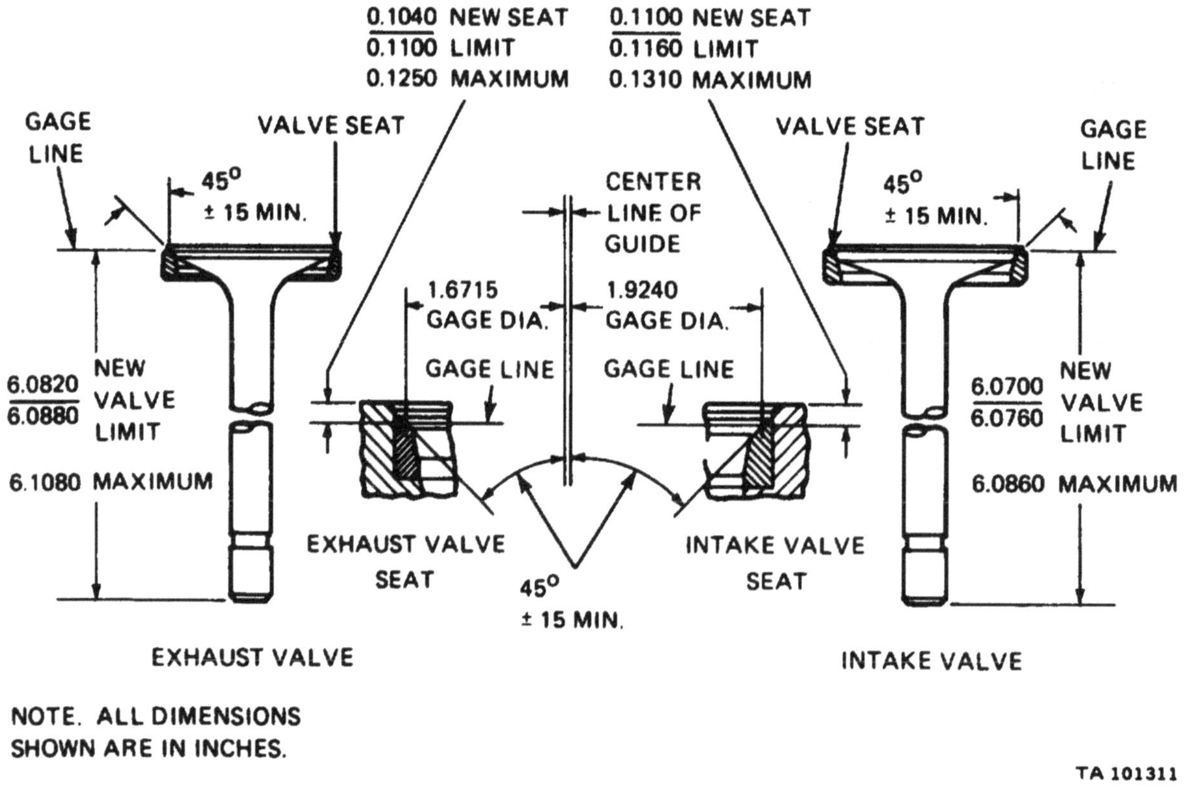

Figure 4-5. Valve Seats and Valves Machining Specifications.

FRAME 18

NOTE

In machine shop, do this frame only if cylinder head is tagged for resurfacing. If cylinder head is not tagged, go to frame 19.

No more than a total of 0.005 inch can be machined off gasket surface of cylinder head (1). If cylinder head has been resurfaced before, amount of metal machined off is stamped or etched on front of cylinder head about 1/2 inch from gasket surface.

1. In machine shop, resurface cylinder head (1) until no scratch or gouge is deeper than 0.001 inch. Resurface cylinder head until it is not out-of-flat lengthwise by more than 0.005 inch or out-of-flat crosswise by more than 0.003 inch. If more than a total of 0.005 inch must be machined off to do this, get a new cylinder head.

NOTE

If old cylinder head (1) can be resurfaced without machining off more than a total of 0.005 inch, do step 2.

2. In machine shop, add amount of metal just machined off cylinder head (1) to amount machined off before, and stamp or etch total on front of cylinder head about 1/2 inch from gasket surface.
3. Measure thickness of cylinder head (1) from gasket surface to bottom of valve spring counterbores (2). If thickness is less than 3.365 inches, get a new cylinder head .

GO TO FRAME 19

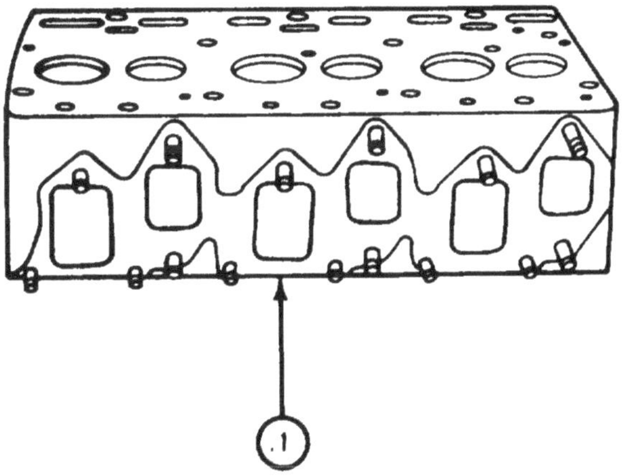

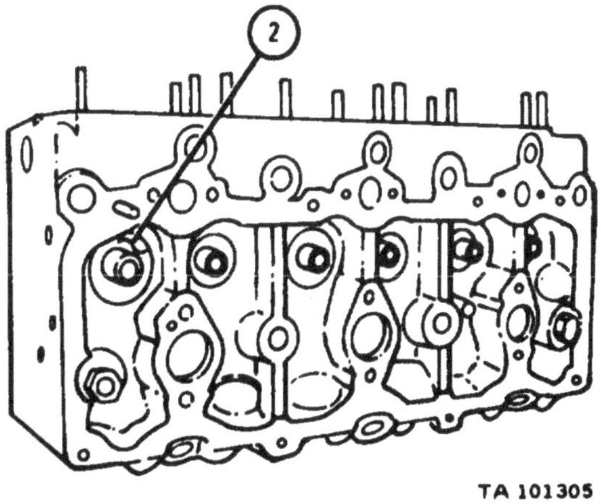

FRAME 19

1. Machine any intake valve seat counterbores (1) tagged for machining to a diameter of 2.0680 to 2.0690 inches.
2. Machine any exhaust valve seat counterbores (2) tagged for machining to a diameter of 1.8180 to 1.8190 inches.

IF CYLINDER HEAD (3) WAS RESURFACED, GO TO FRAME 20.
IF CYLINDER HEAD (3) WAS NOT RESURFACED, GO TO FRAME 21

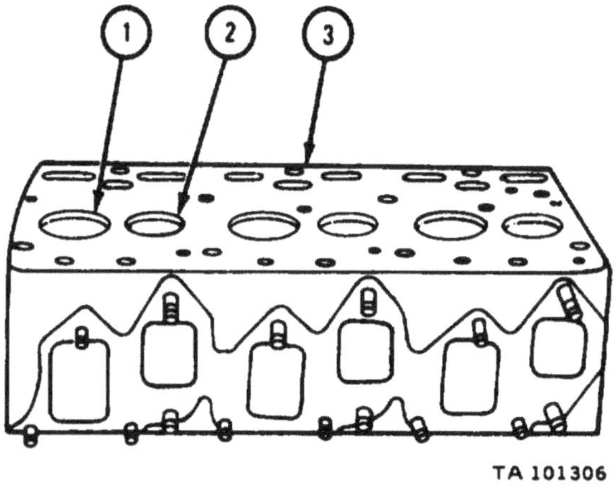

FRAME 2 0

NOTE

Intake valve seat counterbores (1) are bigger than exhaust valve seat counterbores (2) .

1. In machine shop, measure depth of three intake valve seat counterbores (1). Depth must be 0.4155 to 0.4205 inch.
2. In machine shop, if depth of any intake valve seat counterbore (1) is not 0.4155 to 0.4205 inch, machine to correct depth.
3. In machine shop, measure depth of three exhaust valve seat counterbores (2). Depth must be 0.3825 to 0.3875 inch.
4. In machine shop, if depth of any exhaust valve seat counterbore (2) is not 0.3825 to 0.3875 inch, machine to correct depth.
5. Send cylinder head (3) back to mechanic.

GO TO FRAME 21

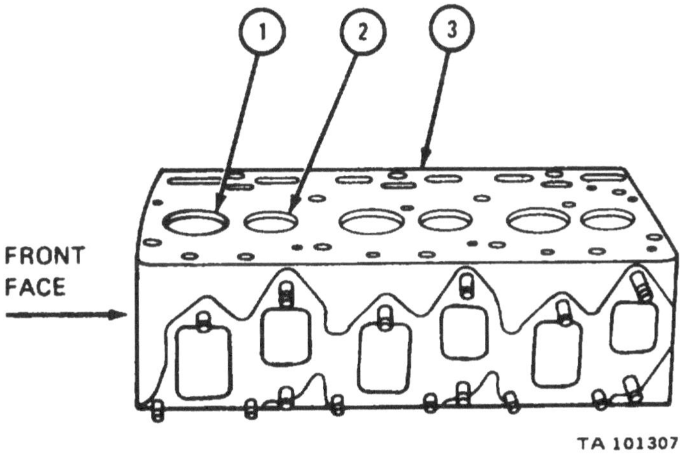

FRAME 21

NOTE

If no intake valve seat inserts (1) were taken out, go to frame 23.

Two sizes of intake valve seat inserts (1) are used in these cylinder heads (2). Cylinder head can use a combination of intake valve seat insert sizes. Measure each intake valve seat counterbore (3) to tell whether to put in a standard size or oversize intake valve seat insert.

Readings must be within limits given in table 4-48. The letter T shows a tight fit. If readings are not within given limits, throw away part and get a new one.

1. Measure and note inside diameter of three intake valve seat counterbores (3). If diameter is 2.0580 to 2.0595 inches, this is a standard size intake valve seat counterbore. If diameter is 2.0680 to 2.0695 inches, this is a 0.0100 oversize intake valve seat counterbore.

NOTE

Used intake valve seat inserts (1) that were not worn were tagged so they can be put back in the same place. Because machine shop may have machined some standard size intake valve seat counterbores (3) into oversize intake valve seat counterbores, some intake valve seat inserts may now be too small.

2. Measure and note outside diameter of each intake valve seat insert (1) to be put in. Standard size intake valve seat inserts go in standard size intake valve seat counterbores (3). For 0.0100-inch oversize intake valve seat counterbores, use 0.0100–inch oversize intake valve seat inserts. Refer to table 4-48 for sizes of intake valve seat inserts.

3. Measure fit of each intake valve seat insert (1) in its intake valve seat counterbore (3).

GO TO FRAME 22

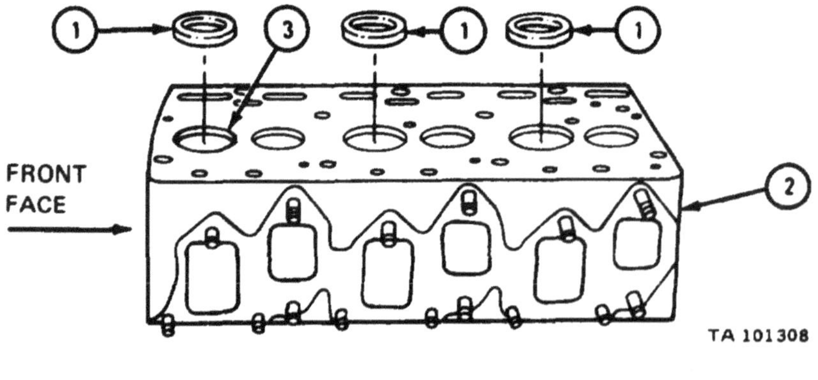

Table 4-48. Intake Valve Seat Inserts and Counterbores Wear Limits

Index Number	Item /Point of Measurement	Size and Fit of New Parts (inches)	Wear Limit (inches)
3	Intake valve seat counterbore inside diameter		
	Standard size	2.0580 to 2.0590	2.0595
	0.0100-inch oversize	2.0680 to 2.0690	2.0695
1	Intake valve seat insert outside diameter		
	Standard size	2.0615 to 2.0625	None
	0.0100-inch oversize	2.0715 to 2.0725	None
3 and 1	Fit of intake valve seat insert in counterbore	0.0025T to 0.0045T	0.0020

FRAME 22

NOTE

Two sizes of intake valve seat inserts (1) are, used in these cylinder heads (2) . Standard size intake valve seat inserts go in standard size intake valve seat counterbores (3) . For 0.0100-inch oversize intake valve seat counterbores, use 0.0100-inch oversize intake valve seat inserts.

1. Put each intake valve seat insert (1) in dry ice for four hours so it will shrink.

NOTE

Intake valve seat insert (1) must be put in intake valve seat counterbore (3) right after it is taken out of dry ice. If intake valve seat insert is left out of ice too long, it will get bigger and will not fit.

2. Using soft-faced hammer and valve seat replacing tool, put intake valve seat insert (1) into its intake valve seat counterbore (3) with beveled edge facing out. Tag new intake valve seat inserts for lapping by machine shop.

NOTE

If no exhaust valve seat inserts (4) were taken out, do step 3.

3. Send cylinder head (2) to machine shop along with intake valves that seat against new intake valve seat inserts (1). Tell machine shop to lap new intake valve seats to valves that seat against them.

GO TO FRAME 23

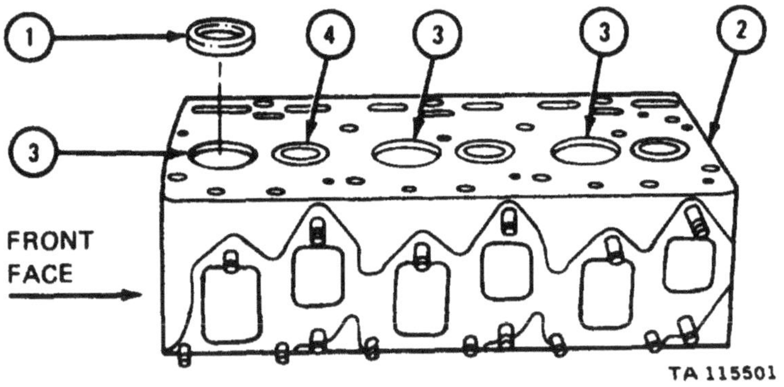

FRAME 2 3

NOTE

If no exhaust valve seat inserts (1) were taken out, go to para 4-19c(2) .

Two sizes of exhaust valve seat inserts (1) are used in cylinder heads (2) . Cylinder head can use a combination of exhaust valve seat insert sizes. Measure each exhaust valve seat counterbore (3) to tell whether to put in a standard size or oversize exhaust valve seat insert.

Readings must be within limits given in table 4-49. The letter T shows a tight fit. If readings are not within given limits, throw away part and get a new one.

1. Measure and note inside diameter of three exhaust valve seat counterbores (3). If diameter is 1.8080 to 1.8095 inches, this is a standard size exhaust valve seat counterbore . If diameter is 1.8180 to 1.8195 inches, this is a 0.0100-inch oversize exhaust valve seat counterbore .

NOTE

Used exhaust valve seat inserts, (1) that were not worn were tagged so they can be put back in the same place. Because machine shop may have machined some standard size exhaust valve seat counterbores (3) into oversize exhaust valve seat counterbores, some exhaust valve seat inserts may be too small.

2. Measure and note outside diameter of each exhaust valve seat insert (1) to be put in. Standard size exhaust valve seat inserts go in standard size exhaust valve seat counterbores (3). For 0.0100-inch oversize exhaust valve seat counterbores, use 0.0100-inch oversize exhaust valve seat inserts . Refer to table 4-49. for exhaust valve seat insert sizes.
3. Measure fit of each exhaust valve seat insert (1) in its exhaust valve seat counterbore (3) .

GO TO FRAME 24

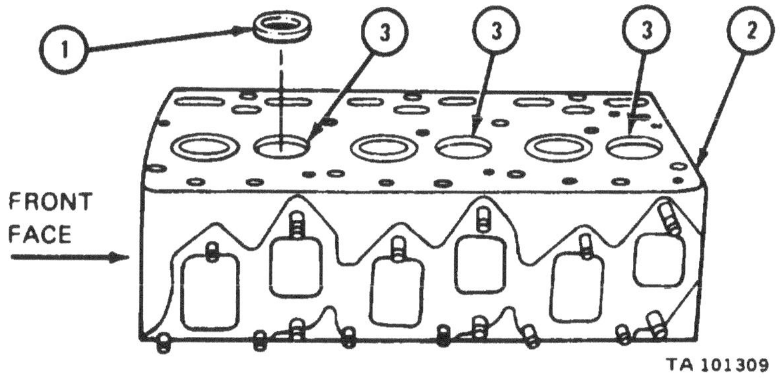

Table 4-49. Exhaustive Valve Seat Inserts and Counterbores Wear Limits

Index Number	Item /Point of Measurement	Size and Fit of New Parts (inches)	Wear Limit (inches)
3	Exhaust valve seat counter-bores inside diameter		
	Standard size	1.8080 to 1.8090	1.8095
	0.0100-inch oversize	1.8180 to 1.8190	1.8195
1	Exhaust valve seat insert outside diameter		
	Standard size	1.8120 to 1.8130	None
	0.0100-inch oversize	1.8220 to 1.8230	None
3 and 1	Fit of exhaust valve seat insert in counterbore	0.0030T to 0.0050T	0.0025T

FRAME 2 4

NOTE

Two sizes of exhaust valve seat inserts (1) are used in these cylinder heads (2). Standard size exhaust valves seat inserts go in standard size exhaust valve seat counterbores (3). For 0.0100-inch oversize exhaust valve seat centerbores, use 0.0100-inch oversize exhaust valve seat inserts.

1. Put each exhaust valve seat insert (1) in dry ice for four hours so it will shrink.

NOTE

Exhaust valve seat insert (1) must be put in its exhaust valve seat counterbore (3) right after it is taken out of dry ice. If exhaust valve seat insert is left out too long, it will get bigger and will not fit.

2. Put exhaust valve seat insert (1) in place on its exhaust valve seat counterbore (3), beveled edge facing out. Using soft-faced hammer and exhaust valve seat replacer, put new exhaust valve seat insert into its exhaust valve seat counterbore. Tag new exhaust valve seat inserts for lapping by machine shop,

3. Send cylinder head (2) to machine shop along with intake and exhaust valves which seat against new intake and exhaust valve seat inserts. Tell machine shop to lap new intake and exhaust valve seats to the valves that seat against them.

END OF TASK

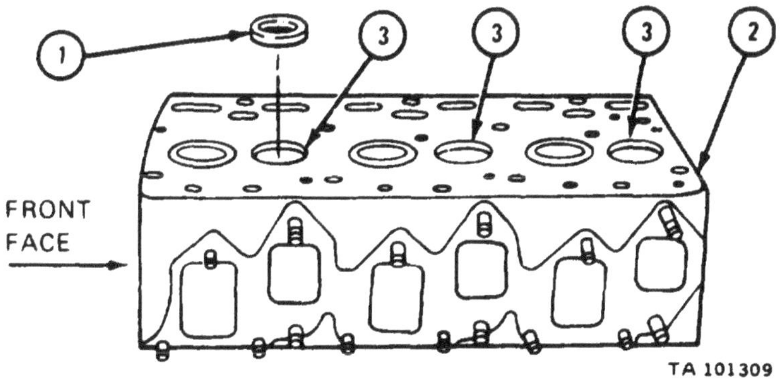

(2) Valve s

CAUTION

It is easy to damage the equipment if you do not know what you are doing. Do not try to do this task unless you are experienced at it or you have an experienced person with you.

FRAME 7

NOTE

Intake valves (1) are bigger than exhaust valves (2).

1. Check that three intake valves (1) and three exhaust valves (2) are not chipped or cracked. Throw away chipped or cracked intake valves and exhaust valves and get new ones.
2. Check that locking grooves in stems of three intake valves (1) and three exhaust valves (2) are not damaged. Throw away intake valves and exhaust valves with damaged locking grooves and get new ones.
3. Check that stems of three intake valves (1) and three exhaust valves (2) are not badly pitted, scored or scratched. **Throw away intake valves and exhaust valves with badly pitted, scored or scratched stems and get new ones.**

GO TO FRAME 2

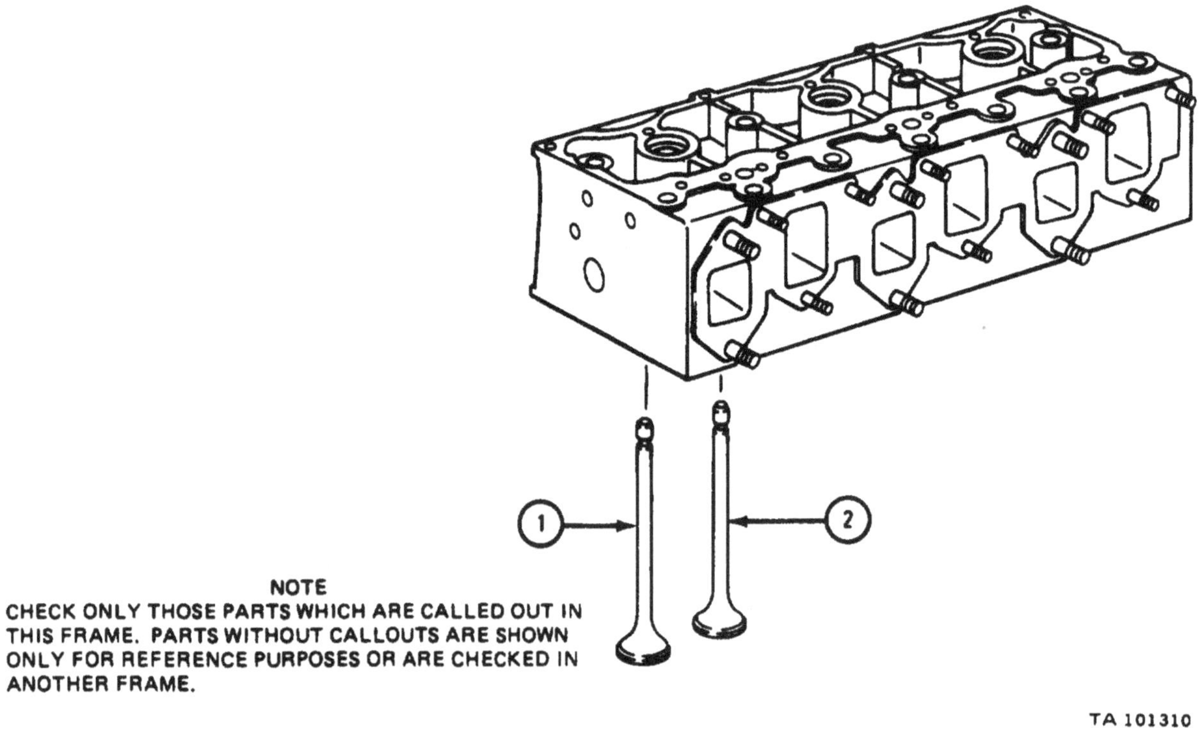

1
2
NOTE
CHECK ONLY THOSE PARTS WHICH ARE CALLED OUT IN THIS FRAME. PARTS WITHOUT CALLOUTS ARE SHOWN ONLY FOR REFERENCE PURPOSES OR ARE CHECKED IN ANOTHER FRAME.
TA 101310

FRAME 2

NOTE

A light frosted appearance or minor discoloration on face of valves (1 and 2) does not mean valve is damaged.

1. Check that faces of three intake valves (1) and three exhaust valves (2) are not badly burned, pitted, worn or heavily discolored from overheating.
2. Throw away damaged or worn intake valves (1) and exhaust valves (2) and get new ones. Send slightly pitted or burned intake valves and exhaust valves to machine shop for grinding. See figure 4-6 .

GO TO FRAME 3

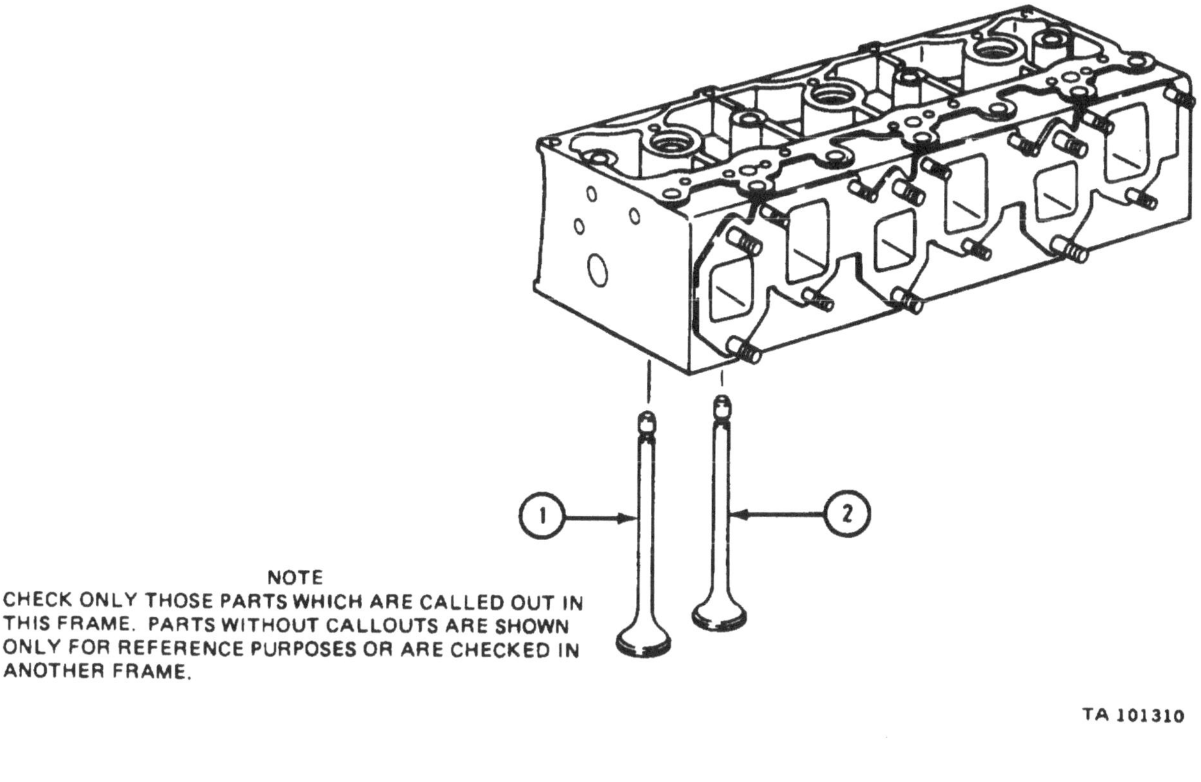
1
2
NOTE
CHECK ONLY THOSE PARTS WHICH ARE CALLED OUT IN THIS FRAME. PARTS WITHOUT CALLOUTS ARE SHOWN ONLY FOR REFERENCE PURPOSES OR ARE CHECKED IN ANOTHER FRAME.

TA 101310

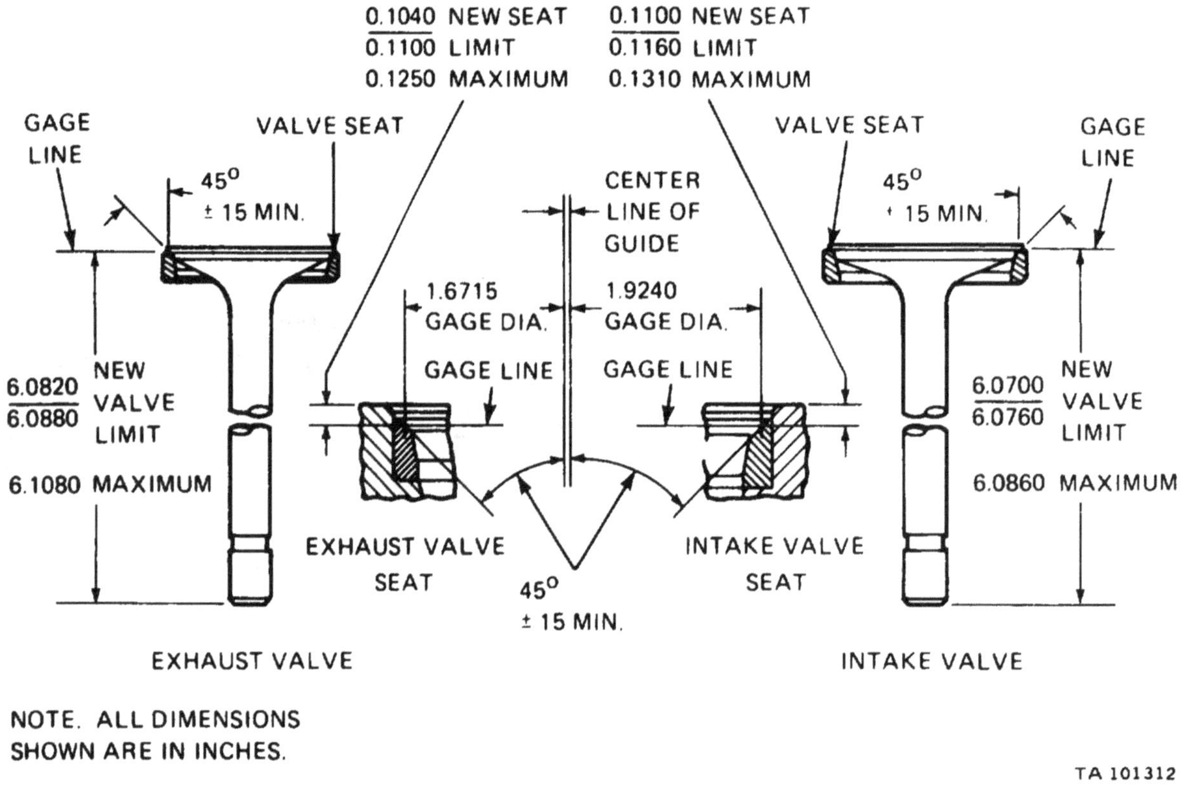

Figure 4-6. Valve Grinding Specifications.

FRAME 3

NOTE

Readings must be within limits given in table 4-50. If readings are not within given limits, throw part away and get a new one.

Intake valves (1) are bigger than exhaust valves (2).

1. Using depth micrometer, measure length of three intake valves (1) from tip to gage line as shown.
2. Using depth micrometer, measure length of three exhaust valves (2) from tip to gage line as shown.

GO TO FRAME 4

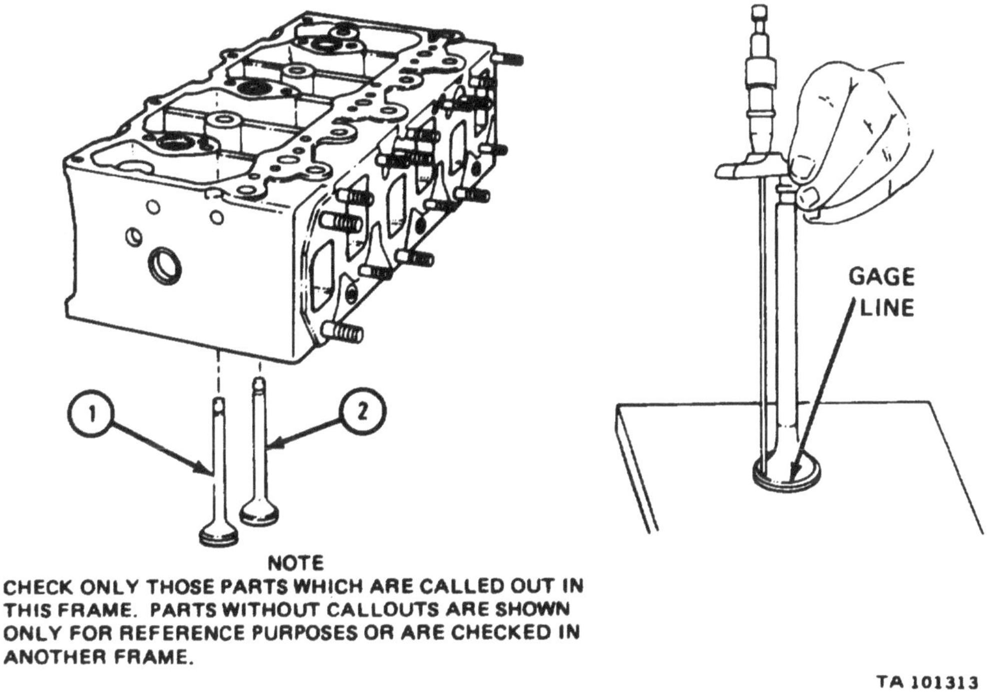

1
2
GAGE LINE
NOTE
CHECK ONLY THOSE PARTS WHICH ARE CALLED OUT IN THIS FRAME. PARTS WITHOUT CALLOUTS ARE SHOWN ONLY FOR REFERENCE PURPOSES OR ARE CHECKED IN ANOTHER FRAME.
TA 101313

Table 4-50. Intake and Exhaust Valve Length Wear Limits

Index Number	Item /Point of Measurement	Size and Fit of New Parts (inches)	Wear Limit (inches)
1	Intake valve length from tip to gage line	6.0820 to 6.0880	6.0980
2	Exhaust valve length from tip to gage line	6.0700 to 6.0760	6.0860

FRAME 4

1. Check whether cylinder head (1) has three expansion plugs (2). If cylinde r head has three expansion plugs, it is late model cylinder head. If cylinder head does not have three expansion plugs, it is early model cylinder head.

IF WORKING ON LATE MODEL CYLINDER HEAD, GO TO FRAME 5.

IF WORKING ON EARLY MODEL CYLINDER HEAD, GO TO FRAME 7

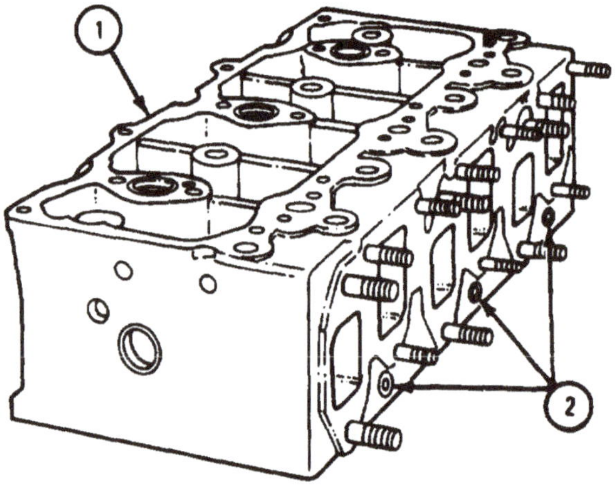

TA 101314

FRAME 5

NOTE

Readings must be within limits given in table 4-51. The letter L shows a loose fit. If readings are not within given limits, throw away part and get a new one.

Intake valves (1) and exhaust valves (2) were tagged so the fit of each valve stem can be measured in valve guide it was taken from.

1. Measure and note outside diameter of stem of three intake valves (1).
2. Measure and note inside diameter of three intake valve guides (3) 1/8 to 1/4 inch from each end of guide. If diameter is different at each end of guide, note biggest diameter.
3. Measure fit of stem of each intake valve (1) in its intake valve guide (3). Do this for all three intake valves and intake valve guides.

GO TO FRAME 6

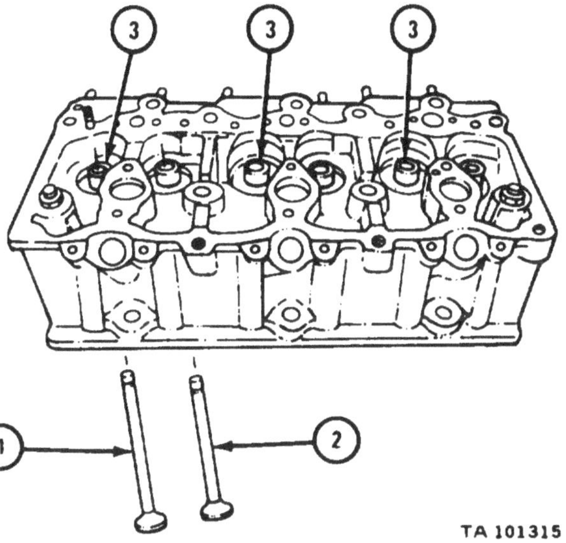

Table 4-51. Intake Valves and Guides Wear Limits

Index Number	Item /Point of Measurement	Size and Fit of New Parts (inches)	Wear Limit (inches)
1	Intake valve stem outside diameter	0.4368 to 0.4373	0.4363
3	Intake valve guide inside diameter	0.4388 to 0.4394	0.4420
1 and 3	Fit of intake valve stem in intake valve guide	0.0015L to 0.0026L	0.0057L

FRAME 6

NOTE

Readings must be within limits given in table 4-52. The letter L shows a loose fit. If readings are not within given limits, throw away part and get a new one.

Intake valves (1) and exhaust valves (2) were tagged so fit of each valve stem can be measured in valve guide it was taken from.

1. Measure and note outside diameter of stem of three exhaust valves (2).
2. Measure and note inside diameter of three exhaust valve guides (3) 1/8 to 1/4 inch from each end of guide. If diameter is different at each end of guide, note biggest diameter.
3. Measure fit of stem of each exhaust valve (2) in its exhaust valve guide (3). Do this for all three exhaust valves and exhaust valve guides.

END OF TASK

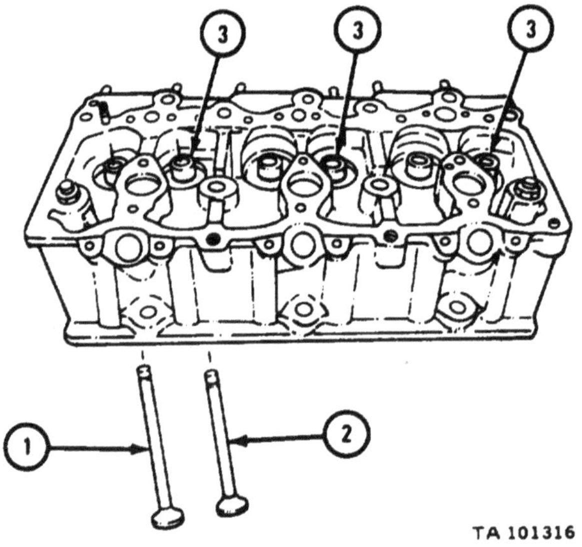

Table 4-52. Exhaust Valves and Guides Wear Limits

Index Number	Item /Point of Measurement	Size and Fit of New Parts (inches)	Wear Limit (inches)
2	Exhaust valve stem outside diameter	0.4363 to 0.4368	0.4358
3	Exhaust valve guide inside diameter	0.4388 to 0.4394	0.4420
2 and 3	Fit of exhaust valve stem in exhaust valve guide	0.0020L to 0.0031L	0.0070L

FRAME 7

NOTE

This frame is for early model cylinder heads (1). Early model cylinder heads used intake valves (2) different from the ones used on late model cylinder heads.

Early model intake valves (2) have a dull shot peened finish. Late model intake valves have a shiny chrome plated finish .

Readings must be within limits given in tables 4-53 and 4-54. The letter L shows a loose fit. If readings are not within given limits, throw away part and get a new one. Be sure to use limits which apply to the type of intake valve (2) you are working on.

Intake valves (2) and exhaust valves (3) were tagged so fit of each valve can be measured in valve guide (4) it was removed from.

1. Measure and note outside diameter of stem of three intake valves (2).
2. Measure and note inside diameter of three intake valve guides (4) 1/8 to 1/4 inch from each end of guide. If diameter is different at each end, note biggest diameter .
3. Measure fit of stem of each intake valve (2) in its intake valve guide (4). Do this for all three intake valves and intake valve guides.
4. Do steps 1, 2, and 3 again for three exhaust valves (3) .

END OF TASK

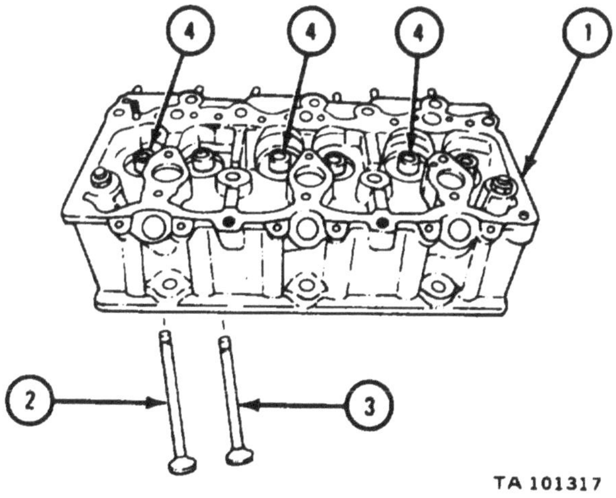

Table 4-53. Early Model Cylinder Head Intake Valves and Guides Wear Limits

Index Number	Item /Point of Measurement	Size and Fit of New Parts (inches)	Wear Limit (inches)
2	Intake valve stem outside diameter		
	Early model shot peened intake valve s	0.4361 to 0.4369	0.4356
	Late model chrome plated intake valve s	0.4368 to 0.4373	0.4363
2 and 4	Fit of intake valve stem in intake valve guide	0.0015L to 0.0003L	0.0070L

Table 4-54. Early Model Cylinder Head Exhaust Valves and Guides Wear Limits

Index Number	Item /Point of Measurement	Size and Fit of New Parts (inches)	Wear Limit (inches)
2	Exhaust valve stem outside diameter		
	Early model shot peened exhaust valve	0.4344 to 0.4352	0.4330
	Late model chrome plated exhaust valve s	0.4363 to 0.4368	0.4358
2 and 4	Fit of exhaust valve stem in exhaust valve guide	0.0032L to 0.0050L	0.0080L

(3) Valve springs, spring retainers, and spring retainer locks .

FRAME 1

NOTE

Readings must be within limits given in table 4-55. If readings are not within given limits, throw away spring and get a new one.

Do this frame for all six inner valve springs (1).

1. Check that inner valve spring (1) is not worn, cracked or damaged. Throw away damaged inner valve springs and get new ones.
2. Measure free length of inner valve spring (1).
3. Using spring tester , measure load needed to squeeze inner valve spring (1) to a length of 1.33 inches.
4. Using spring tester, measure load needed to squeeze inner valve spring (1) to a length of 1.78 inches.
5. Using spring tester, measure solid height of inner valve spring (1).

GO TO FRAME 2

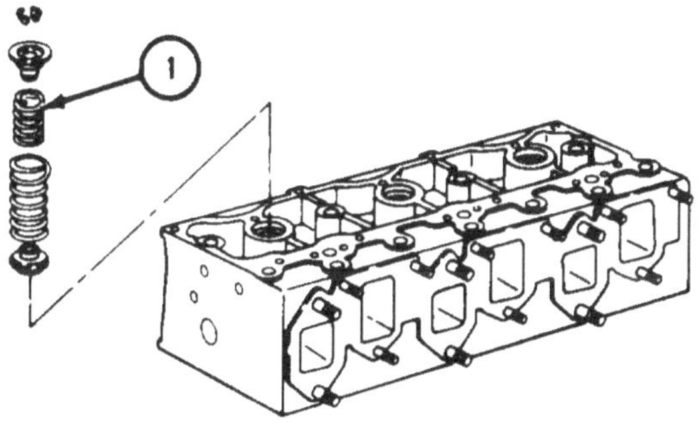

NOTE
CHECK ONLY THOSE PARTS WHICH ARE CALLED OUT IN THIS FRAME. PARTS WITHOUT CALLOUTS ARE SHOWN ONLY FOR REFERENCE PURPOSES OR ARE CHECKED IN ANOTHER FRAME.

TA 101319

Table 4-55. Inner Valve Springs Wear Limits

Index Number	Item /Point of Measurement	Size and Fit of New Parts (inches unles s otherwise noted)	Wear Limit (inches unles s otherwise noted)
1	Inner valve spring free length	2.0100 to 2.2100	None
1	Inner valve spring load at 1.33 inches	81.5 to 91.5 pounds	78 pounds
1	Inner valve spring load at 1.78 inches	34.5 to 38.5 pounds	33 pounds
1	Inner valve spring maximum solid height	1.0700	None

FRAME 2

NOTE

Readings must be within limits given in table 4-56. If readings are not within given limits, throw away spring and get a new one.

Do this frame for all six outer valve springs (1).

1. Check that outer valve spring (1) is not worn, cracked or damaged. Throw away damaged outer valve springs and get new ones.
2. Measure free length of outer valve spring (1).
3. Using spring tester, measure load needed to squeeze outer valve spring (1) to a length of 1.67 inches.
4. Using spring tester, measure load needed to squeeze outer valve spring (1) to a length of 2.12 inches.
5. Using spring tester, measure solid height of outer valve spring (1).

GO TO FRAME 3

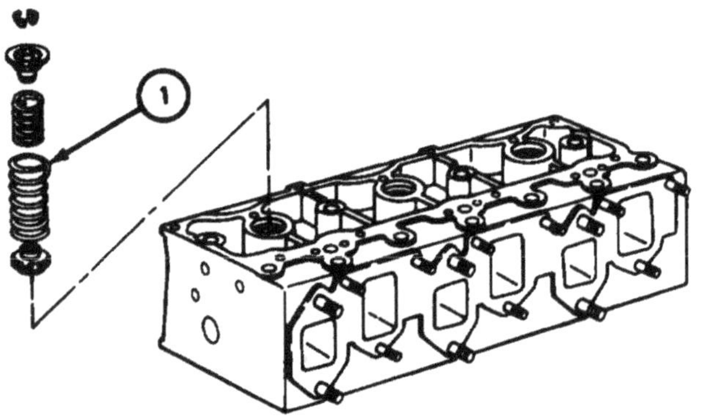

NOTE

CHECK ONLY THOSE PARTS WHICH ARE CALLED OUT IN THIS FRAME. PARTS WITHOUT CALLOUTS ARE SHOWN ONLY FOR REFERENCE PURPOSES OR ARE CHECKED IN ANOTHER FRAME.

TA 101320

Table 4-56. Outer Valve Springs Wear Limits

Index Number	Item /Point of Measurement	Size and Fit of New Parts (inches unless otherwise noted)	Wear Limits (inches unless otherwise noted)
1	Outer valve spring free length	2.5100 to 2.7100	None
1	Outer valve spring load at 1.67 inches	153.7 to 167.7 pounds	147 pounds
1	Outer valve spring load at 2.12 inches	62.7 to 68.7 pounds	63 pounds
1	Outer valve spring maximum solid height	1.5500	None

FRAME 3

1. Check that six valve spring retainers (1) are not worn, chipped or cracked. Throw away worn, chipped or cracked valve spring retainers and get new ones.
2. Check that six pairs of valve spring retainer locks (2) are not chipped or cracked and do not have ridges worn on top. Throw away worn or damaged valve spring retainer locks and get new ones.

END OF TASK

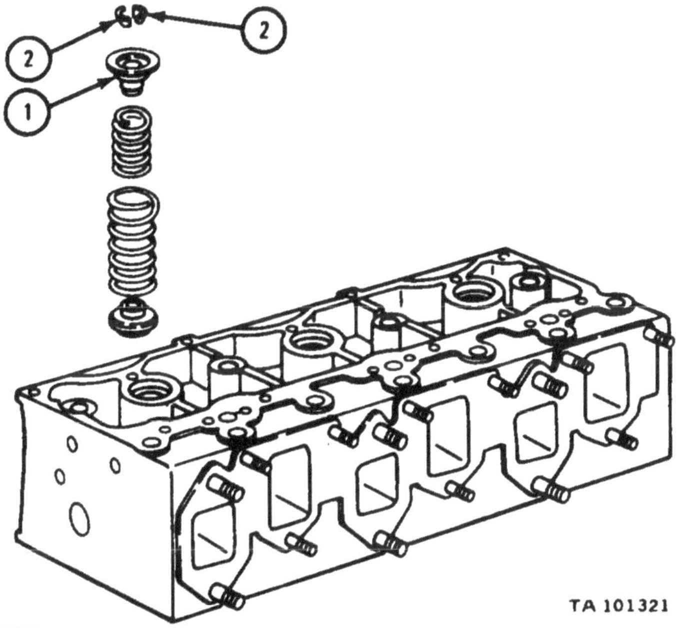

NOTE
CHECK ONLY THOSE PARTS WHICH ARE CALLED OUT IN THIS FRAME. PARTS WITHOUT CALLOUTS ARE SHOWN ONLY FOR REFERENCE PURPOSES OR ARE CHECKED IN ANOTHER FRAME.

(4) Valve rotors .

FRAME 1

NOTE

Three different valve rotors (1) can be used. All are similar in appearance and are tested in the same way. All valve rotors used in engine do not have to be exactly the same.

Do this frame for all six valve rotors (1) in cvlinder head assembly .

1. Check that valve rotors (1) are not worn or cracked. Throw away worn or cracked rotors and get new ones.
2. Cut a sleeve about 1 inch long from 1 1/2-inch inside diameter closed nipple stock.

GO TO FRAME 2

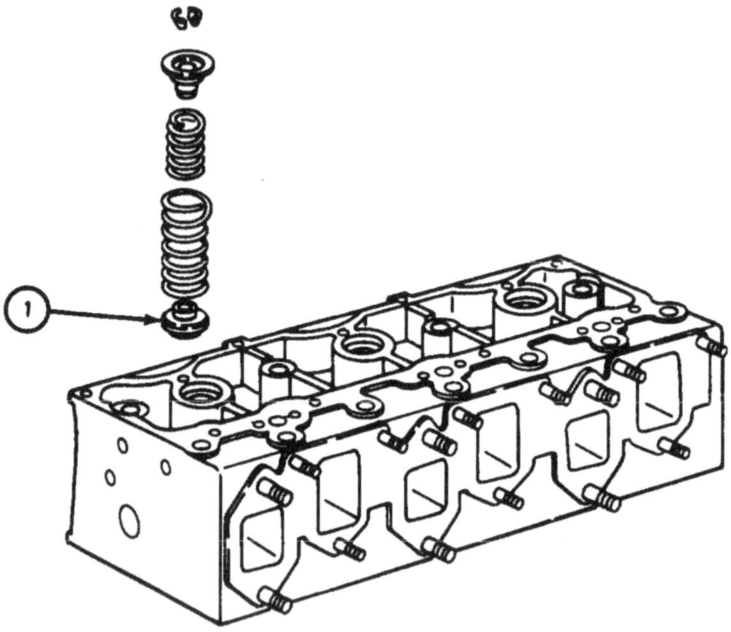

NOTE
CHECK ONLY THOSE PARTS WHICH ARE CALLED OUT IN THIS FRAME. PARTS WITHOUT CALLOUTS ARE SHOWN ONLY FOR REFERENCE PURPOSES OR ARE CHECKED IN ANOTHER FRAME,

TA 101322

FRAME 2

NOTE

Do this frame for all six valve rotors (1) in cylinder head assembly.

1. Draw a chalk mark across inner section (2) and outer section (3) of valve rotor (1) as shown .
2. Put rotor (1) in a container of clean engine oil.
3. Put short steel sleeve (4) on platform of spring tester (5) as shown.
4. Take rotor (1) out of container of oil. Let excess oil drip off. Put rotor (1) on sleeve (4) with valve spring seat side down as shown.
5. Put 3/4-inch hardened steel ball (6) on top of rotor (1) as shown.
6. Work spring tester (5) to put a load of 260 pounds on ball (6) on top of valve rotor (1) . Then let up load to 110 pounds.
7. Do step 6 again 25 to 30 times. Check that marks on inner section (2) and outer section (3) of valve rotor (1) move apart as shown. If marks do not move apart, throw away valve rotor and get a new one.

END OF TASK

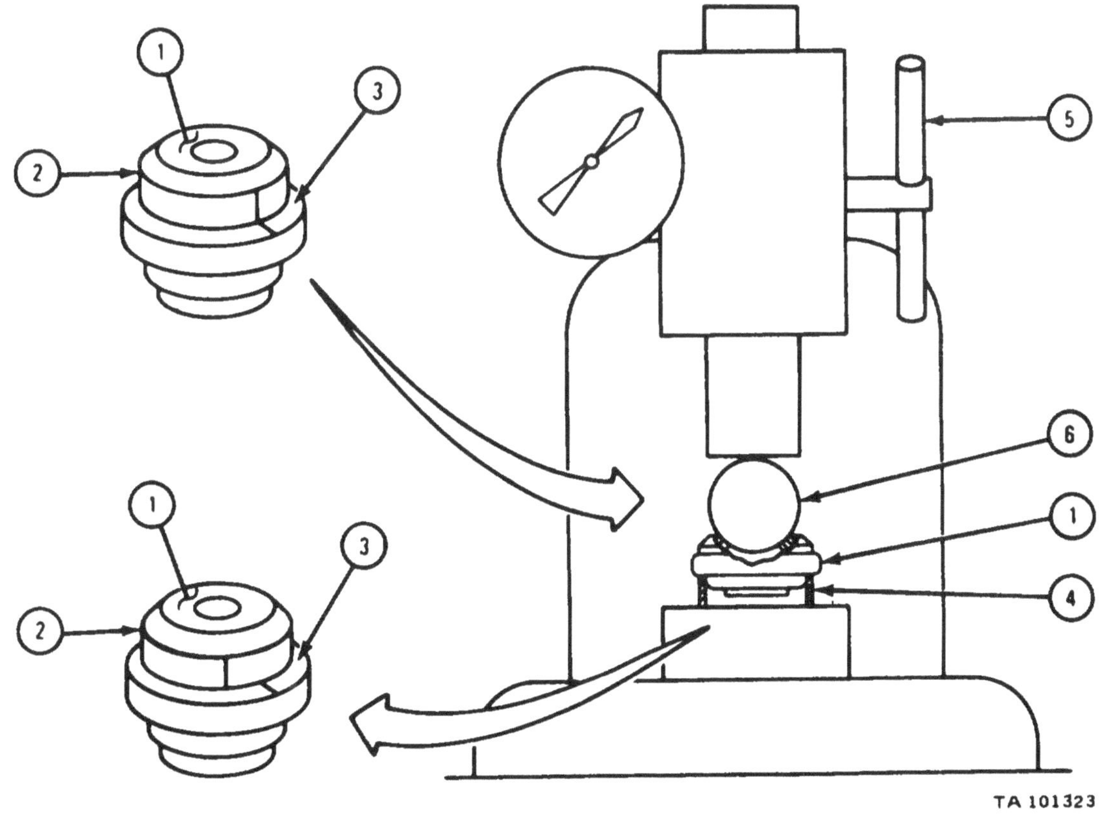

d. Assembly.

FRAME 1

NOTE

Valves must be put back in cylinder head (1) in the same place they were taken from.

Task shown is for an exhaust valve (2). Task is the same for intake valves.

1. Put exhaust valve (2) through exhaust valve guide (3) in cylinder head (1) as shown.
2. Put valve rotor (4) over valve stem (5) and slide it all the way in. Spring seat side of rotor should face outward as shown.

NOTE

Outer valve spring (6) must be put back so that end with tightly wound coils faces cylinder head (1).

3. Put inner valve spring (7) over valve stem (5). Put outer valve spring (6) over inner valve spring .

GO TO FRAME 2

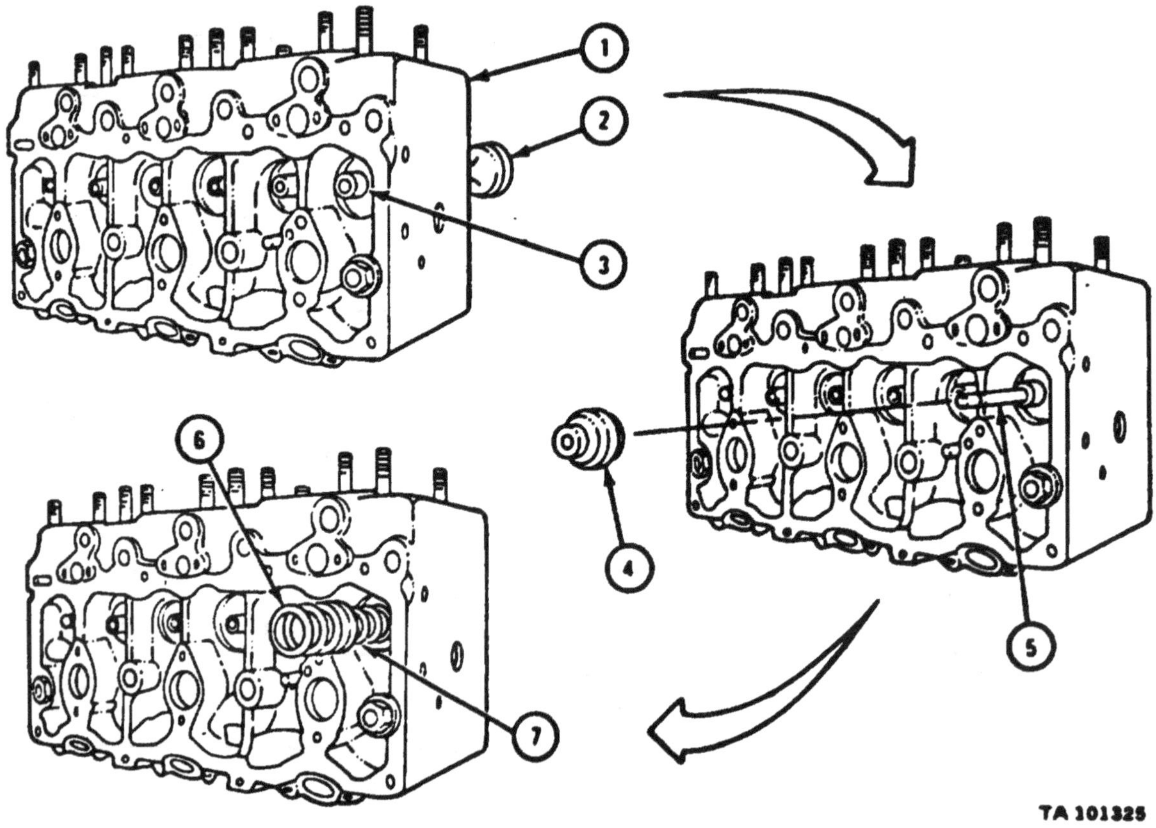

FRAME 2

1. Put spring retainer (1) over outer valve spring (2) as shown. Using valve spring compressor, squeeze spring retainer against springs so that groove (3) in valve stem (4) sticks through hole in spring retainer.
2. Put two spring retainer locks (5) over groove (3) in valve stem (4) and hold retainer locks together .
3. While holding retainer locks (5) together, slowly work valve spring compressor to let all tension off valve springs (2).

CAUTION

> Be sure spring pressure is holding retainer locks (5) in place in groove on valve stem before letting go of retainer locks . If retainer locks are not locked to valve stem, assembly may fly apart when retainer locks are let go and parts will be lost.

4. Let go of retainer locks (5). Take off valve spring compressor .
5. Do frames 1 and 2 again for other two exhaust valves and for three intake valves.

END OF TASK

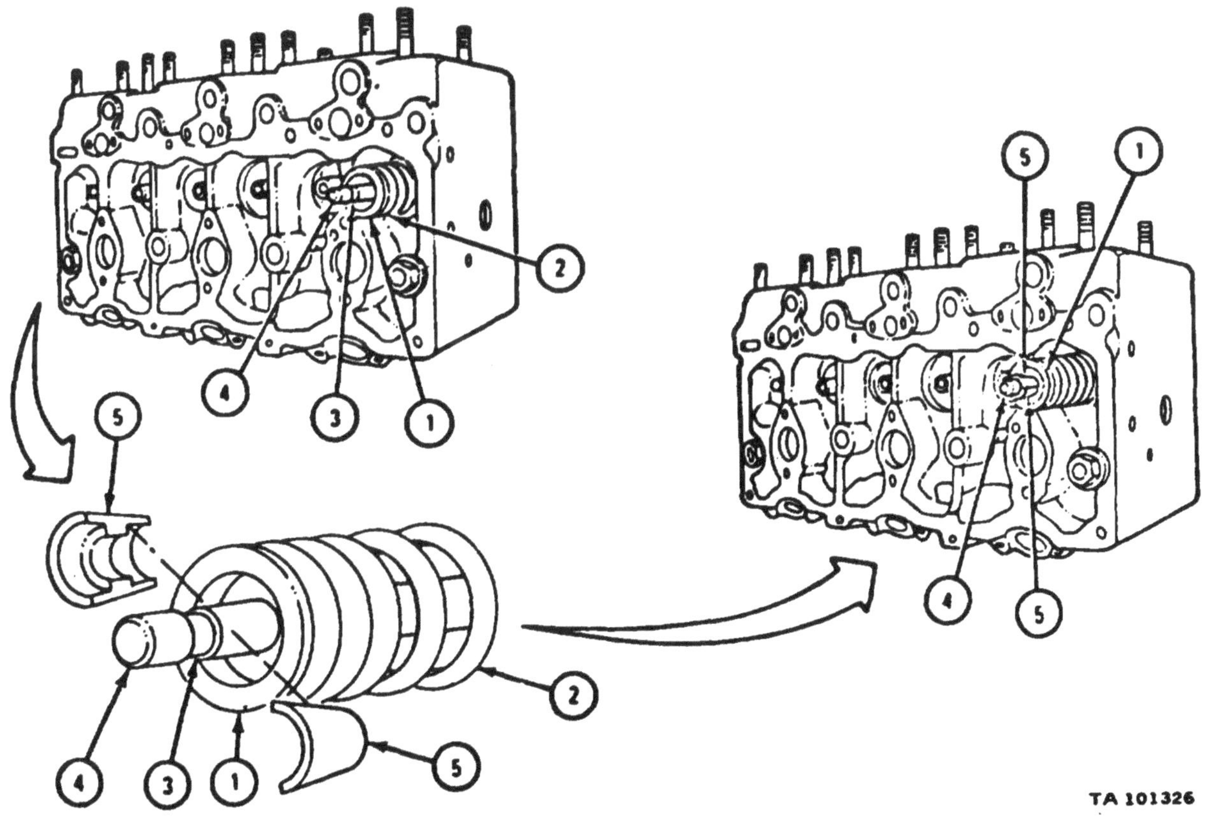

4-20. VALVE ROCKER ARM PUSH RODS.

a. Cleaning. There are no special cleaning procedures needed. Refer to cleaning procedures given in para 4-3.

b. Inspection and Repair.

FRAME 1

1. Check push rod (1) for straightness by rolling on flat surface. If bent, get new push rod.
2. Check that push rod (1) is not out-of-round. If push rod is out-of-round, get new push rod.
3. Do steps 1 and 2 again for all other push rods.

END OF TASK

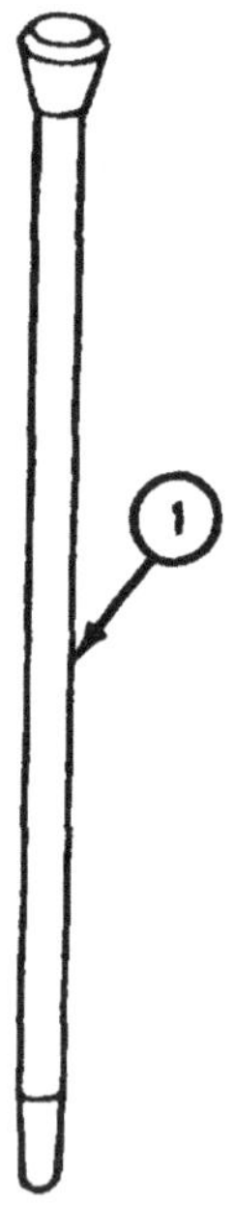

TA 117986

4-21. VALVE ROCKER ARMS AND SHAFTS.

NOTE

The front and rear sets of valve rocker arms and shafts are the same. This task is the same for both sets and must be done for both.

a. Disassembly.

FRAME 1

WARNING

Rocker arm assembly is under spring tension. Be careful when taking off retaining ring (1) or parts of rocker arm assembly will fly off, causing injury to personnel .

1. Take off retaining ring (1) and thrust washer (2). Take off and tag rearmost exhaust valve rocker arm assembly (3) .

GO TO FRAME 2

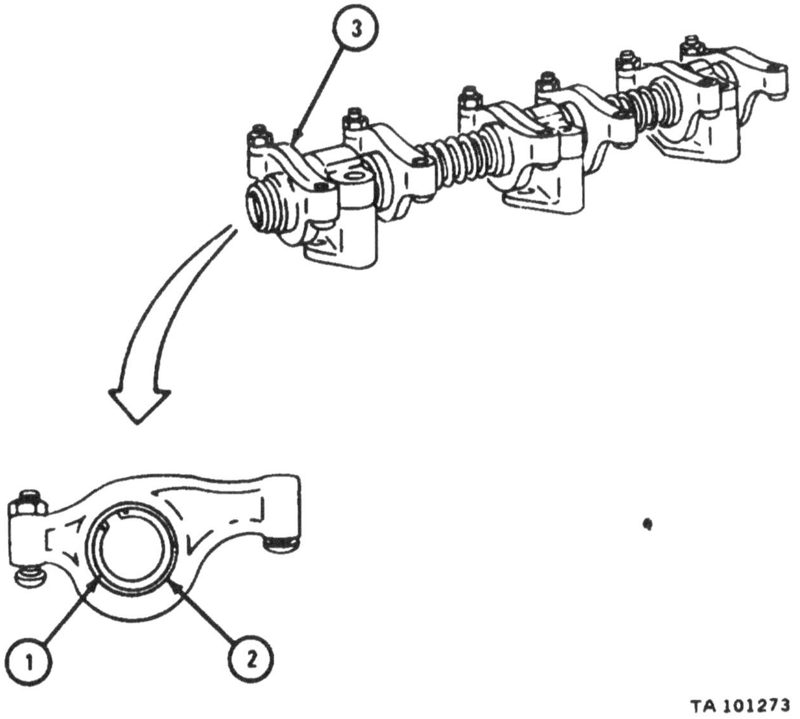

FRAME 2

WARNING

Rocker arm assembly is under spring tension. Be careful when taking off retaining ring (1) or parts of rocker arm assembly will fly off, causing injury to personnel.

1. Take off retaining ring (1) and thrust washer (2). Take off and tag front intake valve rocker arm assembly (3).
2. Take off and tag front rocker arm shaft support (4), front exhaust valve rocker arm assembly (5), front thrust spring (6), and center intake valve rocker arm assembly (7) .

GO TO FRAME 3

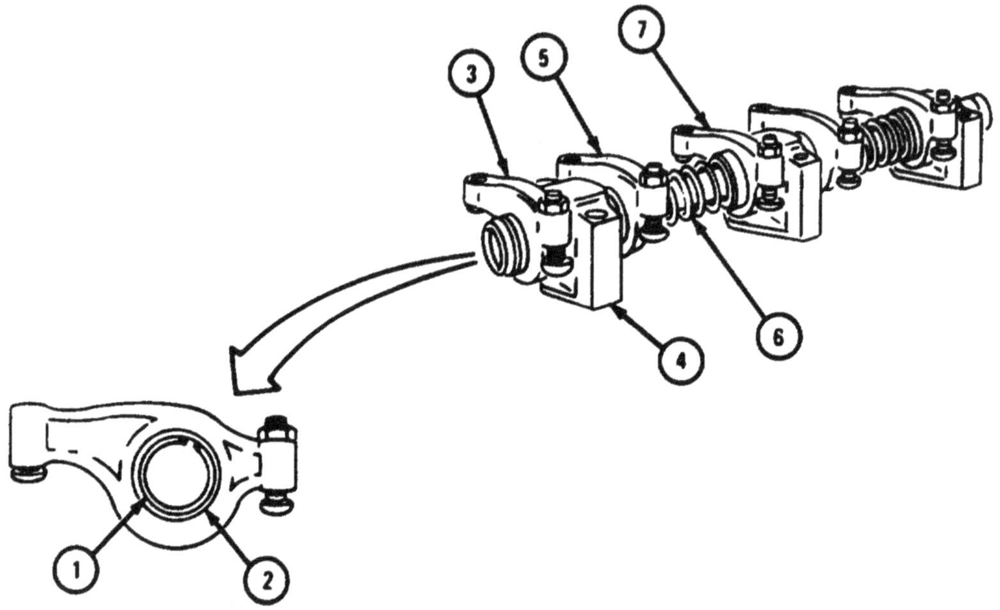

FRAME 3

1. Takeoff and tag center rocker arm shaft support (1) .
2. Takeoff and tag center exhaust valve rocker arm assembly (2) .
3. Take off and tag rear thrust spring (3) and rear intake valve rocker arm assembly (4) .
4. Take off and tag rear rocker arm shaft support (5).
5. Take out two plugs (6) one in each end of rocker arm shaft (7).

GO TO FRAME 4

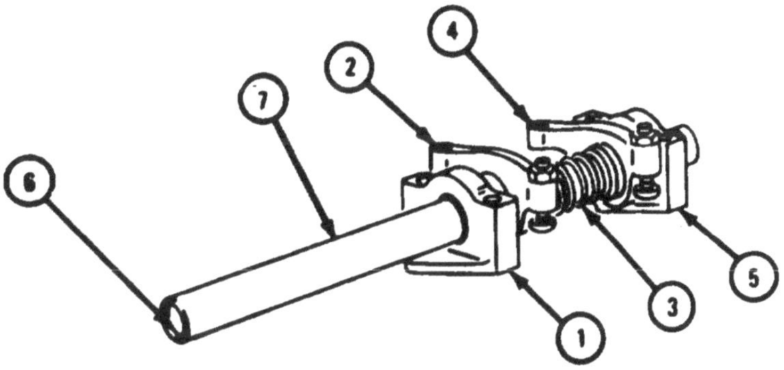

TA 101275

FRAME 4

NOTE

Tag all valve adjusting screws (1) so they can be put back in the same intake or exhaust valve rocker arm assembly. Valve adjusting nuts (2) need not be tagged .

1. Take off valve adjusting screw (1) and valve adjusting nut (2) from all six rocker arm assemblies (3).

END OF TASK

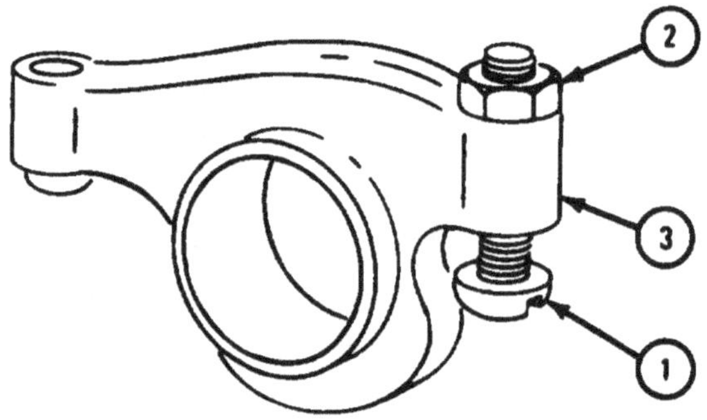

TA 101276

b. Cleaning.

FRAME 1

WARNING

Dry cleaning solvent is flammable. Do not use near an open flame. Keep a fire extinguisher nearby when solvent is used. Use only in well-ventilated places. Failure to do this may result in injury to personnel and damage to equipment.

1. Clean six rocker arms (1), six valve adjusting screws (2), six valve adjusting nuts (3), three rocker arm shaft supports (4), two retaining rings (5), two thrust washers (6), two thrust springs (7), six mounting bolts (8), six lock-washers (9), two plugs (10), and rocker arm shaft (11) in dry cleaning solvent, Clean off sludge and gum deposits with a stiff brush. Dry with clean rags.

WARNING

Eye shields must be worn when using compressed air. Eye injury can occur if eye shields are not used.

2. Clean out all oil passages in six rocker arms (1), front and rear rocker arm shaft supports (4), and rocker arm shaft (11) using brass wire probes. Flus h all oil passages with dry cleaning solvent. Blow all oil passages dry with compressed air .

END OF TASK

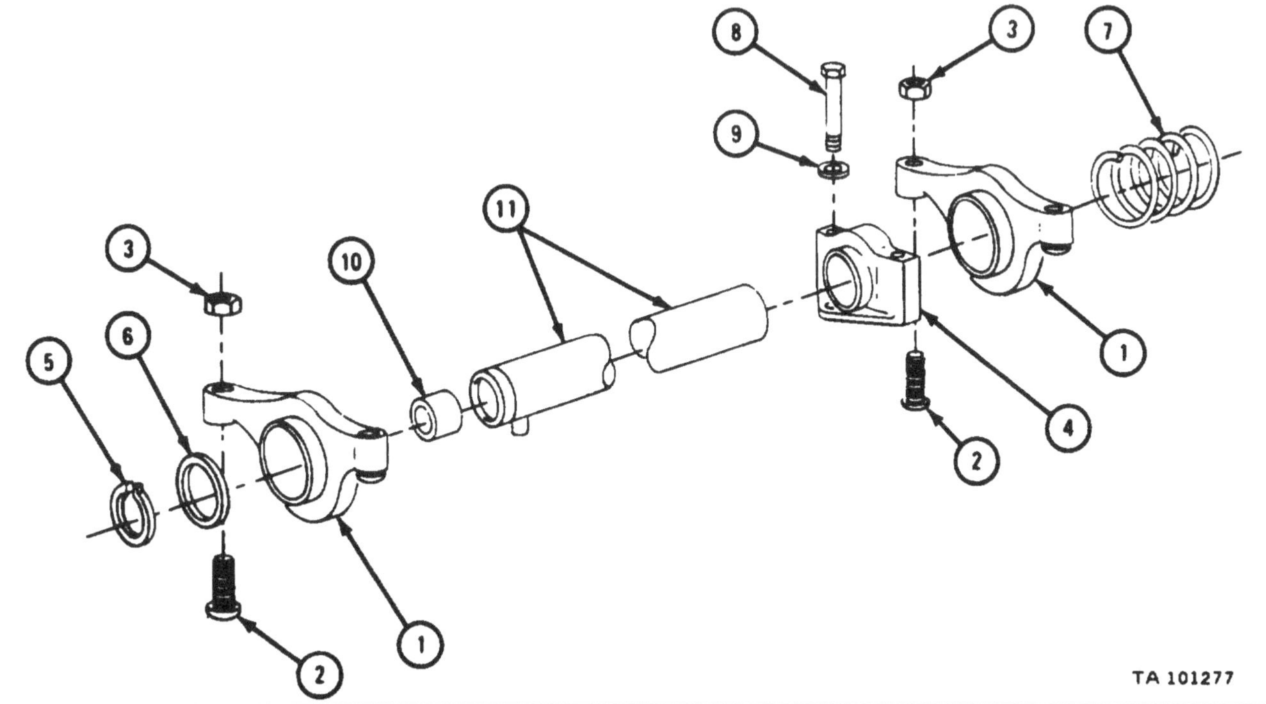

c. Inspection and Repair.

FRAME 1

1. Check that six rocker arms (1) are not cracked. If any rocker arm is cracked, get a new one.

WARNING

Dry cleaning solvent is flammable. Do not use near an open flame. Keep a fire extinguisher nearby when solvent is used. Use only in well-ventilated places. Failure to do this may result in injury to personnel and damage to equipment.

2. Check that six rocker arms (1) have no nicks, scratches or burrs. Fix small nicks, scratches or burrs with a fine mill file or crocus cloth dipped in dry cleaning solvent. If more repair is needed, get a new rocker arm.

3. Check that six bushings (2) are not loose, scored or damaged. If any bushing is loose, scored or damaged, get a new rocker arm.

CAUTION

It is easy to damage the equipment if you do not know what you are doing. Do not try to do this task unless you are experienced at it, or you have an experienced person with you.

4. Check that six valve contact pads (3) are not worn or damaged. If any pad is worn or damaged, get a new rocker arm (1).

5. Check that six valve adjusting screws (4) and nuts (5) are not worn or damaged. Check that screws do not bind in rocker arms (1). If any screw or nut is worn or damaged, get a new one.

GO TO FRAME 2

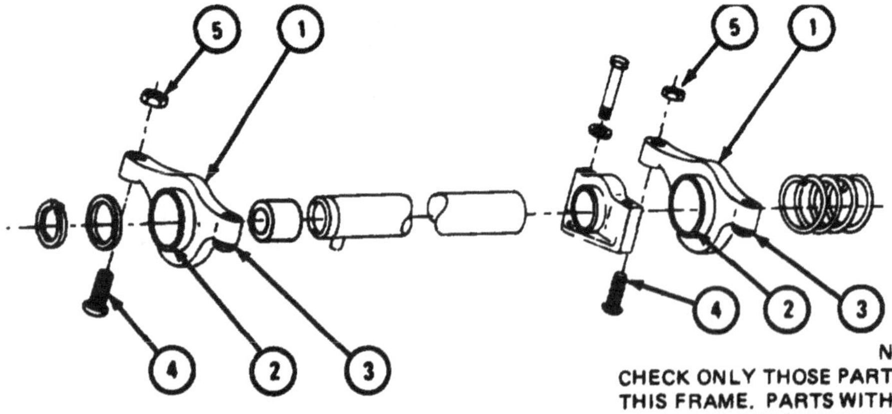

NOTE
CHECK ONLY THOSE PARTS WHICH ARE CALLED OUT IN THIS FRAME. PARTS WITHOUT CALLOUTS ARE SHOWN ONLY FOR REFERENCE PURPOSES OR ARE CHECKED IN ANOTHER FRAME.

TA 101278

FRAME 2

1. Check that two thrust springs (1) are not cracked or broken. If thrust springs are cracked or broken, throw them away and get new ones.
2. Check that two retaining rings (2) are not cracked or bent. If retaining rings are cracked or bent, throw them away and get new ones.
3. Check that two thrust washers (3) are not chipped, cracked, bent or grooved. If thrust washers are chipped, cracked, bent or grooved, throw them away and get new ones.
4. Check that six mounting screws (4) do not have stripped or damaged threads, rounded heads or other damage. If mounting screws are damaged, throw them away and get new ones.
5. Check that six mounting lockwashers (5) and two plugs (6) are not chipped, cracked, or bent. If lockwashers or plugs are damaged, throw them away and get new ones.

GO TO FRAME 3

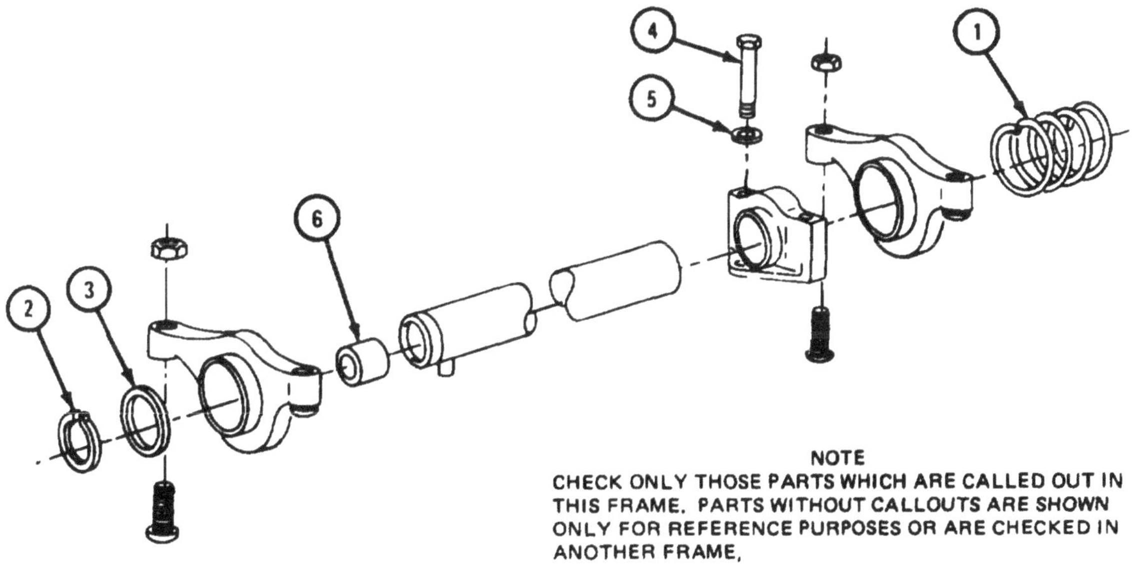

NOTE
CHECK ONLY THOSE PARTS WHICH ARE CALLED OUT IN THIS FRAME. PARTS WITHOUT CALLOUTS ARE SHOWN ONLY FOR REFERENCE PURPOSES OR ARE CHECKED IN ANOTHER FRAME.

TA 101279

FRAME 3

WARNING

Dry cleaning solvent is flammable. Do not use near an open flame. Keep a fire extinguisher nearby when solvent is used. Use only in well-ventilated places. Failure to do this may result in injury to personnel and damage to equipment.

NOTE

Rocker arm shaft supports (1) cannot be changed separately. If any rocker arm shaft support is damaged, rocker arm shaft (2) and all three rocker arm shaft supports must be changed.

1. Check that three rocker arm shaft supports (1) are not cracked, nicked or scored and have no burrs. If rocker arm shaft supports are cracked or badly scored, throw them away and get new ones. Fix small nicks and burrs with a fine mill file or crocus cloth dipped in dry cleaning solvent.

GO TO FRAME 4

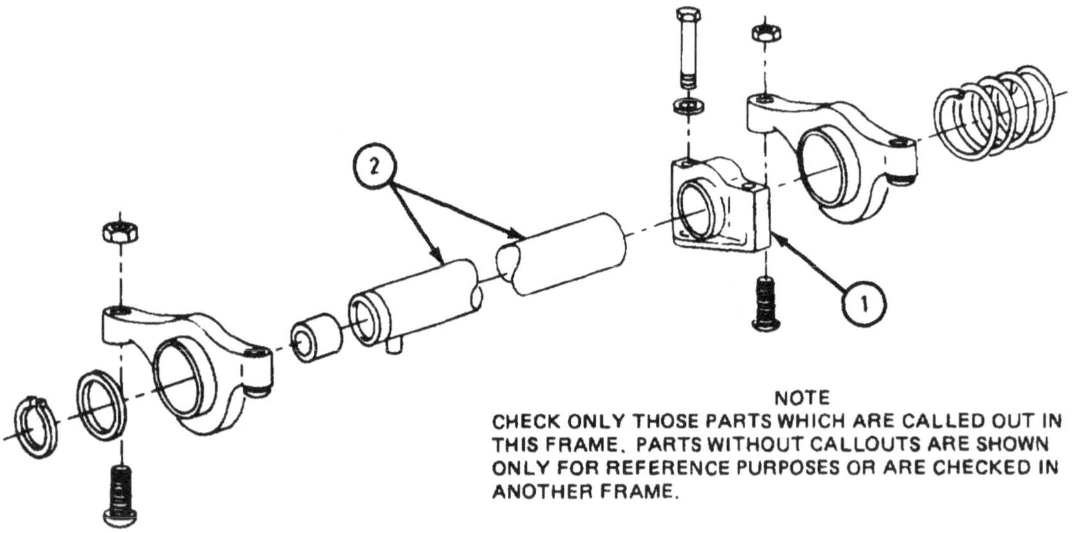
2
1
NOTE
CHECK ONLY THOSE PARTS WHICH ARE CALLED OUT IN THIS FRAME. PARTS WITHOUT CALLOUTS ARE SHOWN ONLY FOR REFERENCE PURPOSES OR ARE CHECKED IN ANOTHER FRAME.
TA 101280

FRAME 4

1. Check that rocker arm shaft (1) is not cracked or broken. If rocker arm shaft is cracked or broken, throw it and rocker arm supports (2) away and get new ones.

CAUTION

It is easy to damage the equipment if you do not know what you are doing. Do not try to do this task unless you are experienced at it, or you have an experienced person with you .

NOTE

Bearing surfaces of rocker arm shaft (1) are surfaces next to oil holes in rocker arm shaft.

2. Check that rocker arm shaft (1) does not have worn bearing surfaces and that rocker arm shaft is not scored. If rocker arm shaft is scored or has worn bearing surfaces, throw it and rocker arm supports (2) away and get new ones.

3. Check that locating dowel pin (3) is not cracked, chipped or loose. If it is damaged or loose, throw away rocker arm shaft (1) and locker arm supports (2) and get new ones.

WARNING

Dry cleaning solvent is flammable. Do not use near an open flame. Keep a fire extinguisher nearby when solvent is used . Use only in well-ventilated places. Failure to do this may result in injury to personnel and damage to equipment.

4. Check that rocker arm shaft (1) has no small nicks, scratches or burrs. Fix small nicks, scratches or burrs with a crocus cloth dipped in dry cleaning solvent.

GO TO FRAME 5

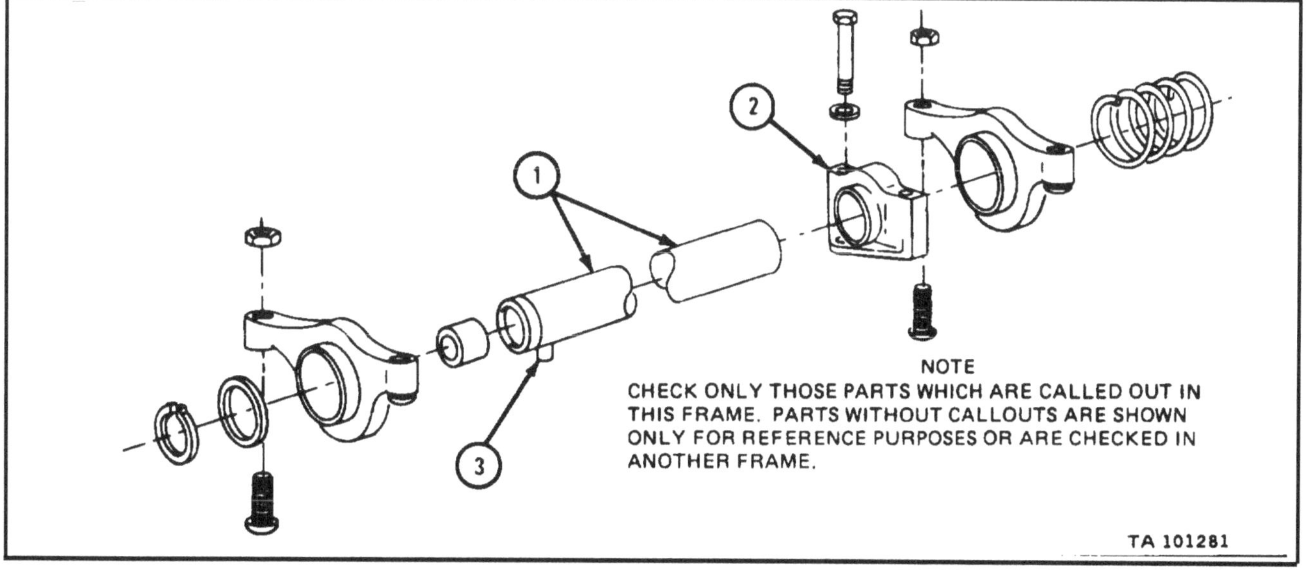
1
2
3
NOTE
CHECK ONLY THOSE PARTS WHICH ARE CALLED OUT IN THIS FRAME. PARTS WITHOUT CALLOUTS ARE SHOWN ONLY FOR REFERENCE PURPOSES OR ARE CHECKED IN ANOTHER FRAME.
TA 101281

FRAME 5

NOTE

Readings must be within limits given in table 4-57. The letter L shows a loose fit. If readings are not within given limits, throw away part and get a new one.

Rear end of rocker arm shaft (1) has a locating dowel pin (2) .

Rocker arm bearing surfaces of rocker arm shaft (1) are surface areas near the first, third, fourth, fifth, sixth, and eighth oil holes in rocker arm shaft, counting from front to rear . Bearing surfaces will look shiny on used rocker arm shafts .

1. Measure inside diameter of six rocker arm bearings (3). Note each measurement and which rocker arm it was taken on.
2. Measure outside diameter of six rocker arm bearing surfaces of rocker arm shaft (1) . Note each measurement and which rocker arm bearing surface it was taken on.
3. Measure fit of each rocker arm bearing (3) on its bearing surface of rocker arm shaft (1) .
4. Measure length A of six valve adjusting screws (4).

GO TO FRAME 6

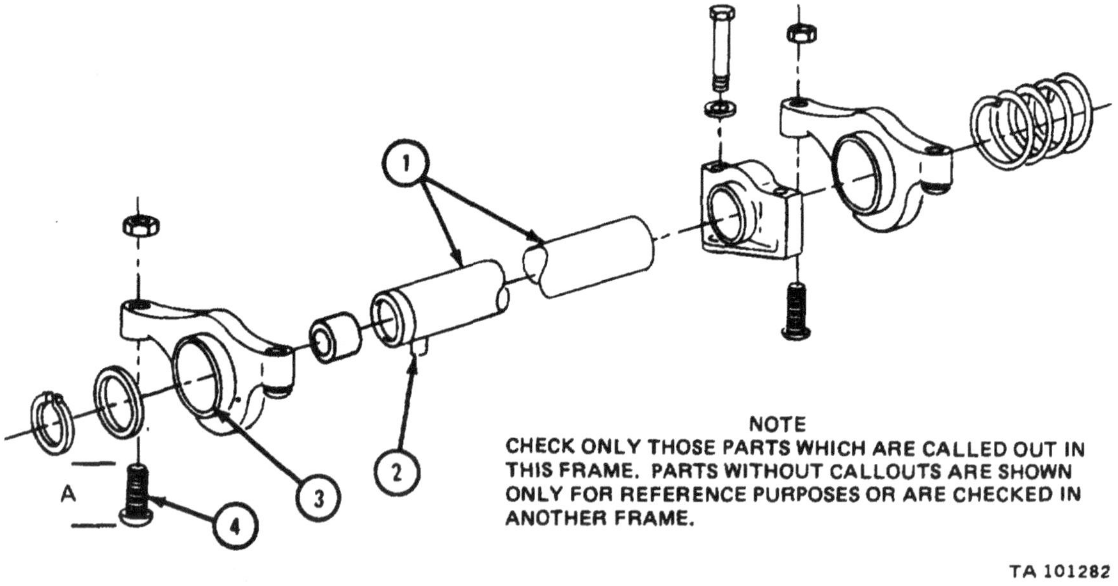
1
2
3
4
A
NOTE
CHECK ONLY THOSE PARTS WHICH ARE CALLED OUT IN THIS FRAME. PARTS WITHOUT CALLOUTS ARE SHOWN ONLY FOR REFERENCE PURPOSES OR ARE CHECKED IN ANOTHER FRAME.
TA 101282

Table 4-57. Valve Rocker Arm Bearings and Shafts Wear Limits

Index Number	Item /Point of Measurement	Size and Fit of New Parts (inches)	Wear Limit (inches)
3	Rocker arm bearing inside diameter		
	Late Model	1.0025 to 1.0038	1.0070
	Early Model	1.0070 to 1.0022	1.0070
1	Rocker arm shaft diameter at rocker arm bearing surface	0.9998 to 1.0005	0.9988
3 and 1	Fit of rocker arm bearing on shaft		
	Late Model	0.0020L to 0.0040L	0.0090L
	Early Model	0.0020L to 0.0024L	0.0090L
4	Valve adjusting screw length A	1.740	None

FRAME 6

NOTE

Readings must be within limits given in table 4-58.
If readings are not within given limits, throw away part and get a new one.

1. Measure free length of two valve rocker arm thrust springs (1).
2. Measure load needed to squeeze two thrust springs (1) down to length of 2 inches.
3. Measure maximum solid height of two thrust springs (1).

END OF TASK

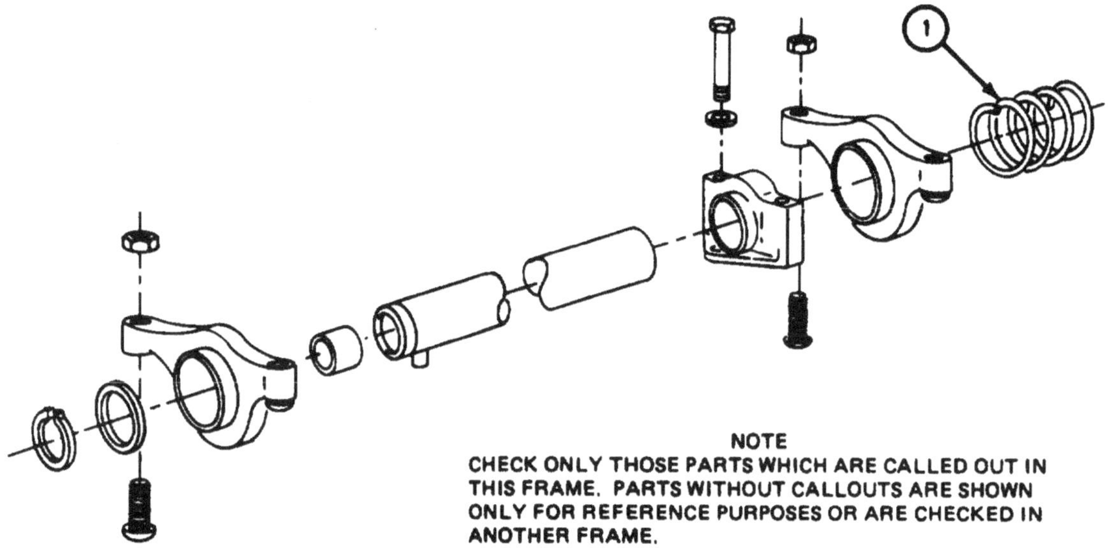
1
NOTE
CHECK ONLY THOSE PARTS WHICH ARE CALLED OUT IN THIS FRAME. PARTS WITHOUT CALLOUTS ARE SHOWN ONLY FOR REFERENCE PURPOSES OR ARE CHECKED IN ANOTHER FRAME.
TA 101283

Table 4-58. Valve Rocker Arm Thrust Spring Wear Limits

Index Number	Item /Point of Measurement	Size and Fit of New Parts “inches unless otherwise noted)	Wear Limit (inches)
1	Thrust spring free lengt h	3.500	None
1	Thrust spring load at 2 inches	8.8 to 9.0 pounds	7 pounds
1	Thrust spring maximum solid height	0.7300	None

d. Assembly.

FRAME 1

1. Put in and hand tighten one valve adjusting screw (1) into each of six rocker arms (2) as tagged . Put screw in as far as it will go.
2. Put on and hand tighten valve adjusting nut (3) on each of six valve adjusting screws (1) .

GO TO FRAME 2

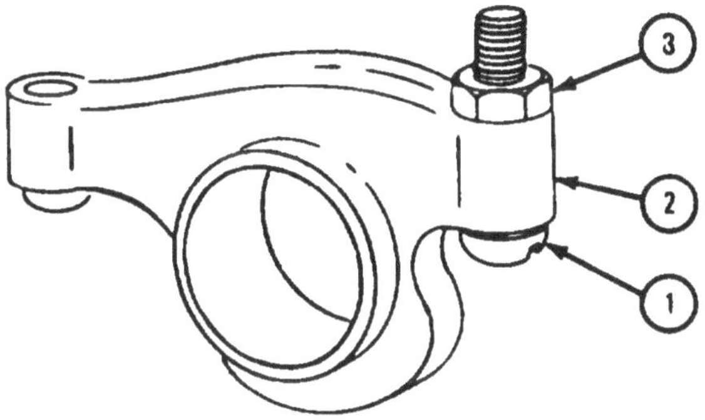

FRAME 2

NOTE

Rear end of rocker arm shaft (1) has locating dowel pin (2) .

Make sure you put rocker arm shaft supports (3 and 4) rocker arm assemblies (5 and 6), and rear thrust spring (7) on rocker arm shaft (1) as tagged.

1. Put in two plugs (8) one in each end of rocker arm shaft (1).
2. Hold rocker arm shaft (1) so that locating dowel pin (2) at rear of rocker arm shaft points down. Slide rear rocker arm shaft support (3) over front end of rocker arm shaft and push it as far back on shaft as it will go. Locating dowel pin must fit in notch of rear rocker arm shaft support.
3. Slide rear intake valve rocker arm assembly (5) over front end of rocker arm shaft (1) and push it all the way back.
4. Slide rear rocker arm thrust spring (7) over front end of rocker arm shaft (1 and push it all the way back as shown.
5. Slide center intake valve rocker arm assembly (6) over front end of rocker arm shaft (1) and push it all the way back.
6. Slide center rocker arm shaft support (4) over front of rocker arm shaft (1) and push it all the way back as shown.

GO TO FRAME 3

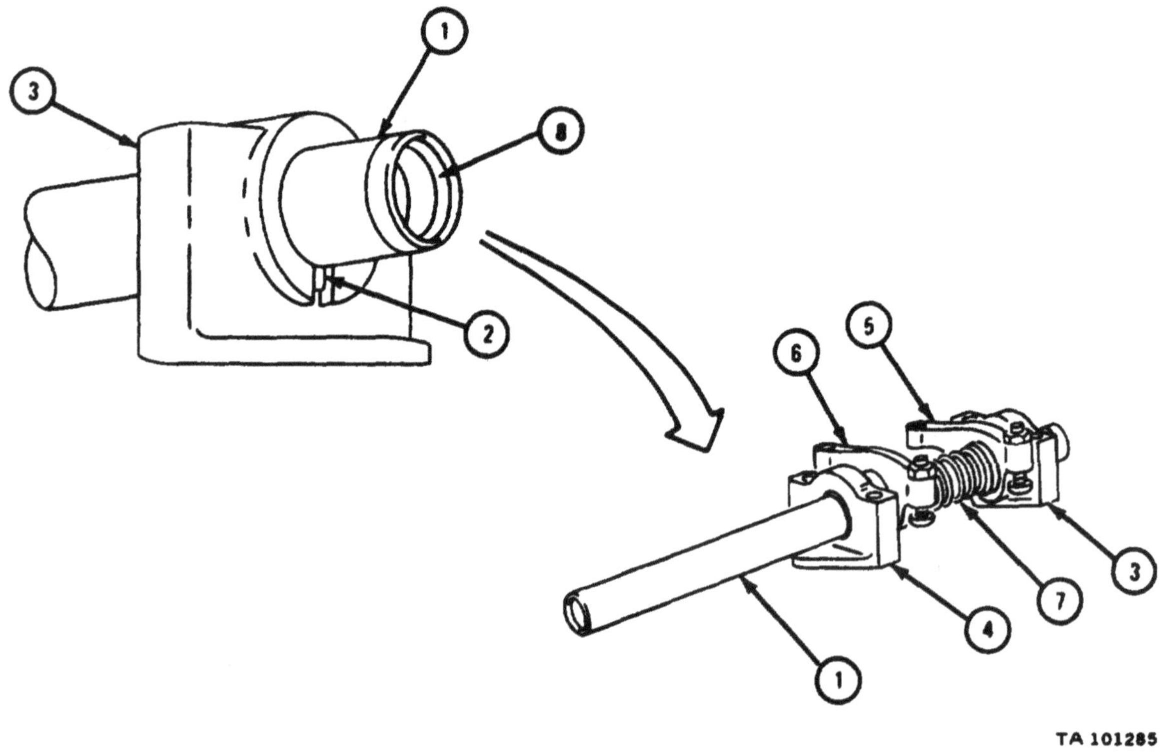

FRAME 3

NOTE

Make sure front rocker arm shaft support (1), rocker arm assemblies (2, 3, and 4), and front thrust spring (5) are put back on rocker arm shaft (6) as tagged.

1. Slide center intake valve rocker arm assembly (2) over front end of rocker arm shaft (6) and push it all the way back.
2. Slide front thrust spring (5) over front of rocker arm shaft (6) and Push it all the way back.
3. Slide front exhaust valve rocker arm assembly (3) over front of rocker arm shaft (6) and push it all the way back.

NOTE

You will have to squeeze down thrust springs (5 and 7) to fit rest of parts on front of rocker arm shaft (6) .

4. Slide front rocker arm shaft support (1) over front of rocker arm shaft (6) and push down thrust springs (5 and 7) so there is enough room at front of rocker arm shaft for front exhaust valve rocker arm assembly (4).
5. Slide front intake valve rocker arm assembly (4) over front of rocker arm shaft (6) and push it down against thrust springs (5 and 7). Put on thrust washer (8) and retaining ring (9) .

GO TO FRAME 4

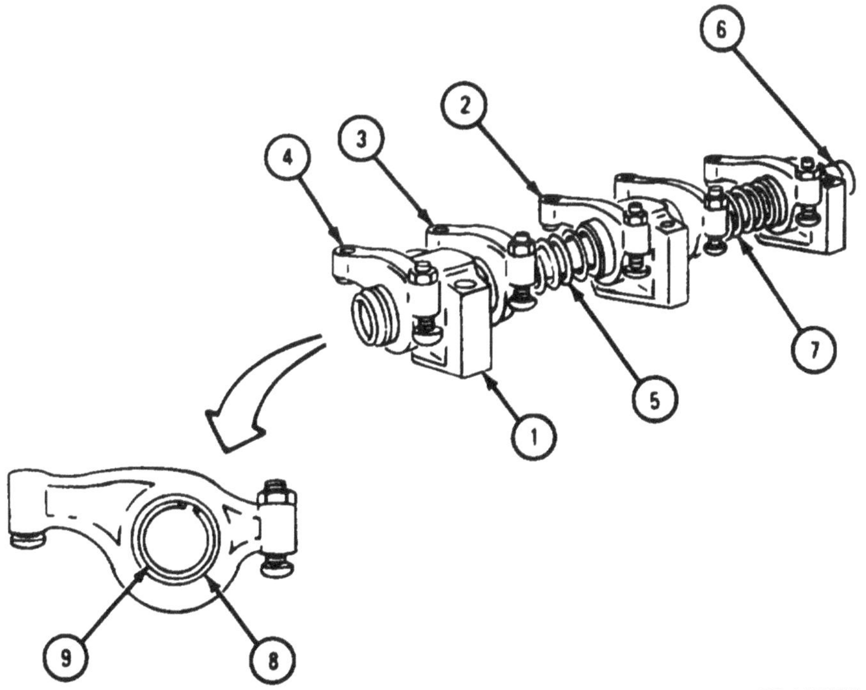

FRAME 4

1. Slide rear exhaust valve rocker arm assembly (1) over rear end of rocker arm shaft (2) as tagged .
2. Put on thrust washer (3) and retaining ring (4).

END OF TASK

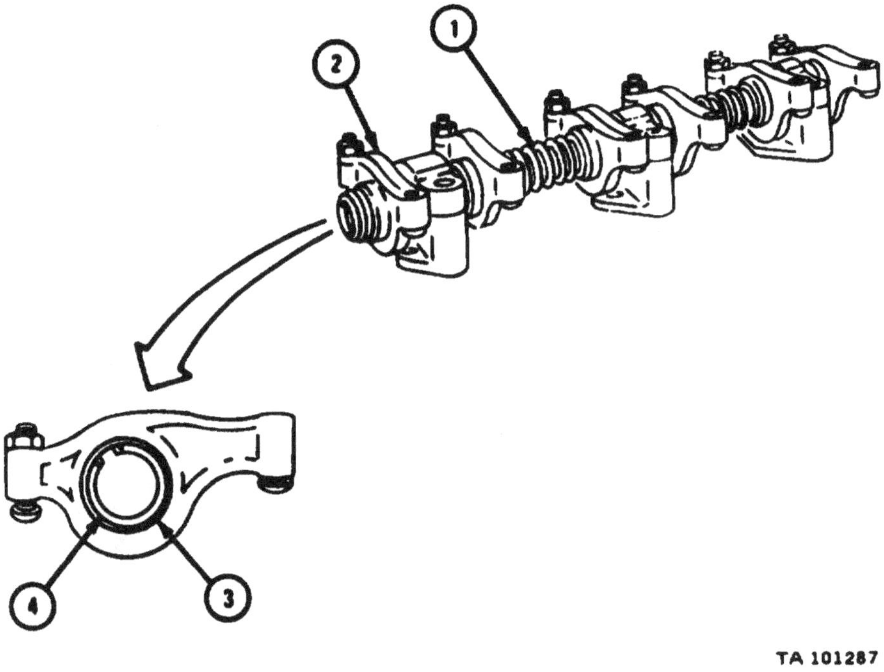

4-22. CYLINDER HEAD COVERS

a. Cleaning.

FRAME 1

1. Clean oil filler cap assembly (1) with warm water and soap.

WARNING

Dry cleaning solvent is flammable. Do not use near an open flame. Keep a fire extinguisher nearby when solvent is used. Use only in well-ventilated places. Failure to do this may result in injury to personnel and damage to equipment.

2. Clean two covers (2) using dry cleaning solvent. Scraper may be used to take off gasket material. Use stiff brush to take off sludge and gum deposits.

END OF TASK

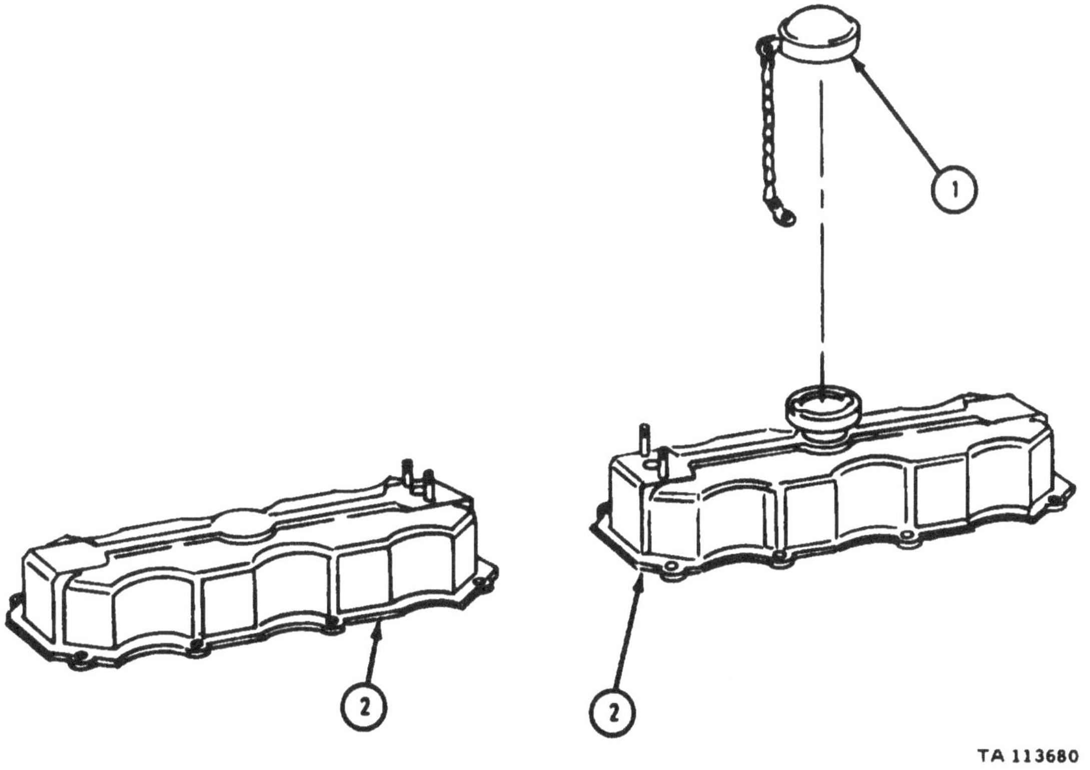

b . Inspection and Repair.

FRAME 1

1. Check that all threaded parts are not stripped or damaged in any way. If threads are damaged, get a new part.
2. Check that covers (1) are not bent or dented. To repair bent or dented covers, refer to FM 43-2. If covers are badly bent or dented, get a new cover.
3. Check that covers (1) are not torn or cracked. To repair tears or cracks, refer to TM 9-237. If covers are badly torn or cracked, get a new cover.

GO TO FRAME 2

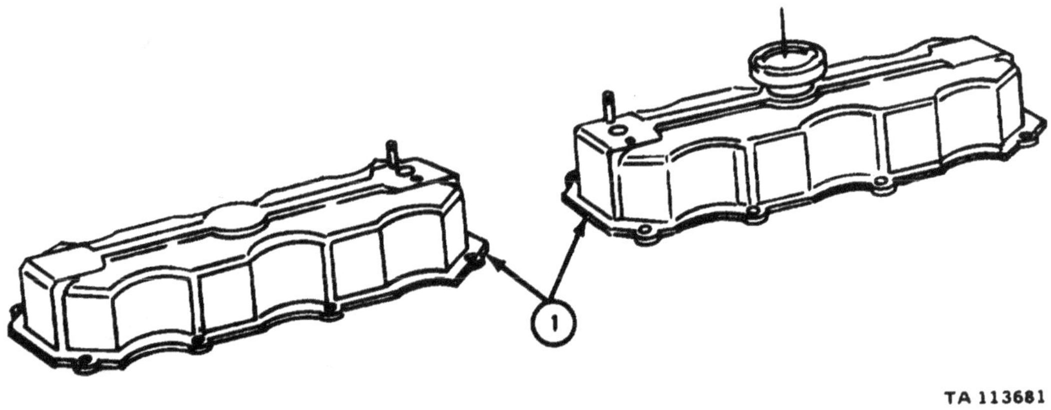

FRAME 2

1. Check that gasket surfaces of covers (1) do not have small nicks or burrs. Take off small nicks and burrs with a fine mill file or crocus cloth.
2. Check that oil filler cap assembly (2) has no cracks, dents or damaged lock tabs (3), and that gasket (4) is not damaged in any way. If oil filler cap assembly is damaged, get a new oil filler cap assembly.

END OF TASK

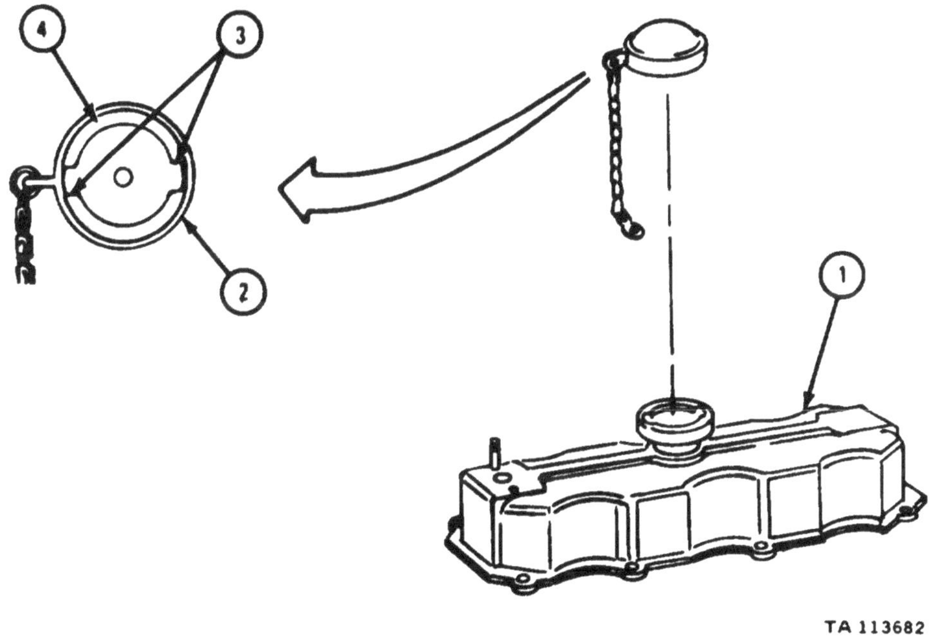

4-23. TACHOMETER ADAPTER.

a. Engine LDS-465-2 .

(1) Cleaning . There are no special cleaning procedures needed. Refer to cleaning procedures given in para 4-3.

(2) Inspection and repair .

FRAME 1

1. Check that tachometer takeoff adapter (1) is not cracked, stripped or damaged in any other way. Fix minor thread damage with thread chaser. If more repair is needed, get a new part.
2. Check that tachometer adapter (2) is not cracked, stripped or damaged in any other way . Fix minor thread damage with thread chaser or tap. If more repair is needed, get a new part.

GO TO FRAME 2

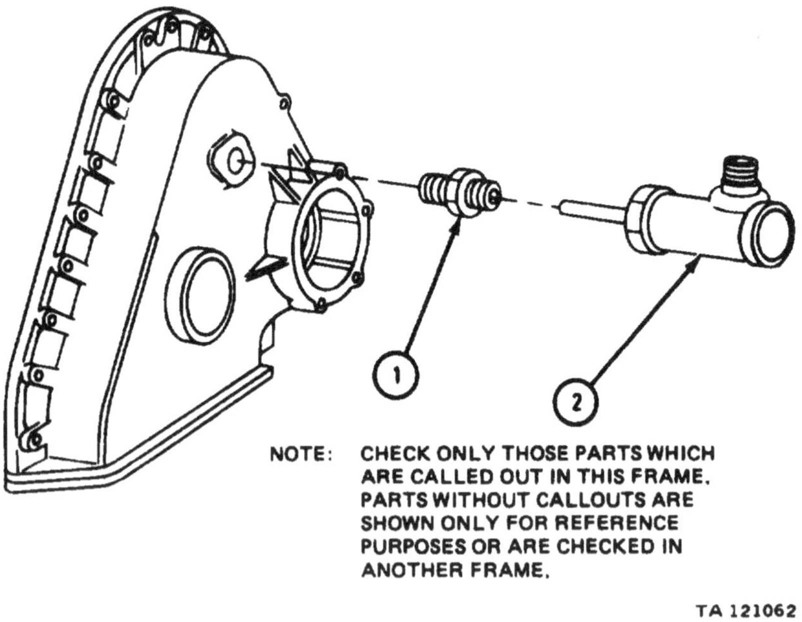

1
2
NOTE: CHECK ONLY THOSE PARTS WHICH ARE CALLED OUT IN THIS FRAME. PARTS WITHOUT CALLOUTS ARE SHOWN ONLY FOR REFERENCE PURPOSES OR ARE CHECKED IN ANOTHER FRAME.
TA 121062

FRAME 2

WARNING

Dry cleaning solvent is flammable. Do not use near an open flame. Keep a fire extinguisher nearby when solvent is used. Use only in well-ventilated places. Failure to do this may result in injury to personnel and damage to equipment.

1. Check that tachometer adapter shaft (1) is not worn, bent or cracked, and has no burrs . Take off small burrs with a crocus cloth dipped in dry cleaning solvent. If more repair is needed, get a new tachometer adapter (2).

2. Turn tachometer adapter shaft (1) . Tachometer adapter shaft must turn freely. If it does not, get a new tachometer adapter (2).

END OF TASK

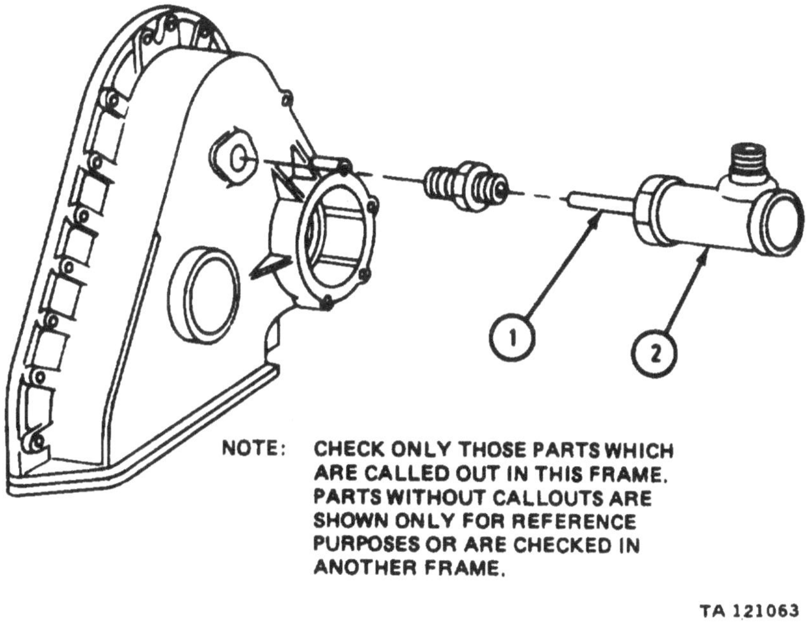
1
2
NOTE: CHECK ONLY THOSE PARTS WHICH ARE CALLED OUT IN THIS FRAME. PARTS WITHOUT CALLOUTS ARE SHOWN ONLY FOR REFERENCE PURPOSES OR ARE CHECKED IN ANOTHER FRAME.
TA 121063

b. All Engines Except LDS-465-2.

(1) Cleaning. There are no special cleaning procedures needed. Refer to cleaning procedures given in para 4-3.

(2) Inspection and repair.

FRAME 1

1. Check that tachometer takeoff adapter (1) is not cracked, stripped or damaged in any other way. Fix minor thread damage with thread chaser. If more repair is needed, get a new part.
2. Check that tachometer adapter (2) is not cracked, stripped or damaged in any other way. Fix minor thread damage with thread chaser or tap. If more repair is needed, get a new part.

GO TO FRAME 2

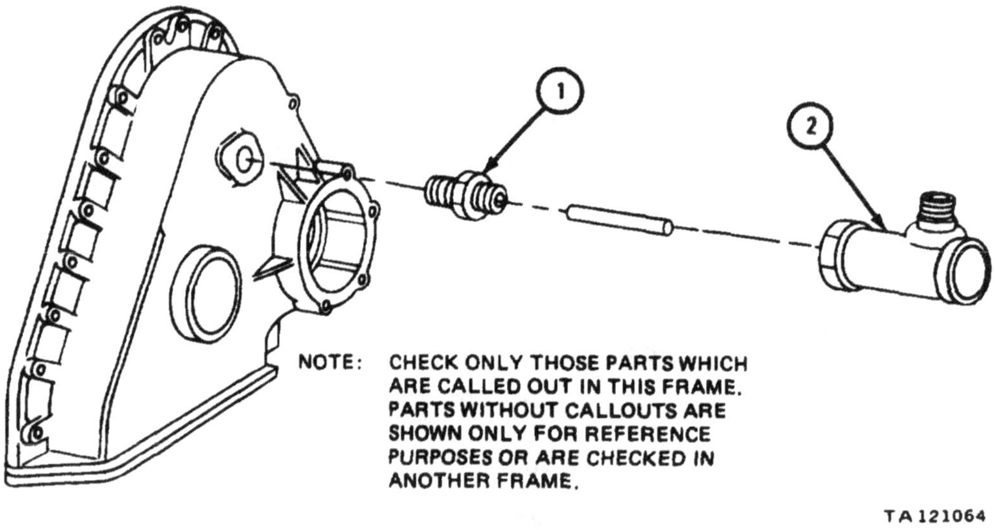
1
2
NOTE: CHECK ONLY THOSE PARTS WHICH ARE CALLED OUT IN THIS FRAME. PARTS WITHOUT CALLOUTS ARE SHOWN ONLY FOR REFERENCE PURPOSES OR ARE CHECKED IN ANOTHER FRAME.
TA 121064

FRAME 2

WARNING

Dry cleaning solvent is flammable. Do not use near an open flame. Keep a fire extinguisher nearby when solvent is used. Use only in well-ventilated places. Failure to do this may result in injury to personnel and damage to equipment.

1. Check that tachometer adapter shaft (1) is not worn, bent or cracked, and has no burrs . Fix small burrs with a crocus cloth dipped in dry cleaning solvent. If more repair is needed, get a new tachometer adapter shaft.
2. Put tachometer shaft (1) in tachometer adapter (2). Turn tachometer adapter shaft (1) . Tachometer adapter shaft must 'turn freely. If it does not, get a new tachometer adapter.

END OF TASK

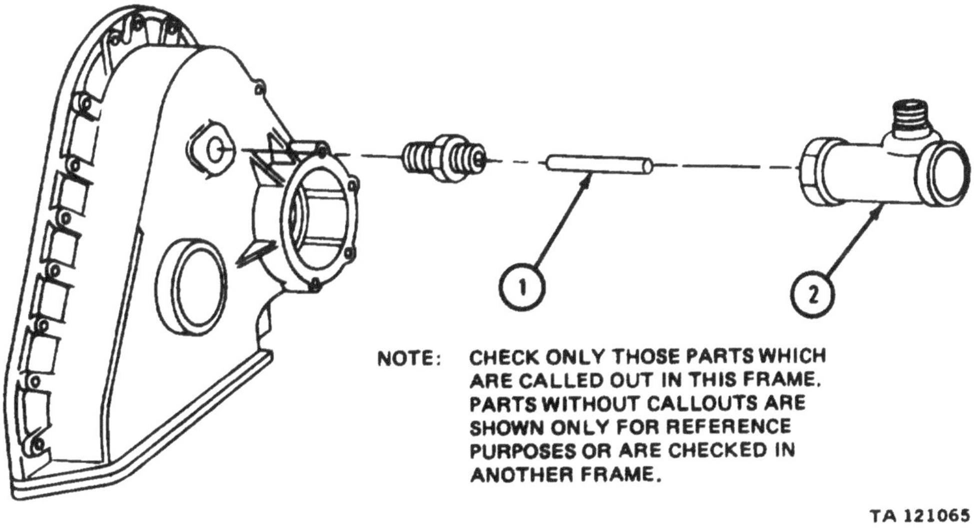
1
2
NOTE: CHECK ONLY THOSE PARTS WHICH ARE CALLED OUT IN THIS FRAME. PARTS WITHOUT CALLOUTS ARE SHOWN ONLY FOR REFERENCE PURPOSES OR ARE CHECKED IN ANOTHER FRAME.
TA 121065

4-24. CRANKCASE BREATHER TUBE ADAPTER.

a. Cleaning There are no special cleaning procedures needed. Refer to cleaning procedures given in para 4-3.

b. Inspection and Repair.

FRAME 1

1. Check that crankcase breather tube adapter (1) is not cracked, bent, dented or damaged in any other way. If crankcase breather tube adapter is damaged in any way, get a new part.

WARNING

Dry cleaning solvent is flammable. Do not use near an open flame. Keep a fire extinguisher nearby when solvent is used . Use only in well-ventilated places. Failure to do this may result in injury to personnel and damage to equipment.

2. Check that crankcase breather tube adapter (1) has no minor nicks, scratches or burrs . To take out minor nicks, scratches or burrs, use a fine mill file or crocus cloth dipped in dry cleaning solvent.

END OF TASK

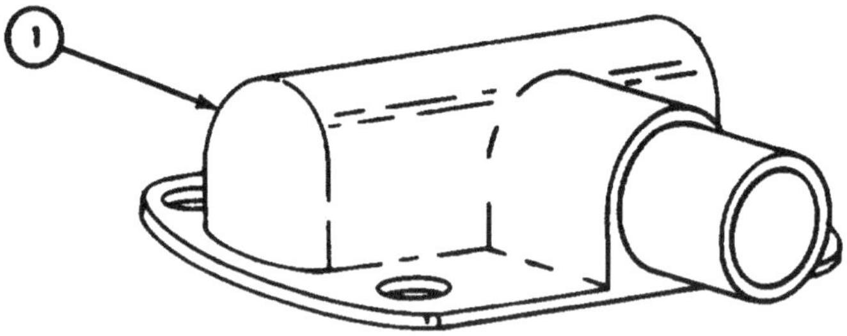

TA 113683

4-25. CRANKCASE BREATHER TUBE (ALL ENGINES EXCEPT LDS-465-2).

a. Cleaning.

FRAME 1

1. Clean rubber hose (1) using warm soapy water. Dry with clean cloth.

WARNING

Dry cleaning solvent is flammable. Do not use near an open flame. Keep a fire extinguisher nearby when solvent is used. Use only in well-ventilated places. Failure to do this may result in injury to personnel and damage to equipment.

2. Clean outside of crankcase breather tube (2) with dry cleaning solvent. Clean off sludge and gum deposits with a stiff brush. Dry with clean cloth.

WARNING

Eye shields must be worn when using compressed air. Eye injury can occur if eye shields are not used.

3. Flush inside of crankcase breather tube (2) with dry cleaning solvent. Blow dry with compressed air.

IF WORKING ON ANY ENGINE EXCEPT LDT-465-1C, GO TO FRAME 2.
IF WORKING ON ENGINE LDT-465-1C, END OF TASK

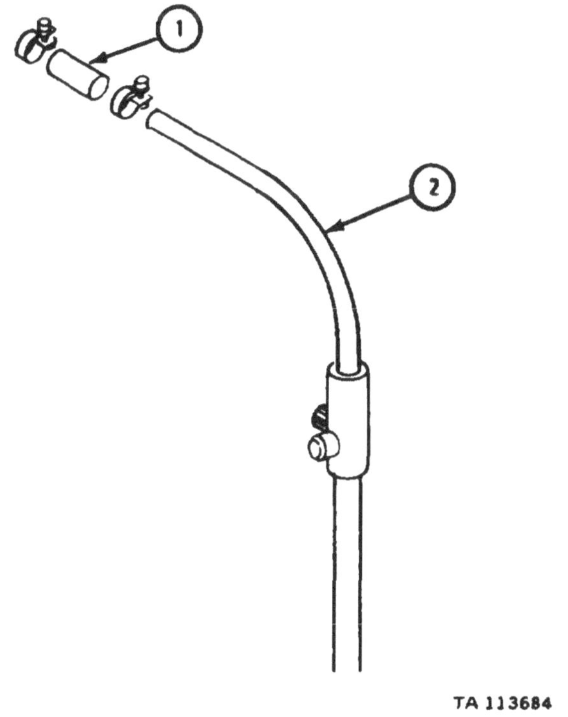

FRAME 2

WARNING

Dry cleaning solvent is flammable. Do not use near an open flame. Keep a fire extinguisher nearby when solvent is used. Use only in well-ventilated places. Failure to do this may result in injury to personnel and damage to equipment.

1. Clean two hose clamps (1), loop clamp (2), screw (3), washer (4), lockwashe r (5), and nut (6) with dry cleaning solvent. Clean off sludge and gum deposits with a stiff brush. Dry with clean cloth.

NOTE

Engines LD-465-1 and LD-465-1C use bracket (6).
Engine LDS-465-1A uses bracket (7) .

2. Clean bracket (6 or 7) with dry cleaning solvent. Clean off sludge and gum deposits with a stiff brush. Dry with clean cloth.

END OF TASK

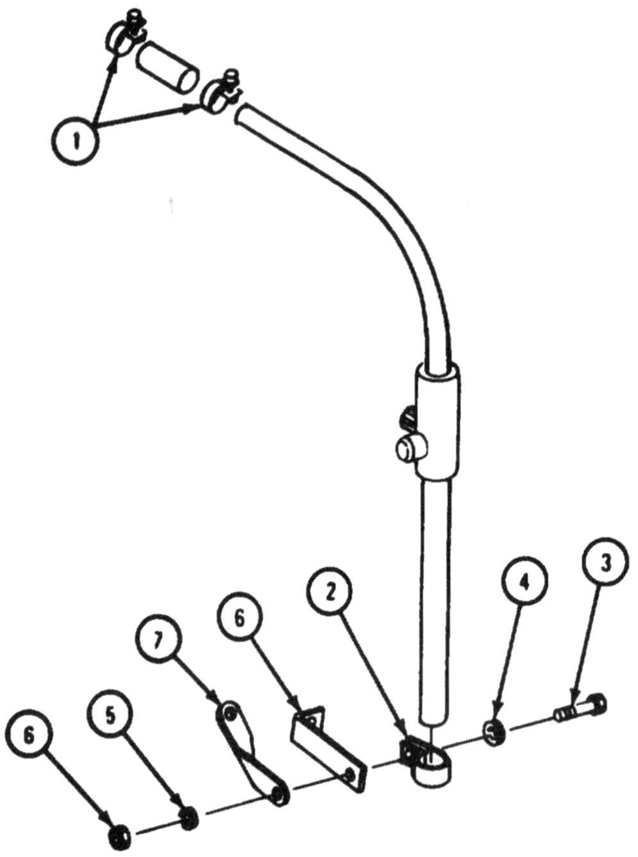

b. Inspection and Repair

FRAME 1

1. Check that all threaded parts are not stripped or damaged in any other way. If threads are damaged, get a new part.
2. Check that two hose clamps (1) and loop clamp (2) are not cracked or bent. Straighten minor bends. If more repair is needed, get a new part.

NOTE

Engines LD-465-1 and LD-465-1C use bracket (3). Engine LDS-465-1A uses bracket (4).

3. Check that bracket (3 or 4) is not cracked or bent. Straighten minor bends. If more repair is needed, get a new bracket.
4. Check that crankcase breather tube (5) is not bent, cracked, dented or damaged in any other way. If crankcase breather tube is damaged, get a new one.

END OF TASK

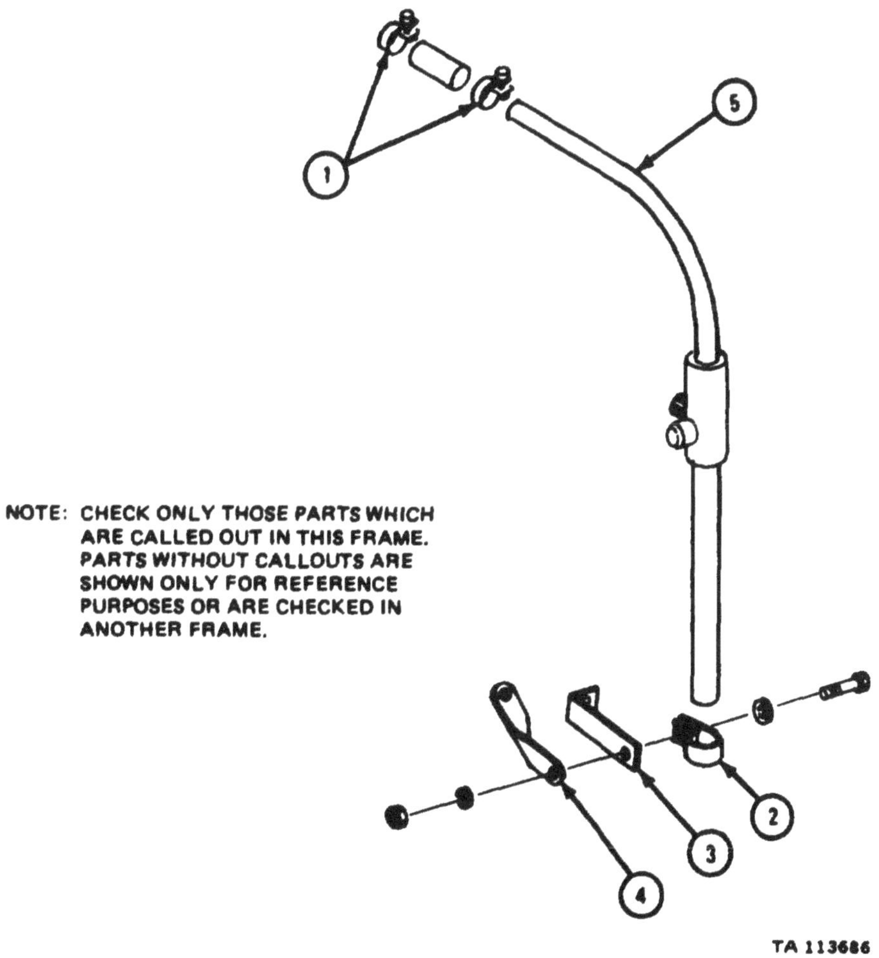
1
5
2
3
4
NOTE: CHECK ONLY THOSE PARTS WHICH ARE CALLED OUT IN THIS FRAME. PARTS WITHOUT CALLOUTS ARE SHOWN ONLY FOR REFERENCE PURPOSES OR ARE CHECKED IN ANOTHER FRAME.
TA 113686

4-26. CRANKCASE BREATHER VALVE ASSEMBLY (ENGINE LDS-465-2).

a. Disassembly.

FRAME 1

1. Take out two screws and lockwashers (1). Take off breather hose adapter (2). Take off and throw away breather tube adapter gasket (3).

2. Take out five screws and lockwashers (4) and take off breather valve spring housing (5). Take out and throw away breather valve spring (6). Take out and throw away diaphragm assembly (7).

GO TO FRAME 2

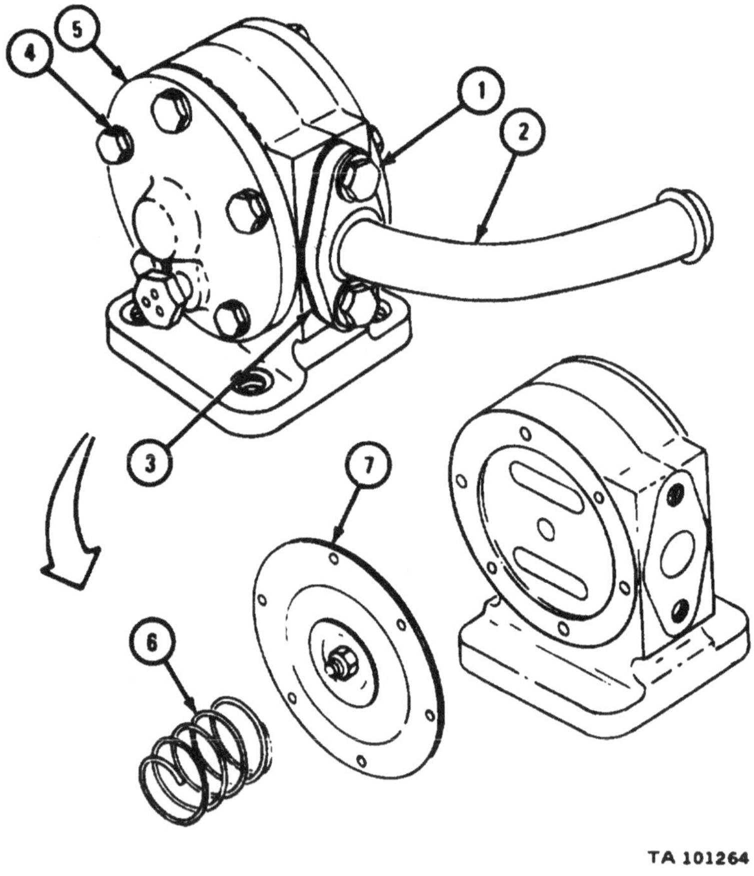

FRAME 2

1. Take out five screws (1) and lockwashers (2) and take off breather valve end cover (3) .
2. Take out and throw away diaphragm assembly (4).

END OF TASK

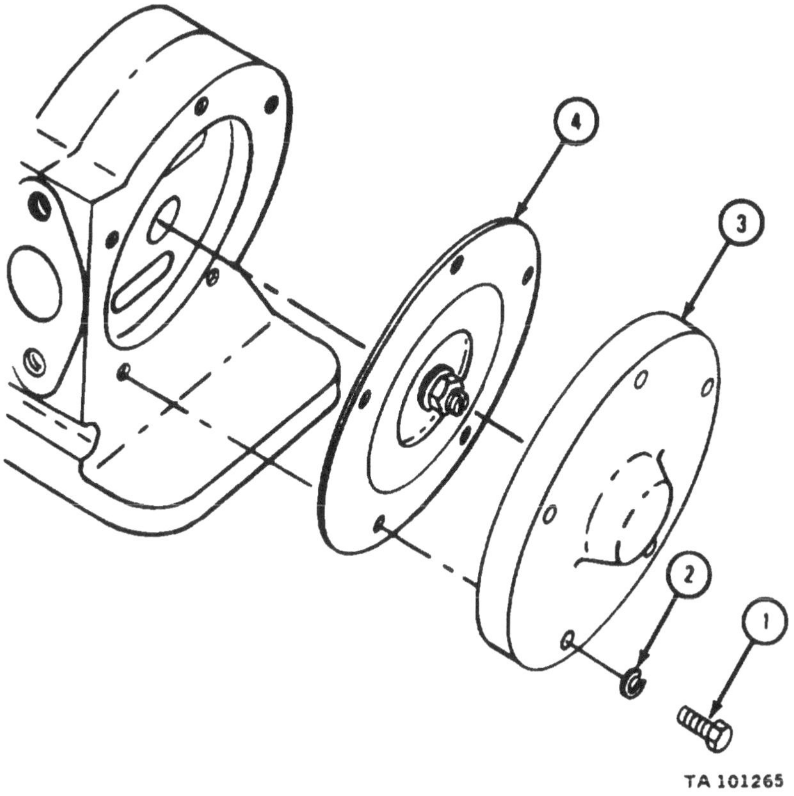

b. Cleaning.

WARNING

Dry cleaning solvent is flammable. Do not use near an open flame. Keep a fire extinguisher nearby when solvent is used. Use only in well-ventilated places . Failure to do this may result in injury to personnel and damage to equipment.

(1) Clean all parts in dry cleaning solvent .

(2) Clean off sludge and gum deposits with a stiff brush.

(3) Dry parts with clean cloth .

c. Inspection and Repair.

FRAME 1

1. Check that breather valve housing (1), breather valve spring housing (2), breather valve end cover (3), and breather tube adapter (4) have no cracks. If any of these parts is cracked, throw away crankcase breather valve assembly and get a new one.

WARNING

Dry cleaning solvent is flammable. Do not use near an open flame. Keep a fire extinguisher nearby when solvent is used. Use only in well-ventilated places. Failure to do this may result in injury to personnel and damage to equipment.

2. Check that breather valve housing (1), breather valve spring housing (2), breather valve end cover (3), and breather tube adapter (4) have no nicks, burrs or raised metal. Fix minor nicks, burrs, and raised metal with a fine mill file or a crocus cloth dipped in dry cleaning solvent.
3. Check that four nuts (5) and all screws (6) have no thread damage. Throw away damaged nuts and screws and get new ones.
4. Check all threaded holes in breather valve housing (1). Fix minor thread damage with a tap.

GO TO FRAME 2

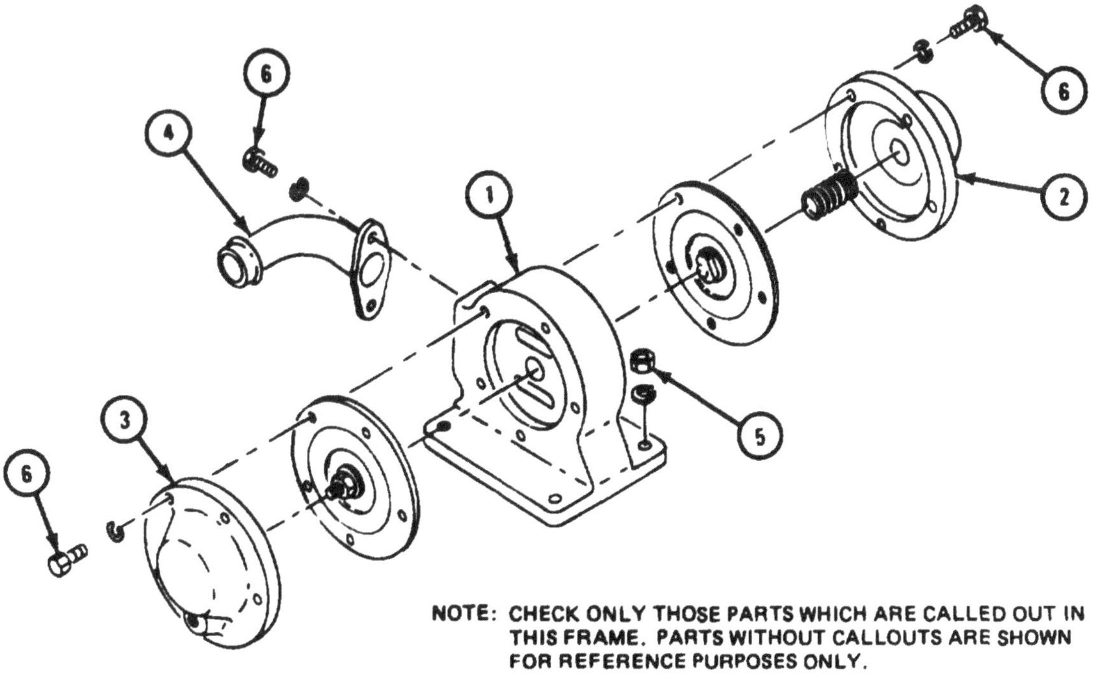

NOTE: CHECK ONLY THOSE PARTS WHICH ARE CALLED OUT IN THIS FRAME. PARTS WITHOUT CALLOUTS ARE SHOWN FOR REFERENCE PURPOSES ONLY.

TA 101266

FRAME 2

1. Check valve seat (1) for proper valve seating. If valve seats properly, an even wear ring (2) will be seen as shown. If valve does not seat properly, wear ring (2) will be broken in places as shown.

NOTE

There are two valve seats (1) in crankcase breather valve housing (3), one on each side of housing.

2. Do step 1 again for valve seat on other side of crankcase breather valve housing (3) .

3. If either valve seat (1) is bad, get a new crankcase breather valve assembly.

END OF TASK

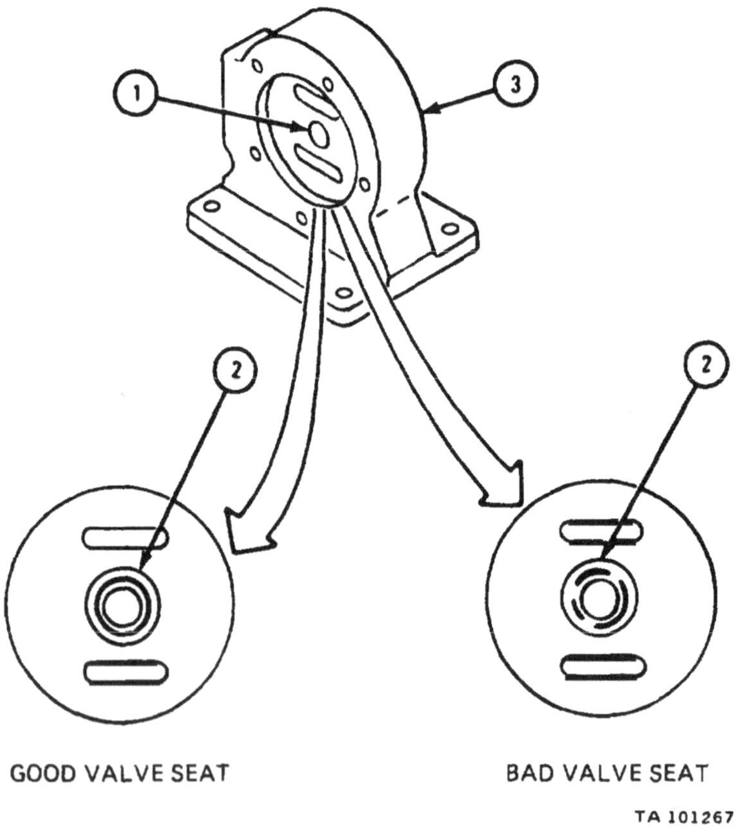

d. Assembly.

FRAME 1

NOTE

One of five holes (1) in diaphragm (2) and one of five screw holes (3) in crankcase breather valve housing (4) are offset . Diaphragm can be put on valve housing only one way.

1. Line up five holes (1) in new diaphragm (2) with five screw holes (3) in crank-case breather valve housing (4). Note which side of diaphragm must face ou t so that holes will line up.

NOTE

Each diaphragm (2) comes with two diaphragm support disks (5) . Both diaphragm support disks are the same.

2. Put diaphragm needle (6) through one diaphragm support disk (5). Curved-in side of support disk must face away from diaphragm as shown.
3. Put diaphragm needle (6) with diaphragm support disk (5) through hole in center of diaphragm (2). Put diaphragm needle in from side of diaphragm which will face in toward valve housing (4).
4. Slide outer diaphragm support disk (5) over end of diaphragm needle (6). Curved-in side of diaphragm support disk must face away from diaphragm (2) as shown.
5. Put on locknut (7) .
6. Do steps 1 through 5 again for other diaphragm (8).

GO TO FRAME 2

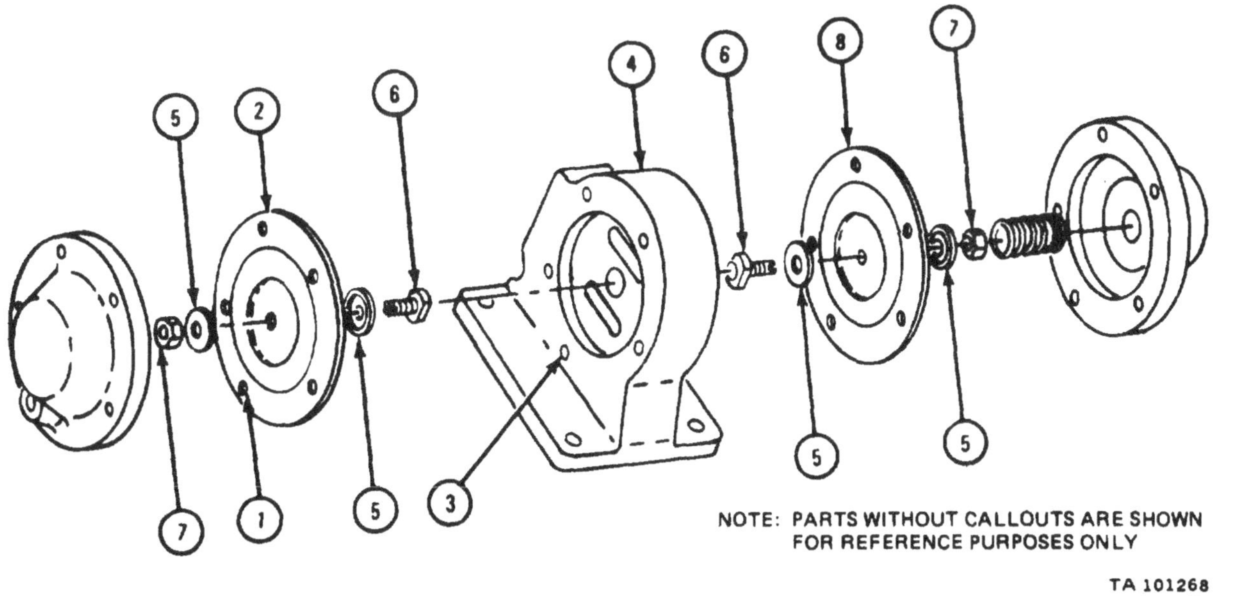

FRAME 2

1. Lay breather valve housing (1) flat on workbench as shown.
2. Put diaphragm assembly (2) on breather valve housing (1), alining holes. Threaded end of diaphragm needle (3) must face up. Valve end of diaphragm needle must rest in valve seat (4).
3. Tap threaded end of diaphragm needle (3) lightly with a short brass hammer to seat diaphragm needle to valve seat (4).
4. Put on diaphragm end cover (5) alining holes. Put in and tighten five screws (6) and lockwashers (7) .

GO TO FRAME 3

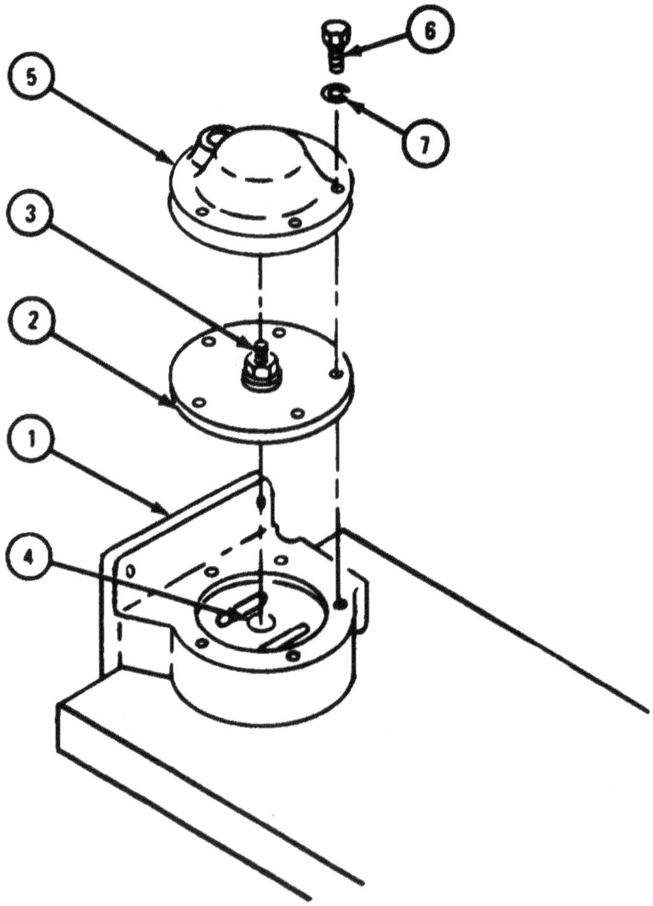

TA 101269

FRAME 3

1. Turn over breather valve housing (1) and lay it flat on workbench as shown.
2. Put diaphragm assembly (2) on breather valve housing (1), alining holes. Threaded end of diaphragm needle (3) must face up. Valve end of diaphragm must rest in valve seat (4).
3. Tap threaded end of diaphragm needle (3) lightly with a short brass hammer to seat diaphragm needle to valve seat (4).
4. Put spring (5) over threaded end of needle valve (3).
5. Put breather valve spring housing (6) on breather valve housing (1), alining holes. Put in five screw; (7) and lockwashers (8) .

GO TO FRAME 4

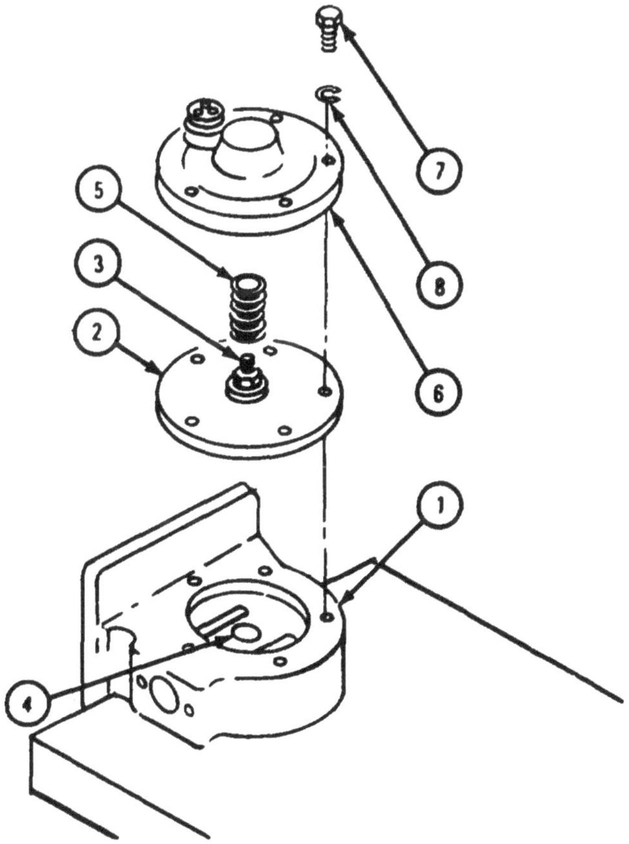

TA 101270

FRAME 4

1. Put gasket (1) on breather valve housing (2), alining holes.
2. Hold tube (3) against gasket (1) and valve housing (2), alining holes.
3. Put in two screws (4) and lock washers (5).

END OF TASK

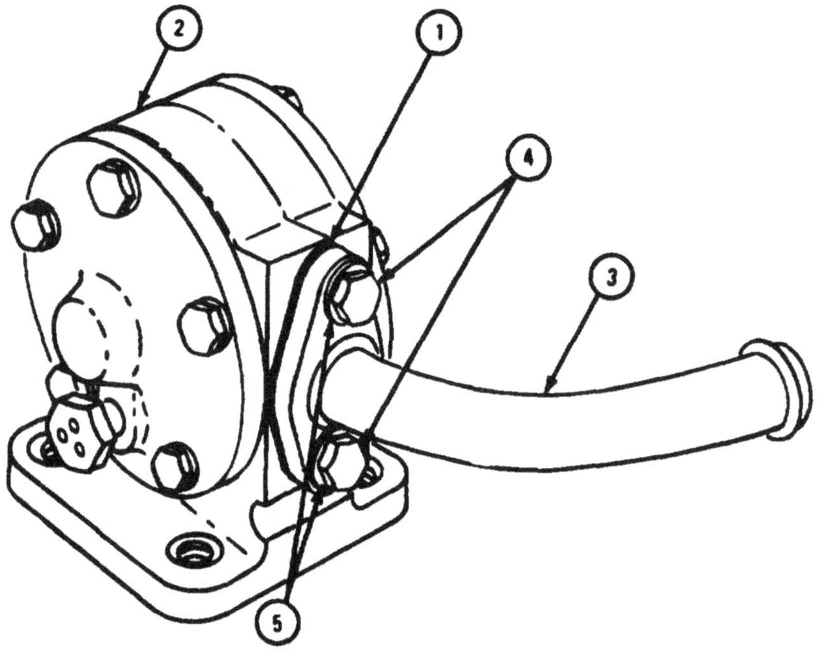

4-27. AIR PRESSURIZATION TUBES AND FITTINGS (ENGINE LDS-465-2).

a. Cleaning.

FRAME 1

WARNING

Dry cleaning solvent is flammable. Do not use near an open flame. Keep a fire extinguisher nearby when solvent is used. Use only in well-ventilated places. Failure to do this may result in injury to personnel and damage to equipment.

Eye shields must be worn when using compressed air. Eye injury can occur if eye shields are not used.

CAUTION

Air pressurization tubes (1, 2 and 3) are copper tubes covered with a nonmetallic sheath. Generator air inlet tube assembly (4) is made of rubber. Do not clean rubber generator air inlet tube assembly with dry cleaning solvent or it will dry out and crack.

1. Clean inside of air pressurization tubes (1, 2 and 3) with brass wire probes. Flush inside of air pressurization tubes (1, 2 and 3) with dry cleaning solvent. Blow dry with compressed air.

2. Clean rubber generator air inlet tube assembly (4) with soapy water. Blow dry inside of tube with compressed air.

GO TO FRAME 2

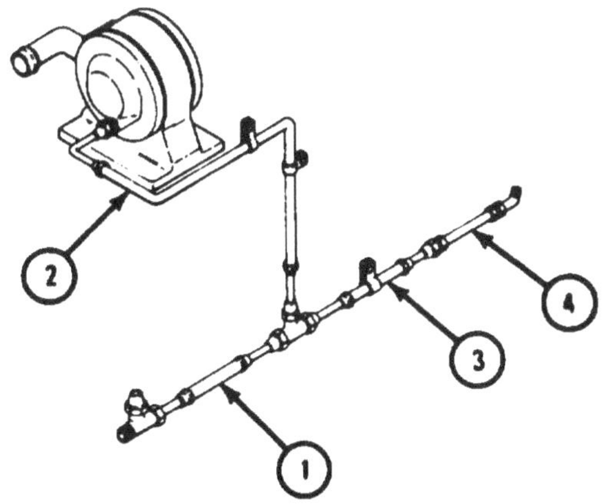

TA 121224

FRAME 2

1. Clean outside of air pressurization tubes (1, 2 and 3) with soapy water. Let tubes air dry .

WARNING

Dry cleaning solvent is flammable. Do not use near an open flame. Keep a fire extinguisher nearby when solvent is used. Use only in well-ventilated places. Failure to do this may result in injury to personnel and damage to equipment.

Eye shields must be worn when using compressed air. Eye injury can occur if eye shields are not used.

2. Clean tee connectors (4 and 5) with dry cleaning solvent. Blow dry with compressed air .

END OF TASK

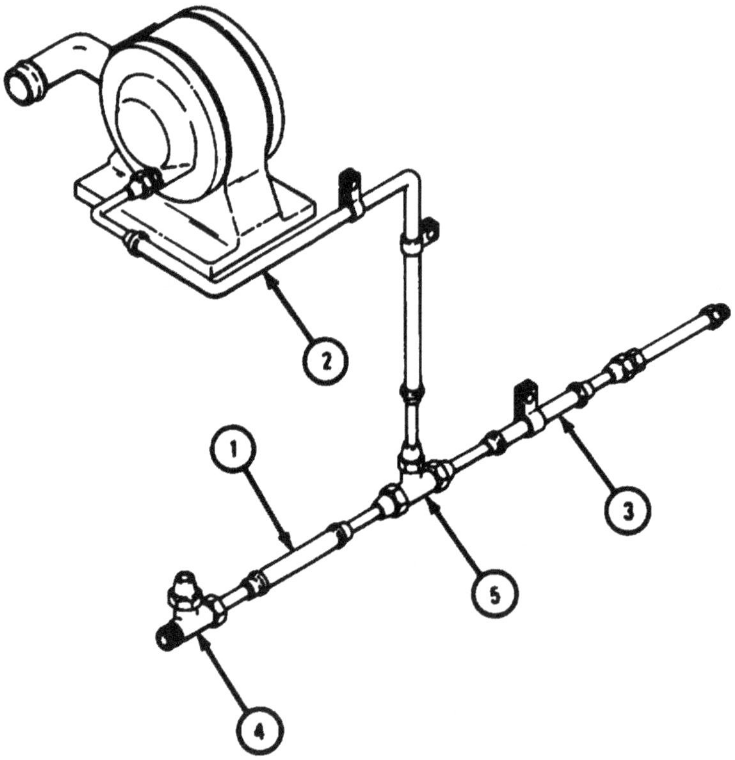

TA 113690

b . Inspection and Repair.

FRAME 1

NOTE

Air pressurization tubes (1, 2, and 3) are not stocked. If you need a new air pressurization tube, cut one the same length as the old tube from seamless copper tube stock. Cut new non-metallic sheath from flexible non-metallic tube stock.

1. Check that six tube nuts (4), two tube connectors (5), and two tee fittings (6 and 7) have no damaged threads. Fix minor thread damage with thread chaser or tap. If more repair is needed, get a new tube or tee fitting.
2. Check that air pressurization tubes (1, 2 and 3) are not cracked, dented or kinked. Check that non-metallic sheaths of air pressurization tubes are not torn, worn away or split. If tubes are damaged, make new ones.
3. Check that rubber generator air inlet tube assembly (8) is not dried out, cracked or split . If tube assembly is damaged, get a new one.

END OF TASK

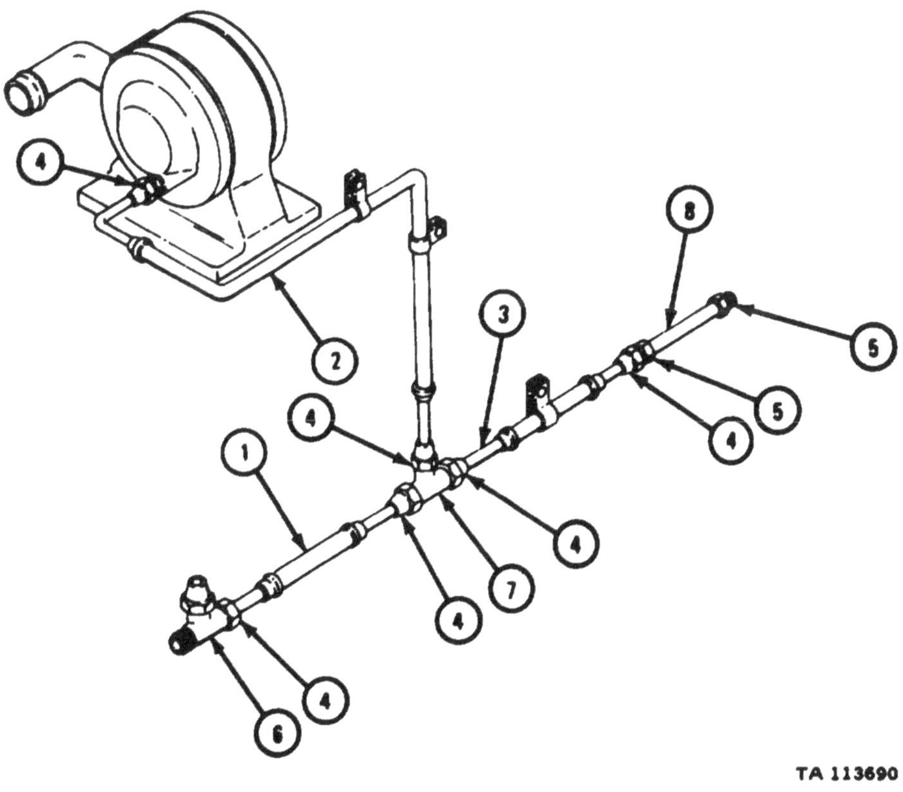

4-28. FLYWHEEL HOUSING.

a. Engine LDS-465-2.

(1) Disassembly .

FRAME 1

1. Take pipe plug (1) and gasket (2) out of flywheel housing (3). Throw away gasket.
2. Take off six screws and washers (4).
3. Takeoff cover (5) and gasket (6). Throw away gasket .
4. Takeoff fitting (7) .

GO TO FRAME 2

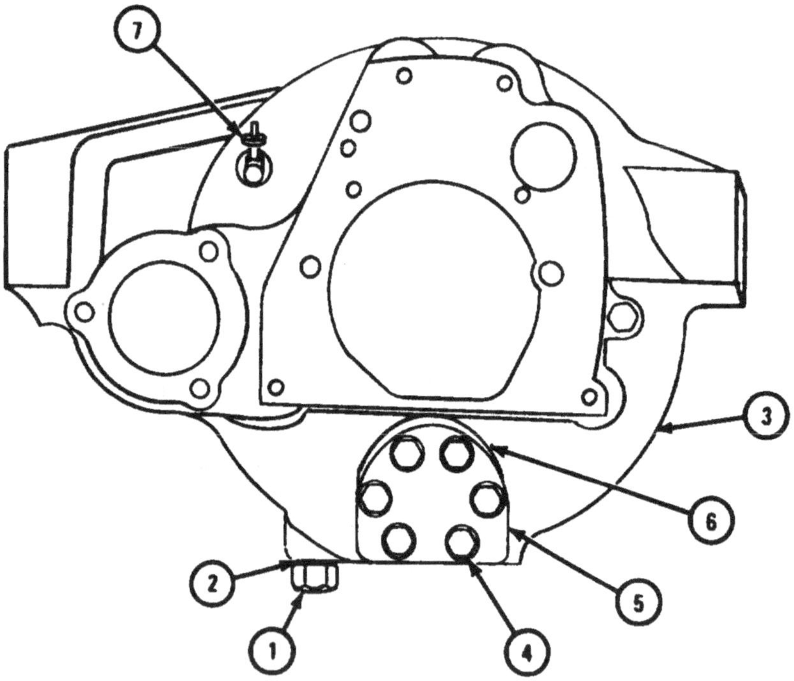

TA 113688

FRAME 2

NOTE

Do not take out three studs (1), six screw thread inserts (3), eight screw thread inserts (4) or 12 screw thread inserts (5) unless they are damaged. Refer to para 4-28a (3) for inspection procedures.

1. Take three studs (1) out of flywheel housing (2).
2. Using screw thread extractor, take out six screw thread inserts (3).
3. Using screw thread extractor, take out eight screw thread inserts (4), four from each side of housing.
4. Using screw thread extractor, take out 12 screw thread inserts (5).

END OF TASK

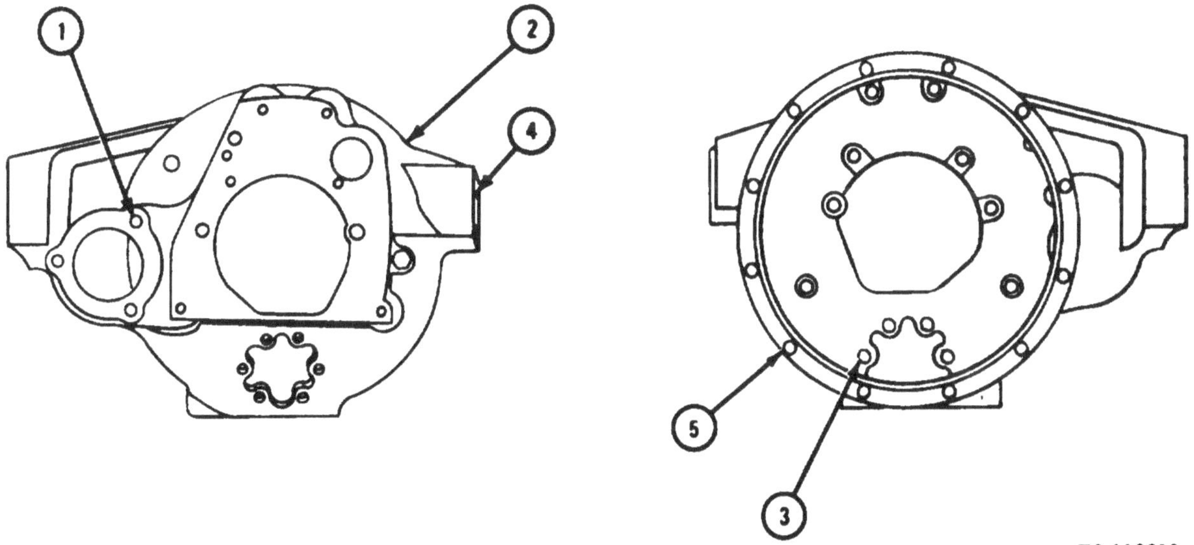

TA 113689

(2) Cleaning.

FRAME 1

WARNING

Dry cleaning solvent is flammable. Do not use near an open flame. Keep a fire extinguisher nearby when solvent is used. Use only in well-ventilated places. Failure to do this may result in injury to personnel and damage to equipment.

1. Clean flywheel housing (1) and cover (2) with dry cleaning solvent. Dry with clean cloth.
2. Clean pipe plug (3), screws (4), studs (5), and inserts (6) with dry cleaning solvent. Let parts air dry.
3. Clean fitting (7) with dry cleaning solvent and wire probe.

END OF TASK

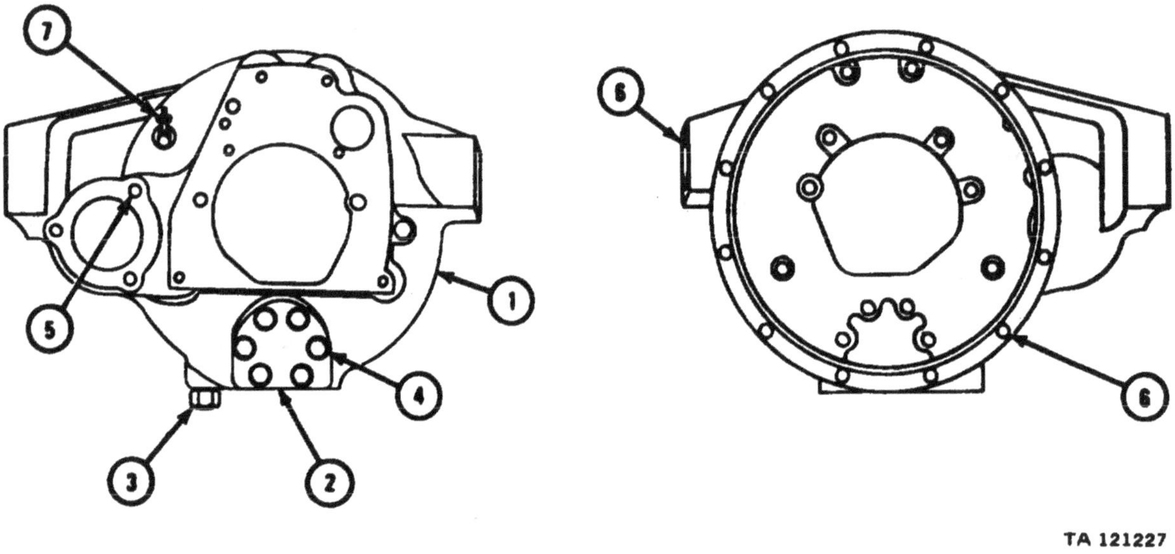

(3) Inspection and repair.

FRAME 1

1. Check that all threaded parts are not stripped or damaged. Repair minor thread damage with a thread chaser or tap. If more repair is needed, get new parts.
2. Check that flywheel housing (1) and cover (2) have no cracks, nicks, burrs or scratches on gasket surfaces. Take off small nicks, burrs or scratches with a fine mill file. If more repair is needed, get a new part.
3. Check fitting (3) for damage. If fitting is damaged, get a new one.

END OF TASK

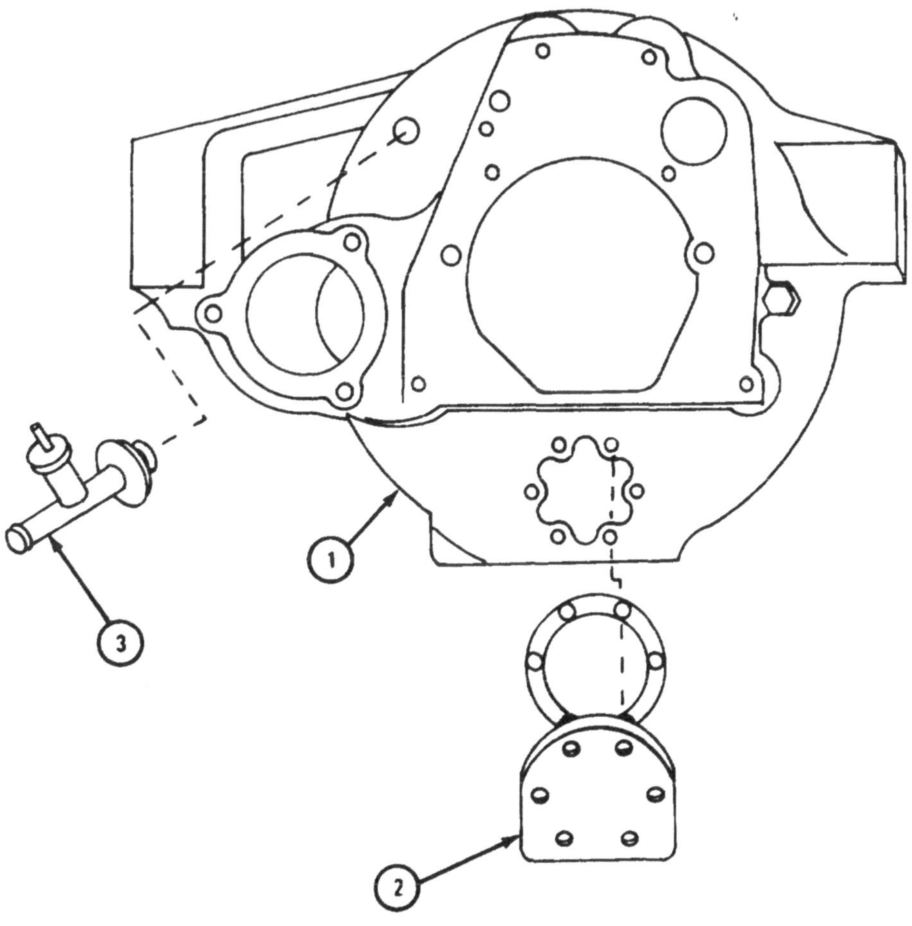

NOTE: CHECK ONLY THOSE PARTS WHICH ARE CALLED OUT. PARTS WITHOUT CALLOUTS ARE SHOWN ONLY FOR REFERENCE PURPOSES.

TA 121226

(4) Assembly.

FRAME 1

NOTE

If any of three studs (1), six screw thread inserts (3), eight screw thread inserts (4) or 12 screw thread inserts (5) were taken out, do steps 1 through 4. If they were not taken out, go to frame 2.

1. Put three studs (1) into flywheel housing (2).
2. Using screw thread inserter, put in six screw thread inserts (3).
3. Using screw thread inserter, put in eight screw thread inserts (4).
4. Using screw thread inserter, put in 12 thread inserts (5).

GO TO FRAME 2

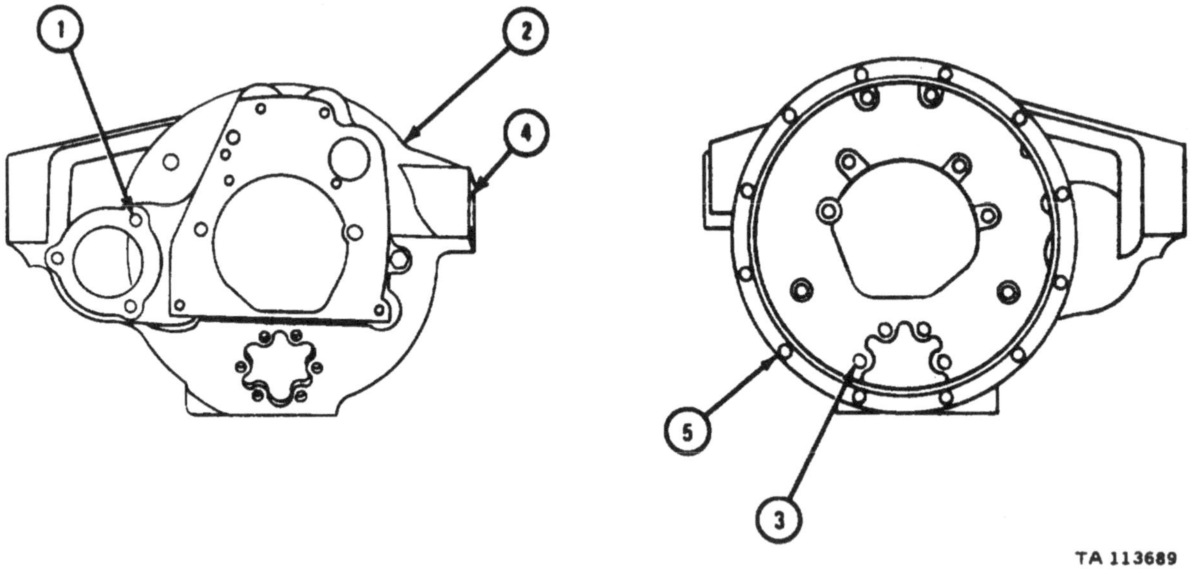

FRAME 2

1. Put gasket (1) on pipe plug (2) and put pipe plug into bottom of flywheel housing (3) .
2. Put gasket (4) on flywheel housing (3) and aline holes.
3. Put cover (5) on gasket (4) and flywheel housing (3). Aline holes and put in six screws (6) and washers (7) .
4. Put in fitting (8).

END OF TASK

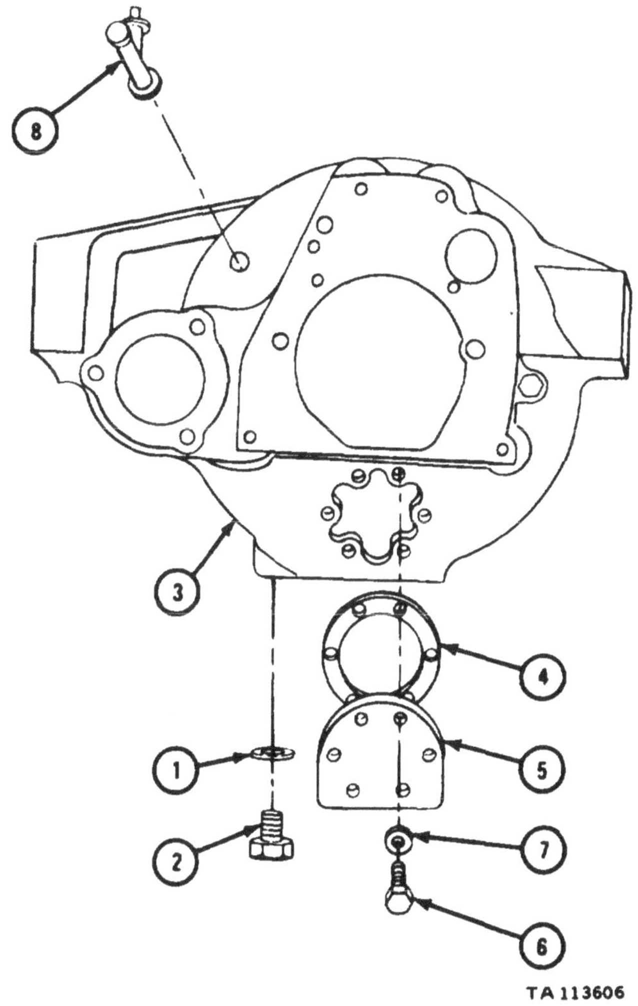

b. Engines LDS-465-1 and LDS-465-1A.

(1) Disassembly.

FRAME 1

1. Take pipe plug (1) out of flywheel housing (2).

NOTE

Do not take out three studs (3) unless they are damaged. Refer to para 4-28b (3) for inspection procedures.

2. Take out three studs (3).

GO TO FRAME 2

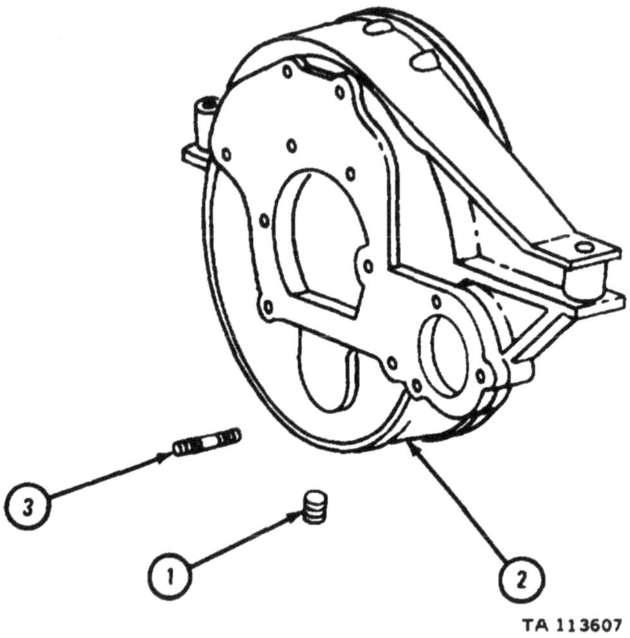

FRAME 2

1. Take pipe plug (1) from flywheel housing (2).

NOTE

Do not take out 12 screw thread inserts (3) unless they are damaged. Refer to para 4-28b (3) for inspection procedures.

2. Using screw thread extractor, take 12 screw thread inserts (3) out of flywheel housing (2).

END OF TASK

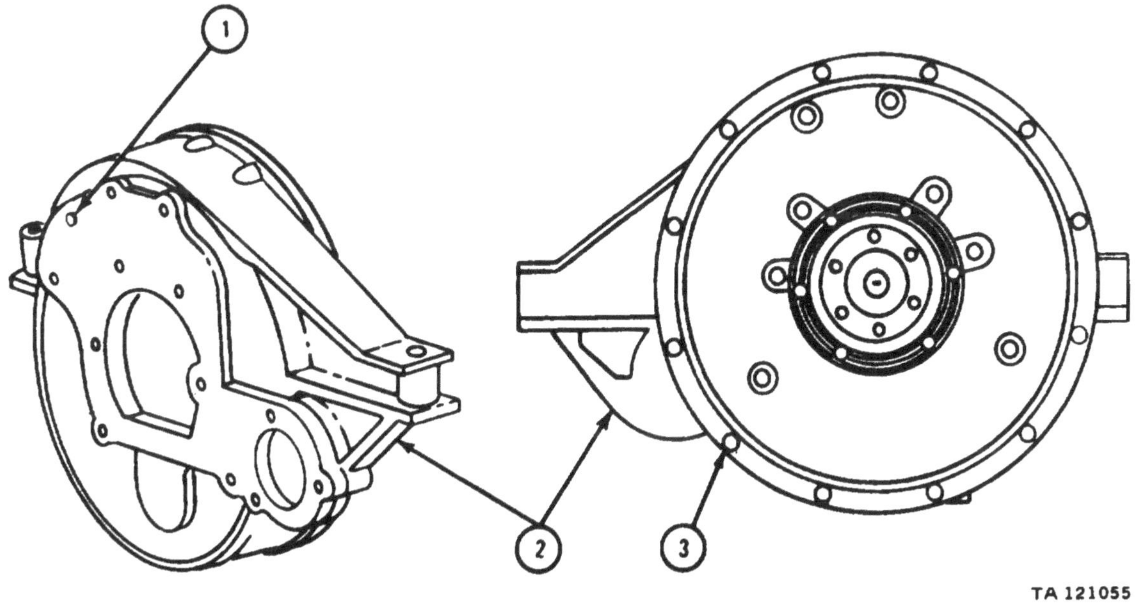

(2) Cleaning .

FRAME 1

WARNING

Dry cleaning solvent is flammable. Do not use near an open flame. Keep a fire extinguisher nearby when solvent is used. Use only in well-ventilated places. Failure to do this may result in injury to personnel and damage to equipment.

1. Clean flywheel housing (1) with dry cleaning solvent. Dry with clean cloth.
2. Clean threaded fittings with dry cleaning solvent. Let fittings air dry.

END OF TASK

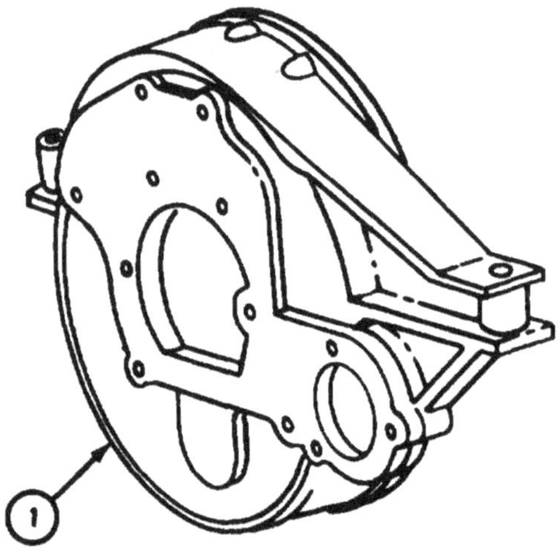

TA121052

(3) Inspection and repair.

FRAME 1

1. Check that three studs (1), two pipe plugs (2), and 12 screw thread inserts (3) are not stripped, crossthreaded or damaged in any other way. If parts are damaged, get a new part.
2. Check that flywheel housing (4) is not cracked, bent or has minor nicks, burrs or scratches on gasket surface (5). Take off small nicks, burrs or scratches with a fine mill file or crocus cloth. If more repair is needed, get a new flywheel housing.

END OF TASK

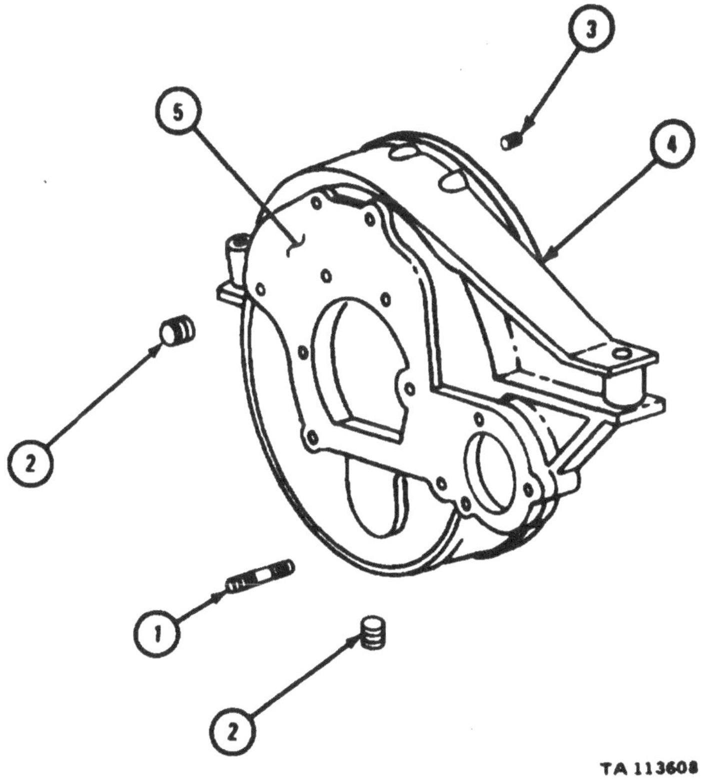

(4) Assembly .

FRAME 1

1. Put two pipe plugs (1) into flywheel housing (2).

NOTE

If any of three studs (3) or 12 screw thread inserts (4) were taken out, do steps 2 and 3. If they were not taken out, skip steps 2 and 3.

2. Put in three studs (3).

3. Using screw thread inserter, put in 12 screw thread inserts (4).

END OF TASK

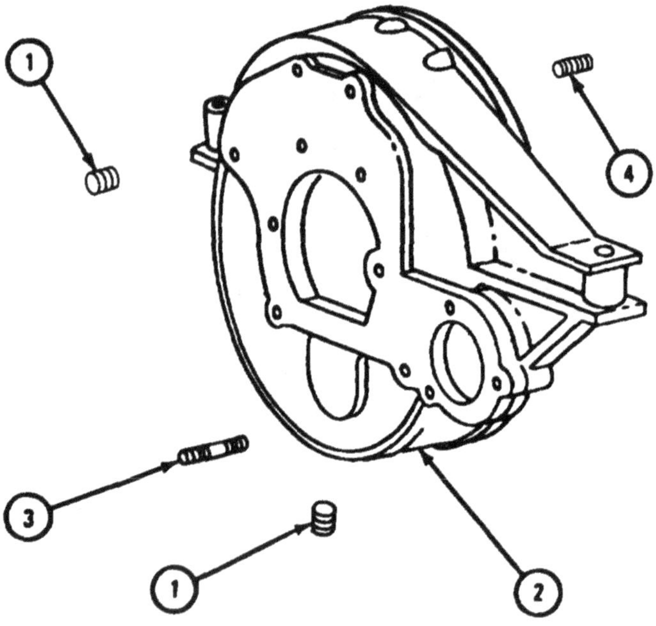

TA 113607

c. Engines LD-465-1, LD-465-1C, and LDT-465-1C.

(1) Disassembly.

FRAME 1

1. Take two pipe plugs (1) out of flywheel housing (2).

NOTE

Do not takeout three studs (3), eight studs (4), or 13 studs (5) unless they are damaged. Refer to para 4-28c (3) for inspection procedures.

2. Take out three studs (3).
3. Take out four studs (4) from each side of housing.
4. Take out 13 studs (5).

END OF TASK

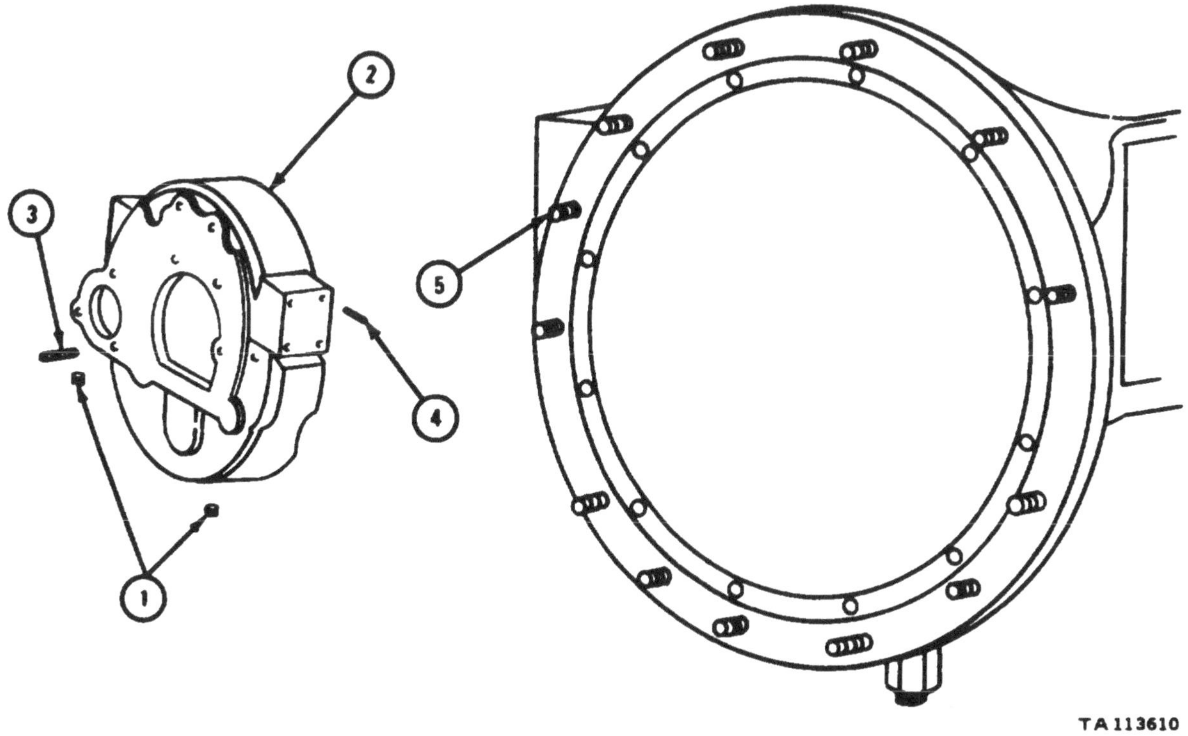

(2) Cleaning .

FRAME 1

WARNING

Dry cleaning solvent is flammable. Do not use near an open flame. Keep a fire extinguisher nearby when solvent is used. Use only in well-ventilated places. Failure to do this may result in injury to personnel and damage to equipment.

1. Clean flywheel housing (1) using dry cleaning solvent. Dry with clean cloth.
2. Clean threaded fittings with dry cleaning solvent. Let fittings air dry.

END OF TASK

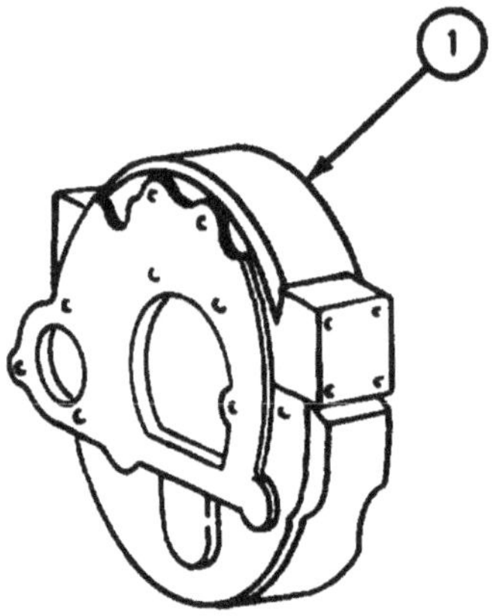

TA 121053

(3) Inspection and repair .

FRAME 1

1. Check that 13 studs (1), eight studs (2), three studs (3), and two pipe plugs (4) are not stripped or damaged in any other way. If parts are damaged, get a new part.
2. Check that flywheel housing (5) is not cracked, bent or has minor nicks, burrs or scratches on gasket surfaces . Take off small nicks, burrs or scratches with a fine mill file or crocus cloth. If more repair is needed, get a new flywheel housing .

END OF TASK

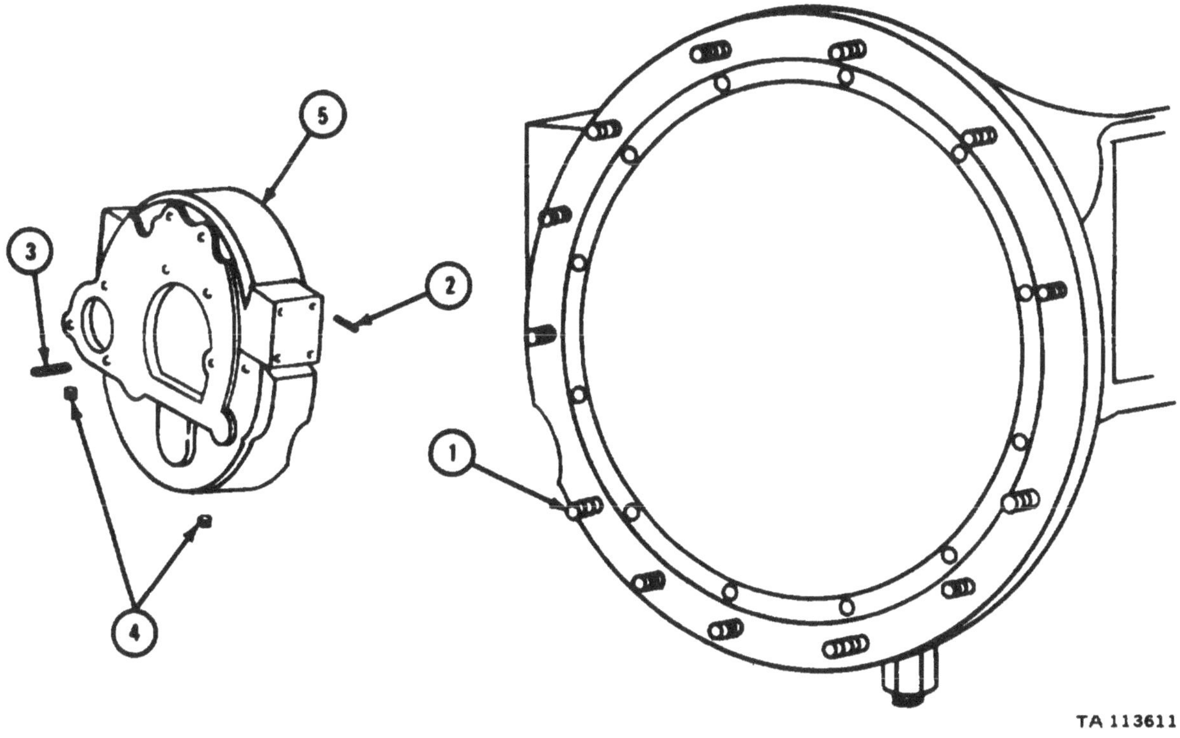

(4) Assembly .

FRAME 1

1. Put two pipe plugs (1) into flywheel housing (2).

NOTE

If any of three studs (3), eight studs (4), or 13 studs (5) were taken out, do steps 2, 3, and 4.

2. Put in three studs (3) .
3. Put in four studs (4) on each side of housing (2).
4. Put in 13 studs (5).

END OF TASK

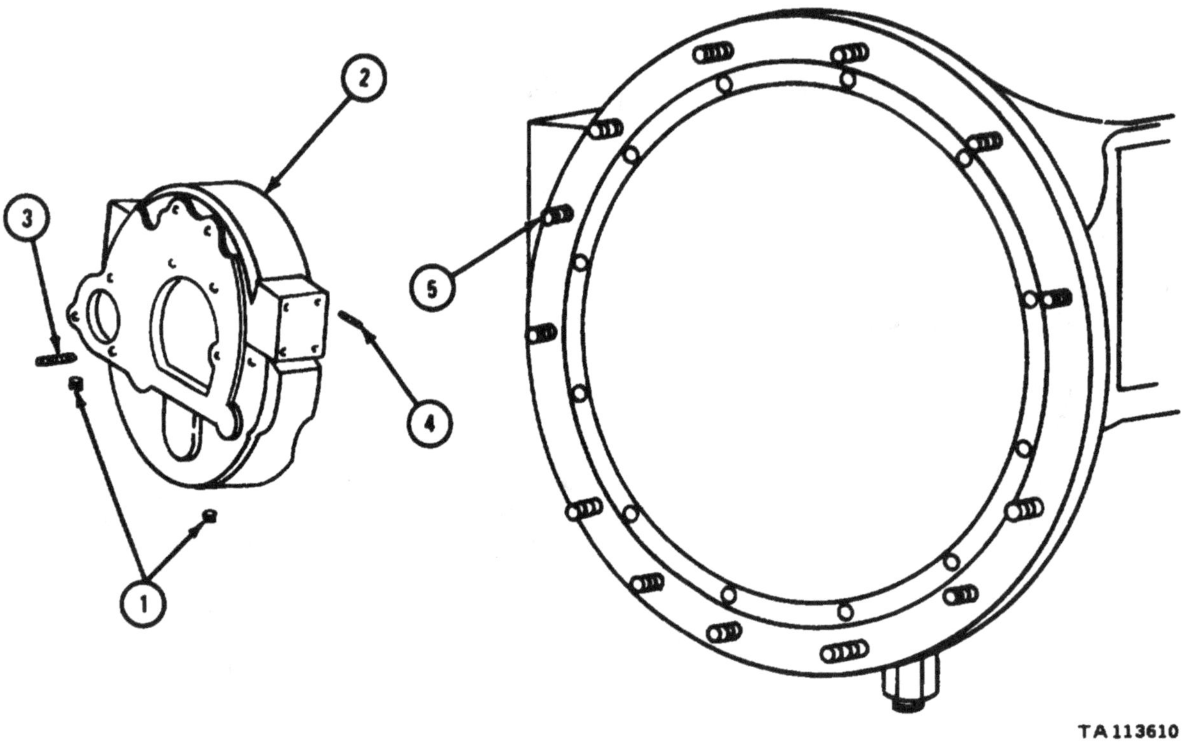

4-29. FLYWHEEL ASSEMBLY.

a. Cleaning. There are no special cleaning procedures needed. Refer to cleaning procedures given in para 4-3.

b. Inspection and Repair.

FRAME 1

1. Check that clutch face (1) of flywheel (2) has no grooving, scuffing, evidence of overheating , warping or cracks .
2. Using straight edge and 0.0040 feeler gage, check flatness of clutch face (1) .

WARNING

Dry cleaning solvent is flammable. Do not use near an open flame. Keep a fire extinguisher nearby when solvent is used. Us-e only in well-ventilated places. Failure to do this may result in injury to personnel and damage to equipment.

3. Fix minor damage to clutch face (1) using crocus cloth dipped in dry cleaning solvent. [f more repair is needed, have clutch face machined or get a new one.

GO TO FRAME 2

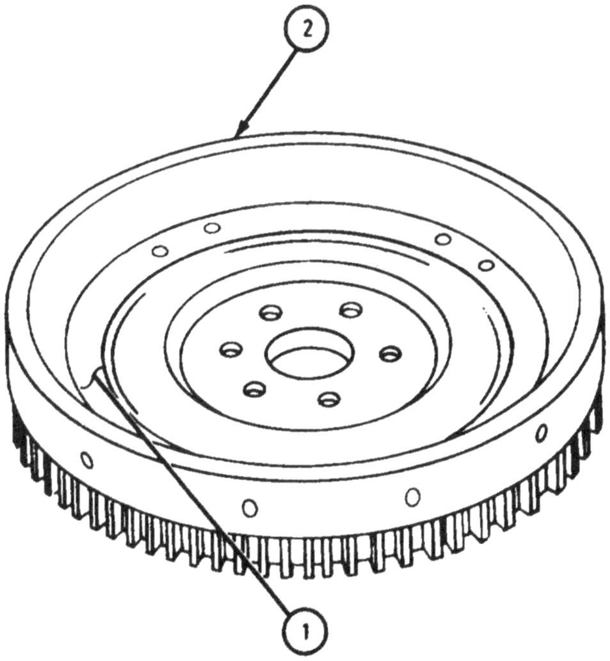

TA 117989

FRAME 2

1. Check that ring gear (1) of flywheel (2) has no wear cracks or damaged or broken teeth . Fix minor damage using a fine mill file. If more repair is needed, put on a new ring gear. Refer to frame 3.
2. Check that pilot bearing (3) is not damaged in any way. Refer to TM 9-214.
3. Check that clutch mounting holes (4) do not have stripped threads or other damage. Fix damaged threads using a thread tap. If damage cannot be fixed, fet a new flywheel.

IF RING GEAR IS DAMAGED. GO TO FRAME 3.
IF RING GEAR IS NOT DAMAGED. END OF TASK

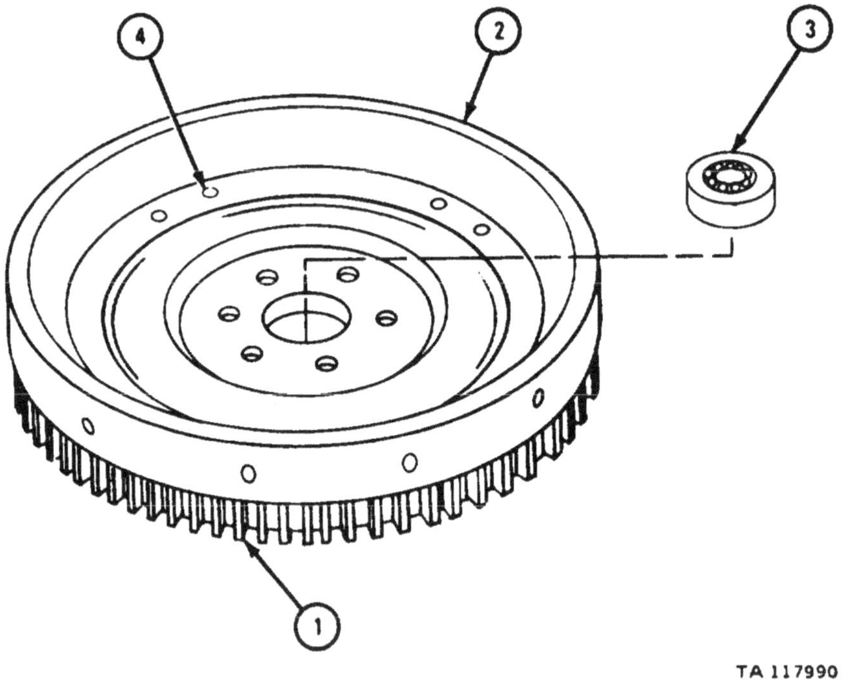

FRAME 2

1. Using hammer and blunt chisel, drive damaged ring gear (1) from flywheel (2).

WARNING

Use welder's gloves when handling hot ring gear (3).

CAUTION

Do not use cutting torch to heat ring gear (3). Use heating torch to avoid damage to gear. Do not heat ring gear to more than 400°F. Temperatures above 400°F will soften gear. Use Tempilstick to measure heat.

2. Using heating torch, heat inner surface of new ring gear (3) until 400°F. Tempilstick melts when touched to side of new ring gear.
3. Using hammer and brass drift, quickly drive heated ring gear (3) onto flywheel (2) .

END OF TASK

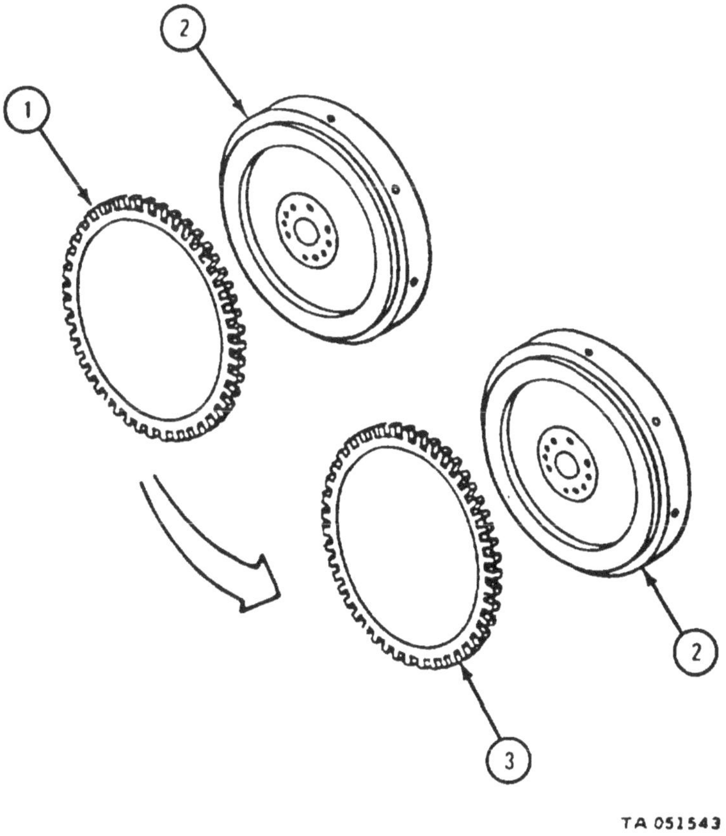

4-30. OIL FILTER BODY ASSEMBLY.

a. Disassembly.

FRAME 1

NOTE

Task is the same for both oil filter body assemblies. Some filters use cotter pin through center post to hold spring (2) and spring cup (3). If used, i t must be taken out.

1. Take out and throw away cotter pin (1), if used.
2. Slide spring (2) and spring cup (3) off post (4) .
3. Take post (4) out of filter body (5) .
4. Take gasket (6) off post (4) and throw away gasket.

END OF TASK

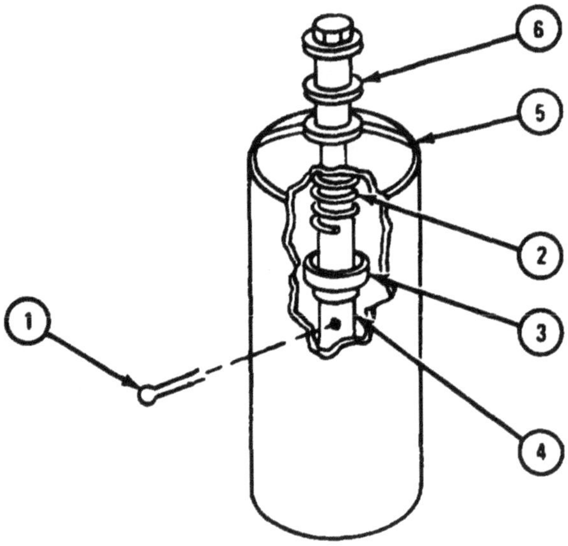

b. Cleaning. There are no special cleaning procedures needed. Refer to cleaning procedure given in para 4-3.

c. Inspection and Repair.

FRAME 1

WARNING

Dry cleaning solvent is flammable. Do not use near open flame. Keep fire extinguisher nearby when solvent is used. Use only in well-ventilated places. Failure to do this may result in injury to personnel and damage to equipment.

1. Check to see that filter body (1) is not damaged. Repair any minor damage with fine mill file or crocus cloth dipped in dry cleaning solvent. If filter body is damaged too much to be repaired, get a new one.
2. Check to see that post (2) is not damaged. Repair minor damage with fine mill file, thread chaser, or crocus cloth dipped in dry cleaning solvent. If post is damaged too much to be repaired, get a new one.
3. Check that spring (3) and spring cup (4) are not damaged. Get a new one if either one is damaged.

END OF TASK

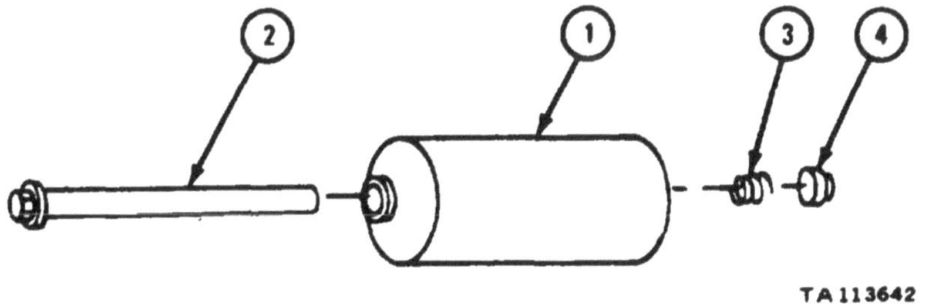

d. Assembly .

FRAME 1

1. Slide gasket (1) on post (2).
2. Put post (2) in through top of filter body (3).
3. Slide spring (4) and spring cup (5) on post (2) .
4. Put in cotter pin (6), if used.

END OF TASK

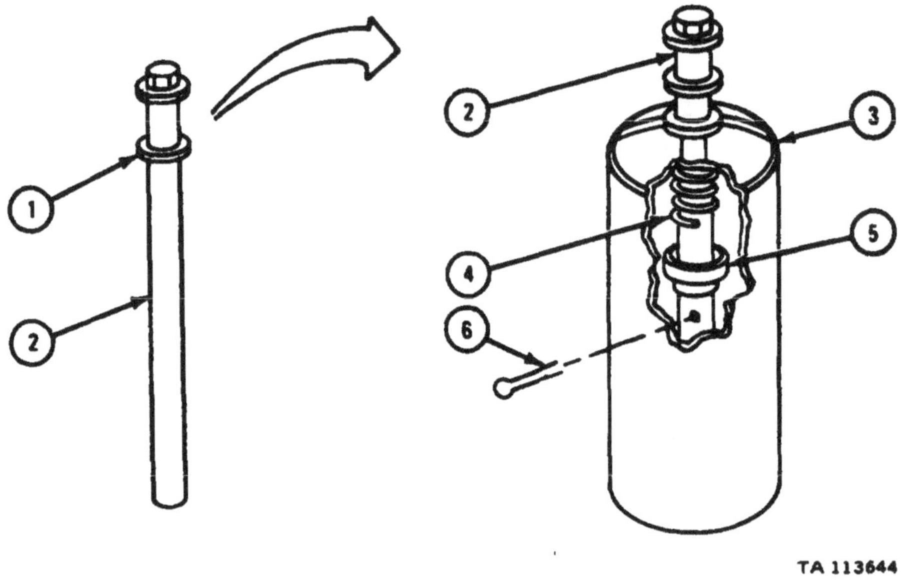

4-31. OIL COOLER AND FILTER HOUSING.

a. Disassembly.

FRAME 1

(1) Take out two plugs (1), two gaskets (2), two springs (3), and two plungers (4). Throw away two gaskets .

(2) Take out two pipe plugs (5) and square-head pipe plug (6).

END OF TASK

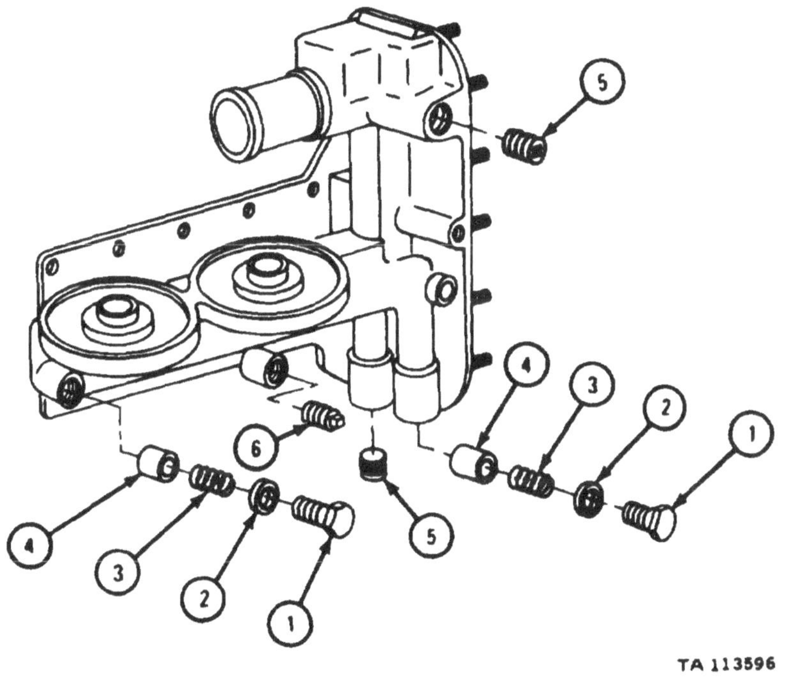

b. Cleaning.

FRAME 1

WARNING

Dry cleaning solvent is flammable. Do not use near an open flame. Keep a fire extinguisher nearby when solvent is used . Use only in well-ventilated places. Failure to do this may result in injury to personnel and damage to equipment.

1. Clean all parts in dry cleaning solvent. Clean off sludge and gum deposits with a stiff brush. Dry oil cooler and filter housing (1) with clean cloth.

WARNING

Eye shields must be worn when using compressed air. Eye injury can occur if eye shields are not used.

2. Clean oil passages in oil cooler and filter housing (1) with dry cleaning solvent. Blow dry with compressed air.

END OF TASK

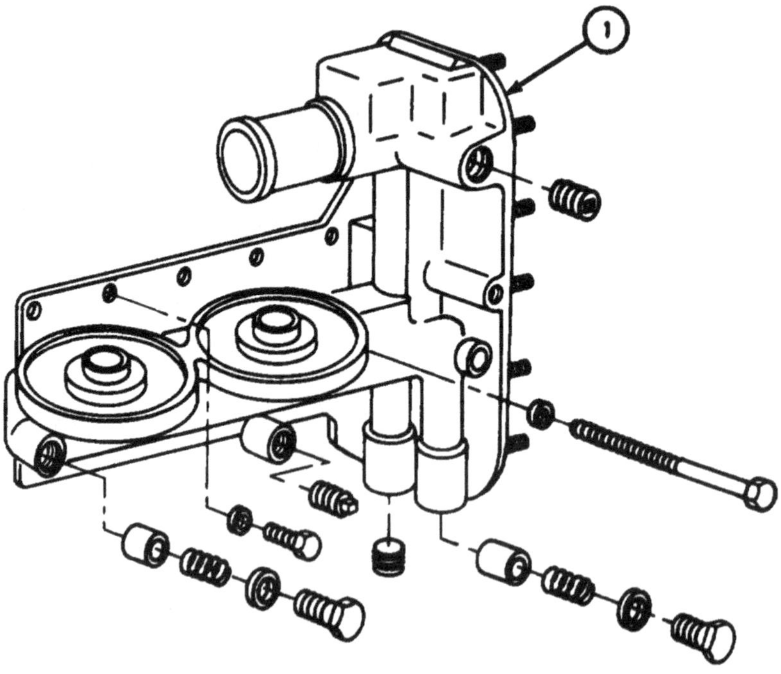

c. Inspection and Repair.

FRAME 1

NOTE

Engine LDS-465-2 uses aluminum oil cooler and filter housing (1) . All other engines use cast iron oil cooler and filter housings.

1. For all engines except LDS-465-2, use strong light and magnifying glass to check that oil cooler and filter housing (1) is not cracked. For engine LDS-465-2, check that oil cooler and filter housing is not cracked using dye penetrant kit .
2. If oil cooler and filter housing (1) is cracked, throw it away and get a new one.

WARNING

Dry cleaning solvent is flammable. Do not use near an open flame. Keep a fire extinguisher nearby when solvent is used. Use only in well-ventilated places. Failure to do this may result in injury to personnel and damage to equipment.

3. Check that oil cooler and filter housing (1) has no nicks or burrs. Fix small nicks or burrs with a fine mill file or crocus cloth dipped in dry cleaning solvent.
4. Check that 12 studs (2) and stud (3) do not have stripped or damaged threads and are not bent or loose. Take out damaged studs and put in new ones.

GO TO FRAME 2

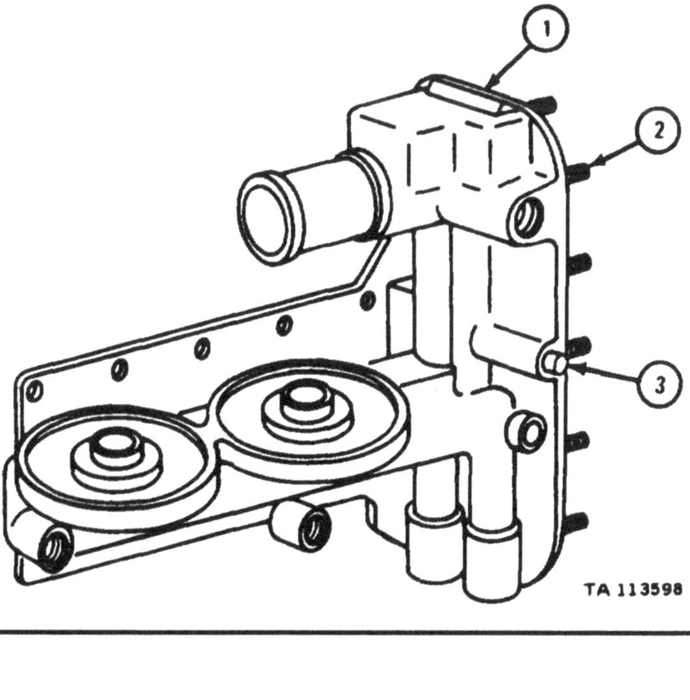

FRAME 2

WARNING

Dry cleaning solvent is flammable. Do not use near an open flame. Keep a fire extinguisher nearby when solvent is used. Use only in well-ventilated places. Failure to do this may result in injury to personnel and damage to equipment.

1. Check that two plunger bores (1) are not scratched. Fix small scratches in plunger bores with crocus cloth dipped in dry cleaning solvent.

NOTE

Some cast iron oil cooler and filter housings (2) do not have threaded inserts (3) or threaded hex-head bushings (4). On these oil cooler and filter housings, threads are tapped directly into two holes (5).

2. Check that all tapped holes in oil cooler and filter housing (2) do not have stripped or damaged threads. Fix minor thread damage with a tap. If threads are stripped or badly damaged, put in a threaded insert of the proper size.
3. On engine LDS-465-2, check that two threaded inserts (3) are not damaged. Take out damaged threaded inserts using screw thread extractor. Put in new threaded inserts using screw thread inserter.
4. On all engines except LDS-465-2, if oil cooler and filter housing (2) has two threaded hex-head bushings (4), check that they are not damaged. Unscrew and take out damaged threaded hex-head bushings. Coat outside of new threaded hex-head bushings with sealant, then screw in and tighten them.

GO TO FRAME 3

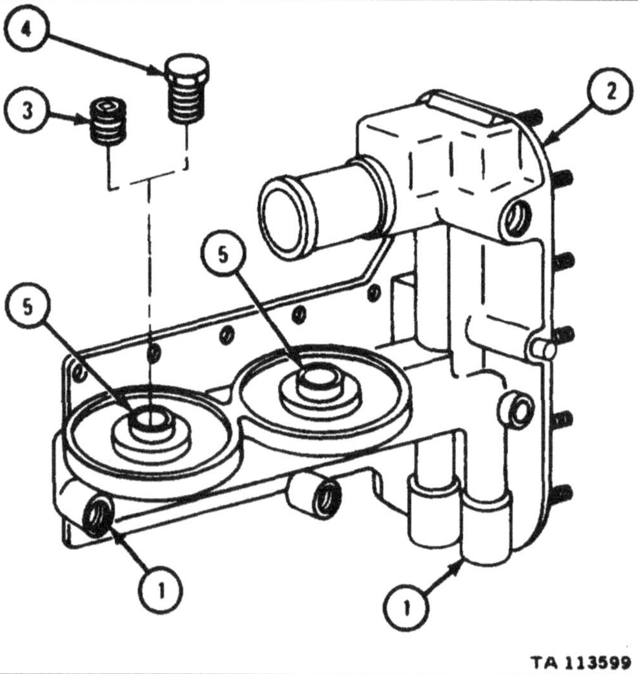

FRAME 3

1. Check that two pipe plugs (1), square head pipe plug (2), and two valve plugs (3) are not cracked and do not have damaged threads. Check that square head of square head pipe plug is not rounded. If pipe plugs, square head pipe plug, or valve plugs are damaged, throw them away and get new ones.
2. Check that 16 mounting screws (4) and long mounting bolt (5) do not have stripped or damaged threads or rounded heads. Check that 16 mounting washers (6) and mounting washer (7) are not cracked or broken. Throw away damaged screws, bolt, and washers and get new ones.
3. Check that two springs (8) are not cracked or squeezed shut. If two springs are cracked or squeezed shut, throw them away and get new ones.

WARNING

Dry cleaning solvent is flammable. Do not use near an open flame. Keep a fire extinguisher nearby when solvent is used. Use only in well-ventilated places. Failure to do this may result in injury to personnel and damage to equipment.

4. Check that two plungers (9) have no scratches or burrs. Fix small burrs and scratches with crocus cloth dipped in dry cleaning solvent. Polish two plungers with crocus cloth dipped in dry cleaning solvent even if they are not scratched.
5. Check that two plungers (9) fit and move freely in bores in oil cooler and filter housing (10). If plungers bind in bores, throw away plungers and get new ones. If new plungers still bind in bores, get a new oil cooler and filter housing.

GO TO FRAME 4

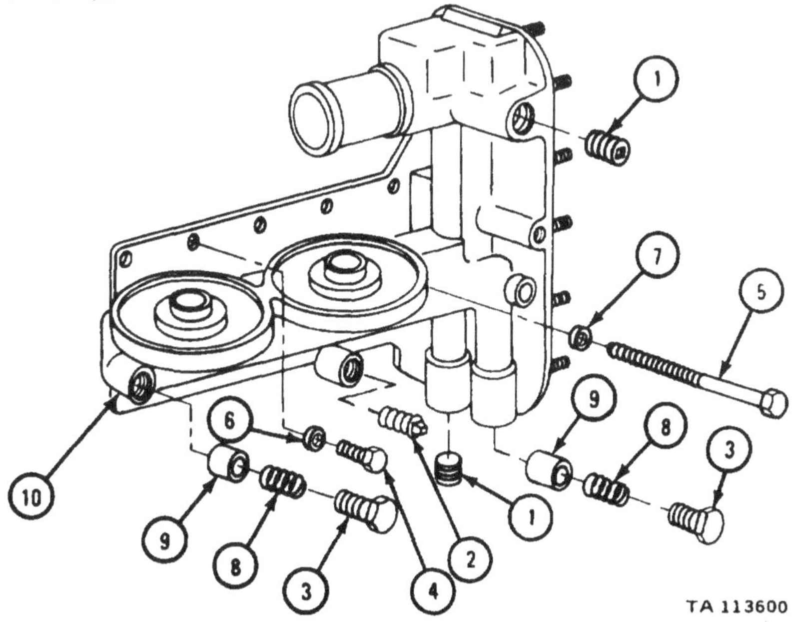

FRAME 4

NOTE

Readings must be within limits given in table 4-59. The letter L shows a loose fit. If readings are not within given limits, throw away part and get a new one.

1. Measure inside diameter of two bypass valve bores (1).
2. Measure out side diameter of two valve plungers (2).
3. Measure fit of each of two valve plungers (2) in its plunger bore (1).

GO TO FRAME 5

NOTE: CHECK ONLY THOSE PARTS WHICH ARE CALLED OUT IN THIS FRAME. PARTS WITHOUT CALLOUTS ARE SHOWN ONLY FOR REFERENCE PURPOSES OR ARE CHECKED IN ANOTHER FRAME.

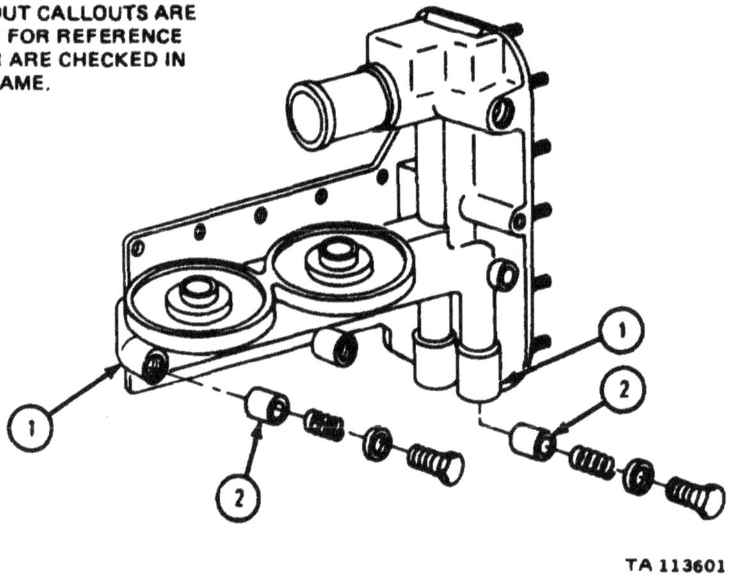

Table 4-59. Oil Cooler and Filter Housing Assembly Wear Limits

Index Number	Item/Point of Measurement	Size and Fit of New Parts (inches)	Wear Limit (inches)
1	Bypass valve bores inside diameter	0.8030 to 0.8040	0.8050
2	Valve plungers out side diameter	0.8000 to 0.8010	0.7990
1 and 2	Fit of plunger in housing bore	0.0020L to 0.0040L	0.0060L

FRAME 5

NOTE

Readings must be within limits given in table 4-60. If readings are not within given limits, throw away part and get a new one.

1. Measure free length of two bypass valve springs (1).
2. Using spring tester , measure load needed to squeeze each of two bypass valve springs (1) down to 1.380 inches.
3. Using spring tester, measure maximum solid height of two bypass valve springs (1).

GO TO FRAME 6

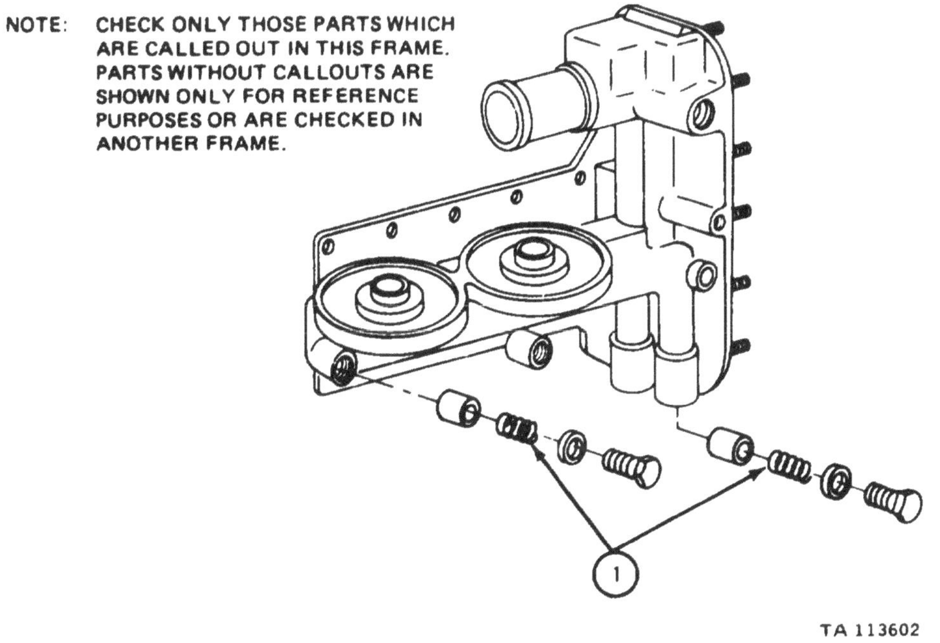
NOTE: CHECK ONLY THOSE PARTS WHICH ARE CALLED OUT IN THIS FRAME. PARTS WITHOUT CALLOUTS ARE SHOWN ONLY FOR REFERENCE PURPOSES OR ARE CHECKED IN ANOTHER FRAME.
1
TA 113602

Table 4-60. Oil Cooler and Filter Bypass Valve Springs Wear Limits

Index Number	Item /Point of Measurement	Size and Fit of New Parts (inches)	Wear Limit (inches)
1	Oil cooler and filter bypass valve spring free lengt h	1.6200	None
1	Oil cooler and filter bypass valve spring load at 1.380 inches	7.58 to 7.72 pounds	None
1	Oil cooler and filter bypass valve spring maximum solid height	0.9520	None

FRAME 6

1. Measure depth of two oil cooler preformed packing counterbores (1), If depth of counterbores is 0.090 inch or more, note that oil cooler and filter housing (2) needs shims put in counterbores.

END OF TASK

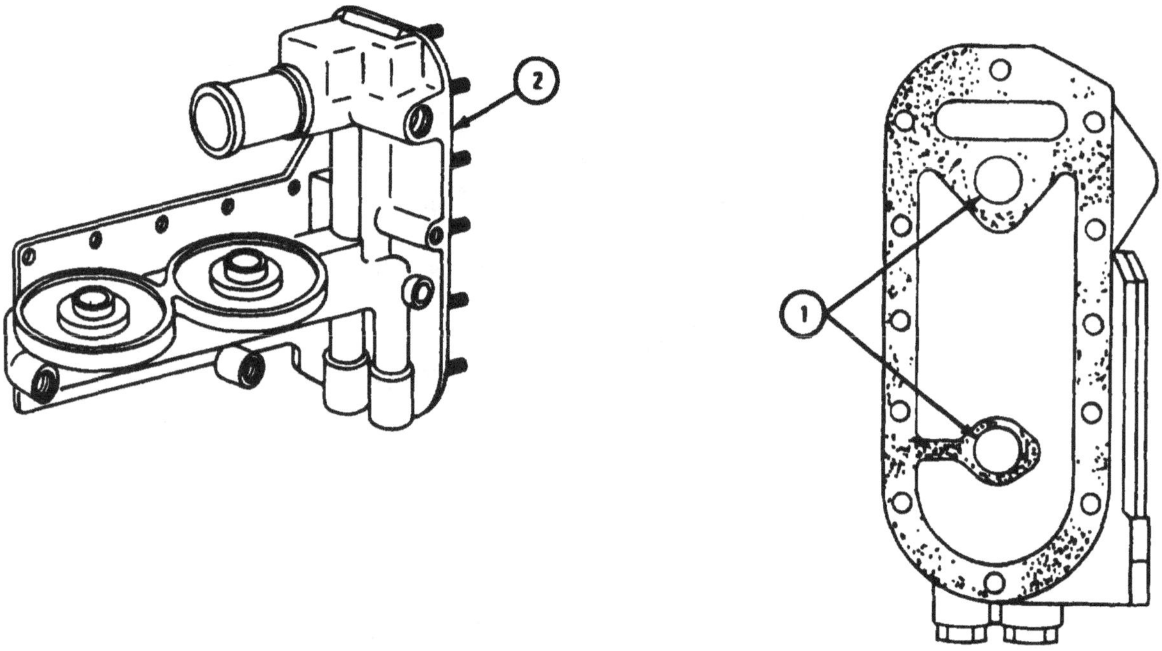

TA 113603

d. Assembly.

FRAME 1

1. Put in two pipe plugs (1).
2. Put in square-head pipe plug (2) .
3. Put in two plungers (3), two springs (4), and two gaskets (5) and two plugs (6) .

END OF TASK

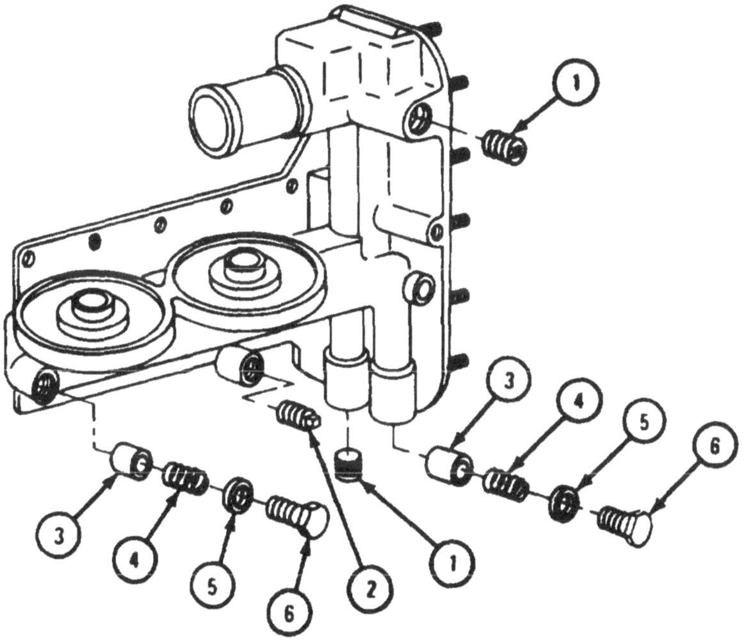

4-32. FUEL INJECTION PUMP OIL HOSE.

a. Cleaning.

FRAME 1

NOTE

Some early model LDS-465-1 fuel injection pumps (1) use two oil hoses (2 and 3). Late model LDS-465-1 fuel injection pumps and fuel injection pumps for all other engines use only one oil hose (2).

1. Clean oil hose (2) or two oil hoses (2 and 3) with soap and water.

WARNING

Dry cleaning solvent is flammable. Do not use near an open flame. Keep a fire extinguisher nearby when solvent is used. Use only in well-ventilated places. Failure to do this may result in injury to personnel and damage to equipment.

2. Clean elbow (4) with dry cleaning solvent. Dry with clean cloth.
3. If working on early model LDS-465-1 with two hoses (2 and 3), clean elbow (5), bracket (6), and loop clamp (7) with dry cleaning solvent. Dry with clean cloth.

END OF TASK

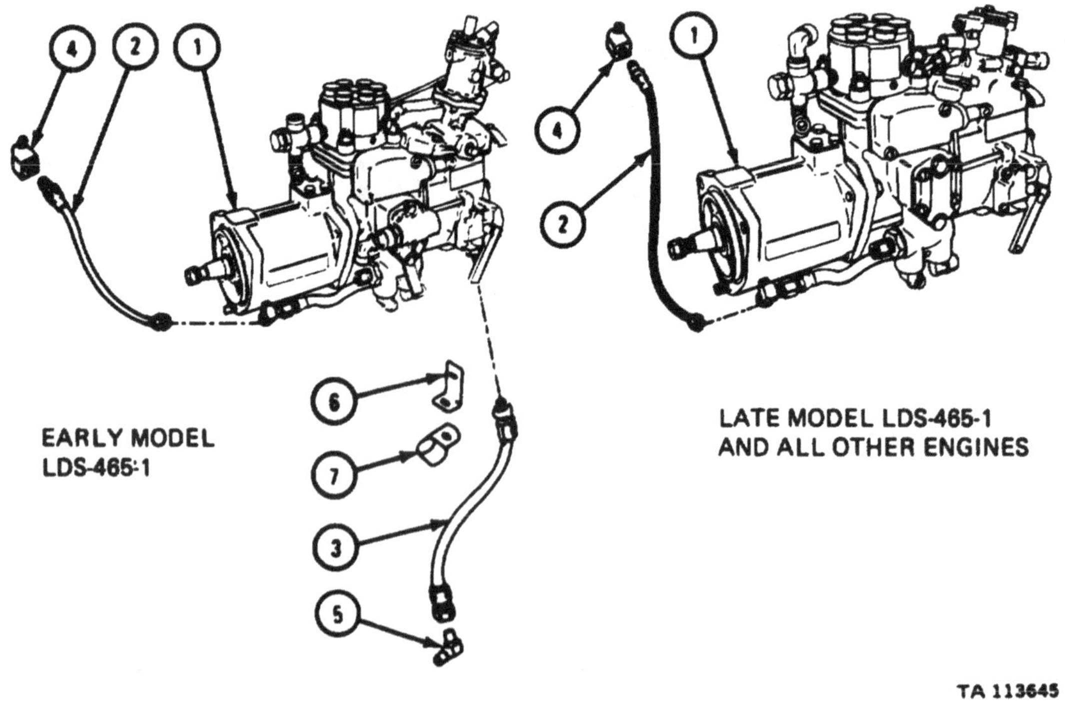

b. Inspection and Repair.

FRAME 1

1. Check that all threaded parts are not stripped or crossthreaded. Take out minor thread damage with thread chaser or tap. If more repair is needed, get a new part.

2. Check that oil hose (1) or two oil hoses (1 and 2) are not cracked, dried out or damaged in any other way. If oil hoses are damaged, get new ones.

END OF TASK

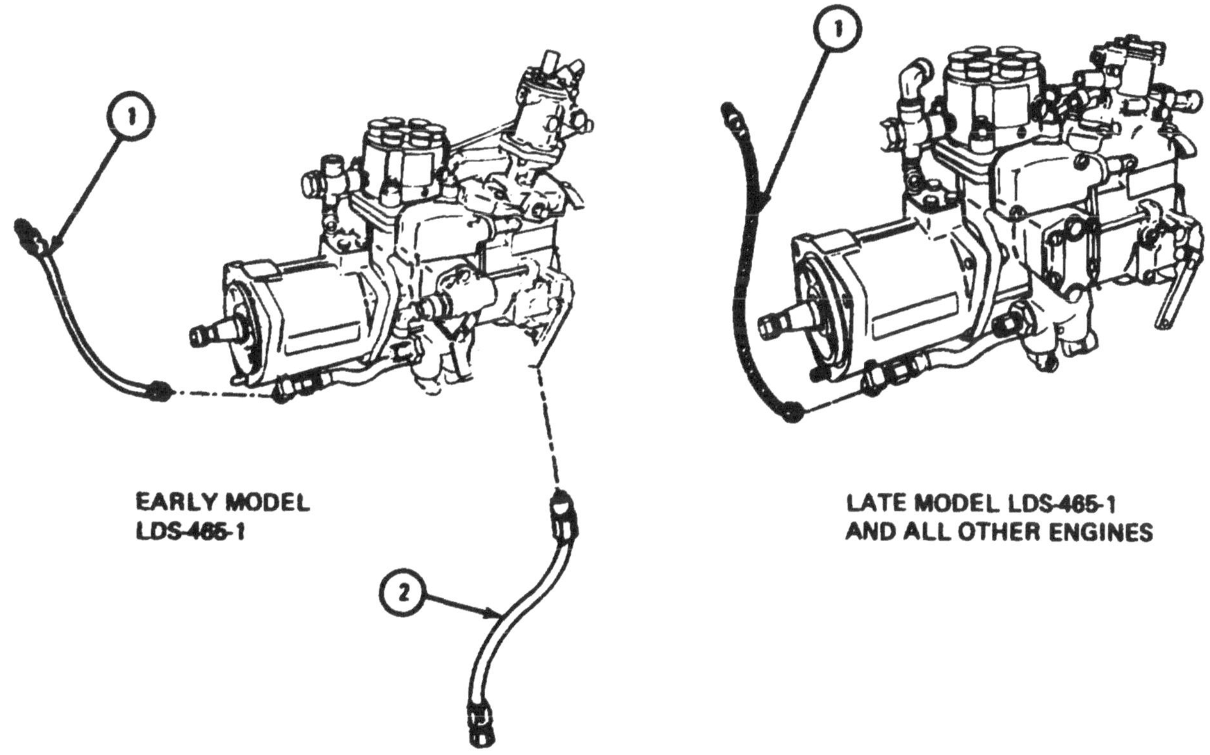

4-33. FUEL FILTER ASSEMBLY.

a. Engines LD-465-1, LD-465-1C, LDT 465-1C, and LDS 465-1A.

(1) Disassembly .

FRAME 1

1. Take two elbows (1) out of fuel filter head (2).
2. Take out tee fitting (3) with elbow (4) and elbow (5). Take off elbows (4 and 5).
3. Take two drain cocks (6) off two fuel filter bodies (7).

GO TO FRAME 2

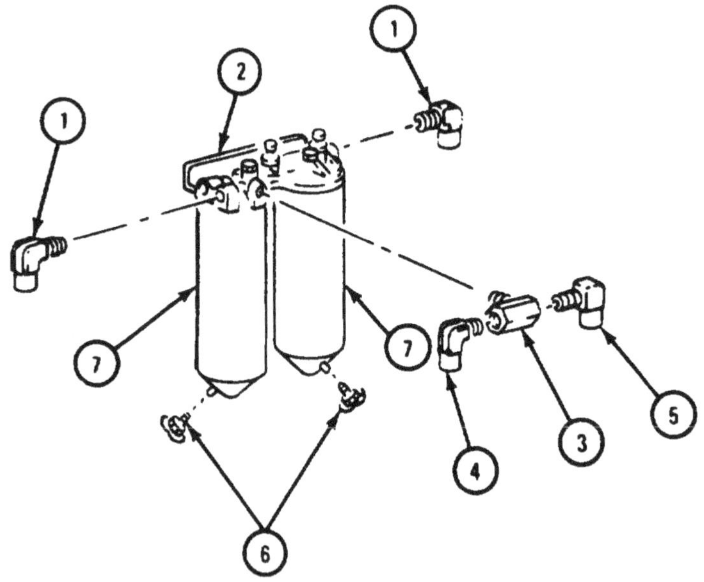

TA 118091

FRAME 2

1. Take two sleeve nuts (1), flat washers (2), and gaskets (3) off filter head (4). Throw away gaskets .
2. Take two filter bodies (5) off filter head (4).
3. Take two gaskets (6) out of fuel filter head (4). Throw away two gaskets.
4. Take pipe plug (7) out of fuel filter head (4).

GO TO FRAME 3

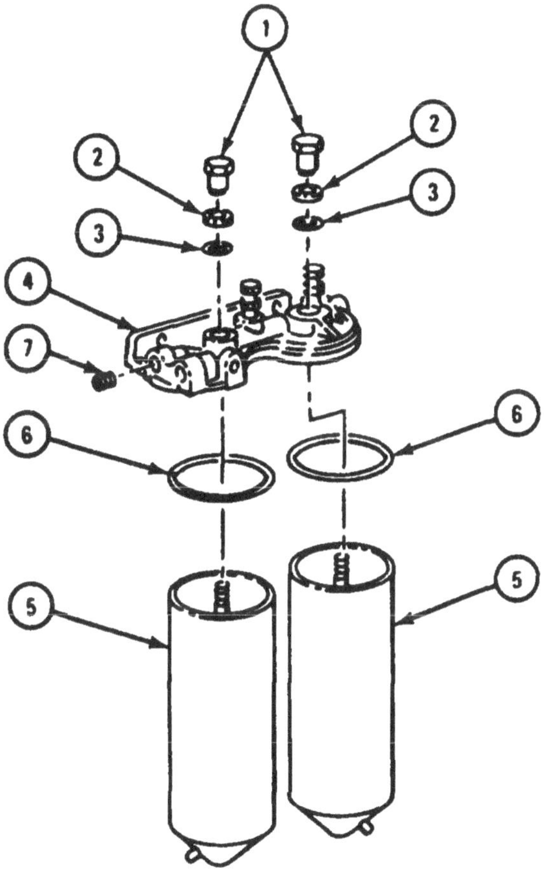

TA 118092

FRAME 3

1. Take two bleeder valves (1) out of fuel filter head (2).

NOTE

Relief valve can have valve, spring, and 3/4-inch hexagon head plug or ball, spring, and 1-inch hexagon plug. Relief valve assemblies cannot be switched in fuel filter heads and individual parts of relief valve assemblies cannot be switched. Early model fuel filter heads have two bleeder valves. Late model fuel filter heads have only one bleeder valve in middle of head.

2. Take out relief valve plug (3 or 4), preformed packing (5) or gasket (6), spring (7 or 8), and relief valve (9) or ball (10) from fuel filter head (2).

3. Throw away preformed packing (5) or gasket (6).

GO TO FRAME 4

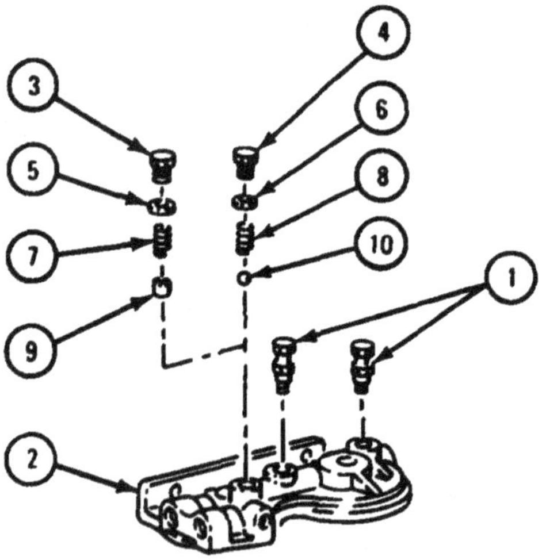

TA 118093

FRAME 4

1. Take cotter pin (1) out of stud (2).
2. Take filter element (3) out of filter body (4).
3. Take cup (5), preformed packing (6), flat washer (7), and spring (8) off stud (2). Throw away preformed packing (6).
4. Do steps 1, 2, and 3 again on filter body (9).

END OF TASK

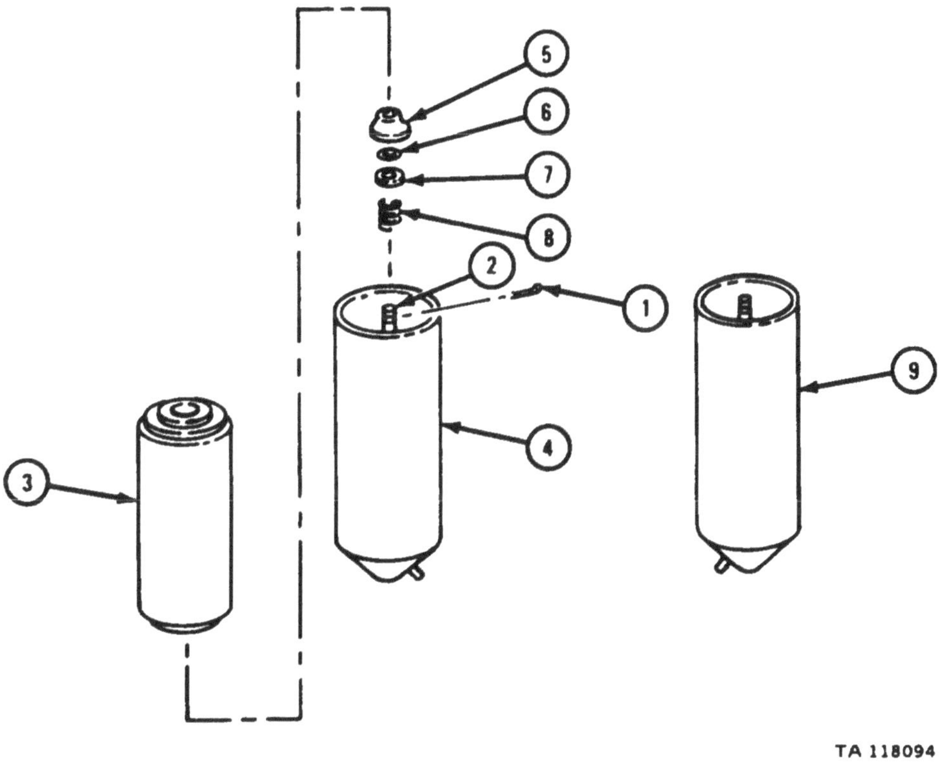

(2) Cleaning .

FRAME 1

WARNING

Dry cleaning solvent is flammable. Do not use near an open flame. Keep a fire extinguisher nearby when solvent is used. Use only in well-ventilated places. Failure to do this may result in injury to personnel and damage to equipment.

1. Wash four elbows (1), tee fitting (2), two drain cocks (3), two filter bodies (4), and two covers (5) with dry cleaning solvent .
2. Wash filter head (6), two relief valve plugs (7), pipe plug (8), two sleeve nuts (9), and two flat washers (10) with dry cleaning solvent.
3. Dry all parts with lint-free cloth.

GO TO FRAME 2

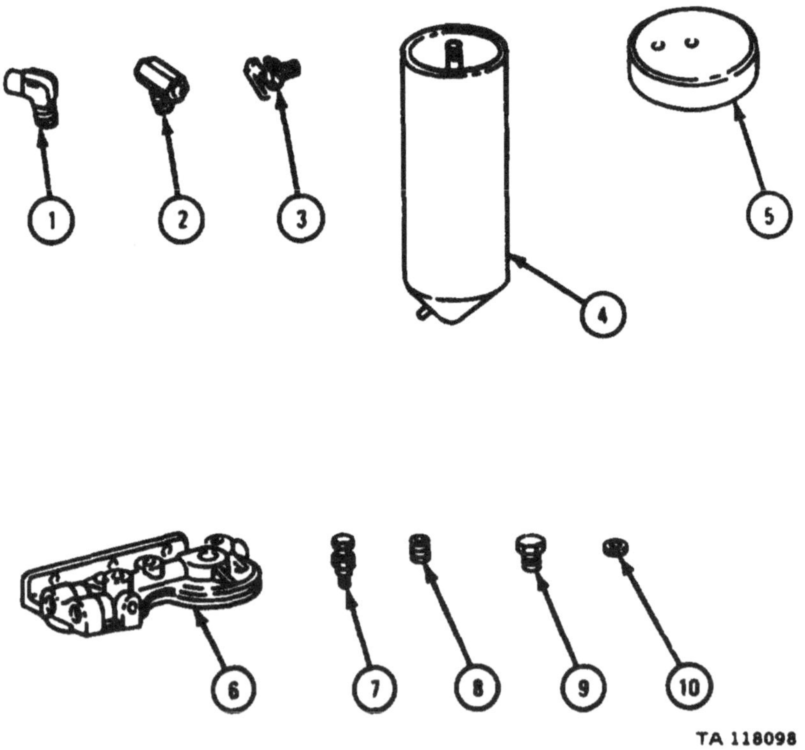

FRAME 2

WARNING

Dry cleaning solvent is flammable. Do not use near an open flame. Keep a fire extinguisher nearby when solvent is used. Use only in well-ventilated places. Failure to do this may result in injury to personnel and damage to equipment.

1. Wash relief plug (1), spring (2), and relief valve (3) or ball (4) with dry cleaning solvent .
2. Dry all parts with lint-free cloth.

END OF TASK

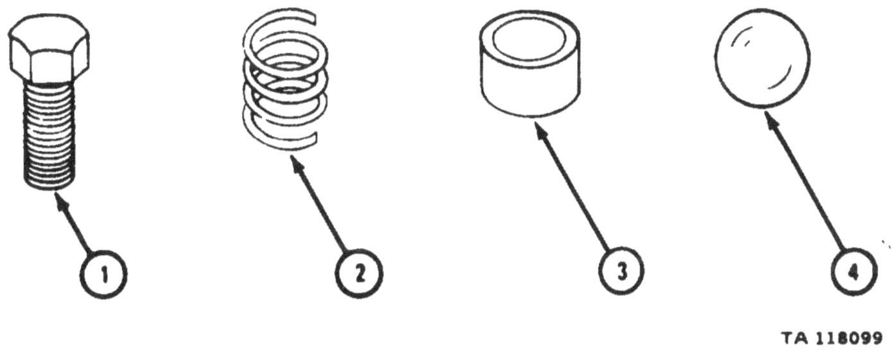

(3) Inspection and repair .

FRAME 1

NOTE

Readings must be within limits given in table 4-61. The letter L indicates a loose fit. If readings are not within given limits, throw away part and get a new one.

1. Measure diameter of pressure relief valve plunger (1).
2. Measure inside diameter of pressure relief valve bore (2) in filter head (3). Get a new filter head if diameter is not within given limits.
3. Measure fit of valve plunger (1) in bore (2). Get a new plunger or plunger and filter head (3) if loose fit is not within given limits.
4. Measure free length, spring load, maximum solid height, and diameter of pressure relief valve spring (4) for plunger (1) type valve and for ball (5) type valve .
5. Check threads in filter head (3) for damage. Repair threads having minor damage with tap. Get a new filter head if damaged threads cannot be repaired with tap.
6. Check pipe plug (6), relief plug (7), two sleeve nuts (8), and two bleeder valves (9) for damaged threads. Repair threads having minor damage with file. Get new parts if damaged threads cannot be repaired with file.

GO TO FRAME 2

NOTE: CHECK ONLY THOSE PARTS WHICH ARE CALLED OUT IN THIS FRAME. PARTS WITHOUT CALLOUTS ARE SHOWN ONLY FOR REFERENCE PURPOSES OR ARE CHECKED IN ANOTHER FRAME.

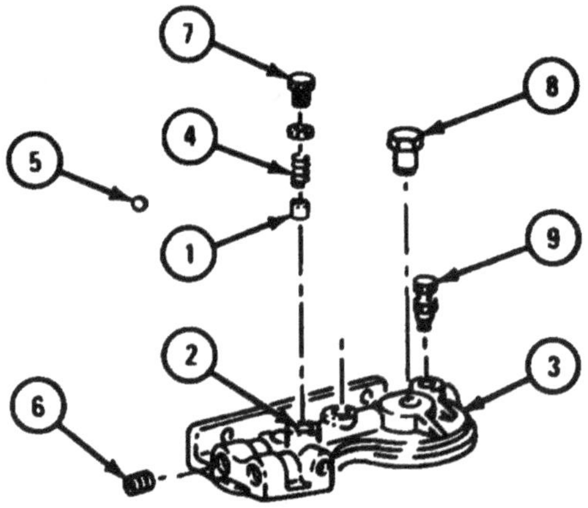

TA 118101

Table 4-61. Fuel Filter Pressure Relief Valve Wear Limits

Index Number	Item/Point of Measurement	Size and Fit of New Parts (inches unless otherwise noted)	Wear Limit (inches)
1	Pressure relief valve plunger diameter	0.4950 to 0.4980	None
2	Pressure relief valve bore diameter	0.5080 to 0.5180	None
1 and 2	Fit of valve plunger in bore	0.0230L	None
4	Pressure relief valve spring free length		
	Plunger-type valve	1.5600	None
	Ball-type valve	1.45	None
4	Pressure relief valve spring load		
	Plunger-type valve (at 2.09 inches)	10.0 to 11.2 pounds	None
	Ball-type valve (at 0.94 inch)	11.0 to 13.0 pounds	None
4	Pressure relief valve spring maximum solid height		
	Plunger-type valve	0.7130	None
	Ball-type valve	0.5400	None
4	Pressure relief valve spring diameter		
	Plunger-type valve	0.3600	None
	Ball-type valve	0.5000	None

FRAME 2

1. Check four elbows (1), tee fitting (2), two drain cocks (3), and two filter bodies (4) for damaged threads. Repair threads having minor damage with file. Get new parts if damaged threads cannot be repaired with file.
2. Check two drain cocks (3) for open fuel passage. If clogged, clean passage with wire. Get a new drain cock if passage cannot be opened.
3. Check two filter bodies (4) for open drain outlets (5). If clogged, clean outlets with wire. Get a new filter body if drain cannot be opened.
4. Check filter body cover (6) for damaged threads. Repair threads having minor damage with tap. Get a new cover if damaged threads cannot be repaired with tap.
5. Check two filter bodies (4) for cracks and dents. If damaged, get new filter bodies (4) .

END OF TASK

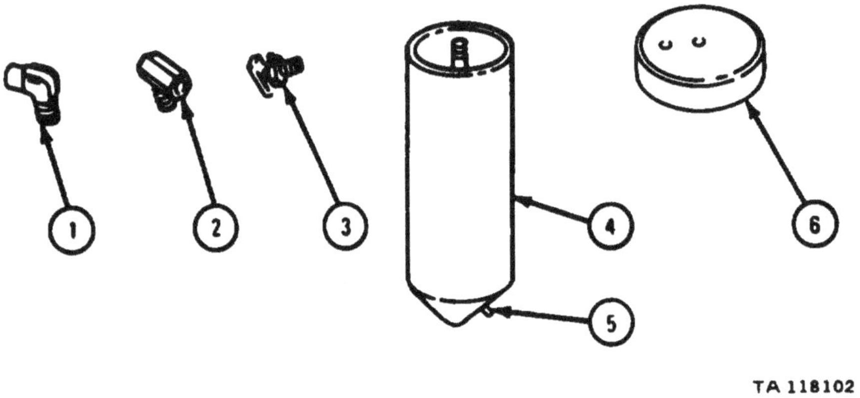

(4) Assembly .

FRAME 1

1. Put two bleeder valves (1) in fuel filter head (2).

NOTE

Relief valve can have valve, spring, and 3/4-inch hexagon head plug or ball, spring, and 1-inch hexagon plug. Relief valve assemblies cannot be switched between fuel filter heads. Early model fuel filter heads have two bleeder valves. Late model fuel filter heads have only one bleeder valve in middle of head.

2. Put in relief valve (3) or ball (4).
3. Put in spring (5) .
4. Put in preformed packing (6) or gasket (7) .
5. Put in relief valve plug (8) .

GO TO FRAME 2

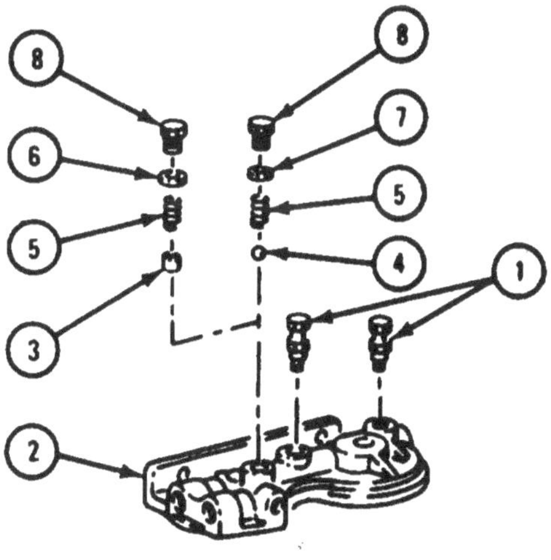

TA 118104

FRAME 2

1. Put spring (1), flat washer (2), gasket (3), and cup (4) on stud (5) .
2. Put cotter pin (6) over cup (4) and through stud (5) .
3. Put fuel element (7) in filter body (8).
4. Do steps 1, 2, and 3 again on filter body (9).

GO TO FRAME 3

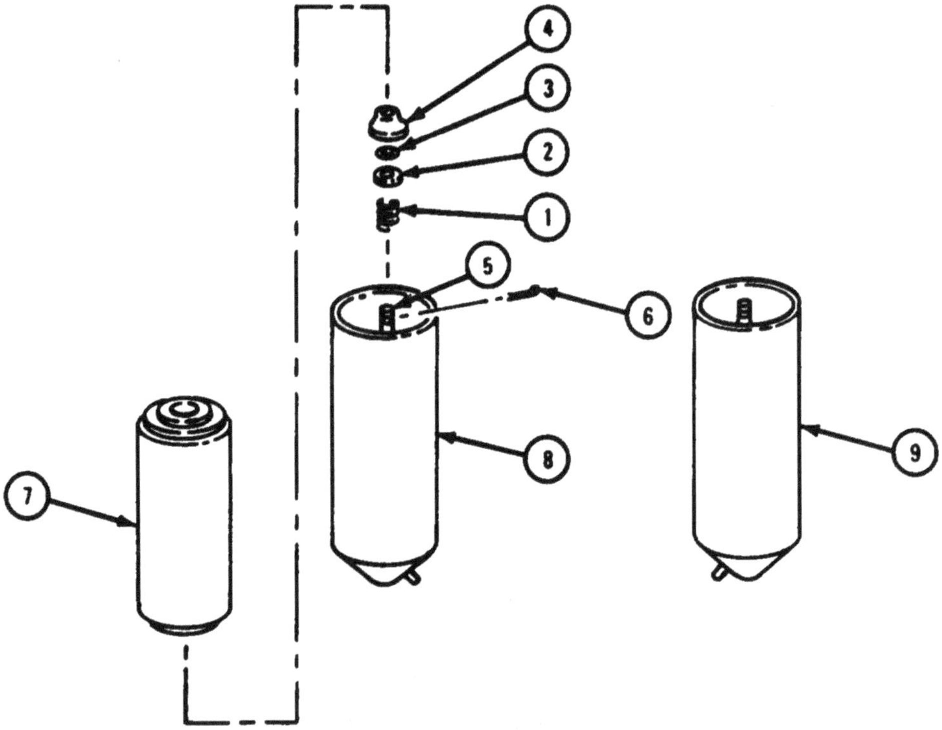

TA 118105

FRAME 3

1. Put filter body gasket (1) on each filter body (2).
2. Put two filter bodies (2) on filter head (3).
3. Put sleeve nut (4) with flat washer (5) and gasket (6) on each of two studs (7). Tighten sleeve nuts (4) to 23 to 27 pound-feet .
4. Put pipe plug (8) in filter head (3).

GO TO FRAME 4

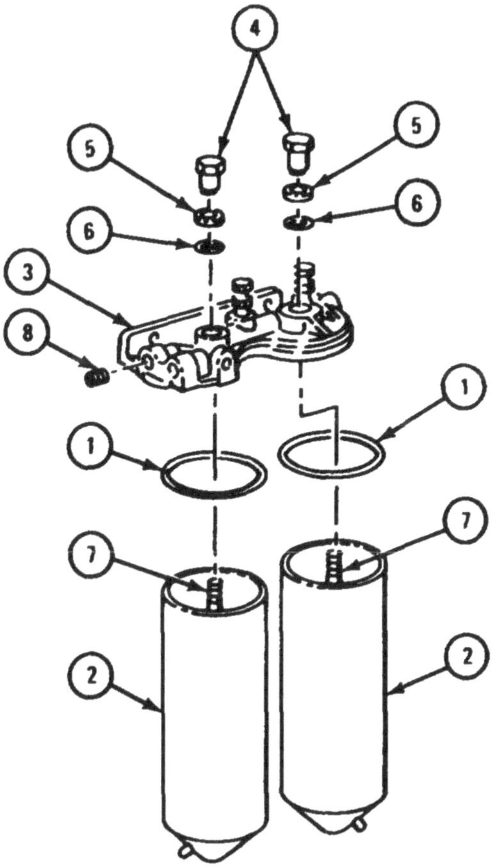

TA 118106

FRAME 4

1. Put two elbows (1) in filter head (2).
2. Put two elbows (3) in tee fitting (4).
3. Put tee fitting (4) in filter head (2).
4. Put drain cock (5) in each of two filter bodies (6).

END OF TASK

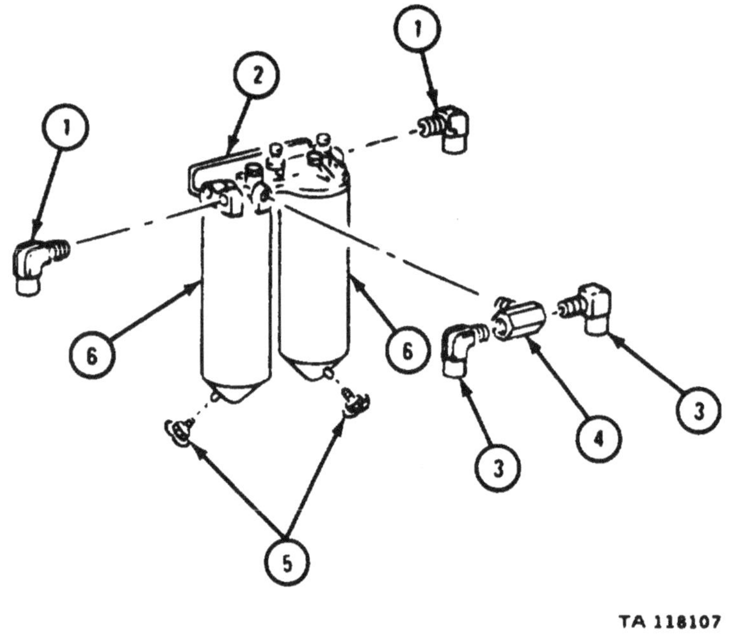

b. Engine LDS-465-2 .

(1) Disassembly .

FRAME 1

1. Take drain cock (1) out of filter body(2) .
2. Take two elbows (3) out of filter body (2).
3. Take two capscrews (4) with lockwashers (5) and nuts (6) out of primary filter bracket (7) .
4* Take rubber bumper (8) out of bracket (7) .
5. Take primary filter body (2) out of bracket (7) .
6. Do steps 1 and 2 again on final filter body (9).
7. Do steps 3 and 4 again on final filter bracket (10).
8. Take final filter body (9) out of bracket (10).

GO TO FRAME 2

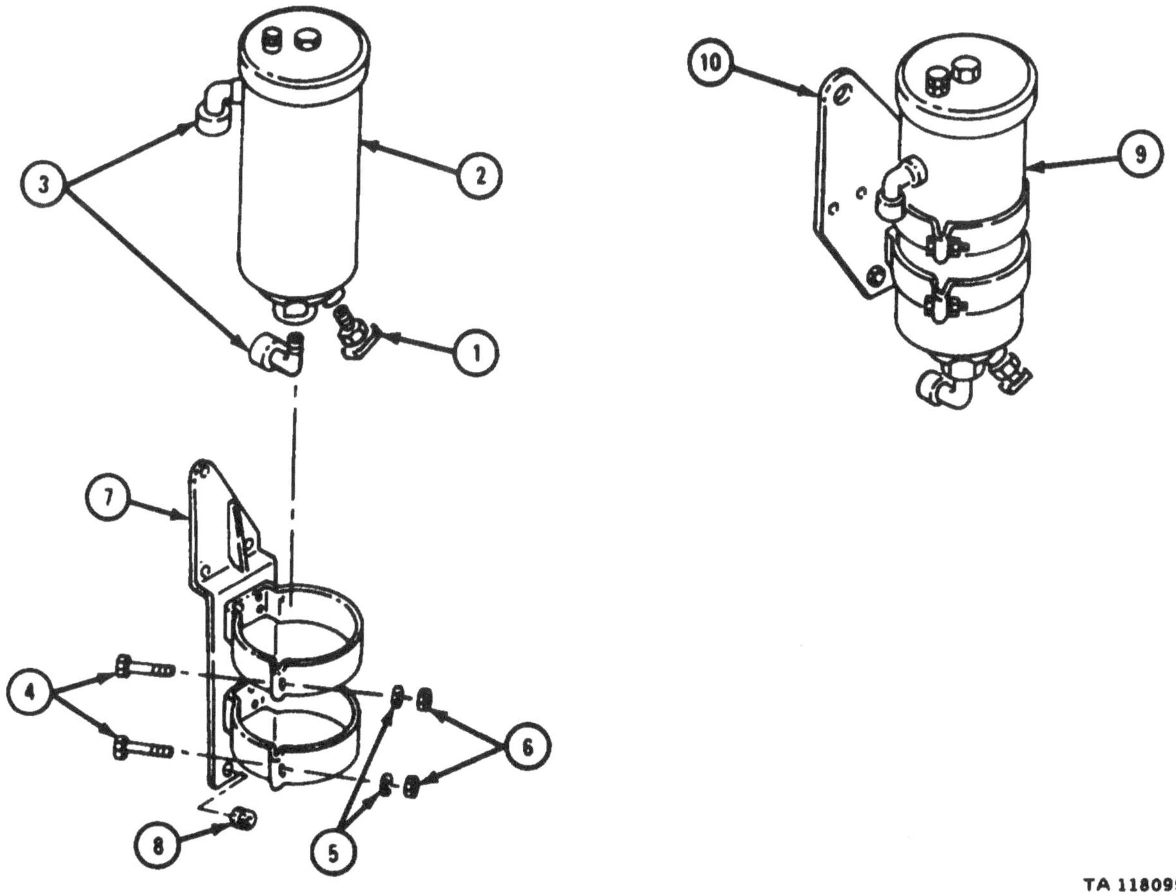

TA 118095

FRAME 2

1. Take bleeder valve (1) out of filter assembly cover (2).
2. Take screw (3) out of cover (2). Take off flat washer (4) and gasket (5). Throw away gasket.
3. Take off cover (2) .
4. Take out and throw away gasket (6) .
5. Take out and throw away filter element (7) .
6. Take off and throw away preformed packing (8).
7. Take out retainer (9) .
8. Take out spring (10) .
9. Do steps 1 through 8 again for other fuel filter assembly.

END OF TASK

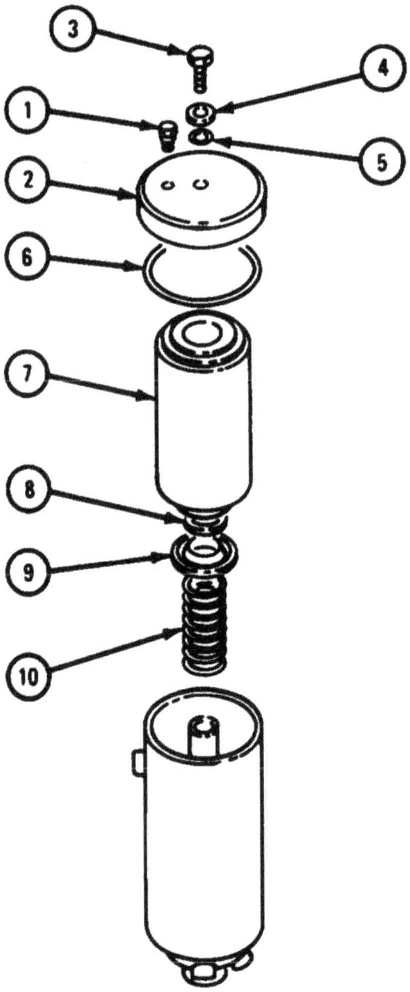

TA 118096

(2) Cleaning .

FRAME 1

WARNING

Dry cleaning solvent is flammable. Do not use near an open flame. Keep a fire extinguisher nearby when solvent is used. Use only in well-ventilated places. Failure to do this may result in injury to personnel and damage to equipment.

1. Wash brackets (1 and 2) in dry cleaning solvent.
2. Wash two elbows (3), two drain cocks (4), and two filter bodies (5) in dry cleaning solvent .
3. Wash two springs (6), two retainers (7), two covers (8), two bleeder valves (9), and two screws (10) in dry cleaning solvent.
4. Dry all parts with lint-free cloth.

END OF TASK

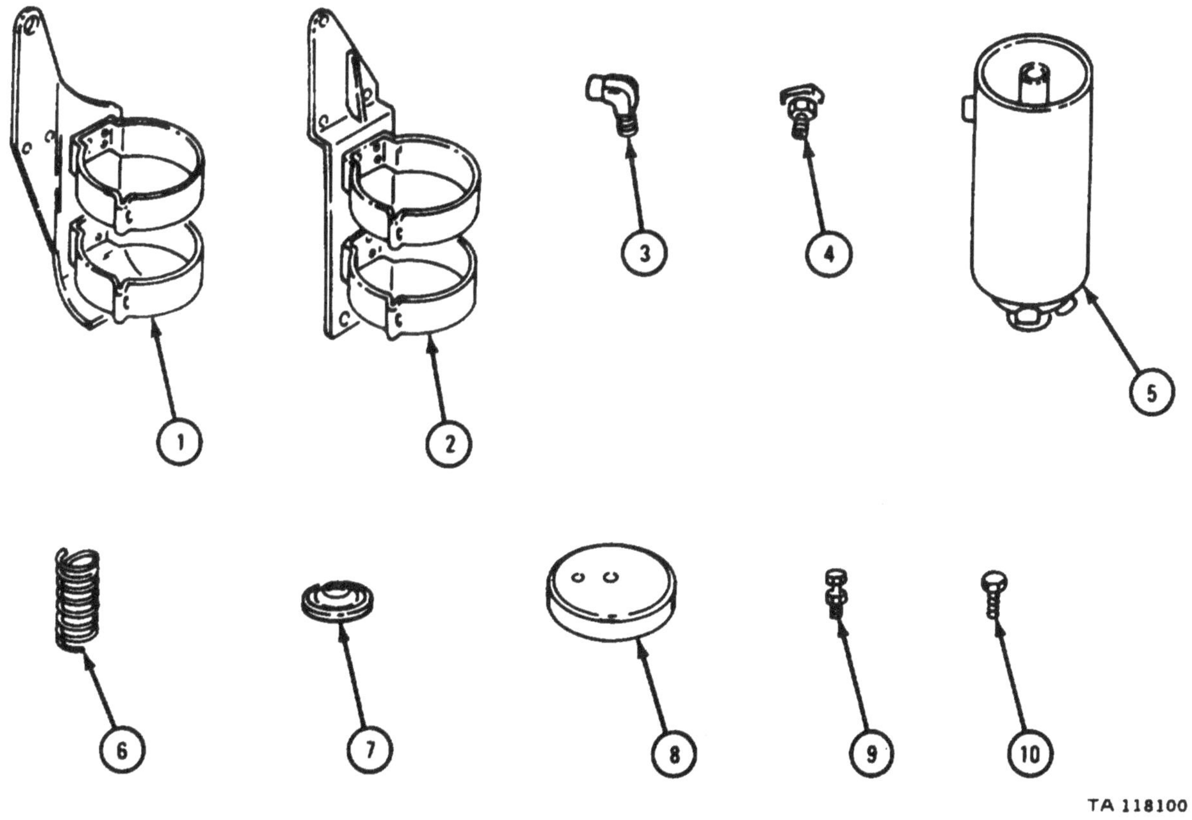

(3) Inspection and repair.

FRAME 1

1. Check that brackets (1 and 2) have no damage. Repair minor damage by straightening or welding. Refer to FM 43-2 and TM 9-237. If brackets have major damage, get new ones.
2. Check that two fuel filter bodies (3) have no dents and cracks. If filter bodies are cracked or dented, get new ones.
3. Check two fuel filter bodies (3), two covers (4), four elbows (5), two drain cocks (6), two bleeder valves (7), and two cover screws (8) for damaged threads. Repair internal threads with tap and repair external threads with file. If parts have damage which cannot be repaired with tap or file, get new ones.
4. Using wire, check that two drain cocks (6) have open fuel passage. If drain cock cannot be cleaned with wire, get a new one.
5. Check that two retainers (9) and two springs (10) have no damage. If damaged, get new retainers or springs .

END OF TASK

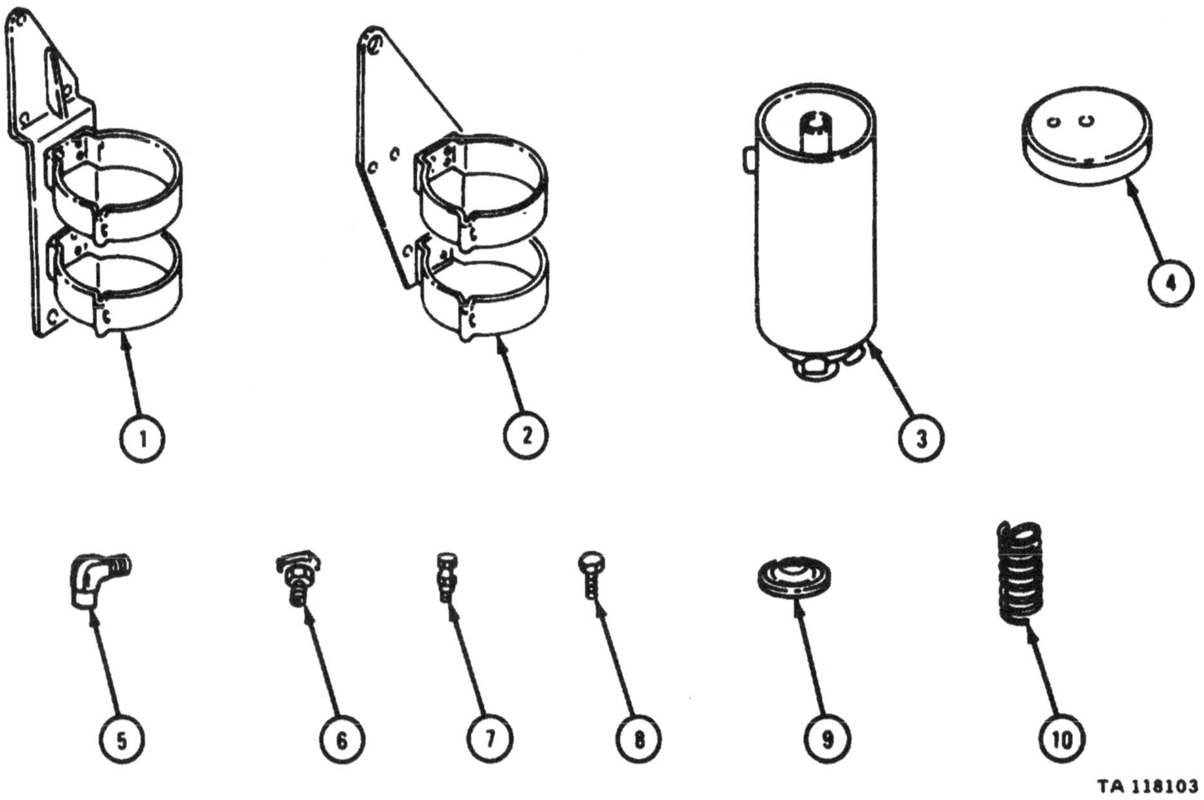

(4) Assembly .

FRAME 1

1. Put spring (1) and retainer (2) in filter body (3) .
2. Put gasket (4) on filter element (5).
3. Put filter element (5) in filter body (3).
4. Put gasket (6) on filter body (3) .
5. Put cover (7) on filter body (3) .
6. Put flat washer (8) and gasket (9) on cover screw (10).
7. Put cover screw (10) in filter body (3) .

GO TO FRAME 2

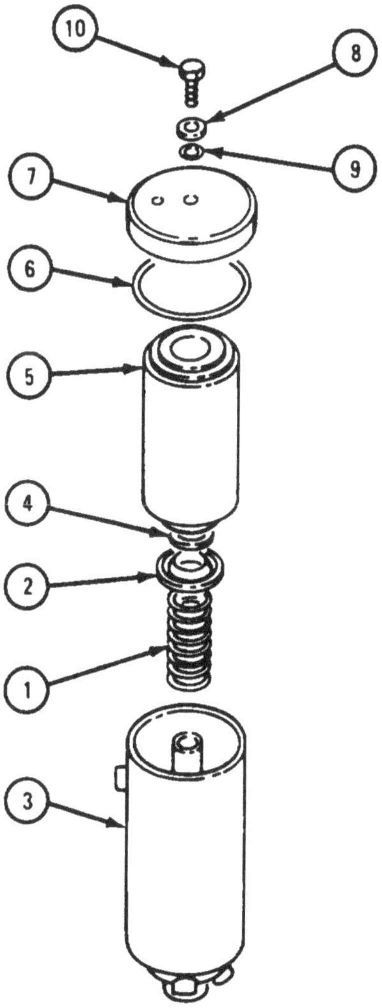

TA 118108

FRAME 2

1. Put pressure relief valve (1) in filter assembly (2).
2. Do steps 1 through 7 of frame 1 again on filter body (3).
3. Put pressure relief valve (4) in filter assembly (5).

GO TO FRAME 3

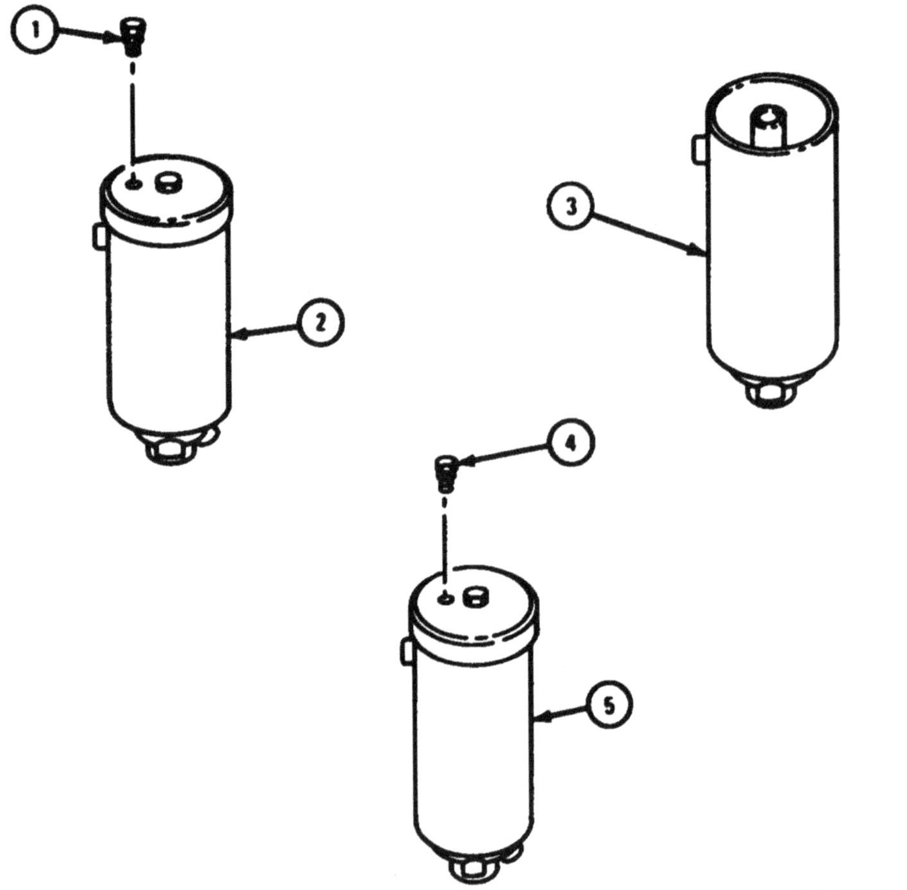

FRAME 3

1. Put drain cock (1) on filter assembly (2).
2. Put two elbows (3) on filter assembly (2).
3. Put filter assembly (2) in primary filter bracket (4).
4. Put in two screws (5). Put on two nuts (6) with washers (7) .
5. Put rubber bumper (8) in bracket (4) .
6. Do steps 1 through 5 again on fuel filter assembly (9) and final filter bracket (10).

END OF TASK

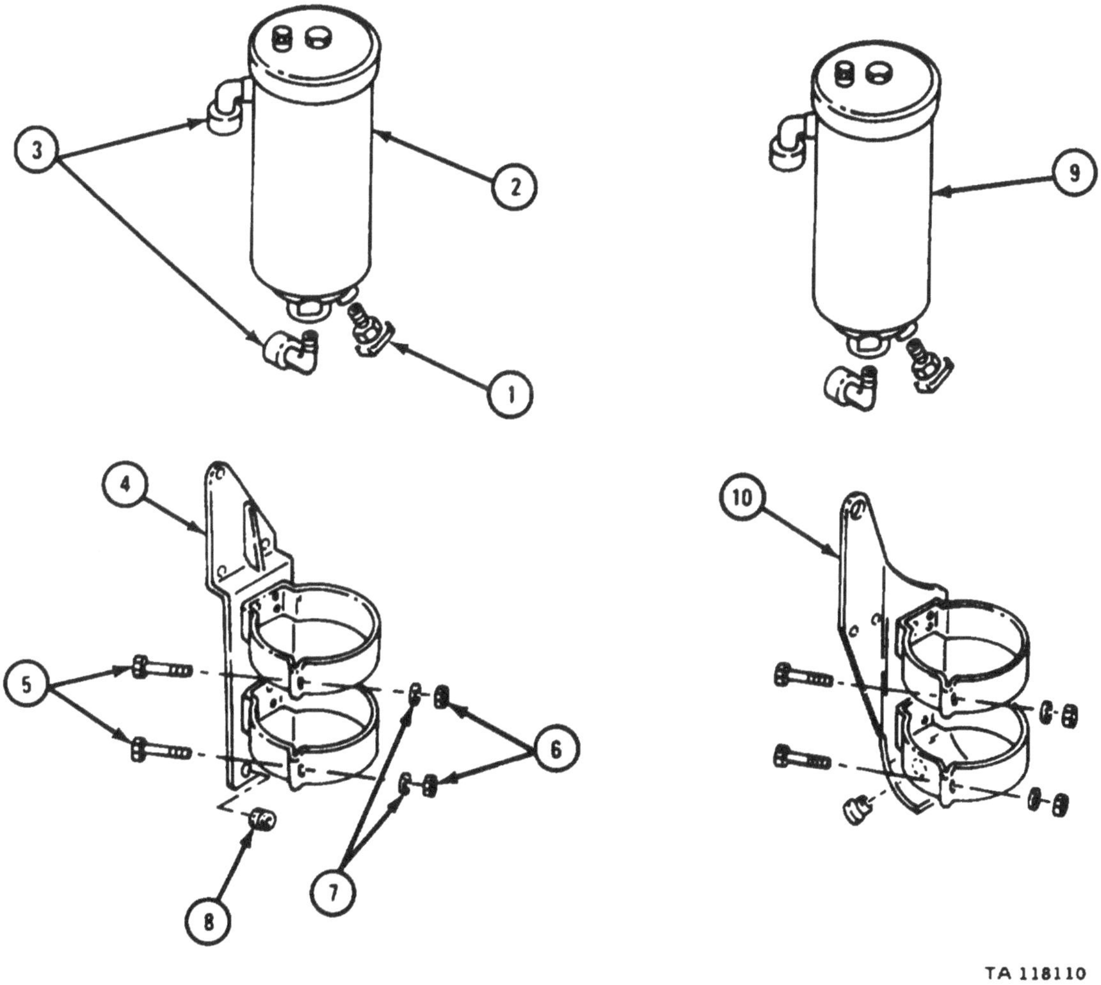

c. Engine LDS-465-1 .

(1) Disassembly .

FRAME 1

1. Take off nut with washer (1) and filter housing cover (2).
2. Take off and throw away gasket (3).
3. Takeout and throw away filter element (4).
4. Takeout drain plug (5) and two elbows (6) .
5. Takeout screw (7) and locknut (8) .
6. Takeoff bracket (9) .

END OF TASK

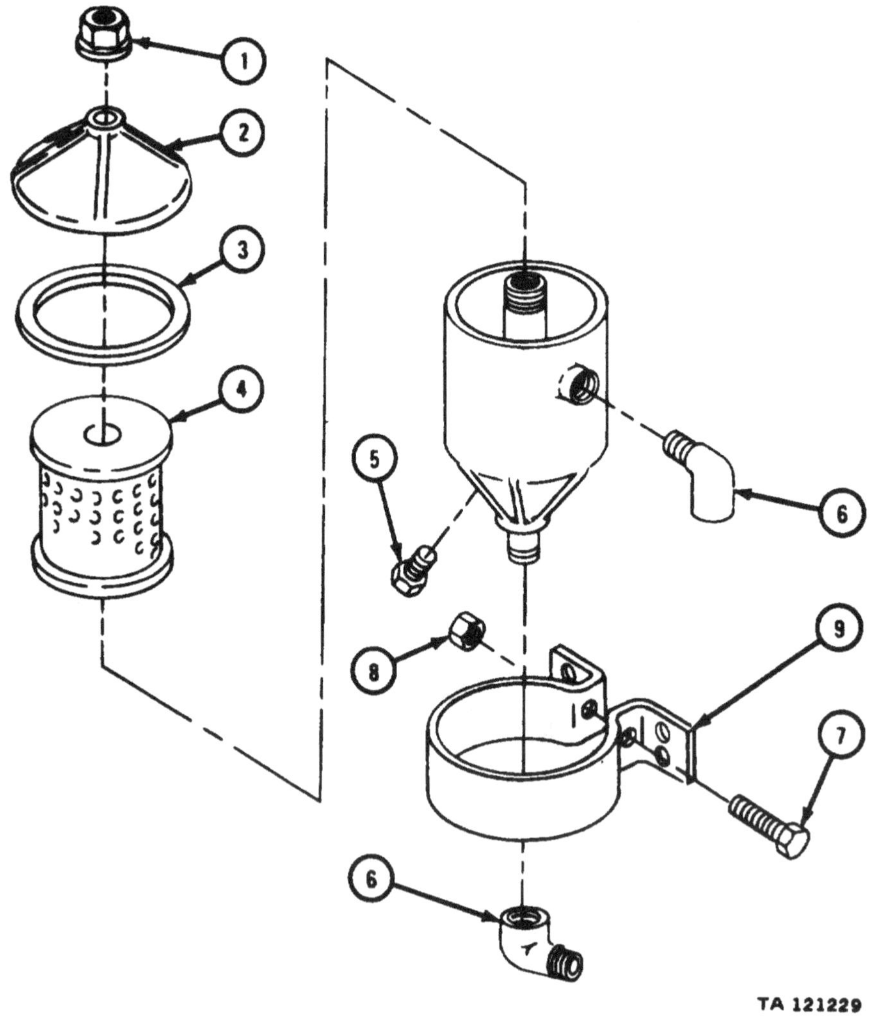

(2) Cleaning .

FRAME 1

WARNING

Dry cleaning solvent is flammable. Do not use near an open flame. Keep a fire extinguisher nearby when solvent is used. Use only in well-ventilated places. Failure to do this may result in injury to personnel and damage to equipment.

1. Wash two elbows (1) and drain plug (2) with dry cleaning solvent.
2. Wash filter housing (3), filter housing cover (4), brackets (5), nut and washer (6) and screw (7) and locknut (8) with dry cleaning solvent.
3. Dry all parts with lint-free cloth.

END OF TASK

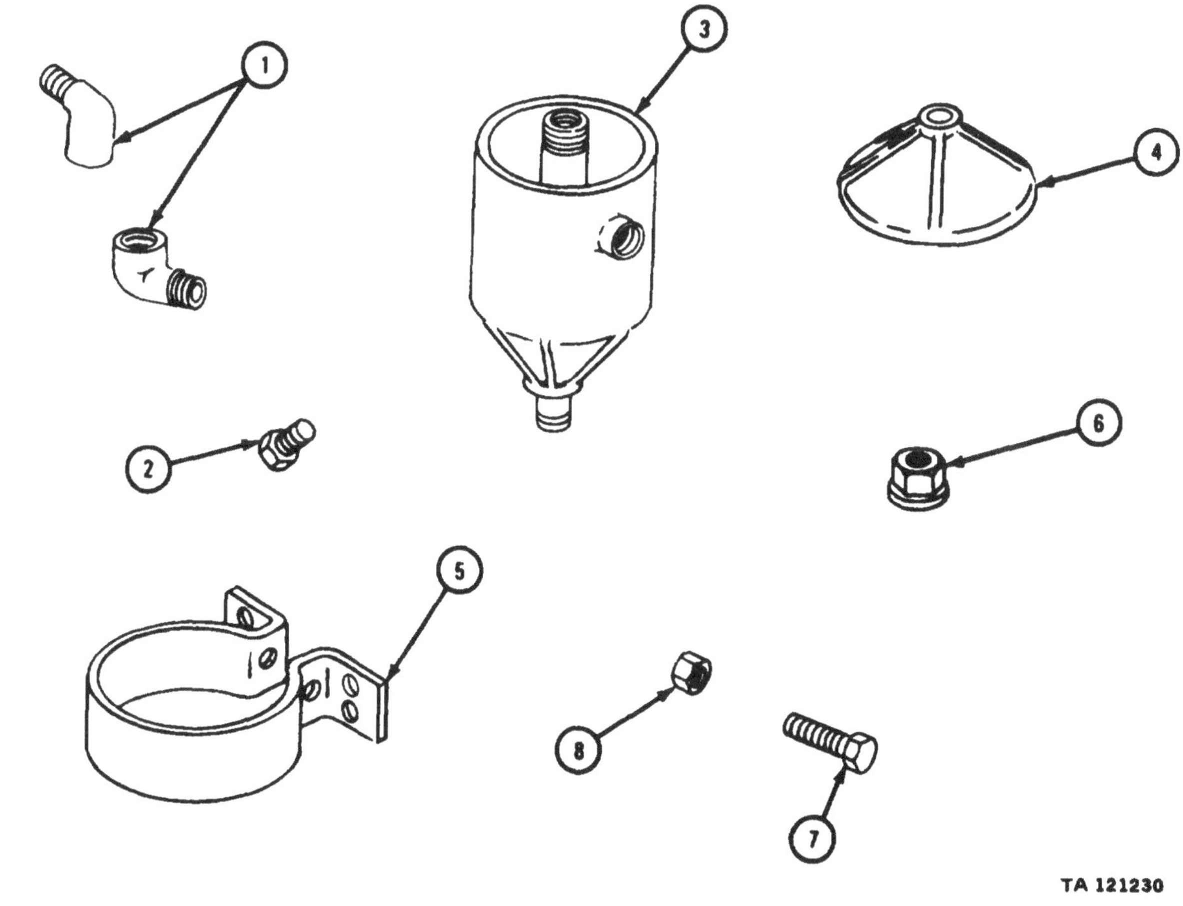

(3) Inspection and repair.

FRAME 1

1. Check that two elbows (1) and drain plug (2) do not have any damaged threads. Repair internal threads with a tap and repair external threads with a file. If parts have damage which cannot be repaired with a tap or file, get new ones.

2. Check that filter housing (3) and filter housing cover (4) have no dents and cracks. If parts are damaged, get new ones.

3. Check that bracket (5) is not damaged. Repair minor damage by straightening or welding. If bracket has major damage, get new one.

4. Check nut (6), screw (7), and locknut (8) for stripped threads. If parts are stripped, get new ones.

END OF TASK

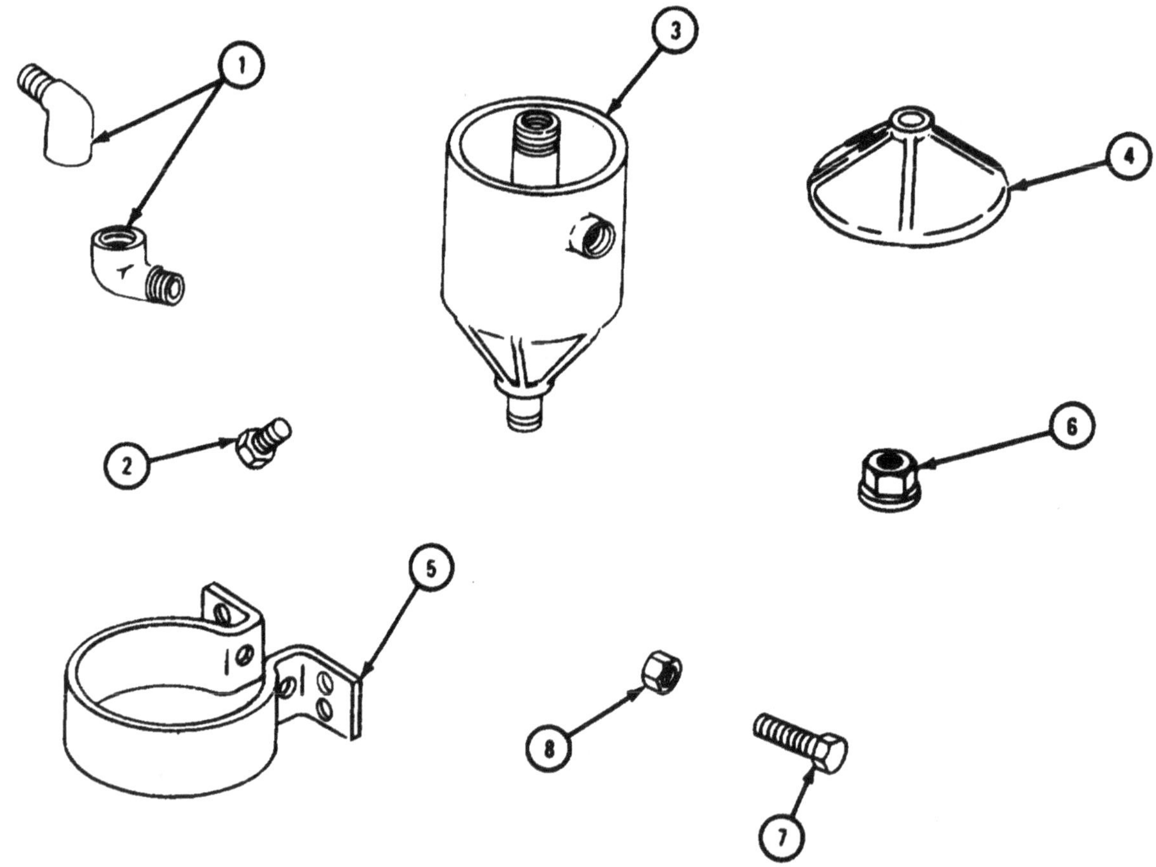

(4) Assembly .

FRAME 1

1. Put on two elbows (1) and put in drain plug (2).
2. Put in new filter element (3).
3. Put new gasket (4) in filter housing cover (5) and put filter housing cover on filter housing (6) .
4. Put on nut and washer (7).

GO TO FRAME 2

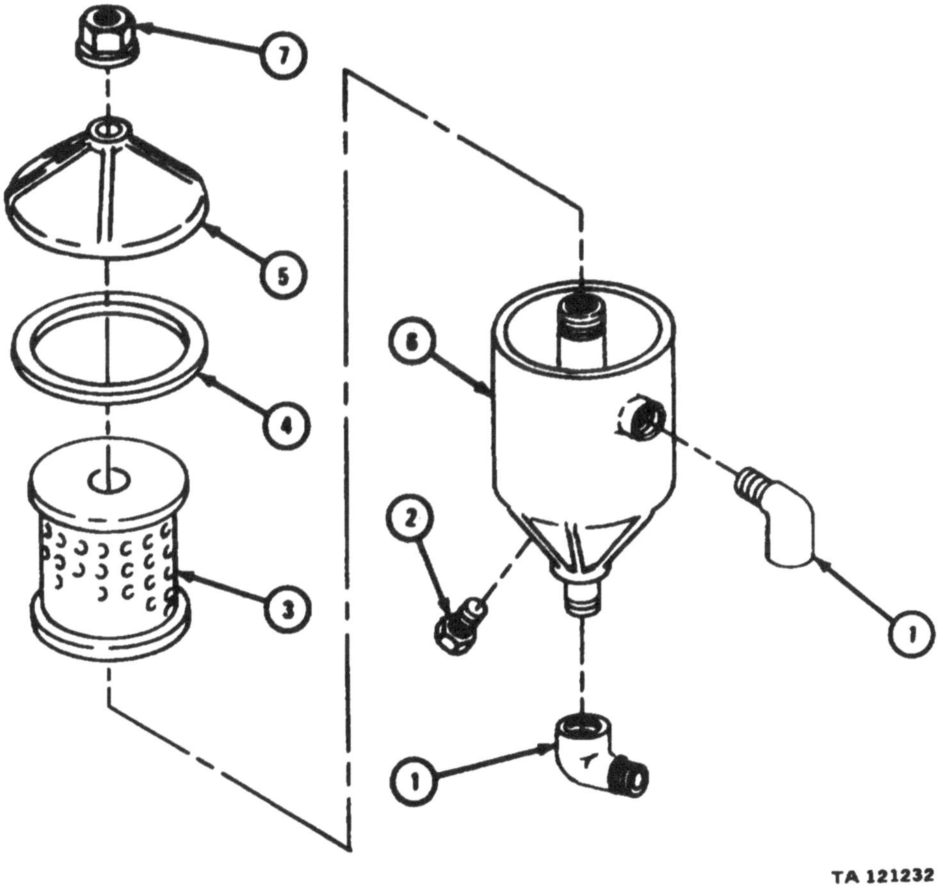

FRAME 2

1. Put filter housing (1) in bracket (2).
2. Put in screw (3) and put *on* locknut (4).

END OF TASK

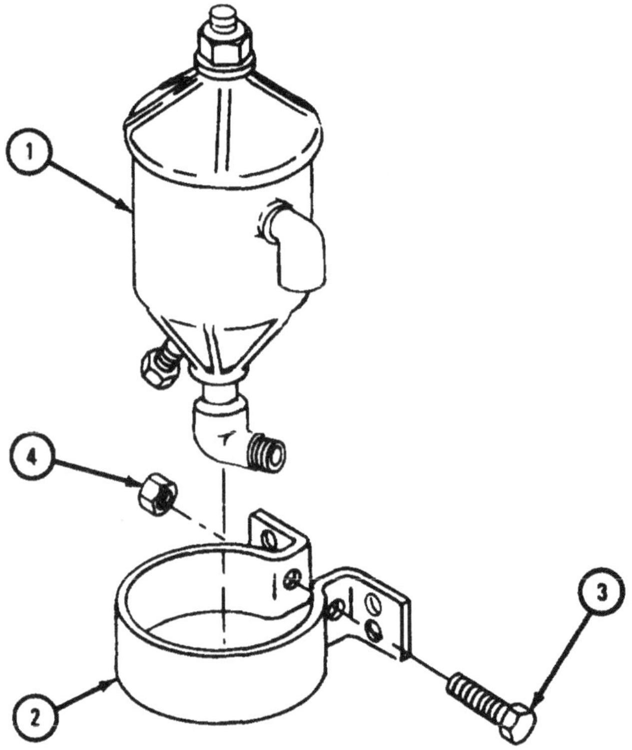

4-34. FUEL FILTER HOSES AND RELATED PARTS.

a. Engine LDS-465-1 .

(1) Cleaning .

FRAME 1

1. Take clamp (1) off hoses (2 and 3).
2. Clean inside and outside of two tubes (2 and 3) with soapy water and dry with clean cloth.

WARNING

Dry cleaning solvent is flammable. Do not use near an open flame. Keep a fire extinguisher nearby when solvent is used. Use only in well-ventilated places. Failure to do this may result in injury to personnel and damage to equipment.

CAUTION

Do not let dry cleaning solvent touch two hoses (2 and 3). Hoses will dry out, rot, and need replacement. If dry cleaning solvent splashes on hoses, dry them with clean cloth.

3. Clean clamp (1) and four tube nuts (4) with dry cleaning solvent. Use a stiff brush to take off sludge or gum deposits. Dry with clean cloth.

END OF TASK

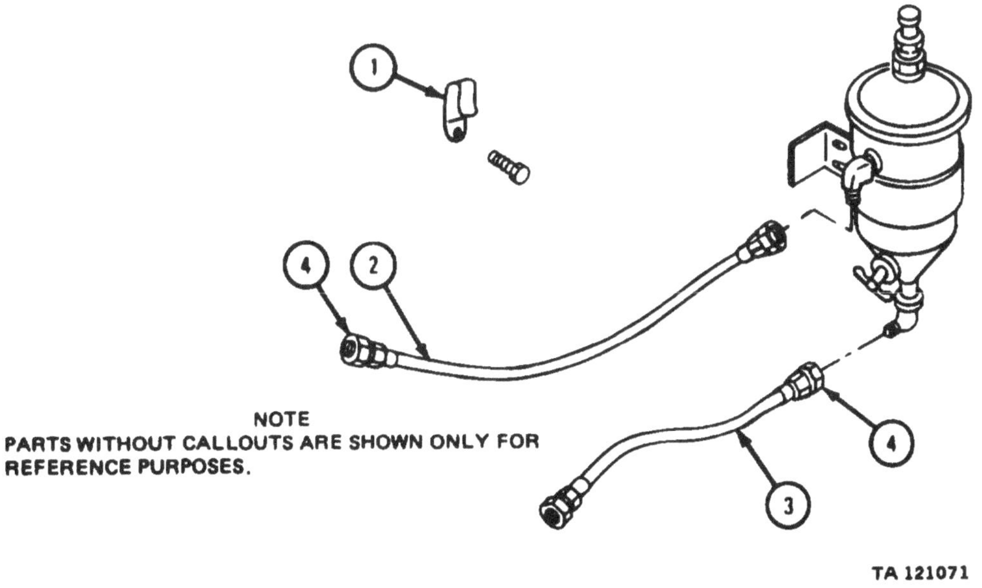
1
4
2
4
3
NOTE
PARTS WITHOUT CALLOUTS ARE SHOWN ONLY FOR REFERENCE PURPOSES.
TA 121071

(2) Inspection and repair .

FRAME 1

1. Check that all threads on tube nuts at end of tubes (1 and 2) are not stripped, crossthreaded or damaged in any other way. Fix minor thread damage with a tap. If more repair is needed, get a new part.
2. Check that two tubes (1 and 2) are not cracked, brittle, split or damaged in any other way. If tube is damaged in any way, get a new part.
3. Check that clamp (3) is not bent, dented or cracked. Take out minor bends or dents. If more repair is needed, get a new part.
4. Put clamp (3) on hoses (1 and 2).

END OF TASK

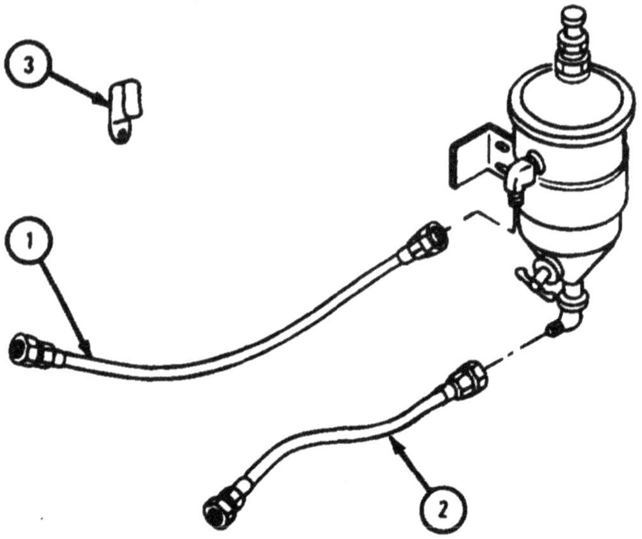

NOTE: CHECK ONLY THOSE PARTS WHICH ARE CALLED OUT. PARTS WITHOUT CALLOUTS ARE SHOWN ONLY FOR REFERENCE PURPOSES'

TA 121072

b. Engine LDS-465-2 .

(1) Cleaning .

FRAME 1

1. Take off four clamps (1).
2. Clean inside and outside of three tubes (2, 3, and 4) with soapy water and dry with clean cloth.

WARNING

Dry cleaning solvent is flammable. Do not use near an open flame. Keep a fire extinguisher nearby when solvent is used. Use only in well-ventilated places. Failure to do this may result in injury to personnel and damage to equipment.

CAUTION

Do not let dry cleaning solvent touch three tubes (2, 3, and 4). Tubes will dry out, rot, and need replacement. If dry cleaning solvent splashes on tubes, dry off with clean cloth.

3. Clean four clamps (1), four screws (5), and six tube nuts (6) with dry cleaning solvent. Use a stiff brush to take off sludge or gum deposits. Dry with clean cloth.

END OF TASK

NOTE
PARTS WITHOUT CALLOUTS ARE SHOWN ONLY FOR REFERENCE PURPOSES.

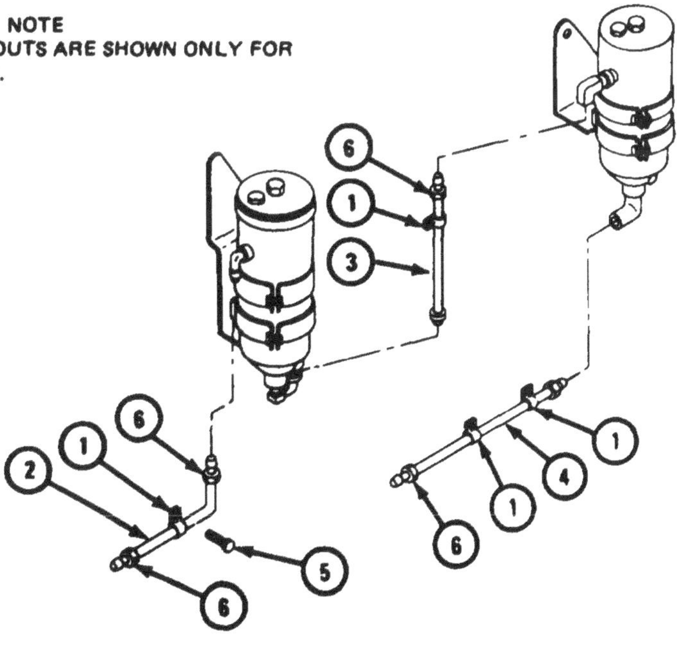

TA 121073

(2) Inspection and repair .

FRAME 1

1. Check that all threaded parts are not stripped, crossthreaded or damaged in any other way. Take out minor thread damage with thread chaser or tap. If more repair is needed, get a new part.
2. Check that four clamps (1) are not bent, dented or cracked. Take out minor bends or dents. If more repair is needed, get a new part.
3. Put on four tube clamps (1).
4. Check that three tubes (2, 3, and 4) are not cracked, brittle, split or damaged in any other way. If tube is damaged in any way, cut a new tube of the same length from plastic tube stock and get two new tube nuts (5). Slide two new tube nuts on tube and tape in place so tube nuts are not lost.

END OF TASK

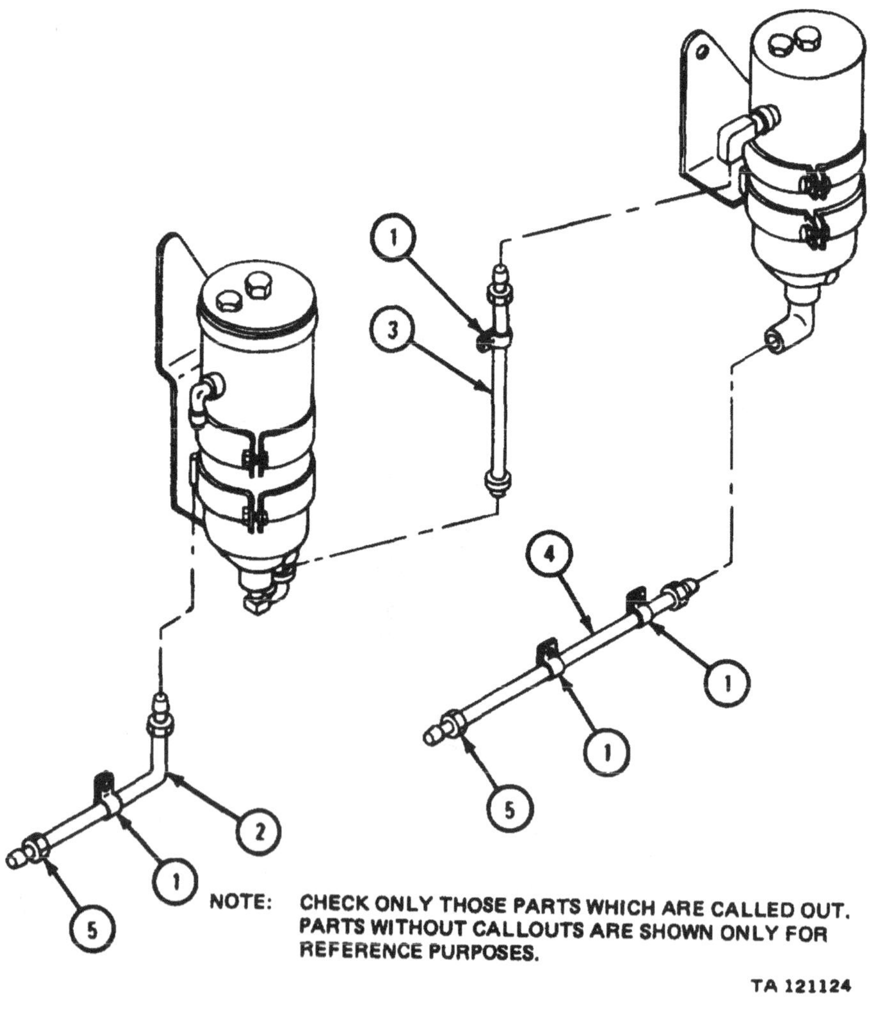
1
3
4
1
1
5
2
1
5
NOTE: CHECK ONLY THOSE PARTS WHICH ARE CALLED OUT. PARTS WITHOUT CALLOUTS ARE SHOWN ONLY FOR REFERENCE PURPOSES.
TA 121124

c. Engines LD-465-1, LD-465-1C, LDT-465-1C, and LDS-465-1A.

(1) Cleaning.

FRAME 1

1. Take off two clamps (1).
2. Clean inside and outside of three tubes (2, 3, and 4) with soap and water and dry with clean cloth.

WARNING

Dry cleaning solvent is flammable. Do not use near an open flame. Keep a fire extinguisher nearby when solvent is used. Use only in well-ventilated places. Failure to do this may result in injury to personnel and damage to equipment.

CAUTION

Do not let dry cleaning solvent touch three tubes (2, 3, and 4). Tubes will dry out, rot, and need replacement. If dry cleaning solvent splashes on tubes, dry them with clean cloth.

3. Clean two clamps (1), two screws (5), and six tube nuts (6) with dry cleaning solvent. Use a stiff brush to take off sludge or gum deposits. Dry with clean cloth.

END OF TASK

NOTE
PARTS WITHOUT CALLOUTS ARE SHOWN ONLY FOR REFERENCE PURPOSES.

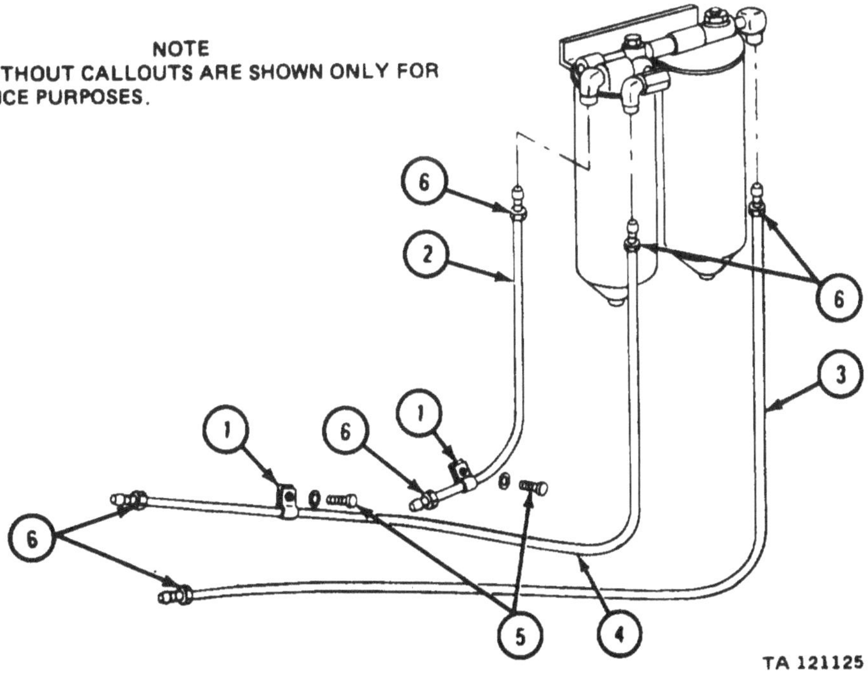

TA 121125

(2) Inspection and repair .

FRAME 1

1. Check that all threaded parts are not stripped, crossthreaded or damaged in any other way. Take out minor thread damage with thread chaser or tap. If more repair is needed, get a new part.
2. Check that two clamps (1) are not bent, dented or cracked. Take out minor bends or dents. If more repair is needed, get a new part.
3. Put on two tube clamps (1).
4. Check that three tubes (2, 3, and 4) are not cracked, brittle, split or damaged in any other way. If tube is damaged in any way, cut a new tube of the same length from plastic tube stock and get two new tube nuts (5). Slide two new tube nuts on tube and tape in place so tube nuts are not lost.

END OF TASK

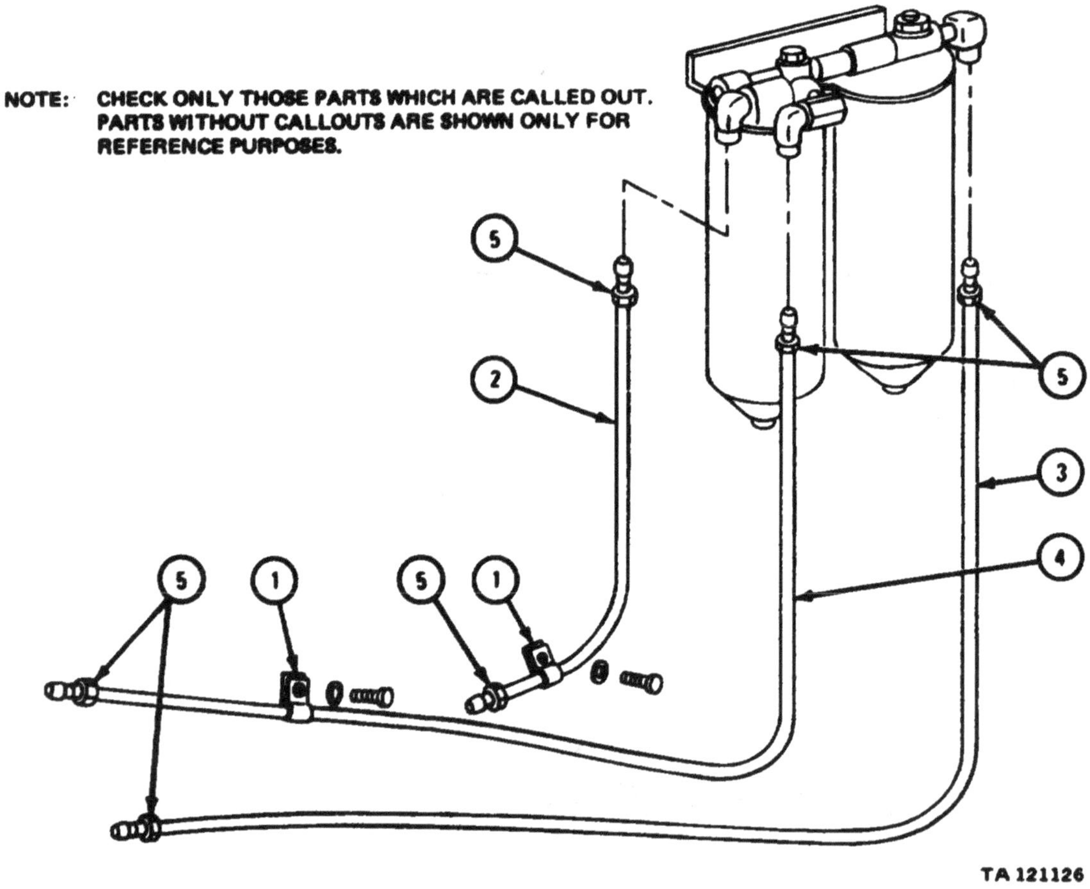
NOTE: CHECK ONLY THOSE PARTS WHICH ARE CALLED OUT. PARTS WITHOUT CALLOUTS ARE SHOWN ONLY FOR REFERENCE PURPOSES.
5
2
5
3
4
5
1
5
1
TA 121126

4-35. FUEL INJECTION PUMP ATTACHING PARTS.

a. Cleaning. There are no special cleaning procedures needed. Refer to cleaning procedures given in para 4-3.

b. Inspection and Repair.

FRAME 1

1. Check that two screws (1), two screws (2), three screws (3), two fue injection pump mounting screws (4), and nut (5) are not stripped, cross-threaded or damaged in any other way. If parts are damaged, get a new part. 1

2. Check that support bracket (6) and lockplate (7) are not bent or cracked. Take out minor bends. If more repair is needed, get a new part.

GO TO FRAME 2

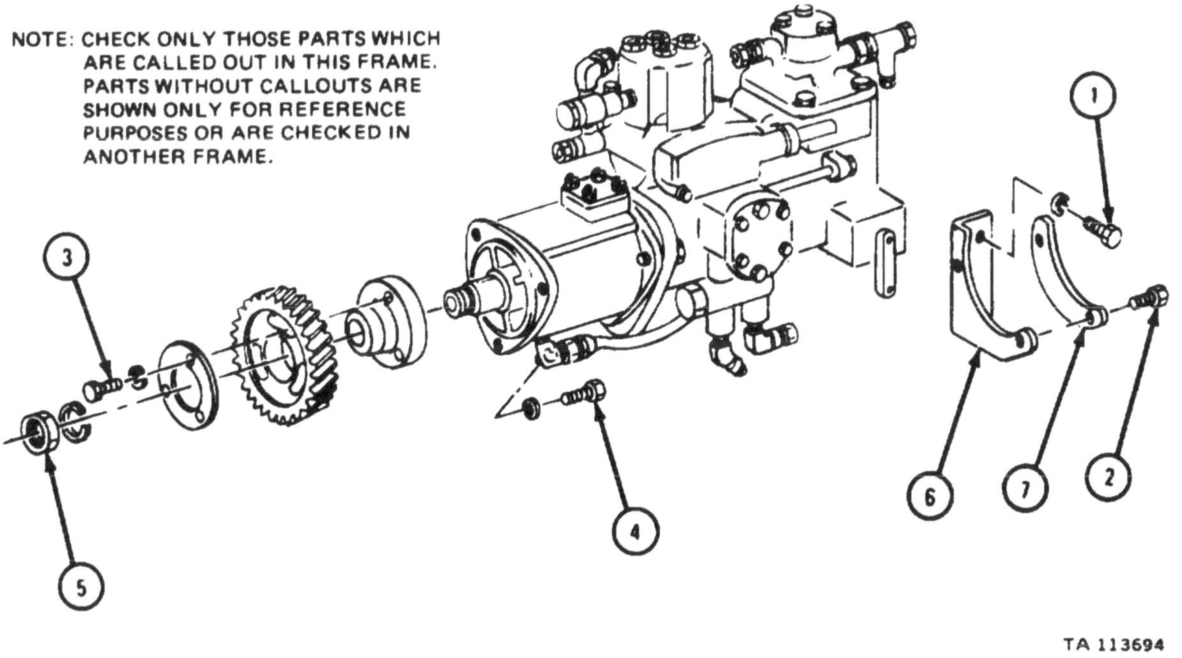
NOTE: CHECK ONLY THOSE PARTS WHICH ARE CALLED OUT IN THIS FRAME. PARTS WITHOUT CALLOUTS ARE SHOWN ONLY FOR REFERENCE PURPOSES OR ARE CHECKED IN ANOTHER FRAME.
1
2
3
4
5
6
7
TA 113694

FRAME 2

1. Check that shaft coupling (1) is not bent, cracked or crossthreaded. Fi x minor thread damage with a tap. If more repair is needed, get a new shaft coupling.
2. Check that retaining plate (2) is not bent, cracked or damaged in any other way. If repair is needed, get a new retaining plate.
3. Check that gear (3) is not cracked, chipped, scratched or badly worn . Take out minor scratches or burrs with a fine mill file. If more repair is needed, get a new gear.

END OF TASK

NOTE: CHECK ONLY THOSE PARTS WHICH ARE CALLED OUT IN THIS FRAME. PARTS WITHOUT CALLOUTS ARE SHOWN ONLY FOR REFERENCE PURPOSES OR ARE CHECKED IN ANOTHER FRAME.

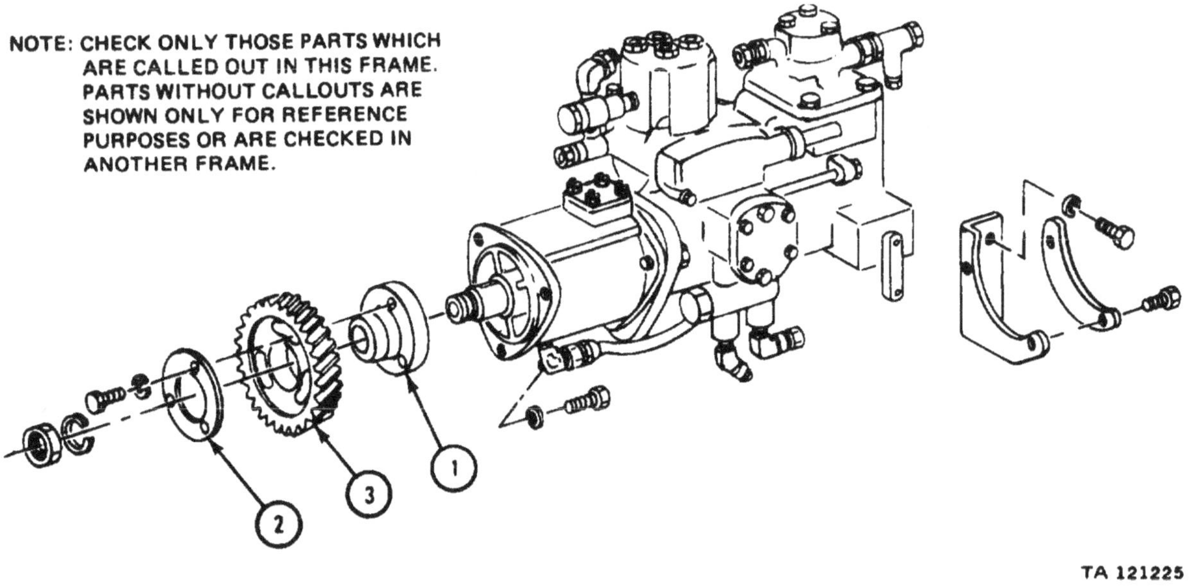

4-36. HYDRAULIC PUMP AND ADAPTER ASSEMBLY (ENGINES LDS-465-1, LDS-465-1A, AND LDS-465-2).

a. Engine LDS-465-2.

(1) Disassembly.

FRAME 1

1. Takeout three sockethead capscrews and lockwashers (1).
2. Takeoff manifold assembly (2) and manifold gasket (3). Throw away gasket.
3. Take off elbow (4), nut (5), and preformed packing (6). Throw away preformed packing.

GO TO FRAME 2

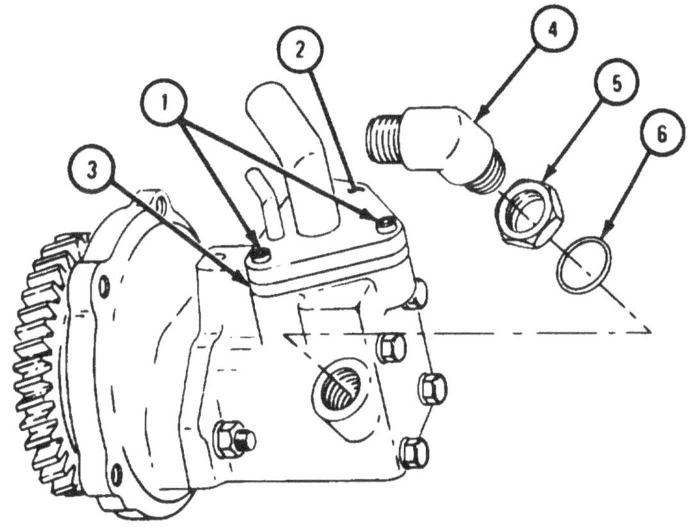

TA 121061

FRAME 2

1. Take off nut (1) and washer (2).
2. Using universal puller, take off drive gear (3).
3. Take woodruff key (4) out of shaft (5).
4. Take off two nuts and washers (6).
5. Take pump adapter (7) apart from pump (8). Take off and throw away pump adapter gasket (9) .

NOTE

Do not take out two studs (10) unless they are damaged.

6. If damaged, take out two studs (10).

END OF TASK

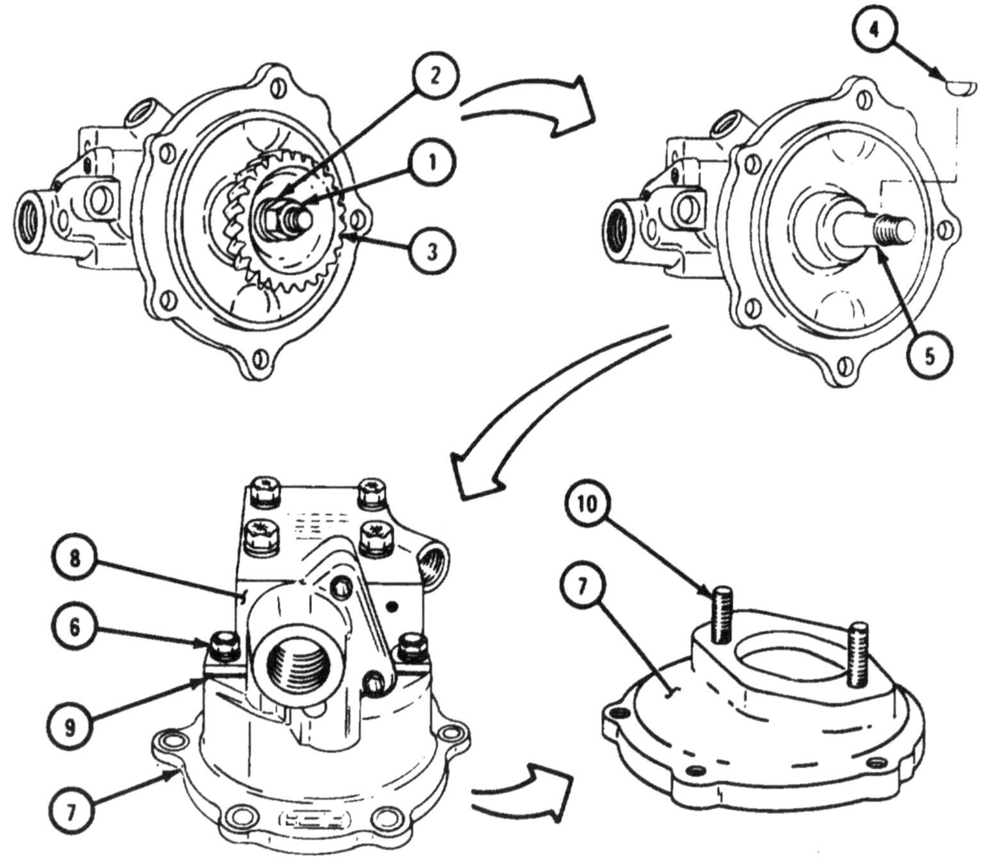

TA 121127

(2) Cleaning .

FRAME 1

WARNING

Dry cleaning solvent is flammable. Do not use near an open flame. Keep a fire extinguisher nearby when solvent is used. Use only in well-ventilated places. Failure to do this may result in injury to personnel and damage to equipment.

CAUTION

Do not put hydraulic pump (1) in dry cleaning solvent. Parts can be damaged and pump will not run properly.

1. Clean hydraulic pump (1) with dry cleaning solvent. Use a stiff brush to take off carbon and sludge. Dry with clean cloth.
2. Clean adapter (2), drive gear (3), manifold assembly (4), elbow (5), three screws (6), nut (7), woodruff key (8), and two nuts (9) in dry cleaning solvent. Use stiff brush to remove carbon and sludge. Dry with clean cloth.

END OF TASK

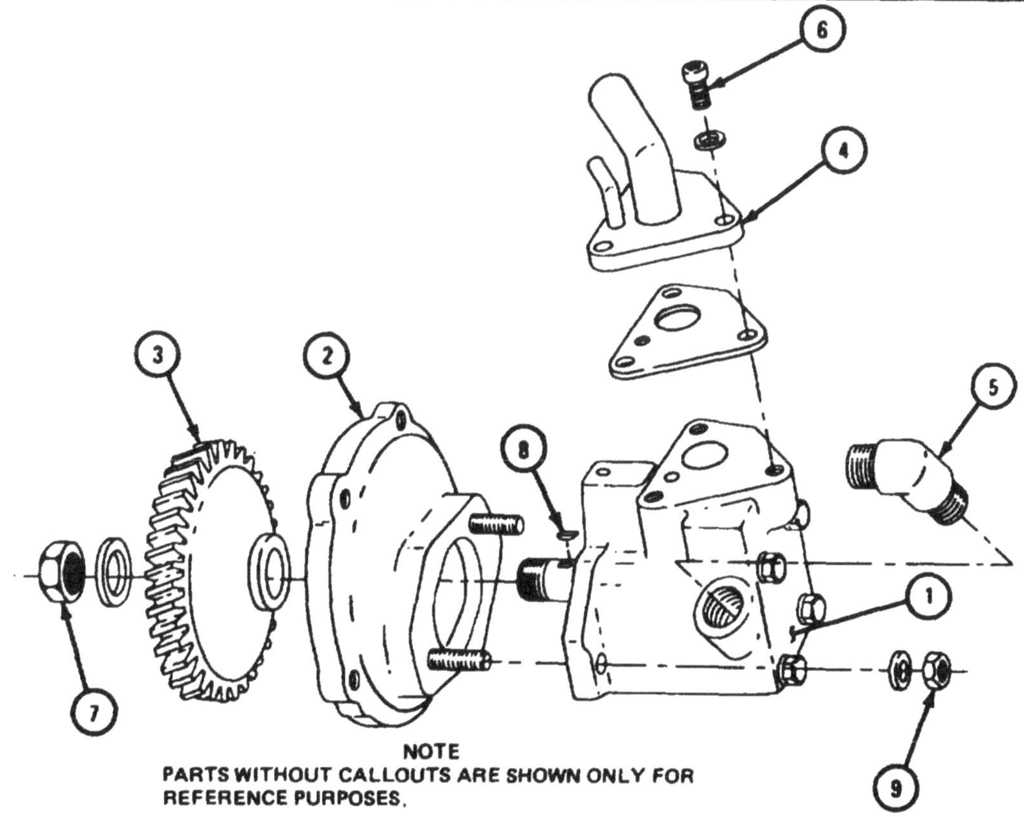

(3) Inspection and repair.

FRAME 1

1. Check that all threaded parts are not stripped, crossthreaded or damaged in any other way. Take out minor thread damage with thread chaser or tap. If more repair is needed, get a new part.

WARNING

Dry cleaning solvent is flammable. Do not use near an open flame. Keep a fire extinguisher nearby when solvent is used. Use only in well-ventilated places. Failure to do this may result in injury to personnel and damage to equipment.

2. Check that hydraulic pump (1) has no cracks, nicks, burrs or scratches on gasket surface (2) . Take out minor nicks, burrs or scratches with a fine mill file or crocus cloth dipped in dry cleaning solvent. If more repair is needed, get a new part.

GO TO FRAME 2

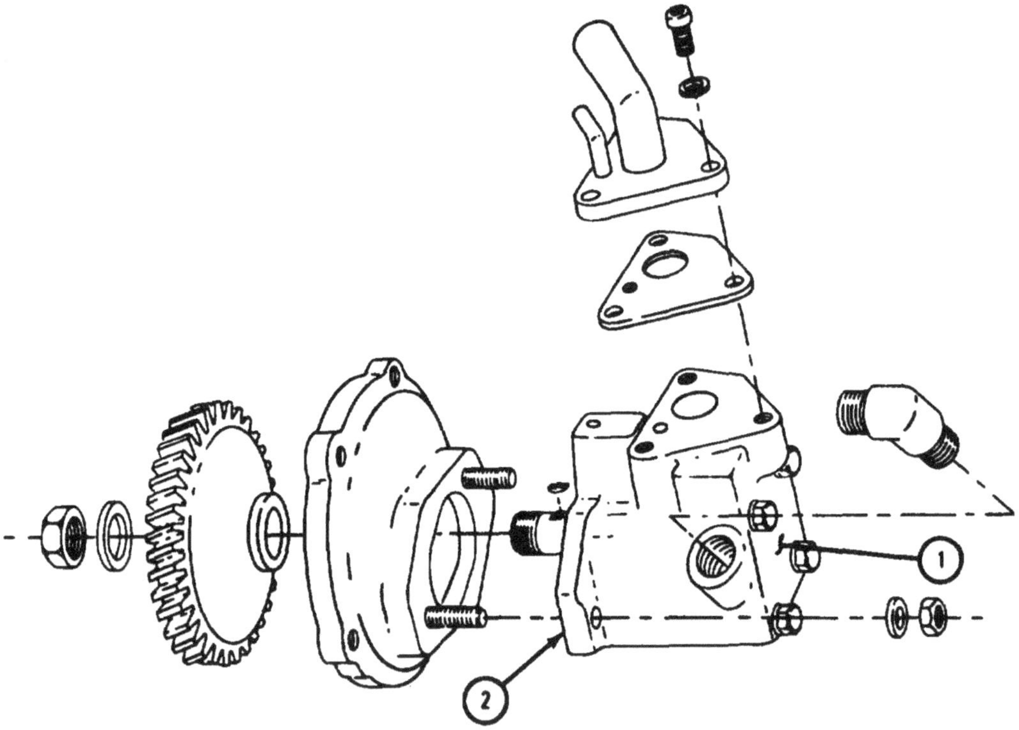

NOTE: CHECK ONLY THOSE PARTS WHICH ARE CALLED OUT IN THIS FRAME. PARTS WITHOUT CALLOUTS ARE SHOWN ONLY FOR REFERENCE PURPOSES OR ARE CHECKED IN ANOTHER FRAME.

TA 121129

FRAME 2

WARNING

Dry cleaning solvent is flammable. Do not use near an open flame. Keep a fire extinguisher nearby when solvent is used. Use only in w en-ventilated places. Failure to do this may result in injury to personnel and damage to equipment.

1. Check that pump adapter (1) and manifold assembly (2) are not cracked, bent, dented and have no nicks, burrs or scratches on gasket surfaces . Take out minor nicks, burrs or scratches with a fine mill file or crocus coth dipped in dry cleaning solvent. If more repair is needed, get a new part.
2. Check that drive gear (3) has no cracks, nicks, burrs, scratches or wor n or chipped gear teeth. Take out minor nicks, burrs or scratches with a fine mill file or crocus cloth dipped in dry cleaning solvent. If more repair is needed, get a new part.

GO TO FRAME 3

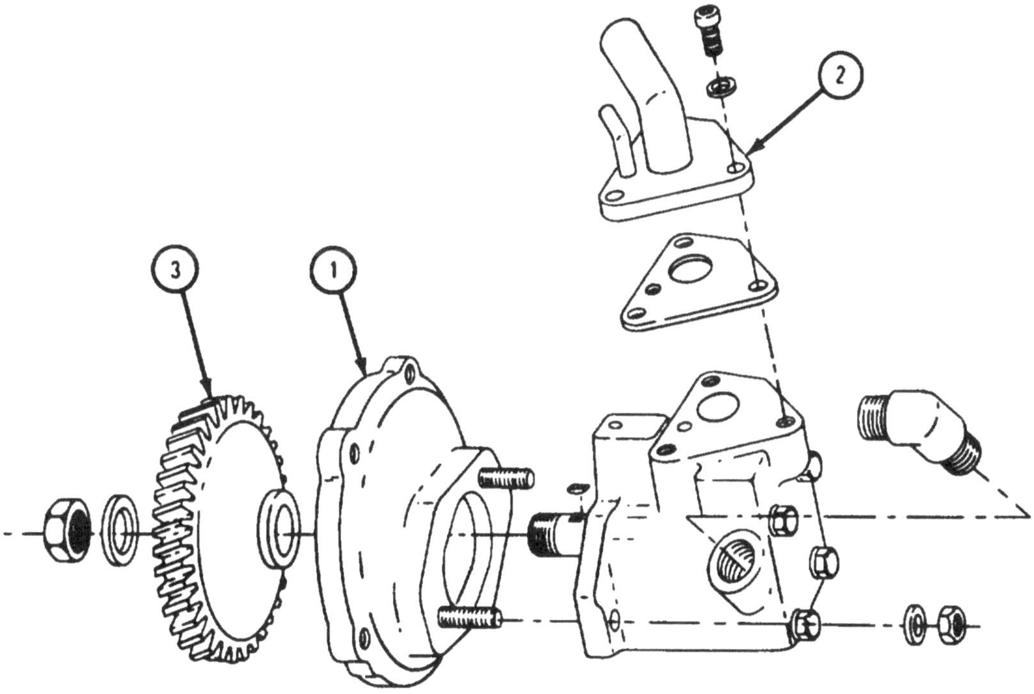

NOTE: CHECK ONLY THOSE PARTS WHICH ARE CALLED OUT IN THIS FRAME. PARTS WITHOUT CALLOUTS ARE SHOWN ONLY FOR REFERENCE PURPOSES OR ARE CHECKED IN ANOTHER FRAME.

TA121130

FRAME 3

1. Check that elbow (1) is not cracked, bent or damaged in any other way. If elbow is damaged, get a new part.

WARNING

Dry cleaning solvent is flammable. Do not use near an open flame. Keep a fire extinguisher nearby when solvent is used. Use only in well en-ventilated places. Failure to do this may result in injury to personnel and damage to equipment.

2. Check that woodruff key (2) is not cracked, chipped, nicked, burred or worn. Take out minor nicks or burrs with fine mill file or crocus cloth dipped in dry cleaning solvent. If more repair is needed, get a new part,

END OF TASK

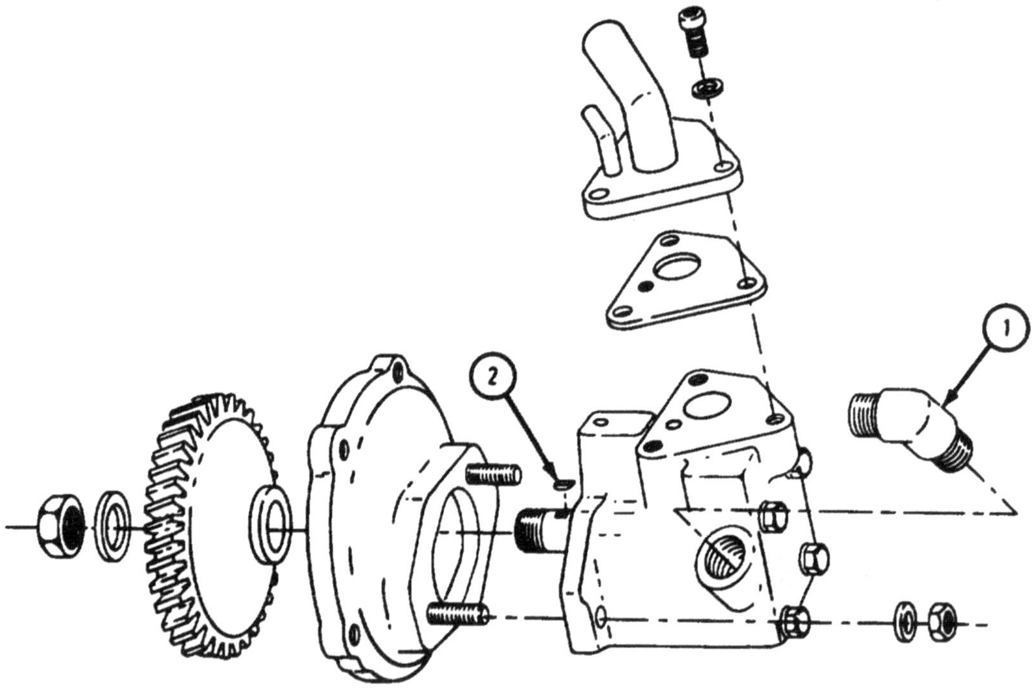

NOTE: CHECK ONLY THOSE PARTS WHICH ARE CALLED OUT IN THIS FRAME. PARTS WITHOUT CALLOUTS ARE SHOWN ONLY FOR REFERENCE PURPOSES OR ARE CHECKED IN ANOTHER FRAME.

TA 121131

(4) Assembly.

FRAME 1

NOTE

If two studs (1) were not taken out, go to step 2.

1. Put in two studs (1).
2. Put gasket (2) on pump adapter (3).
3. Put pump adapter (3) on pump body (4). Two studs (1) on pump adapter must fit through mounting holes in 'pump body.
4. Put on two washers (5) and nuts (6).

GO TO FRAME 2

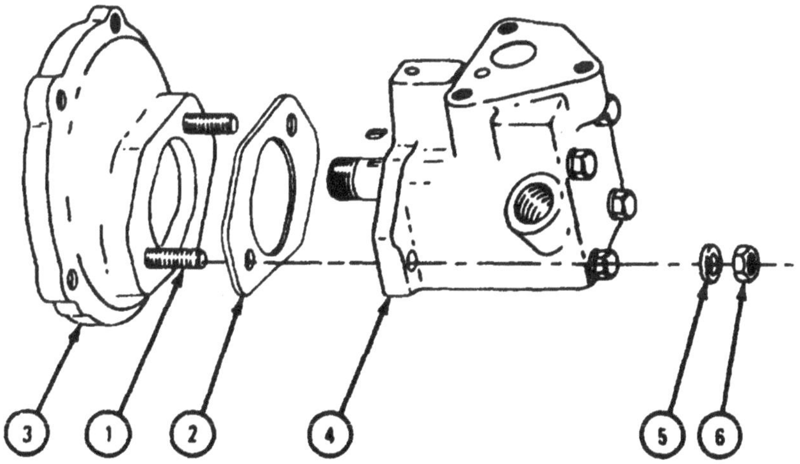

TA 121132

FRAME 2

1. Put woodruff key (1) in shaft (2) .
2. Put drive gear (3) on shaft (2), alining keyway (4) with woodruff key (l) .
3. Using hydraulic pump gear installer, put on drive gear (3).
4. Put on washer (5) and nut (6). Using torque wrench, tighten nut to 45 to 50 pound-feet .

GO TO FRAME 3

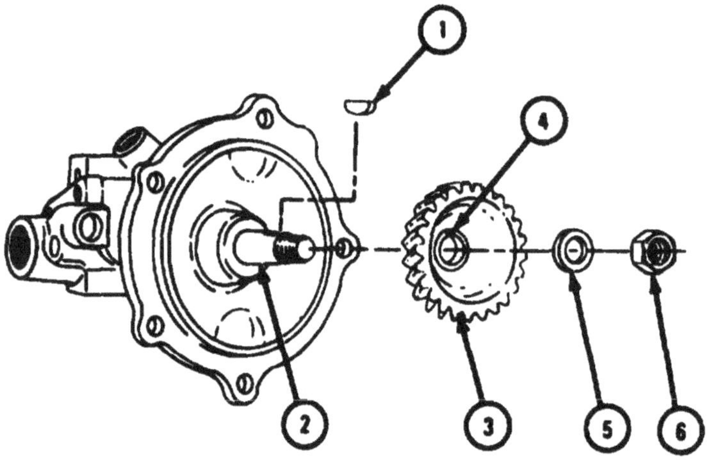

TA 121133

FRAME 3

1. Put hydraulic pump manifold gasket (1) on hydraulic pump (2), dinin g holes.
2. Put hydraulic pump manifold (3) on hydraulic pump (2), alining holes.
3. Put in three screw s (4) and washers (5) .
4. Put preformed packing (6) on elbow (7) .
5. Put elbow (7) in hydraulic pump (2).

END OF TASK

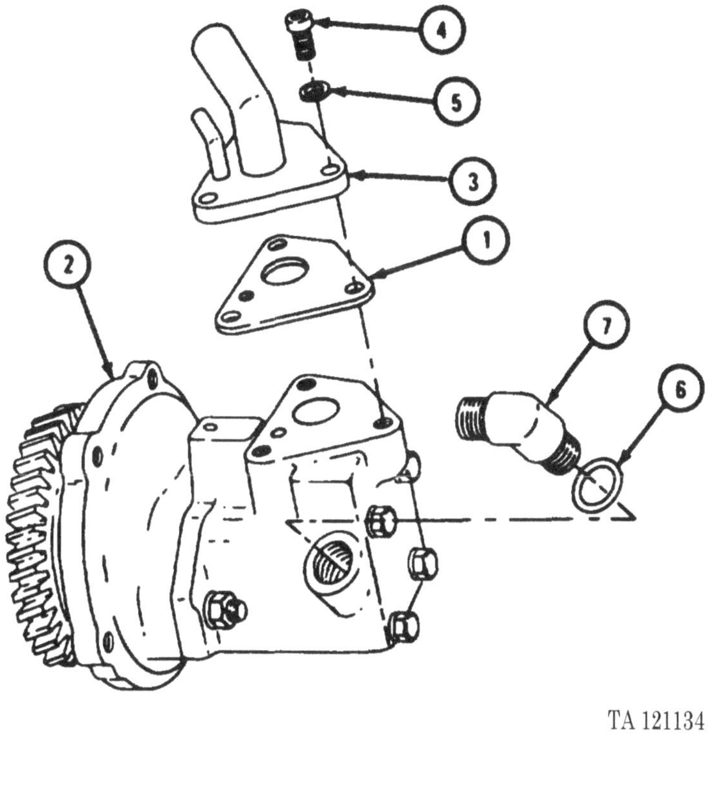

TA 121134

b. Engines LDS-465-1 and LDS-465-1A.

(1) Disassembly.

FRAME 1

1. Take out elbow (1).
2. Take out elbow (2), adapter (3), and preformed packing (4). Throw away preformed packing.
3. Take out elbow (5).

GO TO FRAME 2

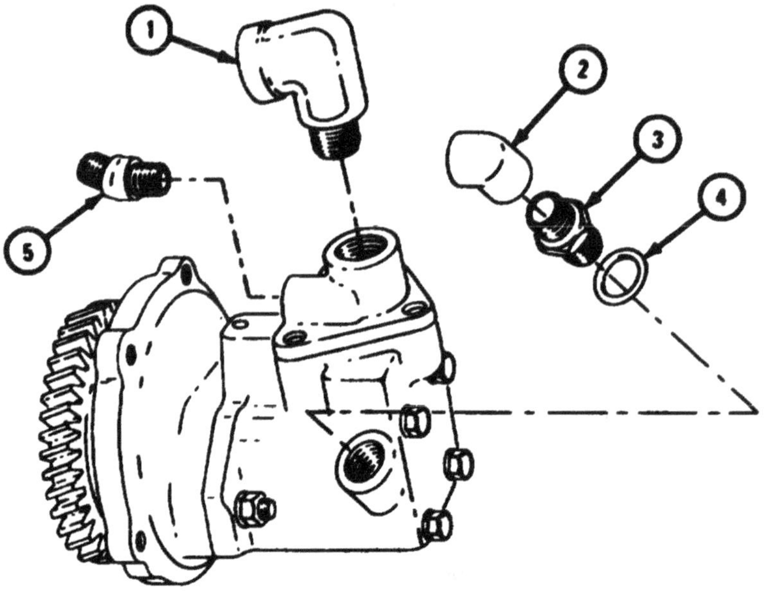

TA 121135

FRAME 2

1. Take off nut (1) and washer (2).
2. Using universal puller, take off drive gear (3).
3. Take woodruff key (4) out of shaft (5) .

GO TO FRAME 3

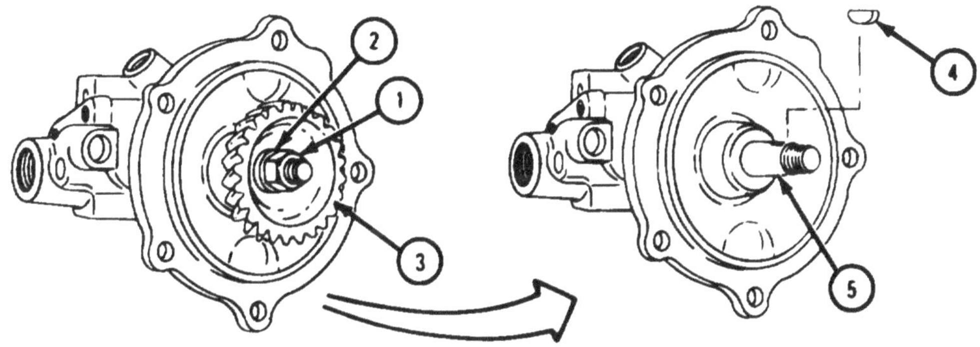

TA 121136

FRAME 3

1. Take off two nuts (1) and washers (2).
2. Take pump adapter (3) apart from pump body (4). Take off and throw away adapter gasket (5) .

NOTE

Do not take out two studs (6) unless they are damaged.

3. If damaged, take out two studs (6).

END OF TASK

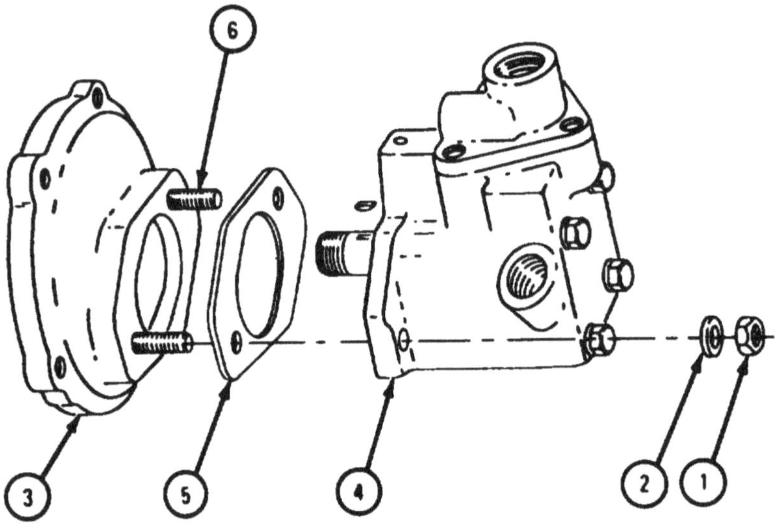

TA 121137

(2) Cleaning.

FRAME 1

WARNING

Dry cleaning solvent is flammable. Do not use near an open flame. Keep a fire extinguisher nearby when solvent is used . Use only in well-ventilated places. Failure to do this may result in injury to personnel and damage to equipment.

CAUTION

Do not put hydraulic pump (1) in dry cleaning solvent, Parts can be damaged and pump will not run properly.

1. Clean hydraulic pump (1) with dry cleaning solvent. Use a stiff brush to take off carbon and sludge. Dry with clean cloth .
2. Clean three elbows (2, 3, and 4) , adapter (5), drive gear (6), pump adapter (7), two nuts (8), woodruff key (9), and nut (10) in dry cleaning solvent . Use a stiff brush to take off carbon and sludge. Dry with clean cloth.

END OF TASK

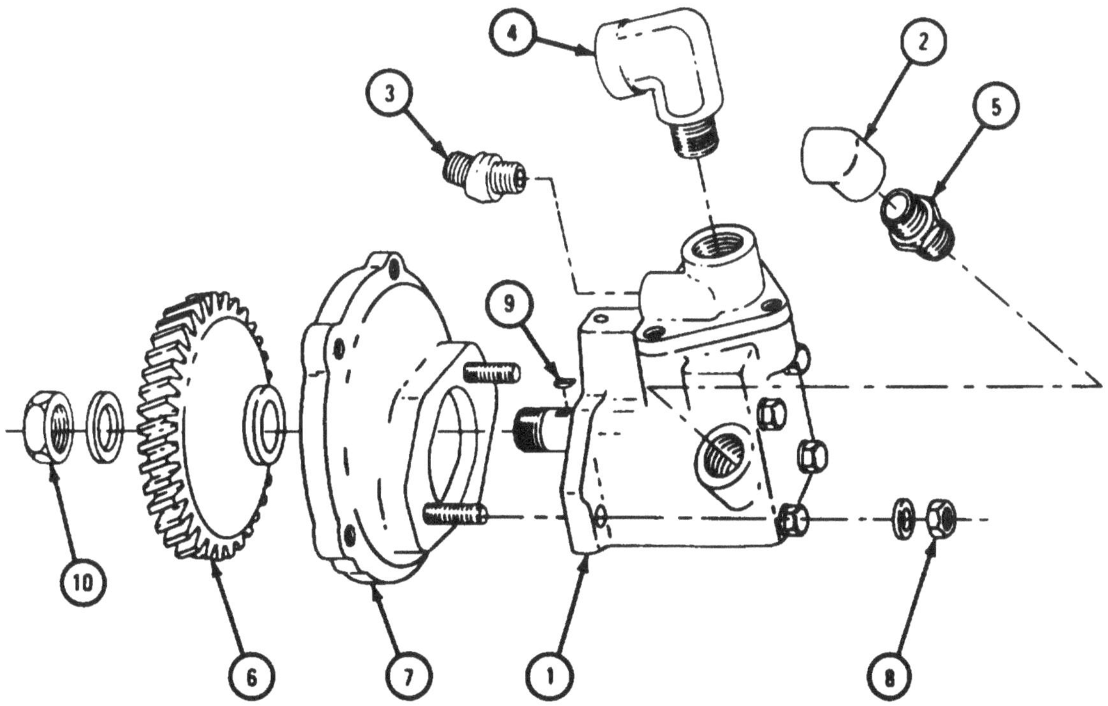

NOTE
PARTS WITHOUT CALLOUTS ARE SHOWN ONLY FOR REFERENCE PURPOSES.

TA 121138

(3) Inspection and repair .

FRAME 1

1. Check that all threaded parts are not stripped, crossthreaded or damaged in any other way. Take out minor thread damage with thread chaser or tap. If more repair is needed, get a new part.

WARNING

Dry cleaning solvent is flammable. Do not use near an open flame. Keep a fire extinguisher nearby when solvent is used. Use only in w en-ventilated places. Failure to do this may result in injury to personnel and damage to equipment.

2. Check that hvdraulic pump (1) has no cracks, nicks, burrs or scratches on gasket surface . Take out minor nicks, burrs or scratches with a fine mill file or crocus cloth dipped in dry cleaning solvent. If more repair is needed, get a new part.

GO TO FRAME 2

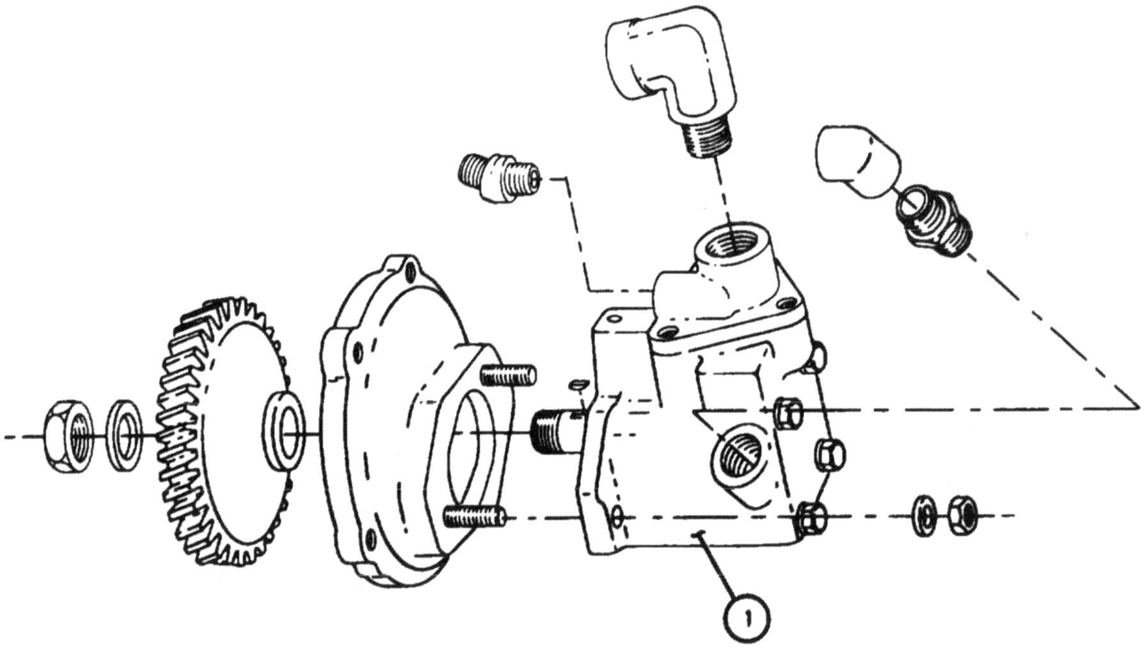

NOTE: CHECK ONLY THOSE PARTS WHICH ARE CALLED OUT IN THIS FRAME. PARTS WITHOUT CALLOUTS ARE SHOWN ONLY FOR REFERENCE PURPOSES OR ARE CHECKED IN ANOTHER FRAME.

TA 121139

FRAME 2

WARNING

Dry cleaning solvent is flammable. Do not use near an open flame. Keep a fire extinguisher nearby when solvent is used. Use only in well-ventilated places. Failure to do this may result in injury to personnel and damage to equipment.

1. Check that pump adapter (1) is not cracked or bent and has no nicks, burrs or scratches on gasket surfaces . Fix minor nicks, burrs or scratches with a fine mill file or crocus cloth dipped in dry cleaning solvent. If more repair is needed, get a new part.

2. Check that drive gear (2) has no cracks, nicks, burrs, scratches or worn or chipped gear teeth . Take out minor nicks, burrs or scratches with a fine mill file or crocus cloth dipped in dry cleaning solvent. If more repair is needed, get a new part.

GO TO FRAME 3

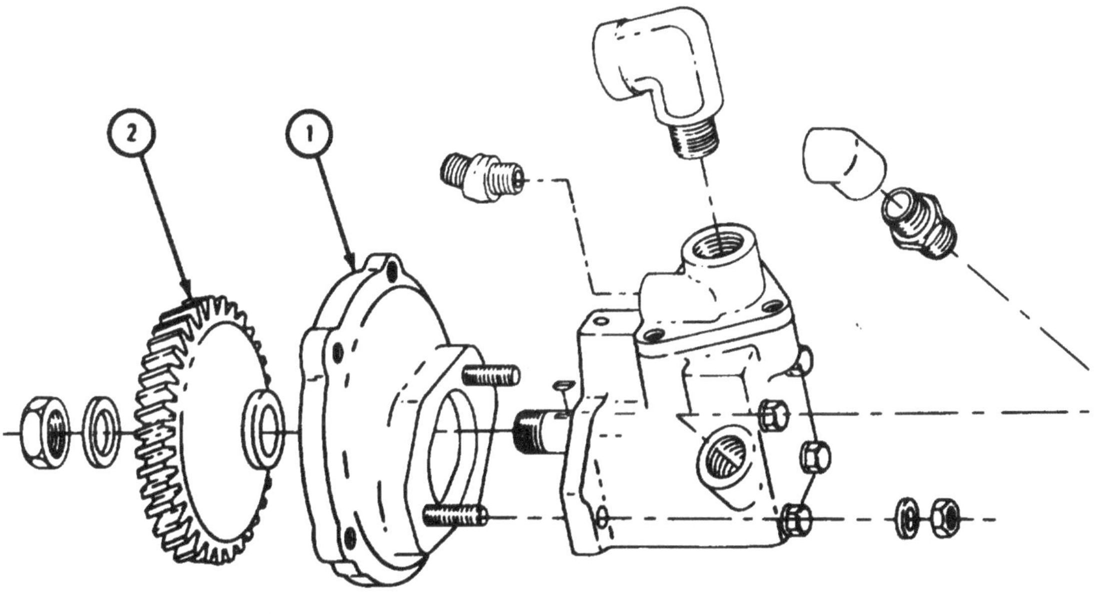

NOTE: CHECK ONLY THOSE PARTS WHICH ARE CALLED OUT IN THIS FRAME. PARTS WITHOUT CAL LOUTS ARE SHOWN ONLY FOR REFERENCE PURPOSES OR ARE CHECKED IN ANOTHER FRAME.

TA 121140

FRAME 3

1. Check that three elbows (1, 2, and 3) and adapter (4) are not cracked, bent or damaged in any other way. If elbows or adapters are damaged, get a new part.

WARNING

Dry cleaning solvent is flammable. Do not use near an open flame. Keep a fire extinguisher nearby when solvent is used. Use only in well-ventilated places. Failure to do this may result in injury to personnel and damage to equipment.

2. Check that woodruff key (5) is not cracked, chipped, nicked or worn. Take out minor nicks or burrs with a fine mill file or crocus cloth dipped in dry cleaning solvent . If more repair is needed, get a new part.

END OF TASK

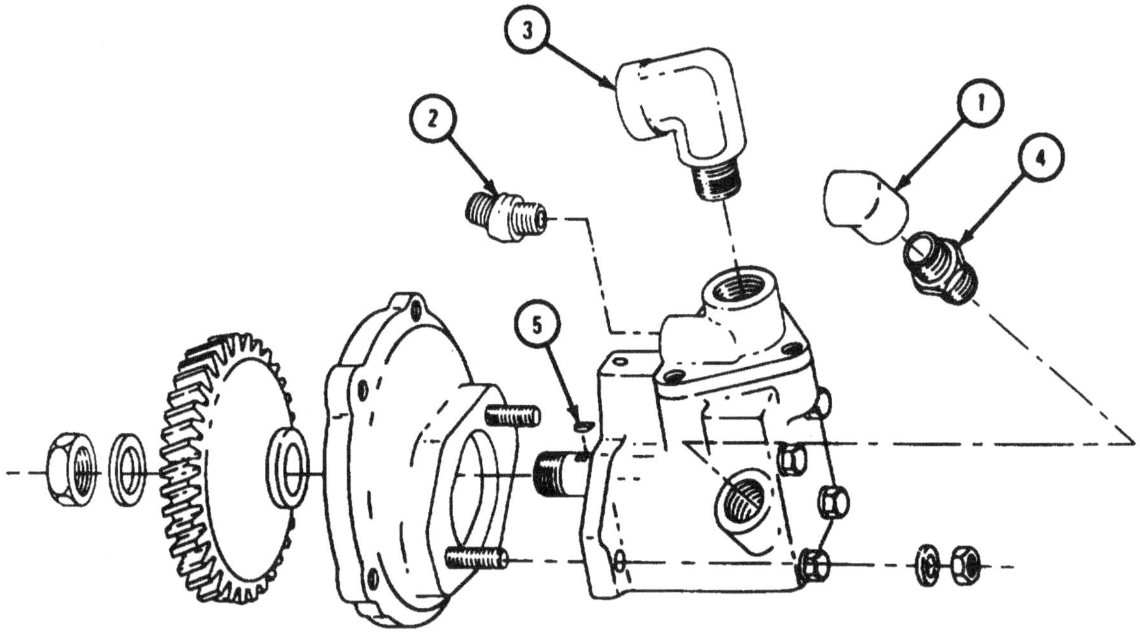

NOTE: CHECK ONLY THOSE PARTS WHICH ARE CALLED OUT IN THIS FRAME. PARTS WITHOUT CALLOUTS ARE SHOWN ONLY FOR REFERENCE PuRPOSES OR ARE CHECKED IN ANOTHER FRAME.

TA 121141

(4) Assembly.

FRAME 1

NOTE

If two studs (1) were not taken out, go to step 2.

1. Put in two studs (1).
2. Put hydraulic pump adapter gasket (2) on hydraulic pump adapter (3).
3. Put hydraulic pump adapter (3) on hydraulic pump (4). Studs in hydraulic pump adapter must fit through mounting holes in hydraulic pump.
4. Put on two washers (5) and nuts (6).

GO TO FRAME 2

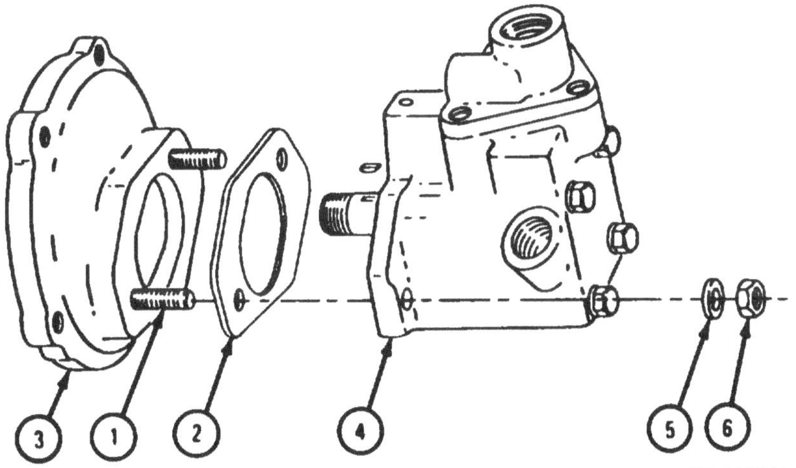

TA 121142

FRAME 2

1. Put woodruff key (1) in shaft (2) .
2. Put drive gear (3) on shaft (2), alining gear keyway with woodruff key (1).
3. Using hydraulic pump gear installer, put on drive gear (3).
4. Put washer (4) and nut (5) on shaft (2). Using torque wrench, tighten nut to 45 to 50 pound-feet.

GO TO FRAME 3

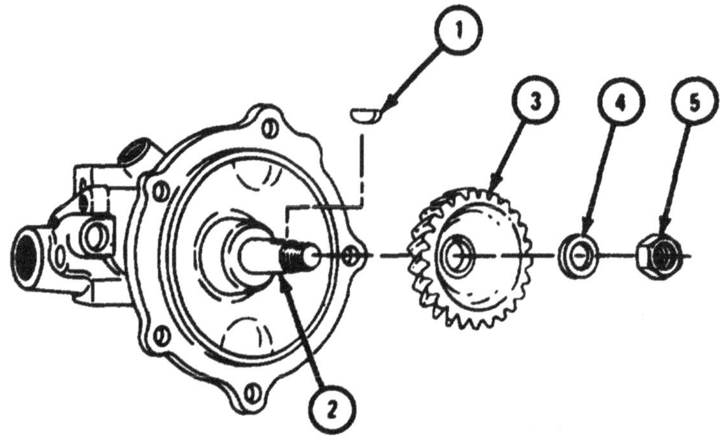

TA 121143

FRAME 3

1. Put preformed packing (1) on adapter (2) .
2. Put adapter (2) in hydraulic pump (3).
3. Put elbow (4) on adapter (2). Elbow must face down as shown.
4. Put in elbow (5) and elbow (6).

END OF TASK

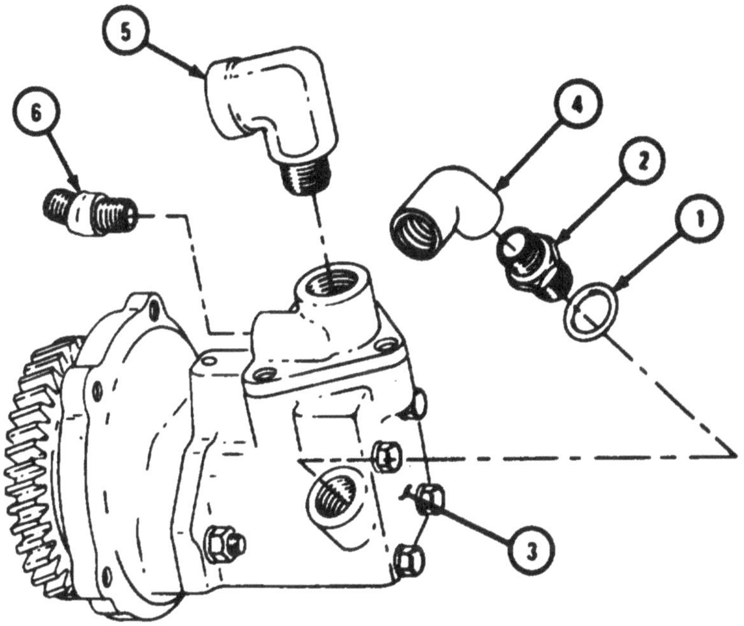

TA 121144

4-37. FUEL INJECTOR TUBE ASSEMBLIES AND ATTACHING PARTS.

a. Engine LDS-465-2.

(1) Cleaning.

FRAME 1

WARNING

Dry cleaning solvent is flammable. Do not use near an open flame. Keep a fire extinguisher nearby when solvent is used. Use only in well-ventilated places. Failure to do this may result in injury to personnel and damage to equipment.

CAUTION

Do not let dry cleaning solvent touch six rubber boots (1). Boots will dry out and crack. If dry cleaning solvent splashes on rubber boots, dry with clean cloth.

1. Slide six rubber boots (1) to one end of six fuel injector tubes (2). Clean outside of six fuel injector tubes with dry cleaning-solvent. Dry with clean cloth.
2. Slide six rubber boots (1) to center of six fuel injector tubes (2). Clean the ends of six fuel injector tubes which were covered by rubber boots in step 1.

WARNING

Eye shields must be worn when using compressed air. Eye injury can occur if eye shields are not used.

3. Flush inside of six fuel injector tubes (2) with dry cleaning solvent. Blow dry with compressed air. Plug ends of six fuel injector tubes to keep out dirt.

END OF TASK

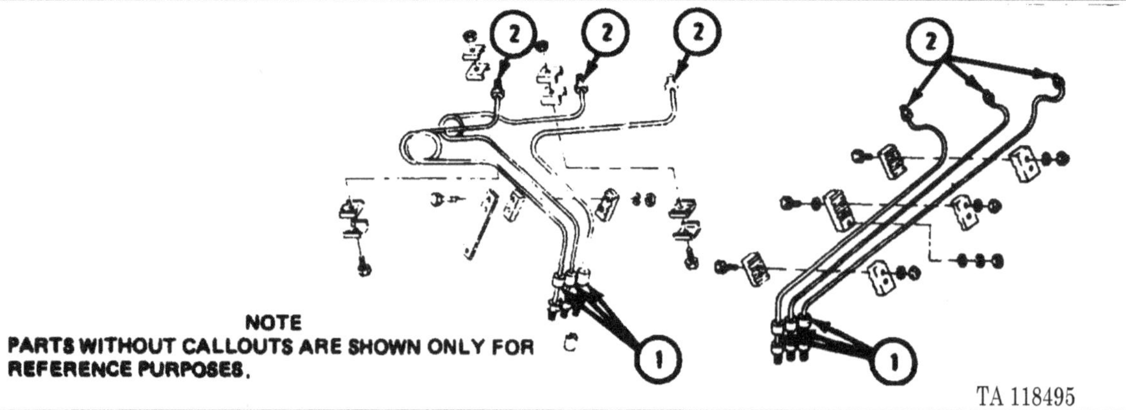
2
2
2
2
1
1
NOTE
PARTS WITHOUT CALLOUTS ARE SHOWN ONLY FOR REFERENCE PURPOSES.
TA 118495

(2) Inspection and repair.

FRAME 1

1. Check that six fuel injector tubes (1) are not cracked, split, dented, kinked or damaged in any other way. If fuel injector tube assembly is damaged, get a new one.
2. Check that 12 coupling nuts (2) are not cracked, split, dented, stripped or damaged in any other way. Use a thread chaser to repair minor thread damage If more repair is needed, get a new fuel injector tube assembly (1).
3. Check that six dust boots (3) are not cracked, torn, dried out or damaged in any other way. If dust boot is damaged in any way, cut off dust boot and get a new dust boot.

IF ANY DUST BOOTS (3) WERE CUT OFF, GO TO FRAME 2.
IF NO DUST BOOTS (3) WERE CUT OFF, GO TO FRAME 3

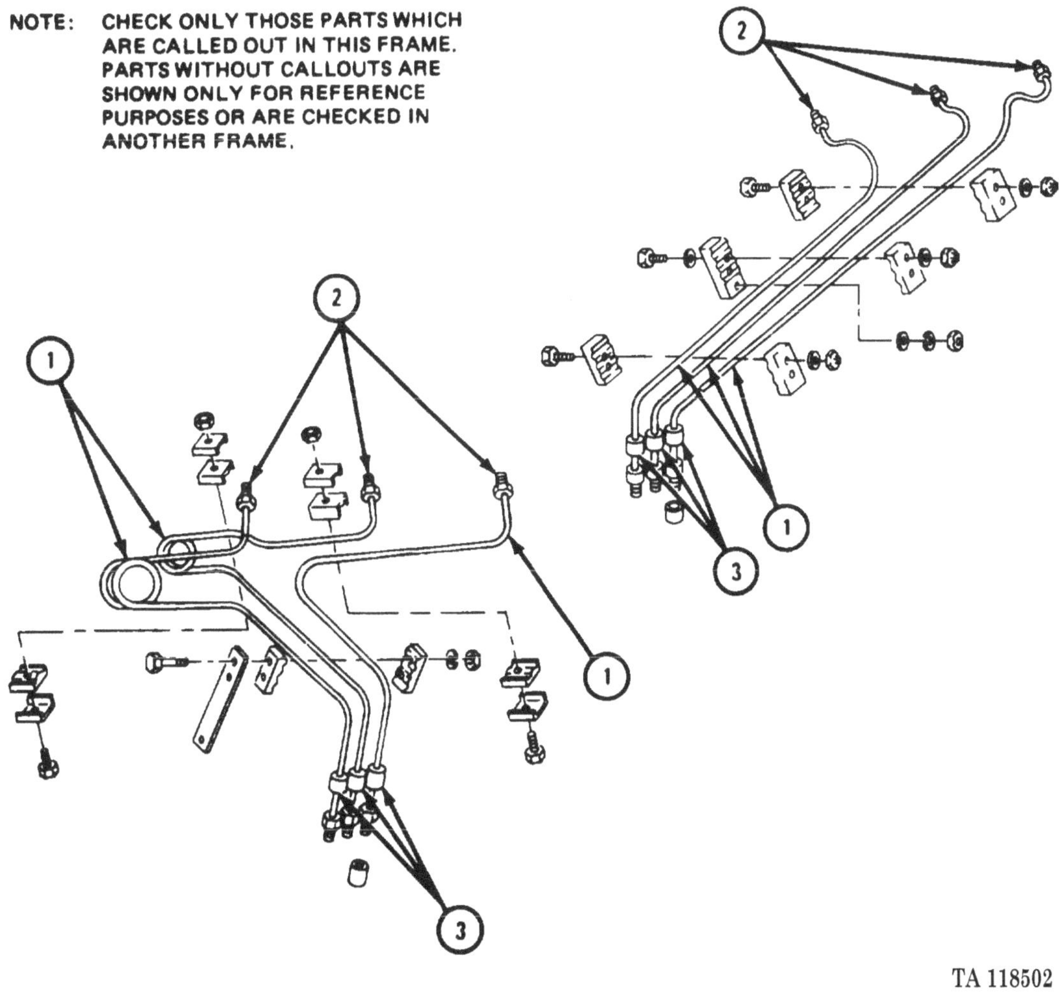
NOTE: CHECK ONLY THOSE PARTS WHICH ARE CALLED OUT IN THIS FRAME. PARTS WITHOUT CALLOUTS ARE SHOWN ONLY FOR REFERENCE PURPOSES OR ARE CHECKED IN ANOTHER FRAME.
1
2
3
TA 118502

FRAME 2

1. Put alight coating of silicone type lubricant on outer surface of dust boot assembly tool (1) and put on dust boot (2).
2. Slide dust boot assembly tool (1) over fuel injector tube coupling nut (3). Slide dust boot (2) over dust boot assembly tool and over coupling nut.

GO TO FRAME 3

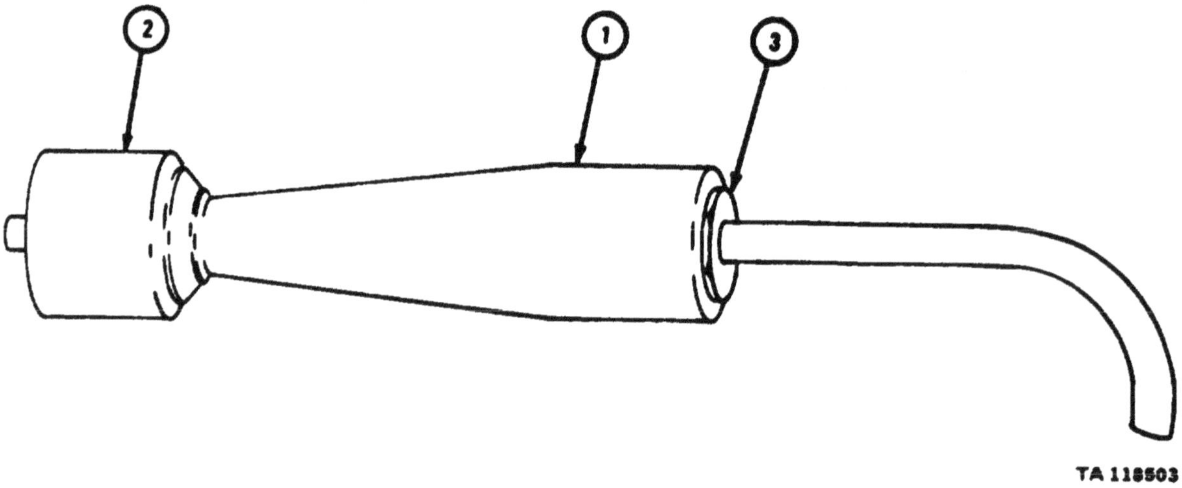

FRAME 3

1. Check that four tube clamp pads (1), seven retaining pads (2), and pad (3) are not torn, cracked, dried out or damaged in any other way. If part is damaged in any way, get a new part.
2. Check that four retaining plates (4) and bracket (5) are not cracked, bent or damaged in any other way. Take out minor bends. If more repair is needed, get a new part.
3. Check that six screws (6) and seven nuts (7) are not cracked, stripped or damaged in any other way. If repair is needed, get a new screw or nut.

END OF TASK

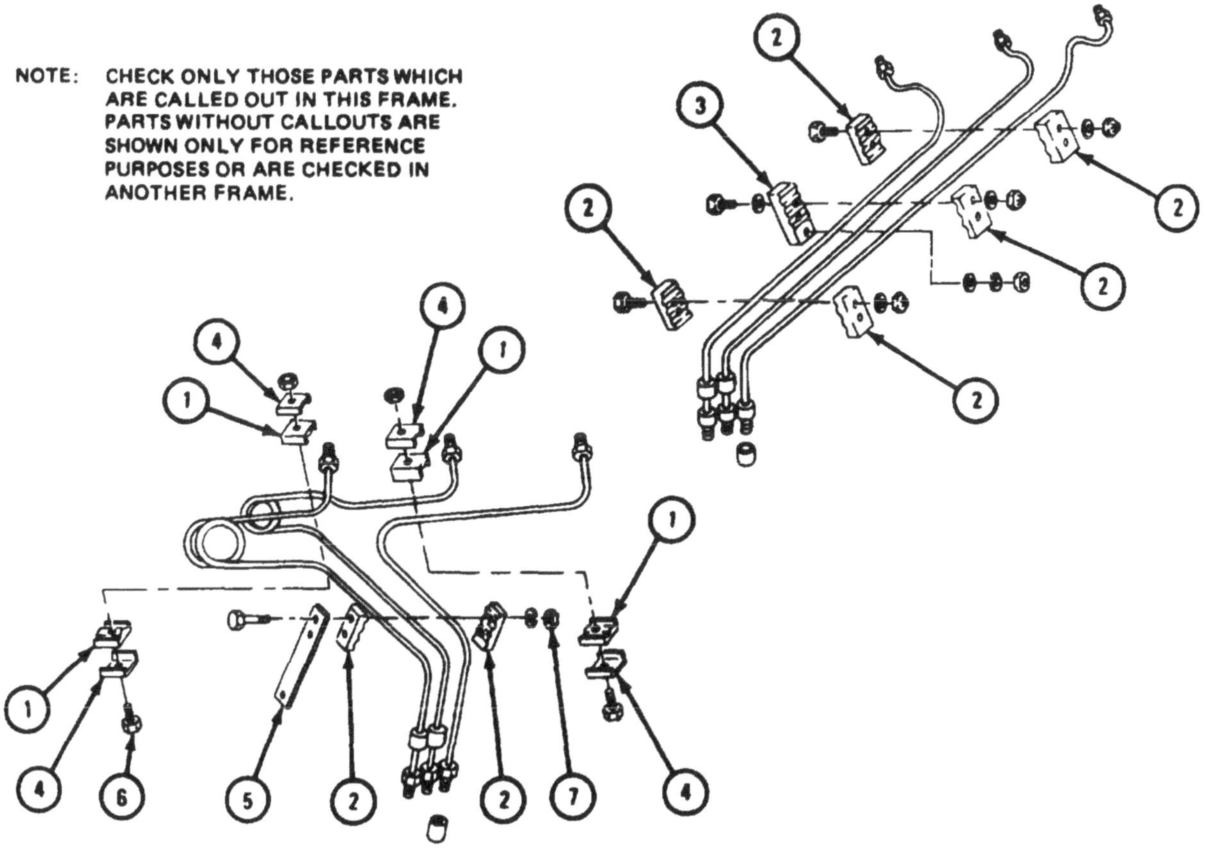
NOTE: CHECK ONLY THOSE PARTS WHICH ARE CALLED OUT IN THIS FRAME. PARTS WITHOUT CALLOUTS ARE SHOWN ONLY FOR REFERENCE PURPOSES OR ARE CHECKED IN ANOTHER FRAME.
1
2
3
4
5
6
7
TA 113700

b. All Engines Except LDS-465-2.

(1) Cleaning.

FRAME 1

WARNING

Dry cleaning solvent is flammable. Do not use near an open flame. Keep a fire extinguisher nearby when solvent is used. Use only in well-ventilated places. Failure to do this may result in injury to personnel and damage to equipment.

CAUTION

Do not let dry cleaning solvent touch six rubber boots (1). Boots will dry out and crack. If dry cleaning solvent splashes on rubber boots, dry with clean cloth.

1. Slide six rubber boots (1) to one end of six fuel injector tubes (2). Clean outside of six fuel injector tubes with dry cleaning solvent. Dry with clean cloth.
2. Slide six rubber boots (1) to center of six fuel injector tubes (2). Clean ends of six fuel injector tubes which were covered by rubber boots in step 1.

WARNING

Eye shields must be worn when using compressed air. Eye injury can occur if eye shields are not used.

3. Flush inside of six fuel injector tubes (2) with dry cleaning solvent. Blow dry with compressed air. Plug ends of six fuel injector tubes to keep out dirt.

GO TO FRAME 2

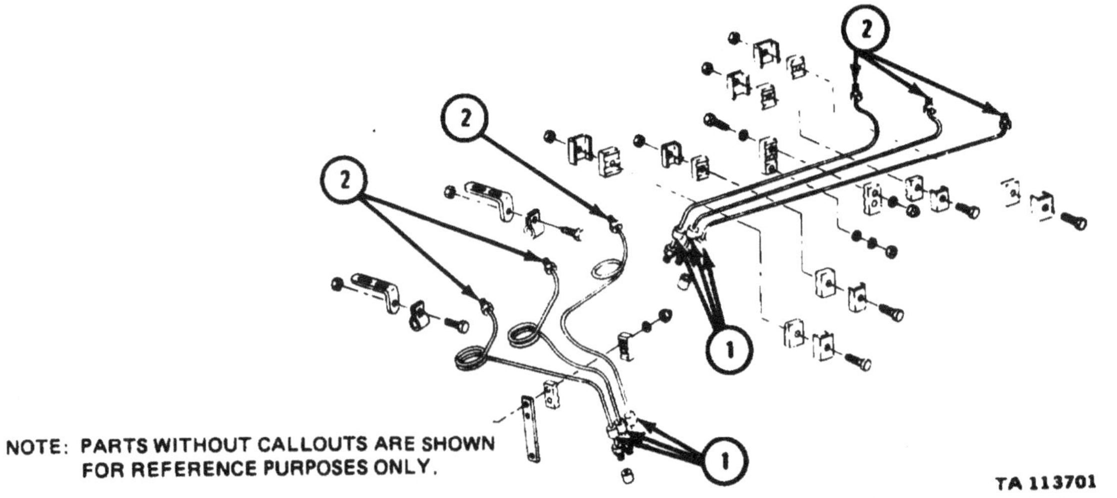
2
2
2
1
1
NOTE: PARTS WITHOUT CALLOUTS ARE SHOWN FOR REFERENCE PURPOSES ONLY.
TA 113701

FRAME 2

1. Clean six boots (1) and 12 tube clamps (2) with soapy water.

WARNING

Dry cleaning solvent is flammable. Do not use near an open flame. Keep a fire extinguisher nearby when solvent is used. Use only in well-ventilated places. Failure to do this may result in injury to personnel and damage to equipment.

2. Clean three brackets (3), two clamps (4), eight retainers (5), eight screws (6), and nine nuts (7) with dry cleaning solvent. Use a stiff brush to take off sludge and gum deposits. Dry with clean cloth.

END OF TASK

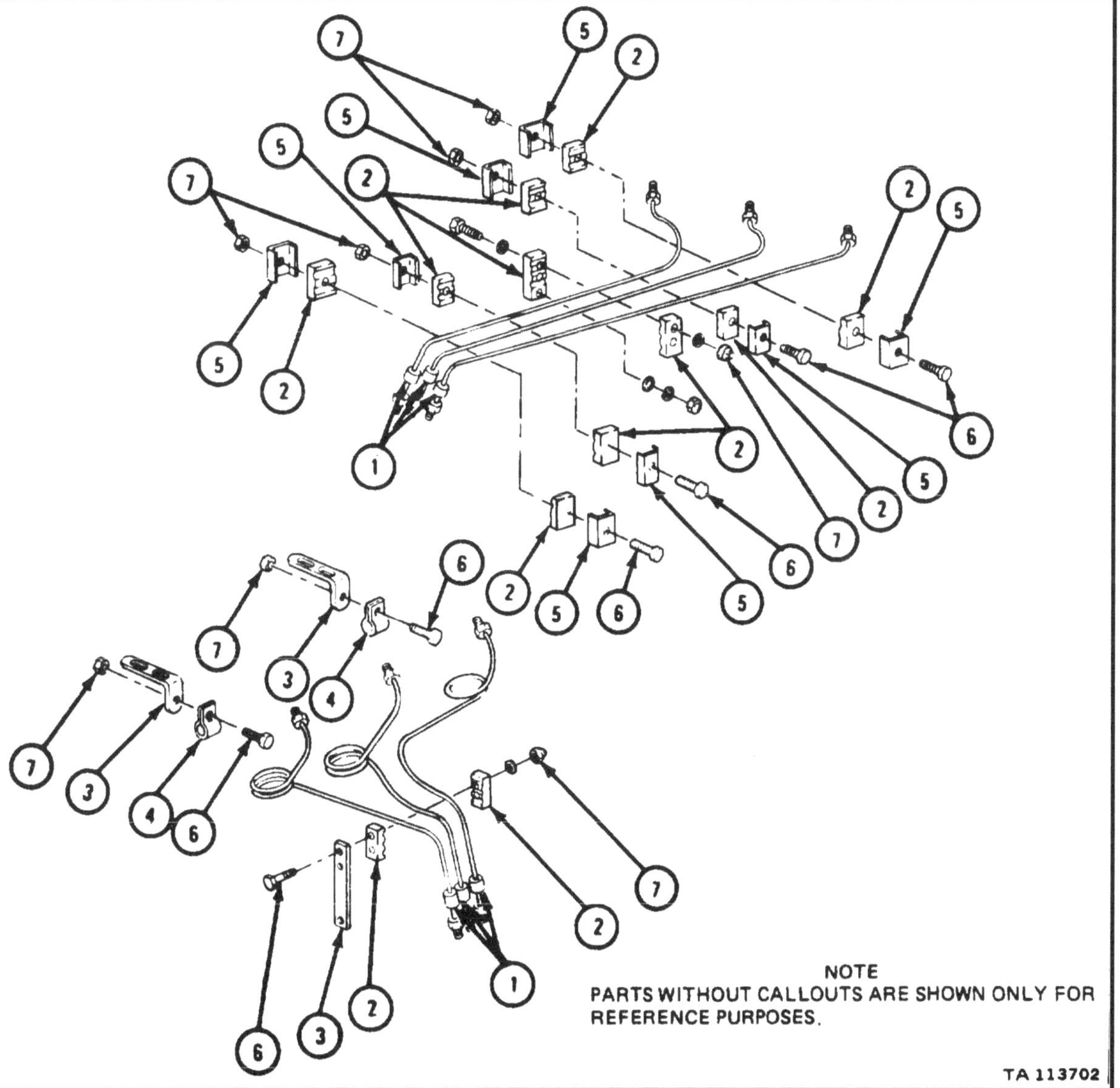

NOTE
PARTS WITHOUT CALLOUTS ARE SHOWN ONLY FOR REFERENCE PURPOSES.
TA 113702

(2) Inspection and repair.

FRAME 1

1. Check that six fuel injector tubes (1) are not cracked, split, dented, kinked or damaged in any other way. If fuel injector tube is damaged in any way, get a new fuel injector tube assembly.

2. Check that 12 coupling nuts (2) are not cracked, split, dented, stripped or damaged in any other way. Use a thread chaser to repair minor thread damage. If more repair is needed, get a new fuel injector tube assembly (1).

3. Check that six dust boots (3) are not cracked, torn, dried out or damaged in any other way. If dust boot is damaged in any way, cut off dust boot and get a new dust boot.

IF ANY DUST BOOTS (3) WERE CUT OFF, GO TO FRAME 2.
IF NO DUST BOOTS (3) WERE CUT OFF, GO TO FRAME 3

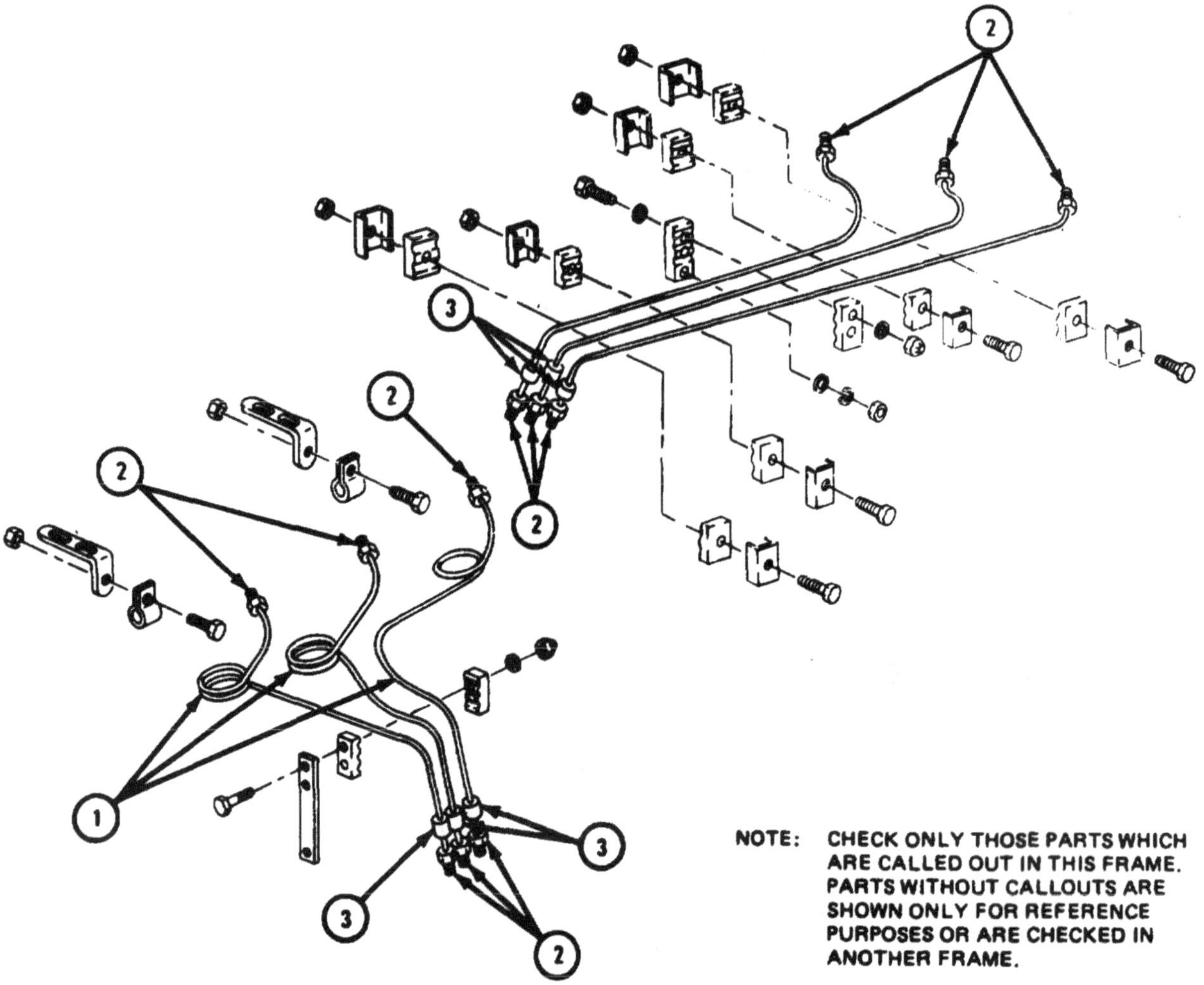

NOTE: CHECK ONLY THOSE PARTS WHICH ARE CALLED OUT IN THIS FRAME. PARTS WITHOUT CALLOUTS ARE SHOWN ONLY FOR REFERENCE PURPOSES OR ARE CHECKED IN ANOTHER FRAME.

FRAME 2

1. Put a light coating of lubricant on outer surface of dust boot assembly tool (1) and put on dust boot (2).
2. Slide dust boot assembly tool (1) over fuel injector tube coupling nut (3) and slide dust boot (2) back over dust boot assembly tool and over coupling nut.

GO TO FRAME 3

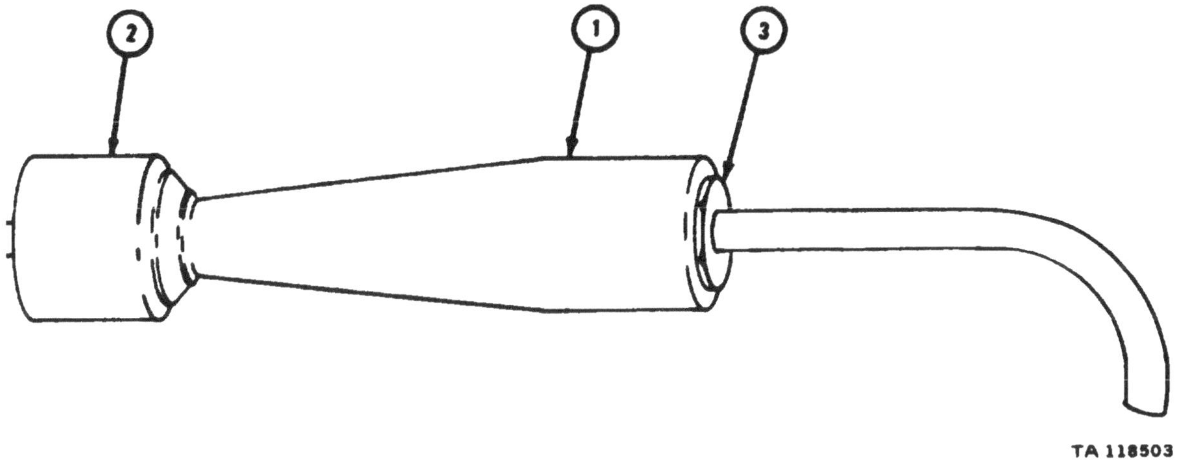

FRAME 3

1. Check that 12 tube clamps (1) are not torn, cracked, dried out or damaged in any other way. If tube clamp is damaged, get a new tube clamp.

2. Check that three brackets (2), two clamps (3), and eight retainers (4) are not cracked, bent or damaged in any other way. Take out minor bends. If more repair is needed, get a new part.

3. Check that eight screws (5) and nine nuts (6) are not cracked, stripped or damaged in any other way. If repair is needed, get a new screw or nut.

END OF TASK

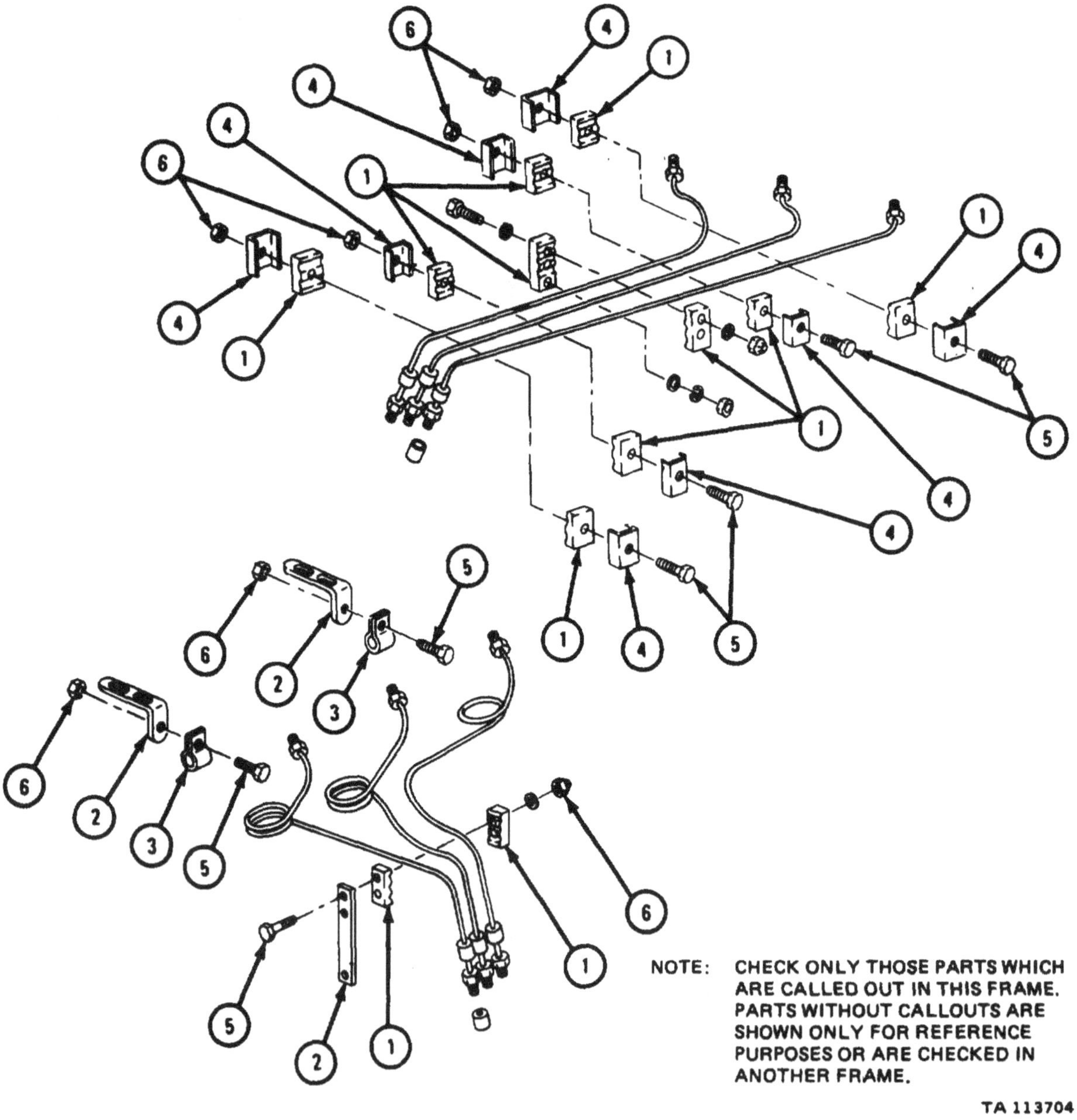

4-38. FUEL INJECTOR FUEL RETURN TUBES.

a. Cleaning.

FRAME 1

1. Clean plastic fuel tubes (1) with soap and water.

WARNING

Dry cleaning solvent is flammable. Do not use near an open flame. Keep a fire extinguisher nearby when solvent is used . Use only in well-ventilated places. Failure to do this may result in injury to personnel and damage to equipment.

Eye shields must be worn when using compressed air. Eye injury can occur if eye shields are not used.

2. Clean clamps (2), connecting tees (3), and elbows (4) with dry cleaning solvent and blow dry with compressed air.

END OF TASK

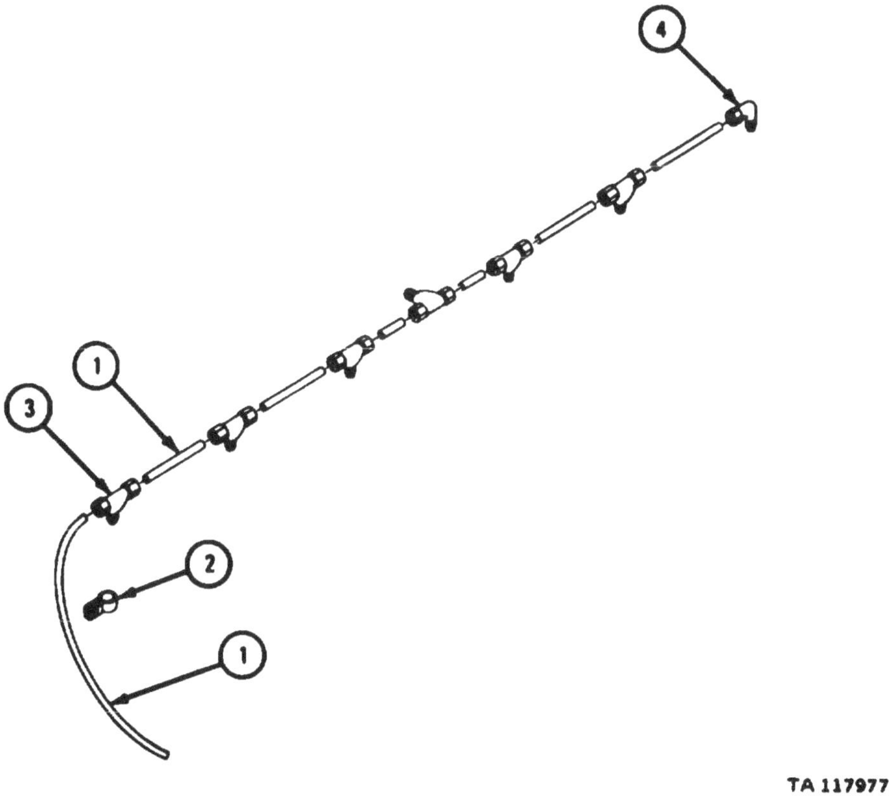

b. Inspection and Repair.

FRAME 1

1. Check that plastic tubes (1) have no cracks, kinks, brittleness or evidence of leaking around tube nuts (2). Throw away tubes if they are cracked or damaged or if tube nuts are stripped or damaged.
2. Check tubes (1) for internal diameter being squeezed closed at tube nut ferrule (3). A No. 28 drill shank (4) may be used as a gage and must pass through the restricted area .

GO TO FRAME 2

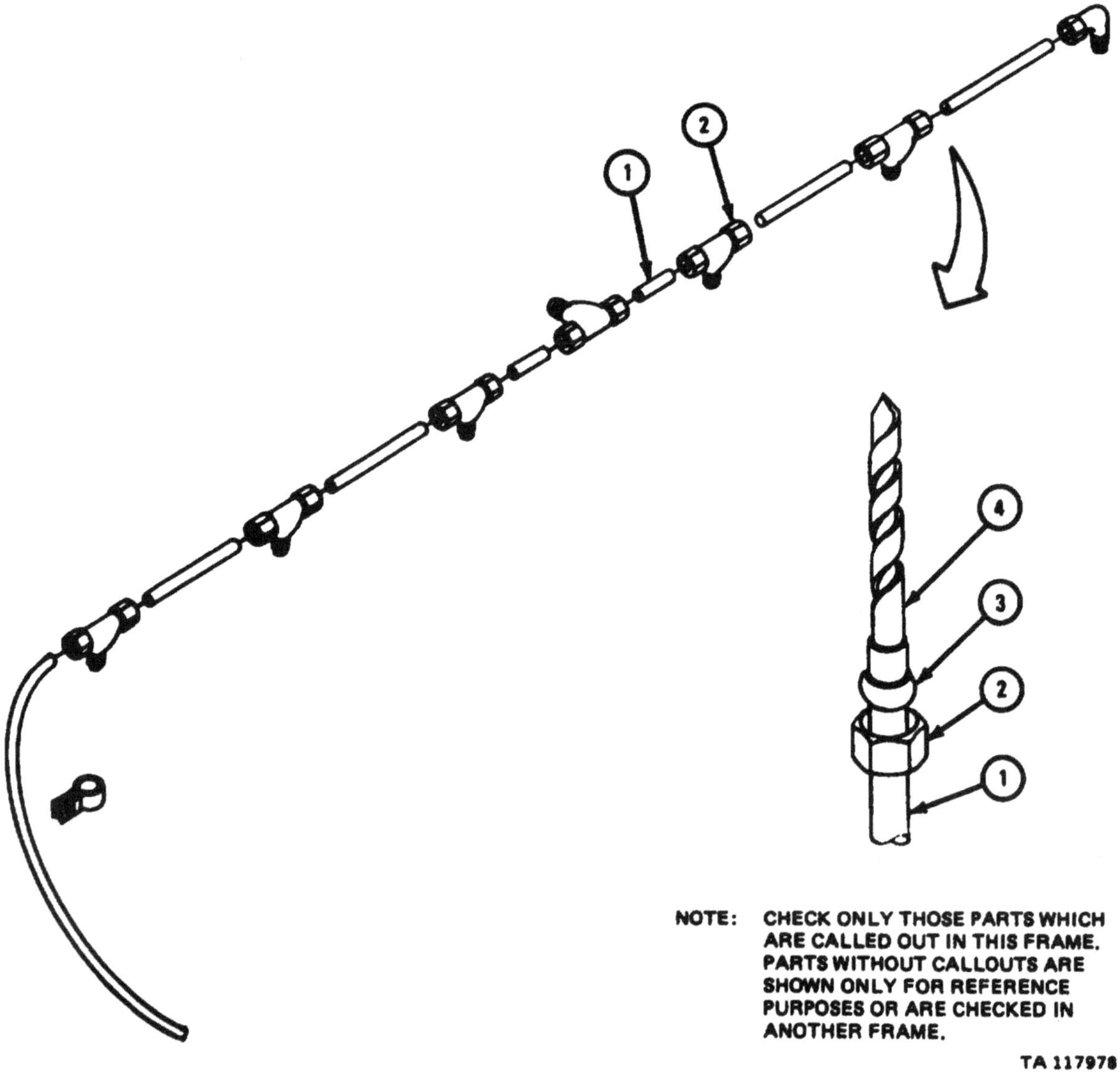
1
2
4
3
2
1
NOTE: CHECK ONLY THOSE PARTS WHICH ARE CALLED OUT IN THIS FRAME. PARTS WITHOUT CALLOUTS ARE SHOWN ONLY FOR REFERENCE PURPOSES OR ARE CHECKED IN ANOTHER FRAME.
TA 117978

FRAME 2

1. Check that tee fittings (1) and elbow (2) have no cracks or stripped or damaged threads . Fix minor thread damage with a small mill file or tap. Throw away badly damaged parts and get new ones.
2. Check that clamp (3) has no cracks, bends in wrong direction or breaks. Straighten minor bends. Throw clamp away and get a new one if it is badly damaged.

END OF TASK

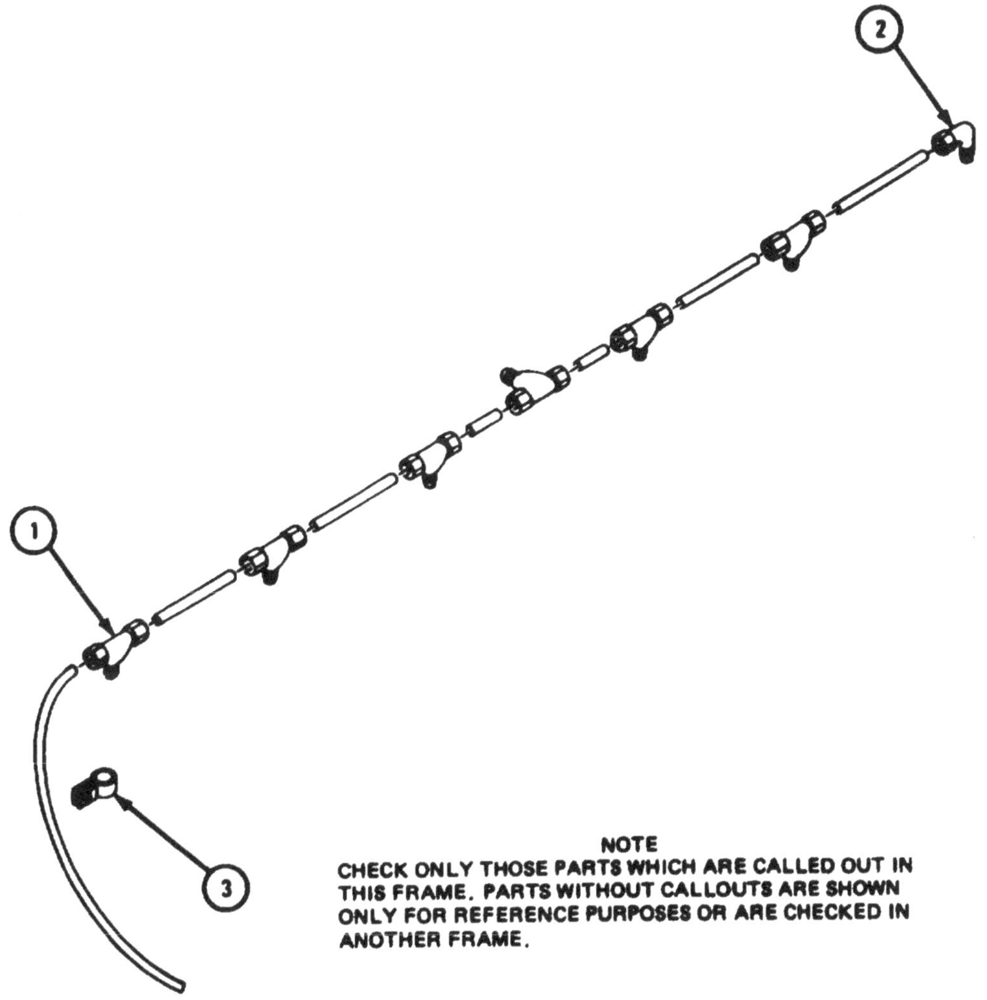

NOTE
CHECK ONLY THOSE PARTS WHICH ARE CALLED OUT IN THIS FRAME. PARTS WITHOUT CALLOUTS ARE SHOWN ONLY FOR REFERENCE PURPOSES OR ARE CHECKED IN ANOTHER FRAME.

TA 117979

4-39. OIL COOLER AND OIL COOLER COVER.

a. Cleaning.

FRAME 1

WARNING

Dry cleaning solvent is flammable. Do not use near an open flame. Keep a fire extinguisher nearby when solvent is used. Use only in well-ventilated places. Failure to do this may result in injury to personnel and damage to equipment.

1. Clean oil cooler cover (1) with dry cleaning solvent. Clean off scale, sludge, and gum with a stiff brush. Dry with clean cloth.

WARNING

Eye shields must be worn when using compressed air. Eye injury can occur if eye shields are not used.

2. Clean scale from outside of oil cooler (2) with a stiff brush. Soak oil cooler in dry cleaning solvent until oil passages (3) are clean. Blow dry oil passages with compressed air.

END OF TASK

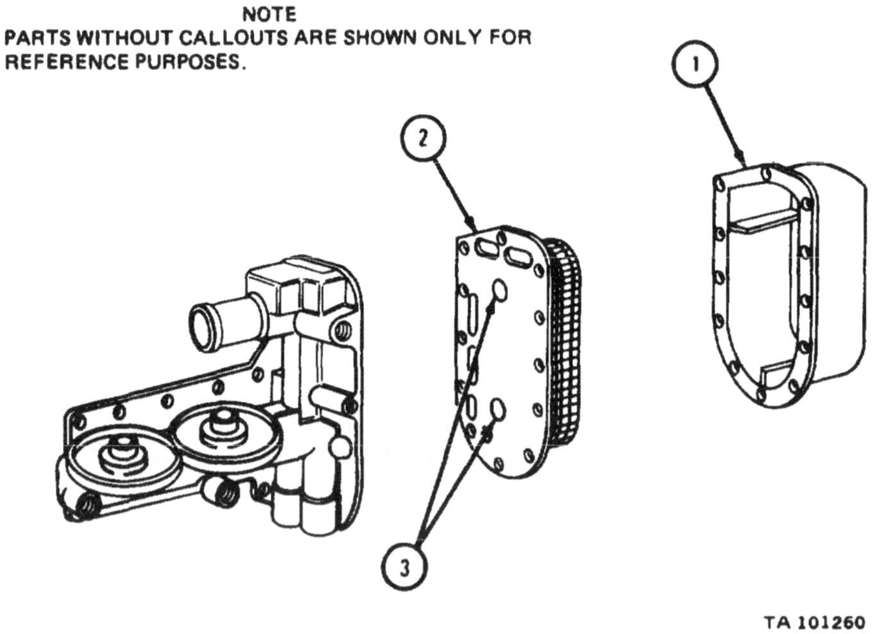
NOTE
PARTS WITHOUT CALLOUTS ARE SHOWN ONLY FOR REFERENCE PURPOSES.
1
2
3
TA 101260

b. Inspection and Repair.

FRAME 1

1. Check that 12 nuts (1) are not stripped, crossthreaded or damaged in any other way. If nut is damaged, get a new nut.
2. Check that oil cooler cover (2) is not bent, dented, cracked or torn. To fix minor bends or dents, refer to FM 43-2 and to fix minor cracks or tears, refer to TM 9-237. If more repair is needed, get a new oil cooler cover.

GO TO FRAME 2

NOTE: CHECK ONLY THOSE PARTS WHICH ARE CALLED OUT IN THIS FRAME. PARTS WITHOUT CALLOUTS ARE SHOWN ONLY FOR REFERENCE PURPOSES OR ARE CHECKED IN ANOTHER FRAME.

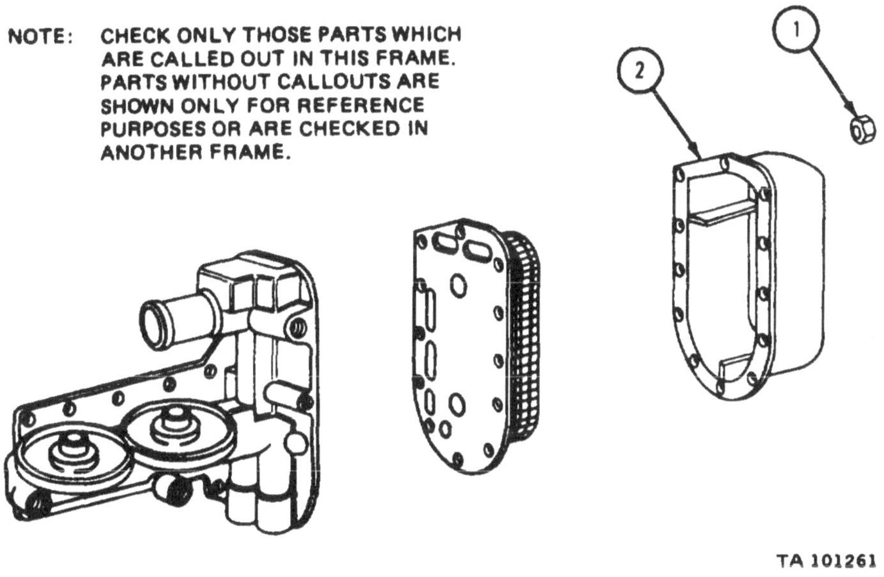

FRAME 2

1. Check that oil cooler (1) has no cracks. If oil cooler is damaged, get a new one.

WARNING

Dry cleaning solvent is flammable. Do not use near an open flame. Keep a fire extinguisher nearby when solvent is used. Use only in well-ventilated places. Failure to do this may result in injury to personnel and damage to equipment.

2. Check that gasket surface (2) of oil cooler (1) and oil cooler cover (3) have no nicks, burrs or scratches. Take out minor nicks, burrs or scratches with a fine mill file or crocus cloth dipped in dry cleaning solvent. If more repair is needed, get a new part.

END OF TASK

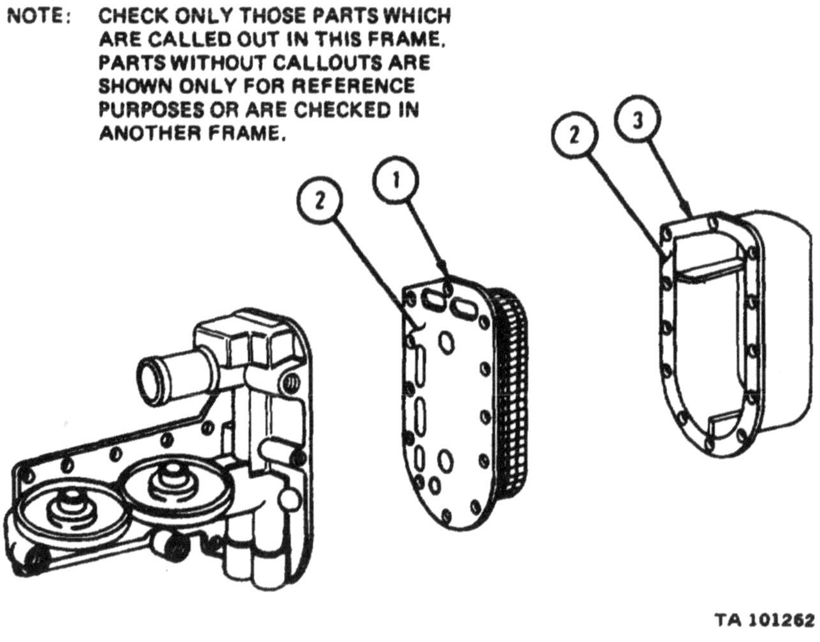
NOTE: CHECK ONLY THOSE PARTS WHICH ARE CALLED OUT IN THIS FRAME. PARTS WITHOUT CALLOUTS ARE SHOWN ONLY FOR REFERENCE PURPOSES OR ARE CHECKED IN ANOTHER FRAME.
1
2
2
3
TA 101262

c. Leakage Test .

FRAME 1

1. Plug one hole (1).

2. Put compressed air hose carrying 150 psi of air pressure in hole (2). Be sure hose seals tightly around hole so no air can get out of oil cooler (3).

NOTE

If oil cooler (3) leaks, air bubbles will be seen in water.

3. Put oil cooler (3) under water and check that it does not leak. If oil cooler leaks, get a new one.

END OF TASK

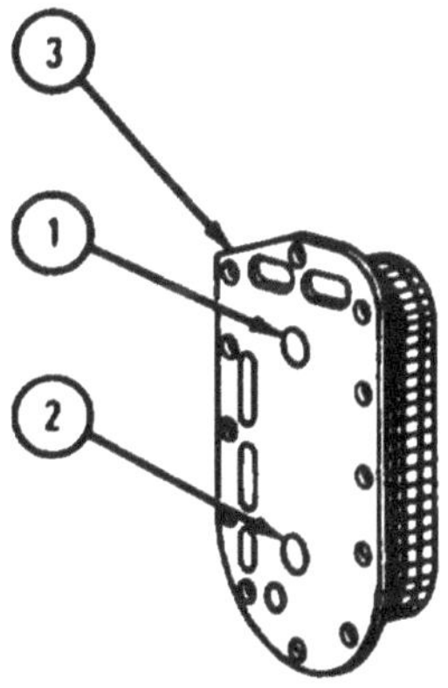

TA113640

4-40. FUEL INJECTOR NOZZLE AND HOLDER ASSEMBLIES.

NOTE

This task is shown for one fuel injector nozzle and holder assembly. This task should be done for all fuel injector nozzle and holder assemblies.

a. Tests and Checks.

WARNING

The high fuel pressure used during these tests and checks can puncture the skin and cause blood poisoning and possible death. Keep hands away from nozzle.

(1) Nozzle opening pressure test.

FRAME 1

WARNING

Carbon removing solvent is flammable. Do not use near an open flame. Keep a fire extinguisher nearby when solvent is used. Use only in well-ventilated places. Failure to do this may result in injury to personnel and damage to equipment.

CAUTION

Make sure no dirt gets in nozzle and holder assembly (1). Dirt can damage equipment.

1. Using carbon removing solvent, clean nozzle area (2).
2. Screw connector tube (3) into nozzle and holder assembly (1).
3. Fill reservoir (4) with fuel. Close pressure valve gate (5). Pump handle (6) about 25 times or until air is cleared from nozzle an-d holder assembly (1).

GO TO FRAME 2

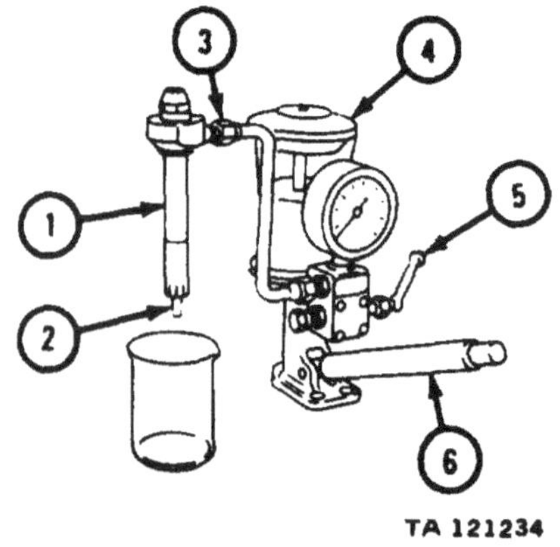

FRAME 2

1. Open pressure gate valve (1) .

NOTE

New nozzles should open between 3050 to 3150 psi. Used nozzles should open at a minimum of 2800 psi. There should not be more than a 300 psi difference between any two nozzles in a complete set.

2. Pump handle (2) and note pressure on gage (3) when nozzle (4) opens and spray comes out. If nozzle does not open within ranges noted above, do nozzle opening pressure adjustment. Refer to para 4-40a (1) .

END OF TASK

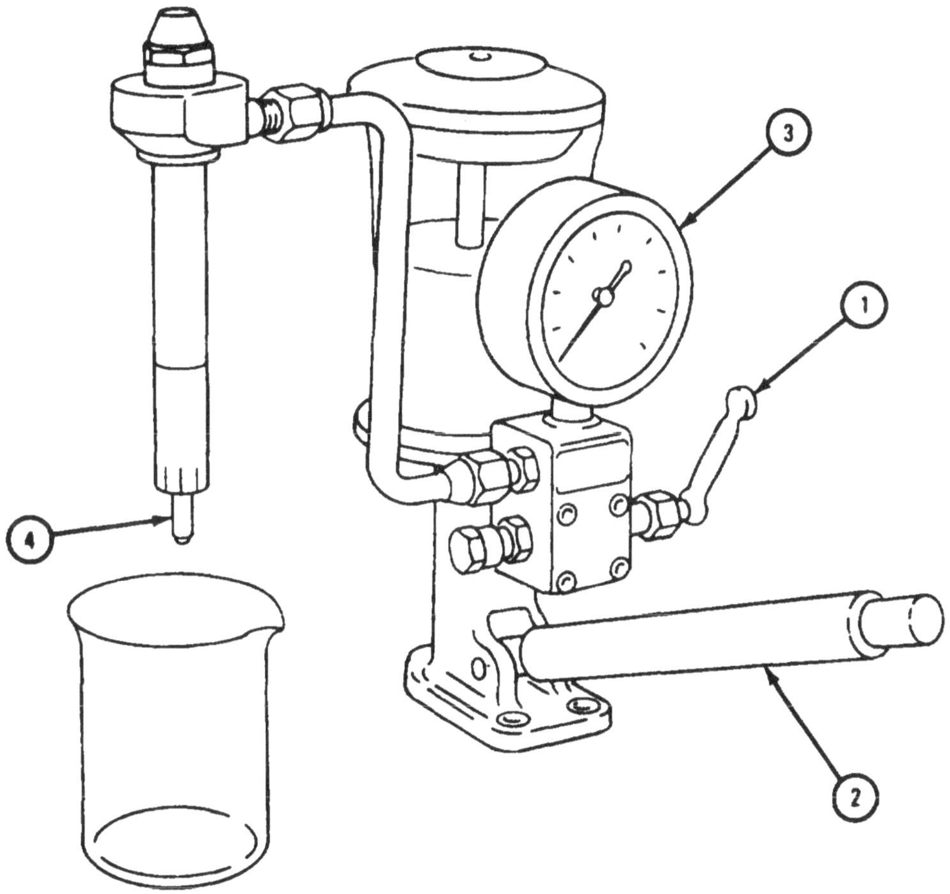

TA 121235

(a) Adjustment of adjusting screw type nozzle .

FRAME 1

1. Unscrew connector tube (1) and take off nozzle and holder assembly (2).
2. Put nozzle and holder assembly (2) in a soft-jawed vise.
3* Hold locknut (3) . Takeoff cap (4) and gasket (5) .

NOTE

Turn adjusting screw (6) to the right to increase pressure, and to the left to decrease pressure. Openin g pressure should be between 3050 to 3150 psi.

4. Turn adjusting screw (6) to the right or to the left to increase or decrease pressure.
5. Tighten locknut (3) . Take nozzle and holder assembly (2) out of vise. Do nozzle opening pressure test. Refer to para 4-40a (1) .
6. Do steps 1 through 5 again until pressure is between 3050 to 3150 psi. If not, disassemble, clean and inspect, and repair nozzle. Refer to para 4-40b, c, a n d .

GO TO FRAME 2

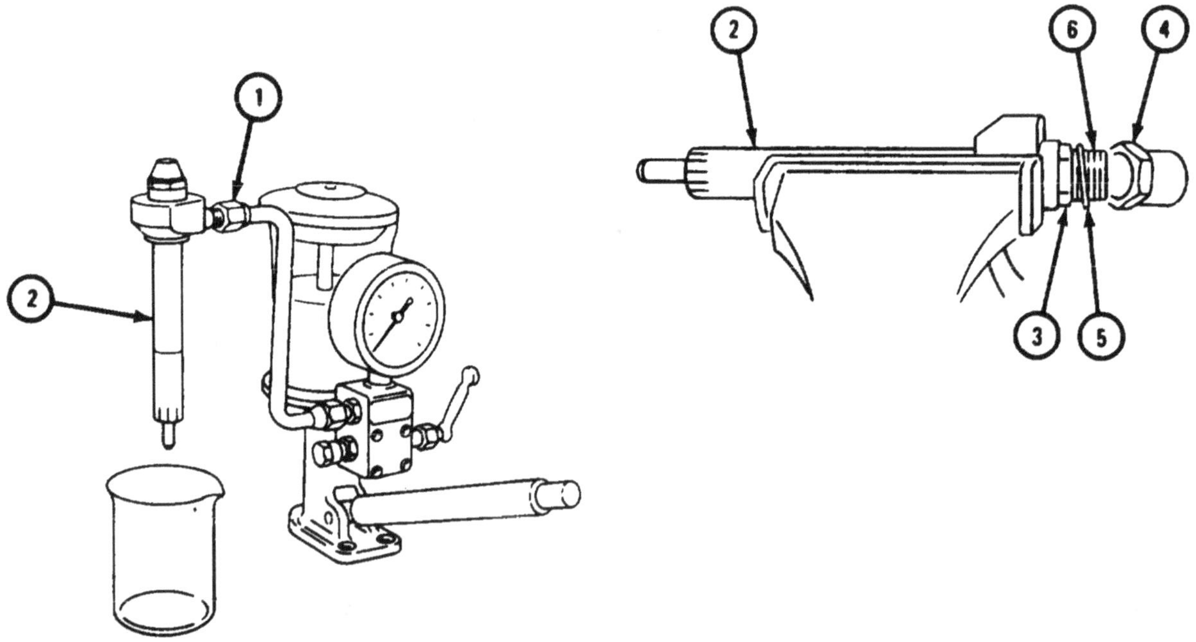

FRAME 2

1. Put nozzle and holder assembly (1) in soft-jawed vise. Tighten locknut (2) to 40 to 45 pound-feet. Take nozzle and holder assembly out of vise.
2. Do nozzle opening pressure test, para 4-40a (1). If adjustment is not right, do frame 1 and step 1 of this frame. If adjustment is still not right, disassemble clean and inspect, and repair nozzle. Refer to para 4-40b, c, and d .
3. Put nozzle and holder assembly (1) in soft-jawed vise. Hold locknut (2) and tighten cap (3) to 40 to 45 pound-feet. Take nozzle and holder assembly out of vise .

END OF TASK

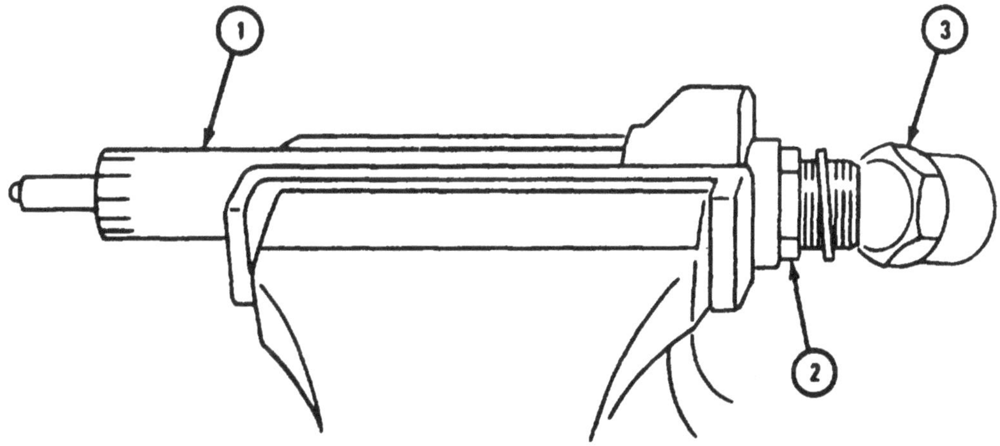

TA 121237

(b) Adjustment of shim type nozzle.

FRAME 1

1. Unscrew connector tube (1) and takeoff nozzle and holder assembly (2).
2. Put nozzle and holder assembly (2) in soft-jawed vise.

NOTE

Add shims to increase pressure and take off shims to decrease pressure. A 0.001-inch shim will change the opening pressure about 150 psi. Pressure opening should be between 3050 to 3150 psi.

3. Take off cap (3). Add or take off shims (4) to increase or decrease pressure.
4. Put on cap (3) and tighten to 40 to 45 pound-feet.
5. Take nozzle and holder assembly (2) out of vise. Do nozzle opening pressure test. Refer to para 4-40a (1).
6. Do steps 1 through 5 again until pressure is right. If not, disassemble, clean and inspect, and repair nozzle. Refer to para 4-40b, c, and d.

END OF TASK

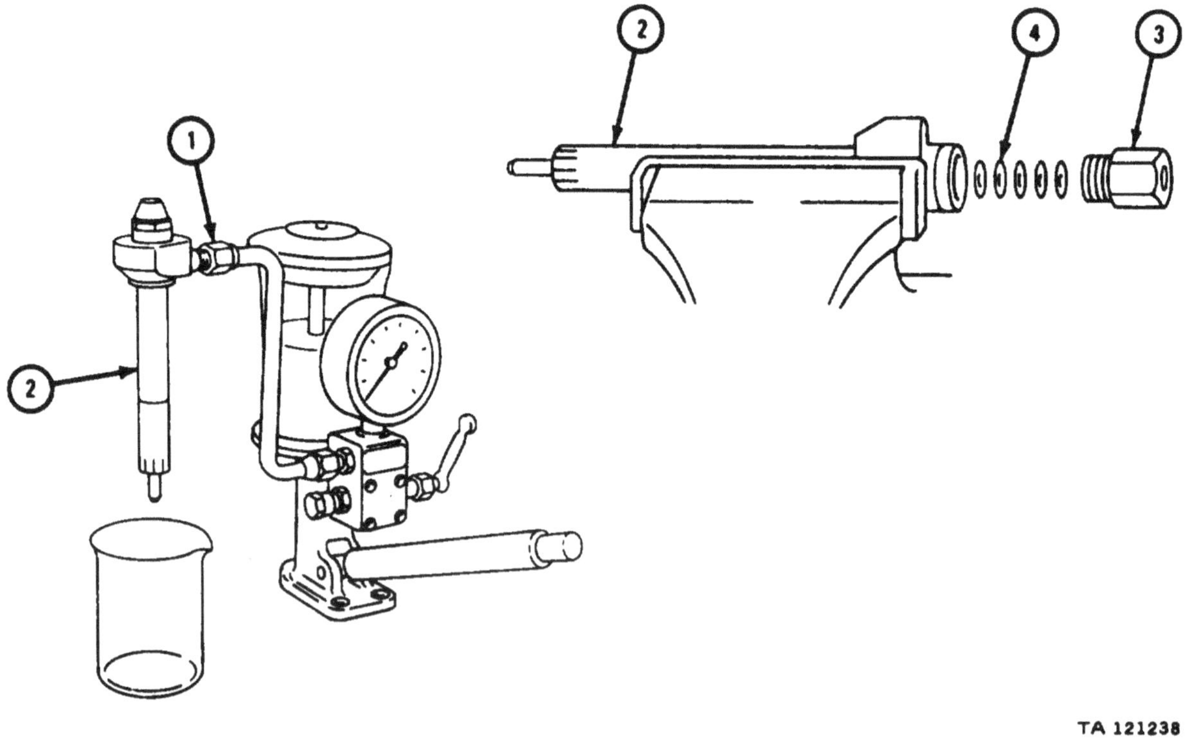

(2) Nozzle leakage check .

FRAME 1

1. Make sure pressure gage valve (1) is open.
2. Pump handle (2) . As pressure comes close to 2800 psi, check that there are no fuel drops on the nozzle (3) or that fuel does not come out as a stream. If either happens, disassemble, clean and inspect, and repair nozzle. Refer to para 4-40b, c, and d .

END OF TASK

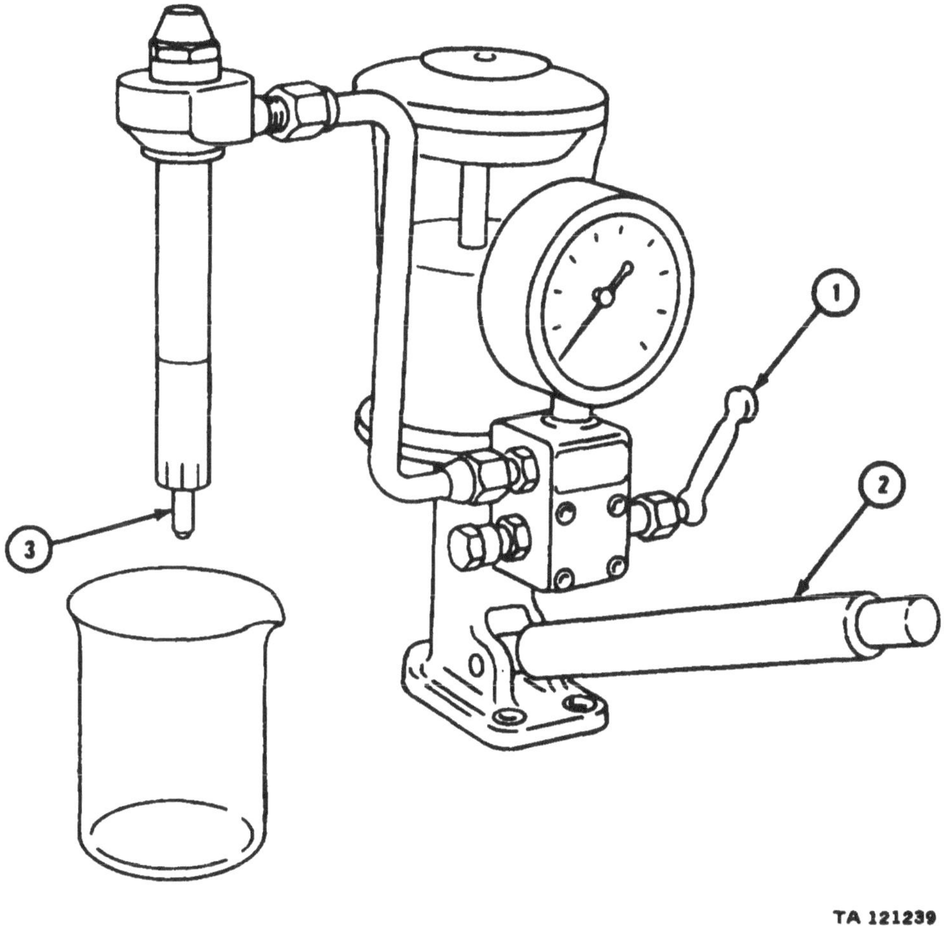

(3) Nozzle spray pattern check.

FRAME 1

1. Close pressure gate valve (1).

NOTE

Nozzles for engines LD-465-1, LD-465-1C, LDT-465-1C, and LDS-465-1 will have dual spray patterns, as shown. Nozzles for engines LDS-465-1A and LDS-465-2 will have single spray patterns, as shown.

2. Pump handle (2) about 15 times per minute. Check that spray patterns are good.

3. If pattern is not good, do steps 1 and 2 several times. If pattern still is not good, disassemble, clean and inspect, and repair nozzle. Refer to para 4-40b, c, and d.

END OF TASK

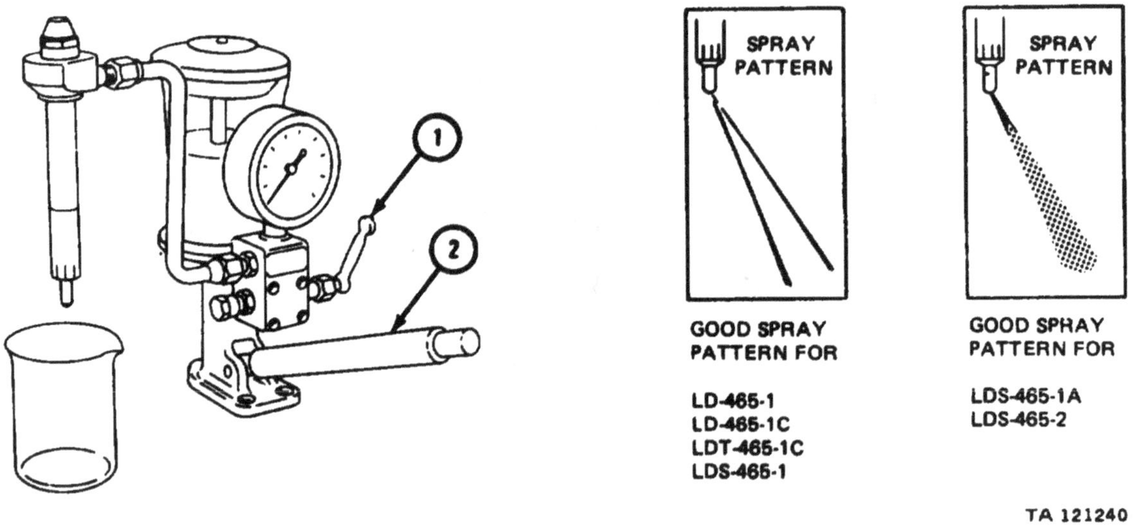

(4) Nozzle chatter check .

FRAME 1

NOTE

Chatter is a normal sound that happens when the valve is open and the valve seat surface is clean.

1. Make sure pressure gate valve (1) is closed.
2. Pump handle (2) and check that a fine spray comes out of nozzle (3) and a chatter sound is heard at each stroke.
3. Do step 2 several times. If spray does not come out or chatter sound only happens sometimes, disassemble, clean and inspect, and repair nozzle. Refer to para 4-40b, c, and d.
4. Unscrew connector tube (4) and take off nozzle and holder assembly.

END OF TASK

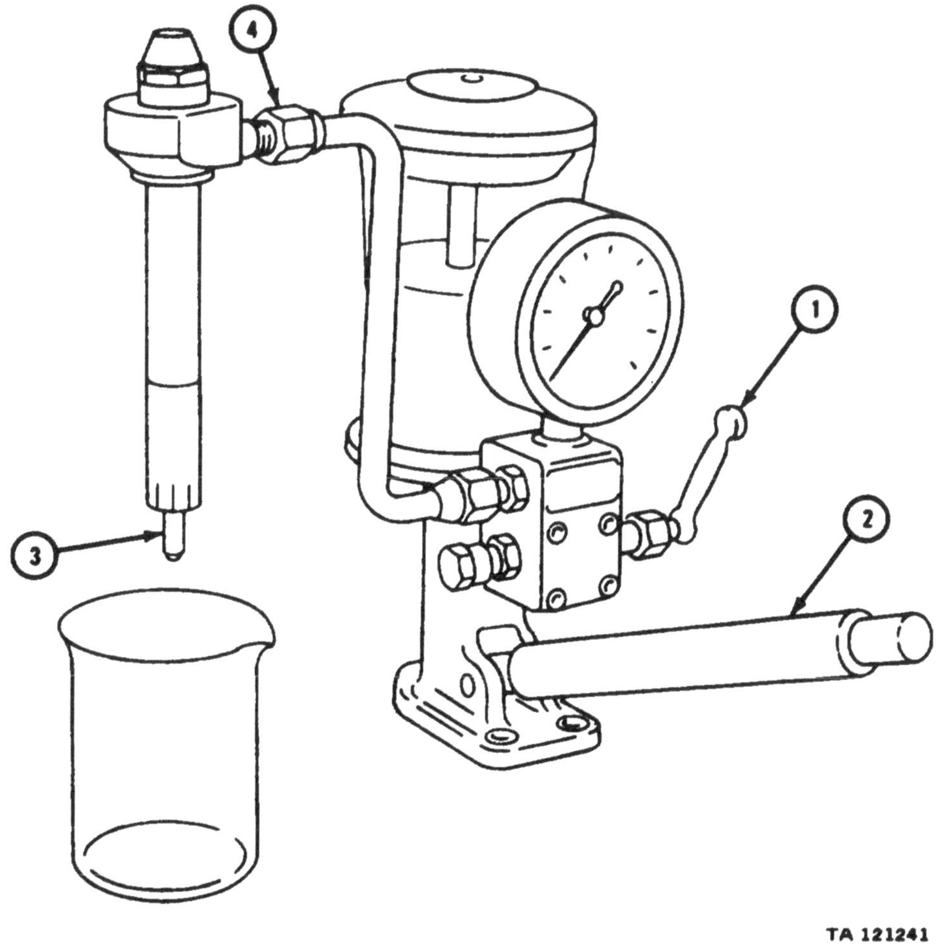

b. Disassembly.

'(1) Adjusting screw type nozzle and holder assembly.

FRAME 1

1. Put nozzle and holder assembly (1) in a soft-jawed vise.
2. Hold locknut (2). Take off cap (3) and gasket (4). Throw away gasket.
3. Loosen locknut (2). Take off adjusting screw (5) and locknut. Take off and throw away locknut gasket (6).
4. Take off spring (7).
5. Take off spindle (8).

GO TO FRAME 2

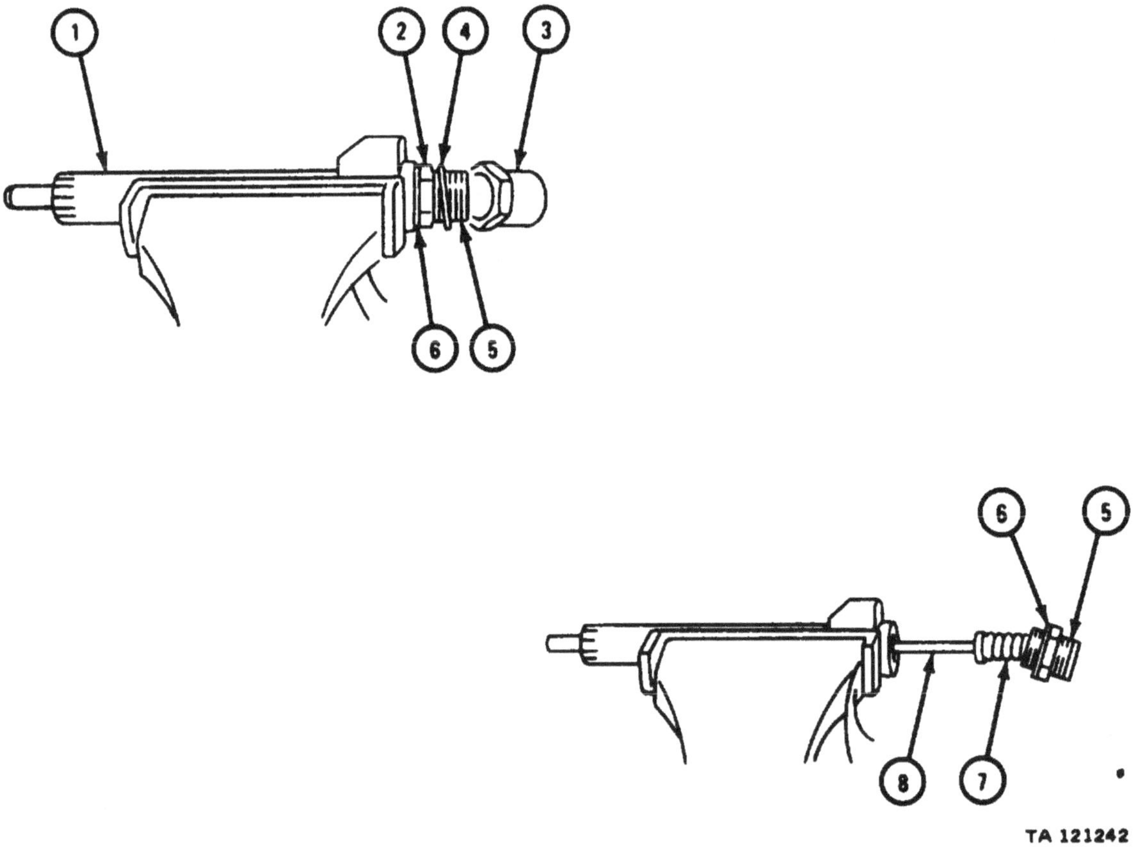

FRAME 2

1. Put nozzle holder (1) in vise, as shown.
2. Take off nozzle capnut (2) .
3. Take off nozzle body (3) and valve (4).

CAUTION

Do not touch valve (4) with hands. Oil may rub off causing a sticking valve which can damage equipment. Do not let polished surface of valve or nozzle body (5) touch any hard surfaces. Nicks or scratches can damage equipment.

4. Hold valve (4) with pliers, as shown, and pull out.

END OF TASK

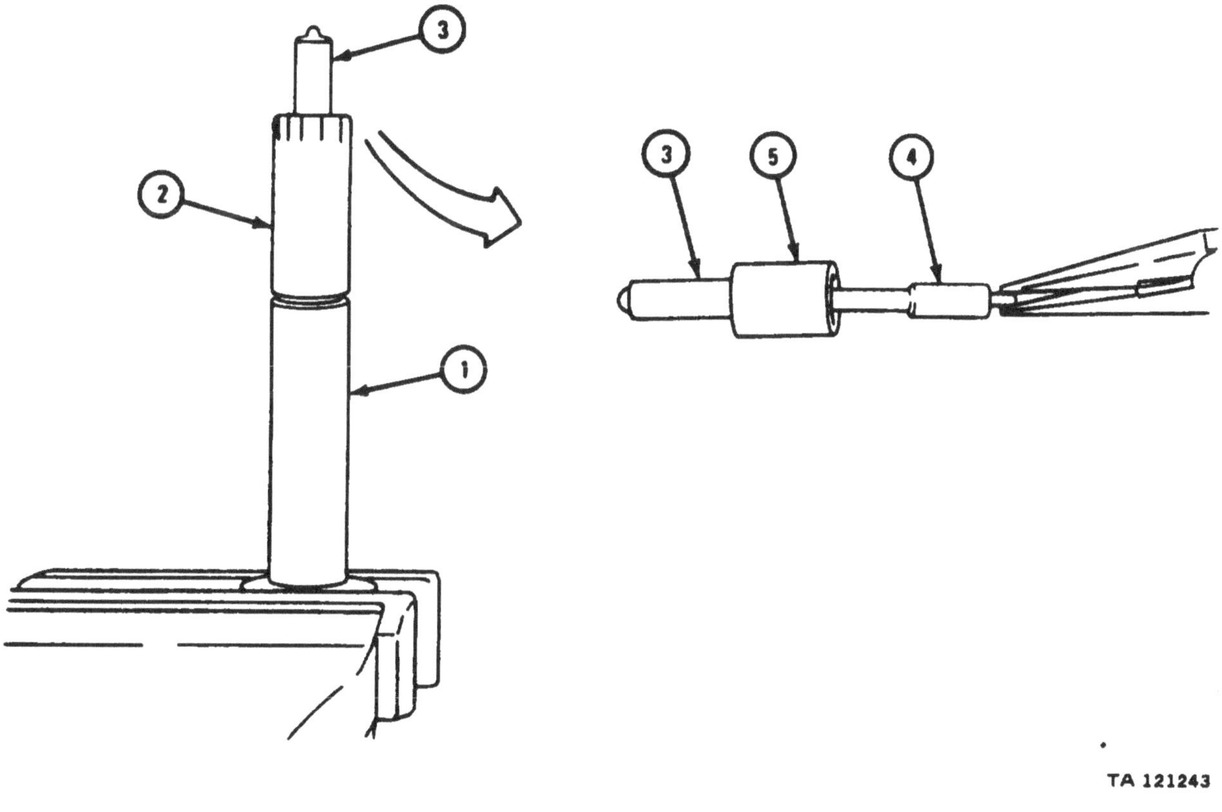

(2) Shim. type nozzle and holder assembly.

FRAME 1

1. Put nozzle and holder assembly (1) in soft-jawed vise.
2. Take off cap (2), shims (3), spring (4), and spindle (5) .
3. Move nozzle and holder assembly (1) in vise, to position shown.
4. Take off nozzle capnut (6) .

GO TO FRAME 2

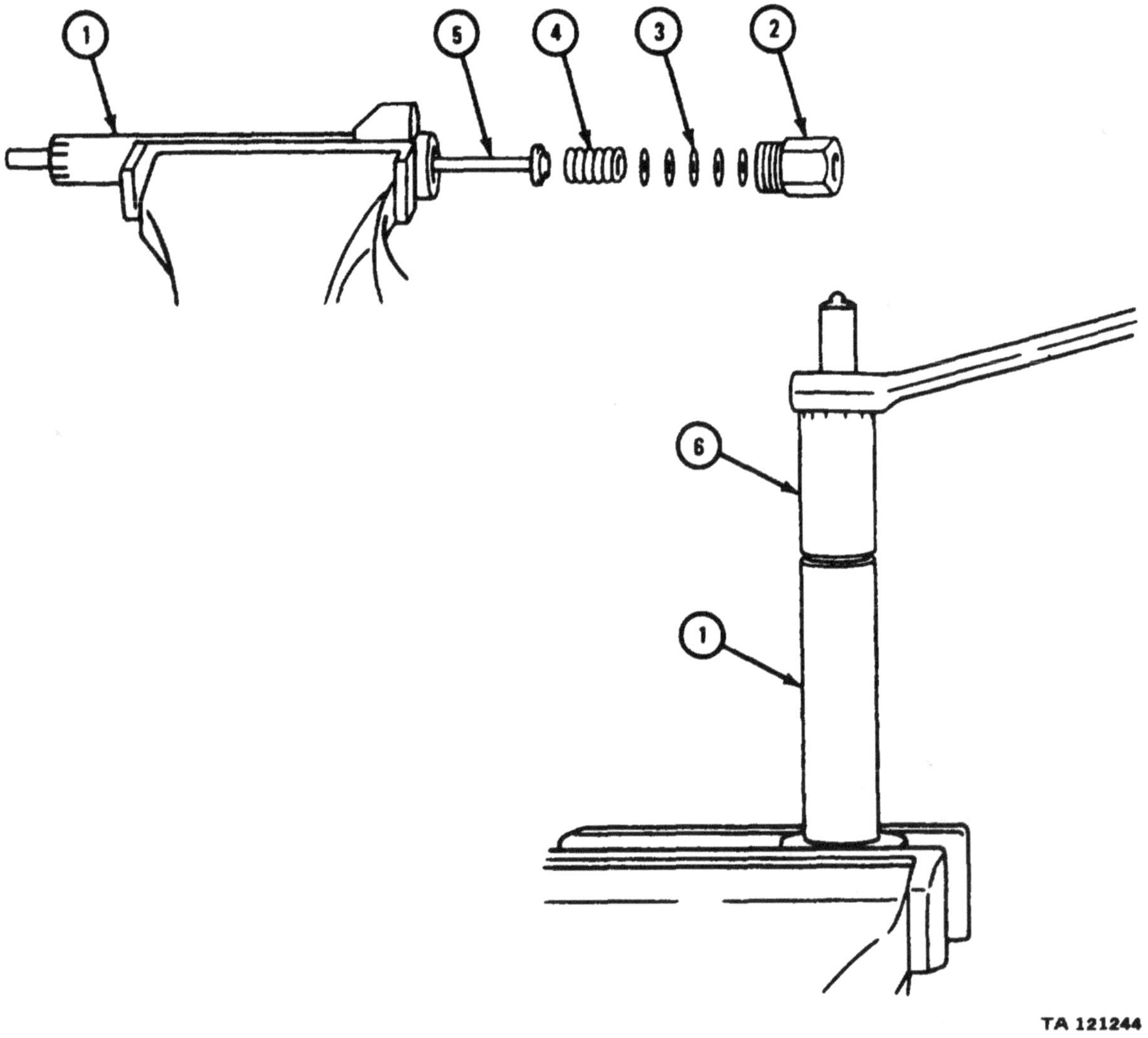

FRAME 2

1. Takeout nozzle body (1) and valve (2) .

CAUTION

Do not touch valve (2) with hands. Oil may rub off causing a sticking valve which can damage equipment. Do not let polished surface of valve or nozzle body (1) touch any hard surfaces. Nicks and scratches ca n damage equipment.

2. Hold valve (2) with pliers, as shown, and pull out.

END OF TASK

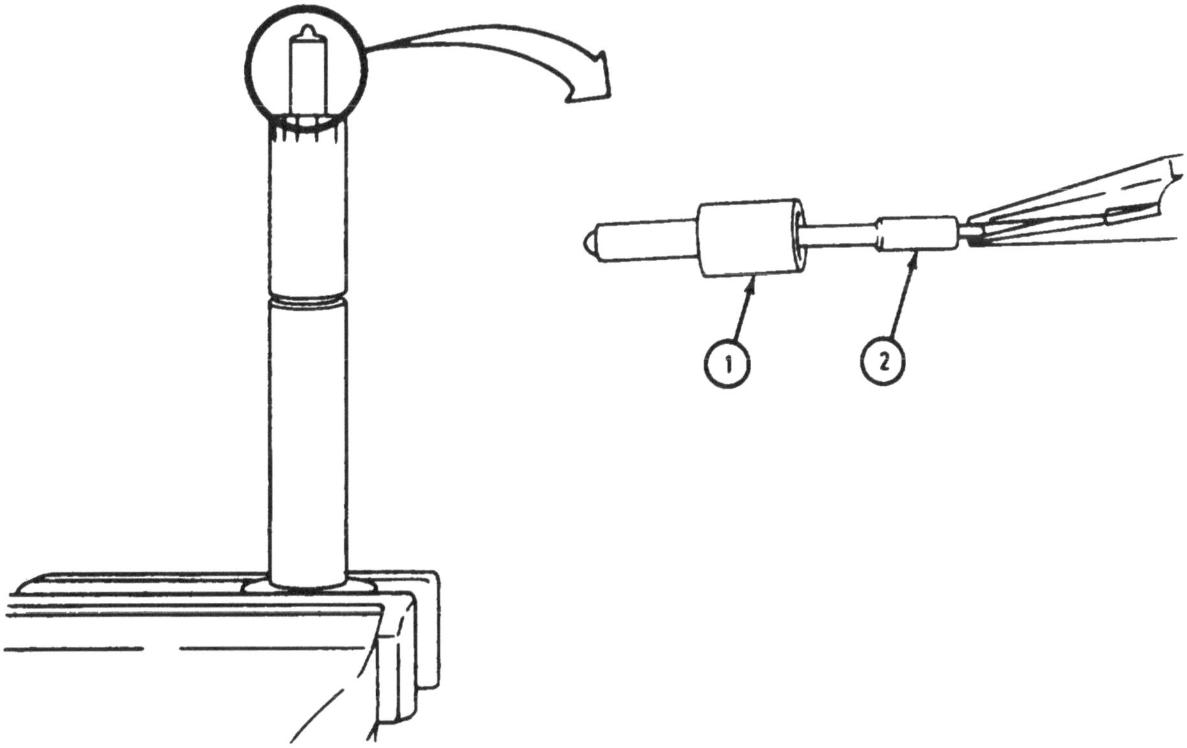

c. Cleaning.

FRAME 1

WARNING

Carbon removing solvent is flammable. Do not use near an open flame. Keep a fire extinguisher nearby when solvent is used. Use only in well-ventilated places. Failure to do this may result in injury to personnel and damage to equipment.

1. Soak valve (1) and nozzle body (2) in carbon removing solvent.

CAUTION

Do not scrape carbon from valve (1) or inner surface of body (2) with any sharp tool, abrasive material or wire brush . Highly polished surfaces may be severely damaged.

2. Using soft cloth and mutton tallow, clean valve (1). Valve (1) may be held by stem (3) in a revolving chuck while cleaning. Clean off carbon with a piece of soft wood well soaked in oil.

CAUTION

Use care while probing with cleaning wire to keep from breaking wire. It may be impossible to take out broken pieces of wire from part being cleaned.

3. Clean carbon from spray nozzle hole (4) in nozzle body (2) by probing with 0.025-inch diameter cleaning wire.
4. Using probe made of soft wood well soaked in oil, clean inside (5) of nozzle body (2) . Point of wood probe should match angle of seat for valve (1).
5. Clean outer surfaces of nozzle body (2) with a soft cloth soaked in carbon solvent.

IF CLEANING ADJUSTING SCREW-TYPE FUEL INJECTOR NOZZLE AND HOLDER ASSEMBLY, GO TO FRAME 2.
IF CLEANING SHIM-TYPE FUEL INJECTOR NOZZLE AND HOLDER ASSEMBLY, GO TO FRAME 3

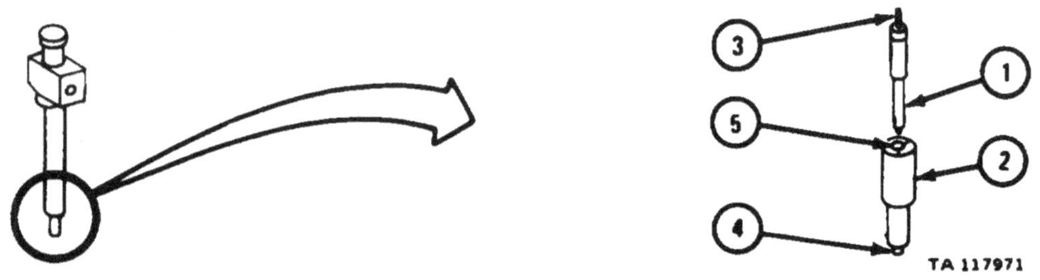

FRAME 2

WARNING

Dry cleaning solvent is flammable. Do not use near an open flame. Keep a fire extinguisher nearby when solvent is used. Use only in well-ventilated places. Failure to do this may result in injury to personnel and damage to equipment.

1. Clean cap (1), locknut (2), adjusting screw (3), spring (4), spindle (5), nozzle holder body (6), and capnut (7) with dry cleaning solvent. Dry with soft cloth.

END OF TASK

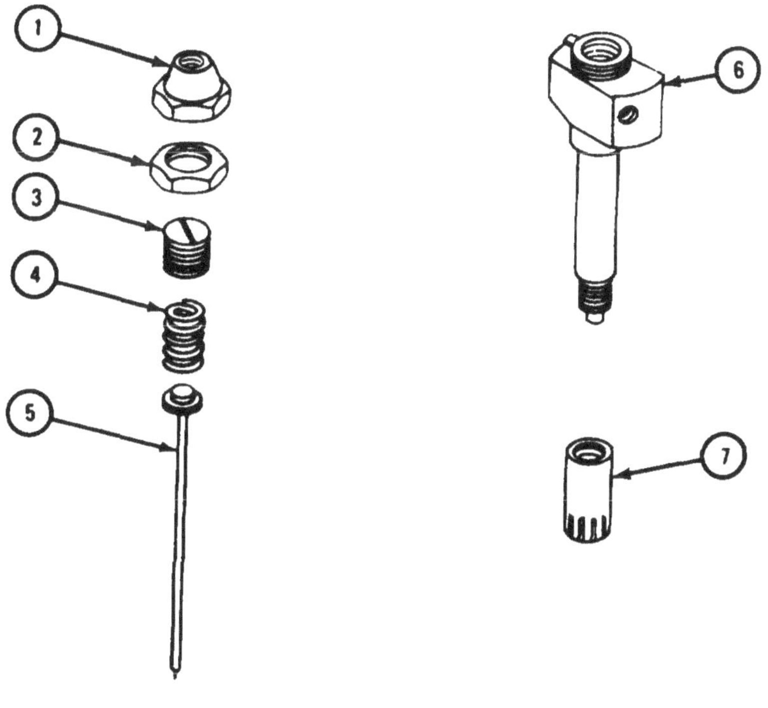

FRAME 3

WARNING

Dry cleaning solvent is flammable. Do not use near an open flame. Keep a fire extinguisher nearby when solvent is used. Use only in well-ventilated places. Failure to do this may result in injury to personnel and damage to equipment.

1. Clean cap (1), shims (2), spring (3), spindle (4), nozzle holder body (5) , and capnut (6), with dry cleaning solvent, Dry with soft cloth.

END OF TASK

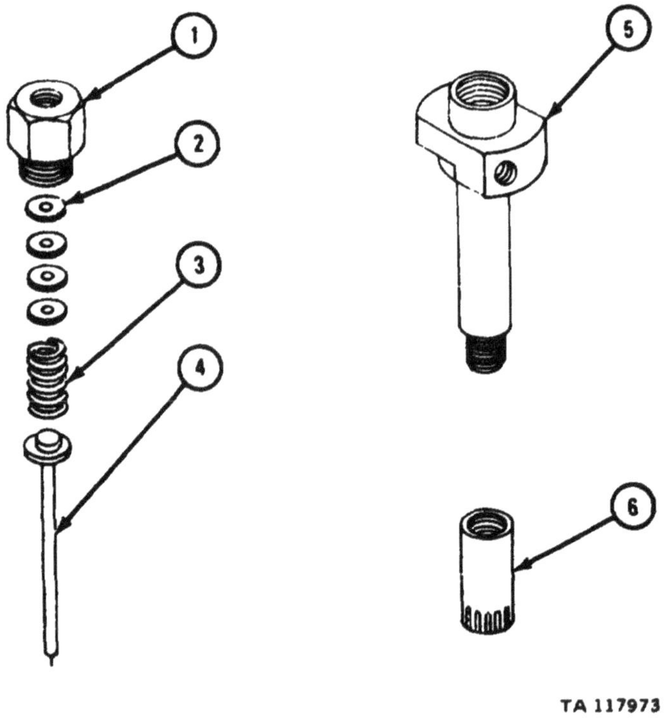

d. Inspection and Repair.

FRAME 1

1. Check that nozzle spring (1) has no cracks and wear. If spring is cracked or worn, get a new one.
2. Using straight edge and strong light, check spindle (2) in at least two positions 90° apart for straightness. If spindle is not straight, get a new one.
3* Check that fuel passage (3) of nozzle holder body (4) has no sludge, gum or other obstructing matter . Get a new nozzle holder body if fuel passage is dirty or clogged .
4. Check that nozzle holder body (4) has no damaged threads at three locations (5). Fix threads which have minor damage with a tap and chaser. If threads have major damage, get new nozzle holder body.
5. Check that nozzle holder body (4) has no loose or damaged locating dowel pins (6) . If dowel pins are loose or damaged, get new nozzle holder body.
6. Check that seat of valve (7) and nozzle body (8) has no pitting, discoloration due to overheating, distortion or any other damage. If pitted, distorted or worn, get a new valve. If valve seat is pitted, distorted or worn, get new nozzle body . Lap valve in nozzle body to take out minor discolorations. If either part has major discoloration, get new valve or nozzle body.
7. Check fit of valve (7) in nozzle body (8). If valve does not slide freely to valve seat of nozzle body (8), lap on a lapping plate or polish to take out minor defects . If defects are not taken out by lapping and polishing, get new valve and nozzle body.
8. Check that nozzle body (8) has no blocked spray holes. If blockage cannot be taken out from spray holes, get new nozzle body.
9. Check that nozzle capnut (9) has no damaged threads. Fix threads which have minor damage with a tap or chaser. If damage cannot be fixed, get new nozzle capnut .

GO TO FRAME 2

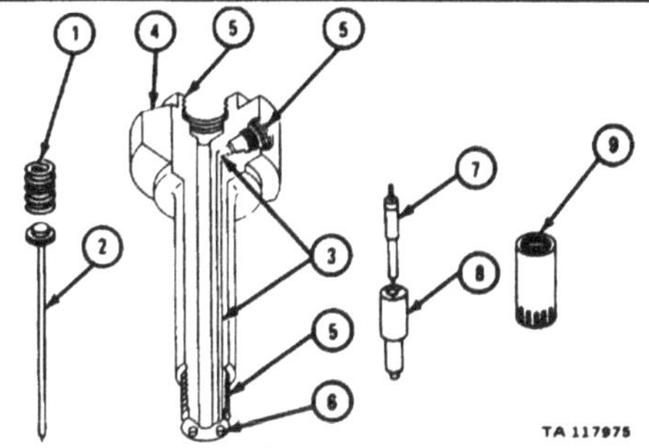

FRAME 2

NOTE

If working on adjusting screw-type fuel injector nozzle and holder assembly, do step 1.

If working on shim-type fuel injector nozzle and holder assembly, do step 2.

1. Check that threads of cap (1), locknut (2), and adjusting screw (3) have no damage. Fix threads which have minor damage with a tap and chaser. If damage cannot be fixed, get new cap, locknut or screw.

2. Check that threads of cap (4) are not damaged. Fix threads which have minor damage with a tap and chaser. If damage cannot be fixed, get new cap.

END OF TASK

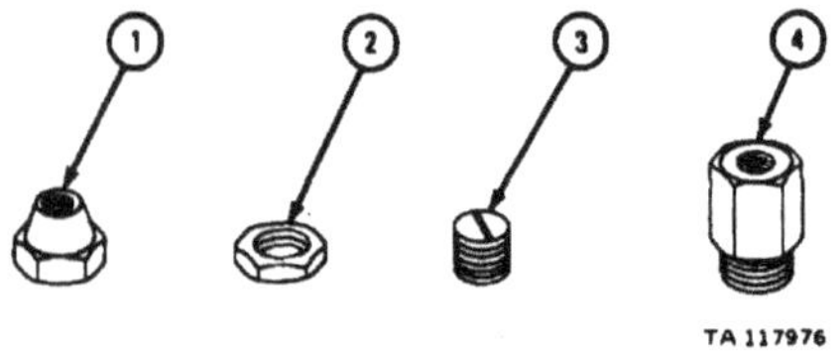

e. Assembly.

(1) Adjusting screw type nozzle and holder assembly.

NOTE

Work area must be kept clean.

FRAME 1

CAUTION

Do not touch valve (1) with hands. Oil may rub off causing a sticking valve which can damage equipment. Do not let polished surface of valve or nozzle body (2) touch any hard surfaces. Nicks or scratches can damage equipment.

1. Hold valve (1) with pliers, as shown, and push into valve body (2).
2. Put valve (1) with valve body (2) into nozzle holder (3). Put nozzle holder into soft-jawed vise, as shown.
3. Put on and tights nozzle capnut (4) to 40 to 45 pound-feet.

GO TO FRAME 2

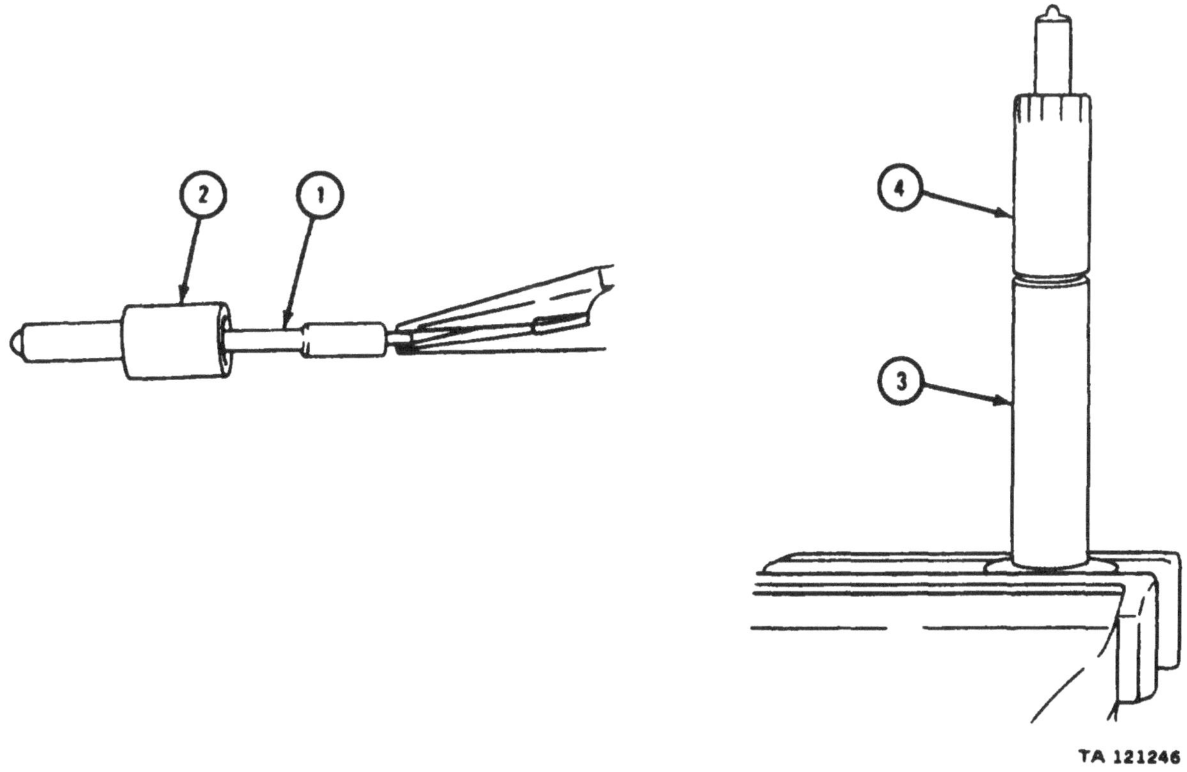

FRAME 2

1. Move nozzle holder (1) in vise to position shown.
2. Put in spindle (2) and spring (3).
3. Put locknut gasket (4) on adjusting screw (5) as shown.
4. Screw in adjusting screw (5) .
5. Tighten locknut (6) to 40 to 45 pound-feet.
6. Put on cap gasket (7) and tighten cap (8) to 40 to 45 pound-feet.
7. Take nozzle and holder assembly out of vise.

END OF TASK

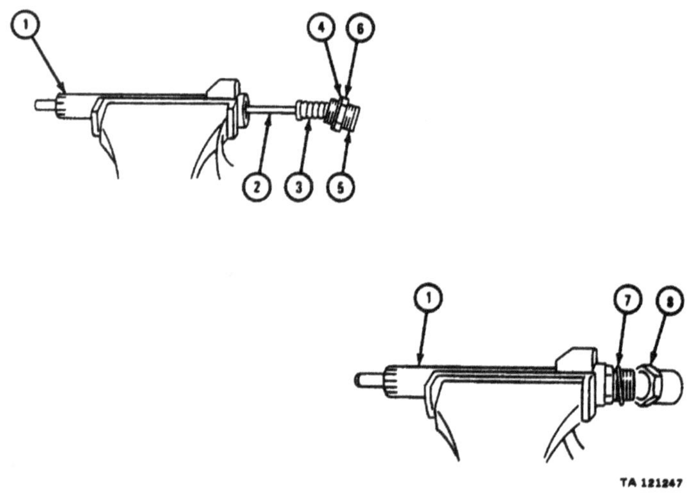

(2) Shim type nozzle and valve holder .

NOTE

Work area must be kept clean.

FRAME 1

CAUTION

Do not touch valve (1) with hands. Oil may rub off causing a sticking valve which can damage equipment. Do not let polished surface of valve or valve body (2) touch any hard surfaces. Nicks or scratches can damage equipment.

1. Hold valve (1) with pliers, as shown, and push into valve body (2).
2. Put valve (1) with valve body (2) into nozzle holder (3). Put nozzle holder into soft-jawed vise as shown.
3. Put on and tighten nozzle capnut (4) to 40 to 45 pound-feet.

GO TO FRAME 2

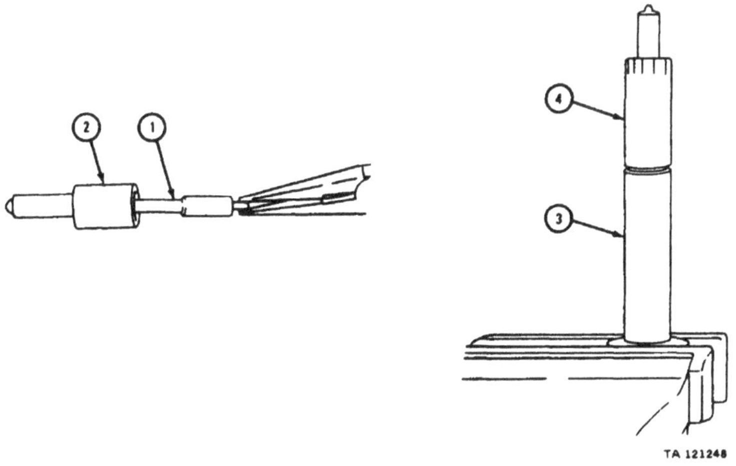

FRAME 2

1. Move nozzle holder (1) in vise to position shown.
2. Put in spindle (2), spring (3), shims (4), and cap (5). Tighten cap to 40 to 45 pound-feet.

END OF TASK

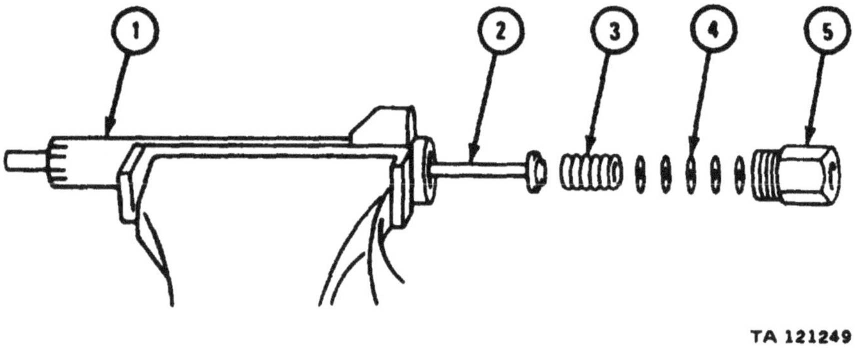

4-41. THERMOSTAT HOUSING

a. Disassembly.

FRAME 1

NOTE

It may be necessary to heat housing (4) to take out pipe plugs (2) .

1. Take out thermostat (1)., two pipe plugs (2), and thermostat seal (3). Throw seal away.

END OF TASK

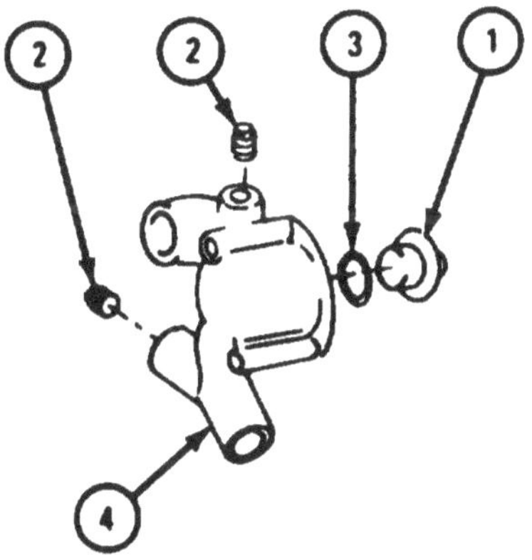

b. Cleaning.

FRAME 1

NOTE

Purpose of agitators to stir cleaning compound while it is boiling . These steps for fabrication of agitator are general and may be altered if necessary.

1. Cut about two feet of copper tube (1) and seal one end (2) by flattening with hammer.
2. Coil tube (1) as shown and drill small holes (3) about 1/2 inch apart along underside of coiled end.
3. Put fitting (4) on other end of tube (1).

GO TO FRAME 2

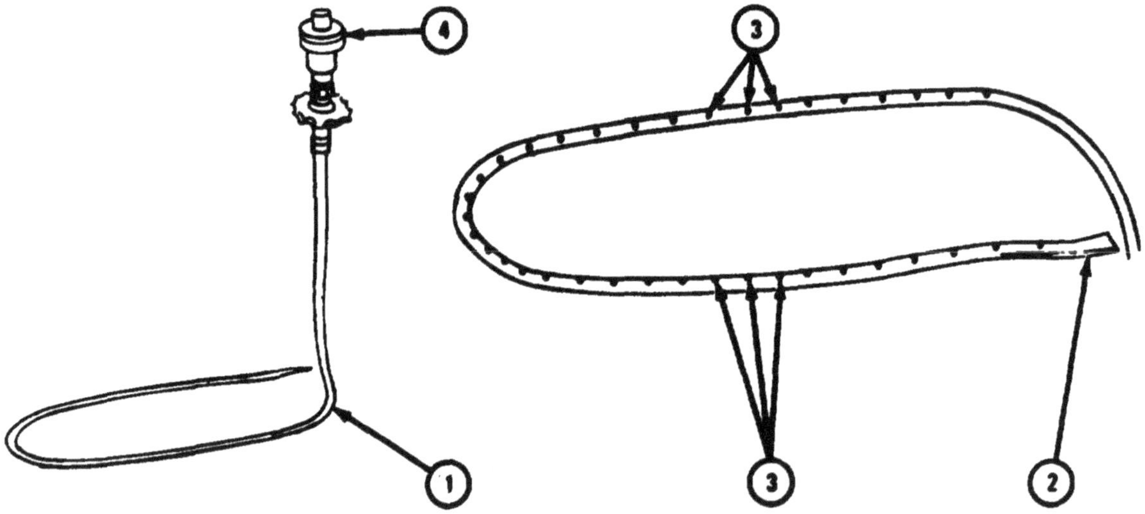

FRAME 2

WARNING

Face shields and gloves must be worn when using heated cleaning compound. Work only in well-ventilated area. Do not breathe in steam. Failure to use proper precautions may result in injury to personnel.

1. Mix cleaning compound in metal container (1). Put agitator (2) in container and put in thermostat housing (3).

NOTE

Set air supply at about 5 psi. Use hot plate or other suitable heating element to boil cleaning compound.

2. Place container (1) on hot plate and hook up air supply to agitator (2).
3. Turn on heat and when cleaning compound starts to boil, turn on air supply. Allow thermostat housing (3) to boil for about 30 minutes.
4. Turn off air supply. Turn off heat . Rinse housing (3) in hot water.

END OF TASK

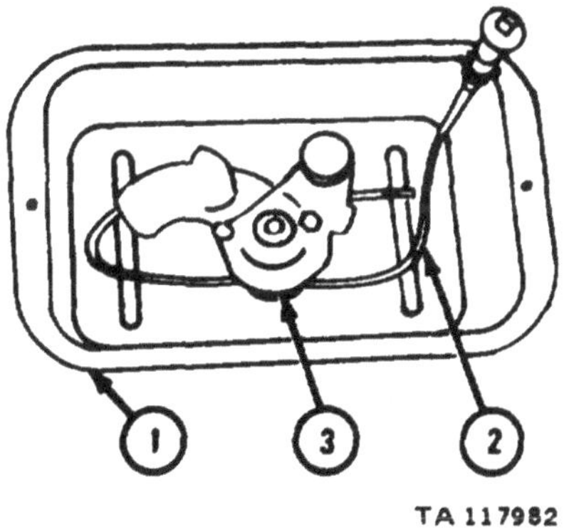

c. Inspection and Repair.

FRAME 1

1. Check that housing (1) has no chips or breaks. Check for damaged threads in two pipe plug holes (2).
2. Using dye penetrant kit, strong light, and a magnifying glass, check that housing (1) has no cracks . Fix small cracks by welding. Refer to TM 9-237.
3. If housing (1) has large cracks, broken out areas or if threads in pipe plug holes (2) are stripped, get new housing .
4. Check that two pipe plugs (3) have no worn threads or rounded heads. If pipe plugs are damaged, get new parts.
5. Check that thermostat (4) is not damaged. If thermostat is damaged, get a new thermostat.

GO TO FRAME 2

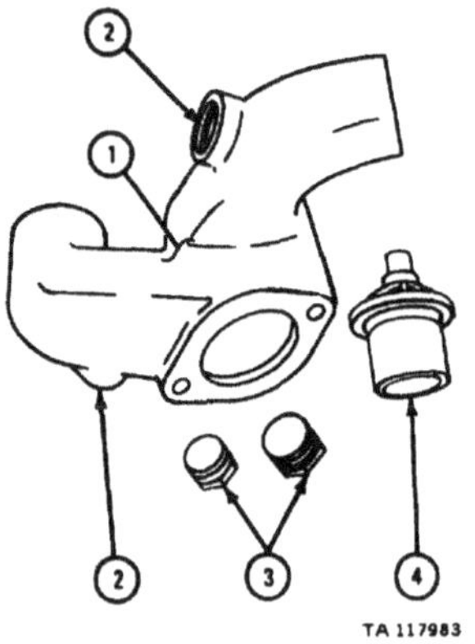

FRAME 2

1. Place thermostat (1) in a container of water. Put thermometer (2) in container.

NOTE

Thermostat (1) for engines LD-465-1, LD-465-1C, and LDT 465- 1C must start to open at 180°F and be fully open at 200°F. Thermostat for engines LDS-465-1A and LDS-465-2 must start opening at 163°F and be fully open at 180°F. All thermostats must open a minimum of 0.310 inch between flange (3) and thermostat spring housing (4) .

2. Slowly raise the water temperature. Check that thermostat (1) opens at correct temperature.

3. If thermostat (1) does not open at correct temperature or does not reach minimum opening, get a new thermostat.

END OF TASK

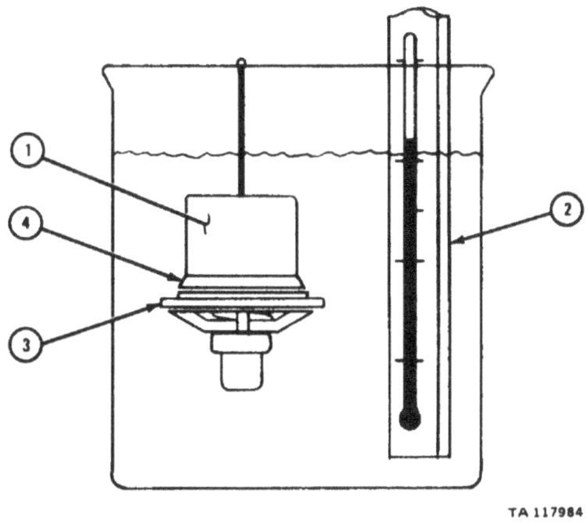

d. Assembly.

FRAME 1

1. Put seal (1) in thermostat housing counterbore (2) with metal edge (3) up.
2. Use driver large enough to touch metal edge of seal (1)., tap driver until seal seats in counterbore (2) .
3. Wrap thread sealant tape around threads of two pipe plugs (4). Put two pipe plugs in housing (5) .
4. Put thermostat (6) in housing (5).

END OF TASK

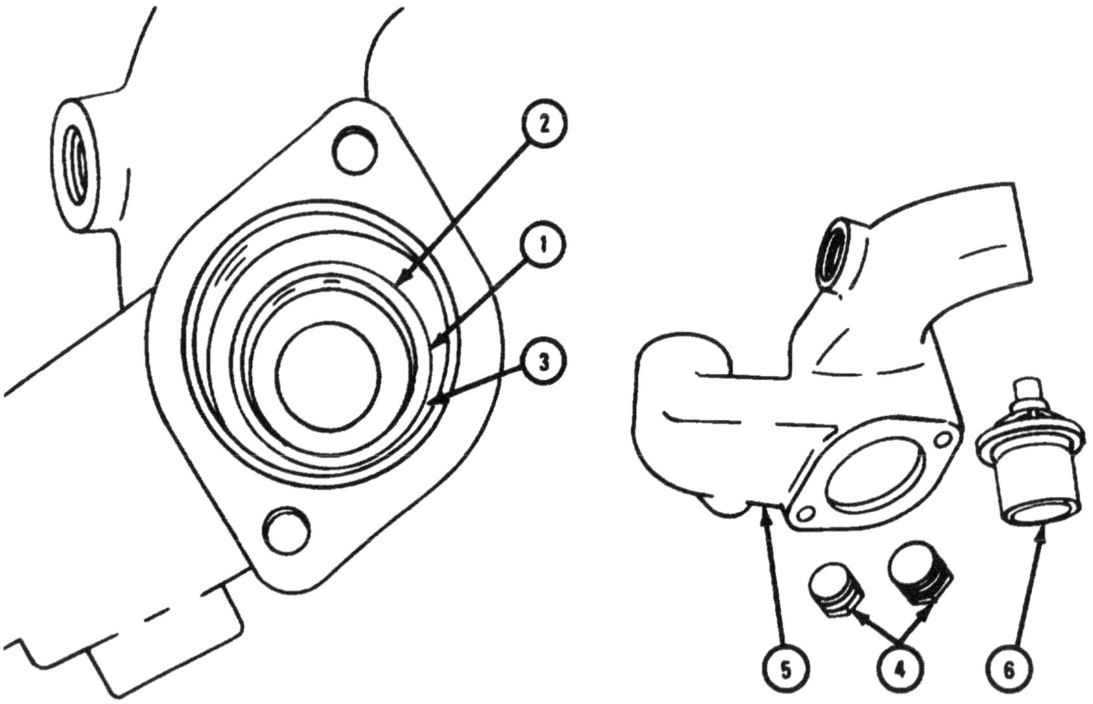

4-42. CYLINDER HEAD WATER OUTLET MANIFOLDS.

Disassembly.

FRAME 1

1. Loosen four hose clamp screws (1) and take off four hose clamps (2) and two hoses (3). Throw away hoses.

END OF TASK

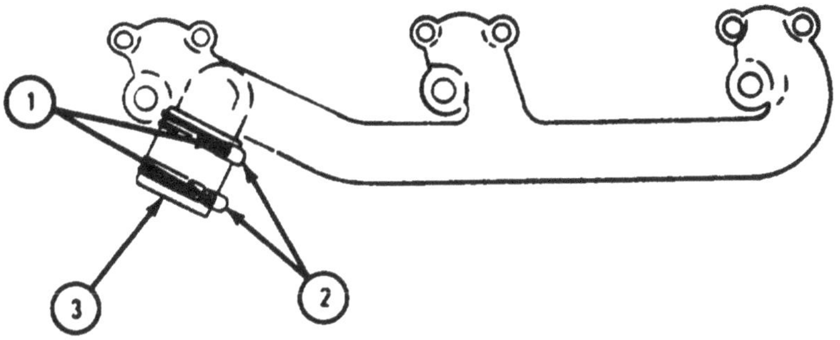

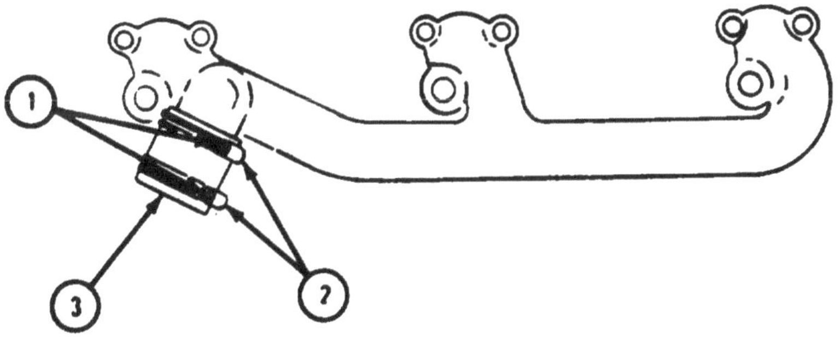

TA 113705

b. There are no special cleaning procedures needed. Refer to cleaning procedures given in para 4-3.

c. Inspection and Repair.

FRAME 1

1. Check that four hose clamps (1) are not cracked or bent. Take out minor bends in hose clamps. If more repair is needed, get a new hose clamp.

2. Check that two cylinder head water outlet manifolds (2) are not cracked, bent or damaged in any other way. To fix cracks, refer to TM 9-237. If more repair is needed, get a new cylinder head water outlet manifold,

END OF TASK

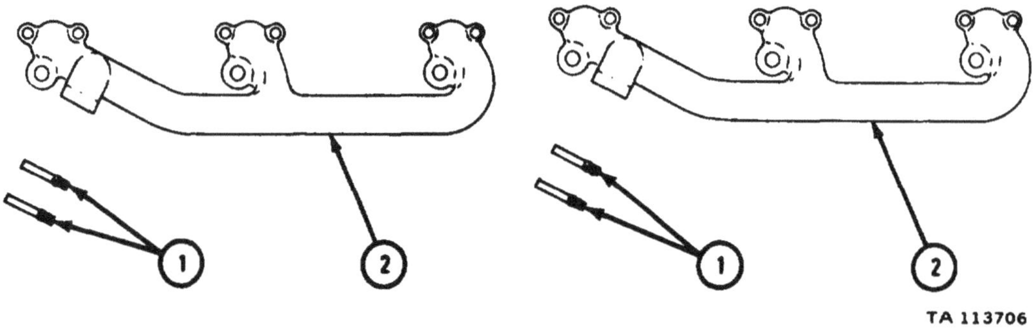

d. Assembly.

FRAME 1

1. Put on two hoses (1).
2. Put on four hose clamps (2).

END OF TASK

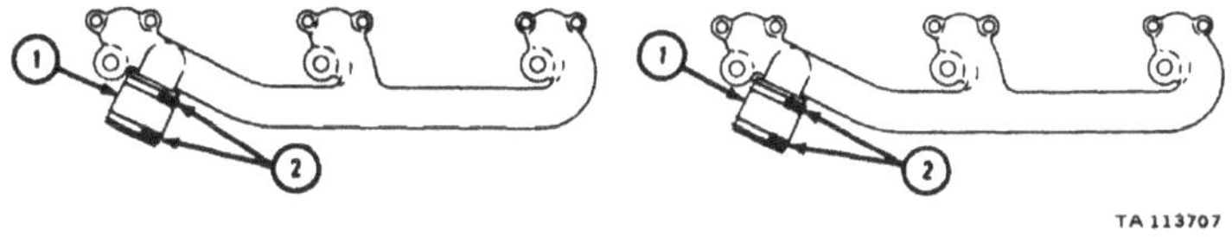

4-43. INTAKE MANIFOLD.

a. Disassembly.

FRAME 1

1. Takeout temperature transmitter (1).
2. Take out three pipe plugs (2, 3, and 4).

NOTE

Do not takeout three expansion plugs (5) or four studs (6) unless they are damaged. Refer to para 4-43c for inspection procedure.

3. Using hammer and punch, take out three expansion plugs (5).
4. Take out four studs (6).

END OF TASK

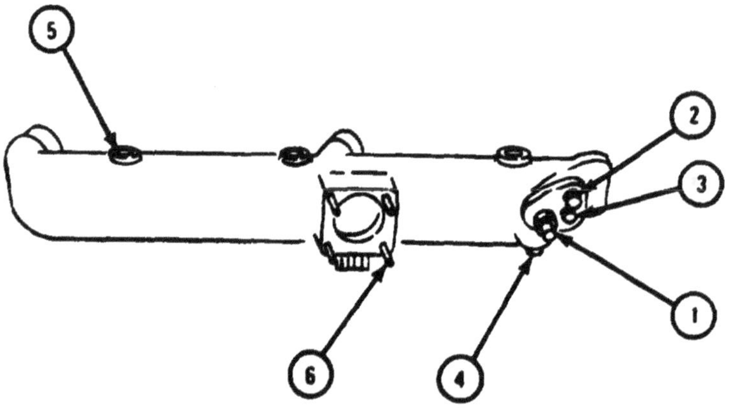

TA 113654

b. There are no special cleaning procedures needed. Refer to cleaning procedures given in para 4-3.

c. Inspection and Repair.

FRAME 1

1. Check that three expansion plugs (1) are not loose, leaking or damaged in any other way. If any expansion plug is damaged, get a new one.
2. Check that four studs (2) are not loose, stripped, crossthreaded or damaged in any other way. Fix minor thread damage with a thread chaser. If more repair is needed, get a new part.

NOTE

To check for cracks, refer to para 4-4.

3. Check that intake manifold (3) is not cracked, bent or damaged in any other way. If repair is needed, get a new part.

WARNING

Dry cleaning solvent is flammable. Do not use near an open flame. Keep a fire extinguisher nearby when solvent is used. Use only in well-ventilated places. Failure to do this may result in injury to personnel and damage to equipment.

4. Check that gasket surfaces (4) have no nicks, burrs or scratches. Take out minor nicks, burrs or scratches with a fine mill file or crocus cloth dipped in dry cleaning solvent. If more repair is needed, get a new intake manifold (3).

GO TO FRAME 2

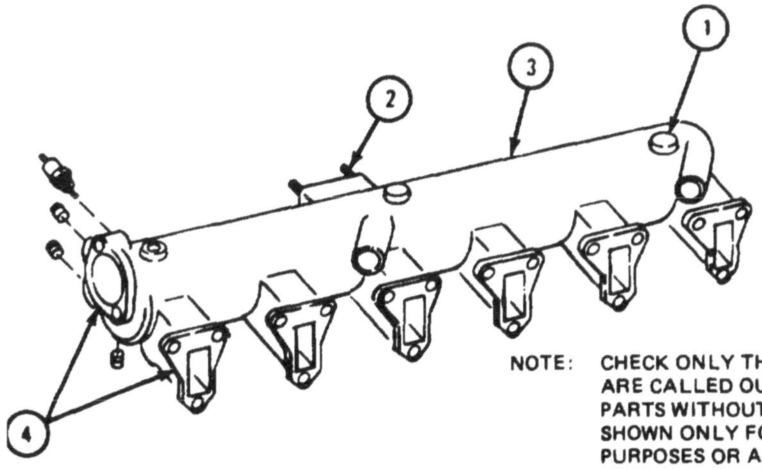

NOTE: CHECK ONLY THOSE PARTS WHICH ARE CALLED OUT IN THIS FRAME. PARTS WITHOUT CALLOUTS ARE SHOWN ONLY FOR REFERENCE PURPOSES OR ARE CHECKED IN ANOTHER FRAME.

TA 113655

FRAME 2

1. Check that three pipe plugs (1, 2 and 3) are not stripped, crossthreaded or damaged in any other way. If pipe plug is damaged in any way, get a new part.
2. Check that temperature transmitter (4) is not dented, cracked, stripped, crossthreaded or damaged in any other way. Fix minor thread damage with a thread chaser.
3. Check that temperature transmitter (4) is not damaged at electrical connector (5). If electrical connector is damaged, get a new temperature transmitter.

END OF TASK

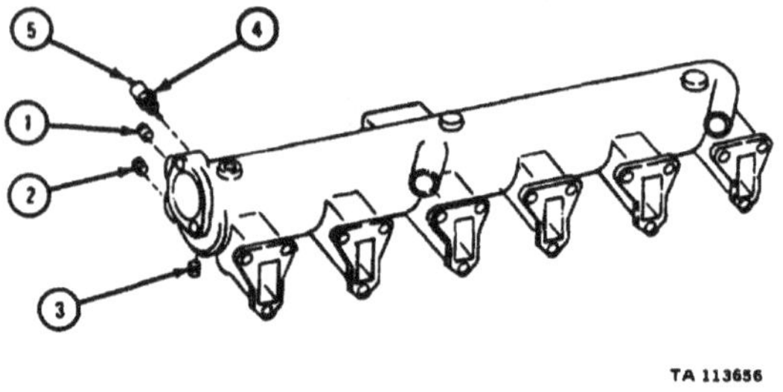

d. Assembly.

FRAME 1

1. Put in temperature transmitter (1),
2. Put in three pipe plugs (2, 3, and 4).

NOTE

If any of three expansion plugs (5) or four studs (6) were taken out, do steps 3 and 4.

3. Using wooden block and hammer, tap in expansion plug (5).
4. Put in stud (6).

END OF TASK

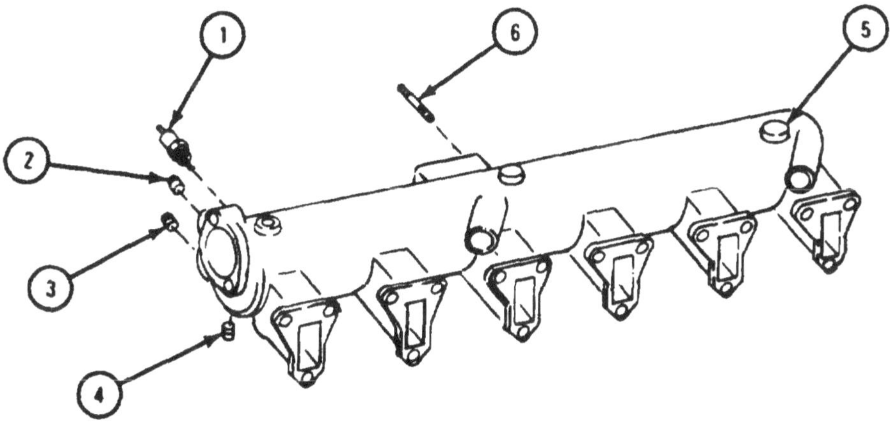

4-44. WATER PUMP AND RELATED PARTS.

a. Cleaning. There are no special cleaning procedures needed. Refer to cleaning procedures given in para 4-3.

b. Inspection and Repair.

FRAME 1

NOTE

If working on water pump from engine LDS-465-1, do steps 1, 3, 4, and 5. For all other engines, do steps 2, 3, 4, and 5.

1. Check that water pump housing (1) has no cracks or other damage. If damaged, get a new water pump assembly.
2. Check that water pump housing (2) has no cracks or other damage. If damaged, get a new water pump assembly.
3. Check that studs (3) have no burrs or other damage to threads. Fix minor damage with thread chaser. If more repair is needed, get new studs.
4. Check that nuts (4) have no damaged threads. If damaged, get new nuts.
5. Check that water pump (5) has no damage or cracks. If damaged, get a new water pump assembly.

END OF TASK

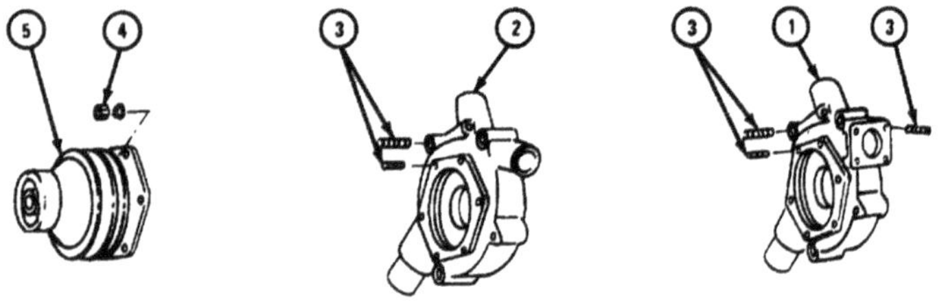

NOTE: CHECK ONLY THOSE PARTS WHICH ARE CALLED OUT. PARTS WITHOUT CALLOUTS ARE SHOWN ONLY FOR REFERENCE PURPOSES.

TA 117987

4-45. EXHAUST MANIFOLD ASSEMBLY.

a. Disassembly.

FRAME 1

WARNING

Wear welder's gloves when handling hot parts. Failure to do so will result in serious burns.

CAUTION

Do not use cutting tip on welding torch or exhaust manifold assembly may be damaged.

1. Put center exhaust manifold section (1) in vise.
2. Using welding torch, heat front exhaust manifold section (2) near joint (3).
3. Pull front exhaust manifold section (2) from center exhaust manifold section (1)
4. Take three seal rings (4) out of groove in front exhaust manifold section (2). Throw away seal rings.
5. Do steps 2, 3, and 4 for rear exhaust manifold section (5).
6. Take center exhaust manifold section (1) out of vise.

END OF TASK

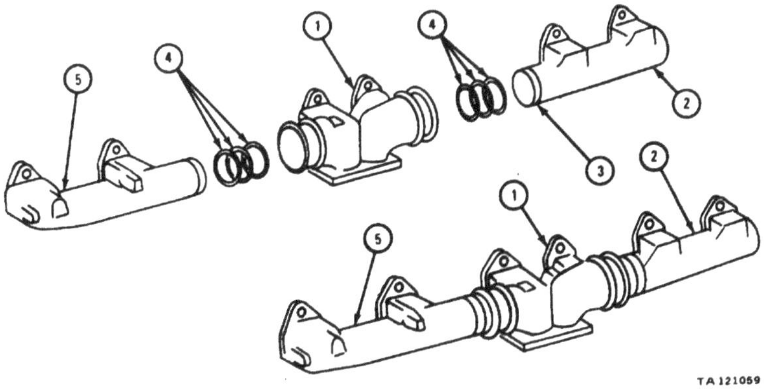

b. Cleaning.

FRAME 1

WARNING

Dry cleaning solvent is flammable. Do not use near an open flame. Keep a fire extinguisher nearby when solvent is used. Use only in well-ventilated places. Failure to do this may result in injury to personnel and damage to equipment.

1. Clean rear exhaust manifold section (1), center exhaust manifold section (2), and front exhaust manifold section (3) with dry cleaning solvent.
2. Using wire brush, clean seal ring grooves (4) and seal ring seating surface (5) Dry with clean cloth.

END OF TASK

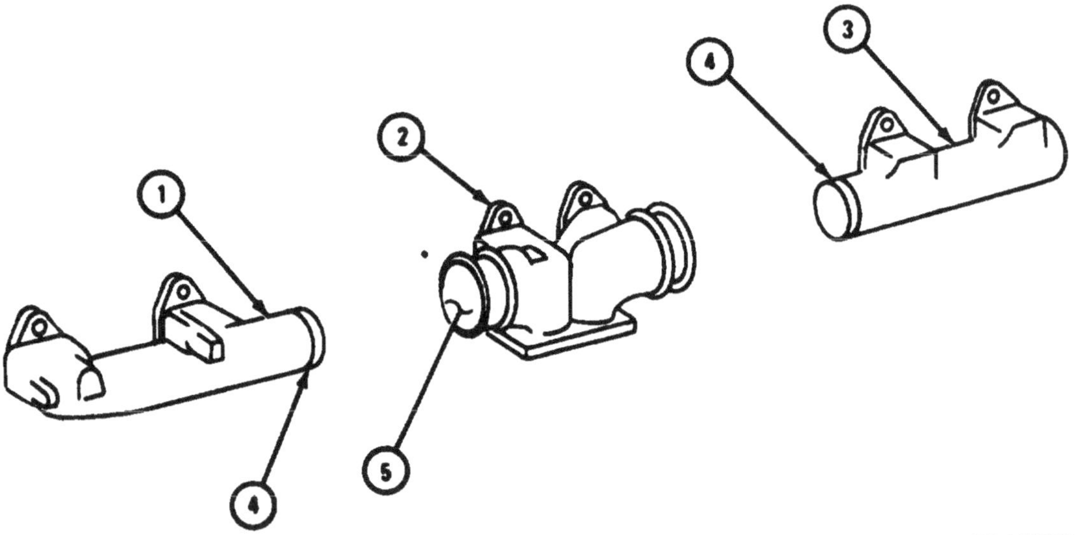

c. Inspection and Repair.

FRAME 1

1. Check that two exhaust manifold sections (1) and center exhaust manifold section (2) are not cracked or bent. Refer to TM 9-237 to fix cracks. Get new parts if cracks cannot be fixed.

END OF TASK

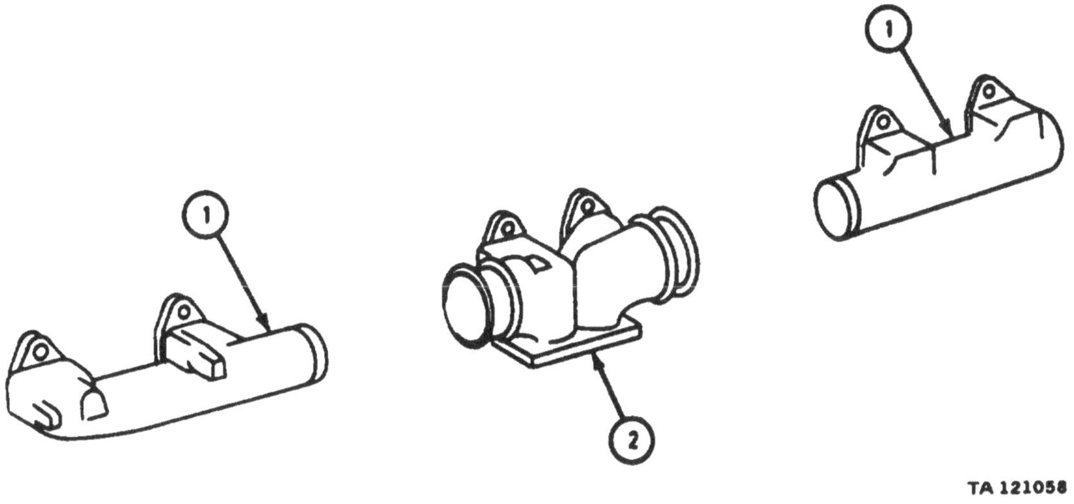

d. Assembly.

FRAME 1

1. Put center exhaust manifold section (1) in vise.
2. Put set of three seal rings (2) in groove in front exhaust manifold section (3). dining gaps in seal rings 120° apart.
3. Press down seal rings (2) and join front exhaust manifold section (3) to center exhaust manifold section (1).
4. Do steps 2 and 3 again for rear exhaust manifold section (4).
5. Take center exhaust manifold section (1) out of vise.

END OF TASK

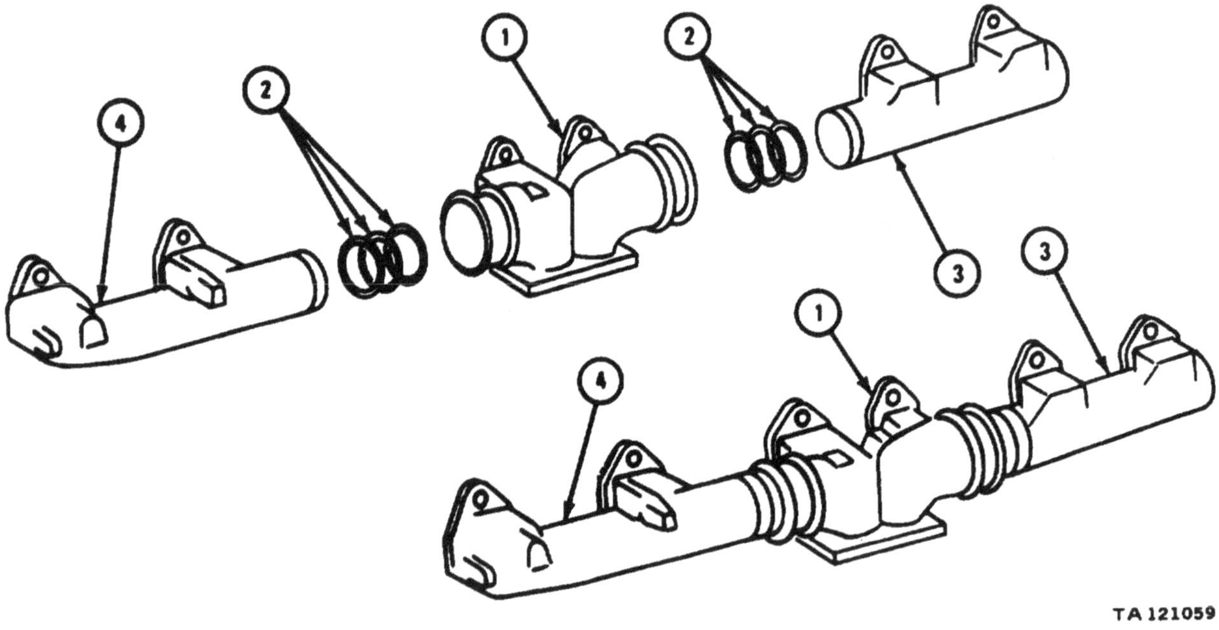

4-46. TURBOCHARGER OIL DRAIN HOSE, OIL DRAIN AND INLET TUBES, AND RELATED PARTS (ENGINES LDT-465-1C, LDS-465-1, LDS-465-1A, AND LDS-465-2).

a. Cleaning.

FRAME 1

1. Take two hose clamps (1) off oil drain hose (2).

CAUTION

Do not let dry cleaning solvent touch oil drain hose (2). Hose will dry out, rot and need replacement. If dry cleaning solvent splashes on hose, dry with clean cloth.

2. Clean oil drain hose (2) with soapy water. Dry with clean cloth.

WARNING

Dry cleaning solvent is flammable. Do not use near an open flame. Keep a fire extinguisher nearby when solvent is used. Use only in well-ventilated places. Failure to do this may result in injury to personnel and damage to equipment.

3. Clean two hose clamps (1), four screws (3), and four lockwashers (4) with dry cleaning solvent. Let parts air dry.

GO TO FRAME 2

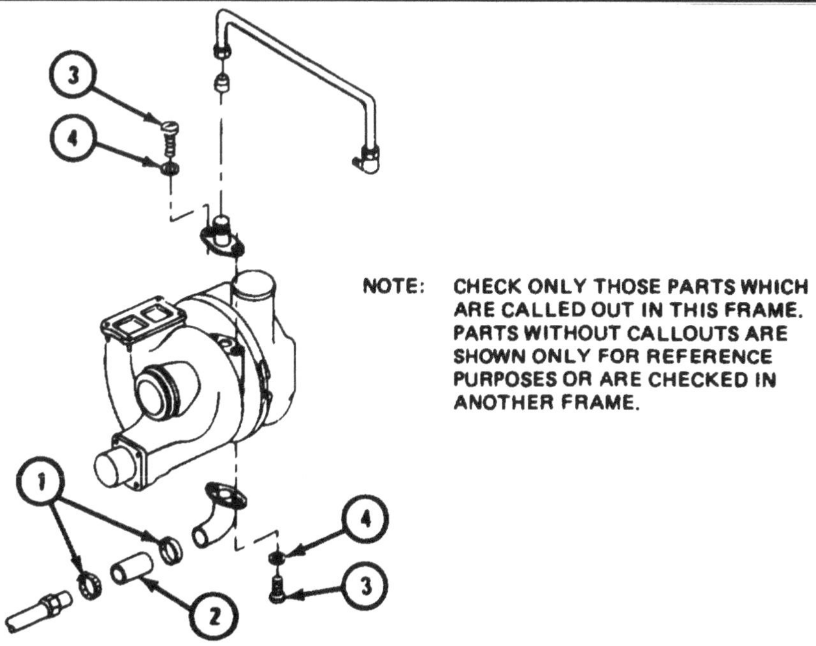
1
2
3
4
4
3
NOTE: CHECK ONLY THOSE PARTS WHICH ARE CALLED OUT IN THIS FRAME. PARTS WITHOUT CALLOUTS ARE SHOWN ONLY FOR REFERENCE PURPOSES OR ARE CHECKED IN ANOTHER FRAME.

TA 121067

FRAME 2

WARNING

Dry cleaning solvent is flammable. Do not use near an open flame. Keep a fire extinguisher nearby when solvent is used. Use only in well-ventilated places. Failure to do this may result in injury to personnel and damage to equipment.

Eye shields must be worn when using compressed air. Eye injury can occur if eye shields are not used.

1. Clean inside of oil inlet tube (1), oil inlet adapter (2), oil drain tube (3), and oil drain hose-to-pipe adapter (4) with brass wire probes. Flush with dry cleaning solvent . Blow dry with compressed air.
2. Clean outside of oil inlet tube (1), oil inlet adapter (2), oil drain tube (3) and oil drain hose-to-pipe adapter (4) with dry cleaning solvent. Dry with clean cloth .

END OF TASK

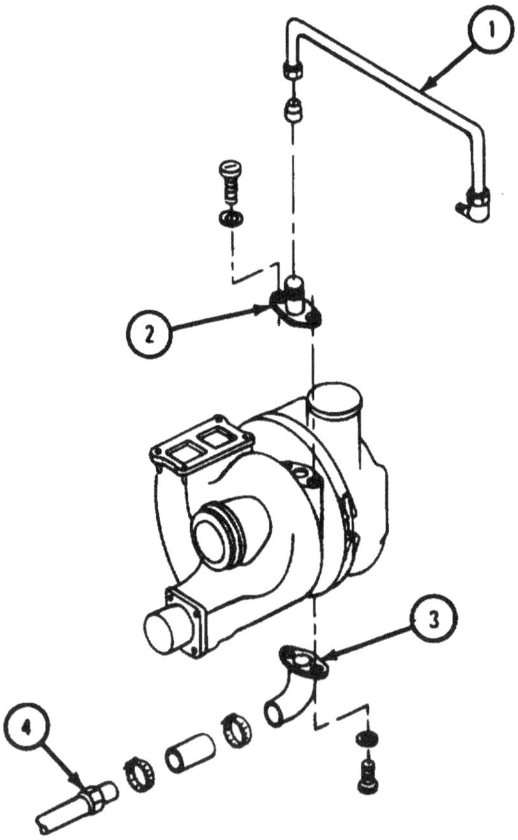

NOTE: CHECK ONLY THOSE PARTS WHICH ARE CALLED OUT IN THIS FRAME. PARTS WITHOUT CALLOUTS ARE SHOWN ONLY FOR REFERENCE PURPOSES OR ARE CHECKED IN ANOTHER FRAME.

TA 121068

b. Inspection and Repair.

FRAME 1

1. Check that oil inlet tube (1) is not bent, dented, cracked or damaged in any other way. If oil inlet tube is damaged, get a new one.
2. Check that two coupling nuts (2) are not cracked, cross threaded or damaged in any other way. If coupling nut is damaged, get a new oil inlet tube (1).
3. Check that two hose clamps (3) are not bent or damaged in any other way. Take out minor bends. If more repair is needed, get a new part.

GO TO FRAME 2

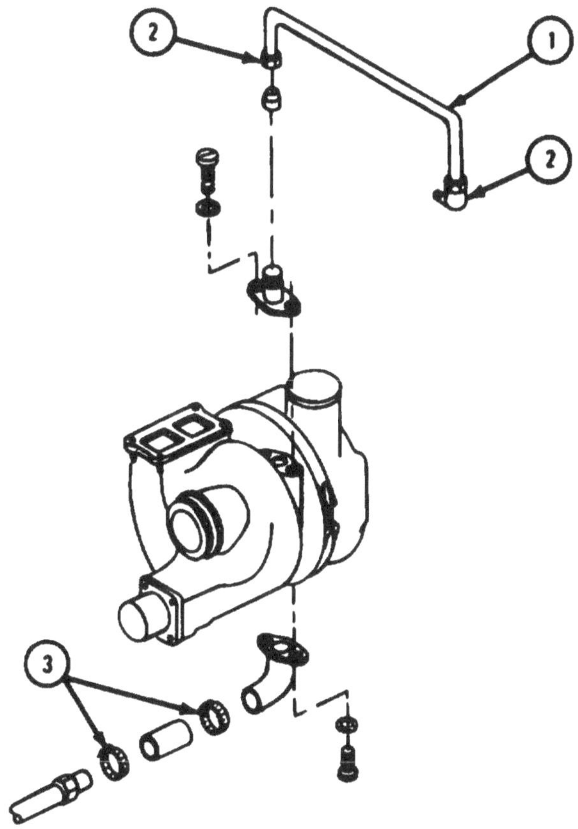

NOTE: CHECK ONLY THOSE PARTS WHICH ARE CALLED OUT IN THIS FRAME. PARTS WITHOUT CALLOUTS ARE SHOWN ONLY FOR REFERENCE PURPOSES OR ARE CHECKED IN ANOTHER FRAME.

TA 121069

FRAME 2

1. Check that oil drain hose (1) is not torn, cracked, dried out or damaged in any way . If oil drain hose is damaged, get a new one.
2. Check that oil drain tube (2) and pipe-to-hose adapter (3) are not bent, dented or cracked and have no minor nicks or burrs. Take out minor nicks or burrs with a fine mill file. If more repair is needed, get a new part.
3. Check that four screws (4) are not stripped, crossthreaded or damaged in any other way. If repair is needed, get a new screw.
4. Check that four lockwashers (5) are not chipped or cracked. If any lock-washer is damaged, get a new one.
5. Check that oil inlet adapter (6) is not bent, cracked or damaged in any other way. If oil inlet adapter is damaged, get a new one.

END OF TASK

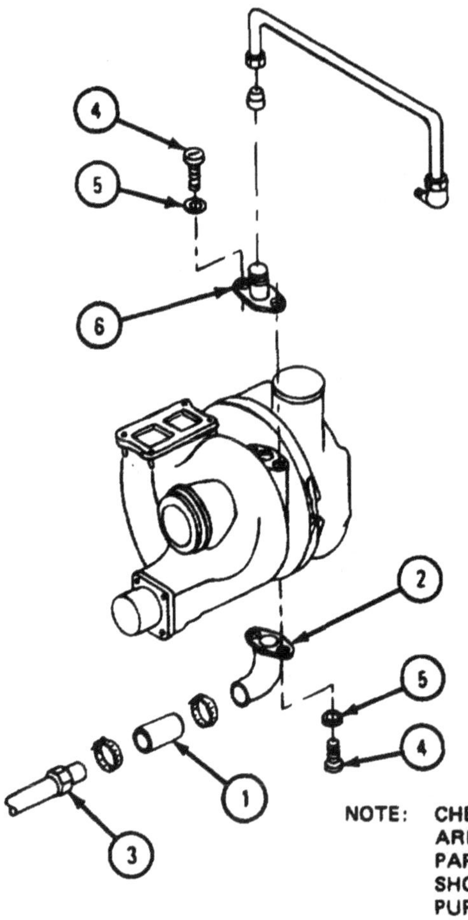

NOTE: CHECK ONLY THOSE PARTS WHICH ARE CALLED OUT IN THIS FRAME. PARTS WITHOUT CALLOUTS ARE SHOWN ONLY FOR REFERENCE PURPOSES OR ARE CHECKED IN ANOTHER FRAME.

TA 121070

4-47. TURBOCHARGER EXHAUST ELBOW OR ADAPTER (ENGINE LDS-465-1).

a. Disassembly.

FRAME 1

NOTE

If working on models with exhaust adapter (1), go to para 4-47b.

1. If working on models with exhaust elbow (2), scribe an alinement mark on exhaust elbow support (3) and exhaust elbow.
2. Take off six nuts (4) and six washers (5).
3. Take out six screws (6) .
4. Take off gasket (7). Throw away gasket .

END OF TASK

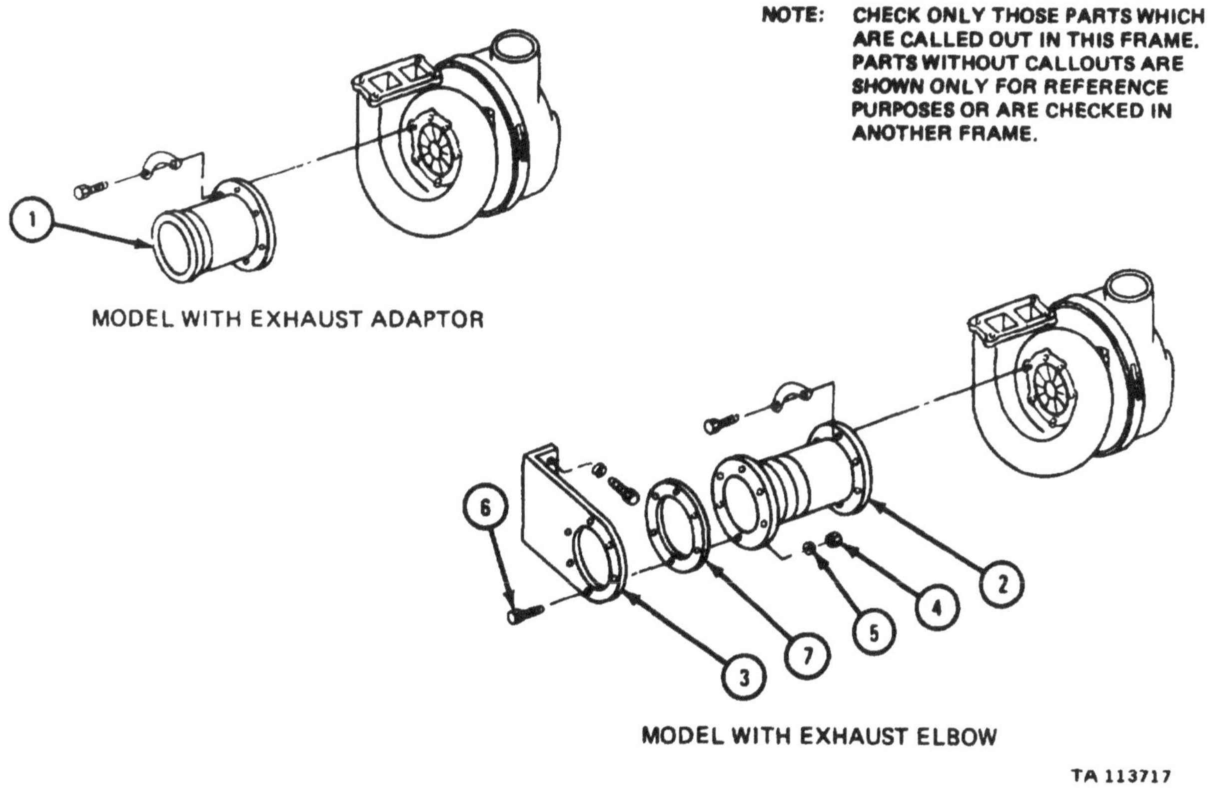
NOTE: CHECK ONLY THOSE PARTS WHICH ARE CALLED OUT IN THIS FRAME. PARTS WITHOUT CALLOUTS ARE SHOWN ONLY FOR REFERENCE PURPOSES OR ARE CHECKED IN ANOTHER FRAME.
1
MODEL WITH EXHAUST ADAPTOR
6
2
4
5
7
3
MODEL WITH EXHAUST ELBOW
TA 113717

b. Cleaning. There are no special cleaning procedures needed. Refer to cleaning procedures given in para 4-3.

c. Inspection and Repair.

FRAME 1

NOTE

If working on model with exhaust elbow (1), do steps 1 through 4 . If working on model with exhaust adapter (2), go to frame 3.

1. Check that exhaust elbow (1) has no cracks, holes or burned areas. If it does, get a new one.
2. Check that flanges of exhaust elbow (1) have no burned areas and are not warped. If flanges are damaged, get a new exhaust elbow.
3. Check that sections of expansion bellows (3) have not come apart. If expansion bellows are damaged, get a new exhaust elbow (1).
4. Check that flanges of exhaust elbow (1) have no nicks or burrs. Take out small nicks or burrs with a fine mill file. If more repair is needed, get a new exhaust elbow .

GO TO FRAME 2

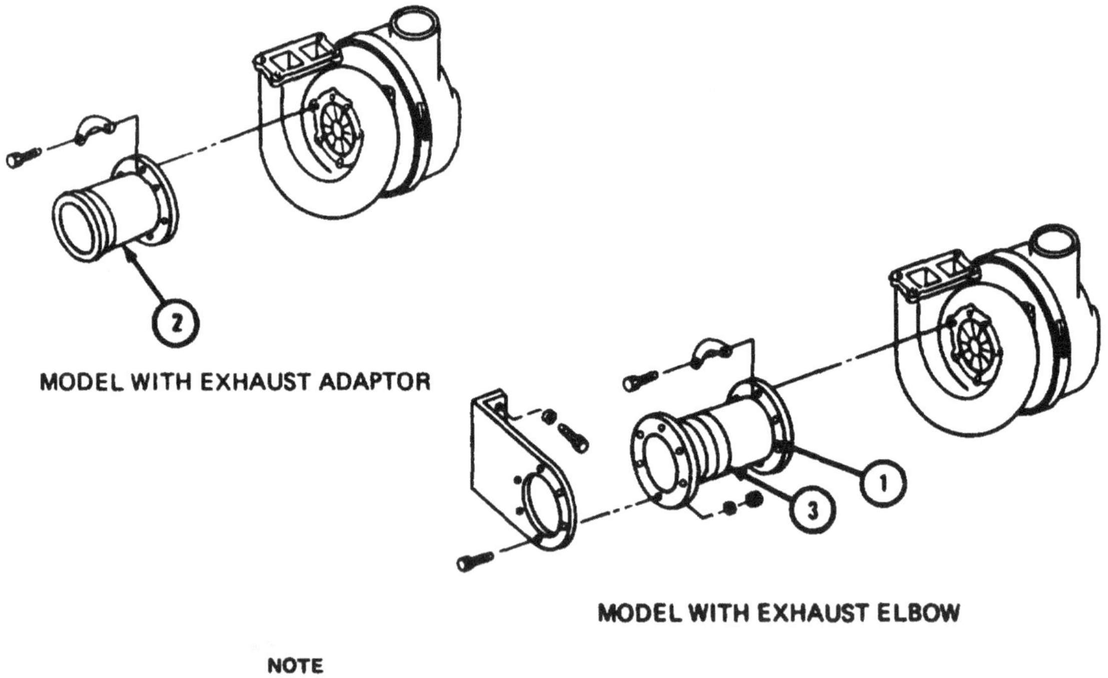
2
MODEL WITH EXHAUST ADAPTOR
1
3
MODEL WITH EXHAUST ELBOW
NOTE
CHECK ONLY THOSE PARTS WHICH ARE CALLED OUT IN THIS FRAME. PARTS WITHOUT CALLOUTS ARE SHOWN ONLY FOR REFERENCE PURPOSES OR ARE CHECKED IN ANOTHER FRAME.
TA 113718

FRAME 2

1. Check that exhaust elbow support (1) is not cracked or bent. Fix small bends. If more repair is needed, get a new elbow support.
2. Check that three lockplates (2) have no dents or cracks. Take out minor dents. If more repair is needed, get new lockplates.
3. Check that 15 screws (3) and six nuts (4) are not stripped or crossthreaded. If screws or nuts are damaged, get new ones.

END OF TASK

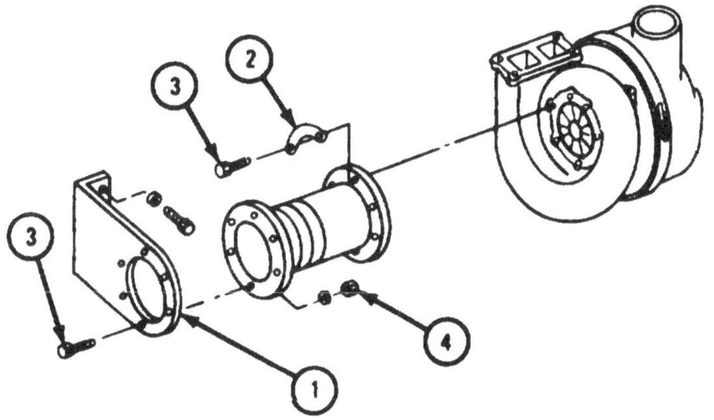

NOTE: CHECK ONLY THOSE PARTS WHICH ARE CALLED OUT IN THIS FRAME. PARTS WITHOUT CALLOUTS ARE SHOWN ONLY FOR REFERENCE PURPOSES OR ARE CHECKED IN ANOTHER FRAME.

TA 113719

FRAME 3

1. Check that exhaust adapter (1) has no cracks, holes or burned areas. If it is cracked or burned, get a new one.
2. Check that exhaust adapter flange (2) is not warped and has no burned areas. If flange is damaged, get a new exhaust adapter (1).
3. Check that exhaust adapter flange (2) has no nicks or burrs. Take out small nicks or burrs with a fine mill file. If more repair is needed, get a new exhaust adapter (1) .
4. Check that three lockplates (3) have no dents or cracks. Take out minor dents. If more repair is needed, get new lockplates.
5. Check that six screws (4) are not stripped or crossthreaded. If screws are damaged, get new ones.

END OF TASK

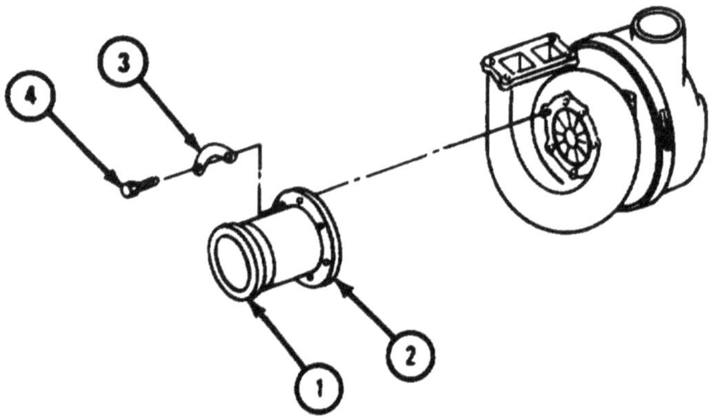

NOTE: CHECK ONLY THOSE PARTS WHICH ARE CALLED OUT IN THIS FRAME. PARTS WITHOUT CALLOUTS ARE SHOWN ONLY FOR REFERENCE PURPOSES OR ARE CHECKED IN ANOTHER FRAME.

TA 113720

d. Assembly.

FRAME 1

1. Aline screw holes of exhaust elbow support (1), gasket (2), and exhaust elbow (3), alining scribe marks .
2. Put in six screws (4).
3. Put on six washers (5) and six nuts (6).

END OF TASK

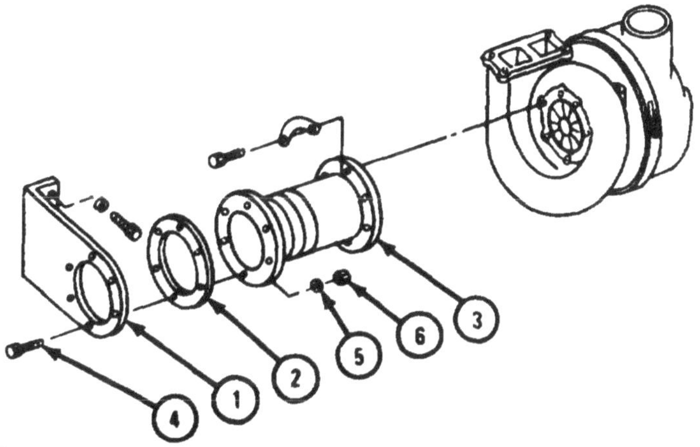

NOTE
PARTS WITHOUT CALLOUTS ARE SHOWN ONLY FOR REFERENCE PURPOSES.

TA 113721

4-48. INTAKE MANIFOLD ELBOW ASSEMBLY.

NOTE

There are two different configurations of intake manifold elbow assemblies. Elbow assembly used on engines LDS-465-1A, LDS-465-2, LD-465-1 and LD-465-1C is shown. This task is the same for elbow assembly used on engine LDT-465-1C.

a. Cleaning.

FRAME 1

WARNING

Dry cleaning solvent is flammable. Do not use near an open flame. Keep a fire extinguisher nearby when solvent is used. Use only in well-ventilated places. Failure to do this may result in injury to personnel and damage to equipment.

1. Wash inner and outer surfaces of intake manifold assembly elbow (1) with dry cleaning solvent . Dry with lint-free cloth.

END OF TASK

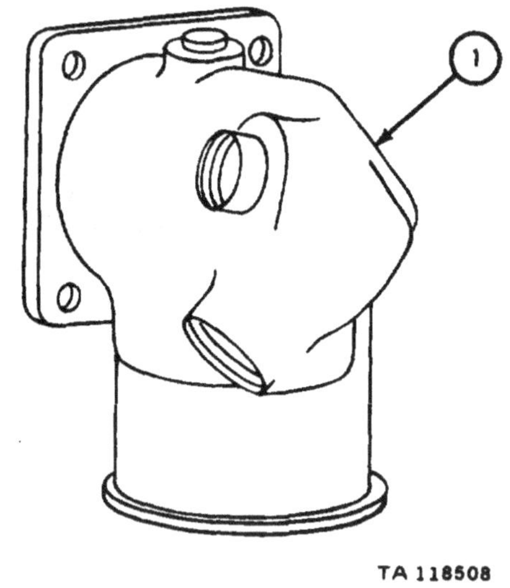

b. Inspection and Repair.

FRAME 1

1. Check intake manifold elbow assembly (1) for cracks using magnetic particle inspection equipment, if available. If magnetic particle inspection equipment is not available, check for cracks using magnifying glass and strong light.
2. Closely check for cracks around four bolt holes (2) and two threaded fittings (3). If cracked, get a new elbow assembly (1) .
3. Check that threads in two fittings (3) have no damage. If damaged, repair threads in fittings with tap or by putting in threaded insert.

END OF TASK

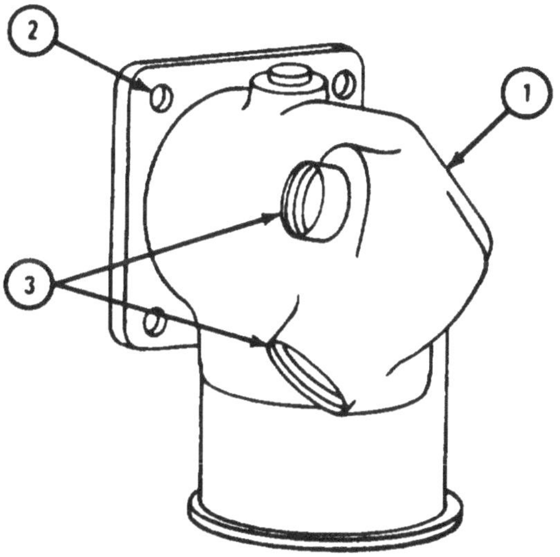

TA 118509

4-49. AIR COMPRESSOR ASSEMBLY.

a. Air Compressor Pulley.

(1) Removal and disassembly .

NOTE

If working on engine LD-465-1, LD-465-1C, LDT-465-1C, LDS-465-1, or LDS-465-1A, go to frame 1. If working on engine LDS-465-2, go to frame 3.

FRAME 1

1. Take out two screws and lockwashers (1).
2. Using air compressor pulley adjusting wrench (2), unscrew and take off pulley adjusting flange (3) .

GO TO FRAME 2

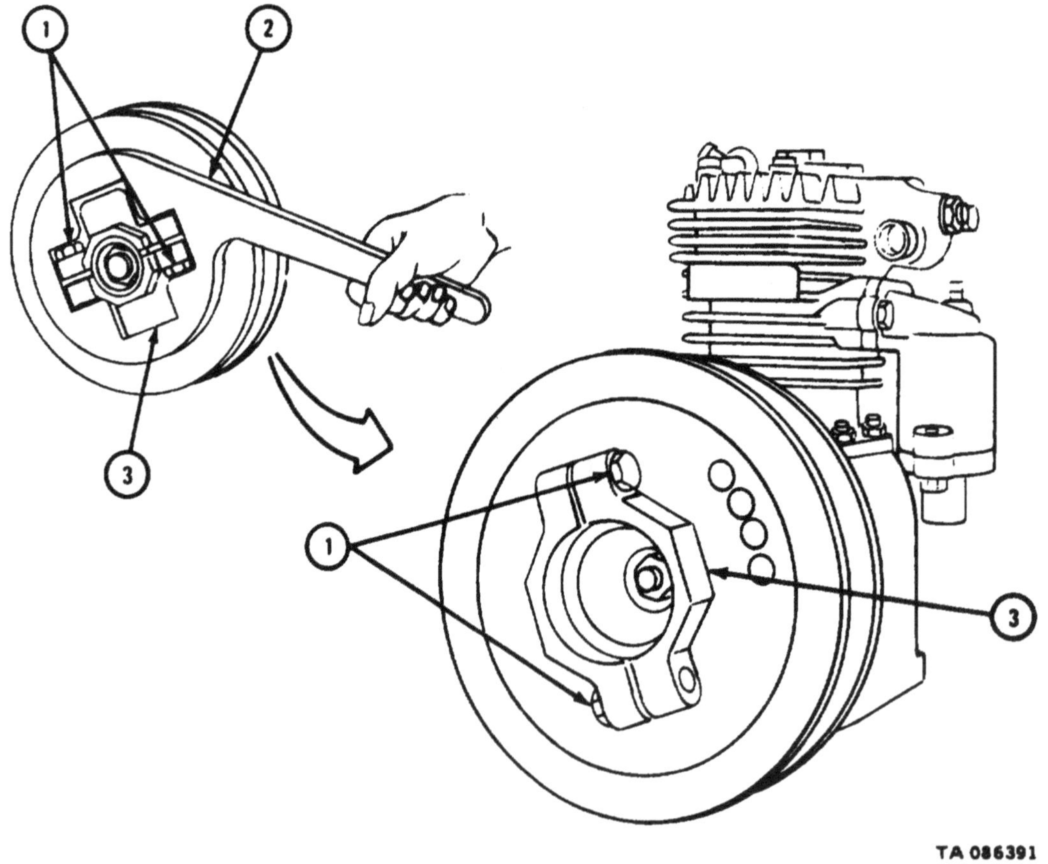

FRAME 2

1. Take off self-locking nut (1) .
2. Using mechanical puller, take pulley half (2) off air compressor shaft (3).

NOTE

Some compressors do not have woodruff key (4).

3. Take woodruff key (4) out of air compressor shaft (3) .

END OF TASK

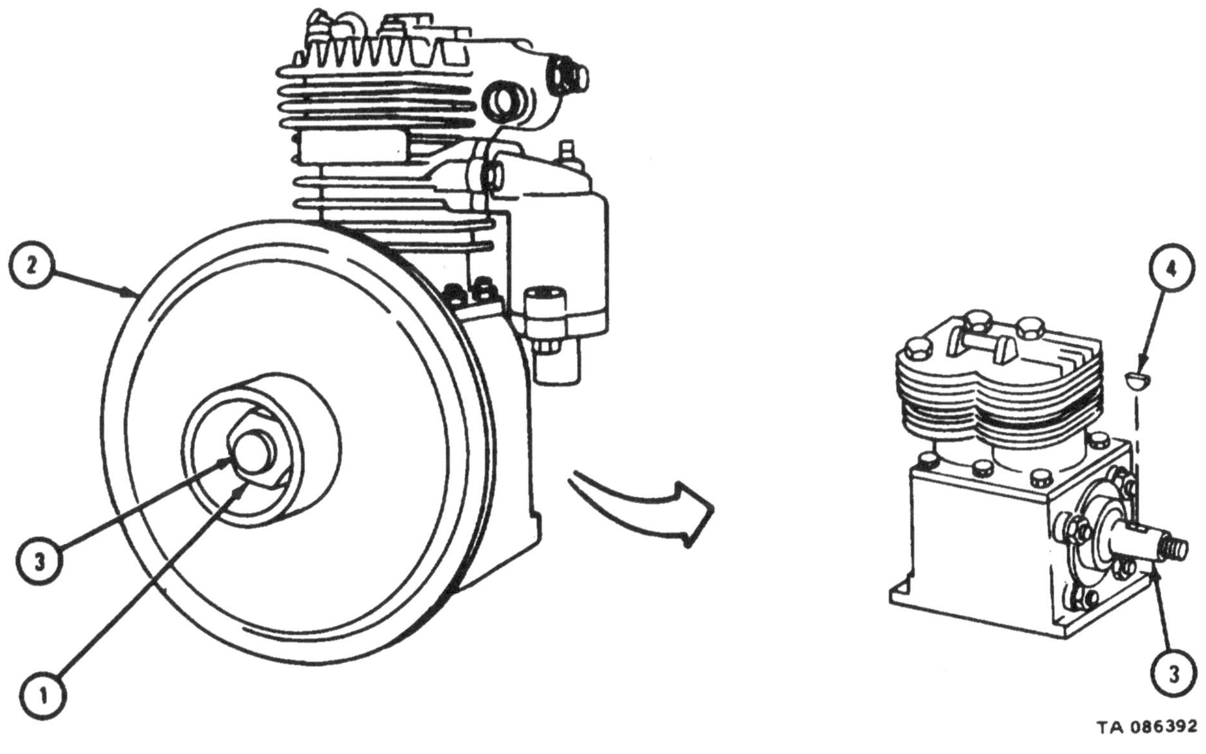

FRAME 3

NOTE

On engine LDS-465-2, air compressor pulley was taken off during engine disassembly .

1. Clamp sir compressor pulley assembly (1) in vise (2) or arbor press as shown.
2. Squeeze pulley assembly (1) together and take out four spring retainer lock rings (3) .

GO TO FRAME 4

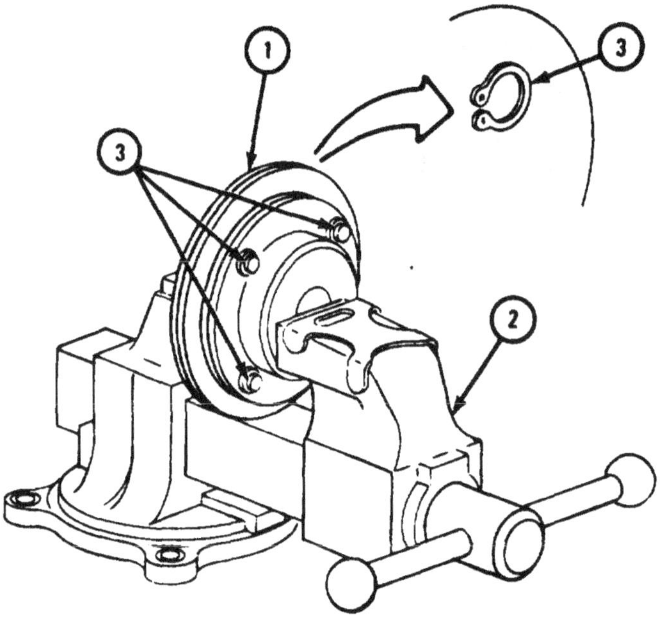

TA 086406

FRAME 4

CAUTION

Pulley assembly (2) has a very strong spring (6) inside. Take off pressure from vise (1) or arbor press slowly or assembly will fly apart and parts will be lost.

1. Work vise (1) or arbor press to let pressure off pulley assembly (2) and take apart inside pulley half (3), outside pulley half (4), spring retainer (5), spring (6), and four retainer pins (7) .
2. Pry out seals (8 and 9) from outside pulley half (4).
3. Using arbor press, press out four sleeve bearings (10) .

END OF TASK

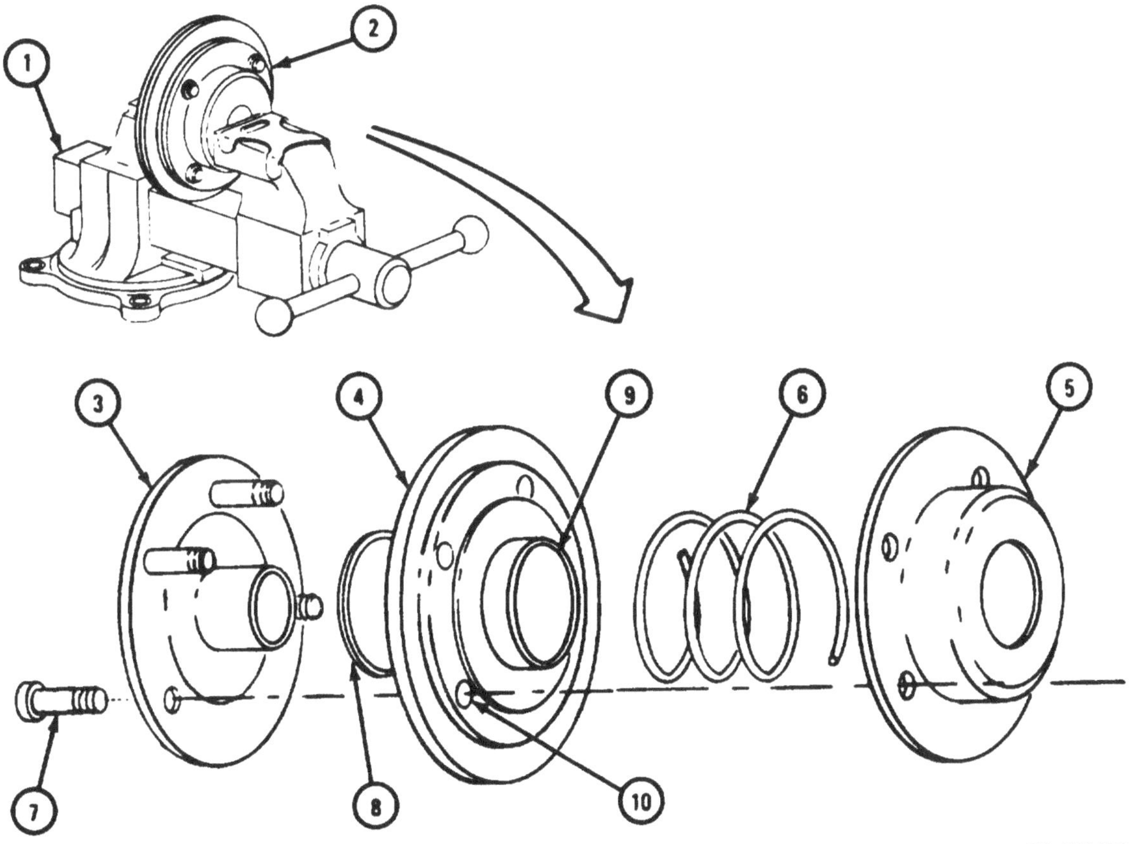

(2) Cleaning. There are no special cleaning procedures needed. Refer to cleaning procedures given in para 4-3.

(3) Inspection and repair.

NOTE

For engines LD-465-1, LD-465-1C, LDT-465-1C, LDS-465-1, and LDS-465-1A, go to frame 1. For engine LDS-465-2, go to frame 2.

FRAME 1

WARNING

Dry cleaning solvent is flammable. Do not use near an open flame. Keep a fire extinguisher nearby when solvent is used. Use only in well-ventilated places. Failure to do this may result in injury to personnel and damage to equipment.

1. Check that pulley flange (1) and pulley half (2) has no cracks, nicks, wear and other damage. Fix minor damage to belt surfaces of pulley flange and pulley half with fine mill file or crocus cloth dipped in dry cleaning solvent.
2. Check that clamping sections (3) of pulley flange (1) have no breaks and bends.
3. Check that threaded holes in clamping section (3) are not damaged. Fix damaged threads with a tap.
4. Check that two screws (4) have no damaged threads. Fix minor thread damage with a file thread restorer.
5. Check that threads on pulley half (2) are not damaged. Fix damaged threads with a file thread restorer.
6. Check that keyway inside pulley half (2) has no damage or wear.
7. If pulley flange (1) or pulley half (2) are damaged beyond repair, throw them away and get new ones.

END OF TASK

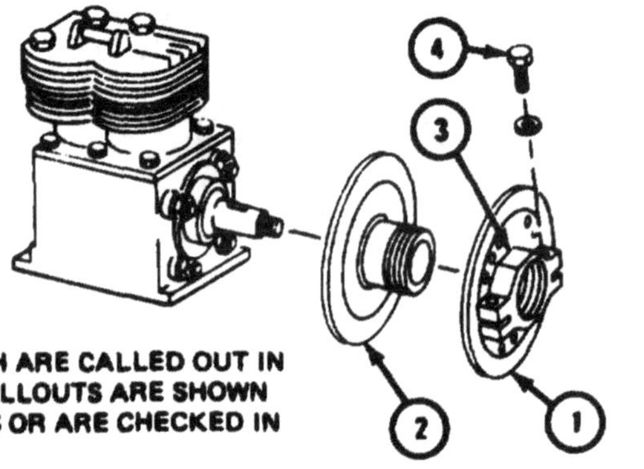

NOTE
CHECK ONLY THOSE PARTS WHICH ARE CALLED OUT IN THIS FRAME. PARTS WITHOUT CALLOUTS ARE SHOWN ONLY FOR REFERENCE PURPOSES OR ARE CHECKED IN ANOTHER FRAME.

TA 086393

FRAME 2

WARNING

Dry cleaning solvent is flammable. Do not use near an open flame. Keep a fire extinguisher nearby when solvent is used. Use only in well-ventilated places. Failure to do this may result in injury to personnel and damage to equipment.

1. Check that inside pulley half (1), outside pulley half (2), and spring retainer (3) have no nicks, burrs, cracks or bends. Fix minor damage with fine mill file or crocus cloth dipped in dry cleaning solvent.
2. Check that oil seals (4 and 5) have no wear, cracks or cuts. If seals are damaged, throw them away and get new ones.

GO TO FRAME 3

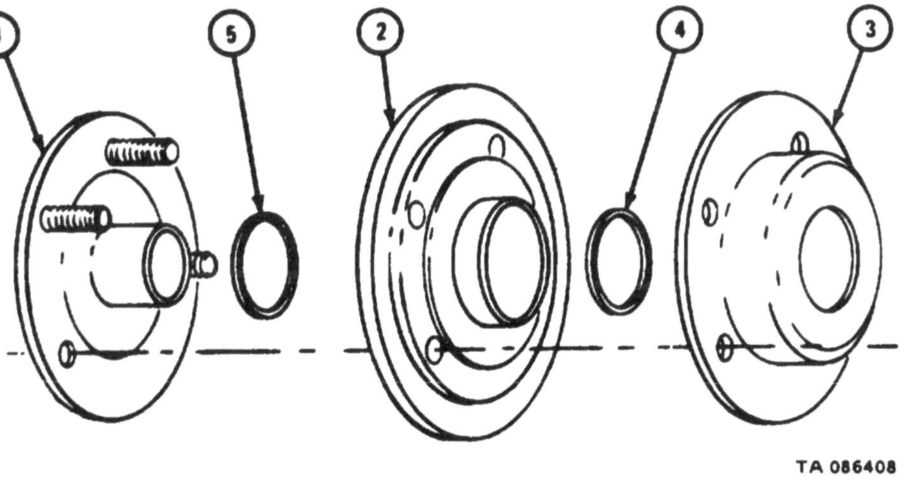

FRAME 3

NOTE

Readings must be within limits given in table 4-62. The letter T shows a tight fit. If readings are not within given limits, throw away part and get a new one.

1. Measure diameter of four retaining pins (1).
2. Measure inside diameter of four retaining pin bores (2) in pulley (3).
3. Measure fit of retaining pin (1) to bore in pulley (3). If fit is not within given limits, throw away item with the most wear and get a new one.
4. Measure inside diameter of bore (4) in pulley (5).
5. Measure outside diameter of four sleeve bearings (6).
6. Measure inside diameter of four sleeve bearings (6) in pulley (5).
7. Measure fit of retaining pin (1) in sleeve bearing (6). If fit is not within given limits, throw away item with the most wear and get a new one.

GO TO FRAME 4

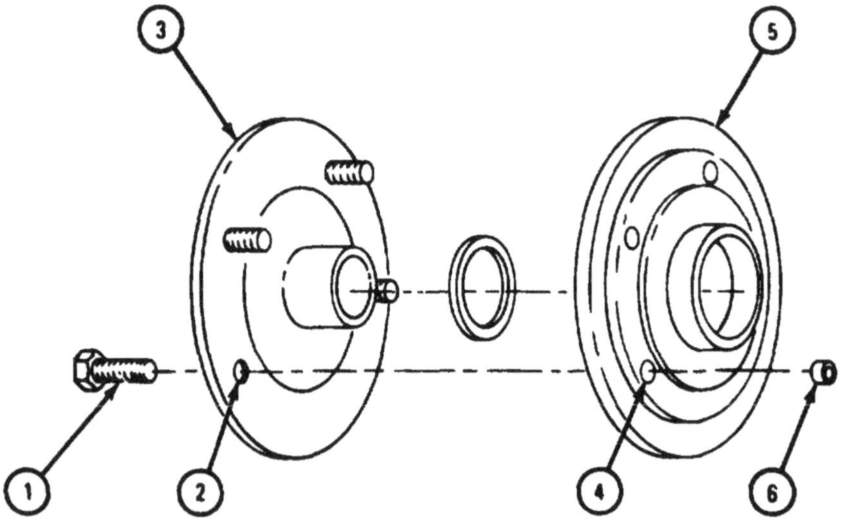

NOTE
CHECK ONLY THOSE PARTS WHICH ARE CALLED OUT IN THIS FRAME. PARTS WITHOUT CALLOUTS ARE SHOWN ONLY FOR REFERENCE PURPOSES OR ARE CHECKED IN ANOTHER FRAME.

TA 086409

Table 4-62. Air Compressor Pulley Pins, Hub, and Sleeve-Type Bearings (Engine LDS-465-2) Wear Limits

Index Number	Item /Point of Measurement	Size and Fit of New Parts (inches)	Wear Limits (inches)
1	Outside diameter of retaining pin	0.4985 to 0.4990	0.4960
2	Inside diameter of retaining pin bore	0.4975 to 0.4980	None
1 and 2	Fit of retaining pin in pulley	0.0005T to 0.0015T	None
4	Inside diameter of bore in pulley	0.5934 to 0.5941	None
6	Outside diameter of sleeve bearing	0.5934 to 0.5941	None
6	Inside diameter of sleeve bearing		0.5060
1 and 6	Fit of retaining pin in bearing sleeve		0.0100

FRAME 4

NOTE

Readings must be within limits given in table 4-63. The letter L indicates a loose fit. If readings are not within given limits, throw away part and get a new one.

1. Measure inside diameter of bore in outside pulley (1).
2. Measure inside diameter of sleeve bearing (2) in outside pulley (1).
3. Measure outside diameter of pulley hub (3).
4. Measure fit of pulley hub (3) into sleeve bearing (2). If fit is not within given limits, throw away item with the most wear and get a new one.
5. Using spring tester, measure free length, maximum solid height, and load pulley spring (4) at 2.000 inches.

END OF TASK

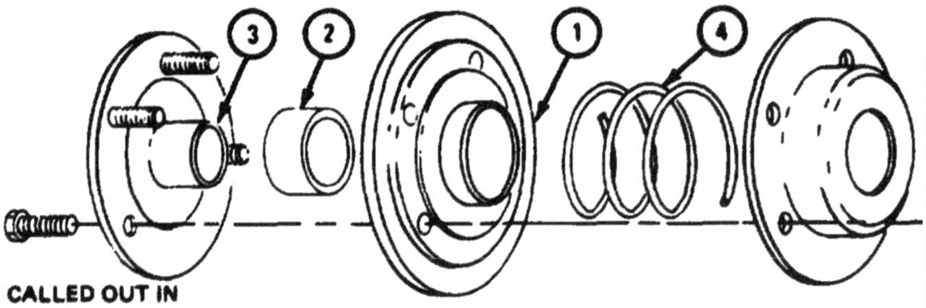
3
2
1
4
NOTE
CHECK ONLY THOSE PARTS WHICH ARE CALLED OUT IN THIS FRAME. PARTS WITHOUT CALLOUTS ARE SHOWN ONLY FOR REFERENCE PURPOSES OR ARE CHECKED IN ANOTHER FRAME.
TA 118086

Table 4-63. Air Compressor Bearing Sleeve and Pulley Spring (Engine LDS-465-2) Wear Limits

Index Number	Item /Point of Measurement	Size and Fit of New Parts (inches unless otherwise noted)	Wear Limits (inches)
1	Inside diameter of bore in pulley	2.1871 to 2.1883	None
2	Inside diameter of sleeve bearing		2.0050
3	Outside diameter of pulley hub	1.9969 to 1.9987	1. 9950
2 and 3	Fit of pulley hub sleeve bearing		0.00100L
4	Air compressor pulley spring :		
	Free length	3.4900	
	Load at 2.000 inches	(179.0 to 209.0 pounds)	
	Maximum solid height	1.7650	

(4) Assembly and replacement .

NOTE

For engine LDS-465-2, go to frame 4. For all other engines, go to frame 1.

FRAME 1

NOTE

Replacement air compressors are fitted with protective cover (3), cover gasket (4), and slotted nut (5) .

If putting on a replacement compressor, do steps 1 and 2. If putting on used compressor, go to frame 2.

1. Take off four push-on nuts (1) and rivets (2) and take off protective cover (3) and gasket (4) . Throw away nuts, rivets, cover, and gasket .
2. Take off and throw away slotted nut (5).

NOTE

Self-locking nut taken off during disassembly will be used instead of slotted nut (5).

GO TO FRAME 2

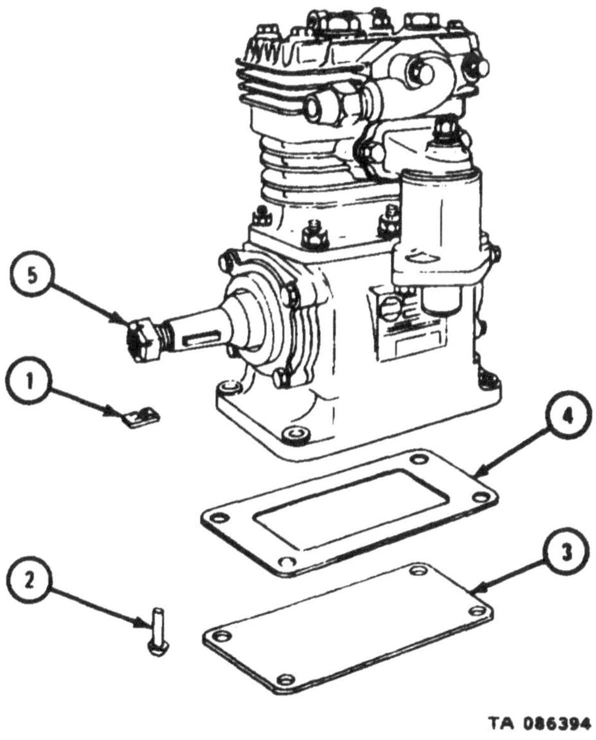

FRAME 2

NOTE

Some air compressor pulley halves (1) have keyway for woodruff key (2) and some do not.

1. Look in bore of air compressor pulley half (1) for keyway for woodruff key (2). If pulley half has keyway, put woodruff key in slot in compressor shaft (3). If it does not, make sure there is no woodruff key in shaft.
2. Slide pulley half (1) onto compressor shaft (3).

NOTE

If pulley half (1) has keyway in pulley half, it must line up with woodruff key (2) in compressor shaft (3).

3. Screw self-locking nut (4) on compressor shaft (3) and using torque wrench tighten to 100 pound-feet.

GO TO FRAME 3

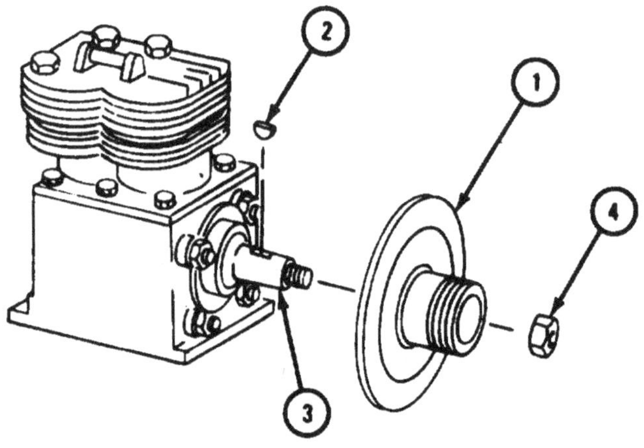

FRAME 3

1. Screw pulley adjusting flange (1) onto pulley half (2).
2. Screw in and tighten two screws and lockwashers (3).

END OF TASK

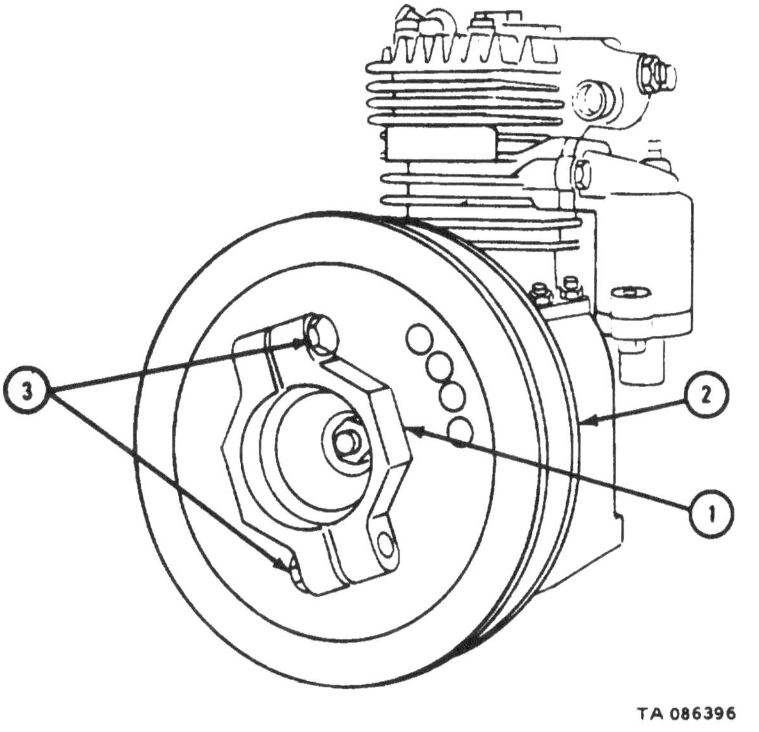

FRAME 4

1. Put several drops of lubricating oil on four retainer pins (1), hub of pulley (2), lips of seals (3) and (4), in four bearing sleeves (5), and in bearing sleeve (6).

GO TO FRAME 5

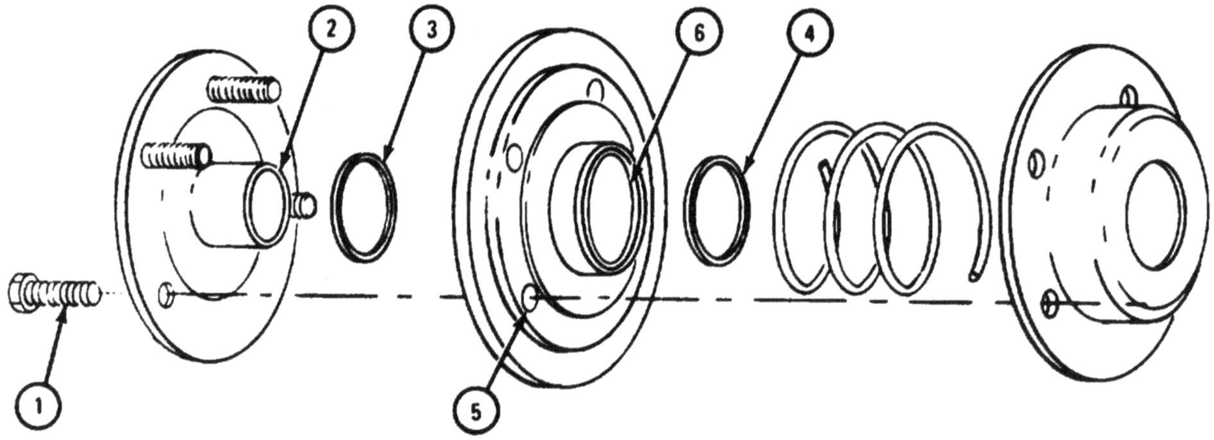

NOTE: PARTS WITHOUT CALLOUTS ARE SHOWN ONLY FOR REFERENCE PURPOSES.

TA 086410

FRAME 5

1. Push four retainer pins (1) through holes in inside pulley half (2).
2. Put inside pulley half (2) on press plates in arbor press (3) so retainer pin (1) points up .
3. Put outside pulley half (4) on top of inside pulley half (2). Retainer pins (1) should fit through holes in outside pulley half.
4. Place spring (5) on top of outside pulley half (4).
5. Place spring retainer (6) on top of spring (5).

GO TO FRAME 6

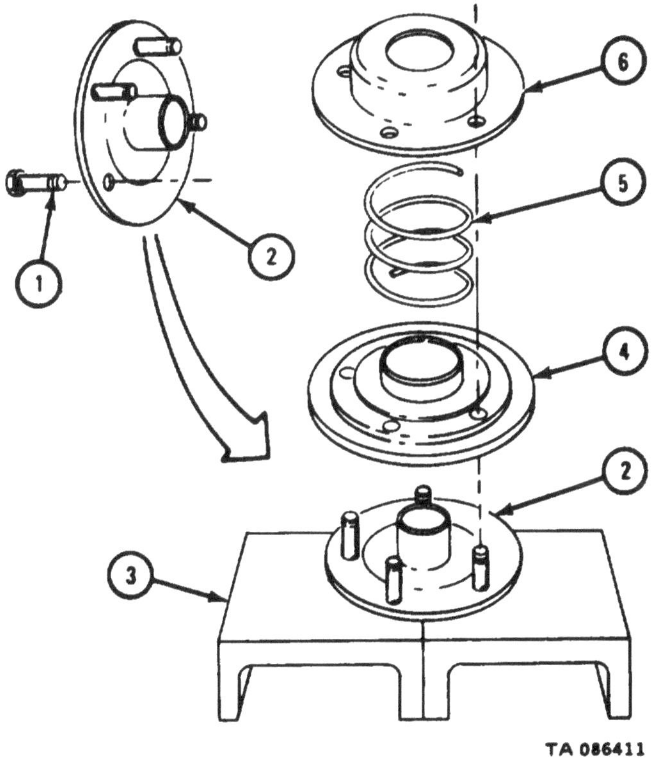

FRAME 6

1. Line up holes in spring retainer (1) with retainer pins (2).
2. Work arbor press (3) to press pulley assembly (4) together so that four retaining pins (2) fit through holes in spring retainer (1).
3. Clip four lock rings (5) around pins (2) to hold pulley assembly (4) together.
4. Take pulley assembly (4) out of arbor press (3).

NOTE

Pulley will be put on compressor after compressor is put on engine.

END OF TASK

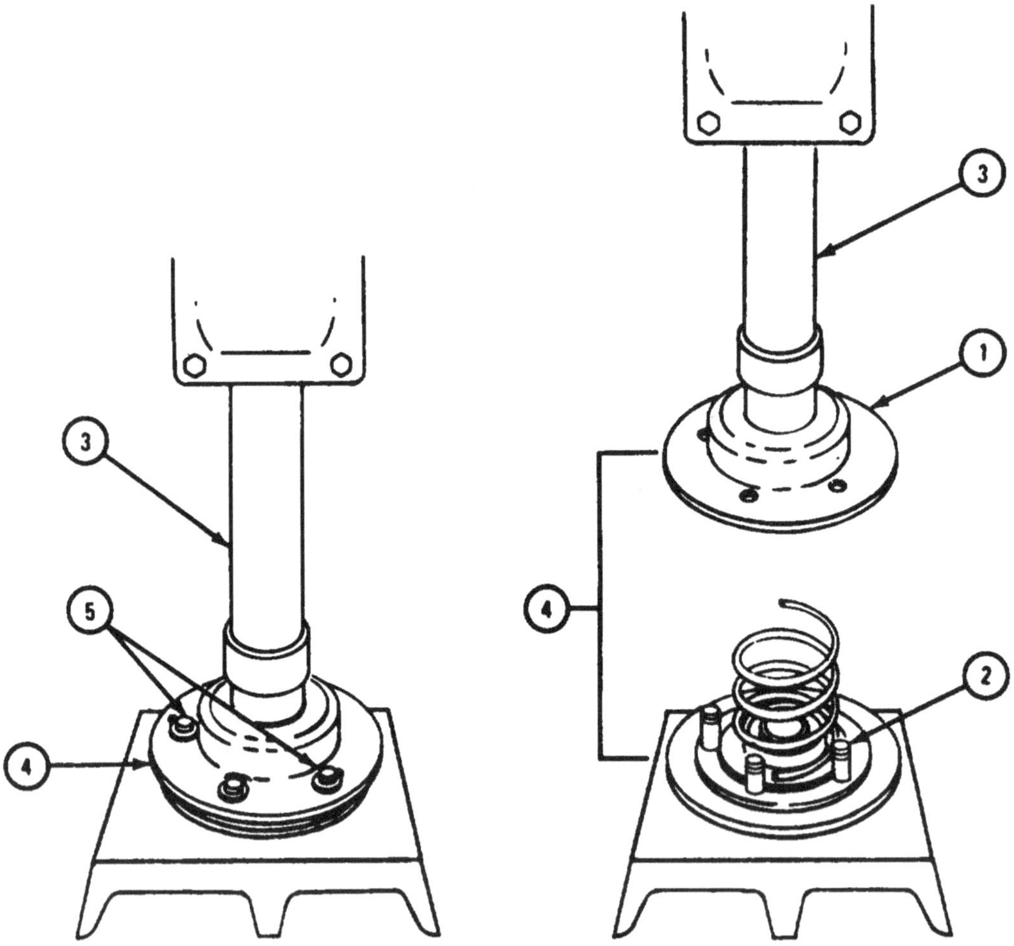

TA 086412

b. Air Compressor Intake Air Cleaner.

(1) Removal.

NOTE

If working on engine LDS-465-2, go to frame 2. For all other engines, go to frame 1.

FRAME 1

NOTE

Air discharge housing (1) may look different from the one shown. Procedure to take off intake air cleaner (3) is same for all compressors.

1. Take out screws and lockwashers (2). Take off air cleaner (3) and throw it away. Take off air cleaner gasket (4) and throw it away.

NOTE

Two different models of intake air cleaner (2) have been used. Early model could be taken apart to change ilter inside. Late model gannet be fixed and must be replaced. Early model is no longer available.

END OF TASK

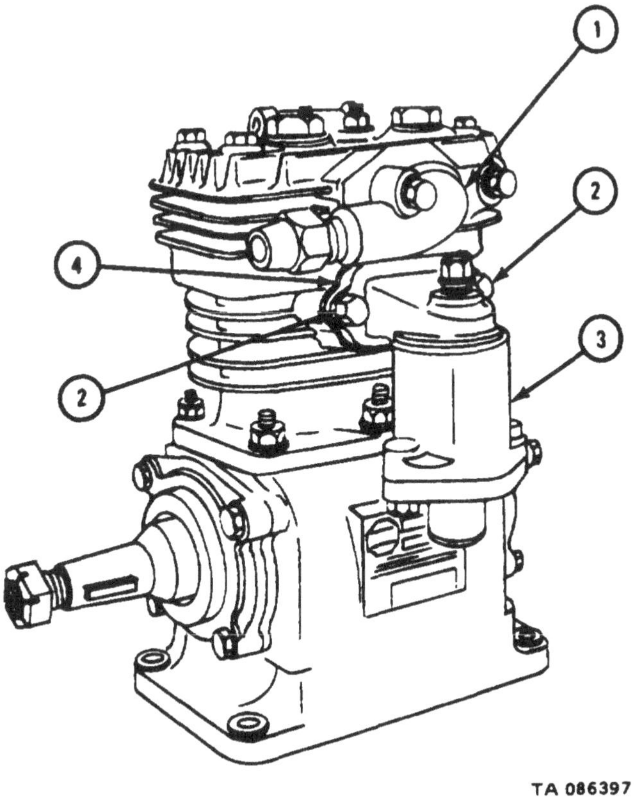

FRAME 2

NOTE

On engine LDS-465-2, air cleaner is separate from air compressor. You only take off intake elbow .

1. Take out two screws and lockwashers (1) and take off intake elbow (2).
2. Take off input elbow gasket (3) and throw it away.

END OF TASK

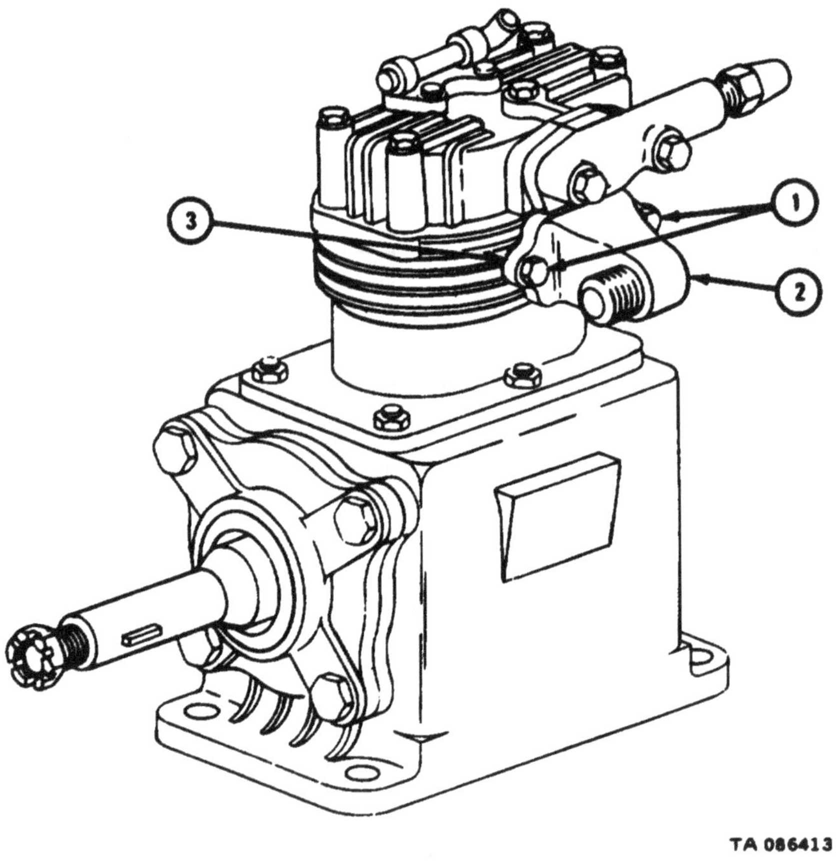

(2) Cleaning .

WARNING

Dry cleaning solvent is flammable. Do not use near an open flame. Keep a fire extinguisher nearby when solvent is used. Use only in well-ventilated places. Failure to do this may result in injury to personnel and damage to equipment.

NOTE

Engines LD-465-1, LD-465-1C, LDT-465-1C , LDS-465-1, and LDS-465-1A all have a throwaway air cleaner and need no cleaning.

Engine LDS-465-2 has a separate air cleaner but intake hookup needs to be cleaned.

(a) Clean intake elbow with dry cleaning solvent.

WARNING

Eye shields must be worn when using compressed air . Eye injury can occur if eye shields are not used.

(b) Blow dry with compressed air .

(3) Inspection and repair .

NOTE

Engine LDS-465-2 is the only one which needs inspection and repair .

FRAME 1

1. Check that intake elbow (1) has no damaged threads, nicks, burrs, and raised metal. Fix minor damage with a fine mill file.
2. Check gasket mounting surface of intake elbow (1) for warpage. If surface is not completely flat, work surface across a sheet of crocus cloth on a flat surface. If surface is badly warped, throw away intake elbow and get a new one.
3. Check that intake elbow (1) has no cracks or other damage. If elbow is cracked or badly damaged, throw it away and get a new one.

END OF TASK

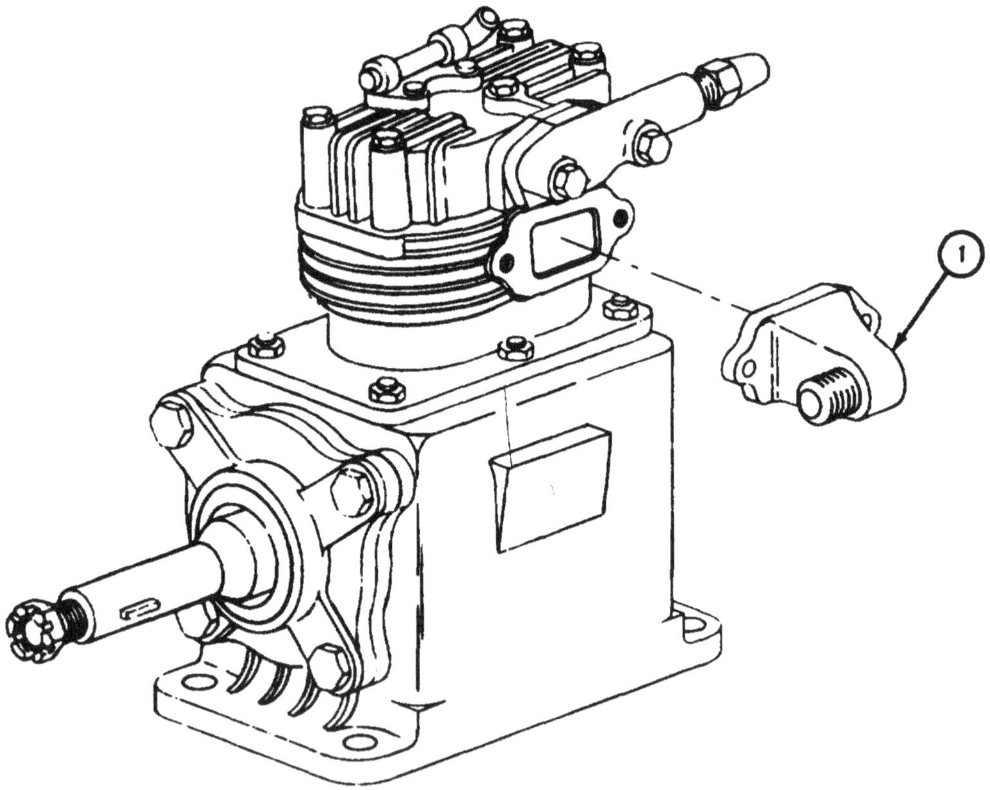

NOTE: CHECK ONLY THOSE PARTS WHICH ARE CALLED OUT. PARTS WITHOUT CALLOUTS ARE SHOWN ONLY FOR REFERENCE PURPOSES.

TA 086414

(4) Replacement .

NOTE

For engine LDS-465-2, go to frame 2. For all other engines, go to frame 1.

FRAME 1

NOTE

Air discharge housing (1) may look different from the one shown . Procedure to put on intake air cleaner (4) is same for all compressors.

1. Put intake air cleaner gasket (2) on air compressor (3).
2. Hold intake air cleaner (4) on sir compressor (3). Screw in and tighten two screws and lockwashers (5) .

END OF TASK

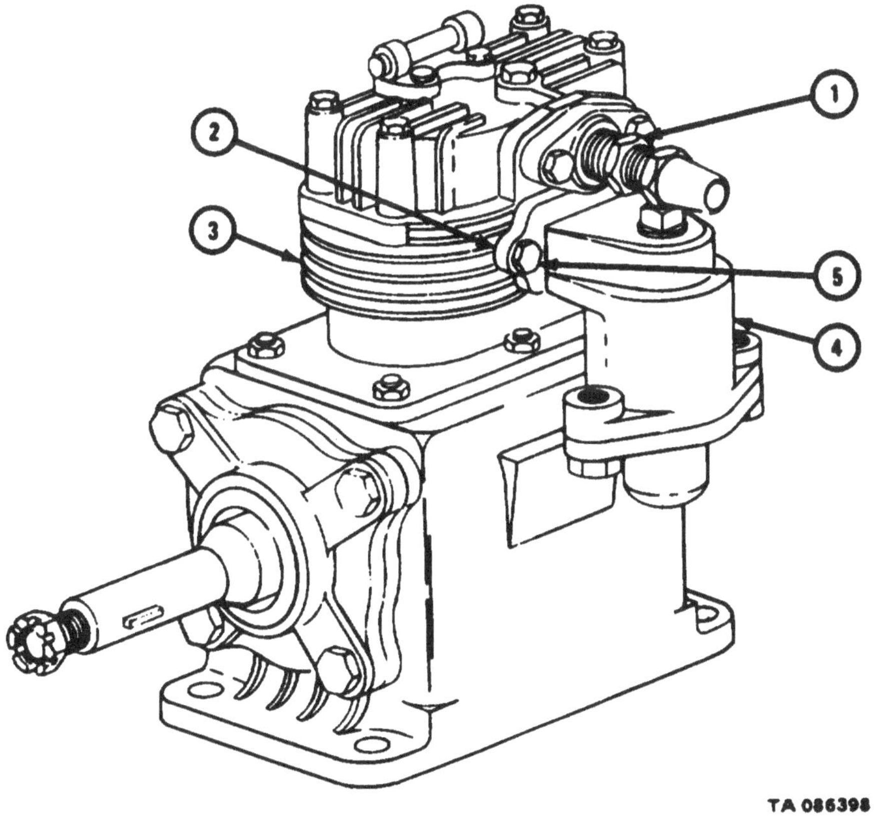

FRAME 2

1. Put intake air cleaner gasket (1) on air compressor (2).
2. Put intake elbow (3) on air compressor (2).
3. Screw in and tighten two screws and lockwashers (4).

END OF TASK

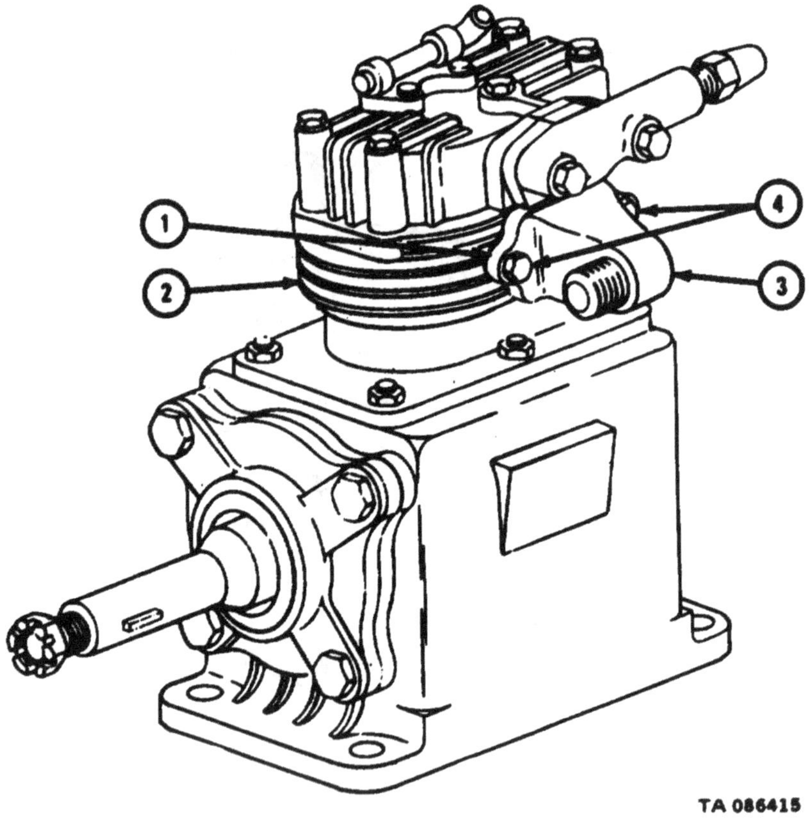

c. Air Compressor Discharge Manifold.

(1) Removal.

FRAME 1

1. Take off two screws and lockwashers (1) and take off air discharge manifold (2).
2. Take off air discharge manifold gasket (3) and throw it away.

END OF TASK

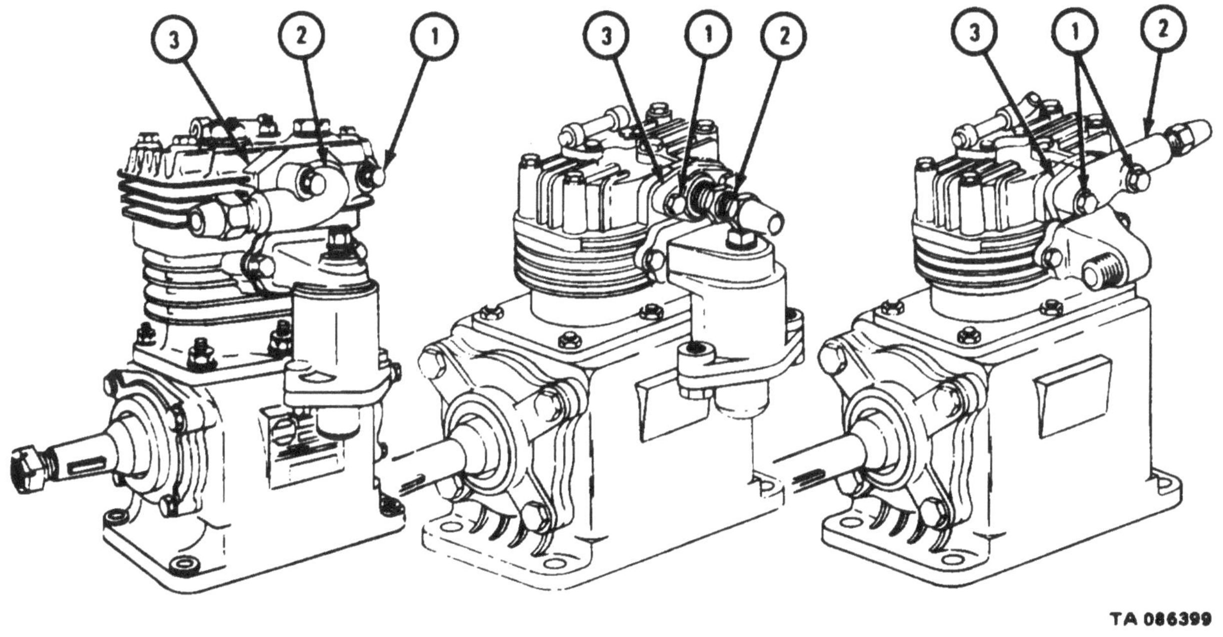

(2) Cleaning. There are no special cleaning procedures needed. Refer to cleaning procedures given in para 4-3.

(3) Inspection and repair.

FRAME 1

1. Check that discharge manifold (1) has no cracks or other damage. Check that machined surfaces have no nicks, burrs or raised metal. Check for stripped threads.
2. Fix burrs or raised metal and minor thread damage with a fine mill file.
3. Check that mating surface of discharge manifold (1) has no warpage. If mating surface is not flat, work surface across a sheet of crocus cloth held tightly to a flat surface.
4. If discharge manifold (1) is cracked or badly damaged, throw it away and get a new one.

END OF TASK

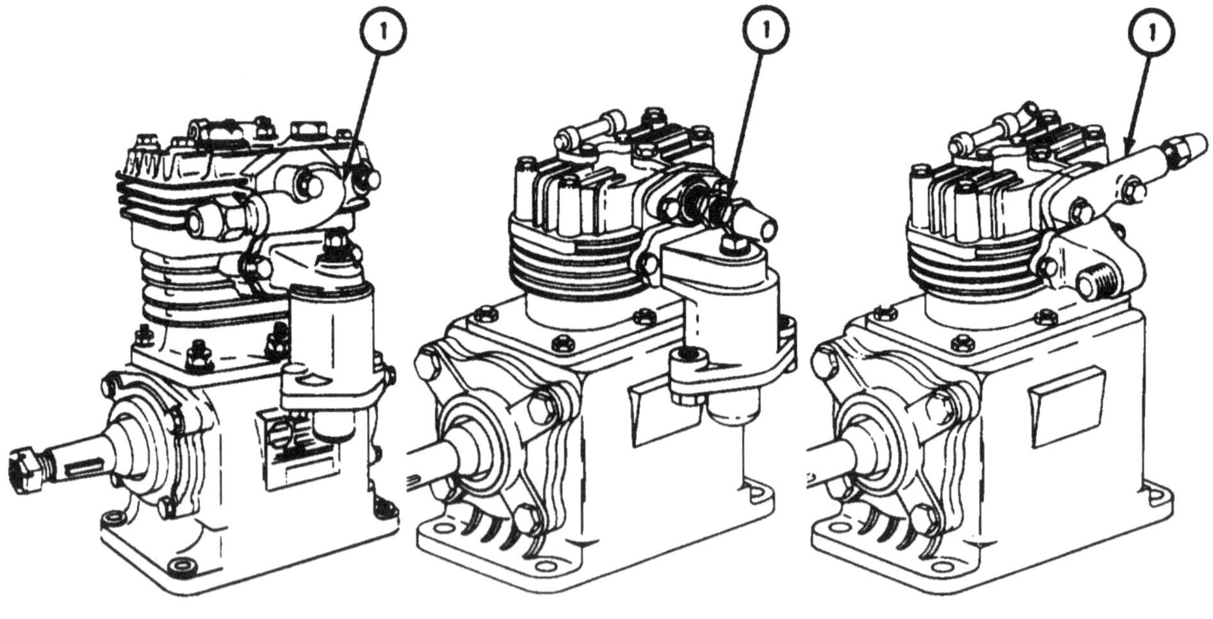

TA 086400

(4) Replacement .

FRAME 1

1. Put air discharge manifold gasket (1) on air compressor (2).
2. Hold air discharge manifold (3) on air compressor (2). Screw in and tighten two screws and lockwashers (4).

END OF TASK

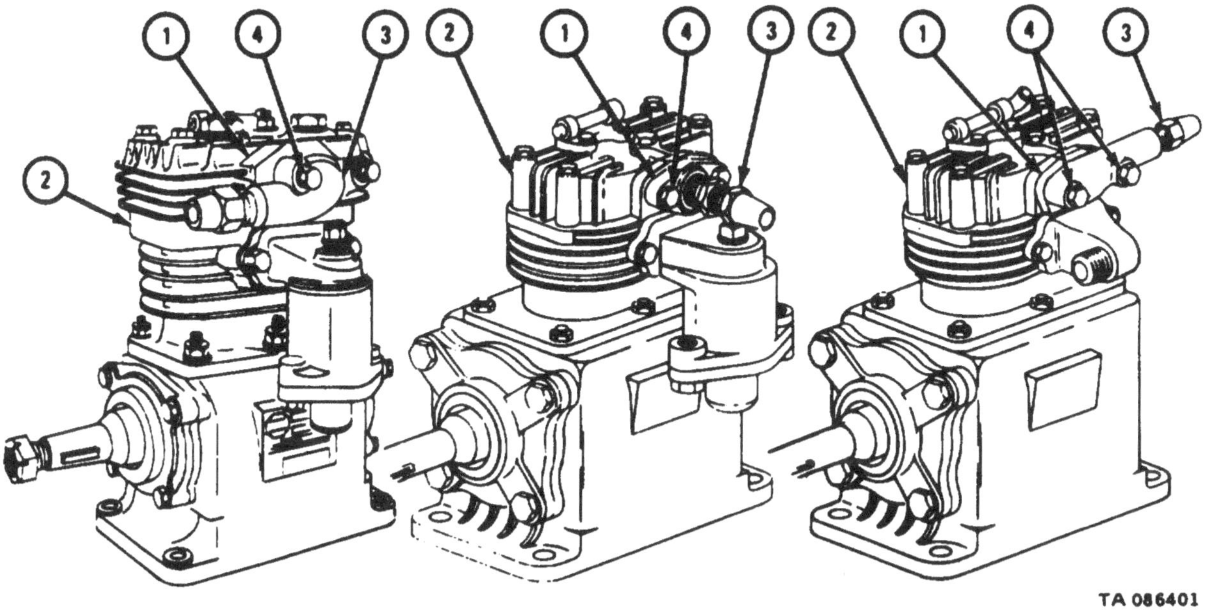

d. Air Compressor Unloader Valve.

(1) Removal.

FRAME 1

NOTE

Air compressor shown is for engine LDS-465-2.
Procedure is same for all other engines.

1. Take out two nuts and lockwashers (1) and take off unloader valve (2). Take off unloader valve gasket (3) and throw it away.

2. Take off unloader valve elbow (4).

END OF TASK

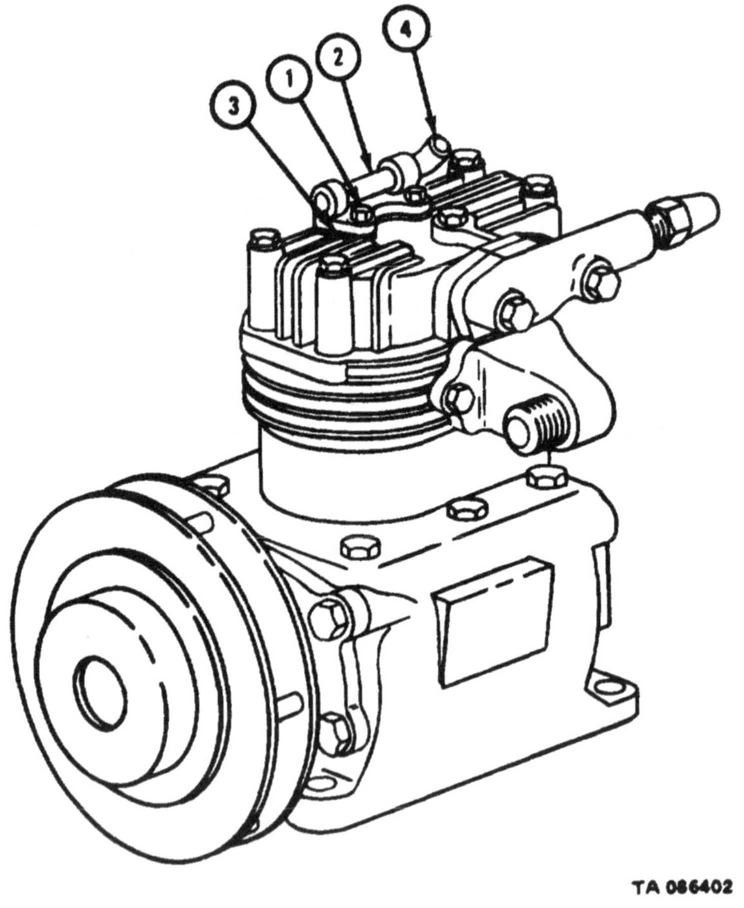

(2) Cleaning.

WARNING

Dry cleaning solvent is flammable. Do not use near an open flame. Keep a fire extinguisher nearby when solvent is used. Use only in well-ventilated places. Failure to do this may result in injury to personnel and damage to equipment.

(a) Clean unloader valve elbow with dry cleaning solvent and wipe dry.

(b) Wipe unloader valve with a cloth wet with dry cleaning solvent and wipe dry.

(3) Inspectio n and repair .

FRAME 1

1. Check that unloader valve (1) and elbow (2) have no cracks, burrs, raised metal or other damage. Fix minor nicks and burrs with a fine mill file.
2. Check that unloader valve (1) and elbow (2) have no stripped threads. If threads are stripped, throw away damaged part and get a new one.
3. If unloader valve (1) or elbow (2) are cracked or badly damaged, throw them away and get new ones.

END OF TASK

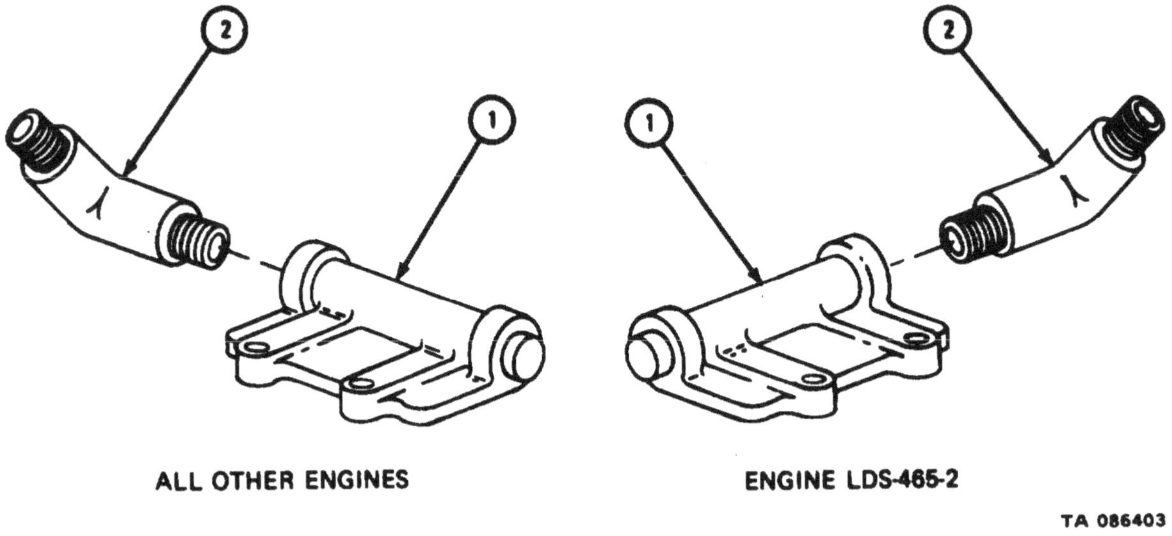

(4) Replacement .

FRAME 1

NOTE

Air compressor shown is for engine LDS-465-2.
Procedure is same for all other engines.

1. Screw unloader valve elbow (1) into unloader valve (2). Tighten elbow so hole in elbow will point up and in towards engine side of compressor as shown.
2. Put unloader valve gasket (3) on air compressor (4).
3. Place unloader valve (2) on air compressor (4) so studs on air compressor fit through holes in valve.
4. Screw on and tighten two nuts and lockwashers (5).

END OF TASK

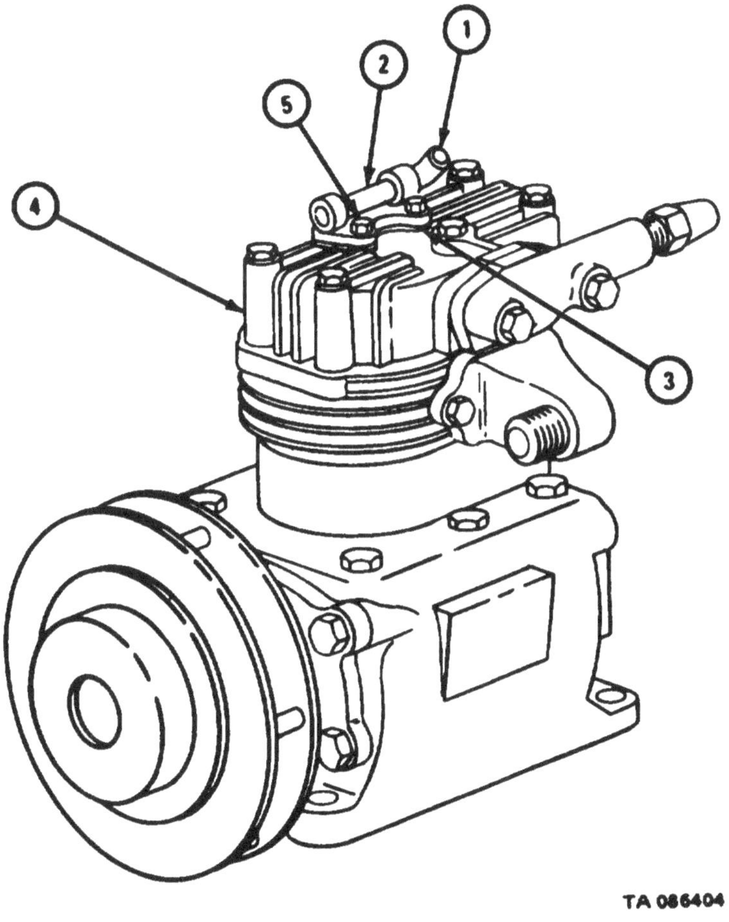

e. Air Compressor.

(1) Cleaning.

WARNING

Dry cleaning solvent is flammable. Do not use near an open flame. Keep a fire extinguisher nearby when solvent is used. Use only in well-ventilated places. Failure to do this may result in injury to personnel and damage to equipment.

Clean air compressor with clean cloth dipped in dry cleaning solvent and wipe dry.

(2) Inspection and repair .

NOTE

Air compressor shown is for engine LDS-465-2. Procedure is same for all engines.

FRAME 1

1. Check that air compressor (1) has no nicks, burrs or raised metal. Fix small nicks, burrs or raised metal with a fine mill file.
2. Check that air compressor (1) has no cracks or other damage. If compressor is cracked or badly damaged, take off pulley assembly (2), intake air cleaner (3), and discharge manifold (4) . Refer to para 4-49a (1), para 4-49b (1), and para 4-49c (1) . Throw away compressor and get a new one.

END OF TASK

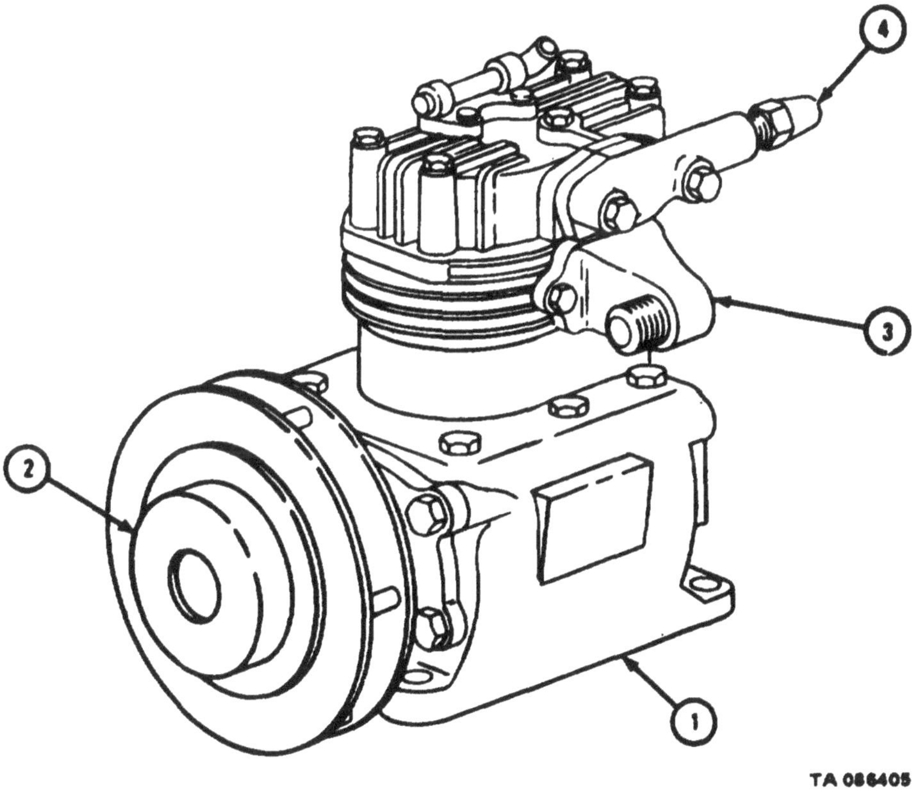

4-50. GENERATOR MOUNTING BRACKET, ADJUSTING STRAP, AND PULLEY.

a. Disassembly.

FRAME 1

1. On engine LDS-465-1, takeout and throw away cotter pin (1). Takeoff nu t (2) and lockwasher (3) .
2. On all engines except LDS-465-1, takeoff locknut (4) and flat washer (5) .
3. Using universal puller, take off generator pulley (6) .
4. Take woodruff key (7) out of shaft (8) .

END OF TASK

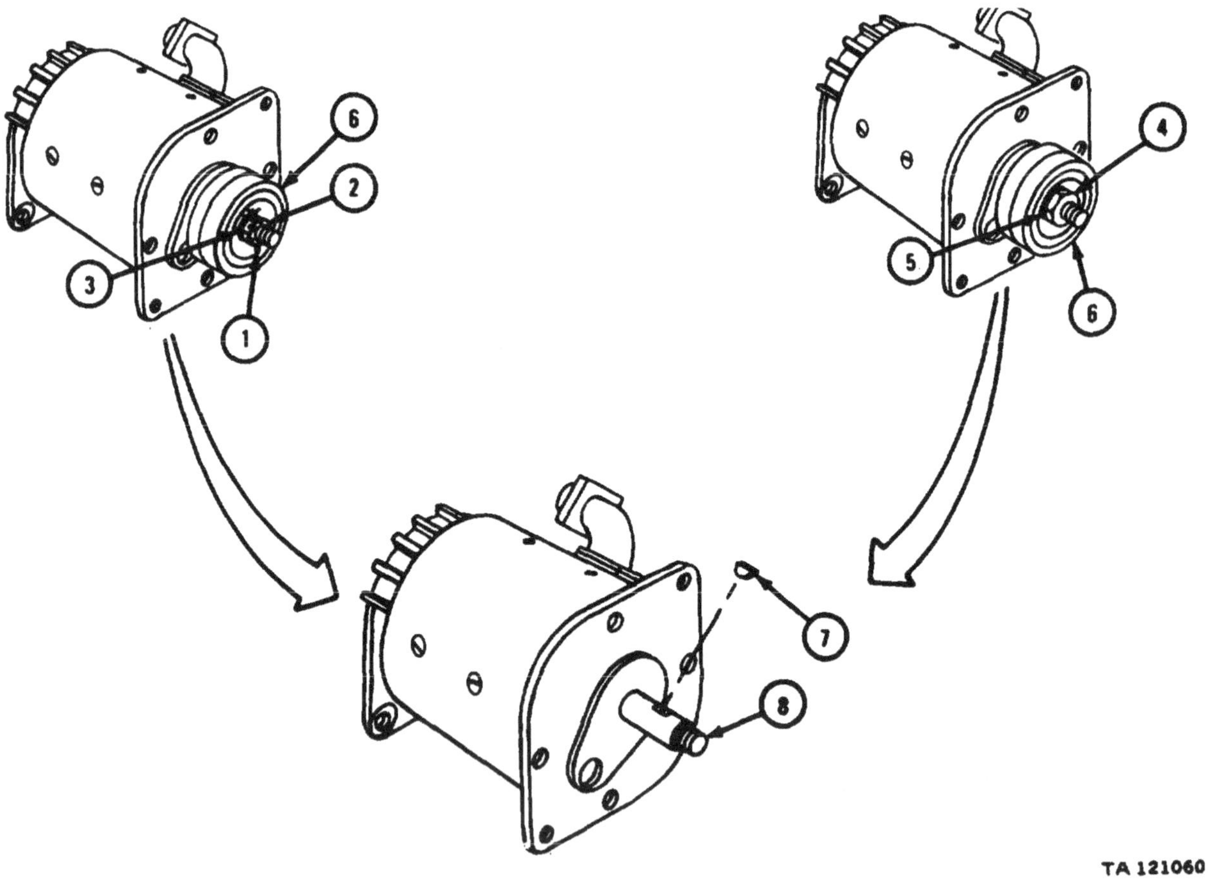

b. Cleaning.

FRAME 1

WARNING

Dry cleaning solvent is flammable. Do not use near an open flame. Keep a fire extinguisher nearby when solvent is used. Use only in well-ventilated places. Failure to do this may result in injury to personnel and damage to equipment.

1. Clean generator mounting bracket (1), adjusting strap (2), and pulley (3) in dry cleaning solvent. Clean off sludge and gum deposits with a stiff brush. Dry with clean cloth.

END OF TASK

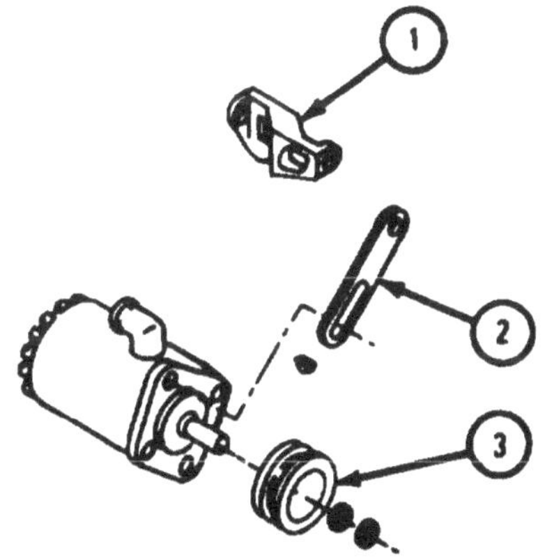

NOTE: PARTS WITHOUT CALLOUTS ARE SHOWN FOR REFERENCE PURPOSES ONLY

TA 113695

c. Inspection and Repair.

FRAME 1

1. Check that mounting bracket (1) and adjusting strap (2) are not cracked or broken. Check that mounting bracket and adjusting strap do not have out-of-round bolt holes. If either part is damaged, get a new one.
2. Check that mating surfaces of mounting bracket (1) and adjusting strap (2) have no nicks or burrs. Fix small nicks or burrs with a fine mill file. If more repair is needed, get a new part.
3. Check that mating surface of mounting bracket (1) is flat. If it is not flat, get a new mounting bracket.
4. Check that adjusting strap (2) is flat. If it is bent, straighten it. If adjusting strap cannot be made flat, get a new one.

GO TO FRAME 2

1
2
NOTE: CHECK ONLY THOSE PARTS WHICH ARE CALLED OUT IN THIS FRAME. PARTS WITHOUT CALLOUTS ARE SHOWN ONLY FOR REFERENCE PURPOSES OR ARE CHECKED IN ANOTHER FRAME.
TA 115428

FRAME 2

1. Check that pulley (1) is not cracked or broken. Check that keyway in pulley is not worn. If keyway is worn or if pulley is damaged, get a new one.

WARNING

Dry cleaning solvent is flammable. Do not use near an open flame. Keep a fire extinguisher nearby when solvent is used. Use only in well-ventilated places. Failure to do this may result in injury to personnel and damage to equipment.

2. Check that belt surface of pulley (1) is not worn and has no nicks, burrs or other damage. Fix small nicks and burrs with a fine mill file or crocus cloth dipped in dry cleaning solvent. If more repair is needed or if belt surface is worn, get a new pulley.
3. Check that woodruff kev (2) is not worn and has no burrs. Fix small burrs with a fine mill file. If more repair is needed or if woodruff key is worn, get a new woodruff key.

END OF TASK

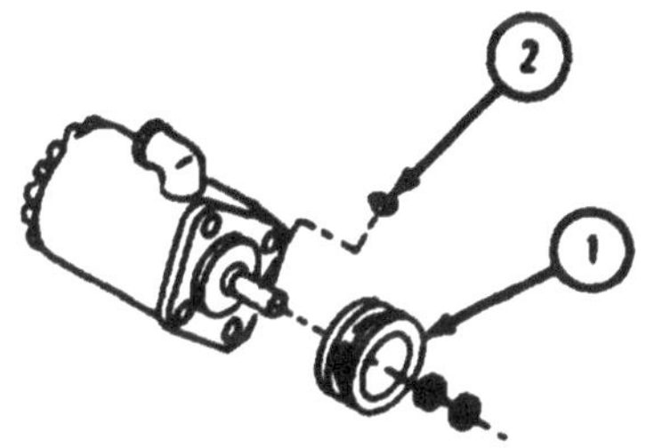

NOTE: CHECK ONLY THOSE PARTS WHICH ARE CALLED OUT IN THIS FRAME. PARTS WITHOUT CALLOUTS ARE SHOWN ONLY FOR REFERENCE PURPOSES OR ARE CHECKED IN ANOTHER FRAME.

TA 115429

d. Assembly.

FRAME 1

1. Put woodruff key (1) in generator shaft (2) .
2. Put pulley (3) on generator shaft (2), and tap gently in place using hollow driver if needed .
3. On engine LDS-465-1, put on lockwasher (4) and nut (5).
4. On all engines except LDS-465-1, put on flat washer (6) and locknut (7).
5. If nut (5 or 7) is 1/2 inch, tighten to 35 pound-feet. If nut is 5/8 inch, tighten to 40 to 50 pound-feet. If nut is 3/4 inch, tighten to 70 to 80 pound-feet.
6. On engine LDS-465-1, put in cotter pin (8).

END OF TASK

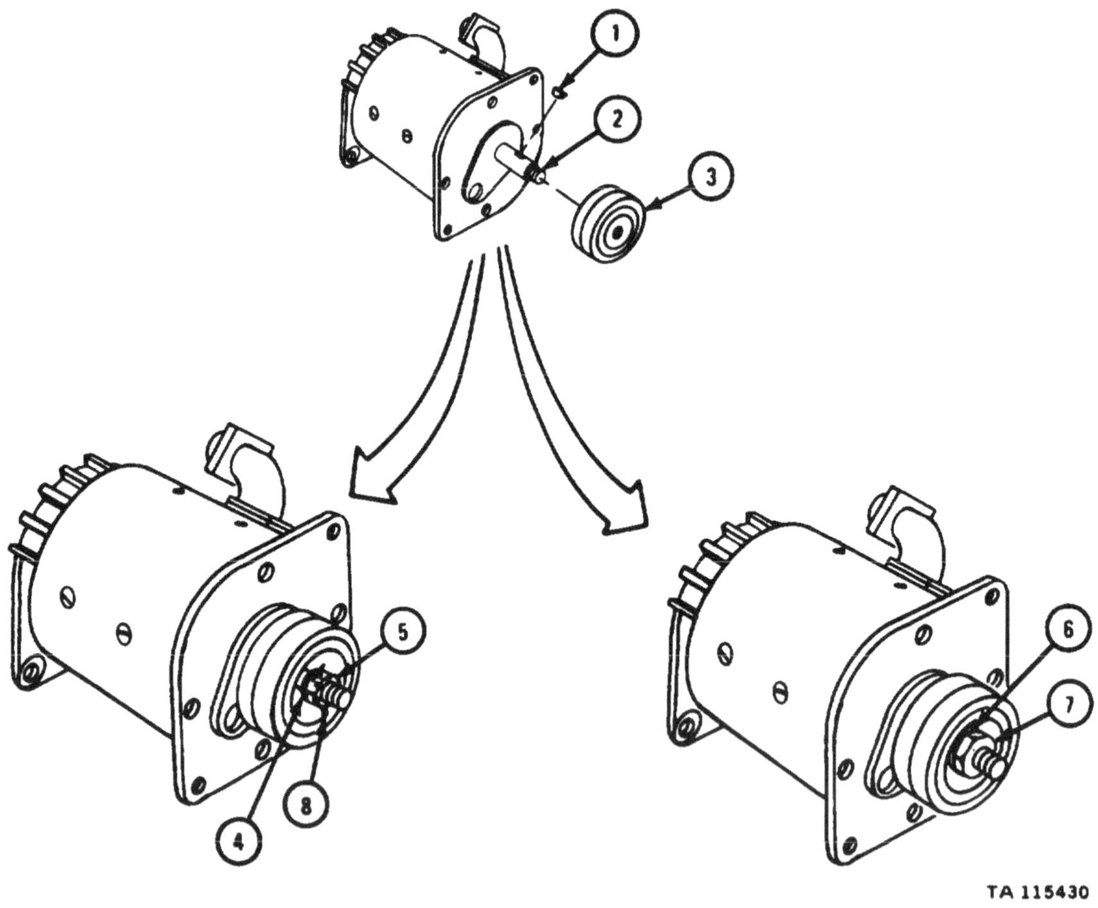

4-51. ENGINE FAN.

a. Cleaning. There are no special cleaning procedures needed. Refer to cleaning procedures given in para 4-3 .

b. Inspection and Repair .

FRAME 1

NOTE

If working on fans with aluminum blades, do frame 1.

If working on fans with steel blades, go to frame 2.

1. Check that fan (1) has no nicks, cracks or any other damage. Take out minor nicks with a fine mill file. If more repair is needed, get a new fan.
2. Check that four bolts (2) have no damaged threads or other damage. If bolts are damaged, get new ones.
3. Check that four lockwashers (3) are not cracked or damaged in any way. If lockwashers are damaged, get new ones.
4. Check that 30 rivets (4) are not loose or damaged. If rivets are loose or damaged, get a new fan (1).

END OF TASK

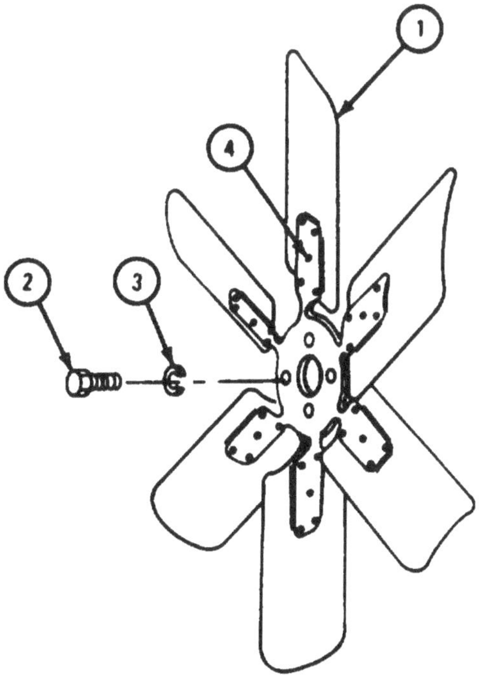

TA 113722

FRAME 2

1. Check that fan (1) has no nicks, cracks or bends. Take out minor nicks with a fine mill file. Take out minor bends on fan blade ends. If more repair is needed, get a new fan.
2. Check that 36 rivets (2) are not loose or damaged. If rivets are loose or damaged, get a new fan (1).
3. Check that four bolts (3) are not crossthreaded, stripped or damaged in any way. If damaged, get a new bolt .
4. Check that four lockwashers (4) are not cracked or damaged in any way. If lockwashers are damaged, get new ones.

END OF TASK

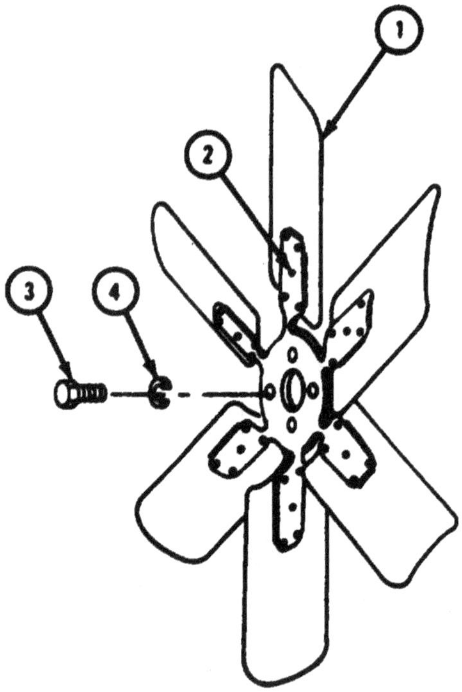

TA 113723

4-52. FAN, GENERATOR, AND AIR COMPRESSOR DRIVE BELTS.

a. Engine LDS-465-2.

(1) cleaning.

FRAME 1

1. Clean three generator and water pump belts (1), two engine fan belts (2), and air compressor belt (3) using warm soapy water.

END OF TASK

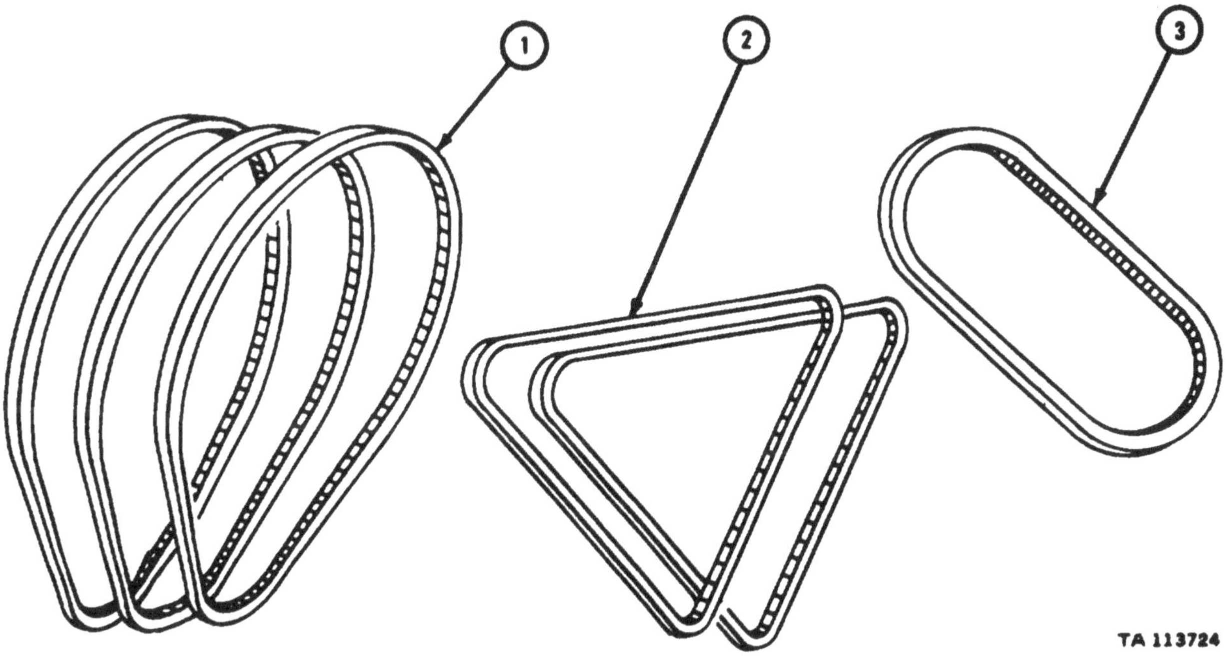

(2) Inspection and repair .

FRAME 1

1. Check that three generator and water pump belts (1) have no nicks, cuts, breaks, hard glazed wear surfaces or any other damage. If any of the three generator and water pump belts are damaged, get a new set of generator and water pump belts.
2. Check that two engine fan belts (2) have no nicks, cuts, breaks, hard glazed wear surfaces or any other damage. If either of the engine fan belts are damaged, get a new set of engine fan belts.
3. Check that air compressor belt (3) has no nicks, cuts, breaks, hard glazed wear surfaces or any other damage. If air compressor belt is damaged, get a new air compressor belt.

END OF TASK

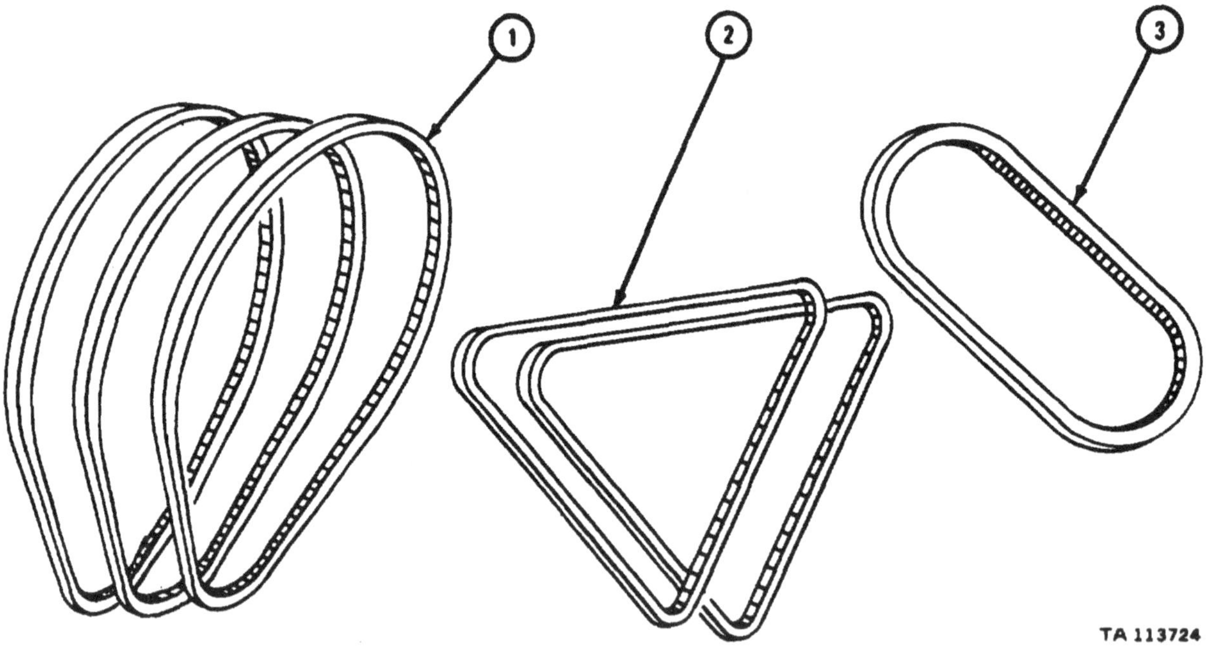

b. All Engines Except LDS-465-2.

(1) Cleaning.

FRAME 1

1. Clean two generator water pump and engine fan belts (1) and air compressor belt (2) using warm soapy water.

END OF TASK

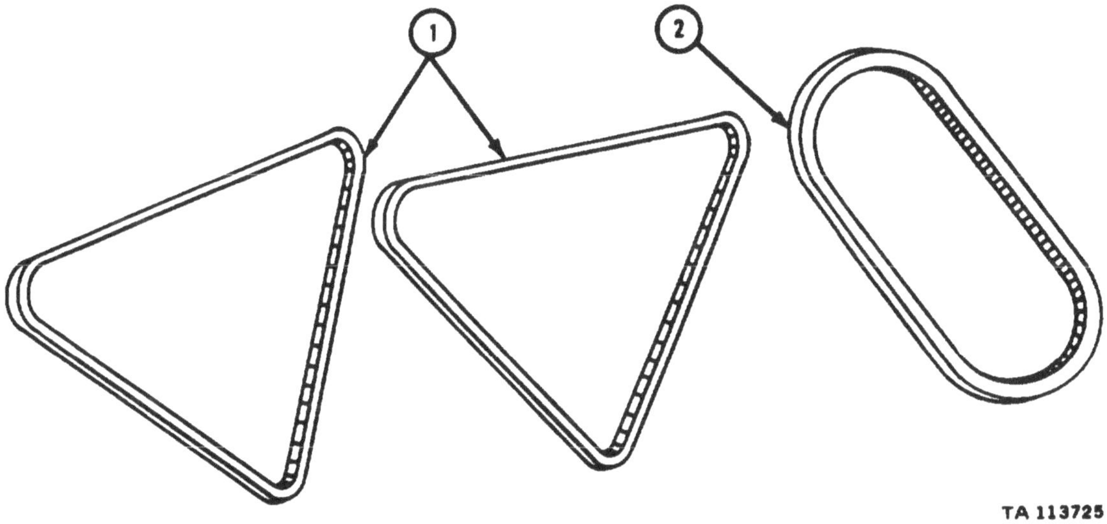

(2) Inspection and repair.

FRAME 1

1. Check that two generator water pump and engine fan belts (1) have no nicks, cuts, breaks, hard glazed wear surfaces or any other damage. If either of the fan belts are damaged, get a new set of fan belts.
2. Check that air compressor belt (2) has no nicks, cuts, breaks, hard glazed wear surfaces or any other damage. If air compressor belt is damaged, get a new air compressor belt.

END OF TASK

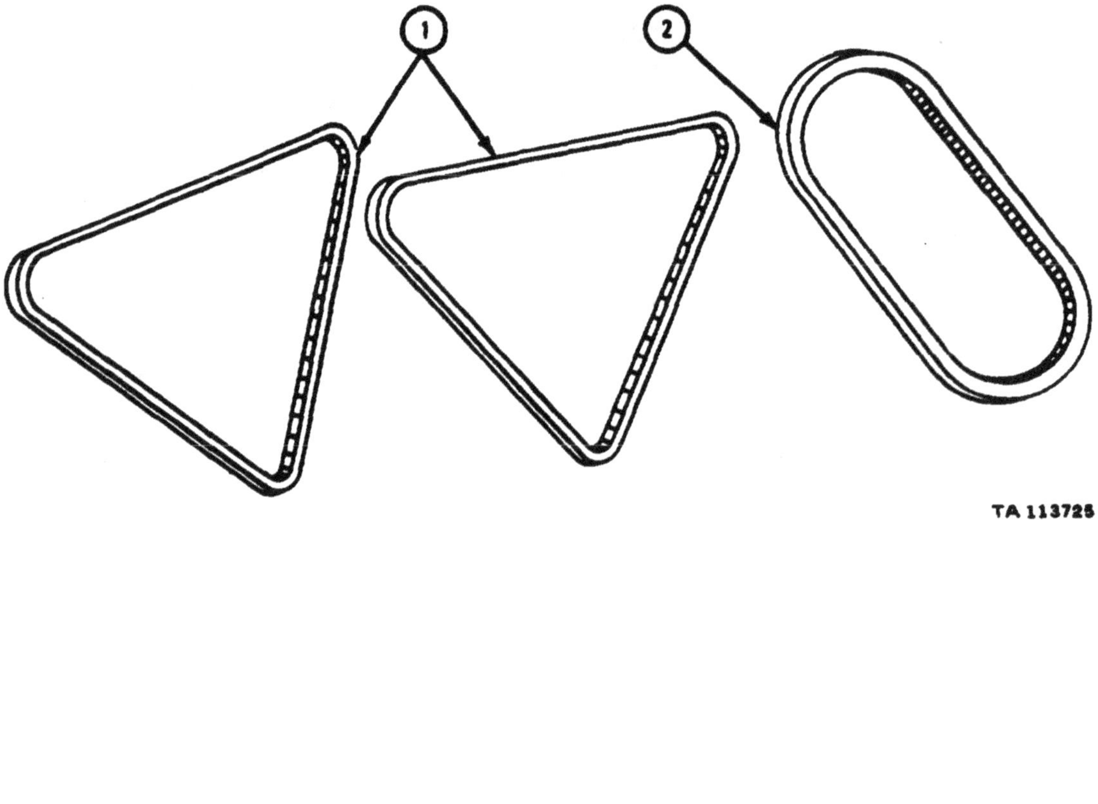

4-53. HYDRAULIC PUMP OIL RESERVOIR (ENGINE LDS-465-2).

a. Disassemble.

FRAME 1

1. Take off oil filler cap (1) and gasket (2). Throw gasket away.
2. Take out dip stick (3) and preformed packing (4). Throw packing away.

WARNING

Cover (7) is spring loaded. Be careful when taking off cover bolt (5) so cover does not fly off. Failure to do so may cause injury to personnel or damage to equipment.

3. Take out cover bolt (5) with washer (6). Take cover (7) and gasket (8) off reservoir body (9). Throw away gasket.

GO TO FRAME 2

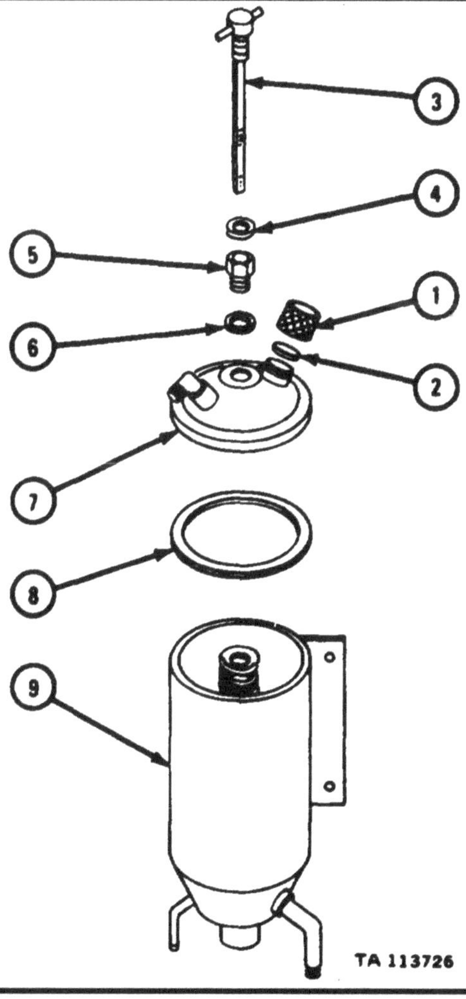

FRAME 2

1. Take spring (1), filter screen (2), and filter element (3) out of reservoir body (4).
2. Take elbow (5) out of cover (6).

END OF TASK

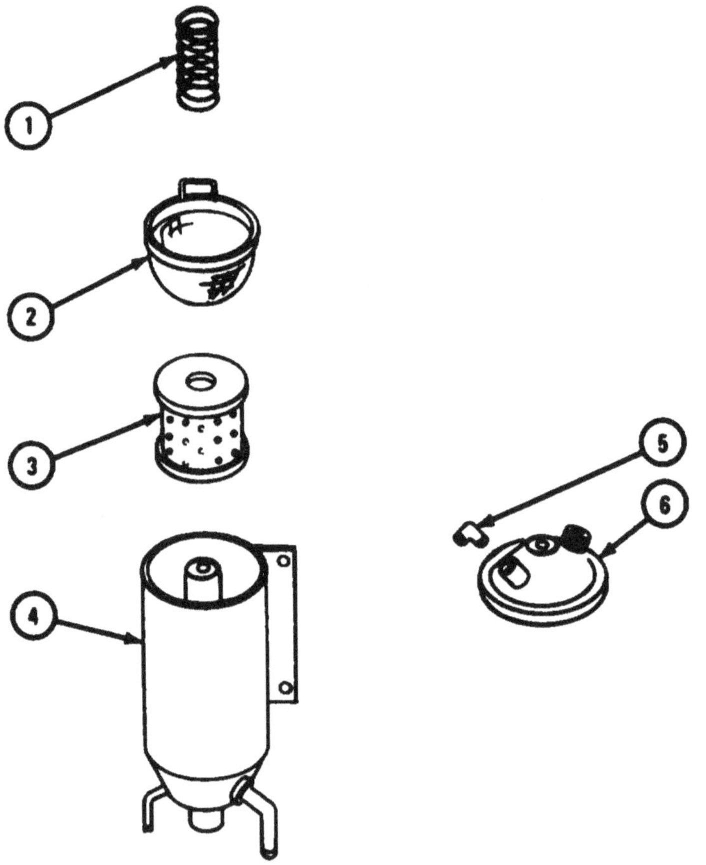

b. Cleaning. There are no special cleaning procedures needed. Refer to cleaning procedures given in para 4-3.

c. Inspection and Repair.

FRAME 1

1. Check that reservoir body (1) and cover (2) have no cracks, dents, burns or other damage. Knock out minor dents and fix nicks or burrs with fine mill file. If more repair is needed, get a new part.
2. Check that cover (2) has no damaged threads. Fix minor thread damage with a tap. If more repair is needed, get a new cover.

GO TO FRAME 2

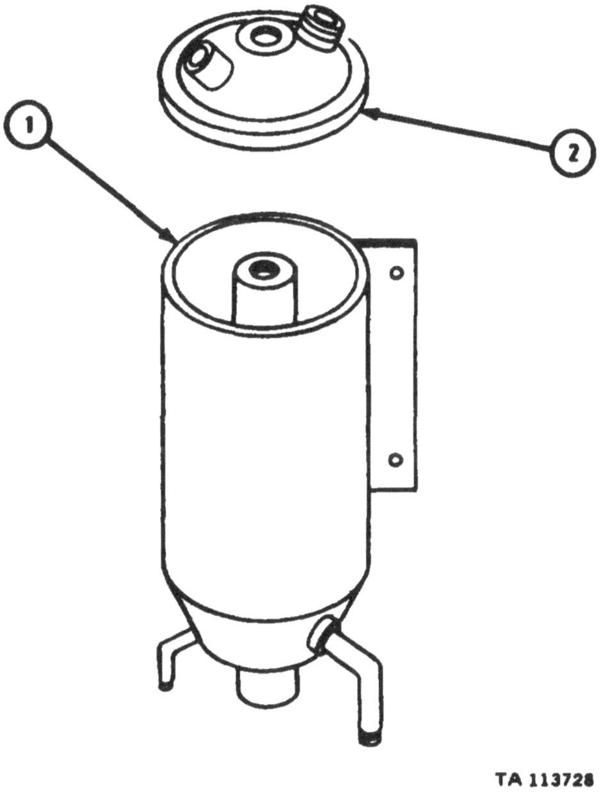

FRAME 2

1. Check that dipstick (1) is not bent, dented or cracked and has no damaged threads. Knock out minor bends and fix minor thread damage with a thread chaser. If more repair is needed, get a new part.
2. Check that cover bolt (2), oil filler cap (3), and elbow (4) are not bent, dented or cracked and have no damaged threads. Fix minor thread damage with a thread chaser or tap. If more repair is needed, get a new part.

GO TO FRAME 3

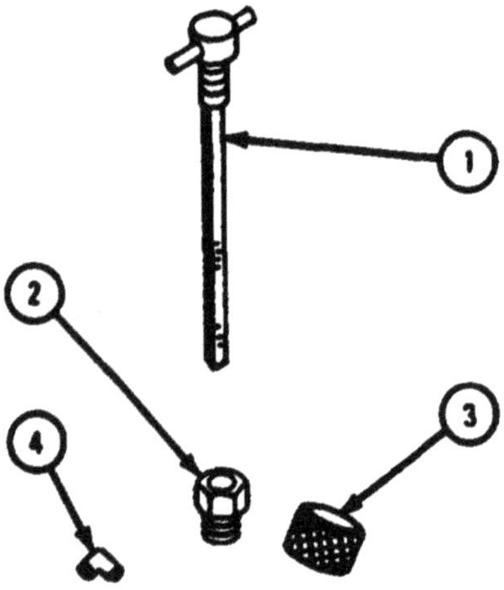

TA 113729

FRAME 3

1. Check that spring (1) is not bent, cracked or damaged in any other way. If spring is damaged, get a new spring.
2. Check that filter screen (2) is not bent, dented, torn or cracked. Knock out minor bends or dents. If more repair is needed, get a new filter screen.

END OF TASK

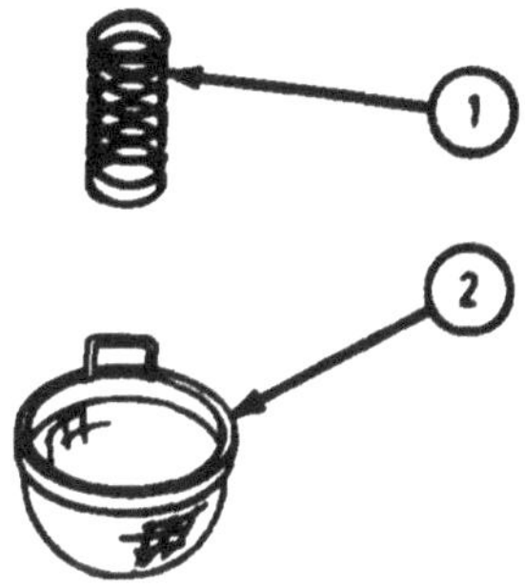

d. Assembly.

FRAME 1

1. Put filter element (1), filter screen (2), and spring (3) in reservoir body (4).

2. Put elbow (5) in cover (6).

GO TO FRAME 2

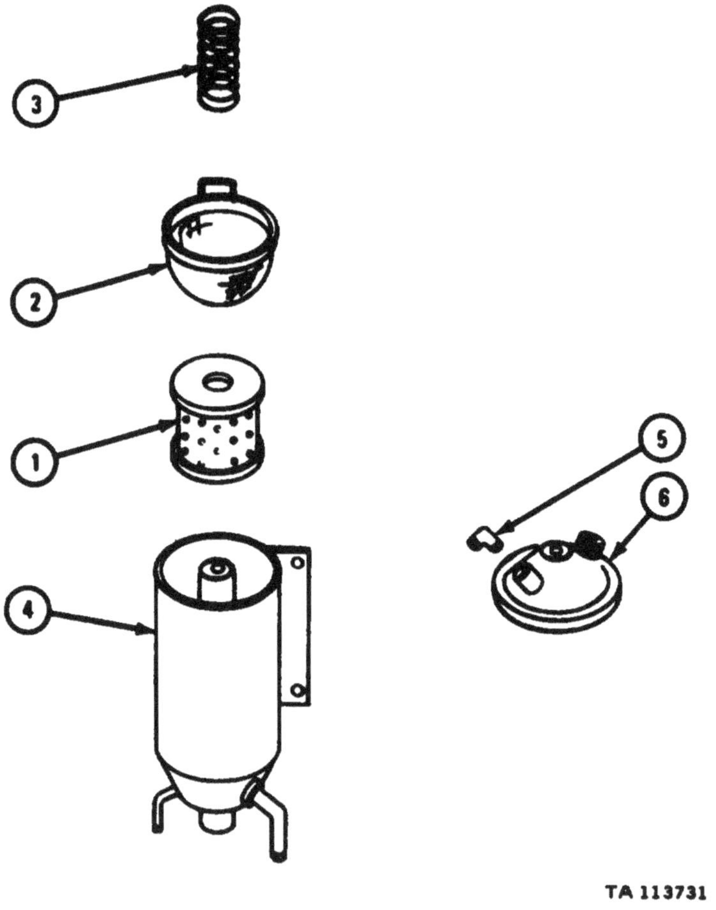

FRAME 2

1. Put cover gasket (1) on reservoir body (2).
2. Put cover (3) on gasket (1) and reservoir body (2).
3. Put on washer (4) and cover bolt (5).
4. Put on oil filler cap gasket (6) and oil filler cap (7).
5. Put preformed packing (8) on dipstick (9) and screw in dipstick.

END OF TASK

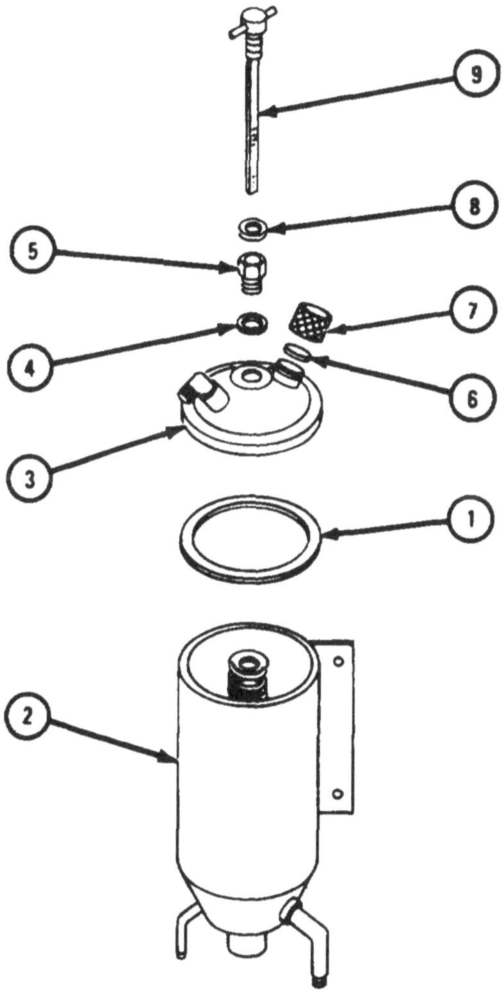

TA 113732

4-54. HYDRAULIC PUMP OIL HOSES (ENGINE LDS-465-2)

a. Cleaning.

FRAME 1

1. Take two hose clamps (1) off hydraulic pump return tube (2). Take two hose clamps (3) off hydraulic pump bypass hose (4) .
2. Take out two screws (5), two nuts (6), and two lockwashers (7). Take two 3/4-inch loop clamps (8) off hydraulic pump return tube (2). Take two 1/2-inch loop clamps (9) off hydraulic pump bypass hose (4) .

WARNING

Eye shields must be worn when using compressed air. Eye injury can occur if eye shields are not used.

3. Clean hydraulic pump return tube (2), bypass hose (4), and supply hose (10) with warm soapy water. Blow dry with compressed sir.

GO TO FRAME 2

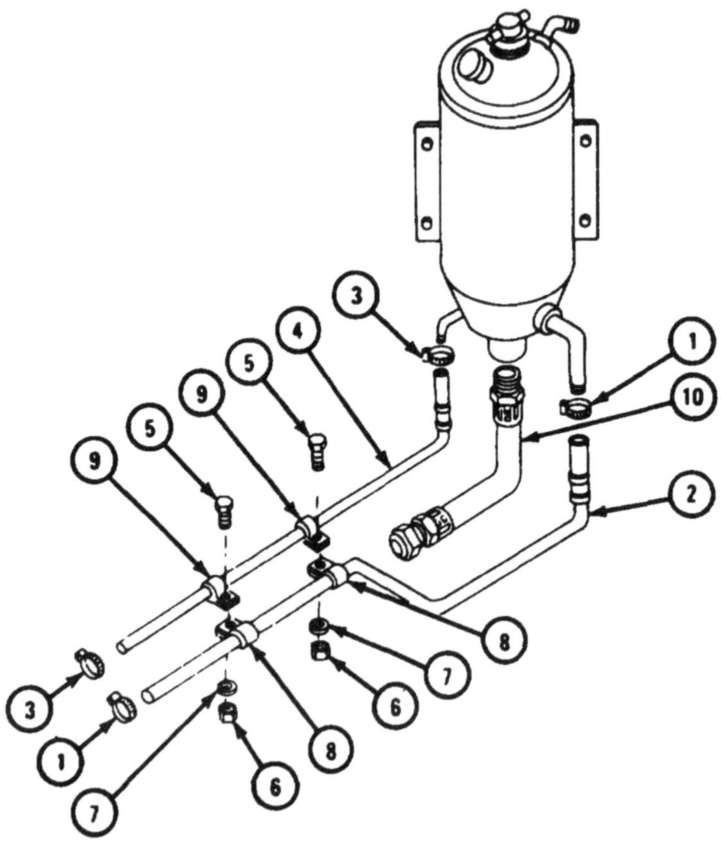

FRAME 2

WARNING

Dry cleaning solvent is flammable. Do not use near an open flame. Keep a fire extinguisher nearby when solvent is used . Use only in well-ventilated places. Failure to do this may result in injury to personnel and damage to equipment.

CAUTION

Do not let dry cleaning solvent touch hydraulic pump return hose (1), bypass hose (2) or supply hose (3) . Hoses will dry out and crack. If dry cleaning solven t splashes on hoses, dry with clean cloth.

1. Clean supply hose fittings (4) with a cloth soaked with dry cleaning solvent. Clean off sludge and gum with a stiff brush. Dry with clean cloth.
2. Clean two screws (5), two nuts (6), two lockwashers (7), four loop clamp s (8), and four hose clamps (9) with dry cleaning solvent. Clean off sludge and gum with a stiff brush. Dry with clean cloth.

END OF TASK

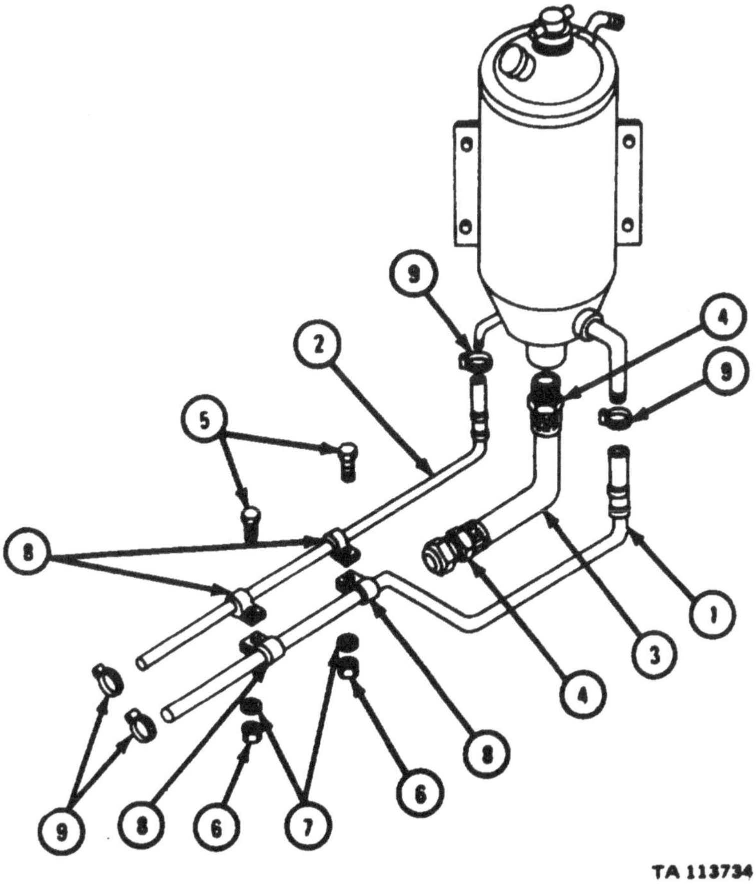

b. Inspection and Repair.

FRAME 1

1. Check that hydraulic pump bypass hose (1) and supply hose (2) are not dried out, cracked or split. If either hose is damaged, get new one.
2. Check that threads on fittings at ends of supply hose (2) are not damaged. Fix small thread damage using thread chaser or tap. If more repair is needed, get new supply hose.
3. Check that rubber ends of hydraulic pump return tube (3) are not dried out, cracked or split. If they are damaged, get a new hydraulic pump return tube.
4. Check that metal part of hydraulic pump return tube (3) is not cracked, split, kinked, dented or damaged in any other way. Straighten small bends. If more repair is needed, get a new hydraulic pump return tube.

GO TO FRAME 2

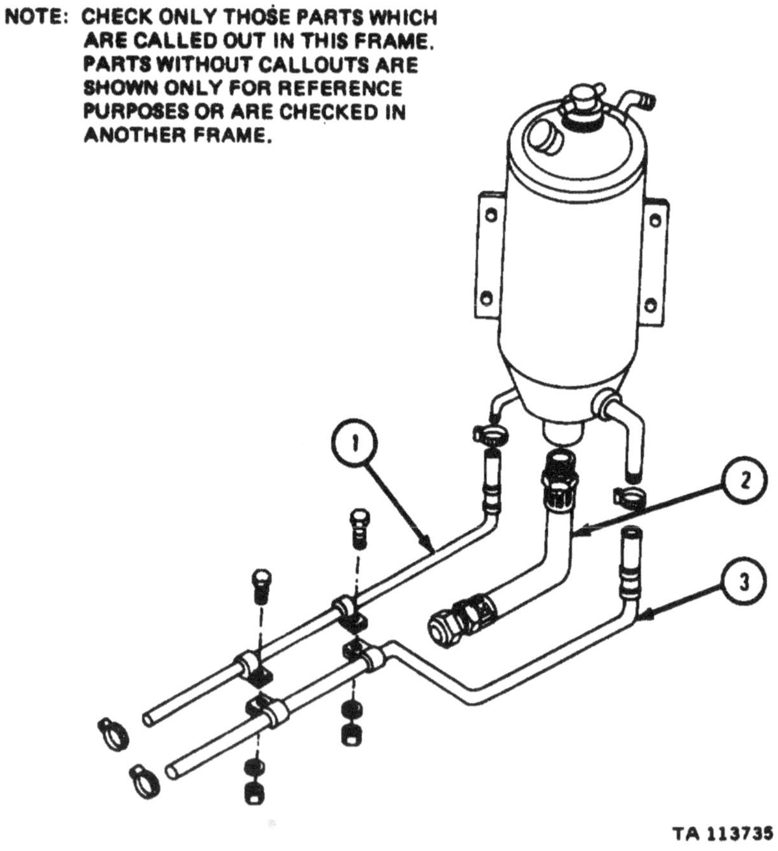
NOTE: CHECK ONLY THOSE PARTS WHICH ARE CALLED OUT IN THIS FRAME. PARTS WITHOUT CALLOUTS ARE SHOWN ONLY FOR REFERENCE PURPOSES OR ARE CHECKED IN ANOTHER FRAME.
1
2
3
TA 113735

FRAME 2

1. Check that two screws (1) and two nuts (2) do not have damaged threads or other damage . If screws or nuts are damaged, get new ones.
2. Check that two lockwashers (3) are not cracked or damaged in any way. If lockwashers are damaged, get new ones.
3. Check that two 3/4-inch loop clamps (4) and two 1/2-inch loop clamps (5) are not cracked, bent or damaged in any way. Fix small bends in loop clamps. If more repair is needed, get new loop clamps.
4. Put two 3/4-inch loop clamps (4) on hydraulic pump return tube (6) as shown. Put two 1/2-inch loop clamps (5) on hydraulic pump bypass tube (7) as shown. Join two loop clamps (4) to two loop clamps (5) using two screws (1), two nuts (2), and two lockwashers (3) as shown .

GO TO FRAME 3

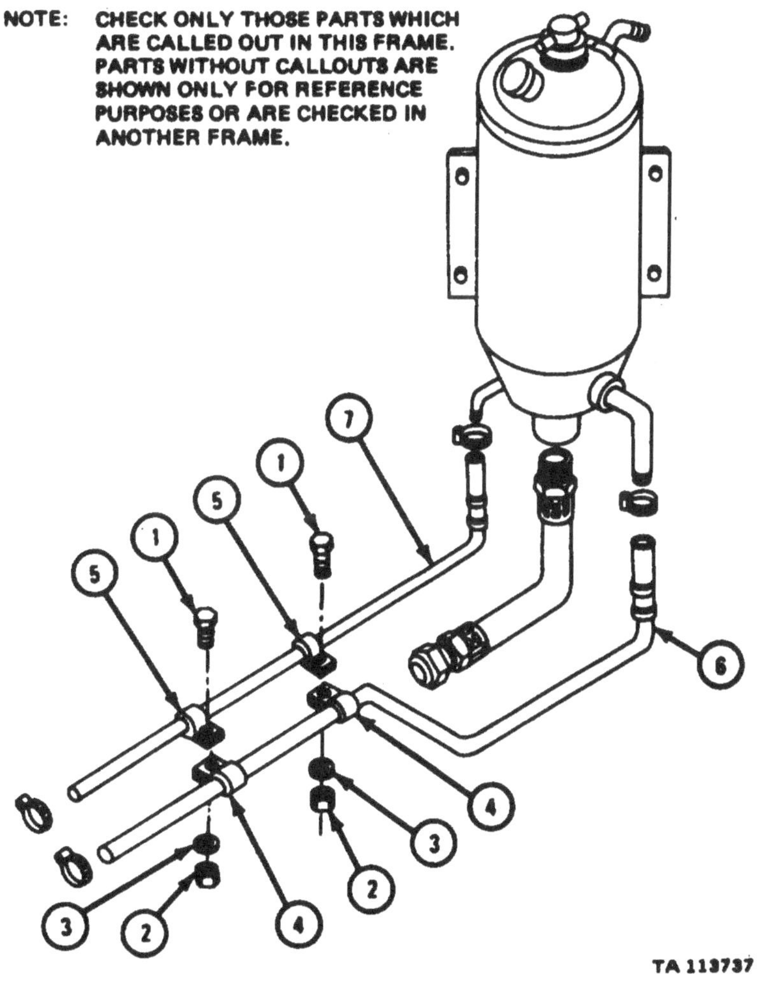
NOTE: CHECK ONLY THOSE PARTS WHICH ARE CALLED OUT IN THIS FRAME. PARTS WITHOUT CALLOUTS ARE SHOWN ONLY FOR REFERENCE PURPOSES OR ARE CHECKED IN ANOTHER FRAME.
7
1
5
1
5
6
4
3
2
3
2
4
TA 113737

FRAME 3

1. Check that two hose clamps (1) and two hose clamps (2) are not cracked, bent or damaged in any other way. Fix small bends and dents with pliers. If more repair is needed, get new hose clamps.
2. Put two hose clamps (1) on hydraulic pump return tube (3) as shown.
3. Put two hose clamps (2) on hydraulic pump bypass hose (4) as shown.

END OF TASK

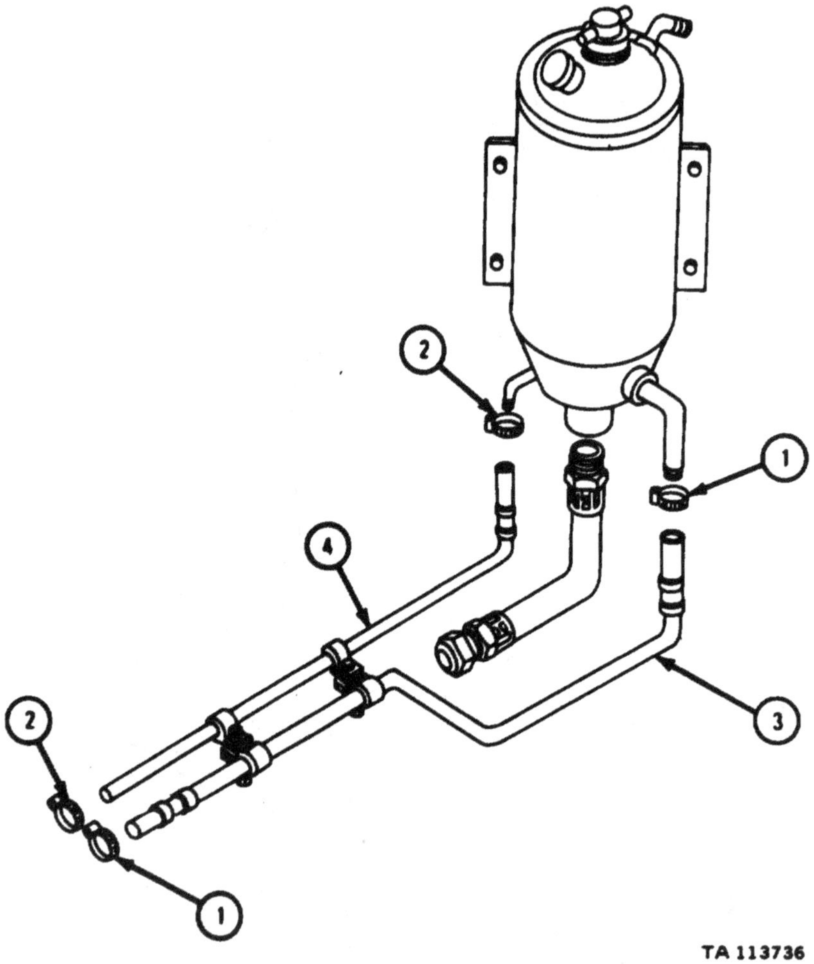

4-55. FAN IDLER PULLEY AND ARM ASSEMBLY AND RELATED PARTS (ENGINE LDS-465-2).

a. Disassembly.

FRAME 1

1. Put grooved pulley (1) and arm assembly (2) in vise, as shown.
2. Using hammer and punch, knock out plug (3).
3. Using retaining ring pliers, take out retaining ring (4).

GO TO FRAME 2

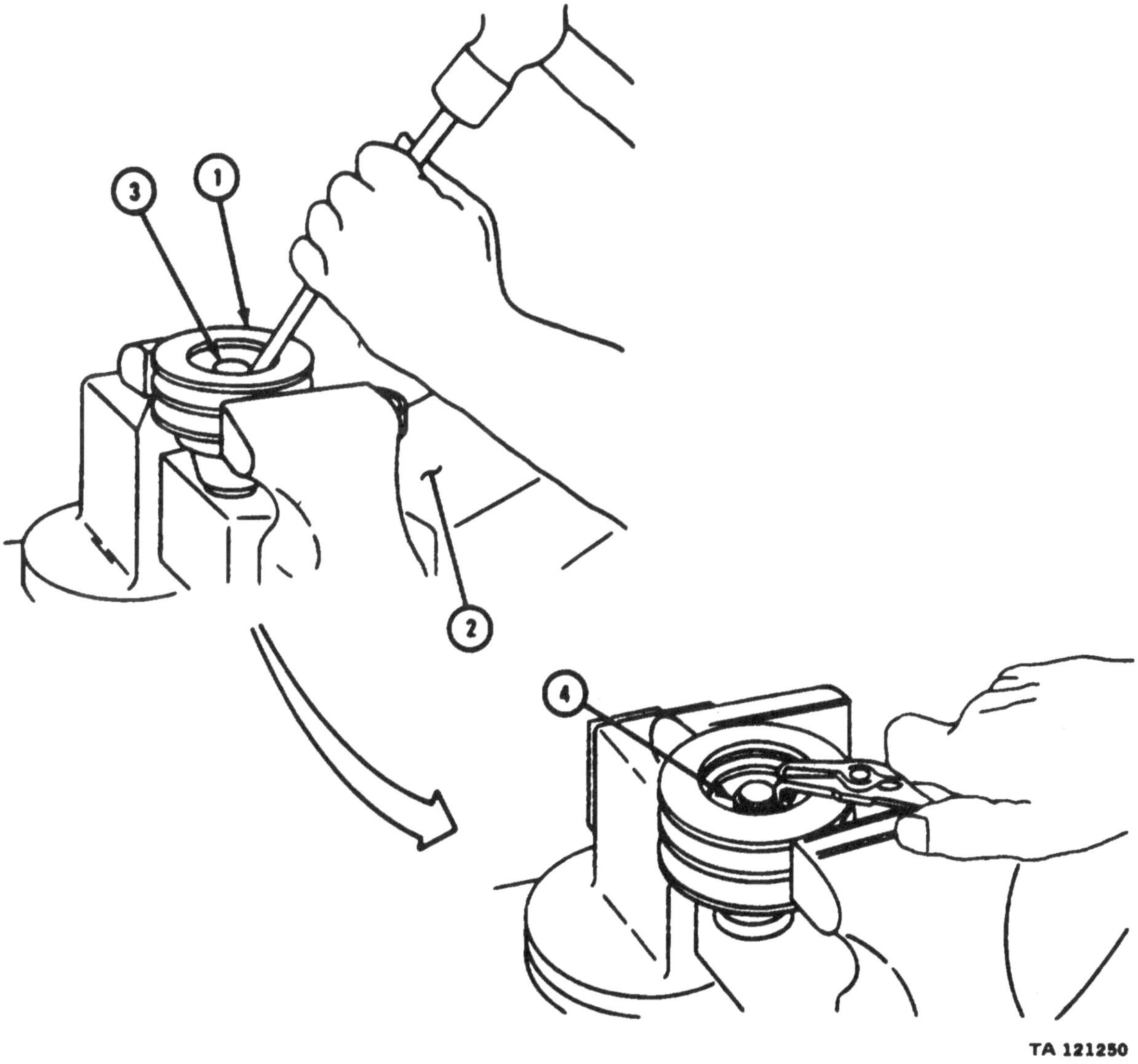

FRAME 2

1. Set up grooved pulley (1) in arbor press, as shown, and press grooved pulley off arm assembly (2).
2. Set up arm assembly (2) in arbor press and press out sleeve bushing (3).

GO TO FRAME 3

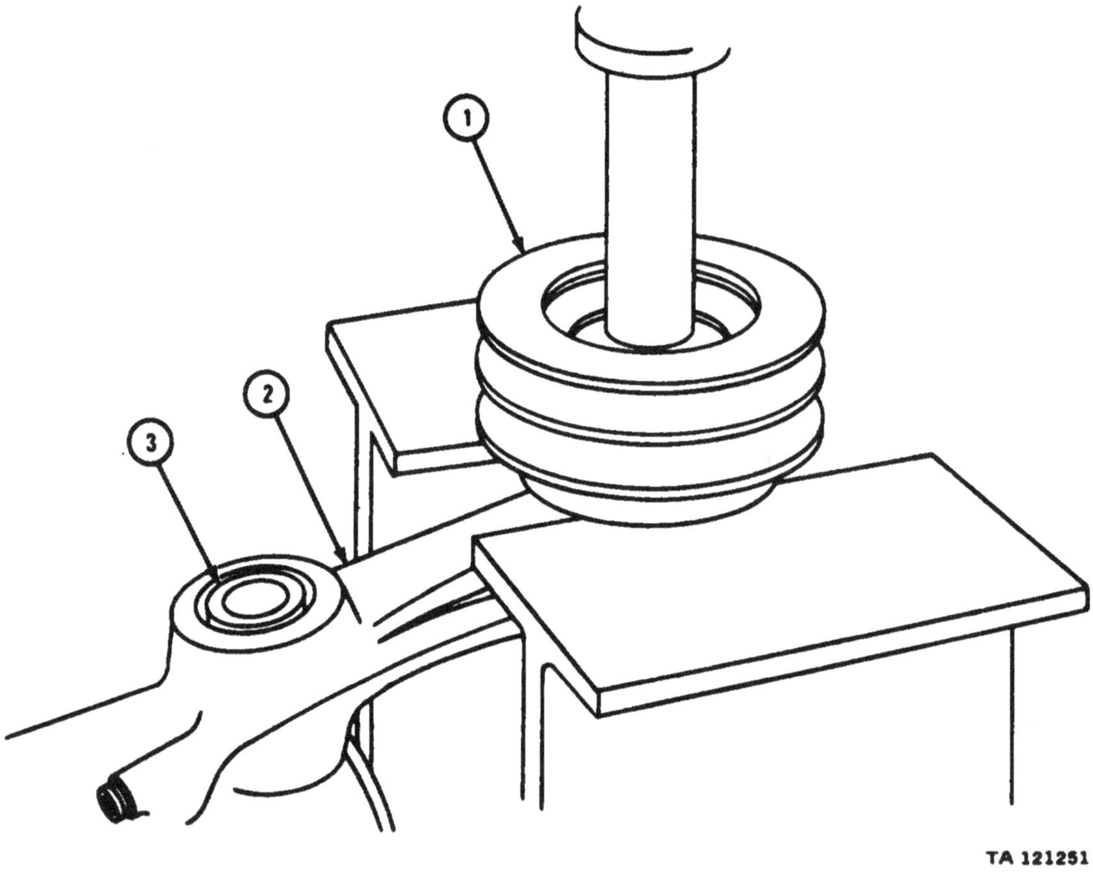

FRAME 3

1. Set up grooved pulley (1) in vise, as shown.
2. Using retaining ring pliers, take out retaining ring (2).
3. Set up grooved pulley (1) in arbor press, as shown, and press out bearing (3).

GO TO FRAME 4

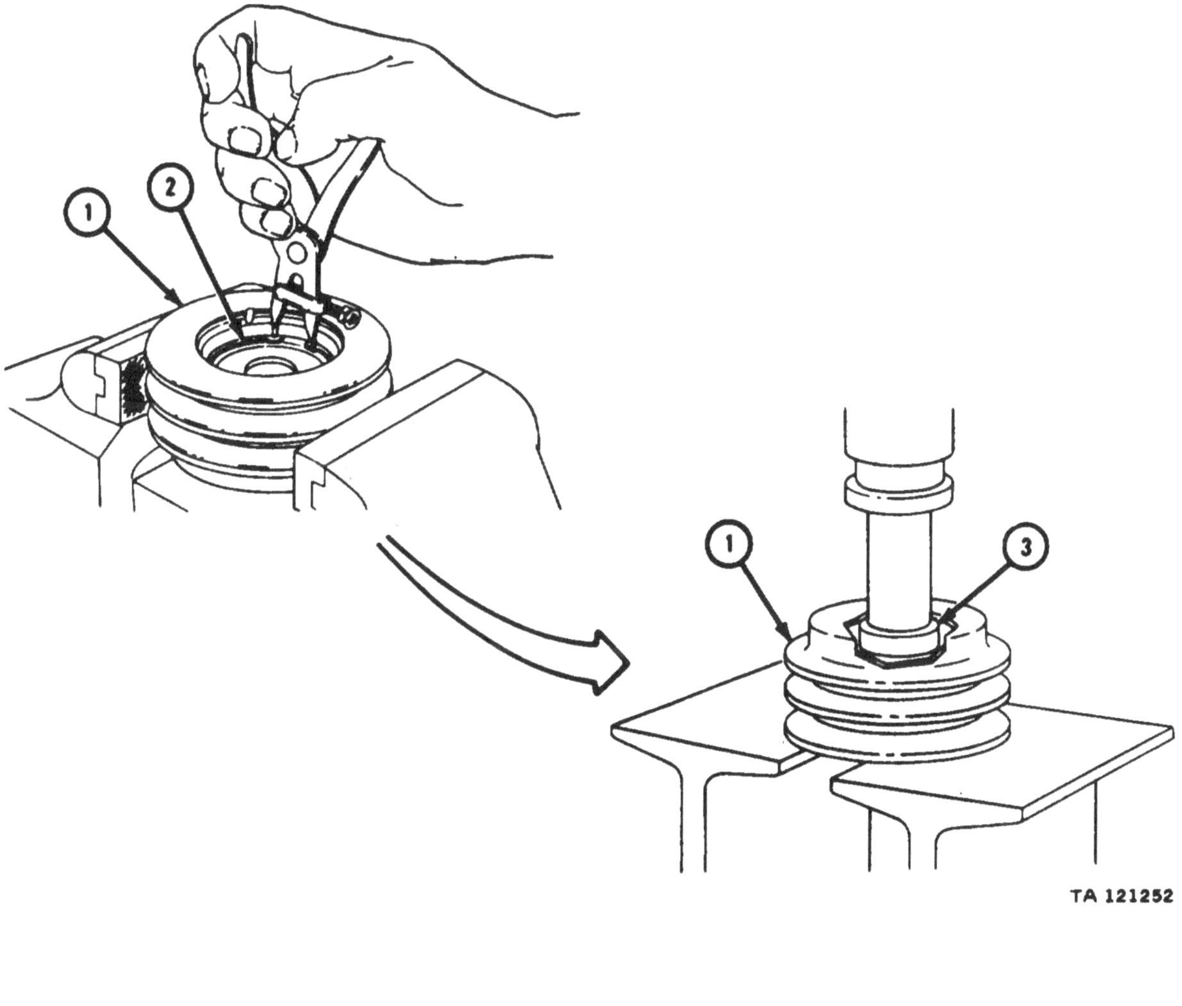

FRAME 4

1. Set up grooved pulley (1) in arbor press, as shown, and press out oil seal (2).
2. Takeout pipe fitting (3) , lubrication fitting (4), and lubricant relief fittin g (5).

END OF TASK

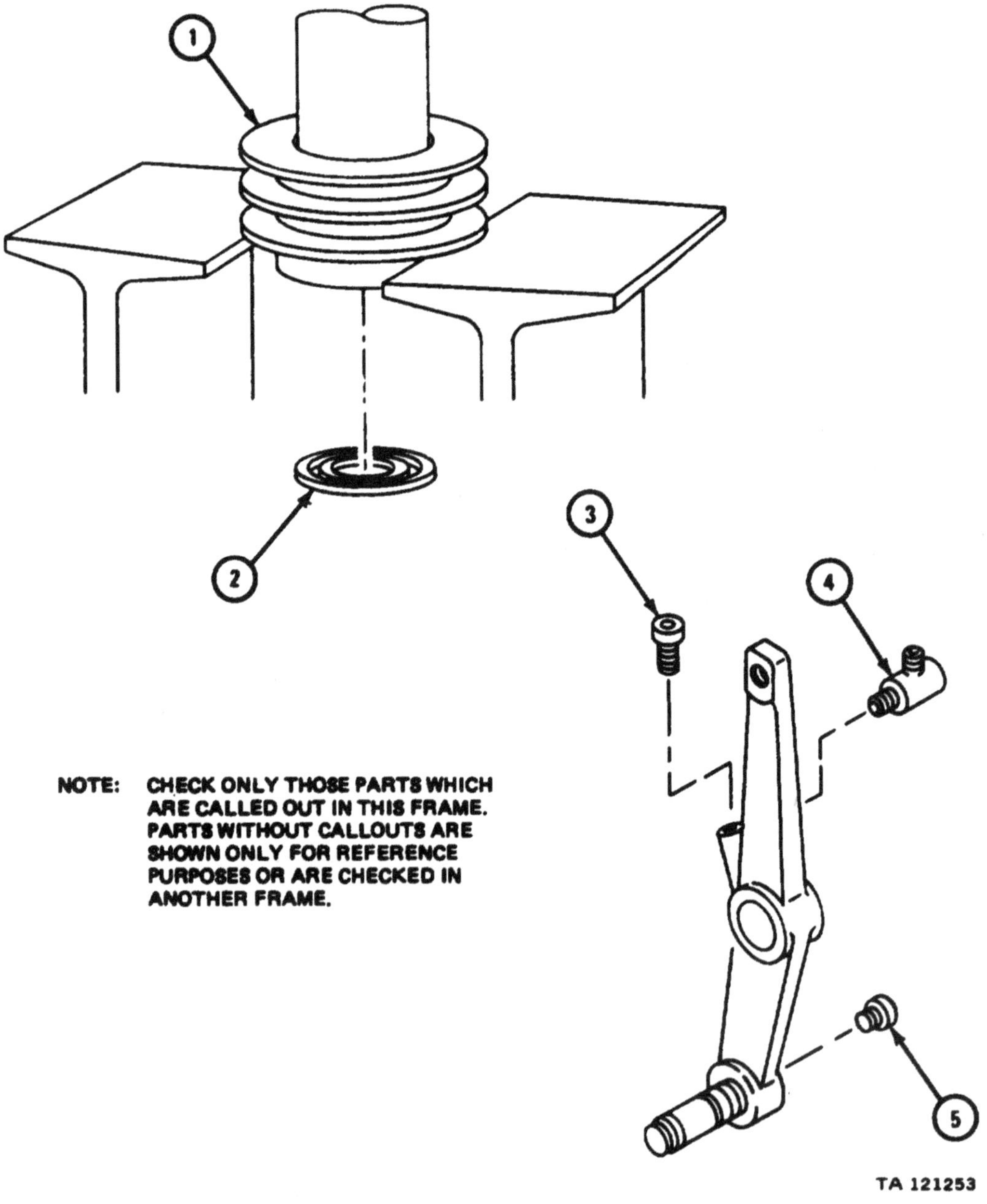
1
2
3
4
5
NOTE: CHECK ONLY THOSE PARTS WHICH ARE CALLED OUT IN THIS FRAME. PARTS WITHOUT CALLOUTS ARE SHOWN ONLY FOR REFERENCE PURPOSES OR ARE CHECKED IN ANOTHER FRAME.
TA 121253

b. Cleaning.

FRAME 1

WARNING

Dry cleaning solvent is flammable. Do not use near an open flame. Keep a fire extinguisher nearby when solvent is used. Use only in well-ventilated places. Failure to do this may result in injury to personnel and damage to equipment.

1. Clean three cap screws (1), four lockwashers (2), three flat washers (3), and nut (4) with dry cleaning solvent. Dry with clean cloth.
2. Clean adjusting arm (5), sliding stay (6), pipe fitting (7) and lubrication fitting (8) with dry cleaning solvent. Dry with clean cloth.
3. Clean two retaining rings (9), lubricant relief fitting (10), and plug (11) with dry cleaning solvent. Dry with clean cloth.

GO TO FRAME 2

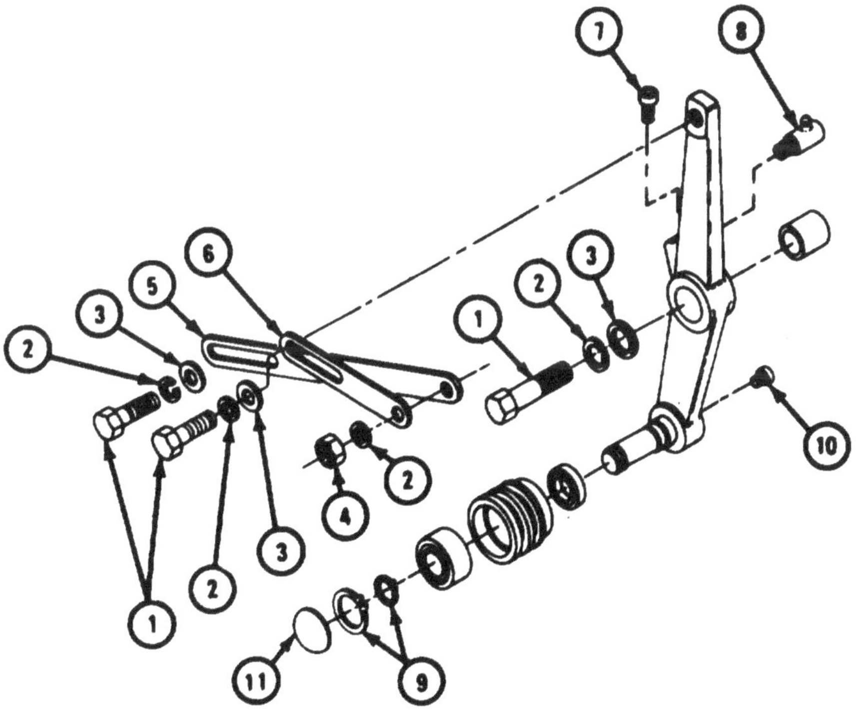

NOTE
CLEAN ONLY THOSE PARTS WHICH ARE CALLED OUT IN THIS FRAME. PARTS WITHOUT CALLOUTS ARE SHOWN ONLY FOR REFERENCE PURPOSES OR ARE CLEANED IN ANOTHER FRAME.

TA 117869

FRAME 2

1. Clean bearing (1). Refer to TM 9-214 .
2. Clean seal (2) using soap and water. Dr y with clean cloth.

WARNING

Dry cleaning solvent is flammable. Do not use near an open flame. Keep a fire extinguisher nearby when solvent is used. Use only in well-ventilated places. Failure to do this may result in injury to personnel and damage to equipment.

3. Clean grooved pulley (3) with dry cleaning solvent. Dry with clean cloth.
4. Clean arm assembly sleeve bushing (4) with dry cleaning solvent. Dry with clean cloth .
5. Clean arm assembly (5) with dry cleaning solvent. Use wire brush probe to clean lubrication passage (6) in arm assembly. Dry with clean cloth-.

END OF TASK

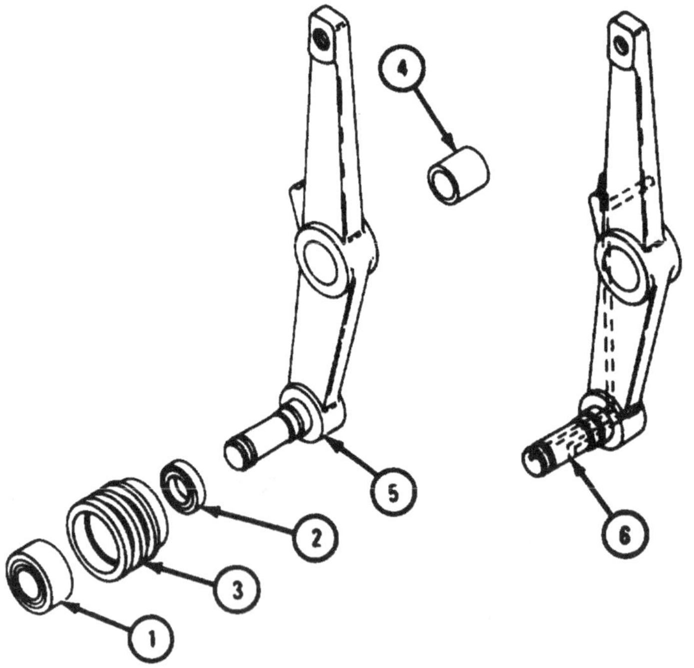

TA 117870

c. Inspection and Repair.

FRAME 1

1. Check that three capscrews (1), four lockwashers (2), three flat washers (3), and nut (4) are not damaged in any way. Repair minor thread damage with thread restorer. If parts are damaged in any other way, get new ones.
2. Check that adjusting arm (5), sliding stay (6), pipe fitting (7), and lubrication fitting (8) are not damaged in any way. If any parts are damaged, get new ones in their place.
3. Check that two retaining rings (9) and lubricant relief fitting (10) and plug (11) are not damaged in any way. If any parts are damaged, get new ones.

GO TO FRAME 2

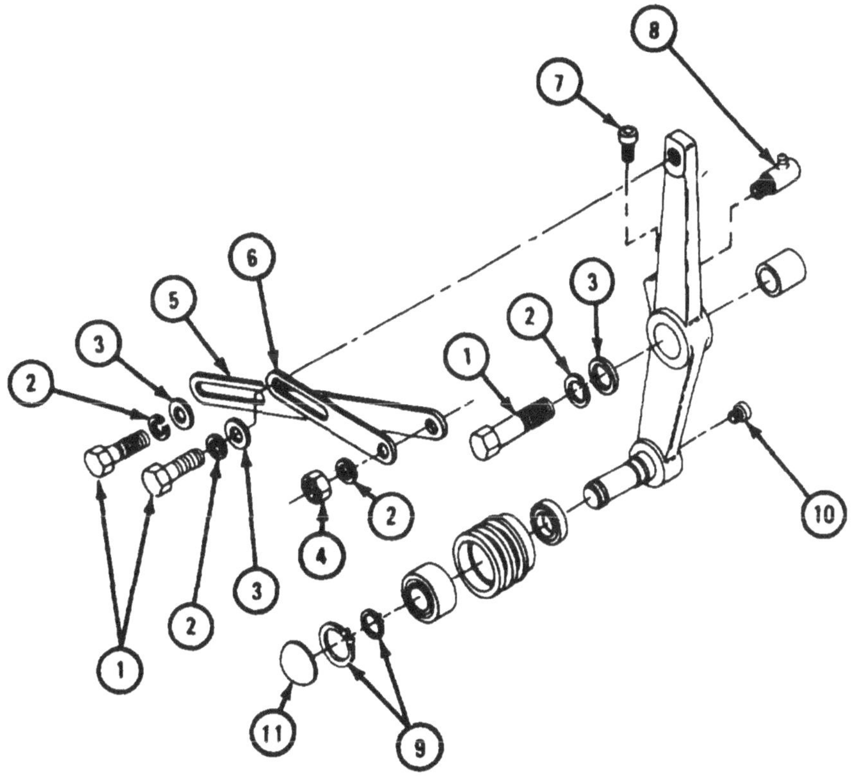

NOTE: CHECK ONLY THOSE PARTS WHICH ARE CALLED OUT IN THIS FRAME. PARTS WITHOUT CALLOUTS ARE SHOWN ONLY FOR REFERENCE PURPOSES OR ARE CHECKED IN ANOTHER FRAME.

TA 117941

FRAME 2

1. Check that bearing (1) turns freely without being too loose or damaged in any way . If bearing is damaged, get a new one.
2. Check that lip of seal (2) is not torn, hard or brittle, and that seal is not damaged in any other way. If seal is damaged, get a new one.

WARNING

Dry cleaning solvent is flammable. Do not use near an open flame. Keep a fire extinguisher nearby when solvent is used . Use only in well-ventilated places. Failure to do this may result in injury to personnel and damage to equipment.

3. Check that grooved pulley (3) has no cracks, breaks or rough grooves. Fi x minor damage with fine mill file or crocus cloth dipped in solvent. If more repair is needed, get a new pulley.
4. Check that arm assembly sleeve bushing (4) has no nicks or burrs and that sleeve bushing fits snugly in arm assembly (5). Fix minor nicks or burrs with fine mill file or crocus cloth dipped in solvent. If more repair is needed or sleeve bushing will not fit snugly in arm assembly, get a new sleeve bushing.
5. Check that arm assembly (5) is not cracked or damaged in any way. If arm assembly is damaged, get a new one.

GO TO FRAME 3

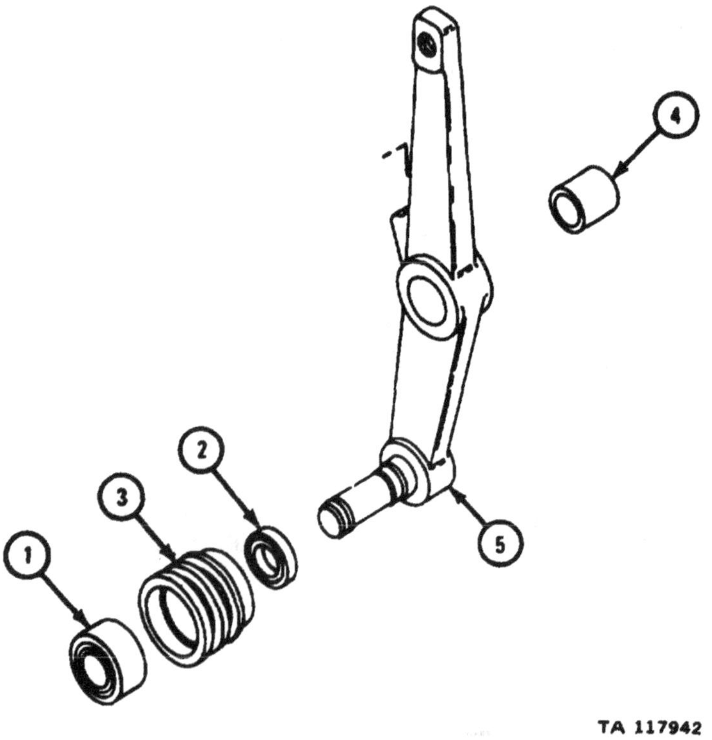

FRAME 3

NOTE

Readings must be within limits given in table 4-64. If readings are not within given limits, throw away part and get a new one.

1. Measure arm assembly bore (1).
2. Measure diameter of fan idler shaft (2).
3. Measure sleeve bushing outside diameter (3) .
4. Measure bearing outside diameter (4) and inside diameter (5).
5. Measure inside diameter of grooved pulley (6) .

GO TO FRAME 4

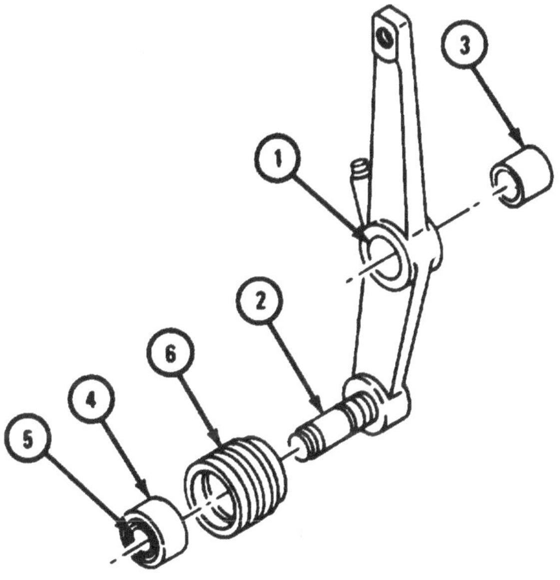

Table 4-64. Fan Idler Pulley and Arm Assembly Wear Limits

Index Number	Item /Point of Measurement	Size and Fit of New Parts (inches)	Wear Limit (inches)
1	Arm assembly bore	1.1220 to 1.1250	1.1280
2	Fan Idler shaft diameter	0.7873 to 0.7877	None
3	Sleeve bushing outside diameter	1.1200 to 1.1210	1.1170
4	Bearing outside diameter	2.0467 to 2.0472	None
5	Bearing inside diameter	0.7870 to 0.7874	None
6	Grooved pulley inside diameter	2.0460 to 2.0472	None

FRAME 4

NOTE

Readings must be within limits given in table 4-65. The letter L shows a loose fit. The letter T shows a tight fit. If readings are not within given limits, throw away part and get a new one.

1. Measure fit of sleeve bushing (1) in bore (2) .
2. Measure fit of bearing inside diameter (3) on idler shaft (4).
3. Measure fit of bearing outside diameter (5) in bore of grooved pulley (6).

END OF TASK

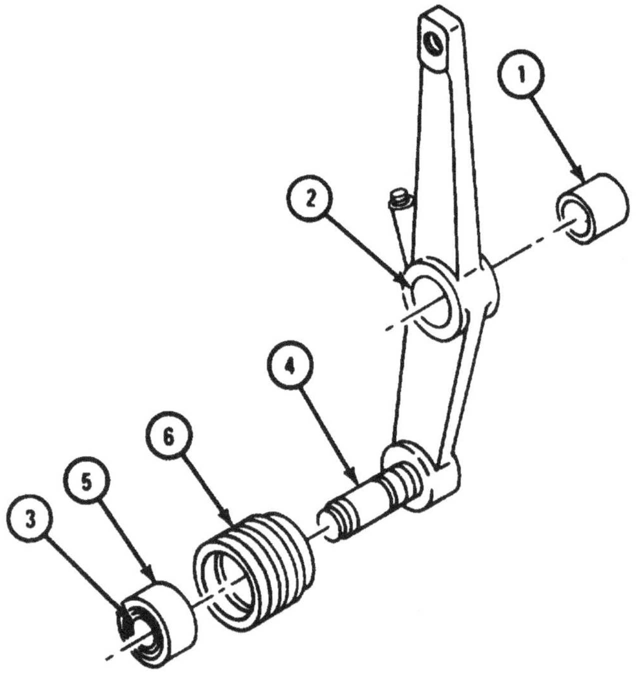

Table 4-65. Fan Idler Pulley and Arm Assembly Fits and Wear Limits

Index Number	Item /Point of Measurement	Size and Fit of New Parts (inches)	Wear Limit (inches)
1 and 2	Fit of sleeve bushing in bore	0.0010L to 0.0050L	0.010L
3 and 4	Fit of bearing inside diameter on idler shaft	0.0007T to 0.0001L	None
5 and 6	Fit of bearing outside diameter in grooved pulley	2.0460 to 2.0472	None

d. Assembly.

FRAME 1

1. Set Up grooved pulley (1) in arbor press and press in bearing (2).
2. Turn over grooved pulley (1) and press in oil seal (3).

GO TO FRAME 2

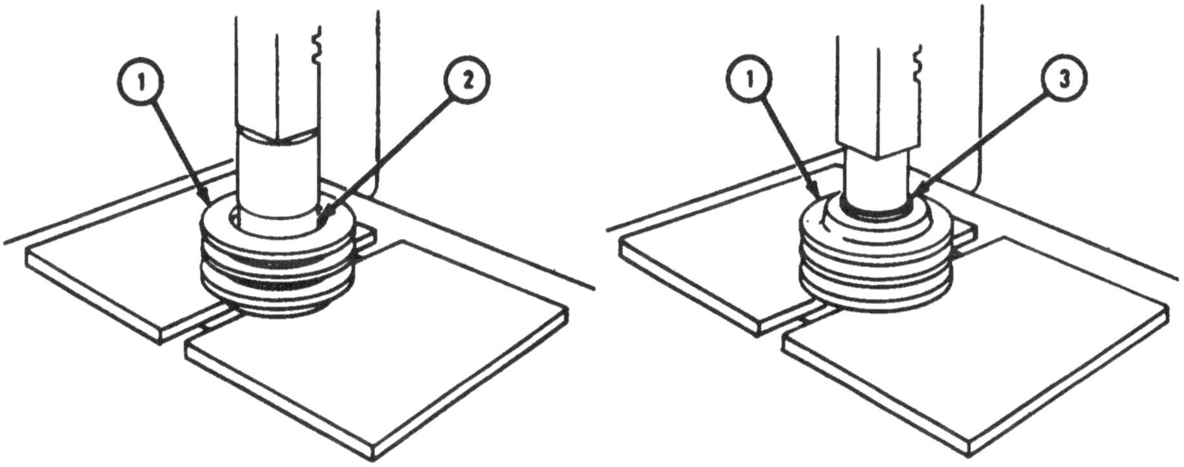

FRAME 2

1. Set up grooved pulley (1) in vise, as shown.
2. Using retaining ring pliers, put in retaining ring (2).

GO TO FRAME 3

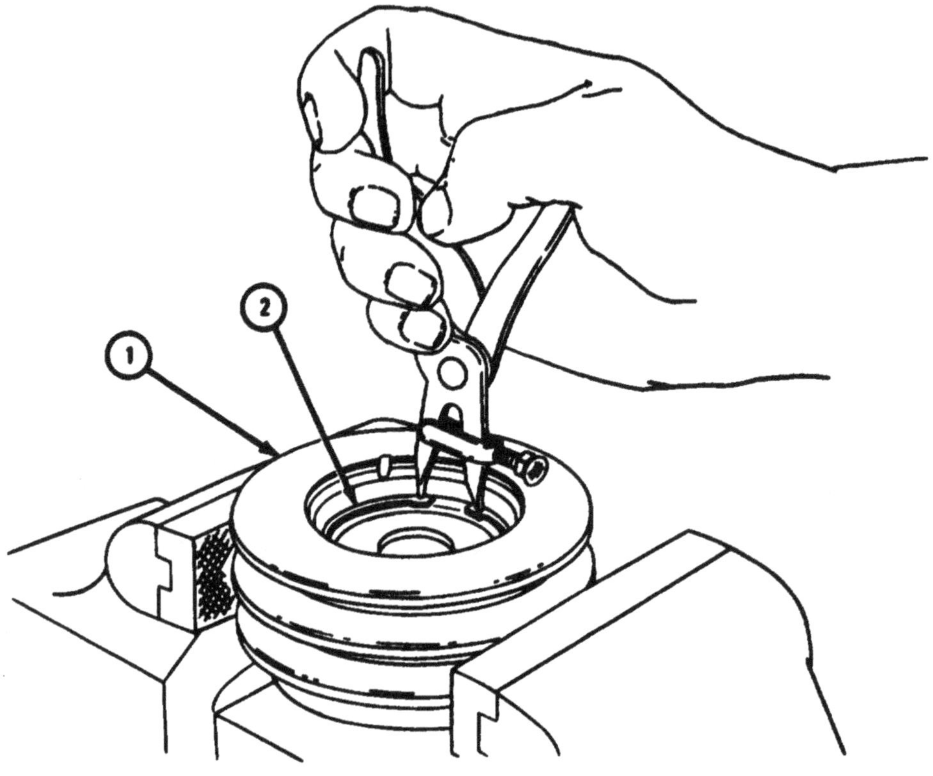

FRAME 3

1. Set up arm assembly (1) with grooved pulley (2) in arbor press, as shown.
2. Press grooved pulley (2) on arm assembly (1) .
3. Set up arm assembly (1) in arbor press and press in sleeve bushing (3).
4. Put in pipe fitting (4), lubrication fitting (5), and lubricant relief fitting (6).

GO TO FRAME 4

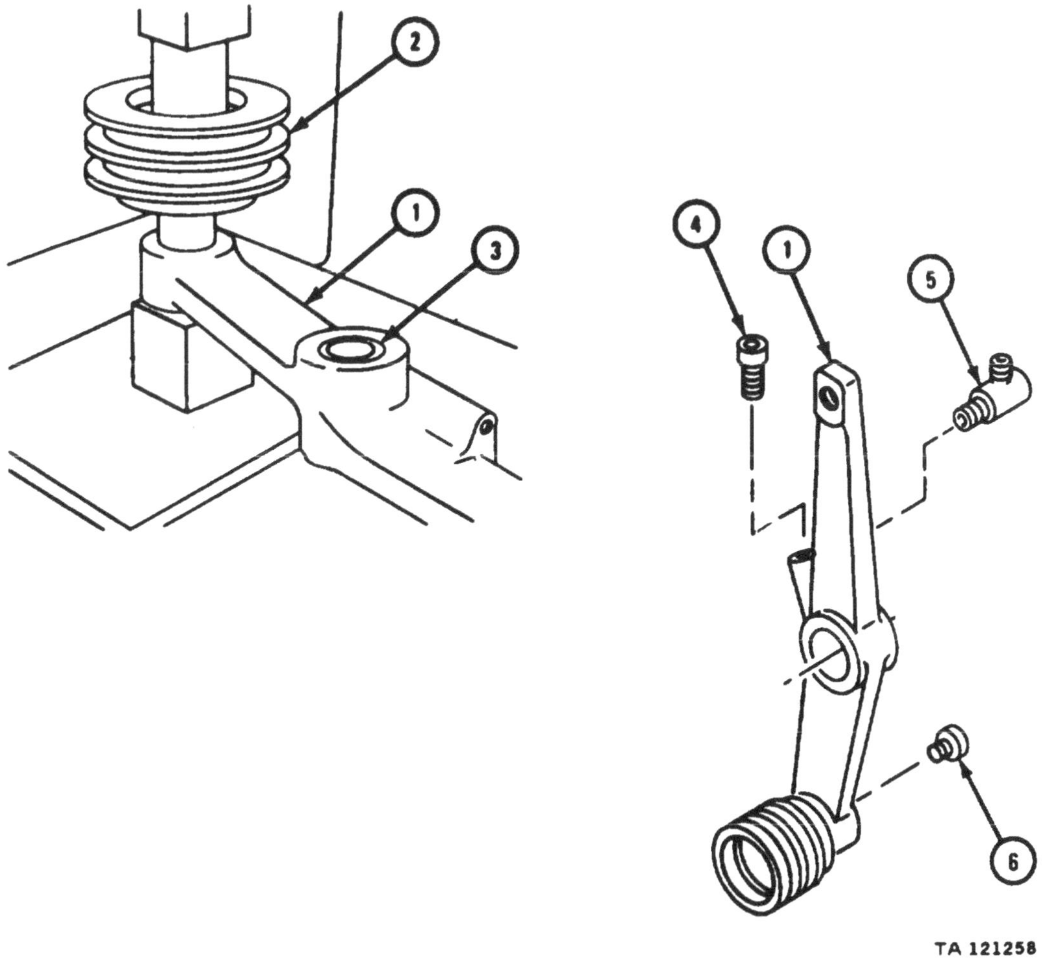

FRAME 4

1. Set up grooved pulley (1) with arm assembly (2) in vise, as shown.
2. Using retaining ring pliers, put on retaining ring (3).
3. Using hammer and punch, punch in plug (4).

END OF TASK

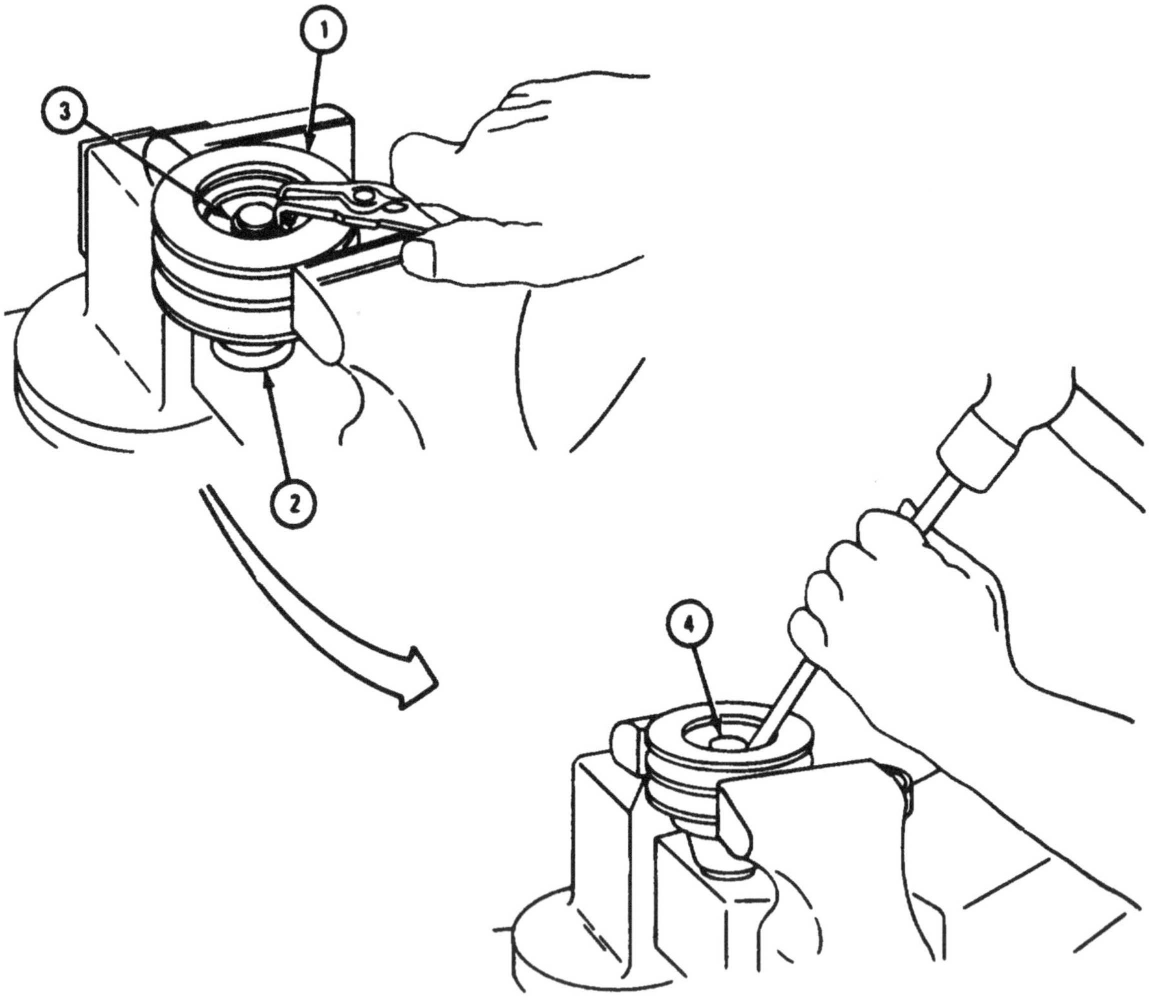

TA 121259

4-56. CLUTCH AND TRANSMISSION DRIVE PLATES.

a. Cleaning.

(1) Engines LD 465-1, LD 465-1C, and LDT 465-1C .

FRAME 1

WARNING

Dry cleaning solvent is flammable. Do not use near an open flame. Keep a fire extinguisher nearby when solvent is used. Use only in well-ventilated places. Failure to do this may result in injury to personnel and damage to equipment.

Eye shields must be worn when using compressed air. Eye injury can occur if eye shields are not used.

1. Clean clutch pressure plate (1) assembly with dry cleaning solvent. Dry passages by blowing them out with dry, compressed air.

CAUTION

Never let dry cleaning solvent, grease or oil of any kind touch clutch disk facings (2) .

2. Clean clutch disk facings (2) with a wire brush. Clean drive hub (3) with a clean rag soaked in dry cleaning solvent.

END OF TASK

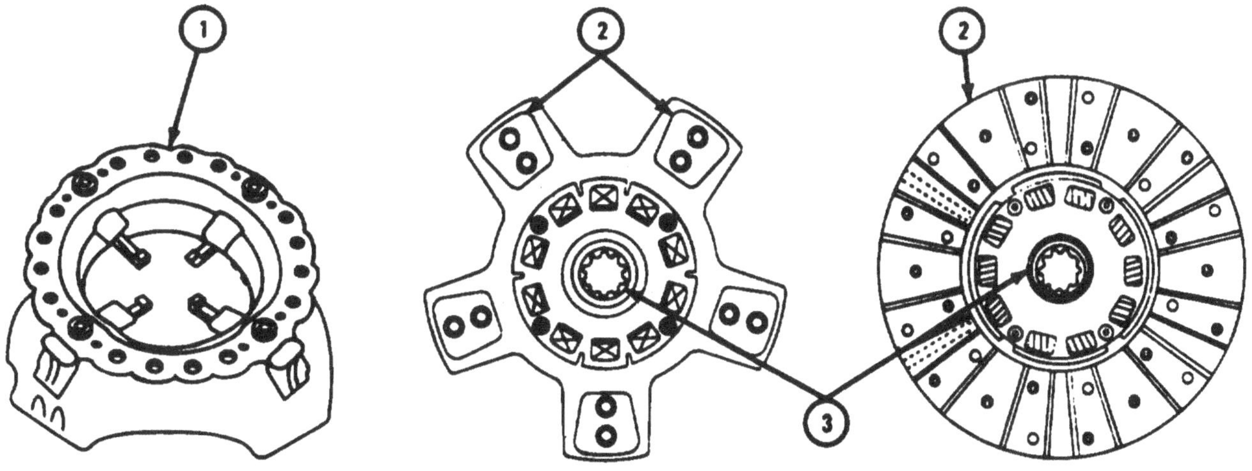

(2) Engines LDS-465-1 and LDS-465-1A .

FRAME 1

WARNING

Dry cleaning solvent is flammable. Do not use near an open flame. Keep a fire extinguisher nearby when solvent is used . Use only in well-ventilated places. Failure to do this may result in injury to personnel and damage to equipment.

Eye shields must be worn when using compressed air. Eye injury can occur if eye shields are not used.

1. Clean clutch pressure plate (1) with dry cleaning solvent. Dry passages by blowing them out with dry, compressed air.

CAUTION

Never let dry cleaning solvent, grease or oil of any kind touch clutch disk facings (2) .

2. Clean clutch disk facings (2) with a wire brush. Clean drive hub (3) with a cloth soaked in dry cleaning solvent and blow dry with dry, compressed air.

END OF TASK

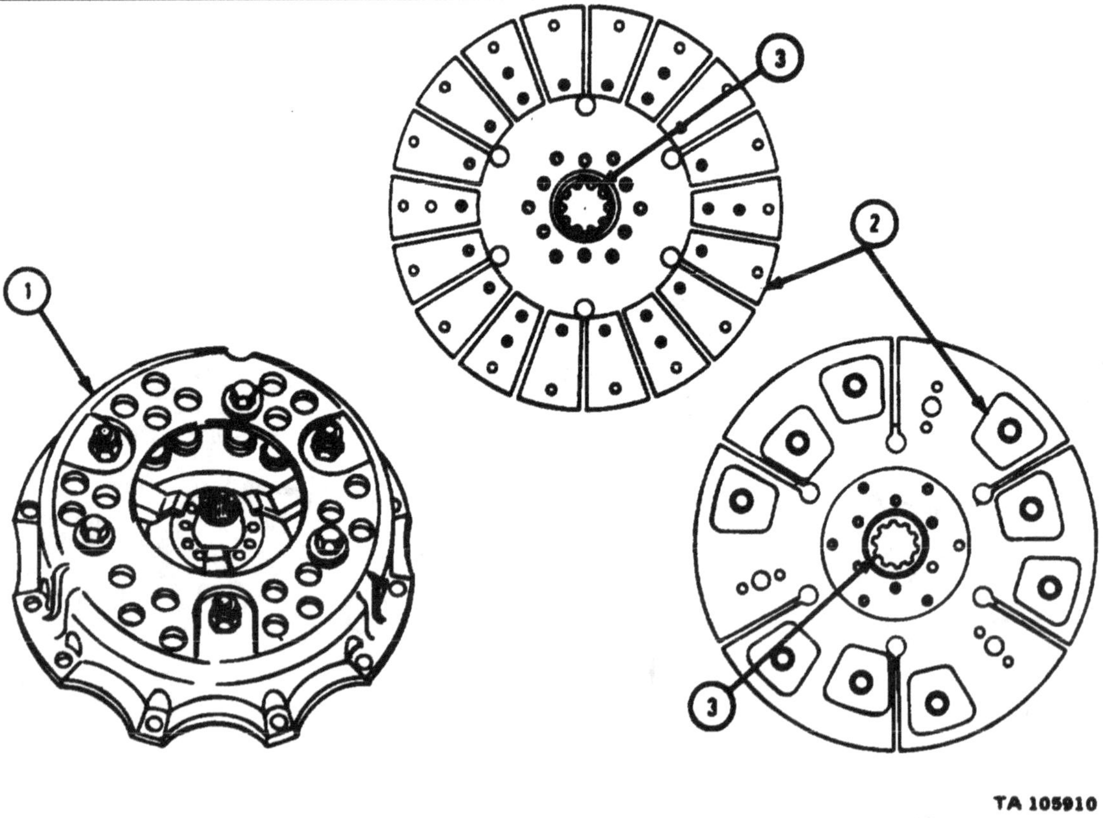

(3) Engine LDS-465-2 .

FRAME 1

WARNING

Dry cleaning solvent is flammable. Do not use near an open flame. Keep a fire extinguisher nearby when solvent is used . Use only in well-ventilated places. Failure to do this may result in injury to personnel and damage to equipment.

Eye shields must be worn when using compressed air. Eye injury can occur if eye shields are not used.

1. Clean all transmission drive plates (1) with dry cleaning solvent and a wire brush.
2. Clean scuff plate (2) and six screws (3) with dry cleaning solvent.
3. Dry all parts with compressed air.

END OF TASK

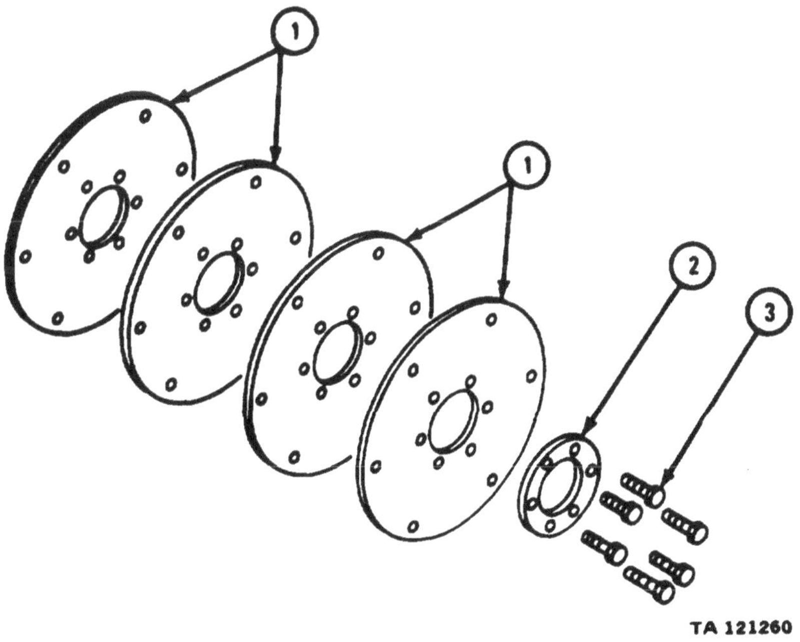

b. Inspection.

NOTE

For engines LD 465-1, LD 465-1A, and LDT 465-1A, go to frame 1. For engines LDS 465-1 and LDS 465-1A, go to frame 2. For engine LDS 465-2 go to frame 5.

FRAME 1

1. Check that clutch disk drive hub (1) has no worn splines and looseness on driven plate (2) .
2. Check that there are no broken or missing damper springs (3).
3. Check that there are no loose pins (4) holding retainer plate (5) to driven plate (2) .
4. Check that clutch disk facing (6) is not burned, loose on driven plates (2) or damaged in any other way.
5. Check that clutch disk (7) is not cracked or worn.
6. If clutch disk (7) is damaged, throw disk away and get a new one.
7. Measure thickness of disk facing (6) at several dates. If thickness is les s than 0.350 inch, throw disk away and get a new one.

GO TO FRAME 3

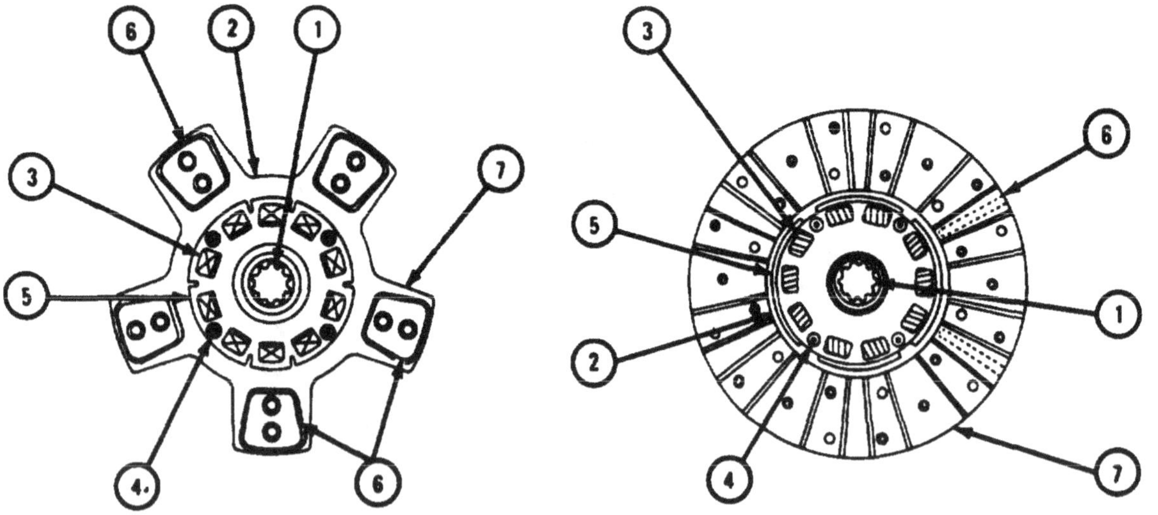

TA 118000

FRAME 2

1. Check that clutch disk drive hub (1) has no worn splines (2) or looseness on driven plate (3) .
2. Check that clutch disk has no loose rivets (4) holding driven plate (3) to drive hub (1) .
3. Check that disk facings (5) are not burned, loose on driven plate (3) or damaged in any other way.
4. Check that disk (6) is not cracked or worn.
5. Check that thickness of disk facing (5) is not less than 0.411 inch.
6. If clutch disk (6) is damaged or worn, put in new clutch disk.

GO TO FRAME 3

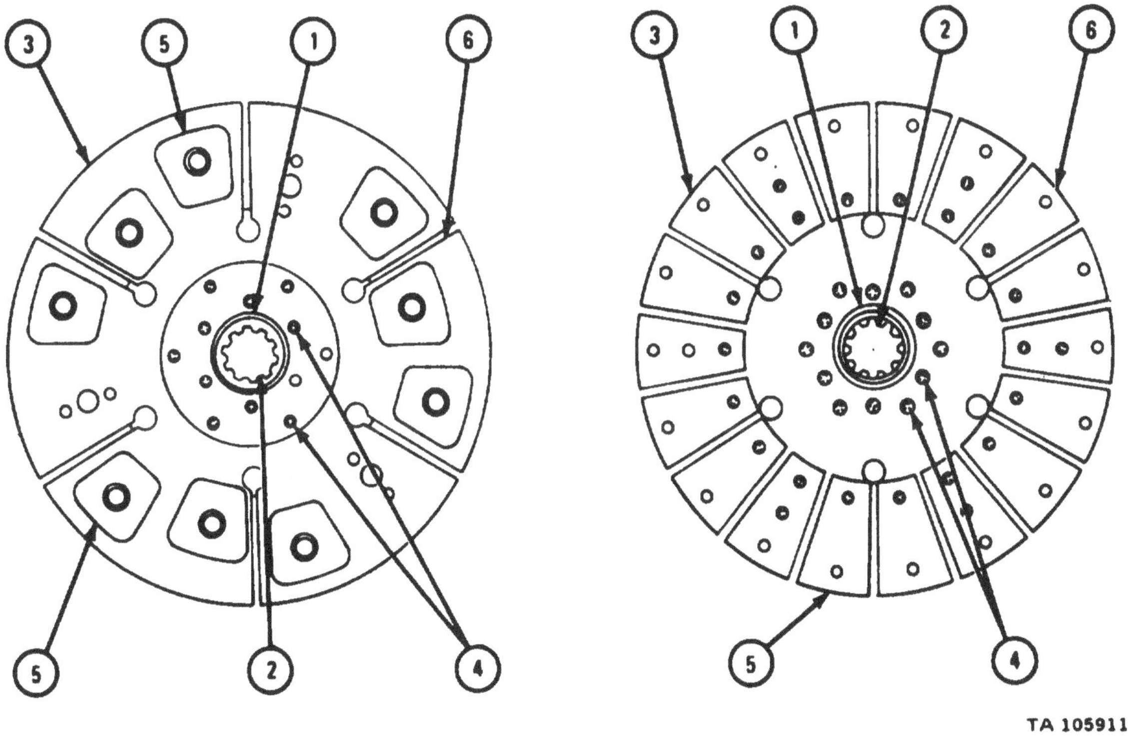

FRAME 3

NOTE

This task is shown for engine LD-465-1. This task is the same for all engines.

1. Stand a new transmission input shaft (1) on gear end on surface plate.
2. Put clutch disk (2) on splined end of input shaft (1) with long end of drive hub toward surface plate.
3. Set up dial indicator (3) as shown.
4. Take measurements at various places around clutch disk (2). Measurements should be made near center, inside, and outside diameters of disk facing.
5. Difference between readings in step 4 should not be more than 0.010 inch. Throw away clutch disk (2) if readings are different by more than given limit.

GO TO FRAME 4

TA 105912

FRAME 4

1. Using 0.010-inch feeler gage (1) and straight edge (2), check that pressure plate (3) is not warped more than 0.010 inch.
2. Check that grooves in pressure plate (3) are not deeper than 0.010 inch.
3. Check that clutch pressure plate (3) has no overheating, rubbing against other metal parts or broken springs.
4. Check that there are no open cracks and heavy heat discoloration on pressure plate (3).
5. Check that there are no closely grouped hairline cracks which go to inner or outer edge of pressure plate (3).
6. If pressure plate (3) is damaged or worn, throw away pressure plate and get a new one.

END OF TASK

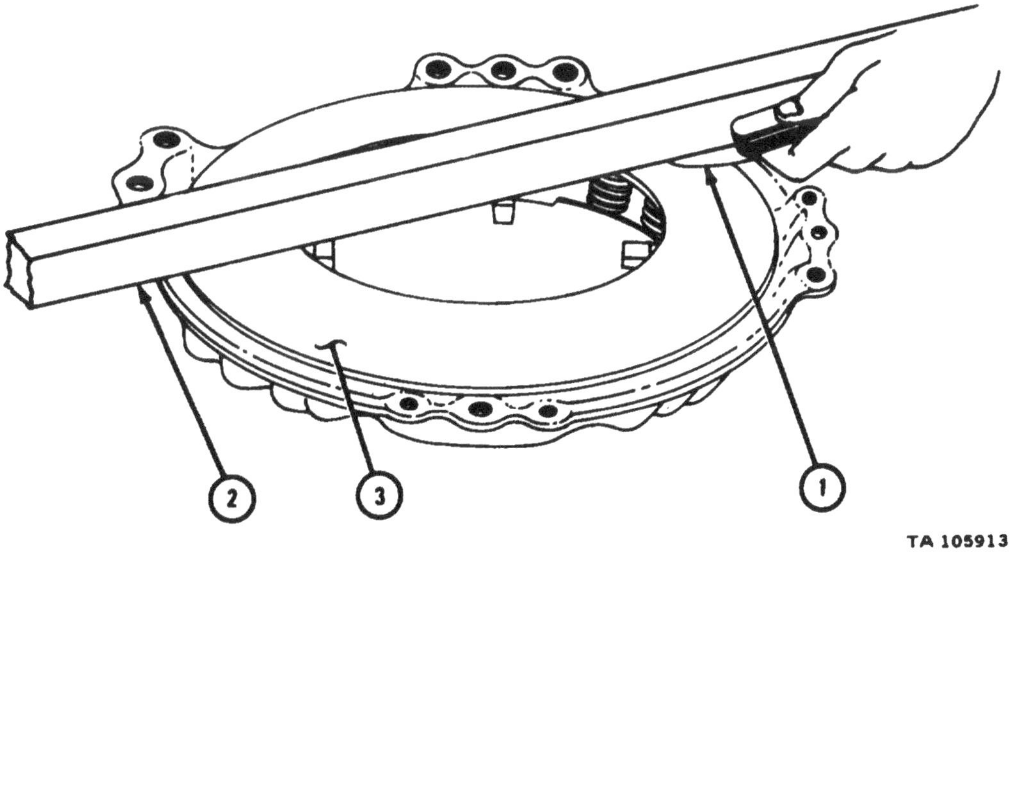

FRAME 5

1. Check that there are no open cracks or burrs on transmission drive plates (1) and scuff plate (2).
2. Check that there are no closely grouped hairline cracks which go to inner or outer edge of transmission drive plates (1) and scuff plate (2).
3. Check that six screws (3) are not stripped. If screws are stripped, throw them away and get new ones.
4. If parts are cracked or broken, throw them away and get new ones.

END OF TASK

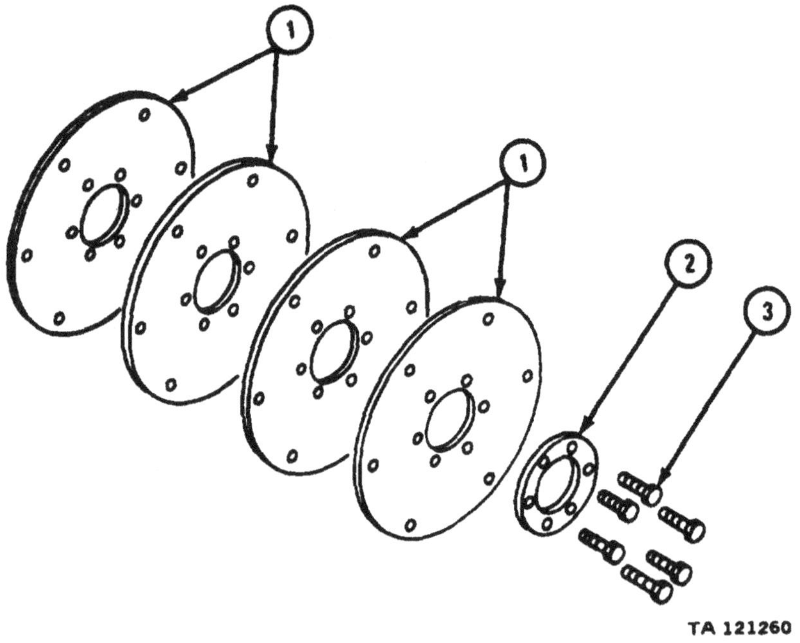

c. Repair.

NOTE

Repair is limited to adjustment only. For engines LD-465-1, LD-465-1C, and LDT-465-1C, go to frame 1. For engines LDS-465-1 and LDS-465-1A, go to frame 4. For engine LDS-465-2 go to frame 6.

FRAME 1

1. Place 0.3749 to 0.3751 inch thick by 13.000 inches dimeter spacer plate (1) into flywheel assembly (2) used for bench testing.
2. Place pressure plate (3) in place and screw in and tighten eight capscrews (4).
3. Using prybar, push in on each clutch finger (5) and take out clutch release lever spacer blocks (6).

GO TO FRAME 2

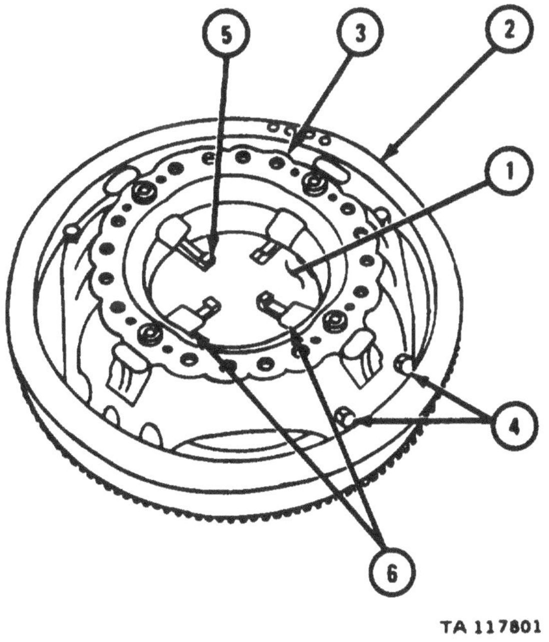

FRAME 2

1. Set depth micrometer (1) to 1.705 inches.
2. Take off adjusting screw lockplates or loosen locknuts (2).
3. Using spanner wrench socket (3), turn release lever adjusting nut (4) until micrometer (1) touches release lever (5) and micrometer anvil (6) touches clutch spacer plate (7) .
4. Do step 3 again for other three release levers (5).
5. When distance in steps 3 and 4 is 1.674 to 1.736 inches, heights of four release levers (5) must be within 0.015 inch of each other.
6. Do steps 3 and 4 again if needed.
7. Put on adjusting screw lockplates or tighten locknuts (2).

GO TO FRAME 3

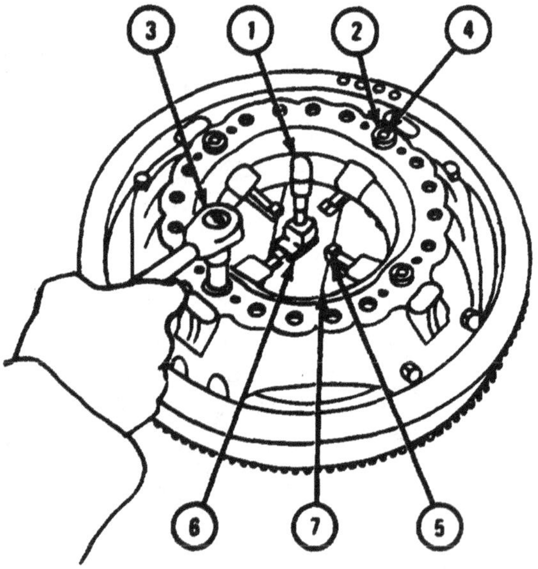

TA 117802

FRAME 3

1. Using prybar (1), push in on each clutch finger (2) and Put in clutch release lever spacer blocks (3) .
2. Take out eight capscrews (4) and pressure plate (5) .
3. Take out spacer plate (6).

END OF TASK

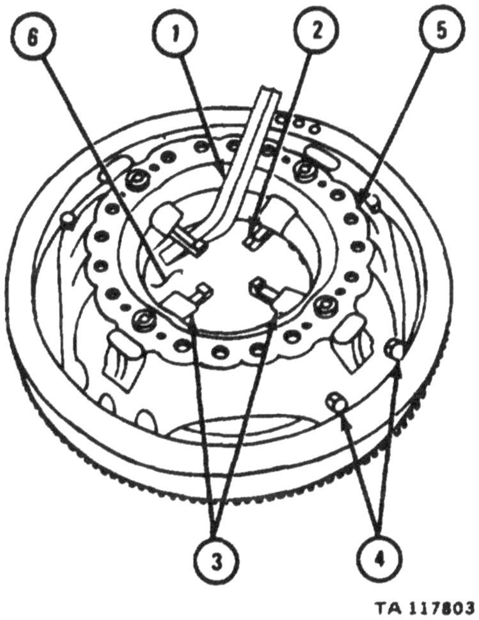

FRAME 4

1. Put clutch pressure plate (1) on flat surface.
2. Using combination square (2), check for 1-9/32 inch clearance A between pressure plate face (3) and pressure rover flange (4).
3. Using 9/16-inch socket (5) with breaker bar (6), set three retaining screws (7) and washers (8) to get 1-9/32 inch clearance A between pressure plate face (3) and pressure cover flange (4).

GO TO FRAME 5

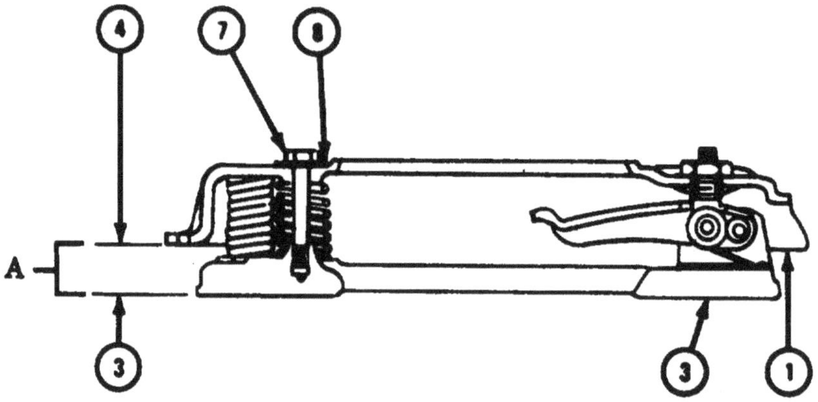

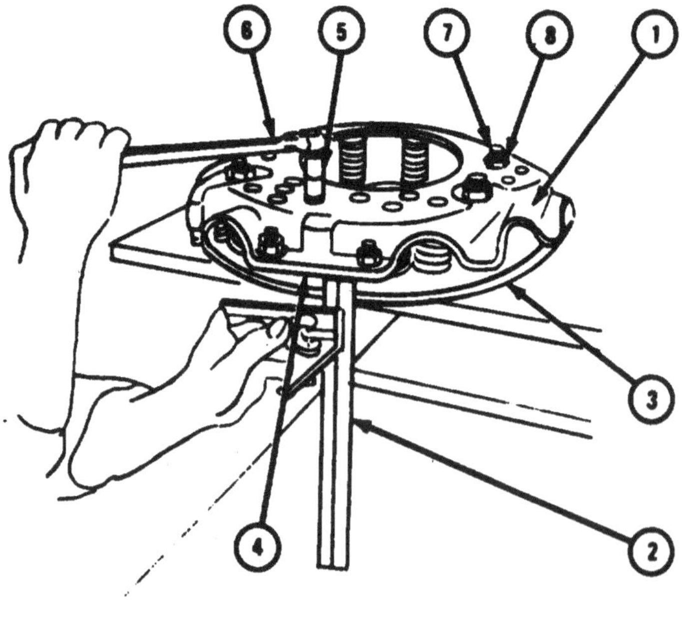

TA 105914

FRAME 5

1. Using combination square (1), check clearance A from three release levers (2) to pressure plate face (3) . Clearance should be 2-5/32 inches.
2. If clearance is not as given in step 1, take off three locknuts (4).
3. Using 1 1/16-inch socket (5) with breaker bar (6), turn three adjusting screws (7) to get clearance A of 2-5/32 inches.
4. Hold adjustment screws (7). Put on locknuts (4) .

END OF TASK

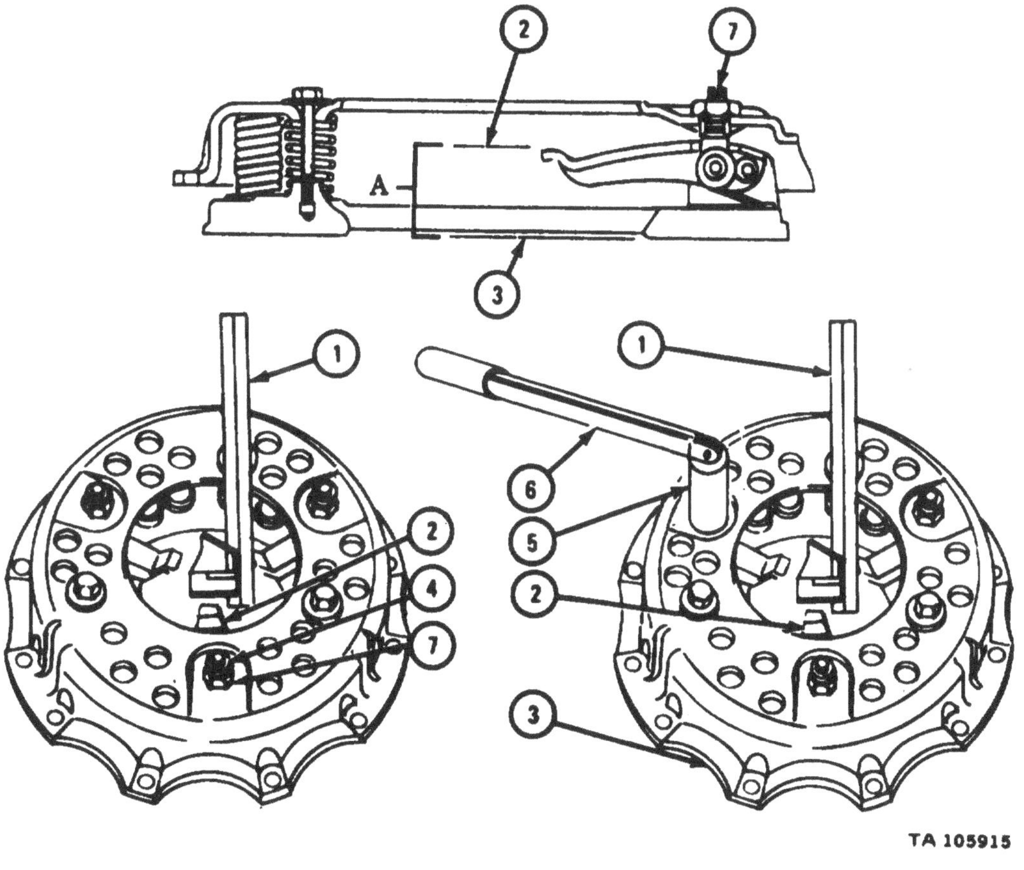

FRAME 6

NOTE

The only repair on transmission drive plates (1) and scuff plate (2) is to take off burrs.

1. Using a fine mill file, take off burrs on transmission drive plates (1) and scuff plate (2) .

END OF TASK

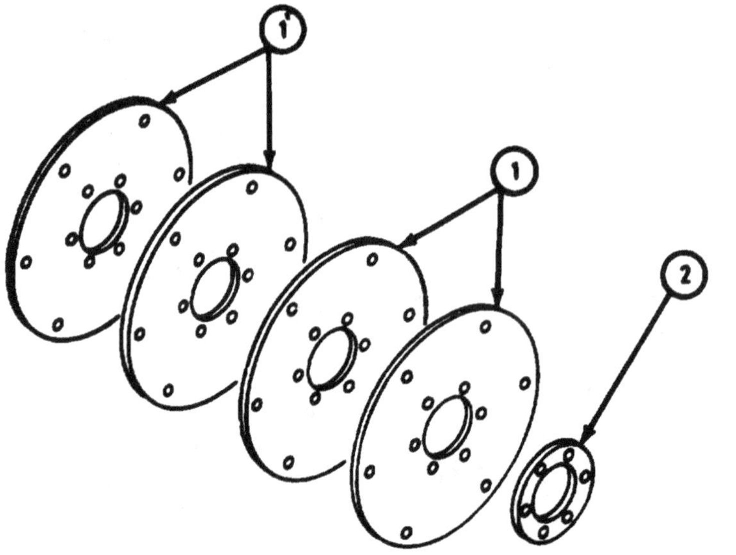

TA 121262

4-57. FUEL INJECTION PUMP DRIVE GEAR ACCESS COVER (ENGINES LD-465-1, LD-465-1C, AND LDT-465-1C).

There are no special cleaning procedures needed. Refer to cleaning procedures given in para 4-3.

b. Inspection and Repair.

FRAME 1

1. Check that five nuts (1) are not stripped, crossthreaded or damaged in any other way. If nuts are damaged, get new ones.

2. Check that access cover (2) is not cracked, bent, dented or damaged in any other way. Take out minor bends or dents. If more repair is needed, get a new access cover.

WARNING

Dry cleaning solvent is flammable. Do not use near an open flame. Keep a fire extinguisher nearby when solvent is used. Use only in well-ventilated places. Failure to do this may result in injury to personnel and damage to equipment.

3. Check that access cover (2) has no nicks, burrs or scratches on gasket surface. Take out minor nicks, burrs or scratches with a fine mill file or crocus cloth dipped in dry cleaning solvent.

END OF TASK

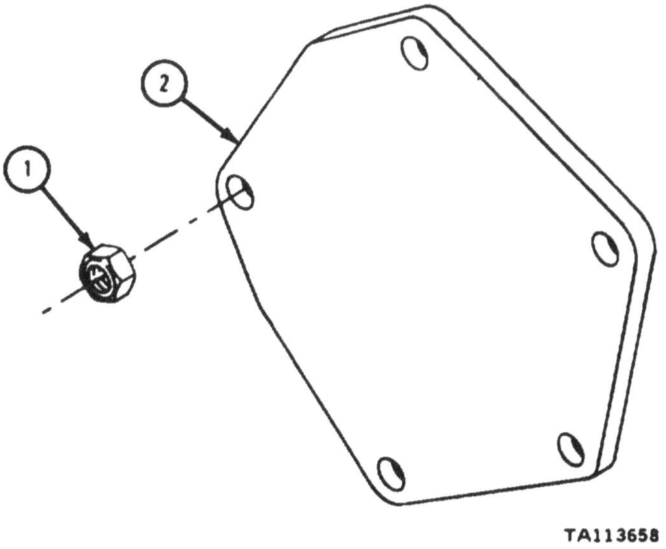

4-58. FLAME HEATER SPARK PLUG (UNCOVERED SYSTEM).

a. Cleaning and Inspection.

FRAME 1

1. Clean spark plug. Refer to TM 9-4910-389-12.
2. Check that spark plug case (1) is not cracked.
3. Check that spark plug electrodes (2) are not cracked or bent.
4. Check that threads (3) are not crossed or burred.

END OF TASK

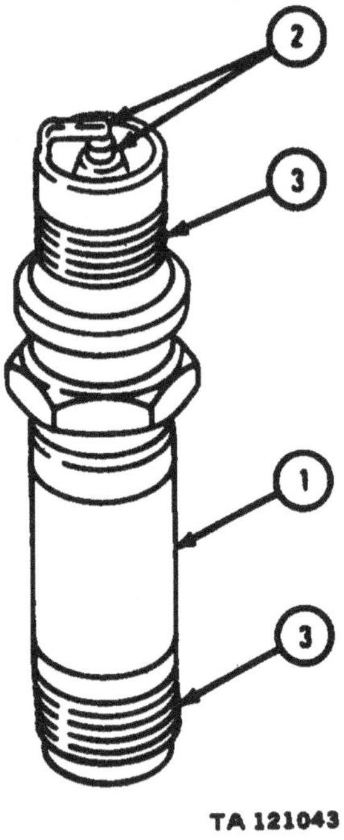

b. Repair.

FRAME 1

1. Spark plug repair is limited to setting spark plug gap.

2. Set gap between pins (1 and 2) to 0.088 to 0.093 inch with feeler gage.

END OF TASK

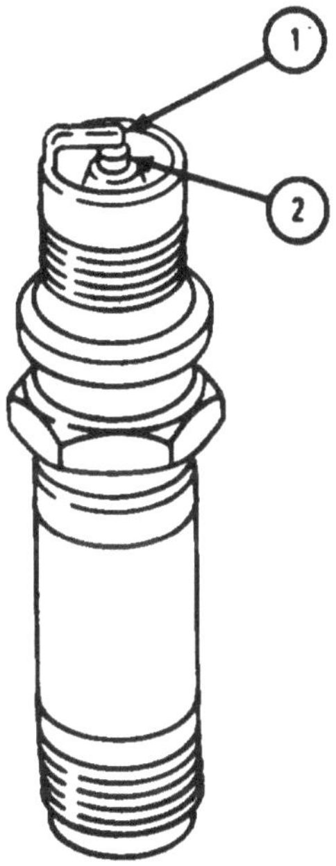

TA 121044

4-59. FLAME HEATER IGNITION UNIT (UNCOVERED SYSTEM).

a. Cleaning.

(1) Cap receptacle openings on ignition unit .

WARNING

Dry cleaning solvent is flammable. Do not use near an open flame. Keep a fire extinguisher nearby when solvent is used. Use only in well-ventilated places . Failure to do this may result in injury to personnel and damage to equipment.

CAUTION

Ignition unit should not be dipped in cleaning solvent.

(2) Clean ignition unit with cloth moistened with dry cleaning solvent.

b. Inspection and Repair .

FRAME 1

WARNING

The voltage output of ignition unit (1) is strong enough to cause a dangerous electrical shock. Do not touch any bare wires or connections during operation. Refer to decal on unit .

1. Check that ignition unit (1) has no cracks or dents. If ignition unit is cracked or dented, get a new one.
2. To check ignition unit (1), join spark plug (2) to output cable assembly (3).
3. Ground pin (4) on spark plug (2) ignition unit (1).
4. Spark plug (2) should spark between pins (4 and 5) when 10 to 12-volt direct current is applied to ignition unit (1). If spark plug does not spark, check spark plug and output cable assembly (3) .
5. Repair minor damage to ignition unit (1) by filing burrs on threads.
6. Throw away ignition unit (1) if spark plug does not spark, and get a new one.

END OF TASK

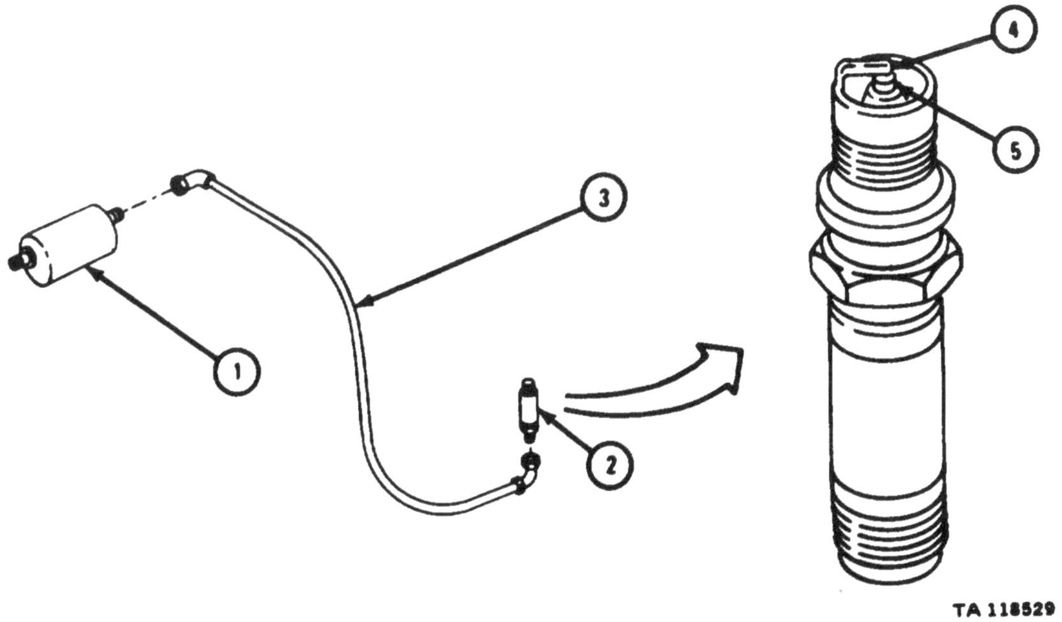

4-60. FLAME HEATER WIRING HARNESS AND ELECTRICAL LEAD ASSEMBLY (UNCOVERED SYSTEM).

a. Cleaning. Clean wiring harness and lead assembly with dry lint-free cloth.

b. Inspection and Repair.

NOTE

Repair is limited to getting anew wiring harness and electrical lead assembly.

FRAME 1

1. Check that wiring harness and electrical lead assembly (1) is not cracked or frayed.
2. Check that threads on receptacle connector (2) are not cracked, cross-threaded or burred.
3. Get new wiring harness and electrical lead assembly (1) if it is damaged.
4. Check that wiring harness and electrical lead assembly (1) has continuity.
5. Get new wiring harness and electrical lead assembly (1) if there is no continuity.

END OF TASK

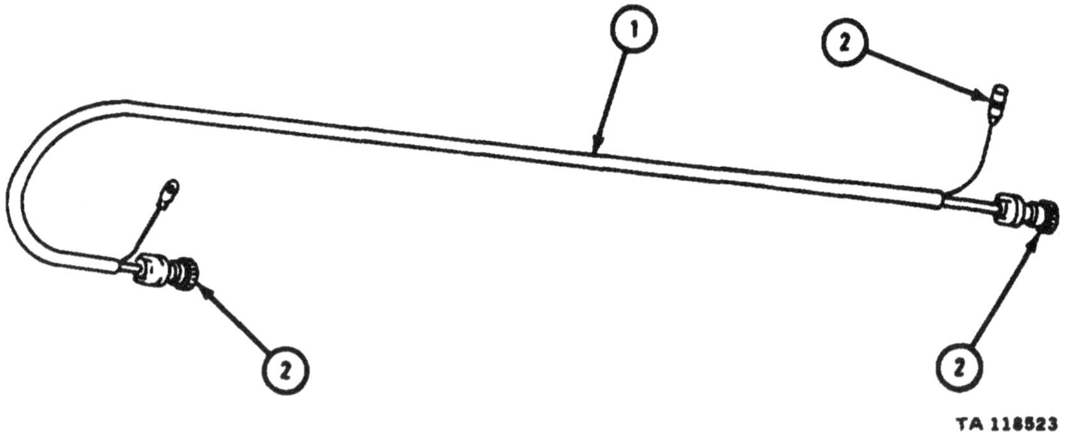

4-61. FLAME HEATER FUEL PUMP (UNCOVERED SYSTEM).

a. Cleaning .

FRAME 1

1. Cap electrical receptacle connection (1) .
2. Plug fuel inlet (2) and outlet (3) ports.

WARNING

Dry cleaning solvent is flammable. Do not use near an open flame. Keep a fire extinguisher nearby when solvent is used . Use only in well-ventilated places. Failure to do this may result in injury to personnel and damage to equipment.

CAUTION

Fuel pump should not be dipped in cleaning solvent. Electrical parts will be damaged.

3. Clean fuel pump with cloth dipped in dry cleaning solvent.

END OF TASK

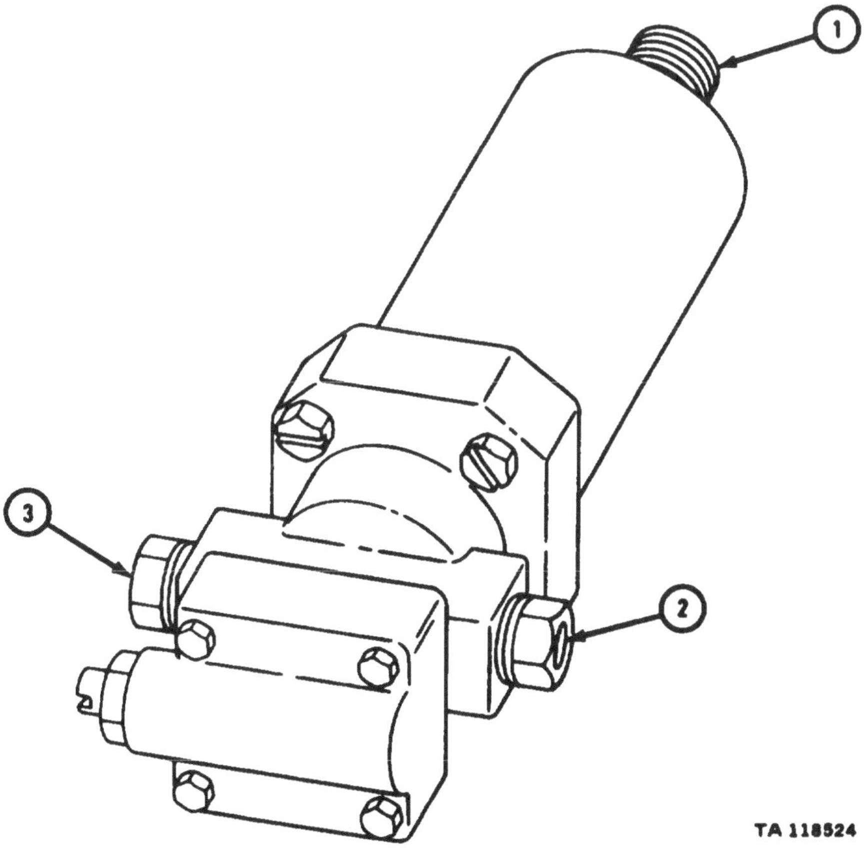

b. Inspection.

FRAME 1

1. Check that fuel pump housing (1) is not dented.
2. Check that receptacle connector (2) is not loose.
3. Check that pins in receptacle connector (2) are not bent or missing.
4. Check that threads on receptacle connector (2) are not crossthreaded or damaged.
5. Check that relief valve (3) and gear housing (4) do not show signs of leakage between housings.
6. Check that fuel inlet port (5) is not cracked.
7. Check that threads in fuel inlet port (5) are not damaged.
8. Check that fuel outlet port (6) is not cracked.
9. Check that threads in fuel outlet port (6) are not damaged.
10. Get a new fuel pump if it is cracked, damaged or shows signs of leakage.

END OF TASK

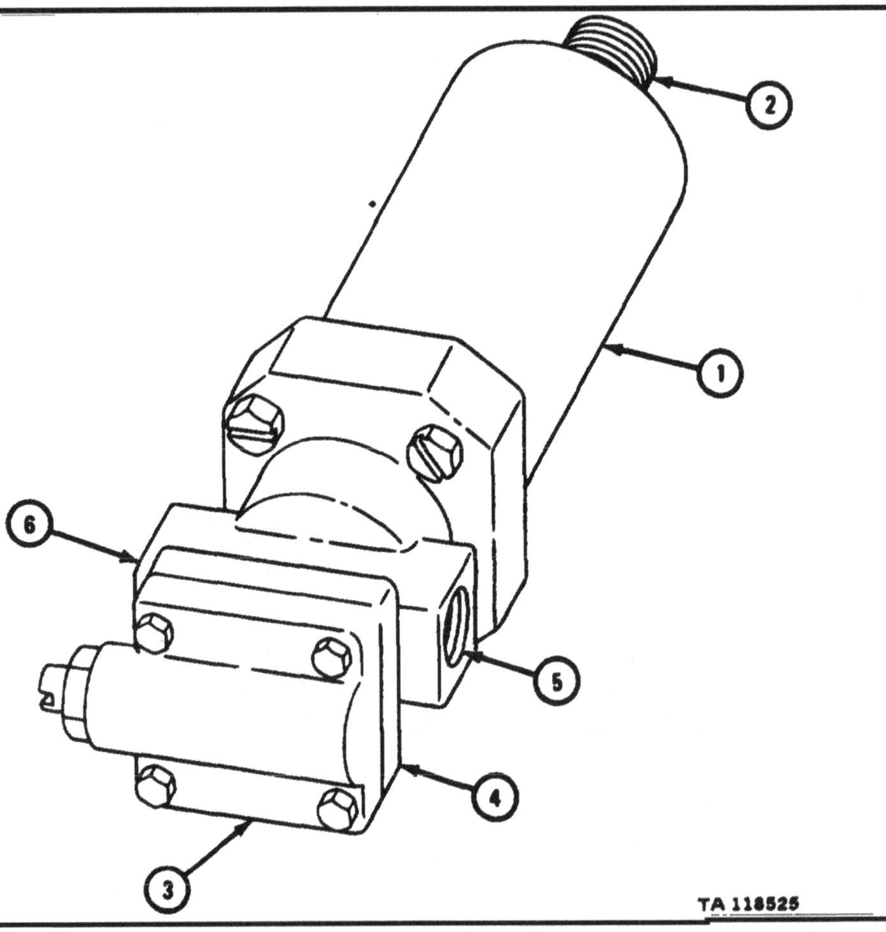

c. Repair. Repair minor damage to fuel pump by filing burrs and threads of receptacle connector.

d. Relief Valve Adjustment.

WARNING

Smoking, sparks or open flames are not allowed within 50 feet of work area during this task. Fire or explosion could occur, causing injury to personnel and damage to equipment.

NOTE

This is an alternate procedure given for reference purposes only. Standard procedure is to adjust relief valve during dynamometer test and adjustment.

FRAME 1

CAUTION

Flame heater fuel pump electrical system is negative grounded. Pin A is positive and pin B is negative. Do not switch polarity of pump during checking or replacement. Changing polarity will permanently damage pump.

1. Hook up hose (1) from supply tank (2) of diesel fuel to pump inlet (3).
2. Hook up return hose (4) from fuel pump outlet (5) back to supply tank (2).
3. Return hose must have 200 psi pressure gage (6) and adjustable valve (7) put on as shown.
4. Hook up 24-volt battery (8) with rheostat (9) to fuel pump receptacle.
5. Hook up voltmeter (10) to fuel pump.

GO TO FRAME 2

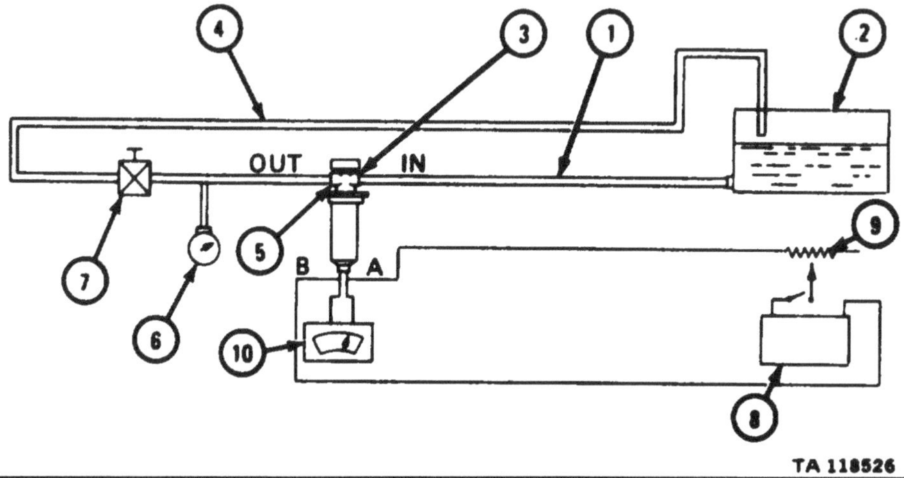

TA 118526

FRAME 2

1. Open adjustable valve (1) and start fuel pump.
2. Keep valve (1) open until system is purged of air.
3. Close valve (1) .
4. Turn rheostat (2) until voltmeter (3) reads 24 volts.
5. Pressure gage (4) should read 90 to 100 psi.
6. To set pressure, if necessary, loosen check valve locknut (5)' with wrench . Turn relief valve adjusting screw (6) with screwdriver until pressure gage reads 90 to 100 psi. Hold relief valve adjusting screw with screwdriver and tighten check valve locknut with wrench.

END OF TASK

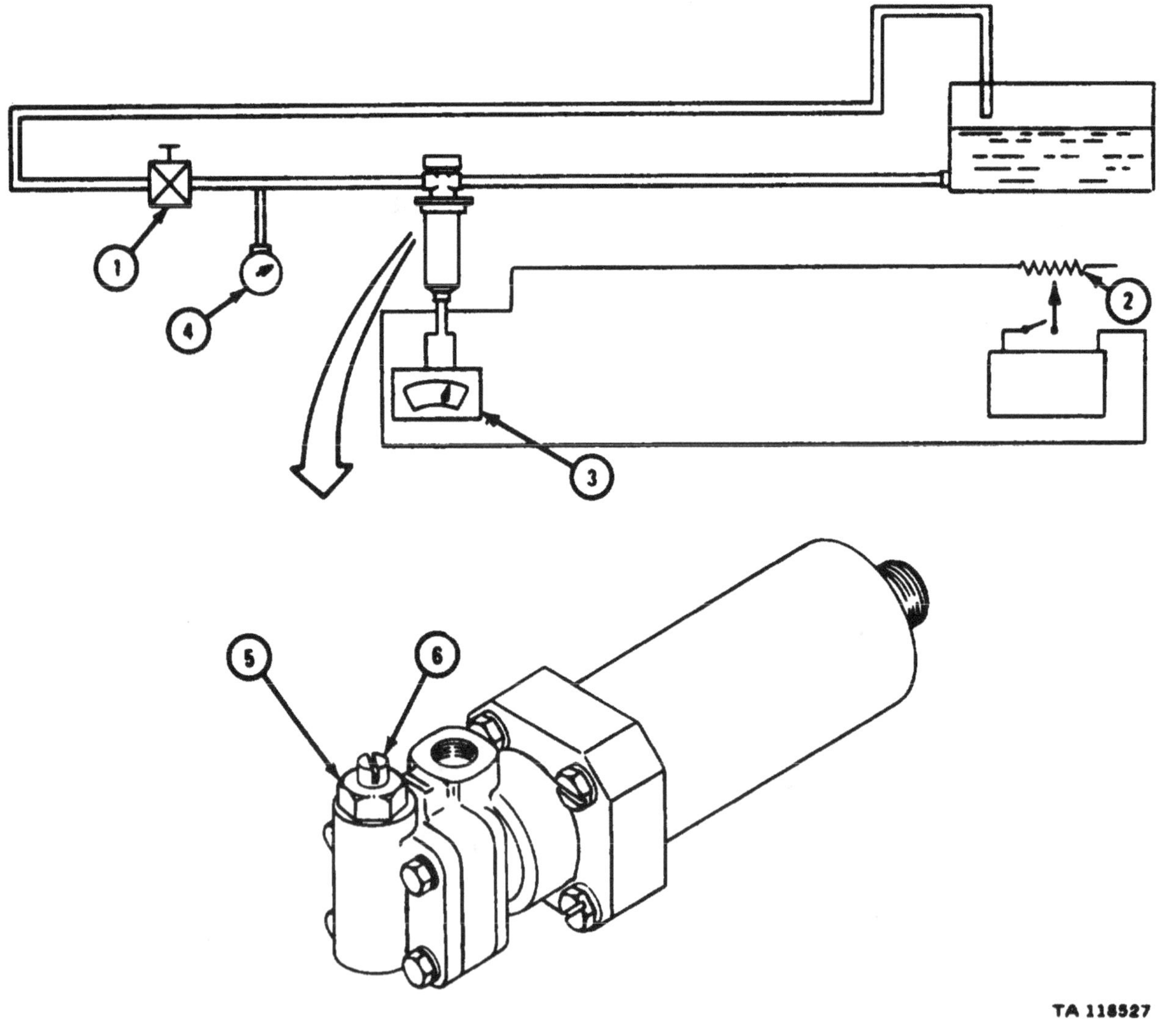

e. Flow Test .

WARNING

Smoking, sparks or open flames are not allowed within 50 feet of work area during this task. Fire or explosion could occur, causing injury to personnel and damage to equipment.

NOTE

This is an alternate procedure included for reference purposes only. Standard procedure is to test fuel pump during dynamometer test and adjustment.

FRAME 1

1. Set rheostat (1) until voltmeter (2) reads 10 volts.
2. Set adjustable valve (3) until pressure gage (4) read 60 psi.
3. Pump (5) should flow a minimum of one pint in 2 minutes and 45 seconds.
4. Turn off rheostat (1) and open adjustable valve (3) to bleed off pressure.
5. Take out pump (5) from test setup.
6. Take apart test setup.

END OF TASK

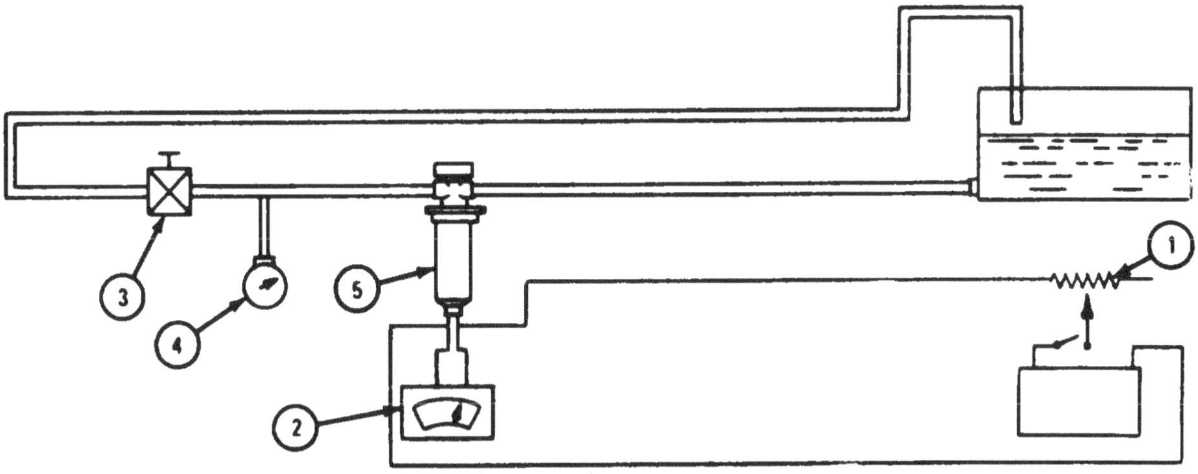

TA 118528

4-62. FLAME HEATER NOZZLE AND VALVE ASSEMBLY (UNCOVERED SYSTEM).

a. Cleaning and Inspection.

FRAME 1

1. Plug nozzle and valve assembly fuel inlet (1) and outlet (2) with pipe plugs.

WARNING

Dry cleaning solvent is flammable. Do not use near an open flame. Keep a fire extinguisher nearby when solvent is used. Use only in well-ventilated places. Failure to do this may result in injury to personnel and damage to equipment.

2. Clean nozzle and valve assembly (3) with dry cleaning solvent. Take off heavy carbon deposits with a stiff brush.
3. Check that fuel inlet (1) and fuel outlet (2) threads are not crossthreaded.
4. Check that nozzle and valve assembly (3) is not cracked or damaged. Get a new nozzle and valve assembly if it is cracked or damaged.

END OF TASK

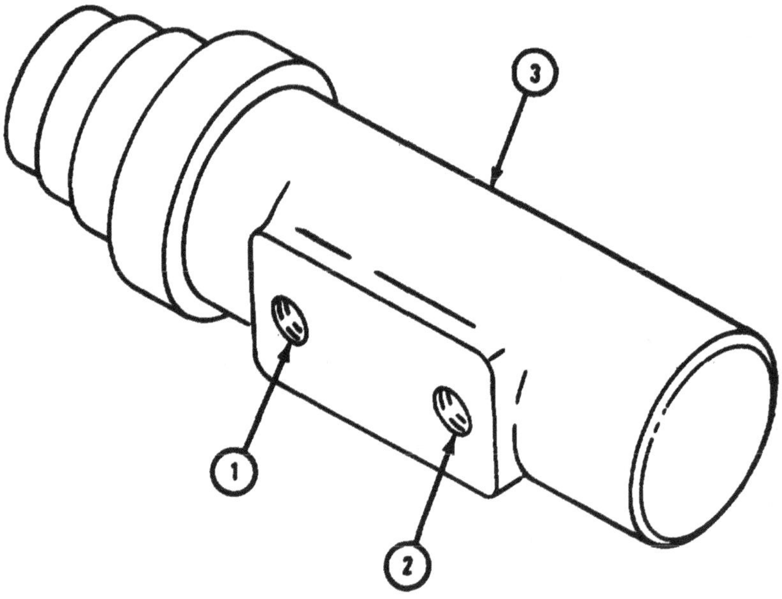

b. Repair.

FRAME 1

1. Repair minor thread damage to fuel inlet (1) and fuel outlet (2) on nozzle and valve assembly (3) with a tap.

END OF TASK

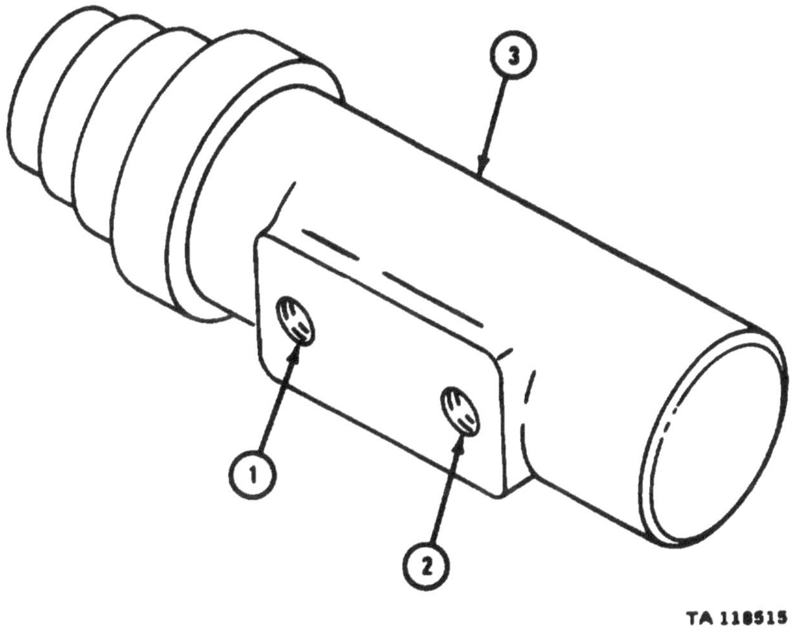

c. Spray Pattern Test.

WARNING

Smoking, sparks or open flames are not allowed within 50 feet of work area during this task. Fire or explosion could occur, causing injury to personnel and damage to equipment.

FRAME 1

1. Take out old fuel line filters (1) from nozzle and check valve assembly (2) if there are any.
2. Put new fuel line filters (1) in fuel inlet (3) and fuel outlet (4).

GO TO FRAME 2

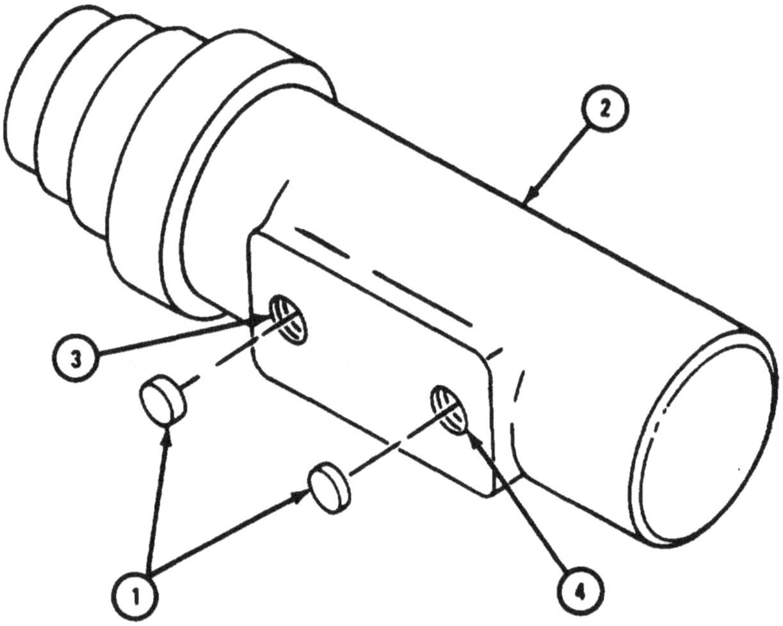

TA 118516

FRAME 2

NOTE

This is an alternate procedure given for reference purposes only . Standard procedure is to test flame heater nozzle during dynamometer test and adjustment.

1. Hook up hose (1) from supply tank (2) of diesel fuel to pump inlet (3).
2. Hook up hose (4) from pump outlet (5) to pressure gage (6).
3. Hook up adjustable valve (7) from pressure gage (6) .
4. Hook UP hose (8) from adjustable valve (7) to nozzle and check valve assembly (9) and inlet opening (10) .

GO TO FRAME 3

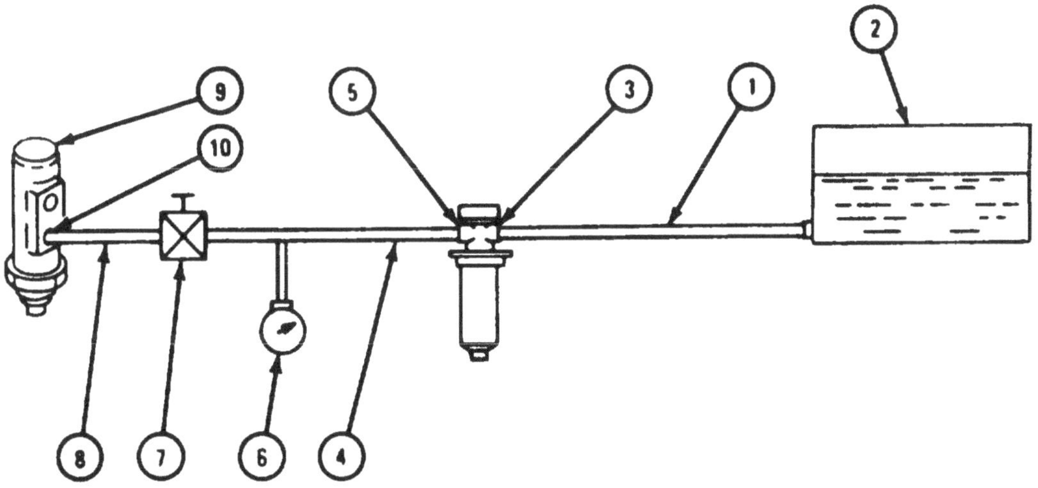

FRAME 3

CAUTION

Flame heater fuel pump electrical system is negative grounded. Pin A is positive and pin B is negative. Do not switch polarity of pump during checking or replacement. Changing polarity will permanently damage pump.

1. Hook up 24-volt battery (1) with rheostat (2) to fuel pump receptacle (3).
2. Hook up voltmeter (4) to fuel pump receptacle (3).

GO TO FRAME 4

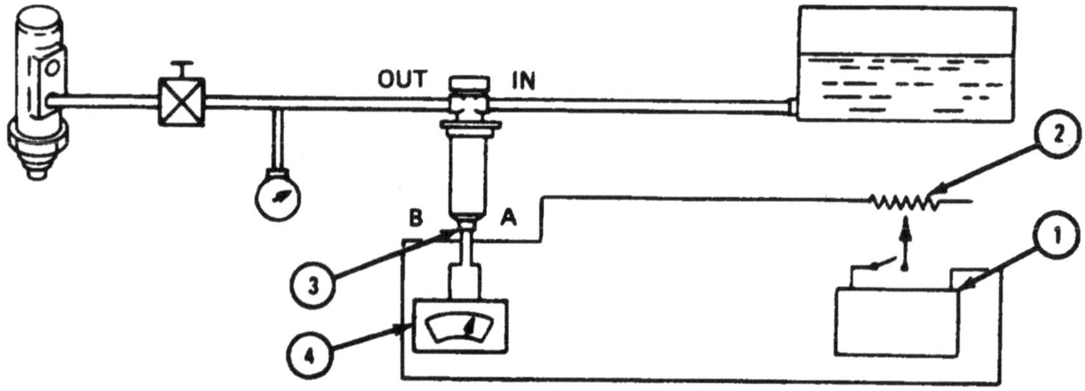

TA 118518

FRAME 4

1. Turn rheostat (1) until voltmeter (2) reads 24 volts.
2. Pressure gage (3) should read 90 to 100 psi.
3. Put bucket (4) under nozzle and check valve assembly (5) to catch discharged fuel from nozzle and check valve assembly (5).
4. Turn adjustable valve (6) to give fuel flow to nozzle and check valve assembly (5).
5. Check that spray pattern (7) is a uniform, fine spray cone with an angle of 60° to 80°.
6. Get a new nozzle and check valve assembly (5) if there is no spray pattern or if spray pattern is coarse or out of shape.
7. Turn off adjustable valve (6).
8. Turn off rheostat (1).
9. Turn on adjustable valve (6) to bleed off pressure.

END OF TASK

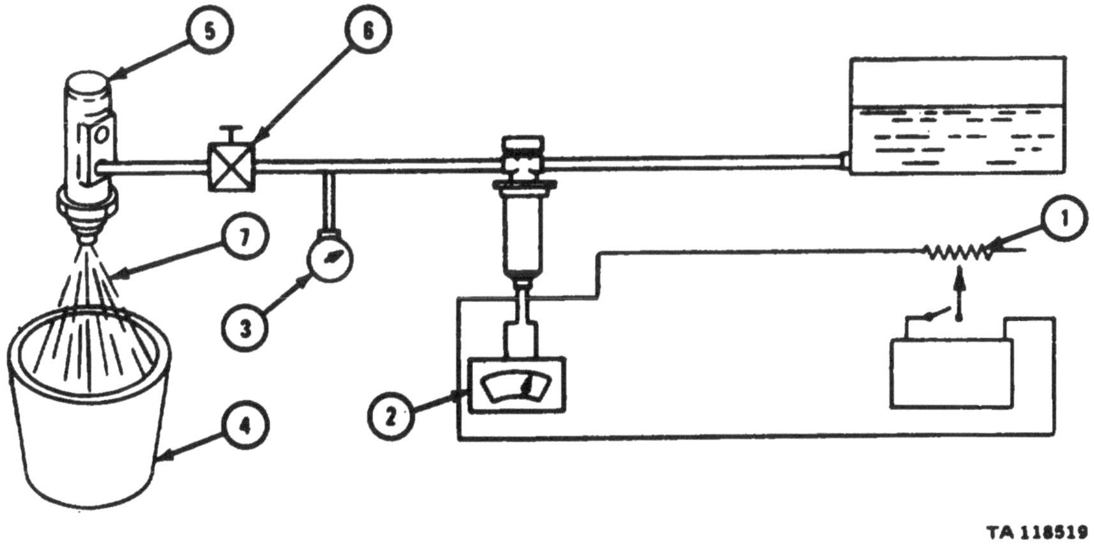

d. Leak Test.

WARNING

Smoking, sparks or open flames are not allowed within 50 feet of work area during this task. Fire or explosion could occur, causing injury to personnel and damage to equipment.

NOTE

This is an alternate procedure used for reference purposes only. Standard procedure is to test for leaks during dynamometer test and adjustment.

FRAME 1

1. Hook up hose (1) from supply tank (2) of diesel fuel to pump inlet (3).
2. Hook up hose (4) from pump outlet (5) to adjustable valve (6).
3. Hook up pressure gage (7) from adjustable valve (6).
4. Hook up hose (8) from pressure gage (7) to nozzle and check valve assembly (9), inlet (10), and outlet (11).

GO TO FRAME 2

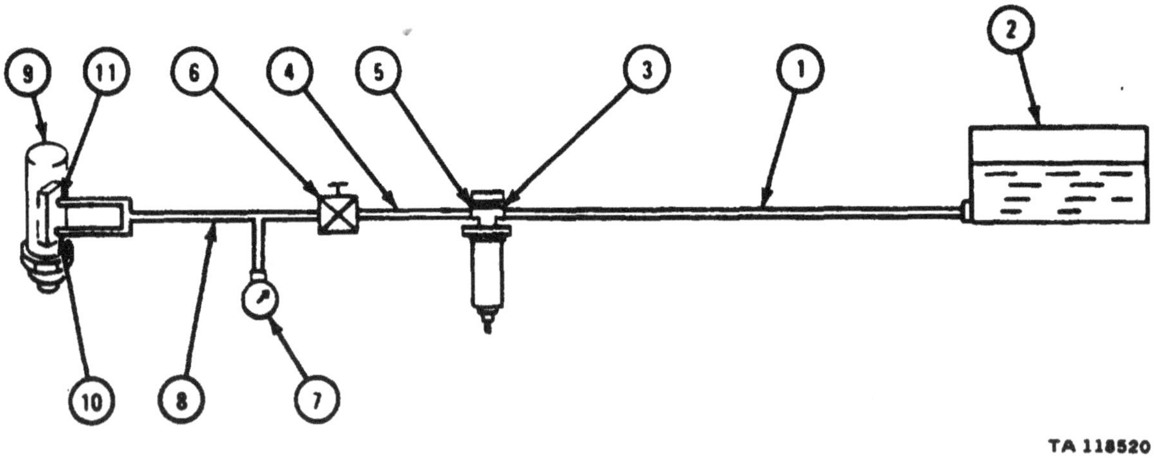

FRAME 2

CAUTION

Flame heater fuel pump electrical system is negatively grounded. Pin A is positive and pin B is negative. Do not switch polarity of pump during checking or replacement. Changing polarity will permanently damage pump.

1. Hook up 24-volt battery (1) with rheostat (2) to fuel pump receptacle (3).
2. Hook up voltmeter (4) to fuel pump receptacle (3).

GO TO FRAME 3

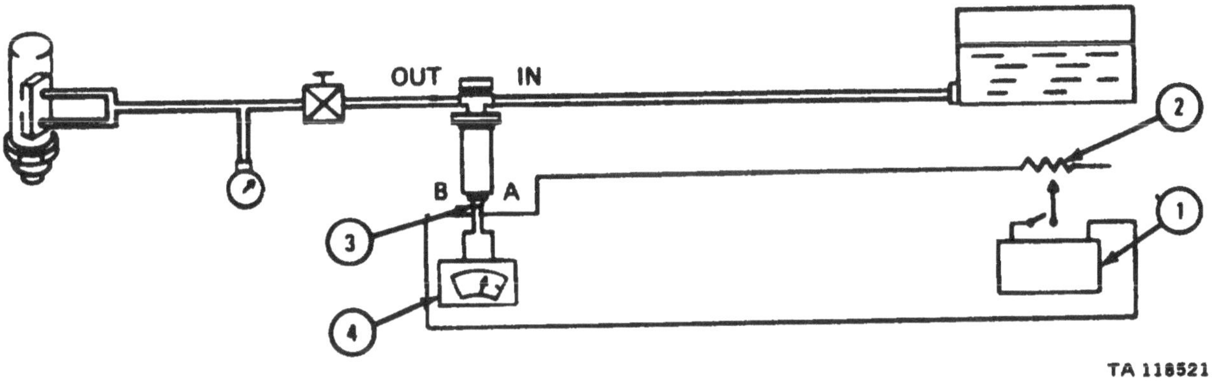

FRAME 3

1. Turn on rheostat (1) until voltmeter (2) reads 24 volts.
2. Turn on adjustable valve (3) until pressure gage (4) reads 5 to 10 psi.
3. Check that nozzle and check valve assembly (5) inlet (6) and outlet (7) do not leak. Get a new nozzle and check valve assembly if inlet or outlet leak.
4. Turn off adjustable valve (3) .
5. Turn off rheostat (1) .
6. Take off nozzle and check valve assembly (5).
7. Turn on adjustable valve (3) to bleed off pressure.

END OF TASK

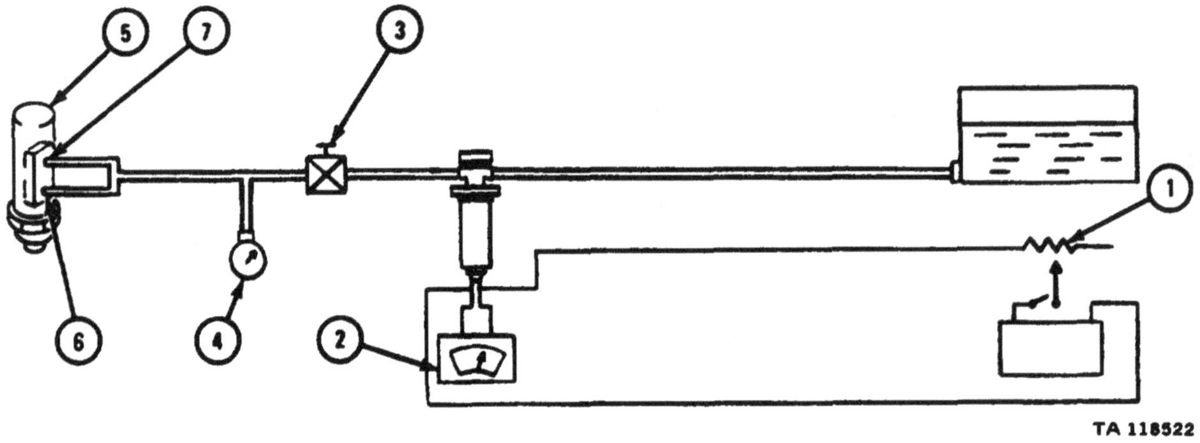

4-63. FLAME HEATER TUBES, CLAMPS, AND FITTINGS (UNCOVERED SYSTEM).

a. Cleaning.

FRAME 1

1. Clean plastic tubing (1) with soap and water.

WARNING

Dry cleaning solvent is flammable. Do not use near an open flame. Keep a fire extinguisher nearby when solvent is used. Use only in well-ventilated places. Failure to do this may result in injury to personnel and damage to equipment.

2. Clean elbows (2) with dry cleaning solvent.
3. Clean connectors (3) with dry cleaning solvent.
4. Clean two loop clamps (4) with soap and water.

END OF TASK

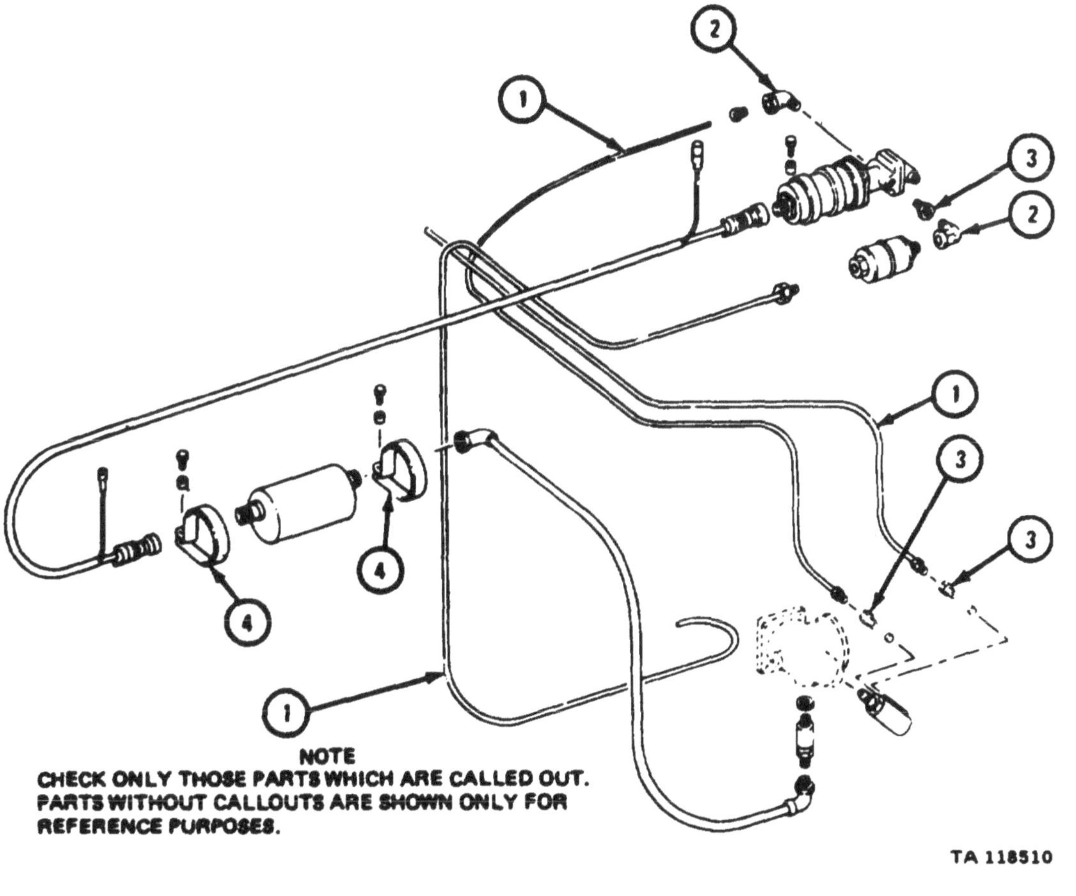
1
2
3
4
NOTE
CHECK ONLY THOSE PARTS WHICH ARE CALLED OUT.
PARTS WITHOUT CALLOUTS ARE SHOWN ONLY FOR
REFERENCE PURPOSES.
TA 118510

b. Inspection and Repair.

FRAME 1

1. Check that plastic tubing (1) is not cracked, kinked or brittle.
2. Check that plastic tubing (1) has no signs of leakage around tube nuts (2).
3. Check that tube nuts (2) are not stripped or damaged.
4. Throw away plastic tubing (1) if cracked, kinked or brittle.
5. Throw away plastic tubing (1) if it shows signs of leakage around tube nuts (2).
6. Throw away plastic tubing (1) if tube nuts (2) are stripped or damaged.

GO TO FRAME 2

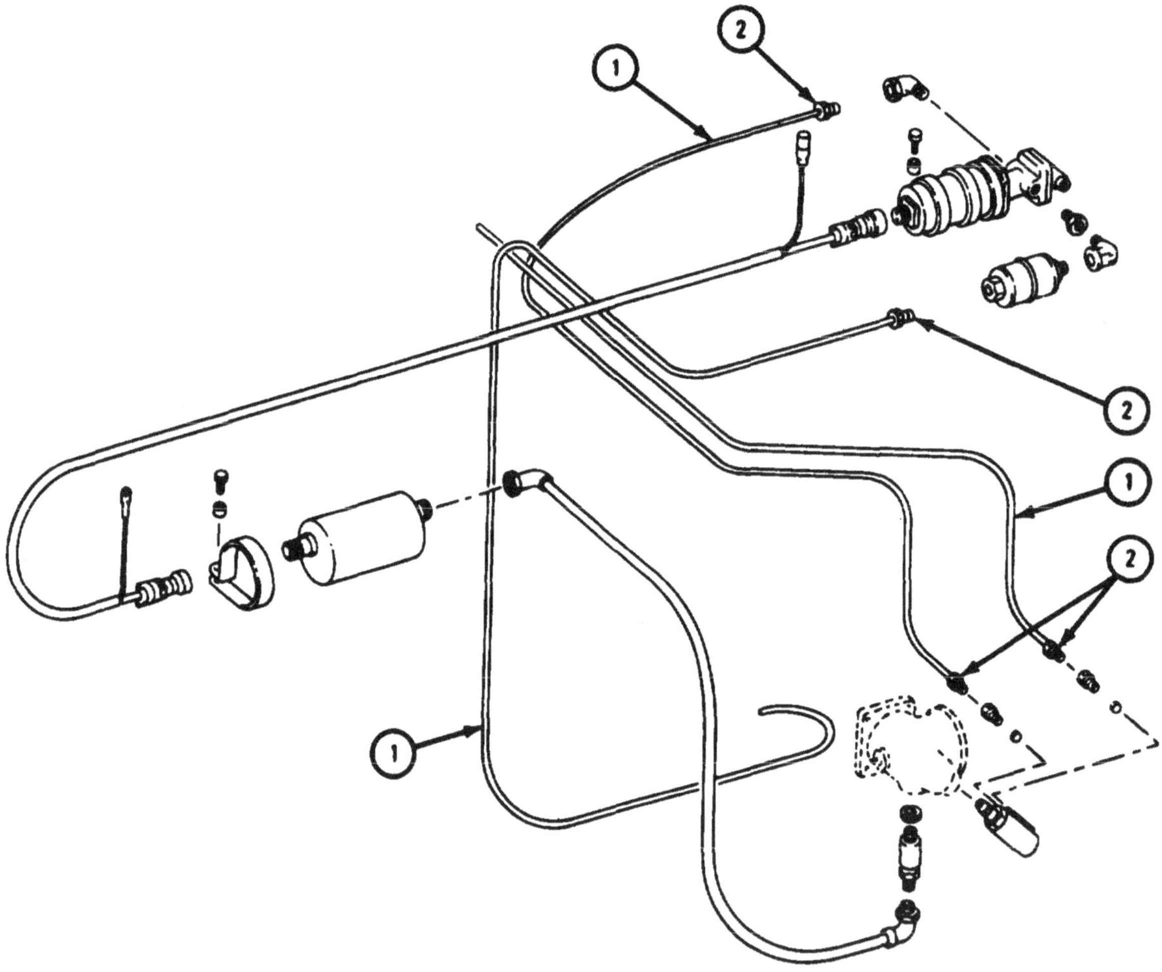

NOTE: CHECK ONLY THOSE PARTS WHICH ARE CALLED OUT IN THIS FRAME. PARTS WITHOUT CALLOUTS ARE SHOWN ONLY FOR REFERENCE PURPOSES OR ARE CHECKED IN ANOTHER FRAME.

TA 118511

FRAME 2

1. Check that 1/4-inch plastic tubing (1) has not shrunk at tube nut ferrules (2).
2. Throw away 1/4-inch plastic tubing (1) if No. 28 drill shank (3) cannot be put in tubing (1) opening .
3. Check that 1/8-inch plastic tubing (4) has not shrunk at tube nut ferrules (5).
4. Get rid of 1/8-inch plastic tubing (4) if a 1/32-inch drill shank (6) cannot be put in tubing opening.

GO TO FRAME 3

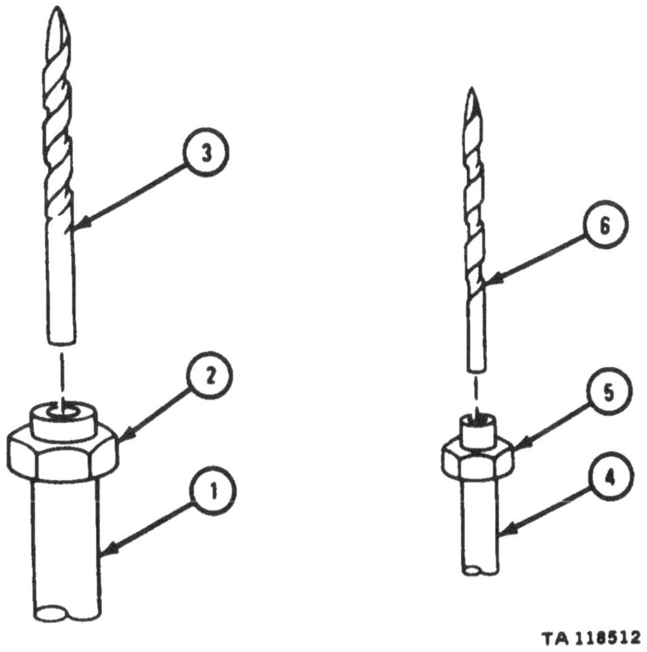

FRAME 3

1. Check that elbows (1) and connectors (2) are not cracked and that threads are not damaged. Throw away elbows and connectors that are cracked or have damaged threads.
2. Check that loop clamp (3) is not cracked or bent and that coating is not worn. Throw away loop clamp (3) if cracked or coating is worn.
3. Straighten minor bends in two loop clamps (3).

END OF TASK

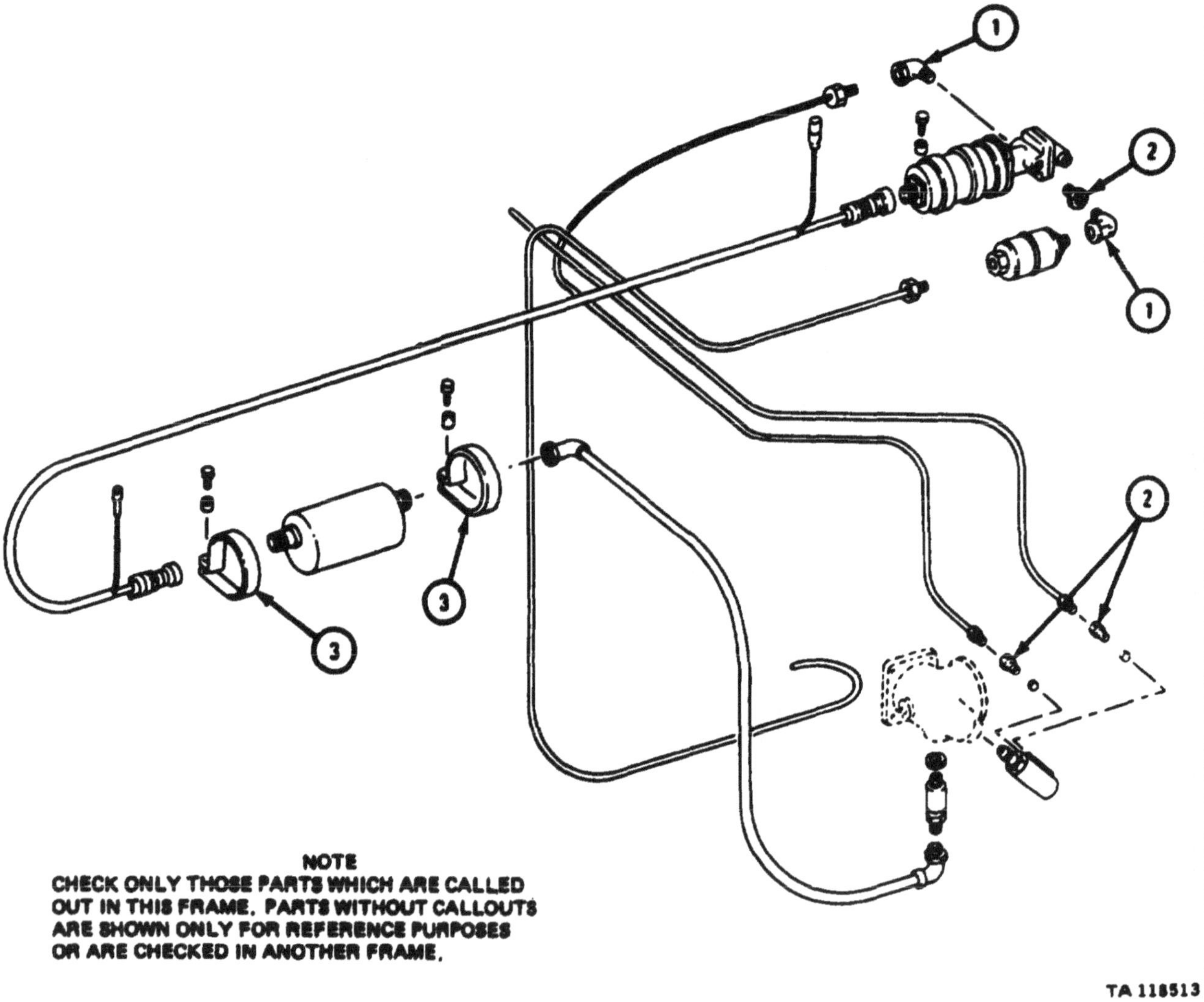
1
2
1
2
3
3
3
NOTE
CHECK ONLY THOSE PARTS WHICH ARE CALLED OUT IN THIS FRAME. PARTS WITHOUT CALLOUTS ARE SHOWN ONLY FOR REFERENCE PURPOSES OR ARE CHECKED IN ANOTHER FRAME.
TA 118513

4-64. FLAME HEATER SPARK PLUG (COVERED SYSTEM).

a. Cleaning and Inspection.

FRAME 1

1. Clean spark plug . Refer to TM 9-4910-389-12.
2. Check that spark plug case (1) is not cracked.
3. Check that spark plug electrodes (2) are not cracked or bent.
4. Check that threads (3) are not crossthreaded or burred .

END OF TASK

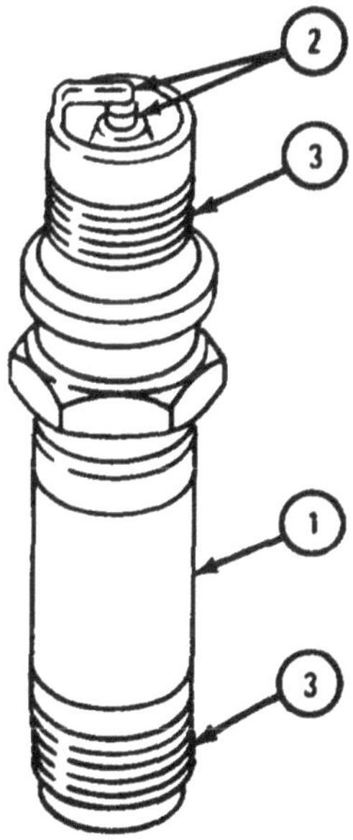

TA 121045

b. Repair.

FRAME 1

NOTE

Spark plug repair is limited to setting sparkplug gap.

1. Set gap between pins (1 and 2) to 0.088 to 0.093 inch with feeler gage.

END OF TASK

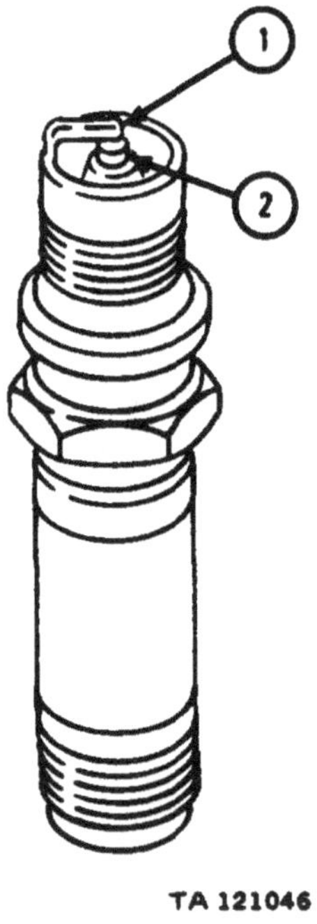

4-65. FLAME HEATER IGNITION UNIT (COVERED SYSTEM).

a. Cleaning.

(1) Cap receptacle openings on ignition unit.

WARNING

Dry cleaning solvent inflammable. Do not use near an open flame. Keep a fire extinguisher nearby when solvent is used. Use only in well-ventilated places. Failure to do this may result in injury to personnel and damage to equipment.

CAUTION

Ignition unit should not be dipped in cleaning solvent. Electrical parts will be damaged.

(2) Clean ignition unit with rag moistened with dry cleaning solvent.

b. Inspection.

FRAME 1

WARNING

Voltage output of ignition unit (1) is strong enough to cause a dangerous electrical shock. Do not touch any bare wires or connections during operation.

1. Check that ignition unit (1) has no cracks or dents. If ignition unit is cracked or dented, get a new one.
2. To check ignition unit (1), join spark plug (2) to output cable assembly (3).
3* Ground pin (4) on spark plug (2) to ignition unit (1).
4. Spark plug (2) should spark between pins (4 and 5) when 10 to 12 volts dc is applied to ignition unit (1). If spark plug does not spark, check spark plug and output cable assembly (3).
5. Do step 4 again. Throw away ignition unit (1) if spark plug does not spark, and get a new one.

END OF TASK

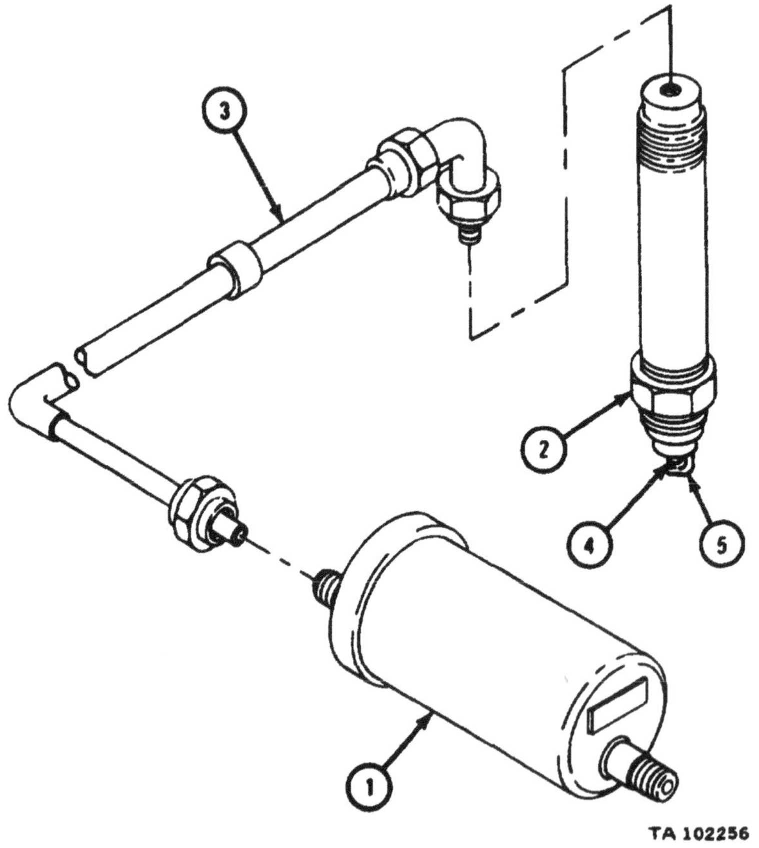

c. Repair.

(1) Repair minor damage to ignition unit by filing burrs on threads.

(2) If damage cannot be repaired, throw unit away and get a new one.

4-66. FLAME HEATER FUEL PUMP (COVERED SYSTEM).

a. Cleaning.

FRAME 1

1. Cap electrical receptacle connection (1).
2. Plug fuel inlet (2) and outlet (3) ports.

WARNING

Dry cleaning solvent is flammable. Do not use near an open flame. Keep a fire extinguisher nearby when solvent is used. Use only in well-ventilated places. Failure to do this may result in injury to personnel and damage to equipment.

CAUTION

Fuel pump should not be dipped in cleaning solvent. Electrical parts can be damaged.

3. Clean fuel pump with cloth dipped in dry cleaning solvent.

END OF TASK

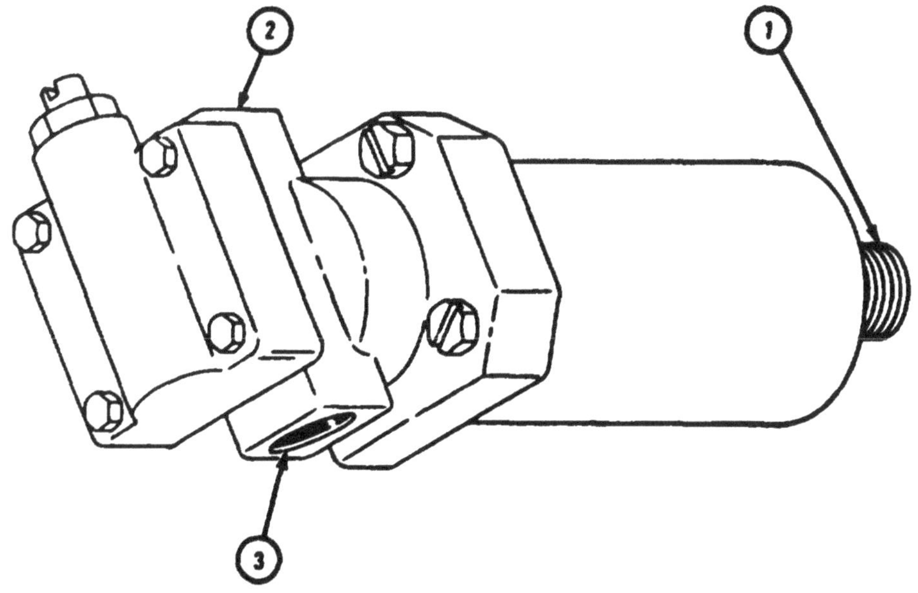

b. Inspection.

FRAME 1

1. Check tha t fuel pump housing (1) is not dented.
2. Check that receptacle connector (2) is not loose .
3. Check that pins in receptacle connector (2) are not bent or missing.
4. Check tha t threads on receptacle connector (2) are not crossthreaded o r damaged.
5. Check tha t relief valve (3) and gear housing (4) do not show signs of leakage. between housings .
6. Check that fuel inlet port (5) is not cracked.
7. Check that threads in fuel inlet port (5) are not damaged.
8. Check that fuel outlet port (6) is not cracked.
9. Check that threads in fuel outlet port (6) are not damaged.
10. Get a new pump if ports are cracked or damaged or if it shows signs of leakage.

END OF TASK

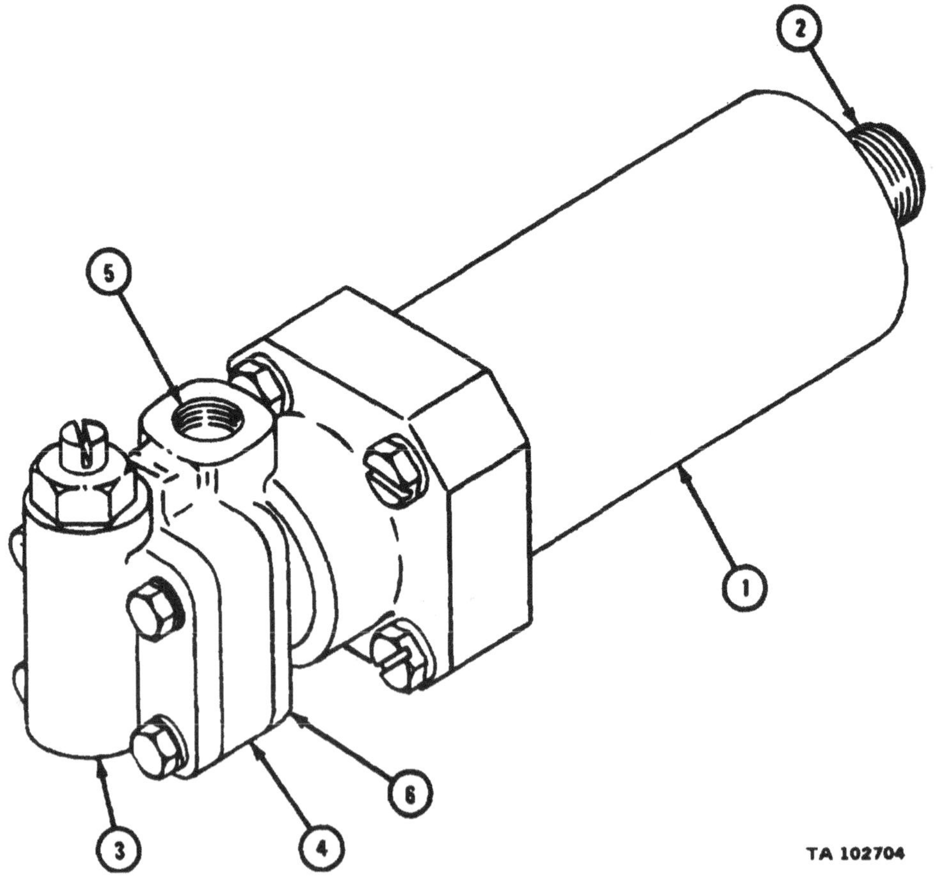

c. Repair. Repair minor damage to fuel pump by filing burrs on threads of receptacle connector .

d. Relief Valve Adjustment.

WARNING

Smoking, sparks or open flames are not allowed within 50 feet of work area during this task. Fire or explosion could occur? causing injur y to personnel and damage to equipment.

NOTE

This is an alternate procedure given for reference purposes only . Standard procedure is to adjust relief valve during dynamometer test and adjustment.

FRAME 1

CAUTION

Flame heater fuel pump electrical system is negatively grounded. Pin A is positive and pin B is negative. Do not switch polarity of pump during checking or replacement. Changing polarity will permanently damage pump.

1. Hook up hose (1) from supply (2) of diesel fuel to pump inlet (3).
2. Hook up return hose (4) from fuel pump outlet (5) back to supply tank (2).
3. Return hose must be 200 psi on pressure gage (6) and adjustable valve (7) must be put on as shown.
4. Hook up 24-volt battery (8) with rheostat (9) to fuel pump receptacle.
5. Hook up voltmeter (10) to fuel pump.

GO TO FRAME 2

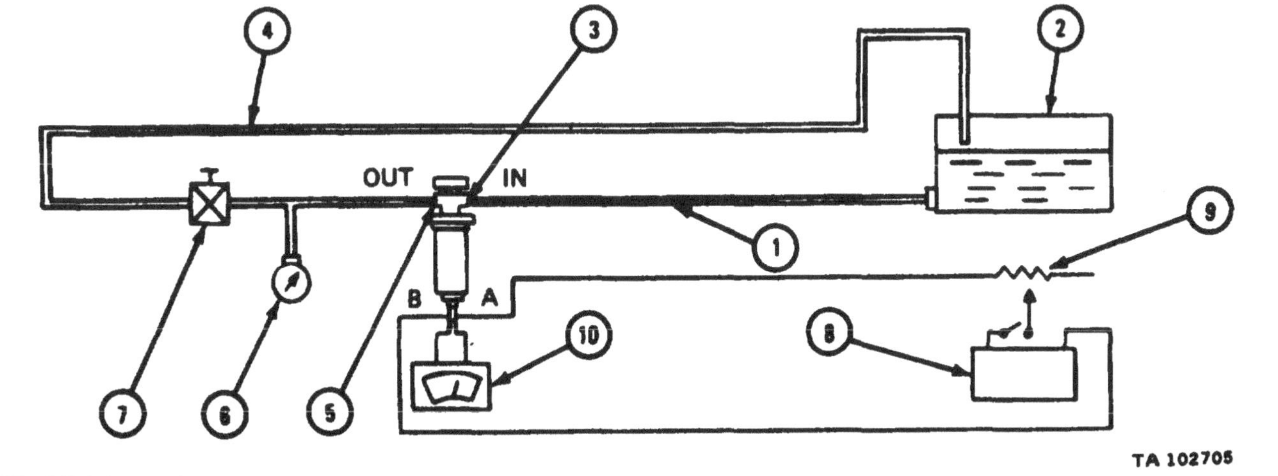

FRAME 2

1. Open adjustable valve (1) and start fuel pump.
2. Keep valve (1) open until system is purged of air.
3. Close valve (1) .
4. Turn rheostat (2) until voltmeter (3) reads 24 volts. Pressure gage (4) should read 90 to 100 psi.
5. To set pressure, if necessary, loosen check valve locknut (5). Turn relief valve adjusting screw (6) until pressure gage (4) reads 90 to 100 psi. Hold relief valve adjusting screw with screwdriver and tighten check valve locknut.

END OF TASK

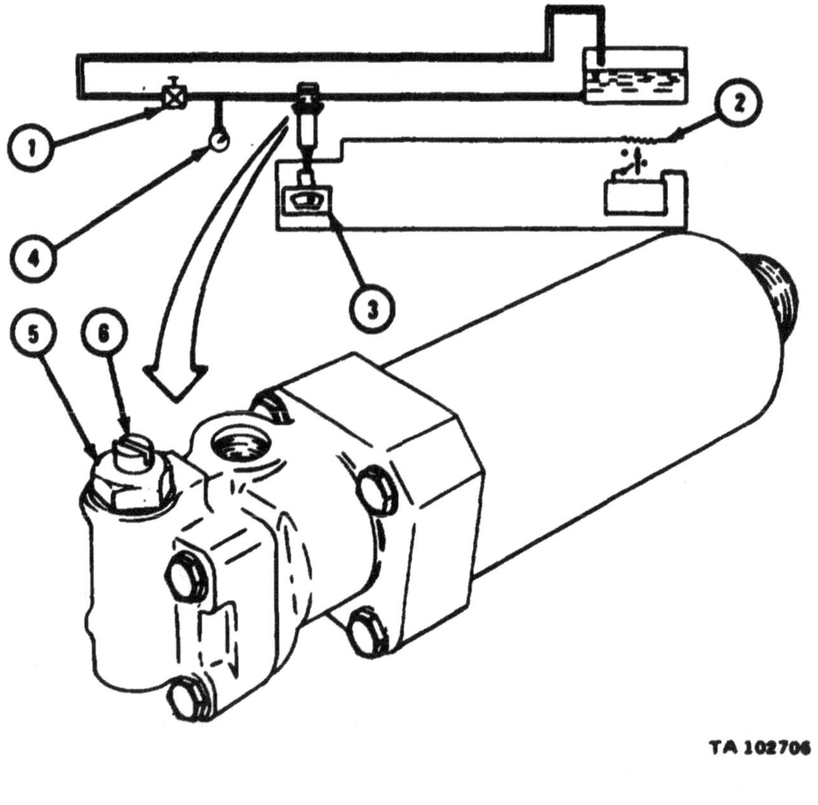

e. Flow test.

WARNING

Smoking, sparks or open flames are not allowed within 50 feet of work area during this task. Fire or explosion could occur, causing injury to personnel and damage to equipment.

NOTE

This is an alternate procedure included for reference purposes only. Standard procedure is to test fuel pump during dynamometer test and adjustment.

FRAME 1

1. Set rheostat (1) until voltmeter (2) reads 10 volts.
2. Set adjustable valve (3) until pressure gage (4) reads 60 psi.
3. Pump (5) should flow a minimum of one pint in 2 minutes and 45 seconds.
4. Turn off rheostat (1) and open adjustable valve (3) to bleed off pressure.
5. Take out pump (5) from test setup.
6. Take apart test setup.

END OF TASK

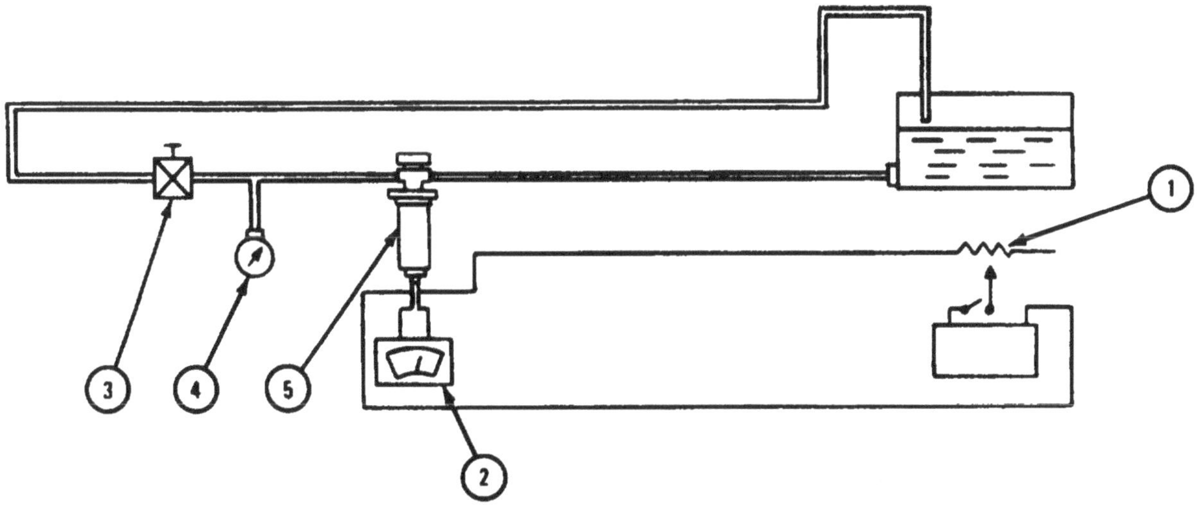

TA 102707

4-67. FLAME HEATER WIRING HARNESS AND ELECTRICAL LEAD ASSEMBLY (COVERED SYSTEM).

a. Cleaning. Clean wiring harness end lead assembly with a dry lint-free cloth.

b. Inspection and Repair.

FRAME 1

1. Check that wiring harness and electrical lead assembly (1) is not cracked or frayed.
2. Check that threads on receptacle connector (2) are not cracked, crossthreaded or burred.
3. Get a new wiring harness and electrical lead assembly (1) if it is cracked or frayed.
4. Check that wiring harness and electrical lead assembly (1) has continuity. Get a new wiring harness and electrical lead assembly if there is no continuity.

END OF TASK

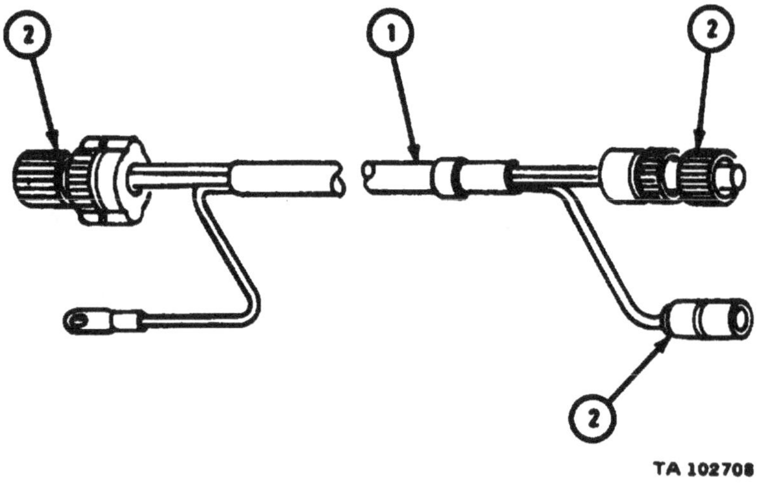

4-68. FLAME HEATER NOZZLE AND VALVE ASSEMBLY (COVERED SYSTEM).

a. Cleaning and Inspection.

FRAME 1

1. Plug fuel inlet (1) and outlet (2) openings with pipe plugs.

WARNING

Dry cleaning solvent is flammable. Do not use near an open flame. Keep a fire extinguisher nearby when solvent is used. Use only in well-ventilated places. Failure to do this may result in injury to personnel and damage to equipment.

2. Clean nozzle and valve assembly (3) with dry cleaning solvent. Take off heavy carbon deposits with a stiff brush.
3. Check that nozzle and valve assembly fuel inlet (1) and fuel outlet (2) threads are not crossthreaded.
4. Check that nozzle and valve assembly (3) is not cracked or damaged. Get a new nozzle and valve assembly if it is cracked or damaged.

END OF TASK

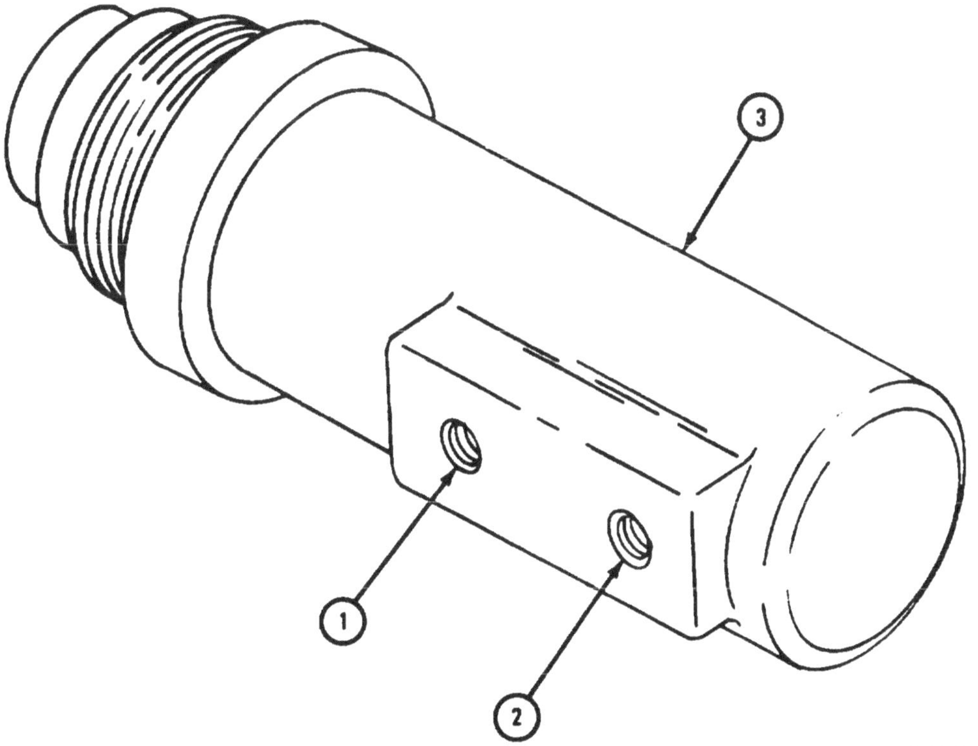

b. Repair.

FRAME 1

1. Repair minor thread damage to fuel inlet (1) and fuel outlet (2) on nozzle and valve assembly (3) with a tap.

END OF TASK

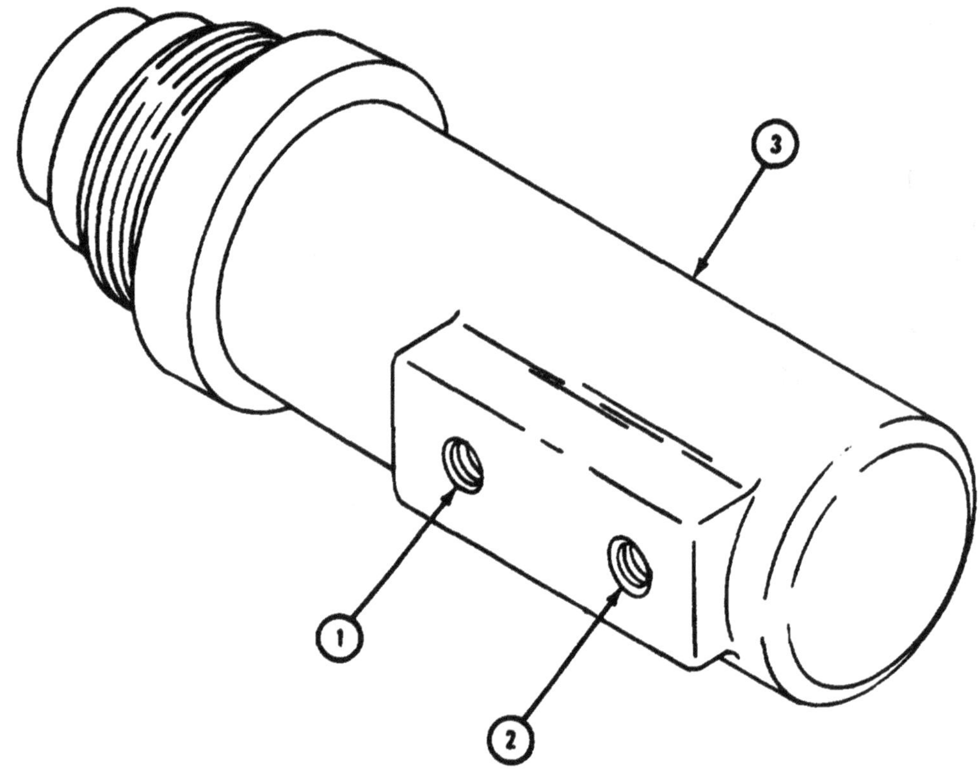

TA 102710

c. Spray Pattern Test.

WARNING

Smoking, sparks or open flames are not allowed within 50 feet of work area during this task. Fire or explosion could occur, causing injury to personnel and damage to equipment.

FRAME 1

1. Take out old fuel line filters (1) from nozzle and check valve assembly (2) if there are any.
2. Put in new fuel line filters (1) in fuel inlet (3) and fuel outlet (4).

GO TO FRAME 2

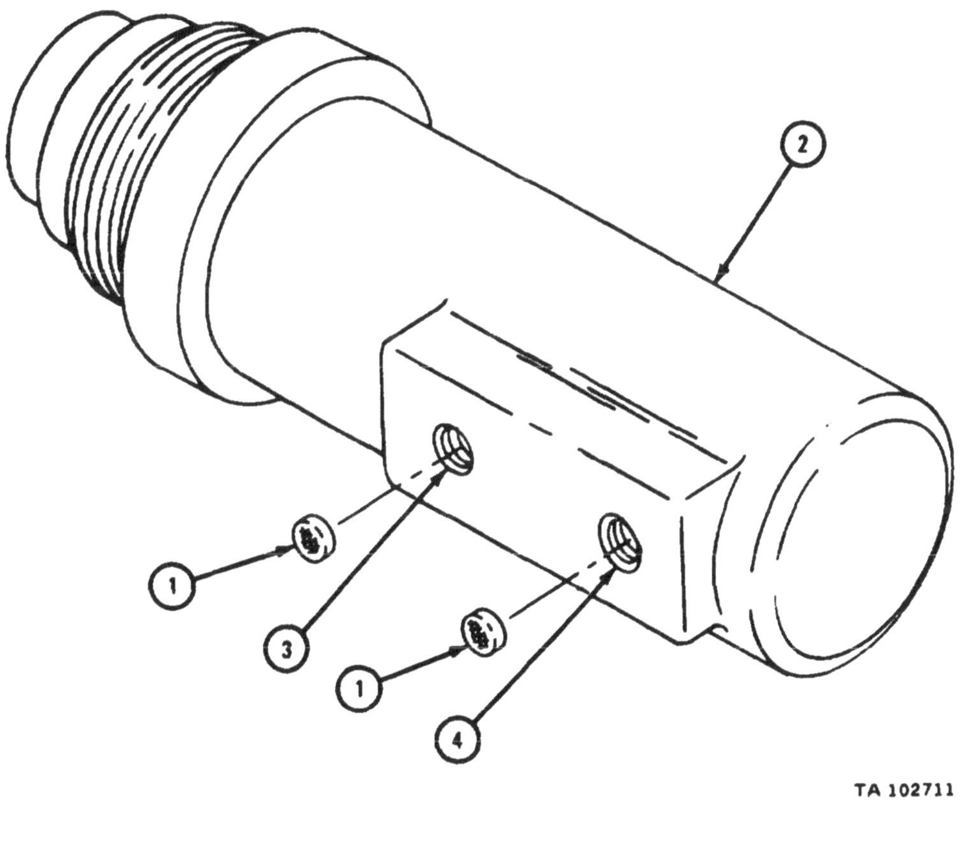

FRAME 2

NOTE

This is an alternate procedure given for reference purposes only . Standard procedure is to test flame heater nozzle during dynamometer test and adjustment.

1. Hook up hose (1) from supply (2) of diesel fuel to pump inlet (3).
2. Hook up hose (4) from pump outlet (5) to pressure gage (6).
3. Hook up adjustable valve (7) from pressure gage (6).
4. Hook up hose (8) from adjustable valve (7) to nozzle and check valve assembly (9) inlet opening (10) .

GO TO FRAME 3

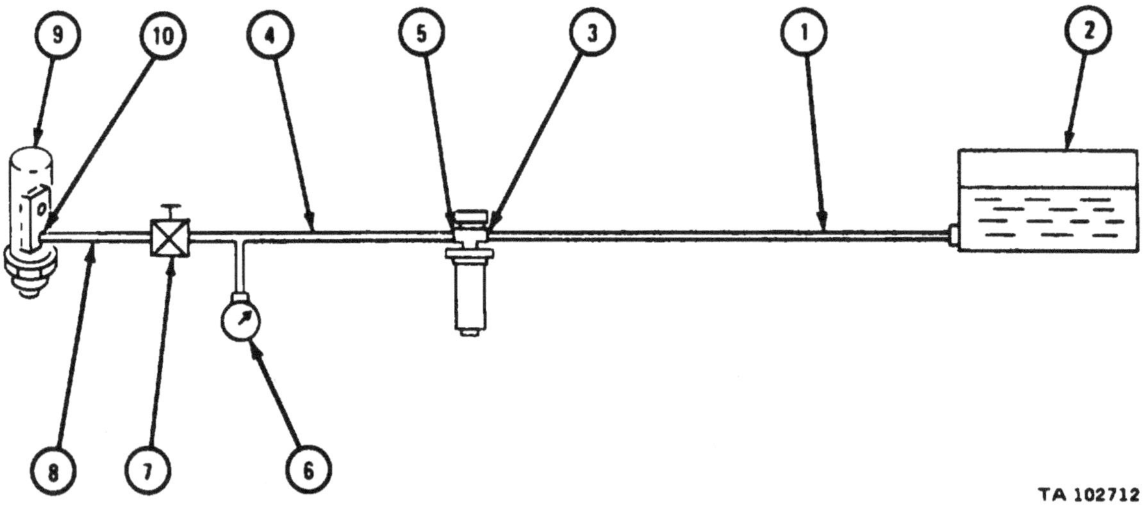

FRAME 3

CAUTION

The flame heater fuel pump electrical system is negatively grounded. Pin A is positive and pin B is negative. Do not switch polarity of pump during checking or replacement. Changing polarity will permanently damage pump.

1. Hook up 24-volt battery (1) with rheostat (2) to fuel pump receptacle (3).
2. Hook up voltmeter (4) to fuel pump receptacle (3) .

GO TO FRAME 4

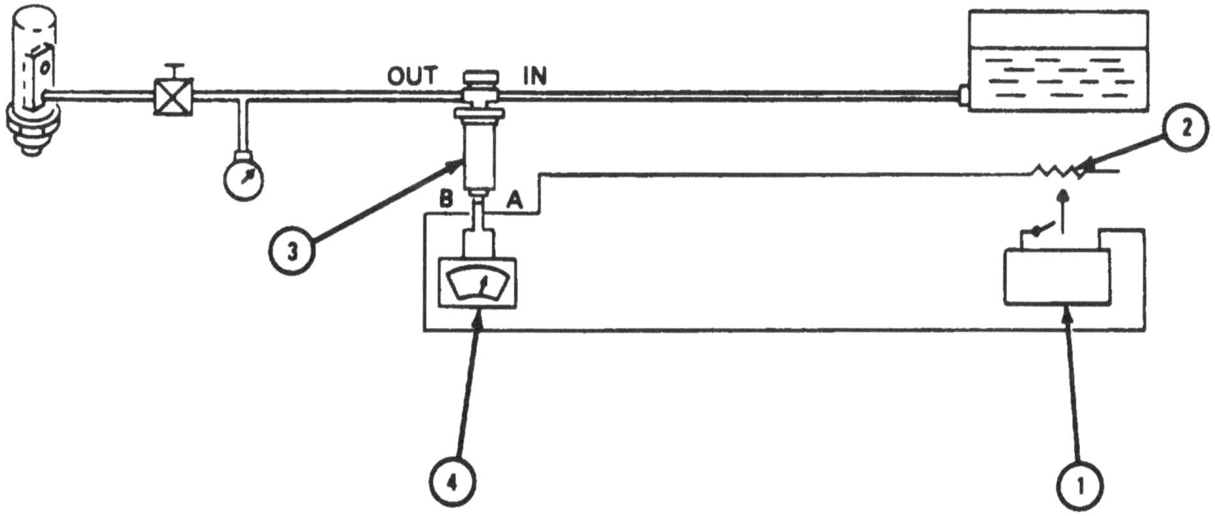

FRAME 4

1. Turn rheostat (1) until voltmeter (2) reads 24 volts. Pressure gage (3) should read **90** to **100** psi.
2. Put bucket (4) under nozzle and check valve assembly (5) to catch discharged fuel.
3. Turn adjustable valve (6) to give fuel flow to nozzle and check valve assembly (5).
4. Check that spray pattern (7) is a uniform, fine spray cone with an angle of 60° to 80°.
5. Get a new nozzle and check valve assembly (5) if there is no spray pattern or if spray pattern is coarse or out of shape.
6. Turn off adjustable valve (6) .
7. Turn off rheostat (1) .
8. Turn on adjustable valve (6) to bleed off pressure.

END OF TASK

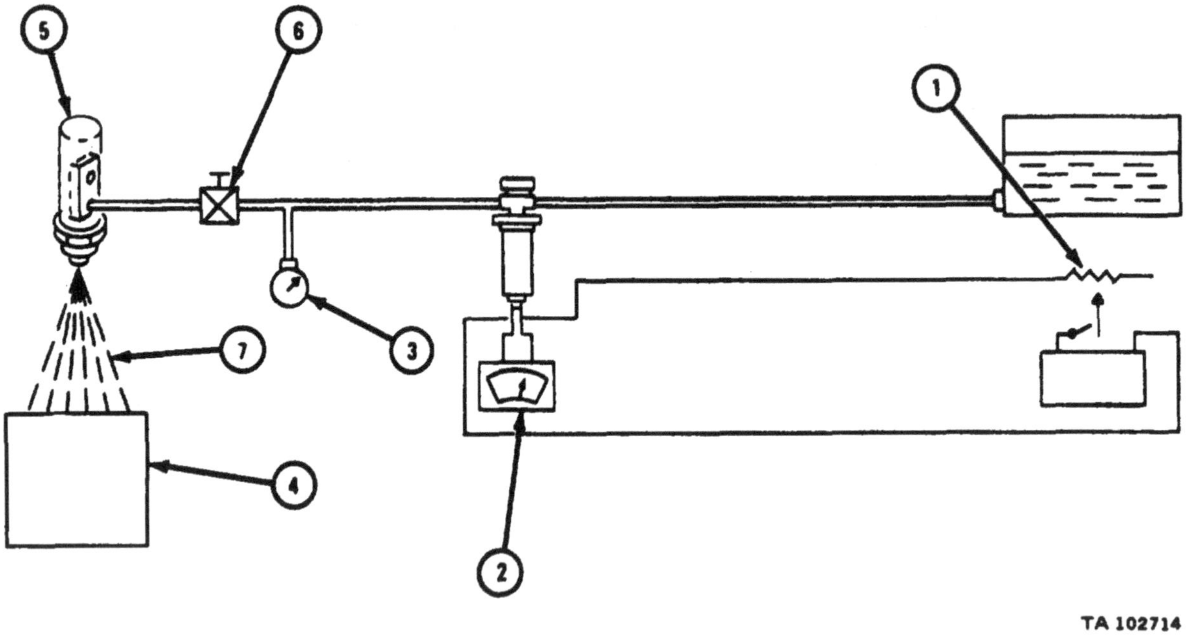

d. Leak Test.

WARNING

Smoking, sparks or open flames are not allowed within 50 feet of work area during this task. Fire or explosion could occur, causing injury to personnel and damage to equipment.

NOTE

This is an alternate procedure given for reference purposes only. Standard procedure is to test for leaks during dynamometer test and adjustment.

FRAME 1

1. Hook up hose (1) from a supply (2) of diesel fuel to pump inlet (3).
2. Hook up hose (4) from pump outlet (5) to adjustable valve (6).
3. Hook up pressure gage (7) from adjustable valve (6).
4. Hook up hose (8) from pressure gage (7) to nozzle and check valve assembly (9), inlet (10), and outlet (11).

GO TO FRAME 2

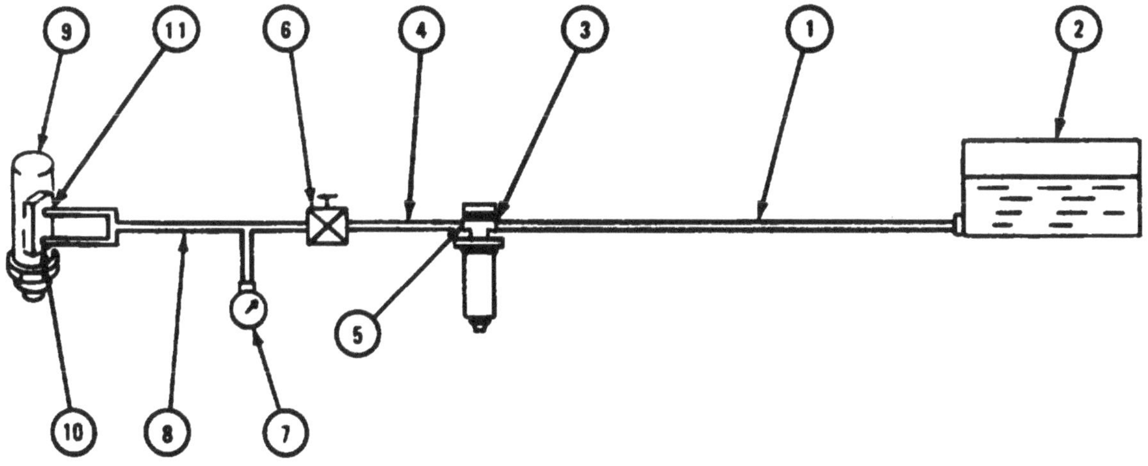

FRAME 2

CAUTION

Flame heater fuel pump electrical system is negatively grounded. Pin A is positive and pin B is negative. Do not switch polarity of pump during checking or replacement. Changing polarity will permanently damage pump.

1. Hook up 24-volt battery (1) with rheostat (2) to fuel pump receptacle (3).
2. Hook up voltmeter (4) to fuel pump receptacle (3).

GO TO FRAME 3

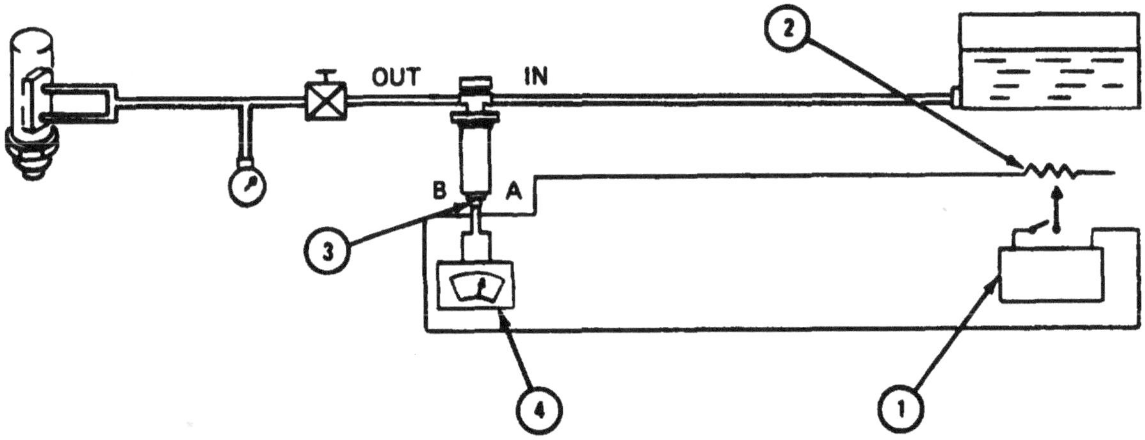

FRAME 3

1. Turn on rheostat (1) until voltmeter (2) reads 24 volts.
2. Turn on adjustable valve (3) until pressure gage (4) reads 5 to 10 psi.
3. Check that nozzle and check valve assembly (5) inlet (6) and outlet (7) openings do not leak. Get a new nozzle and check valve assembly if inlet and outlet openings leak.
4. Turn off adjustable valve (3).
5. Turn off rheostat (1).
6. Take off nozzle and check valve assembly (5).

END OF TASK

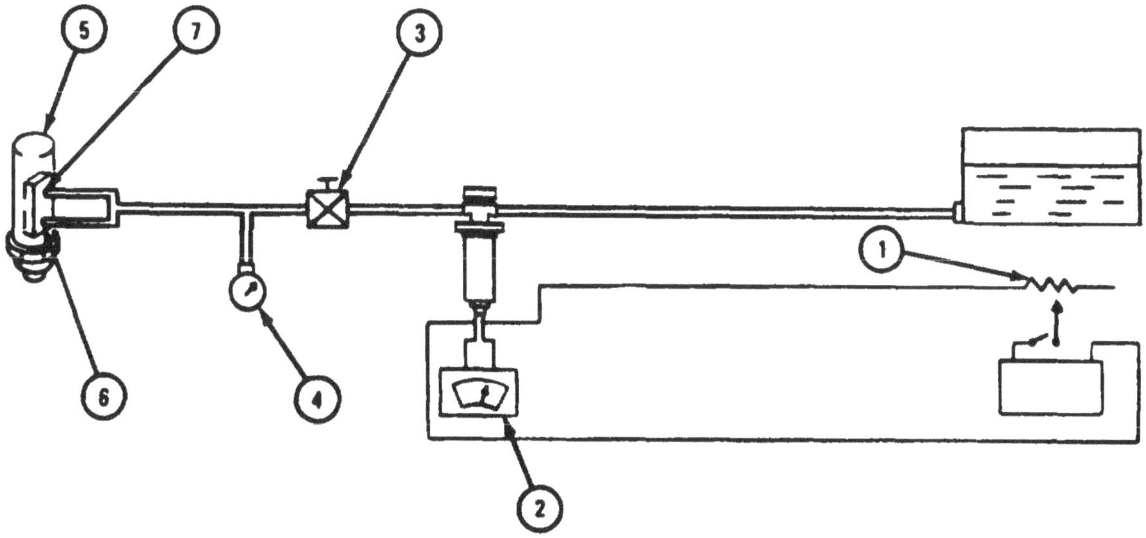

4-69. FLAME HEATER TUBES, CLAMPS, AND FITTINGS (COVERED SYSTEM).

a. Cleaning.

FRAME 1

1. Clean plastic tubing (1) with soap and water.

WARNING

Dry cleaning solvent is flammable. Do not use near an open flame. Keep a fire extinguisher nearby when solvent is used . Use only in well-ventilated places. Failure to do this may result in injury to personnel and damage to equipment.

2. Clean elbows (2) with dry cleaning solvent.
3. Clean connectors (3) with dry cleaning solvent.
4. Clean loop clamps (4) with dry cleaning solvent if they are uncoated or with soap and water if they are coated.
5. Clean spring clamps (5) with dry cleaning solvent if they are uncoated or soap and water if the are coated.

END OF TASK

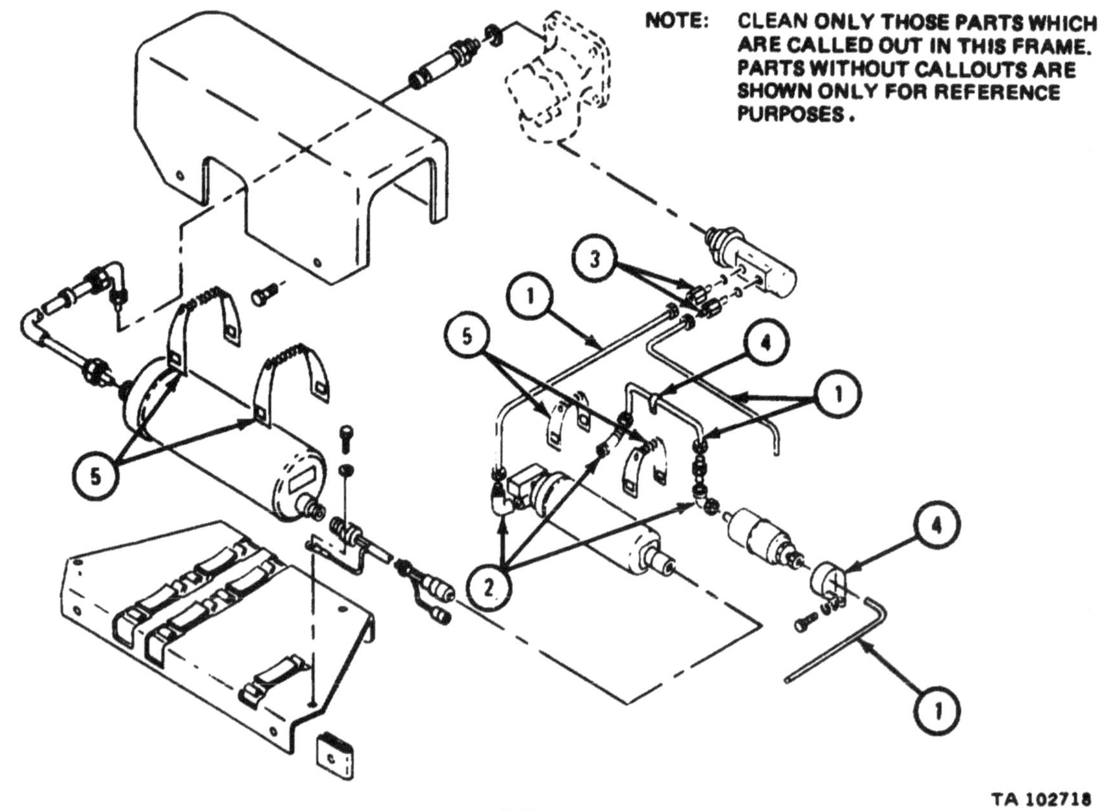
NOTE: CLEAN ONLY THOSE PARTS WHICH ARE CALLED OUT IN THIS FRAME. PARTS WITHOUT CALLOUTS ARE SHOWN ONLY FOR REFERENCE PURPOSES.
1
2
3
4
5
TA 102718

b. Inspection and Repair.

FRAME 1

1. Check that plastic tubing (1) is not cracked, kinked or brittle.
2. Check that plastic tubing (1) has not leaked around tube nuts (2).
3. Throw away plastic tubing (1) if it is cracked or damaged and get a new one.
4. Throw away plastic tubing (1) if tube nuts (2) are stripped or damaged and get a new one.

GO TO FRAME 2

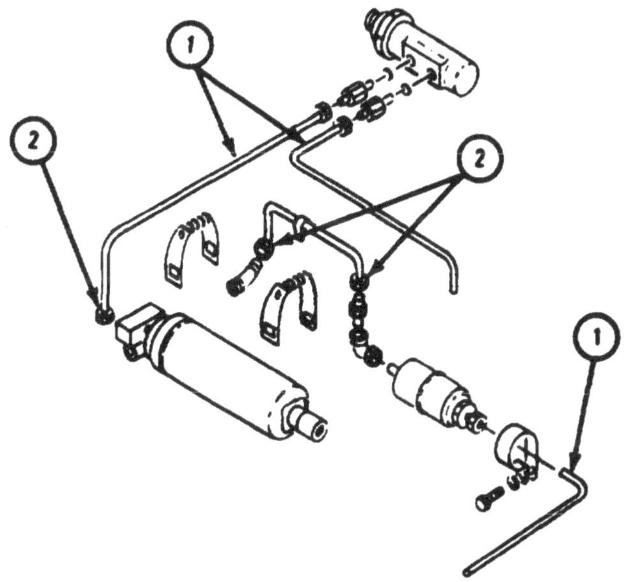

NOTE: CHECK ONLY THOSE PARTS WHICH ARE CALLED OUT IN THIS FRAME. PARTS WITHOUT CALLOUTS ARE SHOWN ONLY FOR REFERENCE PURPOSES OR ARE CHECKED IN ANOTHER FRAME.

TA 102719

FRAME 2

1. Check that 1/4-inch plastic tubing (1) has not shrunk at tube nut ferrules (2).
2. Throw away 1/4-inch plastic tubing (1) if no. 28 drill shank (3) cannot be put in tubing opening.
3. Check that 1/8-inch plastic tubing (4) has not shrunk at tube nut ferrule (5).
4. Throw away 1/8-inch plastic tubing if 1/32-inch drill shank (6) cannot be put in tubing opening.

GO TO FRAME 3

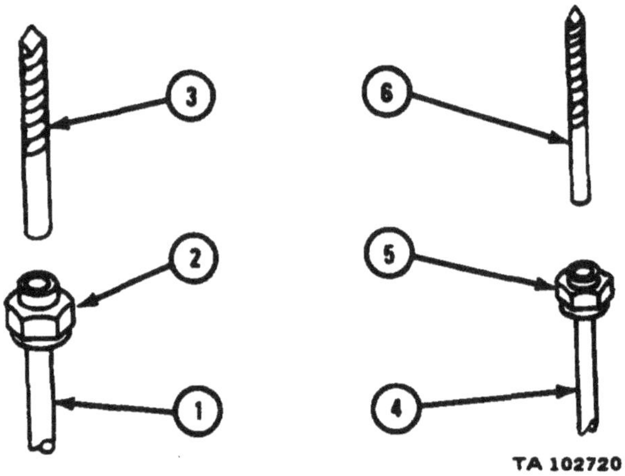

TA 102720

FRAME 3

1. Check that elbows (1) and connectors (2) are not cracked and that threads are not damaged. Throw away elbows and connectors that are cracked or have damaged threads, and get new ones.

GO TO FRAME 4

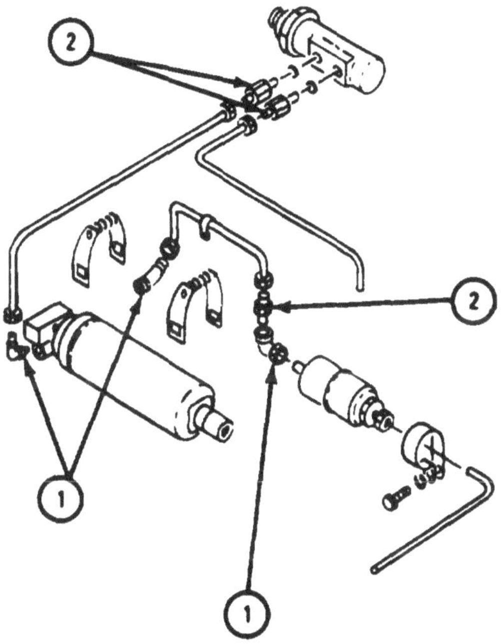

NOTE: CHECK ONLY THOSE PARTS WHICH ARE CALLED OUT IN THIS FRAME. PARTS WITHOUT CALLOUTS ARE SHOWN ONLY FOR REFERENCE PURPOSES OR ARE CHECKED IN ANOTHER FRAME.

TA 102721

FRAME 4

1. Check that loop clamps (1) and spring clamps (2) are not bent or cracked. Throw away loop clamps and spring clamps that are cracked or badly damaged and get new ones.
2. Check that coating on loop clamps (1) is not worn. Throw away 100p clamps that have badly worn coating and get new ones. Straighten minor bends in loop clamps.
3. Check that springs (3) on spring clamps (2) are not stretched or damaged. Throw away springs or spring clamps that are cracked, scratched or damaged, and get new ones.

END OF TASK

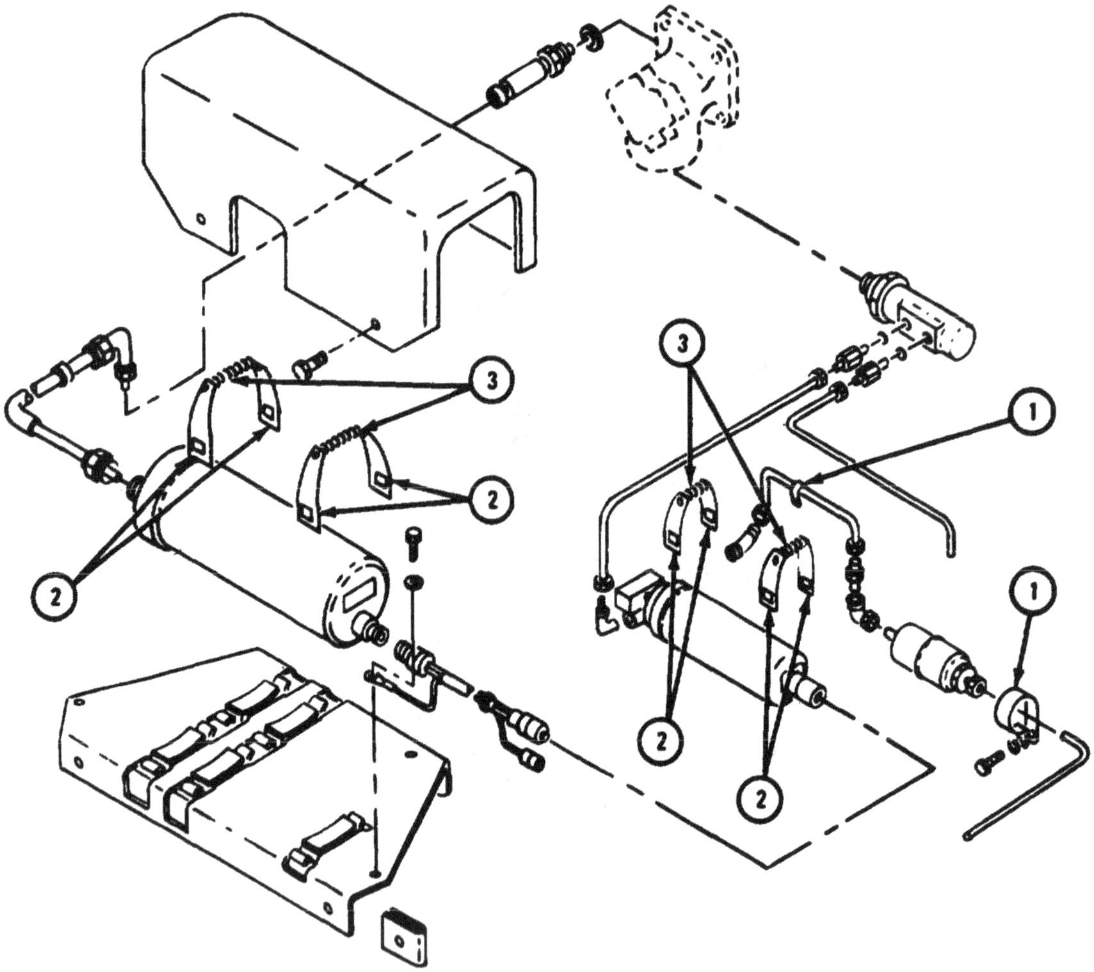

NOTE: CHECK ONLY THOSE PARTS WHICH ARE CALLED OUT IN THIS FRAME. PARTS WITHOUT CALLOUTS ARE SHOWN ONLY FOR REFERENCE PURPOSES OR ARE CHECKED IN ANOTHER FRAME.

TA 102722

4-70. FLAME HEATER COVER AND BRACKET (COVERED SYSTEM).

a. Cleaning and Inspection.

FRAME 1

WARNING

Dry cleaning solvent is flammable. Do not use near an open flame. Keep a fire extinguisher nearby when solvent is used . Use only in w en-ventilated places. Failure to do this may result in injury to personnel and damage to equipment.

1. Clean access cover (1) with dry cleaning solvent .
2. Clean support bracket assembly (2) with soap and water.
3. Check tha t access cover (1) is not cracked, dented or bent .
4. Check that bracket assembly (2) is not cracked or damaged.
5. Check that clip-on nut (3) is not damaged or that threads are not stripped.
6. Check that clip-on nuts (3) fit tightly on bracket (2) .
7. Throw away any part that is cracked or damaged and get a new one.

END OF TASK

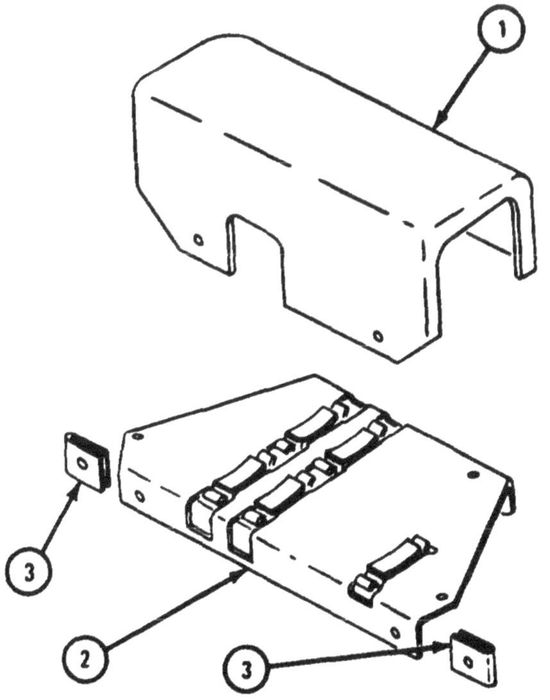

TA 102723

b. Repair.

FRAME 1

1. Repair minor bends in access cover (1) and bracket (2) by straightening. Refer to FM 43-2.
2. File nicks and burrs on access cover (1) and bracket (2) with a fine mill file.

END OF TASK

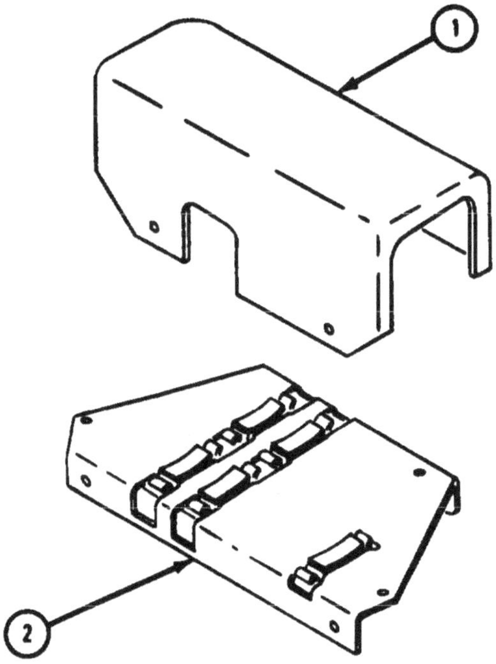

4-71. FLAME HEATER SOLENOID VALVES (SIDE-MOUNTED SYSTEM).

a. Removal.

FRAME 1

1. Unscrew two tube nuts (1) and takeoff fuel line (2).
2. Take out two screws (3) and take fuel inlet solenoid valve (4) off bracket (5). Keep lockwashers (6) on screws and set them aside.
3. Take out two screws (7) and take fuel return solenoid valve (8) off bracket (5). Keep lockwashers (9) on screws and set them aside.

GO TO FRAME 2

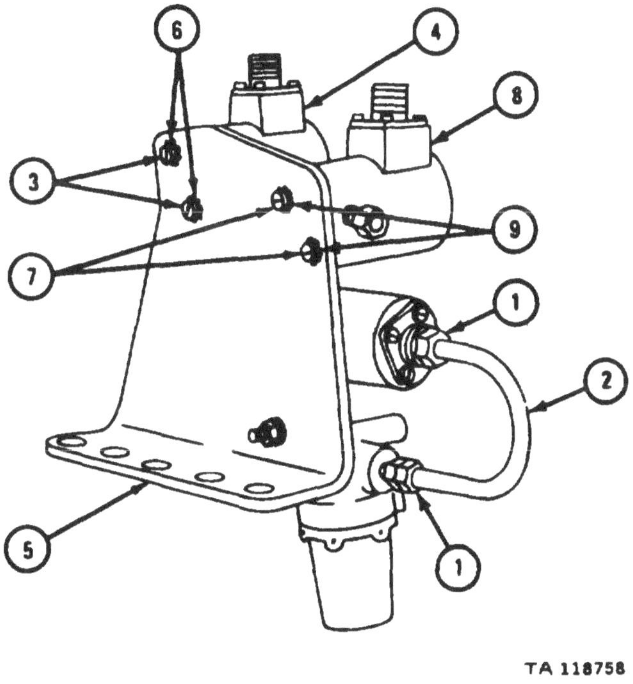

FRAME 2

1. Take out three elbows (1) and tube adapter (2).
2. Plug all openings in two solenoid valves (3).

END OF TASK

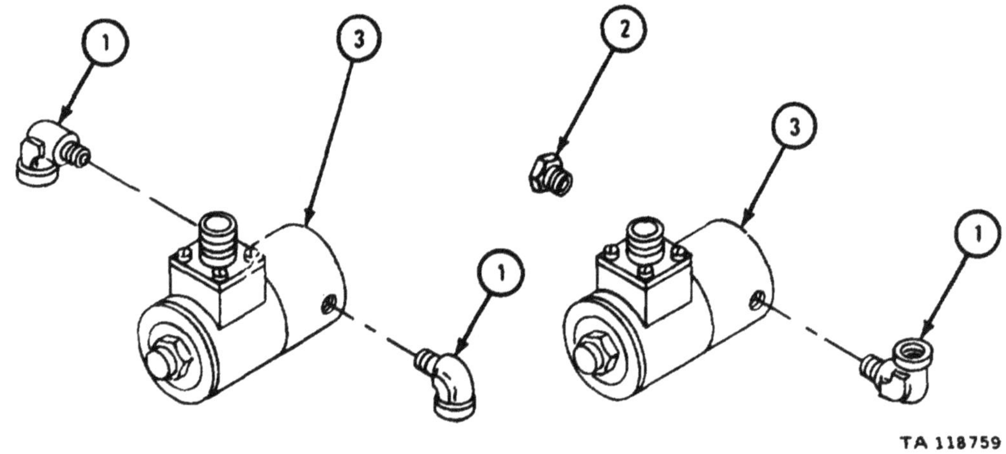

b. Cleaning.

(1) Check that all openings in solenoid valves are plugged.

WARNING

Dry cleaning solvent is flammable. Do not use near an open flame. Keep a fire extinguisher nearby when solvent is used. Use only in well-ventilated places. Failure to do this may result in injury to personnel and damage to equipment.

CAUTION

Do not dip solenoid valves in dry cleaning solvent. Damage to parts inside solenoid valves will result.

(2) Clean outside of solenoid valves using cloth dipped in dry cleaning solvent.

(3) Remove cap plugs from fuel line openings.

c. Inspection.

FRAME 1

1. Check that solenoid valves (1) are not cracked or dented. Get a new valve if cracked or dented.
2. Check that threads in inlet (2) and outlet (3) are not stripped.
3. Check that threads on electrical connector (4) are not stripped.

GO TO FRAME 2

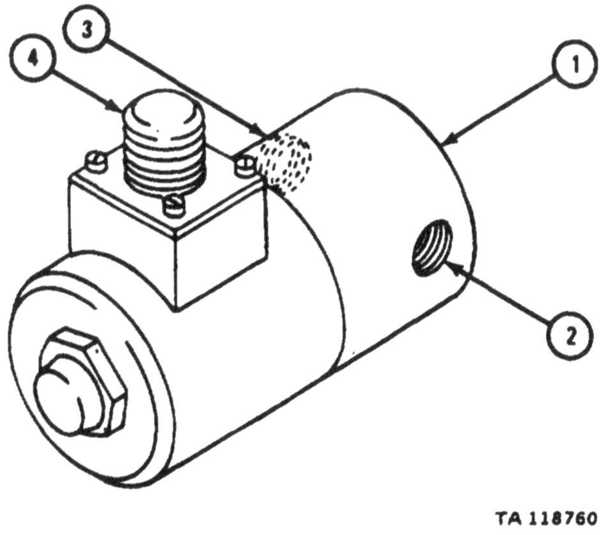

FRAME 2

WARNING

Smoking, sparks or open flames are not allowed within 50 feet of work area during this task. Fire or explosion could occur, causing injury to personnel and damage to equipment.

1. Refer to test setup given in paragraph 4-68d frame 1.
2. Using suitable fitting, hook up solenoid valve inlet opening (1) in place of nozzle and check valve assembly, shown in test setup.
3. Continue with procedures given in paragraph 4-68d frame 2, and frame 3 steps 1 and 2.

CAUTION

The flame heater electrical system is negatively grounded. Pin A is positive and pin B is negative. Do not switch polarity of solenoid valve during checking or replacement. Changing polarity will permanently damage solenoid valve or wiring harness.

4. Join 10 to 12-volt dc power supply to solenoid valve electrical connector (2). Electrical current will open normally closed valve and fuel will flow out of outlet opening (3) .
5. Take off 10 to 12-volt dc power supply. Fuel flow out of outlet opening (3) should stop completely with no leakage. If fuel flow does not stop completely, loosen acorn nut (4) and tighten acorn nut to 40 to 50 pound-inches.
6. Do steps 4 and 5 again.
7. If fuel still leaks from outlet opening (3), get a new solenoid valve.
8. Do steps 1 through 7 again for other solenoid valve.
9. Shut off fuel pressure, refer to paragraph 4-68d frame 3 steps 4 and 5. Take off solenoid valve.

END OF TASK

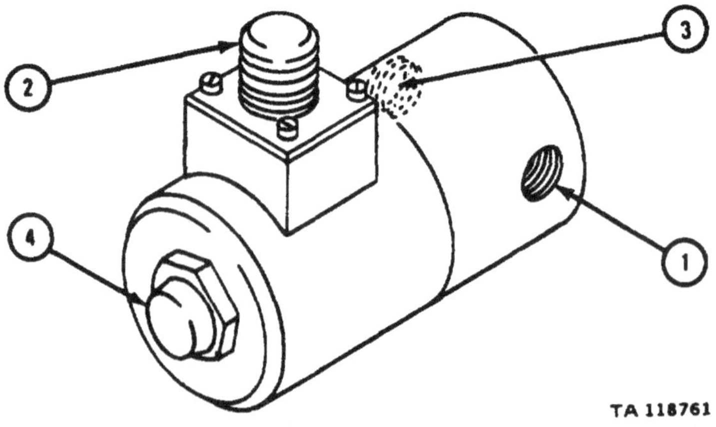

d. Repair. Repair of solenoid valves is limited to taking off burrs from threads of inlet and outlet openings .

e. Replacement.

FRAME 1

1. Screw one 90° elbow (1) into each side of inlet solenoid valve (2).
2. Tighten 90° elbows (1) so openings face down.
3. Screw 90° elbow (3) into return solenoid valve (4).
4. Tighten 90° elbow (3) so opening faces up.
5. Screw tube adapter (5) into other side of return solenoid valve (4).
6. Tighten tube adapter (5) .

GO TO FRAME 2

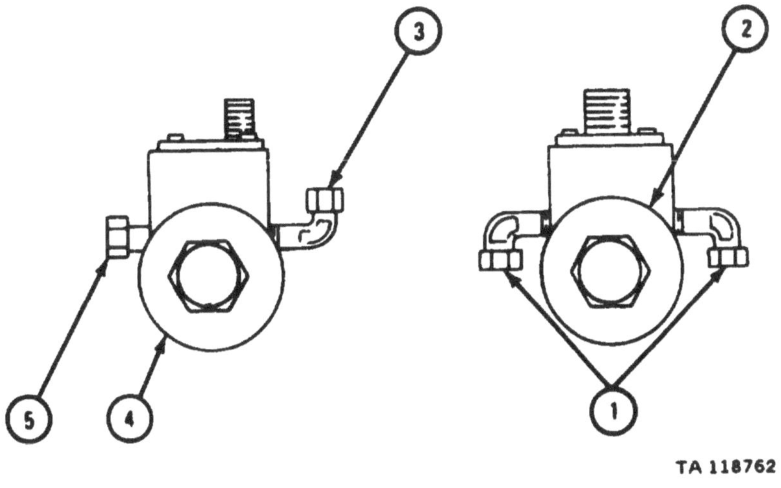

FRAME 2

1. Place inlet solenoid valve (1) on flame heater bracket (2).
2. Aline holes and put in two lockwashers (3) and machine screws (4).
3. Tighten machine screws (4) .
4. Place return solenoid (5) on flame heater bracket (2).
5. Aline holes and put in two lockwashers (6) and machine screws (7).
6. Tighten machine screws (7) .
7. Put fuel tube (8) in place and tighten two tube nuts (9).

END OF TASK

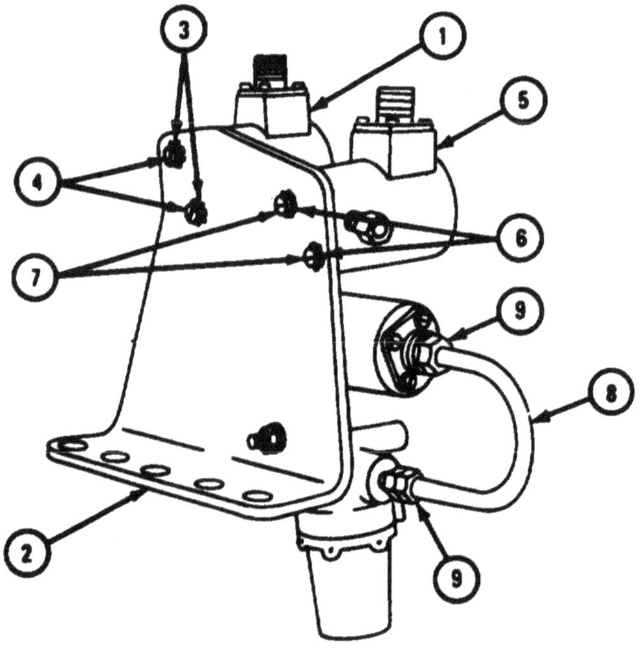

TA 118763

4-72. FLAME HEATER REPLACEABLE ELEMENT FILTER (SIDE-MOUNTED SYSTEM).

a. Removal.

FRAME 1

1. Take off nut (1), lockwasher (2), and machine screw (3) holding filter (4) to flame heater bracket (5).

END OF TASK

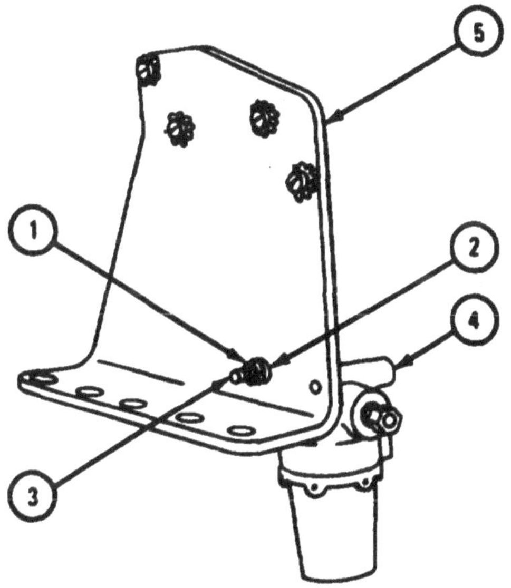

b. Disassembly.

FRAME 1

1. Unscrew filter bowl (1) from filter head (2).
2. Take element assembly (3) out of filter bowl (1).
3. Take out preformed packing (4) from filter bowl (1) and throw it away.

END OF TASK

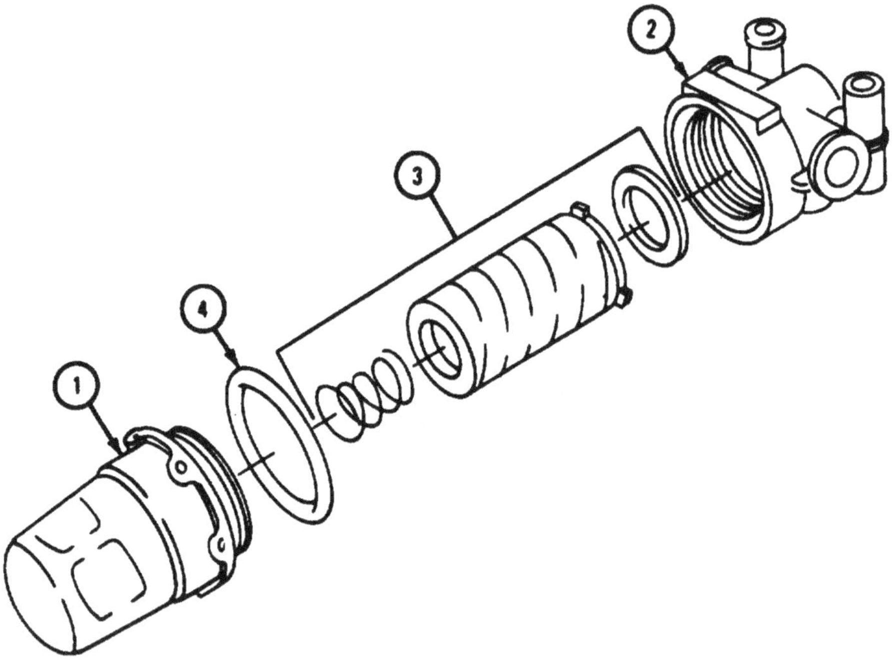

c. Cleaning.

WARNING

Dry cleaning solvent is flammable. Do not use near an open flame. Keep a fire extinguisher nearby when solvent is used. Use only in well-ventilated places. Failure to do this may result in injury to personnel and damage to equipment.

(1) Clean filter bowl and filter head in dry cleaning solvent.

WARNING

Eye shields must be worn when using compressed air. Eye injury can occur if eye shields are not used.

(2) Clean filter element assembly by blowing compressed air through center of element.

(3) Throw away element assembly if it is very dirty or when spring and gasket cannot be used.

d. Inspection and Repair .

FRAME 1

1. Check that filter bowl (1) is not dented, cracked or crossthreaded. Get a new bowl if it is cracked, dented or crossthreaded.
2. Check that filter head (2) is not cracked, and that inlet threads (3) and outlet threads (4) are not burred or crossthreaded. Use a tap to repair inlet threads (3) and outlet threads (4) .
3. Get a complete new filter if head (2) is cracked or if threads (3 and 4) are badly damaged .

END OF TASK

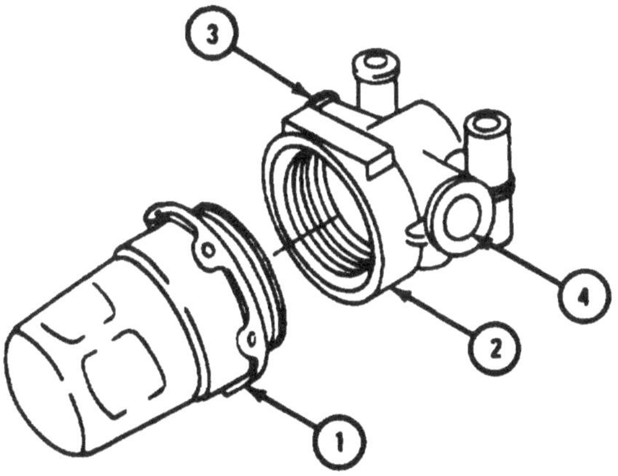

TA 118751

e. Assembly.

FRAME 1

1. Put packing (1) on threaded end of filter bowl (2).
2. Put filter assembly (3) in filter bowl (2).
3. Screw filter bowl into filter head (4).

END OF TASK

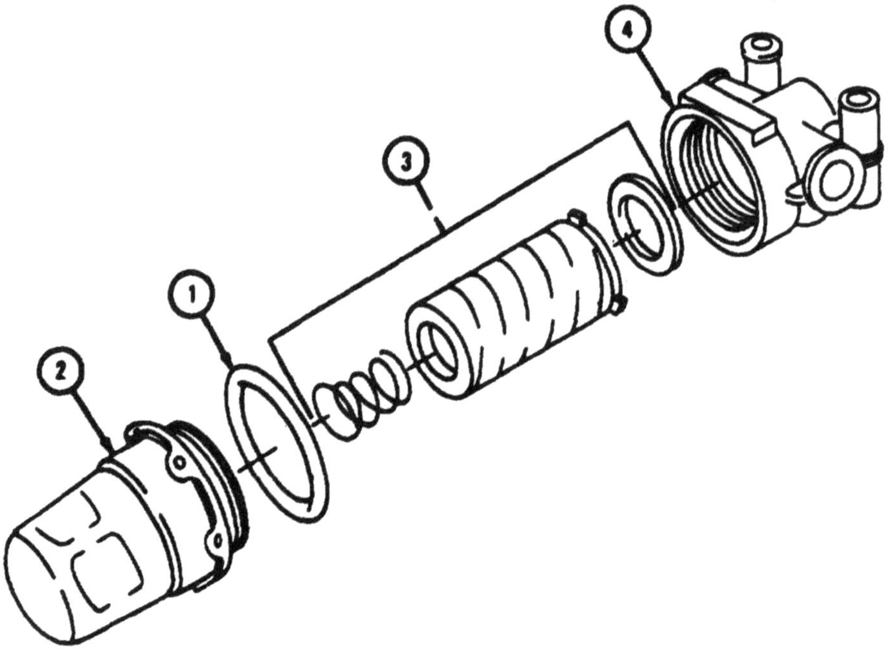

f. Replacement.

FRAME 1

1. Place filter assembly (1) on flame heater bracket (2).
2. Put machine screw (3) in filter assembly (1) and flame heater bracket (2).
3. Put lockwasher (4) and nut (5) on machine screw (3).
4. Tighten nut (5) and machine screw (3).

END OF TASK

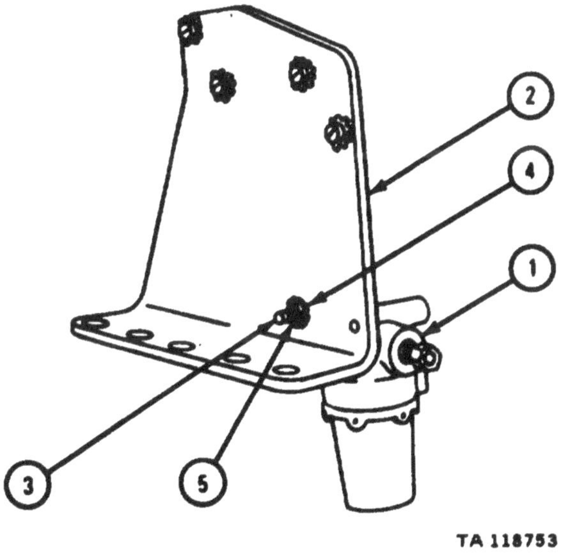

4-73. FLAME HEATER FUEL PUMP (SIDE-MOUNTED SYSTEM).

a. Removal.

FRAME 1

1. Takeoff two nuts (1), lockwashers (2), and machine screws (3) .
2. Take out fuel pump (4) from flame heater bracket (5).

END OF TASK

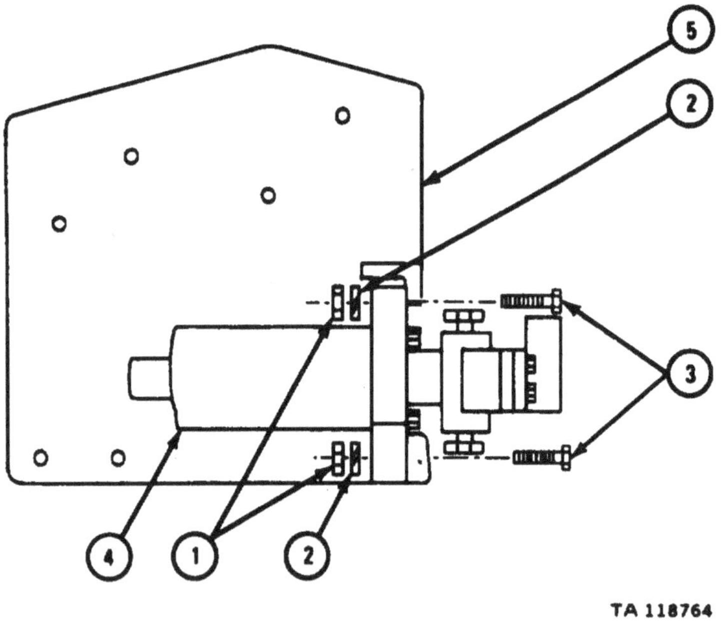

b. Cleaning, Inspection, and Repair. Refer to flame haeater fuel pump (covered system), para 4-66.

c. Replacement .

FRAME 1

1. Place fuel pump (1) in flame heater bracket (2).
2. Aline holes in flame heater bracket (2) and fuel pump (1).
3. Put in two machine screws (3).
4. Place lockwashers (4) and nuts (5) on machine screws (3).
5. Tighten machine screws (3) and nuts (4) .

END OF TASK

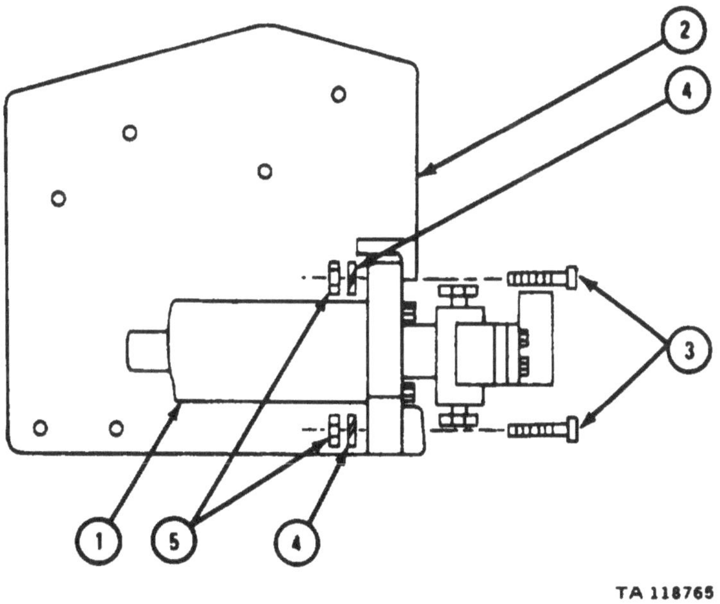

4-74. FLAME HEATER IGNITION UNIT (SIDE-MOUNTED SYSTEM). For procedures to repair the flame heater side-mounted ignition unit, refer to flame heater ignition unit (covered system), para 4-65c .

4-75. FLAME HEATER WIRING HARNESS AND ELECTRICAL LEAD ASSEMBLY (SIDE-MOUNTED SYSTEM). For procedures to repair the side-mounted heater wiring harness, refer to covered flame heater wiring harness, para 4-67b.

4-76. FLAME HEATER NOZZLE AND VALVE ASSEMBLY (SIDE-MOUNTED SYSTEM).

a. Cleaning and Inspection.

FRAME 1

1. Leave felt fuel line filters in fuel inlet (1) and outlet (2) openings in place.
2. Plug nozzle and valve assembly fuel inlet (1) and outlet (2) openings with pipe plugs.

WARNING

Dry cleaning solvent is flammable. Do not use near an open flame. Keep a fire extinguisher nearby when solvent is used. Use only in well-ventilated places . Failure to do this may result in injury to personnel and damage to equipment.

3. Clean nozzle and valve assembly (3) with dry cleaning solvent. Take off heavy carbon deposits with a stiff brush.
4. Check that nozzle and valve assembly fuel inlet (1) and fuel outlet (2) threads are not crossthreaded .
5. Check that nozzle and valve assembly (3) is not cracked or damaged. Get a new nozzle and valve assembly if it is cracked or damaged .

END OF TASK

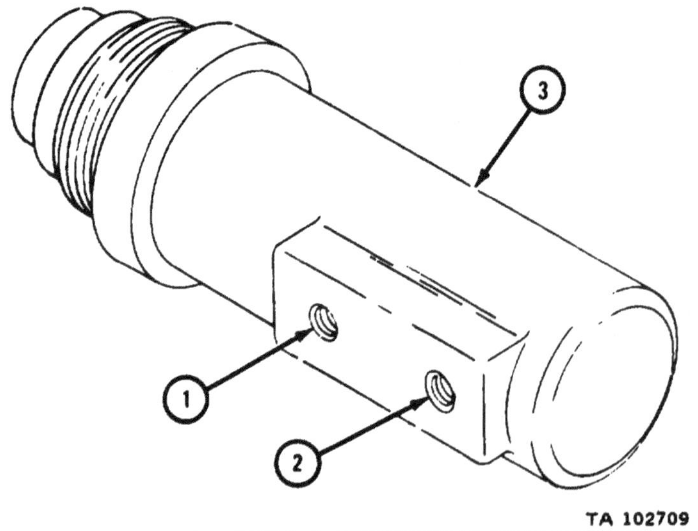

b. Repair.

FRAME 1

1. Repair minor thread damage to fuel inlet (1) and fuel outlet (2) on nozzle and valve assembly (3) with a tap.

GO TO FRAME 2

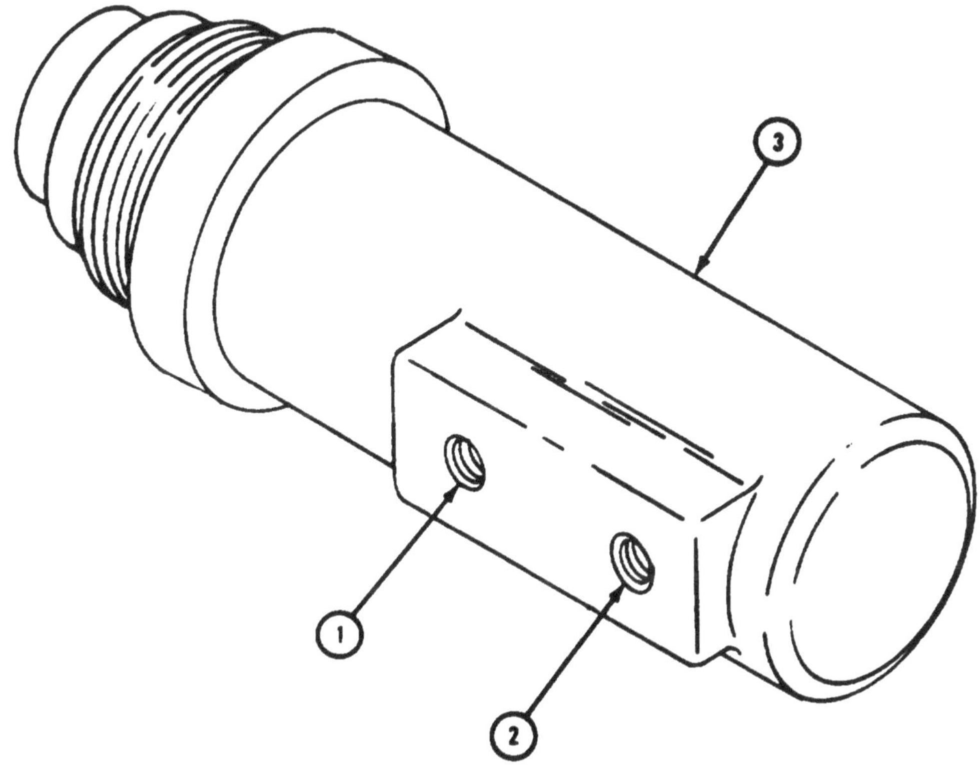

c. Spray Pattern Test .

WARNING

Smoking, sparks or open flames are not allowed within 50 feet of work area during this task. Fire or explosion could occur, causing injury to personnel and damage to equipment.

FRAME 1

1. Take out old fuel line filters (1) from nozzle and valve assembly (2) if there are any .
2. Put in new fuel line filters (1) in fuel inlet (3) and fuel outlet (4).

GO TO FRAME 2

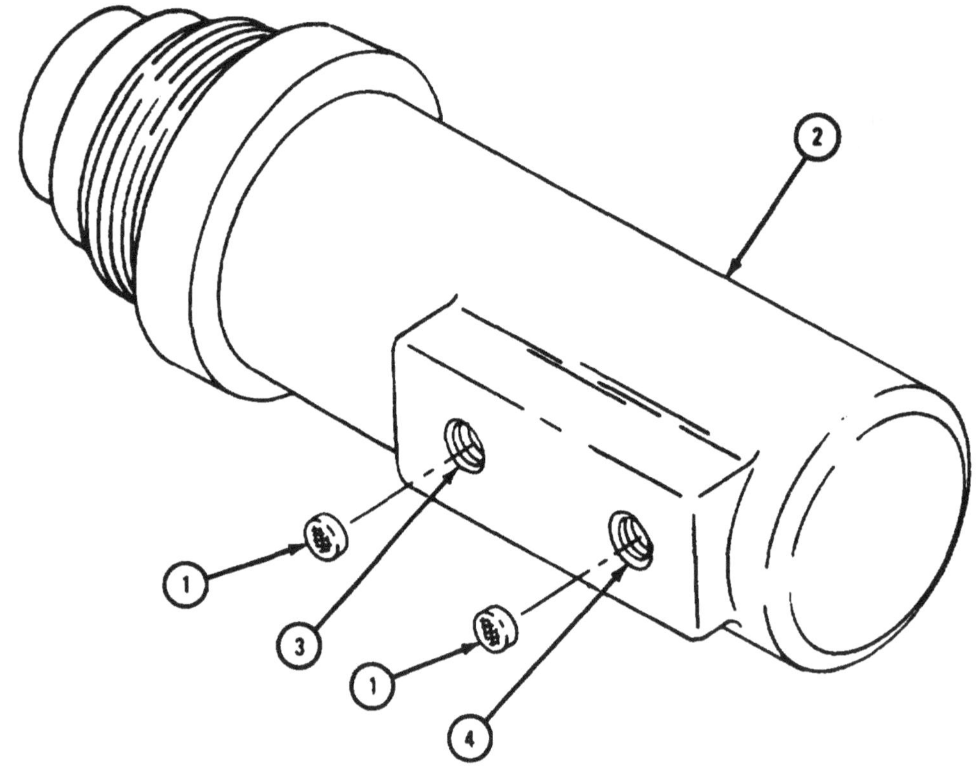

FRAME 2

NOTE

This is an alternate procedure given for reference purposes only. Standard procedure is to test flame heater nozzle during dynamometer test and adjustment.

1. Hook up hose (1) from supply tank (2) of diesel fuel to pump inlet (3).
2. Hook up hose (4) from pump outlet (5) to pressure gage (6).
3. Hook up adjustable valve (7) from pressure gage (6).
4. Hook up hose (8) from adjustable valve (7) to nozzle and valve assembly (9) inlet opening (10).

GO TO FRAME 3

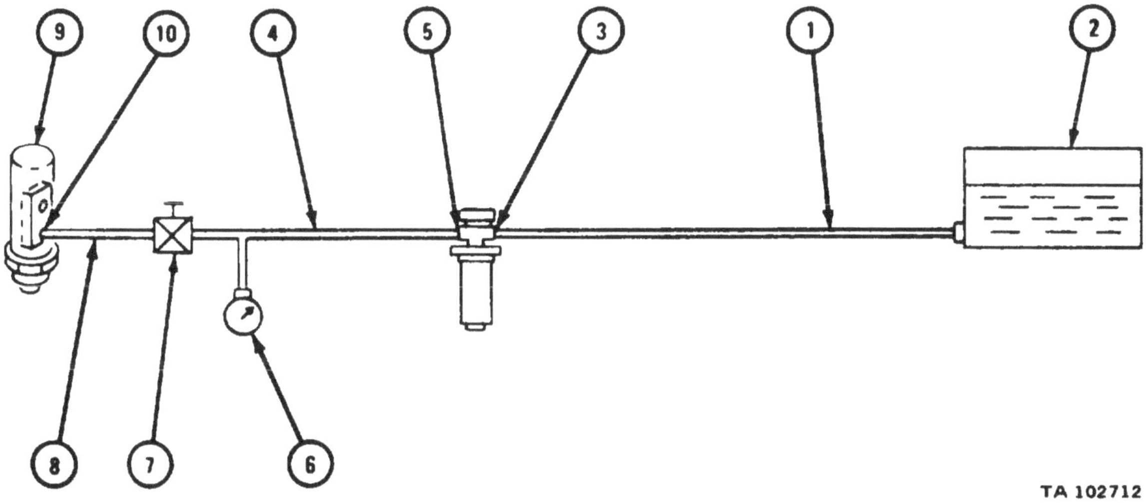

FRAME 3

CAUTION

Flame heater fuel pump electrical system is negative grounded. Pin A positive and pin B is negative. Do not switch polarity of pump during checking or replacement. Changing polarity will permanently damage pump.

1. Hook up 24-volt battery (1) with rheostat (2) to fuel pump receptacle.
2. Hook up voltmeter (4) to fuel pump receptacle (3).

GO TO FRAME 4

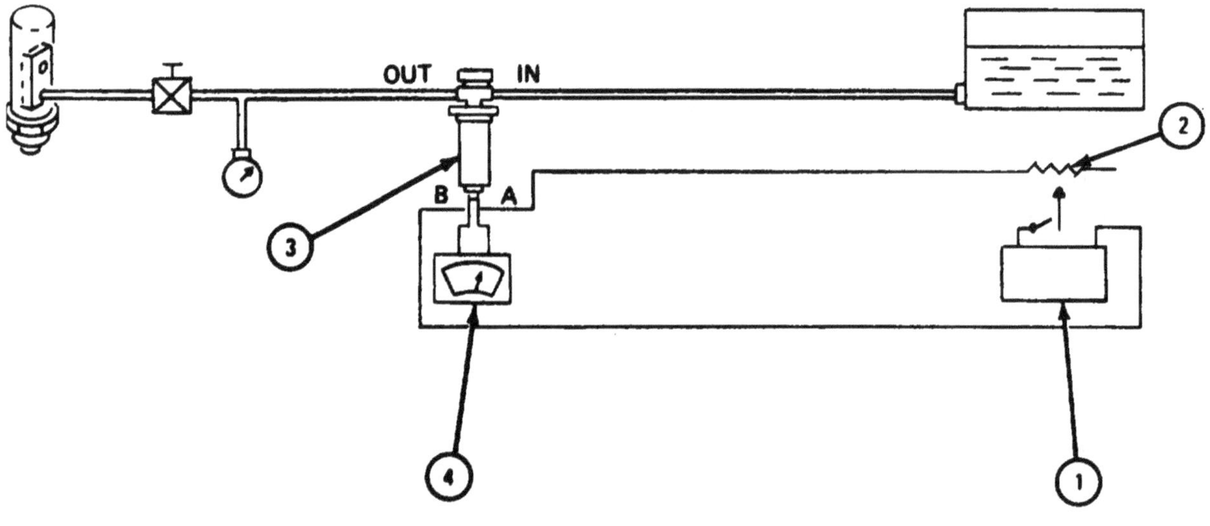

FRAME 4

1. Turn rheostat (1) until voltmeter (2) reads 24 volts. Pressure gage (3) should read 90 to 100 psi.
2. Put bucket (4) under nozzle and valve assembly (5) to catch discharged fuel.
3. Turn adjustable valve (6) to give fuel flow to nozzle and valve assembly (5).
4. Check that spray pattern (7) is a uniform, fine spray cone with an angle of 60° to 80°. Get a new nozzle and valve assembly (5) if there is no spray pattern or if spray pattern is coarse or out of shape.
5. Turn off adjustable valve (6) .
6. Turn off rheostat (1) .
7. Turn on adjustable valve (6) to bleed off pressure.

END OF TASK

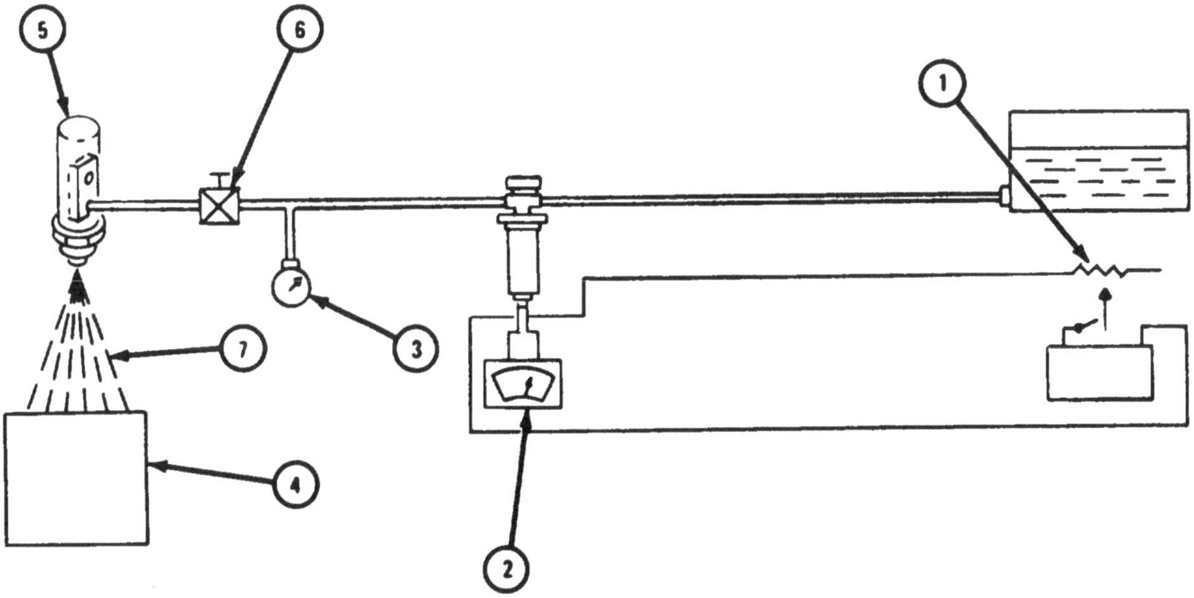

TA 102714

d. Leak Test.

WARNING

Smoking, sparks or open flames are not allowed within 50 feet of work area during this task. Fire or explosion could occur, causing injury to personnel and damage to equipment.

NOTE

This is an alternate procedure given for reference purpose only. Standard procedure is to test for leaks during dynamometer test and adjustment.

FRAME 1

1. Hook up hose (1) from supply tank (2) of diesel fuel to pump inlet (3).
2. Hook up hose (4) from pump outlet (5) to adjustable valve (6).
3. Hook up pressure gage (7) from adjustable valve (6).
4. Hook up hose (8) from pressure gage (7) to nozzle and valve assembly (9), inlet (10) and outlet (11).

GO TO FRAME 2

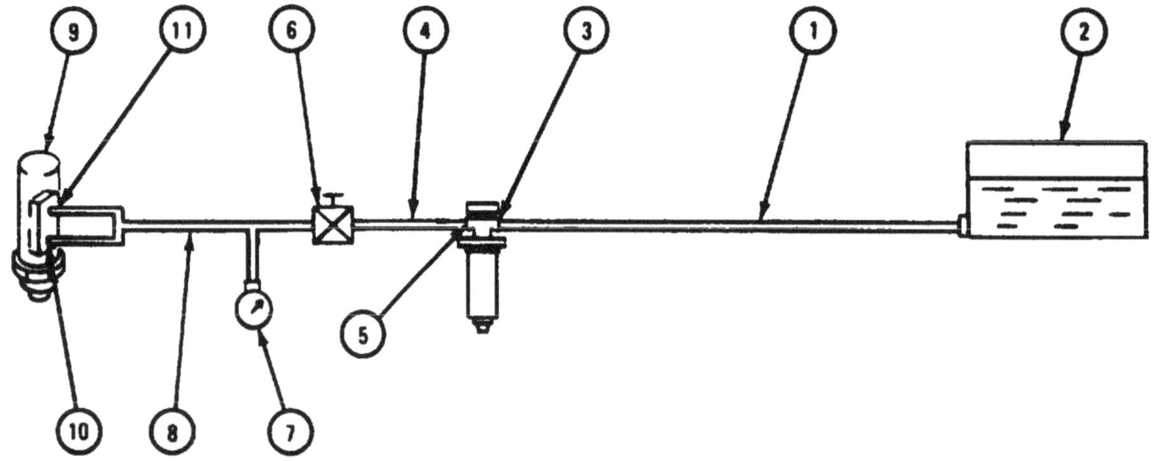

FRAME 2

CAUTION

Flame heater fuel pump electrical system is negatively grounded. Pin A is positive and pin B is negative. Do not switch polarity of pump during checking or replacement. Changing polarity will permanently damage pump.

1. Hook up 24-volt battery (1) with rheostat (2) to fuel pump receptacle (3).
2. Hook up voltmeter (4) to fuel pump receptacle (3).

GO TO FRAME 3

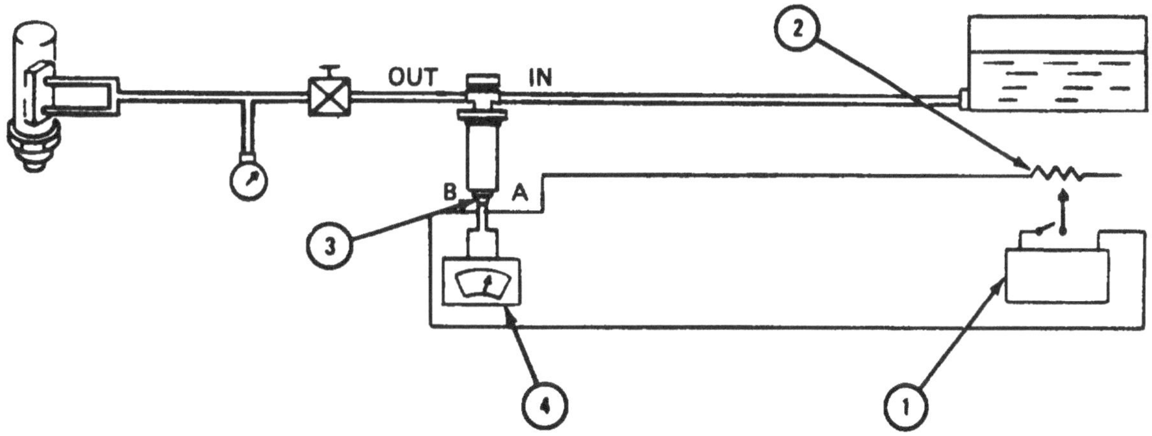

TA 102716

FRAME 3

1. Turn on rheostat (1) until voltmeter (2) reads 24 volts.
2. Turn on adjustable valve (3) until pressure gage (4) reads 5 to 10 psi.
3. Check that nozzle and valve assembly (5) inlet (6) and outlet (7) do not leak. Get a new nozzle and valve assembly if inlet and outlet leak.
4. Turn off adjustable valve (3) .
5. Turn off rheostat (1) .
6. Take off nozzle and valve assembly (5).

END OF TASK

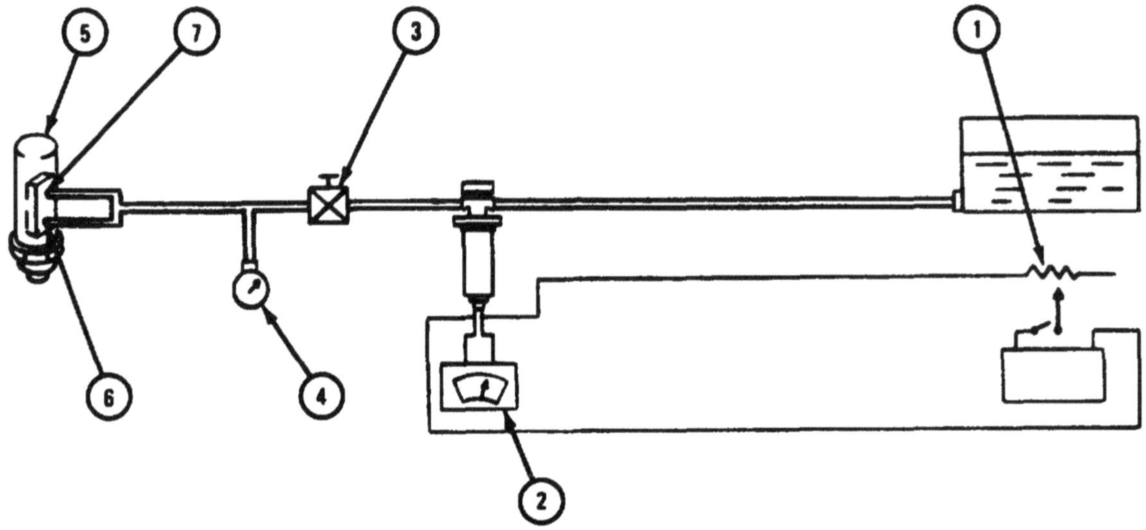

TA 102717

4-77. FLAME HEATER SYSTEM SPARK PLUG (SIDE-MOUNTED SYSTEM).

a. Cleaning and Inspection .

FRAME 1

1. Clean spark plug. Refer to TM 9-4910-389-12.
2. Check that sparkplug case (1) is not cracked .
3. Check that sparkplug electrodes (2) are not cracked or bent .
4. Check that threads (3) are not crossthreaded or burred .

END OF TASK

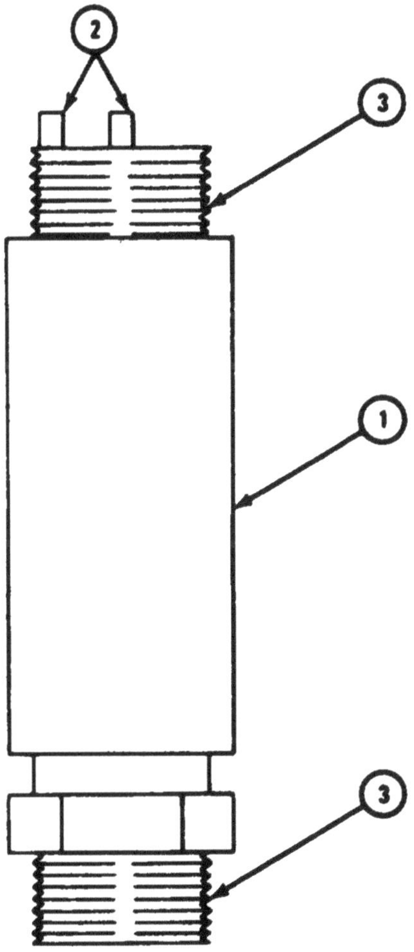

b. Repair.

FRAME 1

NOTE

Spark plug repair is limited to setting spark plug gap .

1. Set gap between pins (1 and 2) to 0.088 to 0.093 inch with feeler gage.

END OF TASK

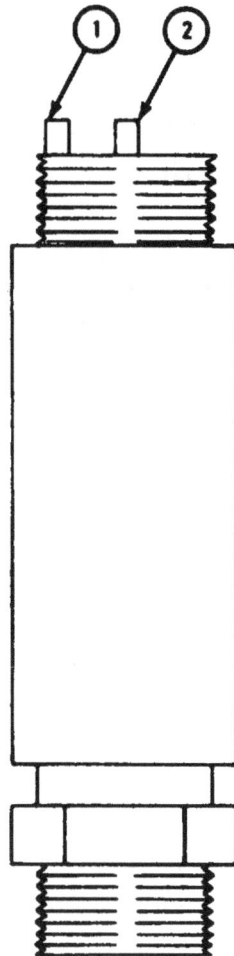

4-78. FLAME HEATER BRACKET (SIDE-MOUNTED SYSTEM).

FRAME 1

WARNING

Dry cleaning solvent is flammable. Do not use near an open flame. Keep a fire extinguisher nearby when solvent is used. Use only in well-ventilated places. Failure to do this may result in injury to personnel and damage to equipment.

1. Clean flame heater bracket (1) in dry cleaning solvent.
2. Check that flame heater bracket (1) has no cracks. Get a new flame heater bracket if it is cracked.
3. Take off nicks and burrs from bracket (1) with a fine mill file.

END OF TASK

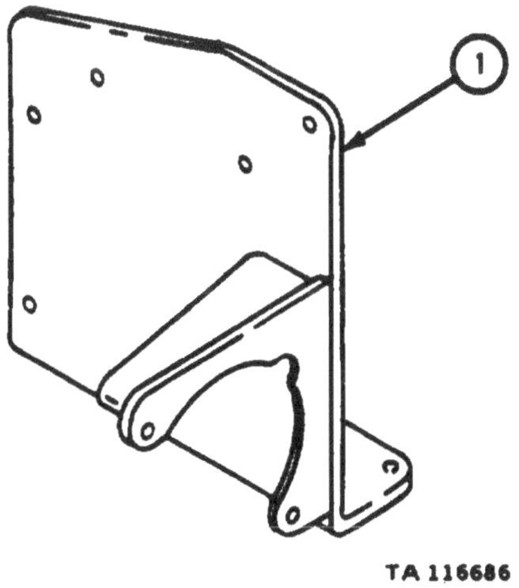

4-79. FLAME HEATER NONREPLACEABLE ELEMENT FILTER (SIDE-MOUNTED SYSTEM).

NOTE

Repair is limited to removal and replacement.

a. Removal.

FRAME 1

1. Take out machine screw (1), lockwasher (2), and machine nut (3) from flame heater bracket (4) .
2. Slip clamp (5) from filter element (6).

END OF TASK

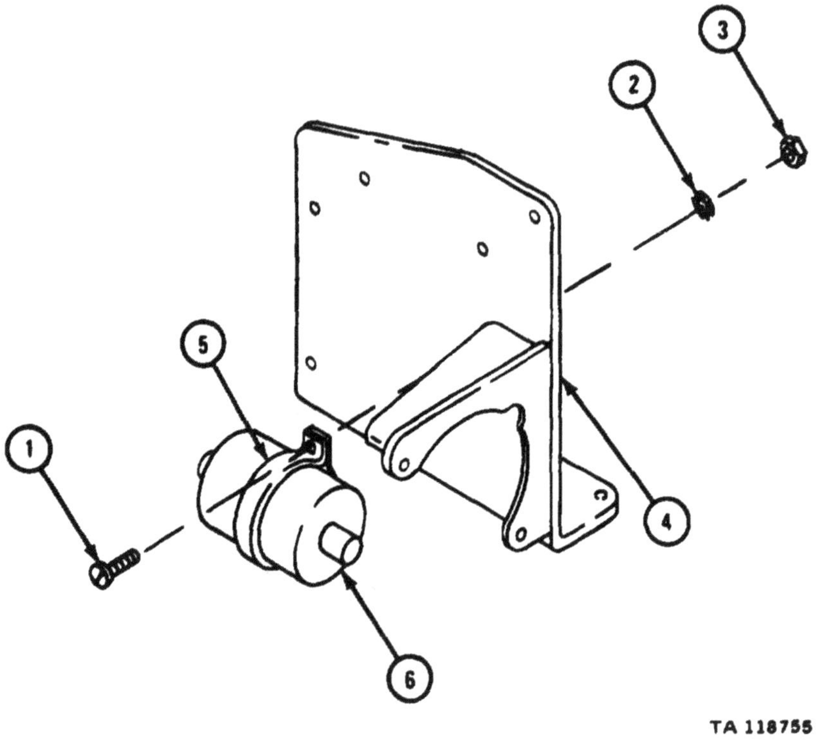

b. Replacement.

FRAME 1

1. Slide loop clamp (1) on filter element (2).
2. Hold filter element (2) in place against bracket (3) as shown.
3. Put in screw (4), lockwasher (5), and nut (6) .

END OF TASK

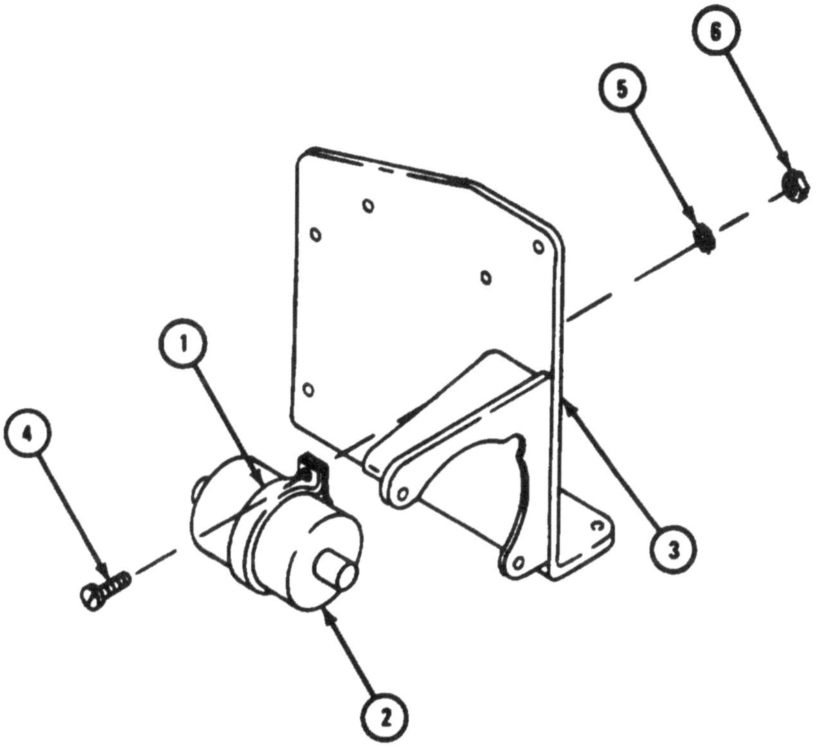

4-80. FLAME HEATER TUBES, CLAMPS, AND FITTINGS (SIDE-MOUNTED SYSTEM).

a. Cleaning.

(1) Clean plastic tubing with soap and water.

WARNING

Dry cleaning solvent is flammable. Do not use near an open flame. Keep a fire extinguisher nearby when solvent is used. Use only in well-ventilated places. Failure to do this may result in injury to personnel and damage to equipment.

(2) Clean elbow fittings and connectors with dry cleaning solvent.

(3) Clean uncoated loop clamps with dry cleaning solvent.

(4) Clean coated loop clamps with soap and water.

b. Inspection and Repair.

NOTE

This task is shown for one tube and one tube nut. It can be used for all flame heater tubes and tube nuts.

FRAME 1

1. Check that plastic tubing (1) is not cracked, kinked or brittle.
2. Check that plastic tubing (1) has not cracked around tube nuts (2).
3. Throw away plastic tubing (1) if it is cracked or damaged and get a new one.
4. Throw away plastic tubing (1) if tube nuts (2) are stripped or damaged and get a new one.
5. Check that 1/4-inch plastic tubing (1) has not shrunk at tube nut ferrules (2). Throw away 1/4-inch plastic tubing if No. 28 drill shank (3) cannot be put in tube opening.
6. Check that 1/8-inch plastic tubing (1) has not shrunk at tube nut ferrule (2). Throw away 1/8-inch plastic tubing if 1/32-inch drill shank (3) cannot be put in tubing opening .

END OF TASK

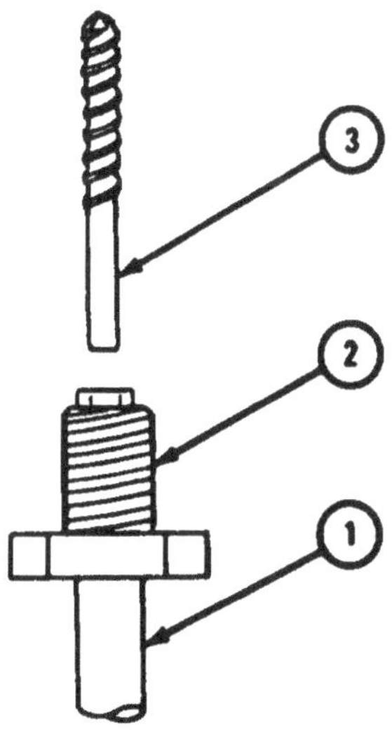

TA 118754

4-81. TIMING GEAR COVER.

a. Cleaning.

(1) Engine LDS-465-2 .

FRAME 1

NOTE

Do not take out dirt and liquid deflector (1) unless it is found to be damaged during inspection.

1. Press two crankshaft front oil seals (2) out of back of timing gear cover (3). Throw away seals.

WARNING

Dry cleaning solvent is flammable. Do not use near an open flame. Keep a fire extinguisher nearby when solvent is used . Use only in well-ventilated places. Failure to do this may result in injury to personnel and damage to equipment.

2. Clean timing gear cover (3) six screws (4), three nuts (5), and nine lockwashers (6) with dry cleaning solvent and stiff brush. Dry with clean cloth .

END OF TASK

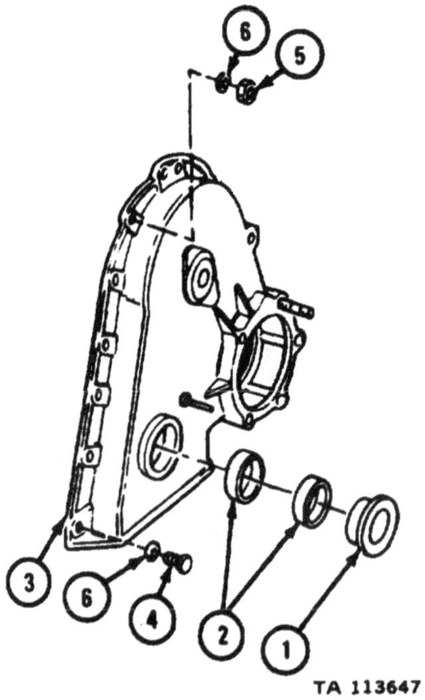

(2) All engines except LDS-465-2 .

FRAME 1

1. Press crankshaft front oil seal (1) out from rear of timing gear cover (2).

WARNING

Dry cleaning solvent is flammable. Do not use near an open flame . Keep a fire extinguisher nearby when solvent is used . Use only in well-ventilated places. Failure to do this may result in injury to personnel and damage to equipment.

2. Clean timing gear cover (2), six screws (3), three nuts (4), and nine lockwashers (5) with dry cleaning solvent and stiff brush. Dry with clean cloth.

END OF TASK

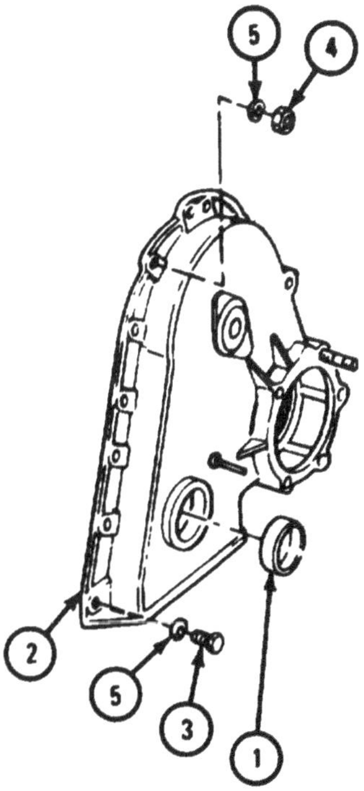

TA 113651

b. Inspection and Repair .

(1) Engine LDS-465-2 .

FRAME 1

1. Check that timing gear cover (1) is not cracked, dented or bent. If timing gear cover is damaged, get a new one.

WARNING

Dry cleaning solvent is flammable. Do not use near an open flame. Keep a fire extinguisher nearby when solvent is used . Use only in well-ventilated places. Failure to do this may result in injury to personnel and damage to equipment.

2. Check that gasket surfaces at bottom and at rear of cover (1) have no nicks, scratches or burrs. Fix minor damage with a fine mill file or crocus cloth dipped in dry cleaning solvent. If more repair is needed, get a new cover.

3. Check that timing pointer pin (2) is not loose, cracked or damaged in any way. If timing pointer pin is damaged, get a new timing gear cover.

4. Check that stud (3) does not have stripped threads and is not damaged in any way. Fix minor thread damage with a thread chaser. If more repair is needed, take out stud and put in a new one.

GO TO FRAME 2

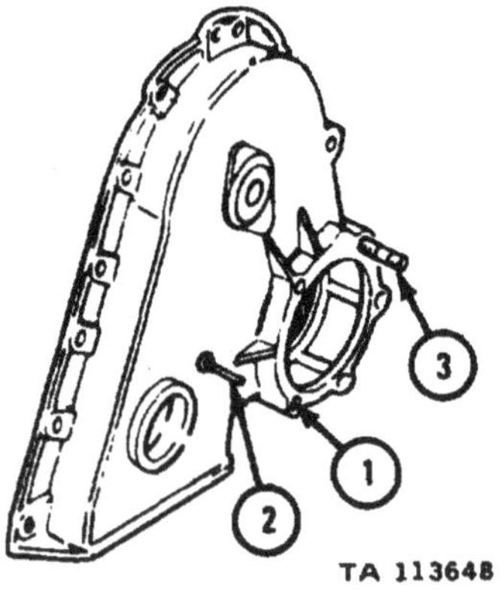

TA 113648

FRAME 2

1. Check that dirt and liquid deflector (1) is not cracked, dented or loose. If it is, press it out and press in a new dirt and liquid deflector.
2. Check that six screws (2), three nuts (3), and nine lockwashers (4) are no t stripped or damaged. If screws, nut or washers are damaged, get new ones.

GO TO FRAME 3

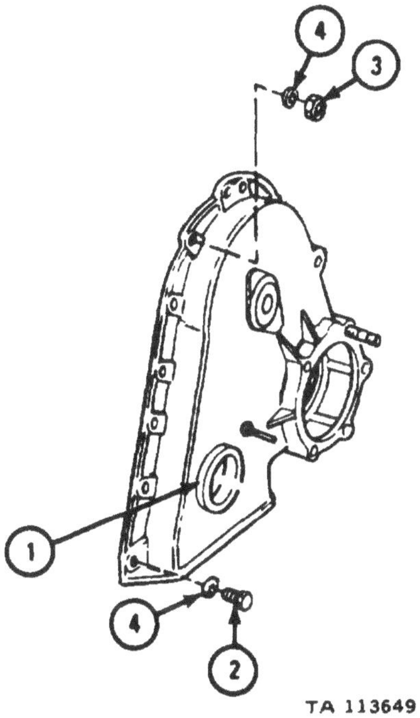

FRAME 3

1. Put a light coat of engine oil on oil seals (1) and seal installer.
2. Using arbor press, press in two oil seals (1) through back of timing gear cover (2) until they seat in recess in cover or seal installer touches cover.

END OF TASK

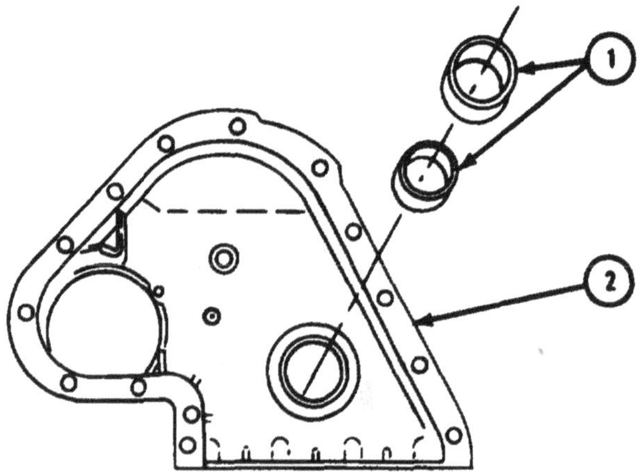

TA 117868

(2) All engines except LDS-465-2 .

FRAME 1

1. Check that timing gear cover (1) is not cracked, dented or burred. If cover is damaged, get a new one.

WARNING

Dry cleaning solvent is flammable. Do not use near an open flame. Keep a fire extinguisher nearby when solvent is used . Use only in w en-ventilated places. Failure to do this may result in injury to personnel and damage to equipment.

2. Check that gasket surfaces at bottom and at rear of cover (1) have no nicks, scratches or burrs . Fix minor damage with a fine mill file or crocus clot h dipped in dry cleaning solvent. If more repair is needed , get a new cover.

3. Check that timing pointer pin (2) is not loose, cracked or damaged in any way If pin is damaged, get a new cover.

4. Check that stud (3) does not have stripped or damaged threads and is not damaged in any other way. Fix minor thread damage with a thread chaser. If more repair is needed, take out stud and put in a new one.

GO TO FRAME 2

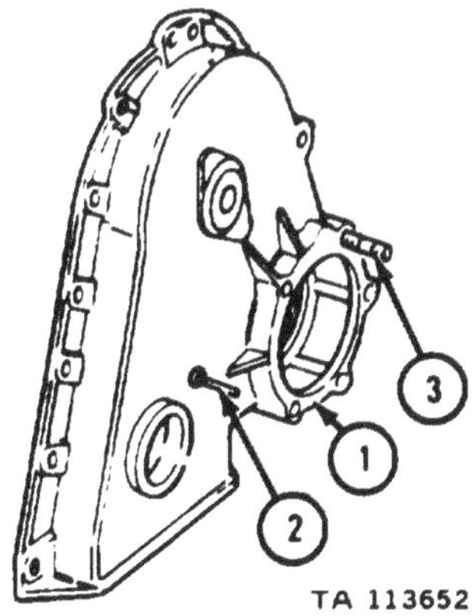

FRAME 2

1. Check to see that six screws (1), three nuts (2), and nine lockwashers (3) are not stripped . If screws, nuts or lockwashers are damaged, throw them away and get new ones.
2. Put a light coat of engine lubricating oil on crankshaft front oil seal (4).
3. Using arbor press and pressing plate, press crankshaft front oil seal (4) in from front of timing gear cover (5). Lip of crankshaft front oil seal must face out . Stop pressing when crankshaft front oil seal is flush with top edge of crankshaft front oil seal bore (6).

END OF TASK

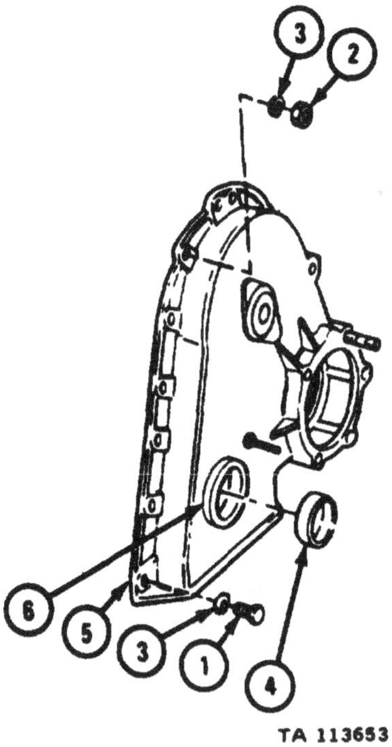

TA 113653

CHAPTER 5

FINAL ASSEMBLY

Section I. SCOPE

5-1. EQUIPMENT ITEMS COVERED. This chapter give s instructions to assemble the engine components into a final assembly.

5-2. EQUIPMENT ITEMS NOT COVERED. All equipment items are covered in this chapter.

Section II. ASSEMBLY OF ENGINE ASSEMBLY

NOTE

This procedure is the same for all engine models except where noted .

TOOLS: Ring compressor, pn C-10899159
-Din-per pulley locator (fabricated locally)
Oil pressure regulator valve adapter (fabricated locally)
Pilot bolt (fabricated locally)
Cylinder head nut wrench, pn D-10951485
Engine transport stand, NSN 4910-00-338-6673
Engine maintenance stand, NSN 4910-00-795-0189

SUPPLIES: Anti-scuff lubricant, Lubrizol 1060 or Texaco TLA111
Sealing compound, type II, MIL-S-45180
Lubricating oil, ICE, OE/HDO 30, MIL-L-2104
Cylinder head gasket (2)
Sealant, MIL-S-7916
Crankshaft rear oil seal
Crankshaft rear oil seal housing gasket
Lubricating oil, ICE, OE/HDO 10, MIL-L-2104
Flywheel housing-to-crankcase gaske t
Intake and exhaust manifold-to-cylinder head gasket (2)
Thermostat housing-to-cylinder head gaske t
Oil pressure regulator valve housing-to-crankcase gaske t
Injector pump adapter-to-crankcase gasket
Oil cooler gasket and preformed packing set
Cylinder head breather tube adapter-to-cylinder head covers gaskets (2)
Hydraulic pump-to-crankcase gaske t
Oil drain gasket
Oil inlet adapter gasket
Turbocharger-to-exhaust manifold gaske t
Intake elbow flame heater spark plug gasket
Starter mounting gasket (engine LDS 465-2)
Starter mounting gasket set

PERSONNEL: Two

EQUIPMENT CONDITION: All engine components cleaned and ready for assembly

5-3. MOUNTING CYLINDER BLOCK ON OVERHAUL STAND.

FRAME 1

NOTE

Cylinder block (1) should be mounted on stand mounting plates (2) with center of balance of cylinder block in center of mounting area.

1. Move chain hoist so cylinder block (1) is between engine stand mounting plates (2) .

GO TO FRAME 2

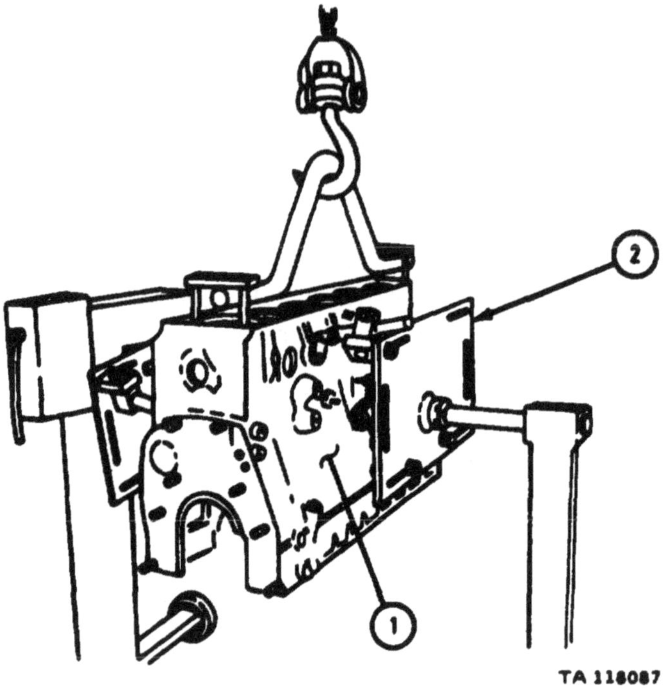

FRAME 2

WARNING

Do not use less than six universal mounting arms. Using less than six arms may cause cylinder block to fall, causing damage to equipment and injury to personnel.

NOTE

Capscrews used to mount cylinder block accessories will be used to mount cylinder block to maintenance stand. Universal mounting arms will be placed as far apart as practical and at various angles. Do not tighten capscrews until all cap-screws are put in.

1. Working on crank side of maintenance stand, put two universal mounting arms (1) in place on mounting plate (2). Put in two cap screws (3).
2. Put two L-shaped brackets (4) in place on cylinder block (5).
3. Put in two capscrews (6).

GO TO FRAME 3

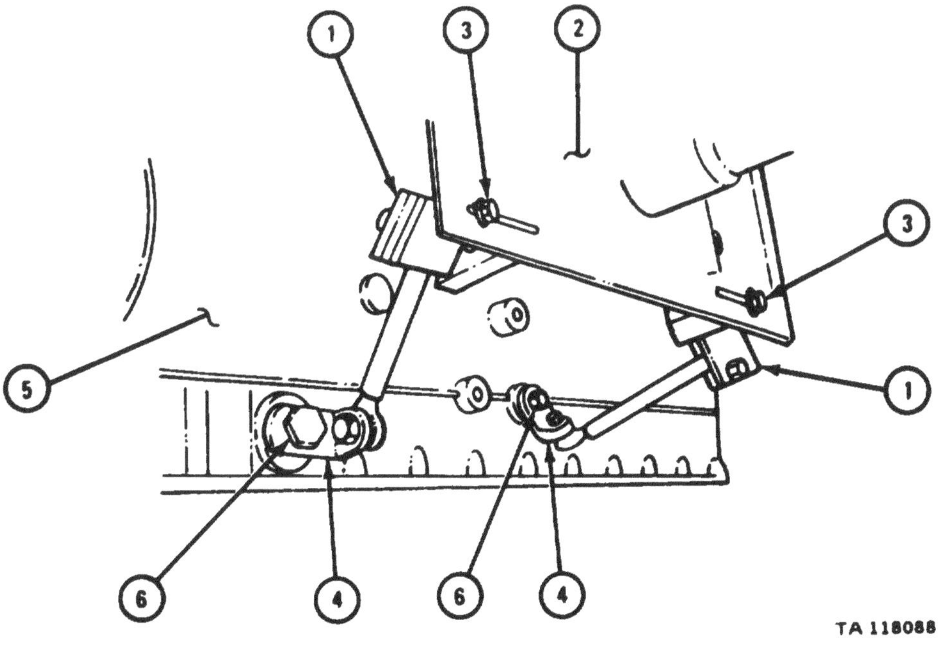

FRAME 3

1. Working on crank side of maintenance stand, put three universal arms (1) in place on mounting plate (2). Put in three capscrews mounting (3).
2. Put three L-shaped brackets (4) in place on cylinder block (5). Put in three capscrews (6) .

GO TO FRAME 4

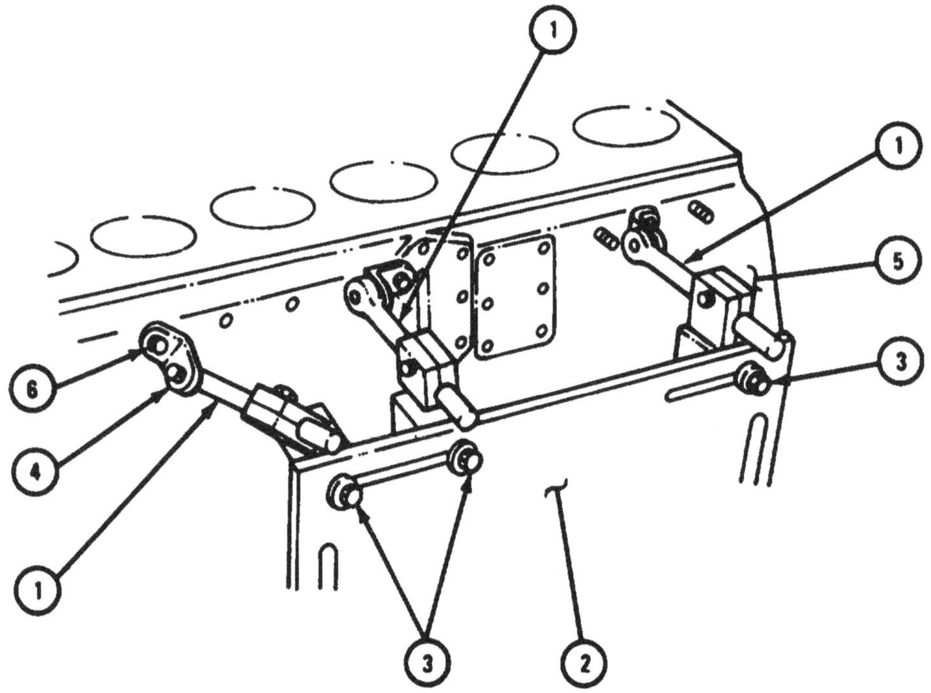

TA 118089

FRAME 4

1. Working on side of stand without crank, put three universal mounting arms (1) in place on mounting plate (2). Put in three capscrews (3) .
2. Put three L-shaped brackets (4) in place on cylinder block (5). Put in three capscrews (6) .
3. Tighten all capscrews .
4. Lower chain hoist and take off lifting sling.

END OF TASK

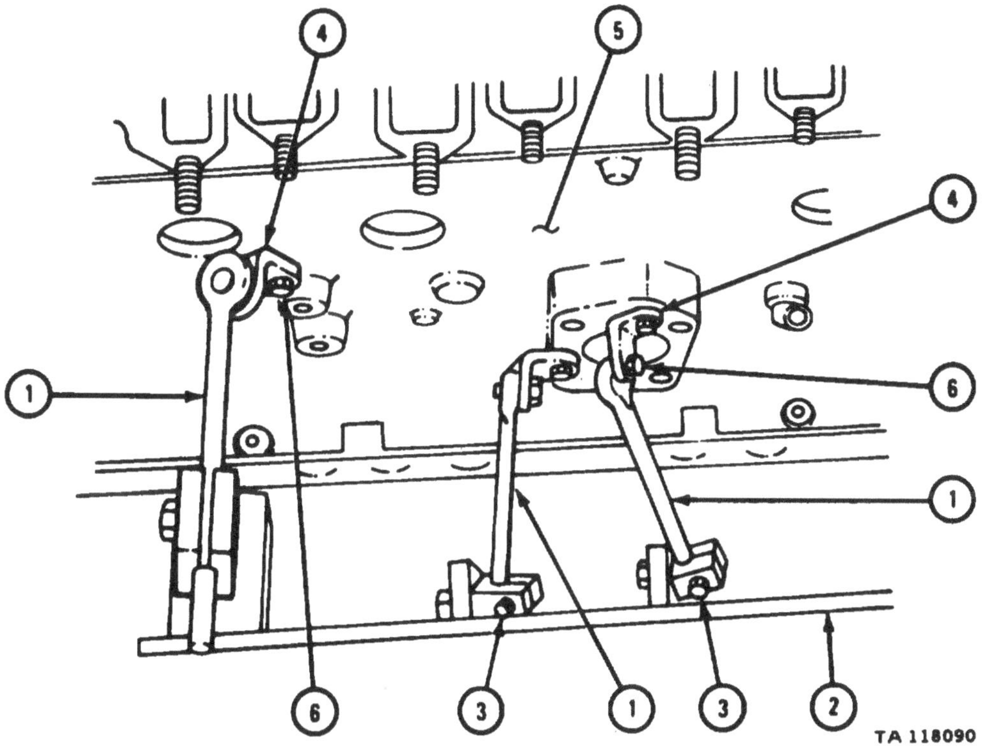

5-4. CAMSHAFT ASSEMBLY, CAMSHAFT REAR BEARING PLUG AND PISTON COOLING NOZZLES.

a. Camshaft Assembly.

FRAME 1

1. Coat 12 lobes (1) of camshaft (2) with anti-scuff lubricant.
2. Push camshaft (2) in crankcase (3).
3. Put camshaft thrust plate (4) over camshaft end (5) and on crankcase (3).
4. Put in and tighten two capscrews (6) with lockwashers (7).

END OF TASK

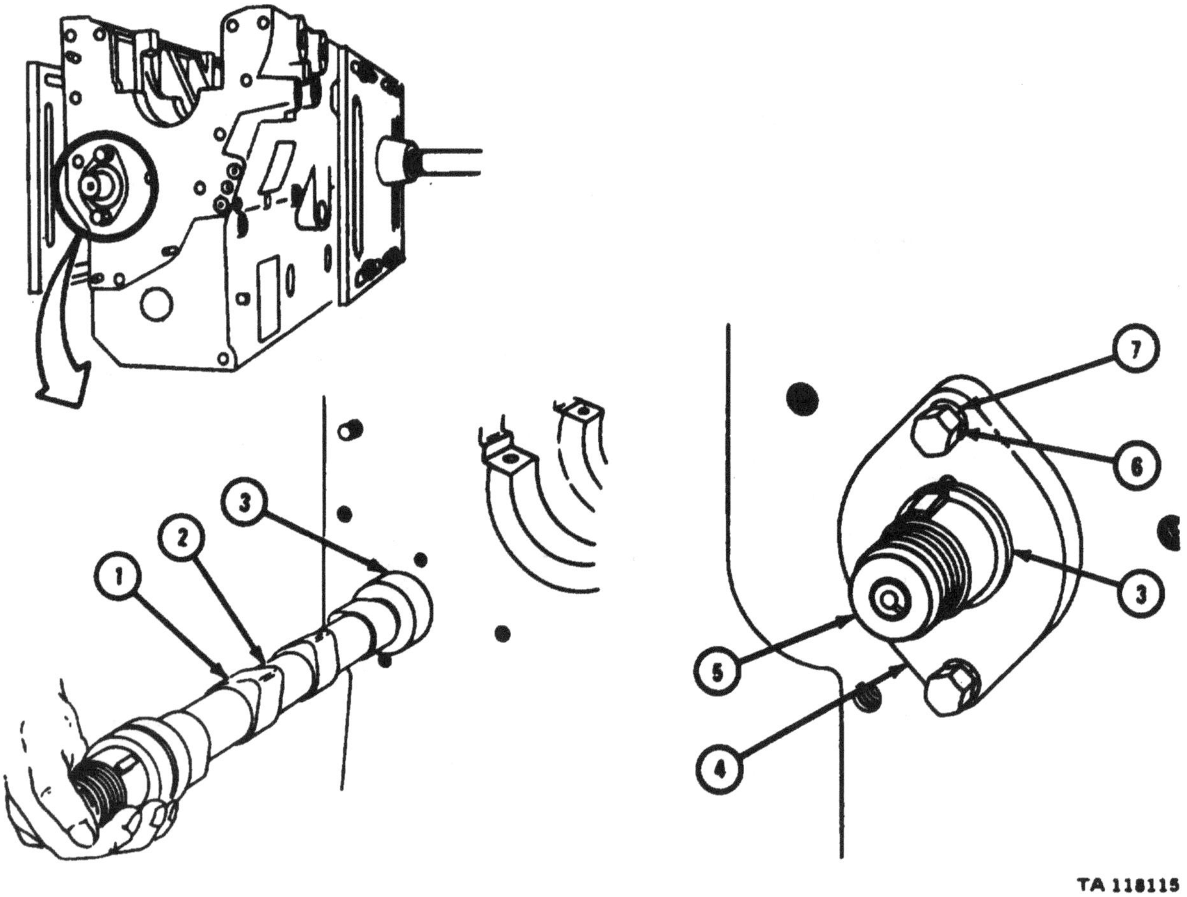

b. Camshaft Rear Bearing Plug.

FRAME 1

1. Coat edge of rear bearing plug (1) with sealant.

NOTE

After putting in plug (1), do not bump camshaft. Bumping could unseat plug and cause an oil leak.

2. Put plug (1) in crankcase (2). Using hammer, tap center of plug to seat plug.

END OF TASK

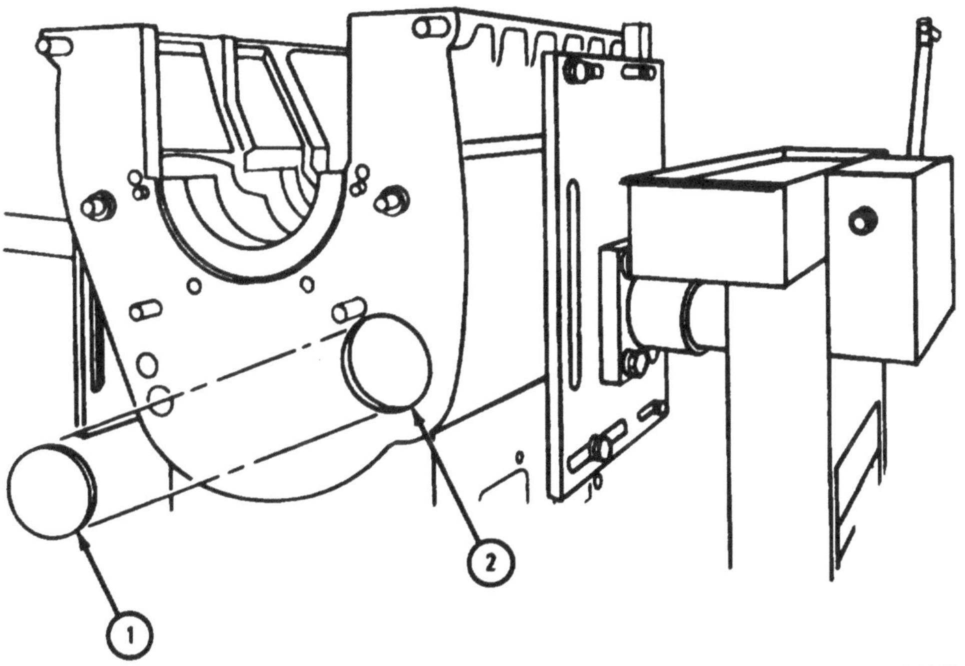

c. Piston Cooling Nozzles.

FRAME 1

NOTE

Piston cooling nozzles (1) must be put on the same cylinder (2) they were taken from.

Cylinders are named by numbers one through six, counting from front to rear of crankcase. Each piston cooling nozzle (1) is tagged with the number of cylinder (2) it was taken from.

1. Put piston cooling nozzle (1) on cylinder block (2) at each of six cylinders (3) as tagged. Take off tags .
2. Put on and tighten six self-locking bolts (4).

END OF TASK

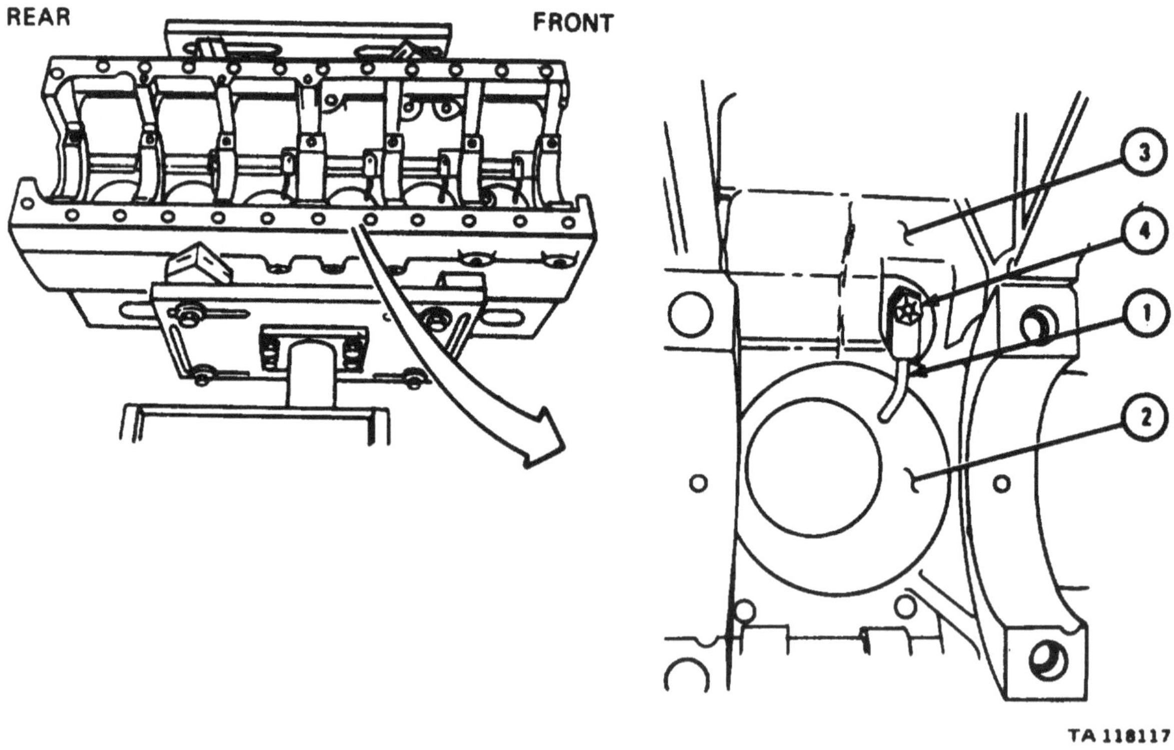

5-5 CRANKSHAFT ASSEMBLY AND ENGINE FRONT PLATE.

FRAME 1

NOTE

Main bearing upper halves were tagged during disassembly and must be put back in the same saddle they were taken from.

1. Oil both sides of six main bearing upper halves (1) and main thrust bearing upper half (2) .
2. Put six main bearing upper halves (1) into six main bearing saddles (3). Put main thrust bearing upper half (2) into center saddle (4).
3. Check that oil hole (5) in main bearing upper halves (1) and main thrust bearing upper half (2) are alined with oil hole (6) in bearing saddles (3 and 4).

GO TO FRAME 2

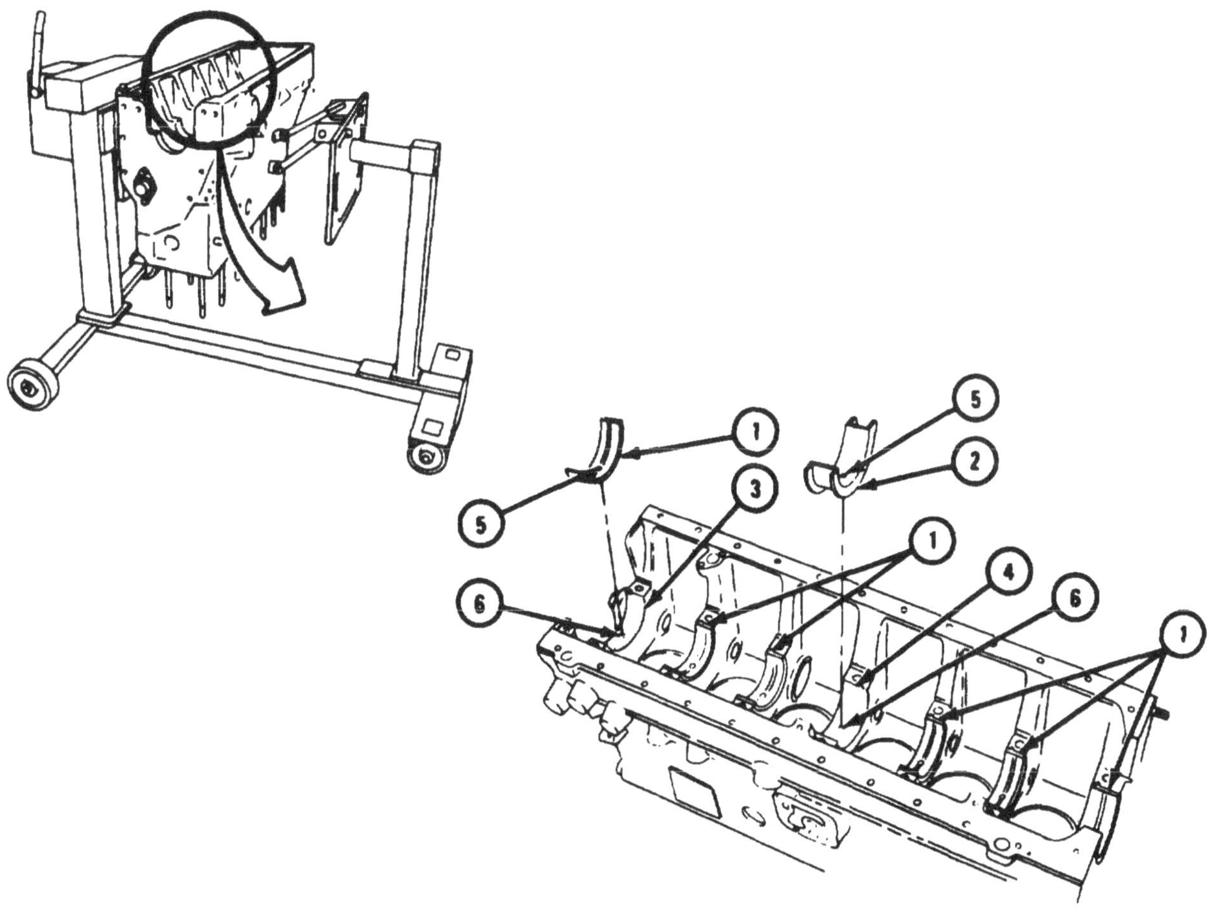

TA 117947

FRAME 2

CAUTION

Use only rope sling to lift crankshaft. Cable or chain sling may damage bearing surfaces of the crankshaft. Bearing surfaces should not be touched with bare hands, Crankshaft bearing surfaces should be wrapped with cheesecloth when handling.

NOTE

Oil all crankshaft bearing surfaces.

1. Put rope sling (1) on crankshaft (2) as shown. Hook rope sling onto hoist hook (3) .
2. Lift crankshaft (2) over engine block (4) .

WARNING

Crankshaft (2) is very heavy. Do not let fingers or hand get between crankshaft and engine block (4). Do not let crankshaft bump engine block. Failure to use proper caution may result in damage to equipment and injury to personnel.

Soldier A 3. Guide crankshaft (2) into engine block (4) .

Soldier B 4. Lower crankshaft (2) into engine block (4) .

GO TO FRAME 3

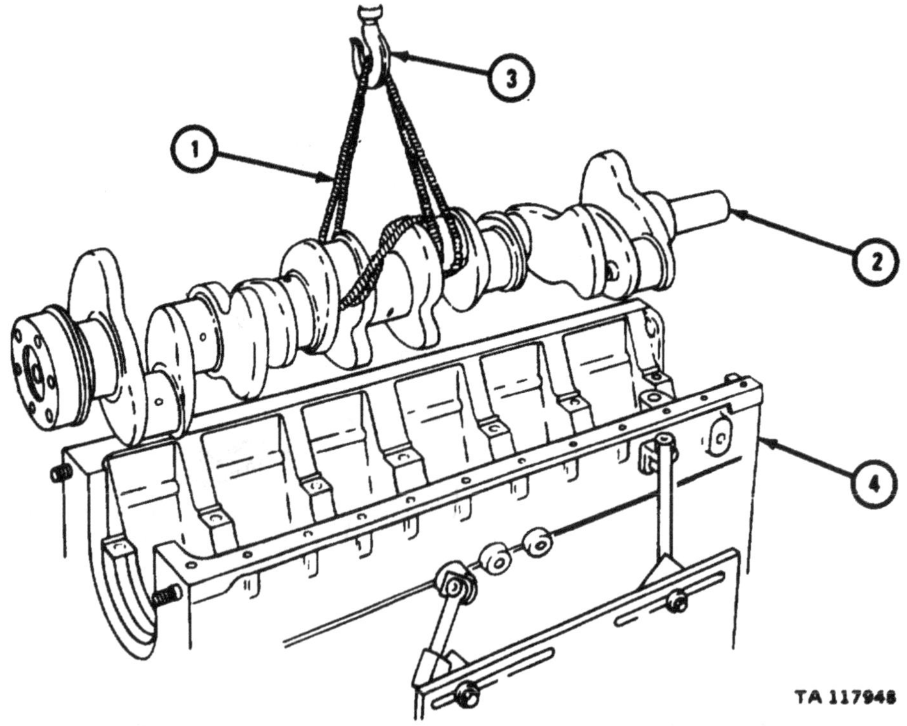

FRAME 3

NOTE

The number four main bearing lower half (1) is called thrust main bearing lower half. It is different from other main bearing lower halves because it has flanges (2).

1. Coat thrust main bearing lower half (1) with oil on both sides.

NOTE

The number four main bearing cap (3) is called thrust main bearing cap. It is different from othe r ing caps because it has a thrust shoulder (4).

2. Put thrust main bearing lower half (1) into thrust main bearing cap (3).
3. Put thrust main bearing cap (3) in place on engine block (5). Put in two cap screws with washers (6). Do not use torque wrench at this time .

GO TO FRAME 4

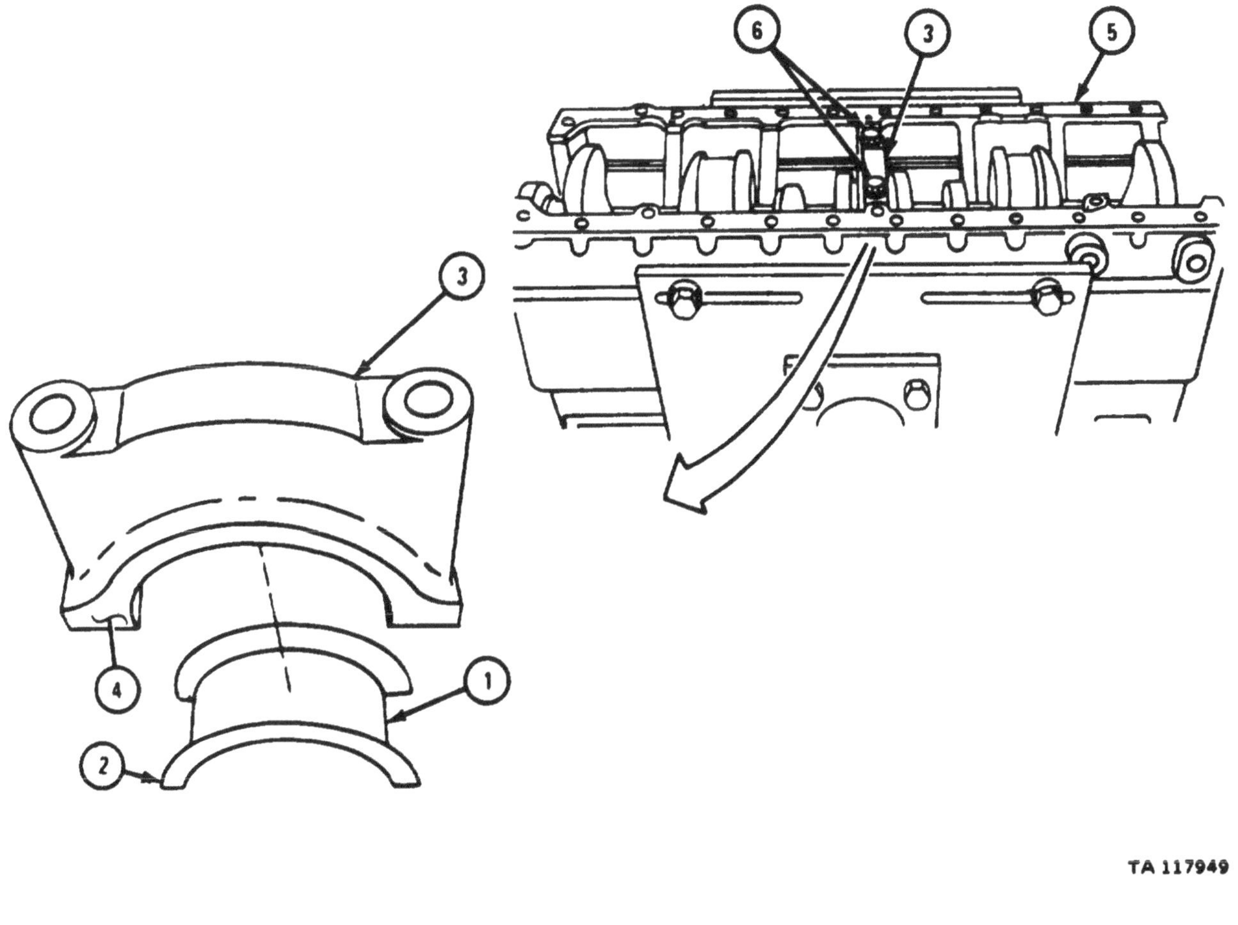

FRAME 4

1. Use prybar (1) between center main bearing web (2) and crankshaft counter-weight (3) . Push crankshaft (4) all the way to the rear of cylinder block (5).
2. Push crankshaft (4) all the way to front of cylinder block (5).
3. Tighten main bearing cap screws (6) to 80 to 90 pound-feet.
4. Tighten main bearing cap screws (6) to 115 to 120 pound-feet.

GO TO FRAME 5

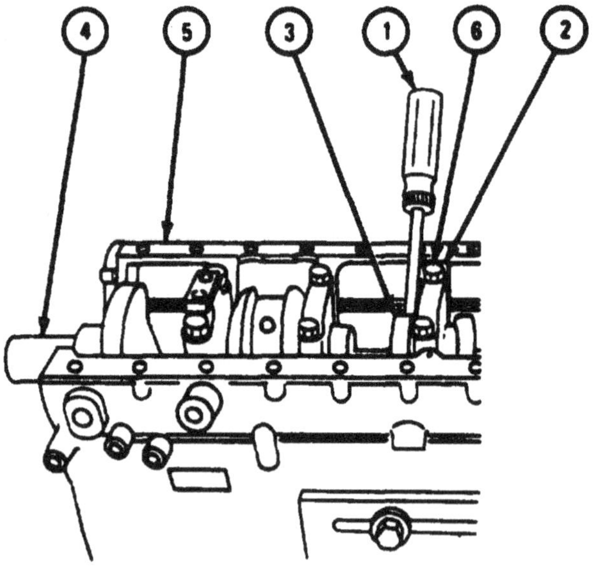

TA 117950

FRAME 5

1. Mount dial indicator (1) on cylinder block (2) so dial indicator point (3) rests on front end of crankshaft (4).
2. Using prybar (5), push crankshaft (4) as far as it will go to rear of cylinder block (2).
3. Set dial indicator (1) to zero and push crankshaft (4) as far as it will go to front of cylinder block (2).
4. Read dial indicator (1). Crankshaft end play must be between 0.008 and 0.022 inch.

GO TO FRAME 6

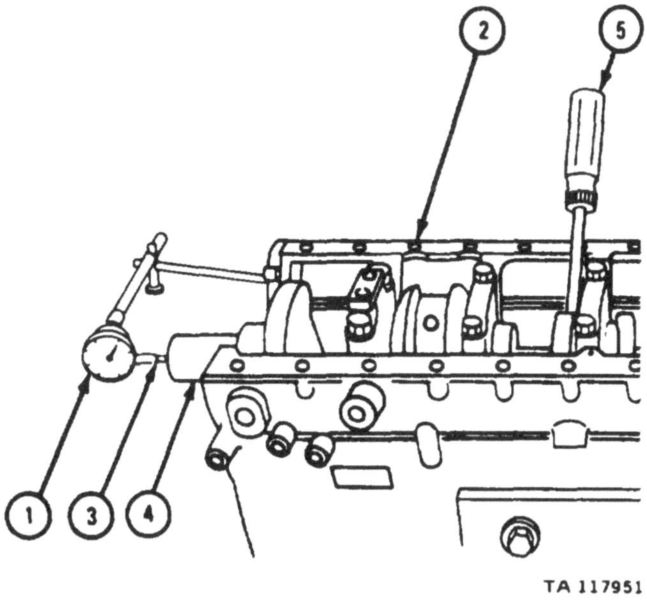

FRAME 6

1. Coat bottom surface of left rear main bearing cap oil seal tab (1) with cement. Let cement set for three minutes.
2. Coat rear main bearing cap oil seal tab left side mating surface (2) with cement. Let cement set for three minutes.
3. Put left rear main bearing cap seal (3) in rear main bearing cap (4) with seal tab (1) flush with bearing cap top.
4. Do steps 1 through 3 again for right rear main bearing cap oil seal (5).
5. Let cement dry for five minutes before putting in rear main bearing cap (4).

GO TO FRAME 7

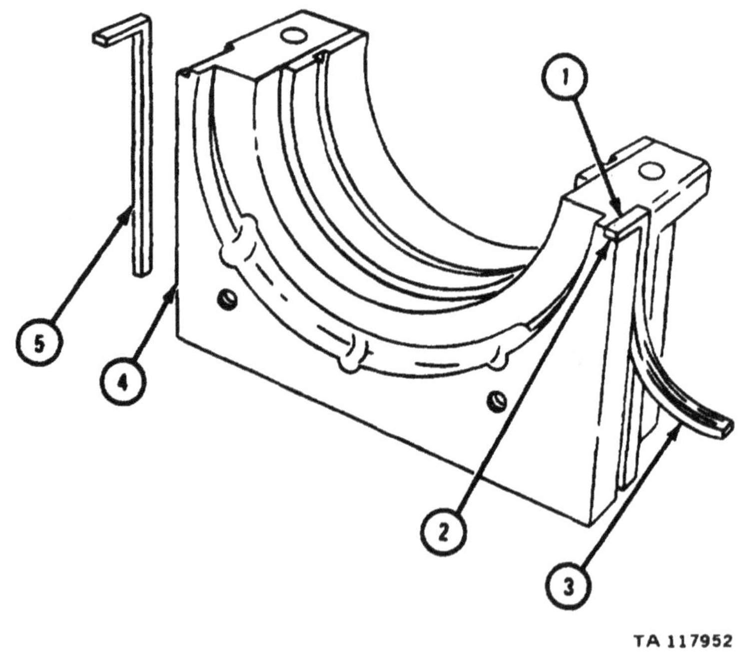

FRAME 7

1. Coat all mating surfaces of rear main bearing cap (1) with oil.
2. Coat both sides of lower rear main bearing half (2) with oil.
3. Put bearing half (2) in bearing cap (1).

NOTE

Be careful when putting in rear main bearing cap (1) so seals (4) are not damaged.

4. Put rear bearing cap (1) in place in cylinder block (3).

GO TO FRAME 8

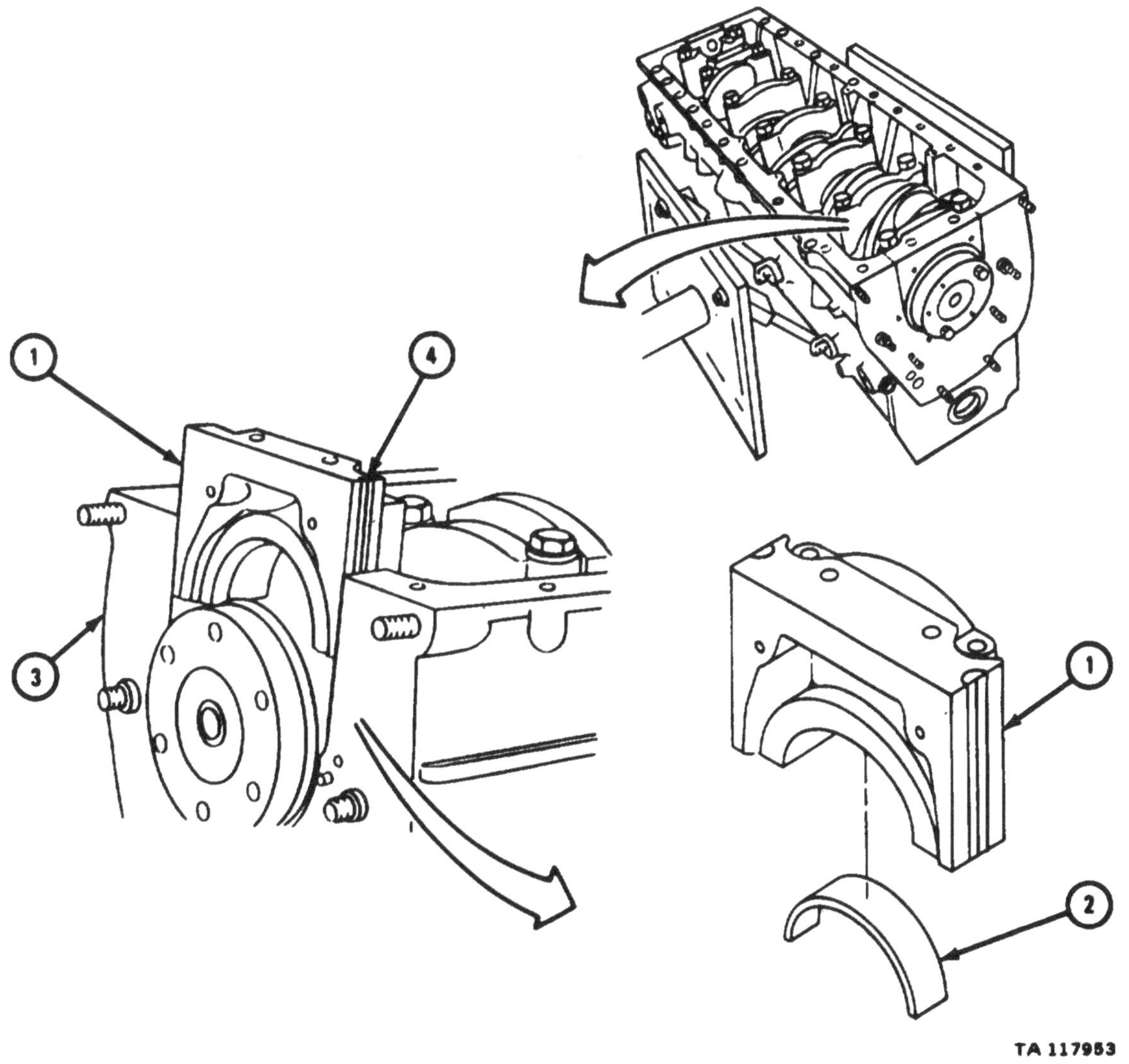

FRAME 8

1. Check that oil seals (1) are flush with top of bearing cap (2) to with in 0.0625 inch.
2. Put in two capscrews with washers (3). Tighten capscrews to 80 to 90 pound-feet.
3. Tighten capscrews (3) to 115 to 120 pound-feet.
4. Using protrusion gage (4), check alinement of main bearing cap (2). Alinement must be within 0.002 inch.

GO TO FRAME 9

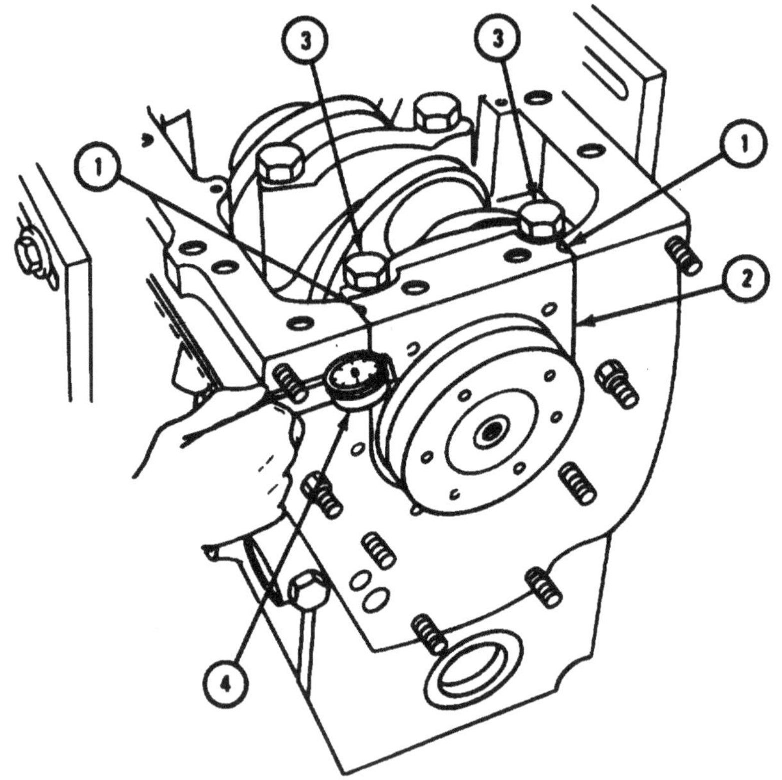

TA 117954

FRAME 9

CAUTION

Main bearing lower halves and main bearing caps were marked during removal. They must be put back in the same position and facing the same way. Failure to do this will result in damage or wear to engine.

1. Coat four main bearing caps (1) and four main bearing lower halves (2) with oil.
2. Put four main bearing lower halves (2) into four main bearing caps (1). Put four main bearing caps into cylinder block (3). Put in eight cap screws with washers (4) .
3. Tighten eight capscrews (4) to 80 to 90 pound-feet. Tighten eight capscrew s to 120 to 125 pound-feet.

GO TO FRAME 10

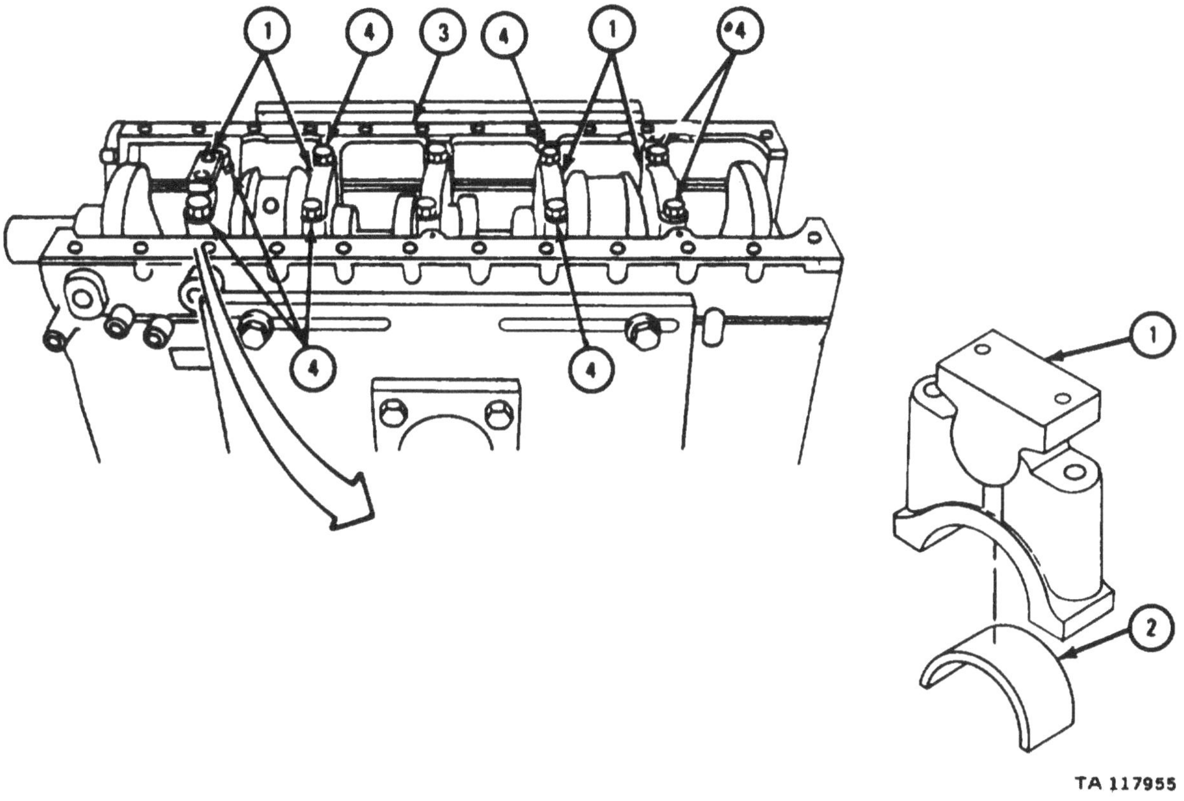

FRAME 10

1. Push crankshaft (1) as far as it will go toward the rear of cylinder block (2).
2. Mount dial indicator (3) so indicator pointer (4) rests on outside edge of front of flywheel mounting flange (5).

NOTE

When turning crankshaft (1), hold it all the way to the rear position.

3. Turn crankshaft (1) one full turn . Dial indicator (3) reading must not be more than 0.002 inch.

GO TO FRAME 11

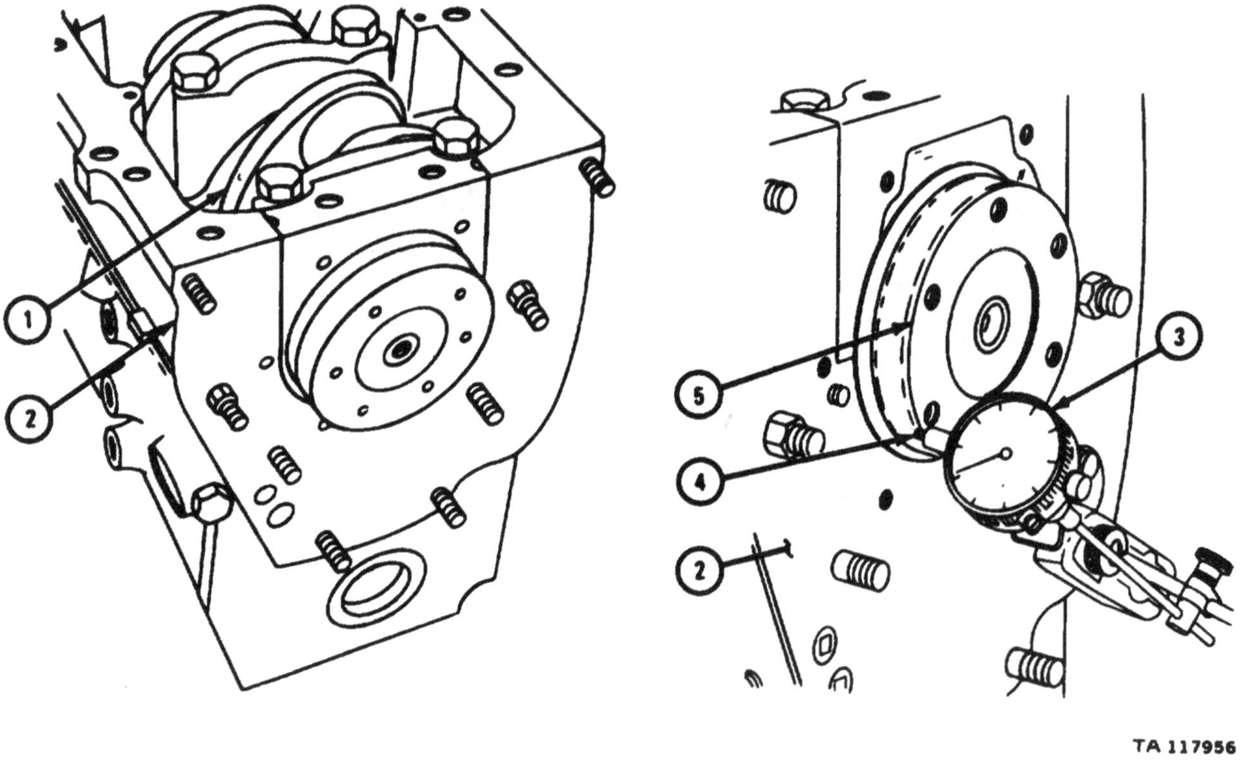

FRAME 11

1. Coat both sides of gasket (1) with sealant. Put gasket in place on cylinde r block (2) .
2. Put engine front plate (3) in place. Put in and tighten six screws with lock-washers (4) .
3. Coat front main bearing cap (5) and lower main bearing half (6) with oil. Put main bearing half into main bearing cap.
4. Put front main bearing cap (5) in place in cylinder block (2). Put in two cap screws with washers (7). Tighten two capscrews to 80 to 90 pound-feet.
5. Tighten two cap screws (7) to 120 to 125 pound-feet.

END OF TASK

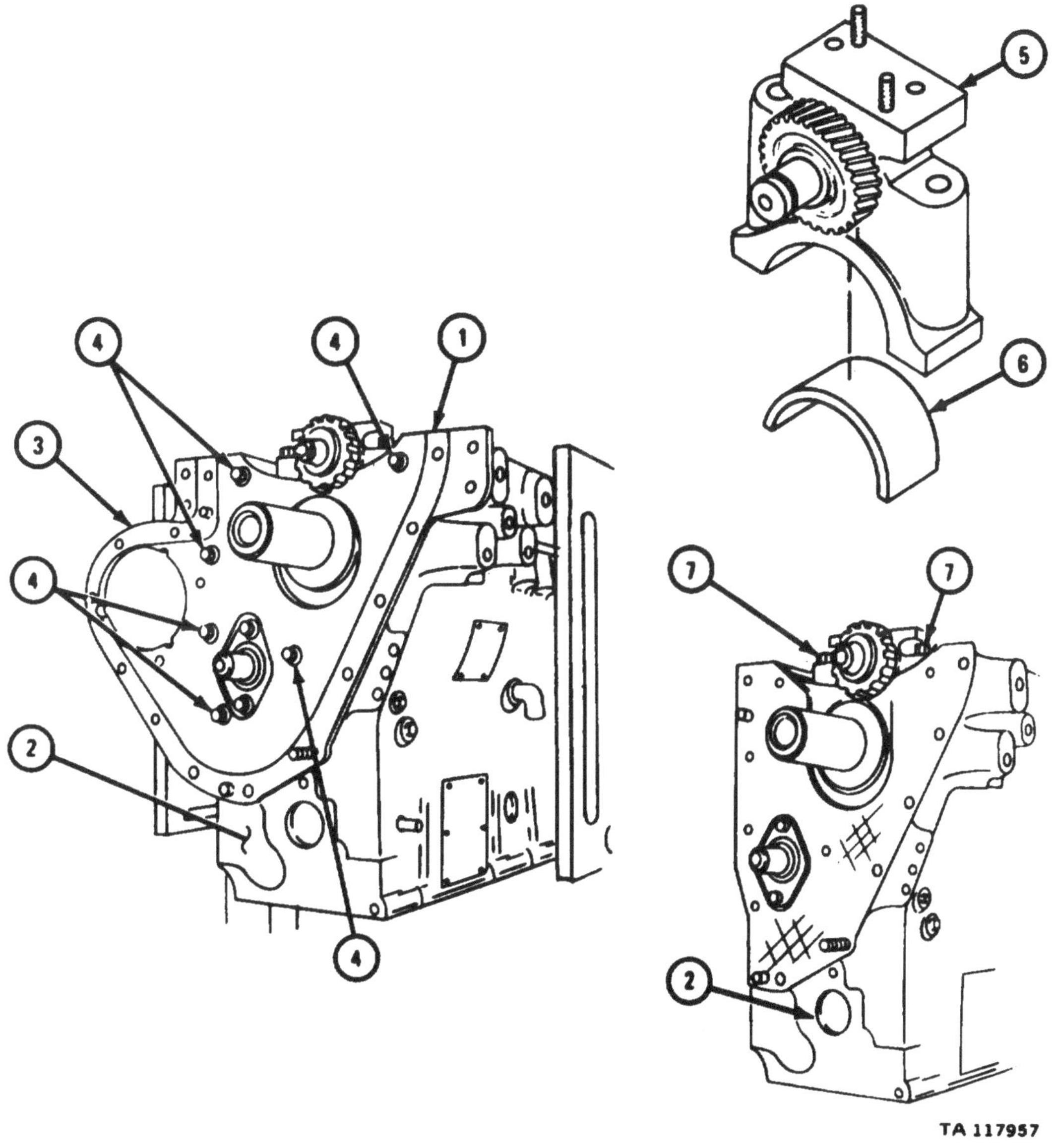

5-6. CRANKSHAFT GEAR AND CAMSHAFT GEAR.

FRAME 1

WARNING

Use welder's gloves to handle heated crankshaft gear (1) and camshaft gear (2). Hot metal can seriously burn personnel.

NOTE

Crankshaft and camshaft gears must be put on, and camshaft gear retaining nut put on and tightened as soon as possible after the gears are heated. Use 250° Tempilstick to check temperature of heated parts.

1. Heat crankshaft gear (1) and camshaft gear (2) to 250°F.
2. Aline keyway in crankshaft gear (1) with key on crankshaft (3) and push crankshaft gear in place on crankshaft, making sure crankshaft gear teeth mesh with teeth on oil pump idler gear (4).
3. As camshaft gear (2) is being put on camshaft (5), aline keyway in camshaft gear with key on camshaft.
4. Aline timing marks (6) on camshaft gear (2) with timing mark (7) on crankshaft gear (1).

GO TO FRAME 2

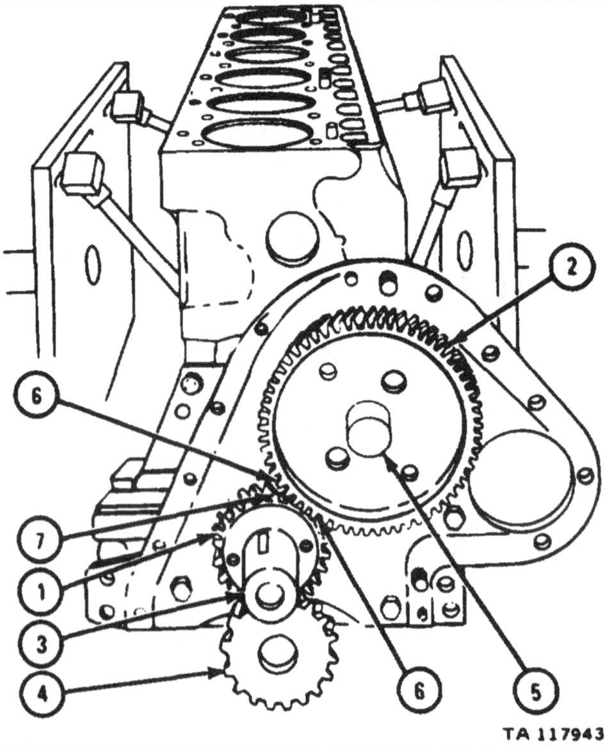

FRAME 2

1. Put on camshaft gear retaining nut (1).
2. Wedge clean rag (2) between camshaft gear (3) and crankshaft gear (4) as shown so gear will not turn.
3. Tighten camshaft gear retaining nut (1) to 325 to 350 pound-feet. Takeou t rag (2) .

GO TO FRAME 3

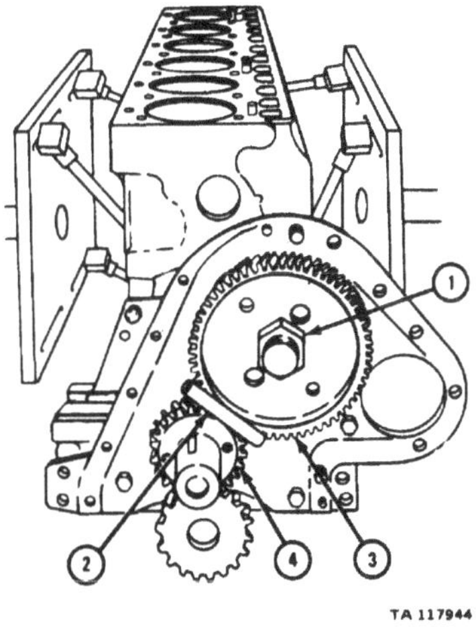

FRAME 3

1. Mount dial indicator (1) so indicator pointer (2) rests on teeth of camshaft gear (3) .
2. Turn gear (3) as far as it will go one way and hold it.
3. Set dial indicator (1) to zero.
4. Turn gear (3) as far as it will go the other way. Read backlash on dial indicator (1) . Camshaft gear backlash must be between 0.003 and 0.009 inch.

GO TO FRAME 4

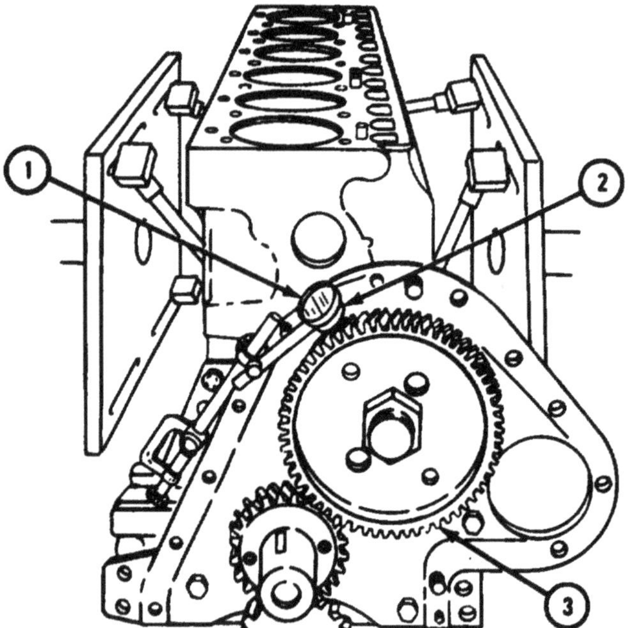

TA 117945

FRAME 4

1. Mount dial indicator (1) so indicator pointer (2) rests on end of camshaft (3).
2. Push camshaft (3) in as far as it will go and set dial indicator (1) to zero.
3. Using heavy screwdriver between camshaft gear (4) and cylinder block (5), pry camshaft (3) out as far as it will go and hold it.
4. Read dial indicator (1). End play of camshaft (3) must be between 0.002 to 0.015 inch.

END OF TASK

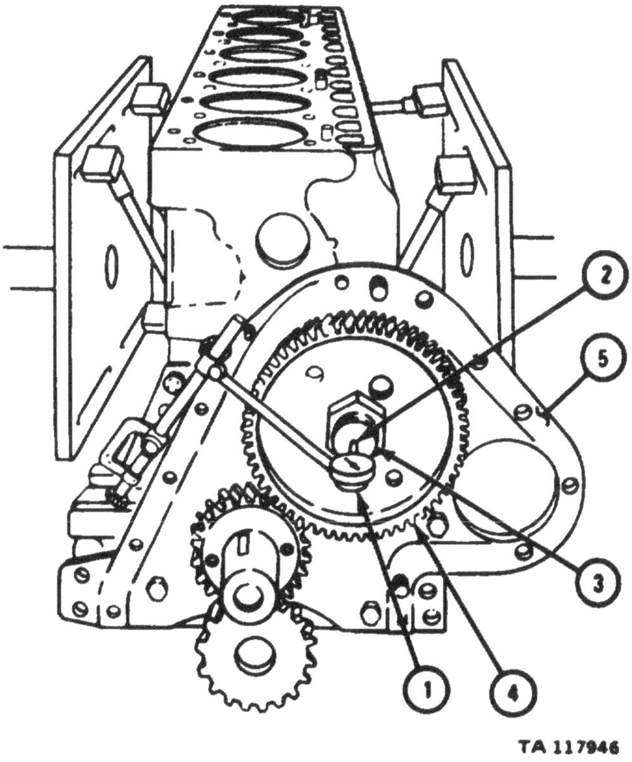

5-7. CRANKSHAFT REAR OIL SEAL AND HOUSING.

FRAME 1

1. Put crankshaft rear oil seal housing (1) on arbor press table (2), Gasket surface of rear oil seal housing should face down as shown.
2. Lightly coat oil seal (3) with lubricating oil.
3. Start oil seal (3) into crankshaft rear oil seal housing (1) by hand. Oil seal lip (4) should face down as shown.

NOTE

Square, flat metal plate may be used instead of round pressing arbor (5) . Plate must be bigger than diameter of oil seal (3).

4. Using pressing arbor (5) with diameter larger than outside diameter of oil seal (3), press oil seal all the way into crankshaft rear oil seal housing (1).

GO TO FRAME 2

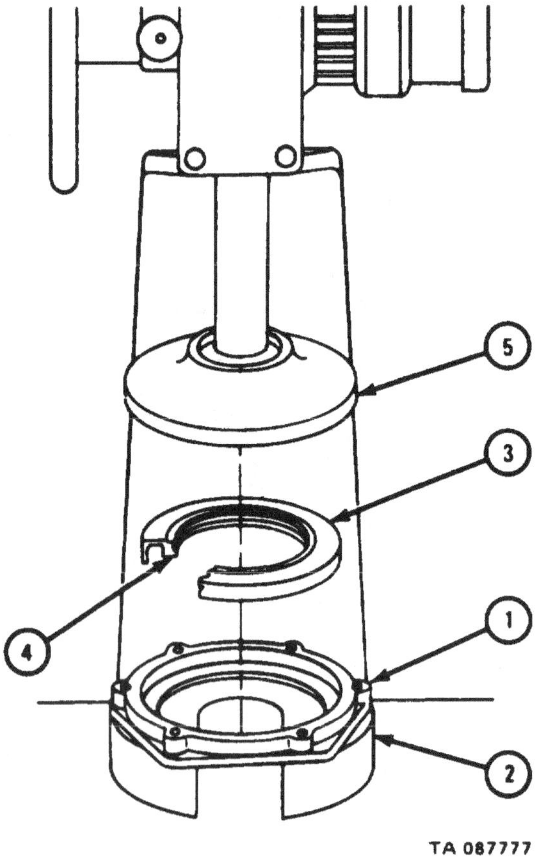

FRAME 2

1. Lightly coat crankshaft hub (1) with lubricating oil.
2. Put crankshaft rear oil seal gasket (2) on rear face of crankcase (3) as shown. Two dowel pins (4) in rear face of crankcase should fit through dowel pin holes in crankshaft rear oil seal gasket.
3. Slide crankshaft rear oil seal housing (5), with rear oil seal inside, over crankshaft hub (1) . Using soft-faced hammer, tap oil seal housing so two dowel pins (4) go through dowel pin holes (6) in oil seal housing.

NOTE

Two lockplates (7) are used only with aluminum crankshaft rear oil seal housings. Do not put on lockplates unless they were taken off.

4. Put on two lockplates (7) and put in six screws (8) and lockwashers (9).
5. If two lockplates (7) were put on, bend up ends of lockplates so screws (8) do not loosen.

END OF TASK

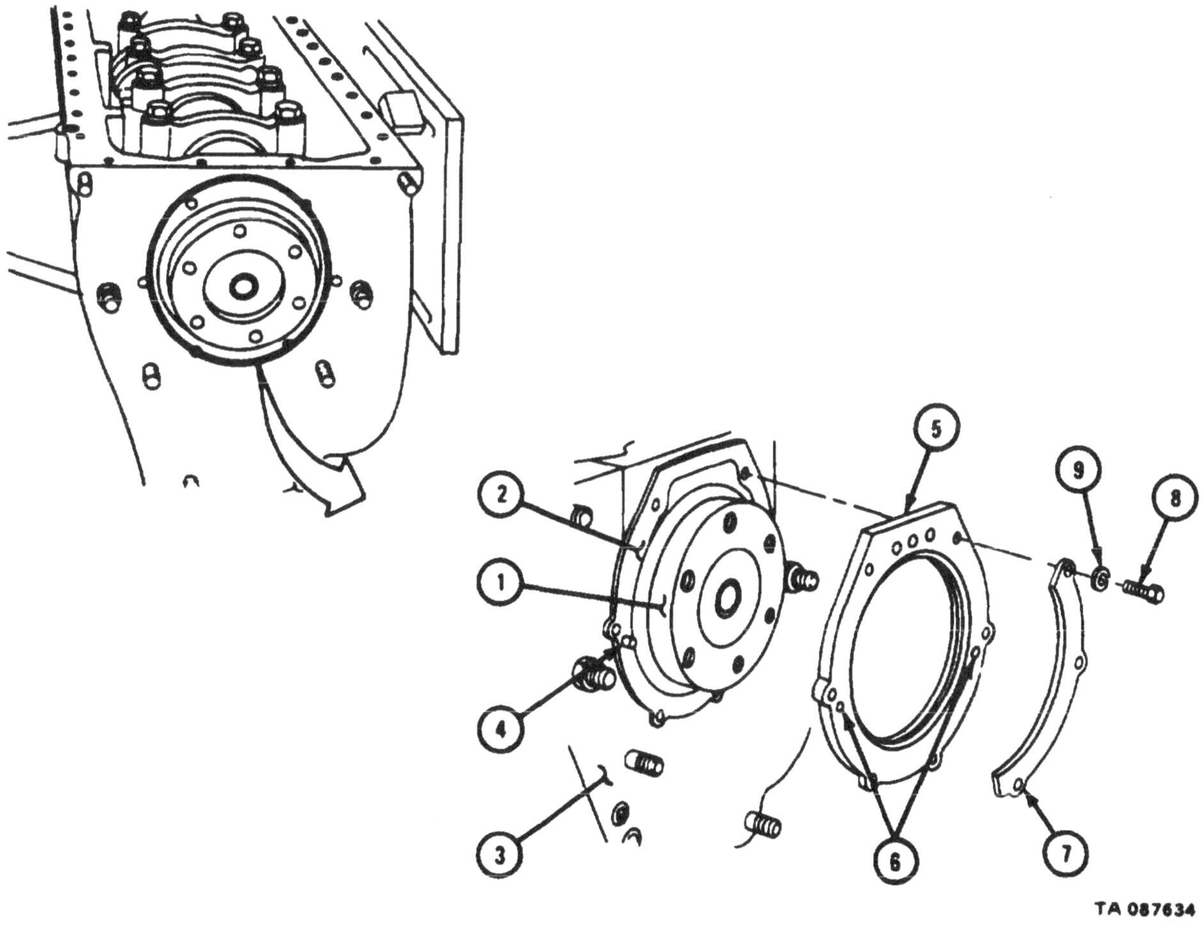

5-8. CRANKSHAFT DIRT AND LIQUID DEFLECTOR TIMING GEAR COVER, TACHOMETERADAPTER, AND CRANKSHAFT DAMPER AND PULLEY ASSEMBLY.

a. Crankshaft Dirt and Liuid Deflector (Engines LD-465-1, LD-465-1C, LDT-465-1C, LDS-465-1, and LDS-465-1A).

FRAME 1

1. With shouldered side of crankshaft dirt and liquid deflector (1) facing crankshaft gear (2), slide deflector over crankshaft (3).

END OF TASK

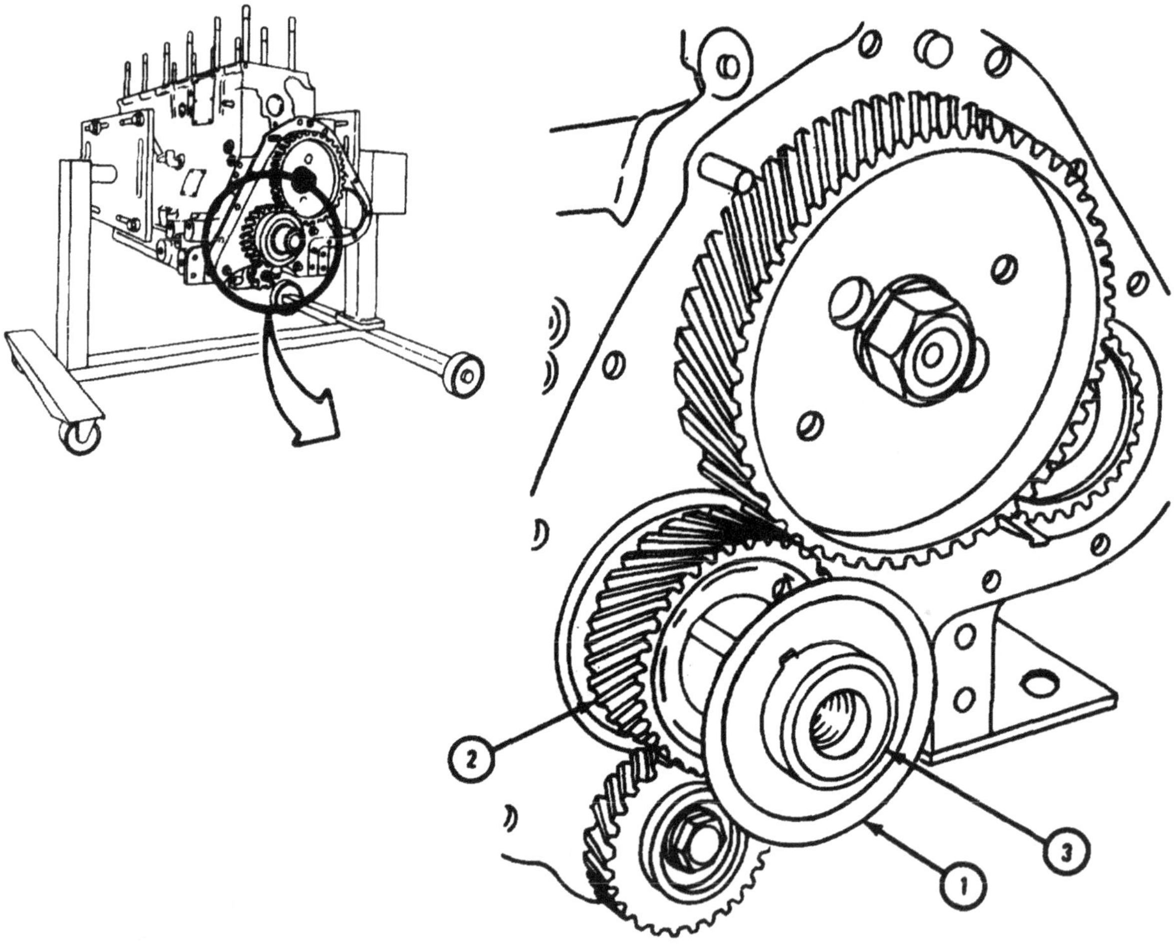

TA 117958

b. Timing Gear Cover.

FRAME 1

1. Put timing gear cover gasket (1) on cylinder block (2).
2. Put timing gear cover (3) on gasket (1).
3. Put in six capscrews with lockwashers (4).
4. Put in capscrew (5) . Put on plain nut (6) with lockwasher (7).
5. Put in capscrew (8) . Put on plain nut (6) with lockwasher (7).
6. Put on plain nut with lockwasher (9).

END OF TASK

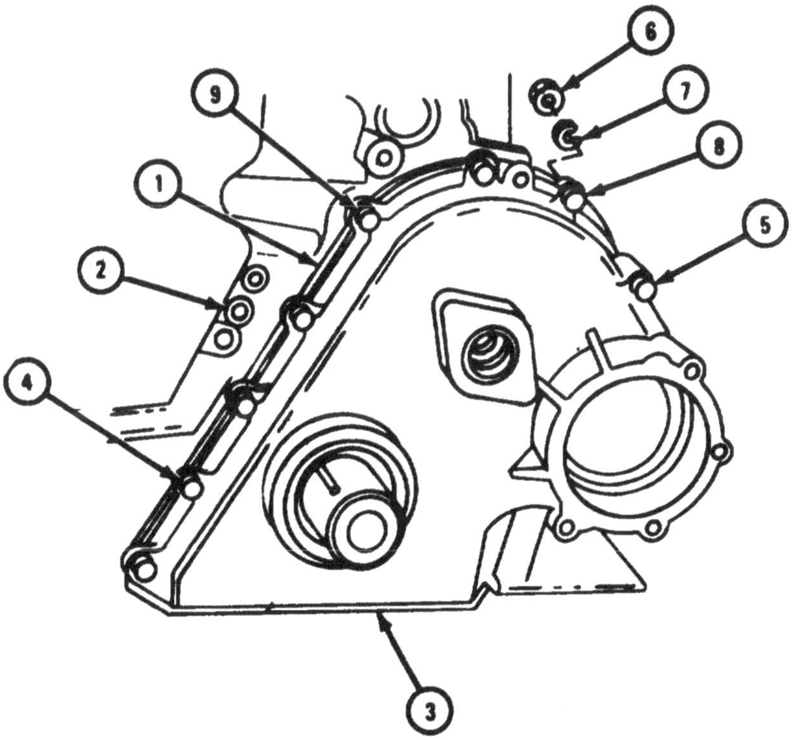

TA 117959

c. Tachometer Adapter.

NOTE

If working on engine LDS-465-2, go to frame 2.

FRAME 1

1. Put tachometer adapter gasket (1) on gear cover (2).
2. Put tachometer takeoff adapter (3) on adapter gasket (1).
3. Put tachometer drive shaft (4) in tachometer drive sleeve of camshaft (5).
4. Put tachometer drive adapter (6) on tachometer takeoff adapter (3).

END OF TASK

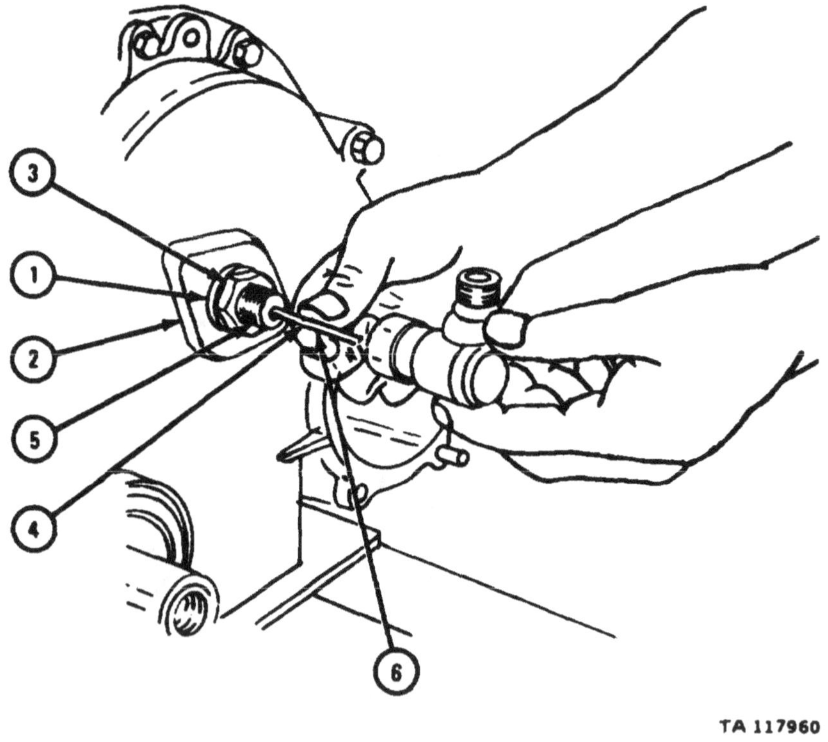

FRAME 2

1. Put tachometer adapter gasket (1) on gear cover (2).
2. Put on and tighten tachometer adapter (3).
3. Put on flat washer (4).
4. Put on and tighten tachometer adapter assembly (5).

END OF TASK

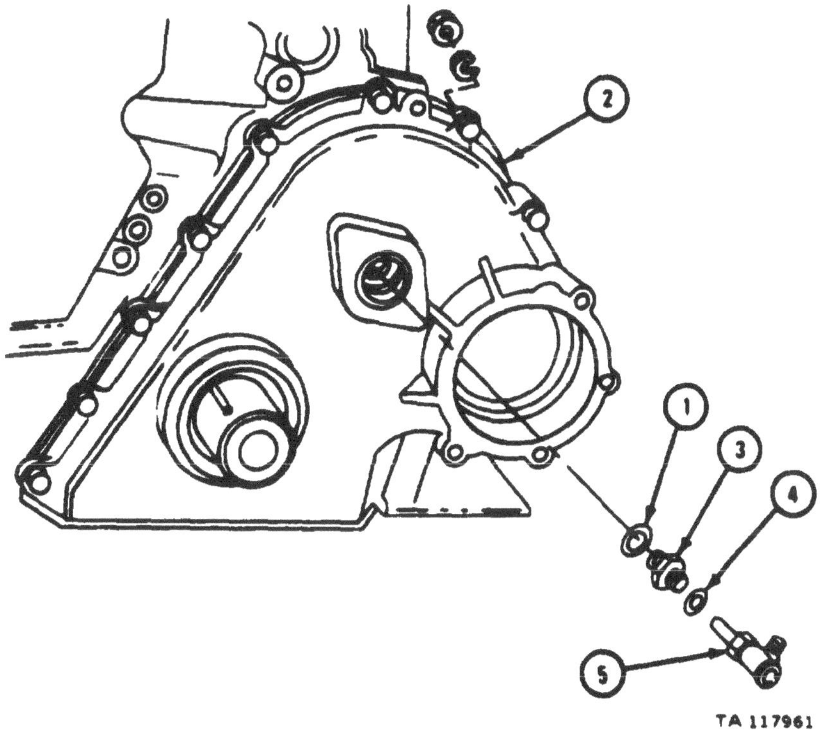

d. Crankshaft Damper and Pulley Assembly.

NOTE

If working on engines other than LDS-465-2, start with frame 1. If working on engine LDS-465-2, go to frame 4.

FRAME 1

1. Coat lip of crankshaft front oil seal (1), crankshaft key (2), and damper pulley assembly bearing area of crankshaft (3) with lubricating oil.

NOTE

Tell machine shop to make damper pulley locator (4). See figure 5-1 .

2. Put on fabricated damper pulley locator (4). Aline locator with crankshaft key (2) .
3. Put in pulley mounting bolt (5).

GO TO FRAME 2

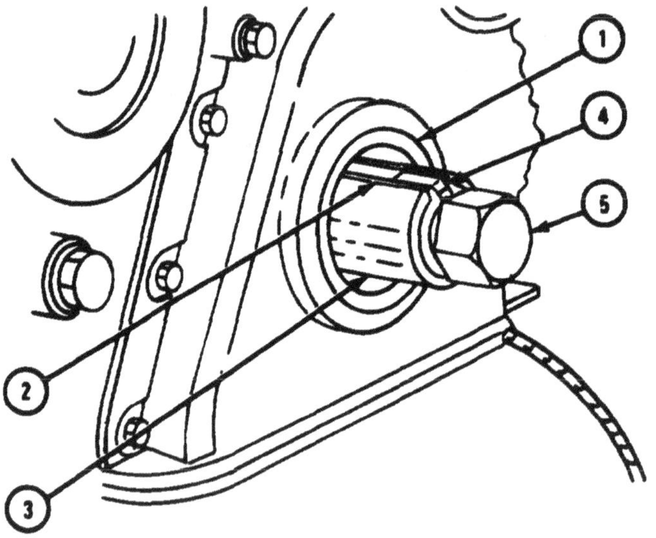

TA 117963

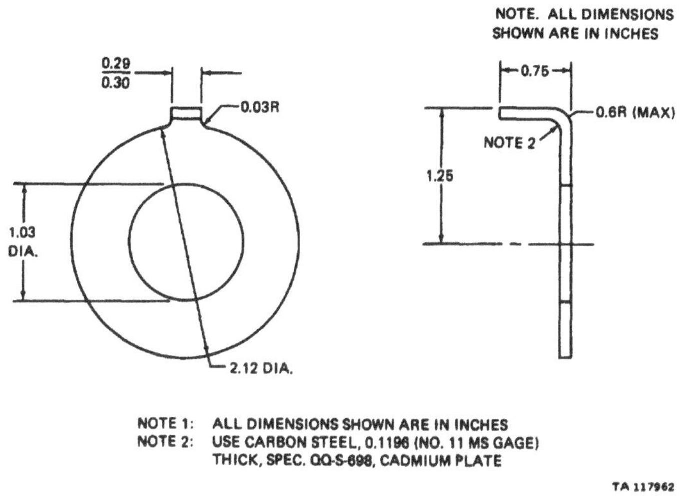

Figure 5-1. Damper Pulley Locator Fabrication Instructions.

FRAME 2

1. Coat surface of crankshaft damper and pulley assembly (1) which touches front oil seal with lubricating oil.

NOTE

To avoid overheating and damage to damper pulley (1), use Tempilstick on rim of damper pulley.

2. Heat damper pulley (1) at 200°F for 30 minutes.

NOTE

To make sure crankshaft damper and pulley assembly is properly seated, put on heated damper and pulley assembly, and put in and tighten retaining screw as soon as possible before temperature of pulley and crankshaft equalize .

3. Aline keyway (2) of damper pulley (1) with damper pulley locator (3).
4. Slide damper pulley (1) on crankshaft (4).
5. Take out damper pulley mounting bolt (5) and damper pulley locator (3).

IF END OF CRANKSHAFT (6) IS FLUSH WITH END OF DAMPER PULLEY BORE (7), GO TO FRAME 7.
IF END OF CRANKSHAFT (6) IS NOT FLUSH WITH END OF DAMPER PULLEY BORE (7) , GO TO FRAME 3

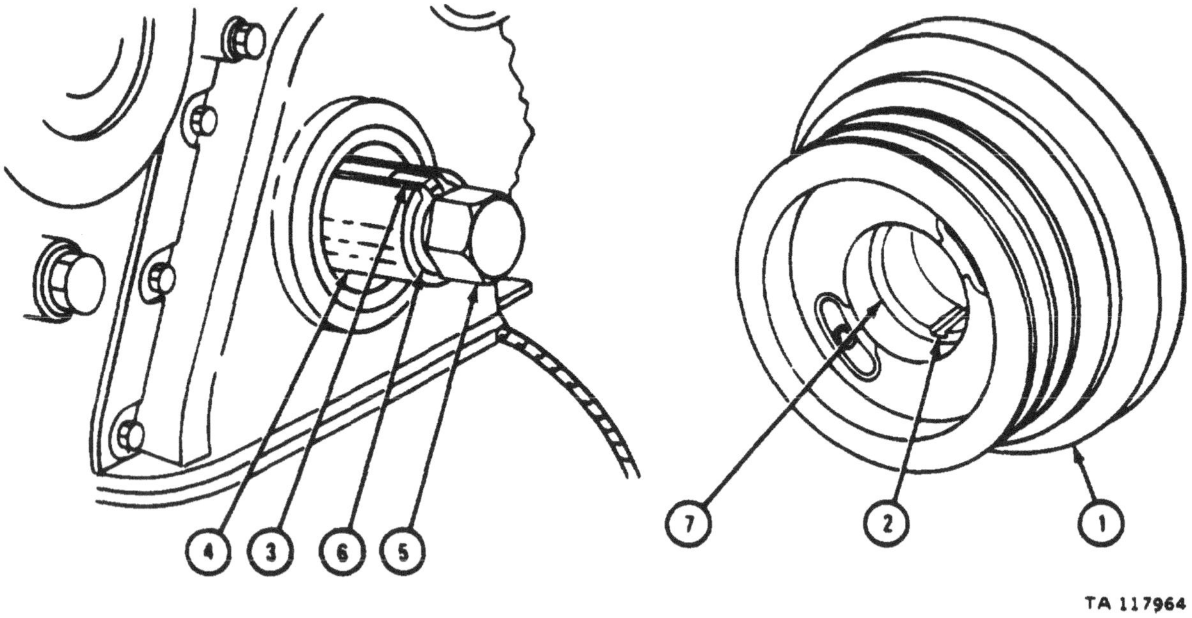

FRAME 3

1. Put crankshaft damper and pulley replacer (1) in end of crankshaft.
2. Hold replacer bolt (2) and turn replacer nut (3) until end of crankshaft is flush with end of damper pulley bore.
3. Take out crankshaft damper and pulley replacer (1).

GO TO FRAME 4

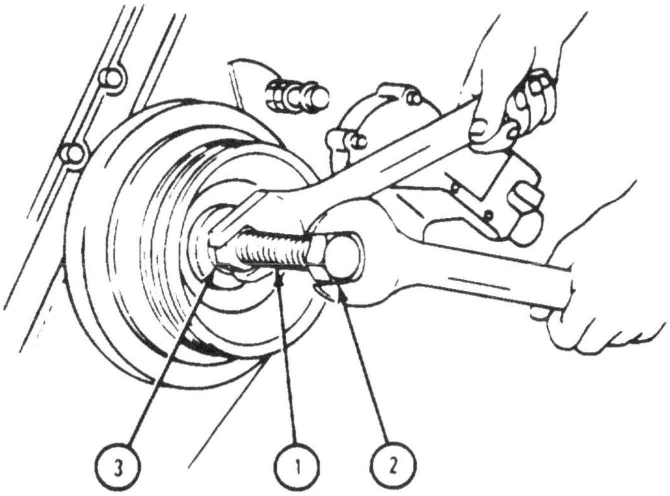

TA 117965

FRAME 4

1. Coat lip of front oil seal (1), crankshaft key (2), and damper pulley bearing area of crankshaft (3) with lubricating oil.
2. Coat crankcase oil seal contact surface of damper pulley (4) with lubricating oil.

GO TO FRAME 5

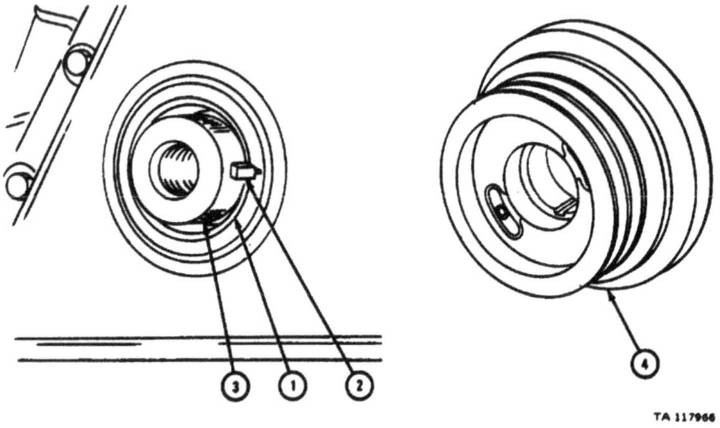

FRAME 5

1. Put crankshaft pulley driver (1) on crankshaft (2) .
2. Using key locator (3), aline key (4) on crankshaft (2) with key (5) on driver (1).
3. Tighten driver shaft (6) and take off key locator (3).

GO TO FRAME 6

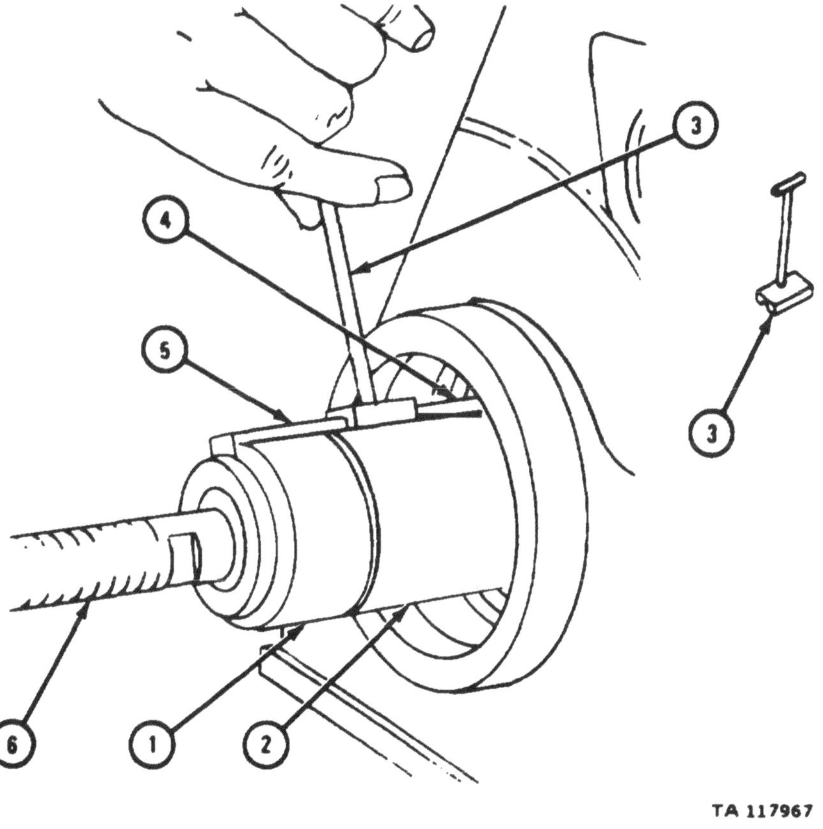

FRAME 6

1. Slide damper and pulley (1) over pulley driver shaft (2).
2. Put pulley driver support (3) and driver nut (4) on driver shaft (2).
3. Tighten nut (4) until end of bore in damper and pulley (1) is even with end of crankshaft .
4. Take off crankshaft pulley driver assembly .

GO TO FRAME 7

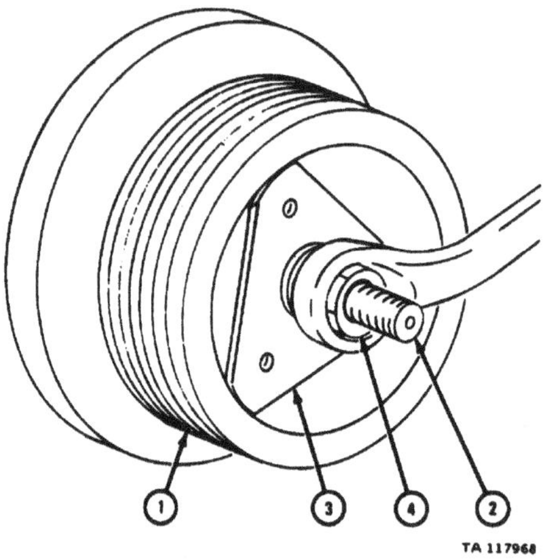

FRAME 7

1. Put in keyways seal (1).
2. Put on retaining washer (2), lockplate (3), and retaining bolt (4). Tighte n retaining bolt to 225 to 250 pound-feet.
3. Put in two capscrews (5) and lockwashers (6).

END OF TASK

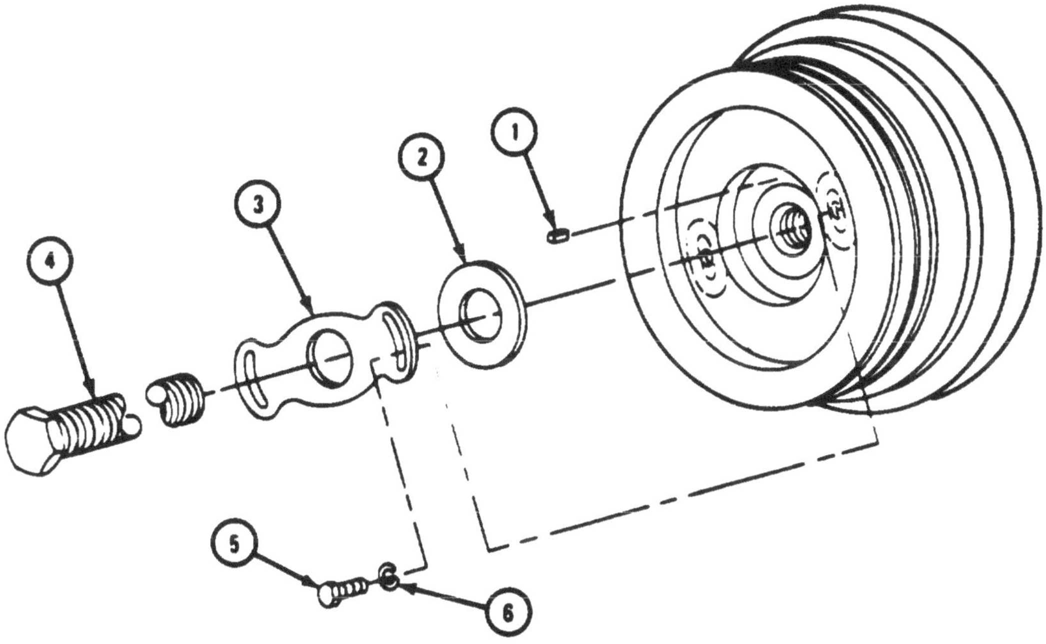

TA 117969

5-9. PISTON AND CONNECTING RODS, AND PISTON COOLING NOZZLES ADJUSTMENT.

a. Piston and Connecting Rods.

FRAME 1

1. Turn up rear end of engine (1) as shown.

NOTE

This task is shown for piston and connecting rod number one . This task is the same for all piston and connecting rods .

2. Turn crankshaft (2) until cylinder number one connecting rod journal (3) is at bottom center.

GO TO FRAME 2

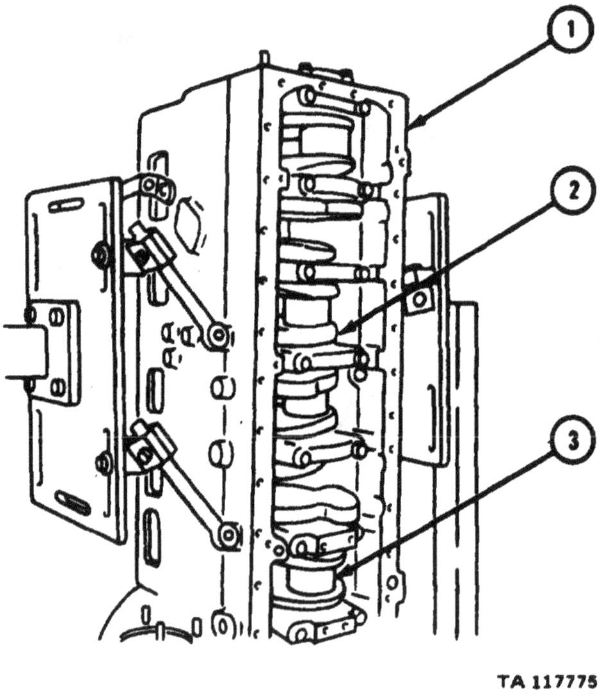

FRAME 2

NOTE

Piston and connecting rod assemblies are numbered. They must be put in cylinder with connecting rod number and piston combustion chamber swirl facing camshaft side of cylinder block. The connecting rod and piston is to be put in. Cylinders are numbered one through six from front to back of cylinder block.

1. Turn four piston rings (1) on piston (2) so gaps (3) are 90° apart,
2. Coat piston (2) with oil, Coat inside of cylinder liner (4) with oil.

GO TO FRAME 3

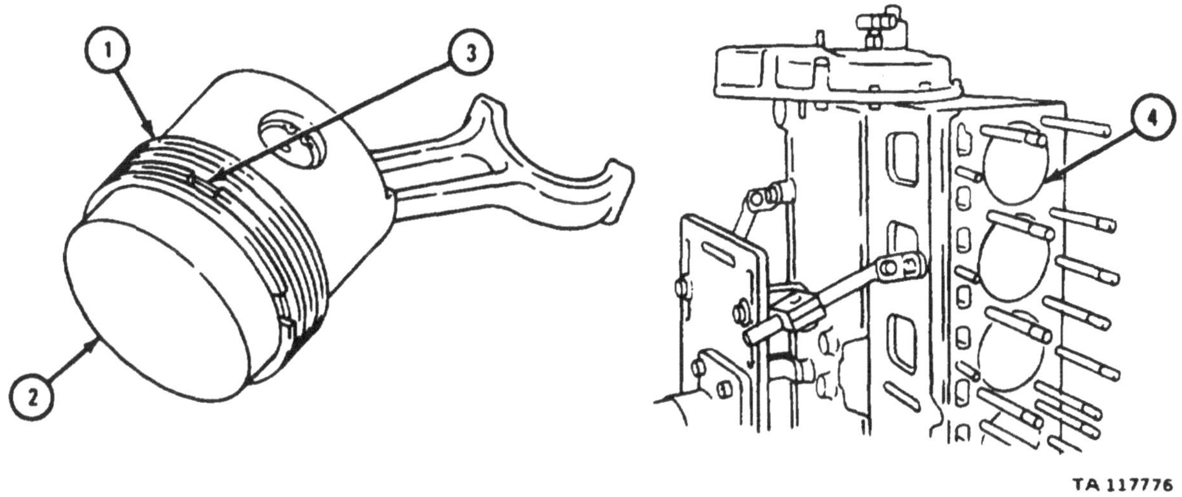

FRAME 3

1. Put ring compressor (1) on piston (2) .

Soldier A 2 . Put connecting rod (3) into cylinder bore (4) .

Soldier B 3 . Guide connecting rod (3) until connecting rod top beating half (5) is in place on crankshaft journal (6).

Soldier A 4 . Using wooden block, push piston (2) into cylinder bore (4).

GO TO FRAME 4

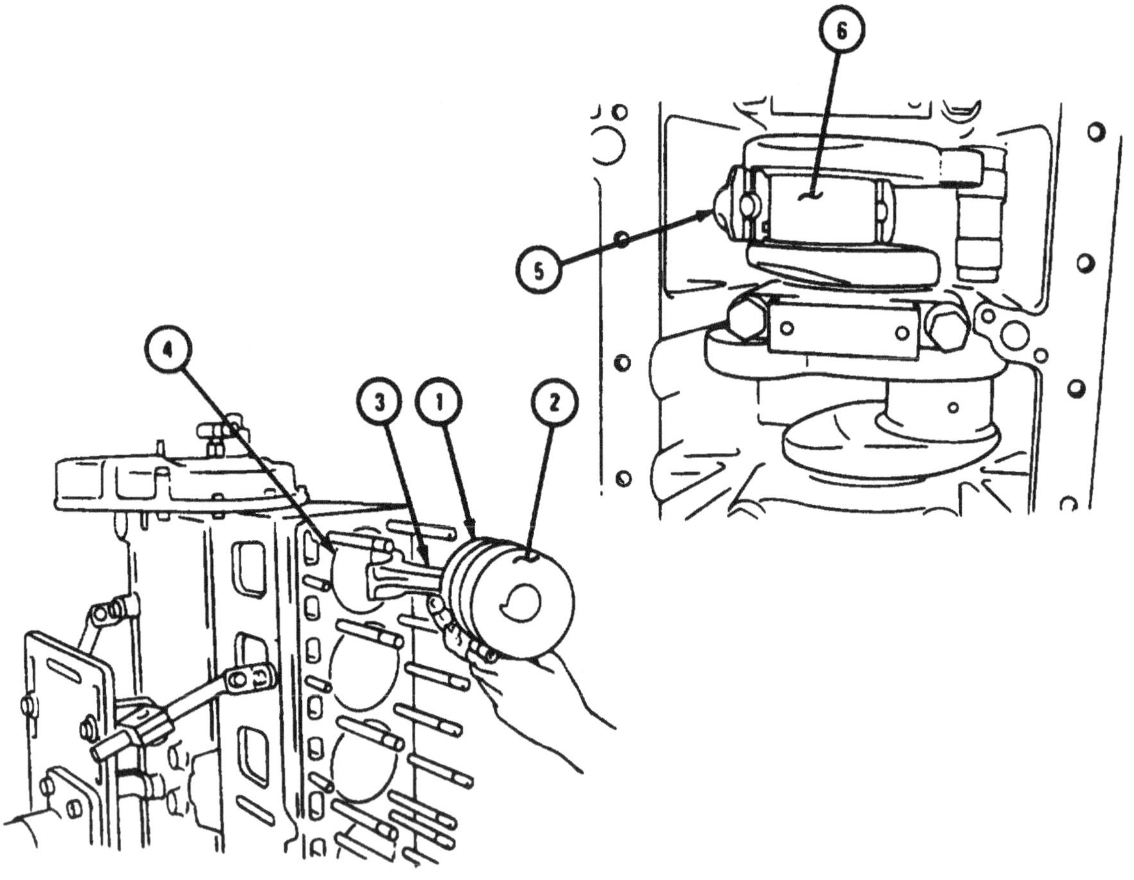

FRAME 4

1. Coat connecting rod bearing lower half (1) and bearing cap (2) with oil.

NOTE

Bearing caps (2) are numbered. Bearing cap number must match and face the same way as connecting rod number.

2. Put bearing cap (2) in place on connecting rod (3). Put in two capscrews (4).
3. Tighten capscrews (4) to 66 to 67 pound-feet.
4. Tighten capscrews (4) to 95 to 100 pound-feet.
5. Do step 2 of frame 1 and frames 2 through 4 again for the other five pistons.

END OF TASK

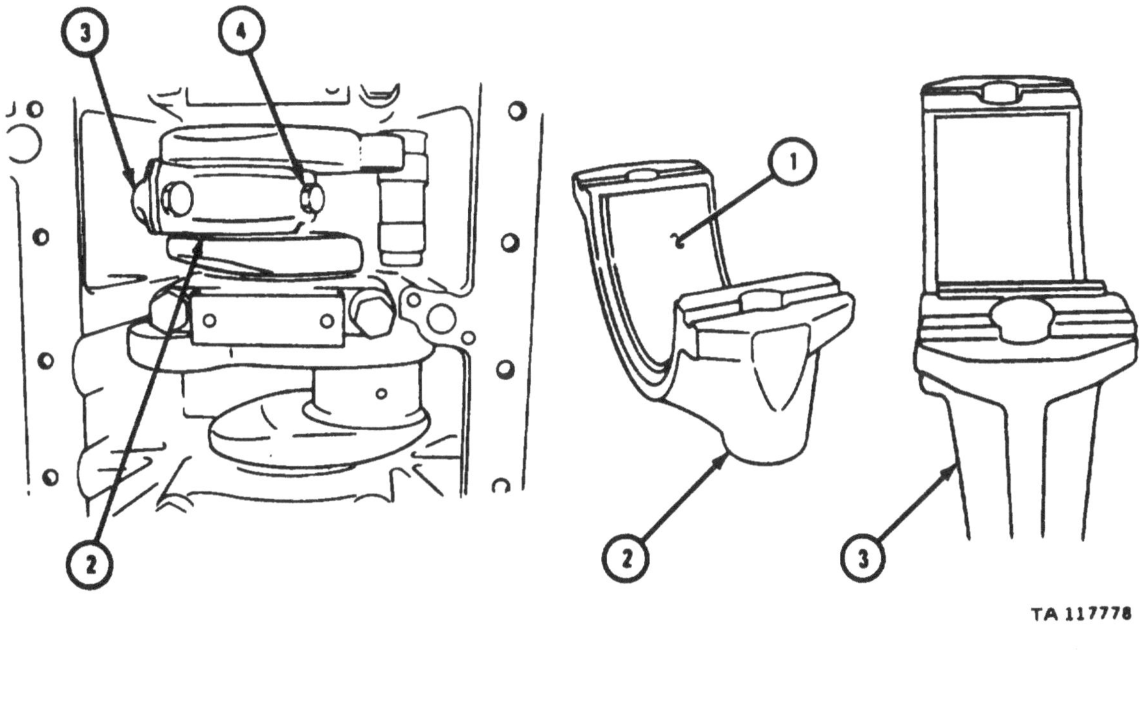

b . Piston Cooling Nozzles Adjustment (Engines LDS-465-1A and LDS-465-2).

FRAME 1

NOTE

Tell machine shop to make adapter with gasket (1). See figure 5-2 .

Use two oil pressure regulator valve adapter capscrews (2) to hold adapter (1) .

1. Put fabricated adapter with gasket (1) in place over nozzle gallery supply hole. Put in two capscrews (2) .
2. Put oil supply line (3) on adapter (1). Put other end of oil supply line on outlet side of oil pressure tank (4).
3. Fill oil pressure tank (4) with OE/HDO 10 engine oil.

NOTE

Air pressure supply must be set to 80 psi.

4. Hook oil pressure tank (4) to air pressure supply.

GO TO FRAME 2

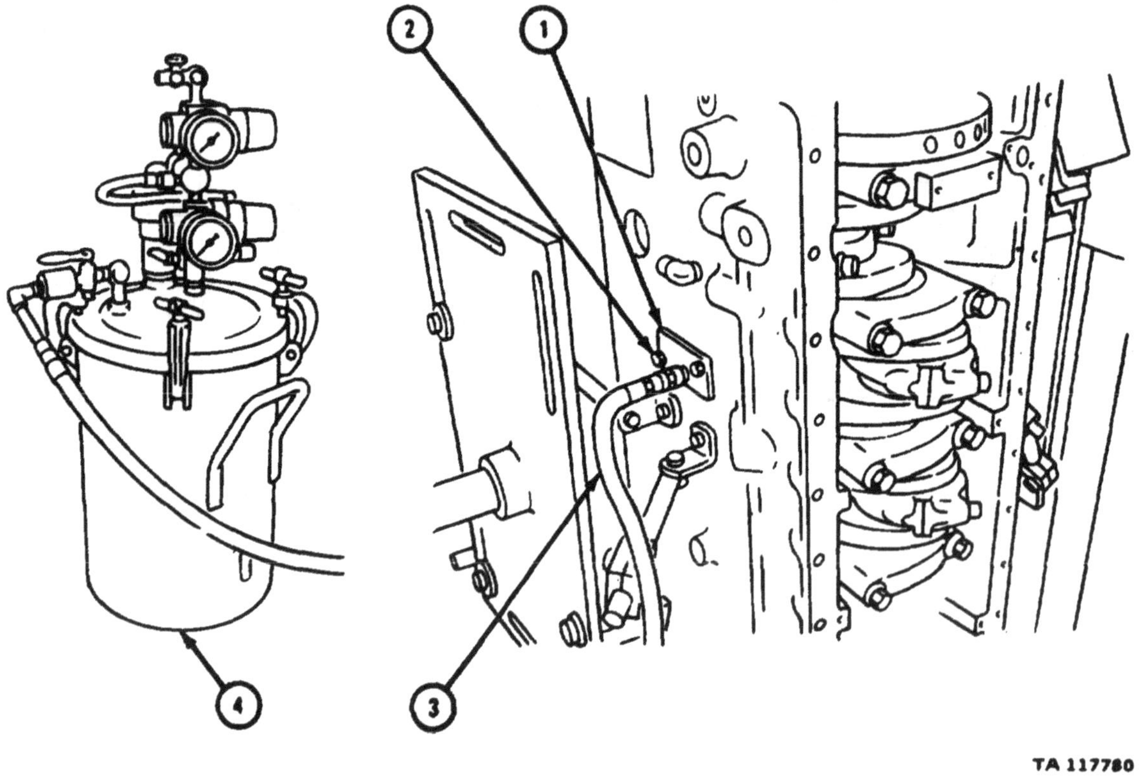

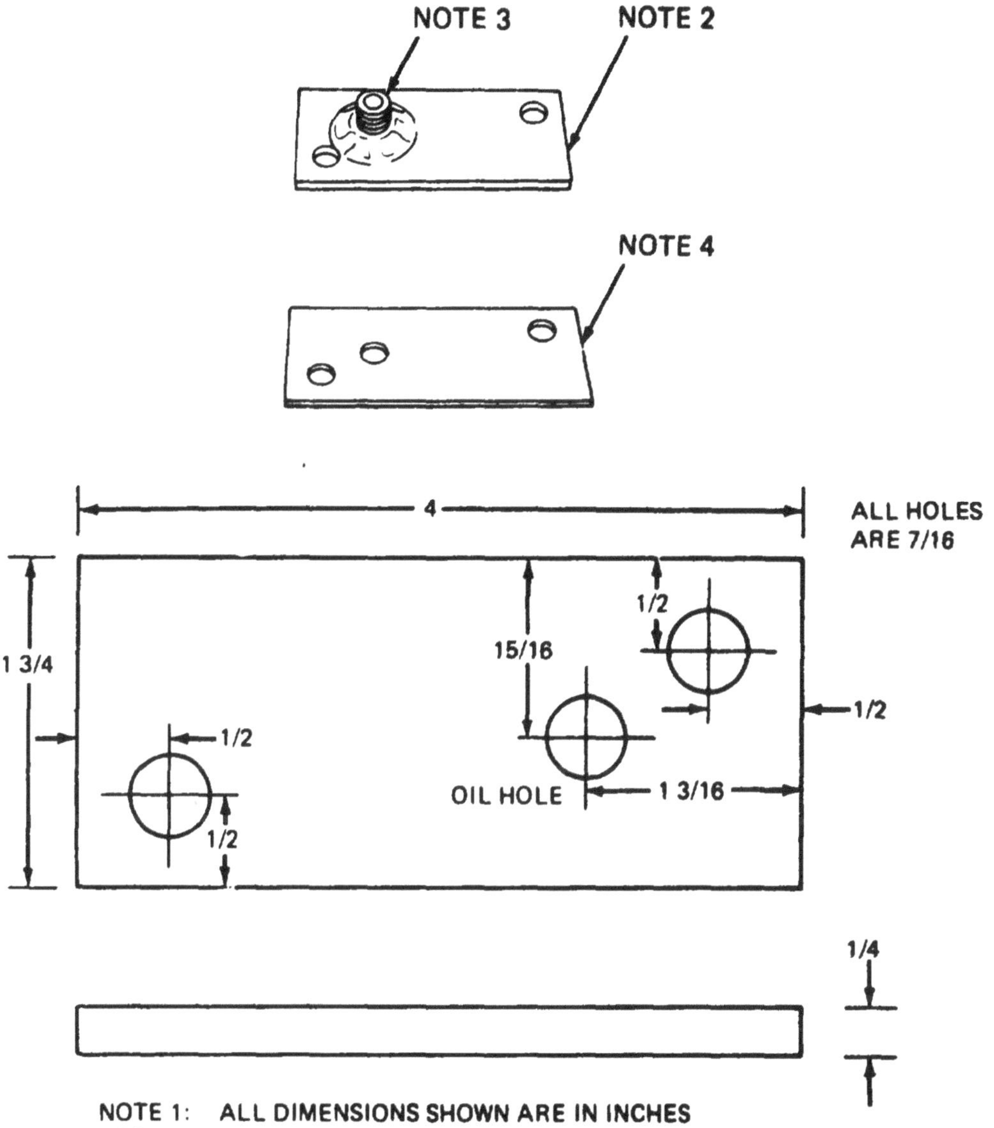
NOTE 3
NOTE 2
NOTE 4
4
ALL HOLES ARE 7/16
1 3/4
1/2
15/16
1/2
1/2
1/2
OIL HOLE
1 3/16
1/2
1/4
NOTE 1: ALL DIMENSIONS SHOWN ARE IN INCHES
NOTE 2: MATERIALS: USE STEEL PLATE, 1/4 INCH THICK
NOTE 3: TAP OUT HOLE AND PUT IN TUBE FITTING
NOTE 4: CUT RUBBER GASKET WITH THREE HOLES TO MATCH HOLES IN ADAPTER
TA 117779

Figure 5-2. Oil Pressure Regulator Valve Adapter Fabrication Instructions.

FRAME 2

NOTE

Piston (2) must be on the down stroke when cooling nozzle adjustment is done.

1. Turn crankshaft (1) until piston (2) is halfway between top and bottom dead center on the down stroke.

GO TO FRAME 3

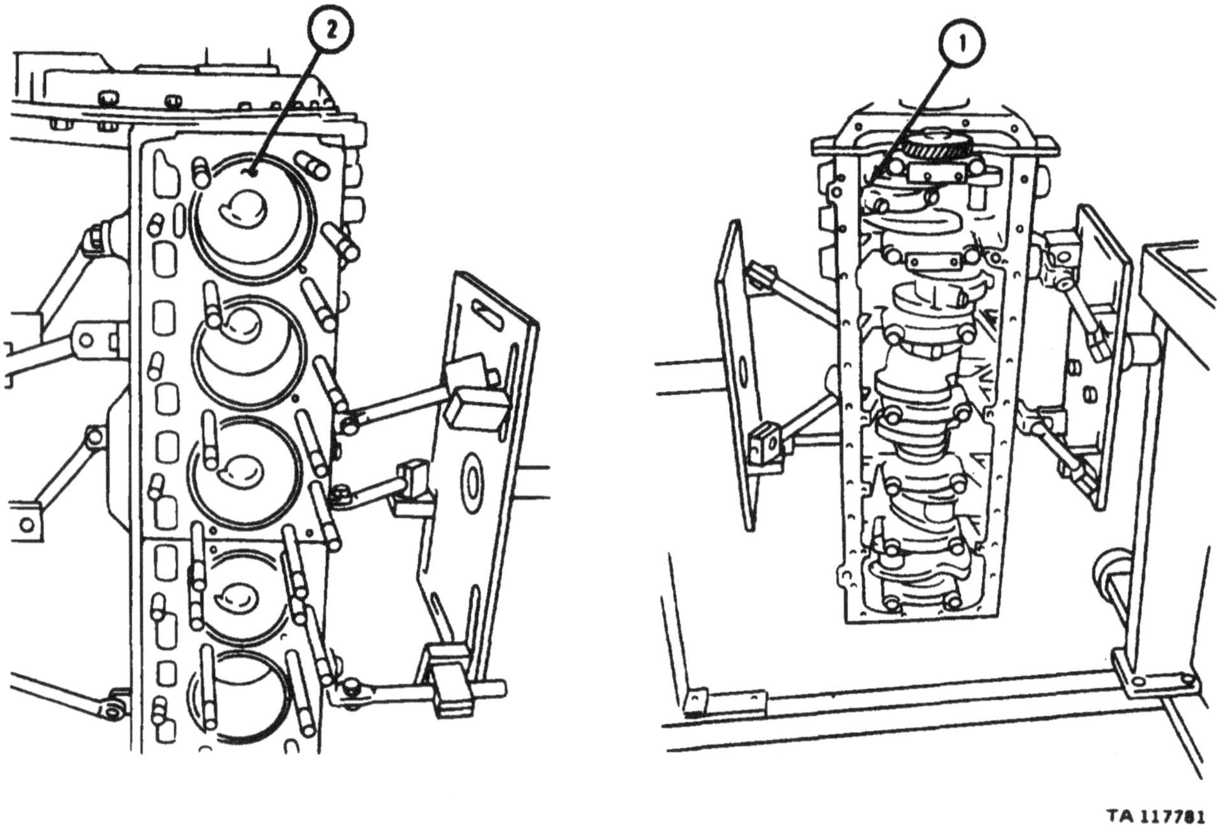

FRAME 3

WARNING

Eye shields must be worn when using compressed air. Eye injury can occur if eye shields are not used.

NOTE

Place drip pan under engine before air supply is turned on. All six nozzles will have an oil stream but only one can be adjusted at a time. Leave air pressure on only as long as needed to check nozzle adjustment.

1. Turn on air supply and check oil stream from piston cooling nozzle (1). Stream must hit center of piston cooling port (2). Turn off air supply .
2. If cooling nozzle (1) adjustment is not correct, aim the nozzle and do step 1 again.
3. Do frames 2 and 3 again for other five cylinders.

GO TO FRAME 4

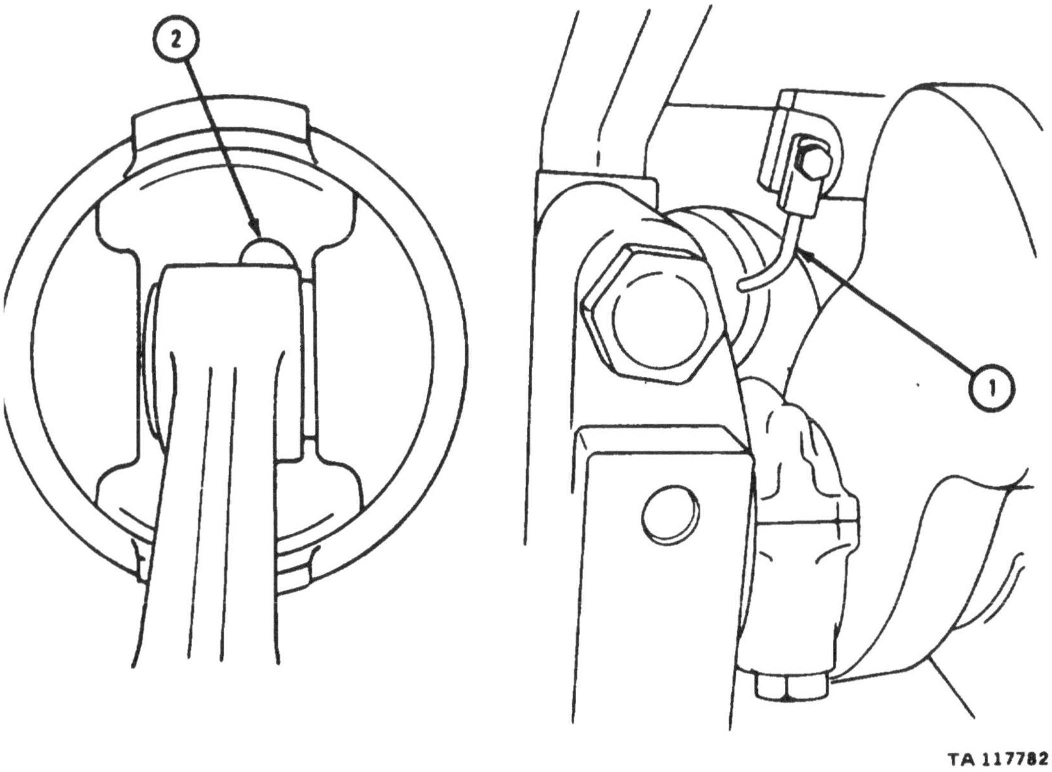

FRAME 4

1. Takeoff oil line (1) . Take out two capscrews (2) and take off adapter with gasket (3) .

END OF TASK

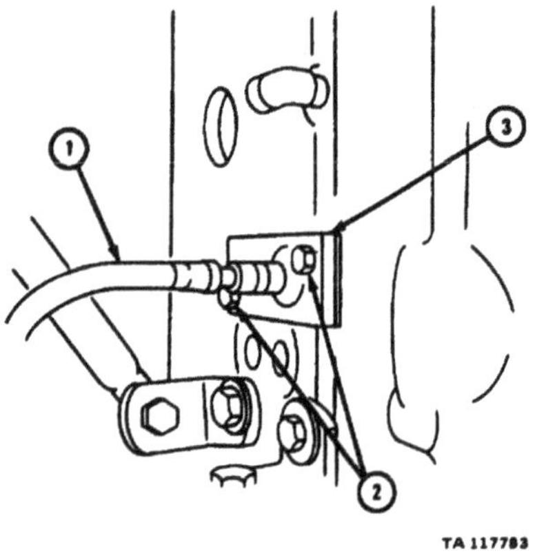

5-10. OIL PUMP ASSEMBLY, OIL PUMP TUBES, AND OIL PAN.

a. Oil Pump Assembly.

NOTE

The oil pump for engine LDS-465-2 looks different than oil pump for other engines. This task is the same for all oil pumps.

FRAME 1

1. Put oil pump (1) on dowels in front bearing cap (2).
2. Mesh gear teeth of oil pump drive gear (3) with teeth of oil pump idler gear (4) .
3. Put in three self-locking bolts (5) . Tighten bolts to 48 to 58 pound-feet.

GO TO FRAME 2

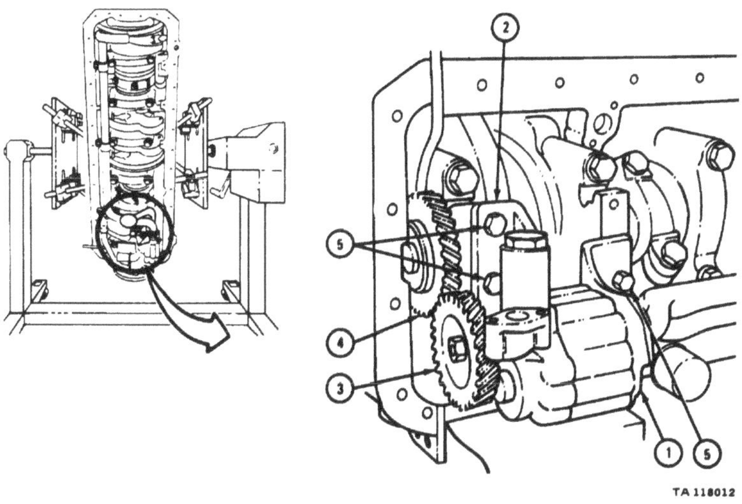

FRAME 2

1. Using dial indicator (1), check backlash of oil pump drive gear (2) and oil pump idler gear (3) .

2. Mount dial indicator (1) on timing gear cover (4). Point of dial indicator (5) should rest against side of a tooth on oil pump drive gear (2).

NOTE

When measuring backlash, make sure that oil pump idler gear (3) does not turn. If gear turns, backlash readings will be wrong.

3. Turn oil pump drive gear (2) as far as you can in one direction. Set dial indicator to read 0.

4. Hold oil pump idler gear (3) so it cannot turn. Turn oil pump drive gear (2) as much as you can in the other direction. Read backlash between oil pump drive gear and oil pump idler gear. Backlash must be between 0.006 and 0.014 inch.

END OF TASK

b. Oil Pump Tubes.

FRAME 1

1. Put end of oil pump outlet tube assembly (1) in oil pump outlet (2). Seat preformed packing (3) on top of oil pump outlet.
2. Put gasket (4) on crankcase oil inlet (5). Put tube assembly (1) on gasket.
3. Put 1 1/4-inch long drilled head bolt (6) in oil pump side of crankcase oil inlet (5) end of tube assembly (1).
4. Put l-inch drilled head bolt (7) in outer side of crankcase end of tube assembly (1).
5. Seat flange (8) on oil pump outlet (2). Put in two cap screws (9) with lock-washers (10) .

GO TO FRAME 2

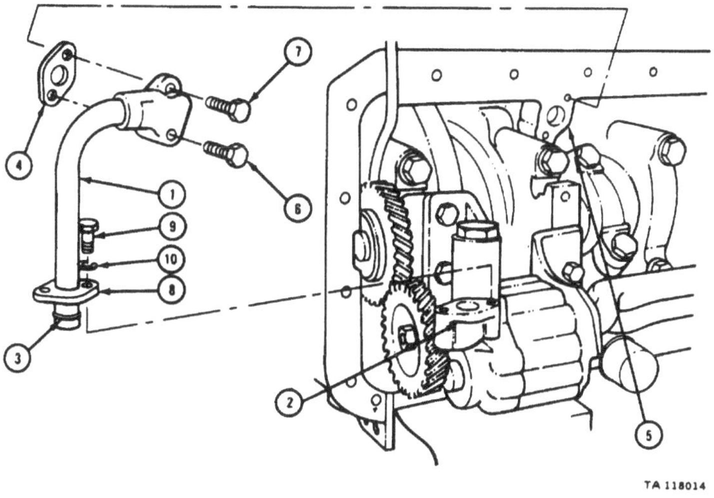

FRAME 2

1. Put one end of nickel copper wire (1) through hole in bolt head (2).
2. Put other end of wire (1) through hole in bolt head (3).
3. Bend ends of wire (1) around bolt heads (2 and 3) to the right. Join ends of wire and twist together.

IF WORKING ON ENGINE LDS-465-1, LDS-465-1A OR LDS-465-2, GO TO FRAME 3.
IF WORKING ON ENGINE LD-465-1, LD-465-1C OR LDT-465-1C, GO TO FRAME 4

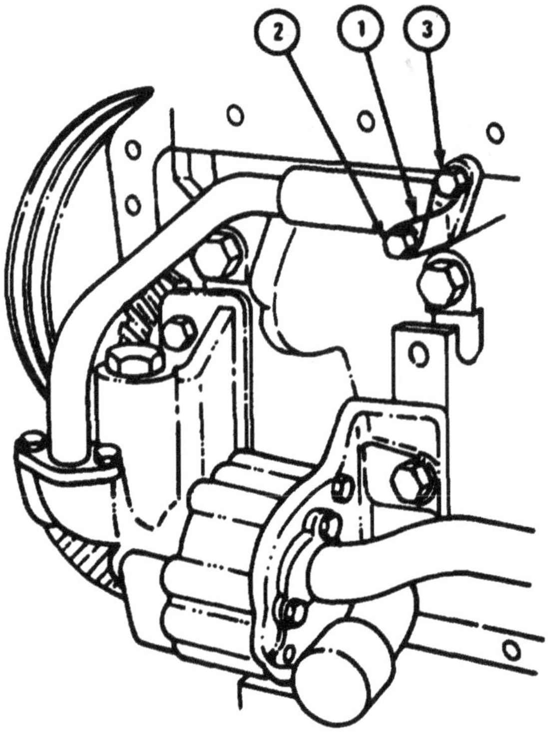

TA 118015

FRAME 3

1. Put machine bolt (1) in tube brace (2) and crankcase (3).
2. Put machine bolt (4) in tube clamp (5) and crankcase (3).

END OF TASK

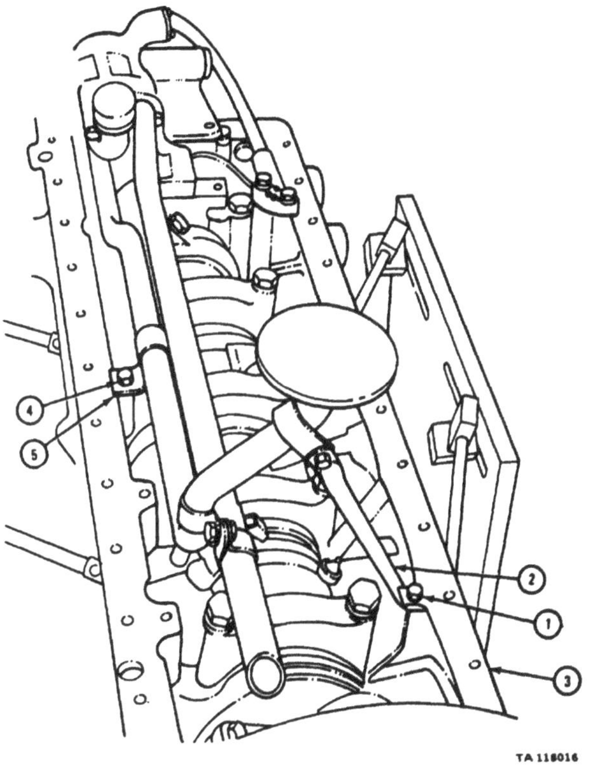

FRAME 4

1. Put machine bolt (1) in tube clamp (2) and crankcase (3).
2. Put machine bolt (4) in tube clamp (5) and crankcase (3).

GO TO FRAME 5

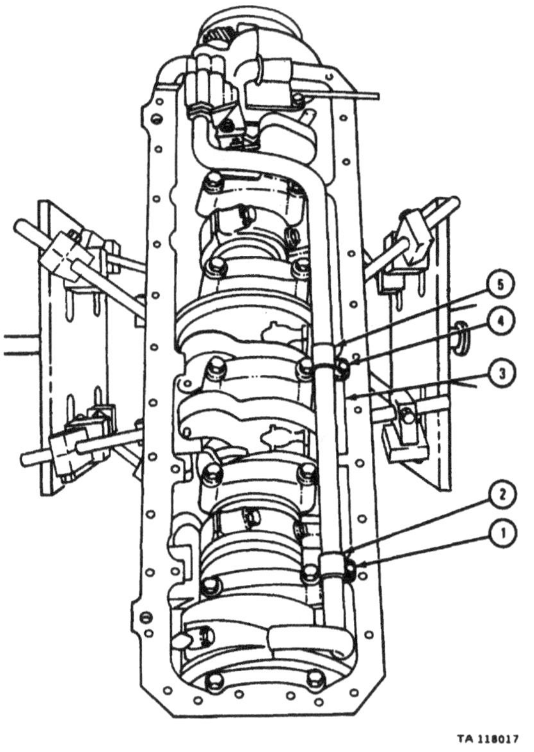

FRAME 5

1. Put oil pump pickup tube gasket (1) on oil pump inlet (2).
2. Put oil pump pickup tube (3) on gasket (1).
3. Put in two cap screws with lockwashers (4).
4. Put cap screw (5) in tube clamp (6) and oil pump bracket (7). Put on nut with lockwasher (8) .

END OF TASK

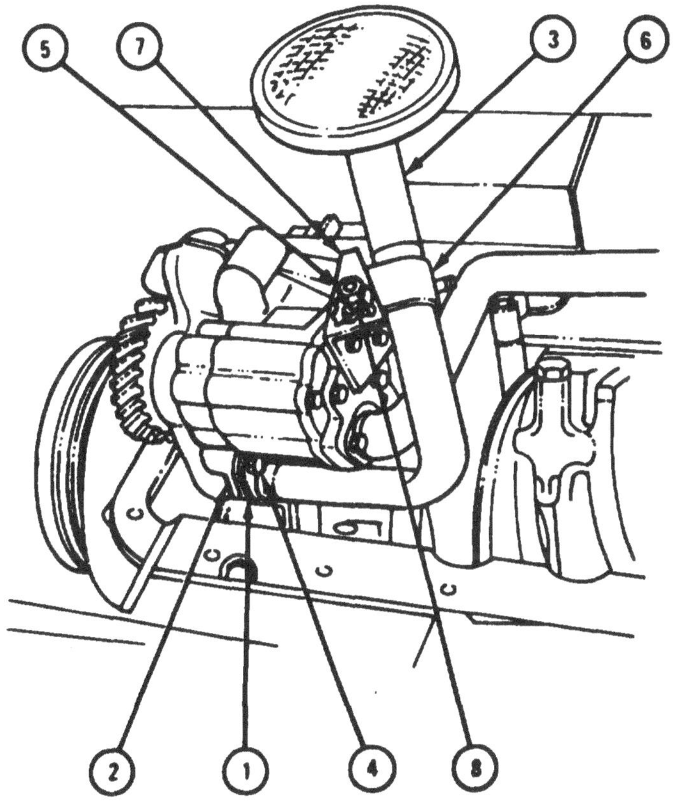

TA 118018

c. Oil Pan.

FRAME 1

1. Coat oil pan gasket (1) on both sides with a thin coat of sealant.
2. Work engine overhaul stand to turn engine upside down so bottom of crankcase (2) faces up .
3. Put gasket (1) on crankcase (2) .

IF WORKING ON ENGINE LD-465-1, LD-465-1C OR LDT-465-1C, GO TO FRAME 2.
IF WORKING ON ENGINE LDS-465-1A, GO TO FRAME 3.
IF WORKING ON ENGINE LDS-465-2, GO TO FRAME 4

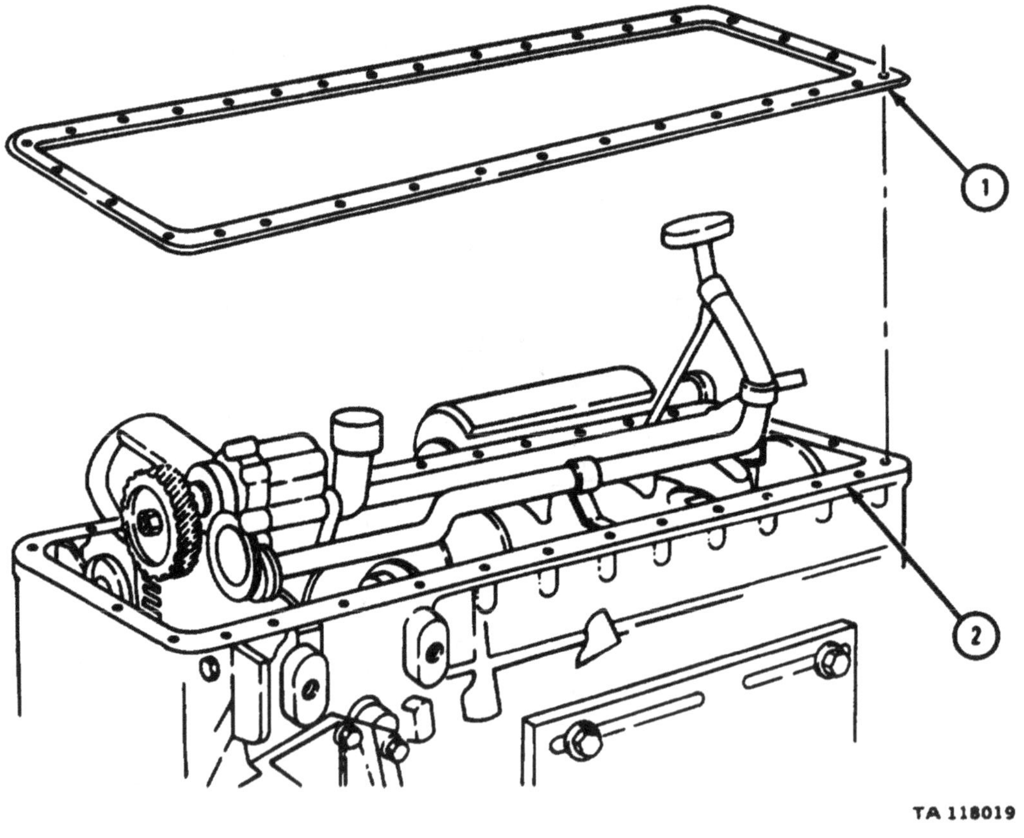

FRAME 2

1. Placing large sump (1) at front of engine (2), put on oil pan (3).
2. Put in 30 capscrews and lockwashers (4).

END OF TASK

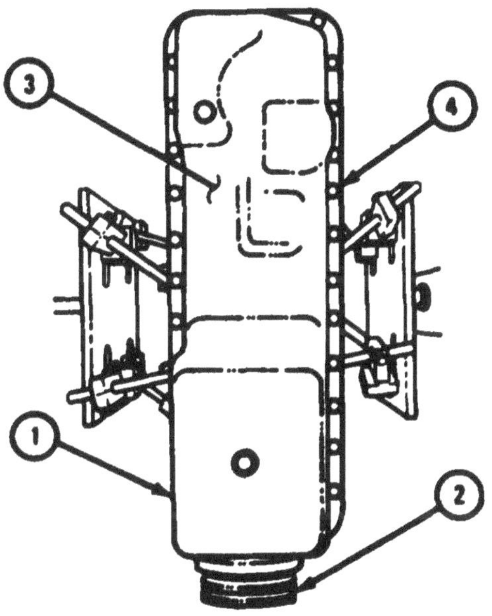

FRAME 3

1. Placing large sump (1) at rear of engine (2), put on oil pan (3).
2. Put on electrical ground lead (4). Put in capscrew and internal tooth lockwasher (5).
3. Put in 30 cap screws and lockwashers (6). Do not put cap screw in cap screw hole (7).

NOTE

Some LDS-465-1A engines have fuel filter drain tube bracket. When put back, drain tube bracket will be mounted on cap screw hole (7).

4. If engine does not have fuel filter drain tube bracket, put capscrew and lockwasher (6) in cap screw hole (7).

END OF TASK

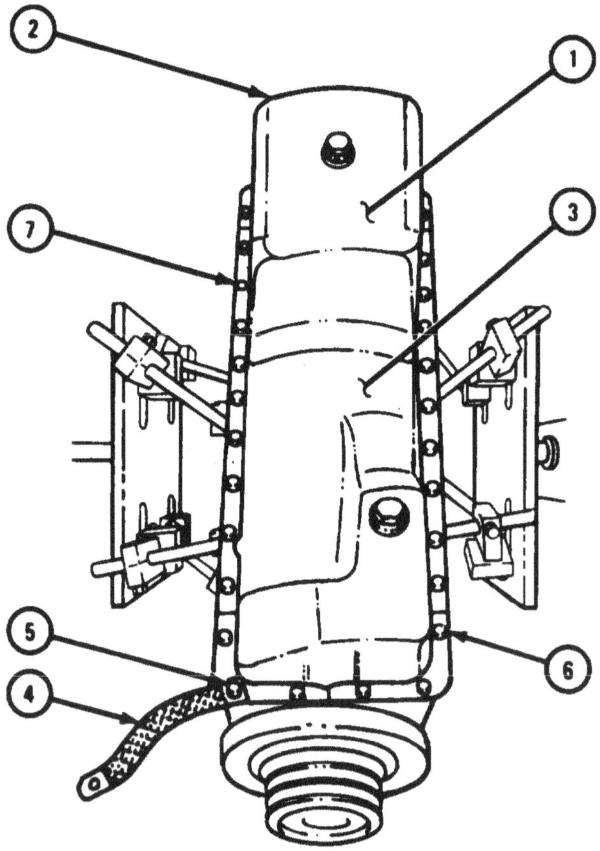

TA 118076

FRAME 4

1. Placing large sump (1) at rear of engine (2), put on oil pan (3).

NOTE

When put back, throttle return spring bracket will be mounted on two cap screw holes (5).

2. Put in 28 capscrews and lockwashers (4). Do not put capscrews in two capscrew holes (5) .

END OF TASK

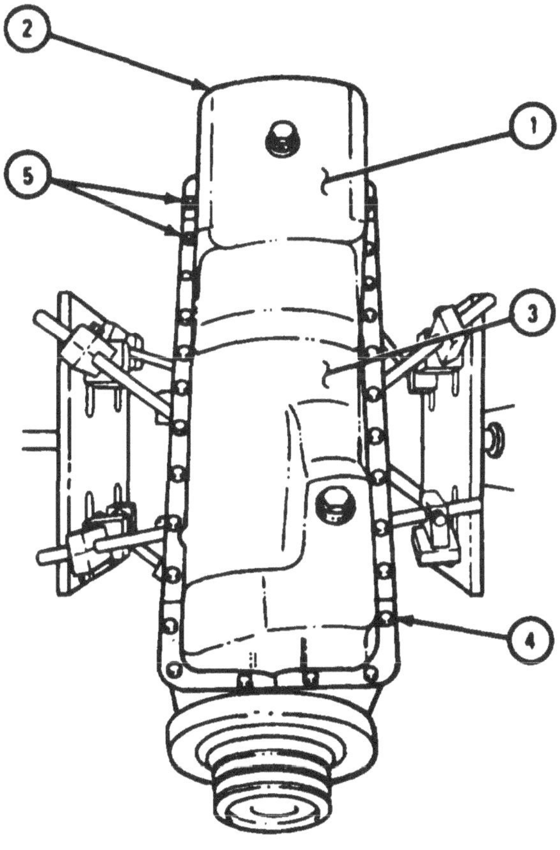

5-11. CYLINDER HEAD.

FRAME 1

1. Put a thin coat of sealant evenly on both sides of two cylinder head gaskets (1) as shown by the shaded area.
2. Put two cylinder head gaskets (1) in place on engine block (2). Put six fire rings (3) in place on engine block.

GO TO FRAME 2

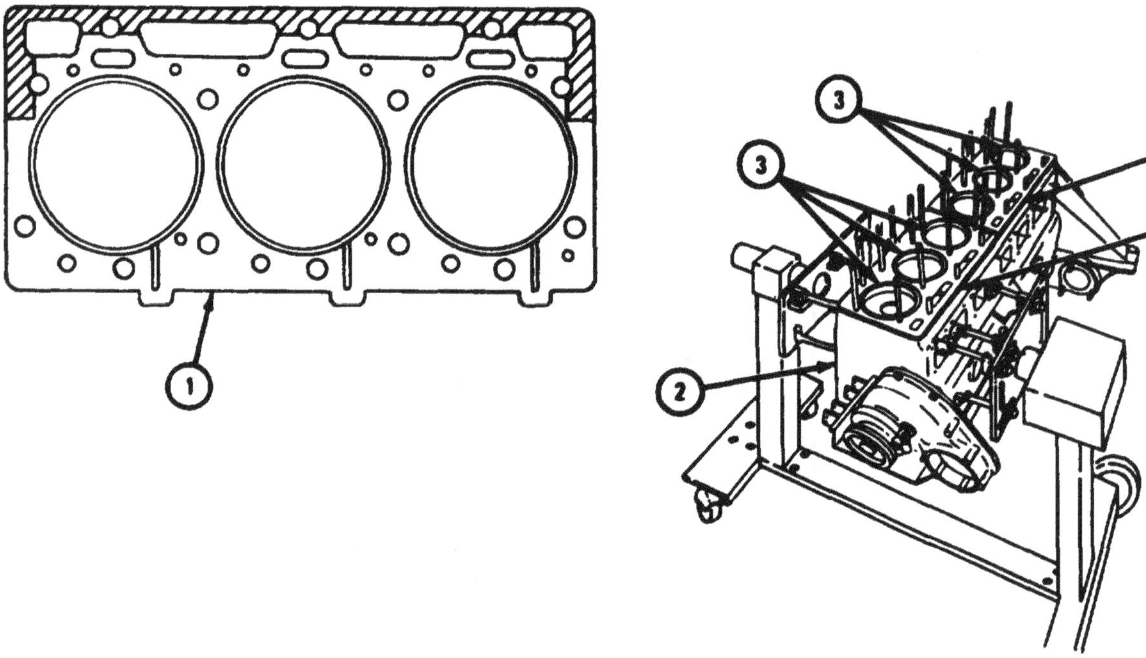

TA 117786

FRAME 2

1. Join lifting sling (1) to front cylinder head assembly (2) using screw and flat washer (3) and nut and flat washer (4) as shown.
2. Put lifting sling (1) on chain hoist hook (5) so cylinder head (2) will hang level when lifted.

GO TO FRAME 3

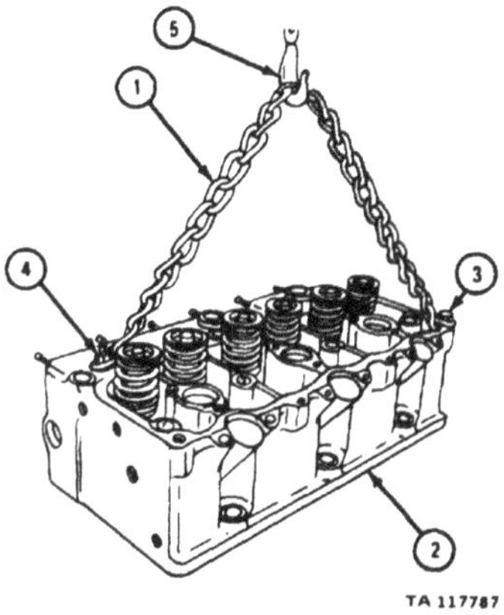

FRAME 3

1. Lift front cylinder head (1) so it hangs over engine block (2).

CAUTION

Make sure front cylinder head (1) is positioned directly above front cylinder head mating surface of engine block (2). Lower cylinder head slowly and evenly onto mating surface and guide cylinder head onto cylinder head studs (3). Failure to use extreme caution may result in damage to cylinder head and cylinder head mounting studs.

Soldier A 2. Guide front cylinder head (1) onto cylinder head mounting studs (3).

Soldier B 3. Lower front cylinder head (1) onto engine block (2).

4. Do frames 2 and 3 again for rear cylinder head (4).

5. Check whether cylinder block (2) has special identification mark (5).

GO TO FRAME 4

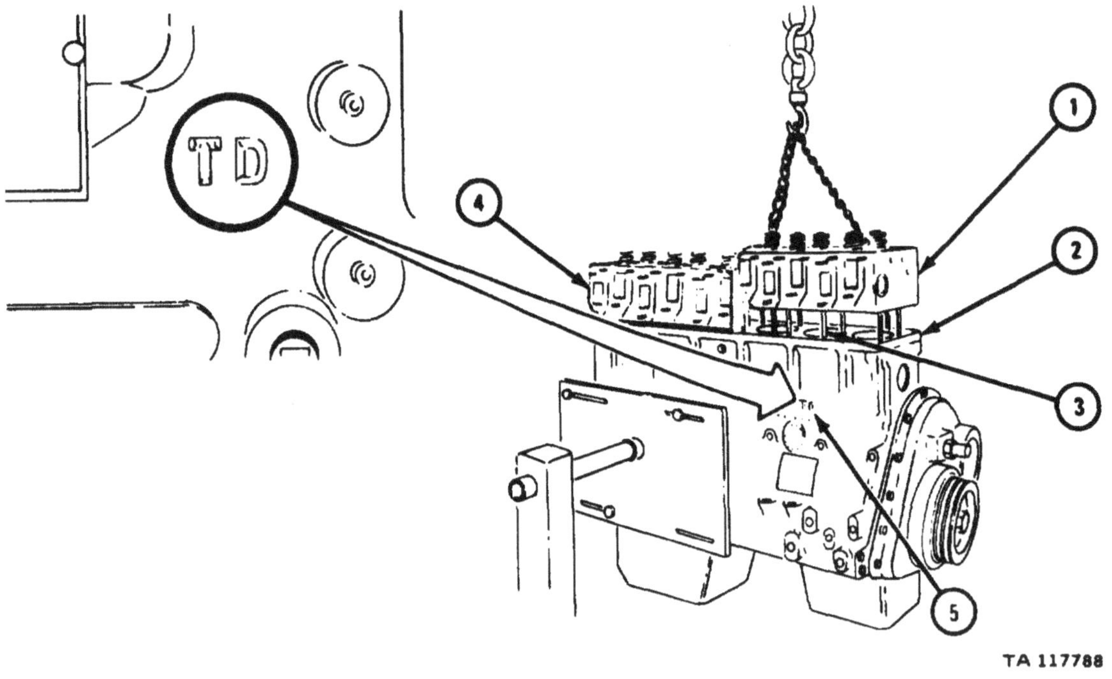

FRAME 4

1. Put a light coat of engine oil on 14 cylinder head studs (1), 14 washers (2) and 14 nuts (3).
2. Put 14 washers (2) and 14 nuts (3) on 14 front cylinder head studs (1).
3. Do steps 1 through 2 again for rear cylinder head (4).

GO TO FRAME 5

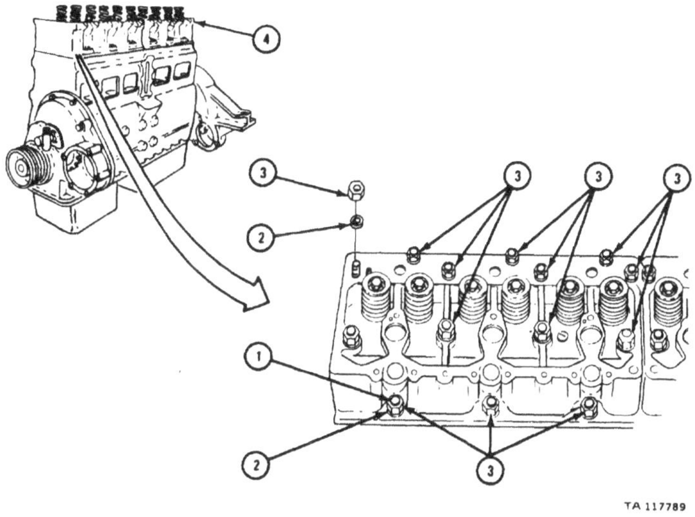

FRAME 5

NOTE

Use cylinder head nut wrench with torque wrench to tighten nuts (1, 8, and 9) .

1. Tighten nuts (1 through 14) to 40 pound-feet in order shown.
2. Tighten nuts (1 through 14) to 80 pound-feet in order shown.
3. Tighten nuts (1 through 14) to 110 pound-feet in order shown.
4. Tighten nuts (1 through 14) to 130 pound-feet in order shown.

NOTE

Do step 5 only if crankcase has "TD" identification. Refer to frame 3.

5. Tighten nuts (1 through 14) to 157 pound-feet in order shown.
6. Do steps 1 through 5 again for other cylinder head.

END OF TASK

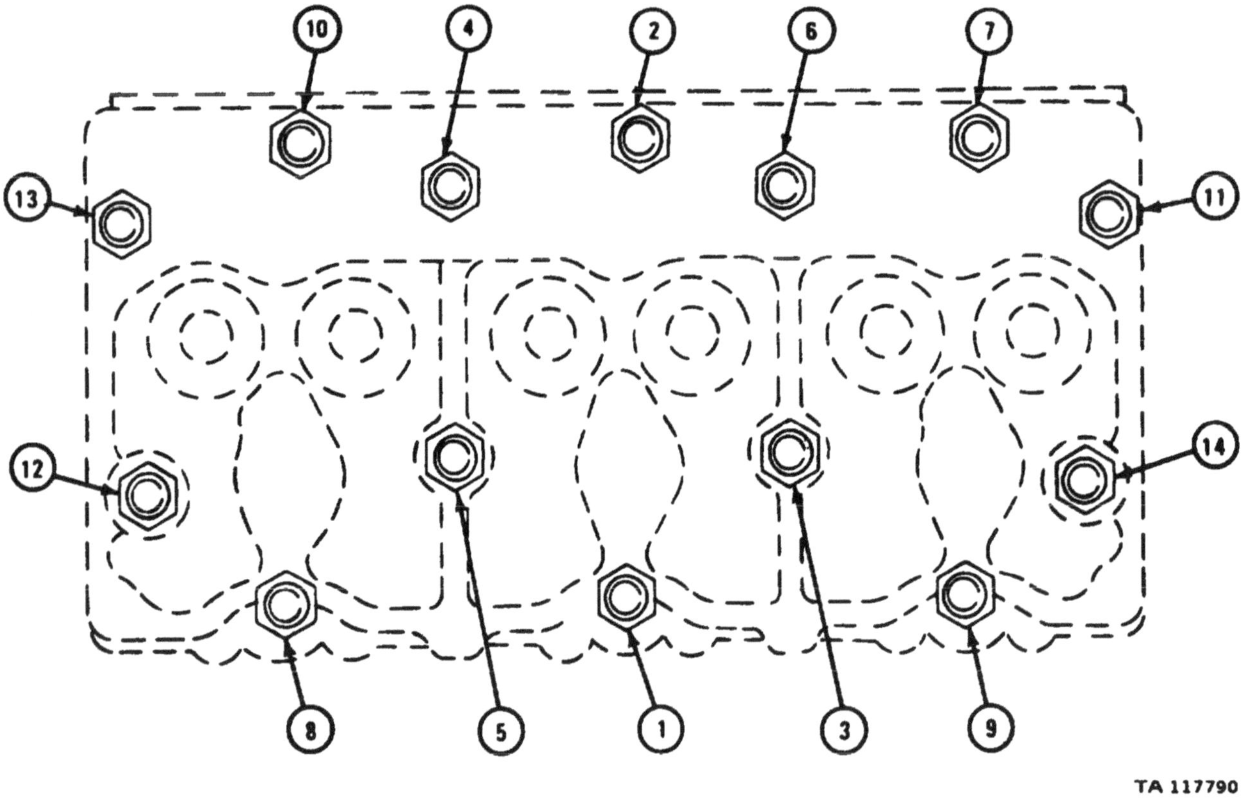

5-12. ENGINE LIFTING BRACKETS.

a. Rear Lifting Bracket .

FRAME 1

1. Put lifting bracket (1) in place on cylinder head (2). Put in two capscrews with lockwashers (3) .

END OF TASK

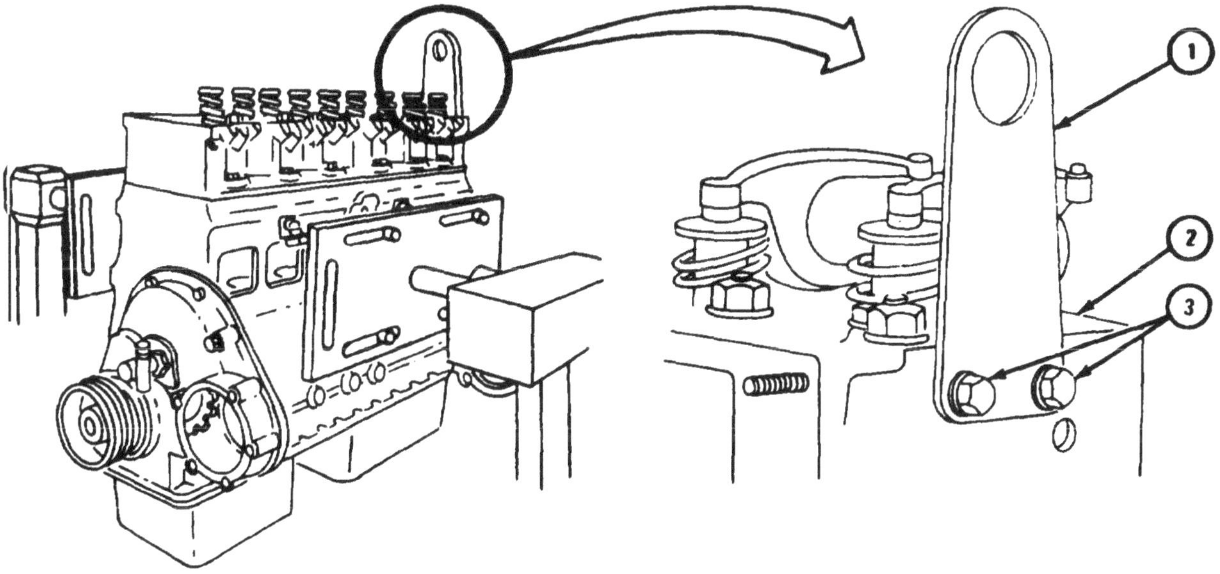

TA 118492

b. Front Lifting Bracket.

(1) All engines except LDS-465-2

FRAME 1

1. Put front lifting bracket (1) in place on cylinder head (2).
2. Put in two capscrews with lockwashers (3).

END OF TASK

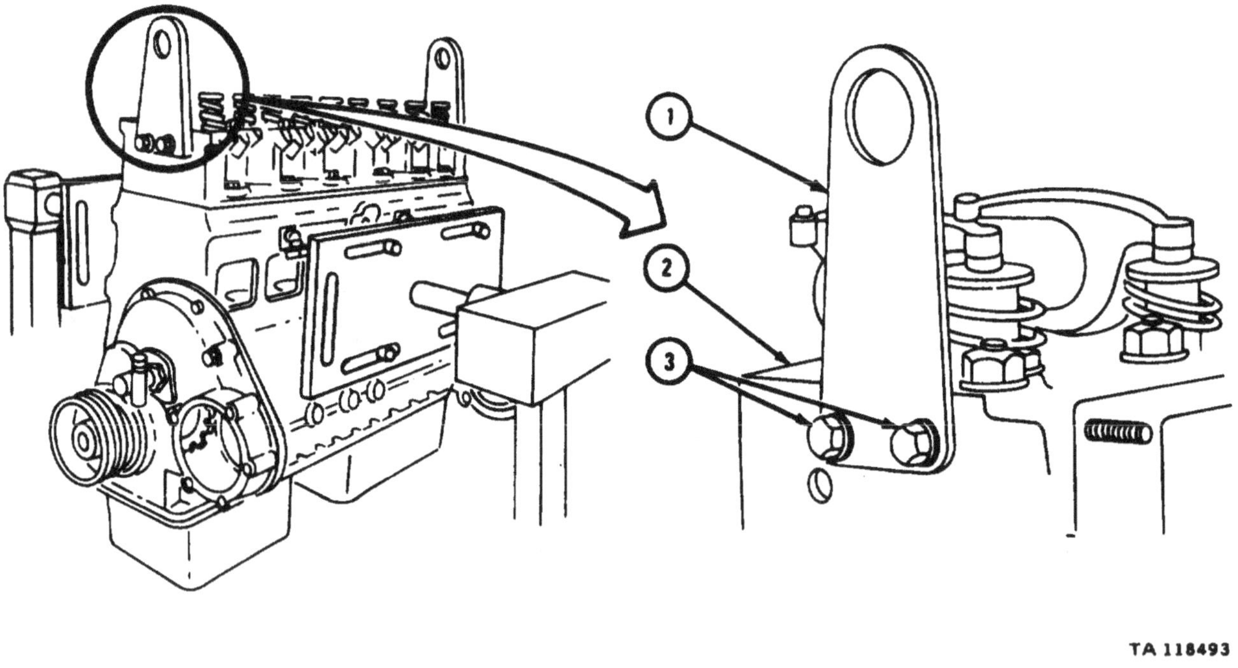

(2) Engine LDS-465-2 .

FRAME 1

1. Put final fuel filter with front lifting bracket (1) in place on cylinder head (2).
2. Put in two capscrews with lockwashers (3).

END OF TASK

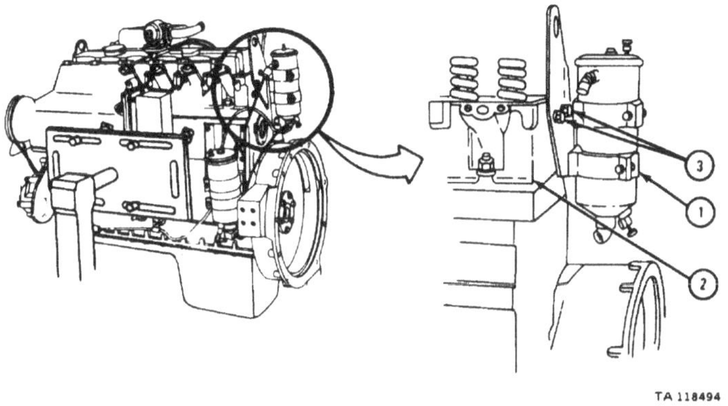

5-13. MOUNTING ENGINE ON TRANSPORT STAND.

FRAME 1

1. Put lifting sling (1) on engine (2) and hook (3) on chain hoist (4). Raise chain hoist to take strain off stand.
2. Working on side without crank, loosen three cap screws (5).
3. Take out three capscrews (6) .
4. Take off three universal mounting arms (7) from mounting plate (8).

GO TO FRAME 2

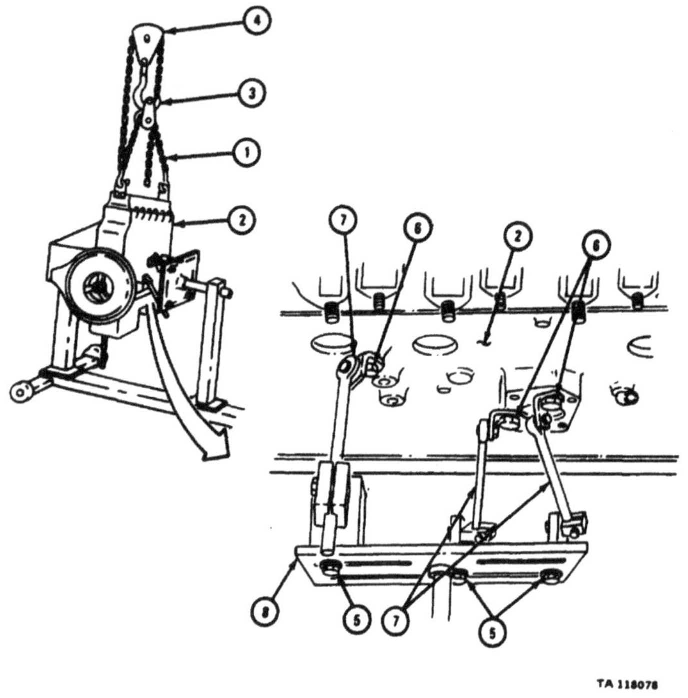

FRAME 2

1. Working on crank side, loosen three capscrews (1).
2. Take out three cap screws (2).
3. Takeout three capscrews (1) and three universal mounting arms (3) .

GO TO FRAME 3

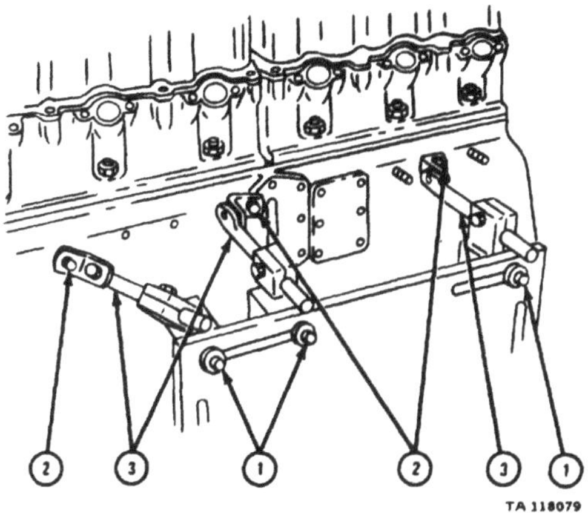

FRAME 3

1. Working on crank side, loosen two screws (1).
2. Take out two capscrews (2) .
3. Take out two capscrews (1).
4. Take off two universal mounting arms (3).

GO TO FRAME 4

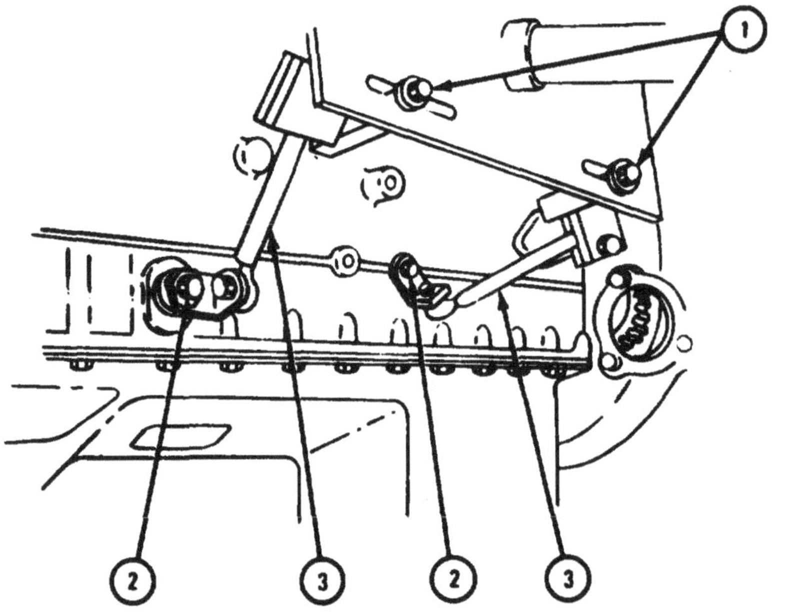

FRAME 4

1. Using hoist (1), lift engine (2) out of engine maintenance stand (3).
2. Move engine maintenance stand (3) out of the way.
3. Loosen four screws (4) . Slide four engine supports (5) out away from center of engine transport stand (6) .
4. Position engine stand (6) so engine (2) hangs between four engine supports (5).

GO TO FRAME 5

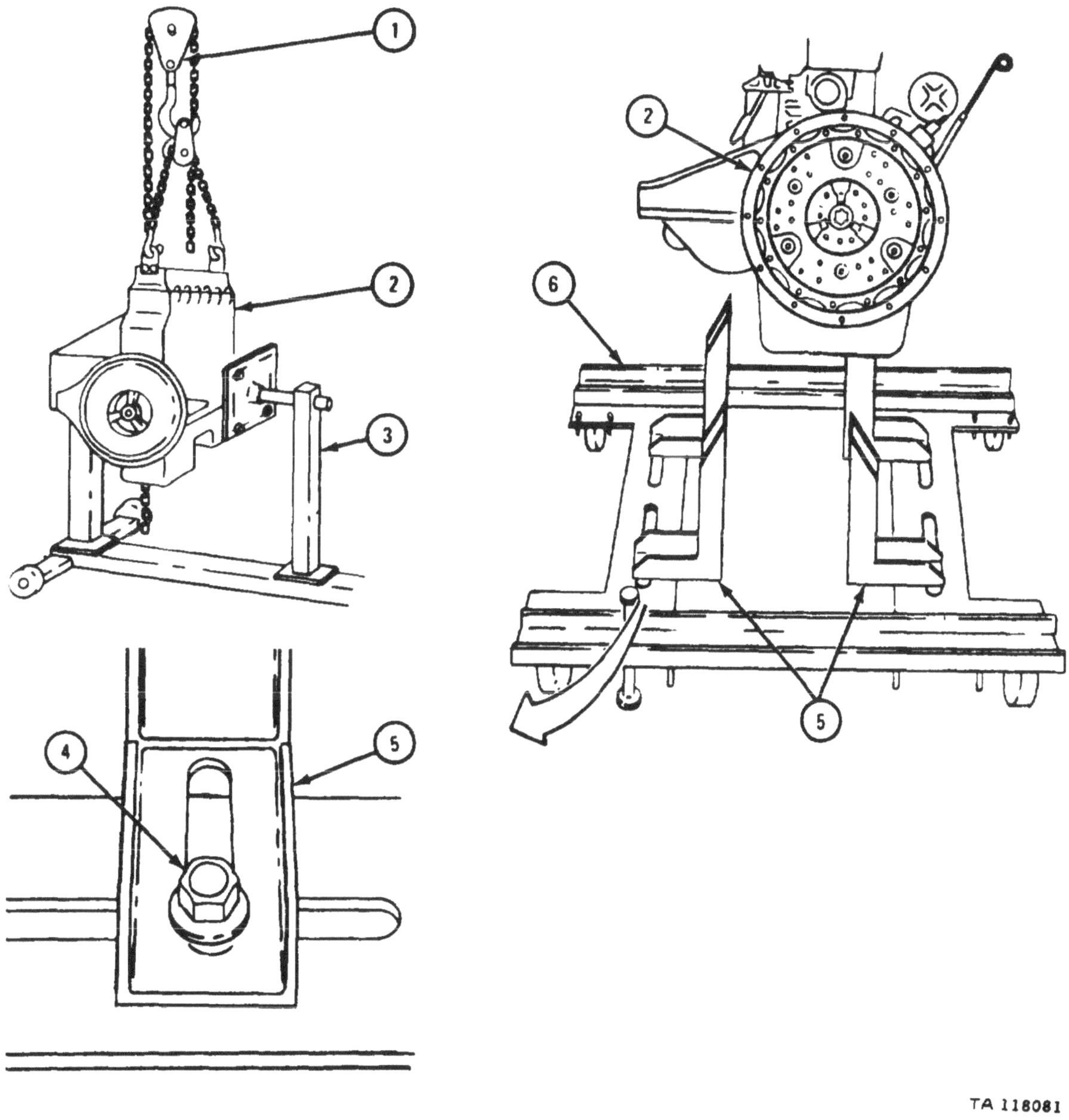

FRAME 5

1. Slide four engine supports (1) in place.

Soldier A 2 . As engine (2) is lowered onto stand (3), guide engine so four engine supports (1) will set on oil pan lip (4) as shown.

Solider B 3. Lower hoist slowly until oil pan lip (4) just touches four engine supports (1) .

Soldier A 4 . Tighten four capscrews (5) .

Soldier B 5. Lower hoist and take off lifting sling.

END OF TASK

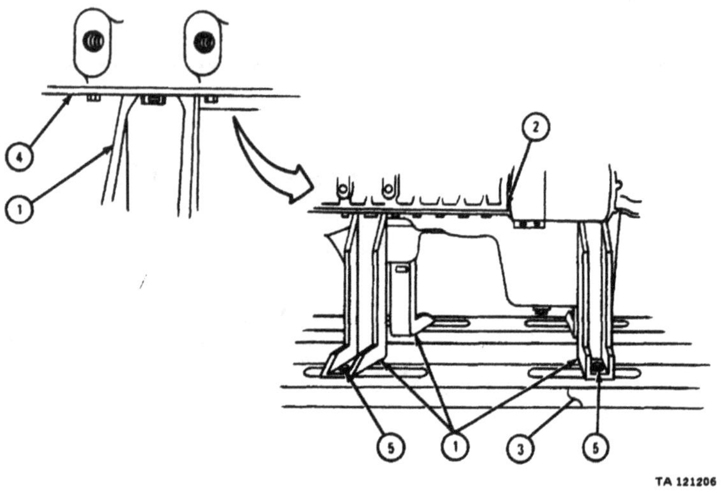

5-14. FLYWHEEL HOUSING.

a. Replacement.

FRAME 1

1. Put gasket (1) on rear of crankcase assembly (2).
2. Put flywheel housing (3) on crankcase assembly (2).
3. Coat threads of two studs (4), threads and seats of two nuts (5), and two 1/8-inch thick flat washers (6) with engine oil.
4. Put nut (5) with 1/8-inch flat washer (6) on each of two studs (4). Hand tighten nuts .
5. Coat threads of six studs (7), threads and seats of six nuts (8), and six 1/16-inch thick flat washers (9) with engine oil.
6. Put nut (8) and 1/16-inch flat washer (9) on each of six studs (7). Hand tighten nuts .

GO TO FRAME 2

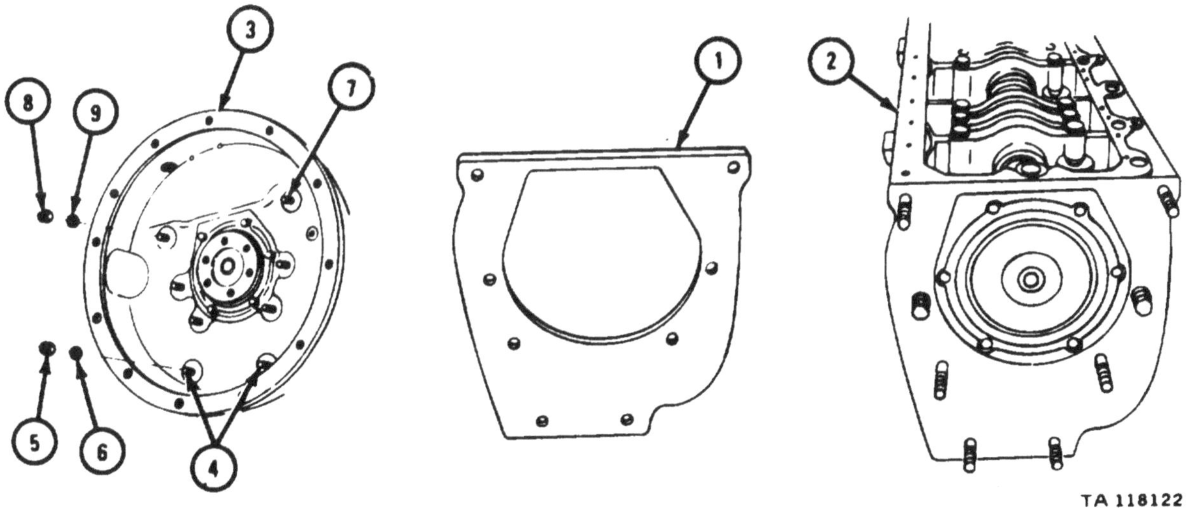

FRAME 2

1. Tighten six nuts (1 through 6) to 20 pound-feet in order shown.

GO TO FRAME 3

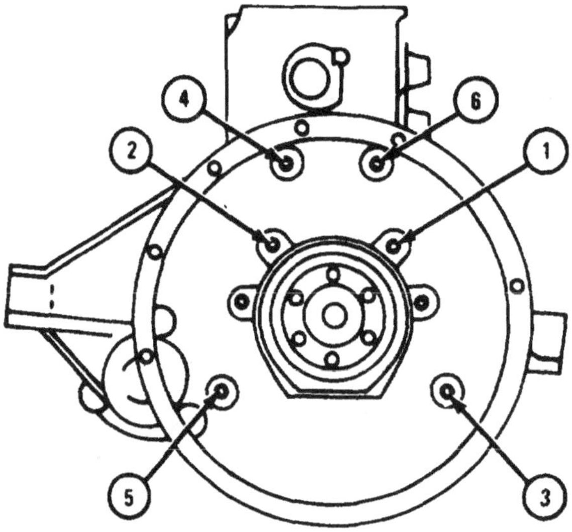

TA 121047

FRAME 3

1. Tighten eight nuts (1 through 8) to 45 pound-feet in order shown.

END OF TASK

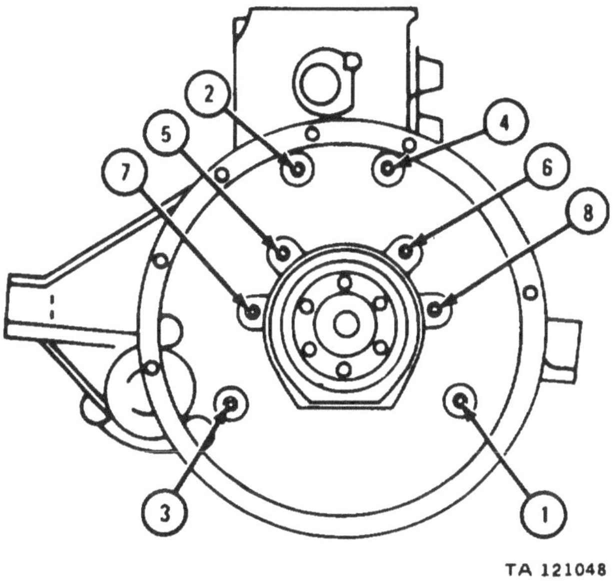

b. Flywheel Housing Face Runout Check.

FRAME 1

1. Mount dial indicator (1) on crankshaft (2) so indicator pointer (3) rests on flywheel housing mounting face (4).
2. Set indicator dial (5) to zero.
3. Turn crankshaft (2) 360° while pushing toward front of engine.
4. Read runout on indicator dial (5) while turning crankshaft (2). Runout cannot be more than 0.012 inch. If runout exceeds 0.012 inch, check to determine cause.

END OF TASK

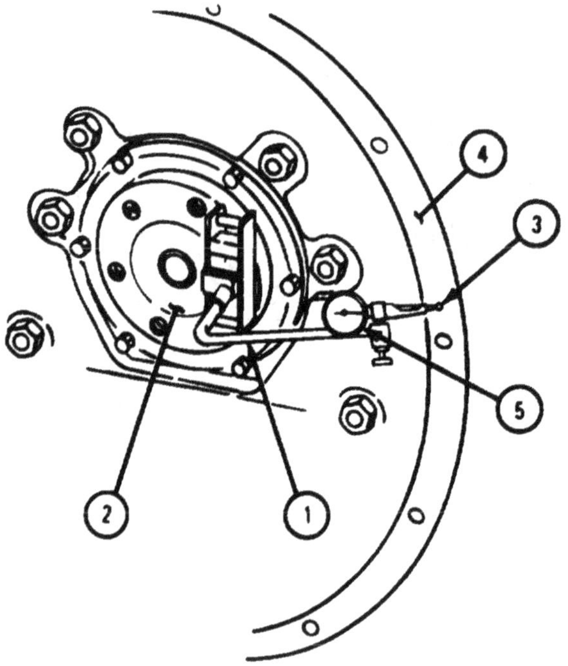

TA 121049

c. Flywheel Housing Rim Out-of-Round Check.

FRAME 1

1. Mount dial indicator (1) on crankshaft (2) so indicator pointer (3) rests on inside rim of flywheel housing (4).
2. Set indicator dial (5) to zero.
3. Turn crankshaft (2) 360° and check that out-of-round reading does not change more than 0.012 inch.

END OF TASK

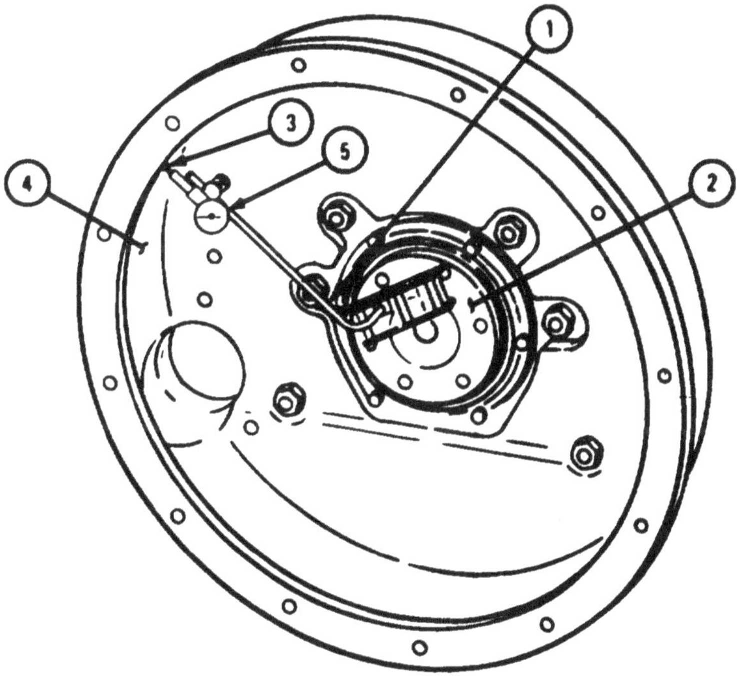

TA 121050

5-15. FLYWHEEL AND CLUTCH.

a. Flywheel.

(1) All engines except LDS-465-2 .

FRAME 1

NOTE

One threaded hole in crankshaft flange (2) is offset 1/16 inch. Because of this, flywheel can only be put on one way. Flange and flywheel were scribed during removal to help alinement during replacement. Pilot bolt (1) was fabricated to take off flywheel and clutch.

1. Put fabricated pilot bolt (1) into crankshaft flange (2).

GO TO FRAME 2

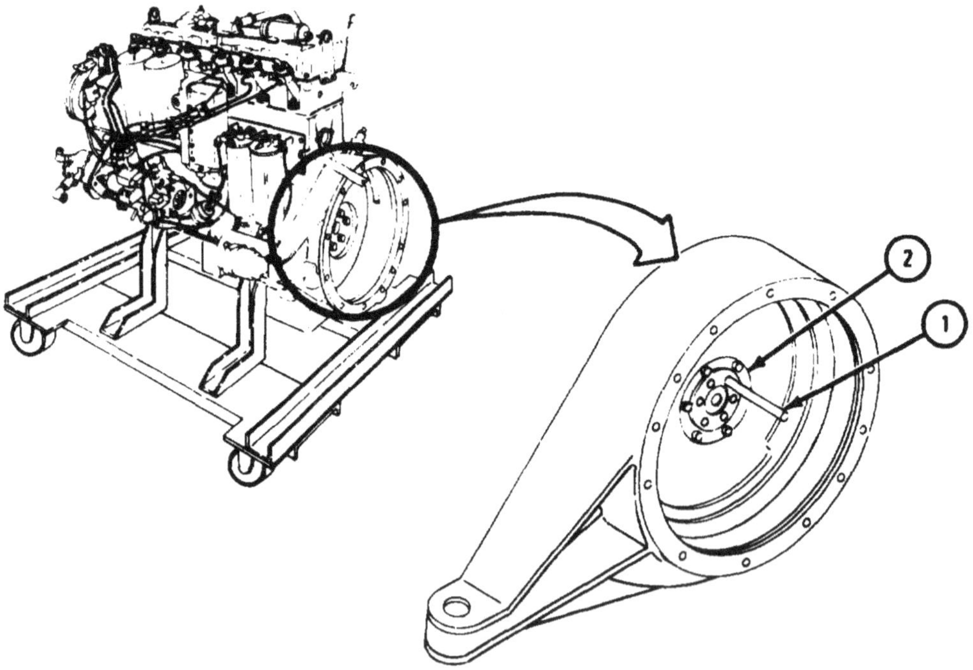

TA 118004

FRAME 2

WARNING

Flywheel (1) is very heavy. It should be lifted into place by two persons. After flywheel is in position, it must be held firmly until mounting bolts are put in. Failure to do so may cause injury to personnel and damage to equipment.

Soldiers A and B	1.	Lift flywheel (1) onto crankshaft flange (2). Aline scribe marks and slide flywheel onto pilot bolt (3).
Soldier A	2.	Hold flywheel (1) in place while soldier B puts in five mounting lockbolts (4).
Soldier B	3.	Coat six lockbolts (4) with oil. Put in and hand tighten five lockbolts.
	4.	Take out pilot bolt (3). put in other lockbolt (4).

GO TO FRAME 3

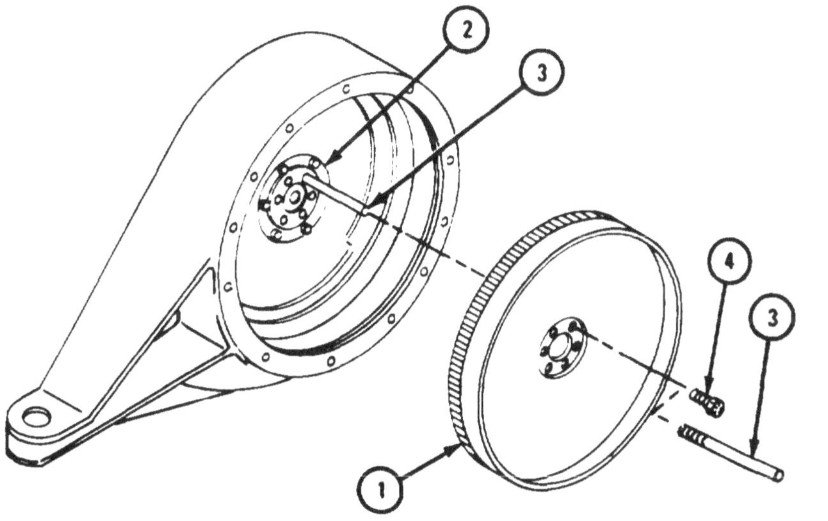

FRAME 3

Soldier A 1. Using engine barring tool, hold crankshaft pulley (1).

Soldier B 2. Tighten six lockbolts (2) to 80 to 90 pound-feet.

3. Tighten six lockbolts (2) to 115 to 120 pound-feet.

GO TO FRAME 4

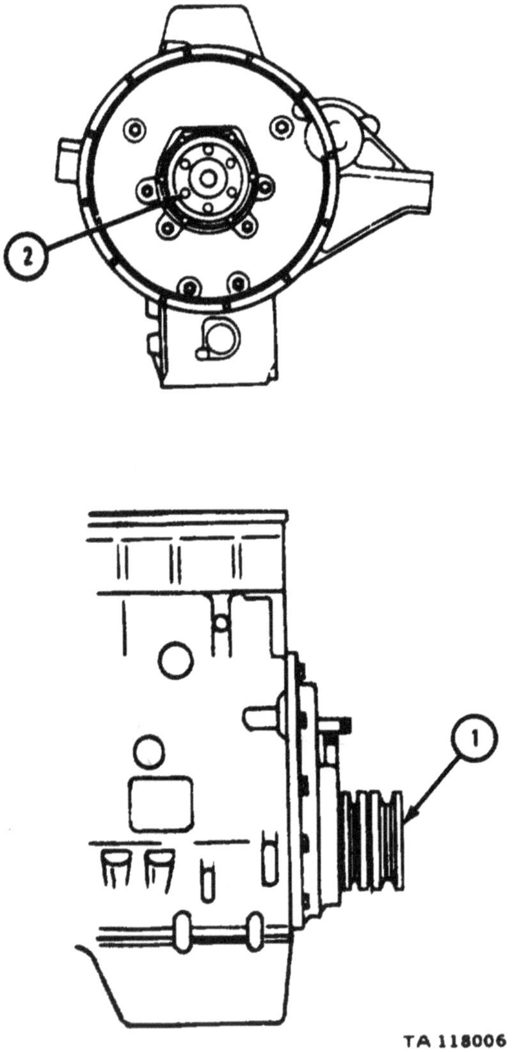

FRAME 4

NOTE

When checking flywheel rim (4) runout, push flywheel forward as far as it will go. Hold forward pressure on flywheel while turning crankshaft pulley (5).

Soldier A 1. Mount dial indicator (1) on flywheel housing (2) so indicator pointer (3) rests on flywheel rim (4) as shown. Set dial indicator to zero.

Soldier B 2. Using engine barring tool, turn crankshaft pulley (5) slowly to the right until soldier A tells you to stop.

Soldier A 3. Read dial indicator (1) as soldier B turns crankshaft pulley (5). When dial indicator is at the highest reading, tell soldier B to stop. Set dial indicator to zero.

Soldier B 4. When soldier A is ready, turn crankshaft pulley (5) slowly one full turn to the right.

NOTE

Dial indicator (1) will now read total flywheel runout. Flywheel runout must not be more than 0.008 inch.

Soldier A 5. Read dial indicator (1) as soldier B turns crankshaft pulley (5).

GO TO FRAME 5

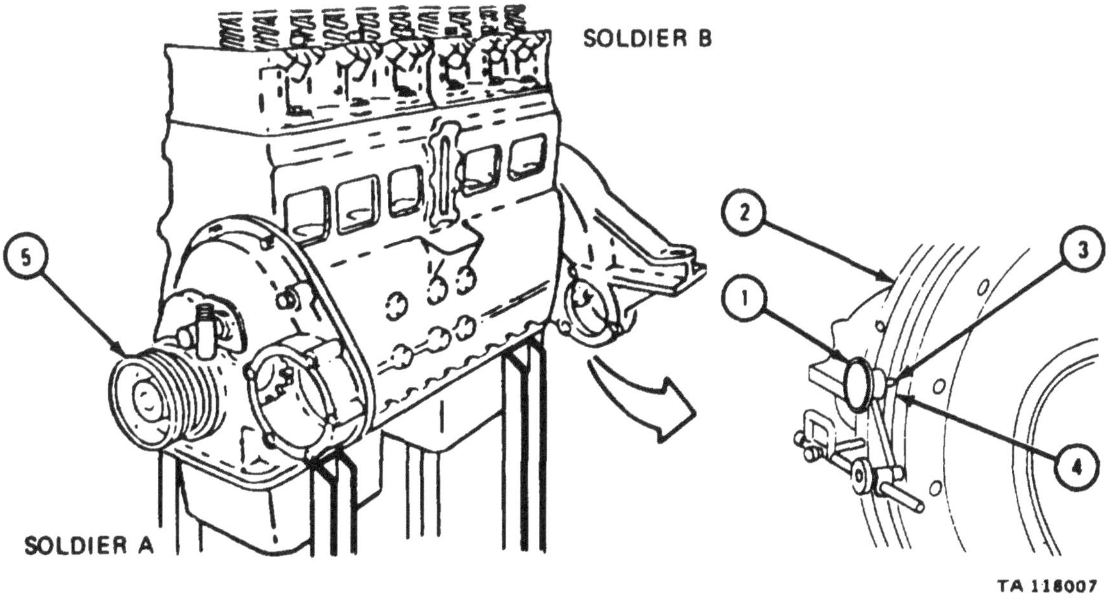

FRAME 5

NOTE

When checking flywheel face (4) runout, push flywheel forward as far as it will go. Hold forward pressure on flywheel while turning crankshaft pulley (5).

Soldier A 1. Mount dial indicator (1) on flywheel housing (2) so indicator pointer (3) rests on flywheel face (4) as shown. Set dial indicator to zero.

Soldier B 2. Using engine barring tool, turn crankshaft pulley (5) slowly to the right until soldier A tells you to stop.

Soldier A 3. Read dial indicator (1) as soldier B turns crankshaft pulley (5). When dial indicator is at the highest reading, tell soldier B to stop. Set dial indicator to zero.

Soldier B 4. When soldier A is ready, turn crankshaft pulley (5) slowly one full turn to the right.

NOTE

Dial indicator (1) will now read total flywheel runout. Flywheel runout must not be more than 0.008 inch.

Soldier A 5. Read dial indicator (1) as soldier B turns crankshaft pulley (5).

END OF TASK

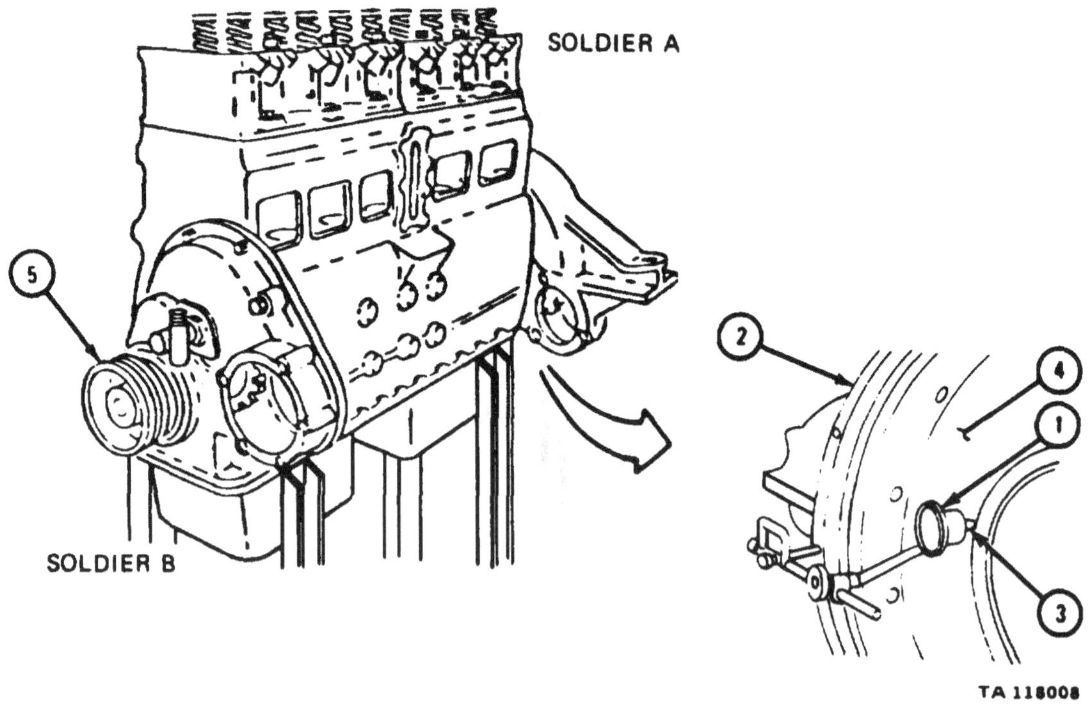

(2) Engine LDS-465-2 .

FRAME 1

NOTE

One threaded hole in crankshaft flange (2) is offset 1/16 inch. Because of this, flywheel can only be put on one way. Flange and flywheel were scribed during removal to help alinement during replacement. Pilot bolt (1) was fabricated to take off flywheel.

1. Put fabricated pilot bolt (1) into crankshaft flange (2).

GO TO FRAME 2

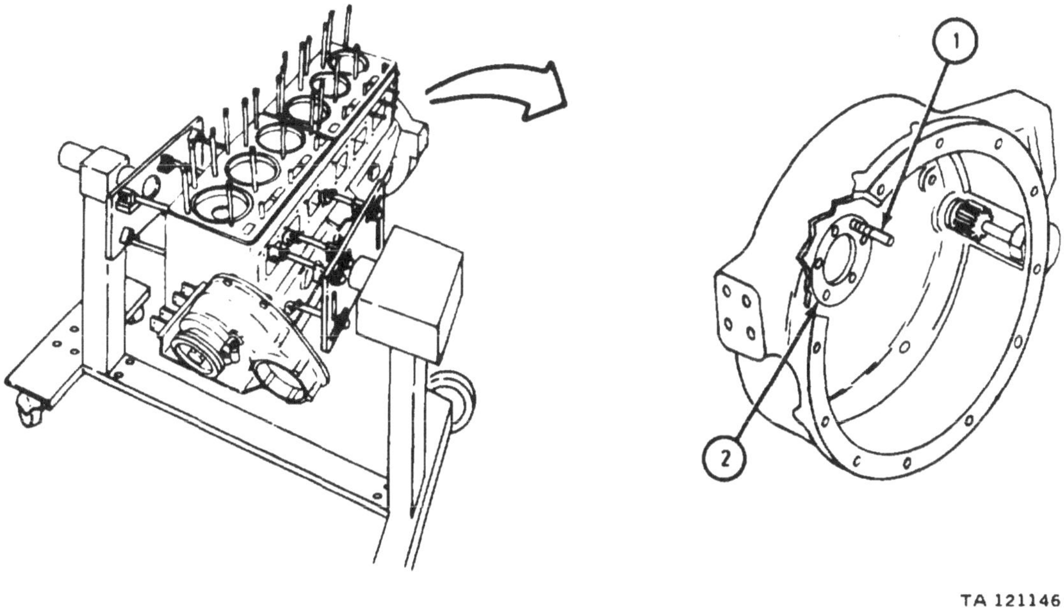

TA 121146

FRAME 2

WARNING

Flywheel (1) is very heavy. It should be lifted into place by two persons. After flywheel is in position, it must be held firmly until mounting bolts are put in. Failure to do so may cause injury to personnel and damage to equipment.

CAUTION

Mating surfaces of crankshaft flange (2) and flywheel (1) must be clean and free of raised metal which could keep flywheel from being properly seated.

Soldiers A and B

1. Lift flywheel (1) and flexible plates (3) up to flywheel housing (4) .
2. Aline scribe marks on flywheel (1) and flexible plates (3) and slide them onto pilot bolt (5).

Soldier A

3. Hold flywheel (1) and flexible plates (3) in place while soldier B puts in five mounting lockbolts (6).

Soldier B

4. Coat six lockbolts (6) with engine oil. Put on scuff plate (7) . Put in and hand tighten five lockbolts.
5. Take out pilot bolt (5). Put in and hand tighten other lockbolt (6).

GO TO FRAME 3

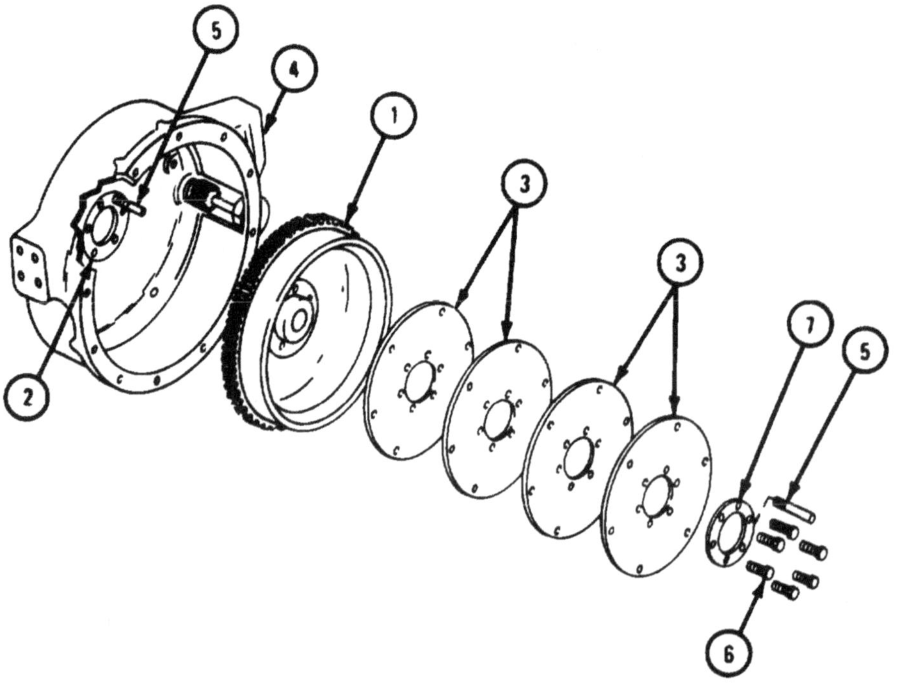

TA 121147

FRAME 3

Soldier A 1. Using engine barring tool, hold crankshaft damper and pulley (1) so it cannot turn to the right.

Soldier B 2. Tighten six lockbolts (2) to 80 to 90 pound-feet.

3. Tighten six lockbolts (2) to 115 to 120 pound-feet.

END OF TASK

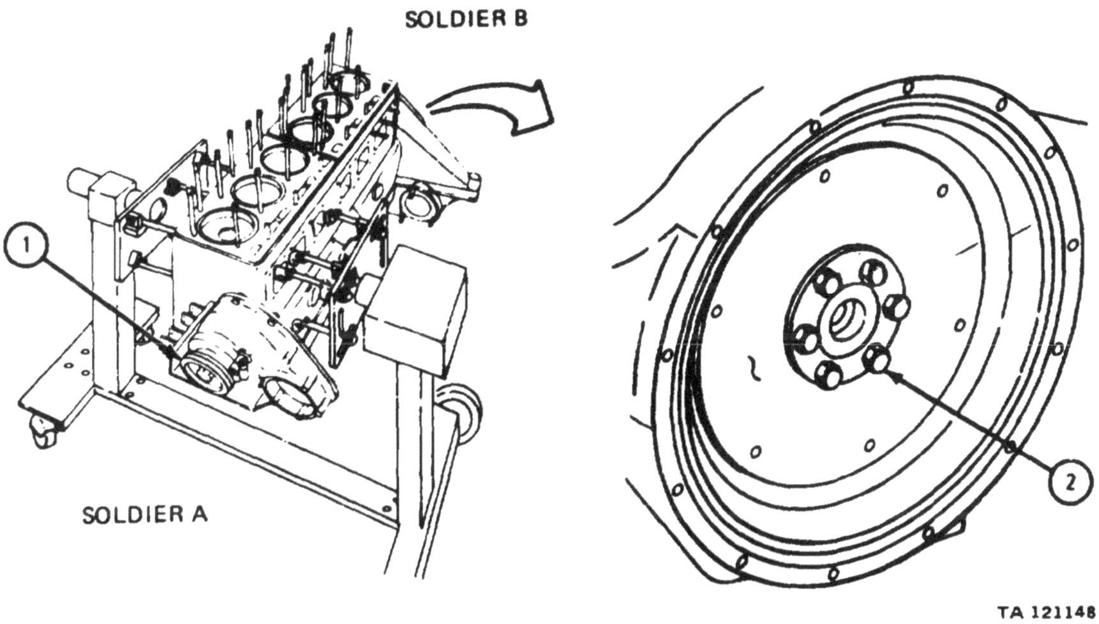

b. Clutch.

(1) Engines LD-465-1, LD-465-1C, and LDT-465-1C.

FRAME 1

WARNING

Clutch disk (1) is heavy. Be careful not to drop it or it can cause injury to personnel and damage to equipment.

1. Put clutch disk (1) in place against face of flywheel (2) with long end of clutch drive hub (3) facing out.

NOTE

Clutch alining tool (4) must be left in place until frame 2 is done.

2. Put clutch alining tool (4) through clutch drive hub (3) and into hole in center of flywheel (2).

GO TO FRAME 2

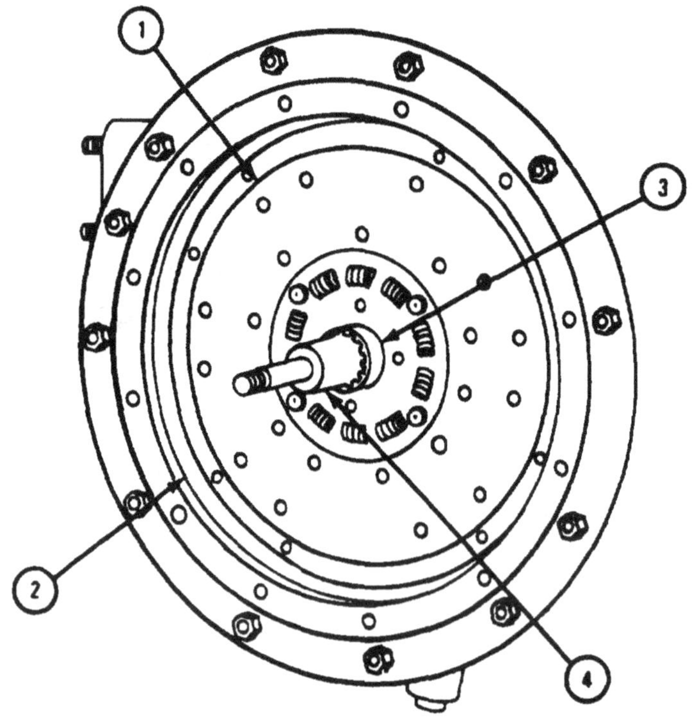

FRAME 2

WARNING

Pressure plate assembly (1) is heavy. It must be held firmly up and against flywheel (2) until mounting screws (4) are put in. If it is not held up, it will fall and cause injury to personnel and damage to equipment.

Soldiers A and B	1.	Put pressure plate assembly (1) in place against flywheel (2) Aline eight holes in pressure plate assembly with holes in clutch disk (3) .
Soldier A	2.	Hold pressure plate assembly (1) firmly up and against clutch disk (3) until eight mounting screws (4) are put in.
Soldier B	3.	Put in eight screws with lockwashers (4).
	4.	Tighten eight screws (4) to 23 to 27 pound-feet.
	5.	Press four release levers (5), one at a time, using prybar, and take out release lever spacer blocks (6).
	6.	Take out clutch alining tool (7).

END OF TASK

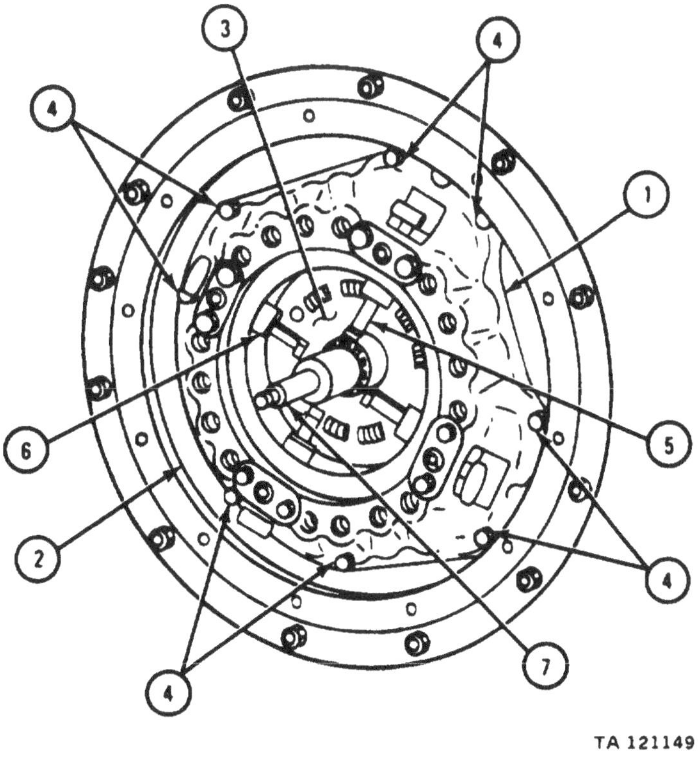

(2) Engines LDS-465-1 and LDS-465-1A .

FRAME 1

NOTE

Three types of clutch disks (1) are used. Replacement procedures are the same for all types.

1. Put clutch disk (1) in place against face of flywheel (2) with long end of clutch drive hub (3) facing out .

NOTE

Clutch alining tool (4) must be left in place until frame 2 is done.

2. Put clutch alining tool (4) through clutch drive hub (3) and into hole in center of flywheel (2) .

GO TO FRAME 2

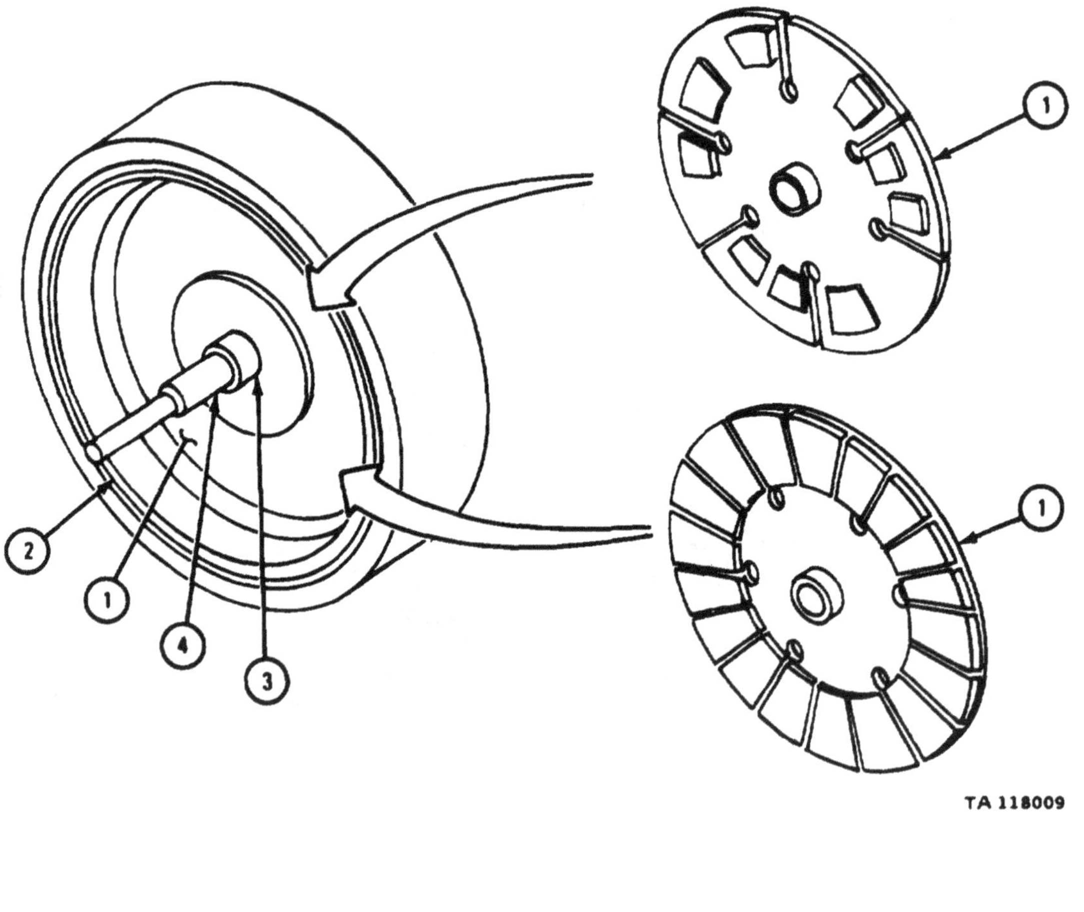

FRAME 2

NOTE

Be sure three retaining screws and washers (1) are screwed tightly in pressure plate assembly (2) before you do step 1. If they are not, put them in. Refe r to para 4-56.

Soldiers A and B

1. Hold pressure plate assembly (2) against rim of flywheel (3). Punch mark (4) on pressure plate assembly should line up with punch mark (5) on rim of flywheel. Twelve mounting holes (6) in pressure plate assembly must line up with threaded holes in rim of flywheel.
2. Join pressure plate assembly (2) to flywheel (3) with 12 screws and lockwashers (7) . Using torque wrench, tighten 12 screws and lockwashers to 23 to 27 pound-feet.
3. Take out three retaining screws and washers (1).
4. Take out clutch alining tool (8).

END OF TASK

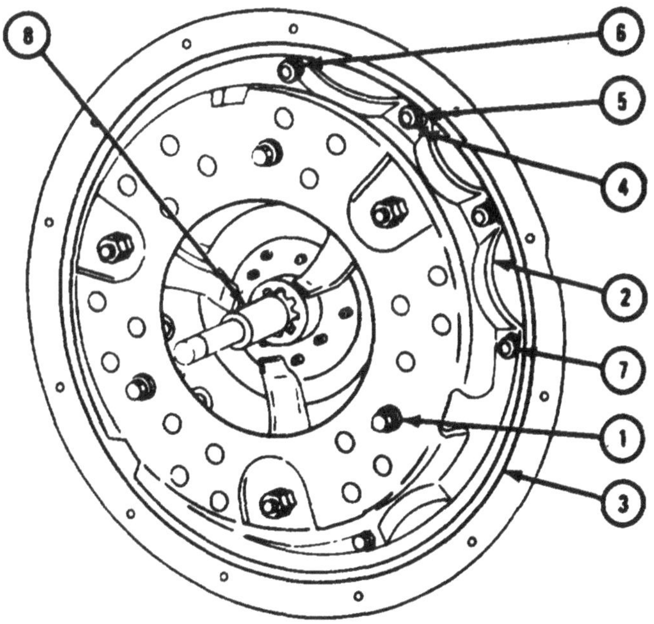

5-16. ROCKER ARMS, ROCKER ARM PUSH RODS, AND VALVE TAPPETS.

FRAME 1

1. Coat 12 tappets (1) with anti-scuff lubricant.

NOTE

If you are using tappets (1) that were taken out, make sure they go in tappet bores (2) they were taken from.

2. Put one tappet (1) into each of 12 tappet bores (2).

GO TO FRAME 2

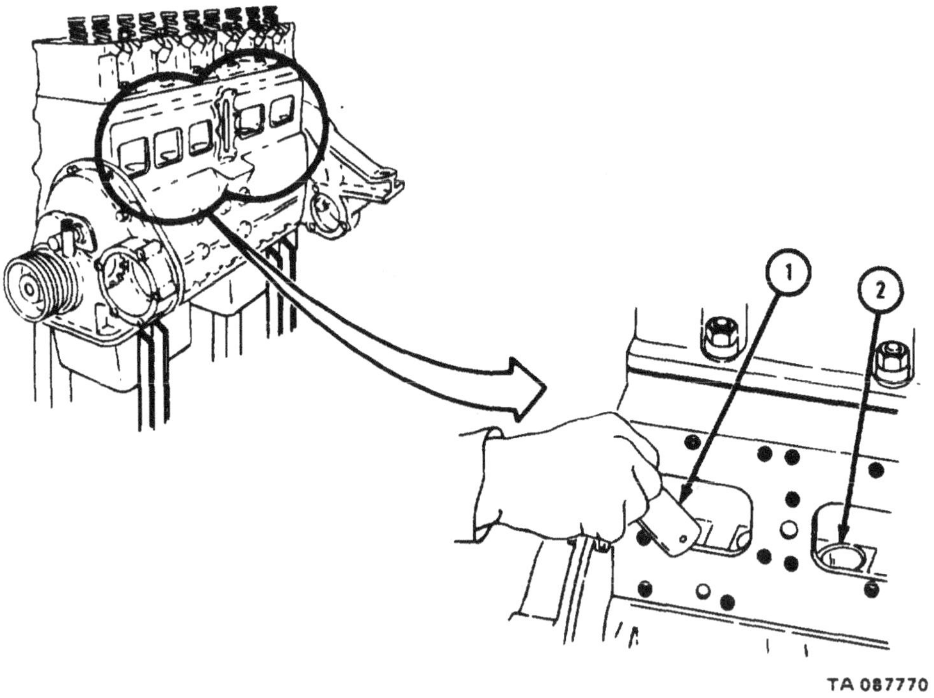

FRAME 2

NOTE

If you are using push rods (1) that were taken out, make sure they go in push rod passage (2) they were taken from .

1. Lower one push rod (1) through each of 12 push rod passages (2) in cylinder heads (3). Bottom end of push rod must fit through tappet bore (4) and rest on tappet (5) inside crankcase (6) as shown.

GO TO FRAME 3

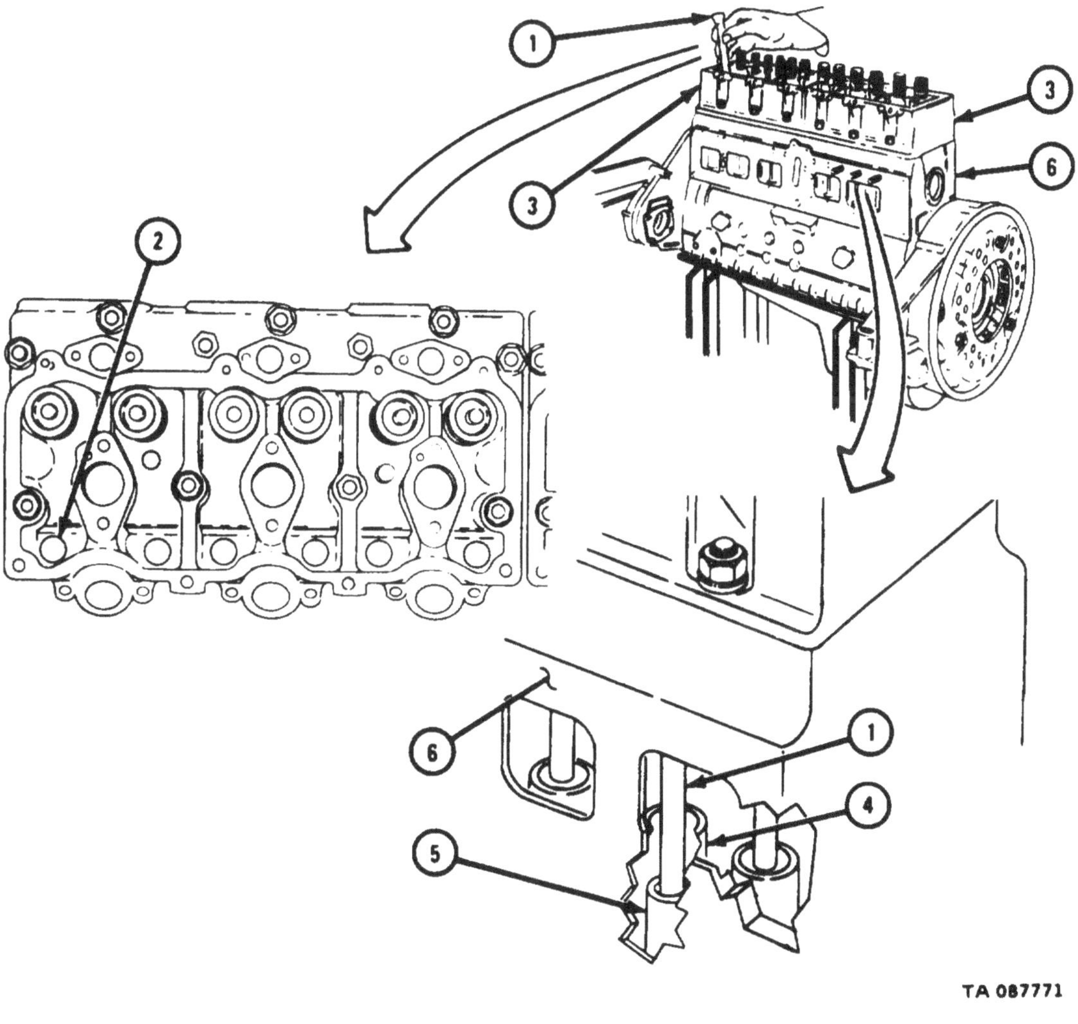

FRAME 3

NOTE

This task is the same for the front and rear rocker arm assemblies. This task is shown for the front rocker arm assembly.

1. Loosen six adjusting screw locknuts (1) on front rocker arm assembly (2). Turn six adjusting screws (3) as far to the left as they will go.
2. Put front rocker arm assembly (2) on front cylinder head (4). Six mounting holes (5) in rocker arm assembly must line up with six holes (6) in cylinder head.
3. Put in and hand tighten two screws and lockwashers (7), Put in and hand tighten two screws and lockwashers (8). Put in and hand tighten two screws and lockwashers (9) .
4. Using torque wrench, tighten two screws (7) to 44 pound-feet. Using torque wrench, tighten two screws (8) to 44 pound-feet. Using torque wrench , tighten two screws (9) to 44 pound-feet.
5. Do steps 1 through 4 again on rear cylinder head (10).

GO TO FRAME 4

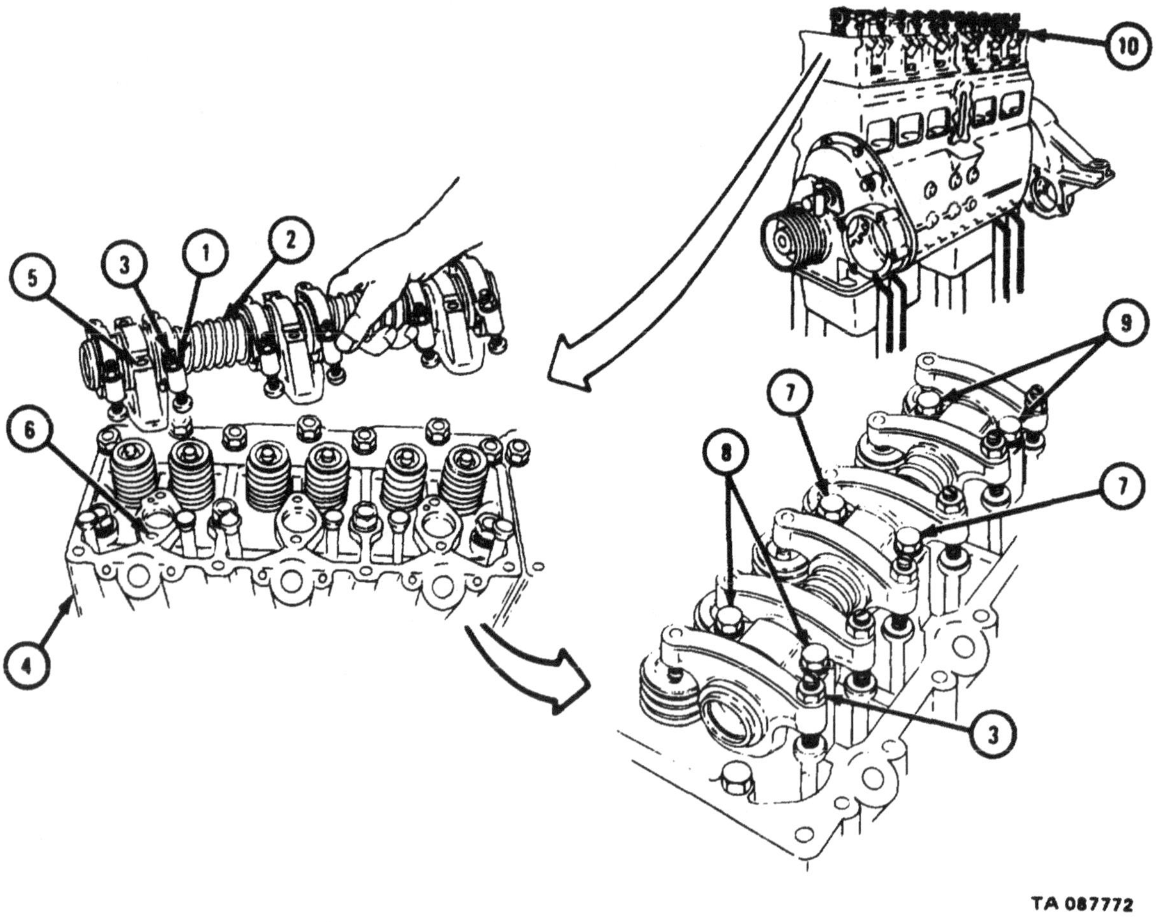

FRAME 4

NOTE

Cylinders are called number one through six, counting from front to rear of engine. Each cylinder has one intake valve and one exhaust valve. Intake valve for each cylinder is toward front of cylinder and exhaust valve is toward rear of cylinder.

1. Turn all 12 adjusting screws (1) in until they just touch push rods (2).

NOTE

While turning in adjusting screws (1), you may have to unscrew locking nuts (3) before adjusting screw can touch push rods (2).

2. Watch push rod (2) which works cylinder number one intake valve. Using engine-barring wrench, turn crankshaft damper and pulley (4) until push rod moves up as high as it will go.
3. Slide 0.010-inch feeler gage between cylinder number two intake rocker arm pad (5) and cylinder number two intake valve stem (6).
4. Turn adjusting screw (1) until feeler gage fits snugly between intake rocker arm pad (5) and intake valve stem (6).
5. Keep adjusting screw (1) from turning and tighten locknut (3).
6. Check that 0.010-inch feeler gage blade still fits snugly between rocker arm pad (5) and intake valve stem (6). If feeler gage does not fit snugly, loosen locknut (3) and do steps 2 through 6 again.
7. Do steps 3 through 6 for cylinder number three and cylinder number six intake valves.

GO TO FRAME 5

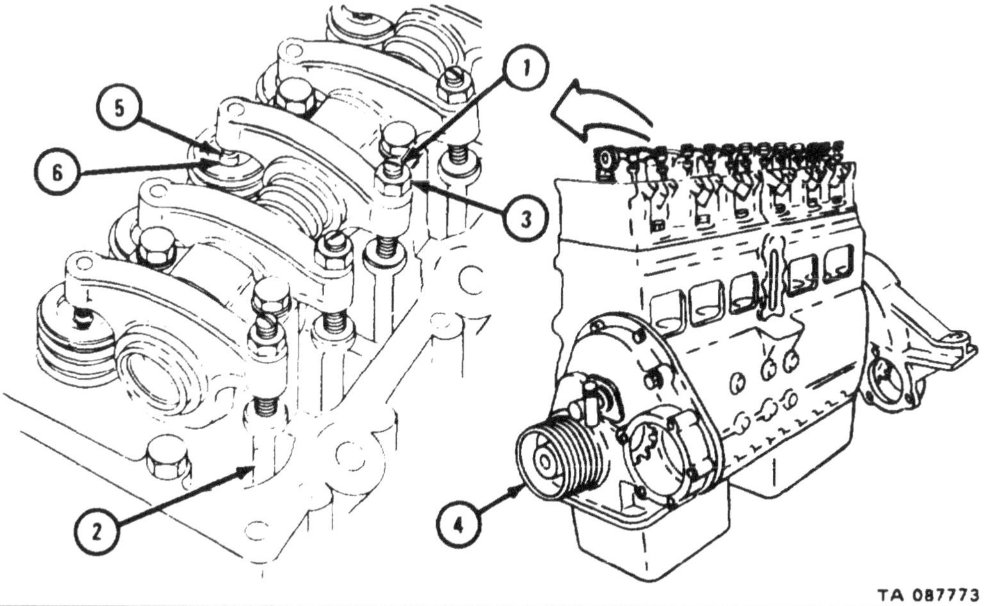

FRAME 5

NOTE

Cylinders are called number one through six counting from front to rear of engine. Each cylinder has one intake valve and one exhaust valve. Intake valve for each cylinder is toward front of cylinder and exhaust valve is toward rear of cylinder.

1. Slide 0.025-inch feeler gage between cylinder number one exhaust rocker arm pad (1) and exhaust valve stem (2).
2. Turn adjusting screw (3) until feeler gage fits snugly between cylinder number one exhaust rocker arm pad (1) and exhaust valve stem (2).
3. Keep adjusting screw (3) from turning and tighten locknut (4).
4. Check that 0.025-inch feeler gage blade still fits snugly between exhaust rocker arm pad (1) and exhaust valve stem (2). If feeler gage does not fit snugly, loosen locknut (4) and do steps 1 through 4 again.
5. Do steps 1 through 4 again for cylinder number two and cylinder number four exhaust valves .

GO TO FRAME 6

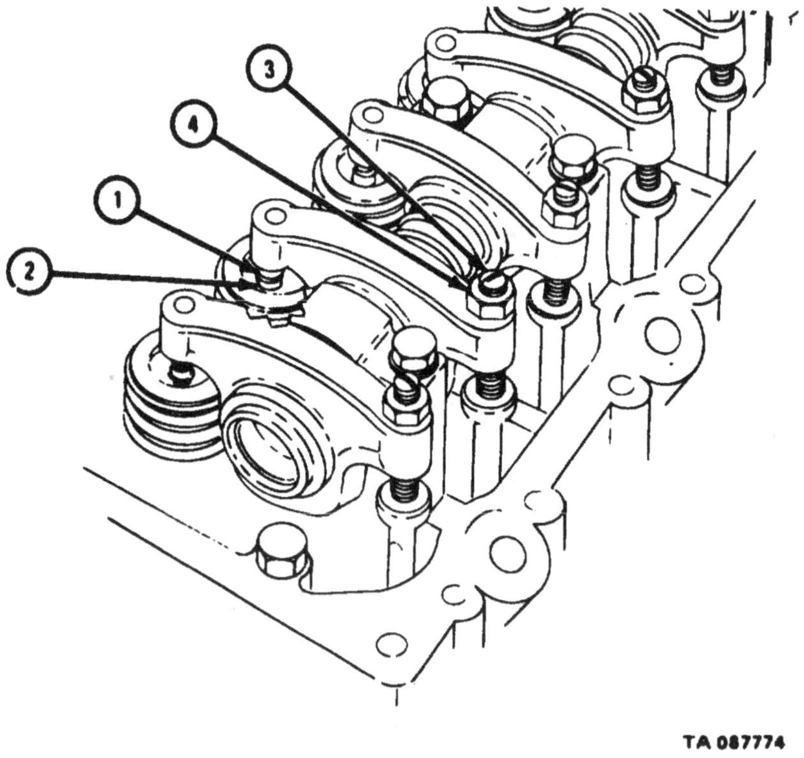

FRAME 6

NOTE

Cylinders are called number one through six counting from front to rear of engine. Each cylinder has one intake valve and one exhaust valve. Intake valve for each cylinder is toward front of cylinder and exhaust valve is toward rear of cylinder.

Crankshaft (1) can be turned back and forth a little to the right or left a little bit until you are sure push rod (2) is as high as it will go.

1. Watch push rod (2) which works cylinder number six intake valve. Turn crankshaft (1) one full circle to the right until push rod (2) moves up as high as it will go.
2. Slide 0.010-inch feeler gage between cylinder number five intake rocker arm pad (3) and intake valve stem (4).
3. Turn adjusting screw (5) until feeler gage fits snugly between cylinder number five intake rocker arm pad (3) and intake valve stem (4).
4. Keep adjusting screw (5) from turning and tighten locknut (6).
5. Check that 0.010-inch feeler gage still fits snugly between rocker arm pad (3) and intake valve stem (4). If feeler gage does not fit snugly, loosen locknut (6) and do steps 2 through 5 again.
6. Do steps 2 through 5 again for cylinder number one and cylinder number four intake valves.

GO TO FRAME 7

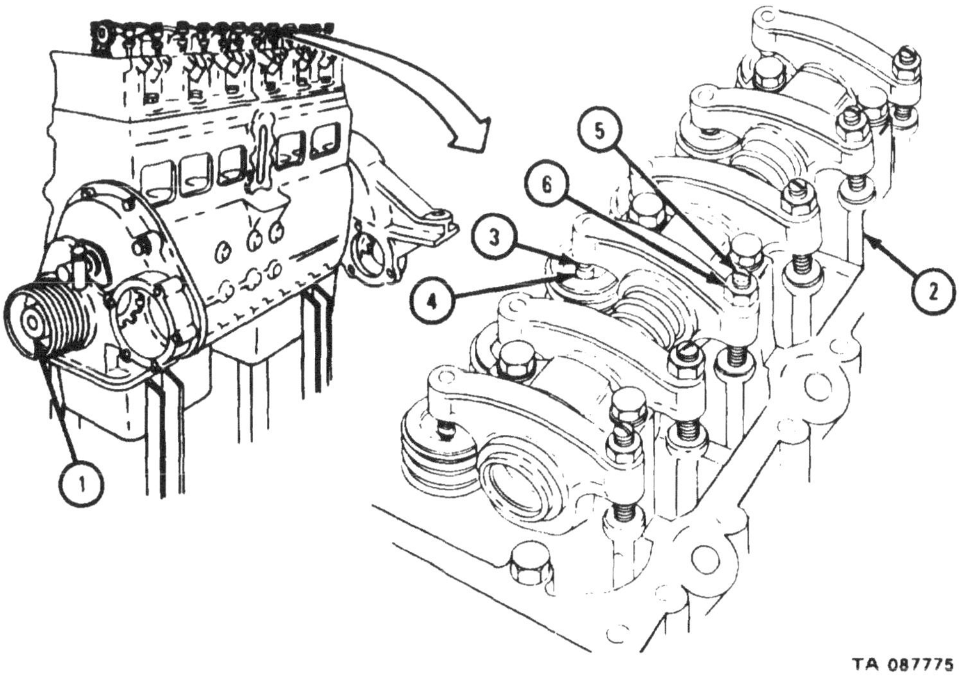

FRAME 7

NOTE

Cylinders are called number one through six counting from front to rear of engine. Each cylinder has one intake valve and one exhaust valve. Intake valve for each cylinder is toward front of cylinder and exhaust valve is toward rear of cylinder.

1. Slide 0.025-inch feeler gage between cylinder number five exhaust rocker arm pad (1) and exhaust valve stem (2).
2. Turn adjusting screw (3) until feeler gage fits snugly between exhaust rocker arm pad (1) and exhaust valve stem (2).
3. Keep adjusting screw (3) from turning and tighten locknut (4).
4. Check that 0.025-inch feeler gage blade still fits snugly between exhaust rocker arm pad (1) and exhaust valve stem (2). If feeler gage does not fit snugly, loosen locknut (4) and do steps 1 through 4 again.
5. Do steps 1 through 4 again for cylinder number three and cylinder number six exhaust valves .

END OF TASK

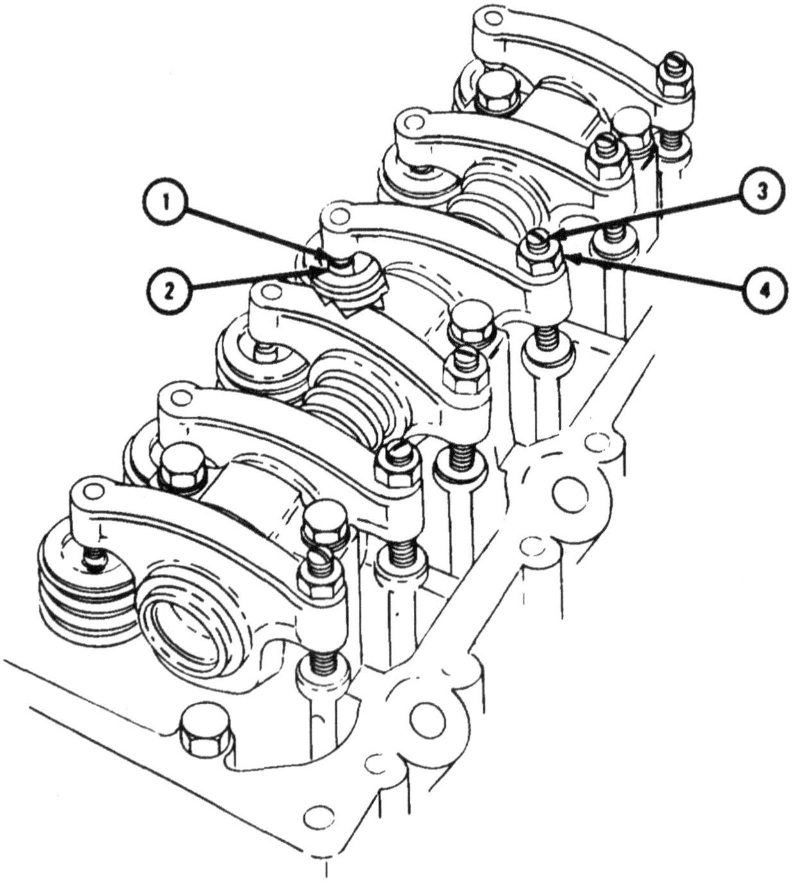

TA 087776

5-17. INTAKE AND EXHAUST MANIFOLD AND OIL LEVEL GAGE SUPPORT TUBE BRACKET.

FRAME 1

1. Put two manifold gaskets (1) in place.
2. Put exhaust manifold (2) in place.

GO TO FRAME 2

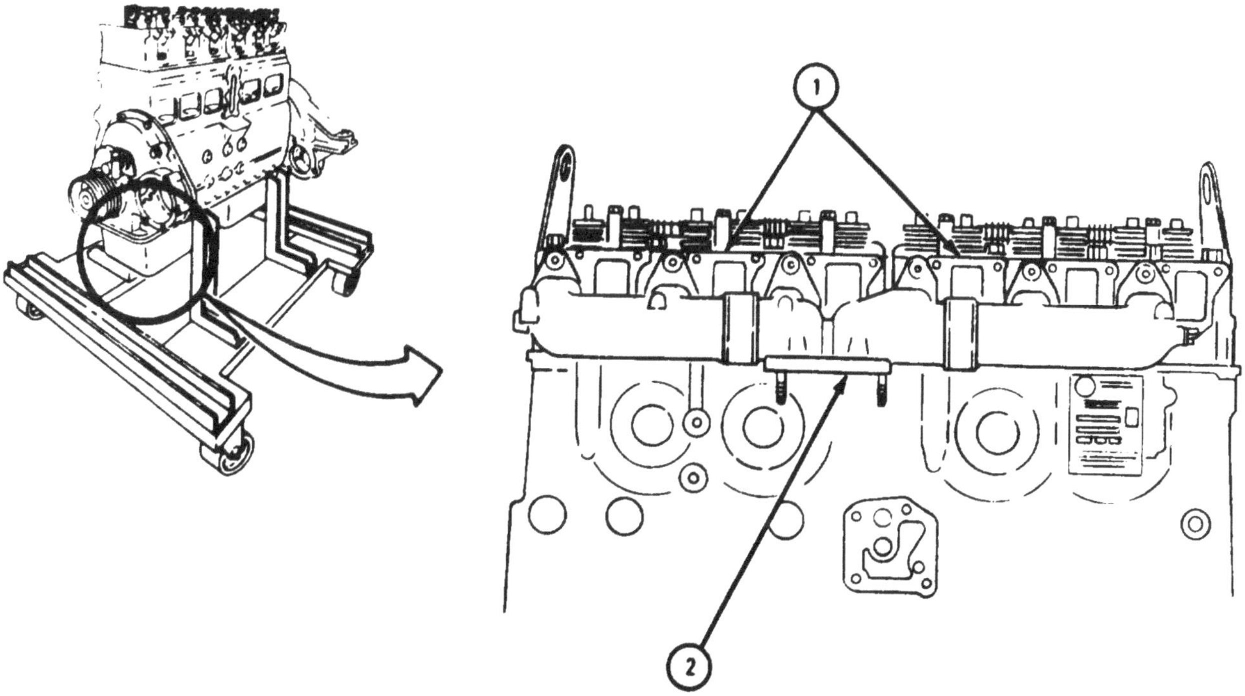

TA 118116

FRAME 2

1. Put intake manifold (1) in place.
2. Put six locknuts and washers (2) on exhaust manifold top flange studs (3).
3. Put 12 nuts and washers (4) on intake manifold top flange studs (5).

IF WORKING ON ENGINE LDS-465-1 OR LDS-465-2, GO TO FRAME 3.
IF WORKING ON ENGINE LD-465-1 OR LDT-465-1C, GO TO FRAME 4

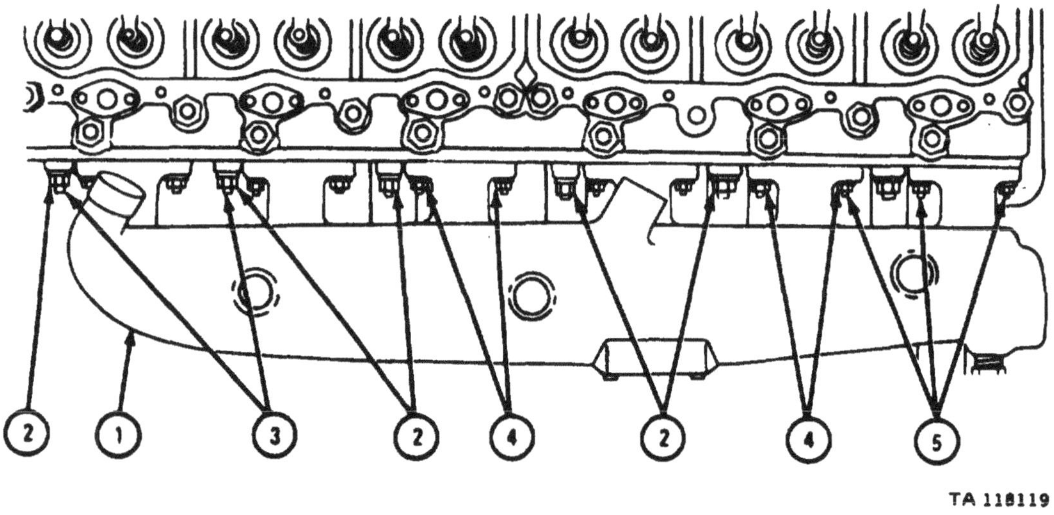

FRAME 3

1. Put six nuts with washers (1) on intake manifold bottom flange studs (2).
2. Put oil level gage support tube bracket (3) in place on rear two exhaust manifold bottom flange studs (4).
3. Put six locknuts with washers (5) on exhaust manifold bottom flange studs (4).

END OF TASK

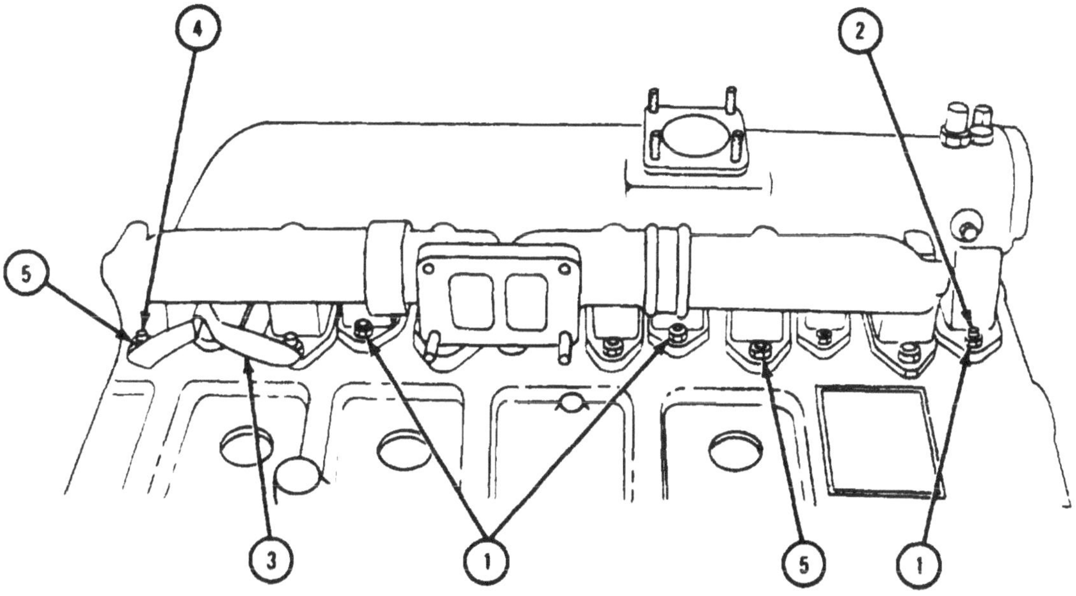

FRAME 4

1. Put oil level gage support tube bracket (1) on the second and third intake manifold bottom flange studs (2).
2. Put six nuts with washers (3) on six intake manifold bottom flange studs (2).
3. Put six locknuts with washers (4) on six exhaust manifold bottom flange studs (5) .

END OF TASK

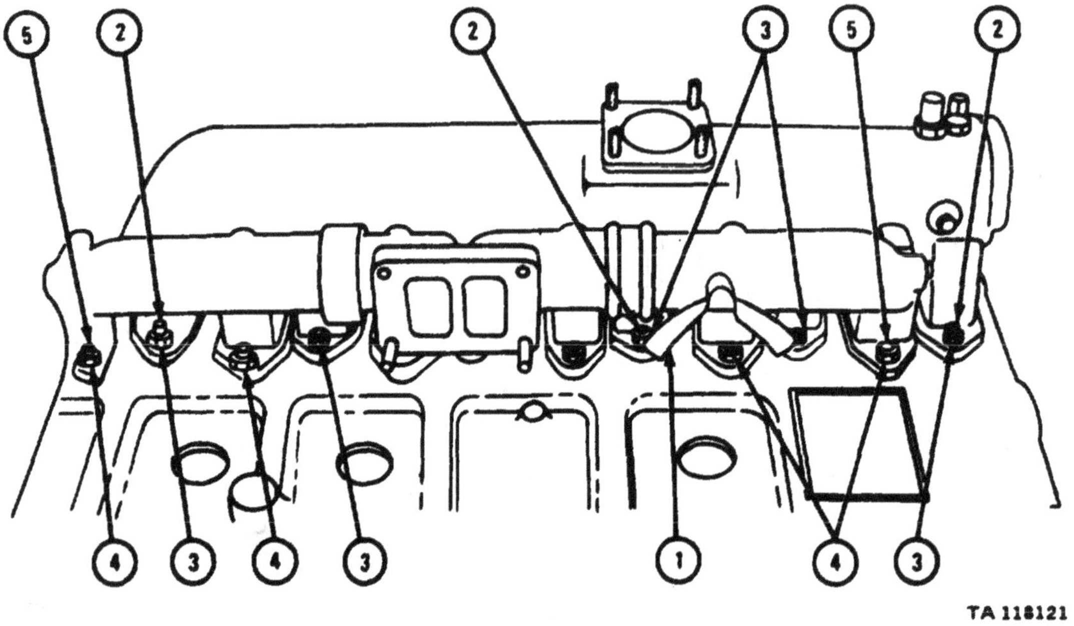

5-18. CYLINDER HEAD WATER OUTLET MANIFOLDS.

FRAME 1

1. Coat six gaskets (1) with sealant. Put six gaskets (1) in place on two cylinder heads (2).
2. Put two hoses (3) on two water manifolds (4). Put on four hose clamps (5).
3. Push two hoses (3) onto two intake manifold water ports (6) and put two water outlet manifolds (4) in place on two cylinder heads (2).

NOTE

It may be necessary to take out three or five cap screws (8) to mount flame heater components.

4. Put oil filler cap retaining chain (7) in place. Put in 12 capscrews with lock-washers (8). Tighten four hose clamps (5).

END OF TASK

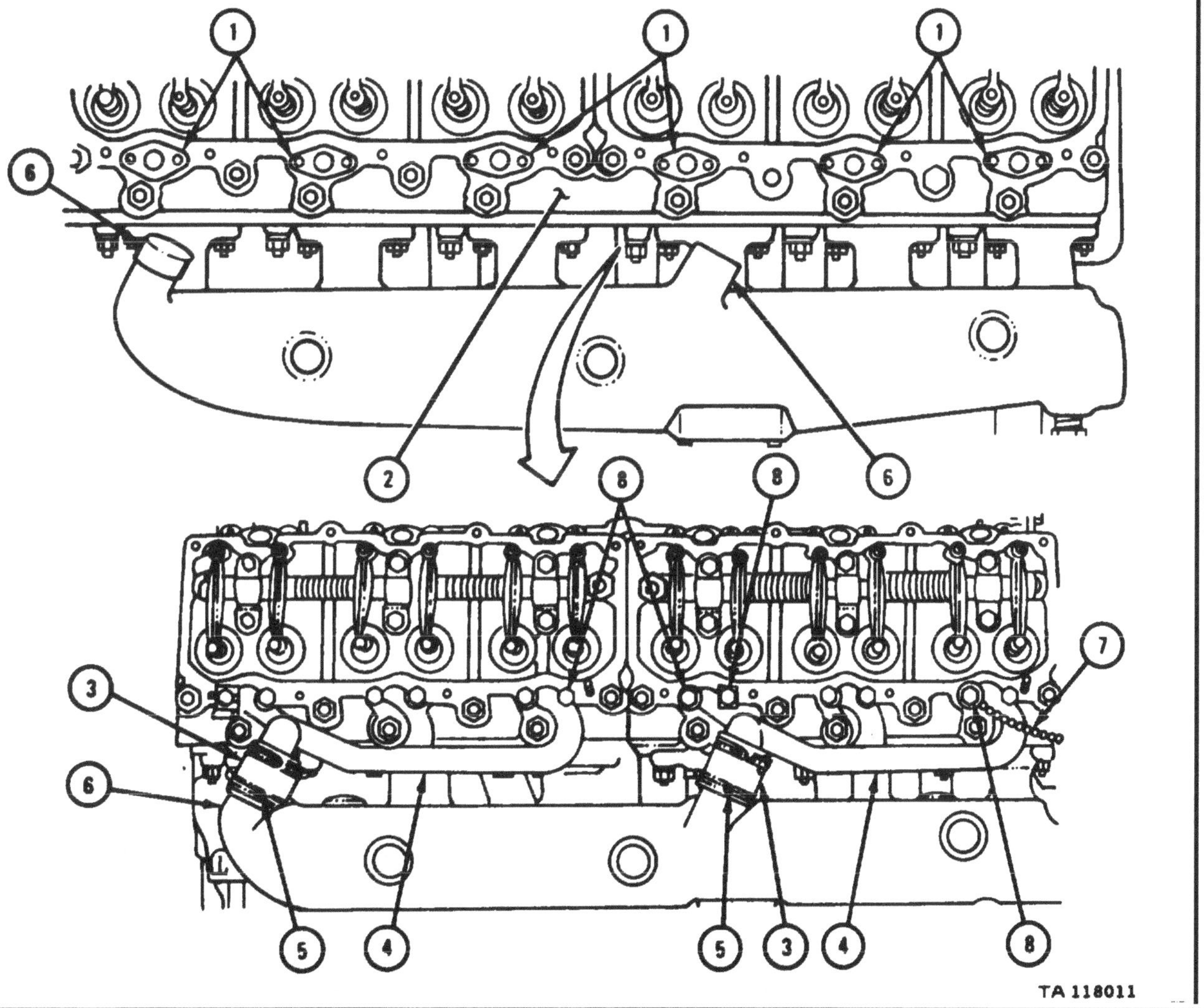

5-19. WATER PUMP AND THERMOSTAT HOUSING.

a. Water Pump.

(1) All engines except LDS-465-2 .

FRAME 1

1. Put water pump housing (1) in place on cylinder block (2).
2. Put in three screws with lockwashers (3).
3. Put on generator adjusting strap (4). Put on but do not tighten nut with lockwasher (5) .
4. Put water pump (6) with new gasket (7) on water pump housing (1).
5. Put on six nuts with lockwashers (8).

END OF TASK

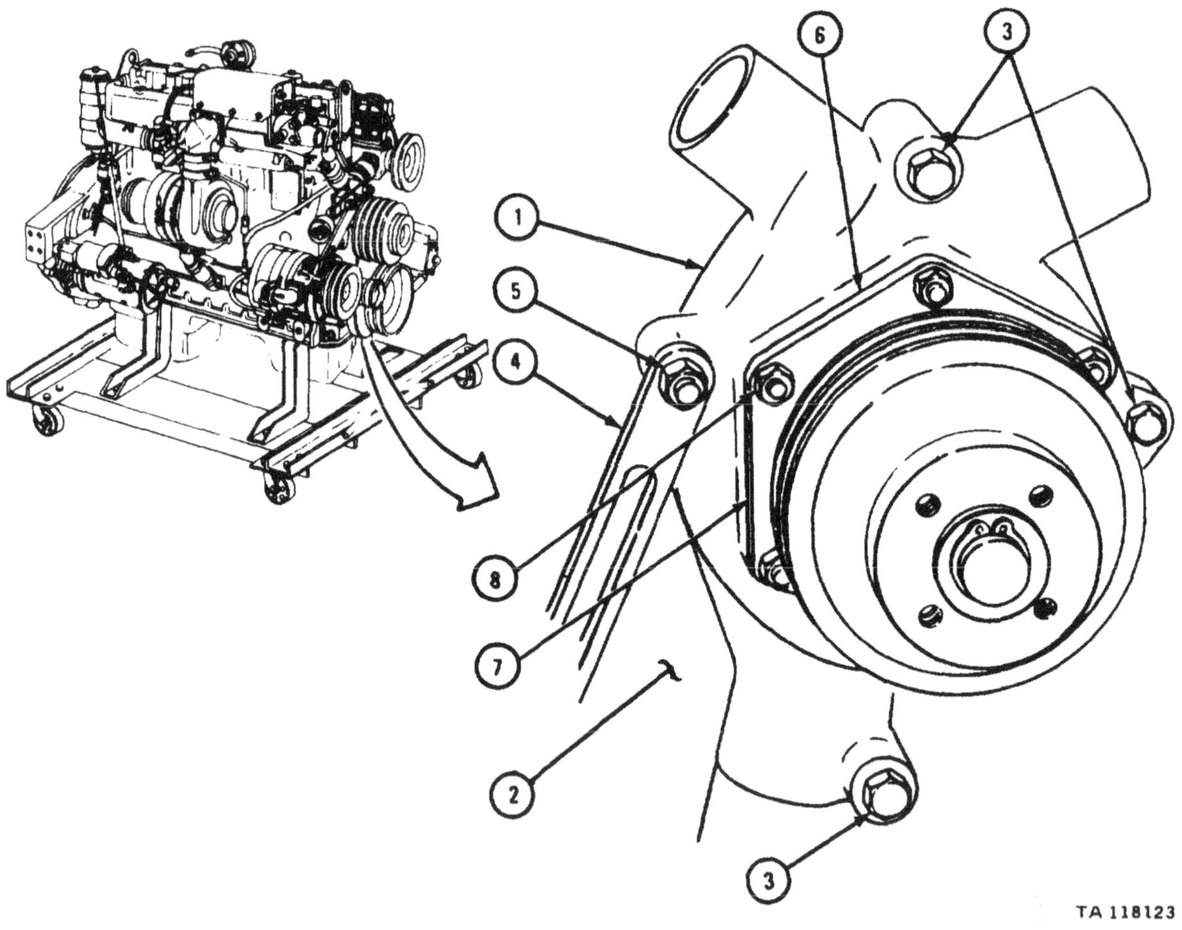

(2) Engine LDS-465-2 .

FRAME 1

1. Put water pump (1) in place on engine block (2).
2. Put in three capscrews with lockwashers (3).

GO TO FRAME 2

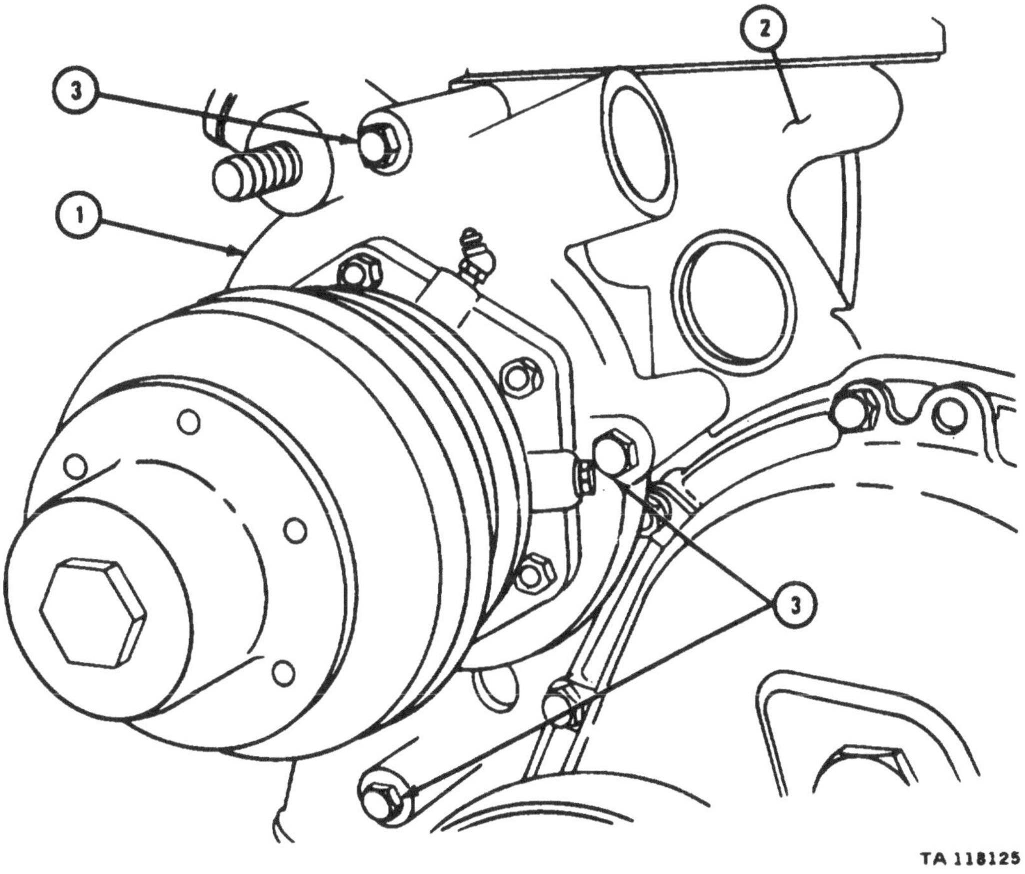

FRAME 2

1. Put idler pulley adjusting arm (1) in place. Put in capscrew (2) with flat washer (3) and lockwasher (4) .
2. Put generator adjusting strap (5) on stud (6). Pu t on idler pulley adjusting strap (7) and put on nut with lockwasher (8).
3. Put in cap screw and lockwasher (9).

END OF TASK

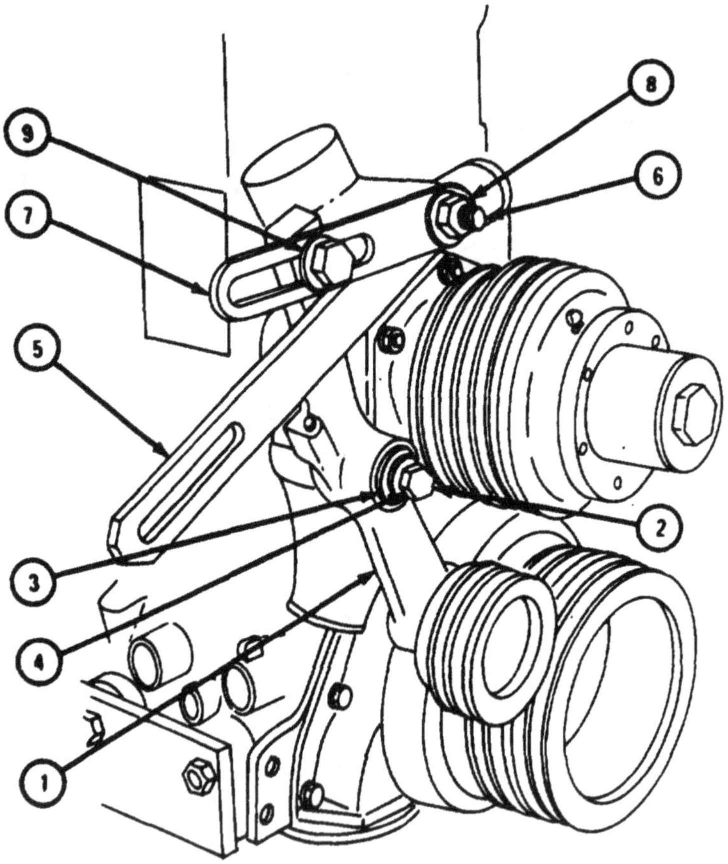

TA 118124

b. Thermostat Housing.

FRAME 1

1. Put water hose (1) on thermostat housing (2). Put on two clamps (3).
2. Put water hose (1) with thermostat housing (2) on water pump (4).
3. Coat both sides of gasket (5) with sealant. Put gasket on thermostat housing (2).

GO TO FRAME 2

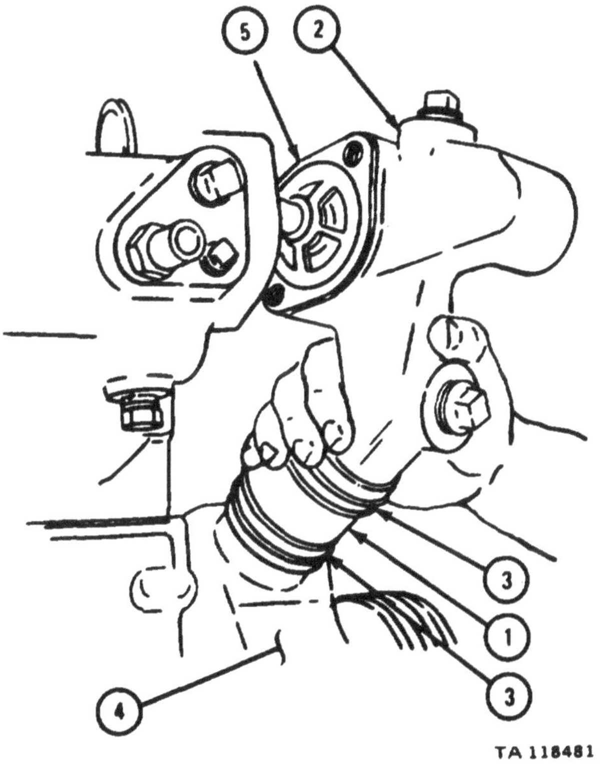

FRAME 2

NOTE

This task is shown for engine LDS-465-1A. This task is the same for all engines.

1. Put thermostat housing (1) in place on intake manifold (2). Put in two capscrews and lockwashers (3) .
2. Tighten two clamps (4).

END OF TASK

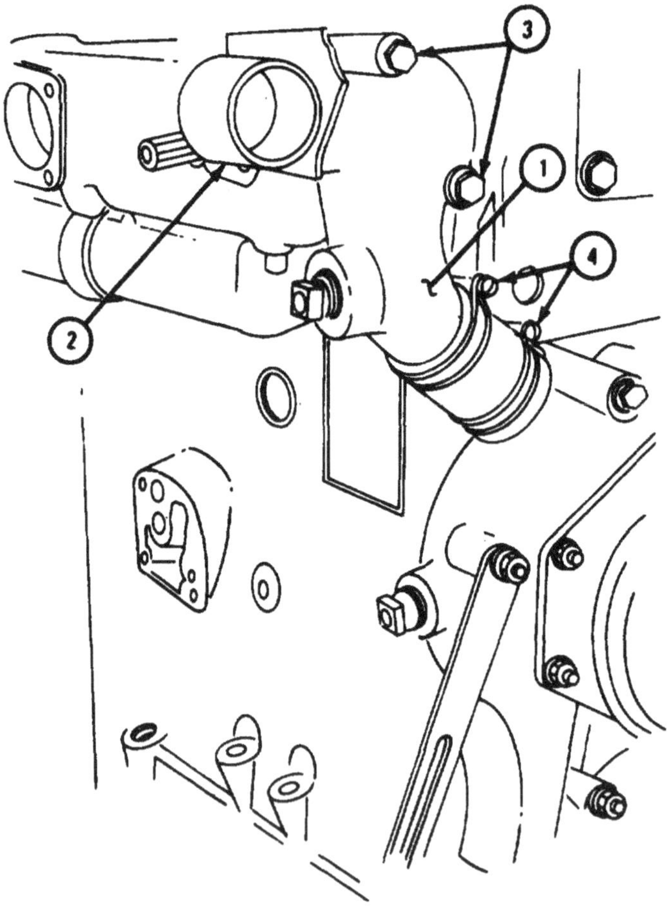

TA 118482

5-20. OIL PRESSURE TRANSMITTER, COOLANT DRAIN COCK, AND OIL LEVEL GAGE AND SUPPORT TUBE.

a. Coolant Drain Cock and Oil Pressure Transmitter.

(1) Engines LDS-465-1, LDS-465-1A, and LDS-465-2 .

FRAME 1

1. Put in elbow (1). Put in coolant drain cock (2).
2. Put in elbow (3), pipe adapter (4), and oil pressure transmitter (5).

END OF TASK

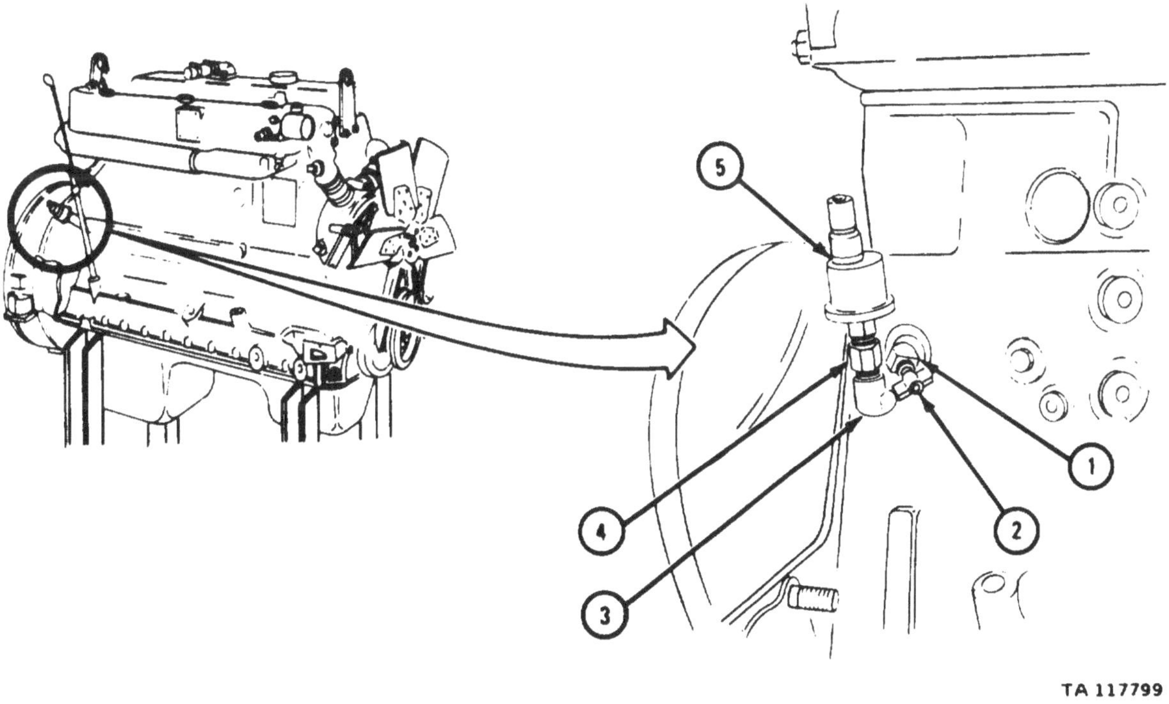

(2) Engines LD-465-1, LD-465-1C, and LDT-465-1C .

FRAME 1

1. Put in elbow (1). Put in coolant drain cock (2).
2. Put in elbow (3), pipe adapter (4), and oil pressure transmitter (5).

END OF TASK

TA 117800

b. Oil Level Gage and Support Tube.

(1) Engines LDS-465-1, LDS-465-1A, and LDS-465-2 .

FRAME 1

1. Put in oil level gage support tube (1). Put in screw with lockwasher and nut (2) .
2. Put in oil level gage (3).

END OF TASK

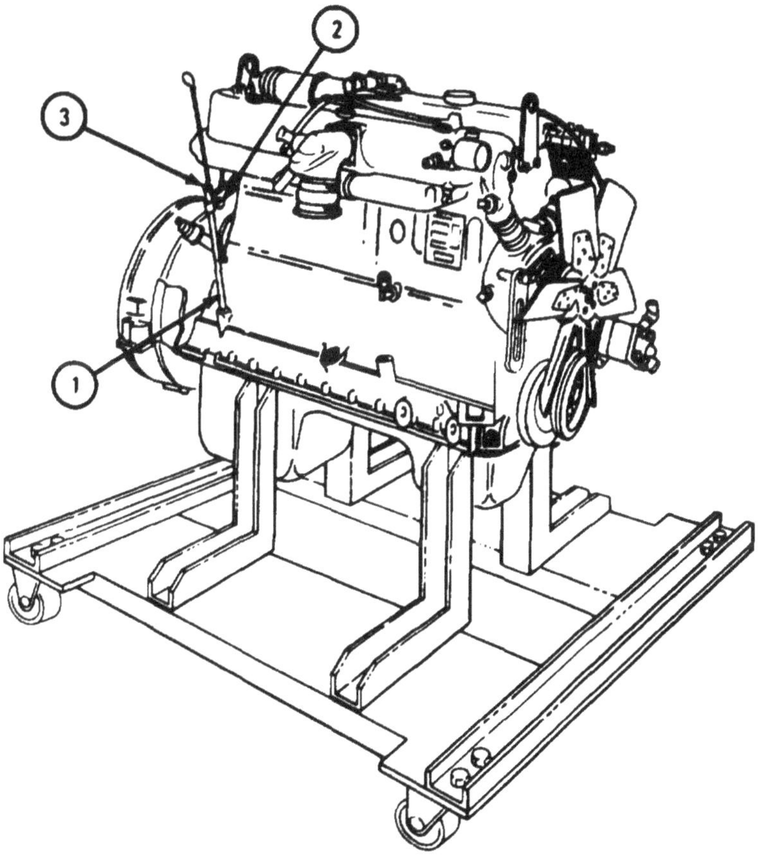

TA 118001

(2) Engines LD-465-1, LD-465-1C, and LDT-465-1C .

FRAME 1

1. Put in oil level gage support tube (1). Put in screw with lockwasher and nut (2) .
2. Put in oil level gage (3).

END OF TASK

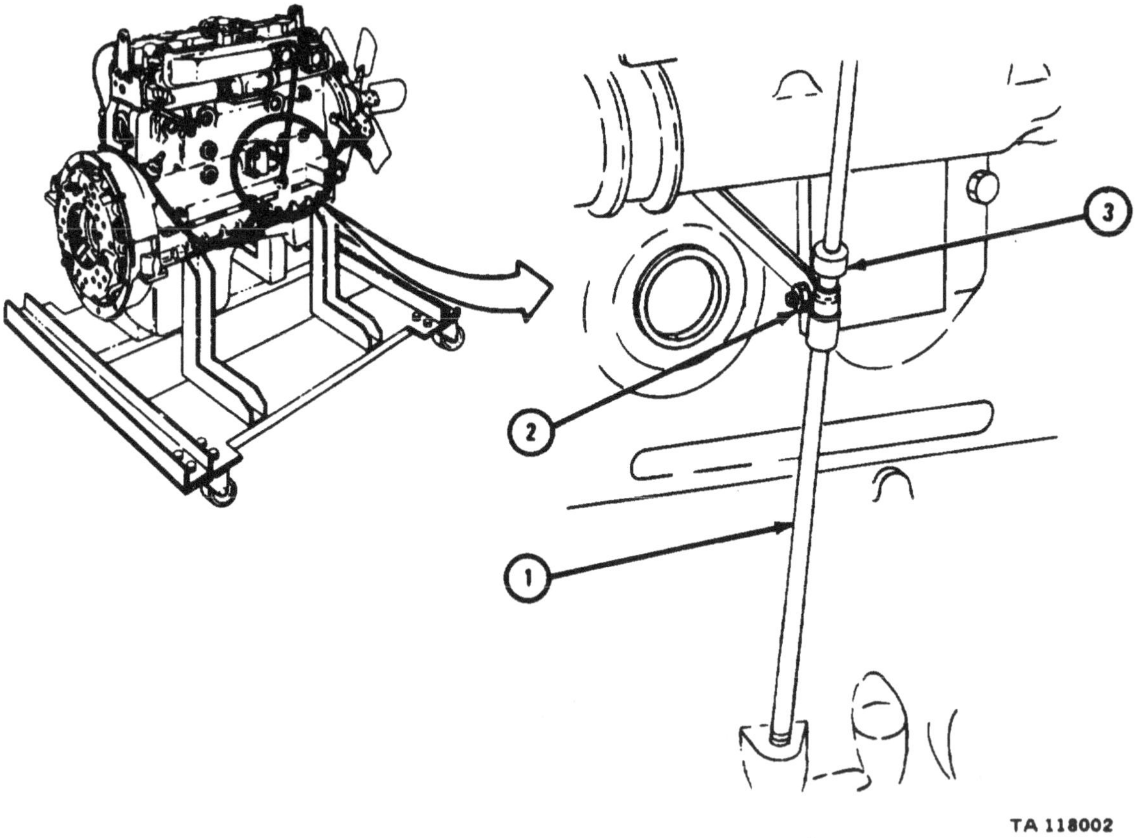

5-21. OIL PRESSURE REGULATOR VALVE HOUSING ASSEMBLY.

FRAME 1

1. Put oil pressure regulator valve gasket (1) in place. Put oil pressure regulator valve (2) in place.

NOTE

Engines LD-465-1 and LD-465-1C have crankcase breather tube support bracket (3) .

2. If crankcase breather tube support bracket (3) is used, put crankcase breather tube support bracket in place .
3. Put in four screws with lockwashers (4).

END OF TASK

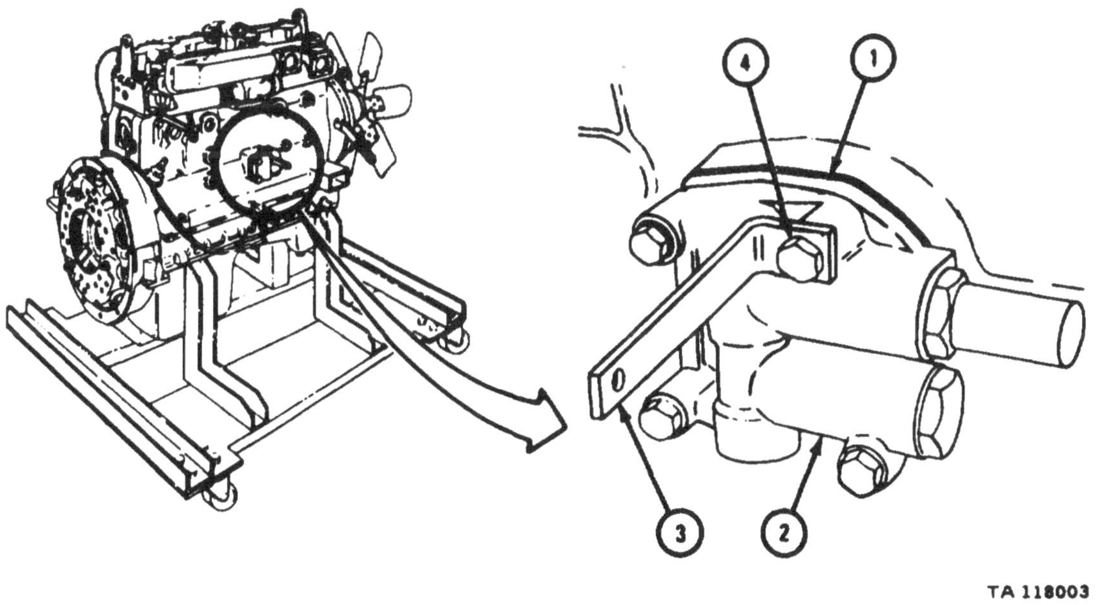

5-22. GENERATOR MOUNTING BRACKET.

FRAME 1

1. For all engines except LDS-465-2, put mounting bracket (1) in place on engine (2) . Put in three screws with starwashers (3).
2. For engine LDS-465-2, put generator mounting bracket (1) with air inlet tube support bracket (4) in place on engine (2). Put in three screws with starwashers (3) .

END OF TASK

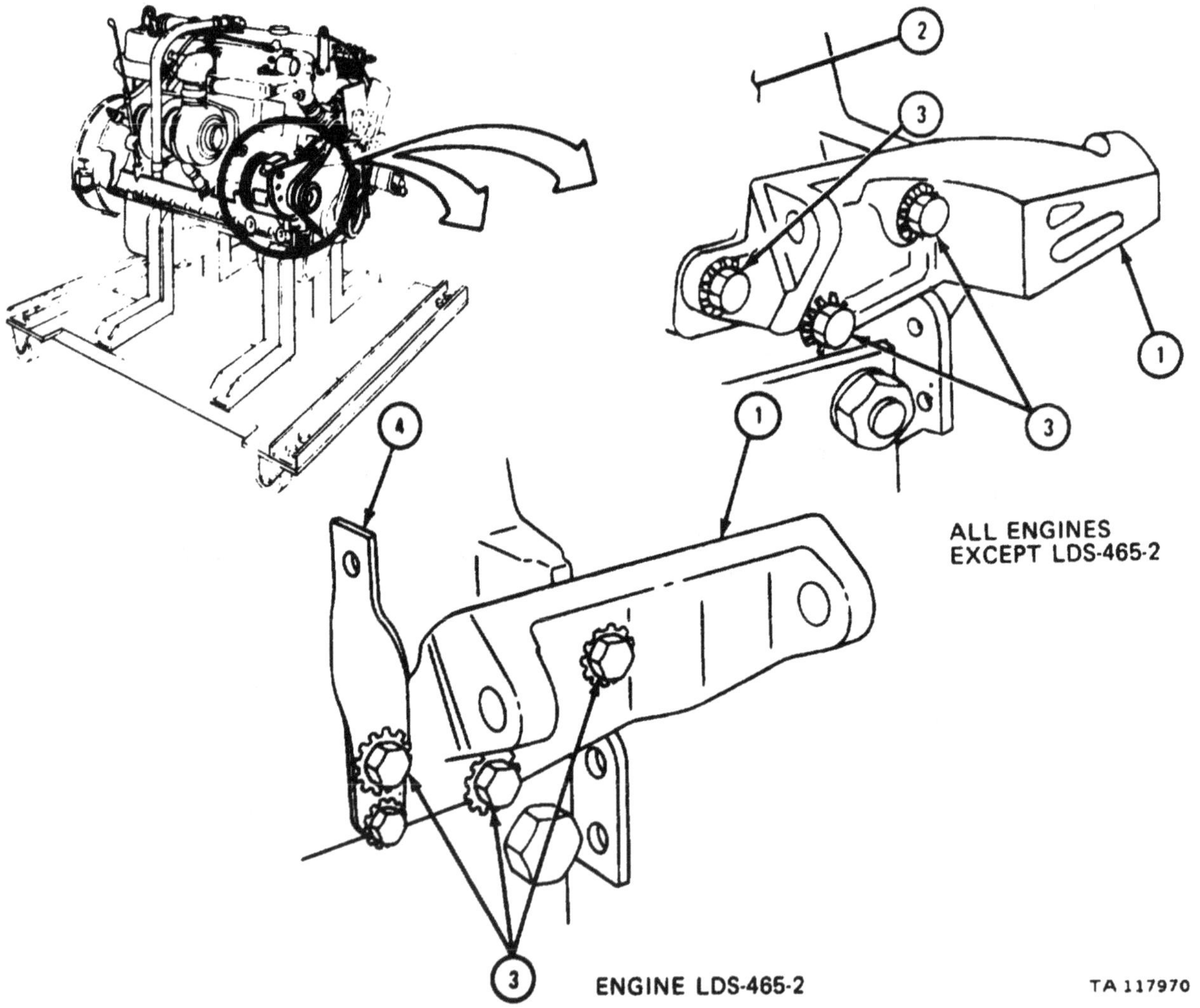

5-23. INTAKE MANIFOLD FLAME HEATER SYSTEM.

a. System Identification.

FRAME 1

1. If engine had cover (1) imposition shown, engine uses top mounted-covered type flame heater system. Go to para 5-23b.
2. If engine had flame heater ignition unit (2) in position shown, engine uses top mounted-uncovered type flame heater system. Go to para 5-23d.
3. If engine had flame heater fuel pump assembly (3) as shown, engine uses side mounted type flame heater system. Go to para 5-23e.

END OF TASK

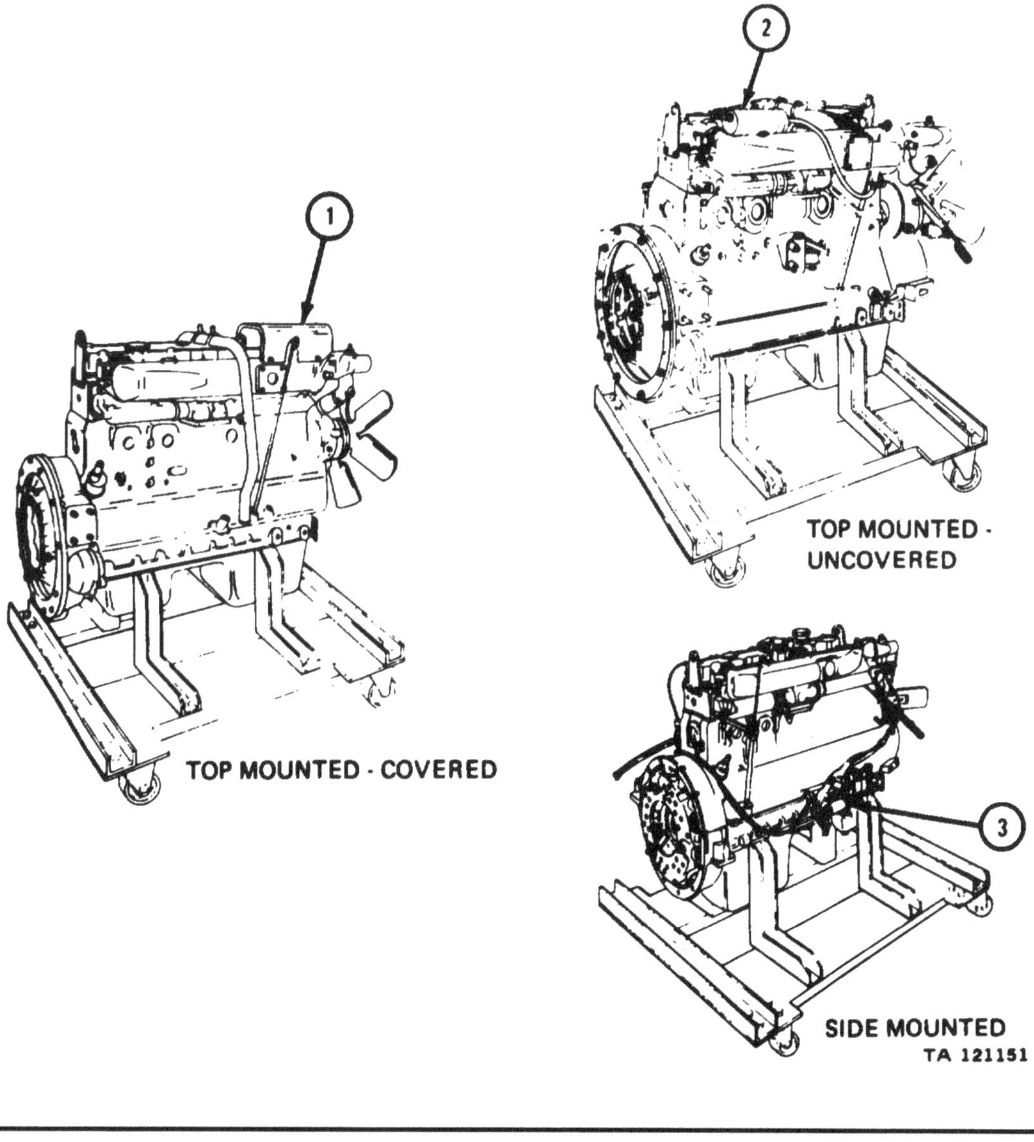

b. Fuel Filter for Top Mounted-Covered Type Flame Heater.

FRAME 1

NOTE

Plugs and tags put on during removal must be taken off during replacement .

1. Put fuel filter with clamp (1) in place. Put in screw with lockwasher (2) .
2. Put fuel filter to flame heater tube (3) in place. Screw in and tighten fitting nut (4) .
3. Put injection pump-to-fuel filter tube (5) in place. Hold adapter (6) and screw in and tighten fitting nut (7).

END OF TASK

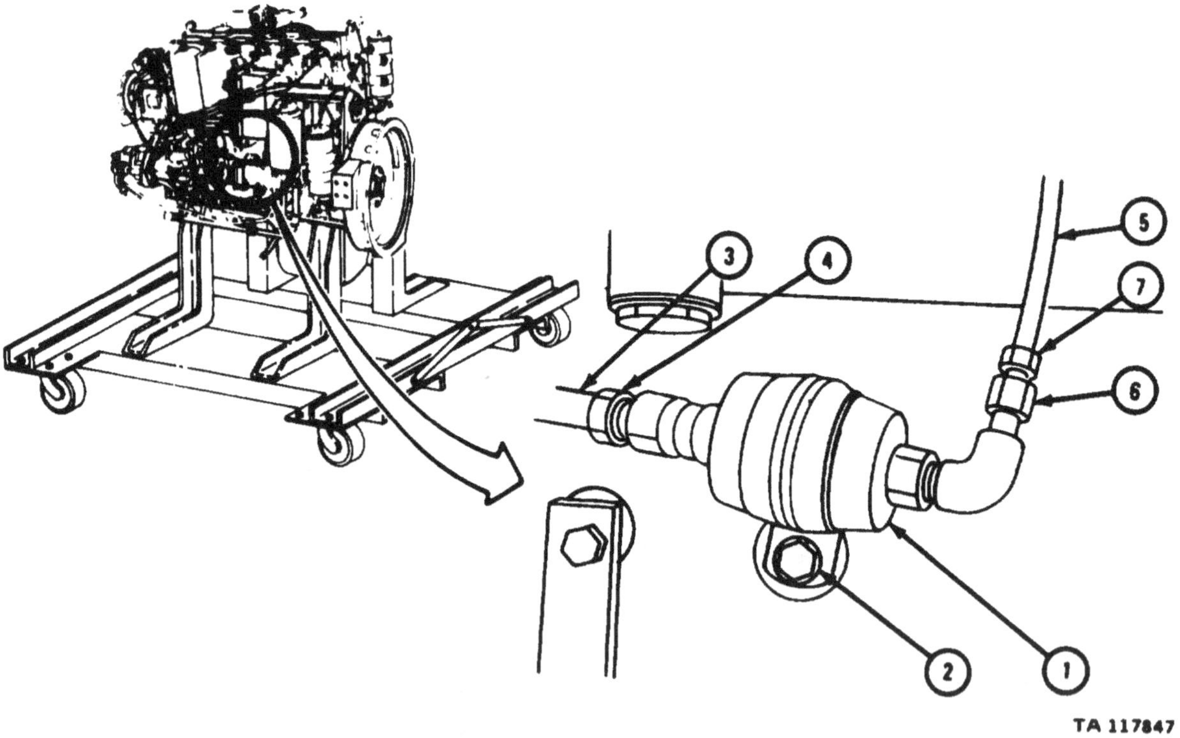

c. Fuel Pump and Ingiition for Top Mounted-Covered Flame Heater .

FRAME 1

1. Put flame heater support bracket (1) in place. Put in two screws with lockwashers (2) .
2. Put flame heater harness (3) in place. Put ground lead (4) in place. put in screw with lockwasher (5) .

GO TO FRAME 2

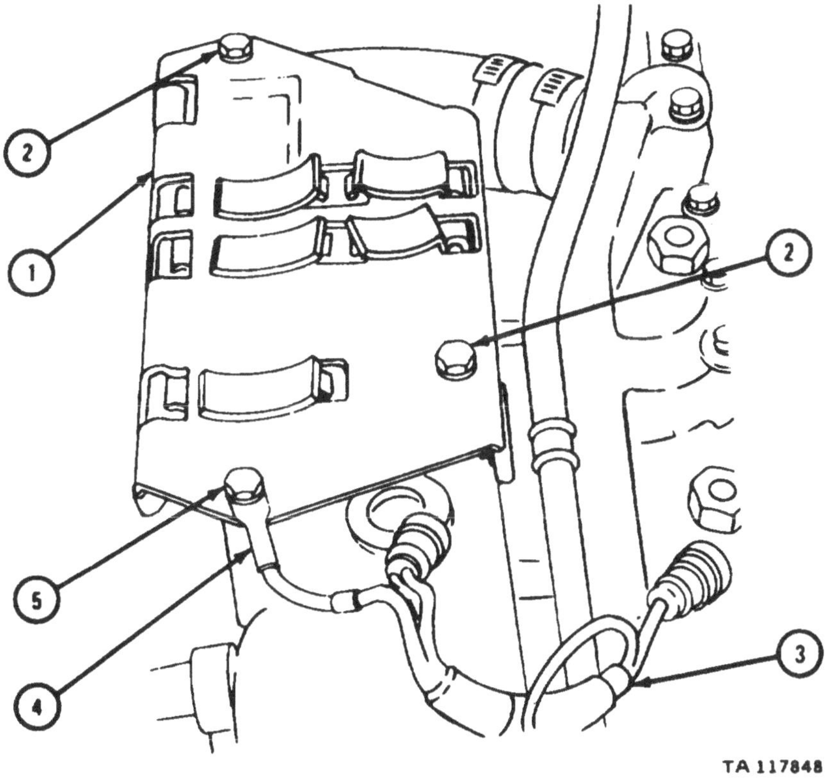

FRAME 2

NOTE

Two types of top mounted-covered flame heater fuel pump and ignition units have been used on these engines. Type A and type B assemblies have the same parts and are put back in the same way.

1. Put 90° elbow (1) and 45° elbow (2) into flame heater fuel pump (3).
2. Put fuel pump (3) in place. Put on two spring clamps (4),
3. Put wiring harness (5) in place. Screw o n and tighten nut (6) .
4. Put flame heater fuel pump-to-flame heater nozzle tube (7) in place. Screw in and tighten fitting 'nut'(8).
5. Put flame heater fuel pump-to-flame heater fuel filter tube (9) in place. Screw in and tighten fitting nut (10).

GO TO FRAME 3

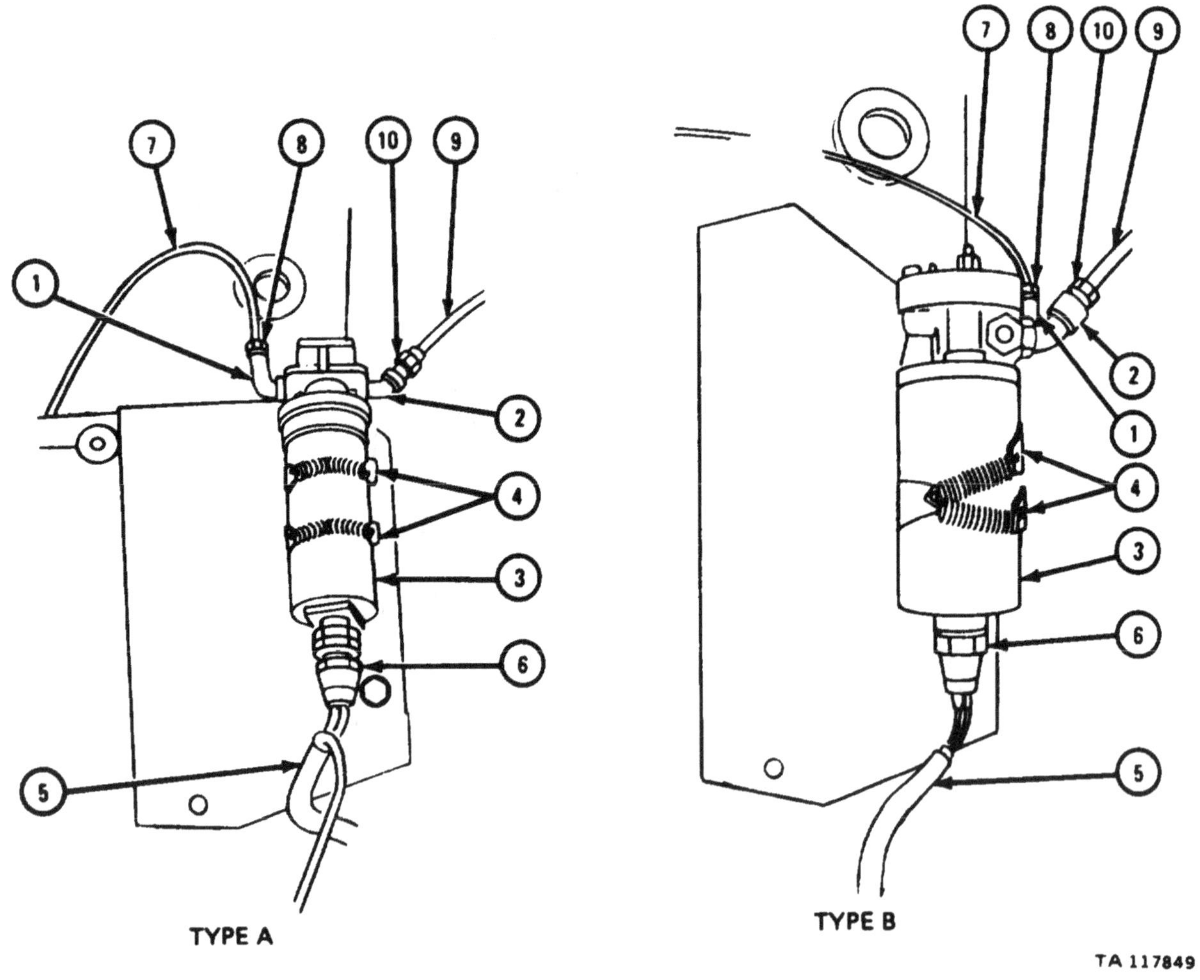

FRAME 3

1. Put ignition unit (1) in place. Put on two spring clamps (2).
2. Put flame heater wiring harness (3) in place. Screw on and tighten nut (4).
3. Put ignition lead (5) in place. Screw on and tighten nut (6).

END OF TASK

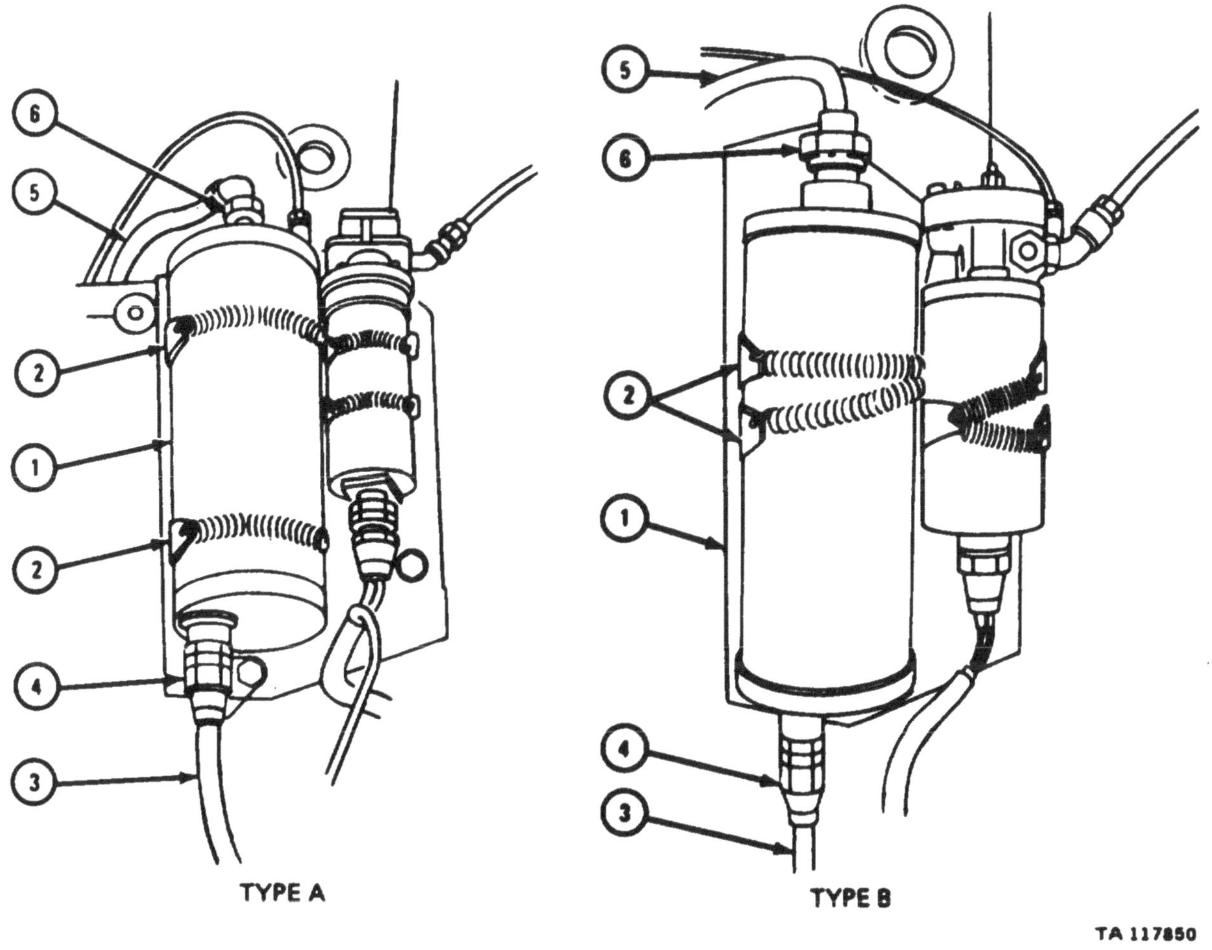

FRAME 4

1. Put flame heater cover (1) in place.

NOTE

Air pressurization tube (2) is used only on engine LDS-465-2.

2. If working on engine LDS-465-2, put on pressurization tube with clamp (2) in place.

3. Put in four screws with lockwashers (3).

END OF TASK

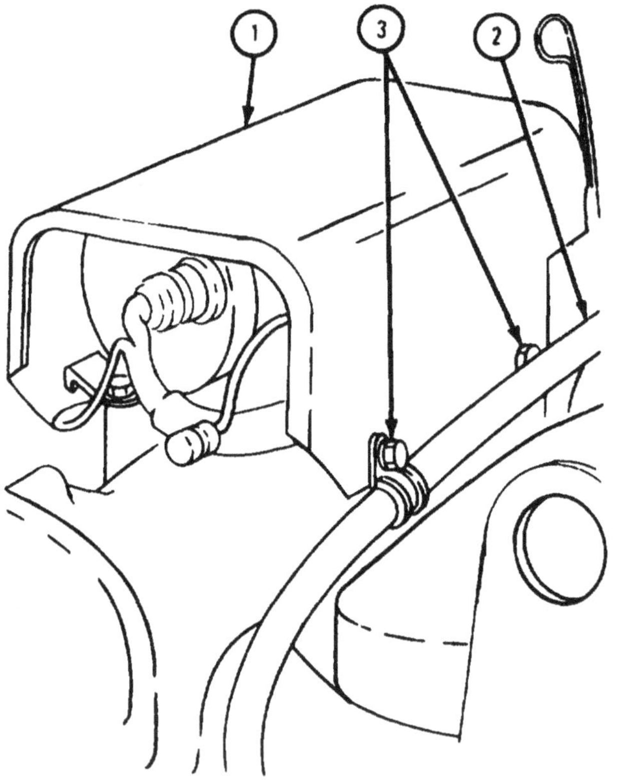

d. Top Mounted-Uncovered Flame Heater Ignition Unit and Fuel Pump.

FRAME 1

1. Put two clamps (1) on ignition unit (2) and put ignition unit in place. Put on ground lead (3) . Put in two screws with lockwashers (4).
2. Put on ignition lead (5). Screw on and tighten nut (6).
3. Put on flame heater harness (7). Screw on and tighten nut (8).

GO TO FRAME 2

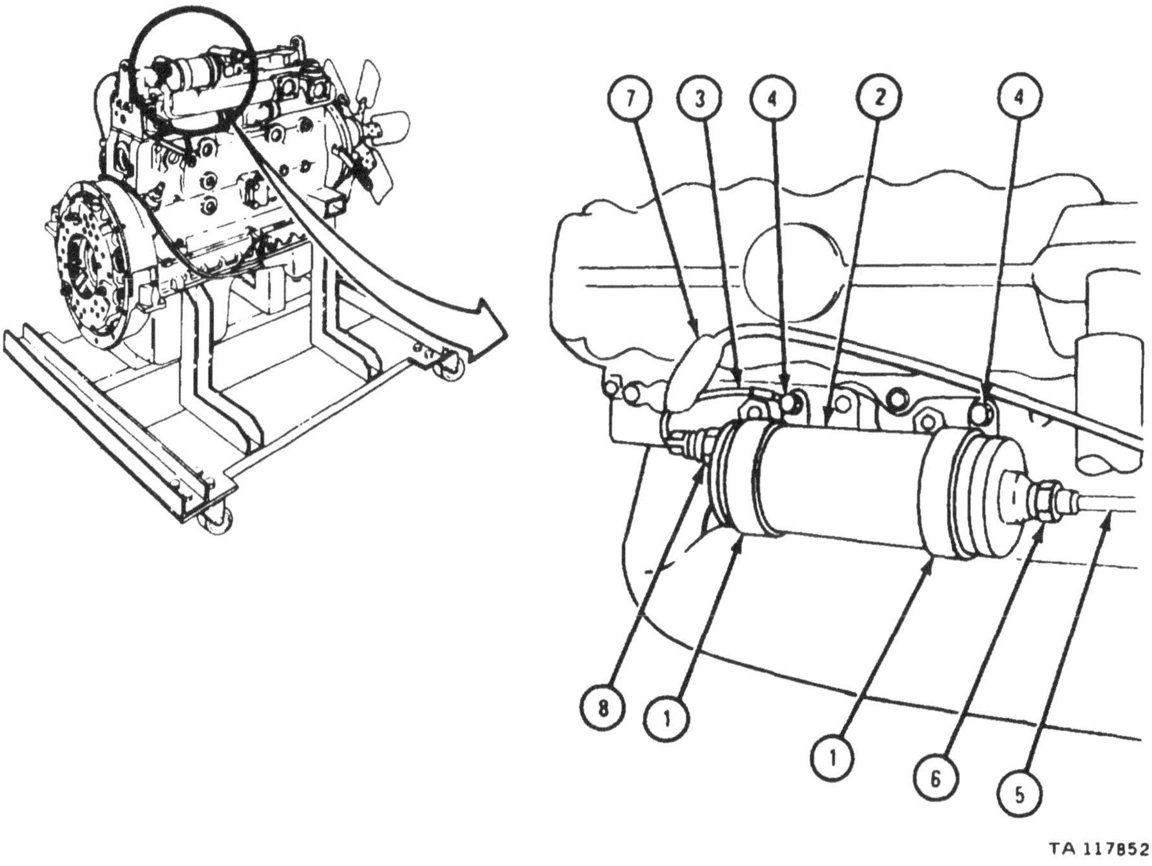

FRAME 2

NOTE

Some pumps (2) may have two clamps (1), Others may have only one clamp.

1. Put two clamps (1) on fuel pump (2). Put in elbow (3) .
2. Put fuel pump (2) in place. Put in two screws with lockwashers (4). Put fuel pump-to-flame heater nozzle tube (5) in place. Screw in and tighten fitting nut (6) .
3. Put flame heater wiring harness (7) in place. Screw on and tighten nut (8).

GO TO FRAME 3

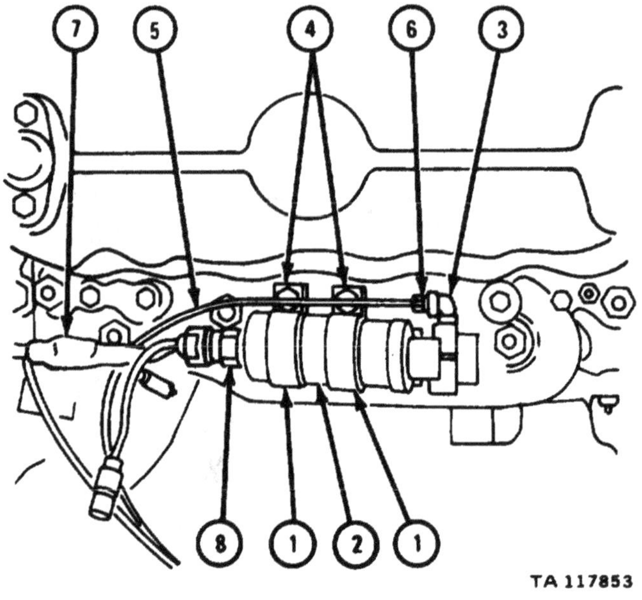

FRAME 3

1. Put pipe fitting (1) into fuel pump (2). Put in elbow (3).
2. Put on fuel filter (4).
3. Put fuel supply pump-to-flame heater fuel filter tube (5) in place. Screw in and tighten fitting nut (6).

END OF TASK

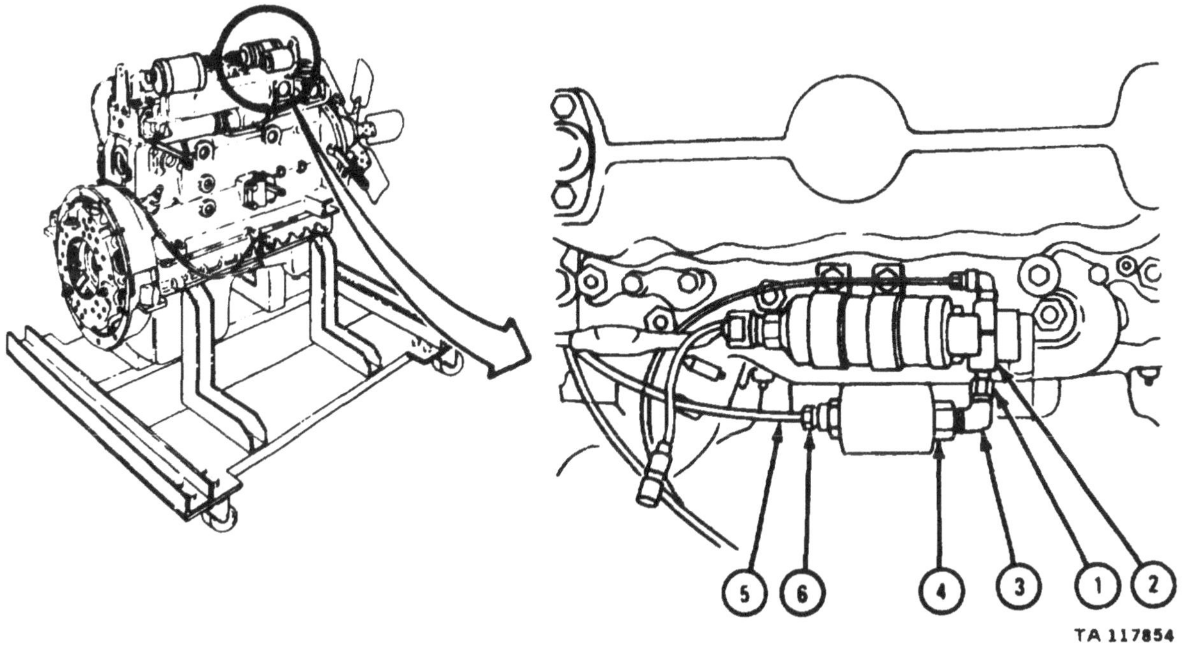

e. Side-Mounted Flame Heater Fuel Pump, Filter, and Solenoid Valve Bracket Assembly.

FRAME 1

1. Put bracket assembly (1) in place on crankcase (2).
2. Put in three bolts with lockwashers (3).

GO TO FRAME 2

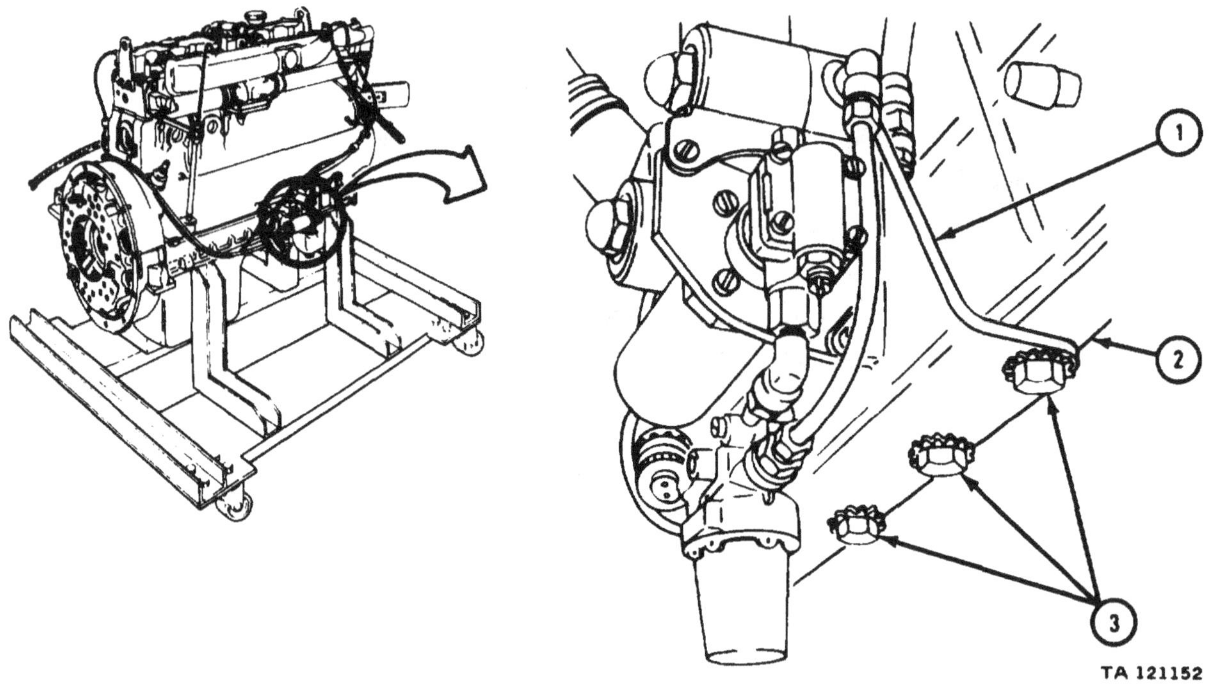

FRAME 2

NOTE

Tags and plugs put on during removal must be taken off during replacement.

1. Put flame heater wiring harness (1) in place. Screw connector (2) onto fuel supply solenoid valve (3).
2. Screw connector (4) onto fuel return solenoid valve (5).
3. Screw connector (6) onto fuel pump (7).
4. Put ground lug (8) in place. Put in screw, nut, and lockwasher (9).

GO TO FRAME 3

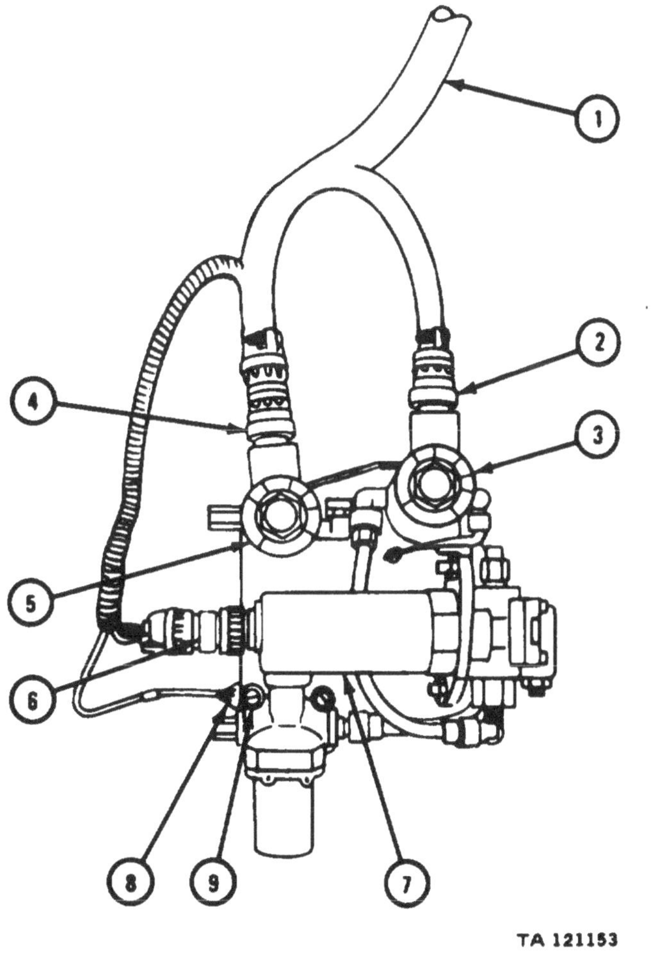

FRAME 3

1. Put fuel return solenoid valve-to-fuel injector nozzle tube (1) in place on solenoid valve (2). Screw on nut (3) .
2. Put fuel injection pump-to-filter tube (4) in place on fuel filter (5). Screw on nut (6) .

GO TO FRAME 4

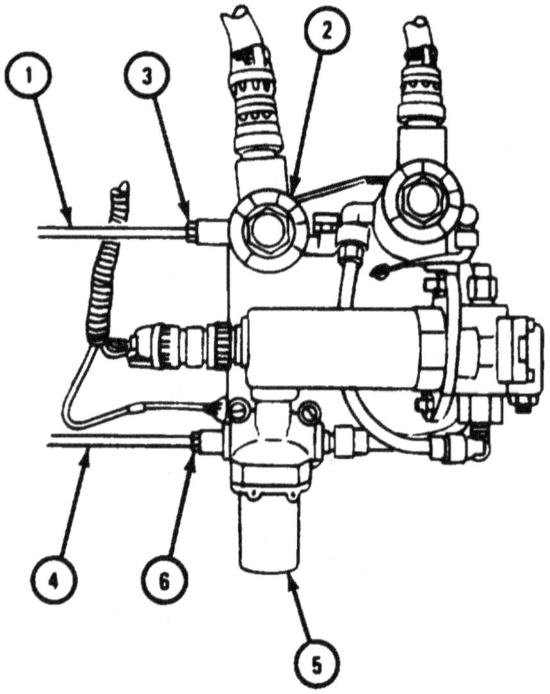

FRAME 4

1. Put flame heater fuel pump-to-flame heater nozzle tube (1) in place on fuel pump (2) . Screw on nut (3) .
2. Put flame heater fuel return-to-solenoid valve tube (4) in place on elbow (5). Screw on nut (6).

END OF TASK

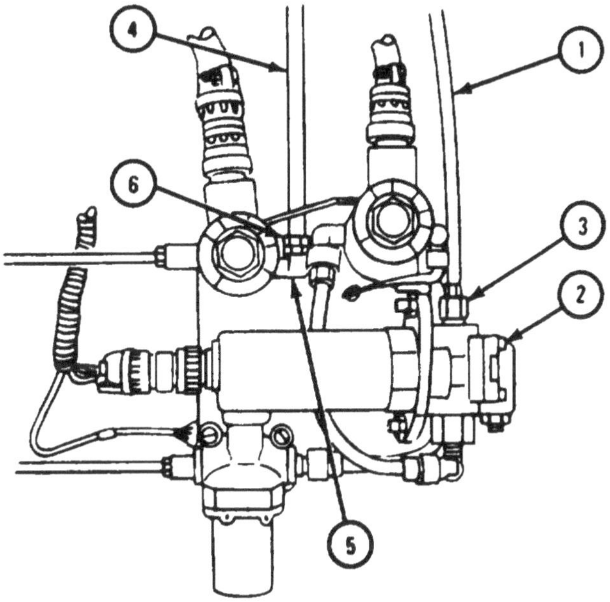

TA 121199

f. Intake Manifold Flame Heater Elbow.

(1) Engines LD-465-1 and LD-465-1C .

FRAME 1

1. Put flame heater gasket (1) in place. Put flame heater elbow (2) in place.

2. Put on four nuts with lockwashers (3).

GO TO FRAME 2

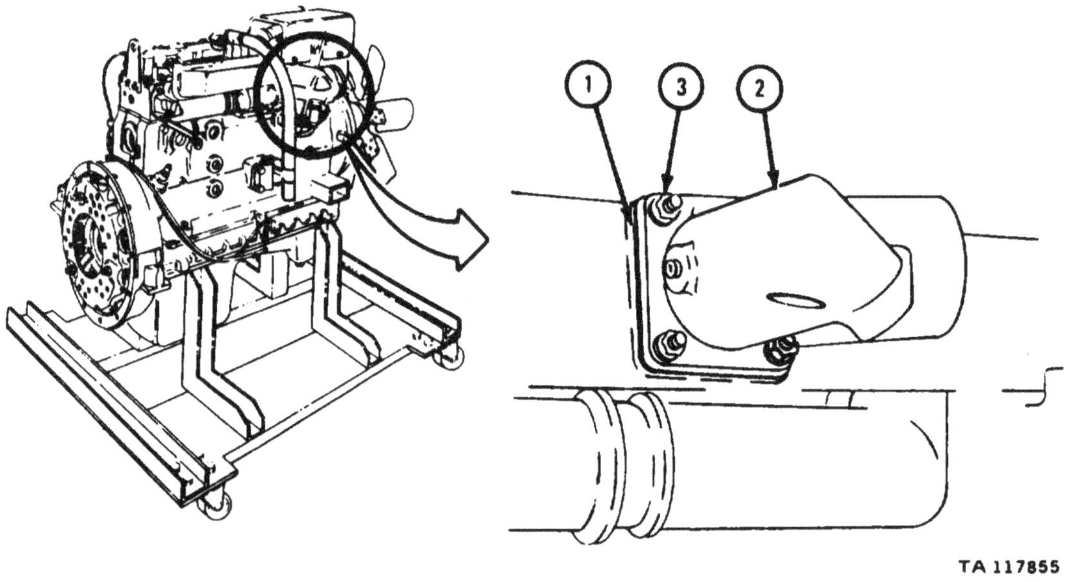

FRAME 2

NOTE

Tags and plugs put on during removal must be taken off during replacement .

1. Put in nozzle (1) and tighten nut (2).

CAUTION

Do not use open end wrench to tighten fitting adapters (3). Use only tubing wrench or box end wrench. Failure to use the correct tool may result in damage to fitting adapters .

2. Put in two fitting adapters (3). Put fuel inlet tube (4) and fuel return tube (5) in place. Put in two fitting nuts (6).

3. Put in spark plug with gasket (7) and tighten to 8 pound-inches. Put in ignition lead (8) .

END OF TASK

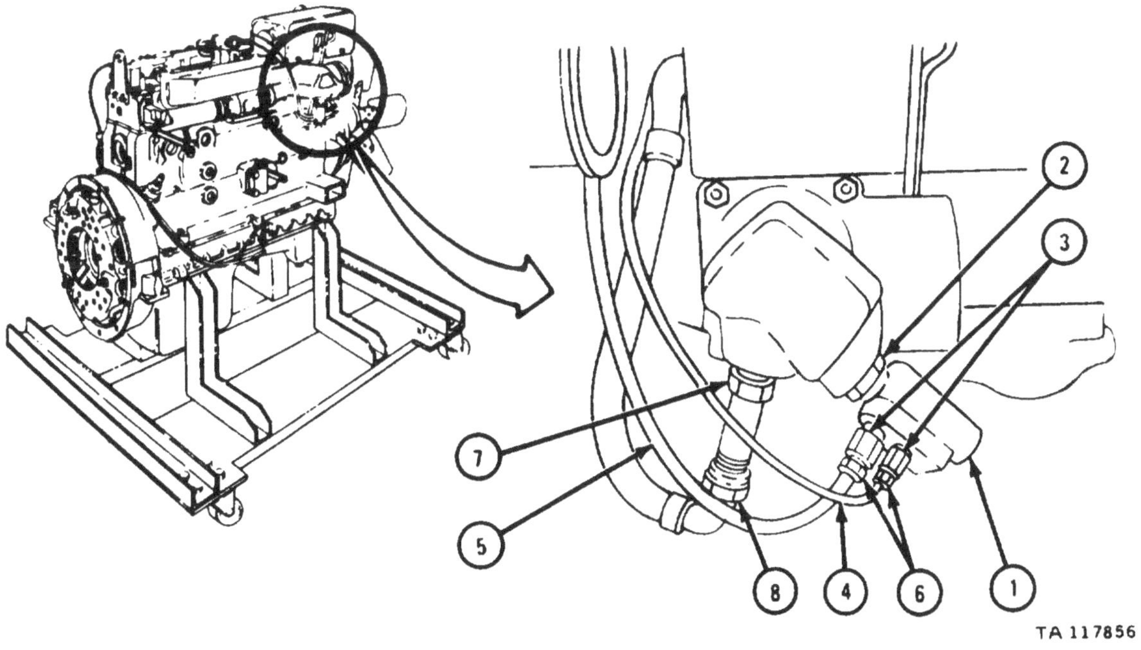

(2) Engines LDS-465-1A and LDS-465-2 .

FRAME 1

1. Put flame heater elbow gasket (1) in place. Put flame heater elbow (2) in place. Put on four nuts with lockwashers (3).
2. Put in nozzle and check valve assembly (4). Tighten locknut (5) .

GO TO FRAME 2

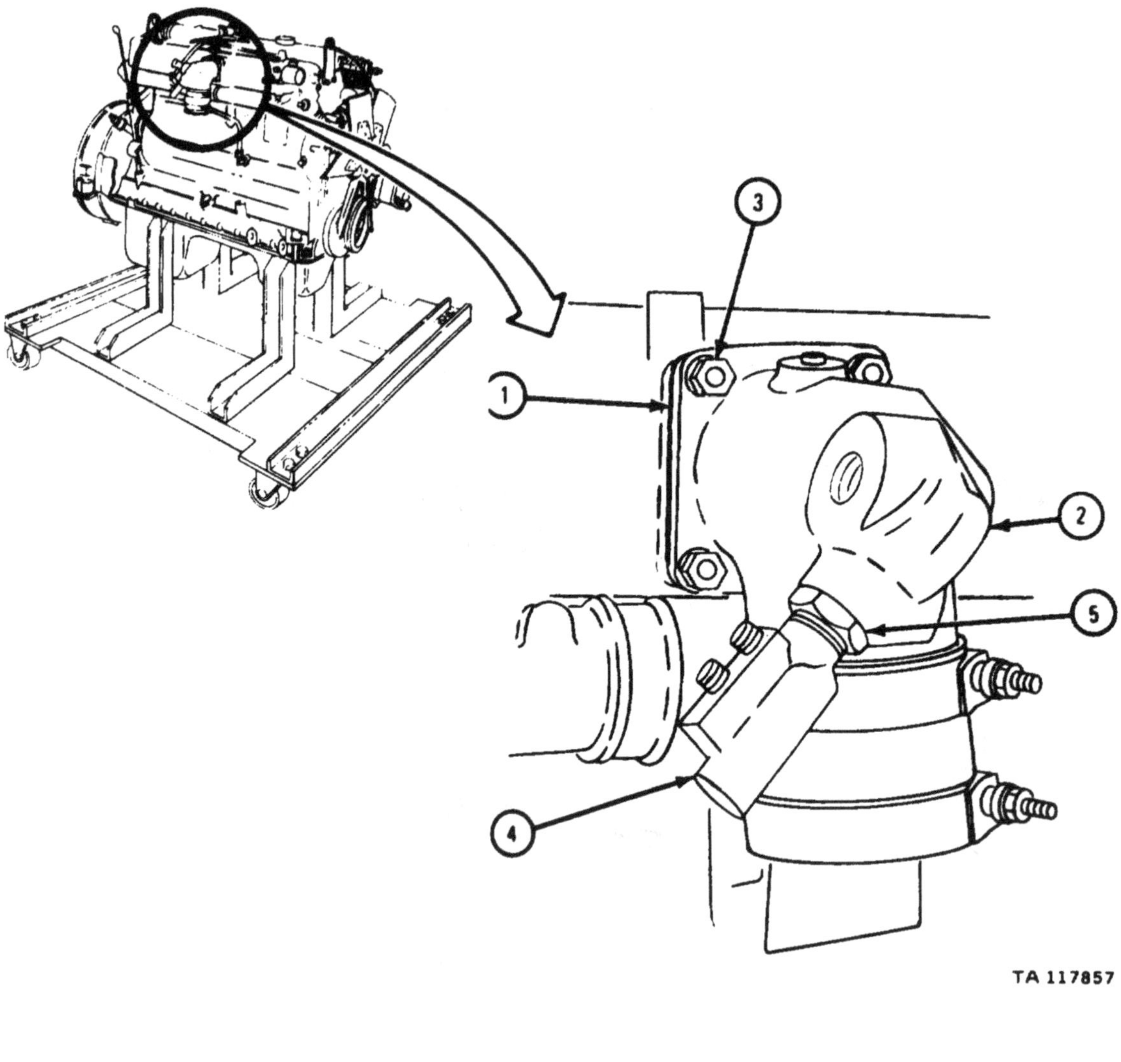

FRAME 2

CAUTION

Do not use open end wrench to tighten fitting adapters (1). Use only tubing wrench or box end wrench. Failure to use the correct tool may result in damage to fitting adapters.

1. Put in two fitting adapters (1).
2. Put fuel return tube (2) in place. Screw in and tighten fitting nut (3).
3. Put fuel inlet tube (4) in place. Screw in and tighten fitting nut (5).
4. Screw in spark plug (6) with gasket (7). Tighten spark plug to 8 pound-inches.
5. Put ignition lead (8) in place. Screw on and tighten nut (9).

END OF TASK

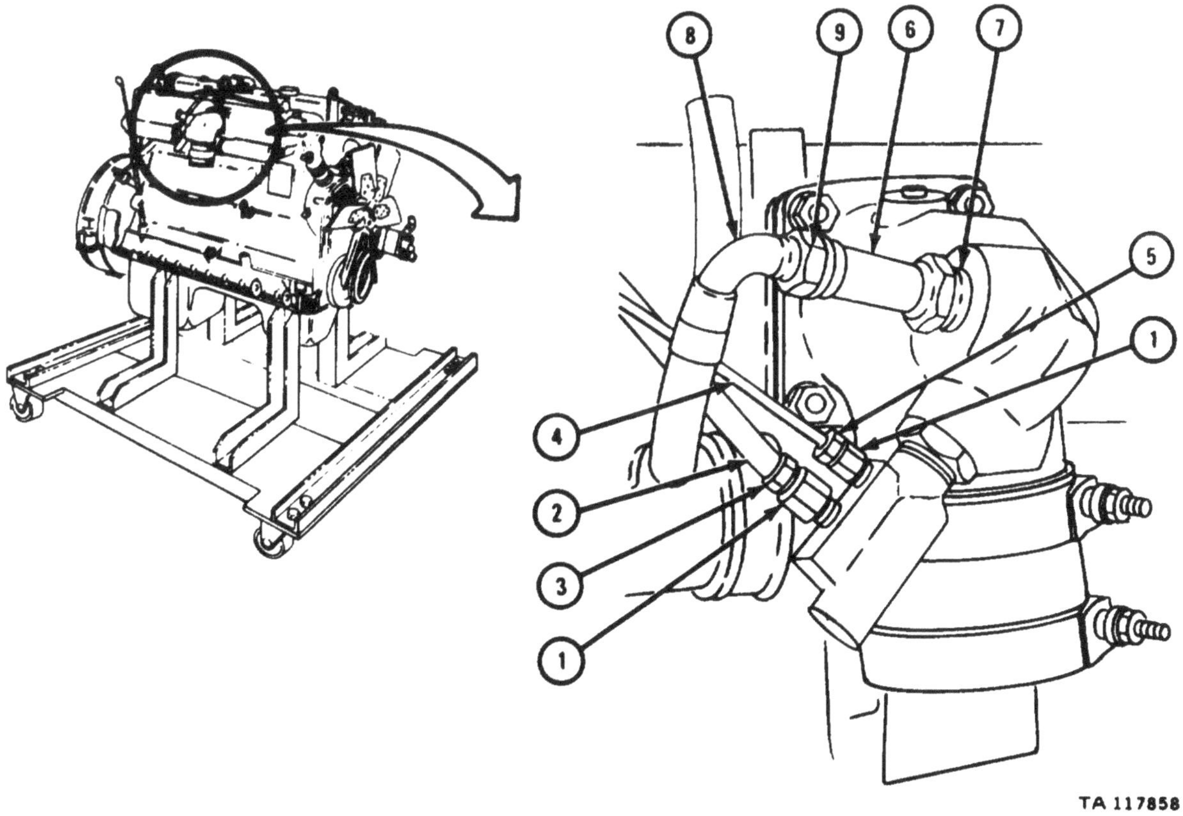

(3) Engine LDS-465-1 .

FRAME 1

NOTE

There are three types of intake manifold flame heater assemblies. All three have nozzle assembly, fuel inlet tube, fuel return tube, spark plug and ignition lead for spark plug. These parts are in different positions on each type of flame heater assembly but are put back the same way except where noted.

If working on flame heater type C, go to frame 4.

1. Put flame heater elbow gasket (1) in place. Put flame heater elbow (2) in place.
2. Put on two nuts with lockwashers (3).
3. Put on hose (4). Put on and tighten two hose clamps (5).

GO TO FRAME 2

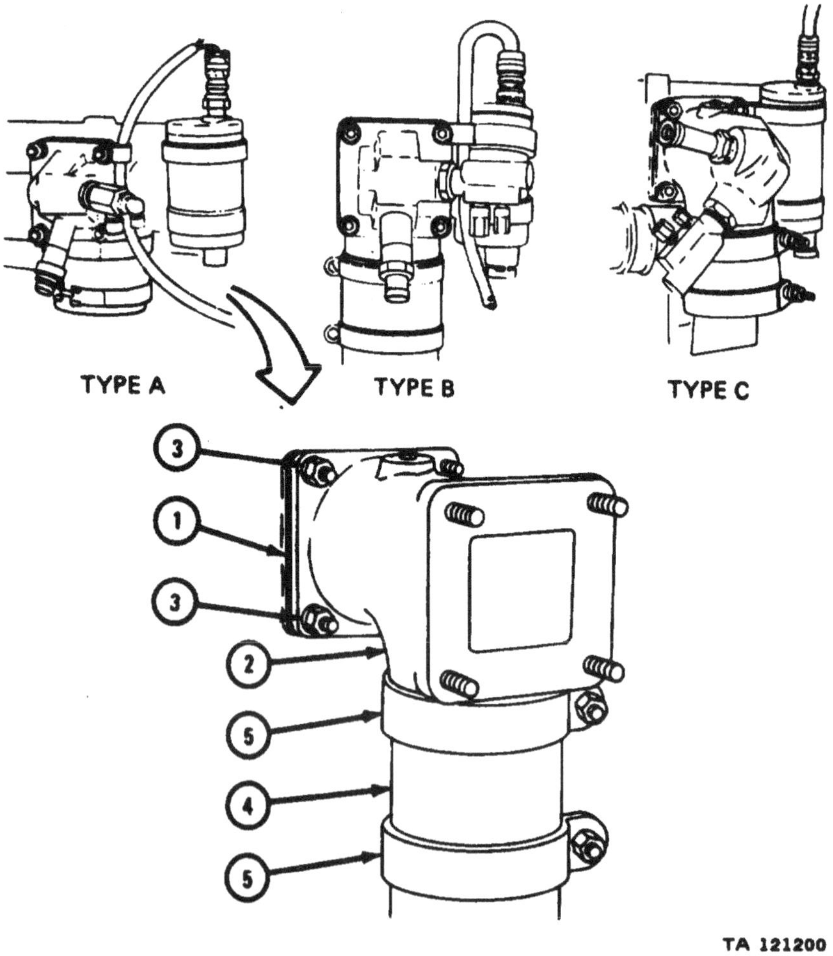

FRAME 2

1. Put flame heater gasket (1) in place. Put flame heater (2) in place.
2. Put on three nuts with lockwashers (3).

GO TO FRAME 3

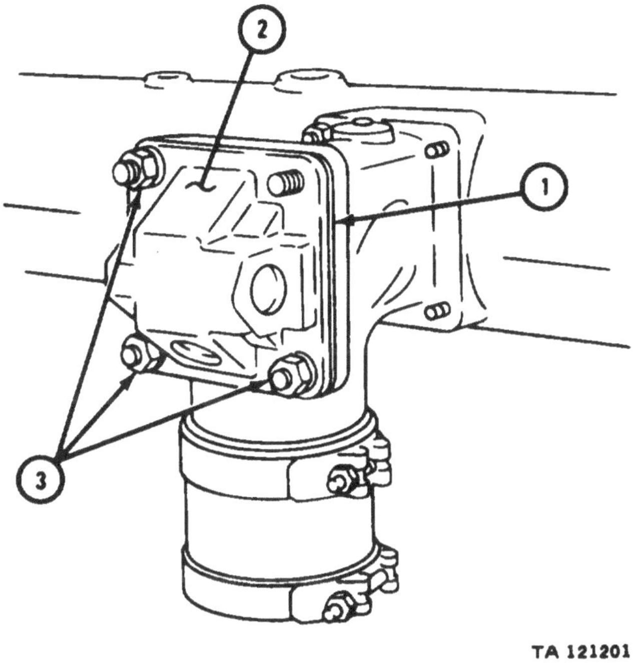

FRAME 3

1. Put nozzle and valve assembly (1) on flame heater assembly (2). Tighten nut (3) .
2. Screw in sparkplug (4) with gasket (5) . Tighten spark plug to 8 pound-inches.

CAUTION

Do not use open end wrench to put on pipe bushing (6) or tube adapter (8) . Use only tubing wrench or box wrench or equipment may be damaged.

3. If working on type A flame heater assembly, put in pipe bushing (6) and tighten. Screw in 90° elbow (7) and tube adapter (8) .

CAUTION

Do not use open end wrench to put on tubing adapters (9). Use only tubing wrench or box wrench or equipment may be damaged.

4. If working on type B flame heater assembly, put in two tube adapters (9).

GO TO FRAME 5

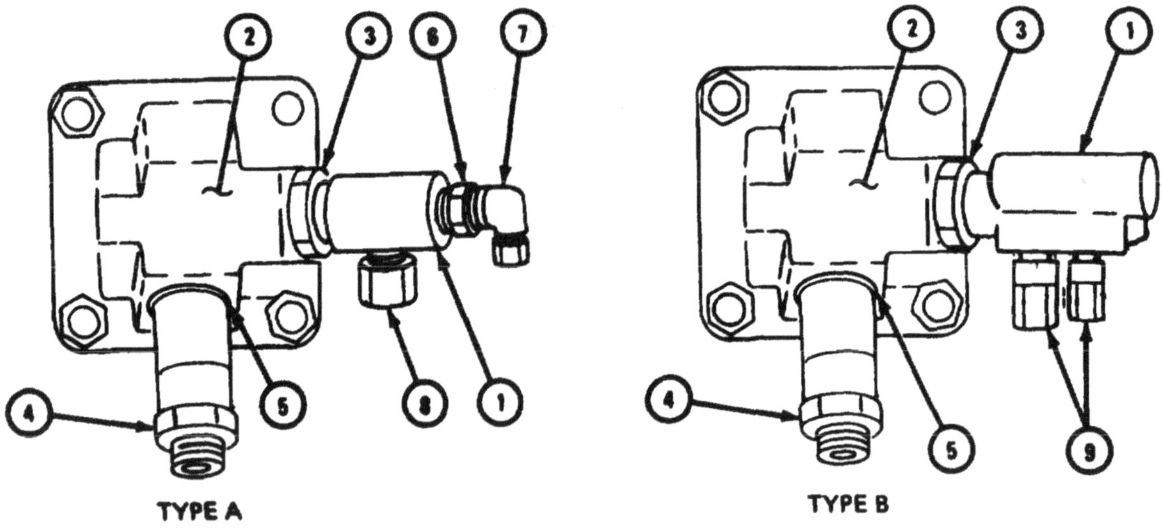

FRAME 4

1. Put flame heater gasket (1) in place. Put flame heater elbow (2) in place.

NOTE

On some engines, ignition coil (3) may be mounted on top of engine instead of next to flame heater elbow as shown.

2. Put ignition coil (3) with two clamps (4) in place.
3. Put on four nuts with lockwashers (5).
4. Screw in spark plug (6) with gasket (7). Tighten spark plug to 8 pound-inches.
5. Screw in nozzle and check valve assembly (8) with locknut (9). Tighten locknut.

CAUTION

Do not use open end wrench to put on tube adapters (10). Use tubing wrench or box wrench or equipment may be damaged.

6. Put on two tube adapters (10).

GO TO FRAME 5

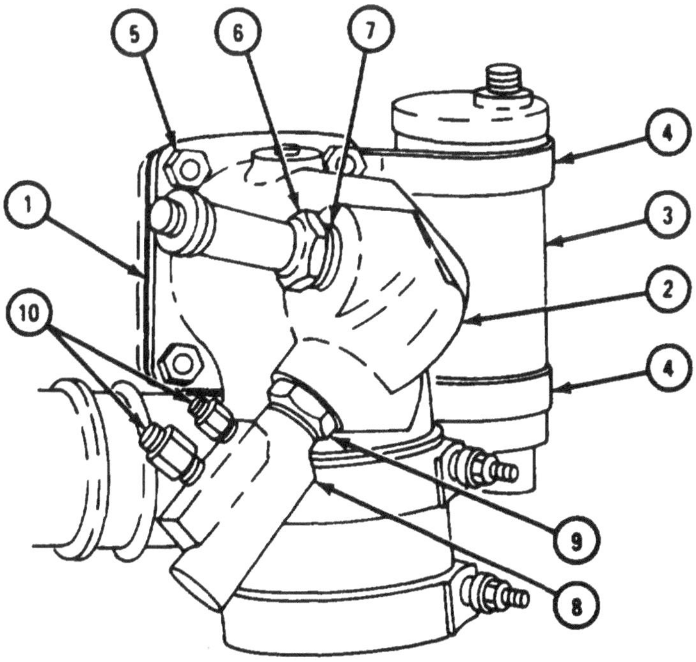

TA 121203

FRAME 5

NOTE

Plugs and tags put on during removal must be taken off during replacement .

1. If working on type A or B, flame heater assembly, put ignition unit (1) with two clamps (2) in place.
2. Put on two nuts with lockwashers (3).
3. Put flame heater cable assembly (4) in place. Tighten nuts (5 and 6).

NOTE

Type C flame heater assembly may not have tube clamp (7).

4. Put wiring harness (8) with tube clamp (7) in place. Tighten nut (9).
5. Put tube clamp (7) in place and put on nut with lockwasher (10).

GO TO FRAME 6

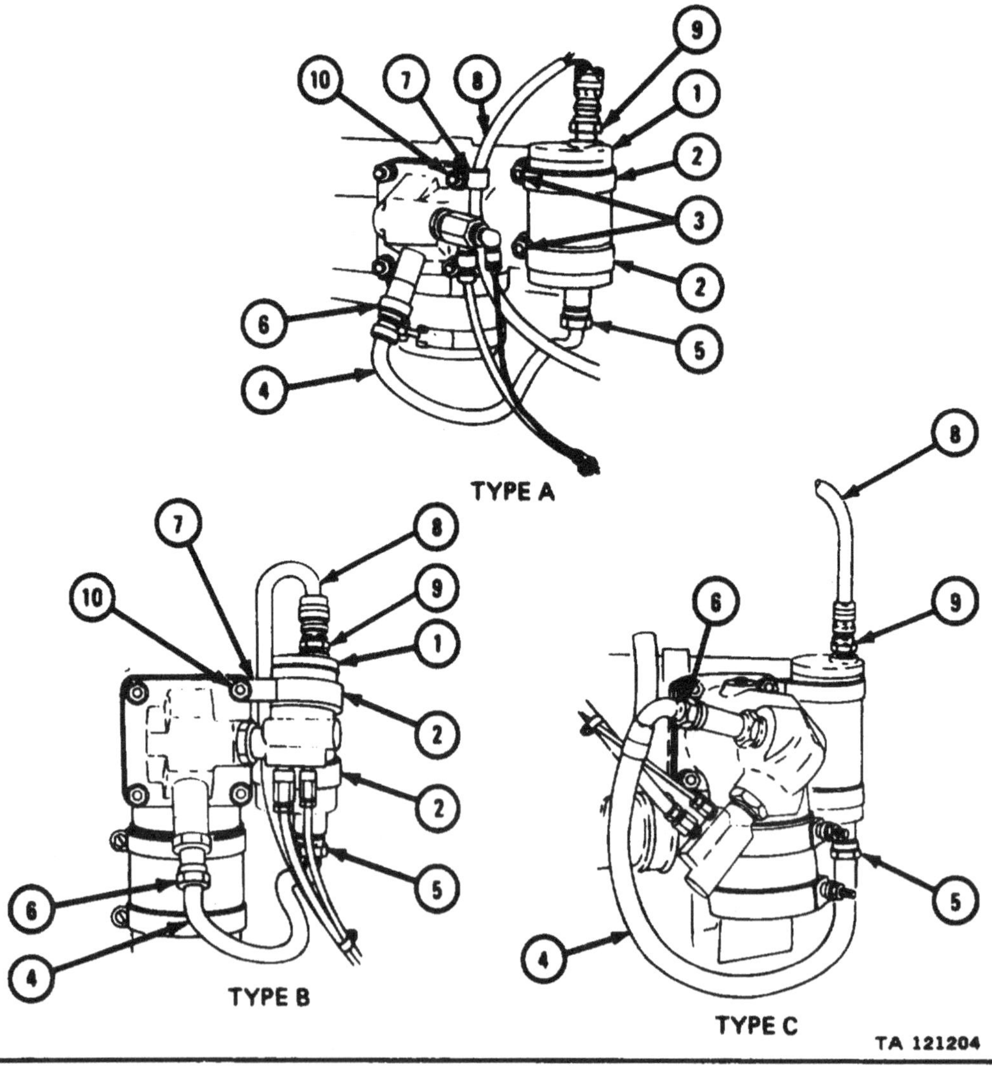

FRAME 6

NOTE

Tags and plugs put on during removal must be taken off during replacement .

1. Put fuel inlet tube (1) and fuel return tube (2) in place. Put on two fitting nuts (3) .
2. Put tube clamp (4) in place. Put in nut, lockwasher, and screw (5).

END OF TASK

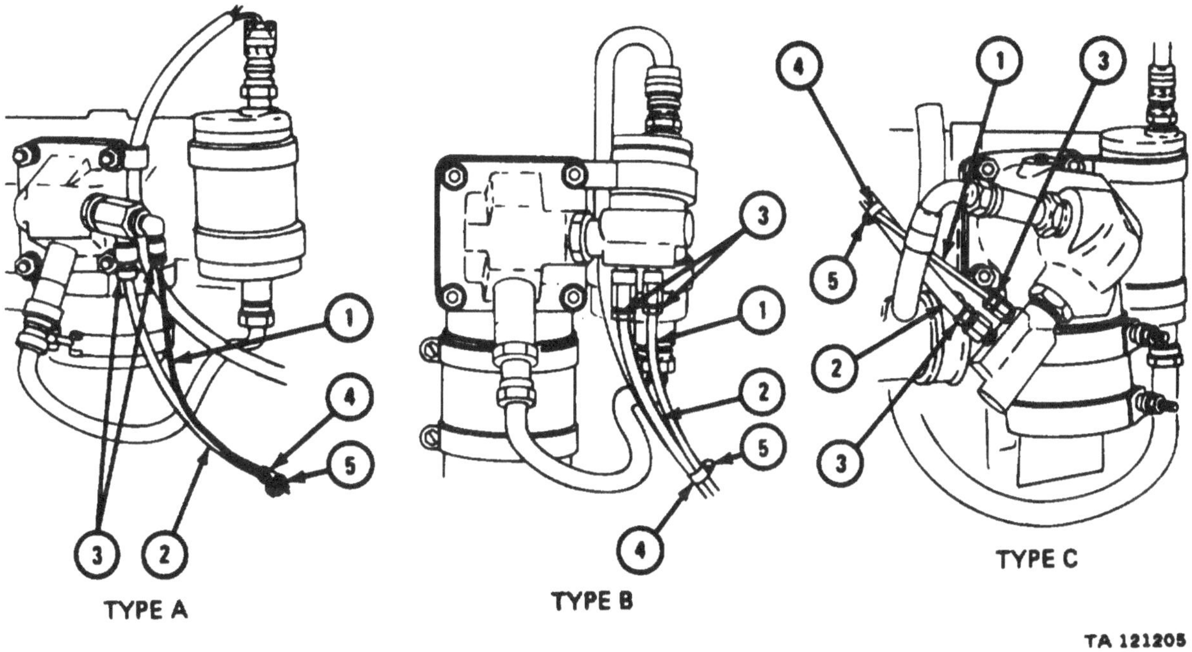

5-24. INJECTION PUMP ADAPTER AND AIR COMPRESSOR SUPPORT.

a. Air Compressor Support .

FRAME 1

1. Put air compressor support with gasket (1) in place on engine (2). Put in seven screws with lockwashers (3) .

END OF TASK

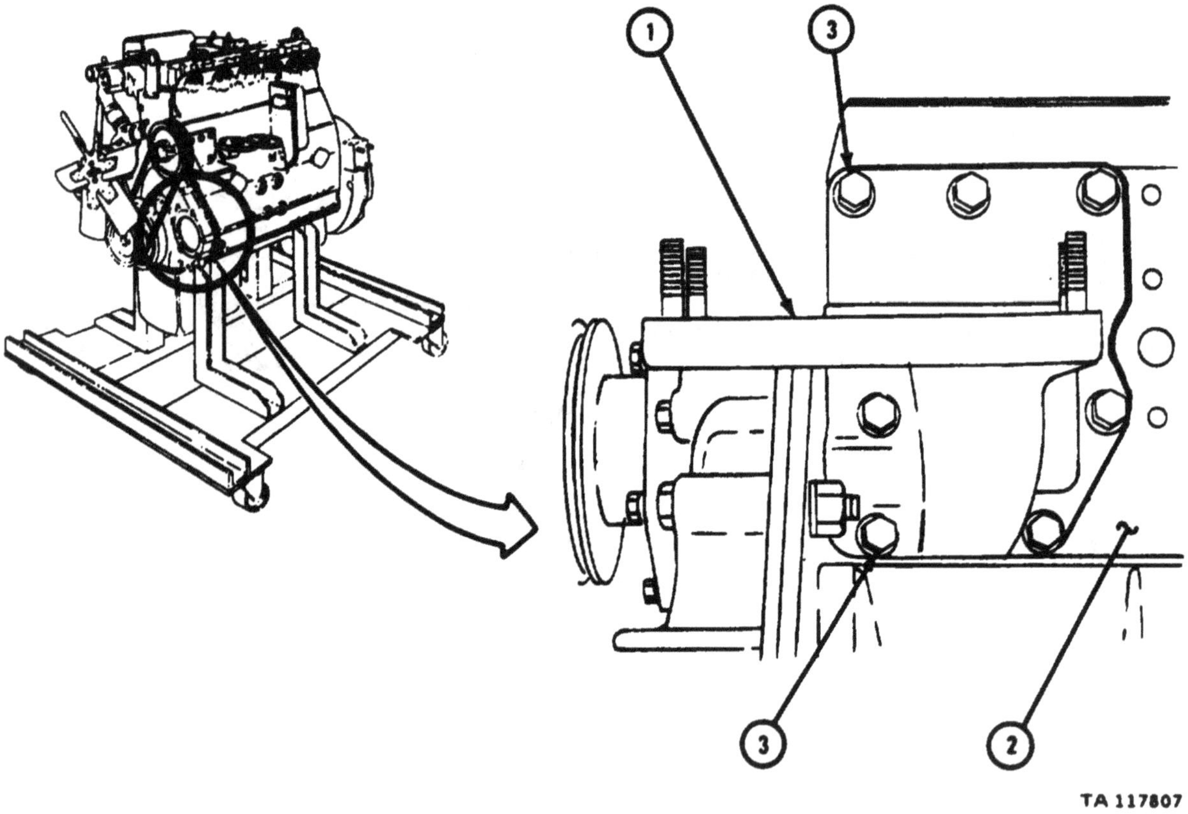

b. Injection Pump Adapter.

FRAME 1

1. Put on injection pump adapter gasket (1). Put on injection pump adapter (2).

NOTE

Some injection pumps have along stud (4). If long stud was taken out, it must be put back in the same hole.

2. Put on two nuts with lockwashers (3). If long stud (4) is used, put in long stud.

END OF TASK

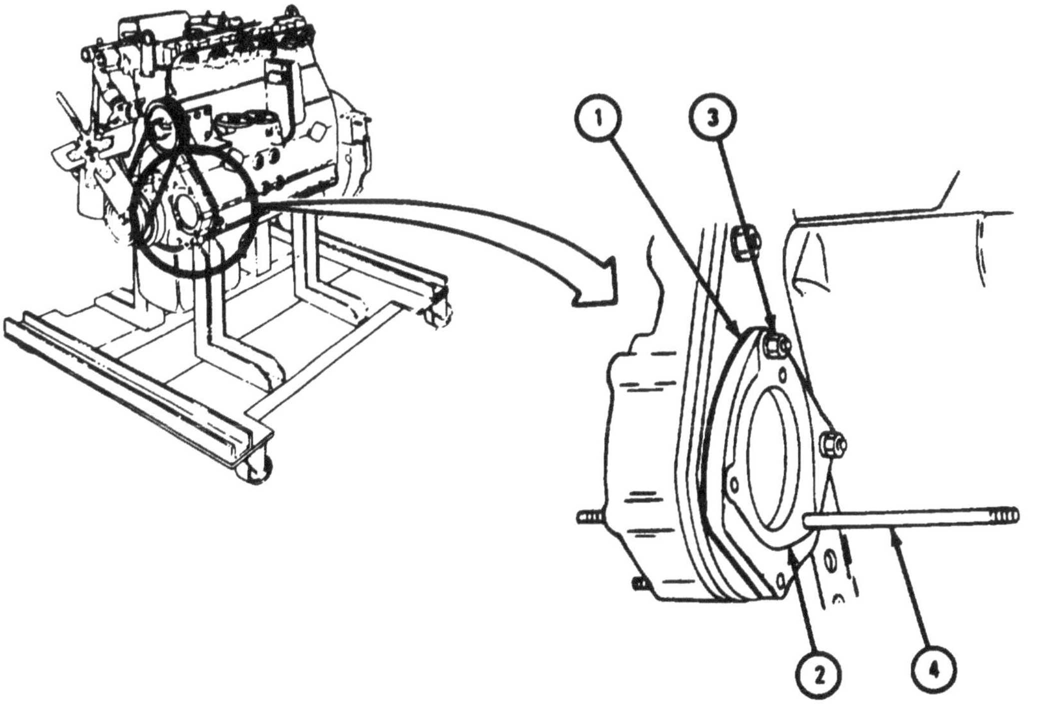

TA 117808

5-25. AIR COMPRESSOR.

a. All Engines Except LDS-465-2.

FRAME 1

1. Put air compressor gasket (1) in place. Put air compressor (2) in place. Put on four nuts with lockwashers (3) and tighten nuts to 25 pound-feet.
2. Loosen two screws with lockwashers (4). Using pulley adjusting wrench (5), turn pulley half (6) to the left and take it off.

NOTE

Refer to TM 9-2320-211-20 for correct drive belt tension.

Hold compressor drive belt (7) while turning pulley half (6) until drive belt moves out to bevel surface of pulley half.

3. Put on drive belt (7) and put on pulley half (6). Using pulley adjusting wrench (5), tighten pulley half until drive belt tension is correct. Put in two screws with lockwashers (4) .

END OF TASK

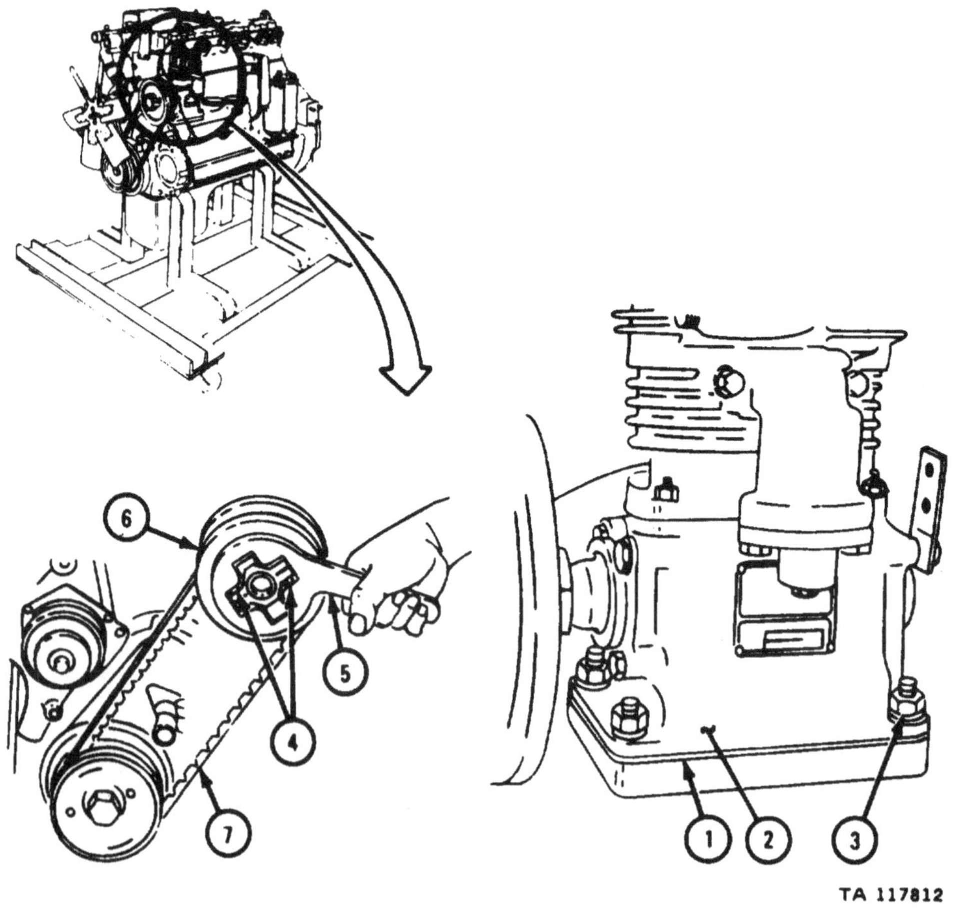

b. Engine LDS-465-2.

FRAME 1

1. Put gasket (1) in place. Put air compressor (2) in place. Put on four nuts with lockwashers (3) and tighten nuts to 25 pound-feet.

GO TO FRAME 2

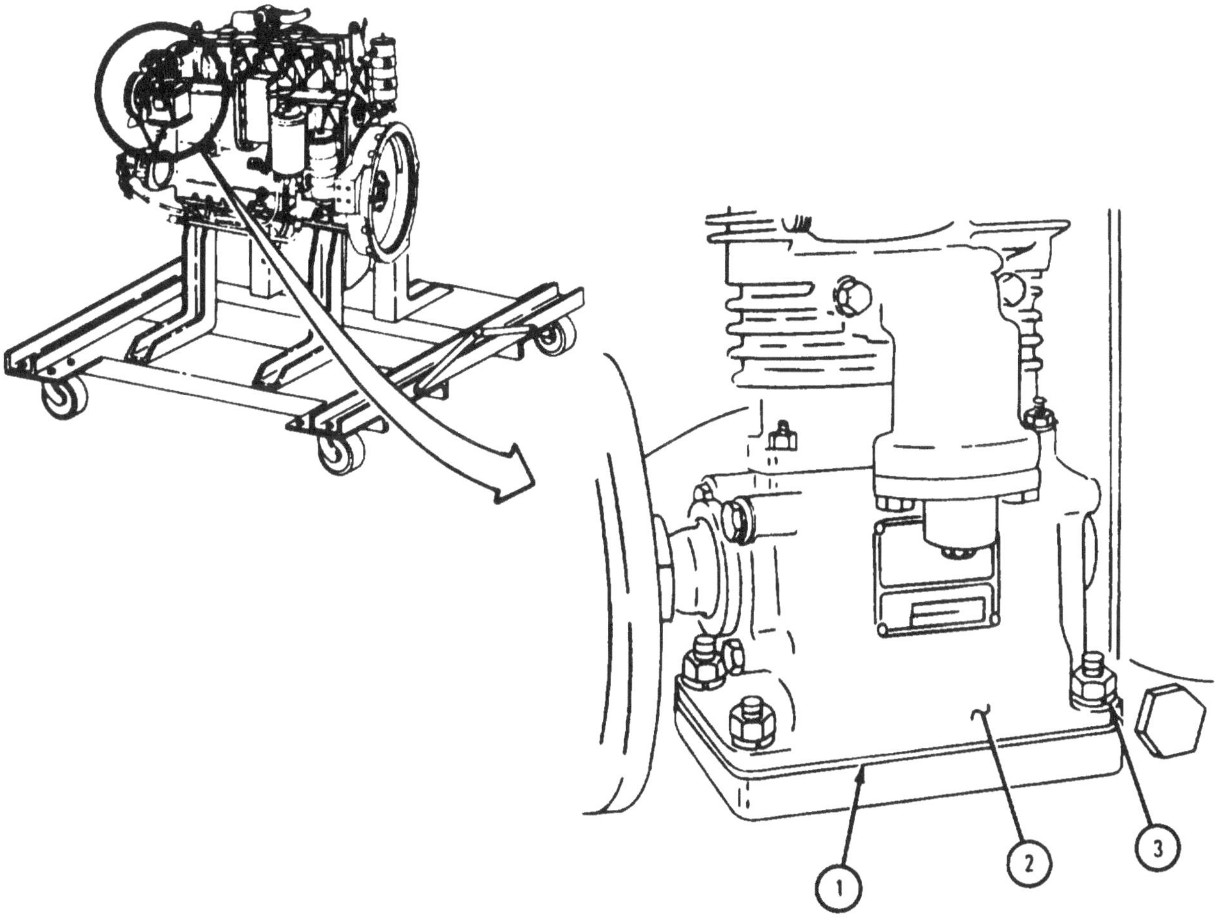

TA 117813

FRAME 2

1. Put compressor drive belt (1) on pulley (2). Put pulley on shaft (3). Pu t on nut (4) and tighten nut to 100 pound-feet.

NOTE

As halves of pulley (2) come together, check that drive belt spreads out evenly.

2. Unscrew and take out two capscrews (5).

END OF TASK

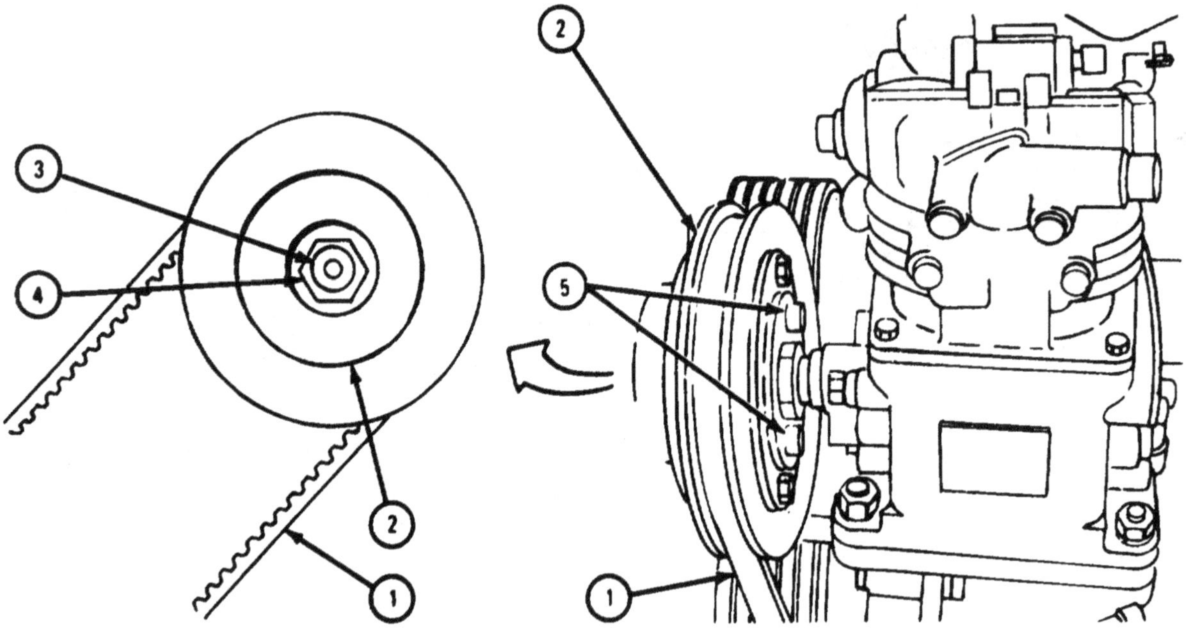

TA 117814

5-26. OIL COOLER AND OIL COOLER AND FILTER HOUSING.

a. Oil Cooler and Filter Housing.

FRAME 1

1. Put gasket (1) in place. Put oil cooler and filter housing (2) in place. Put in 16 screws with flat washers (3).

GO TO FRAME 2

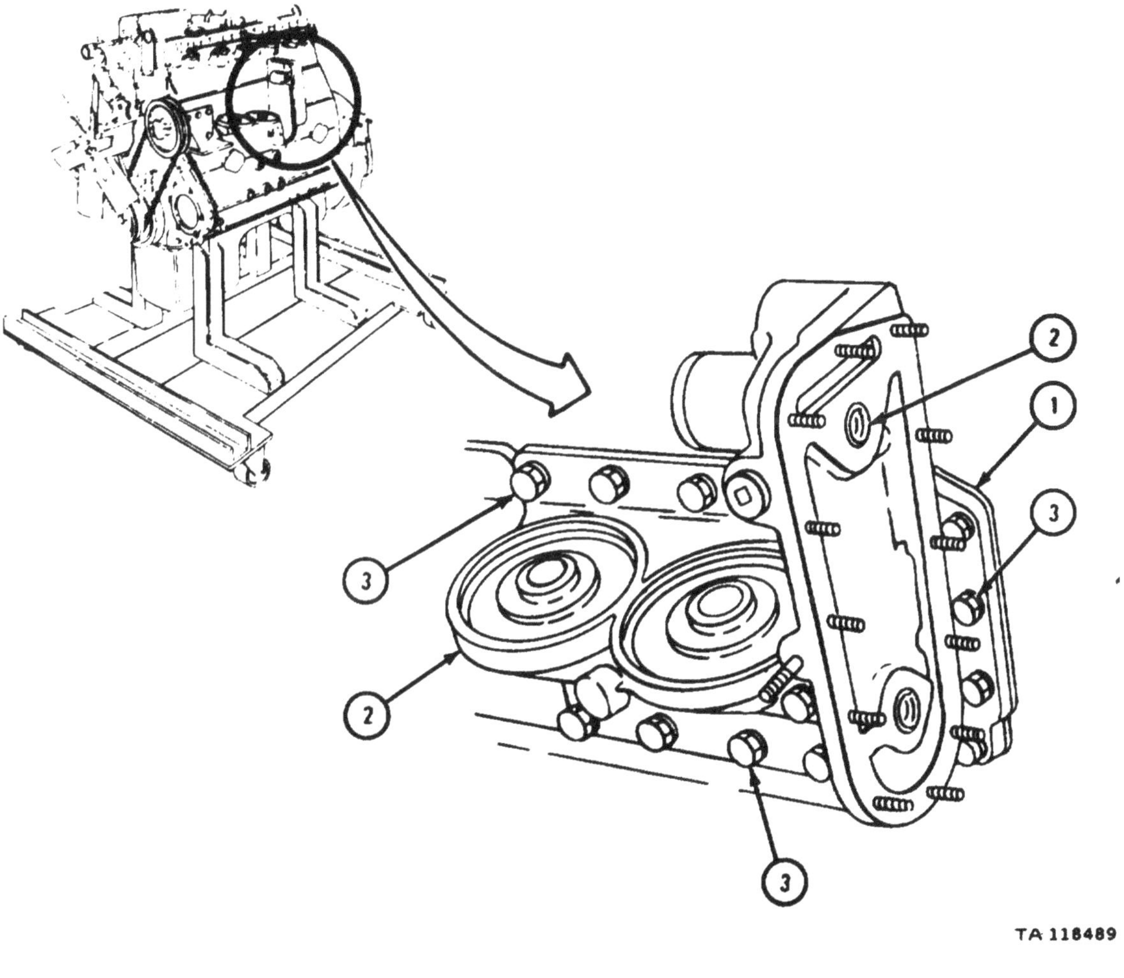

FRAME 2

NOTE

On some engines, oil cooler water inlet tube (1) is joined to water pump (2) with mounting flange (3). On other engines, oil cooler water inlet tube is joined to water pump with hose (4) and clamp (5).

If working on engine with mounting flange (3), do steps 1, 2, and 3. If working on engine without mounting flange (3), do steps 4 and 5.

1. For engines with mounting flange (3), put water inlet tube (1) in place. Slide hose (6) onto oil cooler water inlet (7).
2. Put gasket (8) in place on water pump (2). Aline flange (3) and put in four screws with nuts and lockwashers (9).
3. Tighten two hose clamps (5).
4. For engines without mounting flange (3), put water inlet tube (1) in place. Slide hose (6) onto oil cooler water inlet (7).
5. Slide hose (4) onto water pump (2). Tighten four hose clamps (5).

END OF TASK

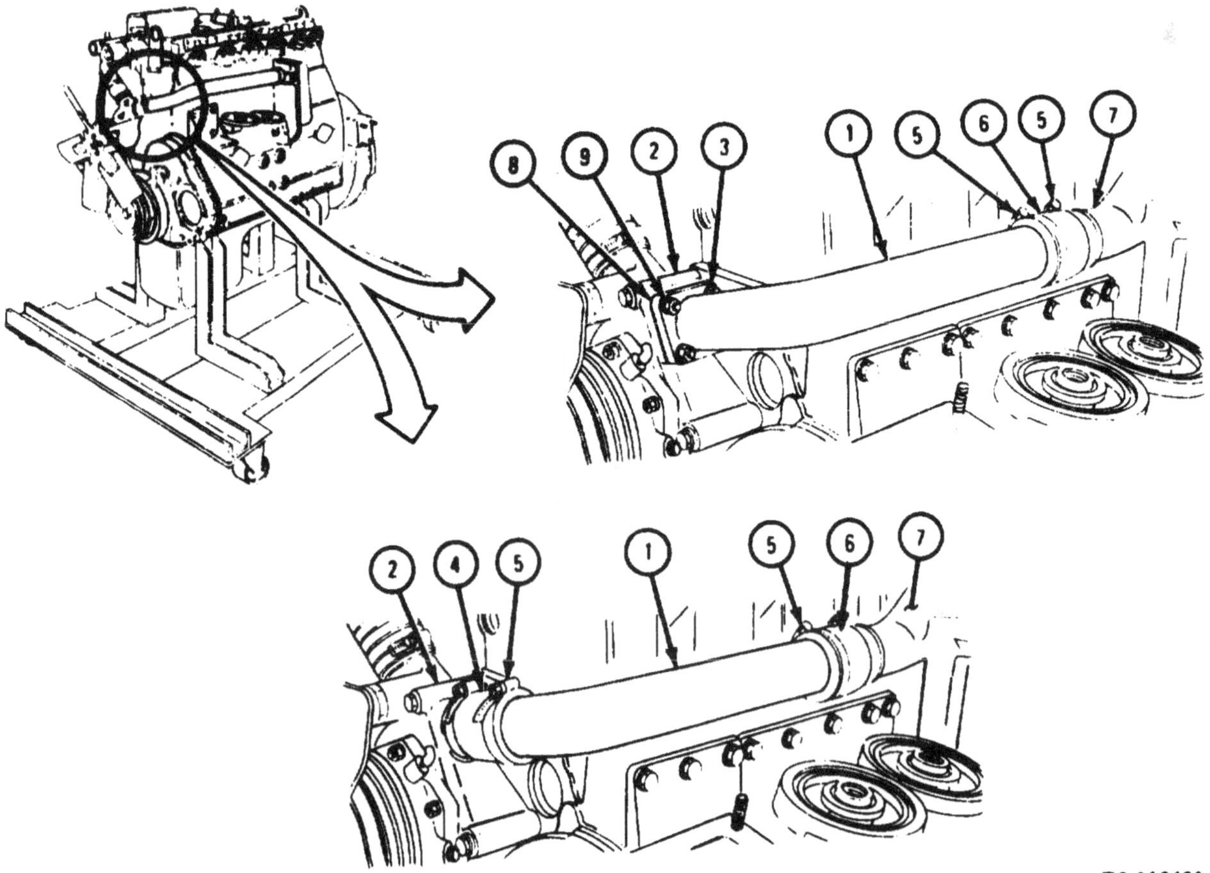

TA 118491

b. Oil Cooler.

FRAME 1

1. Put two preformed packings (1) and gasket (2) in place. Put oil cool elemen t (3) in place. Put gasket (4) in place.
2. Put cover (5) in place.
3. If engine has top mounted-covered type manifold flame heater system, put tube clamp (6) in place.
4. Put on 12 lockwashers (7) and 12 nuts (8).

END OF TASK

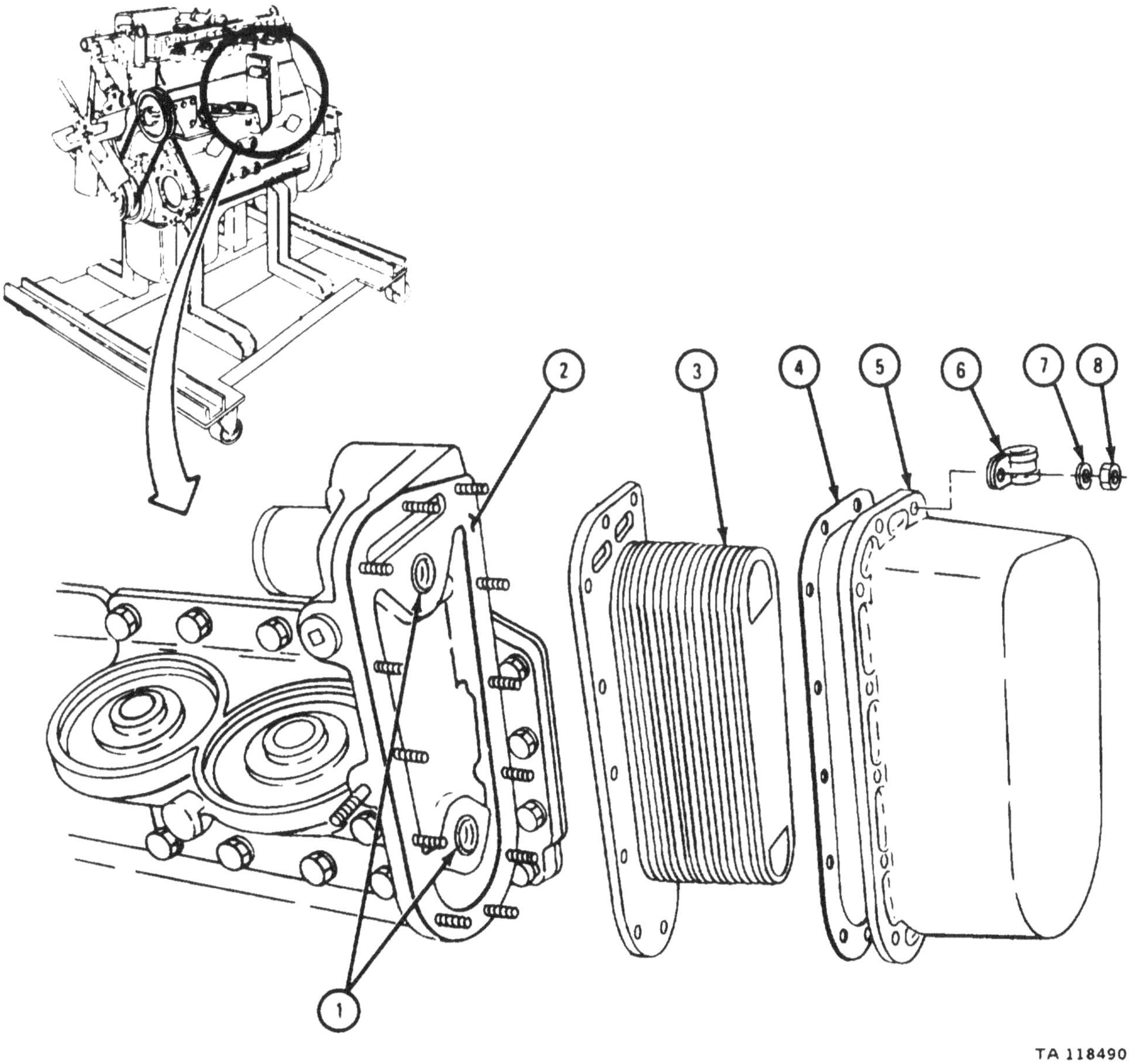

5-27. TAPPET CHAMBER COVER.

a. Engine LDS-465-1.

FRAME 1

1. Put tappet chamber cover with gasket (1) on engine block (2).
2. Put two flame heater tubes (3) and tube clamps (4) in place. Put in two screws with flat washers (5).
3. Put ground strap (6) in place. Put in screw with starwasher (7).

NOTE

Put in screws with flat washers (8) only as shown. Two screw holes (9) will be used to mount fuel filter.

4. Put in six screws with flat washers (8).

END OF TASK

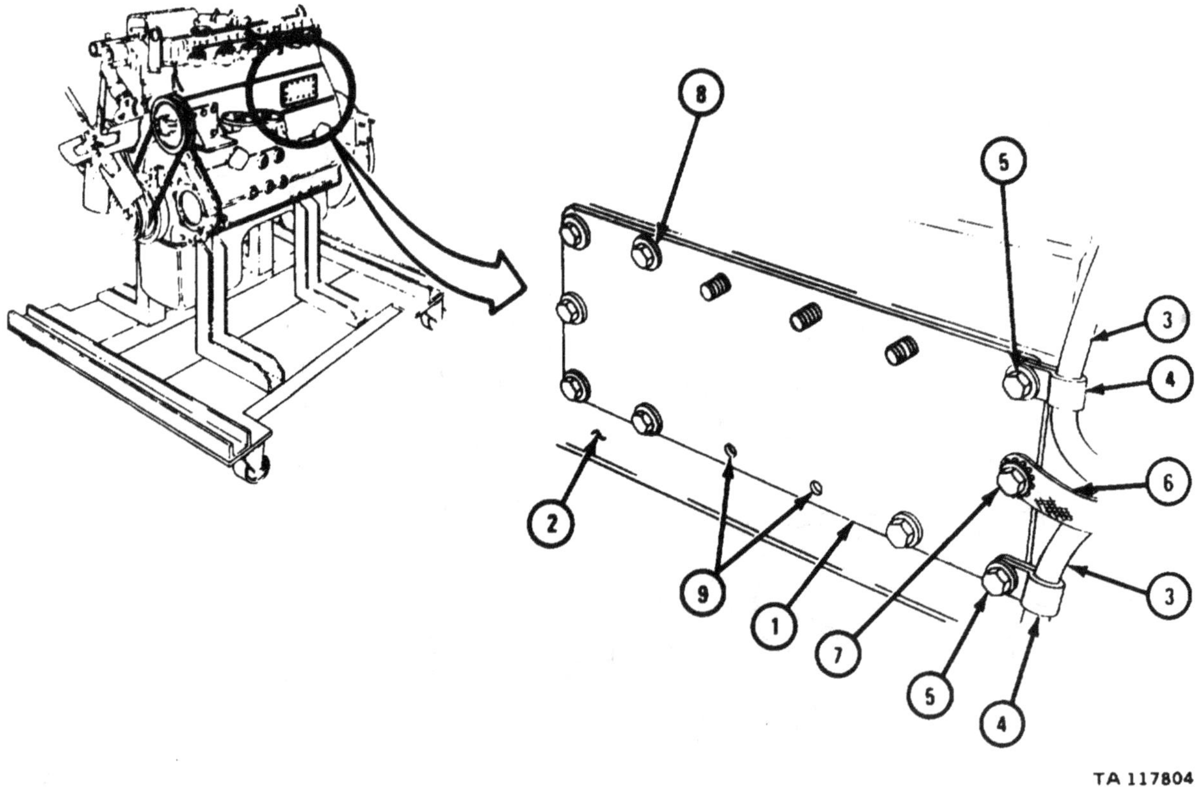

b. Engines LD-465-1, LD-465-1C, LDT-465-1C, and LDS-465-1A.

FRAME 1

1. Put tappet chamber cover with gasket (1) in place.

NOTE

Some engines have a ground strap (2) held on by tappet chamber cover screw (3). If ground strap is used, do steps 2 and 3. If no ground strap is used, do step 4.

2. If engine has ground strap (2), put ground strap in place. Put in screw with flat washer and starwasher (3).

3. Put in 10 screws with flat washers (4).

4. If no ground strap is used, put in 11 screws with flat washers (4).

END OF TASK

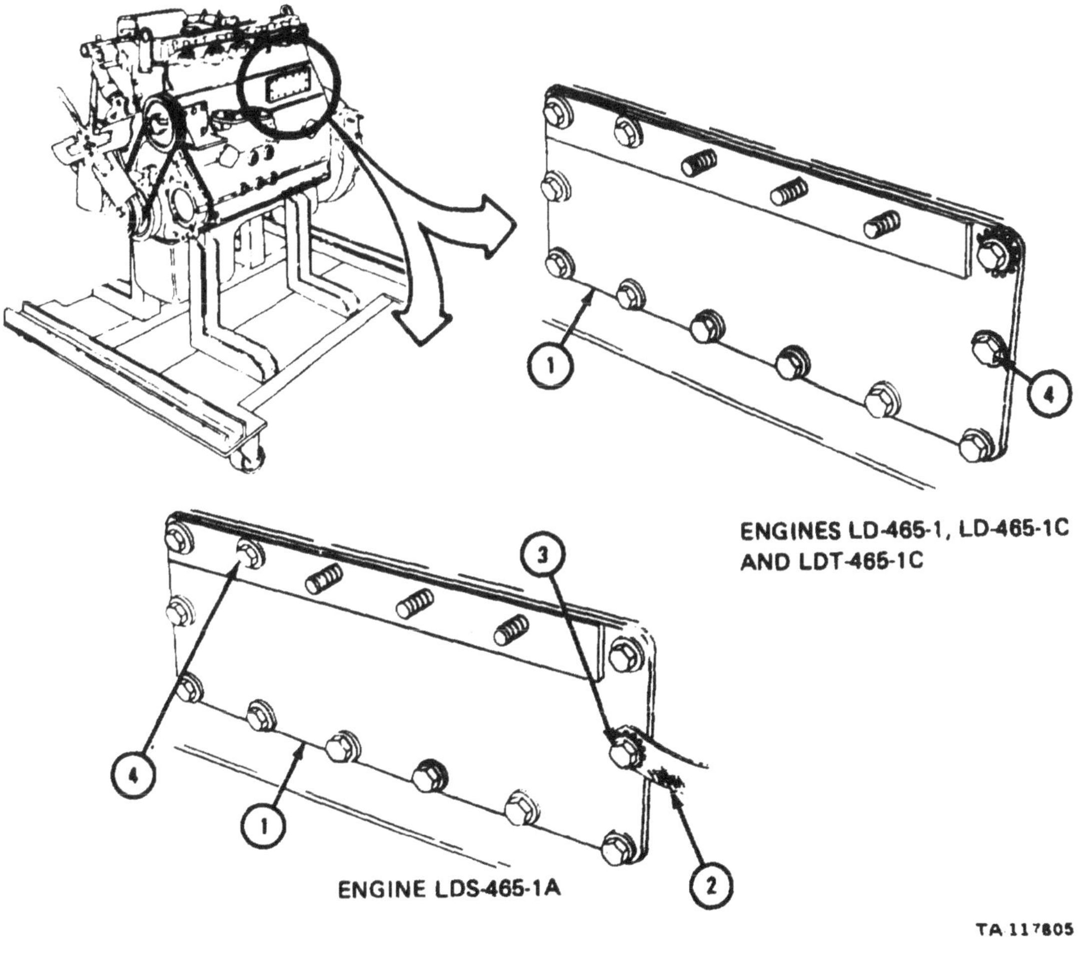

c. Engine LDS-465-2 .

FRAME 1

NOTE

Put in screws (2) only as shown. Other screws will be used to mount engine accessories.

1. Put tappet chamber cover with gasket (1) in place. Put in six screws with flat washers (2) .

END OF TASK

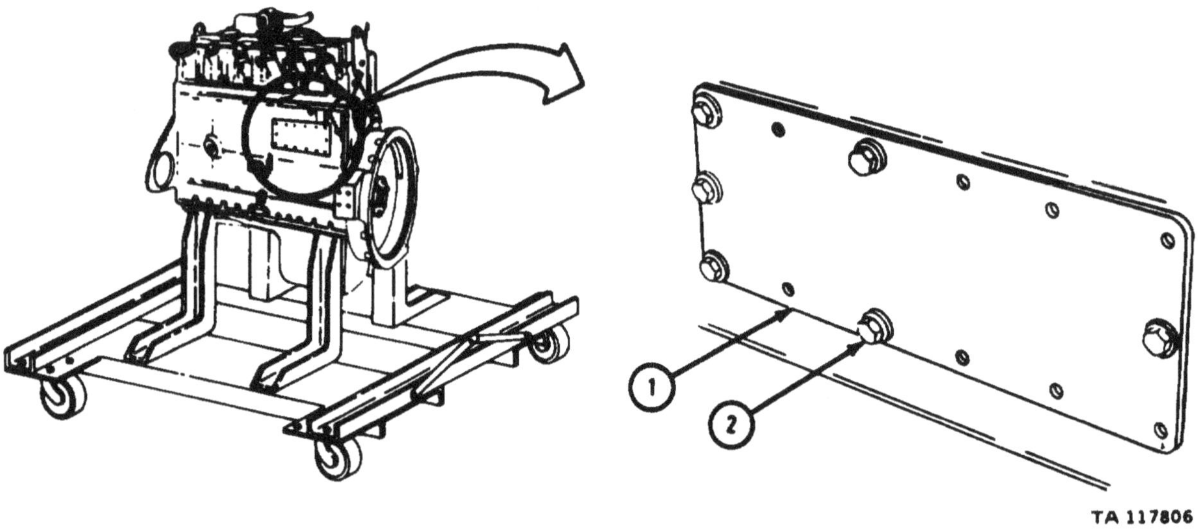

5-28. OIL FILTERS.

FRAME 1

1. Put in filter body gasket (1) and put in filter element (2).
2. Check that spring. and spring retainer (3) are on center post (4) with retainer side facing open end of filter body (5).
3. Put filter body (5) in place. Screw in center post (4) .
4. Do steps 1 through 3 for other filter (6).

END OF TASK

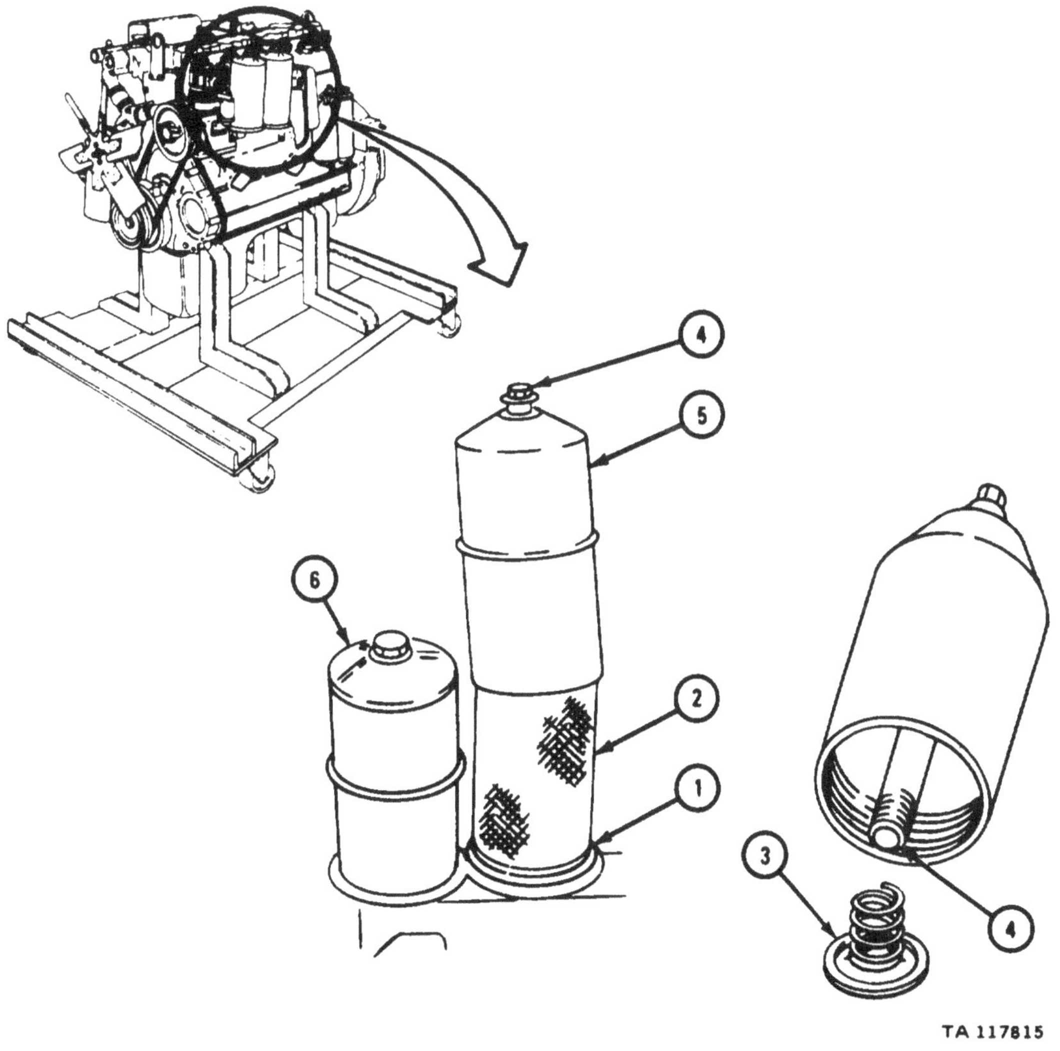

5-29. CYLINDER HEAD COVERS AND CRANKCASE BREATHER ADAPTER.

a. Cylinder Head Covers.

(1) All engines except LDS-465-2.

FRAME 1

1. Put rear cylinder head cover gasket (1) in place. Put on rear cylinder head cover (2).
2. Put tachometer cable support bracket (3) in place. Put in lockbolt with flat washer (4).
3. Put in six lockbolts with flat washers (5). Put on locknut with flat washer (6).

GO TO FRAME 2

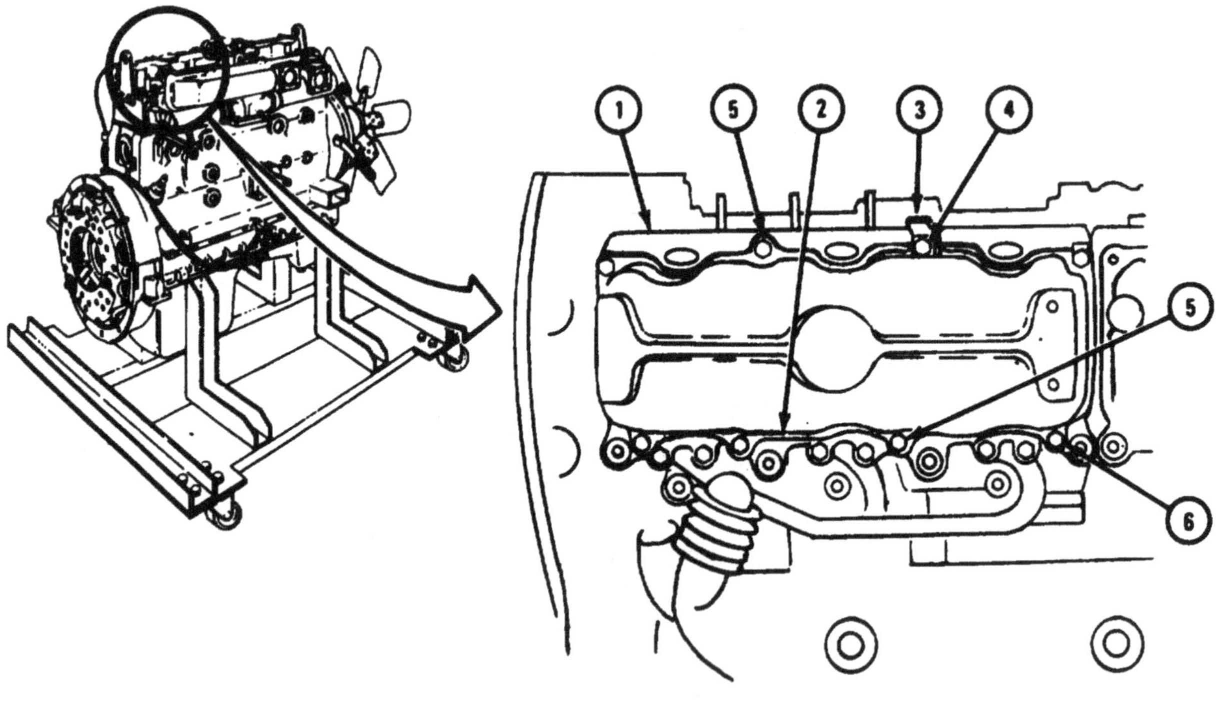

FRAME 2

1. Put cylinder head cover gasket (1) in place. Put on cylinder head cover (2).
2. Put on tachometer cable support bracket (3). Put in lockbolt with flat washer (4).

NOTE

Two lockbolts with flat washers (5) are used to hold injector nozzle tube support brackets and will be put in when injector nozzle tubes are put on.

3. Put in four lockbolts with flat washers (6). Put on locknut with flat washer (7).
4. Put on oil filler cap (8).

END OF TASK

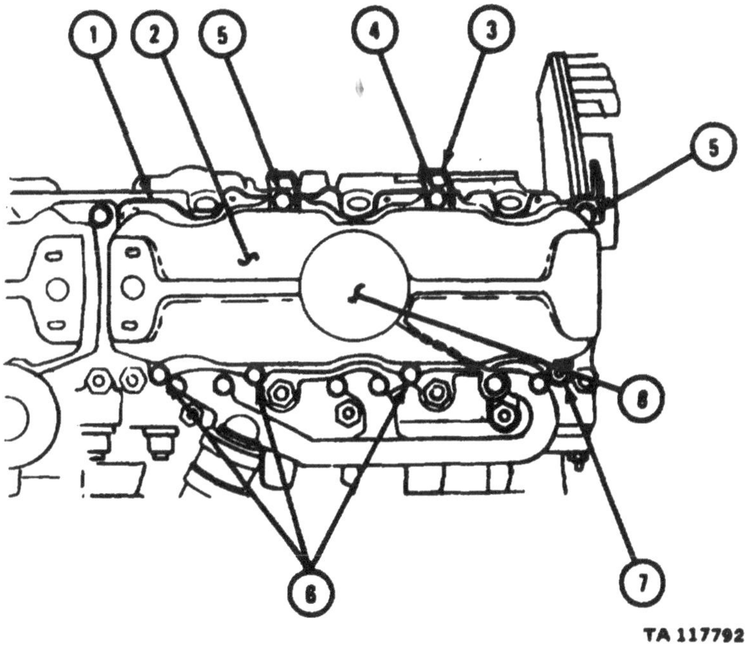

(2) Engine LDS-465-2 .

FRAME 1

1. Put rear cylinder head cover gasket (1) in place. Put on rear cylinder head cover (2) .
2. Put in seven lockbolts with flat washers (3). Pu t on locknut with flat washer (4) .

GO TO FRAME 2

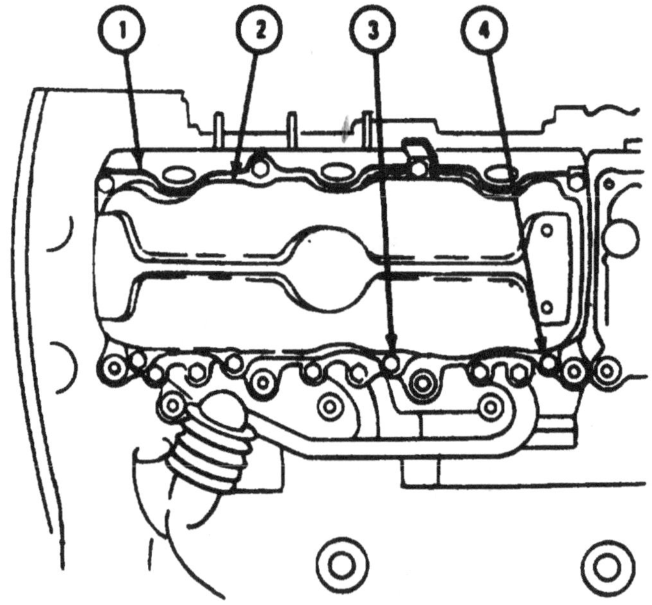

FRAME 2

1. Put front cylinder head cover gasket (1) in place. Put on front cylinder head cover (2).
2. Put in seven lockbolts with flat washers (3). Put on locknut with flat washer (4).
3. Put on oil filler cap (5).

END OF TASK

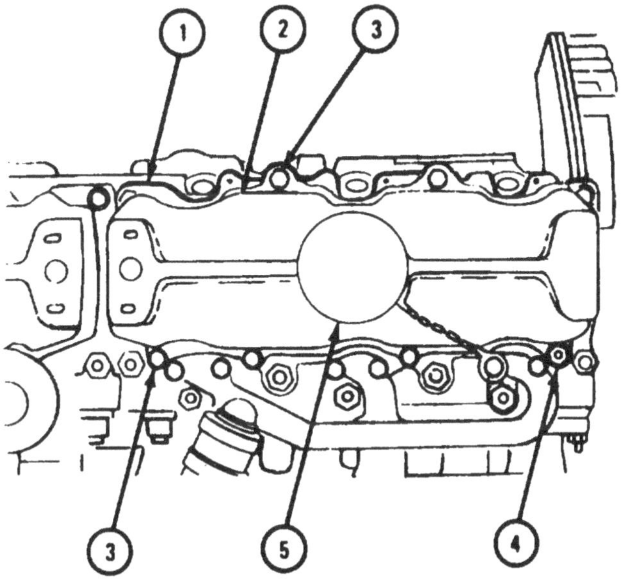

TA 117794

b. Crankcase Breather Adapter.

(1) All engines except LDS-465-2.

FRAME 1

1. Put two crankcase breather adapter gaskets (1) in place. Put on crankcase breather adapter (2) with hose (3) and two hose clamps (4).

2. Put on four locknuts with flat washers (5).

IF WORKING ON ENGINE LD-465-1 OR LDT-465-1C, GO TO FRAME 2.
IF WORKING ON ANY OTHER ENGINE, END OF TASK

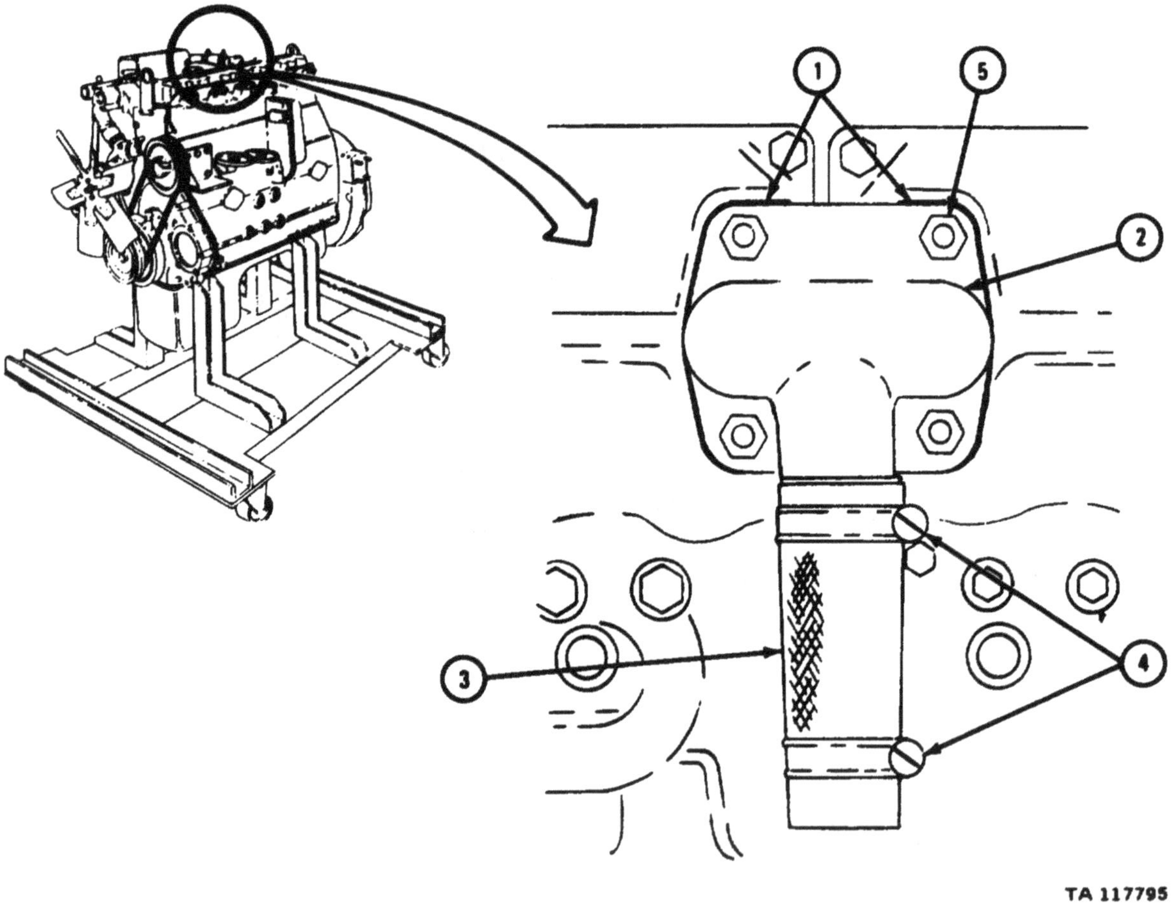

FRAME 2

1. Put crankcase breather tube (1) into crankcase breather adapter hose (2).
2. Aline tube clamp (3) with bracket (4).
3. Put washer (5) on screw (6) and put screw through bracket (4) and clamp (3). Put on lockwasher (7) and nut (8) .
4. Tighten two hose clamps (9).

END OF TASK

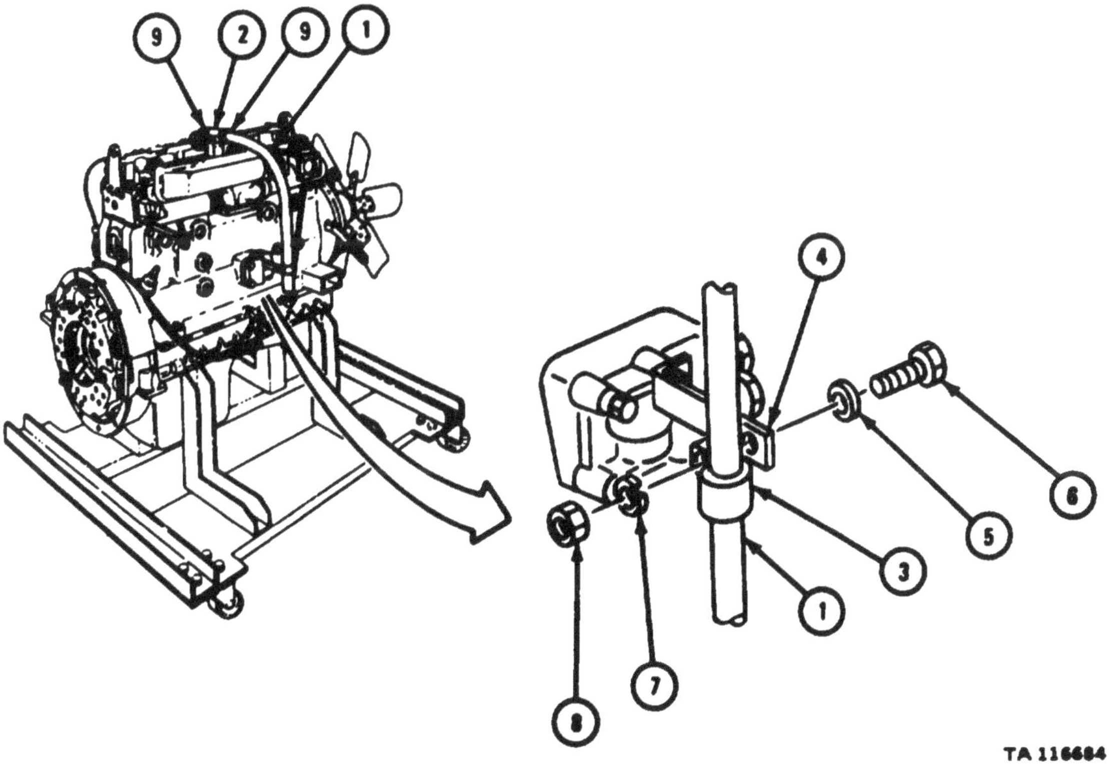

(2) Engine LDS-465-2 .

FRAME 1

1. Put two crankcase breather adapter gaskets (1) in place. Put on crankcase breather adapter (2) .
2. Put on four nuts with lockwashers (3).

GO TO FRAME 2

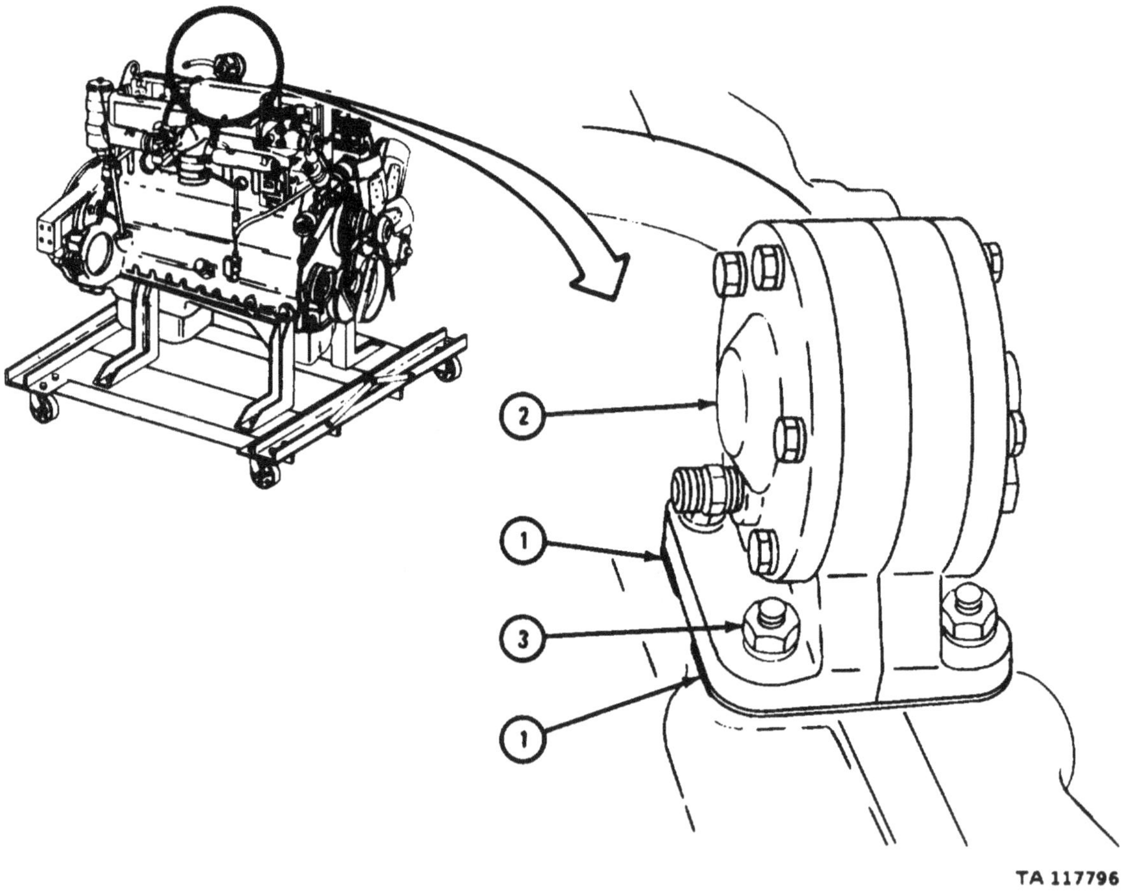

FRAME 2

1. Put air pressurization tube (1) in place. Screw on and tighten fitting nut (2).

NOTE

If intake manifold nut and flat washer (4) has already been put on, take off nut and flat washer. Then do step 2.

2. Put clamp (3) in place. Put on nut with flat washer (4).

GO TO FRAME 3

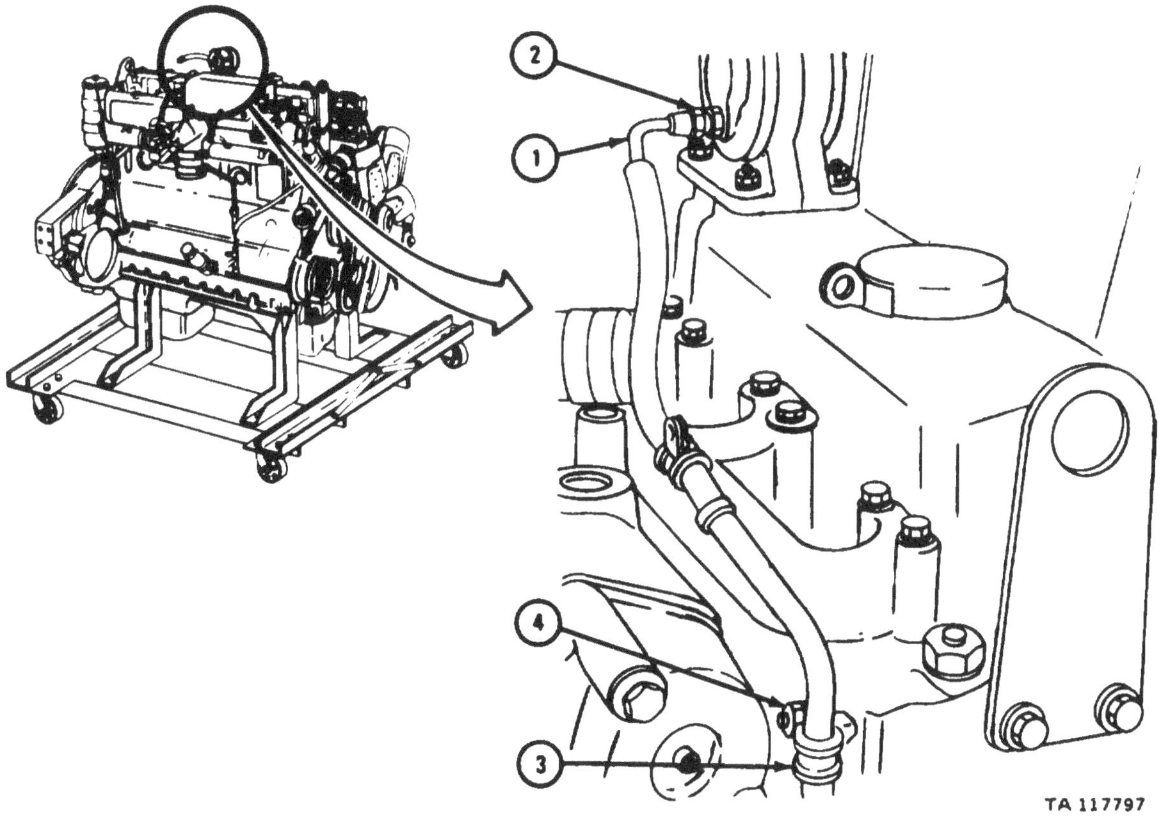

FRAME 3

1. Put air pressurization tube (1) in place. Screw fitting nut (2) onto tee fitting (3) .

2. Screw fitting nut (4) onto tee fitting (5). Put air pressurization tube (6) in place . Screw fitting nut (7) onto tee fitting.

END OF TASK

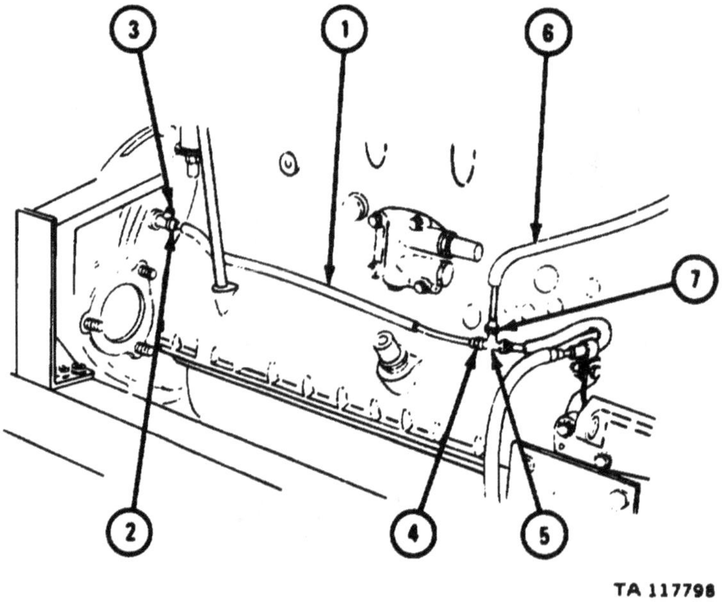

5-30. HYDRAULIC PUMP OIL RESERVOIR.

FRAME 1

NOTE

All tags put on during removal must be taken off during replacement .

1. Put oil reservoir (1) in place. Put in four screws with four washers (2).
2. Put control valve hose (3) in place. Screw in and tighten nut (4).
3. Put on bypass hose (5) and tighten hose clamp (6) .
4. Put hydraulic pump oil return hose (7) in place. Tighten hose clamp (8) .

END OF TASK

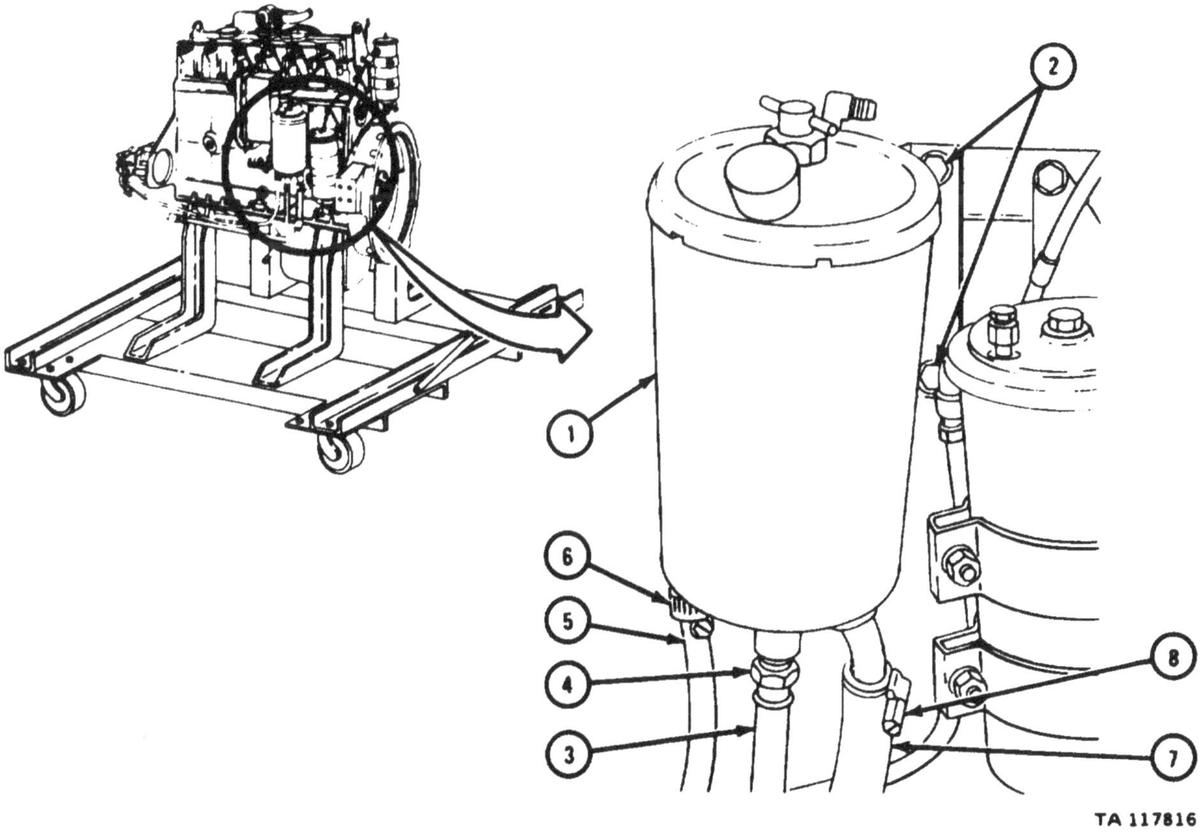

5-31. FUEL FILTER ASSEMBLY, FUEL TUBES, AND OIL HOSE.

a. Engines LD-465-1, LD-465-1C, and LDT-465-1C .

FRAME 1

NOTE

Tags put on during removal must be taken off during replacement.

1. Screw in pipe elbow (1).
2. Put pressure oil hose (2) in place and screw in and tighten fitting nut (3).

GO TO FRAME 2

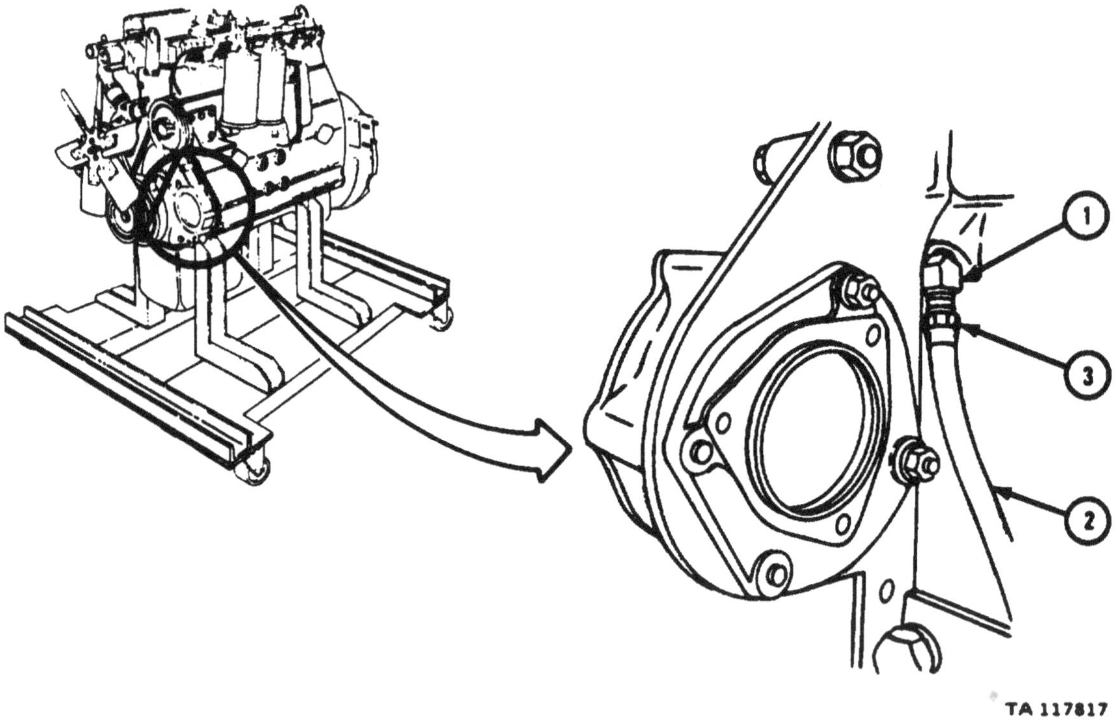

FRAME 2

1. Put fuel filter (1) in place. Put on three lockwashers with three nuts (2).
2. Put fuel filter-to-fuel density compensator tube (3) in place. Screw on and tighten fitting nut (4) .
3. Put fuel inlet tube (5) and fuel return tube (6) in place. Screw on and tighten two fitting nuts (7).
4. Put two drain tubes (8) on fuel filters (1) as shown and put on two clamps (9) .

END OF TASK

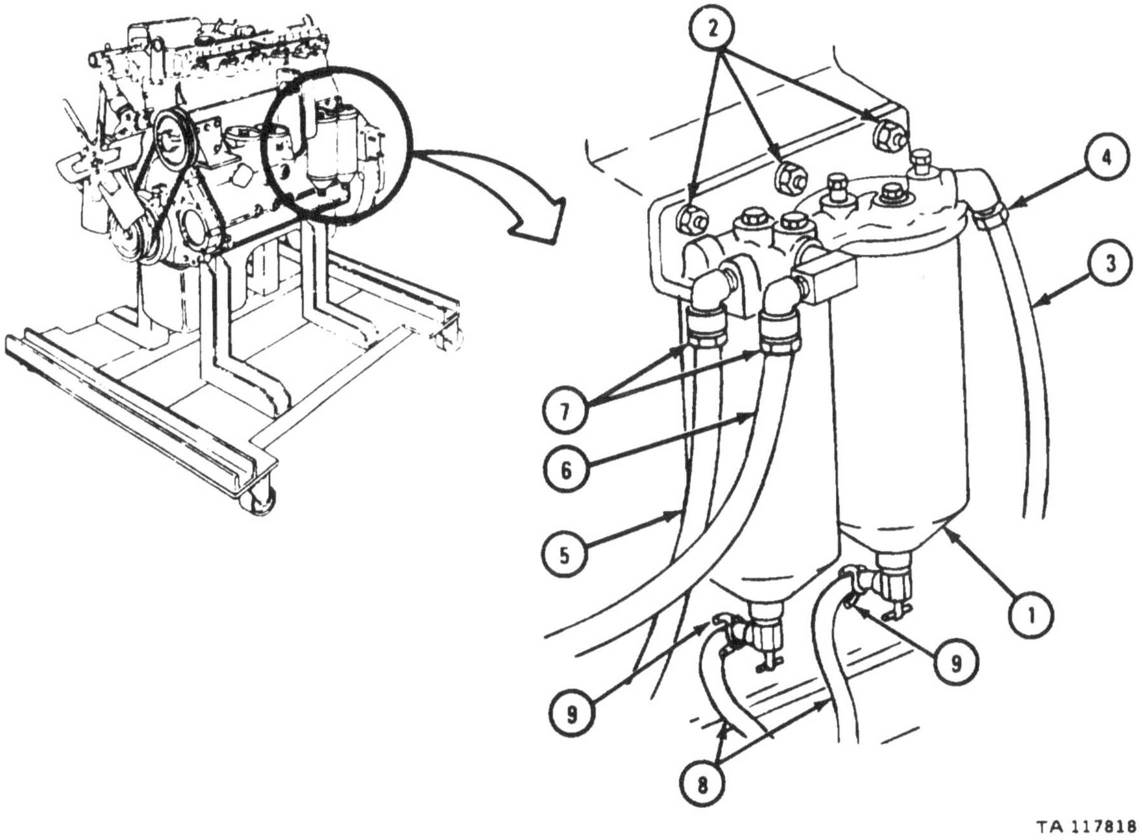

b. Engine LDS-465-1A.

FRAME 1

1. Put fuel filter assembly (1) in place. Put on three lockwashers with three nuts (2).

GO TO FRAME 2

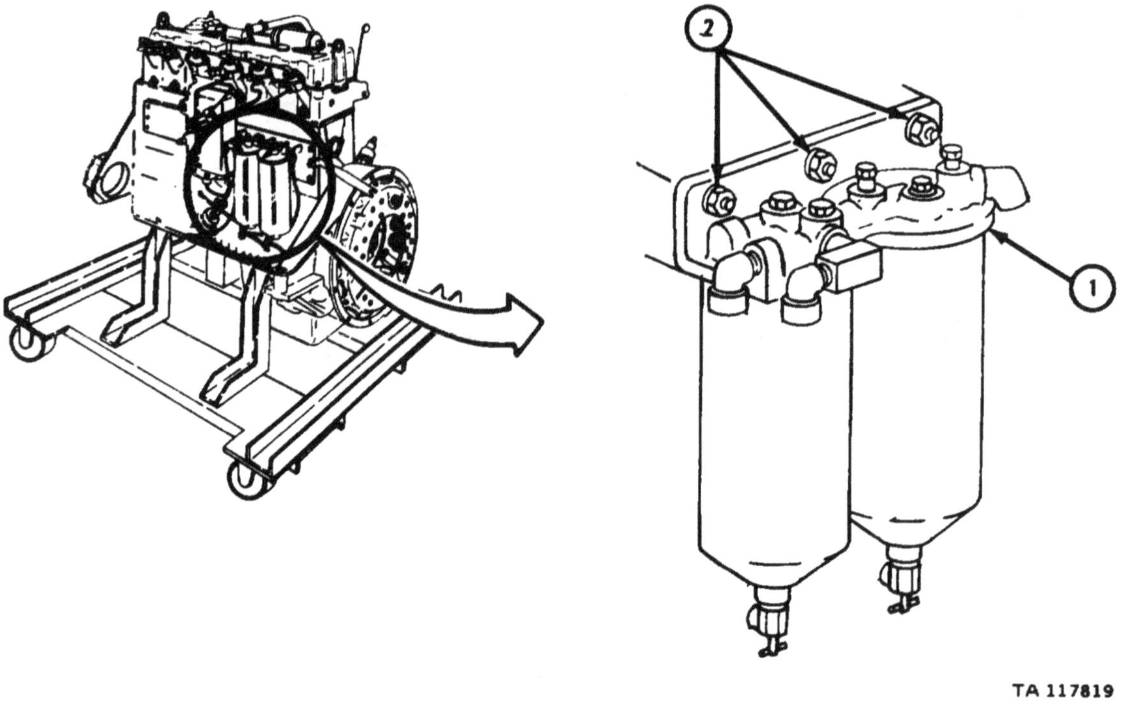

FRAME 2

NOTE

Tags put on during removal must be taken off during replacement.

1. Squeeze hose clamp (1) and put on drain tube (2).
2. Squeeze hose clamp (3) and put on drain tube (4).

NOTE

Some engines do not have tube clamp (5).

3. Put drain tubes (2 and 4) in clamp (5). Put clamp in place and put in screw with lockwasher (6) .

GO TO FRAME 3

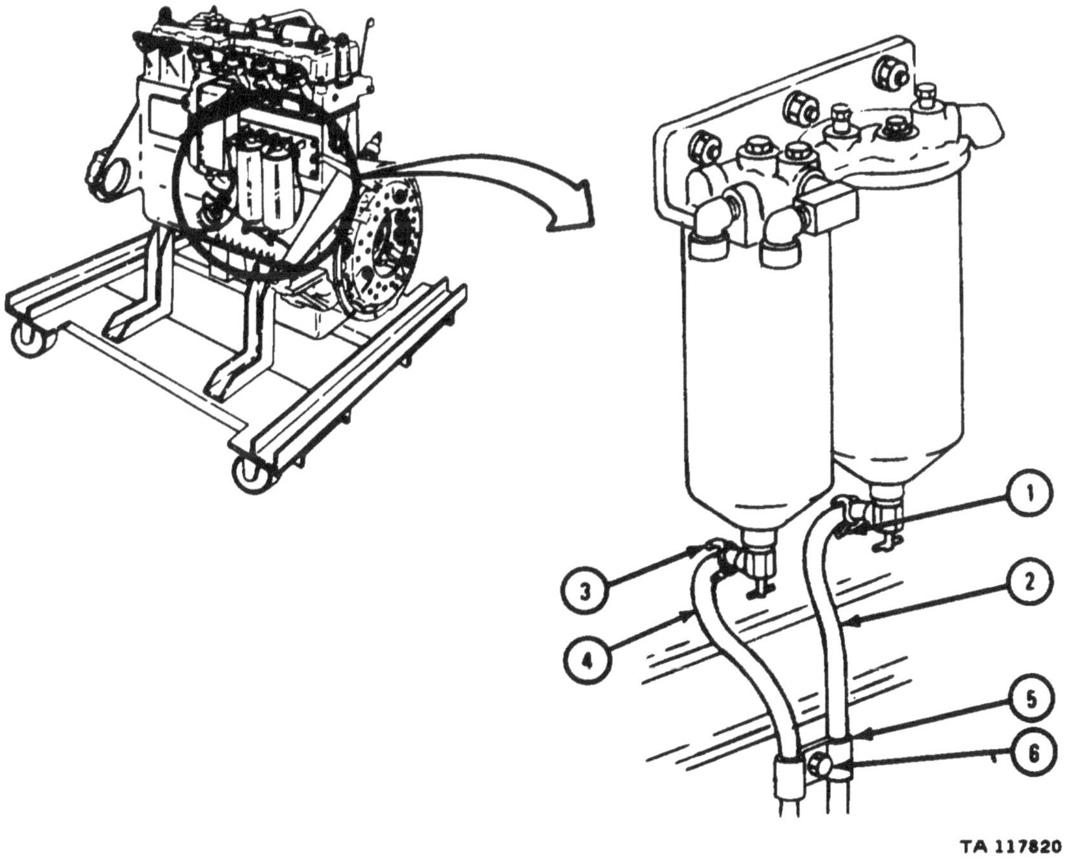

FRAME 3

1. Put fuel filter-to-fuel density compensator tube (1) in place. Screw in and tighten fitting nut (2) .
2. Put fuel inlet tube (3) in place. Screw in and tighten fitting nut (4).
3. Put fuel return tube (5) in place. Screw in and tighten fitting nut (6).
4. Put fuel inlet tube (3) with clamp (7) in place. Put in screw with lockwasher (8).
5. Put fuel return tube (5) with clamp (9) in place. Put in screw with lockwashers (10) .

GO TO FRAME 4

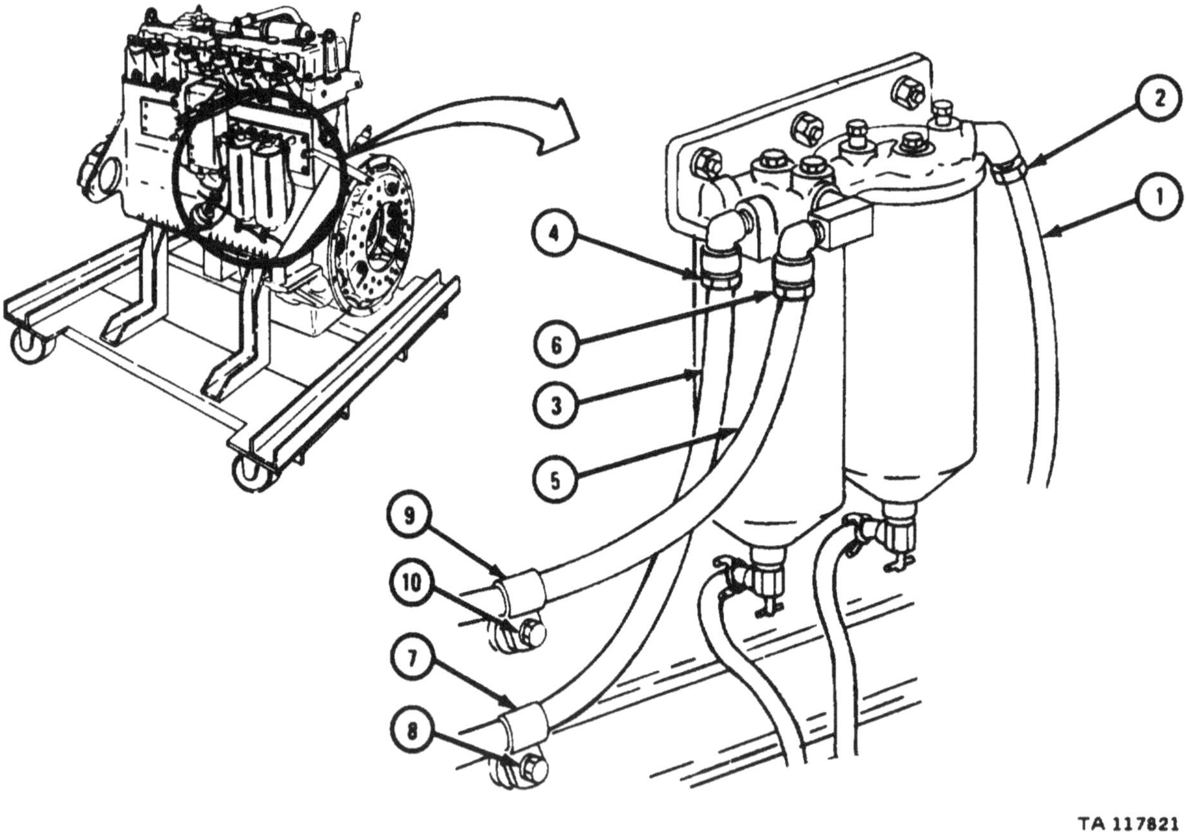

FRAME 4

1. Screw in pipe elbow (1).
2. Put pressure oil hose (2) in place. Screw in and tighten fitting nut (3).

END OF TASK

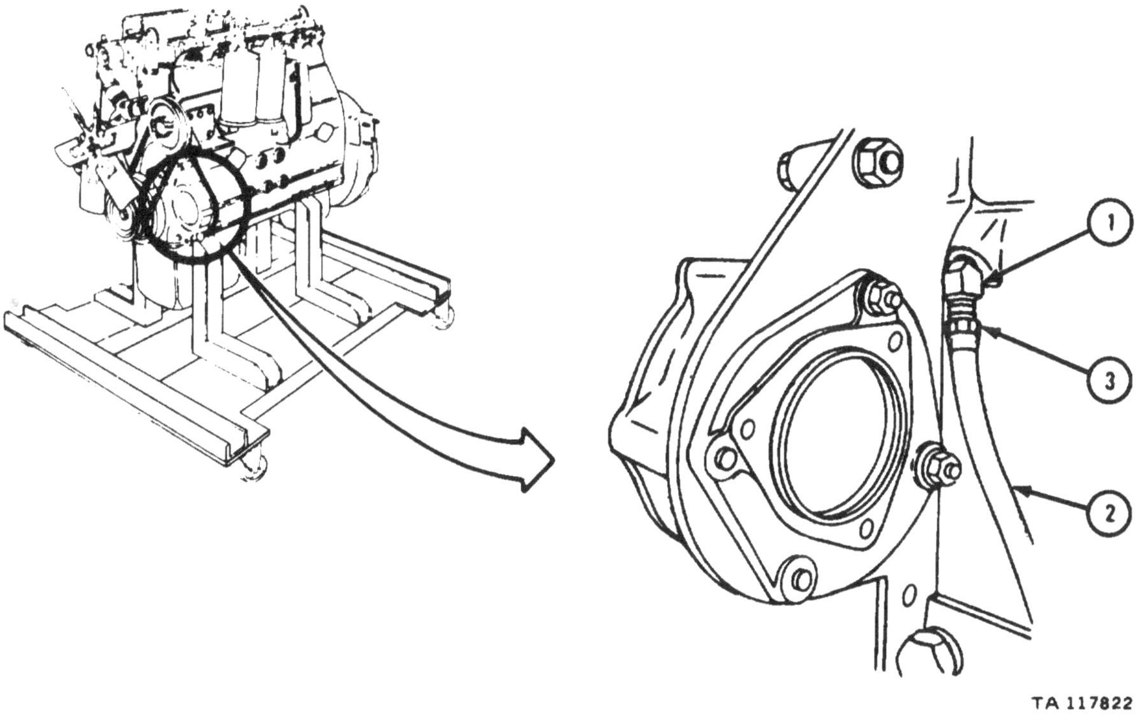

c. Engine LDS-465-1 .

FRAME 1

1. Put fuel filter (1) in place. Put in two screws with washers (2).

GO TO FRAME 2

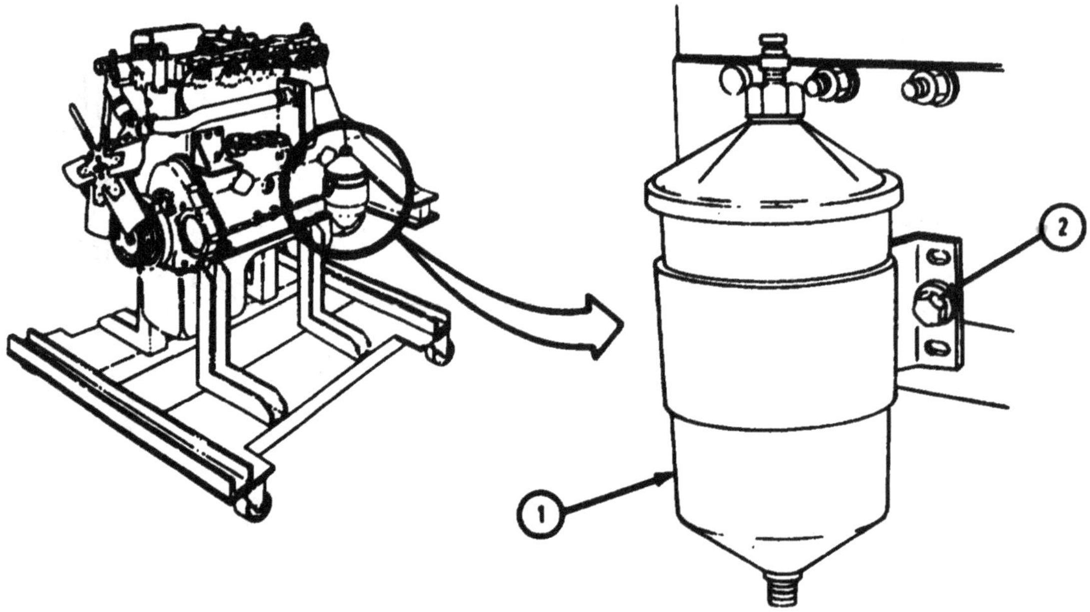

FRAME 2

1. Screw in and tighten fuel outlet elbow (1) and fuel inlet elbow (2).
2. Put fuel outlet hose (3) in place. Screw in and tighten fitting nut (4).
3. Put fuel inlet hose (5) in place. Screw in and tighten fitting nut (6).
4. Put hose clamp (7) in place. Put in screw with lockwasher (8).

GO TO FRAME 3

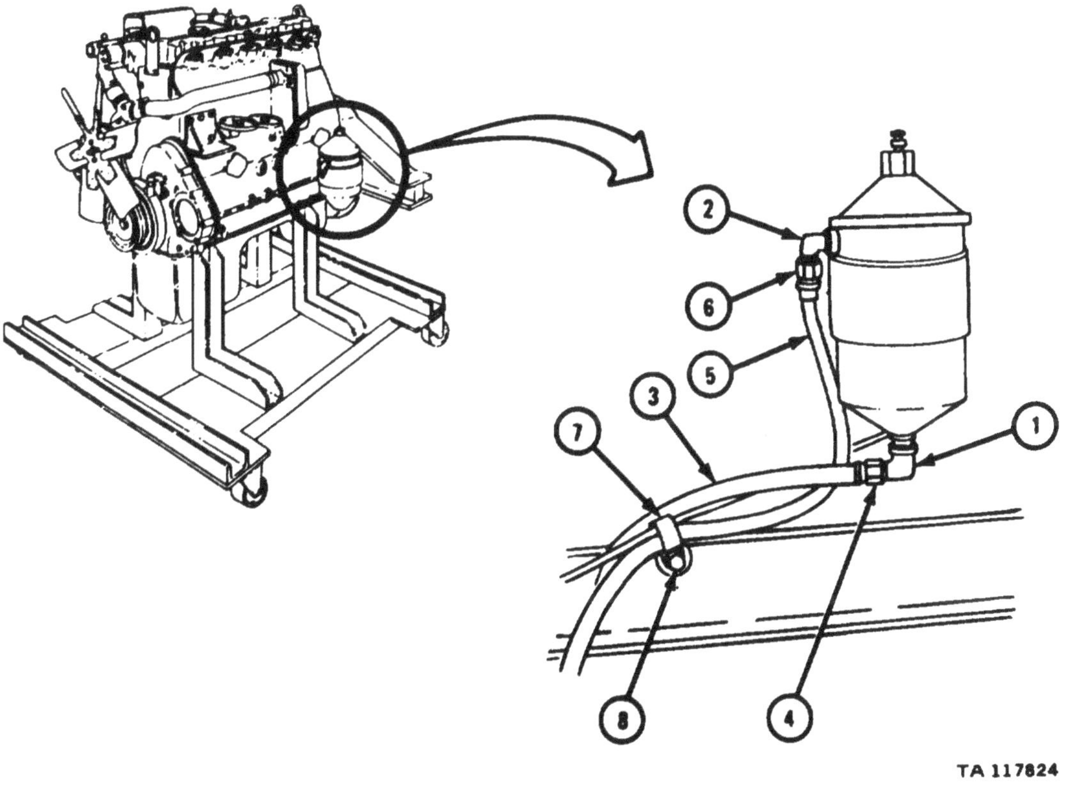

FRAME 3

1. Screw in and tighten elbow (1).
2. Put pressure oil hose (2) in place. Screw in and tighten fitting nut (3).

IF ENGINE HAS FUEL INJECTION PUMP OIL DRAIN HOSE (4), GO TO FRAME 4.
IF ENGINE DOES NOT HAVE FUEL INJECTION PUMP OIL DRAIN HOSE (4),
END OF TASK

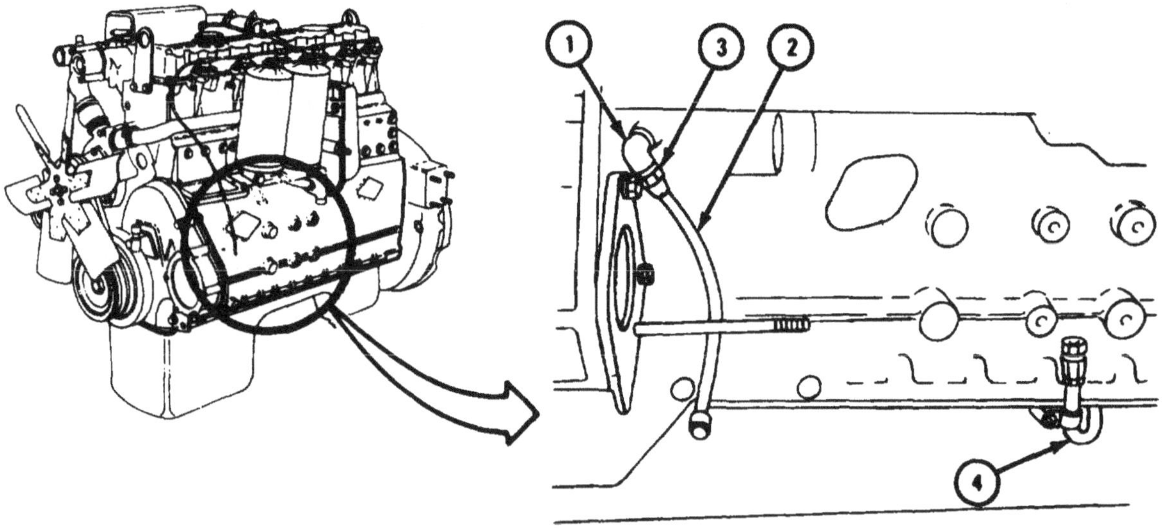

TA 117825

FRAME 4

1. Screw in elbow (1).
2. Put oil drain hose (2) in place. Screw in and tighten fitting nut (3).
3. Put drain hose clamp (4) in place and put in screw and nut (5).

END OF TASK

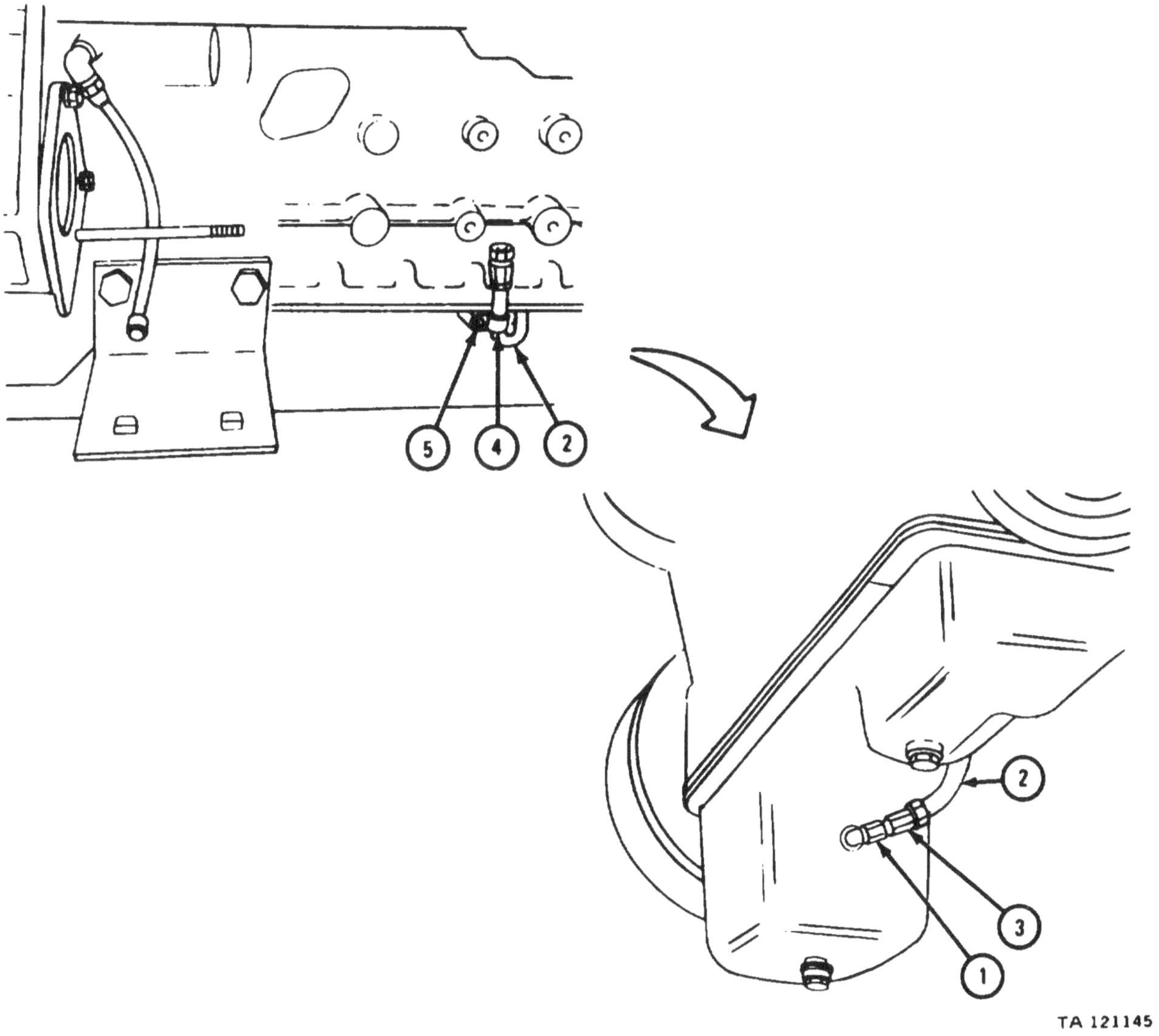

d. Engine LDS-465-2.

FRAME 1

1. Put secondary fuel filter with bracket (1) in place. Put in two screws with washers (2).
2. Put final fuel filter with bracket (3) in place. Put in two screws with lockwashers (4).
3. Put final fuel filter outlet tube (5) in place. Screw in and tighten fitting nut (6).
4. Put final fuel filter outlet tube clamp (7) in place. Put in screw with lockwasher (8).
5. Put drain tube with clamp (9) in place and put in screw with lockwasher (10).

GO TO FRAME 2

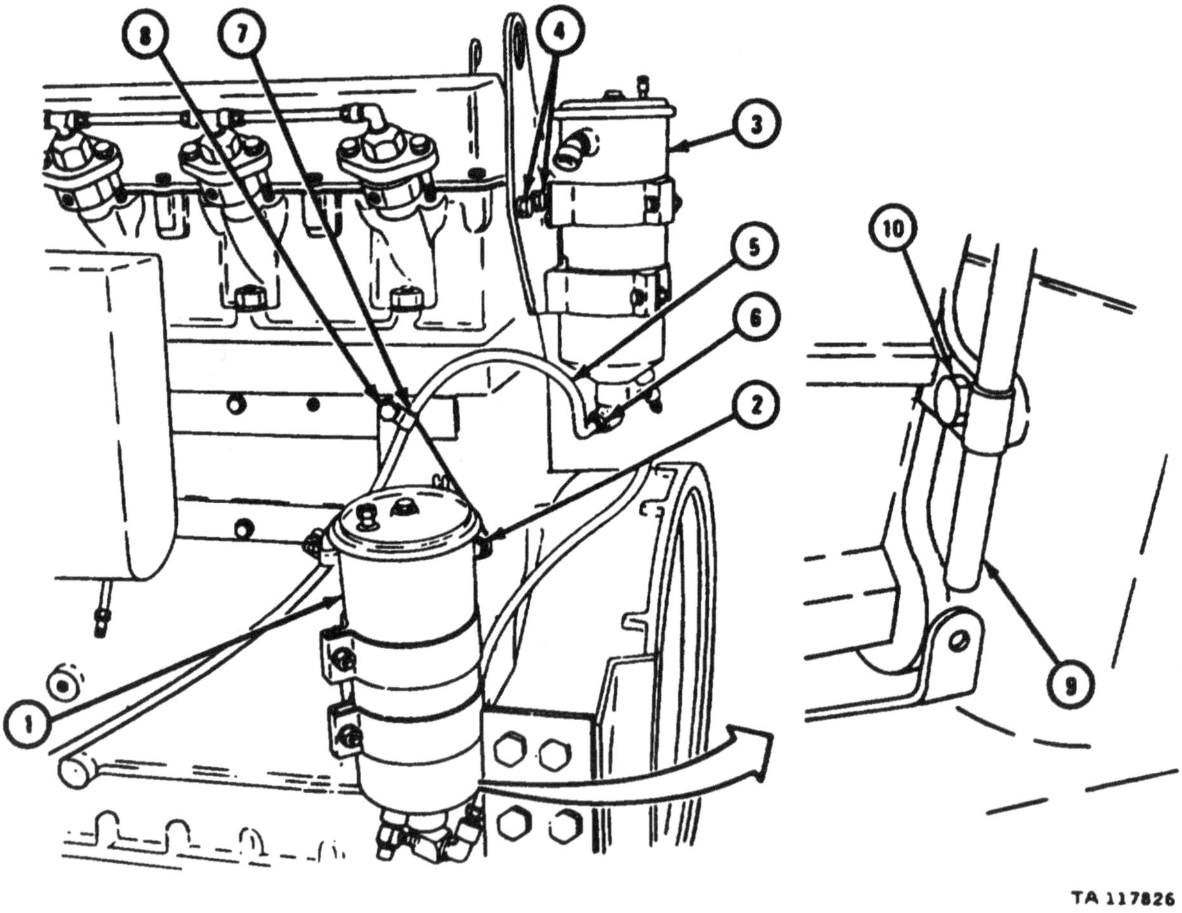

FRAME 2

1. Put final fuel filter-to-secondary fuel filter tube (1) in place. Screw in and tighten two nuts (2).
2. Put hose clamp (3) in place and put in screw with lockwasher (4).
3. Put secondary fuel filter inlet tube (5) in place. Screw in and tighten fitting nut (6) . Put clamp (7) in place and put in screw with lockwasher (8).
4. Push on drain tube (9).

GO TO FRAME 3

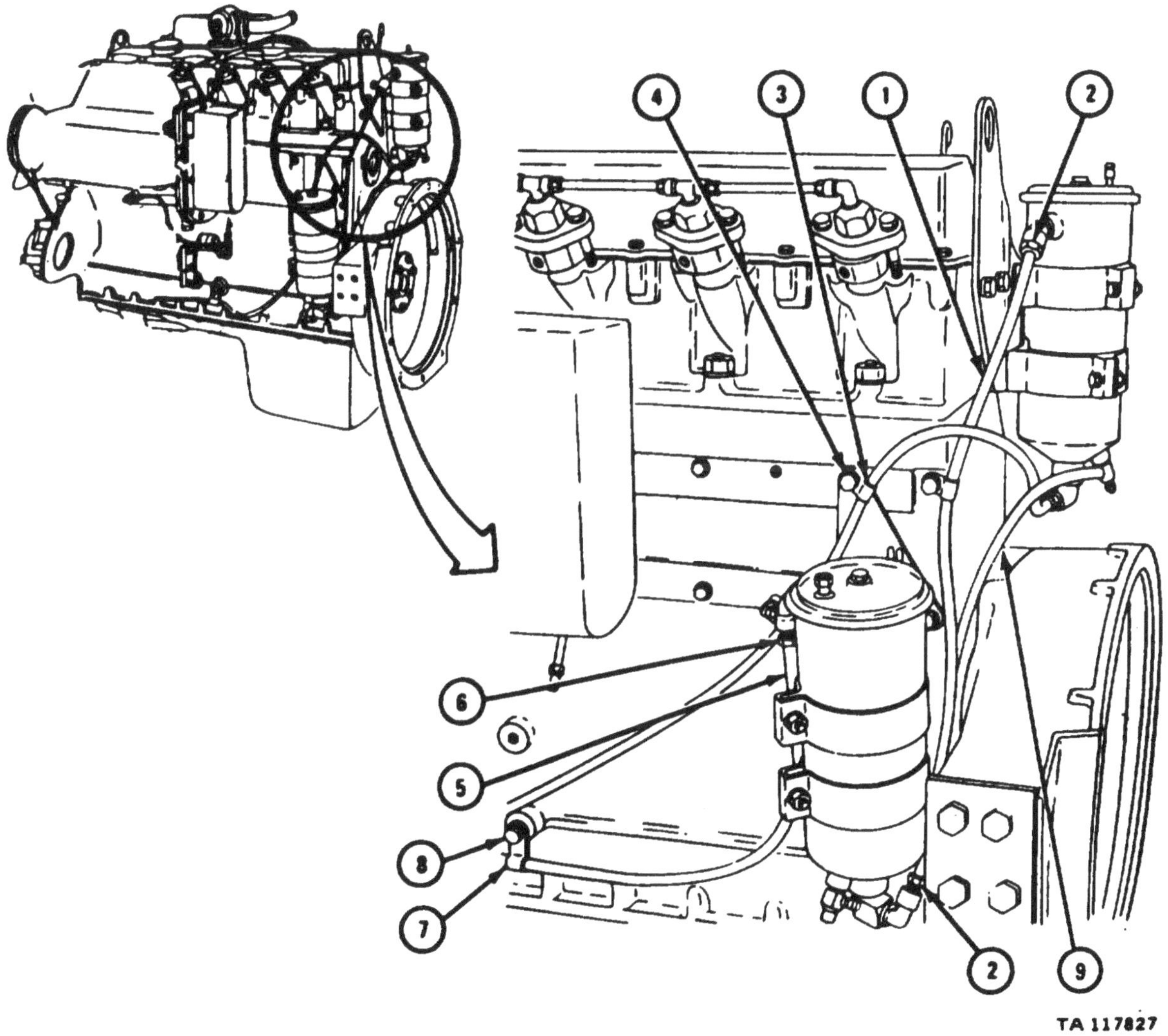

FRAME 3

1. Screw in pipe elbow (1).
2. Put pressure oil hose (2) in place. Screw in and tighten nut (3) .

END OF TASK

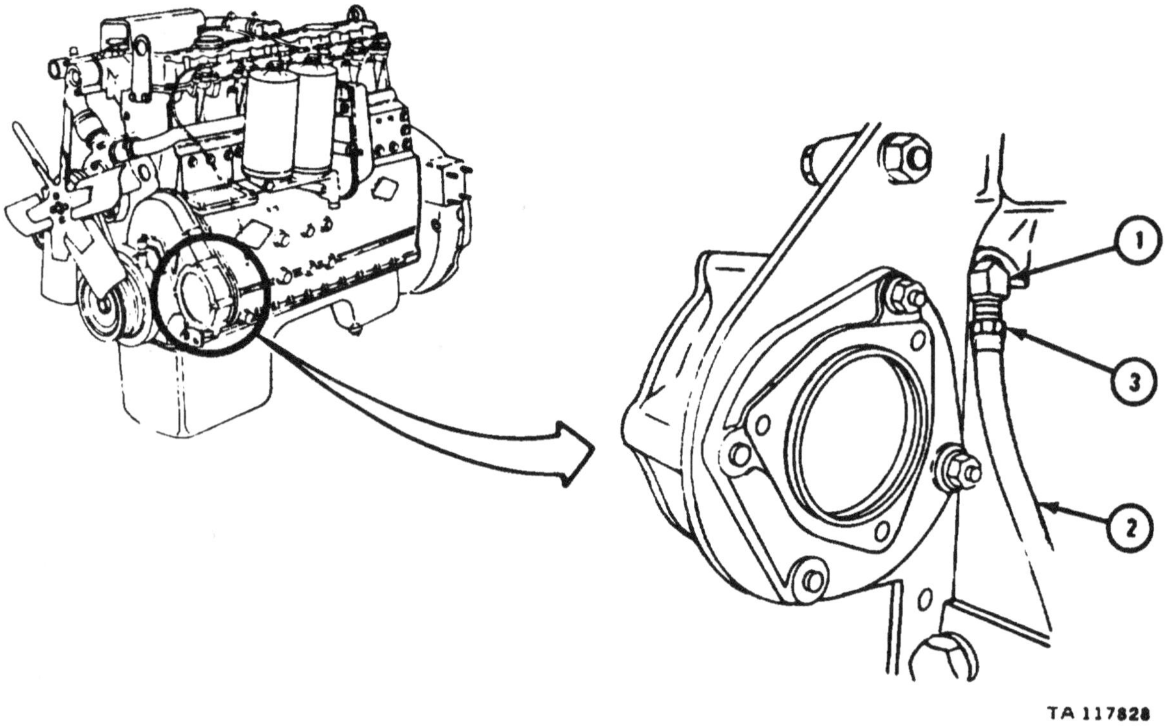

5-32. HYDRAULIC PUMP ASSEMBLY OR FUEL INJECTOR PUMP DRIVE GEAR ACCESS COVER.

a. Engines LDS-465-1, LDS-465-1A, and LDS-465-2.

FRAME 1

1. Put gasket (1) in place. Put hydralic pump (2) in place. Put in two screws (3).
2. Put on three nuts with lockwashers (4).

GO TO FRAME 2

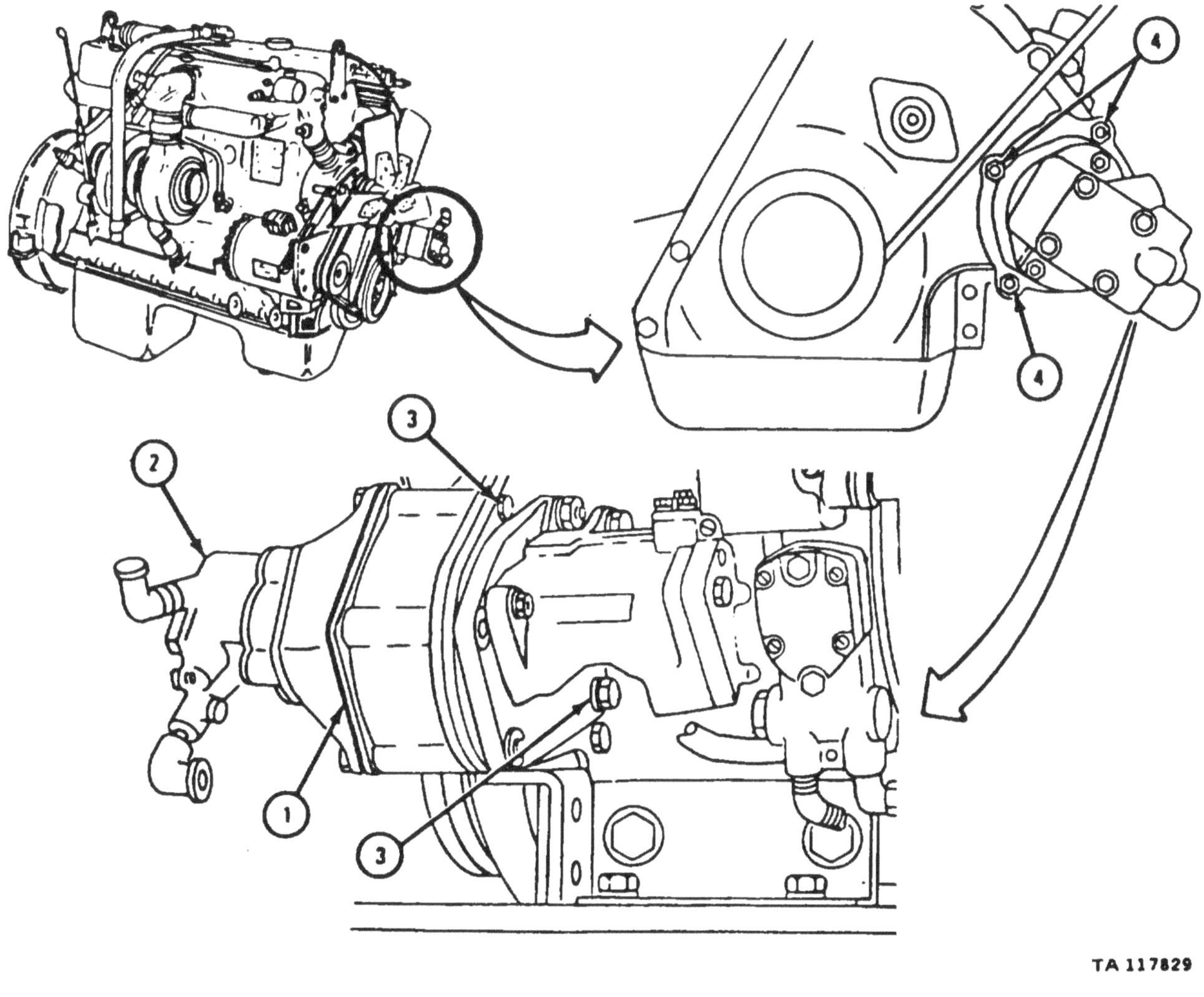

FRAME 2

1. Put on two nuts with lockwashers (1).

NOTE

If working on engine LDS-465-2, do step 2.

2. Put two hoses (2) on two hydraulic tubes (3). Tighten two hose clamps (4).

END OF TASK

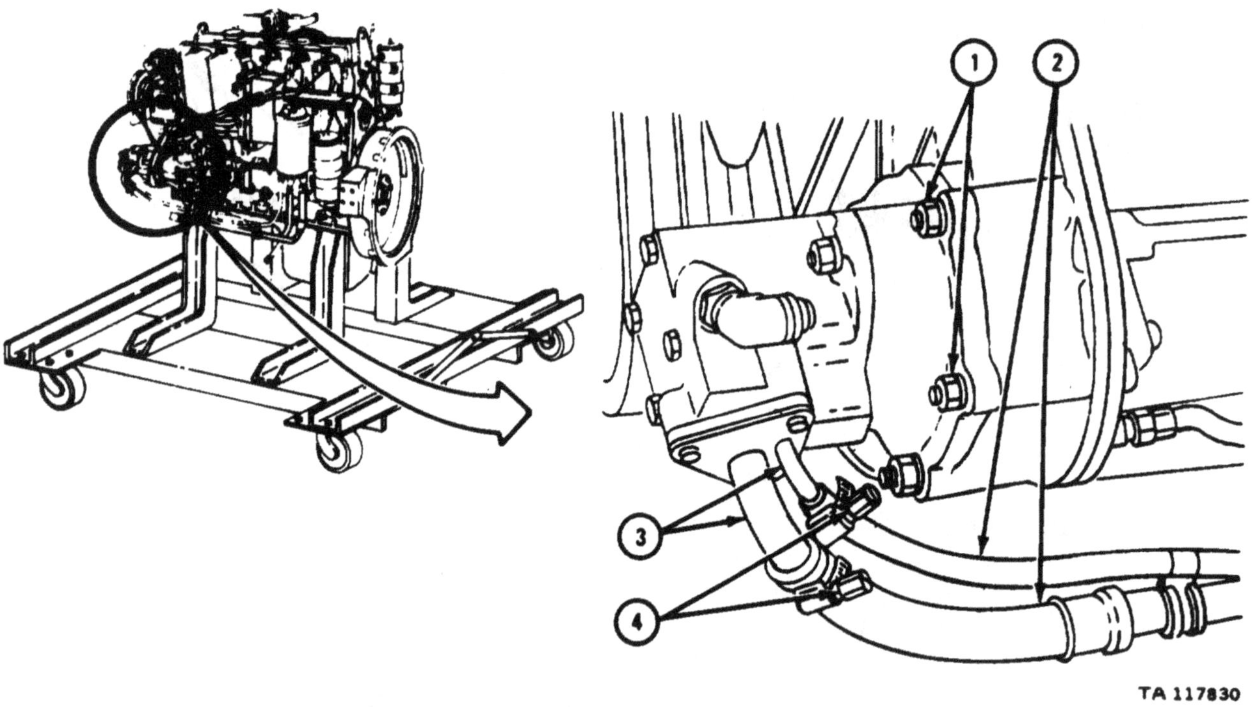

b. Engines LD-465-1, LD-465-1C, and LDT-465-1C.

FRAME 1

1. Put access cover gasket (1) in place. Put access cover (2) in place.
2. Put in two screws (3), and put on five nuts with lockwashers (4).

END OF TASK

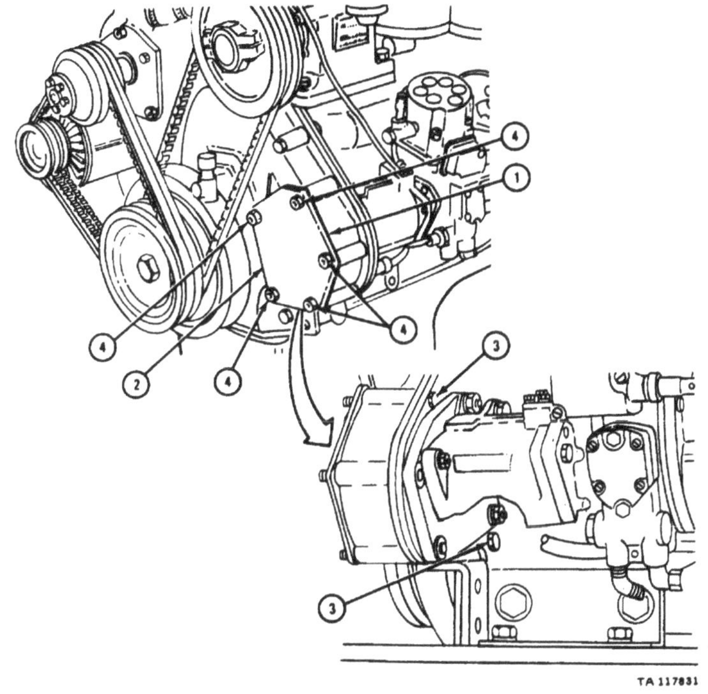

5-33. FUEL INJECTION PUMP ASSEMBLY.

FRAME 1

1. Using engine barring wrench, turn crankshaft damper and pulley (1) to the right until proper timing mark (2) on crankshaft damper and pulley lines up with pointer (3) on timing gear cover (4).

NOTE

There is only one timing mark on crankshaft damper and pulley assembly (2) for engine LDS-465-2.

GO TO FRAME 2

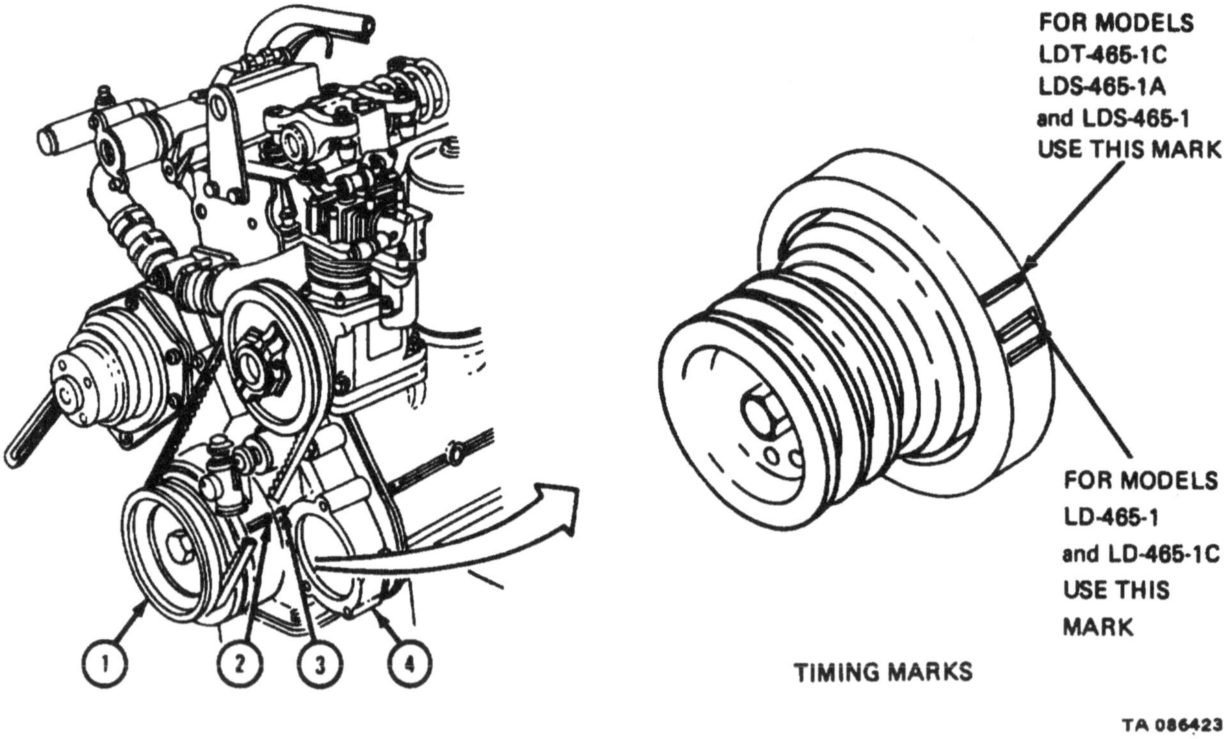

FRAME 2

1. Check that cylinder number one rocker arms (1) have clearance between rocker arm pads (2) and valve stems (3).

2. If there is no clearance between rocker arms (1) and valve stems (3) on cylinder number one, both valves are not closed. Turn crankshaft one ful l turn to the right until proper timing mark and pointer line up again. Refer to frame 1, then do step 1 again.

NOTE

> If there is still no valve clearance, make sure you are using correct timing mark for engine you are working on. Refer to frame 1. If correct timing mark was used and there is no clearance on either valve, refer to para 5-16 and check valve adjustment.

GO TO FRAME 3

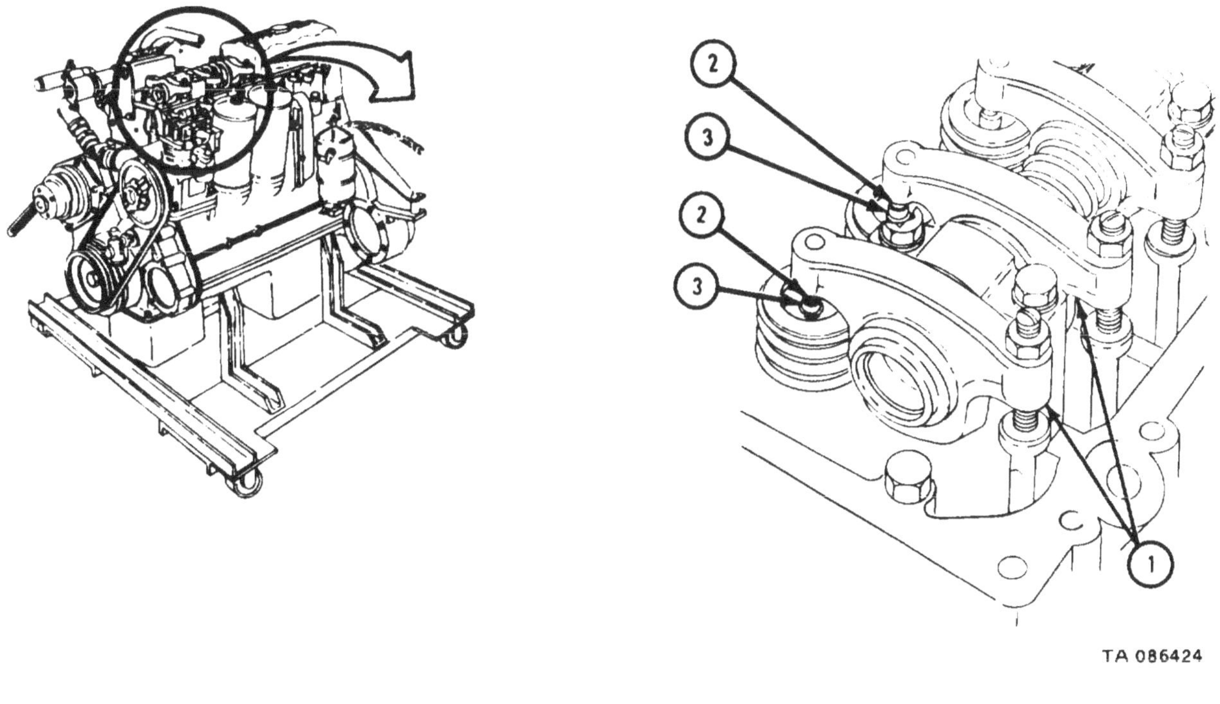

FRAME 3

NOTE

This task is shown for engines LD-465-1, LD-465-1C, and LDT-465-1C. This task is the same for engines LDS-465-1, LDS-465-1A, and LDS-465-2 .

1. Slide fuel injection pump drive hub (1) onto fuel pump shaft (2).
2. Put on and hand tighten nut and lockwasher (3).
3. Join holding wrench (4) to fuel injection pump drive gear hub (1) using two 3/8 x 1 1/4-inch screws (5) as shown. Set fuel injection pump (6) on work-bench as shown . Operating lever (7) must face up.
4. Hold holding wrench (4) so fuel injection pump drive gear hub (1) does not turn. Using torque wrench , tighten nut (3) to 66 to 71 pound-feet.
5. Take out two screws (5) and take off holding wrench (4).

IF WORKING ON ENGINE LDS-465-2, GO TO FRAME 4.
IF WORKING ON ANY OTHER ENGINE, GO TO FRAME 5

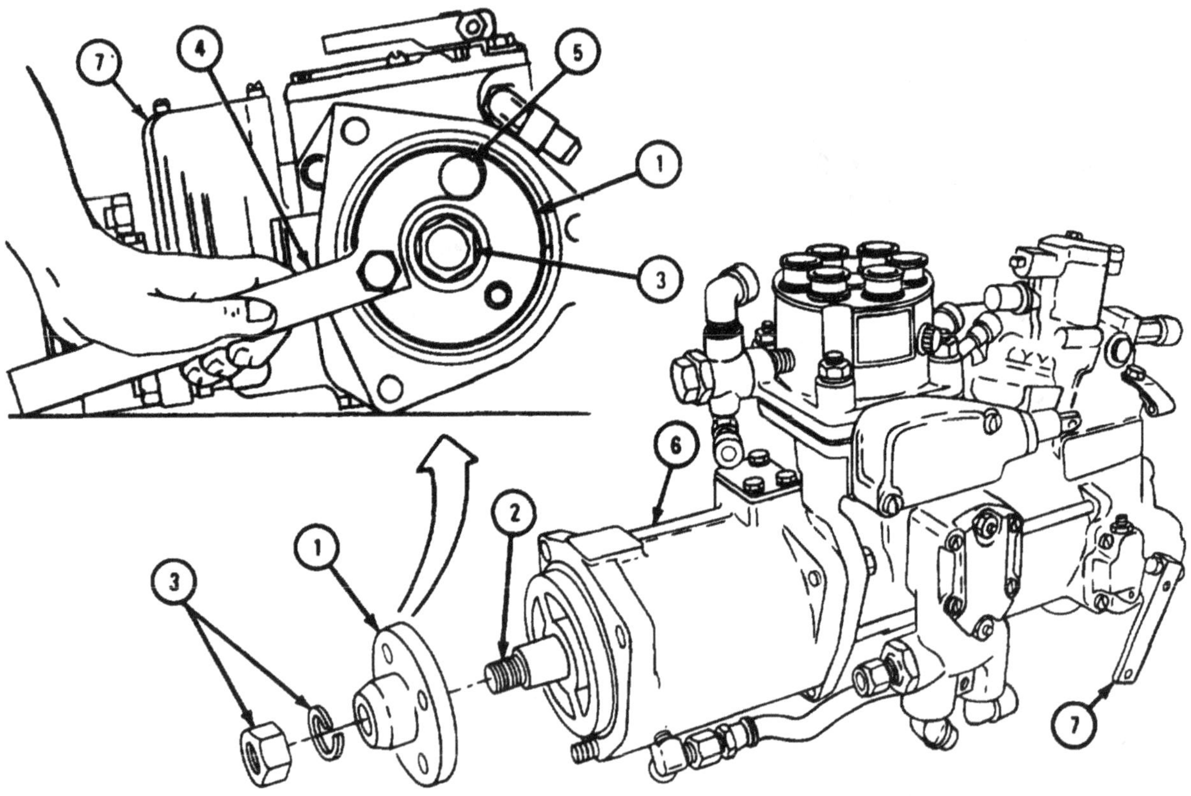

FRAME 4

1. Take out four screws and lockwashers (1) and take off fuel pump automatic timing device cover (2) . Take off automatic timing device cover gasket (3).
2. Take out two sockethead screws and lockwashers (4) and take off fuel shutoff solenoid (5) . Take off fuel shutoff solenoid gasket (6).

GO TO FRAME 6

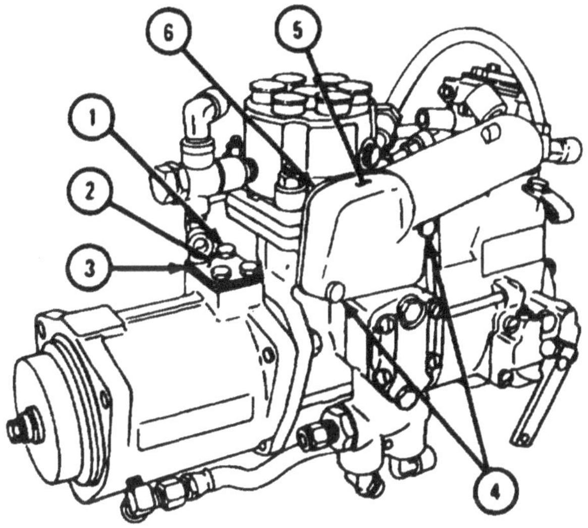

TA 086426

FRAME 5

NOTE

This task is shown for engines LD-465-1, LD-465-1C, and LDT-465-1C . This task is the same for engines LDS-465-1 and LDS-465-1A.

1. Take out four screws and lockwashers (1) and take off fuel pump automatic timing device cover (2) . Take off automatic timing device cover gasket (3).
2. Take out two fillister head screws and lockwashers (4) and take off timing window cover (5) . Take off timing window cover gasket (6).

GO TO FRAME 6

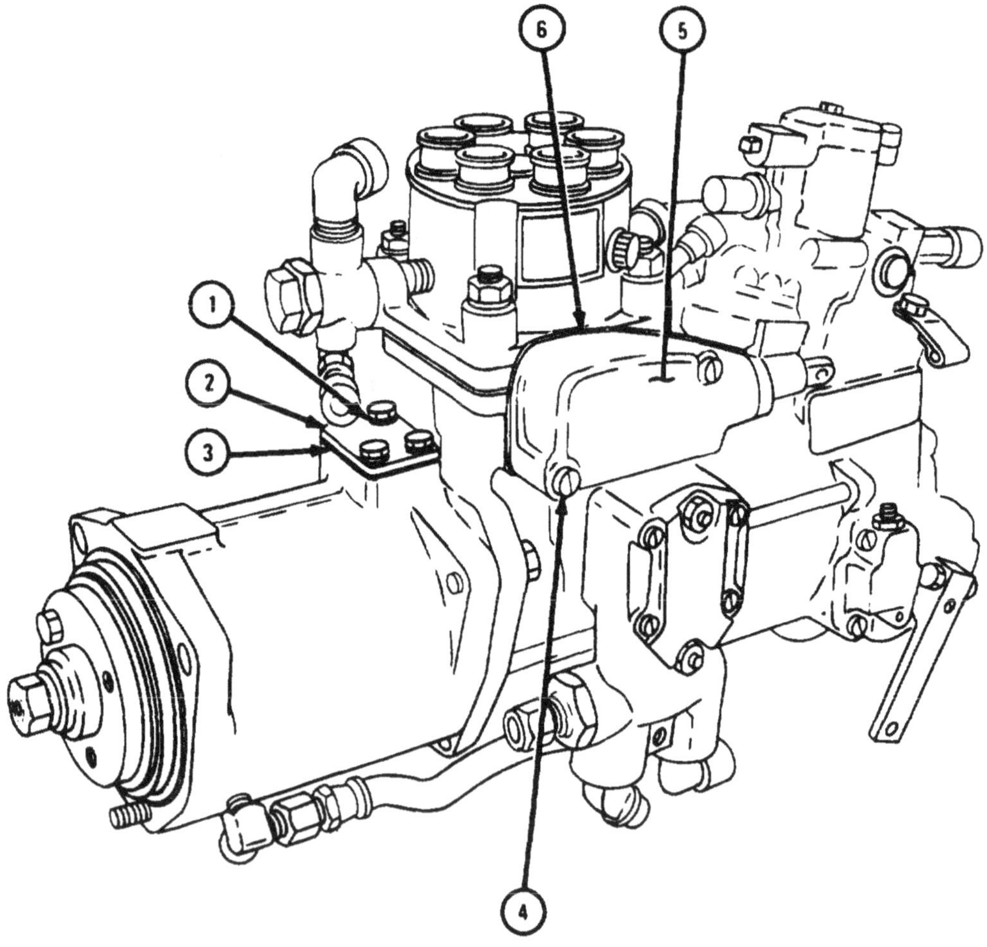

FRAME 6

NOTE

This task is shown for engines LD-465-1, LD-465-1C, and LDT-465-1C . This task is the same for engines LDS-465-1, LDS-465-1A, and LDS-465-2 .

1. Look in window (1). Turn drive gear hub (2) until timing mark (3) lines up with pointer (4) .
2. Look in timing window (5) and check that marked tooth (6) can be seen. If marked tooth (6) cannot be seen, turn drive gear hub (2) exactly one full turn. Marked tooth (6) will now be visible.
3. Look in window 1 again. Check that timing mark (3) is still lined up with pointer (4) . If they are not lined up, turn drive gear hub (2) until they are.
4. Look in timing window (5). Check that marked tooth (6) on plunger gear (7) is one tooth to right of pointer (8).
5. If marked tooth (6) is not one tooth to right of pointer (8), plunger gear (7) must be timed to fuel injection pump timing, device. Refer to TM 9-2910-226-34&p.

IF WORKING ON ENGINE LDS-465-1, GO TO FRAME 7.
IF WORKING ON ANY OTHER ENGINE, GO TO FRAME 9

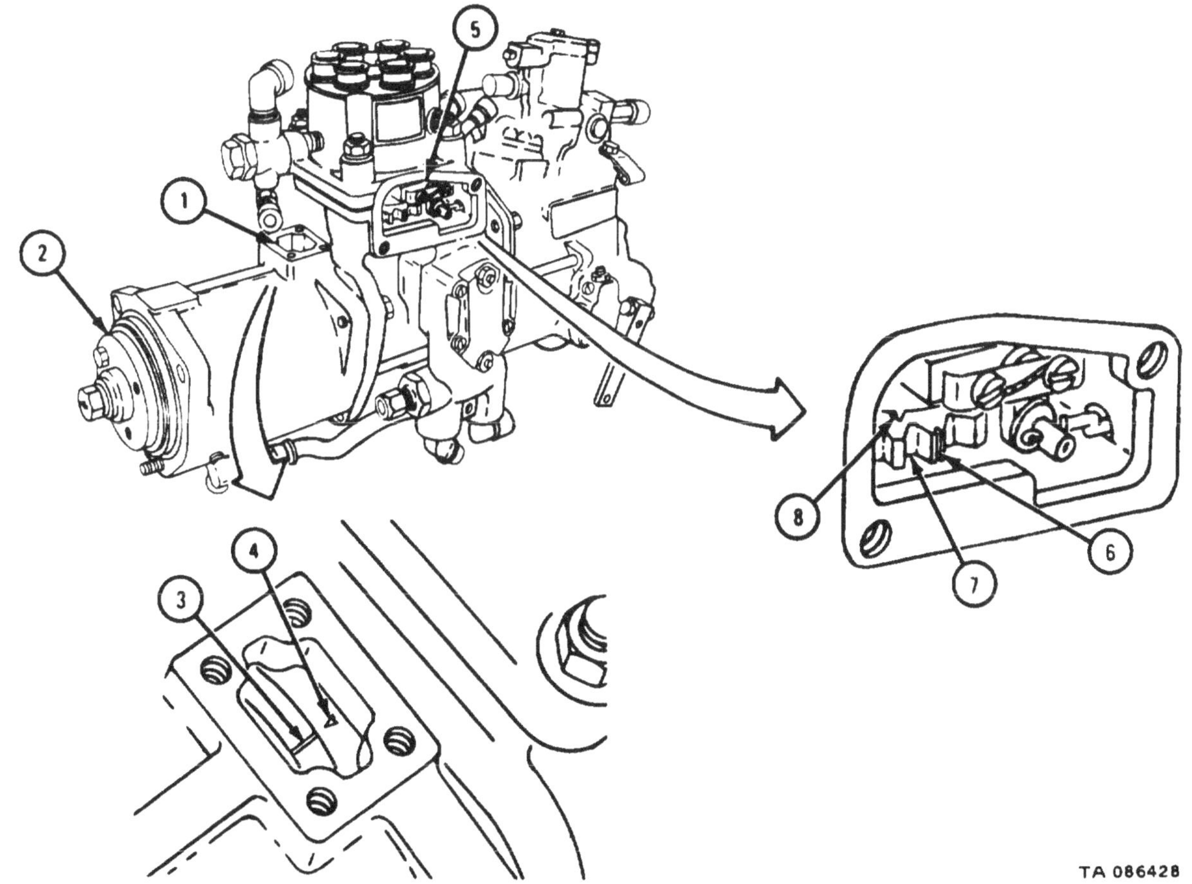

FRAME 7

1. Check whether there is a captive screw (1) in bottom mounting hole of fuel pump (2).
2. Check whether there is a fuel pump mounting stud (3) on engine.
3. If fuel pump (2) has captive screw (1) and engine has fuel pump mounting stud (3), take out mounting stud.

NOTE

New or rebuilt fuel pumps (2) come with new preformed packing (4) .

4. If fuel pump (2) does not have preformed packing (4), put it on.

NOTE

If mounting stud (3) was taken out and thrown away, go to frame 9.

Soldier A 5. If fuel pump (2) does not have captive screw (1) and engine has fuel pu m' mounting stud (3), slide pump over mounting stud and hold it against mounting adapter (5). Mounting stud should fit through bottom mounting hole on fuel pump. Mounting holes (6) on pump should line up with mounting holes (7) on mounting adap- ter.

GO TO FRAME 8

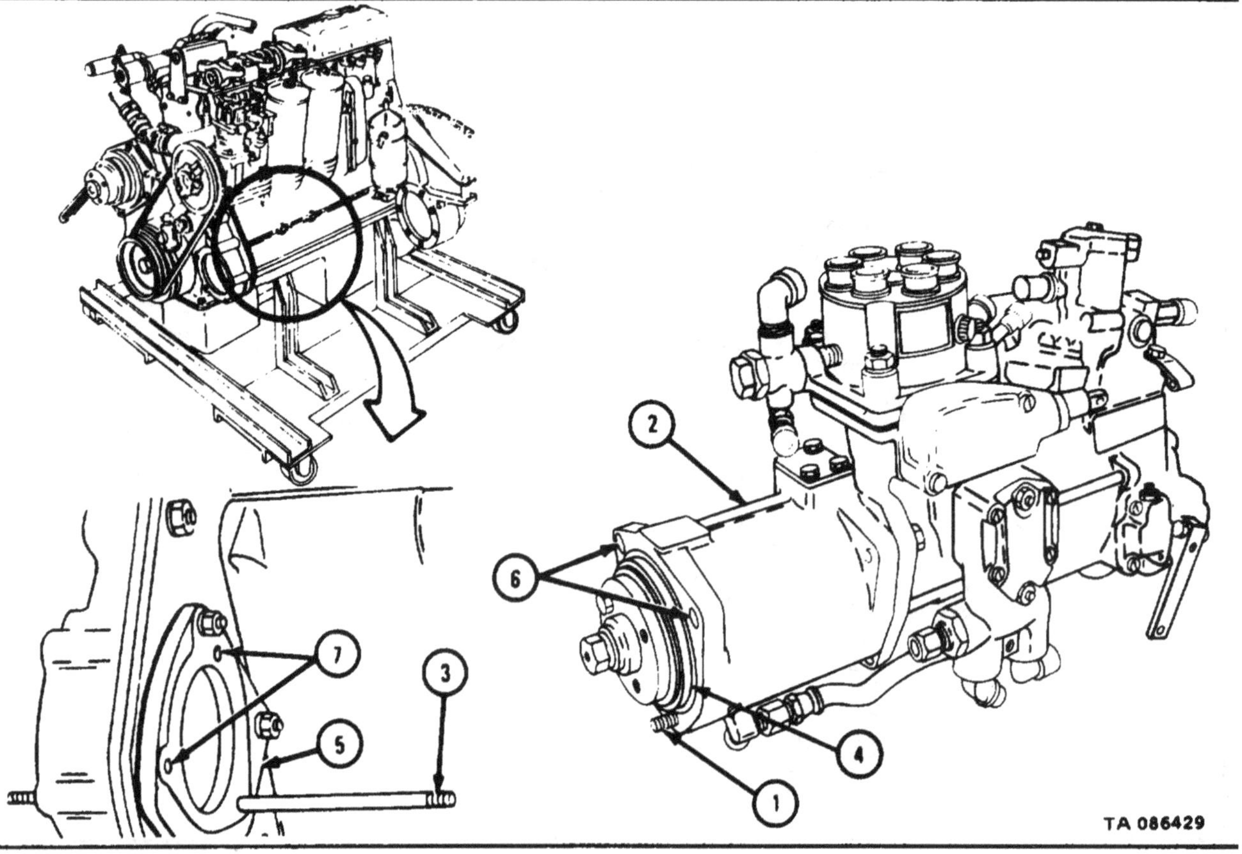

FRAME 8

Soldier B 1 . Put on nut and lockwasher (1).

2. Put in two screws and lockwashers (2).

Soldier A 3 . Let goof pump assembly (3) .

GO TO FRAME 10

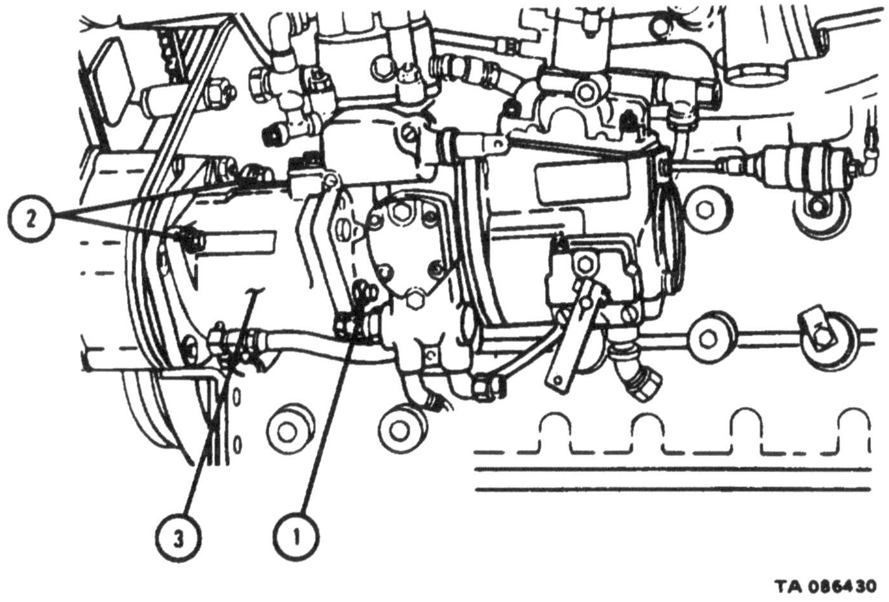

FRAME 9

NOTE

Fuel pump assembly (1) shown is on engines LD-465-1, LD-465-1C, and LDT-465-1C. This task is the same for engines LDS-465-1, LDS-465-1A, and LDS-465-2 .

Soldier A 1. Hold fuel pump assembly (1) against mounting adapter (2) on engine as shown so that mounting holes in fuel pump line up with screw holes in mounting adapter.

Soldier B 2. Tighten captive screw (3) .

3. Put in two screws and lockwashers (4).

Soldier A 4. Let go of fuel pump assembly (1).

GO TO FRAME 10

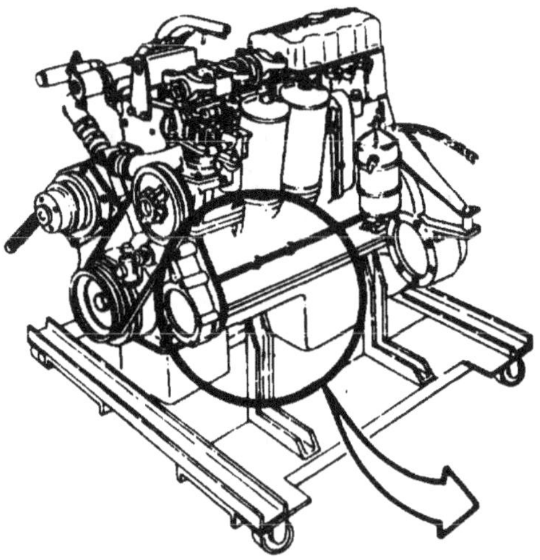

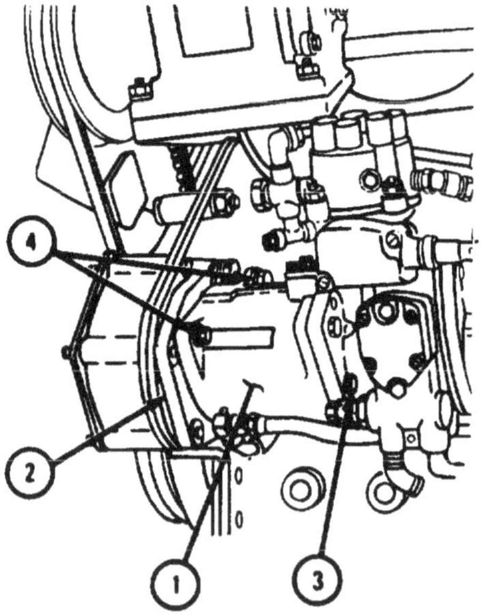

TA 086431

FRAME 10

NOTE

This task is shown for engines LD-465-1, LD-465-1C, and LDT-465-1C . This task is the same for engines LDS-465-1, LDS-465-1A, and LDS-465-2 .

1. Hold fuel pump mounting bracket (1) against engine so that holes in bracket line up with mounting holes on engine.
2. Put in two screws and lockwashers (2).
3. Hold lockplate (3) against mounting bracket (1) as shown.
4. Put in two screws (4).
5. Bend up tabs of lockplate (3) so screws (4) do not loosen.

GO TO FRAME 11

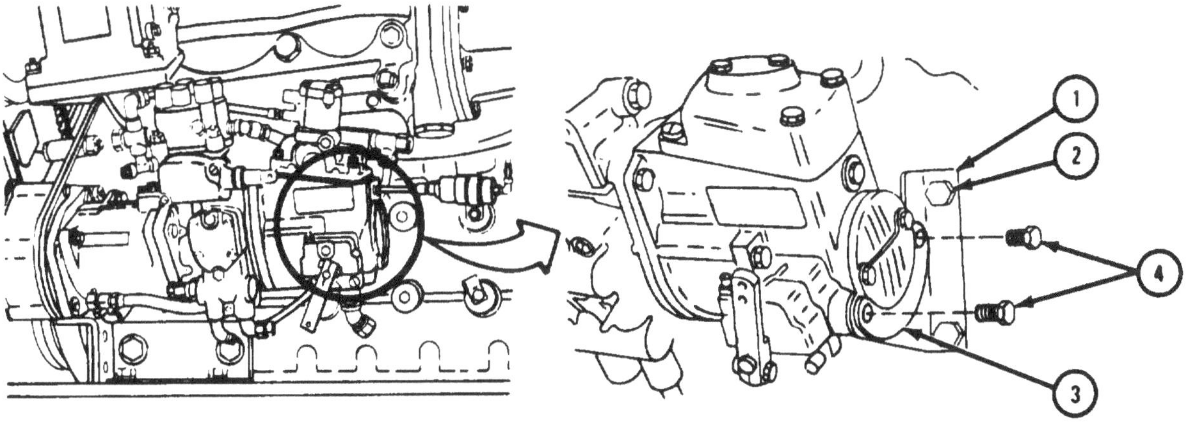

TA 086432

FRAME 11

NOTE

Do not turn fuel pump drive gear hub (1) when putting on fuel pump drive gear (2) and fuel pump drive gear retaining plate (3).

1. Hold fuel pump drive gear (2), on drive gear hub (1). Screw holes (4) in drive gear hub should-be at left end of elongated holes in fuel pump d-rive gear (2) as shown. Fuel pump drive gear must mate with camshaft gear (5) as shown.
2. Hold fuel pump drive gear retaining plate (3) against fuel pump drive gear (2), Holes in retaining plate must line up with screw holes (4) in fuel pump drive gear hub (1) .
3. Put in but do not tighten three screws and lockwashers (6) to hold retaining plate (3) and fuel pump drive gear (2) in place.

GO TO FRAME 12

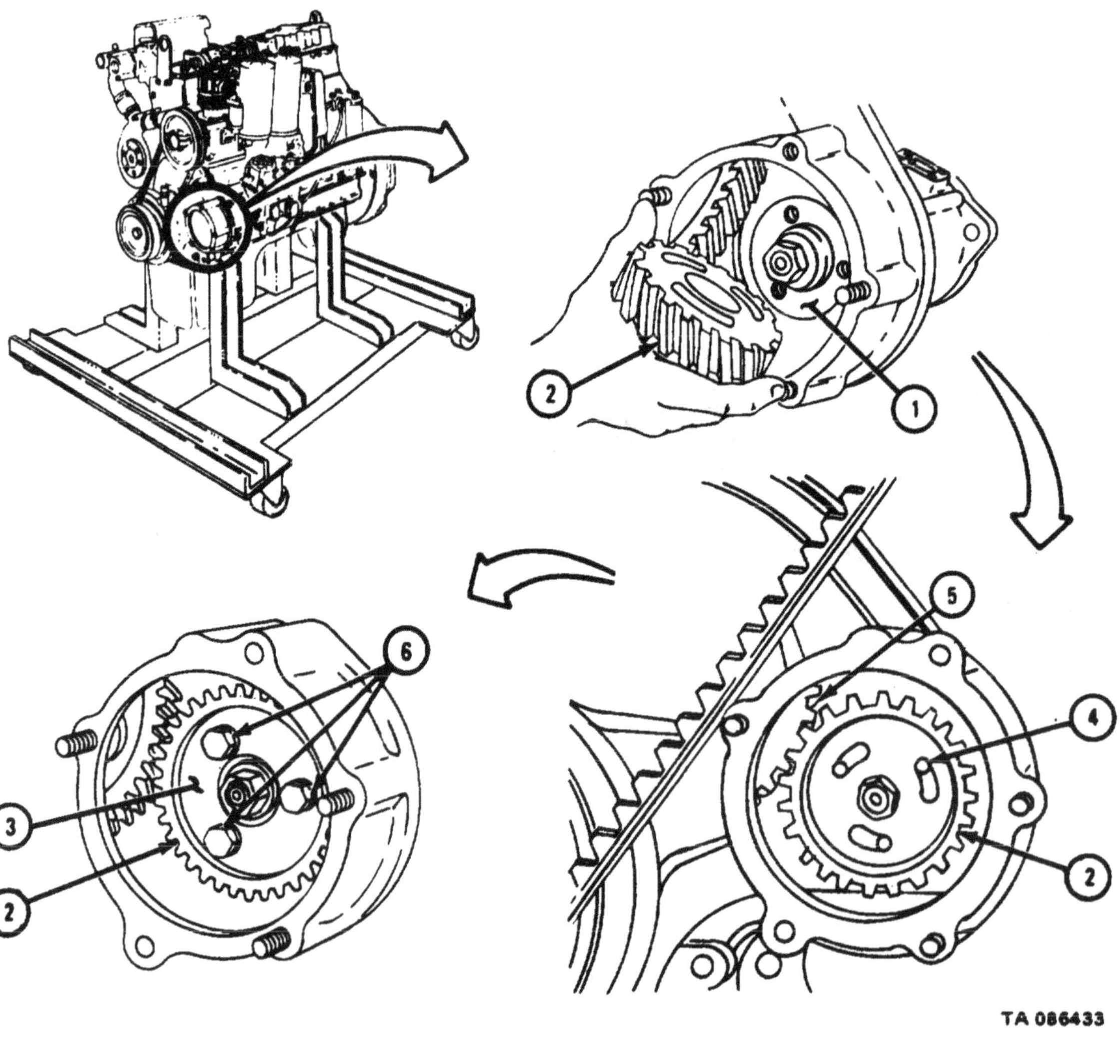

FRAME 12

NOTE

This task is shown for engines LD-465-1, LD-465-1C, and LDT-465-1C . This task is the same for engines LDS-465-1, LDS-465-1A, and LDS-465-2 .

1. Using socket wrench (1) as shown, hold nut at center of fuel pump drive gear (2) .

NOTE

It is hard to keep timing mark (3) lined up with pointer (4) during this task . Engine will not run correctly if timing mark is not kept alined.

2. Look in timing window (5). Check that timing mark (3) is lined up with pointer (4) . If it is not, turn wrench (1) to the right until mark and pointer line up and hold wrench firmly.

NOTE

Elongated holes in fuel pump drive gear (2) will let pump turn about 20° while drive gear is mounted on pump and three screws (7) are loose. If timing mark (3) cannot be lined up with pointer (4) , take off drive gear. Refer to Part 1, para 3-10. Then set pump timing. Refer to frame 6. When pump timing has been set, put back drive gear. Refer to frames 11 and 12.

3. Using torque wrench (6), tighten three screws (7) to 23 to 27 pound-feet while keeping drive gear (2) from turning by holding wrench (1) firmly.

GO TO FRAME 13

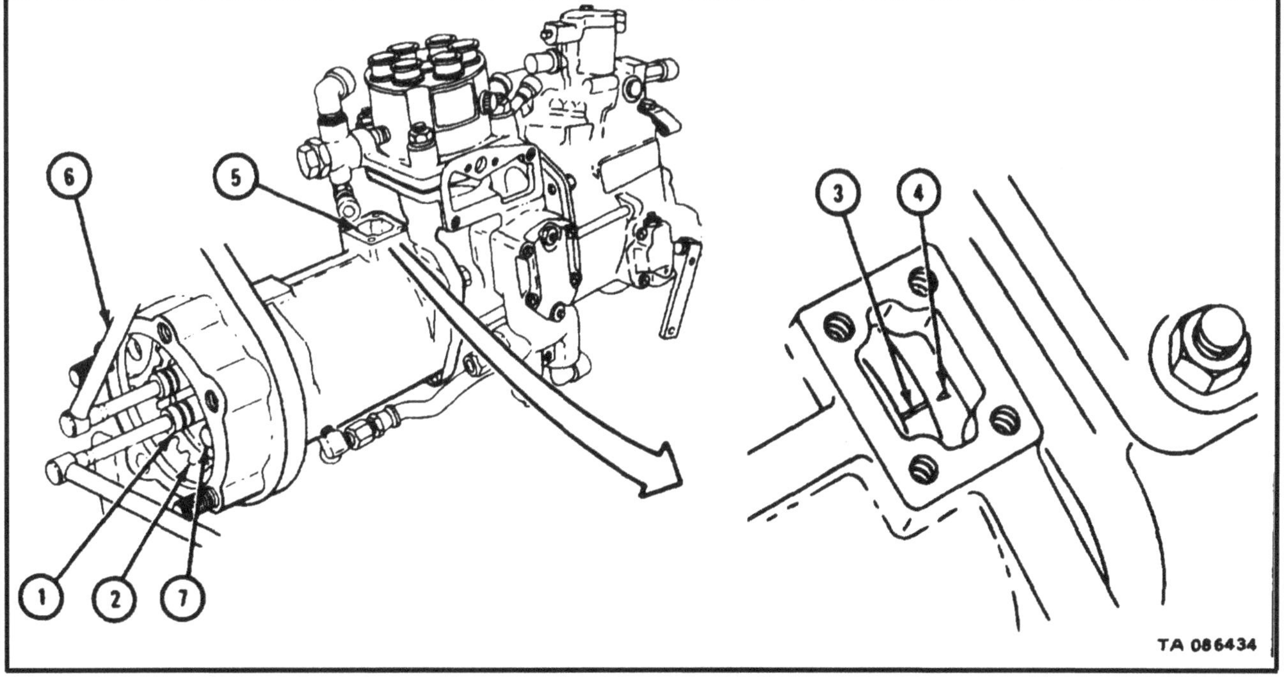

FRAME 1 3

NOTE

Fuel injection pump shown is for engines LD-465-1, LD-465-1C, and LDT-465-1C. The procedure is the same for engines LDS-465-1, LDS-465-1A, and LDS-465-2 .

1. Check that proper timing mark on crankshaft damper and pulley assembly is lined up with pointer on timing gear cover as shown in frame 1. Check that cylinder number one intake and exhaust valves are closed as shown in frame 2.
2. Look in window (1) and check that marked tooth (2) on gear (3) can be seen.
3. If marked tooth (2) on gear (3) cannot be seen through window (1), pump timing is not correct. Take off fuel pump drive gear (4). Refer to Part 1, para 3-10 . Then set and check pump timing. Refer to frame 6 and frames 11 through 13.

GO TO FRAME 14

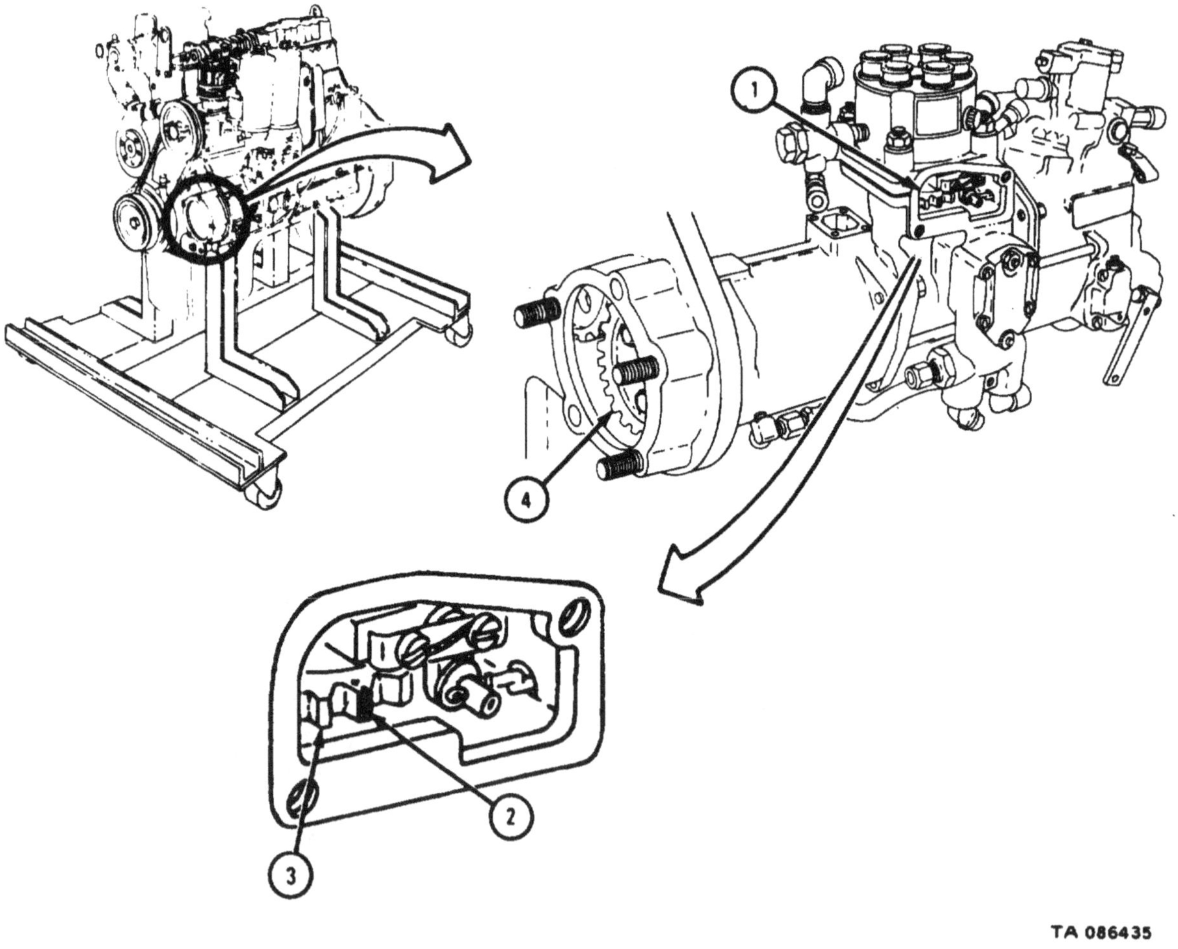

FRAME 14

NOTE

This task is shown for engines LD-465-1, LD-465-1C, and LDT-465-1C . This task is the same for engines LDS-465-1 and LDS-465-1A, and LDS-465-2 .

1. Look in window (1) and check that timing mark (2) lines up with pointer (3). If timing mark lines up with pointer, pump timing is correct. Go to frame 15.
2. Using socket wrench (4) as shown, hold nut at center of fuel pump drive gear (5) . Using socket wrench (6) as shown, loosen but do not take out three screws (7) .
3. Look in window (1). Turn wrench (4) until mark (2) and pointer (3) line up and hold wrench firmly in this position.

NOTE

Elongated holes in fuel pump drive gear (5) will let pump turn about 20° while drive gear is mounted on pump and three screws (7) are loose . If timing mark (2) cannot be lined up with pointer (3), take off drive gear. Refer to Part 1, para 3-10. Then set pump timing. Refer to frame 6 and frames 11 through 14. When pump timing has been set, put back drive gear. Refer to frames 11 and 12.

4. Using torque wrench (6), tighten three screws (7) to 23 to 27 pound-feet while keeping drive gear (5) from turning by holding wrench (4) firmly.
5. Check pump timing again. Refer to frames 13 and 14.

GO TO FRAME 15

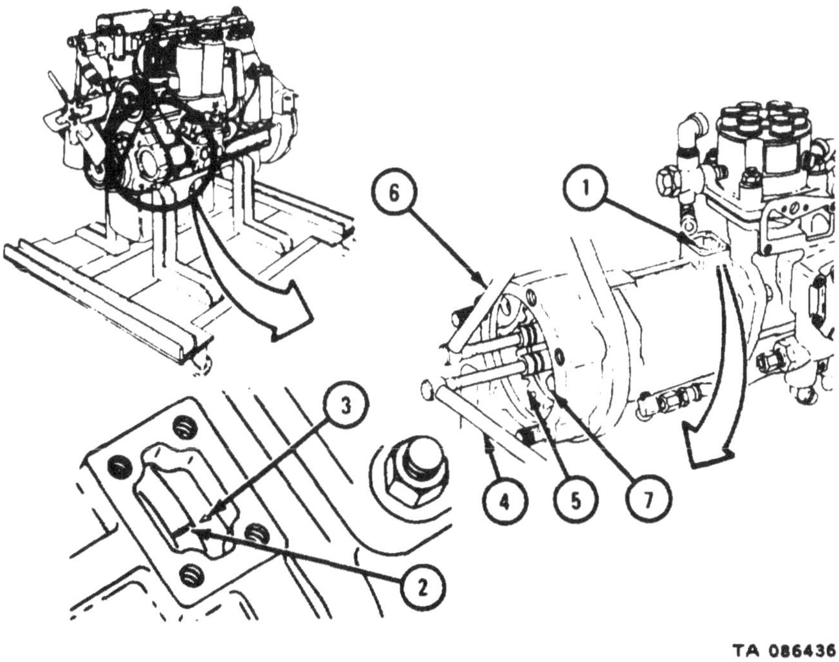

FRAME 1 5

1. Turn crankshaft about 45° to the left as viewed from front of engine. Then turn crankshaft back to the right until proper timing mark on crankshaft vibration damper and pulley lines up with timing pointer on timing gear cover. Refer to frame 1.
2. Check fuel injection pump timing again. Refer to frames 13 and 14.
3. Put automatic timing device cover gasket (1) on automatic timing device window (2) as shown. Put automatic timing device cover (3) on automatic timing device window as shown. Screw in and tighten four screws and lockwashers (4).
4. Put timing window gasket (5) on side of fuel pump (6) as shown.
5. If working on any engine except LDS-465-2, put timing window cover (7) on side of fuel pump (6) as shown. Put in two fillister head screws and lockwashers (8) .
6. If working on engine LDS-465-2, put fuel shutoff solenoid (9) on side of fuel pump (6) . Put in two sockethead screws and lockwashers (10) .

END OF TASK

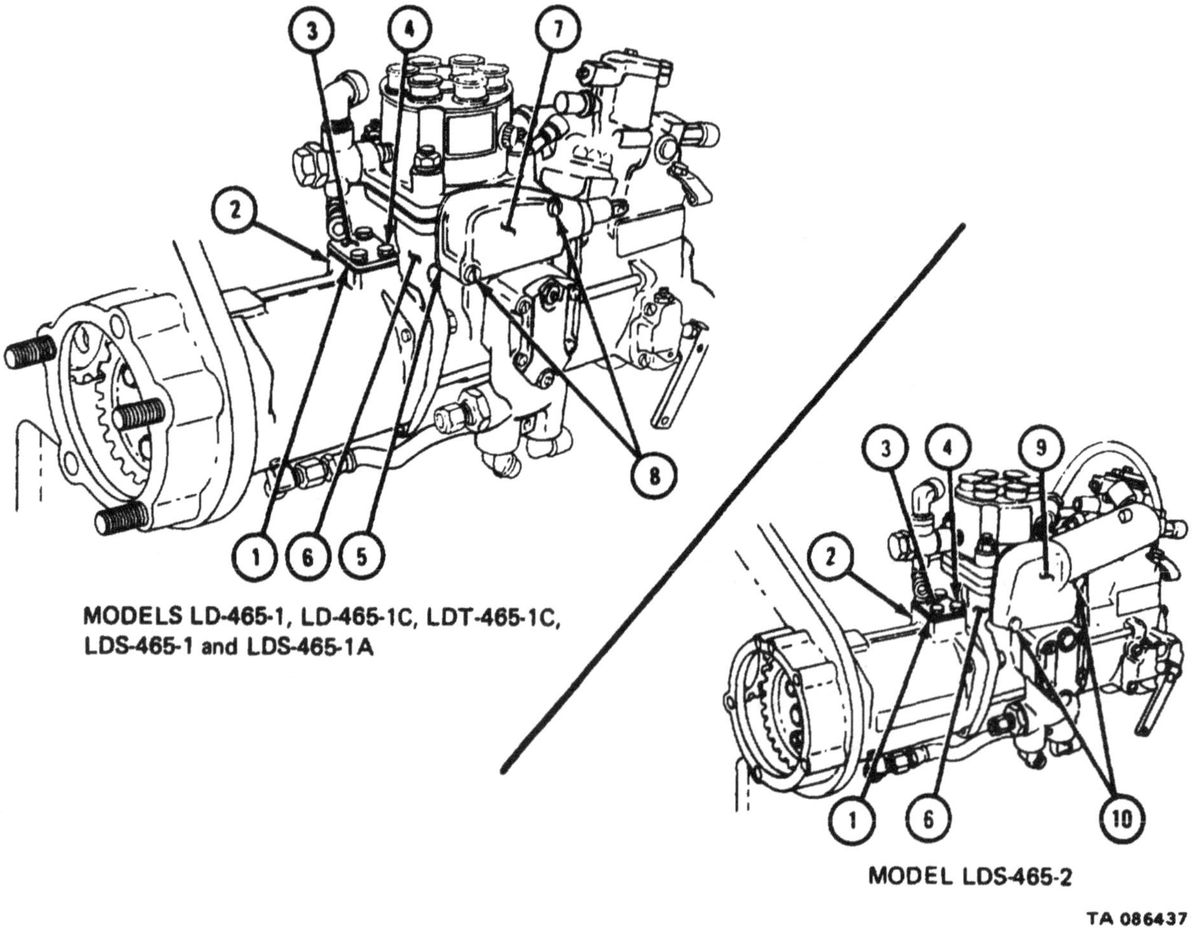

5-34. FUEL INJECTOR NOZZLE, HOLDER ASSEMBLY, AND RETURN TUBE.

FRAME 1

NOTE

Adjusting screw type and shim type fuel injector nozzles can be mixed within the same engine. Plugs put in during removal must be taken out during replacement.

1. Put six gaskets (1) on six injector nozzles (2).
2. Put six injector nozzles (2) into six injector nozzle ports (3).
3. Put six injector nozzle hold down clamps (4) in place. Put in 12 screws with lockwashers (5) . Tighten 12 screws to 150 to 175 pound-feet .
4. Put in five tee fittings (6) as shown.

GO TO FRAME 2

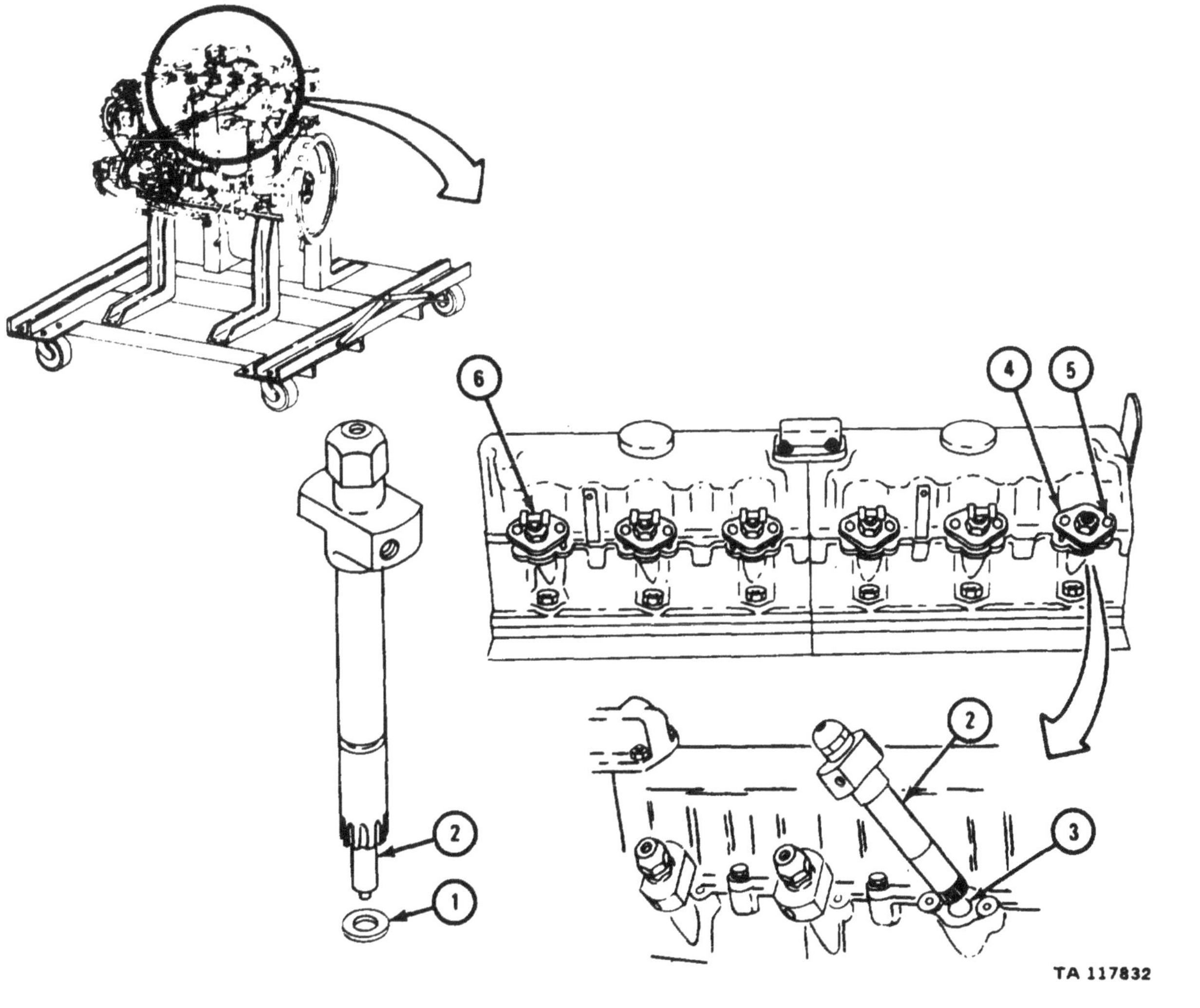

FRAME 2

1. Put in elbow (1) as shown.

CAUTION

Do not turn elbow (1) or tube tee fittings (3) more than 45° or fuel return tubes will break.

2. Turn elbow (1) about 45°. Put one end of fuel return tube (2) into elbow. Turn elbow back straight.

3. Turn tee fitting (3) about 45°. Put other end of fuel return tube (2) into tee fitting. Turn tee fitting back straight and tighten two fitting nuts (4).

4. Do steps 2 and 3 again for fuel return tubes (5, 6, and 7) and tee fittings (8), (9), (10), and (11).

GO TO FRAME 3

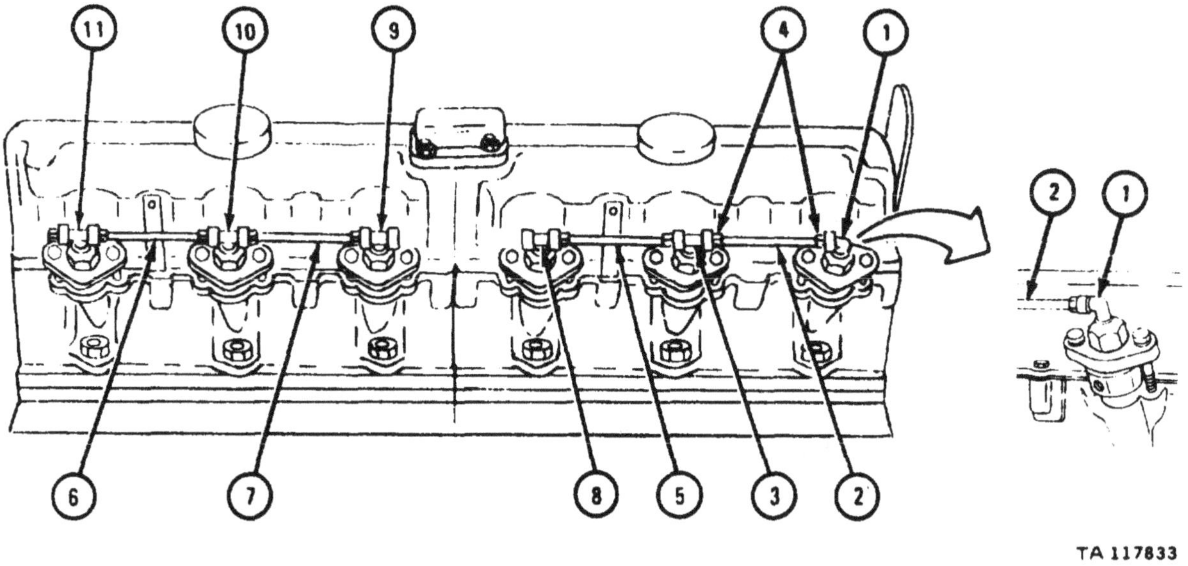

FRAME 3

1. Put two fuel return tubes (1) on tee fitting (2).

CAUTION

Do not turn tube tee fittings (3 and 4) more than 45° or fuel return tubes will break.

2. Turn tee fitting (3) about 45°. Put one fuel return tube free end (1) into tee fitting . Turn tee fitting back straight.
3. Turn tee fitting (4) about 45°. Put other fuel return tube free end (1) into tee fitting. Turn tee fitting back straight.
4. Tighten four fitting nuts (5) .
5. Put flame heater return tube (6) in place. Screw in and tighten fitting nut (7).
6. Put fuel injector to fuel pump return tube (8) in place. Screw in and tighten fitting nut (9) .

END OF TASK

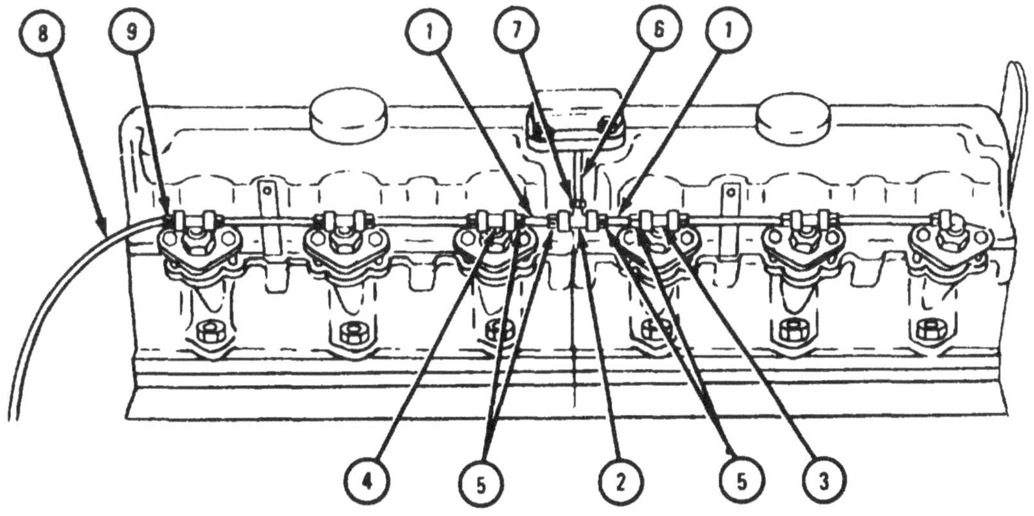

5-35. FUEL INJECTOR TUBE

a. Injector Tube.

FRAME 1

NOTE

Plugs and tags put on injector pumps (1), tubes (2), and injectors (3) during removal must be taken out during replacement.

1. Put six injector tubes (2) in place. Screw in and tighten six injector tube nuts (4) and six injector tube nuts (5).
2. Slide six dust caps (6) down over six injector tube nuts (4).

END OF TASK

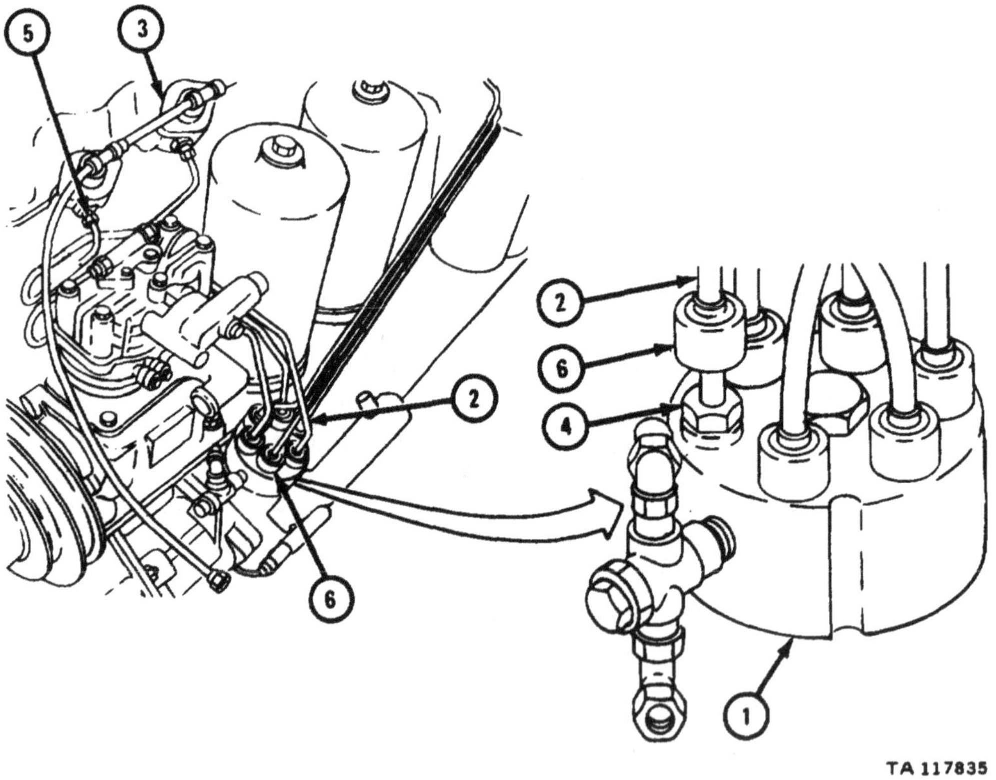

b. Injector Tube Clamp.

FRAME 1

NOTE

Some engines have six tube clamps and some have two tube clamps. This task is shown for engines with six tube clamps. Two screws (1) were put back after tubes were taken out.

1. Take out two screws with flat washers (1). Put two brackets (2) in placse. Put in two screws with flat washers.

NOTE

Some engines do not have bracket (3).

2. Put bracket (3) in place. Put in screw with lockwasher (4). Put clamp (5) in place. Put in two screws with flat washers and locknuts (6).

FOR ALL ENGINES EXCEPT LDS-465-2, GO TO FRAME 2.
FOR ENGINE LDS-465-2, GO TO FRAME 3

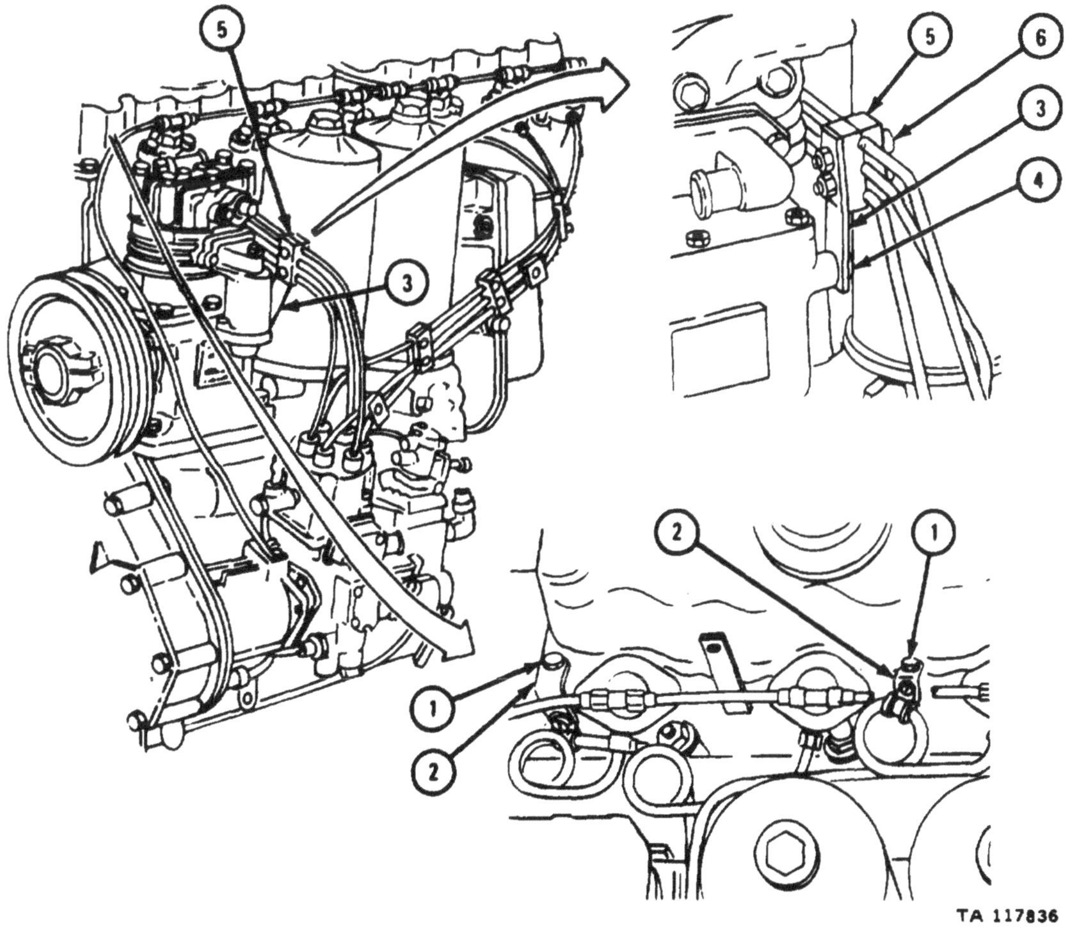

FRAME 2

1. Put clamp (1) in place. Put on nut with lockwasher (2).
2. Put clamp (3) in place. Put in two screws with flat washers and locknuts (4).
3. Put two clamps (5) in place. Put in two screws with locknuts (4).
4. Put two clamps (6) in place. Put in four screws with flat washers and locknuts (4).

END OF TASK

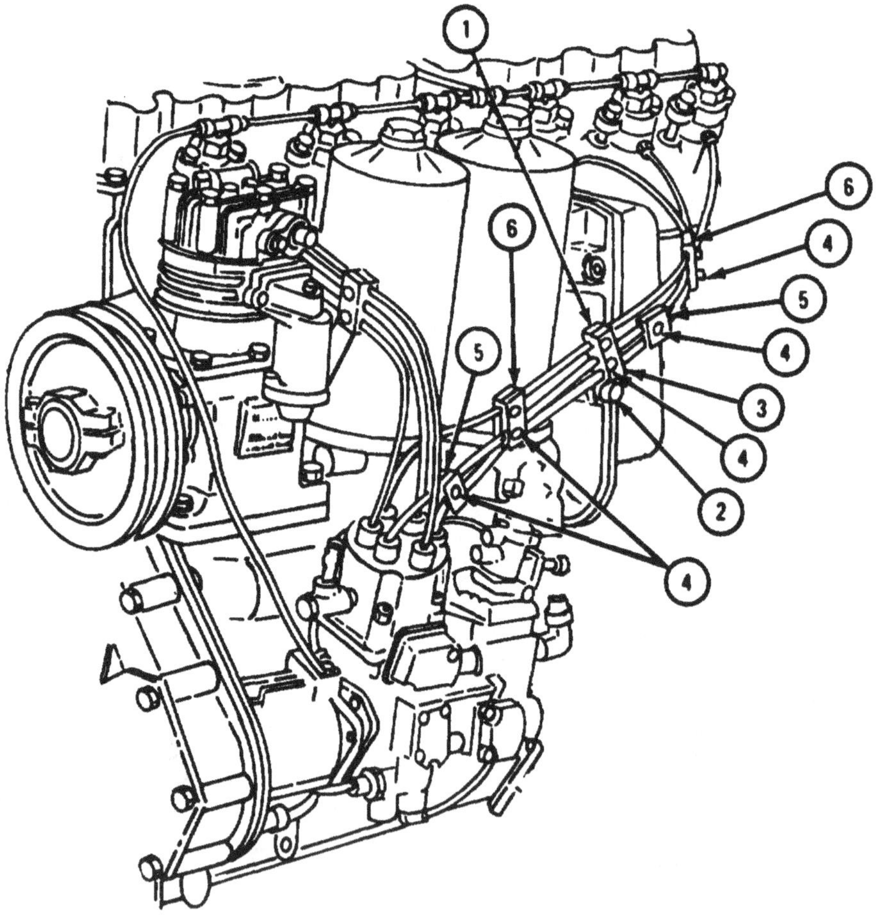

FRAME 3

1. Put two clamps (1) in place. Put in two screws with flat washers and locknuts (2).
2. Put three clamps (3) in place. Put in six screws with flat washers and locknuts (4).

END OF TASK

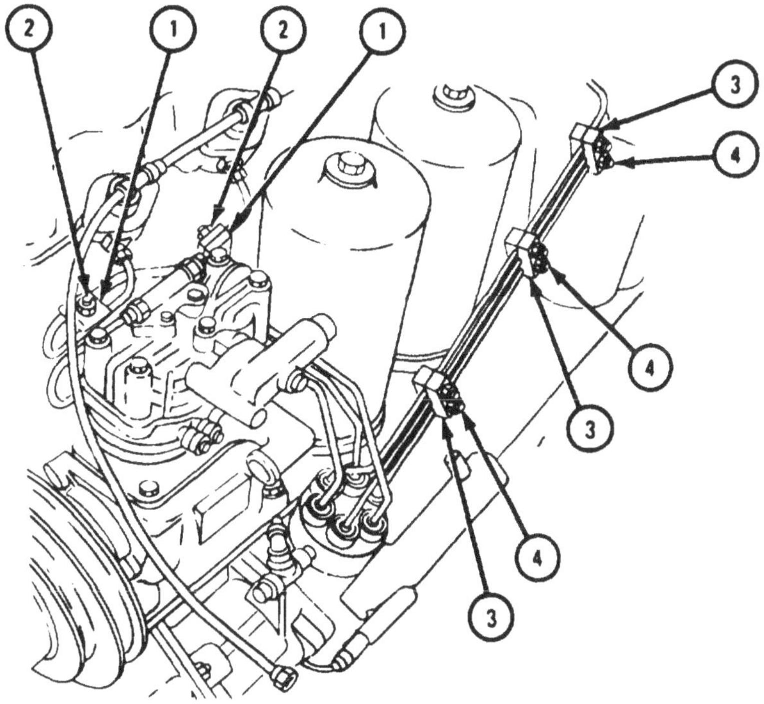

TA 117838

5-36. TURBOCHARGER ASSEMBLY.

a. Engine LDT-465-1C .

FRAME 1

CAUTION

Do not use open end wrench to put flame heater fuel inlet adapter (1) or flame heater fuel return adapter (2) into nozzle and check valve assembly (3). Adapters may be damaged. Use only tubing wrench or box wrench.

1. Put fuel inlet adapter (1) into nozzle and check valve assembly (3).
2. Put fuel return adapter (2) into nozzle and check valve assembly (3).

NOTE

Make sure that locknut (4) is put onto nozzle and check valve assembly (3) before nozzle and check valve assembly is put into intake manifold elbow (5).

3. Hand tighten nozzle and check valve assembly (3) into intake manifold elbow (5). Tighten locknut (4) against intake manifold elbow. Adapters (1 and 2) should face as shown.
4. Put spark plug gasket (6) over end of spark plug (7) as shown. Put in spark plug .

NOTE

Make sure that hose (8) is on end of intake elbow (5) and that both hose clamps (9) are on hose.

GO TO FRAME 2

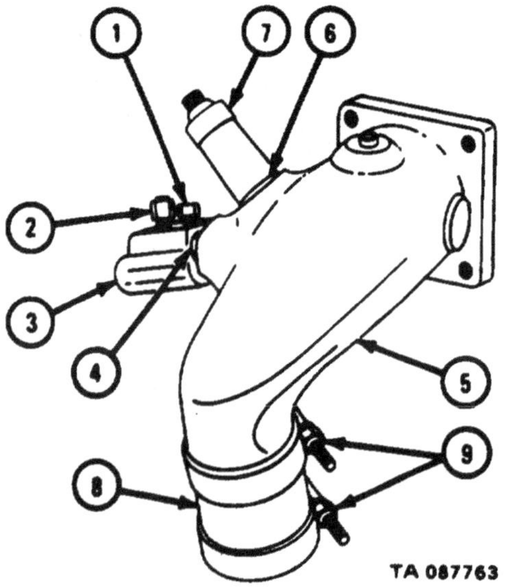

FRAME 2

1. Put turbocharger-to-exhaust manifold gasket (1) on mounting flange of turbocharger (2). Holes in gasket must line up with holes in turbocharger mounting flange.

Soldier A 2. Hold mounting flange of turbocharger (2) against exhaust manifold flange (3). Two studs (4) in exhaust manifold flange must fix through holes in turbocharger-to-exhaust manifold gasket (1) and turbocharger mounting flange .

Soldier B 3. Put in two screws (5) and locknuts (6). put locknut (7) on each of two studs (4) .

NOTE

Make sure that hose clamp (8) is loose.

4. Push crankcase breather tube (9) into breather tube adapter hose (10) . Breather tube must fit between front and rear halves of turbocharger (2) as shown. Tighten hose clamp (8) .

GO TO FRAME 3

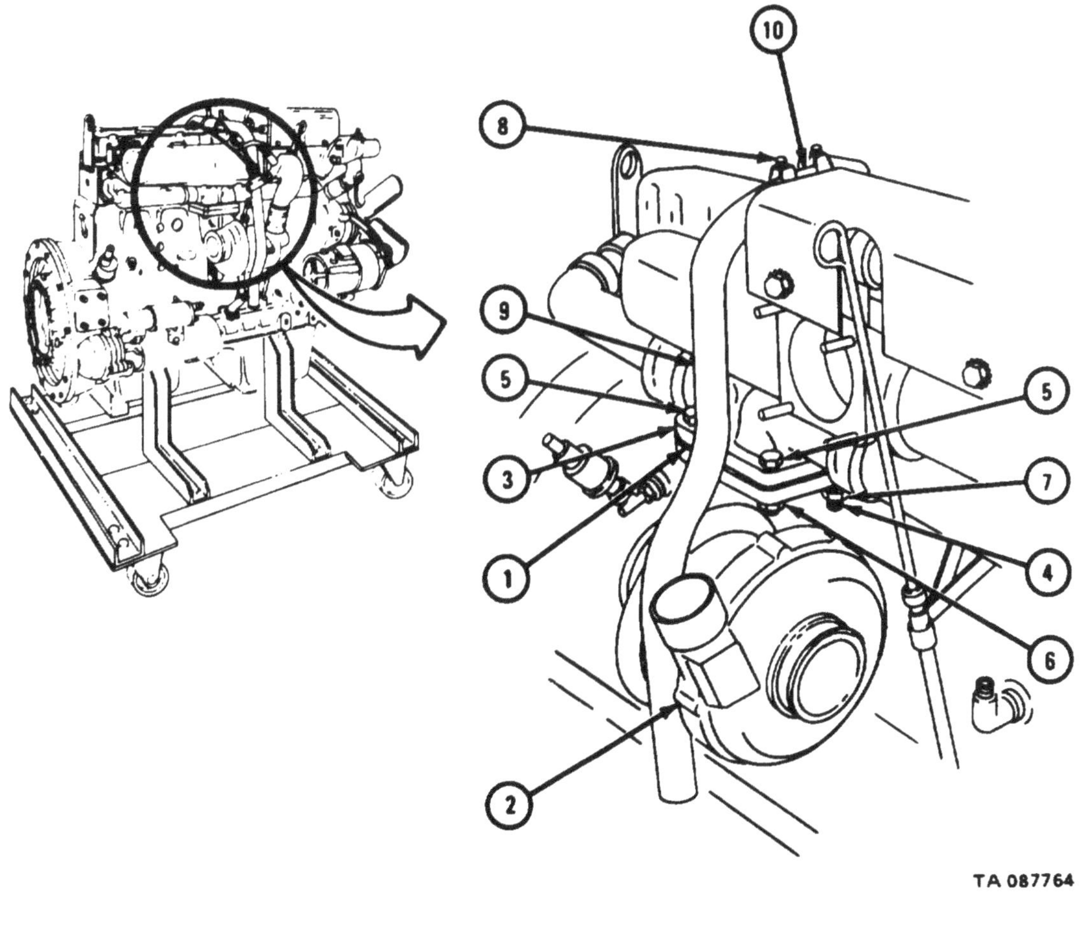

FRAME 3

1. Put in oil drain hose adapter (1).
2. Put oil drain gasket (2) on turbocharger oil drain mounting flange (3).

NOTE

Make sure that two hose clamps (4) are on oil drain hose (5). Hose clamps should be loose and near center of drain hose as shown.

3. Slide oil drain hose (5) onto end of oil drain tube (6).
4. Lift breather tube and bracket (7) up and out of the way of turbocharger oil drain mounting flange (3).
5. Slide oil drain hose (5) over oil drain hose adapter (1). Hold oil drain tube flange (8) against turbocharger oil drain mounting flange (3) so that mounting holes line up.
6. Let breather tube and bracket (7) down so that bracket rests on oil drain tube flange (8) . Hole in bracket should line up with hole in oil drain tube flange.
7. Put in two screws (9) and lockwashers (10).
8. Slide two hose clamps (4) to ends of oil drain hose (5) and tighten clamps.

GO TO FRAME 4

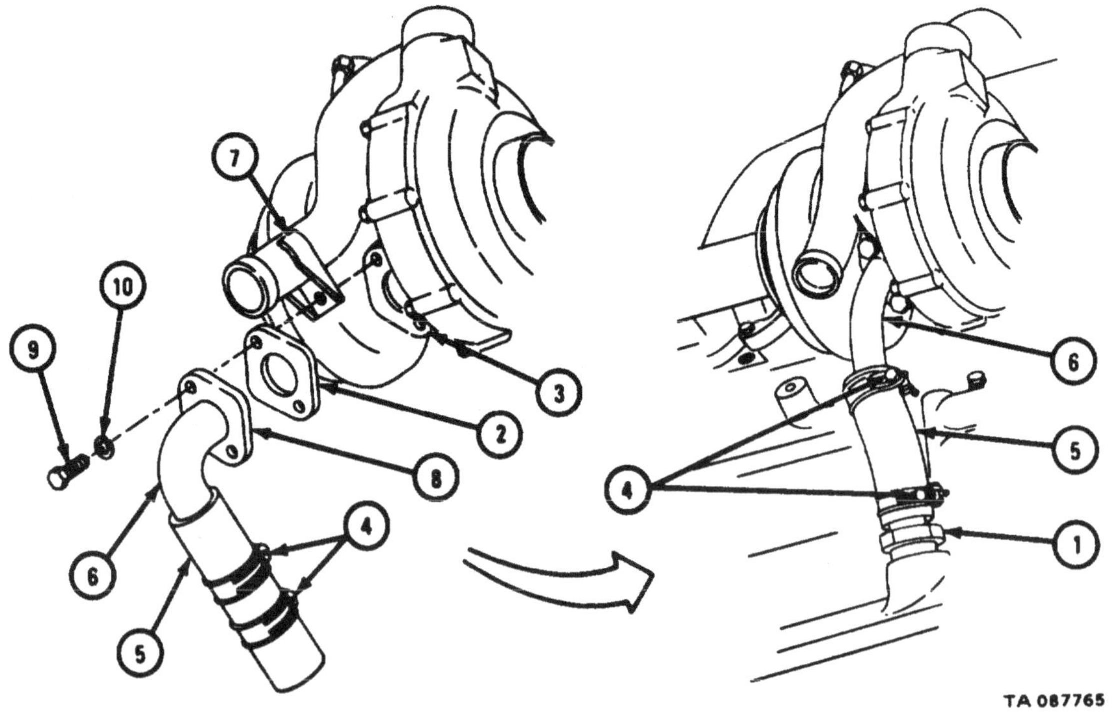

FRAME 4

1. Pour about two ounces of engine lubricating oil into turbocharger oil inlet hole (1) .
2. Put oil inlet adapter flange gasket (2) on oil inlet adapter flange (3).
3. Put oil inlet tube adapter (4) on top of gasket (2) so that mounting holes in adapter, adapter flange (3), and gasket line up. Put in two screws (5) and lockwashers (6) .

GO TO FRAME 5

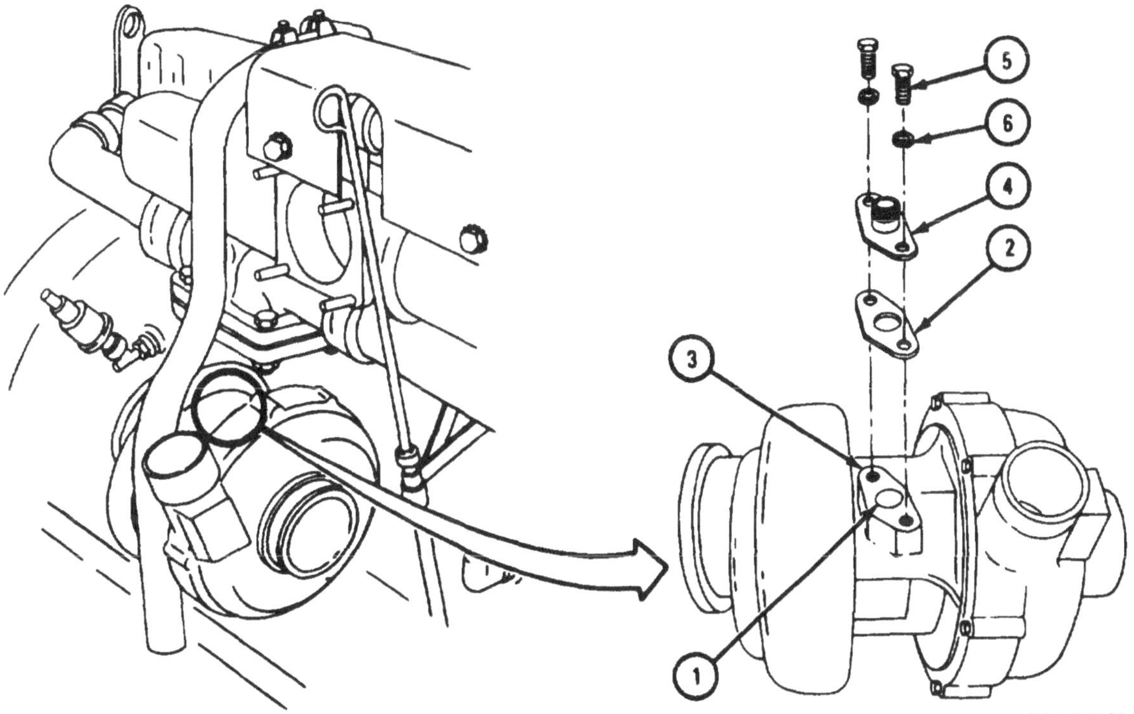

FRAME 5

1. Put oil inlet tube (1) in place as shown and tighten two tube nuts (2).

NOTE

Some engines do not have flame heater cover (3).

2. Take out four screws and starwashers (4) and take off flame heater cover (3).
3. Lift up bracket (5) and put intake elbow gasket (6) on intake manifold elbow mounting flange (7) .

GO TO FRAME 6

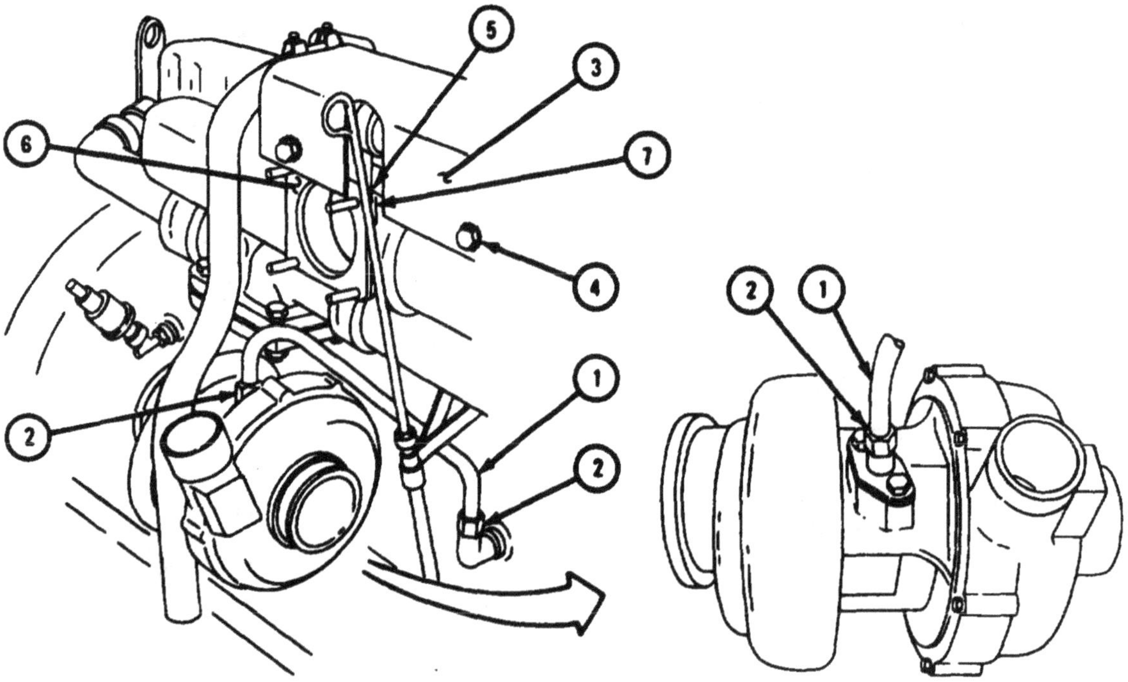

FRAME 6

1. Place intake elbow hose (1) over air outlet tube on turbocharger (2) as shown. Place intake elbow (3) over mounting flange of intake manifold (4). Four studs (5) on mounting flange of intake manifold should fit through holes in mounting flange of intake elbow as shown.
2. Put on four nuts and lockwashers (6). Tighten two hose clamps (7) .
3. Tighten two screws (8) .

NOTE

Some engines do not have flame heater cover (9).

4. Put back flame heater cover (9) as shown. Put in four screws and starwashers (10).

GO TO FRAME 7

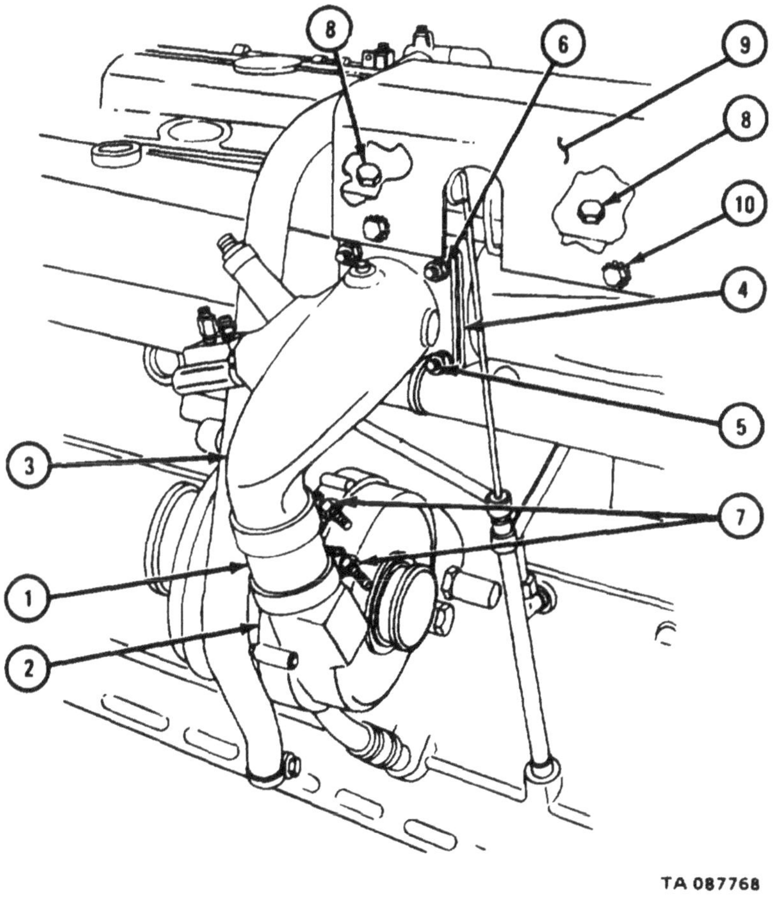

FRAME 7

1. Put fuel inlet tube (1) into adapter (2) and tighten tube nut (3).
2. Put fuel return tube (4) into adapter (5) and tighten tube nut (6).
3. Put ignition lead (7) into spark plug (8) and tighten nut (9).

END OF TASK

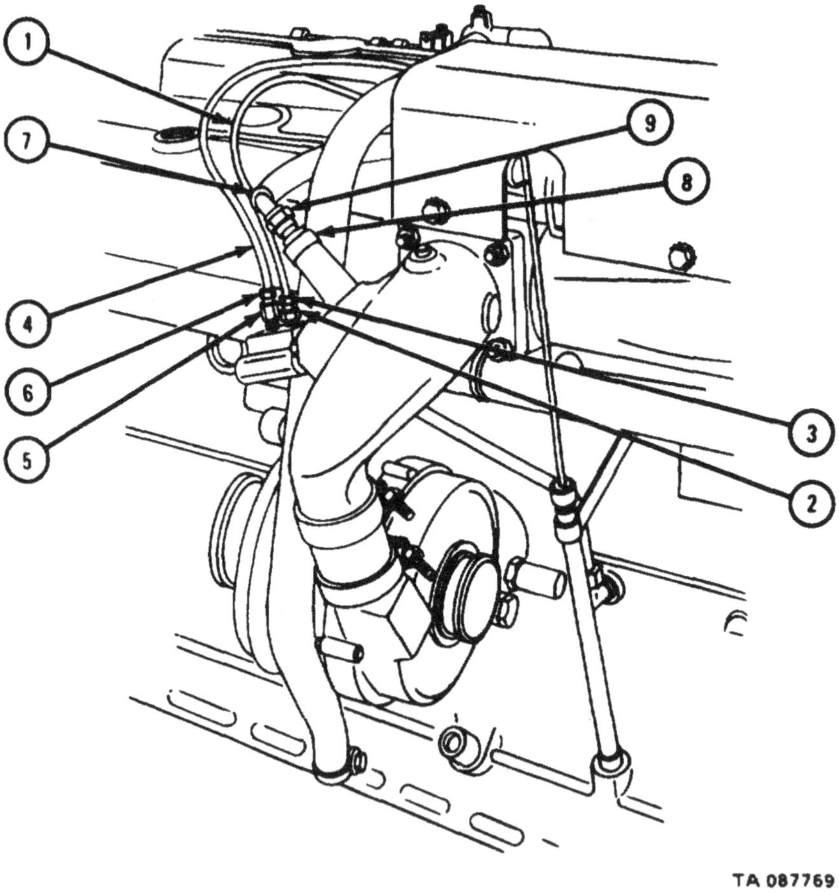

b. Engine LDS-465-1.

FRAME 1

NOTE

Tags, plugs, and tape put on during removal mus be taken off during replacement.

1. Put oil drain tube with gasket (1) in place and put in two screws with lockwashers (2) .
2. Put hose (3) on oil drain tube (1). Slide hose on as far as it will go.

GO TO FRAME 2

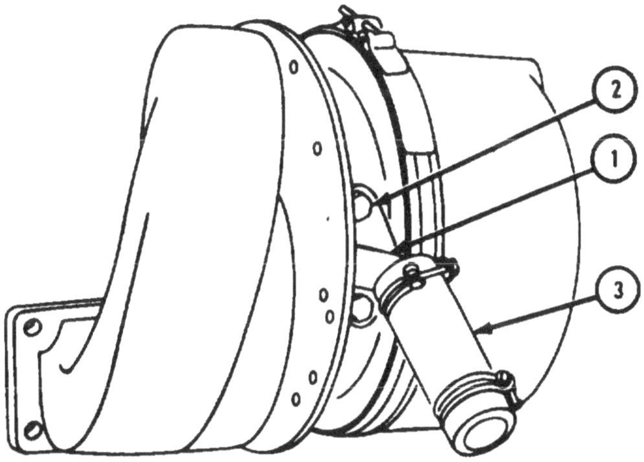

TA 117859

FRAME 2

1. Put in and tighten oil drain adapter (1).
2. Put turbocharger oil inlet tube (2) on engine block fitting (3). Do not tighten.
3. Put oil inlet tube adapter with gasket (4) on turbocharger (5). Put in two screws with lockwashers (6).

GO TO FRAME 3

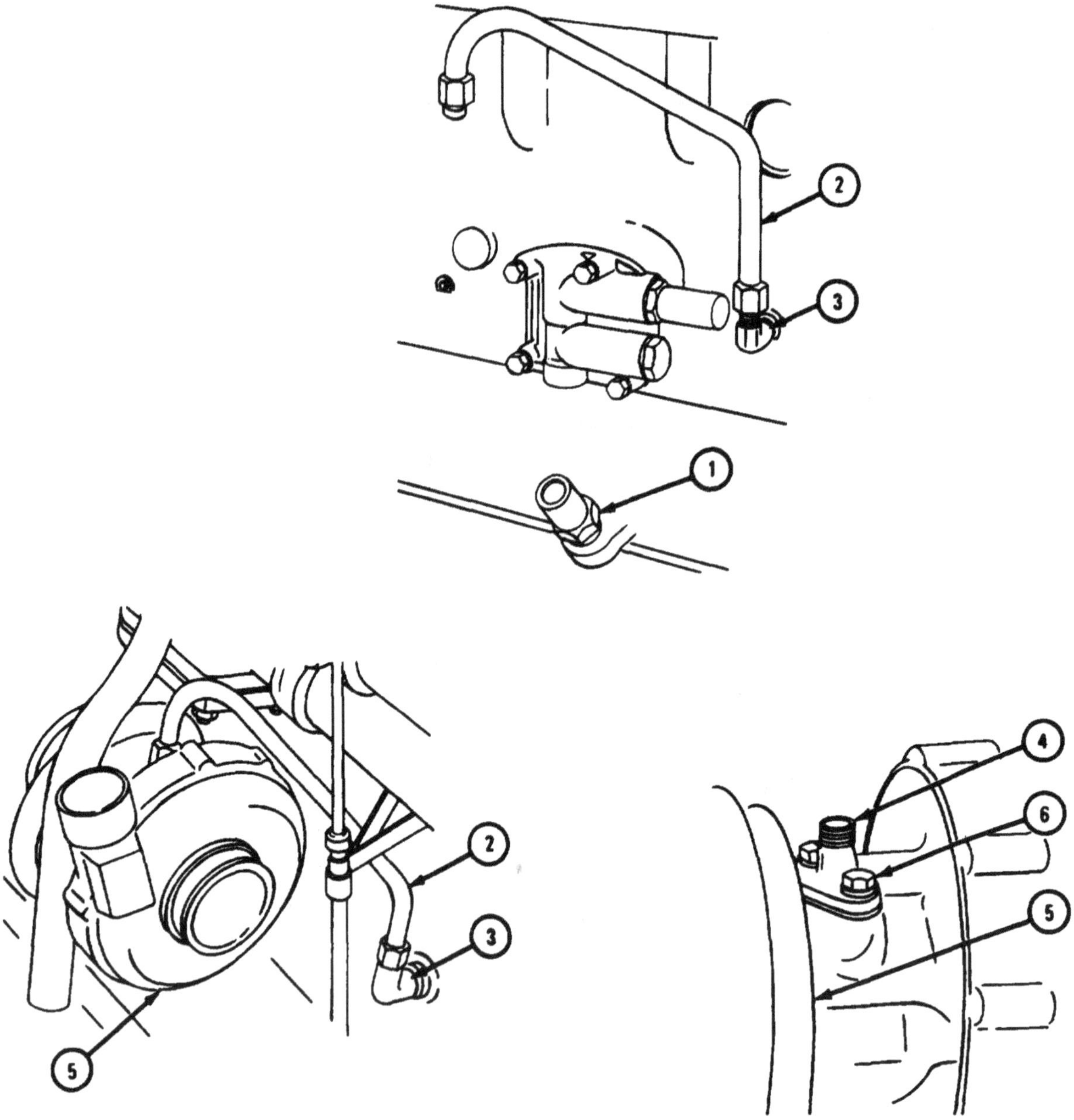

TA 117860

FRAME 3

Soldier A 1. Put turbocharger (1) with gasket (2) in place. Slide hose (3) over turbocharger air outlet (4). Hold turbocharger (1) in place until soldier B puts in mounting screws and nuts.

Soldier B 2. Put in two screws (5). Put on four nuts with lockwashers (6).

3. Slide oil drain hose (7) over oil drain hose adapter (8). Tighten two hose clamps (9).

4. Tighten hose clamp (10).

GO TO FRAME 4

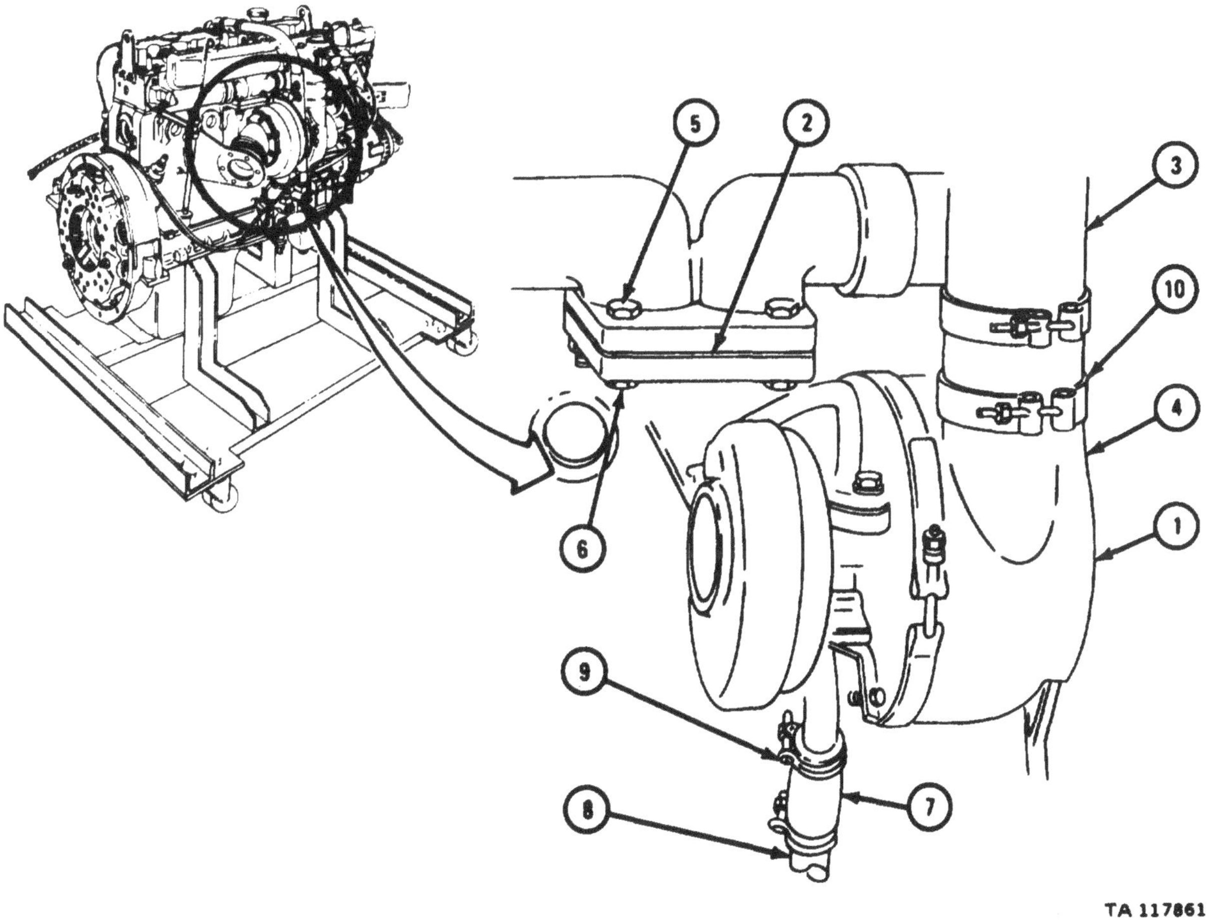

FRAME 4

1. Put turbocharger oil inlet tube (1) on turbocharger oil inlet tube adapter (2). Tighten two fitting nuts (3) .

2. Put crankcase breather tube (4) in place. Tighten hose clamp (5) .

GO TO FRAME 5

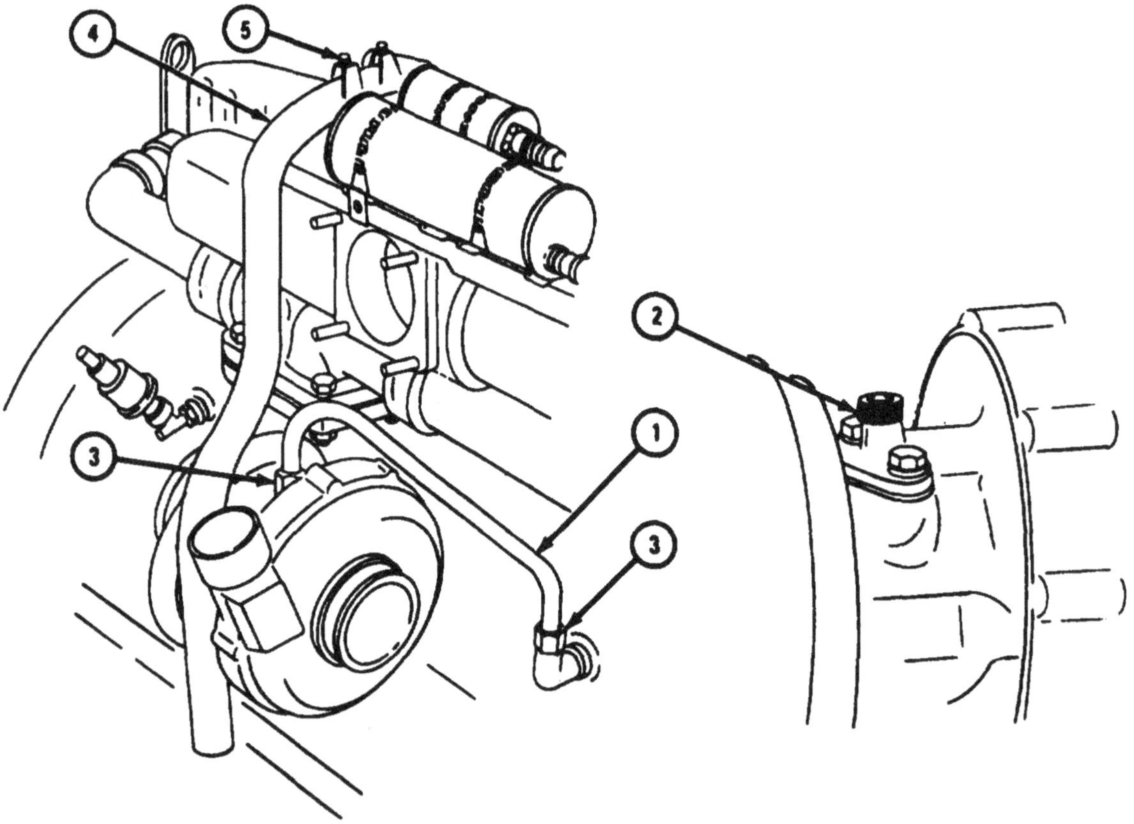

FRAME 5

1. Put in screw (1) with washer (2). Put on lockwasher (3) and nut (4).
2. Put exhaust elbow with bracket and gasket (5) in place. Put on three lockplates (6) and put in six screws with locktabs (7).
3. Put in three screws with lockwashers (8).

END OF TASK

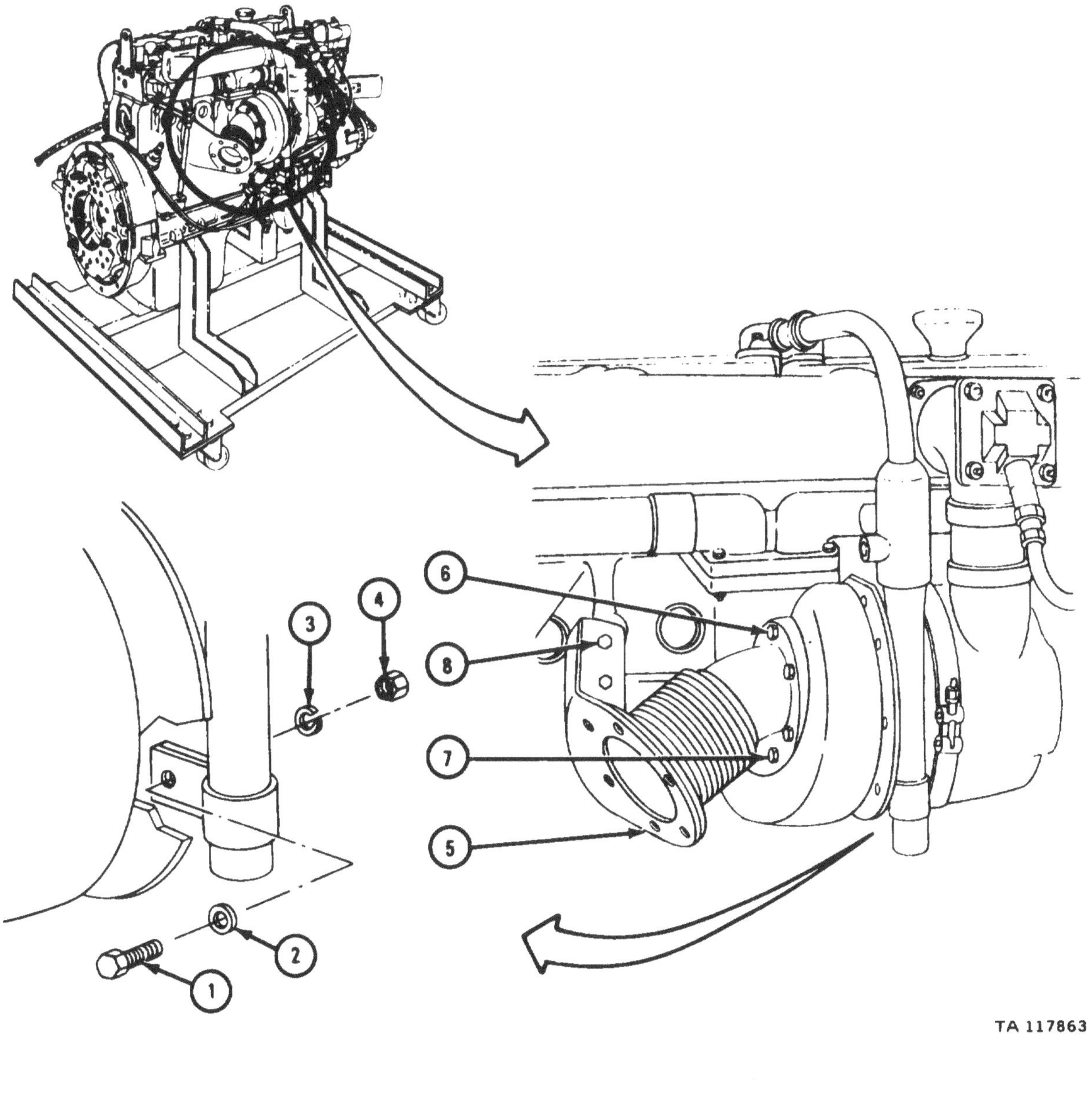

c. Engines LDS-465-1A and LDS-465-2.

FRAME 1

NOTE

Some engines do not have baffle plate (2). If engine has baffle plate, mounting screws (1) were put back in when baffle plate was taken off.

1. Take out two screws with lockwashers (1).
2. Put baffle plate (2) in place and put in two screws with lockwashers (1).

GO TO FRAME 2

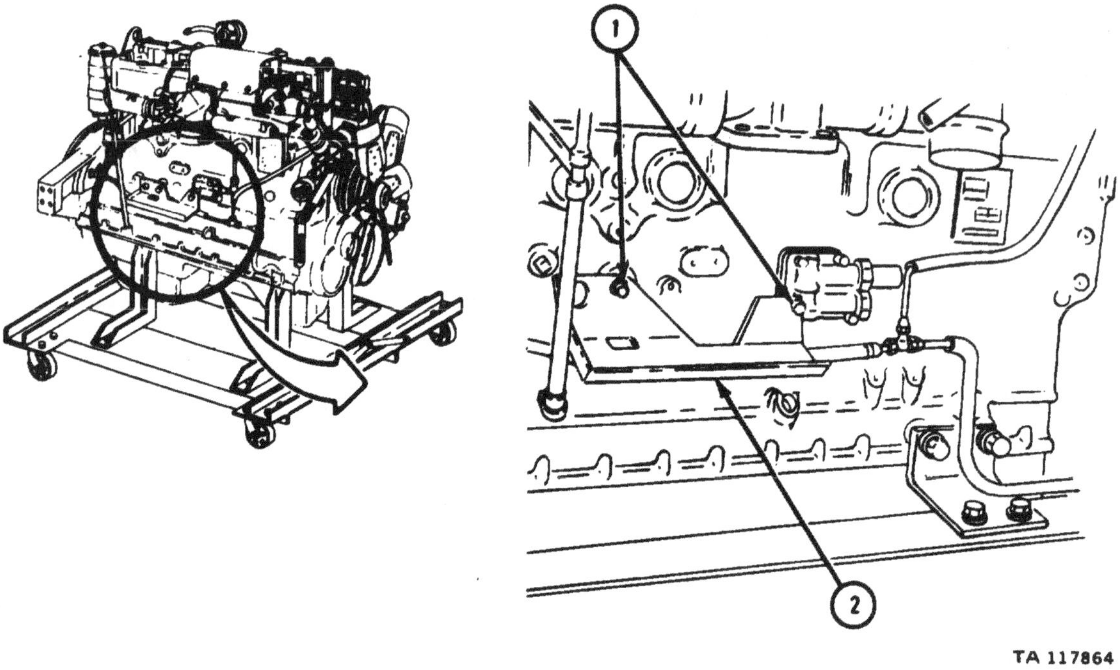

FRAME 2

NOTE

Plugs or tape put on turbocharger during removal must be taken off during replacement.

Soldier A 1. Slide air inlet hose (1) in place and tighten hose clamps (2).

2. Put turbocharger gasket (3) in place. Put turbocharger (4) in place and hold until soldier B puts in mounting screws and nuts.

Soldier B 3. Put in two screws (5). Put on four nuts with lockwashers (6).

4. Put oil inlet adapter with gasket (7) in place. Put in two screws with lockwashers (8).

5. Put oil inlet line (9) in place. Tighten two fitting nuts (10).

GO TO FRAME 3

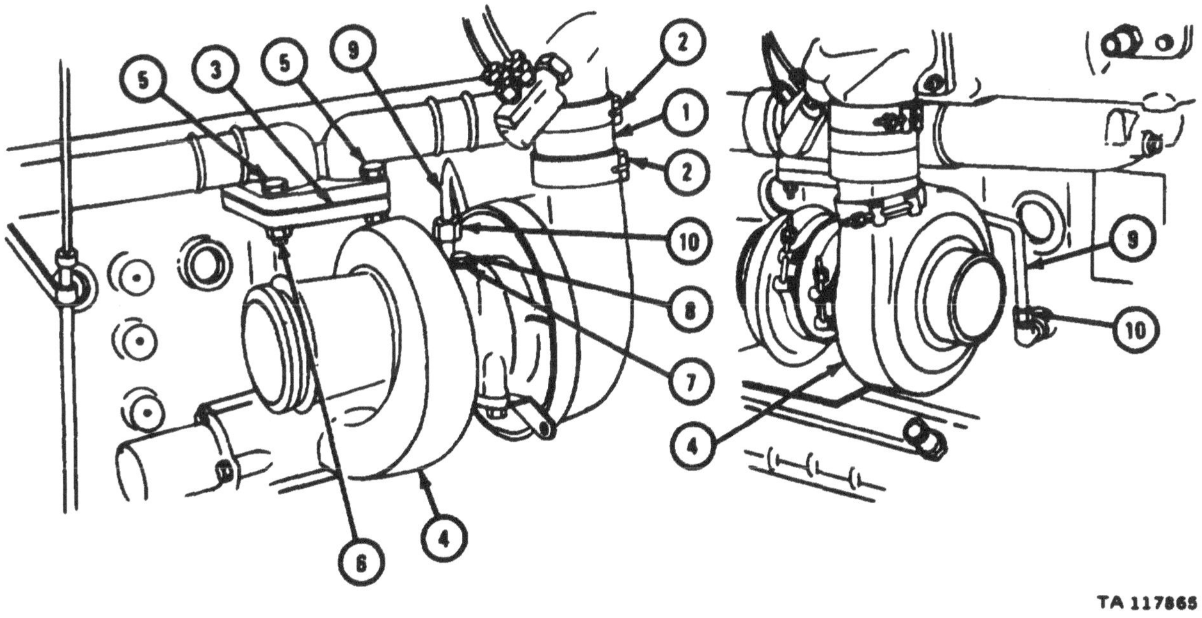

FRAME 3

1. Put in oil drain adapter (1). Slide hose (2) over oil drain adapter and tighten two hose clamps (3).
2. Put oil drain tube with gasket (4) in place. Put crankcase breather tube bracket (5) in place . Put in two screws with lockwashers (6).

GO TO FRAME 4

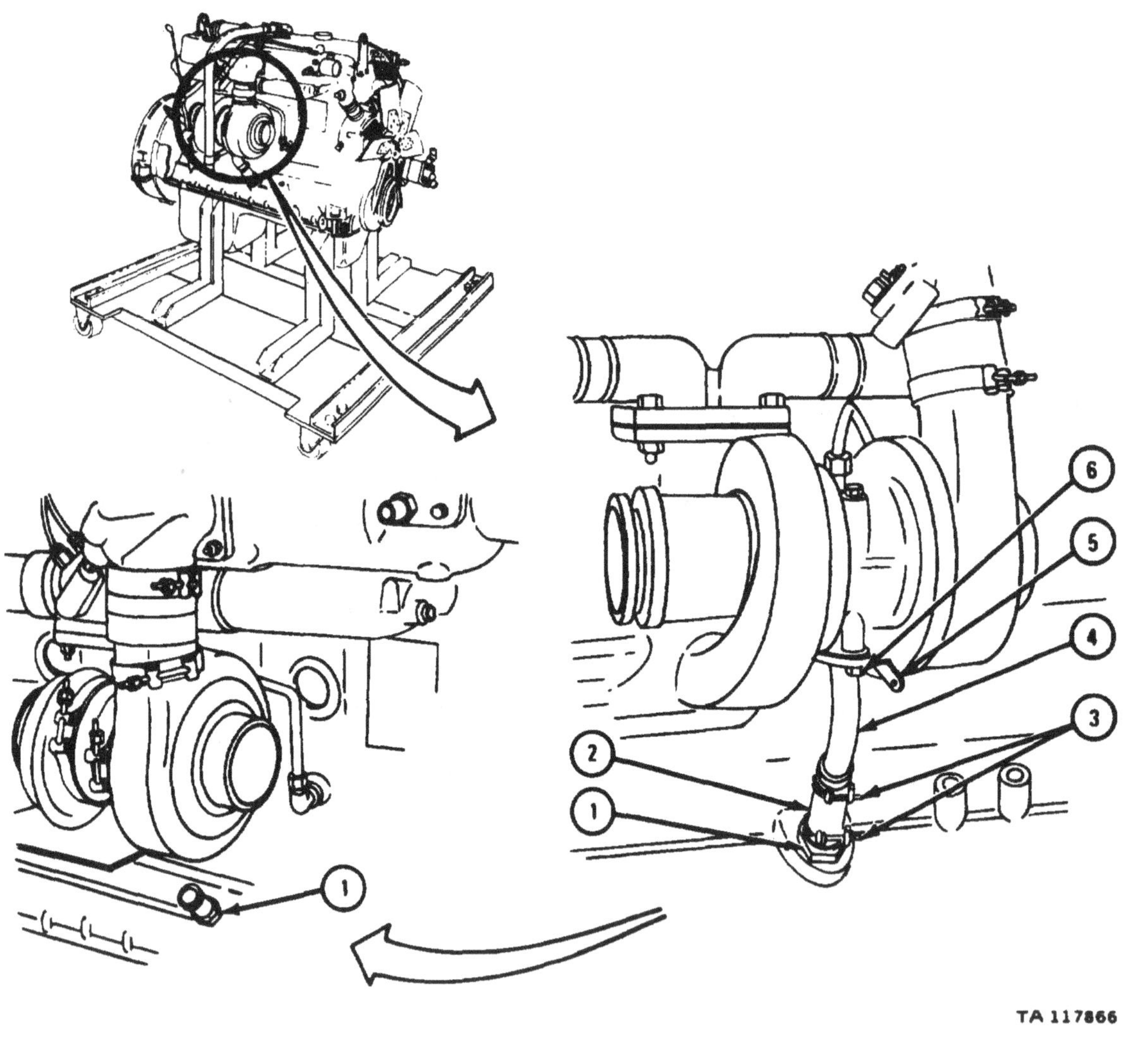

FRAME 4

1. Put crankcase breather tube (1) in place. Tighten hose clamp (2) .
2. Put breather tube clamp (3) against breather tube bracket (4). Put i n screw (5) with washer (6) . Put on lockwasher (7) and nut (8).

END OF TASK

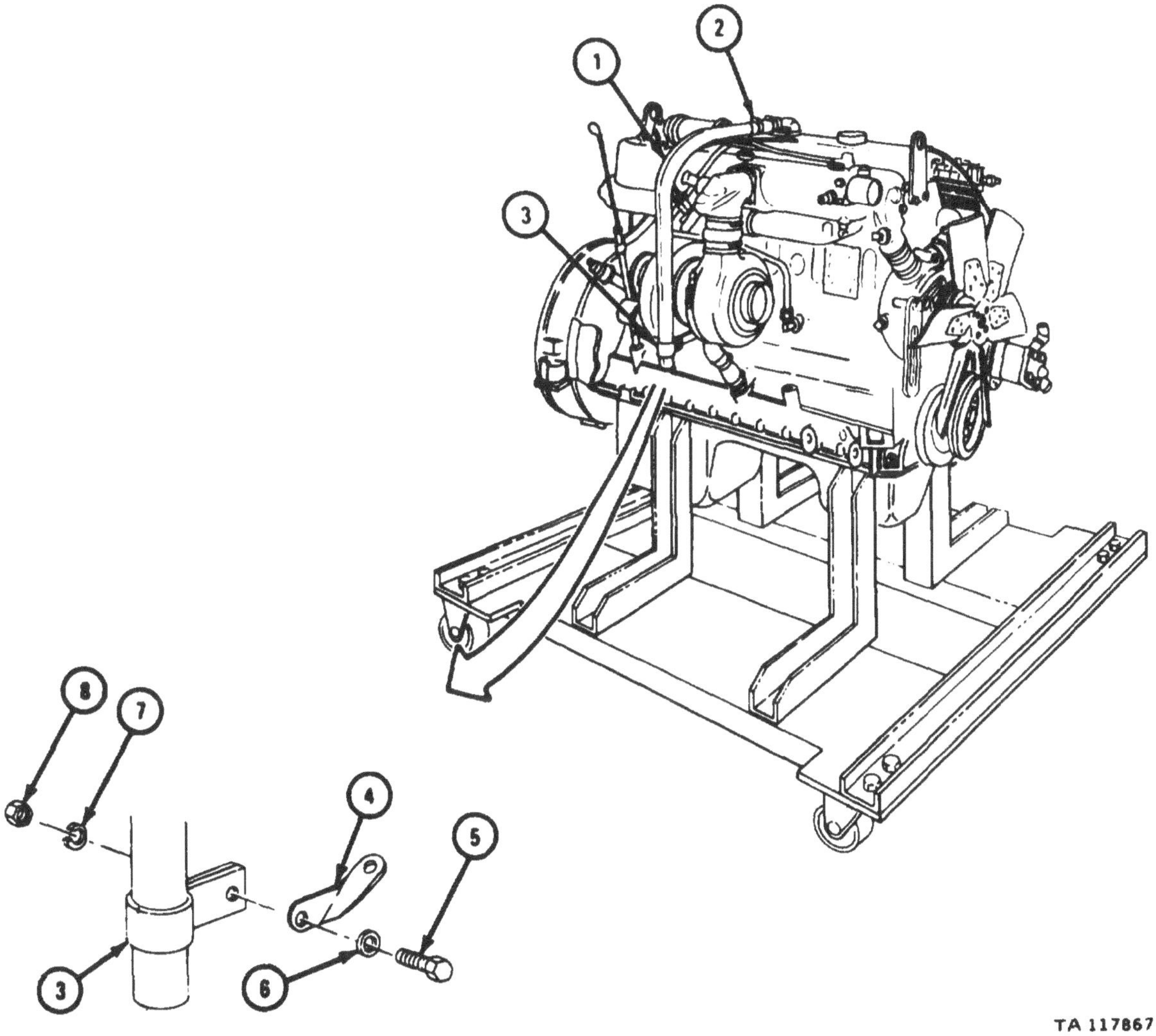

5-37. CONNECTION OF FUEL INJECTION PUMP.

a. Engines LD-465-1, LD-465-1C, and LDT-465-1C.

FRAME 1

NOTE

Plugs or caps put on during removal must be taken off during replacement. When each item is put on, take off tags used for marking.

1. Put fuel pump-to-fuel filter tube (1) in place. Screw on and tighten fitting nut (2).
2. Put fuel filter-to-fuel density compensator tube (3) in place. Screw on and tighten fitting nut (4).
3. Put fuel pump-to-flame heater tube (5) in place. Screw on and tighten fitting nut (6).

GO TO FRAME 2

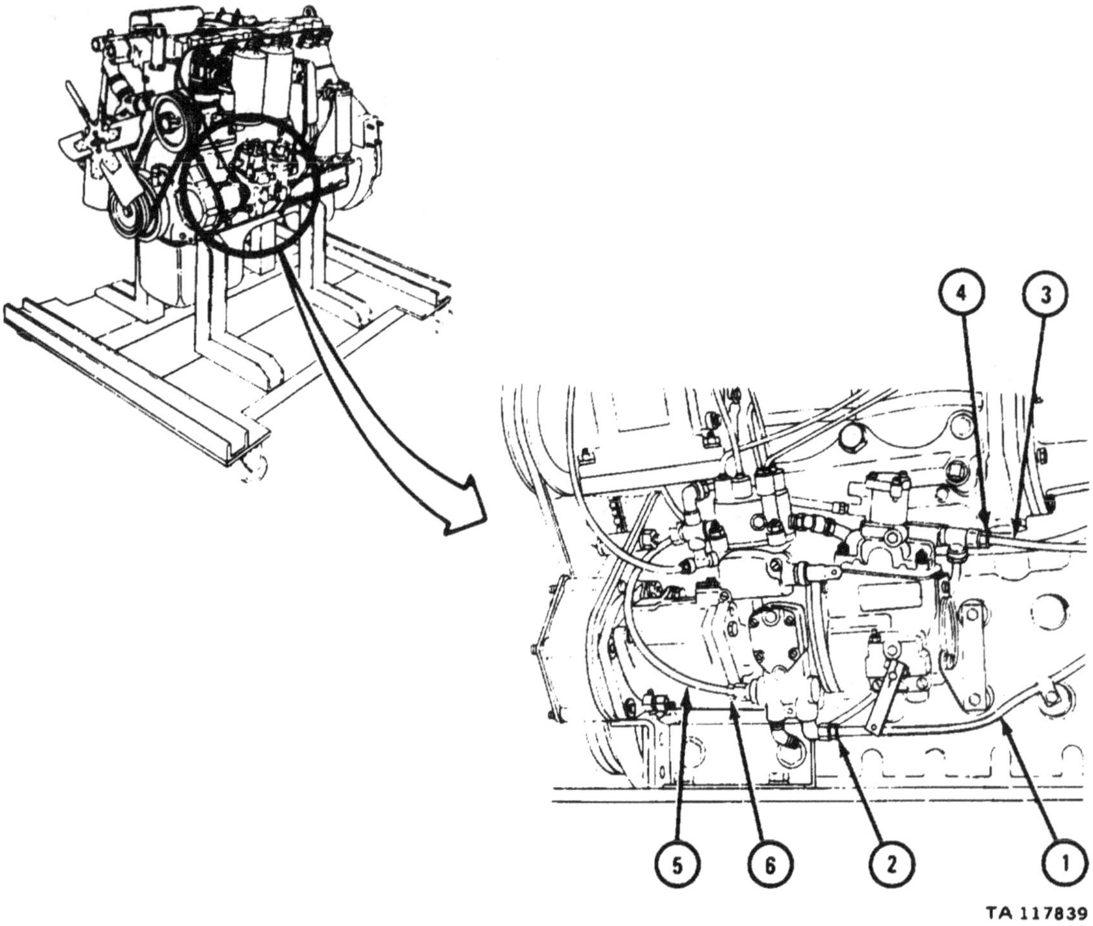

FRAME 2

1. Put fuel pump-to-fuel filter return tube (1) in place. Screw on and tighten fitting nut (2) .
2. Put fuel injector nozzles-to-fuel pump return tube (3) in place. Screw on and tighten fitting nut (4) .
3. Put pressure oil hose (5) in place. Screw on and tighten fitting nut (6) .

END OF TASK

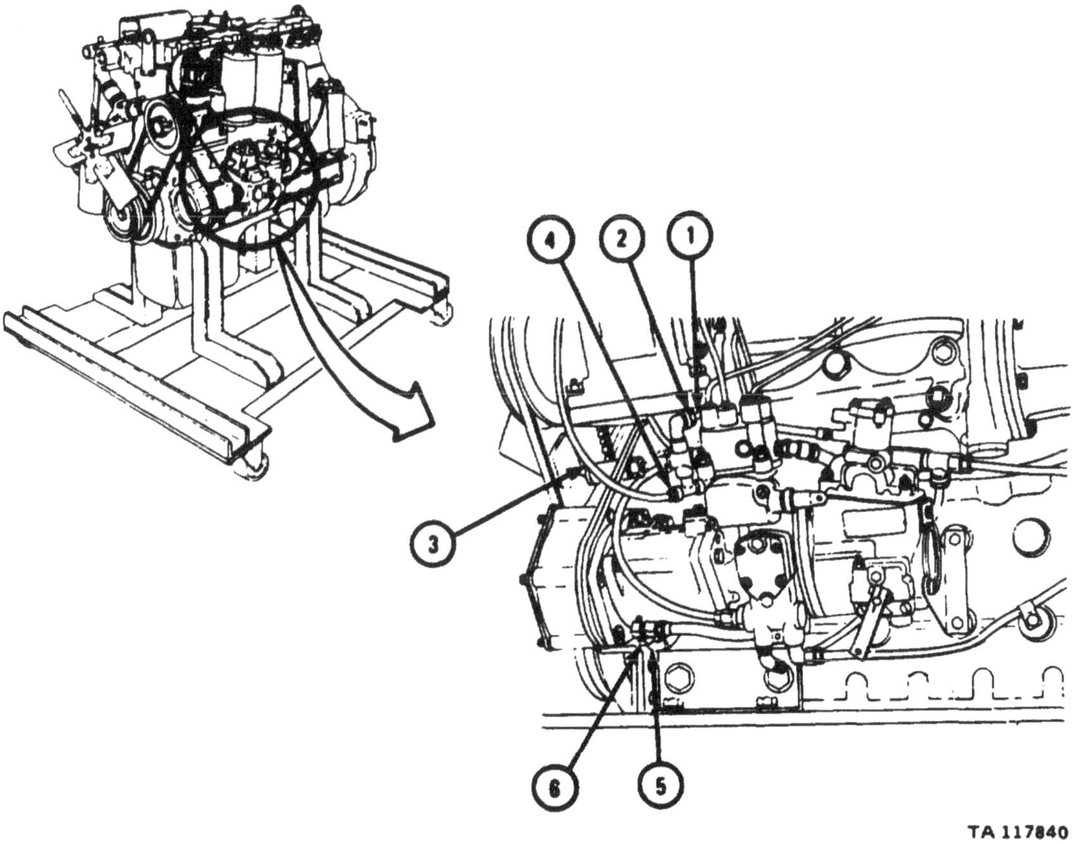

b. Engine LDS-465-1A.

FRAME 1

NOTE

Plugs or caps put on during removal must be taken off during replacement . When each item is put in, take off tags used for marking.

1. Put fuel pump-to-fuel filter tube (1) in place. Screw on and tighten fitting nut (2) .
2. Put fuel filter-to-fuel density compensator tube (3) in place. Screw on and tighten fitting nut (4) .
3. Put fuel pump-to-flame heater tube (5) in place. Screw on and tighten fitting nut (6) .

GO TO FRAME 2

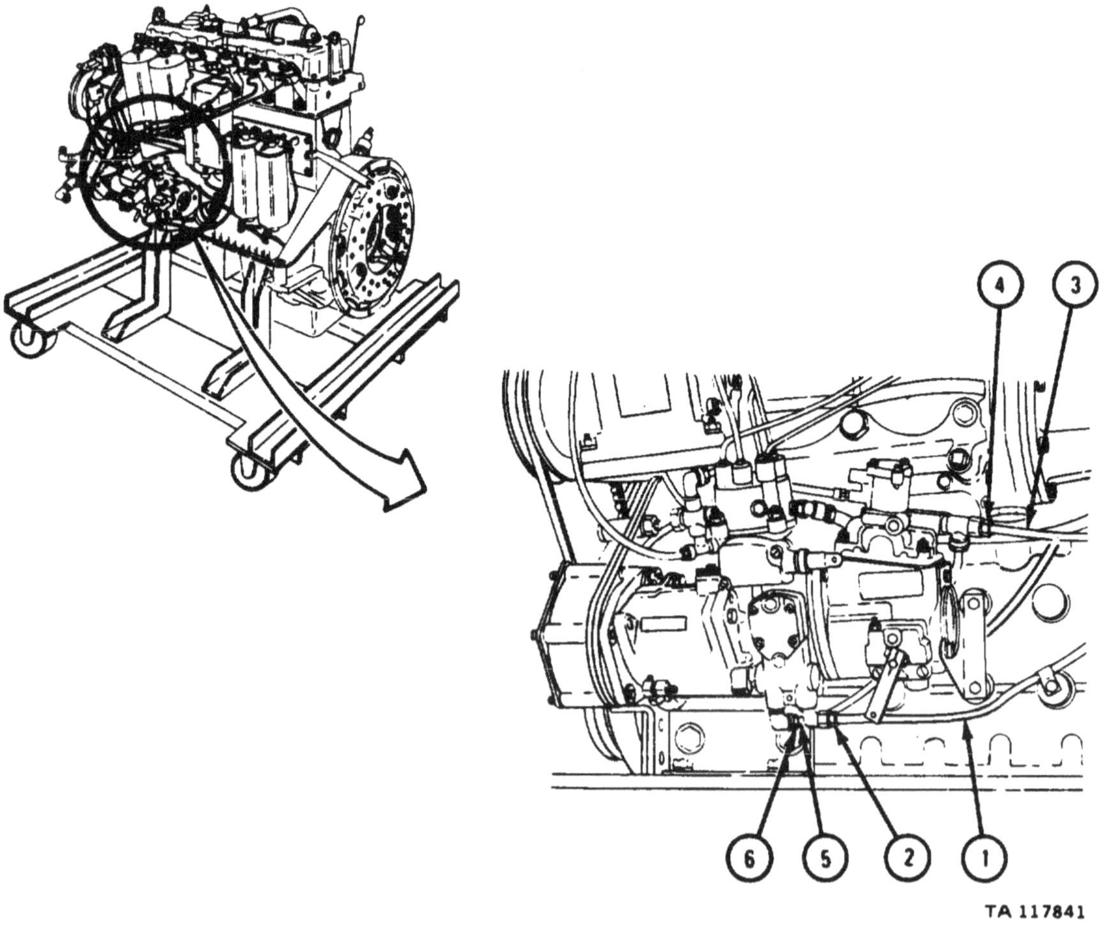

FRAME 2

1. Put fuel pump-to-fuel filter return tube (1) in place. Screw on and tighten fitting nut (2) .
2. Put fuel injector nozzles-to-fuel pump return tube (3) in place. Screw on and tighten fitting nut (4).
3. Put pressure oil hose (5) in place. Screw on and tighten fitting nut (6).

END OF TASK

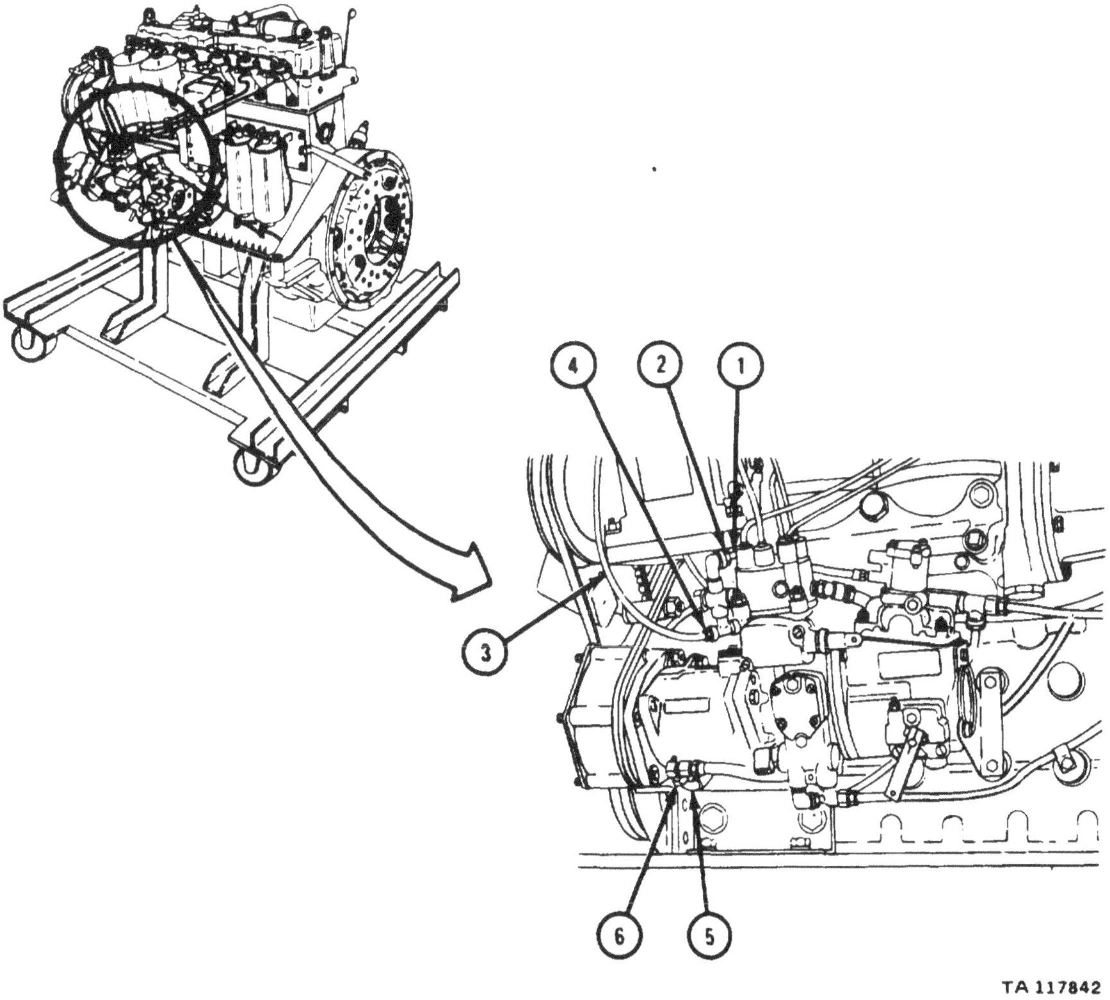

c. Engine LDS-465-1.

FRAME 1

1. Put fuel filter-to-fuel density compensator tube (1) in place. Screw on and tighten fitting nut (2) .
2. Put fuel filter-to-fuel pump tube (3) in place. Screw on and tighten fitting nut (4) .
3. Put fuel pump-to-flame heater tube (5) in place. Screw on and tighten fitting nut (6) .

GO TO FRAME 2

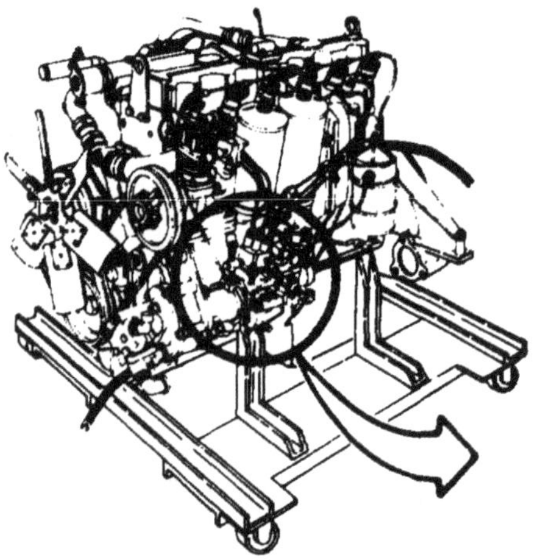

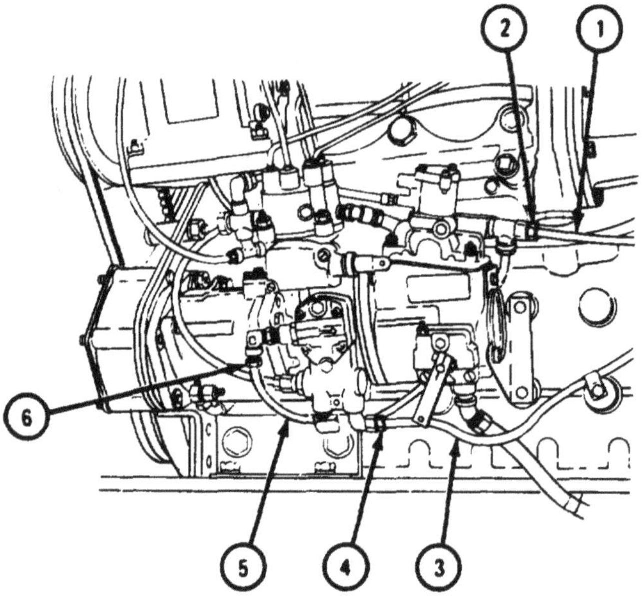

TA 117843

FRAME 2

1. Put fuel pump-to-fuel filter return tube (1) in place. Screw on and tighten fitting nut (2) .
2. Put fuel injector nozzle-to-fuel pump return tube (3) in place. Screw on and tighten fitting nut (4) .
3. Put pressure oil tube (5) in place. Screw on and tighten fitting nut (6).

IF WORKING ON ENGINE WITH OIL DRAIN HOSE (7), GO TO FRAME 3.
IF WORKING ON ENGINE WITHOUT OIL DRAIN HOSE (7), END OF TASK

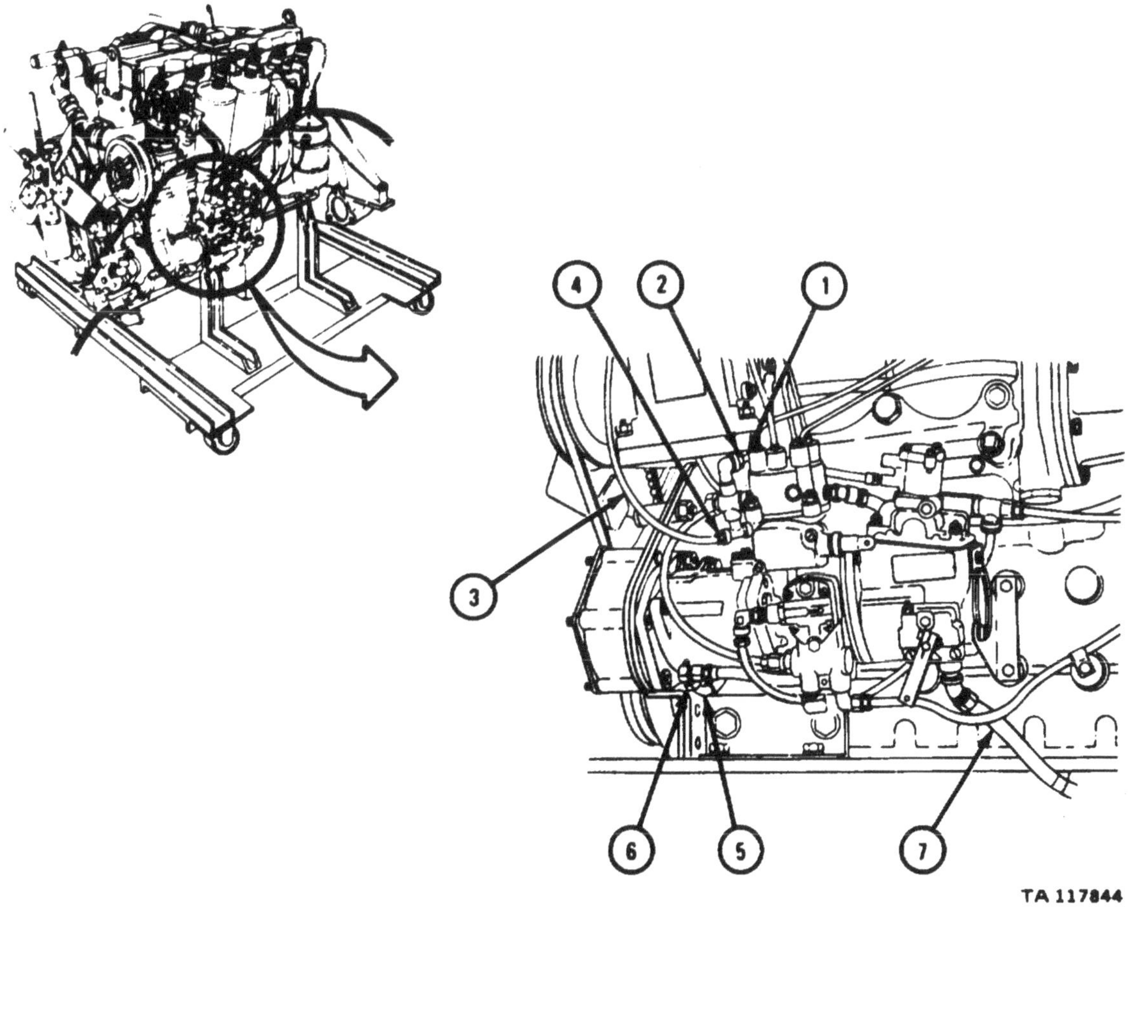

FRAME 3

1. Screw in pipe elbow (1). Screw oil drain tube (2) into pipe elbow.
2. Take out screw with lockwasher (3). Put hose clamp (4) in place and put in screw with lockwasher.
3. Put other end of oil drain tube (2) in place on injector pump (5). Screw in and tighten fitting nut (6).

END OF TASK

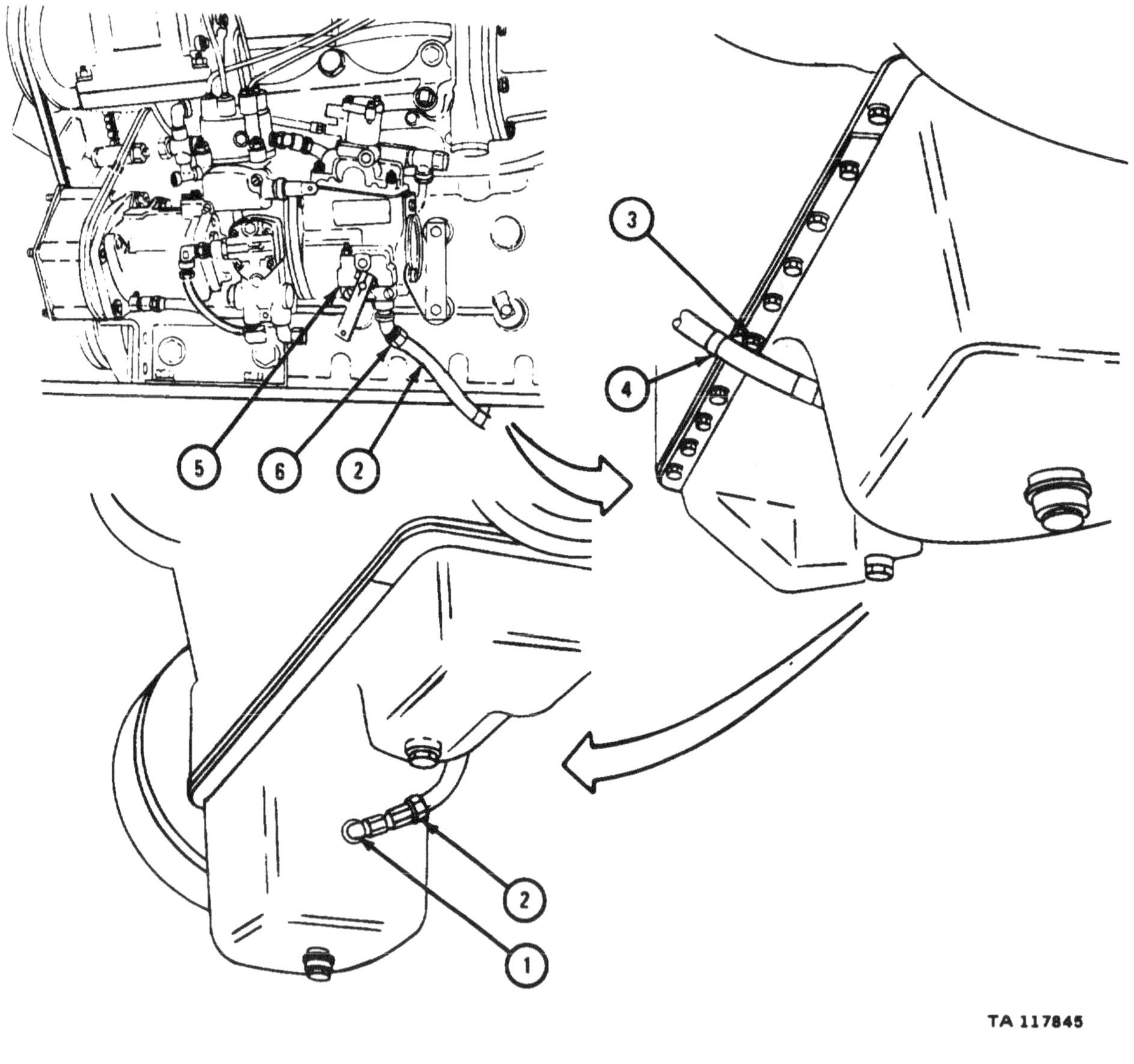

d. Engine LDS-465-2.

FRAME 1

1. Put fuel filter-to-fuel density compensator tube (1) in place. Screw in and tighten fitting nut (2) .
2. Put fuel pump-to-fuel filter tube (3) in place. Screw in and tighten fitting nut (4) .
3. Put fuel pump-to-flame heater tube (5) in place. Screw in and tighten fitting nut (6) .
4. Put fuel injector nozzle-to-fuel pump return tube (7) in place. Screw in and tighten fitting nut (8) .
5. Put pressure oil hose (9) in place. Screw in and tighten fitting nut (10) .

END OF TASK

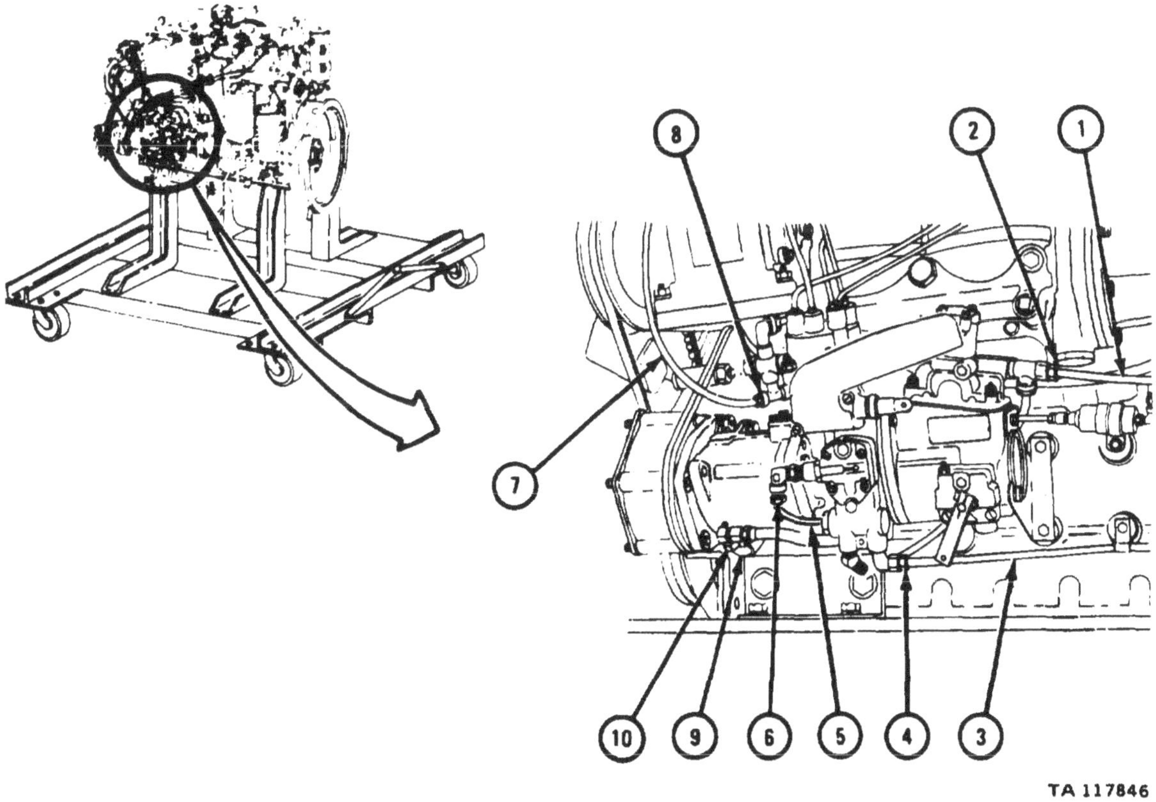

5-38. GENERATOR ASSEMBLY.

a. Engines LD-465-1C and LDT-465-1C.

FRAME 1

NOTE

Some engines have one or more shims (3) between generator (1) and mounting bracket (2). If shims (3) were taken out during removal, they must be put back.

Soldier A 1. Put generator (1) in place on mounting bracket (2). Hold it until soldier B puts in mounting bolts.

Soldier B 2. If shims (3) were used, put in correct number of shims. Put in two bolts (4) with two washers (5). Put on two more washers and two nuts (6) .

NOTE

Some engines use cotter pins (7).

3. If cotter pins (7) were used, put in two cotter pins.

GO TO FRAME 2

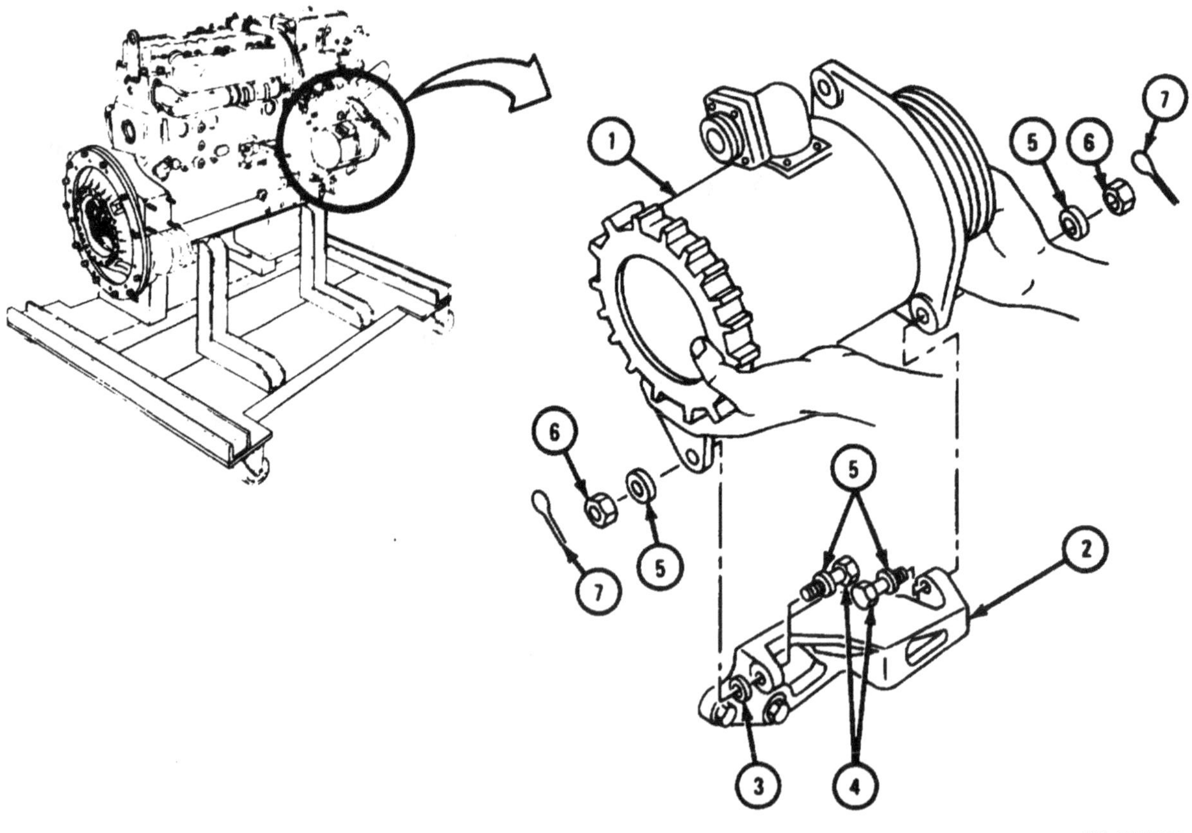

FRAME 2

1. Push generator (1) toward engine (2) . Put two drive belts (3) in place on pulley (4) . Take off tags .
2. Swing adjustment bracket (5) in place and put in screw with lockwasher and step 'washer (6). Do not tighten screw.

NOTE

Refer to TM 9-2320-211-20 for correct drive belt tension.

3. Pry generator (1) away from engine (2) until drive belts (3) are at correct tension. Tighten screw (6) .

END OF TASK

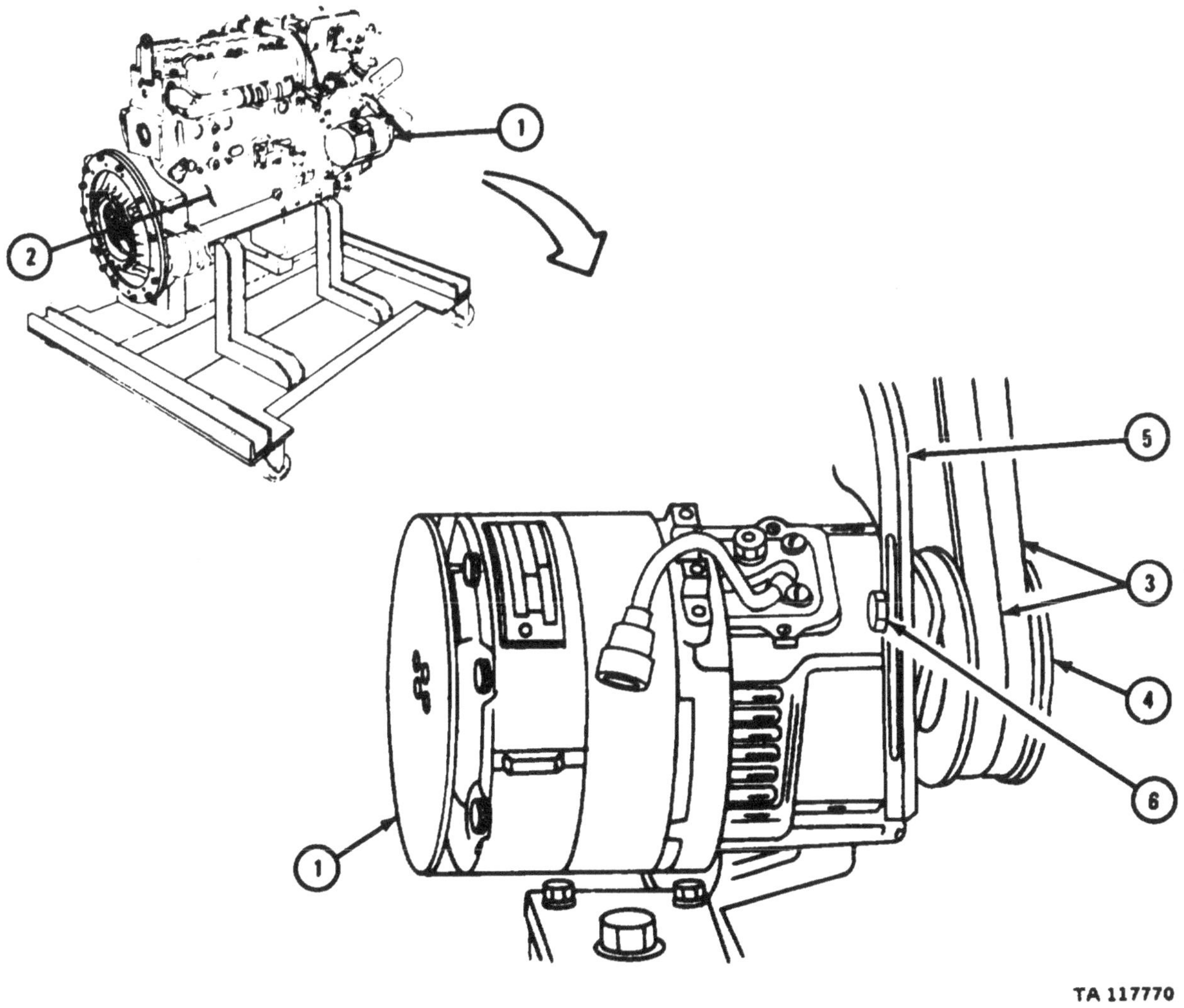

b. Engines LD-465-1, LDS-465-1, and LDS-465-1A.

FRAME 1

NOTE

Some engines have one or more shims (3) between generator (1) and mounting bracket (2). If shims were taken out during removal, they must be put back. Tags that were put on must be taken off during replacement of these parts.

Soldier A 1. Put generator (1) in place on mounting bracket (2). Hold it until soldier B puts in mounting bolts.

Soldier B 2. If shims (3) were used, put in correct number of shims. Put in two bolts (4) with two washers (5). Put on two washers (6) and two nuts (7).

GO TO FRAME 2

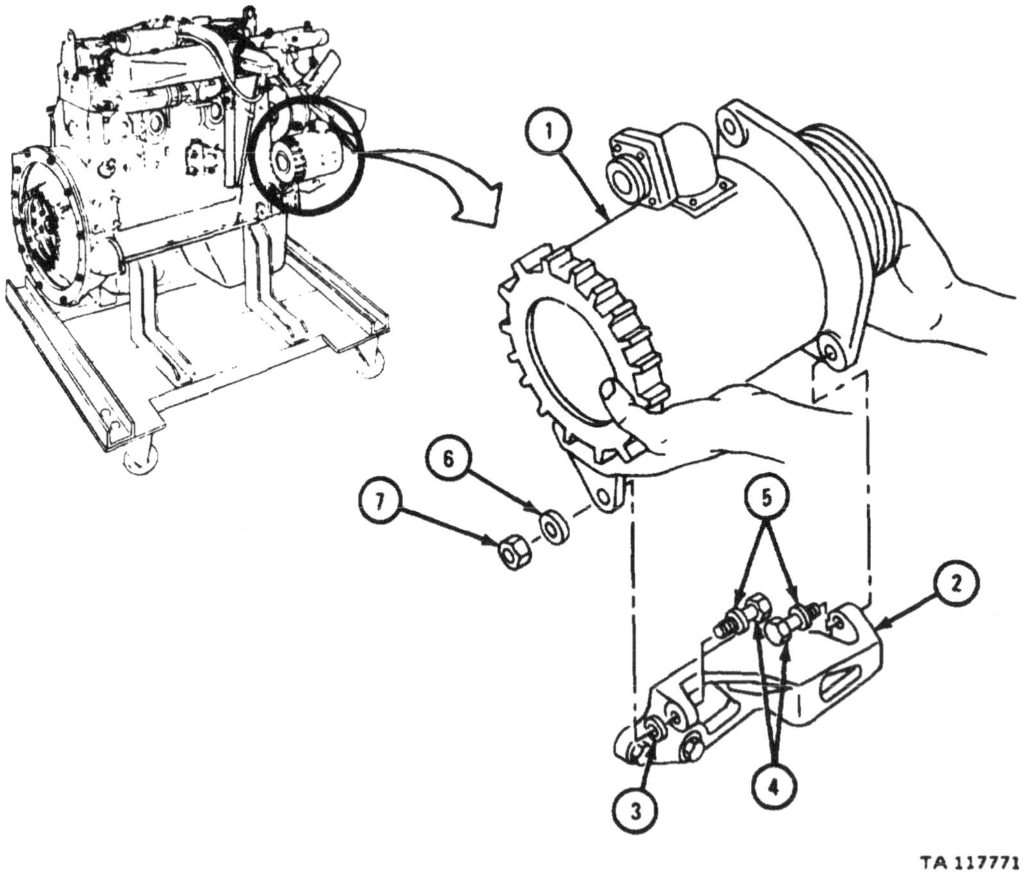

FRAME 2

NOTE

Some engines do not use cotter pins (1).

1. Put in two cotter pins (1).
2. Push generator (2) toward engine. Put on two drive belts (3). Take off tags .
3. Pull generator adjusting bracket (4) into place. Put in screw (5) with lockwasher (6) and step washer (7). Do not tighten screw .

NOTE

Refer to TM 9-2320-211-20 for correct tension of generator drive belts (3) .

4. Pry generator (2) away from engine until tension of drive belts (3) is correct. Tighten screw (5) .

NOTE

Some engines do not have ground strap (8).

5. Put ground strap (8) in place. Put in screw with washer (9).

END OF TASK

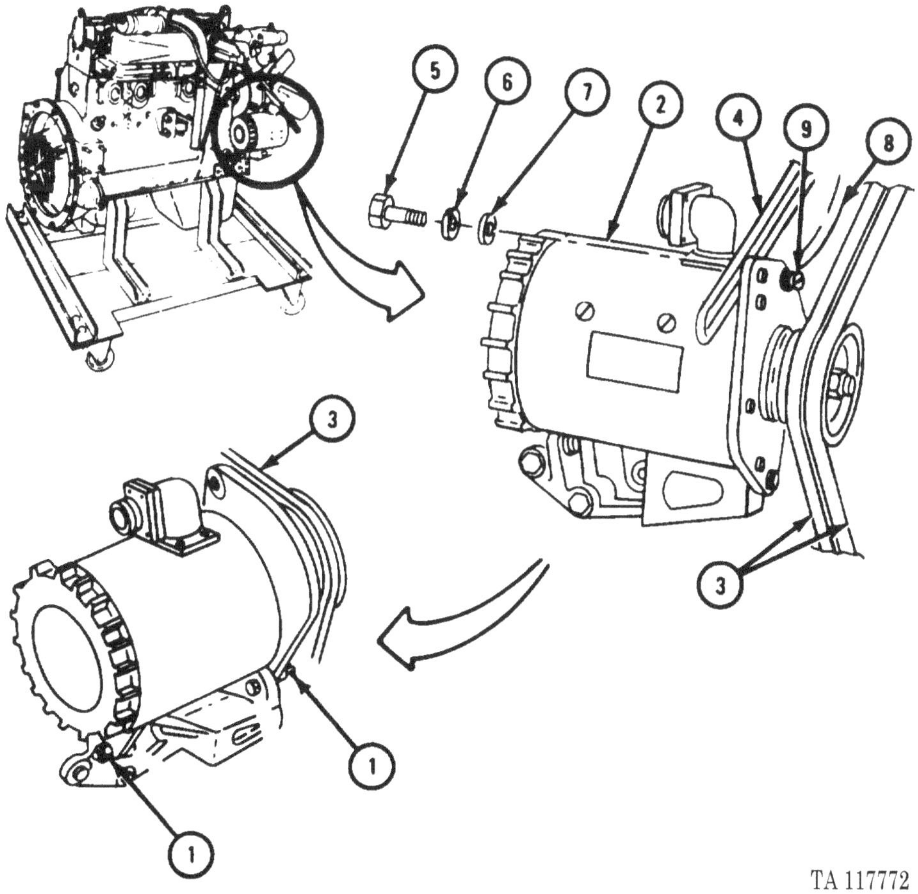

TA 117772

c. Engine LDS-465-2.

FRAME 1

Soldier A 1. Put generator (1) in place on mounting bracket (2). Hold it until soldier B puts in mounting screws.

Soldier B 2. Put in two mounting screws (3) with two washers (4). Put on two more washers and two locknuts (5).

3. Swing generator adjusting bracket (6) in place and put in screw (7) with lockwasher (8) and serrated washer (9). Do not tighten screw.

GO TO FRAME 2

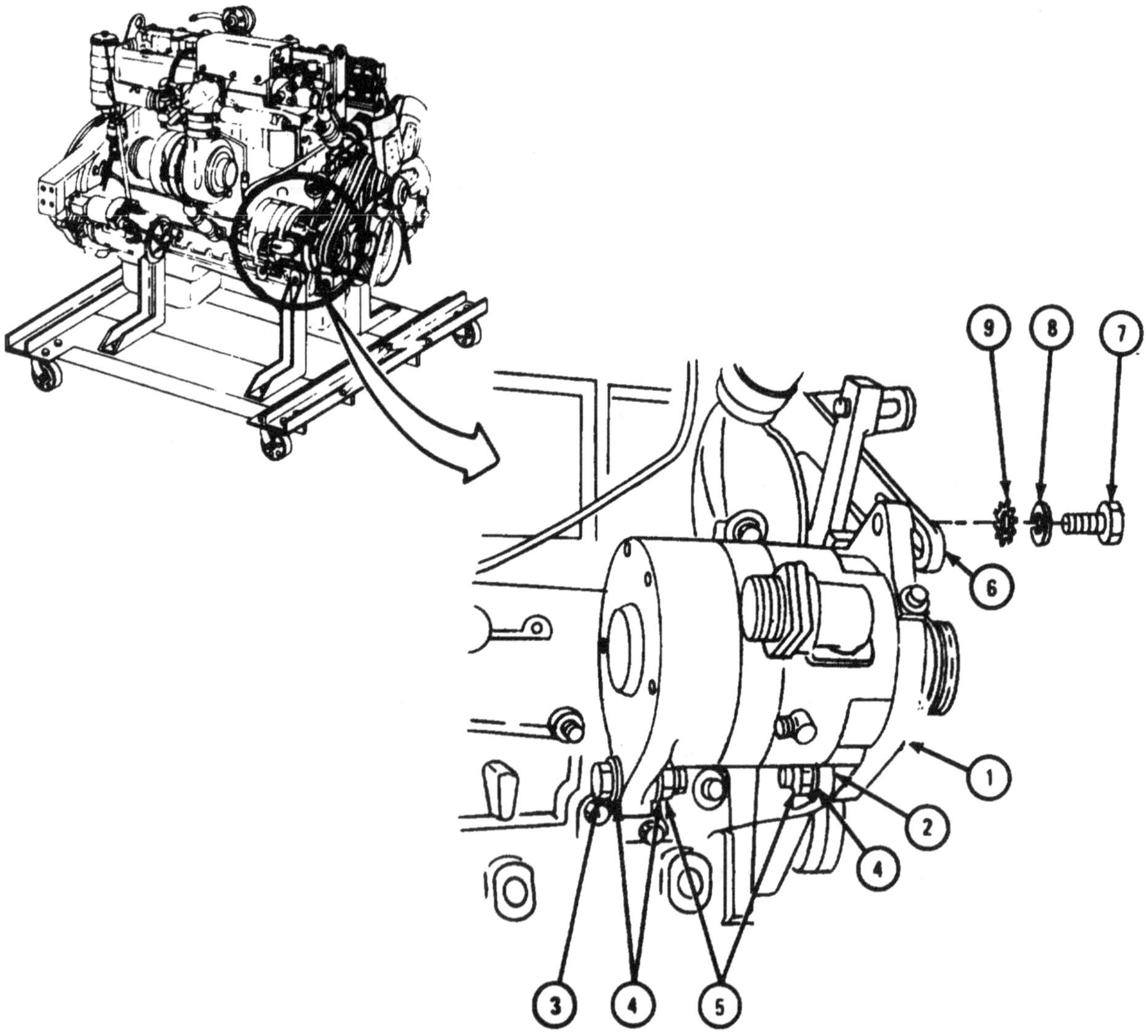

TA 117773

FRAME 2

1. Push generator (1) toward engine and put on two drive belts (2).

NOTE

Refer to TM 9-2320-211-20 for correct tension of drive belts (2).

2. Pry generator (1) away from engine until drive belts (2) are at correct tension. Tighten screw (3).

3. Hook air pressure hose (4) up to fitting (5) and tee fitting (6).

END OF TASK

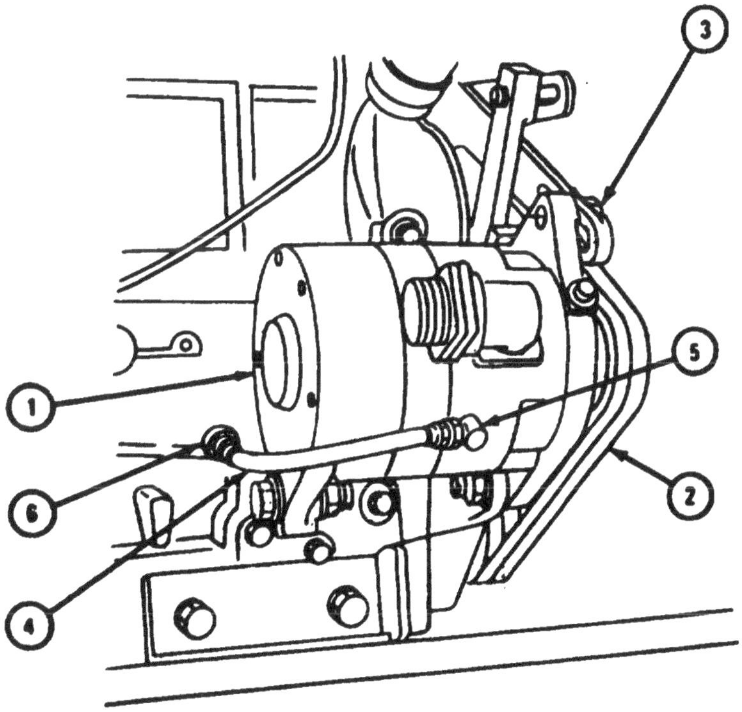

TA 117774

5-39. STARTER ASSEMBLY.

a. Engines LD 465-1, LD 465-1C, and LDT 465-1C.

FRAME 1

Soldier A 1. Put gasket (1) on starter (2). Put on adapter (3) and adapter gasket (4).

WARNING

Starter (2) weighs fifty pounds. Be careful to hold it up firmly when putting it on mounting studs. Starter could fall and cause injury to personnel and damage to equipment.

Soldiers A and B 2. Put starter assembly (2) in place on engine flywheel housing (5). Aline starter mounting holes (6) with flywheel housing studs (7). Push starter assembly straight in.

Soldier A 3. Hold starter assembly (2) until soldier B puts on lockwashers (8) and nuts (9).

Soldier B 4. Put on three lockwashers (8) and three nuts (9). Tighten nuts to 140 to 150 pound-feet.

END OF TASK

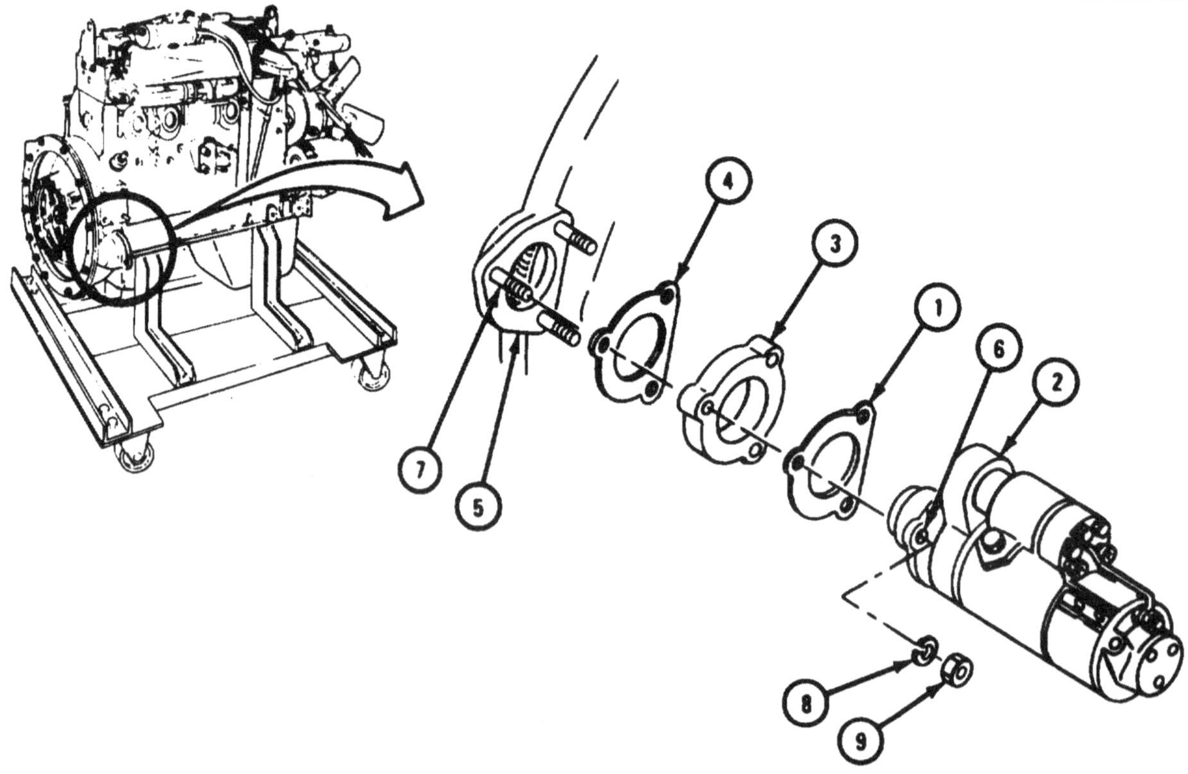

TA 117767

b. Engines LDS 465-1, LDS 465-1A, and LDS 465-2.

FRAME 1

Soldier A 1. Put gasket (1) on flywheel housing (2).

WARNING

Starter (3) weighs fifty pounds. Be careful to hold it up firmly when putting it on mounting studs. Starter could fall and cause injury to personnel and damage to equipment.

Soldiers A and B 2. Hold starter (3) in place up to flywheel housing (2). Aline starter mounting holes (4) with flywheel housing studs (5) and push starter straight into place.

Soldier A 3. Hold starter (3) in place until soldier B puts on lockwashers (6) and nuts (7).

Soldier B 4. Put on three lockwashers (6) and three nuts (7). Tighten three nuts to 140 to 150 pound-feet.

END OF TASK

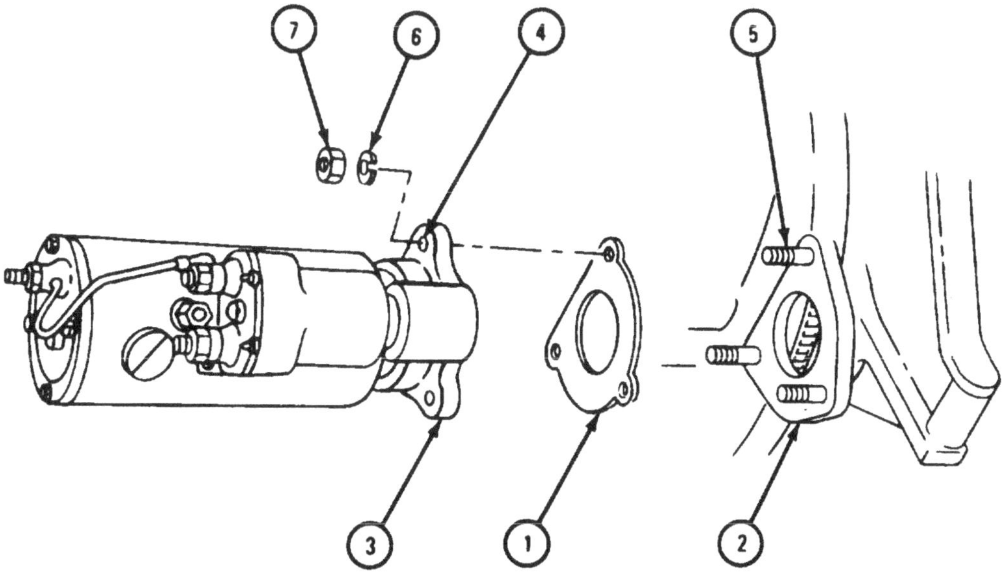

TA 117768

CHAPTER 6

DYNAMOMETER TESTS AND RUN-IN PROCEDURES

Section I. SCOPE

6-1. EQUIPMENT ITEMS COVERED . This chapter gives the tests and run-in procedures that are needed after the engine has been repaired as directed by the maintenance allocation chart.

6-2. EQUIPMENT ITEMS NOT COVERED. All equipment items for which corrective maintenance is authorized at the direct support and general support maintenance levels are covered in this chapter.

Section II. RUN-IN, TESTS, AND ADJUSTMENTS

TOOLS: No special tools required

SUPPLIES: Diesel fuel VV-F-800

PERSONNEL: TWO

EQUIPMENT CONDITION: Engine on engine transport stand.

6-3. MOUNTING ENGINE ON TEST STAND.

a. Installing Engine Mounting Brackets.

FRAME 1

NOTE

Tell machine shop to make two guide bolts (1). See figure 6-1.

1. Take out two bolts (2).
2. Put in two guide bolts (1) in place where two bolts (2) were taken out.
3. Take out four other bolts (2).

GO TO FRAME 2

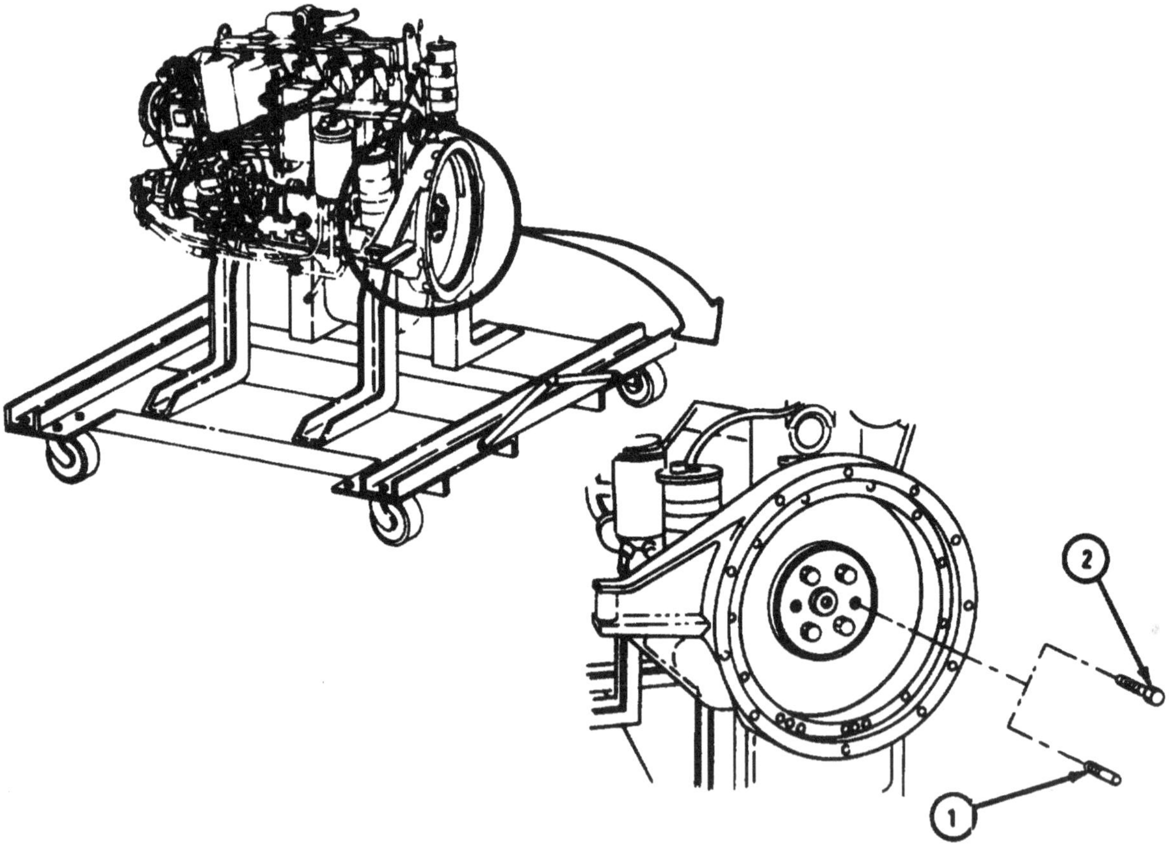

TA 118620

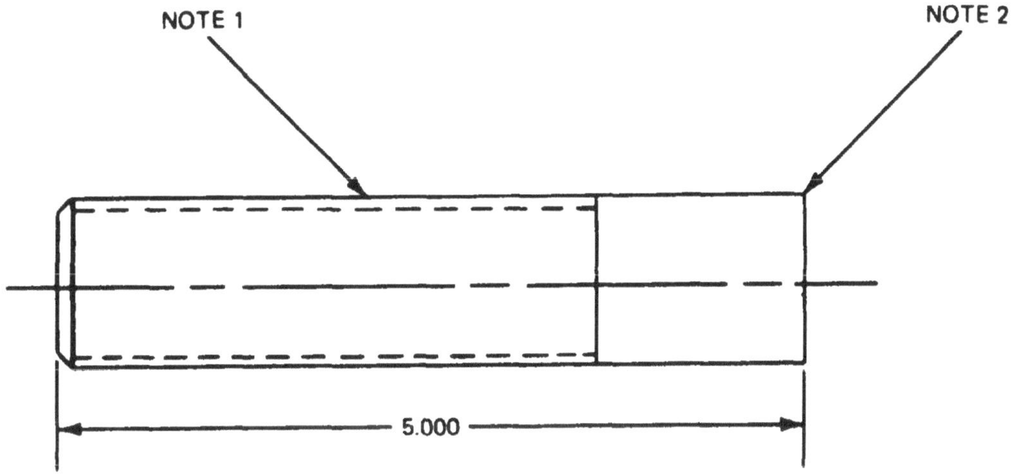

NOTES:
1. MATERIAL - USE 5-INCH LONG BOLT x 1/2-INCH DIAMETER x 20 THREADS PER INCH
2. CUT OFF HEAD IN NOTE 2. GRIND OFF SHARP EDGES

TA 118622

Figure 6-1. Guide Bolt Fabrication Instructions.

FRAME 2

NOTE

Tell machine shop to make flywheel adapter (1). See figure 6-2.

1. Put flywheel adapter (1) over two guide bolts (2) as shown.
2. Put in four bolts (3). Take out two guide bolts (2).

GO TO FRAME 3

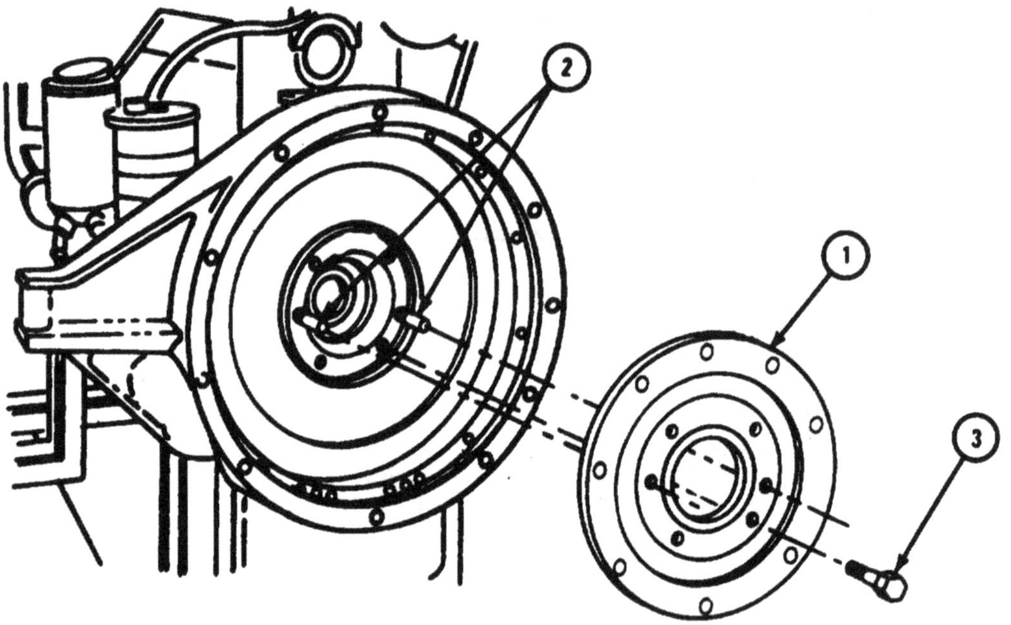

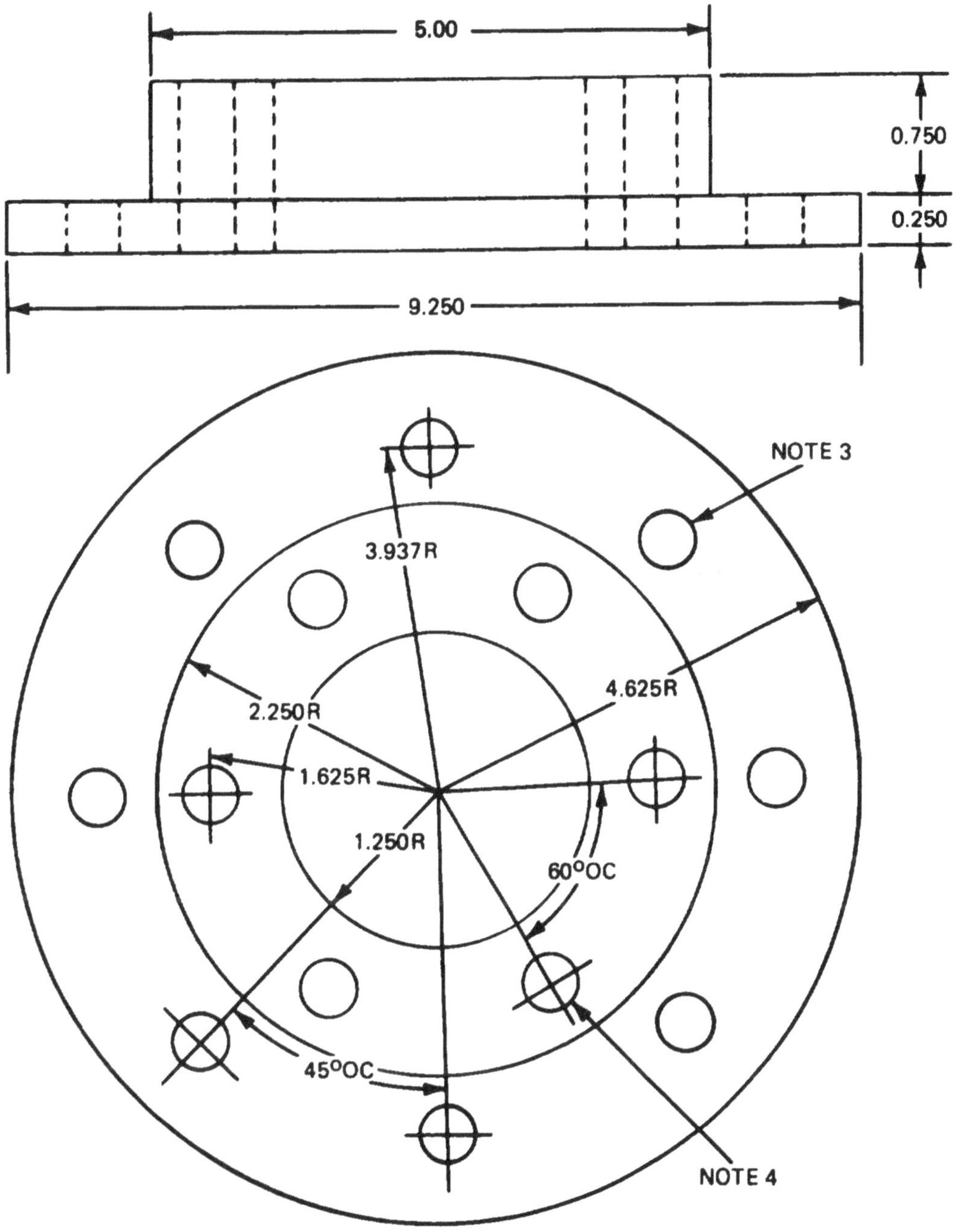

NOTES:
1. ALL DIMENSIONS GIVEN ARE IN INCHES
2. MATERIAL: COLD ROLL STOCK
3. DRILL 8 HOLES 5/16 IN. TAP HOLES USING 3/8 -16 TAP
4. DRILL SIX HOLES 1/2 IN.

TA 118621

Figure 6-2. Flywheel Adapter Fabrication Instructions.

FRAME 3

1. Put in two bolts (1).
2. Tighten six bolts (1) to 80 to 90 pound-feet, and then tighten bolts to 115 to 120 pound-feet .

GO TO FRAME 4

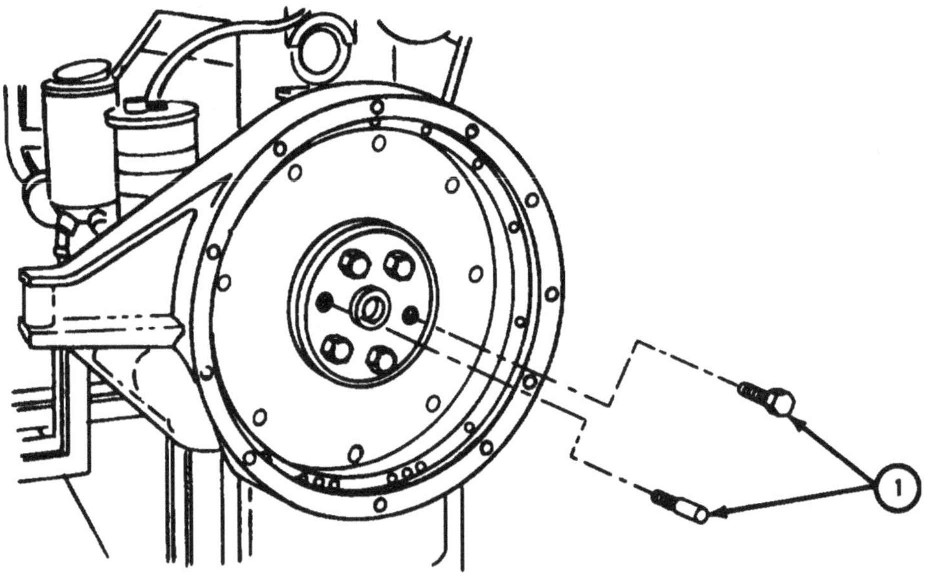

TA 118624

FRAME 4

NOTE

Tell machine shop to make flywheel rear mounting plate (1). See figure 6-3 .

1. Put flywheel rear mounting plate (1) on flywheel housing (2) as shown.
2. Put in 12 cap screws (3) and lockwashers (4).

GO TO FRAME 5

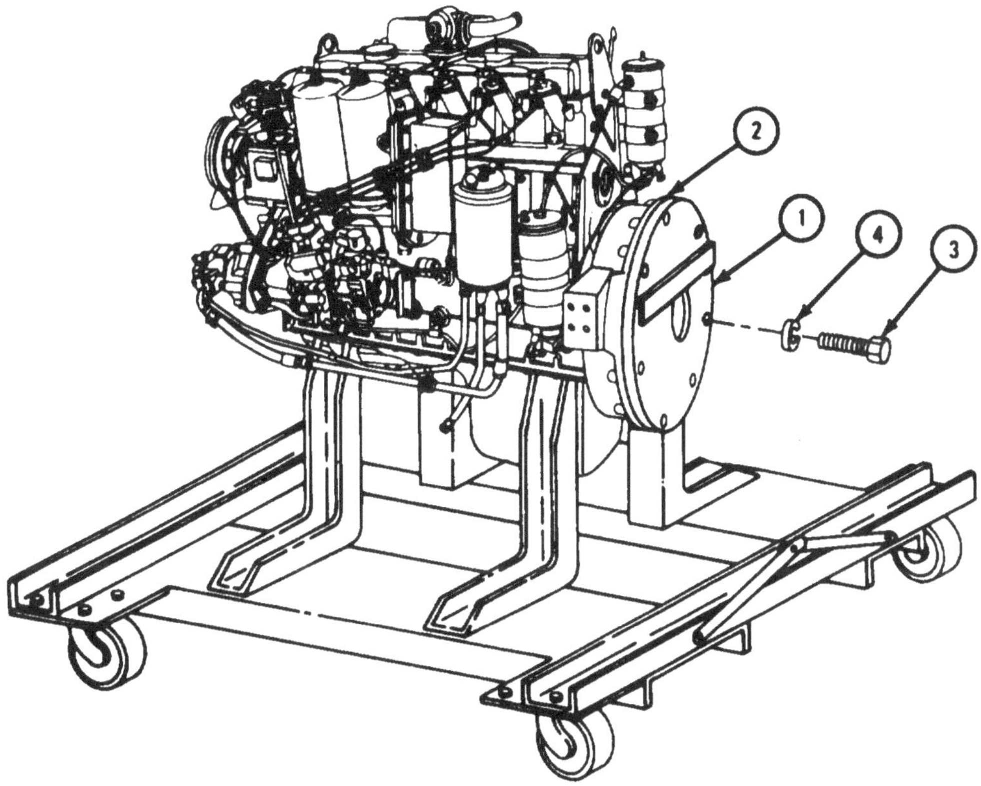

TA 118625

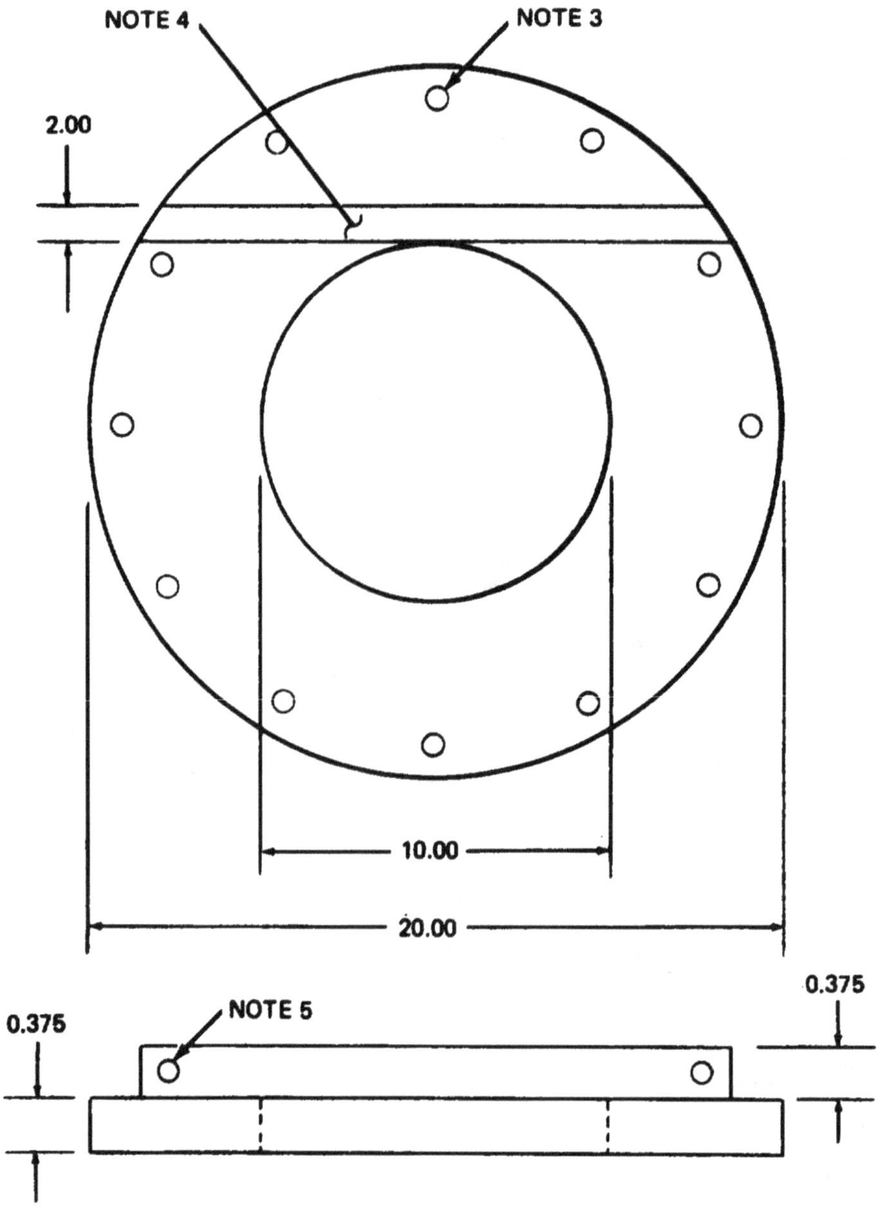

NOTES:

1. ALL DIMENSIONS GIVEN ARE IN INCHES
2. MATERIALS: COLD ROLL STOCK
3. DRILL TWELVE 1/2-IN, HOLES USING FLYWHEEL HOUSING AS A TEMPLATE
4. WELD 3/8 X 2 IN. ANGLE IRON AS SHOWN
5. DRILL TWO 3/4-IN. HOLES USING TEST STAND REAR MOUNTING BRACKET AS A TEMPLATE

Figure 6-3. Flywheel Rear Mounting Plate Fabrication Instructions.

FRAME 5

1. Put front engine adapter (1) in place.
2. Put in four bolts (2) and lockwashers (3).

END OF TASK

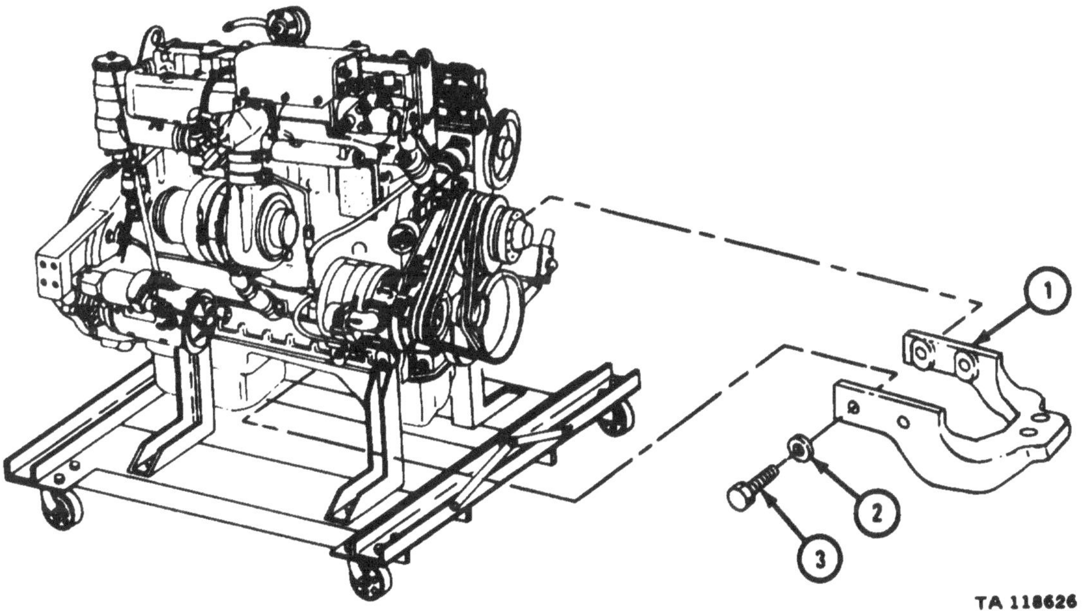

b. Mounting Engine.

WARNING

Make sure all power switches are set to OFF position. Test equipment can cause an electrical shock and injure personnel.

FRAME 1

	1.	Check that main switch (1) is in OFF position.
	2.	Loosen four bolts (2) on two brackets (3).
Soldiers A and B	3.	Slide two mounting brackets (3) toward front of test stand (4) enough for engine to clear drive shaft (5).

GO TO FRAME 2

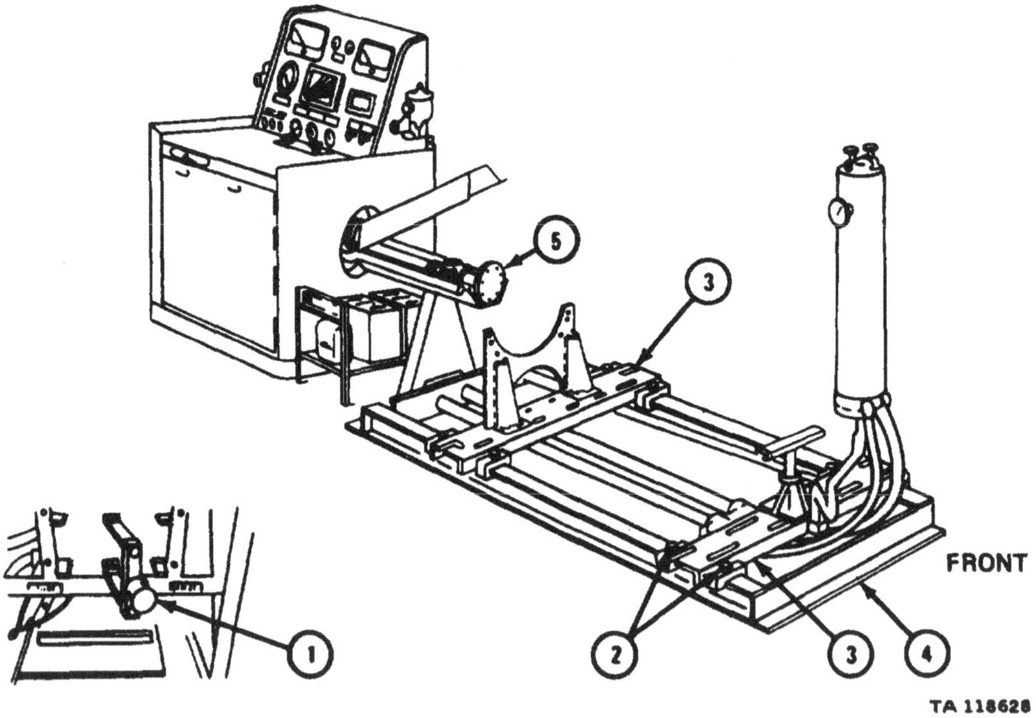

FRAME 2

Soldiers A and B	1.	Move engine transport stand (1) over to dynamometer test stand.
	2.	Hook chain sling (2) to two engine lifting brackets (3) as shown.
	3.	Using hoisting equipment, lift engine assembly (4) off engine transport stand (1) .

GO TO FRAME 3

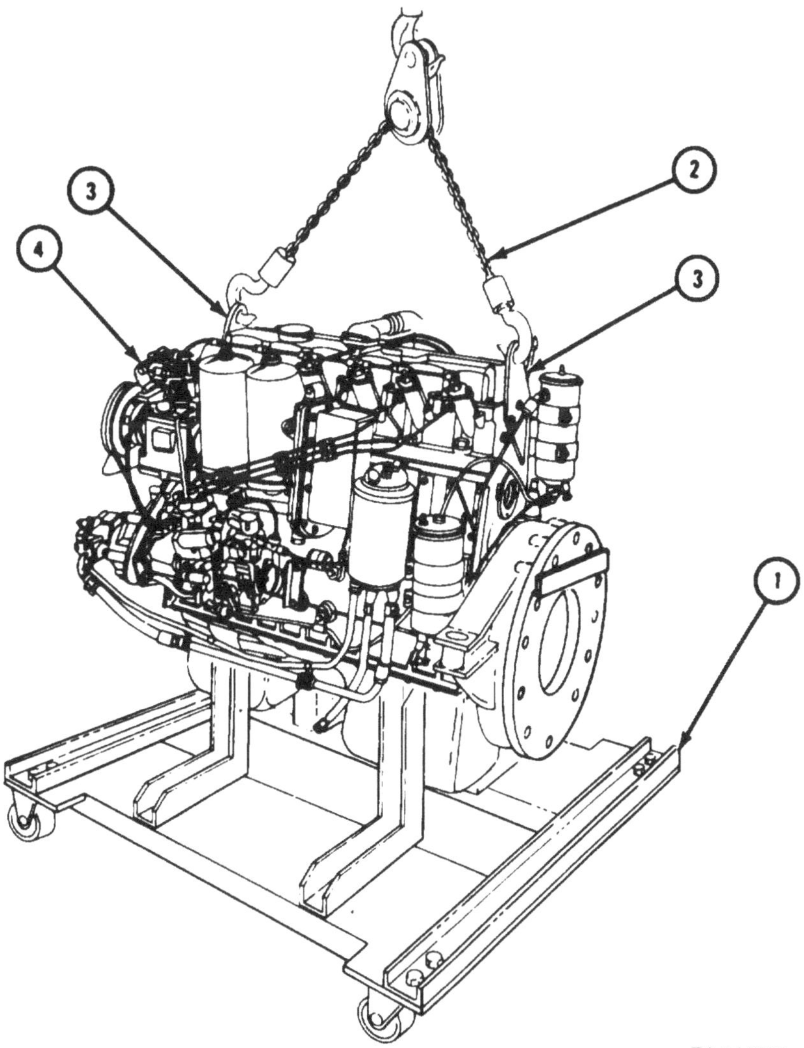

TA 118629

FRAME 3

Soldier A 1. Using hoisting equipment, raise or lower engine assembly (1) as needed when soldier B is ready.

Soldier B 2. Guide engine assembly (1) onto two dynamometer test stand brackets (2 and 3) .

3. Aline two screw holes in bracket (3) with two screw holes in mounting plate (4).

4. Put two rubber washers (5) between bracket (3) and mounting plate (4) as shown. Put in two bolts (6), two washers (7), and two nuts (8) . Do not tighten bolts or nuts.

GO TO FRAME 4

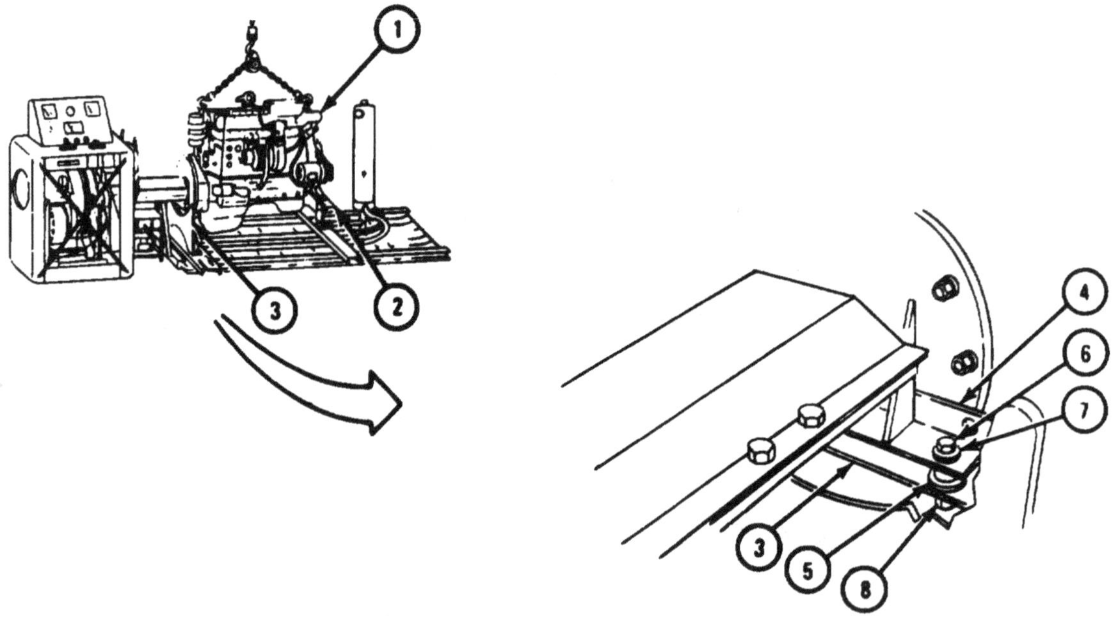

TA 118630

FRAME 4

Soldier B 1. Aline screw holes in front engine adapter (1) with screw holes in bracket (2).

2. Put in two bolts (3), two lockwashers (4), and two nuts (5). Do not tighen nut or bolts.

Soldier A 3. Let engine assembly (6) rest on brackets (2 and 7).

GO TO FRAME 5

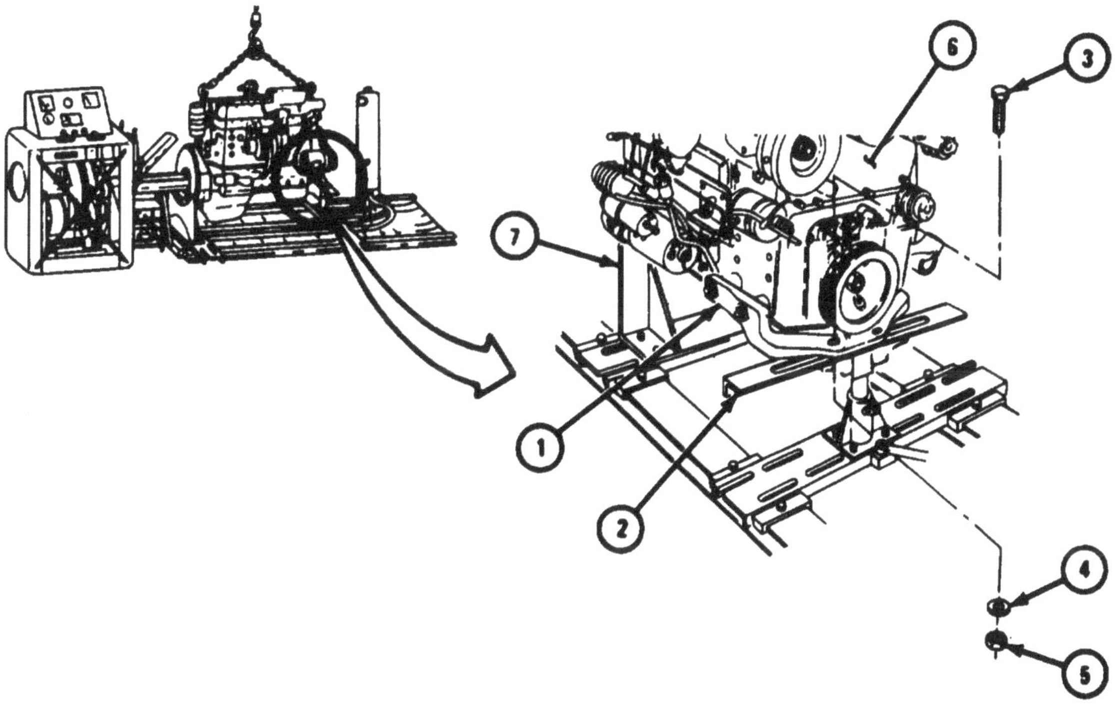

TA 118631

FRAME 5

Soldier A 1 . Loosen four bolts (1), two on each side of cover (2).

Soldier B 2 . Loosen two hinge bolts (3) at rear of cover (2).

Soldier A 3 . Slide cover (2) to the front and lift it up.

Soldier B 4 . Tighten two hinge bolts (3) to hold cover (2) in open position.

GO TO FRAME 6

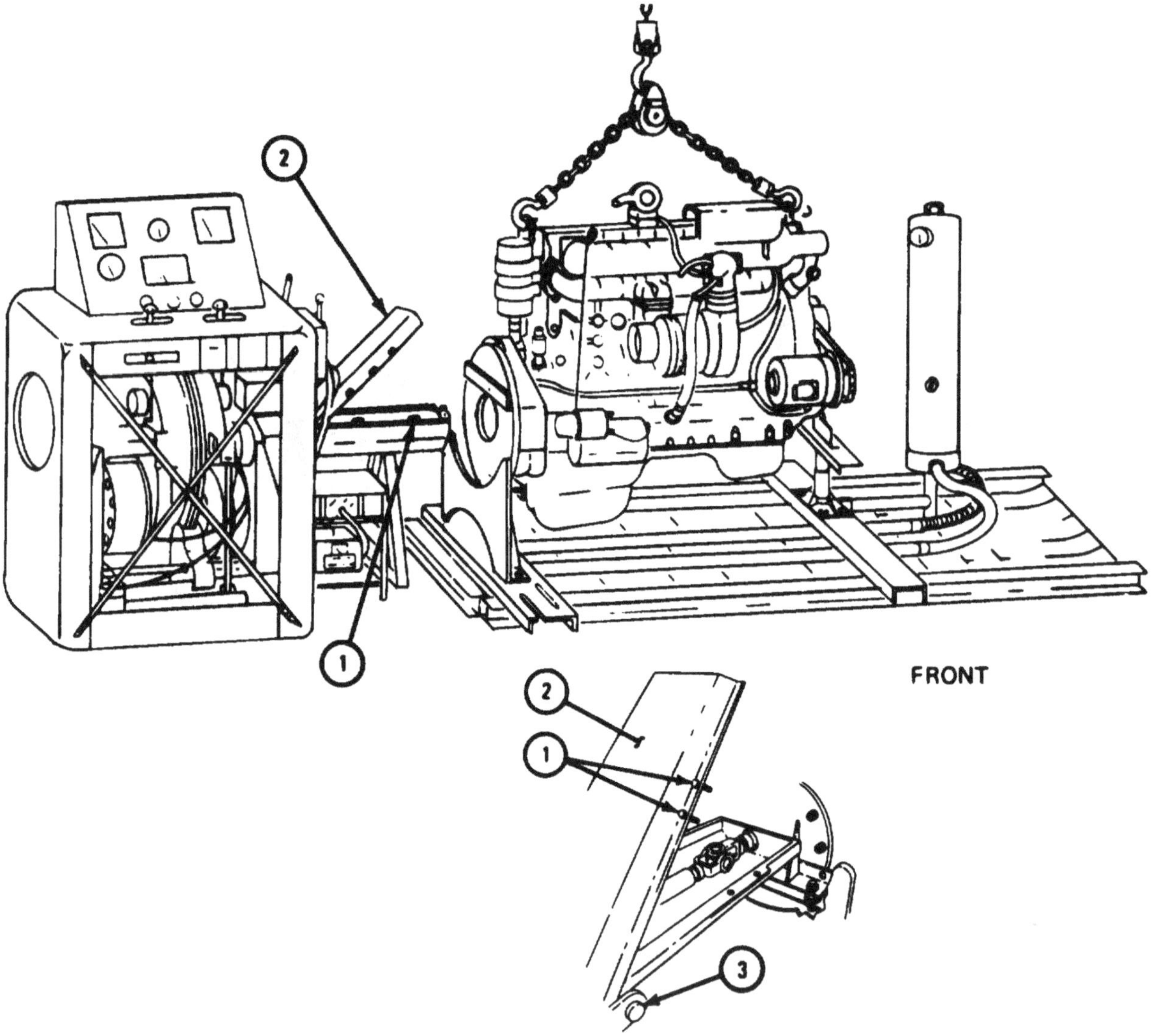

TA 118632

FRAME 6

Soldier A	1.	Using hoisting equipment, raise engine assembly (1) just enough to take weight off dynamometer test stand (2).
Soldiers A and B	2.	Slide engine assembly (1) to the rear until eight screw holes in flywheel adapter (3) are alined with screw holes in drive shaft (4).
Soldier A	3.	Hold drive shaft (4) in place so screw holes stay alined.
Soldier B	4.	Put in eight bolts (5) and lockwashers (6). Do not tighten bolts.
Soldier A	5.	Lower engine assembly (1) all the way down and take off chain sling (7).

GO TO FRAME 7

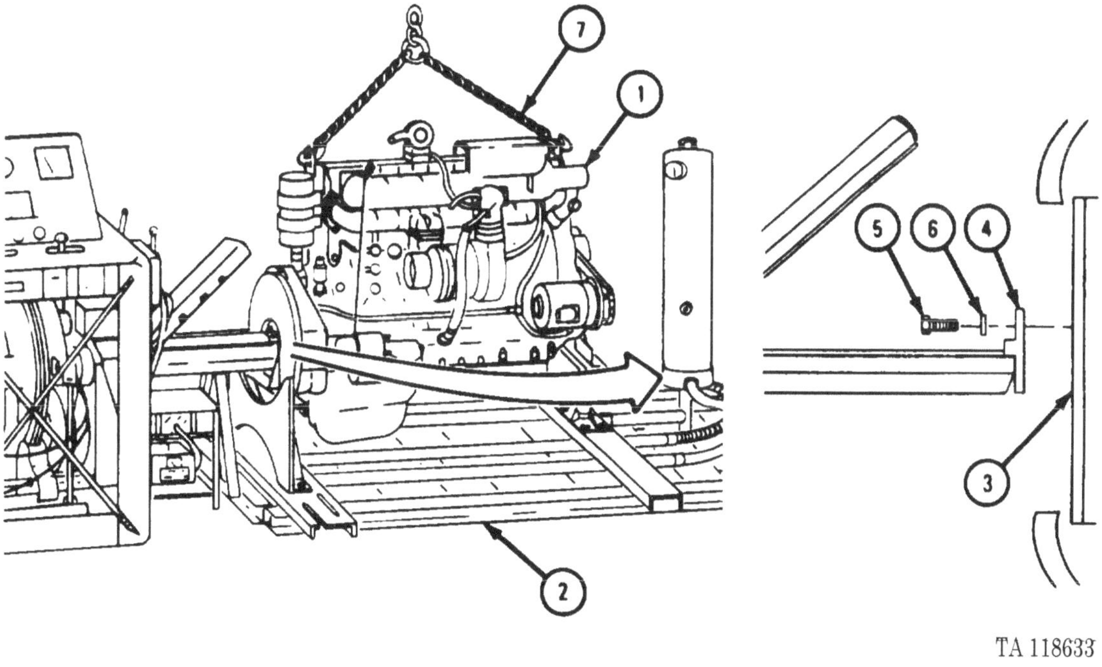

FRAME 7

1. Tighten eight bolts (1) .
2. Tighten two bolts (2) and nuts (3).
3. Tighten two bolts (4) and nuts (5).

GO TO FRAME 8

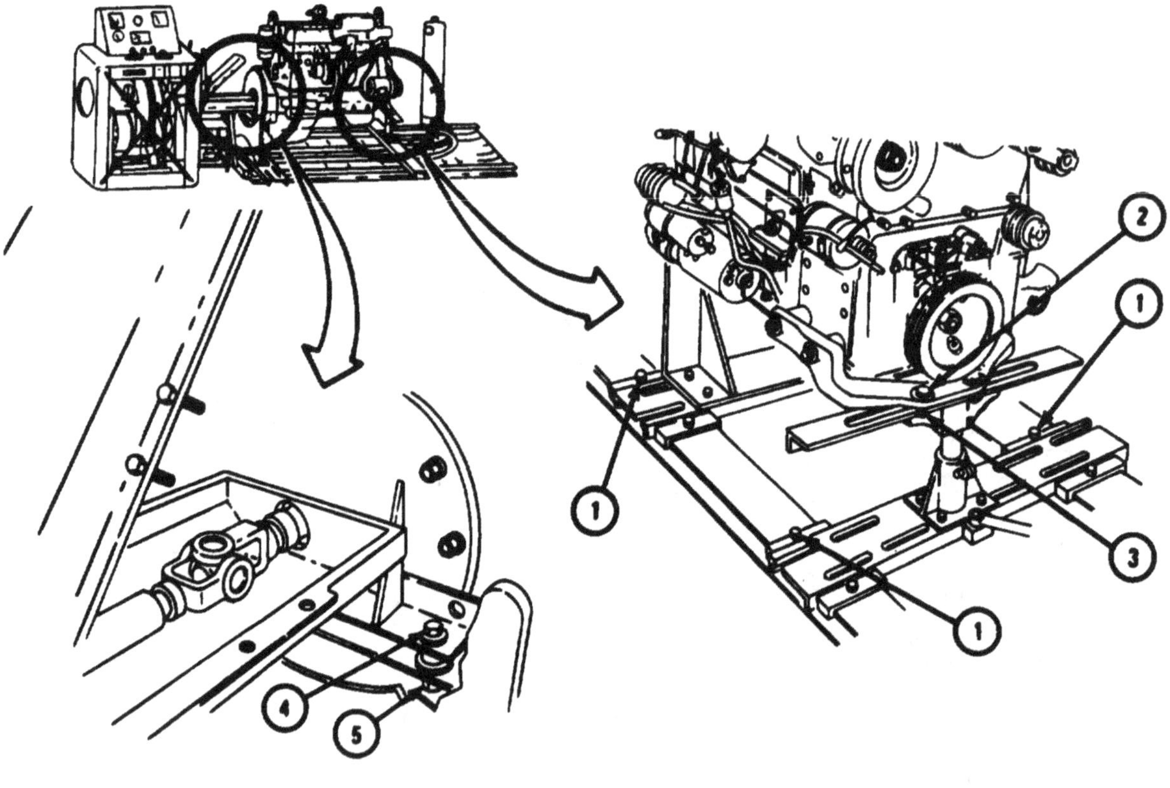

TA 118634

FRAME 8

1. Tighten eight bolts (1).
2. Loosen two hinge bolts (2) and close cover (3). Tighten four bolts (4).
3. Tighten two hinge bolts (2).

GO TO FRAME 9

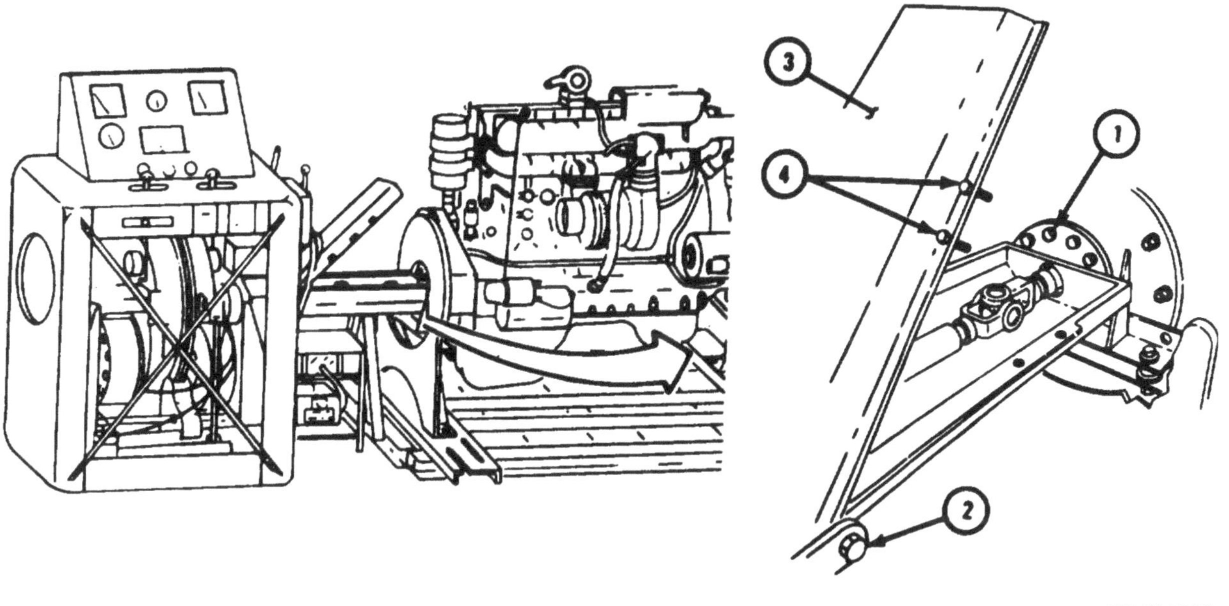

TA 118635

FRAME 9

Soldier A 1.	Loosen locking bolt (1) .
Soldier B 2.	Put carpenter's level on top of rocker arm cover (2). Tel 1 soldier A to raise or lower engine assembly (3) until it is level.
Soldier A 3.	Tighten locking bolt (1) .

END OF TASK

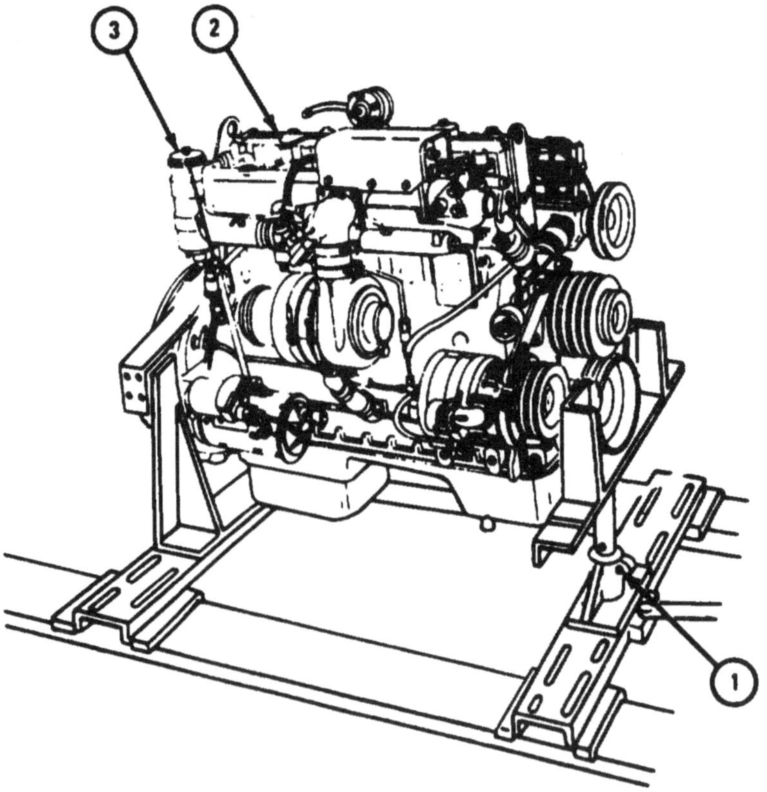

TA 118636

6-4. CONNECTING ENGINE TO DYNAMOMETER,

NOTE

In order to join engine to the dynamometer, it may be necessary to use different types of adapters or fittings other than shown. If fittings are not available, tell machine shop to make them.

FRAME 1

1. Take out oil gage rod (1).
2. Put two hose clamps (2) on outlet hose (3).
3. Put one end of outlet hose (3) on engine water outlet (4) and other end on test stand return tube (5) . Push two hose clamps (2) to ends of outlet hose and tighten.
4. Put two hose clamps (2) on inlet hose (6).
5. Put one end of inlet hose (6) on engine water inlet (7) and other end on test stand supply tube (8) . Push two hose clamps (2) to ends of inlet hose and tighten them.

GO TO FRAME 2

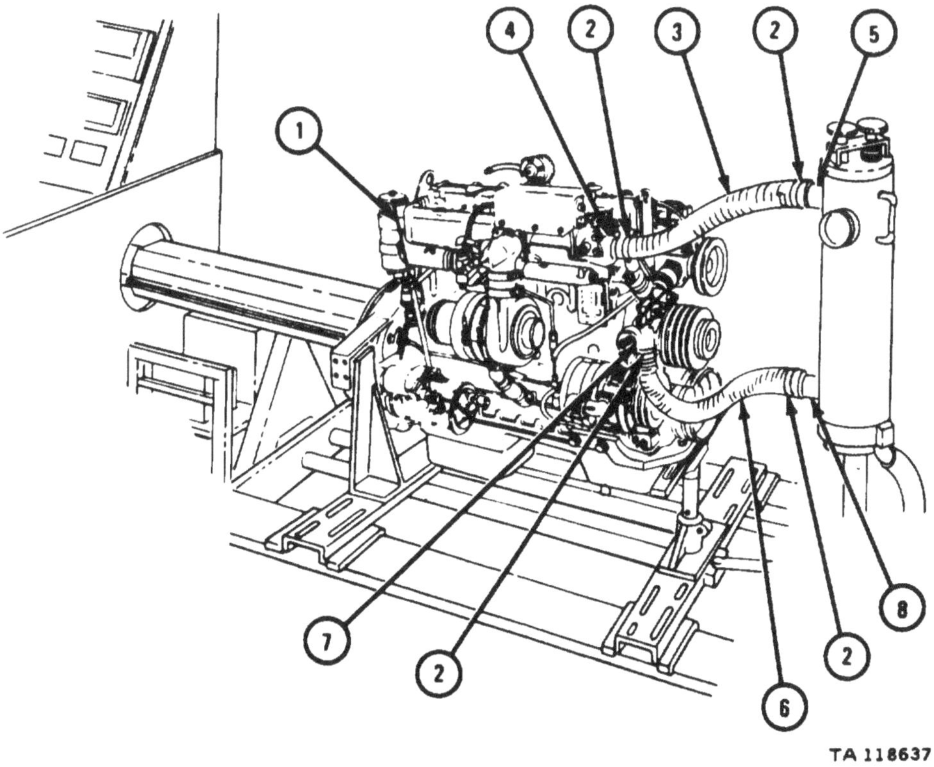

FRAME 2

1. Working on right side of engine, put outside exhaust tube (1) with clamp (2) on engine exhaust manifold outlet (3). Tighten clamp.

GO TO FRAME 3

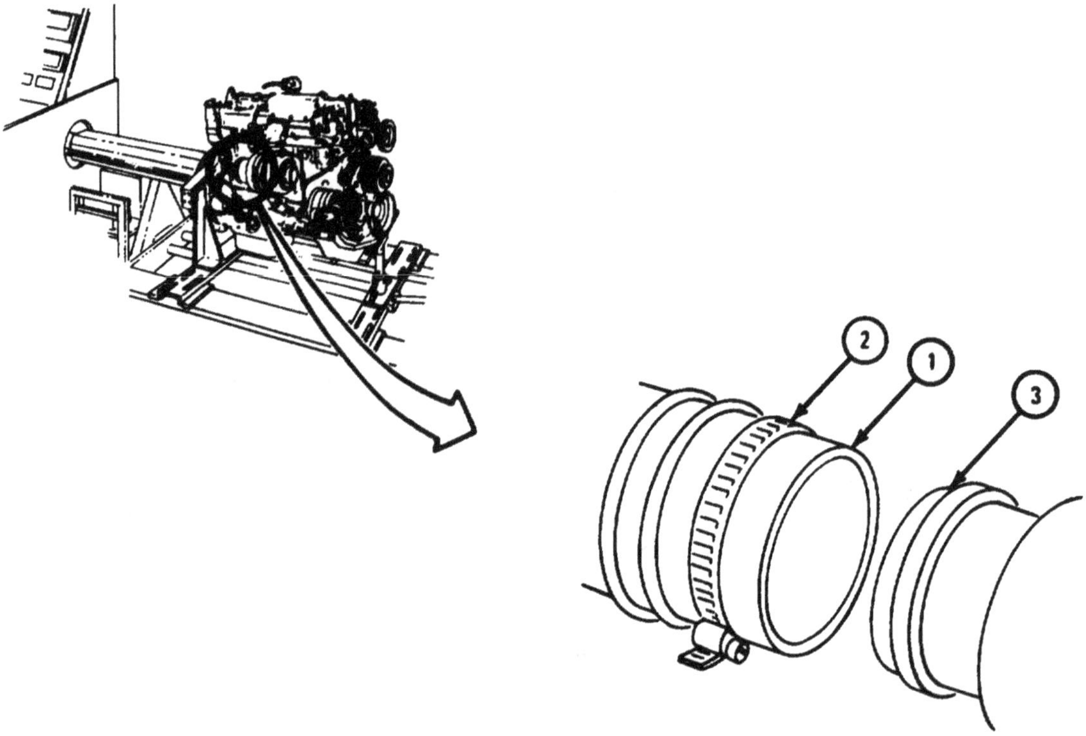

TA 118638

FRAME 3

1. Take off fuel line (1) from fuel pump (2).
2. Using correct adapters and fittings, put in dynamometer fuel line (3) where fuel line (1) was taken out.
3. Take off fuel return line (4) from fuel pump (2).
4. Using correct adapters and fittings, put in dynamometer fuel return line (5) where fuel return line (4) was taken out.

GO TO FRAME 4

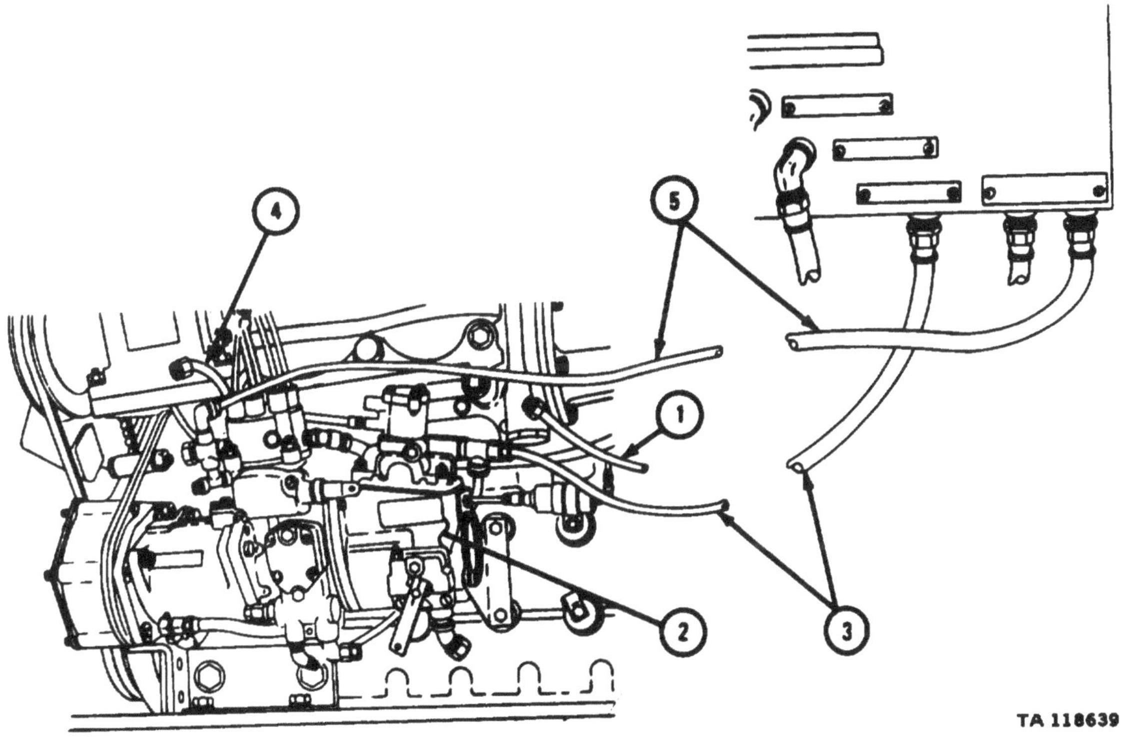

FRAME 4

1. Join dynamometer tachometer drive shaft (1) to tachometer drive (2).

GO TO FRAME 5

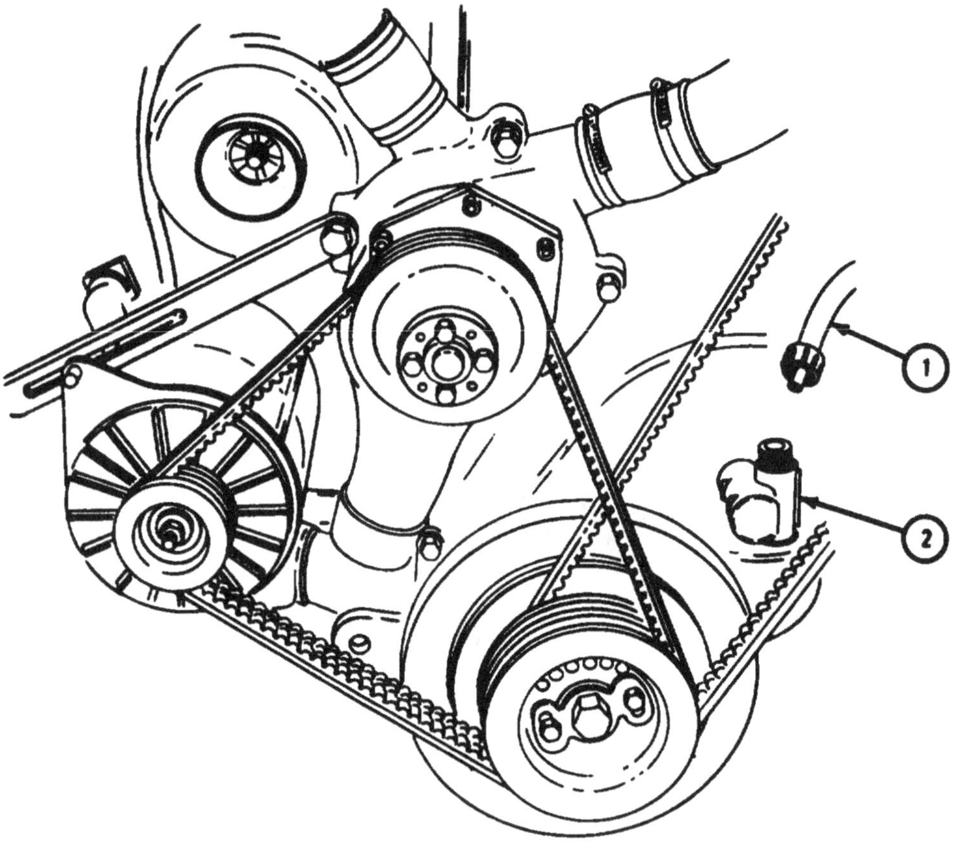

FRAME 5

1. Using correct adapters and fittings, join dynamometer oil pressure line (1) to engine oil pressure fitting (2).
2. Using correct adapters and fittings, join dynamometer manifold pressure line (3) to engine manifold connector (4).

GO TO FRAME 6

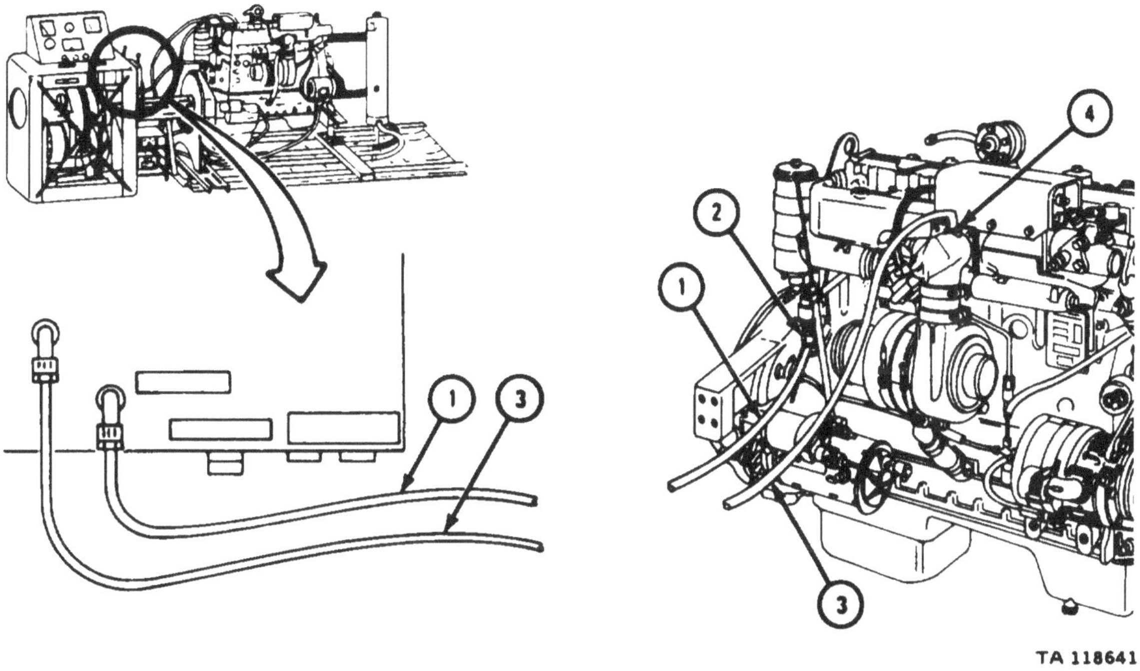

FRAME 6

1. Take off nut (1) and washer (2).
2. Take off nut (3) and washer (4).
3. Put positive (+) cable (5) on starter stud (6) . Put on nut (1) and washer (2).
4. Put starter solenoid wire (7) on starter stud (8). Put on nut (3) and washer (4) .

GO TO FRAME 7

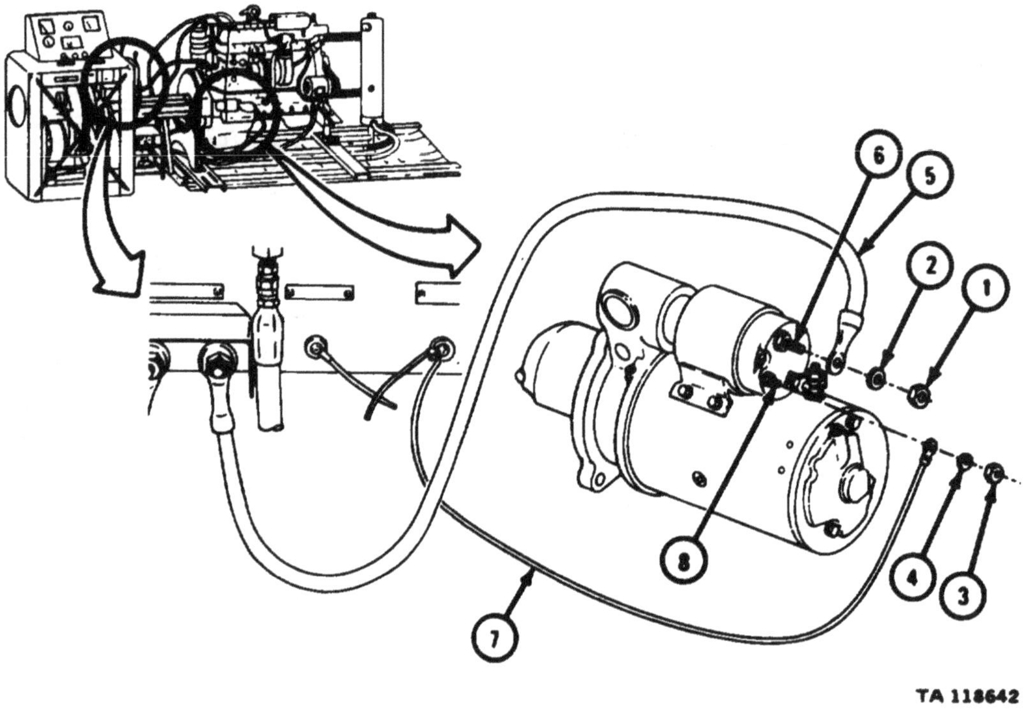

FRAME 7

1. Join dynamometer throttle linkage (1) to engine governor linkage (2).
2. Put manometer tube (3) on crankcase breather tube (4).
3. Put oil temperature gage (5) in oil gage tube (6).

GO TO FRAME 8

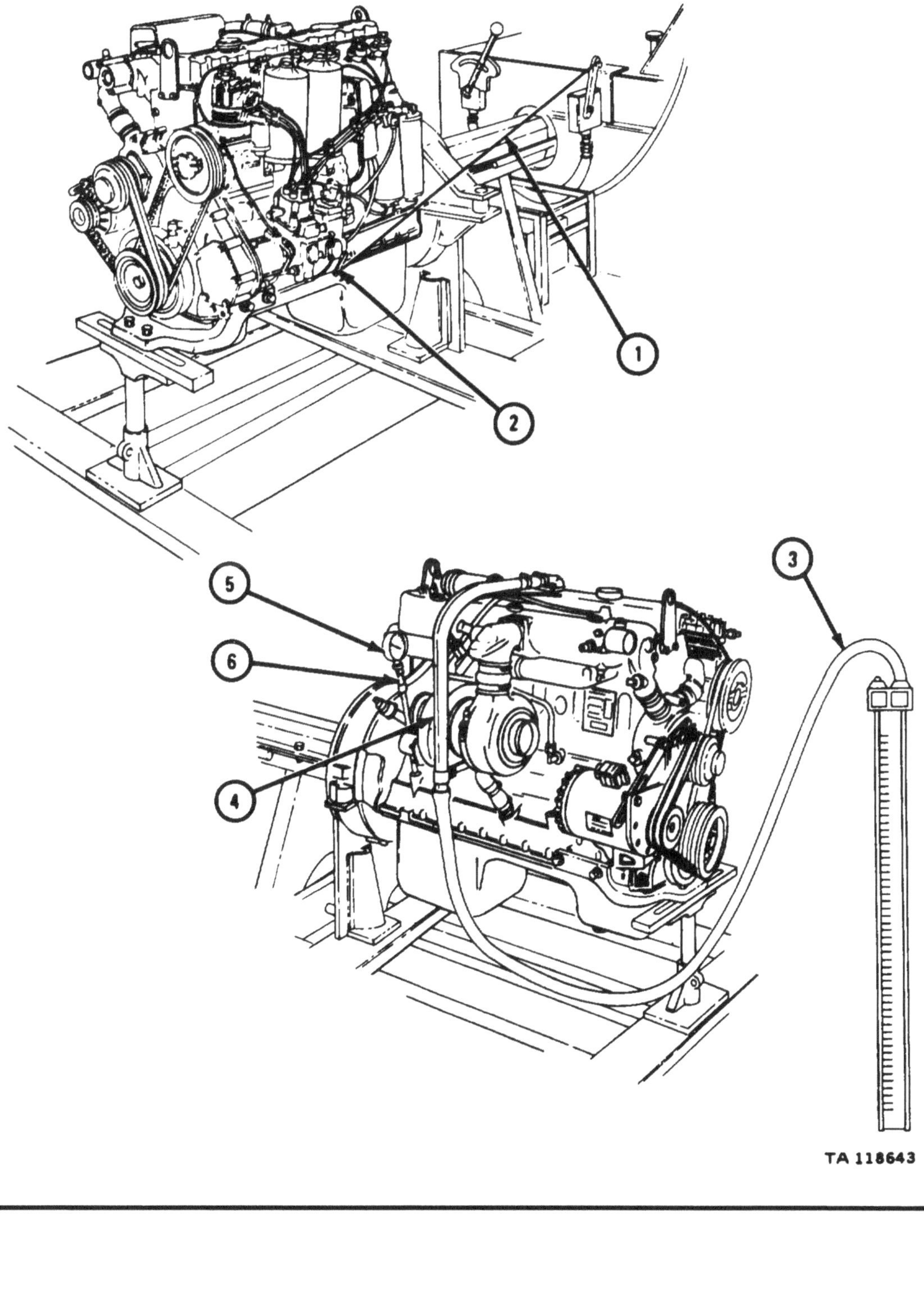

TA 118643

FRAME 8

1. Take out plug (1) from fuel filter (2) on engine (3).
2. Join fuel pressure line (4) on test stand (5) to fuel filter (2) where plug (1) was taken out.

GO TO FRAME 9

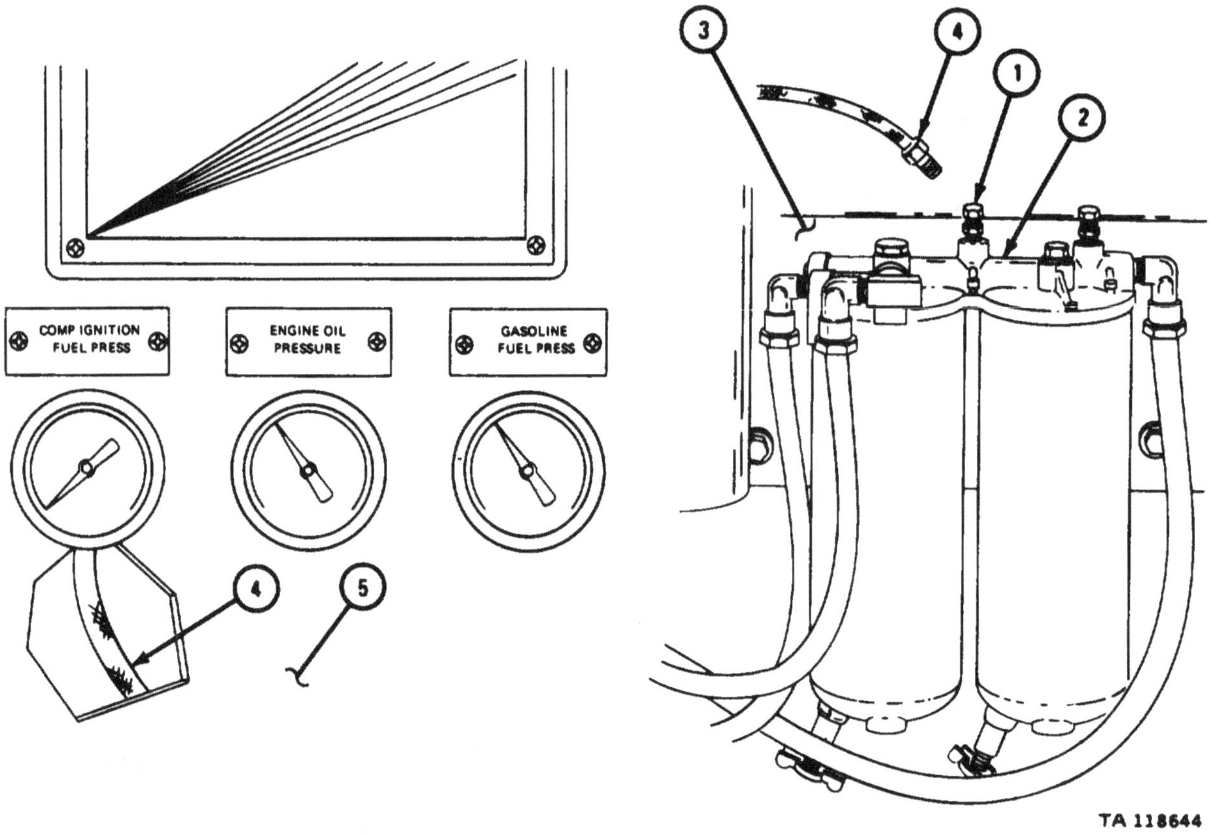

FRAME 9

1. Take oil filler cap (1) from valve cover (2).
2. Using correct adapter (3), put in 0 to 5-inch manometer (4).

GO TO FRAME 10

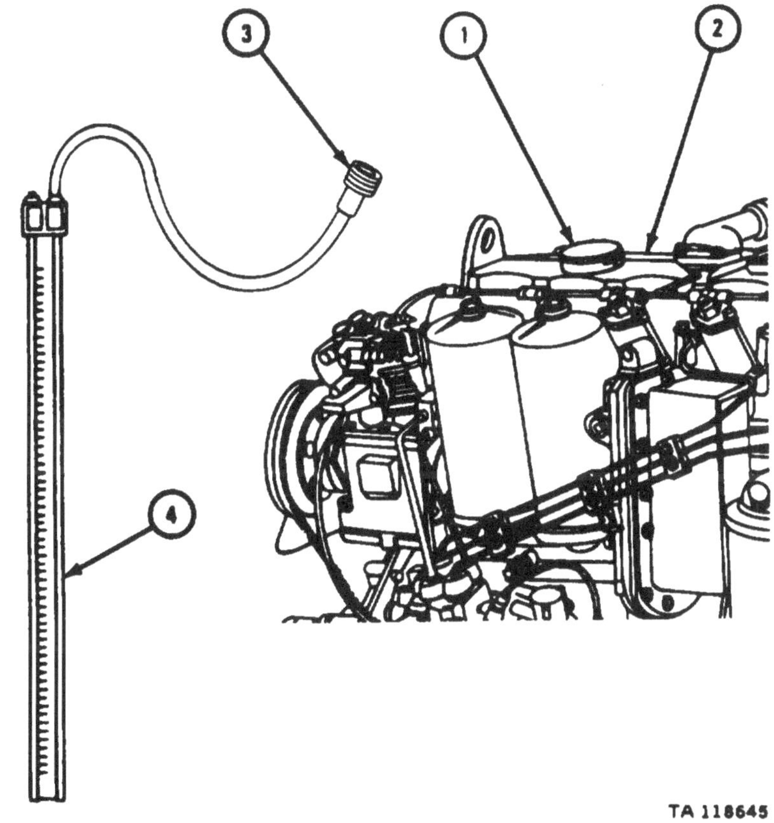

FRAME 10

1. Takeout plug (1) on engine (2) .

2. Put in engine oil pressure line (3) on test stand (4) where plug (1) was taken out .

END OF TASK

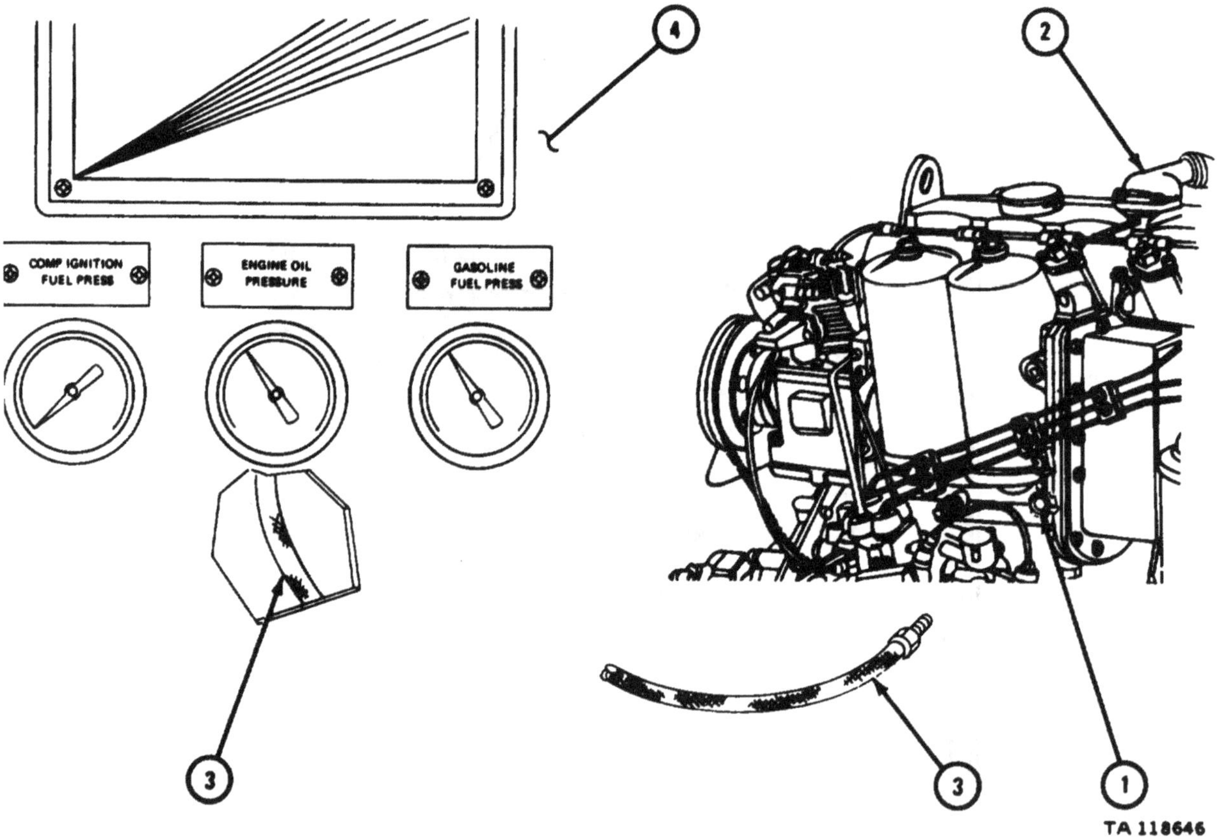

6-5. FLAME HEATER COMPONENT TEST.

NOTE

This task is the same for all intake manifold flame heater systems. The side - mounted flame heater system is shown for this task.

a. Flame Heater Fuel Pump Test.

FRAME 1

1. Take off fuel lines (1) from nozzle (2).
2. Put fuel line fitting ends (3) into container (4).
3. Turn heater to ON position for 15 seconds, then turn heater to OFF position.
4. Check that fuel is in container (4).
5. If no fuel is in container (4), test and repair flame heater system. Refer to para 4-58 through 4-80.
6. If fuel is in container (4), put back fuel lines (1) on nozzle (2).

END OF TASK

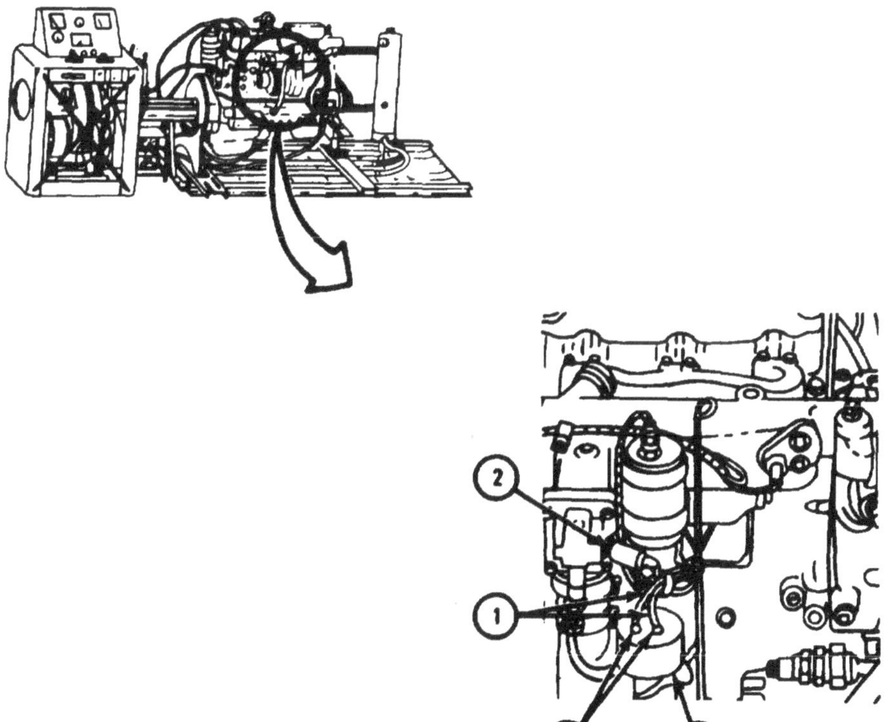

b. Flame Heater Nozzle Test.

FRAME 1

1. Take off two fuel line fittings (1).
2. Takeoff and shake flame heater nozzle (2) .
3. Put back fuel line fittings (1) into nozzle (2).
4. Put container (3) under nozzle (2) .
5. Turn heater to ON position for 15 seconds, then turn heater to OFF position.
6. Check for fuel in container (3).
7. If no fuel is in container (3), test and repair flame heater system. Refer to para 4-58 through 4-80.
8. If fuel is in container (3), put back flame heater nozzle (2).

END OF TASK

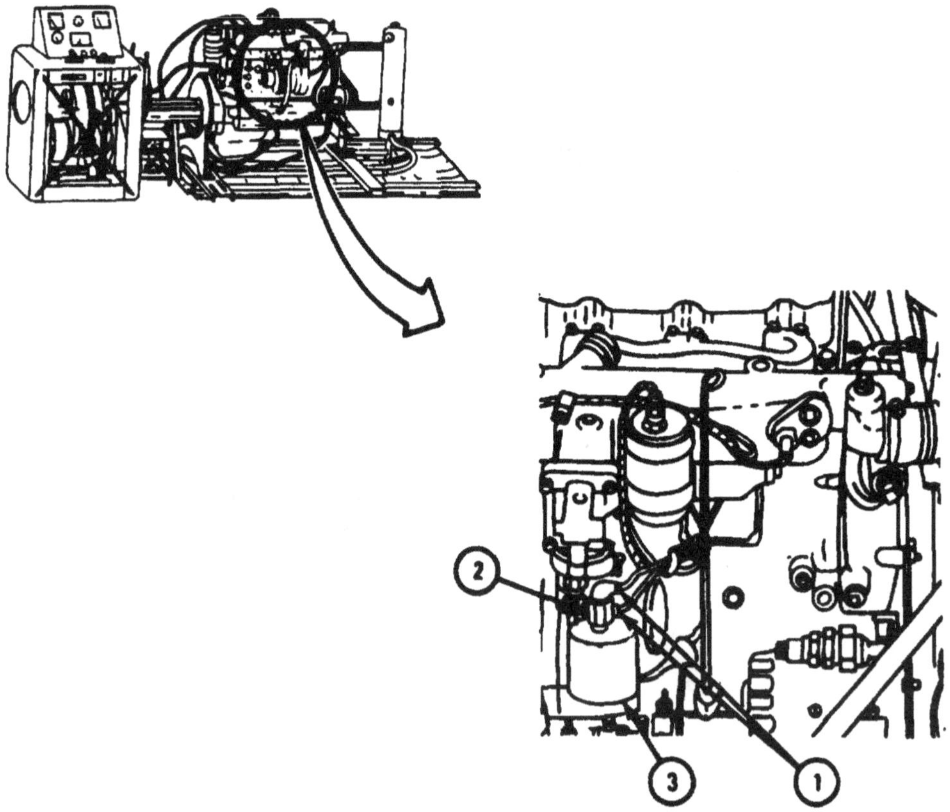

TA 118757

6-6. BASIC ENGINE RUN-IN.

WARNING

Be alert at all times during engine operation for exhaust odors and exposure symptoms. If either are present, immediately ventilate work area. Permanent brain damage or death can result from severe exposure.

Use hearing protection during this test. The test is very noisy and the noise can cause damage to your hearing.

a. Pre-Starting Instructions.

(1) Fill engine crankcase with oil. Refer to LO 9-2320-209-12/1 and LO 9-2320-211-12.

(2) Open dynamometer fuel supply valves.

(3) Open dynamometer water supply valves.

FRAME 1

1. Turn drain control knob (1) to right until finger tight.
2. Turn drain control knob (1) to left one turn.
3. Turn thermostat control knob (2) to right until water can be heard entering cooling column (3).
4. When water level in sight glass (4) no longer changes, turn thermostat control knob (2) to left until it stops.

GO TO FRAME 2

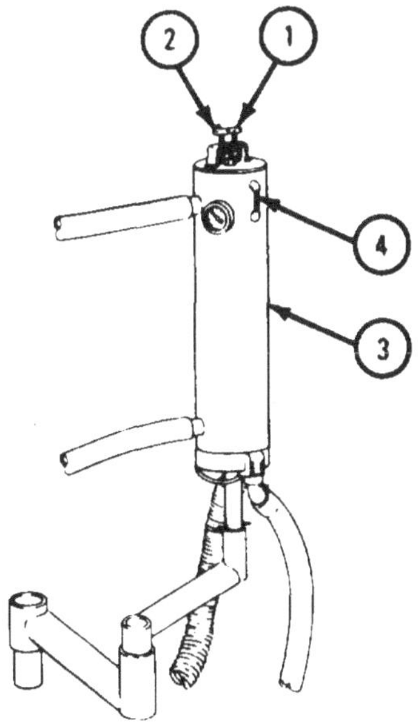

TA 118647

FRAME 2

1. Place main switch (1) in 24 VOLT position (2). Turn knob (3) on OVERSPEED GOVERNOR so that needle (4) points to 2900 RPM. Press RESET button (5) .
2. Place SPEED METER RANGE switch (6) in HI rpm position.
3. Place POWER METER RANGE switoh (7) in LO hp position.
4. Close load control (8) by turning to right until it is finger tight.
5. Open unload control (9) by turning to left.
6. Check and set throttle (10) so it will fully open and fully close engine speed control .

GO TO FRAME 3

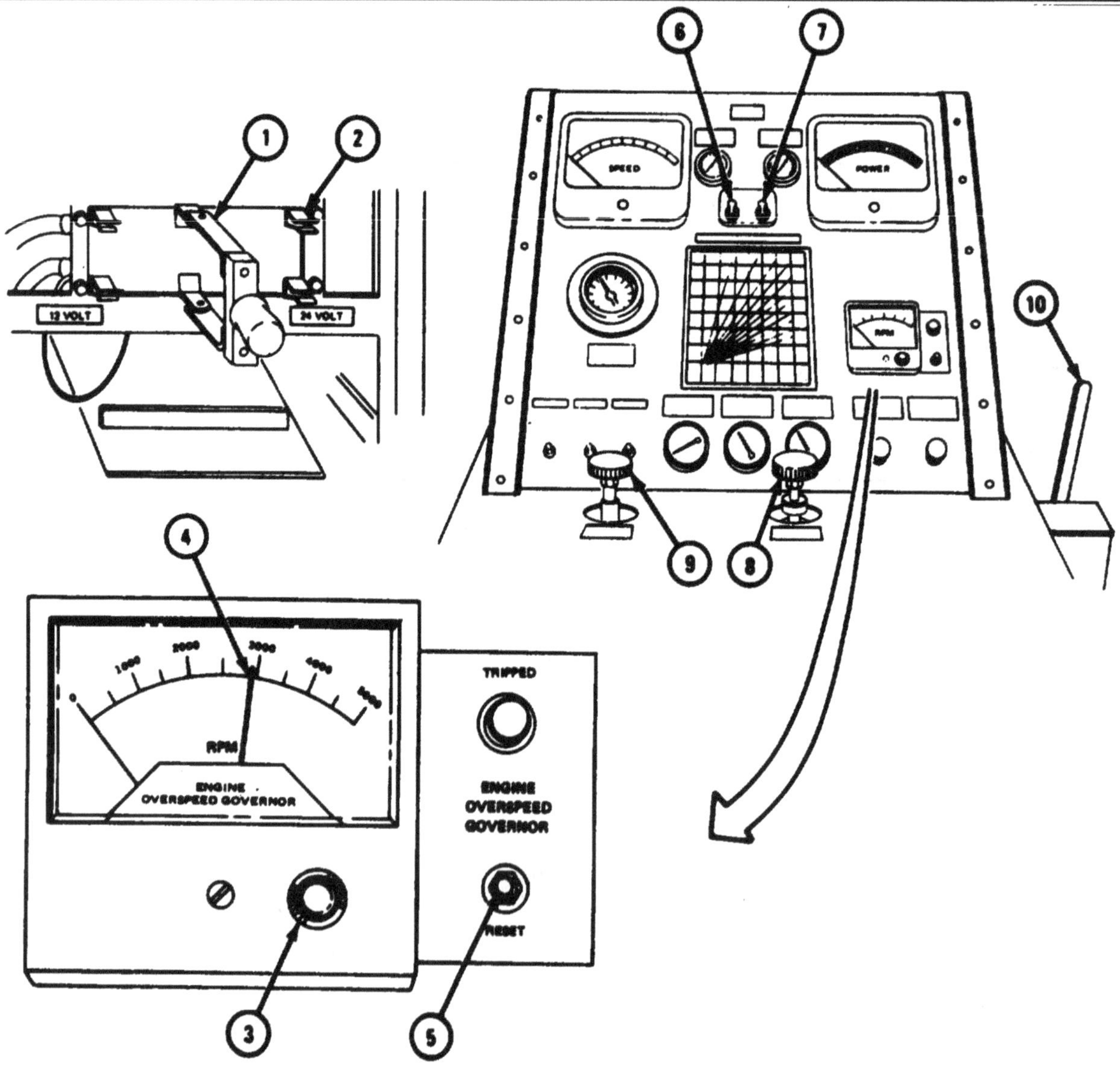

TA 118648

FRAME 3

1. Turn operating lever (1) on flow control valve (2) so it is vertical.
2. Pump fuel primer pump (3) 8 to 10 times to prime engine fuel pump.

END OF TASK

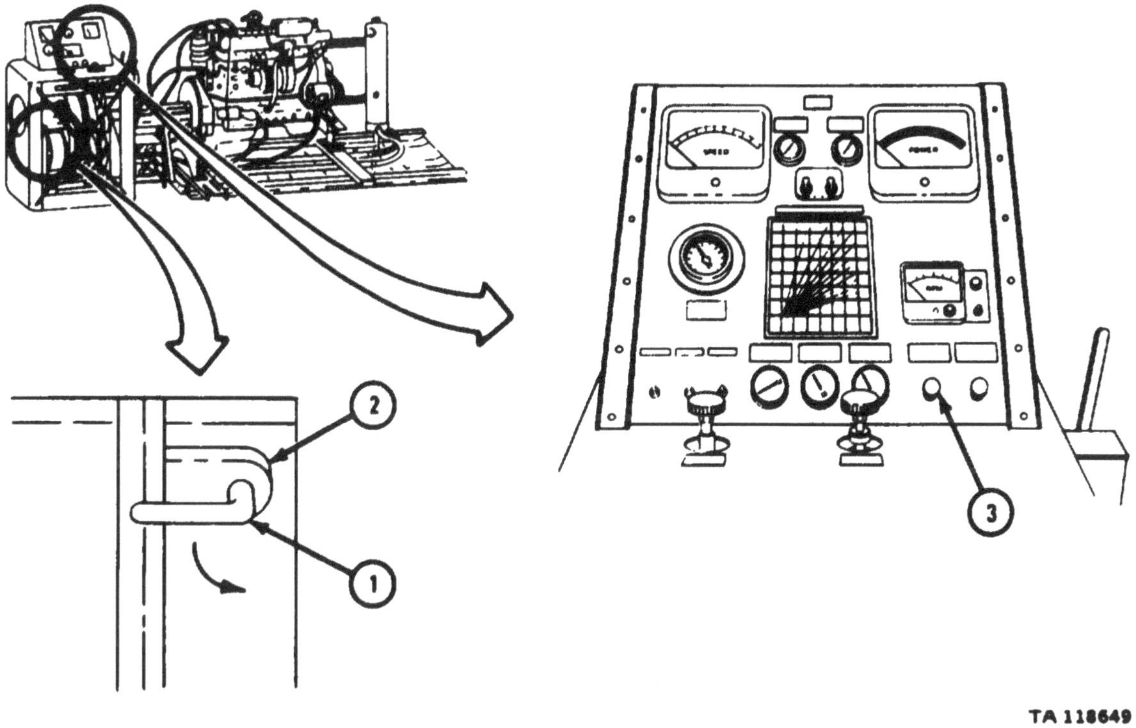

b. Engine Starting and Running Instructions.

FRAME 1

1. Open throttle (1) to idle position.
2. Press ENGINE START button (2).
3. When engine starts, adjust throttle (1) to set engine speed to 500 to 600 RPM on speed gage (3).
4. When water temperature gage (4) reaches 160° to 185°F, check oil temperature gage (5). Oil temperature should be 255° to 265°F. If temperature is too high, engine must be repaired.

END OF TASK

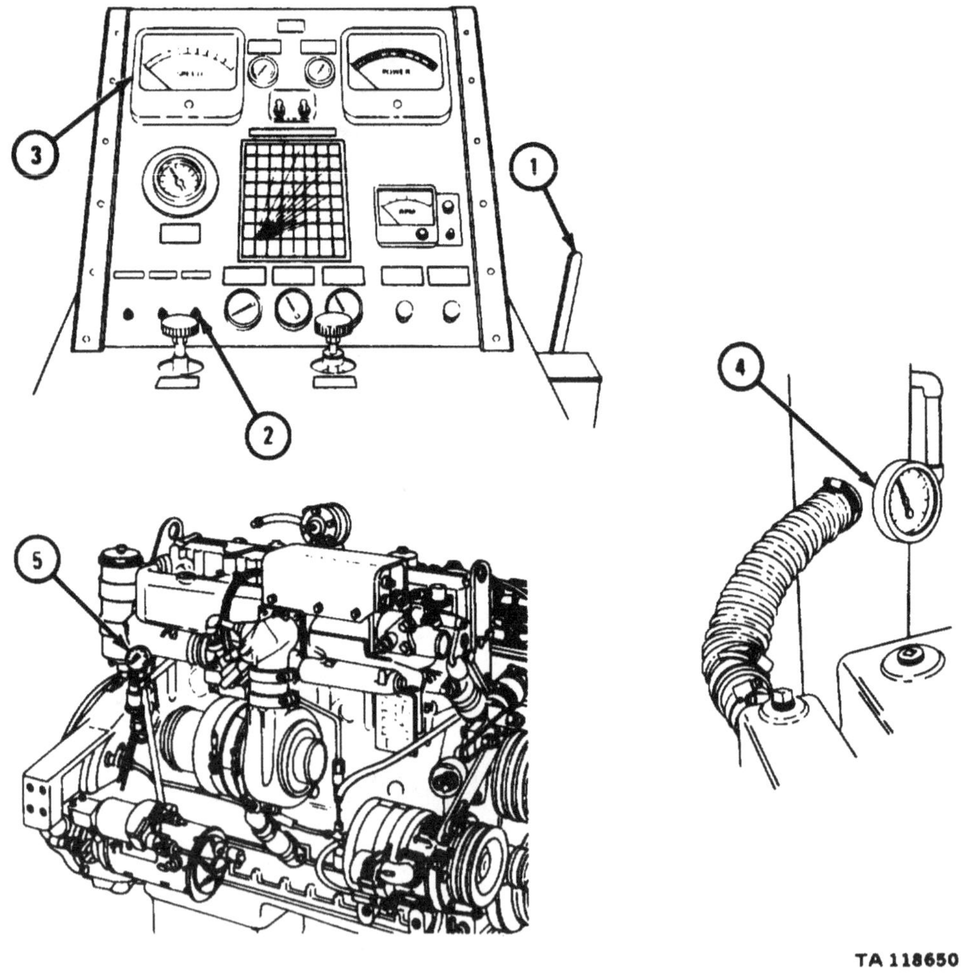

c. Test.

FRAME 1

1. Close unloading valve (1) by turning it to right until it is finger tight.
2. Slowly open loading valve (2) until POWER gage (3) shows 10 to 15 HP. Close loading valve (2) .
3. Set throttle (4) so SPEED gage (5) reads 500 to 600 RPM.
4. If POWER gage (3) reads high, open unloading valve (1). Close unloadin g valve when done. Reset throttle to 2600-2800 RPM.
5. Check ENGINE OIL PRESSURE gage (6). Reading should be 40 PSI. If reading is not within given limits, engine must be repaired.

NOTE

Check all lines and connections for leakage.

GO TO FRAME 2

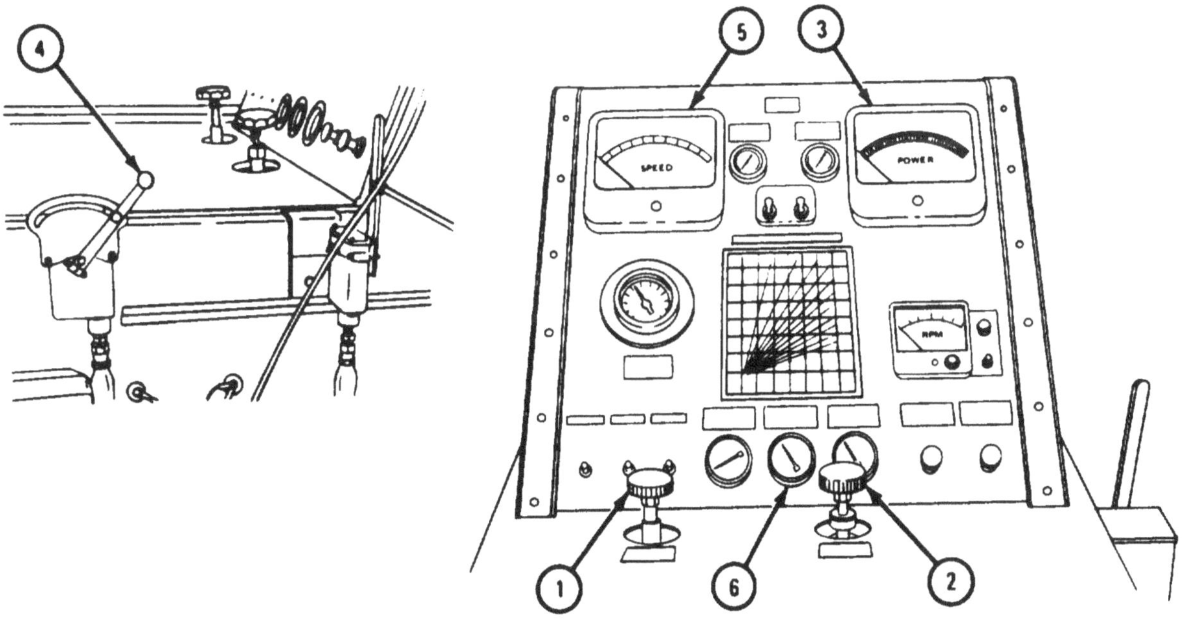

TA 118651

FRAME 2

NOTE

Fill in standard engine test report form used by the installation where this test is being done.

1. Open load control valve (1) until POWER gage (2) reads 15 HP. Close load control valve .
2. Set throttle (3) to 1800 RPM on SPEED gage (4).
3. If POWER gage (2) is above 15 HP, open unload valve (5) to bring it down to 15 HP. Close unload valve .
4. Reset throttle (3), if needed, to keep 1800 rpm.
5. Run engine at 1800 rpm at 15 HP for 10 minutes, then write down lubricating oil pressure and water temperature.
6. Open unload valve (5) until POWER gage (2) reads 1 to 5 HP.
7. Stop engine .

GO TO FRAME 3

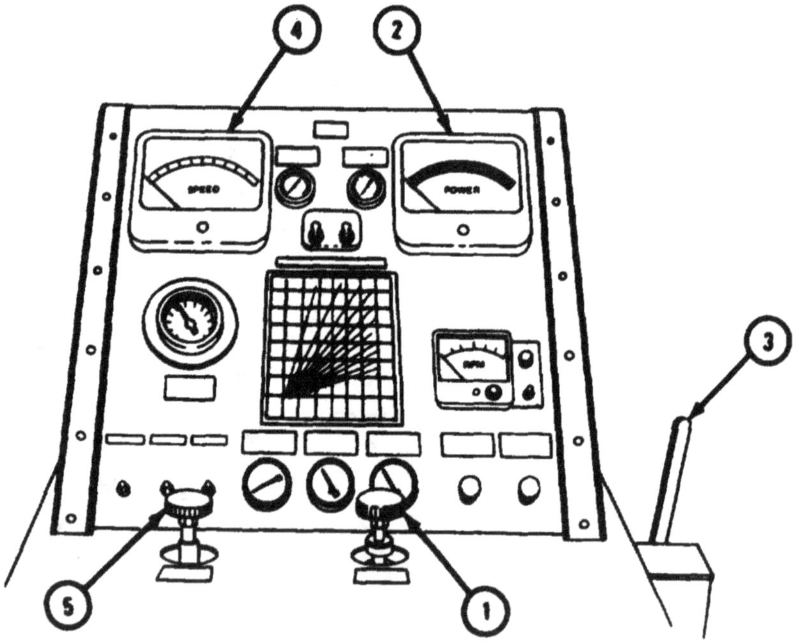

TA 118652

FRAME 3

1. Take off cylinder head cover. Refer to Part 1, para 3-28 .
2. Check that rocker arm mechanism (1) has enough lubricating oil.
3. Check that entire engine has no oil leaks.
4. Check that tube nuts (2) on injectors (3) have no fuel leaks. If leaks are found, tighten nuts .
5. Check around injectors (3), exhaust valves (4), and intake valves (5) for water leaks.
6. If engine has oil, fuel or water leaks, it must be repaired.
7. Replace cylinder head cover. Refer to para 5-29 .
8. Check that all screws and nuts on engine are tight.

END OF TASK

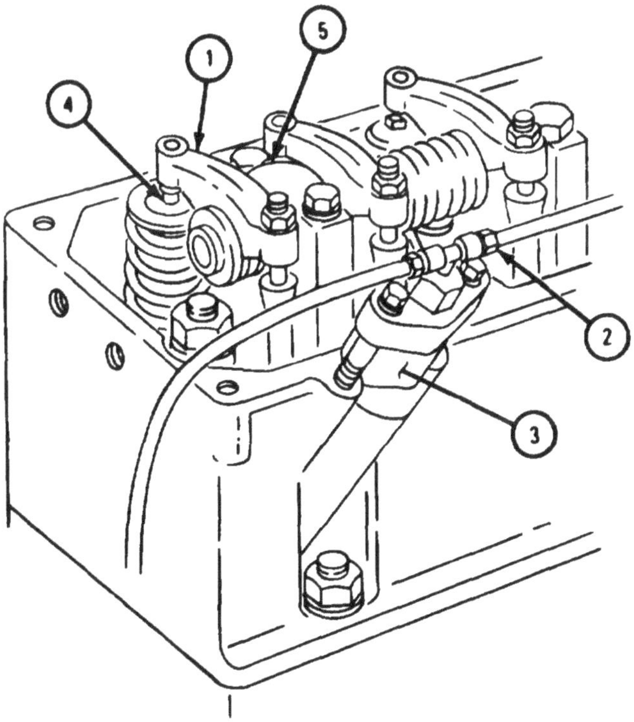

TA 118653

d. Run-In.

CAUTION

If oil pressure drops below 40 psi, stop engine and make the necessary repairs before finishing the tests .

(1) Start engine and check all gage readings. Refer to para 6-6b.

(2) Write down all gage readings taken during each of the following tests on engine test report used at facility you are working.

(a) Run engine for 1/2 hour at idle without a load.

(b) Run engine for 1/2 hour at 2000 rpm at 15 horsepower.

(c) Run engine for 1/2 hour at 2200 rpm at 64 horsepower.

(d) Run engine for 1/2 hour at 2500 rpm at 64 horsepower.

(e) Run engine for 1/2 hour at 2800 rpm at 64 horsepower.

(f) Run engine for 15 minutes at 500 rpm. Take off load a little at a time until engine is unloaded. Stop engine.

e. Inspection After Basic Run-In.

NOTE

Bare engine has air cleaner, but does not have pump, generator, and air compressor drive belts, muffler, cooling fan or any other accessories not essential for operation of the engine connected.

(1) Long run-in schedule:

NOTE

Long run-in schedule is for repaired engines in which new bearings, piston rings or cylinder sleeves have been put in.

Engine Model	Table
LD-465-1 and LD-465-1C	6-1
LDT-465-1C	6-2
LDS-465-1	6-3
LDS-465-1A	6-4
LDS-465-2	6-5

(2) Short run-in schedule :

NOTE

Short run-in schedule is for repaired engines with original bearings, piston rings and cylinder sleeves .

Engine Model	Table
LD-465-1 and LD-465-1C	6-6
LDT-465-1C	6-7
LDS-465-1	6-8
LDS-465-1A	6-9
LDS-465-2	6-10

Table 6-1. Run-In Schedule - Long (Models LD-465-1 and LD-465-1C)

NOTE

During run-in, oil sump temperature shall not be more than 260°F. Oil pressure is measured at engine oil gallery.

Period	Duration (minutes)	Engine Speed (rpm)	Gross Brake Horsepower (bare engine)	Fuel (DF-1 of VV-F-800)
1	1 (minimum)	0 Prelubricate at 25 psi (measured at engine oil gallery)	0	
2	30	1000	15	Diesel
3	15	1200	15	Diesel
4	Retighten cylinder head nuts to 130 pound-feet (157 pound-feet for TD block) and set valves immediately following period 3.			
5	15	1600	33	Diesel
6	30	2000	64	Diesel
7	30	2400	110	Diesel
8	15	2600	120 to 130	Diesel
9	Set high speed adjusting screw to limit maximum speed, no load 2850 to 2900 rpm.			
10	Set full load power	2600	126 to 131 (62 lb/hr max. fuel flow)	Diesel
11	Power check full load	1400	300 pound-feet torque (min)	Diesel
12	Set low idle adjustment	650 to 700	0	Diesel
13	As needed to check for leaks			Diesel

Table 6-2. Run-In Schedule - Long (Model LDT-465-1C)

NOTE

During run-in, oil sump temperature shall not be more than 265°F. Oil pressure is measured at engine oil gallery.

Period	Duration (minutes)	Engine Speed (rpm)	Gross Brake Horsepower (bare engine)	Fuel (DF-1 of VV-F-800)
1	1 (minimum)	0 Prelubricate at 25 psi (measured at engine oil gallery)	0	
2	30	1000	15	Diesel
3	15	1200	15	Diesel
4	Retighten cylinder head nuts to 157 pound-feet and set valves immediately following period 3.			
5	15	1600	33	Diesel
6	30	2000	64	Diesel
7	30	2400	110	Diesel
8	15	2600	125 to 135	Diesel
9	Set high speed adjusting screw to limit maximum speed, no load 2850 to 2900 rpm.			
10	Set full load power	2600	130 to 140 (64 lb/hr max. fuel flow)	Diesel
11	Power check full load	1500	305 pound-feet torque (min) (37.5 lb/hr max. fuel flow)	Diesel
12	Set droop screw	1200	(29 to 30 lb/hr fuel flow)	Diesel
13	Set low idle adjustment	Note 1	0	Diesel
14	As needed to check for leaks			Diesel

Note 1: Check injection pump data plate for recommended low idle adjustment setting.

Table 6-3. Run-In Schedule - Long (Model LDS-465-1)

NOTE

During run-in, oil sump temperature shall not be more than 265°F. Oil pressure is measured at engine oil gallery.

Period	Duration (minutes)	Engine Speed (rpm)	Gross Brake Horsepower (bare engine)	Fuel (DF-1 of VV-F-800)
1	1 (minimum)	0 Prelubricate at 25 psi (measured at engine oil gallery)	0	
2	30	1000	22	Diesel
3	15	1400	40	Diesel
4	Retighten cylinder head nuts to 130 pound-feet (157 pound-feet for TD block) and set valves immediately following period 3.			
5	15	1600	60	Diesel
6	15	2200	120	Diesel
7	30	2600	140	Diesel
8	30	2600	170 to 180	Diesel
9	Set high speed adjusting screw to limit maximum speed, no load 2850 to 2900 rpm.			
10	Set full load power	2600	175 to 185 (83 lb/hr max. fuel flow)	Diesel
11	Adjust droop screw	1200 or 1400	(37 to 39 lb/hr fuel flow) (44 to 45 lb/hr fuel flow)	Diesel
12	Power check full load	2000	425 pound-feet torque (min) (70 lb/hr max. fuel flow)	Diesel
13	Set low idle adjustment	650 to 700	0	Diesel
14	As needed to check for leaks			Diesel

Table 6-4. Run-In Schedule- Long (Model LDS-465-1A)

NOTE

During run-in, oil sump temperature shall not be more than 265°F. Oil pressure is measured at engine oil gallery.

Period	Duration (minutes)	Engine Speed (rpm)	Gross Brake Horsepower (bare engine)	Fuel (DF-1 of VV-F-800)
1	**1** (minimum)	0 Prelubricate at 25 psi (measured at engine oil gallery)	0	
2	30	1000	22	Diesel
3	15	1400	40	Diesel
4	Retighten cylinder head nuts to 130 pound-feet (157 pound-feet for TD block) and set valves immediately following period 3.			
5	15	1600	60	Diesel
6	15	2200	120	Diesel
7	30	2600	140	Diesel
8	30	2600	170 to 180	Diesel
9	Set high speed adjusting screw to limit maximum speed, no load 2850 to 2900 rpm.			
10	Set full load power	2600	175 to 185 (83 lb/hr max. fuel flow)	Diesel
11	Set droop screw	1200 or	(37 to 39 lb/hr fuel flow)	Diesel
		1400	(44 to 45 lb/hr fuel flow)	Diesel
12	Power check full load	2000	425 pound-feet torque (min) (70 lb/hr max. fuel flow)	Diesel
13	Set low idle adjustment	650 to 700	0	Diesel
14	As needed to check for leaks			Diesel

Table 6-5. Run-In Schedule- Long (Model LDS-465-2)

NOTE

During run-in, oil sump temperature shall not be more than 265°F. Oil pressure is measured at engine oil gallery.

Period	Duration (minutes)	Engine Speed (rpm)	Gross Brake Horsepower (bare engine)	Fuel (DF-1 of VV-F-800)
1	**1** (minimum)	0 Prelubricate at 25 psi (measured at engine oil gallery)	0	
2	30	1000	22	Diesel
3	15	1400	40	Diesel
4	Retighten cylinder head nuts to 157 pound-feet and set valves immediately following period 3.			
5	15	1600	60	Diesel
6	15	2200	120	Diesel
7	30	2800	160	Diesel
8	30	2800	190 to 200	Diesel
9	Set high speed adjusting screw to limit maximum speed, no load 3050 to 3100 rpm.			
10	Set full load power	2800	195 to 205 (92 lb/hr max . fuel flow)	Diesel
11	Set droop screw	1200	(37 to 39 lb/ hr fuel flow)	Diesel
		or		
		1400	(44 to 45 lb/ hr fuel flow)	Diesel
12	Power check full load	2000	425 pound-feet torque (min) (70 lb/hr max . fuel flow)	Diesel
13	Set low idle adjust-ment	650 to 700	0	Diesel
14	As needed to check for leaks			Diesel

Table 6-6. Run-In Schedule - Short (Models LD-465-1 and LD-465-1C)

NOTE

During run-in, oil sump temperature shall not be more than 260°F. Oil pressure is measured at engine oil gallery.

Period	Duration (minutes)	Engine Speed (rpm)	Gross Brake Horsepower (bare engine)	Fuel (DF-1 of VV-F-800)
1	1 (minimum)	0 Prelubricate at 25 psi (measured at engine oil gallery)	0	
2	10	1200	15	Diesel
3	15	2000	64	Diesel
4	Retighten cylinder head nuts to 130 pound-feet for TD block and set valves immediate y following period 3.			
5	15	2400	110	Diesel
6	10	2600	120 to 130	Diesel
7	Set high speed adjusting screw to limit maximum speed, no load 2850 to 2900 rpm.			
8	Set full load power	2600	126 to 131 (62 lb/h r max. fuel flow)	Diesel
9	Power check full load	1400	300 pound-feet torque (rein)	Diesel
10	Set low idle adjustment	650 to 700	0	Diesel
11	As needed to check for leaks			Diesel

Table 6-7. Run-In Schedule - Short (Model LDT-465-1C)

NOTE

During run-in, oil sump temperature shall not be more than 265°F. Oil pressure is measured at engine oil gallery.

Period	Duration (minutes)	Engine Speed (rpm)	Gross Brake Horsepower (bare engine)	Fuel (DF-1 of VV-F-800)
1	1 (minimum)	0 Prelubricate at 25 psi (measured at engine oil gallery)	0	
2	10	1200	15	Diesel
3	15	2000	64	Diesel
4	Retighten cylinder head nuts to 157 pound-feet and set valves immediately following period 3.			
5	15	2400	110	Diesel
6	10	2600	125 to 135	Diesel
7	Set high speed adjusting screw to limit maximum speed, no load 2850 to 2900 rpm.			
8	Set full load power	2600	130 to 140 (64 lb/hr max . fuel flow)	Diesel
9	Power check full load	1500	305 pound-feet torque (rein) (37.5 lb/h r max. fuel flow)	Diesel
10	Set droop screw	1200	(29 to 30 lb/ hr fuel flow)	Diesel
11	Set low idle adjust-ment	Note 1	0	Diesel
12	As needed to check for leaks			Diesel

Note 1: Check injection pump data plate for recommended low idle adjustment setting.

Table 6-8. Run-In Schedule - Short (Model LDS-465-1)

NOTE

During run-in, oil sump temperature shall not be more than 265°F. Oil pressure is measured at engine oil gallery.

Period	Duration (minutes)	Engine Speed (rpm)	Gross Brake Horsepower (bare engine)	Fuel (DF-1 of VV-F-800)
1	(minimum)	0 Prelubricate at 25 psi (measured at engine oil gallery)	0	
2	10	1400	40	Diesel
3	15	2200	120	Diesel
4	Retighten cylinder head nuts to 130 pound-feet (157 pound-feet for TD block) and set valves immediately following period 3.			
5	15	2600	140	Diesel
6	10	2600	170 to 180	Diesel
7	Set high speed adjusting screw to limit maximum speed, no load 2850 to 2900 rpm.			
8	Set full load power	2600	175 to 185 (83 lb/hr max. fuel flow)	Diesel
9	Set droop screw	1200 or 1400	(37 to 39 lb/hr fuel flow) (44 to 45 lb/hr fuel flow)	Diesel Diesel
10	Power check full load	2000	425 pound-feet torque (rein) (70 lb/hr max. fuel flow)	Diesel
11	Set low idle adjustment	650 to 700	0	Diesel
12	As needed to check for leaks			Diesel

Table 6-9. Run-In Schedule - Short (Model LDS-465-1A)

NOTE

During run-in, oil sump temperature shall not be more than 265°F. Oil pressure is measured at engine oil gallery.

Period	Duration (minutes)	Engine Speed (rpm)	Gross Brake Horsepower (bare engine)	Fuel (DF-1 of VV-F-800)
1	(minimum)	0 Prelubricate at 25 psi (measured at engine oil gallery)	0	
2	10	1400	40	Diesel
3	15	2200	120	Diesel
4	Retighten cylinder head nuts to 130 pound-feet (157 pound-feet for TD block) and set valves immediately following period 3.			
5	15	2600	140	Diesel
6	10	2600	170 to 180	Diesel
7	Set high speed adjusting screw to limit maximum speed, no load 2850 to 2900 rpm.			
8	Set full load power	2600	175 to 185 (83 lb/h r max. fuel flow)	Diesel
9	Set droop scre w	1200	(37 to 39 lb/hr fuel flow)	Diesel
		o r		
		1400	(44 to 45 lb/hr fuel flow)	Diesel
10	Power check full load	2000	425 pound-feet torque (rein) (70 lb/h r max. fuel flow)	Diesel
11	Set low idle adjust-ment	650 to 700	0	Diesel
12	As needed to check for leaks			Diesel

Table 6-10. Run-In Schedule - Short (Model LDS-465-2)

NOTE

During run-in, oil sump temperature shall not be more than 265°F. Oil pressure is measured at engine oil gallery.

Period	Duration (minutes)	Engine Speed (rpm)	Gross Brake Horsepower (bare engine)	Fuel (DF-1 of VV-F-800)
1	1 (minimum)	0 Prelubricate at 25 psi (measured at engine oil gallery)	0	
2	10	1400	40	Diesel
3	15	1600	60	Diesel
4	Retighten cylinder head nuts to 157 pound-feet and set valves immediately following period 3.			
5	15	2200	120	Diesel
6	10	2800	160 to 190	Diesel
7	Set high speed adjusting screw to limit maximum speed, no load 3050 to 3100 rpm.			
8	Set full load power	2800	195 to 205 (92 lb/h r max. fuel flow)	Diesel
9	Set droop screw	1200	(37 to 39 lb/hr fuel flow)	Diesel
		or		
		1400	(44 to 45 lb/hr fuel flow)	Diesel
10	Power check full load	2000	425 pound-feet torque (rein) (70 lb/h r max. fuel flow)	Diesel
11	Set low idle adjustment	650 to 700	0	Diesel
12	As needed to check for leaks			Diesel

f. Final Run-In.

FRAME 1

1. Start and warm up engine. Refer to para 6-6b.
2. Move throttle (1) until SPEED gage (2) shows 2800 rpm.
3. Open engine load valve (3) until POWER gage (4) shows 160 to 205 HP.
4. Set throttle (1) and load valve (3) to get 160 to 205 hp and 2800 rpm. If hp reading is too high, open unload valve (5).
5. Run engine at 160 to 200 hp and 2800 rpm for 1/2 hour.
6. Write down fuel pressure shown on FUEL PRESS gage (6).

GO TO FRAME 2

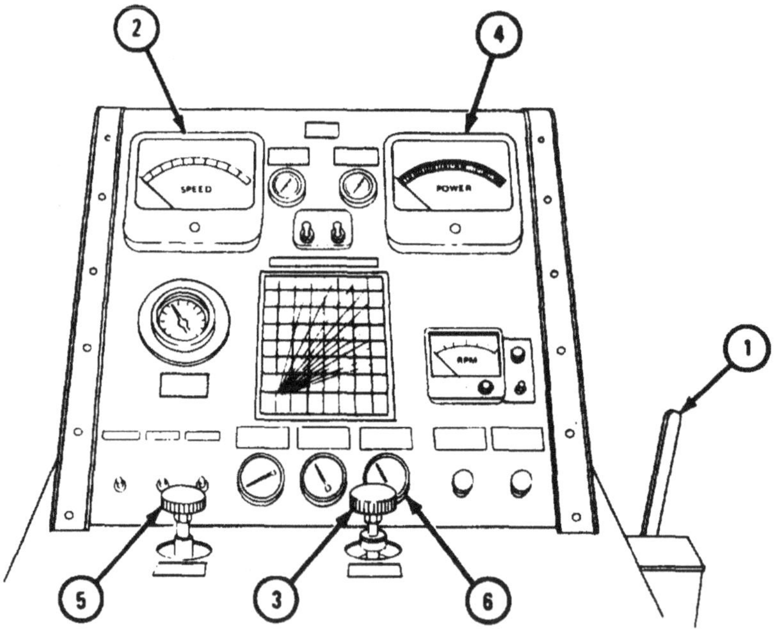

TA 118654

FRAME 2

1. Write down lubricating oil pressure from ENGINE OIL PRESSURE gage (1).
2. Write down lubricating oil temperature from temperature gage (2).
3. Write down water temperature from WATER TEMPERATURE gage (3).
4. Write down crankcase pressure from manometer (4).
5. Write down manifold pressure from ENGINE MANIFOLD PRESS gage (5).

GO TO FRAME 3

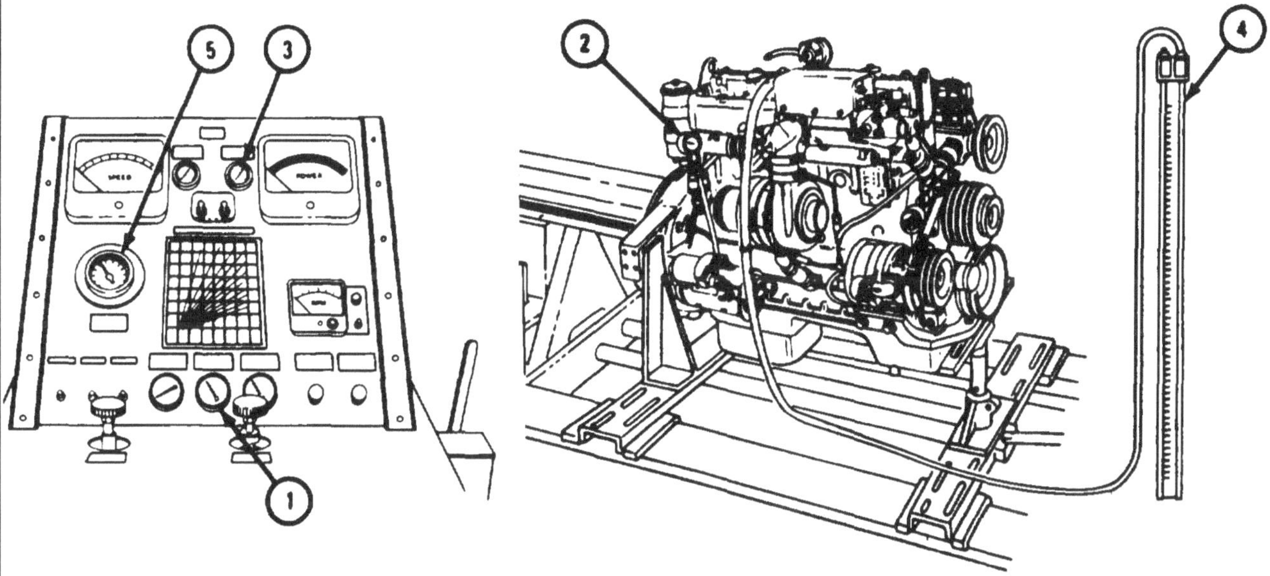

TA 118655

FRAME 3

1. Write down crankcase pressure from crankcase pressure gage (1) .
2. Move throttle (2) back and open unload valve (3) until SPEED gage (4) reads 500 RPM and POWER gage (5) reads 0 to 5 HP.
3. Write down oil pressure from gage (6) and write down idle speed.
4. Put throttle (2) to maximum position. Write down engine rpm from SPEED gage (4) .
5. Slow down engine to 500 rpm for 15 minutes, then stop engine.

GO TO FRAME 4

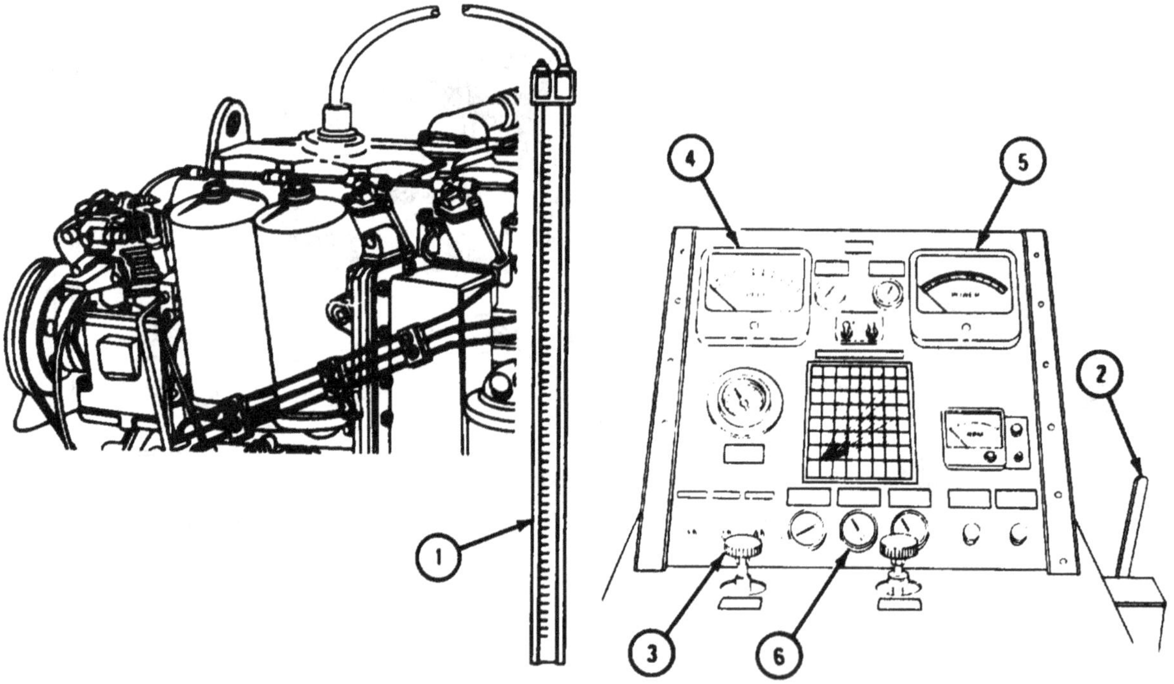

TA 118656

FRAME 4

1. Turn off water supply to test stand.
2. Turn drain control (1) to left until it stops.
3. Close fuel control valve (2) .
4. Turn off engine overspeed governor (3) .
5. Open main switch (4).

END OF TASK

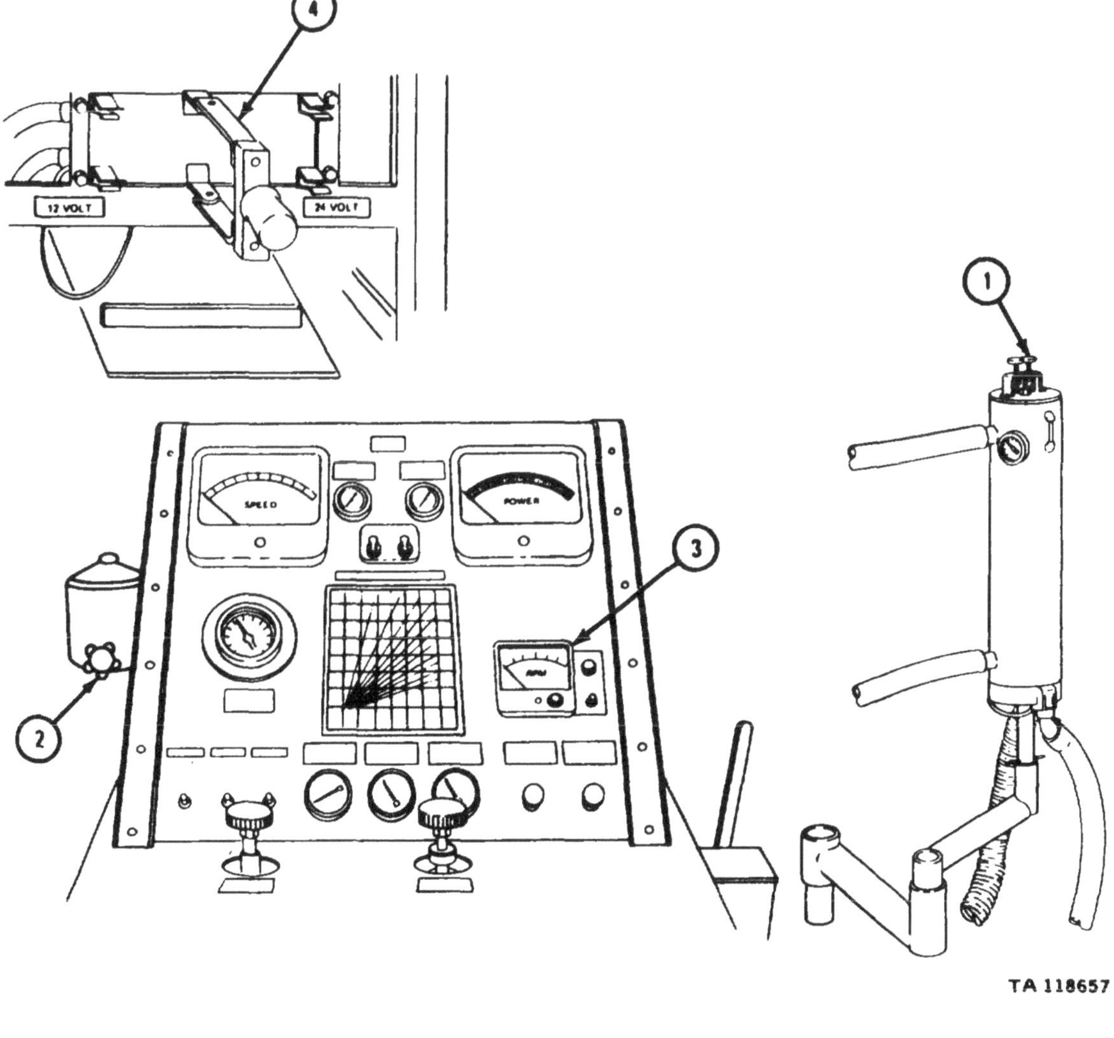

6-7. DISCONNECTING ENGINE FROM TEST STAND.

FRAME 1

1. Take out oil pressure line (1) on engine (2).
2. Put plug (3) in engine (2) .

GO TO FRAME 2

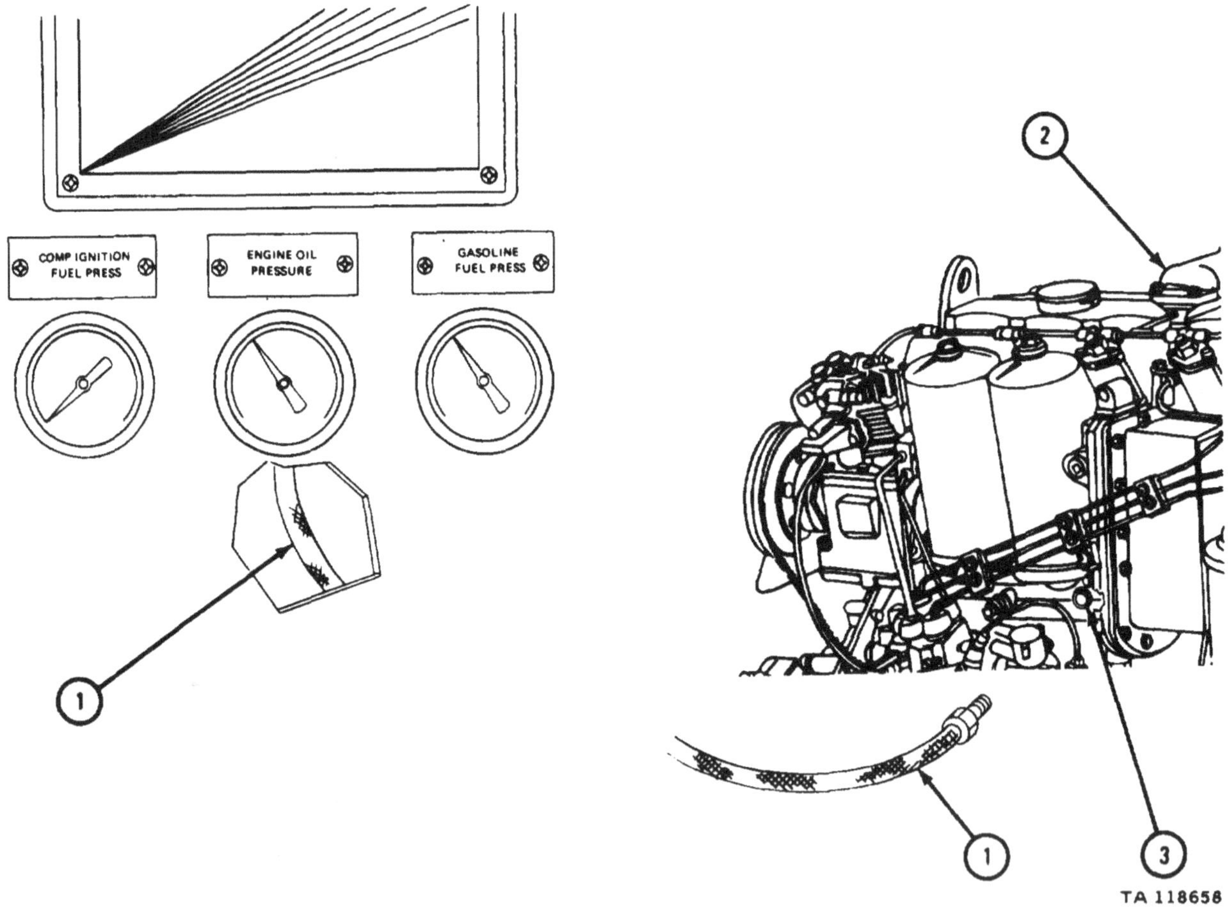

FRAME 2

1. Takeout manometer (1) and adapter (2) .
2. Put oil filler cap (3) on valve cover (4).

GO TO FRAME 3

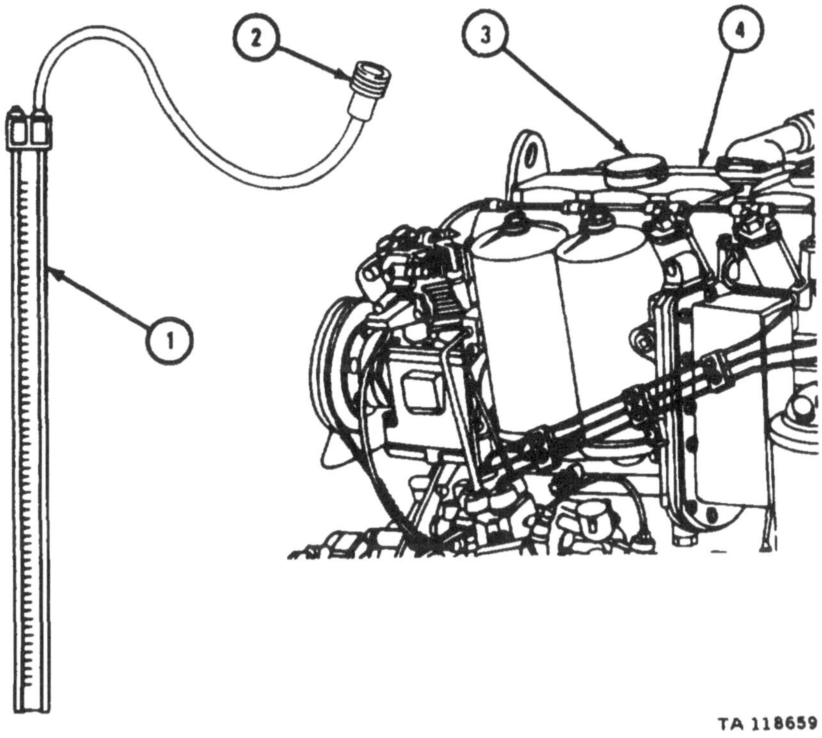

FRAME 3

1. Takeout fuel pressure line (1) from fuel filter (2)
2. Put plug (3) in fuel filter (2).

GO TO FRAME 4

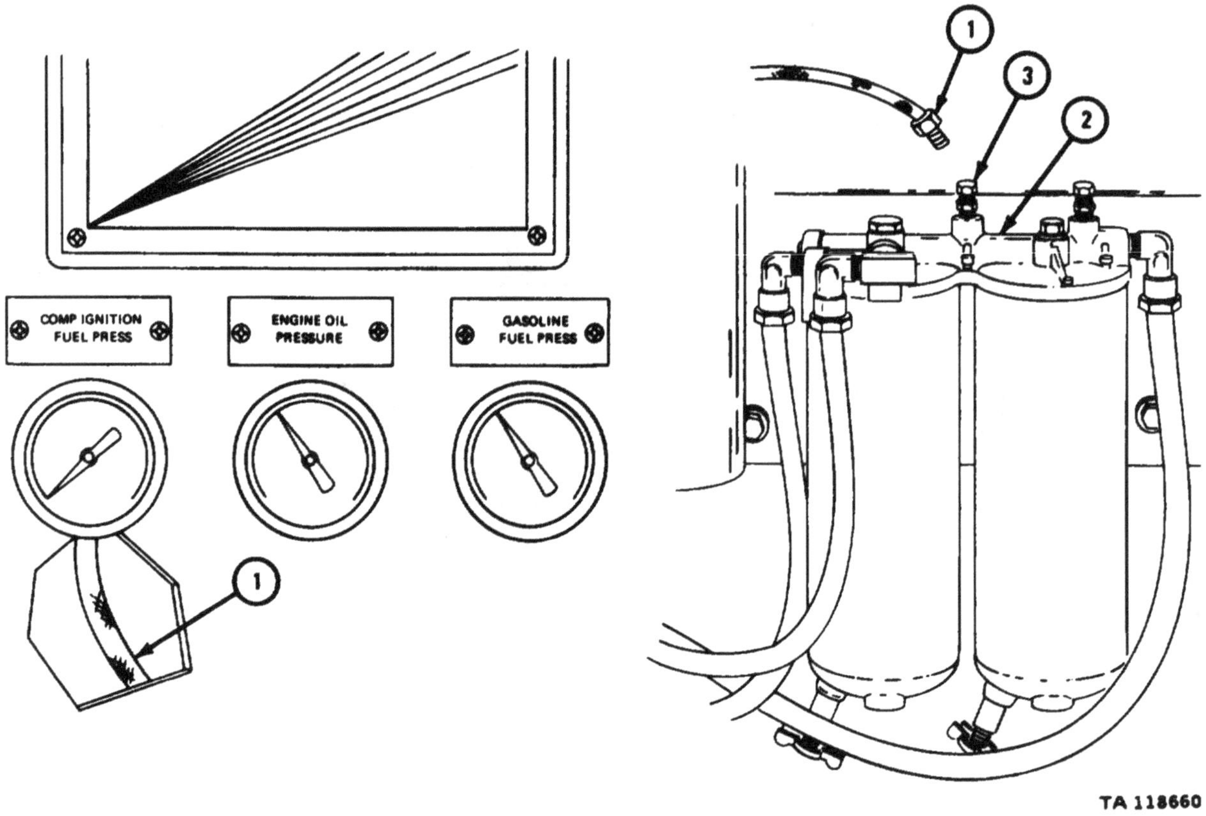

FRAME 4

1. Take out oil temperature gage (1) .
2. Takeoff manometer tube (2) from crankcase breather tube (3) .
3. Take off dynamometer throttle linkage (4) from engine governor linkage (5).

GO TO FRAME 5

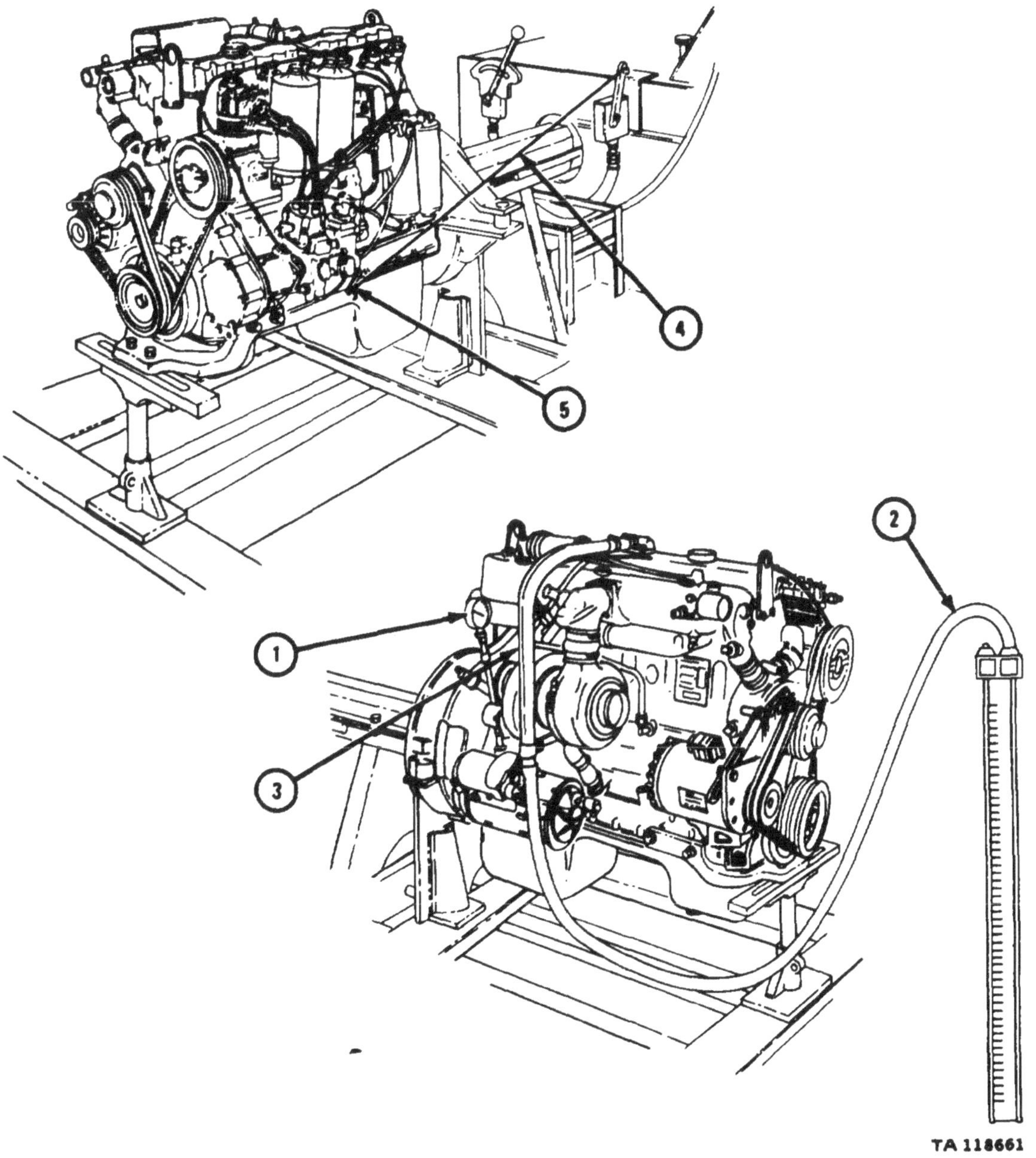

FRAME 5

1. Take off nut (1) and washer (2). Takeoff positive (+) cable (3). Put back nut and washer.
2. Takeoff nut (4) and washer (5). Takeoff starter solenoid wire (6). Put back nut and washer (5).

GO TO FRAME 6

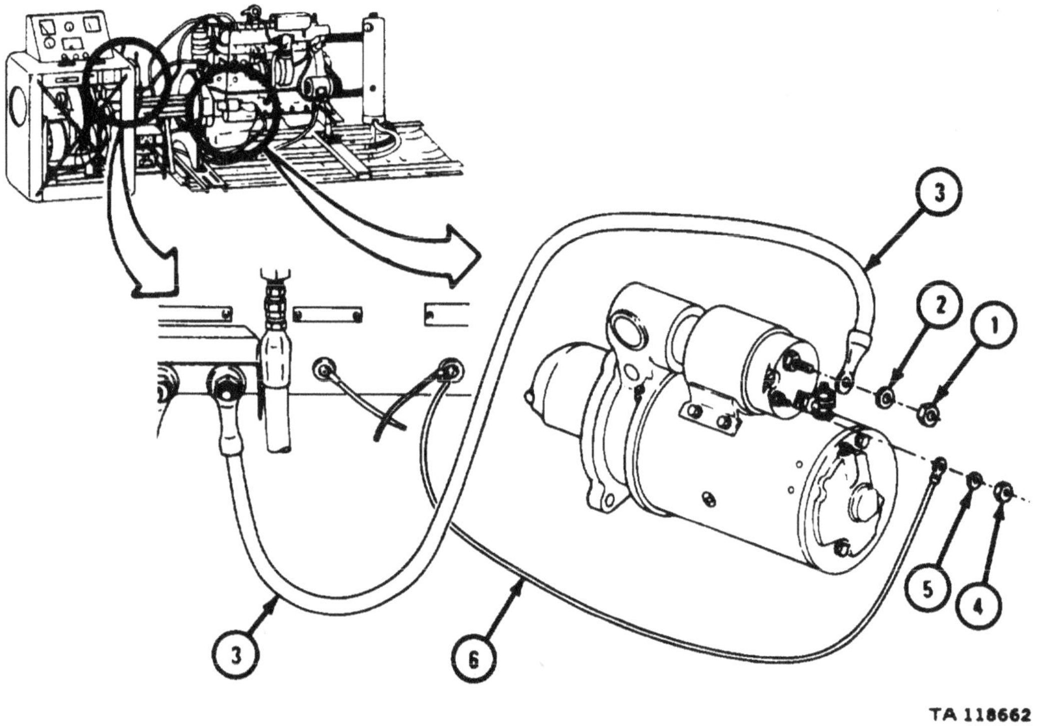

FRAME 6

1. Take off dynamometer manifold pressure line (1) and any adapters or fittings used to join it to the engine.
2. Take off oil pressure line (2) and any adapters or fittings used to join it to the engine .

GO TO FRAME 7

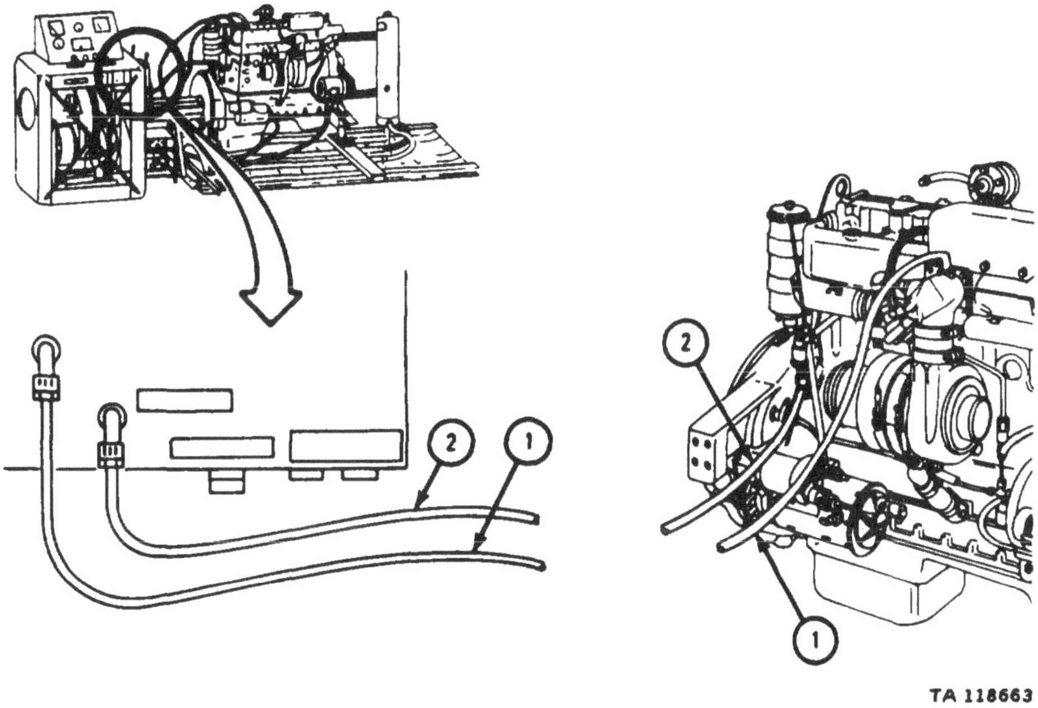

FRAME 7

1. Take off tachometer drive shaft (1) from tachometer drive adapter (2).

GO TO FRAME 8

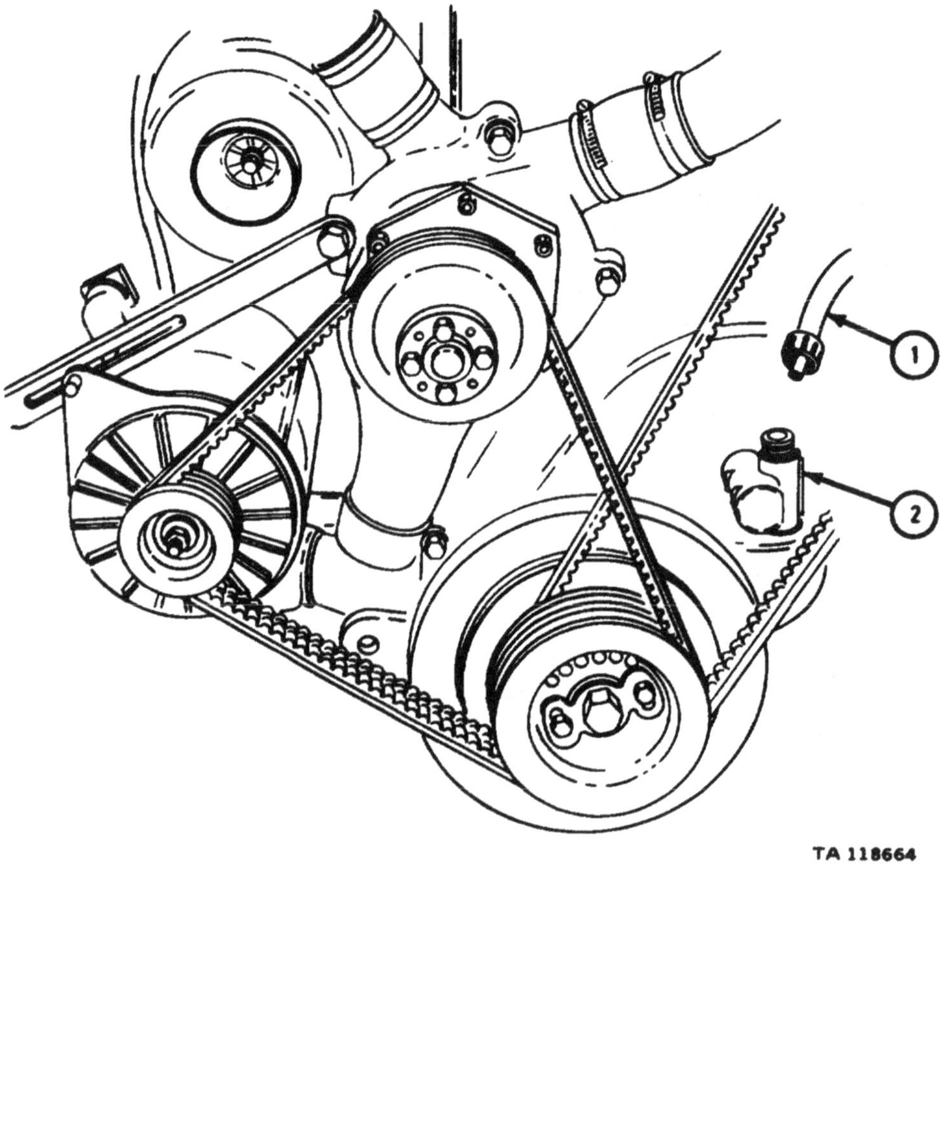

FRAME 8

1. Take off dynamometer fuel return line (1) and any adapters or fittings used to join it to fuel pump (2).
2. Put on fuel return line (3) where dynamometer fuel return line (1) was taken out .
3. Take off dynamometer fuel line (4) and any adapters or fittings used to join it to fuel pump (2).
4. Put on fuel line (5) where dynamometer fuel line (4) was taken out.

GO TO FRAME 9

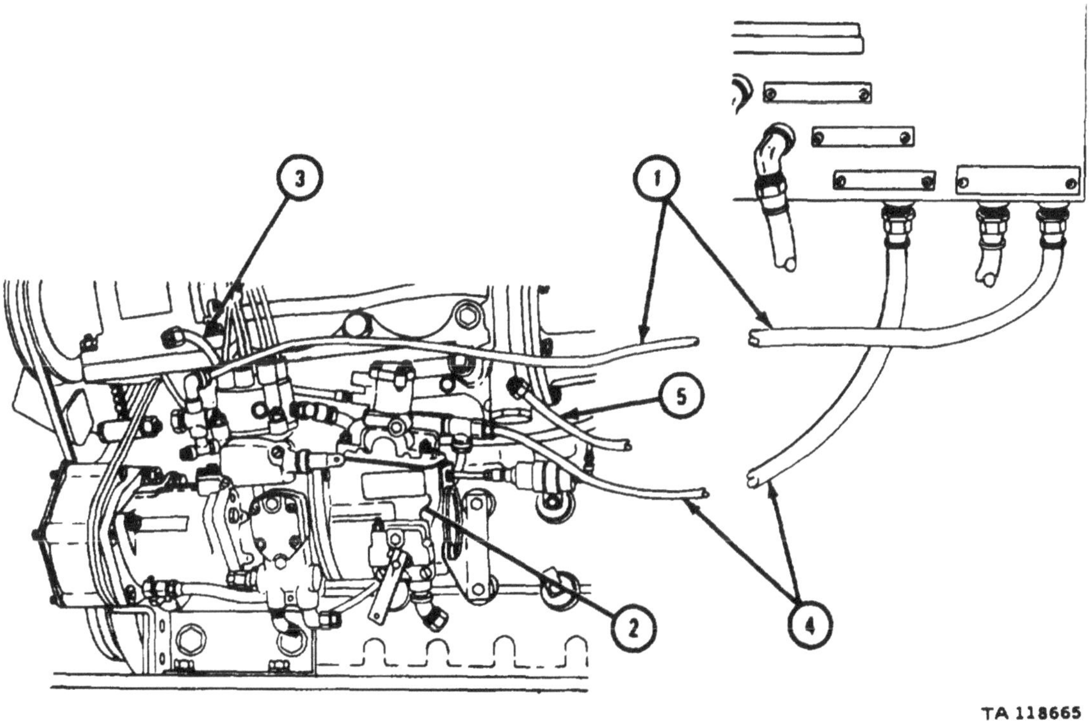

FRAME 9

1. Loosen clamp (1) and take off outside exhaust tube (2) with clamp.

GO TO FRAME 10

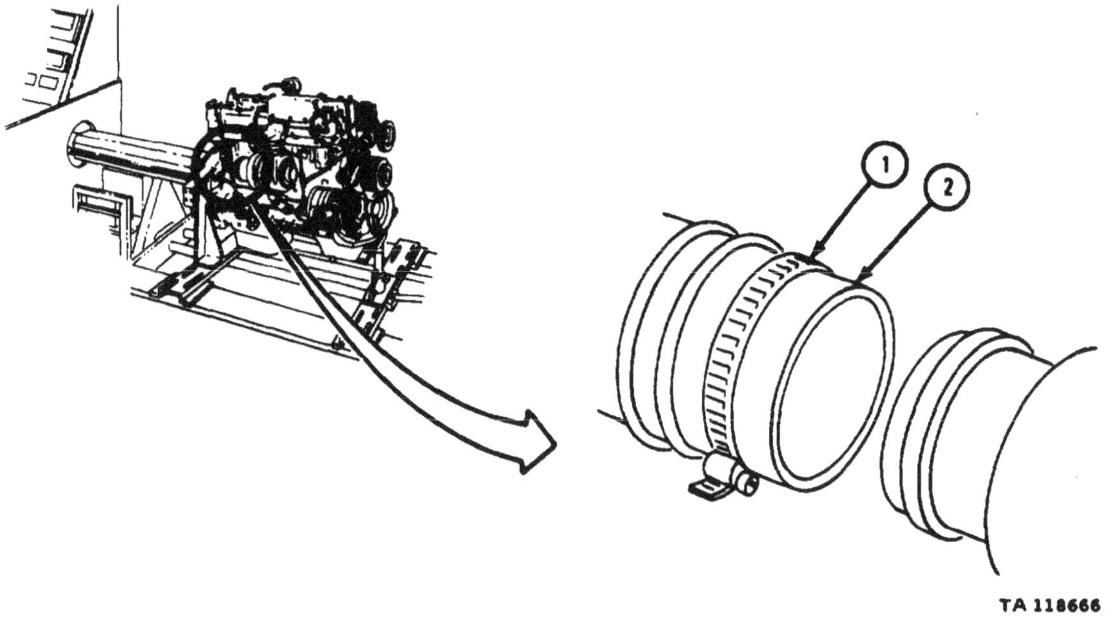

FRAME 10

1. Loosen two hose clamps (1) and takeoff water inlet hose (2).
2. Loosen two hose clamps (3) and takeoff water outlet hose (4).
3. Put oil gage rod (5) oil gage tube (6) .

END OF TASK

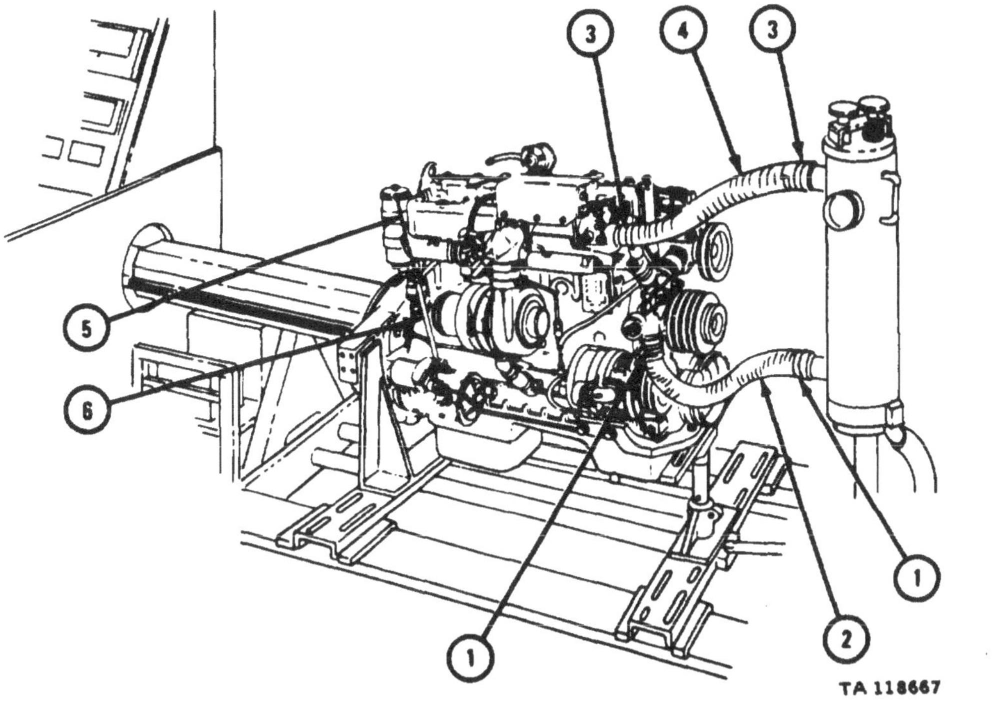

6-8. REMOVING ENGINE.

a. Taking Engine off Test Stand.

FRAME 1

Soldier A 1 . Drain engine oil. Refer to LO 9-2320-209-12/1 or LO 9-2320-211-12.

2. Loosen four bolts (1), two on each side of cover (2).

Soldier B 3 . Loosen two hinge bolts (3) at rear of cover (2).

Soldier A 4 . Slide cover (2) to the front and lift it up.

Soldier B 5 . Tighten two hinge bolts (3) to fasten cover (2) in open position.

Soldier A 6 . Take out eight bolts and lockwashers (4).

GO TO FRAME 2

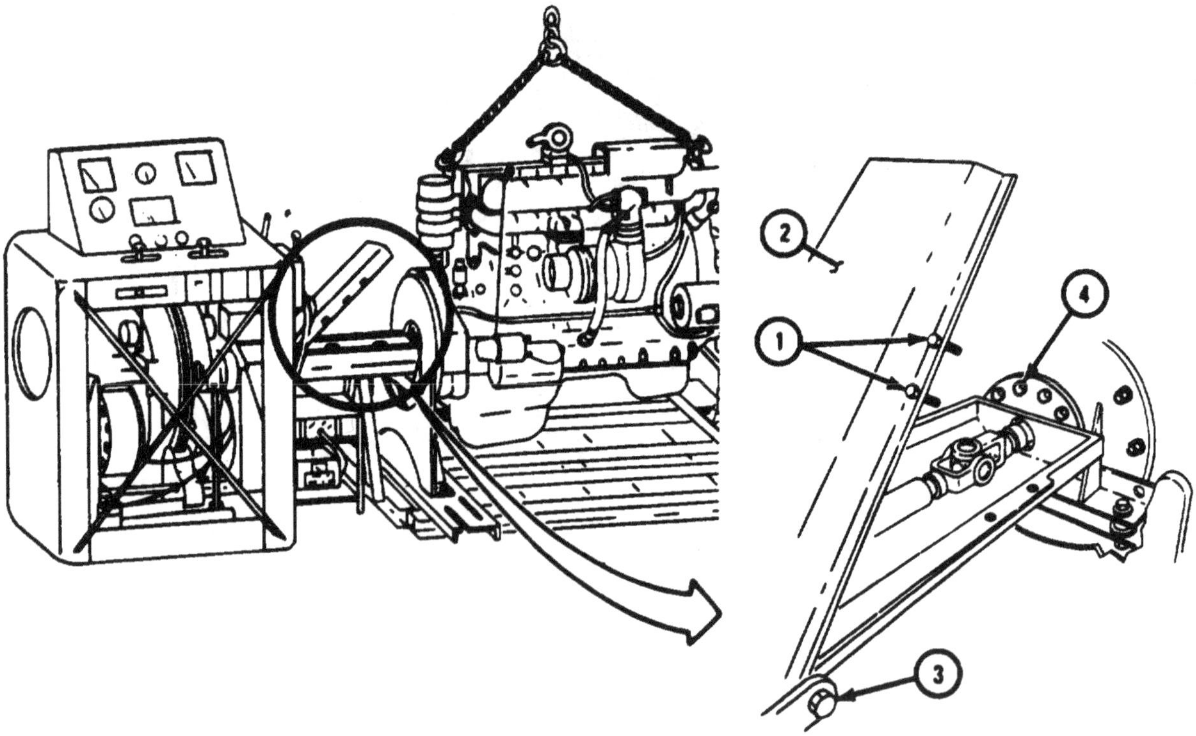

TA 118668

FRAME 2

Soldier A 1. Loosen eight bolts (1).

2. Put chain sling (2) on two engine lifting brackets (3).

3. Using hoisting equipment, lift engine assembly (4) just enough to take weight of engine off test stand (5).

Soldiers A and B 4. Slide engine assembly (4) about one foot to the front.

Soldier A 5. Using hoisting equipment, let engine assembly (4) rest on test stand (5).

GO TO FRAME 3

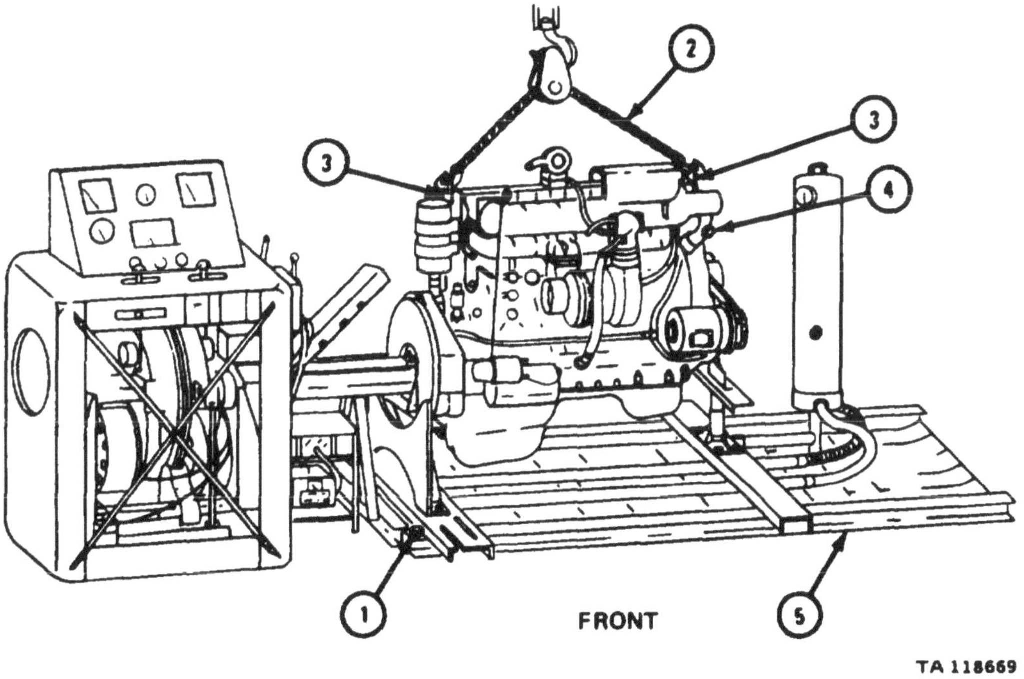

FRAME 3

1. Take out two bolts (1), two lockwashers (2), and two nuts (3).

GO TO FRAME 4

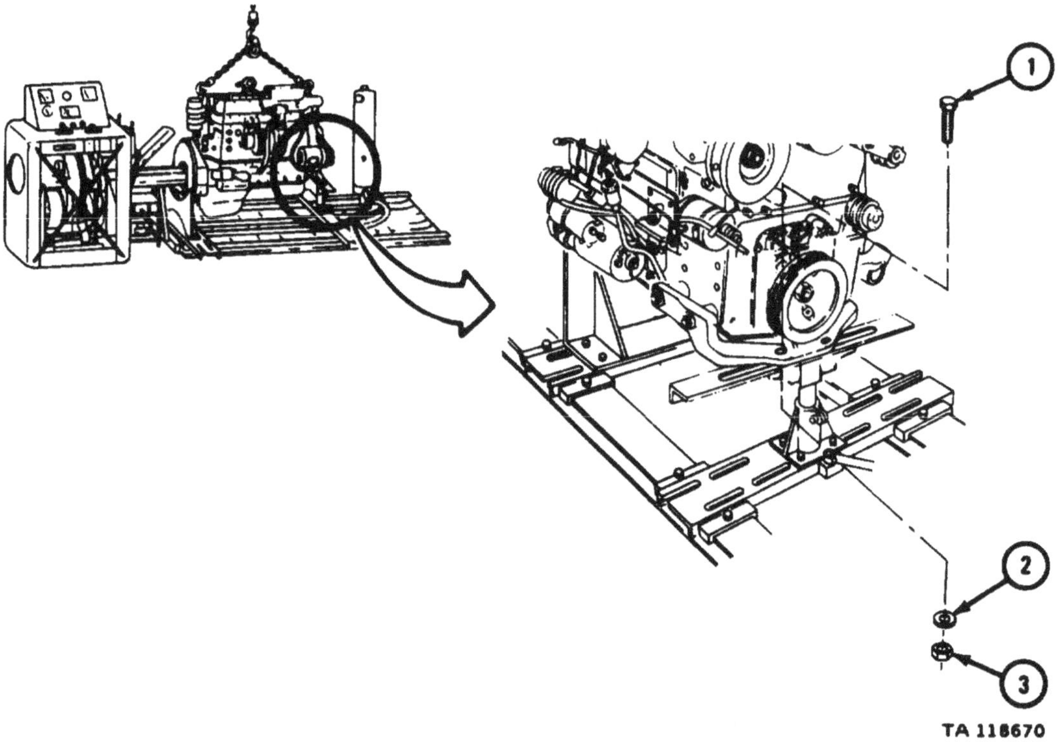

FRAME 4

1. Take out two bolts (1), two washers (2), and two nuts (3). Take off two rubber washers (4) when engine assembly (5) is taken off test stand (6).

GO TO FRAME 5

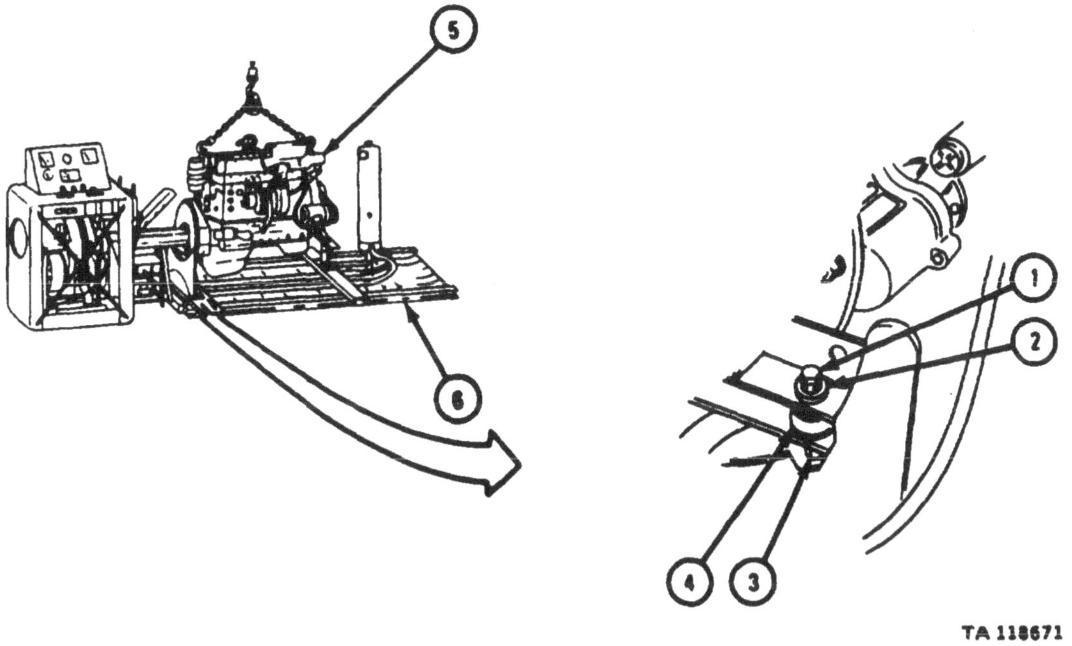

FRAME 5

Soldier A 1. Move engine transport stand (1) over to dynamometer test stand .

Soldier B 2. Using hoisting equipment, raise or lower engine assembly (2) as needed when soldier A is ready.

Soldier A 3. Guide engine assembly (2) off dynamometer test stand onto engine transport stand (1) as shown .

4. Take off chain sling (3) .

END OF TASK

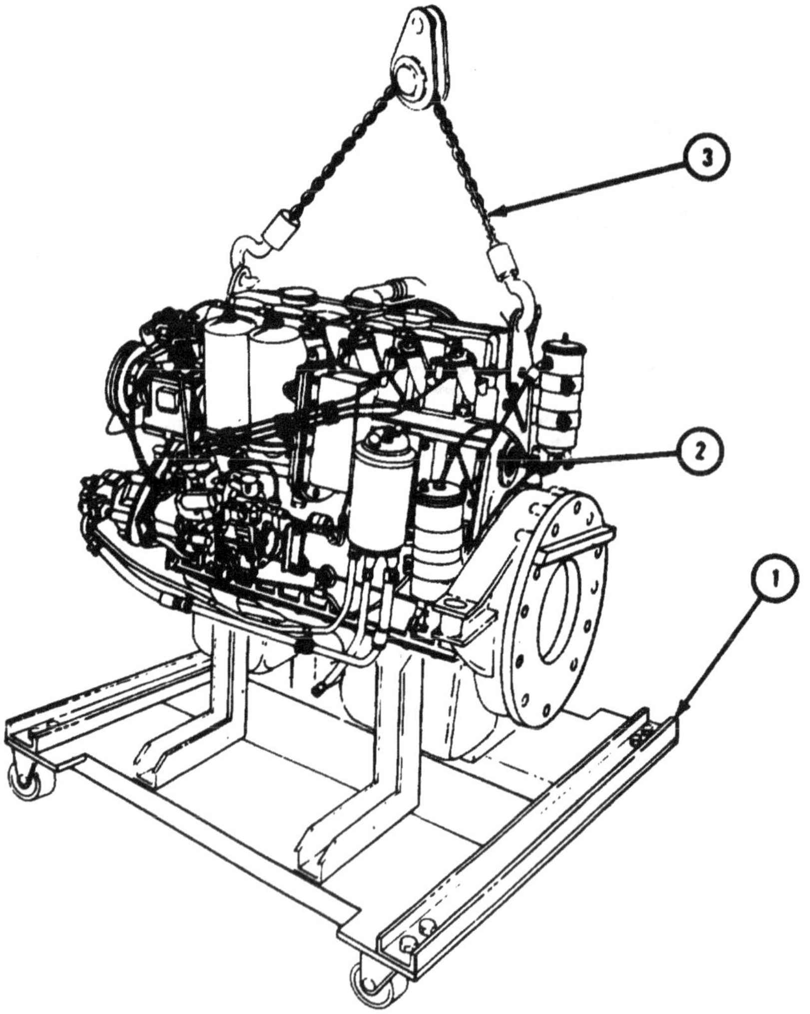

b. Removing Engine Brackets.

FRAME 1

1. Take out four bolts (1) and lockwashers (2).
2. Take off front engine adapter (3).

NOTE

LDS 465-2 engines have six capscrews and lockwashers (5) and safety wire (6) to hold fan (4).

3. Put on fan (4) with four capscrews and lockwashers (5).
4. If fan (4) had safety wire (6), put on safety wire as shown.

GO TO FRAME 2

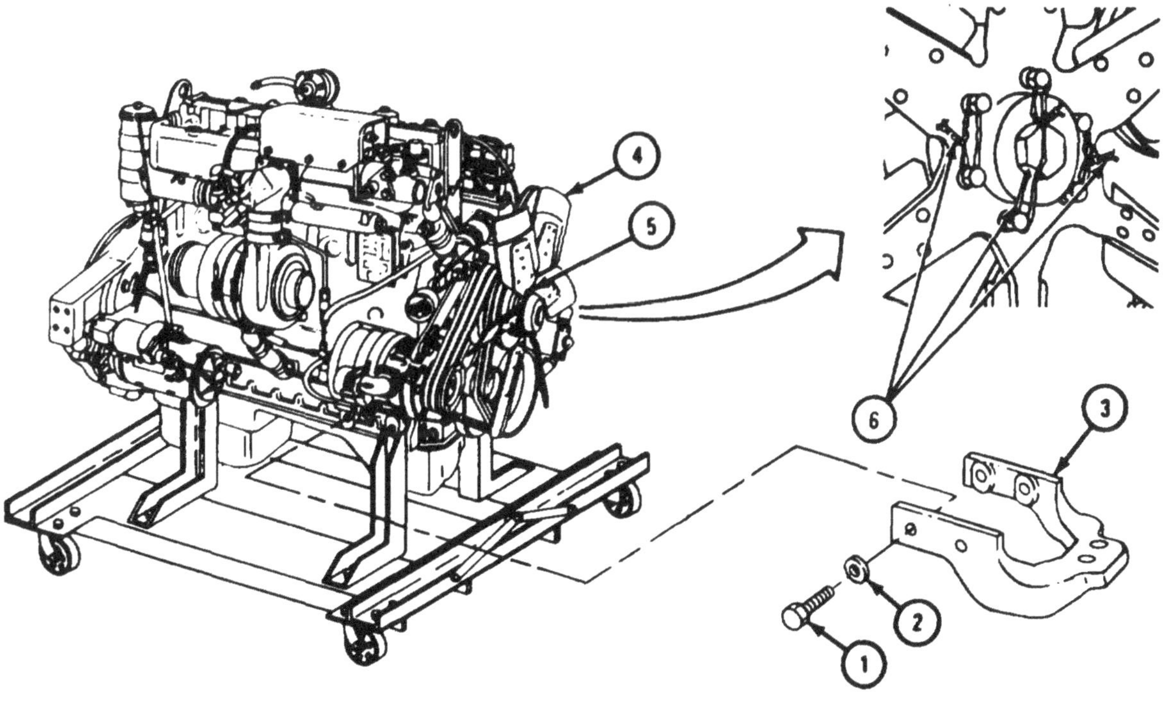

FRAME 2

1. Take out twelve capscrews (1) and lockwashers (2) .

2. Take mounting plate (3) off flywheel housing (4).

GO TO FRAME 3

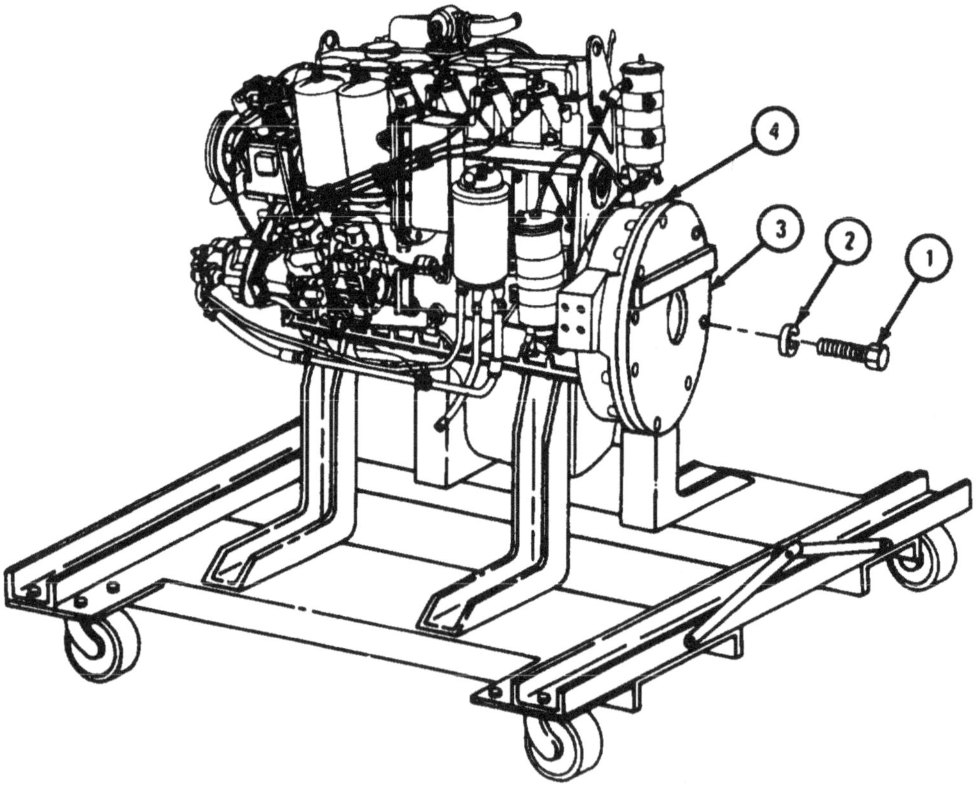

FRAME 3

1. Take out two bolts (1).
2. Put two fabricated guide bolts (2) in place of two bolts (1).

GO TO FRAME 4

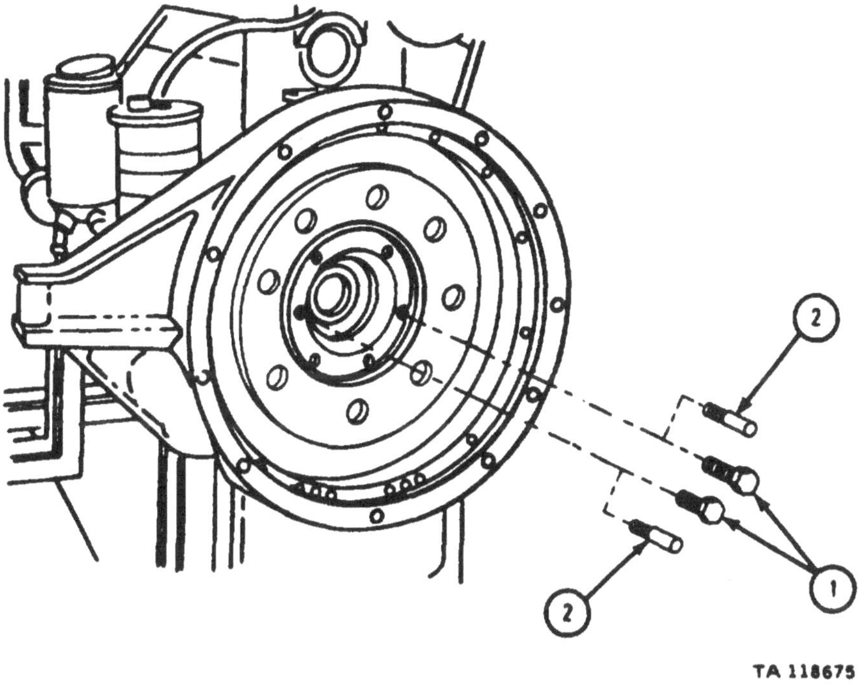

FRAME 4

1. Takeout four bolts (1). Takeoff flywheel adapter (2) .
2. Put in four original bolts taken out.
3. Take out two guide bolts (3) and put in two original bolts taken out.
4. Tighten six bolts to 80 to 90 pound-feet, and then tighten bolts to 115 to 120 pound-feet .

END OF TASK

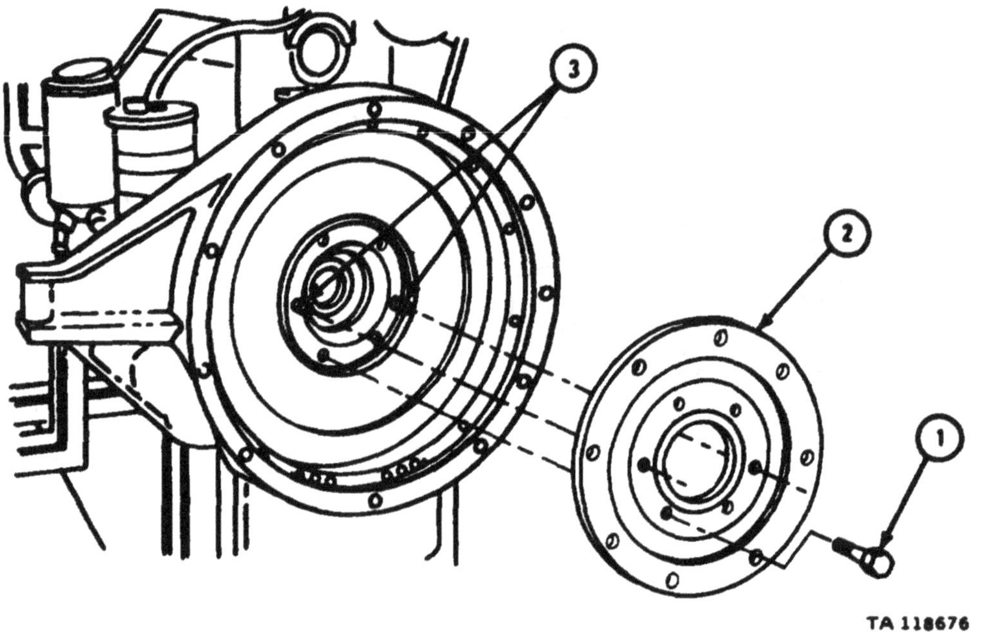

6-9. PLACING ENGINE IN SHIPPING CONTAINER.

FRAME 1

NOTE

Make sure container (2) is clean and dry inside. Chec k that seal (5) and seal groove (6) are clean and not damaged.

1. Using chain hoist, lift engine (1) and move it so it hangs over container (2).
2. Put two front support brackets (3) in place on engine (1). put in four bolt s with lockwashers and nuts (4) .
3. Put box (7) holding clutch parts on engine rear lifting bracket (8).

GO TO FRAME 2

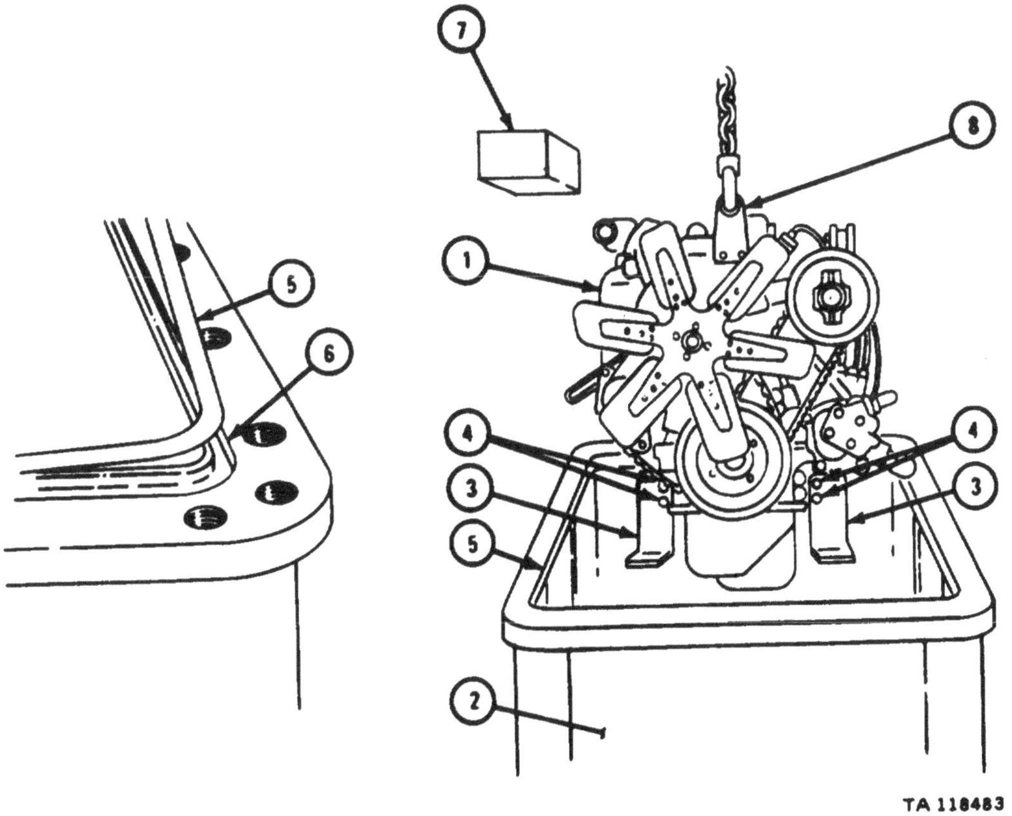

FRAME 2

1. Put mounting bracket (1) on flywheel housing (2). Put in seven capscrews with lockwashers (3).

GO TO FRAME 3

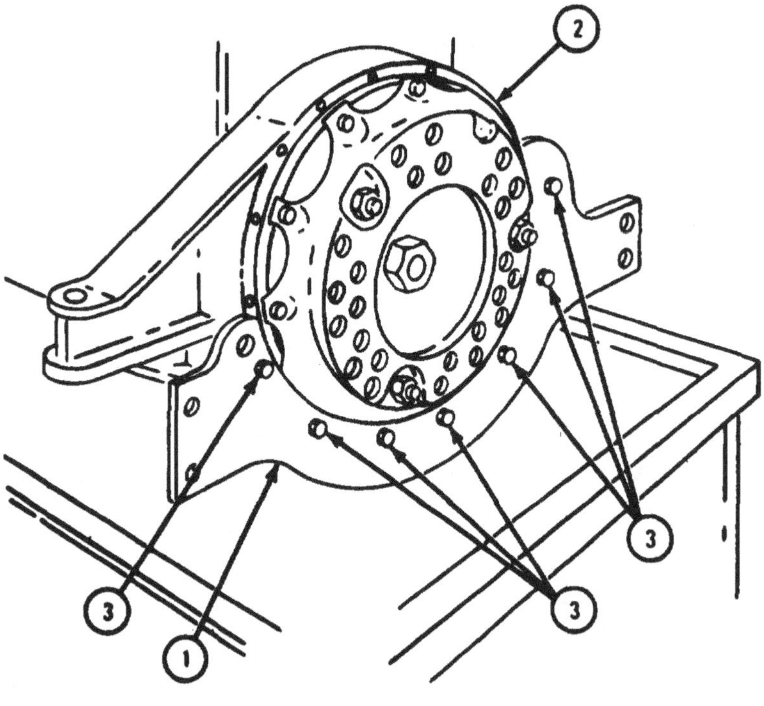

FRAME 3

Soldier A 1. Guide engine (1) into container (2).

Soldier B 2. Lower engine (1) into container (2) until soldier A says to stop.

Soldier A 3. When holes in bracket (3) are alined with holes in container mount (4), tell soldier B to stop.

4. Put in four bolts with lockwashers and nuts (5).

GO TO FRAME 4

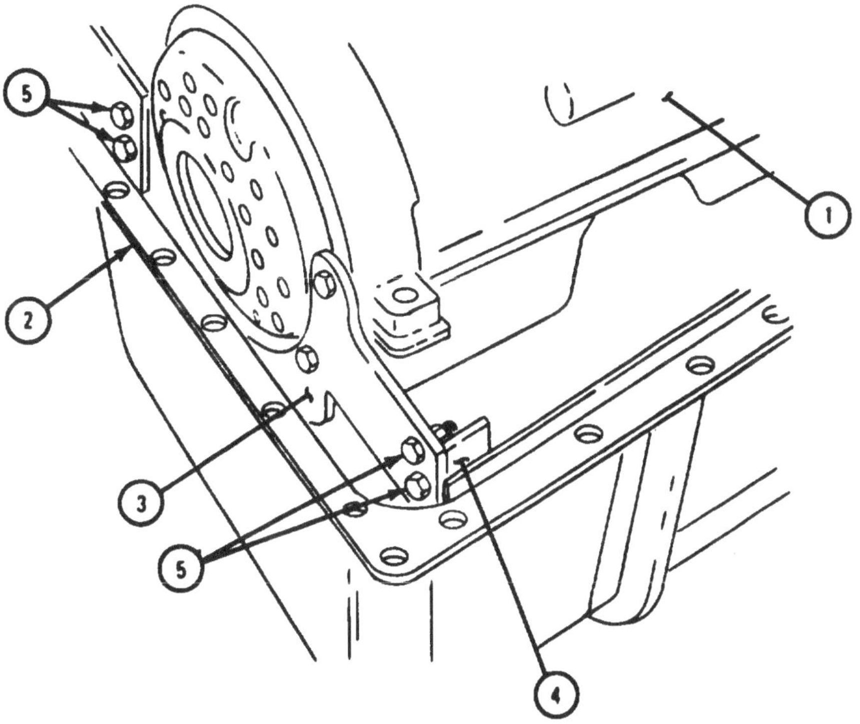

TA 118485

FRAME 4

1. Put four bolts with lockwashers and nuts (1) in two front mounting brackets (2). Remove chain hoist from engine.

GO TO FRAME 5

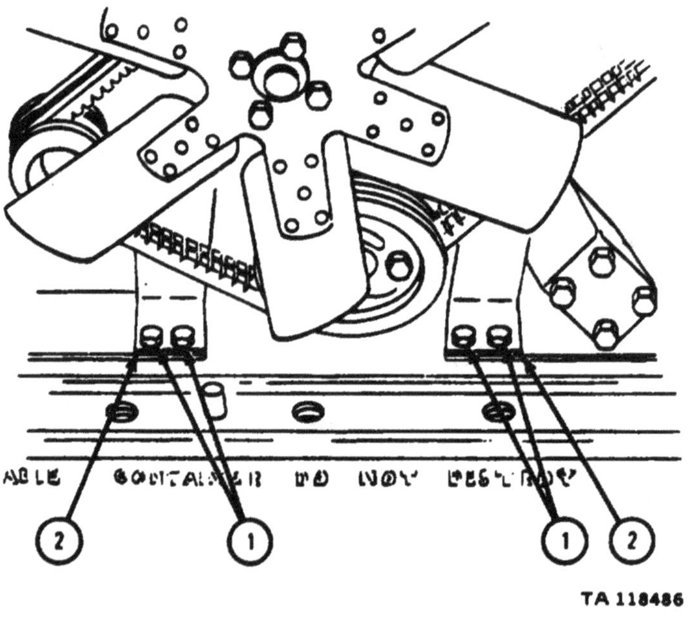

FRAME 5

NOTE

Before cover (1) is put in place, make sure it is clean and dry inside . Place five bags of desicant in container (2) and coat rubber seal with sealing compound.

Soldier B 1 . Using chain hoist, lift cover (1) over container (2) .

Soldier A 2 . Guide cover (1) so dowl pin (3) fits through dowl pin hole (4).

Soldier B 3 . Lower cover (1) onto container (2) .

4. Put in 36 bolts with lockwashers and nuts (5). Tighten bolts (5) to 88 to 115 pound-feet

GO TO FRAME 6

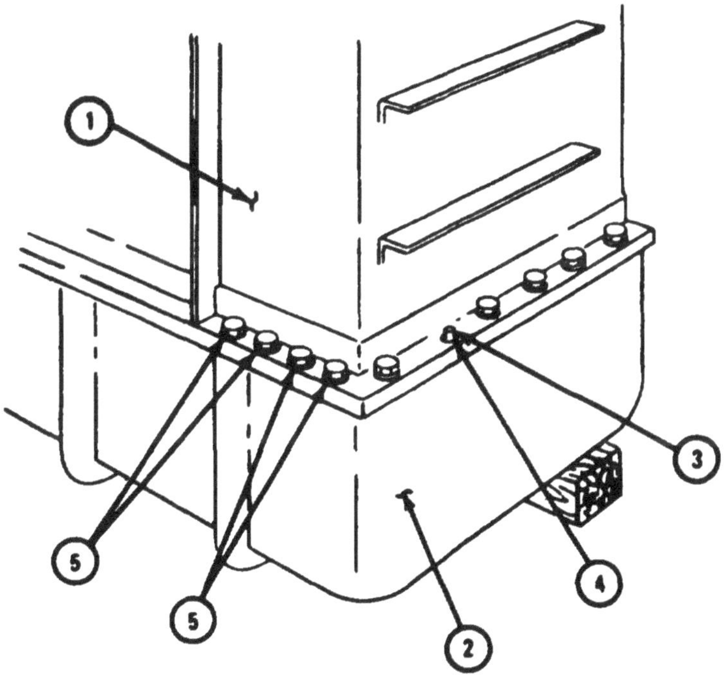

TA 118487

FRAME 6

1. Hook up a source of clean dry air to pneumatic tank valve (1).
2. Pressurize container (2) to 5 psi. After container (2) stands for 12 hours, check that air pressure is still 5 psi.
3. If air pressure is below 5 psi, remove engine from shipping container (part 1, para 2-3), and repair container . Refer to TM 9-2320-209-34 to repair container.

END OF TASK

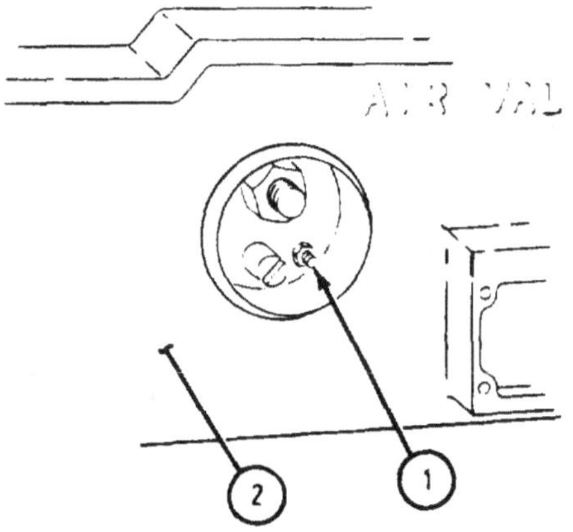

TA 118488

APPENDIX A

REFERENCES

A-1. GENERAL. This appendix contains a list of references which appear in this technical manual.

A-2. PUBLICATION INDEXES.

Index of Technical Publications	DA Pam 310-4

A-3. FORMS .

Quality Deficiency Report	SF 368
Recommended Changes to Equipment Publications	DA Form 2028-2

A-4. TECHNICAL MANUALS.

2 1/2-Ton, 6 x 6 M44A1 and M44A2 Series Trucks (Multifuel): Truck, Cargo: M35A1, M35A2 , M35A2C, M36A2; Truck, Tank, Fuel: M49A1C, M49A2C; Truck, Tank, Water: M50A1, M50A2, M50A3; Truck, Van, Shop: M109A2, M109A3; Truck, Repair Shop : M185A2, M185A3; Truck , Tractor: M275A1, M275A2; Truck, Dump: M342A2; Truck, Maintenance, Pipeline Construction : M765A2; Truck, Maintenance, Earth Boring Machine and Polesetter: M764	TM 9-2320-209-Series
5-Ton, 6 x 6 M39 Series Trucks (Multifuel): Truck, Chassis: M40A2C, M61A2, M63A2; Truck, Cargo: M54A2, M54A2C, M55A2; Truck, Dump: M51A2; Truck, Tractor: M52A2; Truck, Wrecker, Medium : M543A2	TM 9-2320-211-Series
Direct Support and General Support Maintenance Repair Parts and Special Tools Lists (Including Depot Maintenance Repair Parts and Special Tools): Engine Diesel (Multifuel): Turbocharged, Fuel Injected, Water Cooled, 6-Cylinder Assembly (Military Models LD 465-1, 2815-239-5824; LD-465-1C, 2815-134-4830; LDT-465-1C, 2815-103-2642; LDS-465-1, 2815-075-0087; LDS-465-1A, 2815-239-5819; and LDS-465-2, 2815-808-8011 and Clutches)	TM 9-2815-210-34P
Direct Support and General Support Level Maintenance (Repair Parts and Special Tools Lists): Generator Assembly NSN 2920-00-293-4380, (Delco-Remy Model No. 1117495)	TM 9-2920-214-34&P
Direct and General Support Level Maintenance (Including Repair Parts and Special Tools Lists): Generator Assembly NSN 2920-00-903-9534, (Prestolite Model No. GHA 4804JUT); Generator Assembly NSN 2920-00-737-4750, (Autolite Model No. GHA4802UT) ....*..... . . .	TM 9-2920-247-34&P

Direct Support and General Support Level Maintenance (Including Repair Parts and Special Tools List): Pump Fuel, Metering and Distributing, Assembly NSN 2910-00-178-1185, NSN 2910-00-759-5410, NSN 2910-00-908-6320, NSN 2910-00-017-9778, and NSN 2910-00-116-8241	TM 9-2910-226-34&P
Direct Support and General Support Level Maintenance (Including Repair Parts and Special Tools List): Starter, Engine Electrical, Assembly, Military Part Number 10911018-1, NSN 2920-00-226-6545; Military Part Number 10911018, NSN 2920-00-911-5637; Delco-Remy Model 1113944, NSN 2920-00-912-9510; Prestolite Model MFY 6101AUT, NSN 2920-00-999-6216; LEECE-NEVILLE Model M0017072MB, NSN 2920-00-267-9987	TM 9-2920-242-34&P
General Support Level Maintenance (Including Repair Parts and Special Tools List): Turbocharger, Engine Assembly, Schwitzer Model 4-456, NSN 2990-00-074-8930; Schwitzer Model 4D-454C, NSN 2990-00-967-9909; Schwitzer Model 4D-554, NSN 2990-00-974-7583; Schwitzer Model 4LE-354, NSN 2950-00-999-2275; Schwitzer Model 4LE-456, NSN 2950-00-178-1245; Schwitzer Model 3LJ-319, NSN 2950-00-148-7985; Schwitzer Model 3LD-305	TM 9-2990-201-40&P
Inspection, Care and Maintenance of Antifriction Bearings	TM 9-214
Operator's Manual: Welding Theory and Application	TM 9-237
Painting Instructions for Field Use	TM 43-0139
Operator and Organizational Maintenance Manual (Including Repair Parts and Special Tools List) For Cleaner and Tester, Spark Plug; Bench Mounted (Champiom Spark Plug Co., Model 800) (JAT Industries, Model JAT 5000) (The Oiljak Mfg. Co. Inc. , Model B 800M) (The VP Company, Model VP500) (4910-261-5868) Changes 1	TM 9-4910-389-12
Administrative Storage of Equipment	TM 740-90-1
Procedures for Destruction of Tank-Automotive Equipment to Prevent Enemy Use (US Army Tank-Automotive Command)	TM 750-244-6
The Army Maintenance Management System (TAMMS)	TM 38-750
Equipment Improvement Report and Maintenance Summary for TARCOM Equipment	TM 43-0143
Materials Used for Cleaning, Preserving, Abrading, and Cementing Ordnance Materiel and Related Materials Including Chemicals	TM 9-247
Sheet Metal Work: Body, Fender, and Radiator Repairs	FM 43-2

A-5. TECHNICAL BULLETINS.

Equipment Improvement Report and Maintenance Digest: Tank and Automotive Equipment	TB 43-001-39
Safety Inspection and Testing of Lifting Devices	TB 43-0142

Materials Used for Cleaning, Preserving, Abrading, and Cementing Ordnance Materiel and Related Materials Including Chemicals	TM 9-247
Metal Body Repair and Related Operations	FM 43-2

A-6. LUBRICATION ORDERS.

Truck, 2 1/2-Ton, 6 x 6 Cargo, M35A1, M35A2, M35A2C, M36A2; Tank, Fuel, M49A1C , M49A2C; Tank, Water, M50A1, M50A2; Van Shop, M109A2, M109A3; Repair Shop, M185A2, M185A3; Tractor, M275A1, M275A2; Dump, M342A2; Pipeline Construction, M756A2; Earth Boring and Polesetter, M764	LO 9-2320-209-12/1
Truck, Chassis: 5-Ton, 6 x 6, M39, M139, M139A1, M139F, M139A1F, M139A2F, M40, M40A1, M40A1C , M40A2C , M61, M61A1, M61A2, M63, M63A1, M63A1C, M63A2, M63A2C; M63C, Truck, Bolster: M748A1; Truck, Cargo: M54, M54A1, M54A2, M54A1C, M54A2C, M55, M55A2; Truck, Dump: M51, M51A1, M51A2; Truck, Stake, Bridge, Transporting: M328A1; Truck, Tractor: M52, M52A1, M52A2; Truck, Van, Expansible: M291A1, M291A1D; Truck, Tractor, Wrecker: M246, M246A1, M246A2; Truck, Wrecker, Medium: M62, M543, MT543A1, M543A2	LO 9-2320-211-12

INDEX

INDEX-CONT

INDEX-CONT

INDEX-CONT

INDEX-CONT

INDEX - CONT

INDEX - CONT

INDEX - CONT

INDEX - CONT

INDEX - CONT

INDEX - CONT

INDEX - CONT

INDEX-CONT

INDEX-CONT

INDEX-CONT

INDEX-CONT

By Order of the Secretaries of the Army and the Air Force:

E. C. MEYER
General, United States Army
Chief of Staff

Official:

J. C. PENNINGTON
Major General, United States Army
The Adjutant General

LEW ALLEN, JR., General, USAF
Chief of Staff

Official:

VAN L. CRAWFORD, JR., Colonel, USAF
Director of Administration

Distribution:

To be distributed in accordance with DA Form 12-38, Direct and General Support Maintenance requirements for 2-1/2 Ton Truck Cargo; 2-2/1 Ton Truck Van; 5 Ton Truck Chassis; 5 Ton Truck Van; and 5 Ton Truck Cargo, etc.

*U.S. GOVERNMENT PRINTING OFFICE: 1994 300-421/03085

THE METRIC SYSTEM AND EQUIVALENTS

LINEAR MEASURE

1 Centimeter = 10 Millimeters = 0.01 Meters = 0.3937 Inches
1 Meter = 100 Centimeters = 1000 Millimeters = 39.37 Inches
1 Kilometer = 1000 Meters = 0.621 Miles

WEIGHTS

1 Gram = 0.001 Kilograms = 1000 Milligrams = 0.035 Ounces
1 Kilogram = 1000 Grams = 2.2 Lb
1 Metric Ton = 1000 Kilograms = 1 Megagram = 1.1 Short Tons

LIQUID MEASURE

1 Milliliter = 0.001 Liters = 0.0338 Fluid Ounces
1 Liter = 1000 Milliliters = 33.82 Fluid Ounces

SQUARE MEASURE

1 Sq Centimeter = 100 Sq Millimeters = 0.155 Sq Inches
1 Sq Meter = 10,000 Sq Centimeters = 10.76 Sq Feet
1 Sq Kilometer = 1,000,000 Sq Meters = 0.386 Sq Miles

CUBIC MEASURE

1 Cu Centimeter = 1000 Cu Millimeters = 0.06 Cu Inches
1 Cu Meter = 1,000,000 Cu Centimeters = 35.31 Cu Feet

TEMPERATURE

$5/9\,(^{\circ}F - 32) = ^{\circ}C$
212° Fahrenheit is equivalent to 100° Celsius
90° Fahrenheit is equivalent to 32.2° Celsius
32° Fahrenheit is equivalent to 0° Celsius
$9/5\,C^{\circ} + 32 = F^{\circ}$

APPROXIMATE CONVERSION FACTORS

TO CHANGE	TO	MULTIPLY BY
Inches	Centimeters	2.540
Feet	Meters	0.305
Yards	Meters	0.914
Miles	Kilometers	1.609
Square Inches	Square Centimeters	6.451
Square Feet	Square Meters	0.093
Square Yards	Square Meters	0.836
Square Miles	Square Kilometers	2.590
Acres	Square Hectometers	0.405
Cubic Feet	Cubic Meters	0.028
Cubic Yards	Cubic Meters	0.765
Fluid Ounces	Milliliters	29.573
Pints	Liters	0.473
Quarts	Liters	0.946
Gallons	Liters	3.785
Ounces	Grams	28.349
Pounds	Kilograms	0.454
Short Tons	Metric Tons	0.907
Pound-Feet	Newton-Meters	1.356
Pounds per Square Inch	Kilopascals	6.895
Miles per Gallon	Kilometers per Liter	0.425
Miles per Hour	Kilometers per Hour	1.609

TO CHANGE	TO	MULTIPLY BY
Centimeters	Inches	0.394
Meters	Feet	3.280
Meters	Yards	1.094
Kilometers	Miles	0.621
Square Centimeters	Square Inches	0.155
Square Meters	Square Feet	10.764
Square Meters	Square Yards	1.196
Square Kilometers	Square Miles	0.386
Square Hectometers	Acres	2.471
Cubic Meters	Cubic Feet	35.315
Cubic Meters	Cubic Yards	1.308
Milliliters	Fluid Ounces	0.034
Liters	Pints	2.113
Liters	Quarts	1.057
Liters	Gallons	0.264
Grams	Ounces	0.035
Kilograms	Pounds	2.205
Metric Tons	Short Tons	1.102
Newton-Meters	Pound-Feet	0.738
Kilopascals	Pounds per Square Inch	0.145
Kilometers per Liter	Miles per Gallon	2.354
Kilometers per Hour	Miles per Hour	0.621

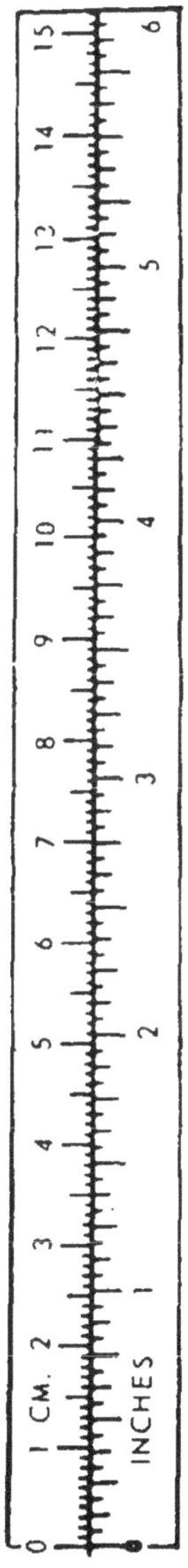

TA089991

PIN: 048858-00 0

www.ingramcontent.com/pod-product-compliance
Lightning Source LLC
Chambersburg PA
CBHW080412030426
42335CB00020B/2430